AF616570

HANDBOOK OF ALGEBRA
VOLUME 1

ELSEVIER
AMSTERDAM · LAUSANNE · NEW YORK · OXFORD · SHANNON · TOKYO

# HANDBOOK OF ALGEBRA

Volume 1

edited by
M. HAZEWINKEL
*CWI, Amsterdam*

1996

ELSEVIER
AMSTERDAM · LAUSANNE · NEW YORK · OXFORD · SHANNON · TOKYO

ELSEVIER SCIENCE B.V.
Sara Burgerhartstraat 25
P.O. Box 211, 1000 AE Amsterdam, The Netherlands

**Library of Congress Cataloging-in-Publication Data**
Handbook of algebra / edited by M. Hazewinkel.
p. cm.
Includes bibliographical references and index.
ISBN 0-444-82212-7 (v. 1 : acid-free paper)
1. Algebra. I. Hazewinkel, Michiel.
QA155.2.H36 1995
512–dc20 95-39024
CIP

ISBN: 0 444 82212 7



This book is printed on acid-free paper.

Printed in The Netherlands

# Preface

## Basic philosophy

Algebra, as we know it today, consists of many different ideas, concepts and results. A reasonable estimate of the number of these different "items" would be somewhere between 50 000 and 200 000. Many of these have been named and many more could (and perhaps should) have a "name" or a convenient designation. Even the nonspecialist is likely to encounter most of these, either somewhere in the literature, disguised as a definition or a theorem or to hear about them and feel the need for more information. If this happens, one should be able to find at least something in this Handbook and hopefully enough to judge if it is worthwhile to pursue the quest. In addition to the primary information, references to relevant articles, books or lecture notes should help the reader to complete his understanding. To make this possible, we have provided an index which is more extensive than usual and not limited to definitions, theorems and the like.

For the purpose of this Handbook, algebra has been defined, more or less arbitrarily as the union of the following areas of the Mathematics Subject Classification Scheme:

- 20 (Group theory)
- 19 ($K$-theory; will be treated at an intermediate level; a separate Handbook of $K$-theory which goes into far more detail than the section planned for this Handbook of Algebra is under consideration)
- 18 (Category theory and homological algebra; including some of the uses of categories in computer science, often classified somewhere in section 68)
- 17 (Nonassociative rings and algebras; especially Lie algebras)
- 16 (Associative rings and algebras)
- 15 (Linear and multilinear algebra, Matrix theory)
- 13 (Commutative rings and algebras; here there is a fine line to tread between commutative algebras and algebraic geometry; algebraic geometry is not a topic that will be dealt with in this Handbook; a separate Handbook on that topic is under consideration)
- 12 (Field theory and polynomials)
- 11 (As far as it used to be classified under old 12 (Algebraic number theory))
- 08 (General algebraic systems)
- 06 (Certain parts; but not topics specific to Boolean algebras as there is a separate three-volume Handbook of Boolean Algebras)

## Planning

Originally, we hoped to cover the whole field in a systematic way. Volume 1 would be devoted to what we now call Section 1 (see below), Volume 2 to Section 2 and so on. A detailed and comprehensive plan was made in terms of topics which needed to be covered and authors to be invited. That turned out to be an inefficient approach. Different authors have different priorities and to wait for the last contribution to a volume, as planned originally, would have resulted in long delays. Therefore, we have opted for a dynamically evolving plan. This means that articles are published as they arrive and that the reader will find in this first volume articles from three different sections. The advantages of this scheme are two-fold: accepted articles will be published quickly and the outline of the series can be allowed to evolve as the various volumes are published. Suggestions from readers both as to topics to be covered and authors to be invited are most welcome and will be taken into serious consideration.

The list of the sections now looks as follows:

Section 1: Linear algebra. Fields. Algebraic number theory
Section 2: Category theory. Homological and homotopical algebra. Methods from logic
Section 3: Commutative and associative rings and algebras
Section 4: Other algebraic structures. Nonassociative rings and algebras. Commutative and associative rings and algebras with extra structure
Section 5: Groups and semigroups
Section 6: Representations and invariant theory
Section 7: Machine computation. Algorithms. Tables
Section 8: Applied algebra
Section 9: History of algebra

For a more detailed plan, the reader is reffered to the Outline of the Series following the Preface.

## The individual chapters

It is not the intention that the handbook as a whole can also be a substitute undergraduate or even graduate, textbook. The treatment of the various topics will be much too dense and professional for that. Basically, the level is graduate and up, and such material as can be found in P.M. Cohn's three-volume textbook "Algebra" (Wiley) will, as a rule, be assumed. An important function of the articles in this Handbook is to provide professional mathematicians working in a different area with sufficient information on the topic in question if and when it is needed.

Each chapter combines some of the features of both a graduate-level textbook and a research-level survey. Not all of the ingredients mentioned below will be appropriate in each case, but authors have been asked to include the following:

- Introduction (including motivation and historical remarks)
- Outline of the chapter

- Basic concepts, definitions, and results (proofs or ideas/sketches of the proofs are given when space permits)
- Comments on the relevance of the results, relations to other results, and applications
- Review of the relevant literature; possibly supplemented with the opinion of the author on recent developments and future directions
- Extensive bibliography (several hundred items will not be exceptional)

**The future**

Of course, ideally, a comprehensive series of books like this should be interactive and have a hypertext structure to make finding material and navigation through it immediate and intuitive. It should also incorporate the various algorithms in implemented form as well as permit a certain amount of dialogue with the reader. Plans for such an interactive, hypertext, CD-Rom-based version certainly exist but the realization is still a nontrivial number of years in the future.

Bussum, September 1995 Michiel Hazewinkel

Kaum nennt man die Dinge beim richtigen Namen,
so verlieren sie ihren gefährlichen Zauber

(You have but to know an object by its proper name
for it to lose its dangerous magic)

E. Canetti

# Outline of the Series

Chapters which have a named author have been written and are ready for publication. The numbers after the title indicate in which volume the chapter either will appear or has appeared. Topics printed in italics already have an author commissioned, and are in the process of being written. For topics printed in roman type no author has been contracted as yet.

No definite plans have been made for Sections 7, 8 and 9 at this stage.

### Section 1. Linear algebra. Fields. Algebraic number theory

*A. Linear Algebra*

**G.P. Egorychev**, Van der Waerden conjecture and applications (1)
**V.L. Girko**, Random matrices (1)
**A.N. Malyshev**, Matrix equations. Factorization of matrix polynomials (1)
**L. Rodman**, Matrix functions (1)
*Linear inequalities (also involving matrices)*
*Orderings (partial and total) on vectors and matrices (including positive matrices)*
*Matrix equations. Factorization of matrices*
*Special kinds of matrices such as Toeplitz and Hankel*
Integral matrices. Matrices over other rings and fields

*B. Linear (In)dependence*

**J.P.S. Kung**, Matroids (1)

*C. Algebras Arising from Vector Spaces*

*Clifford algebras, related algebras, and applications*

*D. Fields, Galois Theory, and Algebraic Number Theory*

(There is an article on ordered fields in Section 4)
**J.K. Denevey and J.N. Mordeson**, Higher derivation Galois theory of inseparable field extensions (1)

**I.B. Fesenko**, Complete discrete valuation fields. Abelian local class field theories (1)
**M. Jarden**, Infinite Galois theory (1)
**R. Lidl and H. Niederreiter**, Finite fields and their applications (1)
**W. Narkiewicz**, Global class field theory (1)
**H. van Tilborg**, Finite fields and error correcting codes (1)
*Skew fields and division rings. Brauer group*
*Topological and valued fields. Valuation theory*
*Zeta and L-functions of fields and related topics*
*Structure of Galois modules*
Constructive Galois theory (realization of groups as Galois groups)

*E. Nonabelian Class Field Theory and the Langlands Program*

(To be arranged in several chapters by Y. Ihara)

*F. Generalizations of Fields and Related Objects*

**U. Hebisch and H.J. Weinert**, Semi-rings and semi-fields (1)
**G.F. Pilz**, Near-rings and near-fields (1)

**Section 2. Category theory. Homological and homotopical algebra. Methods from logic**

*A. Category Theory*

**S. MacLane and I. Moerdijk**, Topos theory (1)
**R.H. Street**, Categorical structures (1)
*Algebraic theories*
*Categories and databases*
*Categories in computer science (in general)*

*B. Homological Algebra. Cohomology. Cohomological Methods in Algebra. Homotopical Algebra*

**J.F. Carlson**, The cohomology of groups (1)
**A.I. Generalov**, Relative homological algebra. Cohomology of categories, posets, and coalgebras (1)
**J.F. Jardine**, Homotopy and homotopical algebra (1)
**B. Keller**, Derived categories and their uses (1)
**A. Helemskii**, Homology for the algebras of analysis (2)
Galois cohomology
Cohomology of commutative and associative algebras
Cohomology of Lie algebras
Cohomology of group schemes

*C. Algebraic $K$-theory*

*Grothendieck groups*
*$K_2$ and symbols*
*Algebraic $K$-theory of $C^*$-algebras, EXT, etc.*
*Hilbert $C^*$-modules*
*Index theory for elliptic operators over $C^*$-algebras*
*Algebraic $K$-theory (including the higher $K_n$)*
*Simplicial algebraic $K$-theory*
*Chern character in algebraic $K$-theory*
*$KK$-theory*
*Noncommutative differential geometry*
$K$-theory of noncommutative rings
Algebraic $L$-theory
Cyclic cohomology

*D. Logic versus Algebra*

*Methods of logic in algebra*
*Logical properties of fields and applications*
*Recursive algebras*
*Logical properties of Boolean algebras*

*E. Rings up to Homotopy*

Rings up to homotopy

**Section 3. Commutative and associative rings and algebras**

*A. Commutative Rings and Algebras*

**J.-P. Lafon**, Ideals and modules (1)
*General theory. Radicals, prime ideals etc. Local rings (general). Finiteness and chain conditions*
*Extensions. Galois theory of rings*
*Modules with quadratic form*
*Finite commutative rings and algebras*
Homological algebra and commutative rings. Ext, Tor, etc. Special properties (p.i.d., factorial, Gorenstein, Cohen–Macauley, Bezout, Fatou, Japanese, Excellent, Ore, Prüfer, Dedekind, ... and their interrelations)
Lifting (Hensel properties) and Artin approximation
Localization. Local–global theory
Rings associated to combinatorial and partial order structures (straightening laws, Hodge algebras, shellability, ...)
Witt rings, real spectra

*B. Associative Rings and Algebras*

**P.M. Cohn**, Polynomial and power series rings. Free algebras, firs and semifirs (1)
**V.K. Kharchenko**, Simple, prime, and semi-prime rings (1)
**V.K. Kharchenko**, Fixed rings and noncommutative invariant theory (2)
**A.R.P. van den Essen**, Algebraic microlocalization and modules with regular singularities over filtered rings (1)
**F. van Oystayen**, Separable algebras (2)
**K. Yamagata**, Frobenius rings (1)
*Classification of Artinian algebras and rings*
*General theory of associative rings and algebras*
*Rings of quotients. Noncommutative localization. Torsion theories*
*von Neumann regular rings*
*Lattices of submodules*
*PI rings*
*Generalized identities*
*Endomorphism rings, rings of linear transformations, matrix rings*
*Homological classification of (noncommutative) rings*
*Group rings and algebras*
*Dimension theory*
*Duality. Morita-duality*
*Groups acting on associative algebras. Noncommutative invariant theory*
*Commutants of differential operators*
Rings of differential operators
Graded and filtered rings and modules (also commutative)
Goldie's theorem, Noetherian rings and related rings

*C. Co-algebras*

*Co-algebras and bi-algebras*

*D. Deformation Theory of Rings and Algebras (Including Lie Algebras)*

Deformation theory of rings and algebras (general)
Deformation theory of Lie algebras

## Section 4. Other algebraic structures. Nonassociative rings and algebras. Commutative and associative rings and algebras with extra structure

*A. Lattices and Partially Ordered Sets*

Lattices and partially ordered sets
Frames and locales

*B. Boolean Algebras*

*C. Universal Algebra*

*D. Varieties of Algebras, Groups, ...*

**V.A. Artamanov**, Varieties of algebras (2)
Varieties of groups
Quasi-varieties
Varieties of semigroups

*E. Lie Algebras*

**Yu.A. Bahturin, M.V. Zaitsev and A.A. Mikhailov**, Infinite dimensional super Lie algebras (2)
General structure theory. Free Lie algebras
Classification theory of semisimple Lie algebras over **R** and **C**
The exceptional Lie algebras
Nilpotent and solvable Lie algebras
Universal envelopping algebras
Modular (ss) Lie algebras (including classification)
Infinite dimensional Lie algebras (general)
Kac–Moody Lie algebras

*F. Jordan Algebras* (finite and infinite dimensional and including their cohomology theory)

*G. Other Nonassociative Algebras* (Malcev, alternative, Lie admissible, ...)

*H. Rings and Algebras with Additional Structure*

*Ordered and lattice-ordered groups, rings and algebras*
*$\lambda$-rings, $\gamma$-rings, ...*
*Difference and differential algebra. Abstract (and p-adic) differential equations. Differential extensions*
*Ordered fields*
Graded and super algebras (commutative, associative and Lie)
Topological rings
Hopf algebras
Quantum groups
Formal groups
Rings and algebras with involution. $C^*$-algebras

*J. The Witt Vectors*

## Section 5. Groups and semigroups

### *A. Groups*

*Simple groups, sporadic groups*
*Abelian groups*
*"Additive" group theory*
Abstract (finite) groups. Structure theory. Special subgroups. Extensions and decompositions
Solvable groups, nilpotent groups, $p$-groups
Infinite soluble groups
Word problems
Burnside problem
Combinatorial group theory
Free groups (including actions on trees)
Formations
Infinite groups. Local properties
Algebraic groups. The classical groups. Chevalley groups
Chevalley groups over rings
The infinite dimensional classical groups
Other groups of matrices. Discrete subgroups
Reflection groups. Coxeter groups
Groups with BN-pair, Tits buildings, ...
Groups and (finite combinatorial) geometry
Probabilistic techniques and results in group theory

### *B. Semigroups*

Semigroup theory. Ideals, radicals, structure theory
Semigroups and automata theory and linguistics

### *C. Algebraic Formal Language Theory*

### *D. Loops, Quasigroups, Heaps, ...*

### *E. Combinatorial Group Theory and Topology*

## Section 6. Representations and invariant theory

### *A. Representations*

**A.U. Klimyk**, Infinite dimensional representations of quantum algebras (2)
Representations of quantum groups
Representation theory of rings, groups, algebras (general)
Modular representation theory (general)

Representation theory of finite groups in characteristic zero
Modular representation theory of finite groups. Blocks
Representation theory of the symmetric groups (both in characteristic zero and modular)
Representation theory of the finite Chevalley groups (both in characteristic zero and modular)
Representation theory of the classical groups. Classical invariant theory
Classical and transcendental invariant theory
Finite dimensional representation theory of the ss Lie algebras (in characteristic zero); structure theory of semi-simple Lie algebras
Infinite dimensional representation theory of ss Lie algebras. Verma modules
Representations of solvable and nilpotent Lie algebras. The Kirillov orbit method
Orbit method, Dixmier map, ... for ss Lie algebras
Modular representation theory of Lie algebras
Representation theory of Kac–Moody algebras
Representations of semigroups
Representations of rings and algebras by sections of sheafs
Representation theory of algebras (Quivers, Auslander–Reiten sequences, almost split sequences, ...)
Invariants of nonlinear representations of Lie groups

*B. Representations, Commutative Algebra and Combinatorics*

*C. Abstract Representation Theory*

**Section 7. Machine computation. Algorithms. Tables**

Some notes on this volume: Besides some general article(s) on machine computation in algebra, this volume should contain specific articles on the computational aspects of the various larger topics occurring in the main volume, as well as the basic corresponding tables. There should also be a general survey on the various available symbolic algebra computation packages.

**Section 8. Applied algebra**

**Section 9. History of algebra**

# Contents

# List of Contributors

Carlson, J.F., *University of Georgia, Athens, GA*
Cohn, P.M., *University College London, London*
Deveney, J.K., *Virginia Commonwealth University, Richmond, VA*
Egorychev, G.P., *Krasnoyarsk State Technical University, Krasnoyarsk*
Fesenko, I.B., *University of Nottingham, Nottingham*
Generalov, A.I., *Skt Petersburg University, Skt Petersburg*
Girko, V.L., *Kiev State University, Kiev*
Hebisch, U., *TU Bergakademie Freiberg, Freiberg*
Jarden, M., *Tel Aviv University, Tel Aviv*
Jardine, J.F., *University of Western Ontario, London, Ont.*
Keller, B., *Université de Paris VII, Paris*
Kharchenko, V.K., *Institute of Mathematics, Novosibirsk*
Kung, J.P.S., *University of North Texas, Denton, TX*
Lafon, J.-P., *Université de Paris XIII, Villetaneuse*
Lidl, R., *University of Tasmania, Hobart, Tasmania*
MacLane, S., *University of Chicago, Chicago, IL*
Malyshev, A.N., *Institute of Mathematics, Novosibirsk*
Moerdijk, I., *Rijksuniversiteit Utrecht, Utrecht*
Mordeson, J.N., *Creighton University, Omaha, NE*
Narkiewicz, W., *Wroclaw University, Wroclaw*
Niederreiter, H., *Austrian Academy of Sciences, Vienna*
Pilz, G.F., *Johannes-Kepler-Universität, Linz*
Rodman, L., *College of William & Mary, Williamsburg, VA*
Street, R.H., *Macquarie University, North Ryde, NSW*
Van den Essen, A.R.P., *Katholieke Universiteit Nijmegen, Nijmegen*
Van Tilborg, H., *Technische Universiteit Eindhoven, Eindhoven*
Weinert, H.J., *TU Clausthal, Clausthal-Zellerfeld*
Yamagata, K., *University of Tsukuba, Tsukuba*

# Section 1A
# Linear Algebra

# Van der Waerden Conjecture and Applications

G.P. Egorychev[1]

*Chair of Software of Discrete Apparatus and Systems, Krasnoyarsk State Technical University, St. Kirenskogo, 26, Krasnoyarsk, 660074, Russia*
*e-mail: gpe@cckr.krasnoyarsk.su*

*Contents*



[1]Jointly supported by RFFI Grant 93-011-1560 and IFS-fSU Grant 1993.

HANDBOOK OF ALGEBRA, VOL. 1
Edited by M. Hazewinkel

**Abstract**

This chapter gives a proof and applications of the well known van der Waerden conjecture about minimum permanents of a doubly stochastic matrix. The conjecture was proved in the early 80s by G.P. Egorychev and the Kiev mathematician D.I. Falikman independently.

## 1. Basic notations and concepts

*Some notation.* $\operatorname{per}(A)$ is the permanent of a matrix $A$ (for a definition see Section 2 below).

$D(A_1, \ldots, A_n)$ is the mixed discriminant of the matrices $A_1, \ldots, A_n$.

$V(K_1, \ldots, K_n)$ is the mixed volume of convex compacts $K_1, \ldots, K_n$ in $\mathbb{R}^n$.

$\Omega_n$ is the set of all doubly stochastic $n \times n$ matrices.

$\Lambda_n^k$ is the set of all $(0, 1)$-matrices of order $n$ which have exactly $k$ units in each row and column.

$P, Q$ are the permutation matrices of order $n$.

$E_{i,j}$ is the matrix having 1 at place $(i, j)$ and all other elements equal to 0.

$e$ is the $n$-row of which each element is 1.

$a_j, \alpha_i$ are the $j$-th column and the $i$-th row of the matrix $A$, respectively.

$O, I_n$ are the null and identity $n \times n$ matrices, respectively.

$J_n$ is the $(n \times n)$ matrix with each element equal to $1/n$.

$A^T, \bar{A}, A^*$ are the transpose, adjoint and complex conjugate matrices to the matrix $A$, respectively.

$A(i/j)$ is the $(n-1) \times (n-1)$ matrix derived from $A$ by deleting the $i$-th row and the $j$-th column.

$$A\begin{pmatrix} i & j \\ x & y \end{pmatrix}$$

is the matrix of order $n$ derived from $A$ by replacing the $i$-th column by the $n$-vector $x$ and the $j$-th column by the $n$-vector $y$.

*Definitions.* A matrix of order $n$ with non-negative elements is called *doubly stochastic* if $eA = e,\ Ae^T = e^T$; a matrix $A \in \Omega_n$ is called *minimizing* if

$$\min_{X \in \Omega_n} \operatorname{per}(X) = \operatorname{per}(A).$$

Let $A = (a_{ij})$ be a non-negative matrix of order $n$. The matrix $A$ is called *fully indecomposable* if it doesn't contain a $k \times (n-k)$ null submatrix for $k = 1, \ldots, n-1$. The matrix $A$ is called *partially decomposable* if it contains $k \times (n-k)$ null submatrix. Matrix $A$ is called nearly decomposable if it is fully indecomposable and such that for each positive element of the matrix $A$ the matrix $A - a_{ij}E_{ij}$ is partially decomposable.

## 2. Introduction: some words about permanents

The concept of the permanent was first introduced independently and practically simultaneously in the well known memoirs of J. Binet (1812) and A. Cauchy (1812). It was at this time that Binet introduced the term "permanent".

*The permanent of a square $n \times n$ matrix $A$* is defined to be the sum

$$\operatorname{per}(A) := \sum_{\sigma \in S_n} a_{1\sigma(1)} \cdots a_{n\sigma(n)}, \tag{2.1}$$

where the sum is taken over all permutations of the set $\{1, \ldots, n\}$ or, which is the same, over all diagonals of the matrix $A$.

The permanent as a matrix function has the following characteristic properties (Egorychev, 1980): *if $f(A)$ is a complex-valued function of $n \times n$-matrix $A$ over the field $\mathbb{C}$, then $f(A) = \operatorname{per} A$, iff:*

(a) *$f(A)$ is a homogeneous polynomial of degree $n$ from the elements of the matrix $A$.*

(b) *$f(A)$ is polyadditive and symmetric with respect to vector-rows and vector-columns of matrix $A$.*

(c) $f(I_n) = 1$.

For the last decade and a half the theory of permanents has been intensively developing, undergoing essential structural modifications. Among the most important achievements of this period are, in our opinion, the proof of the van der Waerden conjecture on permanents (Egorychev, 1980, 1981; Falikman, 1981), the proof of the Tverberg conjecture on permanents (Friedland, 1982); remarkable results by V. Schevelev (1992) and A. Kamenetsky (1990, 1991) concerned with the problem of calculation of the permanents of cyclic matrices and, finally, the fundamental results by A. Razborov (1985) and A. Andreev (1985) about lower estimates of complexity for the permanent of logical matrices.

Many difficult problems of combinatorial analysis, graph theory, linear and polylinear algebra, other areas of mathematics, statistical physics and physical chemistry can be stated and solved in terms of permanents. They basically make use of the major combinatorial properties of permanents of $(0, 1)$-matrices to count the number of systems of different representatives (transversals) of sets.

It is common knowledge that det $A$ can be computed in poly$(n)$ time. On the other hand, the fastest algorithm known for computing per $A$ runs in $n2^{n-1}$ time (Wilf, 1968; Ryser, 1963). Solid grounds for arguing that computing per $A$ even for $(0, 1)$-matrices is an inherently difficult problem were first provided by L. Valiant (1979) who showed that the problem is $P$-complete. One implication of this result is that if $P \neq NP$ then there is no poly$(n)$ time algorithm for computing per $A$.

From there rises the problem of the development of new fundamental algebraic, geometric, theoretical-functional and calculating ideas and methods of computation and estimation of permanents for various classes of matrices. The celebrated van der Waerden conjecture giving a precise lower estimate for the permanents on the convex compact set $\Omega_n$ for a long time occupied a key position in this circle of questions. This problem, in spite of many efforts, remained unresolved for over 50 years.

## 3. Origin of the van der Waerden conjecture

Biographical data about van der Waerden were given by A.I. Borodin and A.S. Bugai (1979, p. 95):

"Barten Leendert van der Waerden is a Dutch mathematician. He was born in Amsterdam on February 2, 1903. He was a professor at the universities of Groningem, Leipzig, Amsterdam, Zürich and was involved in algebra, algebraic geometry, applications of methods of group theory to quantum theory and mathematical statistics (van der Waerden criterion). His work also dealt with the history of mathematics and astronomy in Ancient Egypt, Babylon and Greece. In 1959 his book "Science Awakening" was translated into Russian. The book "Modern Algebra" (1930–1931) marked the culmination of the creative period in "abstract algebra" developed by his teachers E. Noether, E. Steinitz and E. Artin. This book had major influence in the training of specialists in algebra everywhere in the world, defining the character and partially the directions of further research in algebra. An expanded and modernized version titled "Algebra" was published in 1976. B.L. van der Waerden applied the modern algebraic apparatus to strong justification of basic concepts in algebraic geometry. His book "Mathematical Statistics" (1957) is also widely known.

Van Lint, a well-known Dutch mathematician, wrote (1982, p. 72–76) about the previous history of the van der Waerden conjecture:

"Much of the work on permanents is in some way connected to this conjecture and about 75% of the work on permanents is less than 20 years old! ... In 1926 B.L. van der Waerden proposed as a problem (!) to determine the minimal permanent among all doubly stochastic matrices. It was natural to assume that this minimum is $\operatorname{per} J_n = n!n^{-n}$. Let us denote by $\Omega_n$ the set of all doubly stochastic matrices. The assertion

$$(A \in \Omega_n \cap A \neq J_n) \Rightarrow (\operatorname{per} A > \operatorname{per} J_n)$$

became known as the van der Waerden conjecture. Sometimes just showing that $n!n^{-n}$ is the minimal value is referred to as the conjecture.

"This note allows me to save for posterity a humorous experience of the late sixties. Van der Waerden, retired by then, attended a meeting on combinatorics, a field he had never worked in seriously. A young mathematician was desperate to present his thesis in 20 minutes. I was sitting in the front row next to van der Waerden when the famous conjecture was mentioned by the speaker and the alleged author inquired what this famous conjecture stated!! The exasperated speaker spent a few seconds of his precious time to explain and at the end of his talk wandered over to us to read the badge of the person who had asked this inexcusable question. I could foresee what was to happen and yet, I remember how he recoiled. You needn't worry – he recovered and now is a famous combinatorialist. The lesson for the reader is the following. If you did not know of the "conjecture" then it is comforting to realize that it was 40 years old before van der Waerden heard that it had this name.

"What is the origin of the problem? Upon my request van der Waerden went far back in his memory and came up with the following. One day in 1926 during the discussion that took place daily in Hamburg O. Schreier mentioned that G.A. Miller had proved that there is a mutual system of representatives for the right and left cosets of a subgroup $H$ of a finite group $G$. At this moment van der Waerden observed that this was a property of any two partitions of a set of size $\mu n$ into $\mu$ subsets of size $n$. This theorem was published in "Hamburger Abhandlungen" in 1927. In the note, added in the proof, van der Waerden acknowledged that he had rediscovered the theorem which is now known as the König–Hall theorem...

"In the terminology of permanents we can formulate the problem Schreier and van der Waerden were considering as follows. Let $A_i$ $(1 \leqslant i \leqslant \mu)$ and $B_k$ $(1 \leqslant k \leqslant \mu)$ be the subsets in two partitions and let $a_{ik} := |A_i \cap B_k|$. Then, $A = (a_{ik})$ is a matrix with constant line sums $(= n)$. The assertion that there is a mutual system of representatives of the sets $A_i$ respectively of the sets $B_k$ is the same as to say that $\operatorname{per} A > 0$. At this point van der Waerden wondered what the minimal permanent, under the side condition that all line sums are 1, is? He posed this as a problem in Jber. d. D.M.V. **35** and thus the van der Waerden conjecture was born."

## 4. Summary of results

Most essential in proving the van der Waerden conjecture were the results obtained by M. Marcus and M. Newman (1959). They show that:

(a) *If $A$ is a minimizing matrix, then* $\operatorname{per}(A(i/j)) = \operatorname{per}(A)$ *for* $\forall i, j \in \{1, \ldots, n\}$, *where* $a_{i,j} > 0$;

(b) *If $A$ is minimizing then it is fully indecomposable;*

(c) *The permanent has a strong local minimum at point $J_n$;*

(d) *If all elements of the minimizing matrix are positive, then it is equal to $J_n$.*

Using these results D. London (1971) proved that if $A$ is a minimizing matrix, then $\operatorname{per}(A(i/j)) \geqslant \operatorname{per}(A)$ for all $i, j \in \{1, \ldots, n\}$. In 1976 T. Bang announced and in 1979 T. Bang and S. Friedland proved lower bounds for the permanent on $\Omega_n$, which are essentially of the same order as the bound of van der Waerden.

D. König (1916) stated that the permanent of a doubly stochastic matrix $A$ is always positive; if $A \in \Lambda_n^k$, then $\operatorname{per}(A) > k$. G. Frobenius (1917) proved that $\operatorname{per} A$ of a non-negative matrix is equal to 0 if and only if $A$ contains a null submatrix of order $k \times (n - k + 1)$. D. König in his book (1936) devoted to graph theory and its applications mentioned the van der Waerden problem as an unresolved one. Birkhoff in 1946 showed that $\Omega_n$ forms a convex polyhedron with the permutation matrices as vertices. This result implies that $\operatorname{per}(A) \geqslant 1/((n-1)^2 + 1)^{n-1}$ for all $A \in \Omega_n$. M. Marcus and H. Minc improved this bound in 1962 to $n^{-n}$, in 1974 O. Rothaus improved is to $(n!)^{-n}$, and S. Friedland (1979) obtained the substantially better bound of $1/n!$.

M. Marcus and M. Newman (1959) proved the validity of the van der Waerden conjecture for $n = 3$; P. Eberlein and G. Mudholkar (1968) and A. Gleason (1970) proved it for $n = 4$; P. Eberlein (1969) for $n = 5$. In 1962 M. Marcus and M. Newman proved the conjecture for positive semi-definite symmetric matrices, and D. Sasser with M. Slater (1967) extended this theorem to normal matrices of a certain type. The last result was somewhat improved by M. Marcus and H. Minc (1968) and extended to a larger class of matrices by S. Friedland (1974). (For other results related to the van der Waerden conjecture until 1977 see the comprehensive bibliography in the book by Minc (1978).)

A proof of the van der Waerden conjecture appeared in 1980–1981. In his article of 1981 D. Falikman, by a method different from that of G. Egorychev (1980, 1981), obtained the answer to the van der Waerden's question about the minimum of the permanent on $\Omega_n$, but he didn't prove the uniqueness statement of the conjecture. Another attempt to prove this hypothesis was made by V. Reva (1981).

The proof of the Marcus–Newman and van der Waerden conjectures given here is typical for the theory of mixed volumes. It is, on the one hand, connected with the investigation of the structure of the minimizing matrix based on the results of D. König (1916), M. Marcus and M. Newman (1969), D. London (1971). On the other hand, this proof uses geometric inequalities for the permanent as a mixed discriminant (mixed volume) which are a particular case of the well-known Aleksandrov's inequalities for mixed discriminants (1937–1938). This interpretation of the permanent was used in the work of Egorychev (1979–1980) dealing with obtaining a series of polynomial identities and characteristic properties of permanents for plane and space matrices. The role of the geometric inequalities in the theory of permanents was found to be identical to that of Aleksandrov–Fenchel inequalities for mixed volumes, which allow the solving of many important extremal problems and problems of uniqueness for convex bodies in $\mathbb{R}^n$ (see Buseman, 1960; Leichtweiss, 1980; Burago and Zalgaller, 1990; Mitrinovič, Pečarić and Volenec, 1989).

The proof given here of the permanent conjectures is fairly simple and is accessible to any reader familiar with the fundamentals of the linear algebra.

## 5. Structure of the minimizing matrix: the necessary conditions

This section considers several fine assertions on the structure of the minimizing matrix that led D. London to his result (1971). The reader can find the proofs of the well-known facts omitted here in, for example, the work of M. Marcus and H. Minc (1972), A. Marshall, J. Olkin (1983), R. Rockafellar (1970), H. Minc (1978).

THEOREM 5.1 (Birkhoff, 1946). *A set of $n \times n$ doubly stochastic matrices forms a convex polyhedron with permutation matrices as vertices. In other words, if $A \in \Omega_n$, then*

$$A = \sum_{i=1}^{s} \theta_i P_i, \tag{5.1}$$

*where $P_1, \dots, P_s$ are permutation matrices and $\theta_1, \dots, \theta_s \geqslant 0$, $\sum_{i=1}^{s} \theta_i = 1$.*

This theorem has many applications and is one of the main results in the theory of doubly stochastic matrices. The next theorem is one of the most important results in the theory of non-negative matrices.

THEOREM 5.2 (Frobenius–König, 1917). *Let $A$ be an $n \times n$ matrix. A necessary and sufficient condition for every diagonal of $A$ to contain a zero entry is that $A$ contains an $s \times t$ zero submatrix such that $s + t = n + 1$.*

In other words, if $A$ is an $n \times n$ matrix, then $\operatorname{per} A = 0$ iff $A$ contains an $s \times t$ zero submatrix such that $s + t = n + 1$. (For the history of this theorem see (Minc, 1978, pp. 31–34).)

THEOREM 5.3 (König, 1916). *If* $A \in \Omega_n$, *then* $\operatorname{per} A > 0$.

ASSERTION 5.4 (London, 1971). *If* $A = (a_{ij})$ *is a minimizing matrix, then*

$$\operatorname{per}\big(A(i/j)\big) \geqslant \operatorname{per}(A) \quad \textit{for all } i, j \in 1, \ldots, n. \tag{5.2}$$

ASSERTION 5.5. *A non-negative matrix* $A$ *of order* $n$, $n \geqslant 2$, *is fully indecomposable iff* $\operatorname{per}(A(i/j)) > 0$ *for all* $i, j \in \overline{1, n}$.

ASSERTION 5.6 (Marcus and Newman, 1959).

(a) *If* $A$ *is a minimizing matrix, then* $A$ *is fully indecomposable.*

(b) *If* $A = (a_{ij})$ *is a minimizing matrix and some* $a_{ij} > 0$, *then* $\operatorname{per}(A(i/j)) = \operatorname{per}(A)$.

(c) *If the minimizing matrix is positive, then* $A = J_n$.

It is easy to see that Theorem 5.3 is equivalent to the following assertion: *any matrix* $A \in \Omega_n$ *contains at least one diagonal with positive terms, thus, Theorem* 5.3 *follows from the Birkhoff theorem.*

The proof of Assertion 5.6(b) in the work by Marcus and Newman (1959) used to prove Assertion 5.4 adapted the classical Lagrange multipliers method assuming $A \in \Omega_n$.

The result of Assertion 5.5 is a direct corollary of the Frobenius–König theorem.

Assertion 5.4 is proved by taking derivatives in all directions in a neighborhood of a minimizing matrix $A$. For any $n \times n$ permutation matrix $P = (p_{ij})$, $0 \leqslant \theta \leqslant 1$, define the function $f_P(\theta) = \operatorname{per}((1-\theta)A + \theta P)$.

Since $A$ is a minimizing matrix, then $f'_P(0) \geqslant 0$ for any $P$. But

$$\begin{aligned} f'_P(0) &= \sum_{s,t=1}^{n} (-a_{st} + P_{st}) \operatorname{per}\big(A(s/t)\big) \\ &= \sum_{s,t=1}^{n} P_{st} \operatorname{per}\big(A(s/t)\big) - n \operatorname{per}(A) \\ &= \sum_{s=1}^{n} \operatorname{per}\big(A(s/\sigma(s))\big) - n \operatorname{per}(A), \end{aligned}$$

where $\sigma$ is the permutation corresponding to $P$. Hence,

$$\sum_{s=1}^{n} \operatorname{per}\big(A(s/\sigma(s))\big) \geqslant n \operatorname{per} A, \tag{5.3}$$

for every $\sigma$. Since the matrix $A$ must be fully indecomposable (Assertion 5.6(a)), therefore (Assertion 5.5) any entry of $A$ lies on a diagonal all of whose other entries are positive. Thus, for every $(i, j)$ there exists a permutation $\sigma$ such that $j = \sigma(i)$ and $a_{s\sigma(s)} > 0$ for all $s \in 1, \ldots, n$, $s \neq i$.

Now, however, Assertion 5.6(b) ensures that $\operatorname{per}(A(s/\sigma(s))) = \operatorname{per} A$ for the same $s$. Hence, from (5.3) and $j = \sigma(i)$ follows that $\operatorname{per}(A(i/j)) \geqslant \operatorname{per}(A)$.

## 6. Mixed discriminants (volumes) and geometric inequalities for permanents

Consider $m$ quadratic forms

$$f_k = \sum_{i,j=1}^{n} a_{ij}^k x_i x_j$$

in the variables $x_1, \ldots, x_n$. Any linear combination

$$f = \sum_{k=1}^{m} \lambda_k f_k$$

is again a quadratic form with coefficients

$$a_{ij} = \sum_{k=1}^{m} \lambda_k a_{ij}^k.$$

The discriminant of the form $f$ is a homogeneous polynomial of degree $n$ with respect to $\lambda_1, \ldots, \lambda_m$ with coefficients $D(f_{k_1}, \ldots, f_{k_n})$ of $\lambda_{k_1} \ldots \lambda_{k_n}$ chosen not to depend on the order $k_1, \ldots, k_n$. These coefficients, studied by Aleksandrov (1937–1938), are called *mixed discriminants*. They are expressed in terms of coefficients of the given forms as follows

$$D(f_1, \ldots, f_n) = \frac{1}{n!} \sum_{\sigma=(\sigma_1,\ldots,\sigma_n)\in S_n} \det \begin{pmatrix} a_{11}^{\sigma_1} & \ldots & a_{1n}^{\sigma_1} \\ \vdots & & \vdots \\ a_{n1}^{\sigma_n} & \ldots & a_{nn}^{\sigma_n} \end{pmatrix}. \tag{6.1}$$

Because of (6.1) the mixed discriminants are symmetric and polyadditive functions of its arguments and possess a series of other interesting properties (see also (Bapat, 1987)). For us the most important is the following

LEMMA 6.1 (Aleksandrov inequalities for mixed discriminants, 1937–1938).

(a) *Let* $(f_i)$, $i = 1, \ldots, n-1$, *be* $n-1$ *positive definite quadratic forms and let* $\mathfrak{g}$ *be an arbitrary quadratic form. Then,*

$$D^2(\alpha;\ f_{n-1}, \mathfrak{g}) \geqslant D(\alpha;\ f_{n-1}, f_{n-1}) D(\alpha; \mathfrak{g}, \mathfrak{g}), \tag{6.2}$$

*where* $\alpha := (f_1, \ldots, f_{n-2})$.

(b) *Equality holds in* (6.2) *iff*

$$\mathfrak{g} = \lambda f_{n-1}, \quad \lambda \text{ a constant}. \tag{6.3}$$

*Taking suitable limits in* (6.2) *of the coefficients in the forms, assertion* (a) *of Lemma* 6.1 *carries over to the case in which the* $(f_i)$, $i = 1, \ldots, n-1$, *are non-negative definite forms.*

LEMMA 6.2 (Geometric inequalities for permanents, Egorychev, 1980, 1981).

(a) *Let* $(f_i)$, $i = 1, \ldots, n-1$, *be* $n$*-vectors with non-negative components and let* $\mathfrak{g}$ *be an arbitrary* $n$*-vector. Then,*

$$\operatorname{per}^2(\alpha; f_{n-1}, \mathfrak{g}) \geqslant \operatorname{per}(\alpha; f_{n-1}, f_{n-1})\operatorname{per}(\alpha; \mathfrak{g}, \mathfrak{g}). \tag{6.4}$$

(b) *Let* $(f_i)$, $i = 1, \ldots, n-1$, *be* $n$*-vectors with positive components and let* $\mathfrak{g}$ *be an arbitrary* $n$*-vector. Equality holds in* (6.4) *iff*

$$\mathfrak{g} = \lambda f_{n-1}, \quad \lambda \text{ a constant}. \tag{6.5}$$

The results of Lemma 6.2 follow directly from formulas (6.2), (6.3) if we note that

$$\operatorname{per} A = n! D(f_1, \ldots, f_n), \tag{6.6}$$

where $f_i$ is the quadratic form with matrix $A_i = \operatorname{diag}(a_{1i}, \ldots, a_{ni})$, $i = 1, \ldots, n$. Indeed,

$$f = \sum_{k=1}^{n} \lambda_k f_k = \sum_{i=1}^{n} (\lambda_1 a_{i1} + \cdots + \lambda_n a_{in}) x_i^2$$

and

$$\det A = \det\left(\operatorname{diag}\left(\sum_{j=1}^{n} \lambda_j a_{ij}, \ldots, \sum_{j=1}^{n} \lambda_j a_{nj}\right)\right) = \prod_{i=1}^{n}\left(\sum_{j=1}^{n} \lambda_j a_{ij}\right),$$

and it is easy to see that (cf. (6.1)) the coefficient at $\lambda_1, \ldots, \lambda_n$ in the last expression is equal to $\operatorname{per}(A)/n!$.

D. Falikman (1981) obtained an inductive proof of permanent inequalities equivalent to inequalities (6.4).

Papers by A. Panov (1984, 1985) and R. Bapat (1989) investigated the case of equality in (6.2) for non-negative definite forms and D. Knuth (1981) corrected the case of equality in (6.4) by assumptions weaker than in (6.5). R. Bapat also obtained the combinatorial interpretation of mixed discriminants from $(0, 1)$-matrices, and proved a generalization of the well-known König theorem.

Let $\alpha$ be a finite-dimensional linear space over $\mathbb{R}$ and $\varphi$ be a symmetrical bilinear form on $\alpha$. If $\varphi$ has one positive eigenvalue and $n-1$ negative eigenvalues we shall speak of a *Minkovsky* (*Lorentz*) *space*.

ASSERTION 6.3 (van Lint, 1981; Rybnikov, 1985). *If the vector-columns* $a_1, \ldots, a_{n-2}$ *have positive components, then, the quadratic form* $\varphi(x, x) = \operatorname{per}(a_1, \ldots, a_{n-2}, x, x)$, $x \in \mathbb{R}^n$, *has the signature of Minkovsky space. Under the same assumptions the inequalities* (6.4), (6.5) *turn into the Cauchy–Bounjakowsky–Schwarz inequalities in Minkovsky space.*

In conclusion of this section we'll give the representation of the permanent as a mixed volume yielding a family of geometric inequalities for the permanent as a mixed volume.

Let $A+B=\{a+b\colon a\in A,\ b\in B\}$ denote the vector sum (*Minkovsky sum*) of the subsets $A$ and $B$ of Euclidean space $\mathbb{R}^n$, and let $\lambda A=\{\lambda a\colon a\in A\}$ is the result of the homothety of $A$ with coefficients $\lambda$.

THEOREM 6.4 (Minkovsky, 1911). *The volume of the linear combination of nonempty convex compact sets $K_1,\dots,K_s$ ($s\neq n$, in general) with non-negative coefficients $\lambda_1,\dots,\lambda_s$ is a homogeneous polynomial of degree $n$ with respect to $\lambda_1,\dots,\lambda_s$:*

$$V\left(\sum_{i=1}^{n}\lambda_i K_i\right)=\sum_{i_1=1}^{s}\cdots\sum_{i_n=1}^{s}(K_{i_1},\dots,K_{i_n})\lambda_{i_1}\dots\lambda_{i_n}, \tag{6.7}$$

*where it is assumed that for the products of $\lambda_i$ which differ only in the order of the factors the coefficients have the same numerical value. The coefficients $V(K_1,\dots,K_n)$ in the expansion* (6.7) *are called the mixed volumes of convex compact sets $K_1,\dots,K_n$ in $\mathbb{R}^n$.*

Let $A$ be a non-negative matrix of order $n$ and let $K_i,\ i\in 1,\dots,n$, be the family of rectangular parallelopipeds in $\mathbb{R}^n$ induced by it

$$K_i=\{x=(x_1,\dots,x_n)\in\mathbb{R}^n,\ 0\leqslant x_j\leqslant a_{ij},\ j\in 1,\dots,n\}. \tag{6.8}$$

In analogy with the representation (6.6) of the permanent as a mixed discriminant it is easy to see that (Egorychev, 1980–1981)

$$\operatorname{per}(A)=n!V(K_1,\dots,K_n). \tag{6.9}$$

Indeed, the volume of a rectangular parallelopiped

$$K=\sum_{i=1}^{n}\lambda_i K_i=\left\{x=(x_1,\dots,x_n)\in\mathbb{R}^n,\ 0\leqslant x_j\leqslant\sum_{i=1}^{n}\lambda_i a_{ij}\right\}$$

is equal to the product of the lengths of its sides

$$V(K)=\prod_{j=1}^{n}\left(\sum_{j=1}^{n}\lambda_i a_{ij}\right),$$

and the coefficient by $\lambda_1\cdots\lambda_n$ is equal to $\operatorname{per}(A)/n!$.

Formula (6.9) allows one to obtain the following assertion for permanents of a non-negative matrix.

THEOREM 6.5 (cf. Egorychev, 1983). *If $A$ is a non-negative matrix, then, for* $\operatorname{per}(A)$ *there hold the analogs of the Brunn–Minkovsky, Aleksandrov–Fenchel, Shephard, Santalo and other inequalities, including the vector inequalities for mixed volumes in $\mathbb{R}^n$, as well as the corresponding analogs of the results for the various cases of equality* (see Burago and Zalgaller, 1984, Ch. 4).

The papers of G. Ewald (1985), B. Kind and P. Kleinschmidt (1979), P. McMullen and G. Shephard (1971), M. Hochster (1972), R. Stanley (1980, 1981), L. Billera and C. Lee (1981), A. Kouchnirenko (1976), D. Bernstein (1976), B. Teissier (1979, 1982) contain several combinatorial connections of mixed volumes with combinatorial geometry which by virtue of (6.9) are valid for the permanents of non-negative matrices as well.

## 7. Structure of the minimizing matrix: uniqueness of the solution and proof of the conjectures

For the readers' convenience we repeat the assertions used to prove the conjectures.

LEMMA 7.1 (Egorychev, 1980, 1981 – geometrical inequalities for permanents).

(a) *Let* $a_i$, $i = 1, \ldots, n-1$, *be a set of* $n-1$ $n$*-vectors with non-negative components and let* $\mathfrak{g}$ *be an arbitrary* $n$*-vector. Then,*

$$\operatorname{per}^2(\alpha; a_{n-1}, \mathfrak{g}) \geqslant \operatorname{per}(\alpha; a_{n-1}, a_{n-1})\operatorname{per}(\alpha; \mathfrak{g}, \mathfrak{g}), \tag{7.1}$$

*where* $\alpha := (a_1, \ldots, a_{n-2})$.

(b) *Let* $(a_i), i = 1, \ldots, n-1$, *be* $n$*-vectors with positive components and let* $\mathfrak{g}$ *be an arbitrary* $n$*-vector. Equality holds in* (7.1) *iff*

$$\mathfrak{g} = \lambda a_{n-1}, \quad \lambda \text{ a constant.} \tag{7.2}$$

THEOREM 7.2 (König, 1916). *If* $A \in \Omega_n$, *then,* $\operatorname{per}(A) > 0$.

ASSERTION 7.3 (London, 1971). *If* $A$ *is a minimizing matrix, then*

$$\operatorname{per}\bigl(A(i/j)\bigr) \geqslant \operatorname{per} A \quad \text{for all } i, j = 1, \ldots, n. \tag{7.3}$$

THEOREM 7.4 (Egorychev, 1980). (Proof of the Marcus–Newman conjecture on permanents (1965).) *If* $A \in \Omega_n$ *and the inequalities*

$$\operatorname{per}\bigl(A(i/j)\bigr) \geqslant \operatorname{per} A \quad \text{for all } i, j \in \{1, \ldots, n\}, \tag{7.4}$$

*are valid, then,*

$$\operatorname{per}\bigl(A(i/j)\bigr) = \operatorname{per} A \quad \text{for all } i, j \in \{1, \ldots, n\}. \tag{7.5}$$

THEOREM 7.5 (Egorychev, 1980). (Proof of the van der Waerden conjecture on permanents (1926).)

$$\min_{X \in \Omega_n} \bigl(\operatorname{per}(X)\bigr) = n!/n^n, \tag{7.6}$$

$$X \in \Omega_n \quad \text{and} \quad \operatorname{per}(X) = n!/n^n \quad \text{iff} \quad X = J_n. \tag{7.7}$$

PROOF OF THEOREM 7.4. Expanding

$$\operatorname{per}\left(A\begin{pmatrix} a & j \\ a_i & a_i \end{pmatrix}\right) \quad \text{and} \quad \operatorname{per}\left(A\begin{pmatrix} i & j \\ a_j & a_j \end{pmatrix}\right)$$

by the Laplace formula, over the $j$-th and $i$-th columns, respectively, and using (7.1) we get

$$\begin{aligned}\operatorname{per}^2 A &= \operatorname{per}^2\left(A\begin{pmatrix} i & j \\ a_i & a_j\end{pmatrix}\right) \geqslant \operatorname{per}\left(A\begin{pmatrix} i & j \\ a_i & a_i\end{pmatrix}\right)\operatorname{per}\left(A\begin{pmatrix} i & j \\ a_j & a_j\end{pmatrix}\right)\\ &= \Big(\sum_k a_{ki}\operatorname{per}\big(A(k/j)\big)\Big)\Big(\sum_k a_{kj}\operatorname{per}\big(A(k/i)\big)\Big)\\ &\quad \text{for all } i,j \in \{1,\ldots,n\}. \end{aligned} \tag{7.8}$$

Inequalities (7.8) combined with assumption (7.4) give a system of $n^2$ inequalities for the $n^2$ numbers $\operatorname{per}(A(i/j))$, $i,j \in \{1,\ldots,n\}$.

We show that in this case $\operatorname{per}(A(i/j)) = \operatorname{per} A$, $i,j \in \{1,\ldots,n\}$. Assume the opposite, i.e. that there exists a pair $r,s \in \{1,\ldots,n\}$ such that $\operatorname{per}(A(r/s)) > \operatorname{per} A$. Since $A \in \Omega_n$ there is some $t \in \{1,\ldots,n\}$ such that $a_{rt} > 0$. Then, by (7.8),

$$\begin{aligned}\operatorname{per}^2 A &= \operatorname{per}^2\left(A\begin{pmatrix} s & t \\ a_s & a_t\end{pmatrix}\right)\\ &\geqslant \Big(\sum_k a_{ks}\operatorname{per}\big(A(k/t)\big)\Big)\Big(\sum_k a_{kt}\operatorname{per}\big(A(k/s)\big)\Big)\\ &= \Big(\sum_k a_{ks}\operatorname{per}\big(A(k/t)\big)\Big)\Big(a_{rt}\operatorname{per}\big(A(r/s)\big) + \sum_{k\neq r} a_{kt}\operatorname{per}\big(A(k/s)\big)\Big)\\ &> \Big(\sum_k a_{ks}\operatorname{per}(A)\Big)\Big(\sum_k a_{kt}\operatorname{per}(A)\Big) = \operatorname{per}^2 A,\end{aligned}$$

where the strict inequality follows from inequalities (7.4), $a_{rt} > 0$, $\operatorname{per}(A(r/s)) > \operatorname{per}(A)$, $\operatorname{per} A > 0$ (Theorem 7.2), and $A \in \Omega_n$. The contradiction obtained proves the theorem. □

Let $A$ be a minimizing matrix. Then, by London's result, the inequalities (7.4) hold. From Theorem (7.4) follows

LEMMA 7.6 (Egorychev, 1980). *If $A$ is a minimizing matrix, then*

$$\operatorname{per}\big(A(i/j)\big) = \operatorname{per} A, \quad \textit{for all } i,j \in \{1,\ldots,n\}. \tag{7.9}$$

LEMMA 7.7 (Egorychev, 1980). *If $A$ is a minimizing matrix, then for each $i,j \in \{1,\ldots,n\}$*

$$A_\theta = A\begin{pmatrix} i & j \\ \theta a_i + (1-\theta)a_j & (1-\theta)a_i + \theta a_j\end{pmatrix}, \quad 0 \leqslant \theta \leqslant 1,$$

*will be also minimizing matrix. $A_\theta$ is obtained from the matrix $A$ with the help of a "$\theta$-transform" of the $i$-th and $j$-th columns of the matrix $A$.*

PROOF. It is easy to see that $A' \in \Omega_n$. The equality $\mathrm{per}(A_\theta) = \mathrm{per}(A)$ follows immediately from the fact that $\mathrm{per}(A)$ is a multilinear symmetric function of the columns of the matrix $A$, the Laplace formulas, the equalities (7.9), and $A \in \Omega_n$. □

PROOF OF THEOREM 7.5. It is clear that a minimizing matrix cannot be (up to permutation of rows and columns) a matrix of the form

$$A = \begin{pmatrix} 1 & 0 \\ 0 & A_{n-1} \end{pmatrix},$$

where $A_{n-1} \in \Omega_{n-1}$, since in this case (Lemma 7.6) $\mathrm{per}(A(1/1)) = \mathrm{per}(A(1/i)) = 0$ for all $i \in \{2, \ldots, n\}$, i.e. $\mathrm{per}\, A = 0$, a contradiction with Theorem 7.2.

Now we show that if $A$ is a minimizing matrix, then $A = J_n$. Let some column of the matrix $A$, say $a_n$, be different from the $n$-column $e^T/n$. Then, it is easy to see that by Lemma 7.7 for all $i, j \in \{1, \ldots, n-1\}$, $i \neq j$, we can obtain in a finite number of steps a minimizing matrix $B = (b_1, \ldots, b_{n-1}, a_n)$, in which every component of the first $n-1$ columns is positive. This follows from the fact that $A$ does not contain a 1. From the inequalities (7.1), we have

$$\mathrm{per}^2 \left( B \begin{pmatrix} i & n \\ b_i & a_n \end{pmatrix} \right) \geqslant \mathrm{per} \left( B \begin{pmatrix} i & n \\ b_i & b_i \end{pmatrix} \right) \mathrm{per} \left( B \begin{pmatrix} i & n \\ a_n & a_n \end{pmatrix} \right). \tag{7.10}$$

By virtue of equalities (7.9) the minimizing matrix $B$ (Lemmas 7.6 and 7.7) we obtain equality in (7.10). The positivity of the components $b_1, \ldots, b_{n-1}$ allows us to assert immediately (Lemma 7.1, the case of equality) that $a_n = \lambda_i b_i$ for all $i \in \{1, \ldots, n-1\}$. Since the sum of the components of the vector $b_i$ as well as of the vector $a_n$ is equal to 1, we have $a_n = b_i$ for all $i \in \{1, \ldots, n-1\}$. Since $B \in \Omega_n$, we have

$$b_1 = b_2 = \cdots = b_{n-1} = a_n = e^T/n,$$

a contradiction. Thus, $A = J_n$ and the proof of the theorem is complete. □

REMARK 7.8. The reduction from Theorem 7.4 to Theorem 7.5 was well known (see, for example, Minc (1978, p. 101, Problem 18)). Here we gave a simple geometric proof of this reduction.

The proof of conjectures given here, from the necessary conditions to geometric inequalities (7.1), (7.2), is typical for the theory of mixed volumes in the solution of isoperimetric problems. However, specialists usually use such mixed discriminants (volumes) where only two, sometimes three, forms (convex compacts) are different: in our proof all forms are used equally.

## 8. Direct corollaries

As a direct corollary of Theorem 7.5 we obtain the validity of some facts for permanents (see Conjectures 2, 8, 16 and Problem 9 in the list of conjectures and problems in

(Minc, 1978); also a generalization of the van der Waerden conjecture in (Gleason, 1970)). Another direct corollary of Theorem 7.5 is that we obtain lower estimates for some important combinatorial quantities previously expressed by other authors assuming the validity of van der Waerden conjecture. These quantities admit representations in terms of permanents of block $(0,1)$-matrices from $\Lambda_n^k$. Among them are the lower bounds for the number of Latin rectangles and squares (see also (Erdős and Kaplansky, 1946; Yamamoto, 1951, 1956; Gessel 1987; Denes and Keedwell, 1991) etc.) for the number of nonisomorphic Steiner triples (Wilson, 1974) and for the key constant $\lambda_d$ in the $d$-dimensional dimer problem (Hammersley, 1968, 1969; Dubois, 1974; Minc, 1978a, 1980). These bounds are essential improvements of previously known bounds.

The results mentioned above brought about a structural reorganization of sections of combinatorial theory connected with permanents. We should also note that the problem of estimating $\lambda_d$ belongs to an extensive class of mathematical and physical problems connected with finding the number of dimer coverings of the lattice that can be realized by the permanent of special $(0,1)$-matrices (see, for example, the surveys (Percus, 1971) and (Montroll, 1964) about applications of permanents in statistical mechanics and the two-dimensional Ising model in ferromagnetism).

## 9. Further results and new hypotheses

One of the reasons for the interest in computing per$(A)$ is that a $(0,1)$-matrix $A = (a_{ij})$ can be viewed as an adjacency matrix of a bipartite graph, $H = (X, Y, E)$, where $X$ corresponds to the rows in $A, Y$ to the columns in $A$, and $a_{ij} = 1$ if there is an edge between $X_i$ and $Y_j$. The value per$(A)$ is exactly the number of perfect matchings (1-factors) in $H$. This matter finds numerical applications in operations research. A. Schrijver (1982, 1983) published a survey on recent developments of lower and upper bounds for permanents including his interpretation of the van der Waerden conjecture and the well-known Brégman–Minc upper bound (Brégman, 1973; Minc, 1978, Ch. 6). He applied these results to obtain a series of new and hypothetical bounds for the number of perfect matchings, 1-factorizations (edge-colorings), bipartite graphs, and the number of *Eulerian orientations of graphs*. The reader can find other applications of the permanent of (0,1)-matrices in estimating characteristics of various types of graphs in numerous papers (see, for example, (Caianiello, 1953, 1956; Harary, 1969; Hartfiel and Spellman, 1972; Dubois, 1974)).

D. London (1982) used the result of Theorem 7.6 to characterize real zeros of a polynomial defined by the composition of two polynomials and T. Ando (1989) used it to analyze majorization problems.

The van der Waerden conjecture, proved in the early 80s, generated numerous papers (van Lint, 1981, 1982, 1983; Janoś, 1977; Lagaris, 1982; Minc, 1982, 1983; Schrijver, 1982, 1983; Friedland, 1982; Bapat, 1984, 1986; Rybnikov, 1985; London and Minc, 1989) etc., discussing the history of the problem and giving improved and modified versions of the proofs of Egorychev (1980) and Falikman (1981).

The main result and the method used in the work by Egorychev (1980) were essentially used and developed to solve the problem of the permanent minimum at various faces of $\Omega_n$ (see the extensive survey of the current state of the problem in (Seok-Zun Song, 1988, and Minc, 1987; also Knopp and Sinkhorn, 1982; Chang, 1984a; Minc, 1984; Brualdi, 1985; Hwang, 1985; Foregger and Sinkhorn, 1986; Foregger, 1987; London and Minc, 1989) etc.) An elegant conjecture for which the method proposed there didn't work was stated by D. London and H. Minc (1989). Let $\Omega_n^0$ denote a set of $n \times n$ doubly stochastic matrices with zero main diagonal; $J_n^0 = (1 - \sigma_{ij})/(n-1) \in \Omega_n^0$, where $\sigma_{ij}$ is Kronecker symbol. If $A \in \Omega_n^0$, then

$$\operatorname{per}(A) \gg \operatorname{per}(J_n^0) = \frac{n!}{(n-1)^n}\left(1 + \sum_{k=1}^{n}(-1)^k/k!\right). \tag{9.1}$$

What's more, equality in (9.1) holds only for $A = J_n^0$.

Generalizing the previously obtained results, S. Hwang (1985) introduced for a $(0,1)$-matrix $A = (a_{ij})$ of order $n$ the concept of the barycenter matrix $\tilde{A} := (a_{ij}\operatorname{per}(A(i/j)))/\operatorname{per}(A) \in \Omega_n$ and the staircase matrix, and A. Brualdi (1985) stated the problem of the characterization of these classes of matrices and formulated a question of the characterization of faces at $\Omega_n$ where the permanent minimum is achieved on the barycenter matrix (see also (Minc, 1987; Hwang, 1985)).

Finally, the main result and the methods of G. Egorychev (1980), including applications of the geometric inequalities (7.1), (7.2), were efficiently used in numerous articles to prove extremal conjectures for the permanent and other matrix functions on $\Omega_n$ and for other classes of matrices. Most successful in this direction was S. Friedland (1982) proving the Tverberg conjecture (1963) about the minimum of the function $\sigma_k(A)$, $1 \leqslant k \leqslant n$, equal to the sum of the permanents for all $k \times k$ submatrices of a matrix $A$. That article efficiently used a representation of function $\sigma_k(A)$ by the permanent on $\Omega_n$ faces.

The next nontrivial generalization of the van der Waerden conjecture was proposed by E. Dittert (Minc, 1983, Conjecture 28): let $K_n$ be the set of non-negative $n \times n$ matrices the sum of whose entries is $n$. Then,

$$\max\left\{\prod_{i=1}^{n} c_i + \prod_{j=1}^{n} r_j - \operatorname{per}(A) \;\middle|\; A \in K_n\right\} = 2 - n!/n^n, \tag{9.2}$$

where $c_i$ and $r_j$ denote the $i$-th row and the $j$-th column sums of the matrix $A$, respectively. Equality holds in (9.2) iff $A = J_n$. Partial progress in the proof of this conjecture was achieved as the basis of Theorem 7.5 and inequalities (7.1), (7.2) (Sinkhorn, 1984; Hwang, 1986, 1986a, 1987, 1989, 1990; Egorychev, 1994).

Many elegant problems and conjectures for permanents and related matrix functions on various classes of matrices were published recently (Minc, 1978, 1993; Marcus, Minc, 1965; Grone and Merris, 1987; Minc, 1978; book "Permanents: theory and applications", Krasnoyarsk, 1990; Egorychev, 1994). Among them is an explicit representation of a "problem of the century" and a dominant conjecture for Schur functions ("dominance conjecture"). The problems and conjectures for permanents have as a rule a simple formulation but often are very difficult to solve, as is often the case with problems of discrete mathematics. The difficulty in solving them lies not only and not so much in enumerating a lot of variants but is inherent in combinatorial analysis as a whole. The practice of solving them shows that some conjectures appearing almost obvious from their formulation turn out to be invalid. Practice shows also that to solve problems and conjectures it is often necessary to use the apparatus of a series of fundamental models and methods from various sections of modern mathematics.

## 10. Prospects for investigation and conclusion

The analysis of the conjectures and problems for permanents allows us to conclude that research in this field holds much promise in the following interconnected directions:

– to study inequalities for permanents and mixed discriminants, including the case of equality for (non-negative, symmetrical) matrices of arbitrary signature. These questions arise in the analysis of necessary conditions in problems of extremality for matrix combinatorial functions, and turn out to be immediately connected to various definitions of convexity as applied to quadratic functions in $\mathbb{R}^n$ (see Ponstein, 1967; Cottle and Ferland, 1971, 1972; Martos, 1969, 1971; Ferland, 1980). The latter direction emerged, in its turn, from consideration of the problems of quadratic programming and approximation theory (Micchelli, 1986) and finds application in probability theory (Bapat, 1988, 1989a). Making use of several characterizations of the class of subpositive definite quadratic forms (see, for example, (Ferland, 1980)), Bapat (1984) gave some improvement of the result of Theorem 7.5. This research belongs to one of the more promising lines of the matrix combinatorial analysis which has seen rapid progress in recent years (see Maybee, 1988; Johnson, 1988; Brualdi, 1990 etc.);

– to pass to space matrices and the consideration of matrix functions. Analysis of certain problems for permanents shows them to appear more natural when viewed in terms of a problem for a space matrix permanent. This isn't surprising if for no other reason than the representation (6.6) of the permanent as a mixed discriminant from matrices $A_1, \ldots, A_n$ can be considered as a representation of a space matrix with sections $A_1, \ldots, A_n$. A. Gasparyan (1984) and A. Hovansky (1984) found a series of new inequalities for hyperbolic forms (polynomials) for space matrices, while B. Bapat (1986) wrote down analogs of the inequalities (7.1) for "mixed Schur functions";

– it was Gilbert who noted that the relations of the mixed volumes with other fields of mathematics have been settled completely. Aleksandrov (1937–1938) already noted that the coefficients of the characteristic polynomial of a matrix $A$ are mixed discriminants of it and the unit matrix. However, this relationship remained unexploited in research in linear and polylinear algebra. G. Egorychev (1990, 1993, 1994a) develops matrix analysis on numerical fields with nontrivial operations. Introduced and studied are notions and properties for a certain class of $\varphi$-mixed discriminants, polyadditive relative to the operations introduced (for numbers and matrices); there are in particular relations with the operation of parallel summation of matrices and the problem of ordering non-negative definite matrices (Meenakshi, 1987; Anderson and Duffin, 1969; Mitra, 1991);

– to study definitions, properties, and inequalities of permanents and mixed discriminants of matrices over different (partially ordered) algebraic systems. Permanents of a distributive lattice emerged in the work of Skornyakov and Egorova (1984); for permanents with logical variables, see (Razborov, 1985); permanents over finite fields emerged in investigations of the four-color problem (see Shor, 1990); permanents over (partially) ordered fields turned up in the work of Golovanov, Egorychev and Moiseenko (1987). Still more permanents arise in studies of many other authors in the most varied fields of mathematics. Y. Egorychev and Ja. Nuzhin, for example (in press), gave a definition and studied the properties of permanents over noncommutative rings. Studies of $\varphi$-mixed discriminants mentioned in the previous paragraph also allow one, in the opinion of the author, to touch upon different aspects of the theory of matrix functions over different algebraic systems presented in many articles of Vol. 1 of this series "Handbook of Algebra". They also reveal relations with the classical spectral theory of matrices, graph theory and electrical engineering (see, for example, (Egorychev and Moiseenko, 1990a; the survey by Tsvetkovich, 1984, and Strok, 1990, on spectral theory of matrices; Ando and Kubo, 1989, 1990));

– and, finally, of promise for the development of the geometric theory of mixed discriminants (and permanents) is the research making use of their representation as the valuation function in valuation rings (see the survey (MacMillan and Schneider, 1983; Rota, 1973; Barnabei, Brini and Rota, 1986)). This research based on fundamental studies of geometric nature going on for the last two centuries can, in my opinion, essentially enrich the apparatus of enumerative combinatorial analysis as a whole.

The author is grateful to the participants of the seminar on discrete mathematics in Krasnoyarsk for useful remarks concerning this chapter.

## References


Alexandrov, A.D. (1937–1938). *To the theory of mixed volumes of convex bodies*, I–IV. Mat. Sb. **2**, 947–972, 1205–1238; **3**, 27–46, 227–251 (in Russian).

Anderson, W.N. and R.J. Duffin (1969). *Series and parallel addition matrices*, J. Math. Anal. Appl. **26**, 576–594.

Ando, T. (1983). *An inequality between symmetric function means of positive operators*, Acta Sci. Math. **45**, 19–22.

Ando, T. (1989). *Majorization, doubly stochastic matrices, and comparison of eigenvalues*, Linear Algebra Appl. **118**, 163–248.

Ando, T. and F. Kubo (1989). *Some matrix inequalities in multiport network connections*, Oper. Theory Adv. Appl. **40**, 111–131.

Ando, T. and F. Kubo (1990). *Inequalities among operator symmetric function means*, Proc. of Int. Symp. MTNS-89 vol. III, 535–542.

Andreev, A.E. (1985). *About one method of obtaining the lower estimates complexity of individual monotone functions*, Dokl. Akad. Nauk SSSR **282**, 1033–1037 (in Russian).

Bang, T. (1976). *On matrixfunctioner som med et numerisk little dificit viser v.d. Waerdens permanent hypothese*, Proc. Scand. Congr. Turky, 1979: *On matrix functions giving a good approximation to the Van der Waerden permanent conjecture*, Preprint 30, Math. Inst., Kobenhavn.

Bapat, R.B. (1983). *Doubly stochastic matrices with equal subpermanents*, Linear Algebra Appl. **51**, 1–8.

Bapat, R.B. (1984). *A stronger form of the Egorychev–Falikman theorem on permanents*, Linear Algebra Appl. **63**, 95–100.

Bapat, R.B. (1986). *Inequalities for mixed Schur functions*, Linear Algebra Appl. **83**, 143–149.

Bapat, R.B. (1988). *The multinomial distribution and permanents*, Linear Algebra Appl. **104**, 201–205.

Bapat, R.B. (1989). *Mixed discriminants of positive semidefinite matrices*, Linear Algebra Appl. **126**, 107–124.

Bapat, R.B. (1989a). *Multinomial probabilities, permanents and a conjecture of Karlin and Rinott*, Manuscript.

Bapat, R.B. and V.S. Sunder (1985). *On majorization and Schur products*, Linear Algebra Appl. **72**, 107–117.

Bapat, R.B. and V.S. Sunder (1986). *An extremal property of the permanent and the determinant*, Linear Algebra Appl. **76**, 153–163.

Barnabei, M., Brini, A. and J.-C. Rota (1986). *Theory of Möbius functions*, Uspekhi Mat. Nauk **41** (3)(249), 113–157 (in Russian).

Bernstein, D.N. (1976). *The number of roots of a system of equations*, Functional Anal. Appl. **9**, 183–185.

Billera, L.J. and C.W. Lee (1981). *A proof of McMullen's conditions for f-vectors of simplicial convex polytopes*, J. Combin. Theory Ser. A **31**, 237–255.

Binet, J.P.M. (1812). *Mémoire sur un systéme de formules analytiques, et leur application à des considérations géometriques*, J. Éc. Polyt., 9. Cah. **16**, 280–302.

Birkhoff, G. (1946). *Tres observaciones sobre el algebra lineal*, Univ. Nac. Tucumán Rev., Ser A. **5**, 147–151.

Borchardt, C.W. (1855). *Bestimmung der symmetrischen Verbindungen vermittelst ihrer erzeugenden Function*, Monatsb. Akad. Wiss. Berlin, 165–171; Crelle's J. **53**, 193–198; Gesammelte Werke, 97–105.

Borodin, A.I. and A.S. Bugai (1979). *Biographical Dictionary of Activities in the Mathematics*, Radjans'ka Schkola, Kiev (in Russian).

Brégman, L.M. (1973). *Certain properties of non-negative matrices and their permanents*, Dokl. Akad. Nauk SSSR **211**, 27–30 (in Russian).

Brualdi, R.A. (1985). *An interesting face of the polytope of doubly stochastic matrices*, Linear and Multilinear Algebra **17**, 5–18.

Brualdi, R.A. and P.M. Gibson (1977). *The convex polyhedron of doubly stochastic matrices, I. Applications of the permanent function*, J. Combin. Theory Ser. A **22**, 194–230.

Brualdi, R.A. (1990). *The many facets of combinatorial matrix theory*, Proc. Symp. Appl. Math. **40**, 1–35.

Burago, Ju.D. and V.A. Zalgaller (1988). *Geometric Inequalities*, Springer, Berlin (Russian original: Nauka, Leningrad 1980).

Busemann, H. (1958). *Convex Surfaces*. New York, Interscience.

Caianiello, E.R. (1953). *On quantum field theory. I. Explicit solution of Dyson's equation in electrodynamics without use of Fayman graphs*, Nuovo Cimento **10** (9), 1634–1652.

Caianiello, E.R. (1956). Properietá pfaffiani e hafniani, Ricerca, Napoli **7**, 25–31.

Cauchy, A.L. (1812). *Mémoire sur les fonctions qui ne peuvent obtenir que deux valeurs égales et de signes contraires par suite des transpositions operées entre les variables qu'elles renferment*, J. Éc. Polyt., 10. Cah. **17**, 29–112. Oeuvres (2)i.

Chan, N.N. and M.K. Kwong (1985). *Hermitian matrix inequalities and a conjecture*, Amer. Math. Monthly **92**, 533–541.

Chang, D.K. (1983). *Notes on permanents of doubly stochastic matrices*, Linear and Multilinear Algebra **14**, 349–356.

Chang, D.K. (1984). *Minimum permanents of doubly stochastic matrices with one fixed entry*, Linear and Multilinear Algebra **15**, 313–317.

Chang, D.K. (1984a.) *A note on a conjecture of T.H. Foregger*, Linear and Multilinear Algebra **15**, 341–344.

Chang, D.K. (1986). *Permanents of doubly stochastic matrices*, Discrete Math. **62**, 211–213.

Chollet, J. (1982). *Is there a permanental analogue to Oppenheim's inequality?*, Amer. Math. Monthly **89**, 57–58.

Cook, D.J. and H.E. Bez (1984). *Computer Mathematics*, Cambridge Univ. Press, Cambridge.

Cottle, R.W. and J.A. Ferland (1971). *On pseudo-convex functions of non-negative variables*, Math. Programming **1**, 95–101.

Cottle, R.W. and J.A. Ferland (1972). *Matrix-theoretic criteria for the quasi-convexity and pseudo-convexity of quadratic functions*, Linear Algebra Appl. **5**, 123–136.

Dénes, J. and A.D. Keedwell (1991). *Latin Squares and Their Applications*, Publ. House Hungarian Acad. Sciences, Budapest.

Docović, D.Z. (1967). *On a conjecture by van der Waerden*, Mat. Vesnik **4**, 272–276.

Donets, G.A. and N.Z. Shor (1982). *Algebraic Approach to the Problem of Planar Graphs Coloring*, Naukova Dumka, Kiev (in Russian).

Dubois, J. (1974). *A note on the van der Waerden permanent conjecture*, Canad. J. Math. **6**, 352–354.

Eberlein, P.J. (1969). *Remarks on the van der Waerden conjecture, II*, Linear Algebra Appl. **2**, 311–320.

Eberlein, P.J. and G.S. Mudholkar (1968). *Some remarks on the van der Waerden conjecture*, J. Combin. Theory **5**, 386–396.

Egorychev, G.P. (1977). *Integral Representation and the Computation of Combinatorial Sums*, Nauka, Novosibirsk (in Russian). English transl. Transl. Math. Monographs vol. 59, Amer. Math. Soc., Providence, RI, 1984; 2nd ed. in 1989.

Egorychev, G.P. (1979). *Polynomial identity for the permanent*, Math. Zametki. **26**, 961–964 (in Russian).

Egorychev, G.P. (1979a). *Family of identities for the permanent*, Dokl. Akad. Nauk Armenian SSR **69**, 3–7 (in Russian).

Egorychev, G.P. (1980). *The solution of problem Van der Waerden for permanents*, Preprint, Inst. Fiziki Sibirsk. Otd. Akad. Nauk SSSR N 13M, Krasnoyarsk (in Russian). English transl.: Adv. Math. **42**, 299–305.

Egorychev, G.P. (1980a). *New formulas for the permanent*, Dokl. Akad. Nauk SSSR **254**, 784–787 (in Russian).

Egorychev, G.P. (1981). *Proof of van der Waerden conjecture for permanents*, Siberian Math. J. **22**, 65–71. Dokl. Akad. Nauk SSSR **258**, 1041–1044 (in Russian).

Egorychev, G.P. (1982). *History and proof of van der Waerden conjecture about permanents*, Math. Today, Vyschtscha Shkola, Kiev, 44–75 (in Russian).

Egorychev, G.P. (1990). *Mixed discriminants and parallel addition*, Dokl. Akad. Nauk SSSR, **312**, 528–531 (in Russian). English transl.: Soviet Math. Dokl. **41**, 451–455.

Egorychev, G.P. and D.A. Moiseenko (1990a). *Mixed discriminants in graph theory and electrotechnik*, Permanents: Theory and Applications, Mezhvuz. Sb., Krasn. Polytech. Inst., Krasnoyarsk, 138–157 (in Russian).

Egorychev, G.P. (1993). *$\varphi$-mixed discriminants and extremal problems for the permanent*, Conf. Math. Progr. and Appl., Ural Otd. Russian Akad. Nauk, Sverdlovsk, 45 p. (in Russian).

Egorychev, G.P. (1994). *Problems for the permanents and proof of van der Waerden conjecture*, VINITI, 16.12.94, N 2909-B094, 1–208 (in Russian).

Egorychev, G.P. (1994a). *Mixed discriminants on $\varphi$-addition*, Algebraic Systems, Sb. Computer Center of Sibirsk. Otd. Russian Akad. Nauk, Krasnoyarsk, 1–14 (in Russian).

Egorychev, G.P. and Ja.N. Nuzhin, *Permanents over distinct algebraic systems*, In press.

Egorychev, G.P., ed. (1990). *Permanents: Theory and Applications*, Mezhvuz. Sb., Krasn. Polytech. Inst., Krasnoyarsk.

Erdős, P. and I. Kaplansky (1946). *The asymptotic number of Latin rectangles*, Amer. J. Math. **68**, 230–236.

Erickson, K.E. (1959). *A new operation for analyzing series parallel networks*, IEEE Trans. Theory **CT-6**, 124–126.

Ewald, G. (1985). *Convex bodies and algebraic geometry*, Ann. New York Acad. Sci. **440**, 196–204.

Falikman, D.I. (1981). *Proof of van der Waerden conjecture about permanent of doubly stochastic matrix*, Mat. Zametki **29**, 931–938 (in Russian).

Fenchel, W. (1936). *Inégalités quadratiques entre les volumes mixtes des corps convexes*, C. R. Acad. Sci. Paris **203**, 647–650.

Ferland, J.A. (1980). *Positive subdefinite matrices*, Linear Algebra Appl. **31**, 233–244.

Flor, P. (1966). *Research Problem*, Bull. Amer. Math. Soc. **72**, 30.

Foregger, T.H. (1974). *On facial minimizing matrices for the permanent functions*, Notices Amer. Math. Soc. Ser. A **21**, 434.

Foregger, T.H. (1980). *On the minimum value of the permanent of a nearly decomposable doubly stochastic matrix*, Linear Algebra Appl. **32**, 75–85.

Foregger, T.H. (1987). *Minimum permanents of multiplexes*, Linear Algebra Appl. **87**, 197–211.

Foregger, T.H. and R. Sinkhorn (1986). *On matrices minimizing the permanent on faces of the poyhedrom of the doubly stochastic matrices*, Linear and Multilinear Algebra **19**, 395–397.

Formanek, E. (1991). *The polynomial identities and invariants of $n \times n$ matrices*, Notices Amer. Math. Soc. **38**, 150.

Friedland, S. (1974). *Matrices satisfying the van der Waerden conjecture*, Linear Algebra Appl. **8**, 521–528.

Friedland, S. (1978). *A study of the van der Waerden conjecture and its generalization*, Linear and Multilinear Algebra **6**, 123–143.

Friedland, S. (1979). *A lower bound for the permanent of a doubly stochastic matrix*, Ann. Math. **110**, 167–176.

Friedland, S. (1981). *Convex spectral functions*, Linear and Multilinear Algebra **9**, 299–316.

Friedland, S. (1982). *A proof of generalized van der Waerden conjecture on permanents*, Linear and Multilinear Algebra **11**, 107–120.

Frobenius, G. (1917). *Über zerlegbare Determinanten*, Sber. Preuss. Akad. Wiss. 274–277.

Fröberg, C.-E. (1988). *On a combinatorial problem related to permanents*, BIT **28**, 406–411.

Gantmacher, F.R. (1967). *Matrix Theory*, Nauka, Moscow (in Russian).

Garey, M.R. and D.C. Johnson (1979). *Computers and intractability: a guide to the theory of NP-completeness*, Freeman, San Francisco, California.

Gasparjan, A.S. (1983). *About some applications of multidimensional matrices*, Preprint N 18, Computer Center Sibirsk. Otd. Akad. Nauk SSSR, Moscow (in Russian).

Gasparjan, A.S. (1984). *Inequalities for hyperbolic polynoms*, Dokl. Akad. Nauk SSSR **276**, 1294–1296 (in Russian).

Gessel, I. (1987). *Counting Latin rectangles*, Bull. Amer. Math. Soc. **16**, 79–82.

Gibson, P.M. (1980). *Permanental polytopes of doubly stochastic matrices*, Linear Algebra Appl. **12**, 87–111.

Gleason, A.M. (1970). *Remarks on the van der Waerden permanent conjecture*, J. Combin. Theory **8**, 54–64.

Golovanov, M.I., G.P. Egorychev and D.A. Moiseenko (1987). *Geometric inequalities for permanent as mixed discriminant over algebraic systems*, Preprint 39M, Inst. Fiziki Sibirsk. Otd. Akad. Nauk SSSR (in Russian).

Grone, R. and R. Merris (1987). *Conjectures on permanents*, Linear and Multilinear Algebra **21**, 419–427.

Grone, R. and S. Pierce (1988). *Permanental inequalities for correlation matrices*, SIAM J. Matrix Anal. Appl. **9**, 194–201.

Gruber, P.M. and J.M. Wills, eds (1983). *Convexity and Its Applications*, Birkhäuser, Basel.

Günter, E. (1985). *Convex bodies and algebraic geometry. Discrete geometry and convexity*, Ann. New York Acad. Sci. **440**, 196–204.

Gyires, B. (1976). *On permanent inequalities*, Coll. Math. Soc. János Bolyai vol. 18, Combinatorics.

Hammersley, J.M. (1968). *An improved lower bound for the multidimensional dimer problem*, Proc. Cambridge Philos. Soc. **64**, 455–463.

Hammersley, J.M., A. Feuerverger, A. Izenman and S. Makani (1969). *Negative finding for the three-dimensional dimer problem*, J. Math. Phys. **10**, 443–446.

Harary, F. (1969). *Determinants, permanents and bipartite graphs*, Math. Mag. **42**, 146–148.

Hartfiel, D.J. (1970). *A simplified form for nearly reducible and nearly decomposable matrices*, Proc. Amer. Math. Soc. **24**, 388–393.

Hartfiel, D.J. and J.W. Spellman (1972). *A role for doubly stochastic matrices in graph theory*, Proc. Amer. Math. Soc. **36**, 389–394.

Hochster, M. (1972). *Rings of invariants of tori, Cohen–Macaulay rings generated by monomials, and polytopes*, Ann. Math. **96**, 318–337.

Holens, F. (1964). *Two aspects of doubly stochastic matrices: permutation matrices and the minimum of the permanent function* (*Thesis abstract*), Canad. Math. Bull. **7**, 509–510.

Hovansky, A.G. (1984). *Analogs of Alexandrov–Fenchel inequalities for hyperbolic forms*, Dokl. Akad. Nauk SSSR **276**, 1332–1334 (in Russian).

Hwang, S.G. (1985). *Minimum permanent on faces of stair-case type of the polytope of doubly stochastic matrices*, Linear and Multilinear Algebra **18**, 271–306.

Hwang, S.G. (1986). *A note on a conjecture on permanents*, Linear Algebra Appl. **76**, 31–44.

Hwang, S.G. (1986a). *The monotonicity and the Doković conjectures on permanents of doubly stochastic matrices*, Linear Algebra Appl. **79**, 127–151.

Hwang, S.G. (1987). *On a conjecture of E. Dittert*, Linear Algebra Appl. **95**, 161–169.

Hwang, S.G. (1989). *On the monotonicity of the permanent*, Proc. Amer. Math. Soc. **106**, 59–63.

Hwang, S.G., Sohn Mun-gu and Kim Si-ju. (1990). *The Dittert's function on a set of non-negative matrices*, Inst. J. Math. and Math. Sci. **13**, 709–716.

Janos, P. (1977–1981). *Vegues terfogat es a van der Waerden's sejtes*, Mat. Lapok **29**, 46–60.

Johnston, C.R. (1988). *Combinatorial matrix analysis and overview*, Linear Algebra Appl. **107**, 3–15.

Kamenetsky, A.M. (1990). *Permanents and determinants of grouped matrices. Explicit and recurrent formulas for permanents and determinants of arbitrary circulants. The solution of Lemer conjecture on coefficient calculation of circulant determinant*, Publ. Vsesojuz. Proektno-Izysk. NII Ob'edin. Gidroproekt, Moscow (in Russian).

Kamenetsky, A.M. (1991). *Permanents and determinants of grouped matrices. The proof of Mimc H. and Metropolis N., Stein M.L., Stein P.R. conjectures. Solving of generalized Kaplansky–Riordan problem on calculation the rook polynom* $\sum_{i=0}^{t} a_i P_n^i$ *circulant. Solving of generalized problem about matrimonial pair*, VINITI 19.08.91, N 3487-B91, 1–400 (in Russian).

Karlin, S. and Y. Rinott (1981). *Entropy inequalities for classes of probability distributions, II. The multivariate case*, Adv. Appl. Probab. **13**, 325–351.

Karmarkar, N., R. Kapp, R. Lipton, L. Lovász and M. Luby (1993). *A Monte-Carlo algorithm for estimating the permanent*, SIAM J. Comput. **22**, 284–293.

Kind, B. and P. Kleinschmidt (1979). *Schälbare Cohen–Macaulay-Komplexe und ihre Parametriesierung*, Math. Z. **167**, 173–179.

Knopp, P. and R. Sinkhorn (1982). *Minimum permanents of doubly stochastic matrices with at least one zero entry*, Linear and Multilinear Algebra **11**, 351–355.

Knuth, D. (1981). *A permanent inequality*, Amer. Math. Monthly **88**, 731–740.

Kouchnirenko, A.G. (1976). *Polyèdres de Newton et nombres de Milnor*, Invent. Math. **32**, 1–31.

König, D. (1916). *Über Graphen und ihre Anwendung auf Determinantentheorie und Mengelehre*, Math. Ann. **77**, 453–465.

König, D. (1936). *Theorie der Endlichen und Unendlichen Graphen*, Leipzig, Akademie-Verlag.

Kwong, M.K. (1989). *Some results on matrix monotone functions*, Linear Algebra Appl. **118**, 129–153.

Lagaris, J.C. (1982). *The van der Waerden conjecture: two Soviet solutions*, Notices Amer. Math. Soc. **29**, 130–133.

Leichtweiss, K. (1980). *Konvexe Mengen*, VEB Deutscher Verlag der Wissenschaften, Berlin.

Lieb, E.H. (1966). *Proofs of some conjectures on permanents*, J. Math. Mech. **16**, 127–134.

Lint, van J.H. (1981). *Notices on Egoritsjev's proof of the van der Waerden conjecture*, Linear Algebra Appl. **39**, 1–8.

Lint, van J.H. (1982). *The van der Waerden conjecture: two proofs in one year*, Math. Intelligencer **4**, 72–77.

Lint, van J.H. (1983). *The van der Waerden conjecture*, Ann. Discrete Math. **18**, 575–580.

London, D. (1971). *Some notes on the van der Waerden conjecture*, Linear Algebra Appl. **4**, 155–160.

London, D. (1981). *On the Doković conjecture for matrices of rank two*, Linear and Multilinear Algebra **9**, 317–327.

London, D. (1982). *On the van der Waerden conjecture and zeros of polynomials*, Linear Algebra Appl. **45**, 35–41.

London, D. and H. Minc (1989). *On the permanent of doubly stochastic matrices with zero diagonal*, Linear and Multilinear Algebra **24**, 289–300.

Lutwak, E. (1987). *Rotation means of projections*, Israel J. Math. Sci. **58**, 161–169.

Lutwak, E. (1988). *Intersection bodies and dual mixed volumes*, Adv. Math. **71**, 232–261.

Lutwak, E. (1990). *Centroid bodies and dual mixed volumes*, Proc. London Math. Soc. **60**, 1–27.

Marcus, M. and H. Minc (1964). *A Survey of Matrix Theory and Matrix Inequalities*, Boston.

Marcus, M. and H. Minc (1965). *Permanents*, Amer. Math. Monthly **72**, 577–591.

Marcus, M. and H. Minc (1967). *On a conjecture of B.L. van der Waerden*, Proc. Cambridge Philos. Soc. **63**, 305–309.

Marcus, M. and M. Newman (1959). *On the minimum on the permanent of a doubly stochastic matrix*, Duke Math. J. **26**, 61–72.

Marshall, A.W. and J. Olkin (1968). *Scaling of matrices to achieve specified row and column sums*, Num. Math. **12**, 83–90.

Marshall, A.W. and J. Olkin (1979). *Inequalities: Theory of Majorization and Its Applications*, Academic Press, New York.

Martos, B. (1969). *Subdefinite matrices and quadratic forms*, SIAM J. Appl. Math. **17**, 1215–1223.

Martos, B. (1971). *Quadratic programming with a quasi-convex objective function*, Oper. Res. **19**, 87–97.

Maybee, J.S. (1988). *Some possible new directions for combinatorial matrix analysis*, Linear Algebra Appl. **107**, 23–40.

McMullen, P. and G.C. Shepard (1971). *Convex polytopes and the upper bound conjecture*, Cambridge.

McMullen, P. and R. Schneider (1983). *Valuations on convex bodies*, Convexity and Its Applications, 170–247.

Meenakshi, A.R. (1987). *On the partial ordering of parallel summable matrices*, Houston J. Math. **13**, 255–262.

Merris, R. (1987). *The permanental dominance conjecture*, Current Trends in Matrix Theory, Elsevier, New York, 213–223.

Micchelli, C.A. (1986). *Interpolation of scattered data: distance matrices and conditionally positive definite matrices*, Constr. Approx. **2**, 11–22.

Minc, H. (1975). *Subpermanents of doubly stochastic matrices*, Linear and Multilinear Algebra **3**, 91–94.

Minc, H. (1978). *Permanents*, Encyclopedia of Math. and Appl. vol. 6, Addison-Wesley.

Minc, H. (1978a). *An upper bound for the multidimensional dimer problem*, Math. Proc. Cambridge Phil. Soc. **83**, 461–462.

Minc, H. (1980). *An asymptotic solution of the multidimensional dimer problem*, Linear and Multilinear Algebra **8**, 235–239.

Minc, H. (1982). *A note on Egorycev's proof of the van der Waerden conjecture*, Linear and Multilinear Algebra **111**, 367–371.

Minc, H. (1983). *Theory of permanents 1978–1981*, Linear and Multilinear Algebra **12**, 227–263.

Minc, H. (1984). *Minimum permanents of doubly stochastic matrices with prescribed zero entries*, Linear and Multilinear Algebra **15**, 225–243.

Minc, H. (1987). *Theory of permanents 1982–1985*, Linear and Multilinear Algebra **21**, 109–148.

Minc, H. (1988). *Non-negative Matrices*, Wiley, New York.

Minkowsky, H. (1911). *Theorie der convexen Körper, unsbesondere Bergundung ihres Oberflachenbegriffs*, Ges. Abh. B. 2, Leipzig–Berlin, 131–229.

Mitra, S.K. (1991). *Matrix partial orders through generalized inverses: unified theory*, Linear Algebra Appl. **148**, 237–263.

Mitrinović, D.S., J.E. Pečarić and V. Volenec (1989). *Recent Advances in Geometric Inequalities*, Kluwer Academic Press.

Montroll, E.W. (1964). *Lattice statistics*, Applied Combinatorial Mathematics, Wiley, New York, 9–60.

Panov, A.A. (1984). *About signature and kernel of bilinear form connected with permanent*, Uspekhi Mat. Nauk **39**, 202 (in Russian).

Panov, A.A. (1985). *About mixed discriminants connected with positive semidefinite quadratic forms*, Dokl. Akad. Nauk SSSR **282**, 273–276 (in Russian).

Pate, T.H. (1989). *Permanental dominance and the Soules conjecture for certain right ideals in the group algebra*, Linear and Multilinear Algebra **24**, 135–149.

Percus, P. (1971). *Combinatorial Methods*, Springer, New York.

Pierce, S. (1987). *Permanents of correlation matrices*, Current Trends in Matrix Theory, Elsevier, Amsterdam.

Ponstein, J. (1967). *Seven kinds of convexity*, SIAM Rev. **9**, 115–119.

Rämmer, A. (1990). *On even doubly stochastic matrices with minimal even permanent*, Acta Commentat. Universitatis **878**, 103–114.

Razborov, A.A. (1985). *Lower estimates of monotone complexity of logic permanent*, Mat. Zametki **37**, 887–900 (in Russian).

Reva, V.V. (1981). *To van der Waerden conjecture about bistochastic matrices*, Actual Problems of Computers and Programming, Dnepropetrovsk, 121–123 (in Russian).

Rockafellar, R.T. (1970). *Convex Analysis*, Princeton, NJ.

Rota, G.-C. (1973). *The valuation ring of a distributive lattice*, Proc. Univ. Houston, Lattice Th. Conf. Houston, 574–628.

Rothaus, O.S. (1974). *Study of the permanent conjecture and some of its generalizations*, Israel J. Math. **18**, 75–96.

Rybnikov, K.A. (1985). *Introduction to Combinatorial Analysis*, Moscow Univ. Press, Moscow (in Russian).

Ryser, H.J. (1963). *Combinatorial Mathematics*, Math. Assoc. Amer.

Sasser, D.W. and M.L. Slater (1967). *On the inequality* $\sum x_i y_i \geqslant (1/n) \sum x_i \sum y_i$ *and the van der Waerden conjecture*, J. Combin. Theory **3**, 25–33.

Schevelev, V.S. (1992). *Some questions of permutation theory with restricted positions*, Results of Sci. and Techn. Probability Theory. Theor. Cybernetics, Dep. VINITI **30**, 113–177 (in Russian).

Schneider, R. (1966). *On A.D. Alexandrov's inequalities for mixed discriminants*, J. Math. Mech. **15**, 285–290.

Schor, N.Z. (1990). *Permanents and the problem of planar graphs coloring*, Permanents: Theory and Applications, Mezhvuz. Sb., Krasn. Polytechn. Inst., Krasnoyarsk, 135–138 (in Russian).

Schrijver, A. and W.G. Valiant (1980). *On lower bounds for permanents*, Proc. Kon. Ned. Akad. Wet. **A83** (= Indag. Math. **42**), 425–427.
Schrijver, A. (1982). *On the number of edge colourings of regular bipartite graphs*, Discrete Math. **38**, 297–301.
Schrijver, A. (1983). *Bounds on permanents and the number of 1-factors and 1-factorizations of bipartite graphs*, London Math. Soc. Lect. Note Ser. vol. 82, 107–134.
Schur, I. (1918). *Über endliche Gruppen und Hermitische Formen*, Math. Z. **1**, 184–207.
Sheim, D.E. (1974). *The number of edge 3-colorings of a planar cubic graph as a permanent*, Discrete Math. **8**, 377–382.
Sinkhorn, R. (1984). *A problem related to the van der Waerden permanent theorem*, Linear and Multilinear Algebra **16**, 167–173.
Skornjakov, L.A. and D.P. Egorova (1984). *Normal subgroups of complete linear group degree 3 over distributive structure*, Algebra i Logika **23**, 670–683 (in Russian).
Song, S.-Z. (1988). *Minimum permanents of certain faces of matrices containing an identity submatrix*, Linear Algebra Appl. **108**, 263–280.
Stanley, R. (1980). *The number of faces of a simplicial convex polytope*, Adv. Math. **35**, 236–238.
Stanley, R. (1981). *Two combinatorial applications of the Alexandrov–Fenchel inequalities*, J. Combin. Theory Ser. A **31**, 56–65.
Strok, V.V. (1990). *Permanent polynomials and their application in graph theory*, Permanent: Theory and Applications, Mezhvuz. Sb., Krasn. Polytechn. Inst., Krasnoyarsk (in Russian).
Teissier, B. (1979). *Du Theoreme de l'Index de Hodge aux inègalite's isoperimetriques*, C. R. Acad. Sci. Paris **288**, 287–289.
Teissier, B. (1982). *Bonnesen-type inequalities in algebraic geometry, I. Introduction to the problem*, Proceedings, Seminar of Differential Geometry, Princeton, NJ, 85–105.
Tsvetkovich, D., M. Dub and H. Zahs (1984). *Spectrums of Graphs. Theory and Application*, Kiev (in Russian).
Tverberg, H. (1963). *On the permanent of a bistochastic matrix*, Math. Scand. **12**, 25–35.
Valiant, L.G. (1979). *The complexity of computing the permanent*, Theor. Comput. Sci. **8**, 181–201.
Voorhoeve, M. (1979). *A lower bound for the permanents of certain* $(0,1)$*-matrices*, Proc. Kon. Ned. Akad. Wet. **A82** (= Indag. Math. **41**), 83–86.
Waerden, van der B.L. (1926). *Aufgabe 45*, J. der Deutsch. Math. Verein. **35**, 117.
Waerden, van der B.L. (1927). *Ein Satz über Klasseneinteilungen von endlichen Mengen*, Abh. a.d. Math. Seminar Hamburg **5**, 185–188.
Waerden, van der B.L. (1957). *Mathematische Statistic*, Springer, Berlin (Russian transl. in 1960).
Wilf, H.S. (1968). *A mechanical counting method and combinatorial applications*, J. Combin. Theory **4**, 246–258.
Wilson, R.M. (1974). *Nonisomorphic Steiner triple systems*, Math. Z. **135**, 303–313.
Yamamoto, K. (1951). *On the asymptotic number of Latin rectangles*, Japan J. Math. **21**, 113–119.
Yamamoto, K. (1956). *Structure polynomials of Latin rectangles and its application to a combinatorial problem*, Mem. Fac. Sci. Kyushu Univ. A. **10**, 1–13.

# Random Matrices

V.L. Girko

*Department of Mathematics, Kiev State University, 252 017 Kiev, The Ukraine*

*Contents*



HANDBOOK OF ALGEBRA, VOL. 1
Edited by M. Hazewinkel

## Introduction

In this part of the Handbook of Algebra the main results of the theory of random matrices are collected. Distributions of random matrices arise in many applications areas; perhaps the most well-known areas are nuclear physics, multivariate statistics, and test matrices for numerical algorithms. See [1–34] for references to some of these numerous applications.

Note that applications of random matrix theory are not exhausted by the application in physics, in multivariate statistical analysis, and in the theory of nonordered structures. At the moment, the theory of random matrices is also used in the theory of stability of solutions of stochastic systems, in linear stochastic programming, in molecular chemistry, in the theory of experiment planning, and in the theory of ring accelerators.

This part of the Handbook of Algebra is designed for statisticians, mathematicians and physicists, scientists and engineers of different specialities, who use matrix and probability theoretical methods in their work.

## 1. Distribution function of random matrices

A random matrix is a matrix with random entries. Its distribution function is a function of the distributions of all its entries. The majority of formulas for the distribution of some functions of random matrices contain some integrals with respect to invariant measure; therefore, in order to study them, one needs to know properties of invariant measure.

Let $G$ be a separable topological locally compact group, and let $E$ be a space on which the group of transformations $G$ acts. A measure $\mu(A)$ defined on the Borel $\sigma$-algebra $B$ of the space $E$ is called invariant with respect to $G$ if for any $A \in B$ and $s \in G$, such that the $sA$ is measurable, $\mu(sA) = \mu(A)$. Here, $sA$ is the set $\{sg\colon\ g \in A\}$.

If the function $f(p)$, $p \in E$, is measurable with respect to the $\sigma$-algebra $B$ and non-negative, then, under the assumption that at least one of these integrals exists,

$$\int f(p)\,\mu(\mathrm{d}p) = \int f(sp)\,\mu(\mathrm{d}p).$$

We call the measure $\mu$ a left-invariant Haar measure (left Haar measure) if the equations $\mu(sA) = \mu(A)$ and

$$\int f(x)\,\mu(\mathrm{d}x) = \int f(sx)\,\mu(\mathrm{d}x)$$

hold. If $\mu$ is a left Haar measure, then the function $\nu$ defined by the equality $\nu(K) = \mu(K^{-1})$ on a $\sigma$-algebra of measurable subsets of $K$ of $G$ elements is a right Haar measure. By the set $K^{-1}$ we mean $\{K^{-1}:\ k \in E\}$. Obviously, if $\mu$ is a right Haar measure, $\nu$ is a left Haar measure.

We now state the basic results on Haar measures.

A left Haar measure exists on any separable topological locally compact group $T$. If $\mu$ and $\mu'$ are two left Haar measures on $T$, then $\mu' = c\mu$, where $c$ is a positive number.

We shall now give some examples of Haar measures on groups of matrices.

Let $G$ be the locally compact group of invertible real matrices of order $n$, and $B$ the $\sigma$-algebra of Borel sets on it. There are left and right Haar measures defined on the group $G$. The density of a Haar measure on $G$ with respect to Lebesgue measure on $G$ is equal to $c|\det x|^{-n}$, $x \in G$.

Similarly, we consider Haar measure on the group $K$ of complex invertible matrices of order $n$. Here the densities of the left and right Haar measures, with respect to the Lebesgue measure on $K$, are equal to $|\det x|^{-2n}$, $x \in K$, up to a constant positive coefficient. Let $T$ be the group of lower real triangular matrices of order $n$ with positive entries on the main diagonal, and let $B$ be the $\sigma$-algebra of Borel sets on it. The density of the left Haar measure is

$$c\prod_{i=1}^{n} x_{ii}^{-(n+1-i)}, \quad x_{ii} > 0,$$

where $c > 0$ is some constant. The density of the right Haar measure with respect to the Lebesgue measure on $T$, is

$$c\prod_{i=1}^{n} x_{ii}^{-i}, \quad x_{ii} > 0,$$

where $c > 0$ is an arbitrary constant.

## 2. Haar measure on the group of orthogonal matrices

Let $G$ be the group of real orthogonal matrices of order $n$ and let $\mu$ be the invariant normalized Haar measure on it. The entries of a matrix $H \subset G$ satisfy $n(n-1)/2$ equations. Solving these equations, we obtain $n(n-1)/2$ independent parameters of the matrix $H$. The so-called Euler angles are rather convenient parameters of the group $G$. First, the functions by which the entries of the matrix $H$ are expressed in terms of the Euler angles, are almost everywhere differentiable with respect to these angles. Second, the Haar measure expressed in terms of the Euler angles has a simple form.

The Haar measure $\mu$ of the group $G$ of the matrices $H$, defined by means of the Euler angles $\Theta_{ks}$ is absolutely continuous with respect to the Lebesgue measure given on a set of variations of Euler angles $\Theta_{ks}$ with density

$$c_n \prod_{k=1}^{n-1} \prod_{i=1+k}^{n} \sin^{n-i}(\theta_{ki});$$

$$0 < \theta_{kn} < 2\pi, \ 0 < \theta_{ki} < \pi, \ k \in \{1, \dots, n, \}, \ i \in \{k+1, \dots, n-1\},$$

where

$$c_n = 2^{-n+1} \prod_{k=1}^{n-1} \Gamma\big((n-k+1)/2\big)\pi^{-(n-k+1)/2}.$$

## 3. Maximum likelihood estimates of the parameters of a multivariate normal distribution

In this section we consider the mean and covariance matrix formed from a sample from a multivariate normal distribution. A normal distribution of random vectors is

$$(2\pi)^{-m/2} \det R^{-1/2} \exp\left\{-\frac{1}{2}(\vec{x}-\vec{a})^{\mathrm{T}} R^{-1}(\vec{x}-\vec{a})\right\}.$$

We will assume throughout Sections 3–5 that $R$ is a positive definite matrix. Let $\vec{x}_1,\ldots,\vec{x}_n$ be independent $N_m(\vec{a},R)$ observations of a normally $N_m(\vec{a},R)$ distributed random vector. Then the sample mean vector is

$$\vec{\hat{a}} = n^{-1}\sum_{k=1}^{n} \vec{x}_k,$$

and the sample empirical covariance matrix is

$$\hat{R} = (n-1)^{-1}\sum_{k=1}^{n} (\vec{x}_k - \vec{\hat{a}})(x_k - \hat{a})^{\mathrm{T}}.$$

The vector $\vec{\hat{a}}$ and matrix $\hat{R}$ are stochastically independent.

For $n > m$ the maximum likelihood estimates of $\vec{a}$ and $R$ are $\vec{\hat{a}}$ and $(n-1)n^{-1}\hat{R}$.

## 4. The Wishart density $\omega_n(a, R)$

If $\vec{x}_1,\ldots,\vec{x}_n$ are independent observations of normally $N_m(\vec{a},R)$ distributed random vectors and $n > m$ then the Wishart density function of the distribution of the sample covariance matrix

$$\hat{R} = n^{-1}\sum_{i=1}^{n+1} \left(\vec{x}_i - \vec{\hat{a}}\right)\left(\vec{x}_i - \vec{\hat{a}}\right)^{\mathrm{T}}, \quad \text{where } \vec{\hat{a}} = n^{-1}\sum_{k=1}^{n+1} \vec{x}_k$$

is

$$\left[\Gamma_m\left(\frac{n}{2}\right)\det R^{n/2}\right]^{-1}\left(\frac{1}{2}n\right)^{mn/2} \exp\left\{-\frac{1}{2}\operatorname{Tr} nR^{-1}S\right\}(\det S)^{(n-m-1)/2},$$

where $S$ is positive definite matrix of order $m$, and $\Gamma_m(\cdot)$ denotes the multivariate gamma function

$$\Gamma_m(a) = \pi^{m(m-1)/4}\prod_{i=1}^{m}\Gamma\left(a - \frac{1}{2}(i-1)\right), \quad \operatorname{Re} a > \frac{1}{2}(m-1).$$

## 5. Generalized variance

If a multivariate distribution has covariance matrix $R$ then one overall measure of the spread of the distributions is the scalar quantity $\det R$, called the generalized variance by Wilks.

If the matrix $\hat{R}$ has Wishart density $w_n(\vec{a}, R_m)$, where $n \geqslant m$ then $\det \hat{R}/\det R$ has the same distribution as

$$\prod_{i=1}^{m} \chi^2_{n-i+1},$$

where the $\chi^2_{n-i+1}$ for $i = 1, \ldots, m$ denote independent $\chi^2$ random variables with $n-i+1$ degrees of freedom respectively.

This result gives a tidy representation for the distribution of generalized variance, it is not an easy matter to obtain the density function of a product of independent $\chi^2$ random variables. It is, however, easy to obtain an expression for the moments

$$\mathbf{E}\left[n^m \det \hat{R}\right]^k = \det R^k \prod_{i=1}^{m} \left\{2^k \Gamma\left[\frac{n-i+1}{2} + k\right] \Gamma^{-1}\left[\frac{n-i+1}{2}\right]\right\},$$

$$k = 1, 2, \ldots .$$

## 6. Moments of random Vandermonde determinants

The determinant of a matrix $(\xi_i^j)$, $i = 1, \ldots, n$, $j = 0, \ldots, n-1$, where $\xi_i$, $i = 1, 2, \ldots,$ are random variables is called a random Vandermonde determinant.

If the random variables $\xi_i$, $i = 1, \ldots, n$, are independent, identically distributed, and have a $\beta$-distribution with density

$$\left[B(\alpha, \beta)\right]^{-1} x^{\alpha-1}(1-x)^{\beta-1}, \quad 0 < x < 1, \ \alpha > 0, \ \beta > 0,$$

then for $n = 2, 3, \ldots,$

$$\mathbf{E}\left[\det\left(\xi_i^j\right)\right]^{2k} = \mathbf{E}\left[\prod_{1 \leqslant i < j \leqslant n} (\xi_i - \xi_j)\right]^{2k}$$
$$= \prod_{j=1}^{n} \{\Gamma(1+jk)\Gamma(\alpha + (j-1)k)\Gamma(\beta + (j-1)k)$$
$$\times \left[\Gamma(1+k)\Gamma(\alpha + \beta + (n+j-2)k)\right]^{-1}\} B^{-n}(\alpha, \beta),$$

where

$$\operatorname{Re}\alpha > 0, \ \operatorname{Re}\beta > 0, \ \operatorname{Re}k > -\min\left\{n^{-1}, \operatorname{Re}\alpha(n-1)^{-1}, \operatorname{Re}\beta(n-1)^{-1}\right\}.$$

The second moments of complex random Vandermond matrices $(\xi_i^{j-1})$, $i,j = 1,\dots,n$, where $\xi_p = \nu_p + \mathrm{i}\mu_p$, $\nu_p$, $\mu_p$, $p = 1,\dots,n$, are independent random variables distributed according to the standard normal law, are equal to

$$\mathbf{E}\,\big|\det\big(\xi_i^{j-1}\big)\big|^2_{i,j=1,\dots,n} = 2^{n(n-1)/2} n! \prod_{j=1}^{n-1} j!.$$

If the random variables $\theta_i$, $i = 1,\dots,n$, are independent and have a uniform distribution on the interval $(0, 2\pi)$, then for any integer $k \geqslant 0$,

$$\mathbf{E} \prod_{p,l=1,\dots,n,\ p\neq l} \big|\mathrm{e}^{\mathrm{i}\theta_p} - \mathrm{e}^{\mathrm{i}\theta_l}\big|^k = \Gamma\Big(1 + \frac{kn}{2}\Big)\Big[\Gamma\Big(1 + \frac{k}{2}\Big)\Big]^{-n}.$$

## 7. Polar decomposition of random matrices

Let $\Xi = (\xi_{ij})$, $i = 1,\dots,n$, $j = 1,\dots,m$, $m \geqslant n$, be a real rectangular random matrix. We suppose that there is a joint distribution density of the entries $\xi_{ij}$, equal to $p(X)$, where $X$ is a real rectangular matrix. The polar decomposition of $\Xi$ is a representation of $\Xi$ in the form $\Xi = SU$, where $S = \sqrt{\Xi\Xi^{\mathrm{T}}}$ and $U = S^{-1}\Xi$. Let $K_1$ be the set of real $(m \times n)$ matrices, $K_2$ the set of non-negative definite $(n \times n)$ matrices, $K_3$ the set of real orthogonal $(m \times m)$ matrices, $B_1$ and $B_2$ the $\sigma$-algebras of Borel sets in $K_2$ and $K_3$. Let $G$ be the group of real orthogonal $(m \times m)$ matrices and $\mu$ normalized Haar measure on it. Then the joint distribution of $\Xi\Xi^{\mathrm{T}}$ and $(\Xi\Xi^{\mathrm{T}})^{-1/2}\Xi$ is equal to

$$\begin{aligned}&\mathbf{P}\big\{\Xi\Xi^{\mathrm{T}} \in L_1, \big(\Xi\Xi^{\mathrm{T}}\big)^{-1/2}\Xi \in L_2\big\}\\ &\quad = c_{n,m} \int_{Z_n \in L_1,\, H^{(n)} \in L_2} p\big(\sqrt{Z_n} H^{(n)}\big) \det Z_n^{(m-n-1)/2} \mu(\mathrm{d}H)\,\mathrm{d}Z,\end{aligned}$$

where $L_1 \in B_1$, $L_2 \in B_2$, $H = (h_{ij}) \in G$, $H^{(n)} = (h_{ij})$, $i = 1,\dots,n$, $j = 1,\dots,m$,

$$c_{n,m} = \Bigg[\pi^{n(n-1)/4 - nm/2} \prod_{i=1}^{n} \Gamma\Big(\frac{m+1-i}{2}\Big)\Bigg]^{-1}, \qquad \mathrm{d}Z = \prod_{i,j} \mathrm{d}z_{i,j}.$$

If $p(\sqrt{Z_n} H^{(n)}) \equiv q(Z_n)$, in addition to the previous hypotheses, then $\Xi\Xi^{\mathrm{T}}$ and $(\Xi\Xi^{\mathrm{T}})^{-1/2}\Xi$ are stochastically independent and have the distributions

$$\mathbf{P}\big\{\Xi\Xi^{\mathrm{T}} \in M_1\big\} = c_{n,m} \int_{Z_n \in M_1} q(Z_n) \det Z_n^{(m-n-1)/2}\,\mathrm{d}Z_n,$$

$$\mathbf{P}\big\{\big(\Xi\Xi^{\mathrm{T}}\big)^{-1/2}\Xi \in M_2\big\} = \int_{H^{(n)} \in M_2} \mu(\mathrm{d}H).$$

## 8. Symmetric and Hermitian random matrices

Let $\Xi_n$ be a real symmetric $(n \times n)$ random matrix whose entries $\xi_{ij}$, $i \geqslant j$, $i,j = 1,\dots,n$, have joint distribution density, which we denote in what follows by $p(Z_n)$, where $Z_n = (z_{ij})$ is a real symmetric matrix: let $\lambda_1 \geqslant \cdots \geqslant \lambda_n$ be the eigenvalues of $\Xi_n$, and let $\vec{\theta}_i$ be the corresponding eigenvectors whose first nonzero component is positive. If some eigenvalues $\lambda_i$ $(i = 1,\dots,n)$ coincide, then we can choose the $\vec{\theta}_i$ uniquely by fixing in addition some of their components.

The eigenvalues $\lambda_i$ $(i = 1,\dots,n)$ of $\Xi_n$ are distinct with probability 1.

Let $\theta_n$ be the random matrix whose column vectors are equal to $\vec{\theta}_i$, $i = 1,\dots,n$; let $G$ be the group of real $n \times n$ matrices; $B$ the $\sigma$-algebra of Borel sets of orthogonal $n \times n$ matrices on it, and $\mu$ normalized Haar measure on $G$.

If a real symmetric random matrix $\Xi$ has density $p(Z_n)$, then for any subset $E$ of $B$ and any real numbers $\alpha_i$, $\beta_i$ $(i = 1,\dots,n)$

$$\begin{aligned}&\mathbf{P}\{\Theta_n \in E,\ \alpha_i < \lambda_i < \beta_i,\ i = 1,\dots,n\}\\&\quad = c_{1n} \int p\left(X_n Y_n X_n^{\mathrm{T}}\right) \prod_{i>j} (y_i - y_j) \mu(\mathrm{d}X_n)\,\mathrm{d}Y_n,\end{aligned}$$

where the integral is over the domain

$$\begin{aligned}&\{y_1 > y_2 > \cdots > y_n,\ \alpha_i < y_i < \beta_i,\ i = 1,\dots,n,\ x_{1i} > 0,\ i = 1,\dots,n,\\&\quad X_n \in E\},\end{aligned}$$

$$Y_n = (\delta_{ij} y_j), \quad \mathrm{d}Y_n = \prod_{i=1}^{n} \mathrm{d}y_i,$$

$$c_{1n} = 2^n \pi^{n(n+1)/4} \prod_{i=1}^{n} \{\Gamma[(n-i+1)/2]\}^{-1}.$$

If $p(H_n Z_n H_n') \equiv \tilde{p}(Z_n)$ for all $H_n \in G$ and $Z_n$, then $\Theta_n$ is stochastically independent of the eigenvalues of $\Xi$ and has the following distribution:

$$\mathbf{P}\{\Theta_n \in E\} = 2^n \int_{H_n \in E,\ h_{1i} > 0,\, i=1,\dots,n} \mu(\mathrm{d}H_n).$$

The distribution density of the eigenvalues of $\Xi_n$ is

$$2^{-n} c_{1n} p(Y_n) \prod_{i<j} (y_i - y_j), \quad y_1 > y_2 > \cdots > y_n.$$

Let $H_n = (\eta_{ij})$ be a Hermitian $n \times n$ matrix whose entries are complex random variables, and let $X_n$ be a nonrandom Hermitian $n \times n$ matrix. We assume that the real and imaginary parts of entries of the matrix $H_n$ located on the diagonal and above have the joint distribution density $p(X_n)$ (the function $p(X_n)$ depends on the imaginary and real parts of the $X_n$ entries). The eigenvalues $\lambda_1 \geqslant \cdots \geqslant \lambda_n$ of the matrix are real, do not coincide with probability 1, and are random variables. The matrix $\theta = (\theta_{ij})$ whose columns are the eigenvectors of $H_n$ is unitary. The eigenvectors $\vec{\theta}_i$ with probability 1 are defined from the system of equations

$$(H_n - \lambda_i I)\vec{\theta}_i = 0, \quad (\vec{\theta}_i, \bar{\vec{\theta}}_i) = 1.$$

The vectors $\vec{\theta}_i$ can be chosen uniquely by fixing an argument of some nonzero element of each vector $\vec{\theta}_i$. We consider the matrix $\theta_n$ being chosen so that $\arg \theta_{1i} = c_i$, $i = 1, \ldots, n$, where $c_i$ are nonrandom values, $0 \leqslant c_i \leqslant 2\pi$, $i = 1, \ldots, n$.

If a Hermitian random matrix $H_n$ has the distribution density $p(X_n)$, then

$$\begin{aligned} &\mathbf{P}\{\Theta_n \in E,\ \alpha_i < \lambda_i < \beta_i,\ i = 1, \ldots, n\} \\ &\quad = c_{2n} \int p(U_n Y_n U_n^*) \prod_{i>j} (y_i - y_j)^2 \nu(\mathrm{d}U_n \mid \arg u_{1i} = c_i,\ i = 1, \ldots, n)\, \mathrm{d}Y_n, \end{aligned}$$

where the integration is over the domain $y_1 > \cdots > y_n$, $U_n \in E \subset B$, $\alpha_i < y_i < \beta_i$, $i = 1, \ldots, n$; $\nu(U \mid \arg u_{1i} = c_i,\ i = 1, \ldots, n)$ is the regular conditional Haar measure, and

$$c_{2n} = \left[ \pi^{-n^2 + n/2} \prod_{j=0}^{n-1} j! \right]^{-1}.$$

If $p(U_n Y_n U_n^*) \equiv \tilde{p}(Y_n)$ for all unitary matrices $U_n \in \Gamma$, then $\Theta_n$ is stochastically independent of the eigenvalues of $H_n$ and has the following distribution

$$\mathbf{P}\{\Theta_n \in E\} = \int_E \nu(\mathrm{d}U_n \mid \arg u_{1i} = c_i,\ i = 1, \ldots, n).$$

The density of the eigenvalues is

$$c_{2n} p(Y_n) \prod_{p>l} (y_p - y_l)^2, \quad y_1 > \cdots > y_n.$$

An important special case of Hermitian random matrices are the matrices $H_n$ for which the real and imaginary parts of their entries are independent and distributed according to standard normal laws. In this case the density of $H_n$ is

$$p(X_n) = 2^{-n/2} \pi^{-n^2} \exp\left( -2^{-1} \operatorname{Tr} X_n X_n^* \right),$$

and the density of the eigenvalues has the form

$$c'_{2n} \exp\left\{ -2^{-1} \sum_{i=1}^{n} y_i^2 \right\} \prod_{i>j} (y_i - y_j)^2, \quad c'_{2n} = (2\pi)^{-n/2} \left( \prod_{j=1}^{n-1} j! \right)^{-1}$$

## 9. Nonsymmetric random matrices

Let $\Xi_n = (\xi_{ij})$ be a real square random matrix with distribution density $p(X_n)$, where $X_n = (x_{ij})$ is a real $(n \times n)$ matrix. We introduce some notation: $\lambda_k + \mathrm{i}\mu_k$, $\lambda_k - \mathrm{i}\mu_k$ $(k = 1, \ldots, s)$, $\lambda_l$ $(l = s+1, \ldots, n-2s)$ are the eigenvalues of $\Xi_n$, $\vec{z}_k = \vec{x}_k + \mathrm{i}\vec{y}_k$, $\bar{\vec{z}}_k = \vec{x}_k - \mathrm{i}\vec{y}_k$ $(k = 1, \ldots, s)$, $\vec{z}_l$ $(l = s+1, \ldots, n-2s)$ are the eigenvectors of $\Xi_n$. We arrange the complex eigenvalues of $\Xi_n$ in increasing order of their moduli. If some of these complex numbers (among which there are no conjugate pairs) have equal moduli, then we arrange them in increasing order of their arguments. Among pairs of conjugate complex numbers the first is the number with positive imaginary part. The real eigenvalues are arranged in increasing order. The eigenvalues thus chosen are random variables. There are many other ways of ordering the eigenvalues, but we adhere to this one as the most natural. We require that the vectors $\vec{x}_k$, $\vec{y}_k$ $(k = 1, \ldots, s)$, $\vec{x}_l$ $(l = s+1, \ldots, n-2s)$ are of unit length and the first nonzero component of every vector is positive. With probability 1, $\Xi_n$ can be represented in the following form:

$$\Xi_n = T \operatorname{diag}\left\{ \begin{pmatrix} \lambda_1 & \mu_1 \\ -\mu_1 & \lambda_1 \end{pmatrix}, \ldots, \begin{pmatrix} \lambda_s & \mu_s \\ -\mu_s & \lambda_s \end{pmatrix}, \lambda_{s+1}, \ldots, \lambda_{n-2s} \right\} T^{-1},$$

$T$ being a real matrix that is nondegenerate with probability 1 whose column vectors are the $\vec{x}_k$, $\vec{y}_k$ $(k = 1, \ldots, s)$, $\vec{x}_l$ $(l = s+1, \ldots, n-2s)$.

Let $K$ be the group of real nondegenerate $(n \times n)$ matrices, $B$ the $\sigma$-algebra of Borel subsets of $K$, and $\theta_i$ $(i = 1, \ldots, n)$ the eigenvalues of $\Xi_n$ chosen as described above.

If a random matrix $\Xi_n$ has density $p(X_n)$, then for any subset $E \subset B$ and any real numbers $\alpha_i$, $\beta_i$ $(i = 1, \ldots, n)$

$$\begin{aligned} &\mathbf{P}\left\{T_n \in E,\ \operatorname{Re}\theta_i < \alpha_i,\ \operatorname{Im}\theta_i < \beta_i,\ i = 1, \ldots, n\right\} \\ &= \sum_{s=0}^{[n/2]} c_s \int_{K_s} p(X_n Y_s X_n^{-1}) J_s(Y_s) \varphi(Y_s) \, |\det X_n|^{-n} \\ &\quad \times \prod_{i=2}^{n} \left\{ \left(1 - \sum_{j=2}^{n} x_{ji}^2\right)^{-1/2} \right\} \mathrm{d}X_n \, \mathrm{d}Y_s, \end{aligned}$$

where the domain of integration $K_s$ is

$$X_n \in E, \quad x_{1i} = \left[1 - \sum_{j=2}^{n} x_{ji}^2\right]^{1/2}, \quad \sum_{j=2}^{n} x_{ji}^2 \leqslant 1 \ (i = 1, \ldots, n),$$

$$x_1 < \alpha_1,\ y_1 < \beta_1\ , \ldots,\ x_s < \alpha_{2s-1},\ -y_s < \alpha_{2s},\ x_{s+1} < \alpha_{2s+1},$$

$$0 < \beta_{2s+1},\ \ldots,\ x_{n-2s} < \alpha_n,\ 0 < \beta_n,$$

$$Y_s = \operatorname{diag}\left\{ \begin{pmatrix} x_1 & y_1 \\ -y_1 & x_1 \end{pmatrix}, \ldots, \begin{pmatrix} x_s & y_s \\ -y_s & x_s \end{pmatrix}, x_{s+1}, \ldots, x_{n-2s} \right\},$$

$$J_s(Y_s) = \left|\prod_{p\neq l}(q_p - q_l)\right|,$$

$$\mathrm{d}X_n = \prod_{i=1,\dots,n,\ j=1,\dots,n} \mathrm{d}x_{ij}, \quad \mathrm{d}Y_s = \prod \mathrm{d}x_s\,\mathrm{d}y_s,$$

the $q_p$ $(p = 1,\dots,n)$ are the eigenvalues of $Y_s$, and

$$\varphi(Y_s) = \begin{cases} 1 & \text{if the eigenvalues } x_k+\mathrm{i}y_k,\ x_k-\mathrm{i}y_k\ (k=1,\dots,s) \text{ are in increasing order of their moduli and among any two conjugate numbers the first is that with } y_k \geqslant 0, \text{ and the eigenvalues } x_l\ (l = s+1,\dots,n-2s) \text{ are ordered in increasing order;} \\ 0 & \text{otherwise} \end{cases}$$

and the $c_s$ are some constants.

Suppose that $\Xi_n = (\xi_{pl} + \mathrm{i}\eta_pl)$ is a complex random $n \times n$ matrix of random variables $\xi_{pl}$ and $\eta_{pl}$ $(p, l = 1,\dots,n)$ which have a joint distribution density, which is denote by $p(Z_n)$, where $Z_n$ is a complex nonrandom $(n\times n)$ matrix. We assume that the eigenvalues $\lambda_i$ $(i = 1,\dots,n)$ of $\Xi_n$ are ordered in increasing order of their arguments. Let $\vec{\theta}_1,\dots,\vec{\theta}_n$ be the matrix whose column vectors are $\vec{\theta}_i$ $(i = 1,\dots,n)$. With probability 1 we can represent $\Xi_n$ in the form $\Xi_n = \Theta_n \Lambda_n \Theta_n^{-1}$, where $\Lambda_n = (\lambda_i\delta_{ij})$. For $\Theta_n$ to be uniquely determined with probability 1, we require that $(\vec{\theta}_p, \bar{\vec{\theta}}_p) = 1$ and $\arg\theta_{lp} = c_p$ $(p = 1,\dots,n)$, where the $c_p$ $(0 \leqslant c_p \leqslant 2\pi)$ are arbitrary real numbers.

We denote by $K$ the group of nonsingular complex $(n \times n)$ matrices and by $B$ the Borel $\sigma$-algebra of $K$.

If $\Xi_n$ has density $p(Z_n)$, then for any $E \in B$ and complex numbers $\alpha_i$, $\beta_i$ $(i = 1,\dots,n)$

$$\begin{aligned}&\mathbf{P}\{\Theta_n \in E,\ \operatorname{Re}\alpha_k < \operatorname{Re}\lambda_k < \operatorname{Re}\beta_k,\ \operatorname{Im}\alpha_k < \operatorname{Im}\lambda_k < \operatorname{Im}\beta_k,\ k = 1,\dots,n\} \\ &\quad = c_n \int p(X_n Y_n X_n^{-1}) \prod_{i\neq j} |y_i - y_j|^2 \,|\det X_n|^{-2n} \\ &\qquad \times \prod_{i=2,\dots,n,\ j=1,\dots,n} r_{ij}\,\mathrm{d}r_{ij}\,\mathrm{d}\varphi_{ij}\,\mathrm{d}Y_n,\end{aligned}$$

where $X_n = (r_{pl}\mathrm{e}^{\mathrm{i}\varphi_{pl}})$, $\varphi_{1l} = c_l$ $(l = 1,\dots,n)$,

$$r_{1i} = \left[1 - \sum_{j=2}^{n} r_{ji}^2\right]^{1/2}$$

and the integral is over the domain

$$\arg y_1 > \arg y_2 > \cdots > \arg y_n,$$

$$\operatorname{Re}\alpha_k < \operatorname{Re} y_k < \operatorname{Re}\beta_k, \quad \operatorname{Im}\alpha_k < \operatorname{Im} y_k < \operatorname{Im}\beta_k, \quad k = 1, \ldots, n,$$

$$X_n \in E, \quad \sum_{j=2}^{n} r_{ji}^2 \leqslant 1, \quad i = 1, \ldots, n, \; 0 \leqslant \varphi_{ij} \leqslant 2\pi, \; i \neq 1,$$

$$\mathrm{d}Y_n = \prod_{k=1}^{n} \mathrm{d}\operatorname{Re} y_k \, \mathrm{d}\operatorname{Im} Y_k, \quad Y_n = (\delta_{pl} y_l).$$

The constant $c_n$ is determined by the condition that the integral over the domain

$$\varphi_{1l} = c_l, \quad X_n \in K, \quad \arg y_1 > \cdots > \arg y_n, \quad \sum_{j=2}^{n} r_{ji}^2 \leqslant 1$$

equals 1.

## 10. Reduction of random matrices to triangular form

The following theorem of Schur is well known in matrix theory. If $A$ is a complex $(n \times n)$ matrix, then there is a unitary $(n \times n)$ matrix $U_n$ such that $T = U^* A U$ is upper triangular and the entries on the main diagonal of $T$ are the eigenvalues of $A$. If $A$ is real with real eigenvalues, then $U$ can be chosen to be real orthogonal. $A$ is normal if and only if $T$ is diagonal. If the eigenvalues of $A$ are distinct and arranged in any order and the arguments of any nonzero component of every column vector of $U$ are fixed, then the representation $A = UTU^*$ of $A$ is unique.

Suppose that $\Xi_n$ is a complex $(n \times n)$ random matrix and its entries have the distribution density $p(X)$. Let $\Xi_n = USU^*$ be the Schur representation of $\Xi_n$, the diagonal entries $s_{ii}$ $(i = 1, \ldots, n)$ of $S$ being chosen so that their arguments are arranged in nonincreasing order, $\arg u_{1i} = c_i$ $(i = 1, \ldots, n)$, where $c_i$ $(0 \leqslant c_i \leqslant 2\pi)$ are arbitrary real numbers. We denote by $\Gamma$ the group of unitary $(n \times n)$ matrices, by $B$ the Borel $\sigma$-algebra of $\Gamma$, and by $\nu$ the normalized Haar measure on $\Gamma$.

For any $E \in B$ and any measurable set $C$ of complex upper triangular $(n \times n)$ matrices

$$\mathbf{P}\{U \in E, \; S \in C\} = c \int p(HYH^*) \prod_{p \neq l} |y_{pp} - y_{ll}| \, \nu(\mathrm{d}H \mid \arg h_{1p} = c_p, \; p = 1, \ldots, n) \, \mathrm{d}Y,$$

where the integration is over the domain $\arg y_{11} > \cdots > \arg y_{nn}$,

$$Y \in C, \quad H \in E, \quad \mathrm{d}Y = \prod_{i \geqslant j} \mathrm{d}\operatorname{Re} y_{ij} \, \mathrm{d}\operatorname{Im} y_{ij},$$

$$c = \left[ (2\pi)^{-n(n-1)/2} 2^{n(n+1)/2} n! \prod_{j=1}^{n} j! \right]^{-1}.$$

## 11. Gaussian random matrices

Let $\Xi_n = USU^*$ be the Schur representation of $\Xi_n$, the diagonal entries $s_{ii}$ ($i = 1, \dots, n$) of $S$ being chosen so that their arguments are arranged in nonincreasing order, $\arg u_{1i} = c_i$ ($i = 1, \dots, n$), where $c_i$ ($0 \leqslant c_i \leqslant 2\pi$) are arbitrary real numbers.

If the distribution density of the random complex matrix $\Xi_n$ is invariant under a unitary transformation $X = UTU^*$, then the distribution density of $S$ is

$$c'_n p(Y) \prod_{p \neq l} |y_{pp} - y_{ll}|, \quad \arg y_{11} > \cdots > \arg y_{nn},$$

$$c'_n = (2\pi)^{(n-1)n/2} \left[ 2^{n(n+1)/2} \prod_{j=1}^{n} j! \right]^{-1};$$

further, $U$ is stochastically independent of $S$ and has the distribution

$$\mathbf{P}\{U \in E\} = \int_E \nu(\mathrm{d}H \mid \arg h_{1p} = c_p), \quad p = 1, \dots, n.$$

If the real and imaginary parts of the entries of $\Xi_n$ are independent and distributed according to the normal law $N(0,1)$, then the distribution density of the eigenvalues $\lambda_1, \dots, \lambda_n$ of $\Xi_n$ is

$$c''_n \exp\left\{ -\frac{1}{2} \sum_{k=1}^{n} |y_k|^2 \right\} \prod_{p \neq l} |y_{pp} - y_{ll}|, \quad \arg y_{11} > \cdots > \arg y_{nn},$$

$$c''_n = \left[ \pi^{-n} \prod_{j=1}^{n} \Gamma(j) \right]^{-1},$$

the real and imaginary parts of the entries $s_{ij}$, $i > j$, of $S$ are independent, do not depend on $s_{ii}$ and $U$, and are distributed according to the normal law $N(0,1)$.

The eigenvalues $|\lambda_1|, \dots, |\lambda_n|$, $|\lambda_1| > \cdots > |\lambda_n|$, are distributed as corresponding members of order statistics, as obtained from independent random variables $\chi^2_{2i}$, $i = 1, \dots, n$, with $2i$ degrees of freedom.

Let us consider the distribution of the eigenvalues of an asymmetric real random matrix $\Xi_n$ whose entries are independent and distributed according to the standard normal law. The density of such a matrix is

$$p(X) = (2\pi)^{-n^2/2} \exp\left\{ -\frac{1}{2} \operatorname{Sp} XX^{\mathrm{T}} \right\}.$$

Let $\Lambda_n$ be the matrix of the eigenvalues of $\Xi_n$ whose form is given in Section 9. If $E_n$ denotes the mathematical expectation or the number of real eigenvalues of random matrix $\Xi_n$, then

$$\lim_{n\to\infty} \frac{E_n}{\sqrt{n}} = \sqrt{\frac{2}{\pi}}.$$

Let $\lambda_1, \dots, \lambda_k$ be the real eigenvalues of the matrix $\Xi_n$, let $\xi$ be the average real eigenvalue with distribution function

$$F(x) = \mathbf{E} \sum_{i=1}^{k} \chi(\lambda_i < x).$$

If $\xi$ denotes a real eigenvalue of the random matrix $\Xi_n$, then as $n \to \infty$, $\xi/\sqrt{n}$ is uniformly distributed on the interval $[-1, 1]$.

Exact formulas for $E_n$, where $n$ is even:

$$E_n = \sqrt{2} \sum_{k=0}^{n/2-1} \frac{(4k-1)!!}{(4k)!!},$$

while if $n$ is odd,

$$E_n = 1 + \sqrt{2} \sum_{k=1}^{(n-1)/2} \frac{(4k-3)!!}{(4k-2)!!}.$$

The probability that a random matrix $\Xi_n$ has all real eigenvalues is

$$p_{n,n} = 1/2^{n(n-1)/4}.$$

The joint density of the ordered real eigenvalues $\lambda_j$ and ordered complex eigenvalue pairs $x_j \pm \mathrm{i} y_j$, $y_j > 0$, given that $\Xi_n$ has $k$ real eigenvalues is

$$\frac{2^{l-n(n+1)/4}}{\prod_{i=1}^{n} \Gamma(i/2)} \Delta \exp\left( \sum_{i=1}^{\frac{n-k}{2}} (y_i^2 - x_i^2) - \sum_{i=1}^{\frac{n-k}{2}} \lambda_i^2/2 \right) \prod_{i=1}^{\frac{n-k}{2}} \operatorname{erfc}\left(y_i\sqrt{2}\right),$$

where $\Delta$ is the magnitude of the product of the differences of the eigenvalues of $A$. Integrating this formula over the $\lambda_j$, $x_j$ and $y_j > 0$ gives the probability that a matrix $\Xi_n$ has exactly $k$ real eigenvalues.

The density of an average random complex eigenvalue of a normally distributed matrix is

$$p_n(x, y) = \sqrt{2/\pi y\, \mathrm{e}^{y^2-x^2}} \operatorname{erfc}(y\sqrt{2}) \mathrm{e}_{n-2}(x^2 + y^2),$$

where

$$\mathrm{e}_n(z) = \sum_{k=0}^{n} z^k/k!.$$

Consider the generalized eigenvalue problem

$$\det(M_1 - \lambda M_2) = 0,$$

where $M_1$ and $M_2$ are independent random matrices, whose entries are independent and distributed by a standard normal law. If $\lambda$ denotes an average real generalized eigenvalue of this pair of independent random matrices, then its probability density is given by

$$\frac{1}{\pi}\left(1+\lambda^2\right)^{-1},$$

that is, $\lambda$ obeys the standard Cauchy distribution.

## 12. Unitary random matrices

The entries of a unitary matrix $U_n$ can be expressed as almost everywhere continuously differentiable functions of its Euler angles $\varphi_i$ $(i = 1,\dots,n^2)$. The set $D$ of values of the $\varphi_i$ can be split into subsets so that in every measurable subset the angles $\varphi_i$ characterize $\Xi_n$ uniquely. Suppose that the random variables $\varphi_i$ have joint distribution density $p(x_1,\dots,x_{n^2})$. The density $p(x_1,\dots,x_{n^2})$ can be represented as $p(x_1,\dots,x_{n^2}) = \tilde{p}(T_n(x_1,\dots,x_{n^2}))$, where $T_n$ is a unitary matrix determined by the angles $x_i$, since the Euler angles $x_i$ can be expressed in terms of the entries of $U_n$. The distribution of $U_n$ is

$$\mathbf{P}\{U_n \in B\} = \int_{H_n\in B} \tilde{p}(H_n)\,\mathrm{d}H_n,$$

where $B$ is a measurable subset of $\Gamma_n$,

$$\mathrm{d}H_n = \prod_{i=1}^{n^2} \mathrm{d}x_i,$$

and the $x_i$ are the Euler angles of $H_n$. The group $\Gamma_n$ is compact, therefore, there is normalized Haar measure $\mu$ on $\Gamma_n$, which can be represented as follows:

$$\mu(B) = \int_{H_n\in B} q(H_n)\,\mathrm{d}H_n,$$

where $q(H_n)$ is a function of the $x_i$, which we call the density of $\mu$.

Any unitary matrix $U_n$ can be represented as follows: $U_n = H_n \Theta H_n^*$, where $H_n$ is a unitary matrix, $\Theta_n = (\exp(\mathrm{i}\theta_p)\delta_{pl})$, and the $\exp(\mathrm{i}\theta_p)$ are the eigenvalues of $U_n$. We arrange the arguments of the eigenvalues in nonincreasing order $0 \leqslant \theta_1 \leqslant \theta_2 \leqslant \cdots \leqslant \theta_n \leqslant 2\pi$. The eigenvalues thus chosen are random variables. To fix the eigenvectors $\vec{h}_p$ uniquely we require that $\arg h_{ip} = c_p$ $(p = 1, \ldots, n)$, where the $c_p$, $0 \leqslant c_p \leqslant 2\pi$, are nonrandom numbers.

Let $\Gamma$ be the group of unitary $(n \times n)$ matrices, $\nu$ normalized Haar measure on it and $B$ the $\sigma$-algebra of Borel sets of $\Gamma$.

If the Euler angles of a random matrix $U_n$ have the distribution density $p(H_n)$, then for any $E \in B$ and any real numbers $\alpha_i$, $\beta_i$ $(i = 1, \ldots, n)$

$$\begin{aligned}
&\mathbf{P}\{H_n \in E,\ \alpha_k < \theta_k < \beta_k,\ k = 1, \ldots, n\} \\
&\quad = c_n \int \left[\tilde{p}(X_n Y_n X_n^*)/\tilde{q}(X_n Y_n X_n^*)\right] \\
&\quad\quad \times \prod_{k<l} |\mathrm{e}^{\mathrm{i}y_k} - \mathrm{e}^{\mathrm{i}y_l}|^2\, \nu(\mathrm{d}X_n \mid \arg x_{1p} = c_p,\ p = 1, \ldots, n)\, \mathrm{d}Y_n,
\end{aligned}$$

where $q$ is the density of the Haar measure $\nu$,

$$\mathrm{d}Y_n = \prod_{i=1}^{n} \mathrm{d}y_i, \quad Y_n = (\mathrm{e}^{\mathrm{i}y_p}\delta_{pl}),$$

and the integration is over the domain

$$0 < y_1 < y_2 < \cdots < y_n < 2\pi, \quad \alpha_k < y_k < \beta_k, \quad k = 1, \ldots, n.$$

$$X_n \in E, \quad c_n = \left(n!\,(2\pi)^n\right)^{-1}.$$

If the distribution density of the Euler angles of $U_n$ is equal to the density of the Haar measure $\nu$, then the eigenvectors of $U_n$ are stochastically independent of its eigenvalues. The distribution density of the arguments of the eigenvalues of $U_n$ is

$$\left(n!\,(2\pi)^n\right)^{-1} \prod_{k<l} \left|\mathrm{e}^{\mathrm{i}y_k} - \mathrm{e}^{\mathrm{i}y_l}\right|^2, \quad 0 < y_1 < \cdots < y_n < 2\pi.$$

The distribution of $H_n$ is

$$\mathbf{P}\{H_n \in E\} = \int_{X_n \in E} \nu\left(\mathrm{d}X_n \mid \arg x_{1p} = c_p, \quad p = 1, \ldots, n\right).$$

If the distribution of the $U_n$ is absolutely continuous relative to the Haar measure $\nu$ with density $p$ satisfying

$$p(X_n Y_n X_n^*) = \tilde{p}(Y_n), \quad \text{where } X_n \in \Gamma,\ Y_n = (\mathrm{e}^{\mathrm{i}y_p}\delta_{pl}),$$

then the eigenvalues of $U_n$ are stochastically independent of its eigenvectors. The distribution density of the eigenvalues of $U_n$ is

$$\prod_{k>l} \left|e^{iy_k} - e^{iy_l}\right|^2 p(Y_n)\left(n!(2\pi)^n\right)^{-1}, \quad 0 < y_1 < \cdots < y_n < 2\pi.$$

## 13. Distribution of eigenvalues and eigenvectors of orthogonal random matrices

Let $H_n$ be a real orthogonal $(n \times n)$ random matrix. Suppose that there is a joint distribution density $p(x_1, \ldots, x_{n(n-1)/2})$ of its Euler angles $\varphi_i$. For almost all values of $x_i$ we can write $p = \tilde{p}(T_n(x_i,\ i = 1, \ldots, n(n-1)/2))$, since the Euler angles $\varphi_i$ can be expressed in terms of the entries of the orthogonal matrix $T_n(x_i,\ i = 1, \ldots, n(n-1)/2)$.

It is easy to check that when the distribution density of the Euler angles of $H_n$ exists, then the arguments of the eigenvalues of $H_n$ are distinct with probability 1. The eigenvalues of $H_n$ are $\{e^{\pm i\lambda_k},\ k = 1, \ldots, n/2\}$ if $n$ is even and $\{e^{\pm i\lambda_k},\ k = 1, \ldots, (n-1)/2\}$ if $n$ is odd, where the $\lambda_k$ are real numbers with $0 \leqslant \lambda_k \leqslant 2\pi$. Let $\vec{\theta}_k$ be the eigenvectors that correspond to the eigenvalues $e^{\pm i\lambda_k}$. The vectors $\vec{\theta}_k$ corresponding to nonconjugate eigenvalues are orthogonal.

We order the eigenvalues as follows:

$$\left\{e^{i\lambda_1}, e^{-i\lambda_1}, \ldots, e^{i\lambda_{n/2}}, e^{-i\lambda_{n/2}}, 2\pi \geqslant \lambda_1 \geqslant \cdots \geqslant \lambda_{n/2} \geqslant 0\right\}$$

if $n$ is even, and it can happen that some eigenvalues are $\pm 1$. Since the eigenvalues $\lambda_k$ are distinct with probability 1, the case of interest to us is that when two of the eigenvalues $\lambda_k$ are $+1$ and $-1$. In this case we order the eigenvalues as follows:

$$\left\{e^{i\lambda_1}, e^{-i\lambda_1}, \ldots, e^{i\lambda_{(n-2)/2}}, e^{-i\lambda_{(n-2)/2}}, +1, -1, \right.$$
$$\left. 2\pi \geqslant \lambda_1 \geqslant \cdots \geqslant \lambda_{(n-2)/2} \geqslant 0\right\}.$$

For odd $n$ we order them as follows

$$\left\{e^{i\lambda_1}, e^{-i\lambda_1}, \ldots, e^{i\lambda_{(n-1)/2}}, e^{-i\lambda_{(n-1)/2}}, \xi, 2\pi \geqslant \lambda_1 \geqslant \cdots \geqslant \lambda_{(n-1)/2} \geqslant 0\right\},$$

the last eigenvalue $\xi$ is a random variable which takes the values $\pm 1$.

The matrix $H_n$ can be represented almost surely in the following form:

$$H_n = \Theta_n \operatorname{diag}\left\{\begin{pmatrix} \cos\lambda_1 & \sin\lambda_1 \\ -\sin\lambda_1 & \cos\lambda_1 \end{pmatrix}, \ldots, \begin{pmatrix} \cos\lambda_q & \sin\lambda_q \\ -\sin\lambda_q & \cos\lambda_q \end{pmatrix}, +1, -1\right\}$$
$$\times \Theta_n^{\mathrm{T}}$$

for even $n$ and

$$H_n = \Theta_n \operatorname{diag}\left\{\begin{pmatrix} \cos\lambda_1 & \sin\lambda_1 \\ -\sin\lambda_1 & \cos\lambda_1 \end{pmatrix}, \ldots, \begin{pmatrix} \cos\lambda_p & \sin\lambda_p \\ -\sin\lambda_p & \cos\lambda_p \end{pmatrix}, \xi\right\}\Theta_n^{\mathrm{T}},$$

$$p = (n-1)/2,$$

for odd $n$, where $\Theta_n$ is an orthogonal matrix whose column vectors are $\operatorname{Re}\vec{\theta}_k$, $\operatorname{Im}\vec{\theta}_k$. In the first of these equalities there may be no eigenvalues $+1$ or $-1$. However, such a representation is not unique. To make it unique we must fix some entries of $\Theta_n$. Let $\vec{\theta}_p = \vec{x}_p + \mathrm{i}\vec{y}_p$. Then

$$H_n\vec{x}_p = \cos\lambda_p\vec{x}_p - \sin\lambda_p\vec{y}_p, \quad H_n\vec{y}_p = \sin\lambda_p\vec{x}_p + \cos\lambda_p\vec{y}_p.$$

From these equalities we find that $x_p$ and $y_p$ are orthogonal and

$$\left[(H_n - \cos\lambda_p I)^2 + I\sin^2\lambda_p\right]\vec{y}_p = 0.$$

The matrix $(H_n = \cos\lambda_p I)^2$ has real eigenvalues $-\sin^2\lambda_p$ of multiplicity 2. Therefore, we can require that $(\vec{x}_p, \vec{x}_p) = 1$, $(\vec{y}_p, \vec{y}_p) = 1$ and $x_{1p} = c_p$, where $c_p$ is a fixed number with $|c_p| \leqslant 1$.

If $n$ is even and $H_n$ has no eigenvalues $+1$, or $-1$ then we put $x_{1p} = c_p$ $(p = 2, \dots, n)$ if $n$ is even, and if $H_n$ has the eigenvalues $+1$ and $-1$, then we put

$$x_{1p} = c_p \ (p = 2, \dots, n-2), \quad x_{1n-1} \geqslant 0, \quad x_{1n} \geqslant 0,$$

if $n$ is odd, we put $x_{1p} = c_p$ $(p = 2, \dots, n-1)$, $x_{1n} \geqslant 0$.

Let $G$ be the group of real orthogonal $(n\times n)$ matrices, $\mu$ the normalized Haar measure of $G$, $B$ the $\sigma$-algebra of Borel subsets of $G$, and $n$ an odd integer.

If the Euler angles of a random matrix $H_n$ have the distribution density $p$, then for any $E \in B$ and real numbers $\alpha_i$, $\beta_i$ $(i = 1, \dots, (n-1)/2)$ where $0 \leqslant \alpha_i, \beta_i \leqslant 2\pi$,

$$\begin{aligned}
&\mathbf{P}\{\Theta_n \in E,\ \alpha_k < \lambda_k < \beta_k,\ k = 1, \dots, (n-1)/2,\ \xi = \pm 1\} \\
&\quad = c_n^{\pm} \int_{L_1} \tilde{p}(T_n Y_n^{\pm} T_n')\tilde{q}^{-1}(T_n Y_n^{\pm} T_n') \\
&\qquad \times \prod_{s=1}^{(n-1)/2} \left\{\left[\sin^2\frac{x^s}{2}(1\pm 1) + \cos^2\frac{x^s}{2}(1\mp 1)\right]|\sin x_s|\right\} \\
&\qquad \times \prod_{s>m} \sin^2\frac{x_s - x_m}{2}\sin^2\frac{x_s + x_m}{2} \\
&\qquad \times \prod_s \mathrm{d}x_s \mu(\mathrm{d}Y_n \mid t_{1p} = c_p,\ p = 2, \dots, (n-1),\ t_{1n} \geqslant 0),
\end{aligned}$$

$$Y_n^{\pm} = \operatorname{diag}\left\{\begin{pmatrix} \cos x_1 & \sin x_1 \\ -\sin x_1 & \cos x_1 \end{pmatrix}, \dots, \begin{pmatrix} \cos x_{(n-1)/2} & \sin x_{(n-1)/2} \\ -\sin x_{(n-1)/2} & \cos x_{(n-1)/2} \end{pmatrix}, \pm 1\right\}$$

where $c_n^{\pm}$ are some constant.

## 14. Distribution of roots of algebraic equations with random coefficients

In general, the entries of the random matrices can have no distribution density, but in some cases the coefficients of the characteristic equation do have a distribution density. Therefore, it is of interest to find the distribution of roots of random polynomials.

Let $f(t)$: $t^n + \xi_1 t^{n+1} + \cdots + \xi_n = 0$ be the algebraic equation whose coefficients $\xi_i$, $i = 1, \dots, n$, are random variables. Consider the solution of such an equation in the field of complex numbers. It is known from algebra that the equation $f(t) = 0$ has $n$ roots $\nu_i$, $i = 1, \dots, n$, and the roots $\nu_i$, $i = 1, \dots, n$, are continuous functions of the coefficients $\xi_i$, $i = 1, \dots, n$. Therefore, the roots $\nu_i$, $i = 1, \dots, n$, can be selected in such a way that they will be random variables.

The roots of the equation $f(t) = 0$ have the following form:

$$\begin{aligned} &\nu_1 = \lambda_1 + \mathrm{i}\mu_1, \quad \nu_2 = \lambda_1 - \mathrm{i}\mu_1, \quad \dots, \quad \nu_{2k-1} = \lambda_k + \mathrm{i}\mu_k, \\ &\nu_{2k} = \lambda_k - \mathrm{i}\mu_k, \quad \nu_{2k+1} = \tau_1, \quad \dots, \quad \nu_n = \tau_n - 2k, \end{aligned}$$

where $\lambda_i$, $\mu_i$, $i = 1, \dots, n$, $\tau_j$, $j = 1, \dots, n - 2k$, are real variables, and the index $k$ is a random variable taking values from 0 to $[n/2]$.

Arrange the complex roots in increasing order of their moduli. If the moduli of the complex roots coincide, then we arrange them in increasing order of their arguments; among the conjugate pairs of roots, the one with negative imaginary part comes first. Real roots are arranged in increasing order. Roots selected in such a way are random variables. Of course, there are many other ways of ordering eigenvalues, but we adhere to this one as the most natural procedure.

If the random coefficients $\xi_i$, $i = 1, \dots, n$, of the equation $f(t) = 0$ have a joint distribution density $p(x_1, \dots, x_n)$, then for any real numbers $\alpha_i$, $\beta_i$, $i = 1, \dots, n$,

$$\begin{aligned} &\mathbf{P}\{\operatorname{Re}\nu_i < \alpha_i,\ \operatorname{Im}\nu_i < \beta_i,\ i = 1, \dots, n\} \\ &\quad = \sum_{s=0}^{[n/2]} 2^s \int_{L_s} p(\Delta_1, \Delta_2, \dots, \Delta_n) \cdot \varphi(z_1, z_2, \dots, z_n) \Big| \prod_{i>j} (z_i - z_j) \Big| \\ &\qquad \times \prod_{i=1}^{s} \mathrm{d}x_i\, \mathrm{d}y_i \prod_{i=2s+1}^{n} \mathrm{d}z_i \end{aligned}$$

where $z_{2p-1} = x_p + \mathrm{i}y_p$, $z_{2p} = x_p - \mathrm{i}y_p$, $p = 1, \dots, s$; $z_l$, $l = 2s + 1, \dots, n$, are real variables, the domain of integration $L_s$ is equal to

$$\begin{aligned} &\{x_i, y_i, z_i\colon x_1 < \alpha_1,\ y_1 < \beta_1, \dots, x_s < \alpha_{2s},\ -y_s < \beta_{2s}, \\ &\quad z_{2s+1} < \alpha_{2s+1},\ 0 < \beta_{2s+1}, \dots, z_n < \alpha_n,\ 0 < \beta_n\}, \end{aligned}$$

and the

$$\Delta_k = (-1)^k \sum_{i_1 < i_2 < \cdots < i_k} z_{i_1}, z_{i_2}, \dots, z_{i_k}$$

are the symmetric functions of the variables $z_i$, $i = 1, \dots, n$; $\varphi(z_1, \dots, z_n) = 1$, if the values $z_i$, $i = 1, \dots, n$, are ordered as described above, and $\varphi(z_1, \dots, z_n) = 0$ otherwise.

The probability that the equation $f(t) = 0$ has exactly $s$ pairs of conjugate complex roots is

$$2^s \int_{R^n} p(\Delta_1, \ldots, \Delta_n)\varphi(z_1, \ldots, z_n) \left| \prod_{i>j} (z_i - z_j) \right| \prod_{i=1}^{s} \mathrm{d}x_i \, \mathrm{d}y_i \prod_{i=2s+1}^{n} \mathrm{d}z_i.$$

Let $C$ be some measurable set of the complex plane whose Lebesgue measure given on this plane is equal to zero and such that the linear measure of the Lebesgue set which is equal to the intersection of the set $C$ with the real line is also zero. Then the roots of the equation $f(t) = 0$ get into $C$ with zero probability. The probability that the roots of the equation $f(t) = 0$ are on the real line is

$$\int_{z_1 > \cdots > z_n} p(\Delta_1, \ldots, \Delta_n) \prod_{i>j} (z_i - z_j) \prod_{i=1}^{n} \mathrm{d}z_i.$$

Suppose that the coefficients of the equation $f(t) = 0$ are complex random variables. The roots of such an equation $\nu_i$, $i = 1, \ldots, n$, will be complex. Order the roots $\nu_i$, $i = 1, \ldots, n$, in increasing order of their arguments. The density of the real and imaginary parts of coefficients $\xi_i$, $i = 1, \ldots, n$, will be denoted by $p(\operatorname{Re} z_i, \operatorname{Im} z_i, i = 1, \ldots, n)$, where $z_i$, $i = 1, \ldots, n$, are complex variables. Then the density of the roots $\nu_i$, $i = 1, \ldots, n$, is

$$p(\operatorname{Re} \Delta_i, \operatorname{Im} \Delta_i) \prod_{i>j} |z_i - z_j|^2, \quad \arg z_1 > \cdots > \arg z_r,$$

where the $\Delta_i$, $i = 1, \ldots, n$, are the elementary symmetric functions of the complex variables $z_i$, $i = 1, \ldots, n$.

Let the real function $f(t)$ be continuous on the segment $\alpha \leqslant t \leqslant b$, let it have continuous derivatives be on the interval $\alpha < t < b$, and have a finite number of points in which the derivative $f(t)$ vanishes. Then the number of zeroes of the function $f(t)$ on the interval $(a, b)$ is equal to

$$n(a,b) = (2\pi)^{-1} \int \mathrm{d}y \int_a^b \cos\left[yf(t)\right] \left|f'(t)\right| \mathrm{d}t.$$

Moreover, a multiple zero is counted once but a zero coinciding with $a$ or $b$ gives the contribution in $n(a,b)$, equal to 1/2. With the help of this formula the following result was obtained. If $E_n$ denotes the mathematical expectation of the number of real roots of an algebraic equation with independent normally $N(0,1)$ distributed coefficients, then $\lim_{n\to\infty} E_n / \ln n = 2\pi^{-1}$.

## 15. The logarithmic law

For each $n$ let the random elements $\xi_{ij}^{(n)}$, $i,j=1,\dots,n$, of a matrix $\Xi_n$ be independent, $\mathbf{E}\,\xi_{ij}^{(n)}=0$, $\mathbf{V}\xi_{ij}^{(n)}=1$, $\mathbf{E}\,[\xi_{ij}^{(n)}]^4=3$, and for some $\delta>0$,

$$\sup_n \sup_{i,j=1,\dots,n} \mathbf{E}\,\bigl|\xi_{ij}^{(n)}\bigr|^{4+\delta}<\infty.$$

Then

$$\lim_{n\to\infty}\mathbf{P}\bigl\{\bigl[\ln\det\Xi_n^2-\ln(n-1)!\bigr](2\ln n)^{-1/2}<x\bigr\}$$
$$=(2\pi)^{-1/2}\int_{-\infty}^{x}\exp\bigl(-y^2/2\bigr)\,\mathrm{d}y,$$

$$\lim_{n\to\infty}\mathbf{P}\,\{\operatorname{sign}\det\Xi_n=+1\}=1/2,$$

$$\lim_{n\to\infty}\mathbf{P}\{\operatorname{sign}\det\Xi_n=-1\}=1/2.$$

## 16. Limit theorems for random determinants

Let us call a set of random variables $\xi_{ij}^{(n)}$, $i,j=1,\dots,n$, asymptotically constant if there can be found nonrandom numbers $a_{ij}^{(n)}$, such that for all $\varepsilon>0$,

$$\lim_{n\to\infty}\sup_{k,l=1,\dots,n}\mathbf{P}\bigl\{\bigl|\xi_{kl}^{(n)}-a_{kl}^{(n)}\bigr|\geqslant\varepsilon\bigr\}=0.$$

The random vectors $\vec{\xi}_{nk}$, $k=1,\dots,n$, $n=1,2,\dots$, are called asymptotically constant if there are constant vectors $\vec{a}_{nk}$, $k=1,\dots,n$, such that for all $\varepsilon>0$,

$$\lim_{n\to\infty}\sup_{k=1,\dots,n}\mathbf{P}\bigl\{\bigl(\vec{\xi}_{nk}-\vec{a}_{nk},\,\vec{\xi}_{nk}-\vec{a}_{nk}\bigr)\geqslant\varepsilon\bigr\}=0.$$

Let us consider the random variables $\nu_{ij}^{(n)}=\xi_{ij}^{(n)}-a_{ij}^{(n)}-p_{ij}^{(n)}$, where

$$p_{ij}^{(n)}=\int_{|x|<\tau}x\,\mathrm{d}F_{ij}\bigl(x+a_{ij}^{(n)}\bigr),$$

$\tau>0$ is an arbitrary constant, and $F_{ij}(x)=\mathbf{P}\,\{\xi_{ij}^{(n)}<x\}$. The square matrix $B_n:=(b_{ij}^{(n)})$ is composed of the $b_{ij}^{(n)}:=p_{ij}^{(n)}+a_{ij}^{(n)}$.

If for each $n$, the vectors $(\xi_{ij}^{(n)}, \xi_{ji}^{(n)})$, $i \geqslant j$, $i, j = 1, \ldots, n$, are independent and the column vectors and row vectors of the matrix $\Xi_n = (\xi_{ij}^{(n)})$ are asymptotically constant,

$$\lim_{h \to \infty} \lim_{n \to \infty} \mathbf{P}\left\{\left|\sum_{i=1}^{n} \nu_{ii}^{(n)}\right| + \sum_{i,j=1}^{n} (\nu_{ij}^{(n)})^2 \geqslant h\right\} = 0;$$

$$\sup_n \left[ |\mathrm{Tr}\, B_n| + \mathrm{Tr}\, B_n B_n^{\mathrm{T}} \right] < \infty,$$

then

$$\det(I + \Xi_n) \sim \det(I + B_n) \prod_{i>j,\ i,j=1}^{n} (1 - \nu_{ij}^{(n)} \nu_{ji}^{(n)}) \prod_{i=1}^{n} (1 + \nu_{ii}^{(n)}),$$

where the symbol "$\sim$" means that for any sequences of random variables $\xi_n$ and $\eta_n$ and for almost all $x$

$$\lim_{n \to \infty} \left[ \mathbf{P}\{\xi_n < x\} - \mathbf{P}\{\eta_n < x\} \right] = 0.$$

## 17. The spectral function of random matrices

Let $\Xi_n$ be a complex random matrix. Denote its eigenvalues by $\lambda_i$, $i = 1, \ldots, n$. By the spectral function of the matrix $\Xi_n$ is meant the expression

$$\mu_n(x, y) = c_n^{-1} \sum_{k=1}^{n} F(x - \mathrm{Re}(\lambda_k / b_n)) F(y - \mathrm{Im}(\lambda_k / b_n)),$$

where $F(x) = 0$ as $x < 0$, and $F(x) = 1$ as $x \geqslant 0$, and $c_n$, $b_n$ are certain nonrandom numbers. If $c_n = n$, then $\mu_n(x, y)$ is the normalized spectral function of the matrix $\Xi_n b_n^{-1}$. If the eigenvalues $\lambda_i$ of the matrix $\Xi_n$ are real, then the normalized spectral function of the matrix $\Xi_n$ takes the form

$$\mu_n(x) = n^{-1} \sum_{i=1}^{n} F(x - \lambda_i).$$

Obviously, the distribution functions are realizations of the function $\mu_n(x)$.

A random determinant can be represented in the form

$$c_n^{-1} \ln \det \left( \Xi_n b_n^{-1} \right) = \iint \ln(x + \mathrm{i}y)\, \mathrm{d}\mu_n(x, y),$$

assuming that the integral on the right-hand side of the equation exists.

Let $\Xi_n$ be a real random matrix, $\lambda_i$, $i = 1, \ldots, n$, the eigenvalues of the matrix $\Xi_n \Xi_n' \exp(-2a_n)$. Denote the spectral function by

$$\nu_n(x) = (2b_n)^{-1} \sum_{i=1}^{n} F(x - \lambda_i)\varphi(\lambda_i),$$

where $\varphi(x)$ is a continuous function on $(-\infty, \infty)$.

Then

$$\big[\ln|\det \Xi_n| - a_n\big] b_n^{-1} = \int_0^\infty \ln x \big(\varphi(x)\big)^{-1} \mathrm{d}\nu_n(x),$$

under the condition that the integral on the right-hand side of the equation exists.

By the notation $\mu_n(x) \Rightarrow \mu(x)$ $\{\mu_n(x) \simeq\!\!\rightarrow \mu(x)\}$ is meant the convergence of the finite-dimensional distributions of the random spectral functions $\mu_n(x)$ to the corresponding finite-dimensional distributions of random spectral function $\mu(x)$ (at the points of the stochastic continuity of the latter function).

Let the function $f(x)$ $(-\infty < x < \infty)$ be continuous and bounded on the whole real line $R_1$, let the $\mu_n(x)$ be the normalized spectral functions of the symmetric random matrices $\Xi_n$, $\mu_n(x) \Rightarrow \mu(x)$ on some everywhere dense set $C$ of the real line $R_1$, $\mu_n(-\infty) \Rightarrow \mu(-\infty)$, $\mu_n(+\infty) \Rightarrow \mu(+\infty)$, where $\mu(x)$ is a random distribution function. Then

$$\int f(x)\,\mathrm{d}\mu_n(x) \Rightarrow \int f(x)\,\mathrm{d}\mu(x).$$

Let the function $f(x)$ be continuous on the real line $R_1$; $\mu_n(x) \Rightarrow \mu(x)$ on some everywhere dense set $C$ of the real line $R_1$; $\mu_n(-\infty) \Rightarrow \mu(-\infty)$, $\mu_n(+\infty) \Rightarrow \mu(+\infty)$ for some $\alpha > 0$,

$$\sup_n \mathbf{E} \int \left|f(x)\right|^{1+\alpha} \mathrm{d}\mu_n(x) < \infty.$$

Then $\int f\,\mathrm{d}\mu_n \Rightarrow \int f\,\mathrm{d}\mu$.

Let the $\Xi_n$ be symmetric random matrices, $\mu_n(x)$ their normalized spectral functions, $\mu_n(x) \Rightarrow \mu(x)$ on some everywhere dense set $C$ of the real line $R_1$, $\mu_n(-\infty) \Rightarrow \mu(-\infty)$, $\mu_n(+\infty) \Rightarrow \mu(+\infty)$ for some $\alpha > 0$,

$$\sup_n n^{-1} \mathbf{E}\,\mathrm{Tr} \left| \ln|\Xi_n| \right|^{1+\alpha} < \infty.$$

Then

$$n^{-1} \ln|\det \Xi_n| \Rightarrow \int \ln|x|\,\mathrm{d}\mu(x).$$

It is convenient to prove limit theorems for $\mu_n(x)$, $\nu_n(x)$ with the help of the Stieltjes transformation:

$$\int (x-z)^{-1}\,\mathrm{d}\mu_n(x) = n^{-1}\,\mathrm{Tr}(\Xi - zI)^{-1},$$

where $z$ is a complex number, $\mathrm{Im}\, z \neq 0$, $\Xi$ is a symmetric random matrix, and $\mu_n(x)$ is its normalized spectral function.

Write

$$\eta_n(t) = \int (1+\mathrm{i}tx)^{-1}\,\mathrm{d}\mu_n(x), \quad \xi_n(z) = \int (x-z)^{-1}\,\mathrm{d}\mu_n(x), \quad \mathrm{Im}\, z \neq 0.$$

The inversion formula at points of stochastic continuity $x_1$ and $x_2$ of the function $\mu_n(x)$ has the form

$$\mathbf{P}\{\mu_n(x_2) - \mu_n(x_1) < u\} = \lim_{\varepsilon\to 0}\mathbf{P}\Big\{\pi^{-1}\int_{x_2}^{x_1}\mathrm{Im}\,\xi_n(y+\mathrm{i}\varepsilon)\,\mathrm{d}y < u\Big\}.$$

Analogously, we obtain the inversion formula for finite dimensional distributions of the function $\mu_n(x)$

$$\mathbf{P}\{\mu_n(x_2^k) - \mu_n(x_1^k) < u_k,\ k=1,\dots,m\}$$
$$= \lim_{\varepsilon\to 0}\mathbf{P}\Big\{\pi^{-1}\int_{x_2^k}^{x_1^k}\mathrm{Im}\,\xi_n(y+\mathrm{i}\varepsilon)\,\mathrm{d}y < u_k,\ k=1,\dots,m\Big\},$$

where $x_1^k$, $x_2^k$, $k=1,\dots,m$, are points of stochastic continuity of the function $\mu_n(x)$.

Let $\mu_n(x)$ be a sequence of the random spectral functions and with probability 1,

$$\lim_{h\to-\infty}\sup_n \mathbf{E}\,\mu_n(h) = 0.$$

Then, in order that $\mu_n(x) \simeq\rightarrow \mu(x)$, where $\mu(x)$ is some random spectral function, it is necessary and sufficient that $\xi_n(z) \Rightarrow \xi(z)$, $\mathrm{Im}\ z \neq 0$.

Let $\mu_n(x)$ be a sequence of the random spectral functions and

$$\lim_{h\to-\infty}\sup_n \mathbf{E}\,\mu_n(h) = 0.$$

Then

a) in order that $\mu_n(x) \simeq\rightarrow \mu(x)$, where $\mu(x)$ is some random spectral function, it is necessary and sufficient that $\eta_n(t) \Rightarrow \eta(t)$;

b) in order that at every point of the continuity of the nonrandom distribution function $\mu(x) p\lim_{n\to\infty}\mu_n(x) = \mu(x)$, it is necessary and sufficient that for every $t$, $p\lim_{n\to\infty}\eta_n(t) = \eta(t)$ where

$$\eta(t) = \int (1+itx)^{-1}\,\mathrm{d}\mu(x).$$

Let $\mu_n(x)$ and $\lambda_m(x)$ be sequences of random spectral functions given on a common probability space, and with probability 1

$$\lim_{h\to-\infty}\sup_n \mu_n(h) = 0, \qquad \lim_{h\to-\infty}\sup_n \lambda_n(h) = 0,$$

$$m_n(t) = \int (1+\mathrm{i}tx)^{-1}\,\mathrm{d}\mu_n, \quad p_n(t) = \int (1+\mathrm{i}tx)^{-1}\,\mathrm{d}\lambda_n.$$

Then

a) in order that $\mu_n(x) \sim \lambda_n(x)$ on some everywhere dense set $C$, it is necessary and sufficient that $m_n(t) \sim p_n(t)$, $-\infty < t < \infty$;

b) in order that $p\lim_{n\to\infty}[\mu_n(x) - \lambda_n(x)] = 0$ for all $x$ from some everywhere dense set $C$, it is necessary and sufficient that for each $t$, $p\lim_{n\to\infty}[m_n(t) - p_n(t)] = 0$.

Let $\mu_n(x)$ and $\lambda_n(x)$ be the sequences of random spectral functions and with probability 1, $\lim_{h\to-\infty}\sup_n \mathbf{E}\mu_n(h) = 0$. In order that at every point of continuity of some nonrandom distribution function $\mu(x)$ whose Stieltjes transformation equals

$$\eta(t) = \int (1+\mathrm{i}tx)^{-1}\,\mathrm{d}\mu(x), \qquad \lim_{n\to\infty}\mu_n(x) = \mu(x),$$

it is necessary and sufficient that with probability 1 for every $t$, $\lim_{n\to\infty}\eta_n(t) = \eta(t)$.

## 18. Canonical spectral equation

A peculiar feature of the normalized spectral functions of a symmetric random matrix with independent entries on the diagonal and above is their convergence to some nonrandom function of distribution under the condition that the dimension of the matrices is increasing. Let $\Xi_n = (\xi_{ij}^{(n)})$ be a symmetric random matrix and $\mu_n(x)$ its normalized spectral function.

If for every $n$ the vectors $\vec{\xi}_i = (\xi_{ii}^{(n)}, \xi_{ii+1}^{(n)}, \ldots, \xi_{in}^{(n)})$, $i = 1, \ldots, n$, are given by independent, random values $\xi_{ij}^{(n)}$, $i, j, n = 1, 2, \ldots$, on a common probability space, and there exists a limit

$$\lim_{n\to\infty} n^{-1}\mathbf{E}\,\mathrm{Tr}(I + \mathrm{i}t\Xi_n)^{-1} = m(t)$$

and the function $m(t)$ is continuous at zero, then with probability 1, $\lim_{n\to\infty}\mu_n(x) = \mu(x)$ at every point of continuity of the nonrandom function $\mu(x)$, whose Stieltjes transformation is equal to

$$\int (1+\mathrm{i}tx)^{-1}\,\mathrm{d}\mu_n = m(t).$$

For every $n$, let the random entries $\xi_{ij}^{n}$, $i \geqslant j$, $i,j = 1,\dots,n$, of the matrix $\Xi_n = (\xi_{ij}^{(n)} - a_{ij}^{(n)})$ be independent, infinitesimal,

$$a_{ij} = \int_{|x|<\tau} x\,\mathrm{d}\mathbf{P}\{\xi_{ij} < x\}, \quad \tau > 0, \text{ is an arbitrary constant,}$$

$K_n(u,v,z) \Rightarrow K(u,v,z)$, where

$$K_n(u,v,z) = n\int_{-\infty}^{z} y^2\left(1+y^2\right)^{-1}\mathrm{d}\mathbf{P}\{\xi_{ij} - a_{ij} < y\};$$

$$i/n \leqslant u < (i+1)/n;\ j/n \leqslant v < (j+1)/n,$$

with the $K(u,v,z)$ a nondecreasing function with bounded variation on $z$ and continuous on $u$ and $v$ in the domain $0 \leqslant u,\ v \leqslant 1$. Then with probability 1 for almost all $x$

$$\lim_{n\to\infty} \mu_n(x) = F(x),$$

where $F(x)$ is a distribution function whose Stieltjes transformation equals

$$\int (1+\mathrm{i}tx)^{-1}\,\mathrm{d}F(x) = \lim_{\alpha\downarrow 0}\int_0^1\int_0^1 x\,\mathrm{d}G_\alpha(x,z,t)\,\mathrm{d}z,$$

where $G_\alpha(x,z,t)$ is a distribution function on $x$ $(0 \leqslant x \leqslant 1,\ 0 \leqslant z \leqslant 1,\ -\infty < t < \infty)$, satisfying the canonical spectral equation at the points of continuity

$$G_\alpha(x,z,t) = \mathbf{P}\left\{\left[1 + t^2\xi_\alpha\left(G_\alpha(\cdot,\cdot,t),z\right)\right]^{-1} < x\right\},$$

and $\xi_\alpha(G_\alpha(\cdot,\cdot,t),z)$ is a random functional whose Laplace transformation of one dimensional distributions equals

$$\begin{aligned}
&\mathbf{E}\exp\left\{-s\xi\left(G(\cdot,\cdot,t),z\right)\right\}\\
&\quad = \exp\left\{\int_0^1\int_0^1\left[\int_0^\infty \left(\exp\left\{-syx^2\left(1+\alpha|x|\right)^{-2}\right\} - 1\right)\right.\right.\\
&\qquad \left.\left.\times\left(1+x^{-2}\right)\mathrm{d}K(v,z,x)\right]\mathrm{d}G(y,v,t)\,\mathrm{d}v\right\}, \quad \alpha > 0,\ s \geqslant 0.
\end{aligned}$$

The solution of the canonical spectral equation exists and is unique in the class $L$ of the functions $G(x,z,t)$, which are distribution functions on $x$ $(0 \leqslant x \leqslant 1)$ for any fixed $0 \leqslant z \leqslant 1$, $-\infty < t < \infty$ and such that for any integer $k > 0$ and $z$ the functions $\int x_k dG_\alpha(x,z,t)$ are analytical on $t$ (excluding, perhaps, the point zero).

## 19. The Wigner semicircle law

Let $\Xi_n = (\xi_{ij}^{(n)})_{i,j=1}^n$, $n = 1, 2, \ldots$, be symmetric matrices,

$$\mu_n(x) = n^{-1} \sum_{\lambda_{in} < x} 1,$$

where $\lambda_{in}$, $i = 1, \ldots, n$, are the eigenvalues of $\Xi_n$, and the random variables $\xi_{ij}^{(n)}$, $i, j = 1, \ldots, n$, $n = 1, 2, \ldots$, given on a common probability space.

A semicircle law is any assertion which states that the normalized spectral function $\mu_n(x)$ converges, with probability 1 or in probability, to a nonrandom spectral function $\mu(x)$ whose density has the semicircle form:

$$\mu'(x) = \begin{cases} (2\pi\sigma^2)^{-1}\sqrt{4\sigma^2 - x^2}, & |x| \leqslant 2\sigma, \\ 0, & |x| > 2\sigma, \ \sigma > 0. \end{cases}$$

If the random variables $\xi_{ij}^{(n)}$, $i \geqslant j$, $i, j = 1, \ldots, n$, are independent for each $n$, $\mathbf{E}\,\xi_{ij}^{(n)} = 0$, and $\mathbf{Var}\,\xi_{ij}^{(n)} = \sigma^2/n$, and $0 < \sigma^2 < \infty$, then $\lim_{n\to\infty} \mu_n(x) = \mu(x)$ with probability 1 if and only if for every $\tau > 0$

$$\lim_{n\to\infty} n^{-1} \sum_{i,j=1}^{n} \mathbf{E}\big[\xi_{ij}^{(n)}\big]^2 \chi\big(|\xi_{ij}^{(n)}| > \tau\big) = 0.$$

## 20. Limit theorems for determinants of random Jacobi matrices

Many problems of theoretical physics and numerical analysis can be reduced to the determination of the distribution function $F(x)$ of the eigenvalues of a random Jacobi matrix

$$\Xi_n = (\xi_i \delta_{ij} + \eta_i \delta_{ij-1} + \zeta_i \delta_{ij+1}),$$

where

$$\delta_{ij} = \begin{cases} 0, & i \neq j, \\ 1, & i = j, \end{cases}$$

is the Kronecker symbol.

In particular limit theorems for $\det \Xi_n$ are very important.

For each $n$, let the random 3-vectors $(\xi_k, \eta_k, \zeta_k)$, $k = 1, \ldots, n$, be independent and with probability 1,

$$\xi_k - |\eta_k \zeta_k| \geqslant 1, \quad |\eta_{k-1}\zeta_{k-1}| \leqslant 1, \quad \xi_n \geqslant 1, \ k = 1, \ldots, n, \ n = 1, 2, \ldots;$$

$$\sup_n \sup_{k=1,\dots,n} \mathbf{E} \ln^2 \left(\xi_k \pm |\eta_k \zeta_k| \pm |\eta_{k-1}\zeta_{k-1}|\right) < \infty.$$

Then

$$p \lim_{n\to\infty} n^{-1}(\ln \det \Xi_n - \mathbf{E} \ln \det \Xi_n) = 0.$$

## 21. The Dyson equation

Let the random variables $\xi_i$, $i = 1, 2, \dots$, of the matrices

$$\Xi_n = \left((2 + \xi_i)\delta_{ij} - \delta_{ij-1} - \delta_{ij+1}\right)$$

be independent, non-negative, and identically distributed, let the sequence of sums

$$\sum_{k=1}^{n} \xi_k, \quad n = 1, 2, \dots,$$

tend in probability to infinity, and let for some $\delta > 0$,

$$\mathbf{E}|\ln \xi_1|^{1+\delta} < \infty.$$

Then

$$p \lim_{n\to\infty} n^{-1} \ln \det \Xi_n = \int_1^\infty \ln x \, \mathrm{d}F(x),$$

where the distribution function $F(x)$ satisfies the Dyson integral equation

$$F(x) = \iint_{2+y-z^{-1}<x} \mathrm{d}F(z)\, \mathrm{d}\mathbf{P}\{\xi_1 < y\}.$$

## 22. The stochastic Sturm–Liouville problem

Let us study the distribution of eigenvalues of the differential equation

$$u''(t) + \left(\xi(t) + \lambda\right)u(t) = 0; \quad u(0) = u(1) = 0,$$

where $\xi(t)$ is a real, continuous and bounded from below random process defined on $[0, L]$.

Sometimes, instead of boundary conditions, we use the following conditions

$$u(0)\cos\alpha - u'(0)\sin\alpha = 0,$$

$$u(L)\cos\beta - u'(L)\sin\beta = 0.$$

In the case when such differential equation can be approximately reduced to a difference equation in order to solve the stochastic Sturm–Liouville problem, it is necessary to use limit theorems for determinants of random Jacobi matrices

$$\Xi_n(\lambda) = \left\{\delta_{ij}\left(2 + n^2\left(\xi\left(\frac{i}{n}\right) + \lambda\right)\right) - \delta_{ij-1} - \delta_{ij+1}\right\}.$$

The matrix $\Xi_n$ is a non-negative-positive definite matrix. Consider the random process

$$\lambda_n(x) = \sum_{i=1}^{n} \lambda_{in}^{-1} F(x - \lambda_{in}),$$

where $\lambda_{1n} \geqslant \lambda_{2n} \geqslant \cdots \geqslant \lambda_{nn}$ are the eigenvalues of the matrix $\Xi_n(0)$. It is obvious that

$$\int_0^\infty (1 + \lambda x)^{-1}\, \mathrm{d}\lambda_n(x) = \frac{\mathrm{d}}{\mathrm{d}\lambda} \ln\det \Xi_n(\lambda).$$

Let $\xi(t)$ be a measurable process on $[0, L]$ such that

$$\mathbf{P}\left\{\inf_{0\leqslant t\leqslant L} \xi(t) > 0\right\} = 1,$$

$$\lim_{h\to\infty} \mathbf{P}\left\{\sup_{0\leqslant t\leqslant L} \xi(t) \geqslant h\right\} = 0.$$

Then for all $\lambda \geqslant 0$,

$$n^{-1} \ln\det \Xi_n(\lambda) \Rightarrow \int_0^L \left\{\mathbf{E}\left[\exp\left\{-\frac{1}{2}\int_0^t (\xi(x) + \lambda)\omega^2(x)\, \mathrm{d}x\right\}/\sigma\right]\right\}^{-2} \mathrm{d}t,$$

$$\lambda_n(x) \simeq\!\!\to \lambda(x), \quad 0 \leqslant x < \infty,$$

where $\lambda(x)$ is a nondecreasing, random process, bounded with probability 1, whose Stieltjes transform is

$$\int_0^\infty (1 + tx)^{-1} \mathrm{d}\lambda(x)$$
$$= \frac{\mathrm{d}}{\mathrm{d}t} \ln \int_0^L \left\{\mathbf{E}\left[\exp\left\{-\frac{1}{2}\int_0^t (\xi(x) + \lambda)\omega^2(x)\, \mathrm{d}x\right\}/\sigma\right]\right\}^{-2} \mathrm{d}t.$$

as $n \to \infty$, where $\omega(x)$ is a Brownian motion process which is independent of $\xi(t)$, and $\sigma$ is the minimal $\sigma$-algebra, with respect to which the process $\xi(x)$ is measurable, $x \in [0, L]$.

In order to find the limiting spectral functions of random Jacobi matrices, we have to invert their Stieltjes transform, which is the solution of the Dyson equation. Note that such an inversion, in general, is a very difficult task. The Sturm oscillation theorem makes it possible to avoid this operation in some cases. We shall give one of its generalizations.

Let $A_n$ be a symmetric real matrix of order $n$ and let $\det A_i$, $i = 1, \ldots, n$, $(\det A_0 = 1)$ be the sequence of its main minors, $\det A_i \neq 0,\ i = 0, \ldots, n$.

Then the number of negative eigenvalues of matrix $A$ is equal to the number of changes of the sign in the sequence $\det A_i,\ i = 0, \ldots, n$.

Let $\Xi_n = (\xi_{ij}^{(n)})$ be a random real symmetric matrix of order $n$, let $\det \Xi_0 = 1$, $\det \Xi_i,\ i = 0, \ldots, n$, be its main minors, $\mu_n(x)$ the normalized spectral function of $\Xi_n$.

If the random entries $\xi_{ij}^{(n)},\ i \geqslant j,\ i, j = 1, \ldots, n$, are independent and have continuous distributions for every $n$,

$$\lim_{h \to -\infty} \sup_n \mathbf{E}\, \mu_n(h) = 0,$$

then with probability 1 for almost all $x$,

$$\lim_{n \to \infty} \big(\mu_n(x) - \mathbf{E}\, \mu(x)\big) = 0,$$

where

$$\mathbf{E}\, \mu_n(x) = n^{-1} \sum_{i=1}^{n} \mathbf{P}\big\{ \det(\Xi_{i-1} - Ix)\big(\det(\Xi_i - Ix)\big) < 0\big\}.$$

## 23. The central limit theorem for determinants of random Jacobi matrices

Let $\Xi_n = (\xi_i \delta_{ij} + \eta_i \delta_{ij-1} + \zeta_j \delta_{ij+1})$, and let $\sigma_k$ be the minimal $\sigma$-algebra with respect to which the random variables $\xi_l\, \eta_l,\ \zeta_l,\ l = k+1, \ldots, n$, are measurable. Suppose that $\mathbf{E} \ln^2 |\det \Xi_n|$ exists. Then

$$\begin{aligned} &\ln|\det \Xi_n| - \mathbf{E} \ln|\det \Xi_n| \\ &\quad = \sum_{k=1}^{n} \Big\{ \mathbf{E}\Big[ \ln\Big| - \eta_{k-1}\zeta_{k-1} \frac{d_{k-2}}{d_{k-1}} + \xi_k - \eta_k \zeta_k \frac{b_{n-(k+2)}}{b_{n-(k+1)}} \Big| \Big/ \sigma_{k-1} \Big] \\ &\qquad - \mathbf{E}\left[ \ln| - \eta_{k-1}\zeta_{k-1} d_{k-2} d_{k-1}^{-1} + \xi_k - \eta_k \zeta_k b_{n-(k+2)} b_{n-(k+1)}^{-1} | / \sigma_k \right] \Big\}, \end{aligned}$$

where

$$d_k = \det(\xi_i \delta_{ij} + \eta_i \delta_{ij-1} + \zeta_j \delta_{ij+1}), \quad i, j = 1, \ldots, k,$$

$$b_{n-k} = \det(\xi_i \delta_{ij} + \eta_i \delta_{ij-1} + \zeta_j \delta_{ij+1}), \quad i, j = k, \ldots, n.$$

On the basis of such representation the following assertion holds.

Let the random variables $\xi_i$, $i = 1, 2, \ldots$, of the matrix

$$\Xi_n = \big((2+\xi_i)\delta_{ij} - \delta_{ij+1} - \delta_{ij-1}\big)$$

be independent, non-negative, identically distributed, and suppose $\mathbf{E} \ln^2 \xi_1 < \infty$, and $\sigma^2 > 0$, where

$$\sigma^2 = \int_1^\infty \int_0^\infty \Big[ \int_1^\infty \ln(-z^{-1} + 2 + u - x^{-1})\, \mathrm{d}G(x) - \int_0^\infty \int_1^\infty \ln\big(-z^{-1} + 2 + u - x^{-1}\big)\, \mathrm{d}G(x)\, \mathrm{d}F(u) \Big]^2 \mathrm{d}F(u)\, \mathrm{d}G(z),$$

with the distribution function $G(z)$ satisfying the integral equation

$$G(x) = \iint_{2+y-z^{-1}<x,\, z \geqslant 1} \mathrm{d}G(z)\, \mathrm{d}F(y), \quad F(y) = \mathbf{P}\{\xi_1 < y\}.$$

Then

$$\lim_{n\to\infty} \mathbf{P}\big\{n^{-1/2}\sigma^{-1}[\ln \det \Xi_n - \mathbf{E} \ln \det \Xi_n] < x\big\} = (2\pi)^{-1/2} \int_{-\infty}^{x} \exp\big(-y^2/2\big)\, \mathrm{d}y.$$

## 24. The Fredholm random determinants

Let $\Xi_n$ be a square random matrix. We call the random function $\det(I + t\Xi_n)$, where $t$ is a real or complex variable, the Fredholm random determinant of the matrix $\Xi_n$. Fredholm random determinants carry important information about random matrices. With their help, the limiting distributions for eigenvalues of the random matrices can be found. In this section, on the basis of limit theorems for Fredholm random determinants, limit theorems for the eigenvalues of symmetric and nonsymmetric random matrices are given.

Let $\Xi_n = (\xi_{ij}^{(n)})$ be square random matrices of the order $n$. Arrange the eigenvalues of the matrix $\Xi_n \Xi_n^{\mathrm{T}}$ in nonincreasing order $\lambda_{1n} \geqslant \lambda_{2n} \geqslant \cdots \geqslant \lambda_{nn}$. Consider the random process $\lambda_n(x)$ equal to the sum of eigenvalues belonging to the semi-interval $[0, x)$. If with probability 1, $\operatorname{Tr} \Xi_n \Xi_n^{\mathrm{T}} < \infty$, then

$$\int_0^\infty (1+tx)^{-1} \mathrm{d}\lambda_n(x) = \mathrm{d} \ln \det\big(I + t\Xi_n \Xi_n'\big)/\mathrm{d}t := \eta_n(t).$$

If

$$\lim_{h\to\infty} \overline{\operatorname{Lim}}_{n\to\infty} \mathbf{P}\big\{\lambda_n(+\infty) \geqslant h\big\} = 0,$$

then in order that $\lambda_n(x) \simeq\rightarrow \lambda(x)$, $x \geqslant 0$, where $\lambda(x)$ is a random function, nondecreasing and of bounded variation with probability 1, it is necessary and sufficient that $\eta_n(t) \Rightarrow \eta(t)$, $t \geqslant 0$, where $\eta(t)$ is some random function.

## 25. Limit theorems for eigenvalues of random matrices

Let $\Xi_n = (\xi_{ij}^{(n)})$ be square random matrices of the order $n$. Arrange the eigenvalues of the matrix $\Xi_n \Xi_n^{\mathrm{T}}$ in nonincreasing order $\lambda_{1n} \geqslant \lambda_{2n} \geqslant \cdots \geqslant \lambda_{nn}$. See notation in Section 16.

For every $n$ let the random entries $\xi_{ij}^{(n)}$, $i, j = 1, \ldots, n$, of the matrix $\Xi_n$ be independent; let the vector rows and vector columns of the matrix $\Xi_n$ be asymptotically constant,

$$\lim_{n\to\infty} \operatorname{Tr} B_n B_n^{\mathrm{T}} = 0,$$

$$\sum_{i,j=1}^{n} \left[1 - F_{ij}^{(n)}(z)\right] \Rightarrow K(z), \quad z \geqslant 0,$$

where $F_{ij}^{(n)}(z) = \mathbf{P}\{\nu_{ij}^2 < z\}$, and the function $K(z)$ is continuous and bounded for all $z > 0$.

Then for all integers $k_1 > k_2 > \cdots > k_m > 0$ and real numbers $x_m \geqslant x_{m-1} \geqslant \cdots \geqslant x_1 > 0$,

$$\begin{aligned}
&\lim_{n\to\infty} \mathbf{P}\{\lambda_{k_1 n} < x_1, \ldots, \lambda_{k_m n} < x_m\} \\
&\quad = (-1)^m \left[(k_{m-1})!\right]^{-1} \int_0^{x_1} \exp\left(-K(z_1)\right) \mathrm{d}K(z_1) \\
&\qquad \times \prod_{i=1}^{m-1} \left\{ \left[(k_i - k_{i+1} - 1)!\right]^{-1} \right. \\
&\qquad \times \left. \int_{z_i}^{x_{i+1}} \left[K(z_i) - K(z_{i+1})\right]^{k_i - k_{i+1} - 1} \mathrm{d}K(z_{i+1}) \right\} \left[K(z_m)\right]^{k_m - 1}.
\end{aligned}$$

From this formula it follows that

$$\lim_{n\to\infty} \mathbf{P}\{\lambda_{kn} < x\} = -\left[(k-1)!\right] \int_0^x \exp\left[-K(z)\right] \left[K(z)\right]^{k-1} \mathrm{d}K(z), \quad x > 0,$$

and for every $k > m$, $0 < x < y$,

$$\begin{aligned}
&\lim_{n\to\infty} \mathbf{P}\{\lambda_{kn} < x,\ \lambda_{mn} < y\} \\
&\quad = \left[(m-1)!(k-m-1)!\right]^{-1} \int_0^x \exp\left[-K(z_1)\right] \mathrm{d}K(z_1)
\end{aligned}$$

$$\times \int_{z_1}^{y} \left[K(z_1) - K(Z_2)\right]^{k-m-1} \left[K(z_2)\right]^{m-1} \mathrm{d}K(z_2).$$

If the function $K(z)$ is differentiable, then

$$\lim_{n\to\infty} \mathbf{P}\{\lambda_{kn} < x\} = \exp\left(-K(x)\right) \sum_{m=0}^{k-1} (m!)^{-1} K^m(x).$$

Let $\lambda_{m_n} < \cdots < \lambda_1$ be the eigenvalues of the covariance matrix $R_{m_n}$ and let the vectors $\vec{x}_1, \ldots, \vec{x}_n$ be independent observations of a random vector $\vec{\xi}$ distributed according to the normal law $N(\vec{a}, R_{m_n})$, and $\hat{R}_{m_n}$ is the empirical covariance matrix:

$$\hat{R}_{m_n} = n^{-1} \sum_{k=1}^{n+1} \left(\vec{x}_k - \hat{\vec{a}}\right)\left(\vec{x}_k - \hat{\vec{a}}\right)^{\mathrm{T}}$$

and $\hat{\vec{a}}$ is the empirical expectation:

$$\hat{\vec{a}} = n^{-1} \sum_{k=1}^{n} \vec{x}_k.$$

Assume that the conditions $\limsup_{n\to\infty} m_n n^{-1} < 1$, $\lambda_k(R_{m_n}) \leqslant c < \infty$ hold. Then

$$p \lim_{n\to\infty} [\lambda_1 - \alpha_2] = 0, \quad p \lim_{n\to\infty} [\lambda_m - \alpha_1] = 0,$$

where

$$\alpha_i = \nu_i(1-\gamma) - \frac{1}{m} \sum_{k=1}^{m} \gamma \nu_i^2 (\lambda_k - \nu_i)^{-1}, \qquad \gamma = m_n n^{-1},\ i = 1, 2,$$

$\nu_1 = \min\{y_i\}$, $\nu_2 = \max\{y_i\}$ and the $y_i$ are the real solutions of the equation

$$1 - \gamma - \frac{2}{m} \sum_{k=1}^{m} \gamma y_i (\lambda_k - y_i)^{-1} = \frac{1}{m} \sum_{k=1}^{m} \gamma y_i^2 (\lambda_k - y_i)^{-2}.$$

If, in addition, $\lambda_k(R_{m_n}) = 1$, $k = 1, \ldots, m$, then

$$p \lim_{n\to\infty} \left[\lambda_m(\hat{R}_{m_n}) - (1-\sqrt{\gamma})^2\right] = 0,\ p \lim_{n\to\infty} \left[\lambda_1(\hat{R}_{m_n}) - (1+\sqrt{\gamma})^2\right] = 0,$$

$$p \lim_{n\to\infty} \left[\lambda_k(\hat{R}_{m_n}) - c_k\right] = 0, \quad k \neq 1, m,$$

where $c_k$ is the unique real solution of the equation

$$(k-1)m^{-1} = (2\pi\gamma)^{-1} \int_{(1-\sqrt{\gamma})^2}^{c_k} y^{-1} \{ \left[y - (1-\sqrt{\gamma})^2\right]$$

$$\times \left[ (1+\sqrt{\gamma})^2 - y \right] \Big\}^{1/2} \mathrm{d}y, \quad k = 2, \ldots, m_n - 1.$$

Consider the sequence of symmetric random matrices $\Xi_n = (\xi_{ij}^{(n)})_{i,j=1}^n$, $n = 1, 2, \ldots$, whose entries $\xi_{ij}^{(n)}$, $i \geqslant j$, $i, j = 1, \ldots, n$, are independent for every $n$, and let

$$\mathbf{E}\, \xi_{ij}^{(n)} = a_i^{(n)} \delta_{ij}, \quad \mathbf{E} \left[ \xi_{ij}^{(n)} \right]^2 = \sigma^2 n^{-1}, \quad 0 < \sigma^2 < \infty,$$

and for some $\beta > 0$

$$\sup_n \sup_{i,j=1,\ldots,n} \mathbf{E} \left| \xi_{ij}^{(n)} n^{1/2} \right|^{4+\beta} < \infty,$$

and $\sup_n \sup_{i=1,\ldots,n} |a_i^{(n)}| < \infty$. Let $\lambda_1 \geqslant \cdots \geqslant \lambda_n$ be the eigenvalues of this random matrix. Then

$$\operatorname*{p\,lim}_{n\to\infty} (\lambda_1 - r_1) = 0, \quad \operatorname*{p\,lim}_{n\to\infty} (\lambda_n - r_2) = 0,$$

where

$$r_i = y_i + \frac{1}{n} \sum_{k=1}^n \sigma^2 \left( y_i - a_i^{(n)} \right)^{-1},$$

$y_1 = \max\{\nu_i\}$, $y_2 = \min\{\nu_i\}$, and $\nu_i$ are the real solutions of the equation

$$n^{-1} \sum_{k=1}^n \left( x - a_k^{(n)} \right)^{-2} \sigma^2 = 1.$$

If in addition $a_i^{(n)} \equiv 0$, $i = 1, \ldots, n$, $\sigma^2 = 1$, then

$$\operatorname*{p\,lim}_{n\to\infty} \lambda_1 = 2, \qquad \operatorname*{p\,lim}_{n\to\infty} \lambda_n = -2,$$

$$\operatorname*{p\,lim}_{n\to\infty} |\lambda_k - b_k| = 0, \quad k \neq n,$$

where the value $b_k$, $k = 1, \ldots, n-1$, is the unique real solution of the equation

$$(2\pi)^{-1} b_k \left( 1 - b_k^2/4 \right)^{1/2} + \pi^{-1} \arcsin(b_k/2) + 1/2 = kn^{-1}.$$

## 26. The systems of linear algebraic equations with random coefficients

By a system of linear random algebraic equations we mean a equality $\Xi \vec{x}(\omega) = \vec{\eta}(\omega)$, where $\Xi = (\xi_{ij})$ is a random matrix, $\vec{\eta}(\omega)$ is a random vector, and $\vec{x}(\omega)$ is a desired vector from some set $D$ of random vectors whose dimension is the same as that of the vector $\vec{x}(\omega)$.

We call the system of equations $\Xi\vec{x}(\omega) = \vec{\eta}(\omega)$ normal if the entries of matrix $\Xi$ or of the vector $\vec{\eta}$ or if the entries of both of them are distributed according to a joint normal law.

The equation $\Xi\vec{x}(\omega) = \vec{\eta}(\omega)$ has a unique solution if $\Xi_n$ is a square matrix and $\mathbf{P}\{\det \Xi = 0\} = 0$. If the square matrix $\Xi_n$ and the vector $\vec{\eta}$ have the joint distribution density $p(Z_n, x)$ then the distribution density of the solution of equations $\Xi\vec{x} = \vec{\eta}$ is equal to

$$\int p(Z_n, Z_n\vec{y})|\det Z_n|\,\mathrm{d}Z_n,$$

on the assumption that this integral exists.

If the vector $\vec{\eta}_n$ does not depend on the matrix $\Xi_n$ and is distributed according to a nondegenerate normal law with parameters $\vec{a}_n$, $T_n$ and $\mathbf{P}\{\det \Xi_n = 0\} = 0$, then the distribution density of the solution $\vec{x}_n(\omega)$ of the system $\Xi_n\vec{x}_n(\omega) = \vec{\eta}_n(\omega)$ is equal to

$$\begin{aligned} p(\vec{y}_n) := {} & (2\pi)^{-n/2}\det T_n^{-1/2}\mathbf{E}|\det \Xi_n| \\ & \times \exp\big\{-0.5\big(T_n^{-1}(\Xi_n\vec{y}_n - \vec{a}_n), (\Xi_n\vec{y}_n - \vec{a}_n)\big)\big\}. \end{aligned}$$

In this formula, we suppose the distribution of matrix $\Xi_n$ to be such that the density $p(\vec{y}_n)$ exists. For example, we can require that

$$\mathbf{E}\,|\det \Xi_n| < \infty.$$

In particular the density of the solutions of some systems of normal linear algebraic equations has an explicit form:

Let the vector $\eta_n$ be normally distributed $N(0,1)$, the column vectors $\vec{\xi}_i = (\xi_{i1}, \ldots, \xi_{in})$, $i = 1, \ldots, n$, of matrix $\Xi_n$ be independent, not depending on the vector $\vec{\eta}$, and let they be normally $N(0, R_n)$ distributed (the matrix $R_n$ nondegenerate). Then the distribution density of the solution $\vec{x}_n$ of the system of equations $\Xi_n\vec{x}_n = \vec{\eta}_n$ is equal to

$$p(\vec{y}_n) = \Gamma\big((n+1)/2\big)\det R_n^{1/2}\pi^{-(n+1)/2}\big[1 + (R_n\vec{y}_n, \vec{y}_n)\big]^{-(n+1)/2}.$$

## 27. The arctangent law

Let us consider systems of linear algebraic equations $\Xi_n\vec{x}_n = \vec{\eta}_n$, where $\Xi_n = (\xi_{ij}^{(n)})$ is a real random square matrix of order $n$, and $\vec{\eta}_n = (\eta_1, \ldots, \eta_n)$ is a random vector. If $\det \Xi \neq 0$, then the solution of this system exists and equals $\vec{x}_n = \Xi_n^{-1}\vec{\eta}_n$; if $\det \Xi_n = 0$, then the solution cannot exist. Suppose, that the components $x_k^{(n)}$ of the vector $x_n$ are equal to some constant if $\det(\Xi_n) = 0$.

For every $n$, let the random variables $\xi_{ij}$, $\eta_i$, $i,j=1,\dots,n$, be independent, $\mathbf{E}\,\xi_{ij}=0$, $\mathbf{E}\,\eta_i=0$, $\mathbf{Var}\,\xi_{ij}=\mathbf{Var}\,\eta_i=\sigma^2$, $0<\sigma^2<\infty$, $i,j=1,\dots,n$, for some $\delta>0$

$$\sup_{n,i,j}\mathbf{E}\left[|\xi_{ij}|^{4+\delta}+|\eta_i|^{4+\delta}\right]<\infty.$$

Then for any $k\neq l,\ k,l=1,\dots,n$,

$$\lim_{n\to\infty}\mathbf{P}\{x_k^{(n)}<z\}=\lim_{n\to\infty}\mathbf{P}\{x_k^{(n)}/x_l^{(n)}<z\}=1/2+\pi^{-1}\operatorname{arctg} z,$$

$$\begin{aligned}&\lim_{n\to\infty}\mathbf{P}\{x_{i_1}^{(n)}<y_1,\dots,x_{ik}^{(n)}<y_k\}\\&\quad=\pi^{-(k+1)/2}\Gamma\big((k+1)/2\big)\\&\qquad\times\int_{-\infty}^{y_1}\cdots\int_{-\infty}^{y_k}\left(1+z_1^2+\cdots+z_k^2\right)^{-(k+1)/2}\prod_{i=1}^{k}\mathrm{d}z_1,\end{aligned}$$

where $i_1,\dots,i_k$ are any distinct integer numbers from 1 to $n$.

## 28. The circle law

For every $n$, let the random entries $\xi_{pl}^{(n)}$, $l,p=1,\dots,n$, of a complex matrix $H_n=(\xi_{pl}^{(n)}n^{-1/2})$ be independent, $\mathbf{E}\,\xi_{pl}^{(n)}=0$, $\mathbf{E}\,|\xi_{pl}^{(n)}|^2=\sigma^2$, $0<\sigma<\infty$ and let the quantities $\operatorname{Re}\xi_{kl}^{(n)}$, $\operatorname{Im}\xi_{kl}^{(n)}$ have distribution densities $p_{kl}(x)$ and $q_{ik}(x)$ satisfying the condition: for some $\beta>1$,

$$\sup_n\sup_{k,l=1,\dots,n}\int\left[p_{kl}^{\beta}(x)+q_{kl}^{\beta}(x)\right]\mathrm{d}x<\infty,$$

and for some $\delta>0$,

$$\sup_n\sup_{k,l=1,\dots,n}\mathbf{E}\left|\xi_{pl}^{(n)}\right|^{2+\delta}<\infty.$$

Then for any $x$ and $y$,

$$p\lim_{n\to\infty}\nu_n(x,y)=\nu(x,y),$$

where

$$\left[\partial^2\nu(x,y)/\partial x\partial y\right]=\sigma^{-2}\pi^{-1}$$

for $x^2+y^2<\sigma^2$, $=0$, $x^2+y^2\geqslant\sigma^2$,

$$\nu_n(x,y)=n^{-1}\sum_{k=1}^{n}\chi(\operatorname{Re}\lambda_k<x)\chi(\operatorname{Im}\lambda_k<y),$$

$\lambda_k$ are the eigenvalues of the matrix $H_n$.

## 29. The elliptic law

Assume that the random entries $\xi_{pl}^{(n)}$ and $\xi_{lp}^{(n)}$ of the matrix $H_n=(\xi_{pl}^{(n)})$ are dependent.

Suppose that for every $m=1,2,\ldots$ the random vectors $(\xi_{pl}^{(n)},\xi_{lp}^{(n)})$, $p\geqslant l$ $(p,l=1,\ldots,n)$ are stochastically independent,

$$\mathbf{E}|\xi_{pl}^{(n)}|^2=n^{-1},\quad \mathbf{E}\,\xi_{pl}^{(n)}\xi_{lp}^{(n)}=\rho/n,\quad 0\leqslant|\rho|<1,\ p\neq l,$$

and that the real and imaginary parts of random elements $\xi_{pl}^{(n)},\xi_{lp}^{(n)}$ have distribution densities $q_{pl}(x_1,x_2,y_1,y_2)$ satisfying the condition

$$\sup_n\ \sup_{p,l=1,\ldots,n}\ \sup_{y,x} q_{pl}(x,y)<\infty,$$

where

$$q_{pl}(x,y)=\iint q(x,x_1,y,y_1)\,\mathrm{d}x_1\,\mathrm{d}y_1.$$

Then

$$p\lim_{n\to\infty}\nu_n(x,y)=\lambda(x,y),$$

where

$$\begin{aligned}(\partial^2/\partial x\partial y)\lambda(x,y)=\pi^{-1}&\big[1-\big(a^2+b^2\big)^2\big]^{-1}\\ &\times\chi[(bx-ay)^2(1-a^2-b^2)^{-2}(a^2+b^2)^{-1}+(ax+by)^2\\ &\times\big(1+a^2+b^2\big)^{-2}\big(a^2+b^2\big)^{-1}<1],\end{aligned}$$

$$a=\operatorname{Re}\rho^{1/2},\quad b=\operatorname{Im}\rho^{1/2}.$$

## 30. The unimodal law

Suppose that for every $n$ the entries of the random matrices $A_n = (\xi_{ij}^{(n)})_{i,j=1}^n$, $B_n = (\eta_{ij}^{(n)})_{i,j=1}^n$ are independent, and

$$\mathbf{E}\,\xi_{ij}^{(n)} = \mathbf{E}\,\eta_{ij}^{(n)} = 0,$$

$$\mathbf{E}\left[\xi_{ij}^{(n)}\right]^2 = \delta_{in}^2, \qquad \mathbf{E}\left[\eta_{ij}^{(n)}\right]^2 = \sigma_{in}^2,$$

$$0 < c_1 \leqslant \sigma_{in}^2 \leqslant c_2 < \infty, \quad 0 < c_1 \leqslant \delta_{in}^2 \leqslant c_2 < \infty,$$

and that for the random entries $\xi_{pl}^{(n)}$, $\eta_{ij}^{(n)}$, the condition

$$\sup_n \sup_{i,j=1,\dots,n} \mathbf{E}\left[\left|\xi_{ij}^{(n)}\right|^{4+\delta} + \left|\eta_{ij}^{(n)}\right|^{4+\delta}\right] < \infty$$

is satisfied, $\delta > 0$.

Then for almost all $x, y$,

$$p\lim_{n\to\infty}\left[\mu_n(x,y) - \int_{-\infty}^{x}\int_{-\infty}^{y} p_n(u,v)\,\mathrm{d}u\,\mathrm{d}v\right] = 0,$$

where

$$p_n(u,v) = n^{-1}\sum_{k=1}^{n}\left[\sigma_{kn}^2 + (u^2+v^2)\delta_{kn}^2\right]^{-2}\alpha,$$

$$\alpha = \left(\pi n^{-1}\sum_{k=1}^{n}\sigma_k^{-2}\delta_k^{-2}\right)^{-1},$$

$$\mu_n(x,y) = n^{-1}\sum_{k=1}^{n}\chi(\operatorname{Re}\lambda_{kn} < x)\chi(\operatorname{Im}\lambda_k < y),$$

and the $\lambda_{kn}$ are the eigenvalues of the matrix $A_n^{-1}B_n$.

## 31. The distribution of eigenvalues and eigenvectors of random matrix-valued processes

Let $\Xi_n(t)$ be a random symmetric square matrix of order $n$, with real random processes $\xi_{ij}(t)$, $t \geqslant 0$, be as its elements. Let $\lambda_1(t), \dots, \lambda_n(t)$, $\vec{\theta}_1(t), \dots, \vec{\theta}_n(t)$, be respectively the eigenvalues and eigenvectors of the matrix $\Xi_n(t)$. We arrange the eigenvalues so that $\lambda_1(t) \geqslant \cdots \geqslant \lambda_n(t)$ for every $t$ and choose the eigenvectors $\vec{\theta}_i(t)$, $i = 1, \dots, n$, in such a way that their first nonzero component be positive. Assume that the finite-dimensional $k$-th dimensional distributions of the random process $\Xi_n(t)$ have the densities $p(t_1, \dots, t_k, X_1, \dots, X_k)$, where $t_1, \dots, t_k$ are some values of the time parameter, and the $X_i$, $i = 1, \dots, k$, are real symmetric matrices of order $n$. The density of the finite-dimensional $k$-dimensional distributions of the eigenvectors $\{\lambda_1(t), \dots, \lambda_n(t)\}$ is equal to

$$
\begin{aligned}
&q(t_1, \dots, t_k, Y_1, \dots, Y_k) \\
&\quad := c^k \int \cdots \int_{y_{1s} > \cdots > y_{ns},\ h_{in}^{(s)} > 0,\ s=1,\dots,k} p(t_1, \dots, t_k, H_1 Y_1 H_1', \dots, H_k Y_k H_k') \\
&\quad \times \prod_{s=1}^{k} \prod_{i>j} (y_{is} - y_{js}) \prod_{s=1}^{k} \mu(\mathrm{d}H_s),
\end{aligned}
$$

where $Y_s = (\delta_{ij} y_{is})$, $H_s = (h_{ij}^{(s)})$ are real orthogonal matrices of order $n$,

$$
c = \pi^{n(n+1)/4} \prod_{i=1}^{n} \{\Gamma((n-i+1)/2)\}^{-1}.
$$

## 32. Perturbation formulas

If we assume the existence of a distribution density of a random matrix, its eigenvalues with probability 1 will be different. Therefore, we need formulas for the perturbations of different eigenvalues of matrices.

Let $\lambda_i(A + \varepsilon B)$ and $l_j(A + \varepsilon B)$ be the eigenvalues and eigenvectors of the matrix $A$, where $\varepsilon$ is some arbitrary real parameter, with the eigenvalues $\lambda_j(A + \varepsilon B)$, satisfying the relation $\lim_{\varepsilon \to 0} \lambda_j(A + \varepsilon B) = \lambda_j$. The coefficients of the characteristic equation for the matrix $A + \varepsilon B$ are analytical functions of $\varepsilon$. Therefore, the eigenvalues of such a matrix are analytical functions of $\varepsilon$ having only algebraic singularities. Then for $\varepsilon$ sufficiently small, the following expansions hold:

$$
\lambda_j(A + \varepsilon B) = \sum_{m=0}^{\infty} \lambda_j^{(m)} \varepsilon^{(m)}, \quad \lambda_j^{(0)} = \lambda_j,
$$

$$
l_j(A + \varepsilon B) = \sum_{m=0}^{\infty} l_j^{(m)} \varepsilon^{m}, \quad l_j^{(0)} = l_j.
$$

If $\lambda_i \neq \lambda_j$, $i \neq j$, then

$$
\begin{aligned}
\lambda_j^{(m)} = &\sum_{s_1+s_2+s_3=m-2,\ s_i \geqslant 0} \operatorname{Tr} (S_j B)^{s_1} E_j B (S_j B)^{s_2} E_j B (S_j B)^{s_3} \\
&+ \operatorname{Tr} E_j B (S_j B)^{m-1}, \quad m \geqslant 2,
\end{aligned}
$$

$$\lambda_j^{(1)} = \operatorname{Tr} E_j B,$$

$$E_j^{(m)} = \sum_{s_1+s_2=m-1,\, s\geqslant 0} (S_jB)^{s_1} E_j B (S_jB)^{s_2} S_j + (S_jB)^{m-1} E_j,$$

where

$$S_j = \sum_{k\neq j,\, k=1,\dots,n} E_k(\lambda_k - \lambda_j)^{-1}, \quad E_k = l_k l_k^{\mathrm{T}}.$$

## 33. Forward and backward spectral Kolmogorov equations for distribution densities of eigenvalues of random matrix processes with independent increments

Let $w_n(t)$ be the symmetric matrix process of Brownian motion of order $n$, i.e. the elements of the matrix $w_n(t)$ are random processes of the form $\delta_{ij}\mu_i + w_{ij}(t)(1+\delta_{ij})/2$, where the $w_{ij}(t)$, $i \geqslant j$, are independent processes of Brownian motion, and the $\mu_1 > \mu_2 > \cdots > \mu_n$ are arbitrary real nonrandom values. For the Markov process $\Lambda(t) = \{\lambda_1(t), \dots, \lambda_n(t)\, ;\, \lambda_1(t) \geqslant \lambda_2(t) \geqslant \cdots \geqslant \lambda_n(t)\}$ the transition probability density exists,

$$p(s, \vec{x}, t, \vec{y}), \quad \vec{x} = (x_1, \dots, x_n),\ \vec{y} = (y_1, \dots, y_n).$$

Let $f(\vec{x})$ be continuous and bounded,

$$u(s, \vec{x}) := \int_{R^n} f(\vec{y}) p(s, \vec{x}, t, \vec{y})\, \mathrm{d}\vec{y},$$

$$M^n := \{\vec{x}\colon\ x_i \neq x_j,\ i \neq j\}.$$

Then $u(s, \vec{x})$ for $\vec{x} \in M^n$, $s \in (0, t)$ satisfies the equation

$$-\frac{\partial u(s,\vec{x})}{\partial s} = \sum_{i=1}^{n} a_i(\vec{x})/2 \frac{\partial u(s,\vec{x})}{\partial x_i} + \frac{1}{2}\sum_{i=1}^{n} \frac{\partial^2 u(s,\vec{x})}{\partial x_i^2},$$

where

$$a_i(\vec{x}) = \sum_{k\neq i} [1/(x_i - x_k)]$$

and the boundary condition $\lim_{s\uparrow t} u(s, \vec{x}) = f(\vec{x})$.

The function $p(s, \vec{x}, t, \vec{y})$ in the domain $s \in (t, T)$, $\vec{x} \in R_n$, $\vec{y} \in M^n$ satisfies the equation

$$\frac{\partial p(s,\vec{x},t,\vec{y})}{\partial t} = -2^{-1}\sum_{i=1}^{n} \frac{\partial}{\partial y_i}\left[a_i(\vec{y}) p(s,\vec{x},t,\vec{y})\right] + 2^{-1}\sum_{i=1}^{n} \frac{\partial}{\partial y_i^2} p(s,\vec{x},t,\vec{y}).$$

This is the forward Kolmogorov equation for $p(s,\vec{x},t,\vec{y})$ which is also called the Fokker–Planck or Einstein–Smoluchowski equation.

For any fixed $s$, the solution of this equation exists and is unique for all initial functions $p(s,\vec{x},t,\vec{y})$ belonging to the class of functions that are everywhere compact in the metric of uniform convergence on the space of all continuous and differentiable functions $p(s,\vec{x},t,\vec{y})$, once with respect to $s$ and twice with respect to $y_i$, $i=1,\dots,n$.

## 34. Spectral stochastic differential equations for random symmetric matrix processes with independent increments

Let $W_n(t)$ be a symmetric matrix process of Brownian motion. The eigenvalues $\lambda_k(t)$ satisfy a system of spectral stochastic differential equations

$$\mathrm{d}\lambda_k(t) = \frac{1}{2}\sum_{m\neq k}\bigl(\lambda_k(t)-\lambda_m(t)\bigr)^{-1}\mathrm{d}t + \mathrm{d}w_k(t),$$

$$\lambda_k(0)=\mu_k,\quad k=1,\dots,n,$$

where $w_k(t)$ are independent random processes of Brownian motion. A weak solution of the system of this equations exists and is unique in a strong sense.

## 35. Spectral stochastic differential equations for random matrix-valued processes with multiplicative independent increments

Let $w_0^t$ be a random matrix-valued process of dimension $m\times m$ satisfying the following conditions; for any $0<t_1<t_2<\dots<t_k<s$,

$$w_0^s = w_0^{t_1}w_{t_1}^{t_2}\dots w_{t_k}^s,\quad w_0^0=A,\ w_t^t=I,$$

the random matrices $w_{t_1}^{t_i+1}$, $i=1,2,\dots$, are independent, and their distributions depend only on the difference $t_{i+1}-t_i$, $A$ is a real deterministic matrix, the eigenvalues $\alpha_i$ of the matrix $AA'$ are different, $\alpha_1>\alpha_2>\dots>\alpha_n$,

$$\lim_{\Delta t\downarrow 0}\mathbf{E}w_t^{t+\Delta t}=A$$

for any vectors $\vec{x}$ and $\vec{y}$ of dimension $m$, and

$$\lim_{\Delta t\to 0}(\Delta t)^{-1}\mathbf{E}\bigl[\bigl(\bigl(w_0^{t+\Delta t}-w_0^t\bigr)\vec{x},\vec{y}\bigr)^2/w_0^t\bigr] = (\vec{x},\vec{x})\bigl(\vec{y},w_0^t\bigl(w_0^t\bigr)^*\vec{y}\bigr).$$

Let $\lambda_k(t)$ and $\vec{l}_k(t)$, respectively, be the eigenvalues and eigenvectors of the process $\xi(t):=w_0^t(w_0^t)^*$.

Let $L = \{X := \lambda_i(X) \neq \lambda_j(X),\ i \neq j\}$, where $X$ are non-negatively definite matrices of dimension $m \times m$,

$$\nu = \inf\{t \geqslant 0\colon\ w(t) \in \bar{L}\}.$$

By using the perturbation formulas for the eigenvalues, we have that the eigenvalues $\lambda_k(t)$ satisfy the following system of stochastic differential equations as $t < \nu$,

$$\mathrm{d}\lambda_k(t) = \sum_{s \neq k} \left(\lambda_s(t) + \lambda_k(t)\right)\left(\lambda_k(t) - \lambda_s(t)\right)^{-1} \mathrm{d}t + n\,\mathrm{d}t + \mathrm{d}w_k(t),$$

$$\lambda_k(0) = \alpha_k, \quad k = 1, \ldots, n,$$

where the $w_k(t)$ are independent processes of Brownian motion.

## 36. The stochastic Ljapunov problem for systems of stationary linear differential equations

Let $\vec{x}'(t) = A\vec{x}(t),\ \vec{x}(0) = \vec{x}_0,\ \vec{x}_0 \neq \vec{0}$ be a system of linear differential equations with a random matrix of coefficients $A$. The stochastic Ljapunov problem for such systems is that of finding the probability of the event

$$\{w\colon\ \vec{x}(t) \to \vec{0},\ t \to \infty\}.$$

Let us consider a system $(\mathrm{d}/\mathrm{d}t)\vec{x}'(t) = A\vec{x}(t),\ \vec{x}(0) = \vec{c}$ of linear differential equations with constant real coefficients, where $A$ is a square matrix of order $n$, and $\vec{x}$ and $\vec{c}$ are $n$-vectors. The solution of such an equation converges to the null vector, as $t \to \infty$, for any vector $\vec{c} \neq \vec{0}$ if and only if $\operatorname{Re}\lambda_i < 0$ where the $\lambda_i$ are the eigenvalues of $A$. A matrix $A$ for which $\operatorname{Re}\lambda_i < 0$ will be said to be *stable*. To prove the stability of $A$ we can use Ljapunov's theorem: $A$ is stable if and only if the matrix $Y$ determined by the equation $A'Y + YA = -I$ is positive-definite. However, if $A$ is a random matrix, this stability criterion is inefficient.

Let $\Xi$ be a random symmetric matrix of order $n$ with probability density $p(X)$ and let $\lambda_i$ be its eigenvalues. Then,

$$\mathbf{P}\{\lambda_i < 0,\ i = 1, \ldots, n\} = c \int p\left(-Z_{n\times(n+1)} Z^{\mathrm{T}}_{n\times(n+1)}\right) \mathrm{d}Z_{n\times(n+1)},$$

where $Z_{n\times(n+1)}$ is an $n \times (n+1)$ real matrix, and

$$c = \pi^{-n(n+3)/4} \prod_{i=1}^{n} \Gamma\left[(i+1)/2\right], \quad \mathrm{d}Z_{n\times(n+1)} = \prod_{i=1,\ldots,n,\ j=1,\ldots,n+1} \mathrm{d}Z_{ij}.$$

If the entries $\xi_{ij}$, $i \geqslant j$, of the symmetric matrix $\Xi$ are independent and have $N(a_{ij}, \sigma_{ij}^2)$ distributions ($\sigma_{ij}^2 \neq 0$), then

$$\mathbf{P}\{\lambda_i < 0,\ i = 1, \dots, n\}$$
$$= (2\pi)^{-n(n+1)/2} \prod_{i \geqslant j} \sigma_{ij}^{-1} c \int \cdots \int \exp\left\{ -\frac{1}{2} \right.$$
$$\times \sum_{i>j} \sigma_{ij}^{-2} \left( a_{ij} - \sum_{k=1}^{n+1} z_{ik} z_{jk} \right)^2 - \frac{1}{2} \sum_{i=1}^{n} \sigma_{ii}^{-2} \left( a_{ii} - \sum_{k=1}^{n+1} z_{ik}^2 \right)^2 \Bigg\}$$
$$\times \prod_{i=1,\dots,n,\ j=1,\dots,n+1} \mathrm{d}z_{ij}.$$

## 37. Equation for the resolvent of empirical covariance matrices if the Lindeberg condition holds

Let $R_{m_n}$ be the covariance matrix of the $m_n$-dimensional random vector $\vec{\xi}$, $\mathbf{E}\vec{\xi} = \vec{a}$. The expression

$$\mu_{m_n}(x, R) = m_n^{-1} \sum_{k=1}^{m_n} F(x - \lambda_k)$$

is called the normalized spectral function of the matrix $R_{m_n}$, where $F(x - \lambda_k) = 1$ if $\lambda_k < x$ and $F(x - \lambda_k) = 0$ if $\lambda_k \geqslant x$, and $\lambda_k$ are roots of the characteristic equation

$$\det(Iz - R) = 0.$$

Let $\vec{x}_1, \vec{x}_2, \dots, \vec{x}_n$ be the observations of the $m$-dimensional random vector $\vec{\xi}$. Suppose this vector has covariance matrix $R_{m_n}$,

$$\xi_i = (\xi_{ij},\ j = 1, \dots, m)^{\mathrm{T}} = H R_{m_n}^{-1/2} (\vec{x}_i - \vec{a}),$$

where $H = (h_1, \dots, h_m)$, $h_p$ is an eigenvector corresponding to the eigenvalue $\lambda_p$, and the random variables $\xi_{ij}$ are independent,

$$0 < c_1 \leqslant \lambda_k \leqslant c_2 < \infty,$$

$$\lim_{n \to \infty} m_n n^{-1} = c, \quad 0 < c < 1.$$

Then, in order that for every $t > 0$,

$$p \lim_{n \to \infty} \left[ \int_0^\infty (t + x)^{-1}\, \mathrm{d}\mu(x, \hat{R}_{m_n}) - a_{m_n}(t) \right] = 0,$$

where

$$\hat{R}_m = (n-1)^{-1}\sum_{k=1}^{n}\left(\vec{x}_k - \hat{\vec{x}}\right)\left(\vec{x}_k - \hat{\vec{x}}\right)^{\mathrm{T}}, \quad \hat{\vec{x}} = n^{-1}\sum_{k=1}^{n}\vec{x}_k,$$

and the $a_{m_n}(t)$ are non-negative analytical functions satisfying the equation

$$a_{m_n}(t) = \int_0^\infty \left\{t + \left[(1 - m_n n^{-1} + m_n n^{-1} t a_{m_n}(t)\right] x\right\}^{-1} \mathrm{d}\mu_{m_n}(x, R_{m_n}),$$
$$t \geqslant 0,$$

it is sufficient, and in the case of symmetric variables $\xi_{ij}$, it is also necessary, that the Lindeberg condition holds, i.e. for every $\tau > 0$,

$$\lim_{n\to\infty} m_n^{-1}\sum_{i=1}^{m_n}\sum_{j=1}^{n}\mathbf{E}\left|\xi_{ij}(n-1)^{-1/2}\right|^2\chi\left(|\xi_{ij}|(n-1)^{-1/2} > \tau\right) = 0.$$

## 38. Equation for the Stieltjes transformation of normalized spectral functions of the empirical covariance matrix pencil

Let $R_1$ and $R_2$ be nonsingular covariance matrices of two independent $m$-dimensional random vectors $\vec{\xi}_1$ and the $\vec{\xi}_2$, $\vec{a}_1 = \mathbf{E}\,\vec{\xi}_1$, $\vec{a}_2 = \mathbf{E}\,\vec{\xi}_2$.

The expression

$$\mu_n(x_1, R_1, R_2) = m_{-1}\sum_{k=1}^{m} F(x - \lambda_k)$$

is called the normalized spectral function of the $R_1$ and $R_2$ covariance pencil, where $F(x - \lambda_k) = 1$ if $\lambda_k < x$, and $F(x - \lambda_k) = 0$ if $\lambda_k \geqslant x$; $\lambda_k$ are roots of the characteristic equation

$$\det(R_1 z - R_2) = 0, \quad 0 < d_1 \leqslant \lambda_k \leqslant d_2 < \infty.$$

Let $x_1, \ldots, x_{n_1}, y_1, \ldots, y_{n_2}$ be observations of the random vectors $\vec{\xi}_1$ and $\vec{\xi}_2$,

$$\xi_i = (\xi_{ij}, j = 1, \ldots, m)^{\mathrm{T}} = R_1^{-1/2}(\vec{x}_i - \vec{a}_1),$$

$$\eta_i = (\eta_{ij}, j = 1, \ldots, m)^{\mathrm{T}} = R_2^{-1/2}(\vec{y}_i - \vec{a}_2),$$

let the random variables $\xi_{ij}, \eta_{ij}, i, j = 1, 2, \ldots,$ be independent

$$\lim_{m\to\infty}\frac{m}{n_1} = c_1, \quad \lim_{n\to\infty}\frac{m}{n_2} = c_2, \quad c_1^{-1} + c_2^{-1} < 1, \ c_1 \neq 1,$$

and let the Lindeberg condition be fulfilled, i.e. we have

$$\lim_{n\to\infty}\left[\sum_{i=1}^{m}\mathbf{E}\left|\xi_{i1}(n_1-1)^{-1/2}\right|^2\chi\left(|\xi_{i1}|(n_1-1)^{-1/2}>\tau\right)\right.$$
$$\left.+m_n^{-1}\sum_{i=1}^{m}\mathbf{E}\left|\eta_{i1}(n_2-1)^{-1/2}\right|^2\chi\left(|\eta_{i1}|(n_2-1)^{-1/2}>\tau\right)\right]=0,$$

for every $\tau>0$. Then

$$p\lim_{m\to\infty}\left[\int_0^{\infty}(t+x)^{-1}\,\mathrm{d}\mu_m(x,\hat{R}_1,\hat{R}_2)-a_m(t)\right]=0,\quad t>0,$$

where the function $a_m(t)$, $t>0$, is equal to

$$a_m(t)=-\int_0^{\infty}\left(\frac{\partial}{\partial t}\right)b_m(t,x)\,\mathrm{d}x,$$

and the function $b_m(t,x)$ satisfies the equation

$$b_m(t,\alpha)=\int_0^{\infty}\left[\alpha+t\left(1+tc_1b_m(t,\alpha)\right)\right)^{-1}+x\left[1-c_2+\alpha c_2\right.$$
$$\times\, b_m(t,\alpha)\left(\alpha+tc_1b_m(t,\alpha)\right)\right)^{-1}\mathrm{d}\mu_m(x,R_1,R_2).$$

## 39. Consistent estimates of generalized variance

Let the independent observations $\vec{x}_1,\dots,\vec{x}_m$ of the $m_n$-dimensional random vector $\vec{\xi}$, $n>m_n$, be given,

$$\hat{R}:=(n-1)^{-1}\sum_{k=1}^{n}\left(\vec{x}_k-\hat{\vec{x}}\right)\left(\vec{x}_k-\hat{\vec{x}}\right)^{\mathrm{T}},\quad \hat{\vec{x}}=n^{-1}\sum_{k=1}^{n}\vec{x}_k.$$

The expression $\det R$ is called a generalized variance. If the vectors $x_i$, $i=1,\dots,n$, are independent and distributed according to the multidimensional normal law $N(a,R)$, then

$$\det\hat{R}\approx\det R(n-1)^{-m}\prod_{i=n-m}^{n-1}\chi_i^2,$$

where $\chi_i^2$ are independent random variables distributed according to the $\chi^2$-law with $i$ degrees of freedom. In the general case, the distribution of $\det\hat{R}$ is difficult to find, and therefore finding consistent estimates for $\det R$ is a very complicated problem. It is proved, that under certain conditions the $G$-estimates for the variables $c_n^{-1} \ln \det R$, where $c_n$ is a sequence of constants such that

$$\lim_{n\to\infty} c_n^{-2} \ln n(n-m_n)^{-1} = 0,$$

can be represented in the form

$$G_1(\hat{R}) := c_n^{-1}\{ \ln \det \hat{R} + \ln \left[(n-1)^m \left(A_{n-1}^m\right)^{-1} n(n-m_n)^{-1}\right\},$$

where $A_{n-1}^m = (n-1)\cdots(n-m)$.

For every value $n > m_n$, let the random $m_n$-dimensional vectors $x_1^{(n)}, \dots, x_n^{(n)}$ be independent and identically distributed with a mean vector $\vec{a}$ and nondegenerate covariance matrices $R_{m_n}$, such that for a certain $\delta > 0$

$$\sup_n \sup_{i=1,\dots,n,\ j=1,\dots,m_n} \mathbf{E}\left|\tilde{x}_{ij}^{(n)}\right|^{4+\delta} < \infty,$$

where the $\tilde{x}_{ij}^{(n)}$ are the vector components of $\tilde{x}_i = R_{m_n}^{-1/2}(x_i^{(n)} - a)$,

$$\lim_{n\to\infty} (n-m_n) = \infty, \qquad \lim_{n\to\infty} n m_n^{-1} \geqslant 1;$$

and for each value of $n > m_n$, the random variables $x_{ij}^{(n)}$, $i = 1,\dots,n$, $j = 1,\dots,m_n$, are independent.

Then

$$p \lim_{n\to\infty} \left[G_1\left(\hat{R}_{m_n}\right) - c_n^{-1} \ln \det R_{m_n}\right] = 0.$$

If, in addition to the previous condition,

$$\mathbf{E}\left(\tilde{x}_{ij}^{(n)}\right)^4 = 3, \quad i = 1,\dots,n,\ j = 1,\dots,m_n,$$

then

$$\lim_{n\to\infty} \mathbf{P}\left\{\left(c_n C_1\left(\hat{R}_{m_n}\right) - \ln \det R_{m_n}\right)\left(-2\ln\left(1 - m_n n^{-1}\right)\right)^{-1/2} < x\right\}$$
$$= (2\pi)^{-1/2} \int_{-\infty}^{x} \mathrm{e}^{-y^2/2}\, \mathrm{d}y.$$

## 40. Consistent estimates of the Stieltjes transform of the normalized spectral function of covariance matrices

Consider the main problem of statistical analysis of observations of large dimension: the estimation of the Stieltjes transforms of normalized spectral functions

$$\mu_{m_n}(x) = m_n^{-1} \sum_{k=1}^{m_n} \chi(\lambda_k < x)$$

of the covariance matrices $R_{m_n}$ from observations of the random vector $\vec{\xi}$ with covariance matrix $R_{m_n}$, where the $\lambda_k$ are the eigenvalues of the matrix $R_{m_n}$. Note that many analytic functions of the covariance matrices that are used in multivariate statistical analysis can be expressed through the spectral functions $\mu_{m_n}(x)$. For example,

$$m_n^{-1} \mathrm{Tr}\, f(R_{m_n}) = \int_0^\infty f(x)\, \mathrm{d}\mu_{m_n}(x),$$

where $f(x)$ is an analytic function.

The expression

$$\varphi(t, R_{m_n}) = \int_0^\infty (1+tx)^{-1}\, \mathrm{d}\mu_{m_n} = m_n^{-1} \mathrm{Tr}(I + tR_{m_n})^{-1}$$

is called the Stieltjes transform of the function $\mu_{m_n}(x)$. The consistent estimate of the Stieltjes' transform $\varphi(t, R_{m_n})$ is by definition the following expression: $G_2(t, \hat{R}_{m_n}) = \varphi(\hat{\theta}_n(t), \hat{R}_{m_n})$, where $\hat{\theta}_n(t)$ is the positive solution of the equation

$$\theta(1 - m_n(n-1)^{-1} + m_n(n-1)^{-1}\varphi(\theta, \hat{R}_{m_n})) = t, \quad t \geqslant 0.$$

It is obvious that the positive solution of this equation exists and is unique as $t \geqslant 0$, $m_n(n-1)^{-1} < 1$.

Let the independent observations $\vec{x}_1, \ldots, \vec{x}_n$ of the $m_n$-dimensional random vector $\vec{\xi}$ be given, and let the $G$-condition be fulfilled:

$$\limsup_{n\to\infty} m_n n^{-1} < 1, \quad 0 < c_1 \leqslant \lambda_i \leqslant c_2 < \infty, \; i = 1, \ldots, m_n,$$

let the components of the vector $(\eta_{1k}, \ldots, \eta_{m_n k})^{\mathrm{T}} = R_{m_n}^{-1/2}(\vec{\xi} - \mathbf{E}\vec{\xi})$ be independent, and

$$\sup_n \sup_{k=1,\ldots,n} \sup_{i=1,\ldots,m_n} \mathbf{E}|\eta_{ik}|^{4+\delta} < \infty, \quad \delta > 0.$$

Then

$$\lim_{n\to\infty} \mathbf{P}\Big\{ \big[G_2 - \varphi(t, R_{m_n})\big] \sqrt{(n-1)m_n a_n(t) + c_n(t)} < x \Big\}$$

$$= (2\pi)^{-1/2} \int_{-\infty}^{x} \mathrm{e}^{-y^2/2}\,\mathrm{d}y,$$

for $t > 0$, where $a_n(t)$ and $c_n(t)$ are some bounded functions.

## 41. Stochastic condition of complete controllability

The basis of the theory of linear stationary controllable systems is the following mathematical model,

$$\frac{\mathrm{d}\vec{x}}{\mathrm{d}t} = A\vec{x} + B\vec{u}; \quad \vec{z} = C\vec{x},$$

where $\vec{x}(t) = (x_1(t), \ldots, x_n(t))^{\mathrm{T}}$ is the state vector; $\vec{u}(t) = (u_1(t), \ldots, u_r(t))^{\mathrm{T}}$ is the input of the system, or control; $\vec{z}(t) = (z_1(t), \ldots, z_m(t))^{\mathrm{T}}$ is the output of the system, or observation; $A = (a_{ij})$ is an $n \times n$ matrix; $B = (b_{ij})$ is an $n \times r$ matrix; and $C = (c_{ij})$ is an $m \times n$ matrix.

This system is "in general position" if and only if

$$\det\left(\vec{b}_i, A\vec{b}_i, \ldots, A^{n-1}\vec{b}_i\right) \neq 0, \quad i = 1, 2, \ldots, r,$$

where $\vec{b}_i$ is the $i$-th column of the matrix $B$.

If $B = \vec{b}$ is a vector, i.e. $r = 1$, then the condition

$$\det\left(\vec{b}A\vec{b}\ldots A^{n-1}\vec{b}\right) \neq 0,$$

is a criterion for the system's complete controllability.

Thus the quantity $\det(\vec{b}, A\vec{b}, \ldots, A^{n-1}\vec{b}) = D_n$ plays a central role in the theory of stationary controllable systems.

However, in real systems, because of a number of always existing factors (obstacles, noise, inaccuracies in measurings, wrong information), the elements of the matrix $A$ and of the vector $\vec{b}$ cannot be regarded as deterministic quantities.

Obviously, if the elements of the matrix $A$ and the vector $\vec{b}$ are continuous random variables, then $\mathbf{P}\{D_n \neq 0\} = 1$. A much more interesting (but rather complicated) problem is to find the probability

$$\mathbf{P}\left\{|D_n| > \varepsilon\right\}, \quad \varepsilon > 0.$$

This problem, in general, is far from being solved. We shall consider only one particular important case.

Let $A$ be a random symmetric matrix which does not depend on the vector $\vec{b}$, and such that the entries on the main diagonal and above it, are independent and distributed according to the normal law $N(0, 1)$; and such that the components of the vector $\vec{b}$ are also independent and distributed according to the standard normal law. Then for every $k = 0, 1, 2, \dots$

$$\mathbf{E}\,|D_n|^k = \frac{2^{n(k+2)/2}\,\pi^{-n(n+1)/4}}{n!\,(k+1)^n}\prod_{j=1}^{n}\frac{\Gamma\left(1+\frac{(k+1)j}{2}\right)}{\Gamma(j/2)}.$$

## 42. Random matrices in physics: Spacing of eigenvalues

Energy levels of heavy atomic nuclei under high energy are arranged rather close to one another, and it is practically impossible to find them even if the Hamiltonian of system is known. In this connection, Wigner proposed a statistical model of highly excited states of heavy atomic nuclei in which complex nuclei are regarded as some "black cavity", and the particles, which make up a nucleus, interact according to an unknown random law. The central problem is the choice of a mathematical model for such systems. As a model of an ensemble of complex nuclei, Wigner chose a Hermitian matrix of large dimension, the elements of which were independent random variables with zero mean, identical variances, and bounded moments.

Wigner proposed the following hypothesis: In a sequence of a great numbers of levels, on the average separated by distance $D$ from one another and having identical values for all quantum numbers considered for identification such as moments of the amount of movement and even parity, the probability to find two levels at a distance between $t$ and $t + \Delta t$ is equal to

$$Q(t)\,\mathrm{d}t := (2D)^{-1}\pi t \exp\left(-\pi t^2 4^{-1} D^{-2}\right)\mathrm{d}t.$$

As the mathematical model for checking this Hypothesis, again Wigner's model was chosen. Mehta and Gaudin obtained the density $p(t)$ of the distance between two neighboring eigenvalues of a random Hermitian matrix, of which the elements on the diagonal and above it were independent and distributed according to the standard normal law. This density differs from the value $Q(t)$ but not too much, $|Q(t) - p(t)| \leqslant 0.0162$.

Let $\lambda_1 \geqslant \lambda_2 \geqslant \cdots \geqslant \lambda_n$ be the eigenvalues of a symmetric random matrix $\Xi$ and suppose $\Xi$ is such that the random variables $\lambda_i$ have density $p(x_1, \dots, x_n)$, $x_1 > \cdots > x_n$, and the function $p$ is symmetric, i.e. $p$ is invariant under a simultaneous permutation of the variables. We consider the spectral function for spacing of the eigenvalues of the random matrix $\Xi$

$$\theta_n(x) = n^{-1}\sum_{i=1}^{n-1}\mathbf{E}\,F\big(x - (\lambda_i - \lambda_{i+1})\big),$$

where

$$F(x) = \begin{cases} 1, & x > 0, \\ 0, & x \leqslant 0. \end{cases}$$

Then

$$\theta_n(x) = n^{-1} \int_0^x \int_{-\infty}^{\infty} \left( \frac{\partial^2}{\partial u \partial v} \right) p(u, v)_{u=y,\, v=z+y} \, dy \, dz$$

where $p(u, v)$ is the probability that all eigenvalues lie outside the interval $(u, v)$.

If $\Xi$ is a Gaussian Hermitian random matrix, then the limit of the probability that the eigenvalues $\lambda_i$ of $\Xi$ do not lie in the interval $(\alpha n^{-1/2}, \beta n^{-1/2})$ is equal to

$$\left[ \mathbf{E} \exp \left\{ \int_\alpha^\beta \eta^2(z) \, dz \right\} \right]^{-2} = \prod_{i=1}^{\infty} (1 - \mu_i),$$

where $\mu_i$ are the eigenvalues of the integral equation

$$\pi^{-1} \int_\alpha^\beta (x - y)^{-1} \sin(x - y) t(y) \, dy = \lambda t(x), \quad x \in (\alpha, \beta).$$

## 43. Band random matrices

The study of the characteristic properties of quantum systems whose classical limit is chaotic has led to increased interest in Random Matrix Theory. A lot of interesting problems in physics are described by matrices with a band-like structure. For example, the quantum kicked rotator, which is the most popular model in quantum chaos, is described by a unitary matrix which is very close to a band symmetric matrix: $\Xi = (\xi_{ij}^{(n)})_{i,j=1}^n$ where $\xi_{ij}^{(n)} \equiv 0$ if $|i - j| > m$, $m = 1, 2, \ldots$.

Let $\xi_{ij}^{(n)}$, $|i-j| \leqslant m$, $i \geqslant j$, be independent random variables, $\mathbf{E}\, \xi_{ij}^{(n)} = 0$, $\mathbf{Var}\, \xi_{ij}^{(n)} = \sigma_{ij}^2$, $|i - j| > m$, for every $i$

$$\sum_{|i-j| \leqslant m} \sigma_{ij}^2 = 1$$

and let the following Lindeberg conditions be fulfilled: for every $\tau > 0$

$$\lim_{n \to \infty} \frac{1}{n} \sum_{|i-j| \leqslant m,\ i,j=1,\ldots,n} \mathbf{E}\, \xi_{ij}^2 \chi\big(|\xi_{ij}| > \tau\big) = 0.$$

Then with probability 1 for every $x$

$$\lim_{n \to \infty} n^{-1} \sum_{k=1}^{n} \chi(\lambda_k < x) = \frac{1}{2\pi} \int_{(y<x) \cap (|y|<2)} \sqrt{4 - y^2} \, dy,$$

where the $\lambda_k$ are the eigenvalues of the matrix $\Xi_n$.

## References


[1] T.W. Anderson, *An Introduction to Multivariate Statistical Analysis*, Wiley, New York, London (1958).

[2] F.A. Berezin, *Some remarks on the Wigner distribution*, Theoret. and Math. Phys. **17** (1973), 1163–1175. English transl. from the Soviet journal Teoret. Mat. Fiz. **17** (1973), 305–318.

[3] A.T. Bharucha-Reid, *Probabilistic Method in Applied Mathematics, Vol. 2*, Academic Press, New York (1970).

[4] T.A. Brody, J. Flores, J.B. French, P.A. Mello, A. Pandey and S. Wong, *Random-matrix physics: Spectrum and strength fluctuations*, Rev. Modern Phys. **53** (1981), 385.

[5] H. Cramer, *Mathematical Methods in Statistics*, Princeton Univ. Press, Princeton, NJ (1946).

[6] F.J. Dyson, *The dynamics of disordered linear chain*, Phys. Rev. **92**(6) (1953), 1331–1338.

[7] F.J. Dyson, *Statistical theory of the energy levels of complex systems, I–V*, J. Math. Phys. **3** (1963), 701–712, 713–719 (Parts IV and V coauthored with Madan Lab Mehta).

[8] F.J. Dyson, *A Brownian-motion model for the eigenvalues of a random matrix*, J. Math. Phys. **3**(6) (1962), 31–60.

[9] A. Edelman, E. Kostlan and M. Shub, *How many eigenvalues of a random matrix are real?*, J. Amer. Math. Soc. (1994) (to appear).

[10] R. Fortet, *Random determinants*, J. Research Nat. Bur. Standards **47** (1951), 465–470.

[11] V.L. Girko, *On the distribution of solutions of systems of linear equations with random coefficients*, Theory Probab. Math. Statist. no. 2 (1974), 41–44. English transl. from the Soviet journal Teor. Verojatnost. i Mat. Statist. 2.

[12] V.L. Girko, *Random Matrices*, Vishcha Shkola (Izdat. Kiev. Univ.), Kiev (1975) (in Russian).

[13] V.L. Girko, *A limit theorem for products of random matrices*, Theory Probab. Appl. **21** (1973). English transl. from the Soviet journal Teor. Veroyatnost. i Primenen. **21**, 201–202.

[14] V.L. Girko, *The distribution of the eigenvalues and eigenvectors of hermitian random matrices*, Ukrainian Math. J. **31** (1979), 533–537. English transl. from the Soviet journal Ukrainian Mat. Zh. **31**.

[15] V.L. Girko, *The central limit theorem for random determinants*, Theory Probab. Appl. **24** (1979), 729–740.

[16] V.L. Girko, *Theory of Random Determinants*, Kluwer Academic Publishers (1990).

[17] V.L. Girko, *Distribution of eigenvalues and eigenvectors of unitary random matrices*, Theory Probab. Math. Statist. **25** (1981), 13–16. English transl. from the Soviet journal Teor. Veroyatnost. i Mat. Statist. **25**.

[18] V.L. Girko, *The elliptic law*, Teor. Verojatnost. i Primenen. **30**(4) (1985), 640–651 (in Russian).

[19] N.R. Goodman, *Distribution of the determinant of a complex Wishart distributed matrix*, Ann. Math. Statistics **34**(1) (1963), 178–180.

[20] U. Grenander, *Probabilities on Algebraic Structures*, Wiley, New York, London (1963).

[21] M. Kac, *Probability and Related Topics in Physical Sciences*, Interscience Publishers, London, New York (1957).

[22] T. Kato, *Perturbation Theory for Linear Operators*, Springer, Berlin (1966).

[23] J. Komlos, *On the determinant of random matrices*, Studia Sci. Math. Hungar. **3**(4) (1968), 387–399.

[24] E. Kostlan, *On the spectra of Gaussian matrices*, Linear Algebra Appl. **162–164** (1992), 385–388.

[25] I.M. Lifshits, *On the structure of the energy spectrum and quantum state of disordered condenser systems*, Soviet Physics Uspekhi **7** (1965), 549–573. English transl. from the Soviet journal Uspekhi Fiz. Nauk **83**, 617–655.

[26] V.A. Marchenko and L.A. Pastur, *Distribution of the eigenvalues in certain sets of random matrices*, Math. USSR-Sb. **1** (1967), 457–483. English transl. from the Soviet journal Mat. Sb. **72** (1968), 507–536.

[27] M.L. Mehta, *Random Matrices and the Statistical Theory of Energy Levels*, Academic Press, New York, London (1967).

[28] F.D. Murnaghan, *The Theory of Group Representations*, Johns Hopkins Press, Baltimore, MD (1938). Reprint Dover, New York (1963).

[29] H. Nyguist, S. Rice and J. Riordan, *The distribution of random determinants*, Quart. Appl. Math. **12**(2) (1954), 97–104.

[30] L.A. Pastur, *The spectra of random self-adjoint operators*, Russian Math. Surveys **28**(1) (1973), 1–67. English transl. from the Soviet journal Uspekhi Mat. Nauk **28**(1) 4–63.

[31] C.E. Porter, *Statistical Theories of Spectra: Fluctuations*, Academic Press, New York, London (1965).
[32] A. Prekopa, *On random determinants, I*, Studia Sci. Math. Hungar. **2**(1–2) (1967), 125–132.
[33] E.P. Wigner, *Random matrices in physics*, SIAM Rev. **9**(1) (1967), 1–23.
[34] E.P. Wigner, *On the distribution of the roots of certain symmetric matrices*, Ann. Math. **67**(2) (1968), 325–327.

# Matrix Equations. Factorization of Matrix Polynomials

A.N. Malyshev

*Institute of Mathematics, Novosibirsk, 630090, Russia*
*e-mail: malyshev@math.nsk.su*

*Contents*



HANDBOOK OF ALGEBRA, VOL. 1
Edited by M. Hazewinkel

This chapter is devoted to the main results and methods in the theory of matrix equations. The problem of factorization of matrix polynomials is essentially a problem concerning special systems of matrix equations. It is not easy to describe the scope of the theory of matrix equations. Rather the subject consists of a set of concrete equations and algorithms to solve them. These algorithms are often useful for the theoretical study of matrix equations and properties of their solutions.

The theory of such highly structured matrix equations as the Lyapunov and Riccati equations, which are widely used in applications, represents an advanced part of matrix analysis. A great many books and papers on this subject have already been published, but the bibliography is still rapidly growing, mainly in publications on automatic control and scientific computation.

Our attention here will be primarily directed to algebraic aspects of the theory of matrix equations and matrix polynomials. As a rule, our main tool will be the use of invariant subspaces of certain associated matrices and linear matrix pencils, thus reducing the problem of solving matrix equations to certain spectral matrix problems. Nevertheless, it is worthwhile to point out that more efficient approaches in one or other sense may be applied to some highly structured problems.

This chapter is organized as follows. Section 1 prepares the ground for the sections that follow. Mostly attention is paid to the definitions and properties of invariant subspaces of regular linear matrix pencils, which are a straightforward generalization of invariant subspaces of matrices.

Linear matrix equations are treated in Section 2. A brief survey of the theory of linear matrix equations is followed by a summary of the main results on the Lyapunov equations.

Nonlinear matrix equations are studied in Section 3. First the approaches based on perturbation theory and iterative methods are discussed. Then the methods of reduction of quadratic and other nonlinear equations to spectral problems for linear matrix pencils are presented in detail. Considerable attention is paid to the algebraic Riccati equations.

Section 4 deals with the theory of factorization of matrix polynomials. It includes a description of necessary and sufficient conditions for the existence of divisors and a theoretical algorithm for computation of right polynomial divisors. Then the theorems on symmetric factorization of self-adjoint matrix polynomials are discussed.

All scalars, vectors, and matrices in this chapter are complex.

## 1. Spectral characteristics of a regular linear matrix pencil

A regular linear matrix pencil is a $\lambda$-matrix $\lambda B - A$ with $N \times N$ matrices $A$ and $B$ provided that the scalar polynomial $p(\lambda) = \det(\lambda B - A)$ is not identically zero.

THEOREM 1 (The Weierstrass canonical form). *For any regular matrix pencil $\lambda B - A$ there exist nondegenerate matrices $U$ and $V$ such that*

$$\lambda B - A = V \begin{bmatrix} \lambda I - J_F & 0 \\ 0 & \lambda J_\infty - I \end{bmatrix} U^{-1}, \tag{1}$$

*where $J_F$, $J_\infty$ are in the Jordan canonical form and $J_\infty$ is nilpotent, i.e. it has only zero eigenvalues.*

THEOREM 2 (The Schur canonical form). *For any regular matrix pencil $\lambda B - A$ there exist unitary matrices $U$ and $V$ such that*

$$\lambda B - A = V(\lambda T_B - T_A)U^{-1}, \tag{2}$$

*where $T_A$, $T_B$ are a pair either of upper triangular matrices or of lower triangular ones.*

The roots of the polynomial $p(\lambda) = \det(\lambda B - A)$ are called the finite eigenvalues of a pencil $\lambda B - A$. The set of all finite eigenvalues of a pencil $\lambda B - A$ will be referred to as the finite spectrum of $\lambda B - A$. It is clear from (1) that the finite spectrum of $\lambda B - A$ is precisely the set of all eigenvalues of $J_F$. The Schur form (2) also yields finite eigenvalues of the pencil $\lambda B - A$: let $\alpha_i$ and $\beta_i$ be equally located diagonal elements of the triangular matrices $T_A$ and $T_B$ respectively, then the finite spectrum of $\lambda B - A$ consists of all numbers $\alpha_i/\beta_i$ with $\beta_i \neq 0$.

If the degree of the polynomial $p(\lambda)$ is less than $N$ then the pencil $\lambda B - A$ is said to have an infinite eigenvalue. Strictly speaking, the zero eigenvalues of the pencil $B - \mu A$ are the infinite eigenvalues of $\lambda B - A$ and vice versa. Henceforth the matrix pencil $B - \mu A$ is referred to as the dual pencil to $\lambda B - A$. From (1) we derive that $J_\infty$ corresponds to the infinite spectrum of $\lambda B - A$, i.e. all infinite eigenvalues of the pencil $\lambda B - A$. The set of all eigenvalues of a pencil $\lambda B - A$ is called the spectrum of $\lambda B - A$ and consists of the finite and infinite spectra of the pencil.

Let $\lambda_0$ be a finite eigenvalue of a regular matrix pencil $\lambda B - A$. A sequence of vectors $x_0, x_1, \ldots, x_{l-1} \in C^N$, $x_0 \neq 0$, forms a right Jordan chain of $\lambda B - A$ corresponding to $\lambda_0$ if

$$(\lambda_0 B - A)x_0 = 0, \qquad (\lambda_0 B - A)x_i + Bx_{i-1} = 0, \quad i = 1, 2, \ldots, l-1.$$

Analogously, a sequence of vectors $x_0, x_1, \ldots, x_{l-1} \in C^N$, $x_0 \neq 0$, forms a right Jordan chain of $\lambda B - A$ corresponding to the infinite eigenvalue if

$$Bx_0 = 0, \qquad Bx_i - Ax_{i-1} = 0, \quad i = 1, 2, \ldots, l-1.$$

Left Jordan chains are defined in a similar way.

It is not hard to verify that the columns of the matrix $U$ in (1) are composed of right Jordan chains of $\lambda B - A$, and the columns of $V$ consist of left Jordan chains.

Now we introduce the important notion of invariant subspace of a regular pencil $\lambda B - A$. A linear subspace $\mathcal{L} \subset \mathbf{C}^N$ is called a right invariant subspace of $\lambda B - A$ if

$$\dim(A\mathcal{L} + B\mathcal{L}) \leqslant \dim \mathcal{L}, \tag{3}$$

where $A\mathcal{L} + B\mathcal{L} = \{x \in \mathbf{C}^N \mid x = Ay + Bz,\ y \in \mathcal{L},\ z \in \mathcal{L}\}$. One can show that by virtue of the regularity of the pencil $\lambda B - A$ only the equality sign occurs in (3). We note that a right invariant subspace $\mathcal{L}$ of $\lambda B - A$ is an invariant subspace of the matrix $A$ if $B = I$, and is invariant for the matrix $B$ if $A = I$. Moreover, if $Q$ is a nondegenerate matrix then the pencils $\lambda B - A$ and $\lambda QB - QA$ have the same invariant subspaces.

Left invariant subspaces of a regular pencil $\lambda B - A$ are defined in a similar way. In what follows the word "right" before "invariant" will be often omitted.

We shall say that an $N \times d$ matrix $X$ of rank $d$ is a basis matrix for a $d$-dimensional subspace $\mathcal{L} \subset \mathbf{C}^N$ if the linear span of the columns of $X$ equals $\mathcal{L}$.

Let $X$ be a basis matrix for a $d$-dimensional invariant subspace $\mathcal{L}$ of a pencil $\lambda B - A$ and let the linear span of the columns of an $N \times d$ matrix $Y$ contain the subspace $A\mathcal{L} + B\mathcal{L}$. Since $A\mathcal{L} \subset A\mathcal{L} + B\mathcal{L}$ and $B\mathcal{L} \subset A\mathcal{L} + B\mathcal{L}$, there exist $d \times d$ matrices $A_{\mathcal{L}}$ and $B_{\mathcal{L}}$ such that

$$AX = YA_{\mathcal{L}}, \qquad BX = YB_{\mathcal{L}}. \tag{4}$$

Conversely, if (4) holds and $\operatorname{rank}(X) = d$, then for the subspace $\mathcal{L} \subset C^N$ with basis matrix $X$ we obtain that $\dim(A\mathcal{L} + B\mathcal{L}) \leqslant \operatorname{rank} Y \leqslant \dim \mathcal{L}$. Thus, (4) with a matrix $X$ of full rank is shown to be an equivalent definition of the invariant subspace $\mathcal{L}$ of the pencil $\lambda B - A$ with basis matrix $X$.

The spectral characteristics of the $d \times d$ matrix pencil $\lambda B_{\mathcal{L}} - A_{\mathcal{L}}$ are restrictions of the spectral characteristics of the regular pencil $\lambda B - A$. Indeed, the pencil $\lambda B_{\mathcal{L}} - A_{\mathcal{L}}$ is regular, the eigenvalues of $\lambda B_{\mathcal{L}} - A_{\mathcal{L}}$ are eigenvalues of the pencil $\lambda B - A$, i.e. the spectrum of $\lambda B_{\mathcal{L}} - A_{\mathcal{L}}$ is a subset of the spectrum $\lambda B - A$. If $z_0, z_1, \ldots, z_{l-1}$ is a Jordan chain of the pencil $\lambda B_{\mathcal{L}} - A_{\mathcal{L}}$ corresponding to an eigenvalue $\lambda_0$, then $Xz_0$, $Xz_1, \ldots, Xz_{l-1}$ is a Jordan chain of $\lambda B - A$ corresponding to the eigenvalue $\lambda_0$.

It is clear that full information on an invariant subspace $\mathcal{L}$ and the spectral properties of the original regular pencil $\lambda B - A$ restricted to $\mathcal{L}$ is given by the four matrices $(X, Y, A_{\mathcal{L}}, B_{\mathcal{L}})$. But this structure is too inconvenient for our purposes, and we introduce the following

DEFINITION. A pair $(X;T)$ consisting of an $N \times d$ matrix $X$ of rank $d$ and a $d \times d$ matrix $T$ is a monic block eigenpair of dimension $d$ for a regular matrix pencil $\lambda B - A$ if $AX = BXT$.

This definition implies that $AX = YT$, $BX = Y$, i.e. $A_{\mathcal{L}} = T$, $B_{\mathcal{L}} = I$. Therefore, $X$ is a basis matrix of an invariant subspace $\mathcal{L}$ of the pencil $\lambda B - A$. It is obvious that the invariant subspace $\mathcal{L}$ can be represented by a monic block eigenpair $(X;T)$ if and only if the matrix $B_{\mathcal{L}}$ of the four matrices $(X, Y, A_{\mathcal{L}}, B_{\mathcal{L}})$ satisfying (4) is nondegenerate.

It is important to observe the fact that if two pairs $(X;T)$ and $(X',T')$ generate the same invariant subspace $\mathcal{L}$ then there exists a nondegenerate matrix $Q$ such that $X' = XQ$, $T' = Q^{-1}TQ$. As a consequence, monic block eigenpairs $(X;T)$ and $(X';T')$ are called similar when $X' = XQ$, $T' = Q^{-1}TQ$ with some nondegenerate matrix $Q$.

DEFINITION. A pair $(X;T)$ consisting of an $N \times d$ matrix $X$ of rank $d$ and a $d \times d$ matrix $T$ is referred to as a comonic block eigenpair of dimension $d$ for a regular matrix pencil $\lambda B - A$ if $AXT = BX$.

In this case, $A_{\mathcal{L}} = I$, $B_{\mathcal{L}} = T$, $Y = AX$. A comonic block eigenpair $(X;T)$ can be used if and only if the matrix $A_{\mathcal{L}}$ of the four matrices $(X, Y, A_{\mathcal{L}}, B_{\mathcal{L}})$ is nondegenerate. Two comonic block eigenpairs $(X;T)$ and $(X';T')$ are called similar when there exists a nondegenerate matrix $Q$ such that $X' = XQ$, $T' = Q^{-1}TQ$. It is also obvious that if the matrix $T$ of a comonic block eigenpair $(X;T)$ is nondegenerate then the pair $(X;T^{-1})$ is a monic block eigenpair for the dual pencil $B - \mu A$.

DEFINITION. If the matrix $T$ of a monic (or comonic) block eigenpair $(X;T)$ for a pencil $\lambda B - A$ is a Jordan matrix, then such a matrix pair $(X;T)$ will be called a monic (comonic) block Jordan pair.

Let $(X;T)$ be a monic block Jordan pair of a regular matrix pencil $\lambda B - A$ and $T = \text{block diag}[T_1, T_2, \ldots, T_k]$ be the partition of $T$ into the Jordan blocks $T_j$ of order $l_j$, $j = 1, 2, \ldots, k$, respectively. Let us also partition $X = [X_1 X_2 \ldots X_k]$ into blocks consistent with the partition of $T$, i.e. $X_j$ will be a matrix of size $N \times l_j$. Then the columns of the matrix $X_j$ form a Jordan chain corresponding to an eigenvalue of the matrix $T_j$. The structure of a comonic block Jordan pair is analogous.

Yet it is not possible to represent all invariant subspaces of an arbitrary regular matrix pencil $\lambda B - A$ by monic and comonic block eigenpairs only. It is instructive here to exhibit the structure of an arbitrary invariant subspace $\mathcal{L}$ of a pencil $\lambda B - A$. Let $X$ be a basis matrix of $\mathcal{L}$ and $AX = YA_{\mathcal{L}}$, $BX = YB_{\mathcal{L}}$. We make use of the canonical Weierstrass decomposition of $\lambda B_{\mathcal{L}} - A_{\mathcal{L}}$:

$$\lambda B_{\mathcal{L}} - A_{\mathcal{L}} = V \begin{bmatrix} \lambda I - J_F & 0 \\ 0 & \lambda J_\infty - I \end{bmatrix} U^{-1}.$$

Then

$$AXU = YV \begin{pmatrix} J_F & 0 \\ 0 & I \end{pmatrix}, \qquad BXU = YV \begin{pmatrix} I & 0 \\ 0 & J_\infty \end{pmatrix}.$$

The matrices $XU$ and $YV$ are partitioned consistently with the sizes of the two blocks $J_F$ and $J_\infty$: $XU = [X_F X_\infty]$, $YV = [Y_F Y_\infty]$. Hence, $AX_F = Y_F J_F$, $BX_F = Y_F$, $AX_\infty = Y_\infty$, $BX_\infty = Y_\infty J_\infty$. Thus, we have defined a monic block Jordan pair $(X_F; J_F)$ and a comonic block Jordan pair $(X_\infty; J_\infty)$ of the pencil $\lambda B - A$, with the columns of the composed matrix $[X_F\ X_\infty]$ being a basis of $\mathcal{L}$. If $J_F \neq 0$ and $J_\infty \neq 0$ then, obviously, the subspace $\mathcal{L}$ cannot be described only by means of monic or comonic block eigenpairs.

DEFINITION. A pair $(X_1, X_2; T_1, T_2)$ consisting of an $N \times d_1$ matrix $X_1$, an $N \times d_2$ matrix $X_2$, a $d_1 \times d_1$ matrix $T_1$ and a $d_2 \times d_2$ matrix $T_2$ is referred to as a decomposable block eigenpair of dimension $d = d_1 + d_2$ for a regular linear matrix pencil $\lambda B - A$ if $\text{rank}[X_1\ X_2] = d$ and $AX_1 = BX_1T_1$, $AX_2T_2 = BX_2$.

As was shown above, every invariant subspace of a regular pencil $\lambda B - A$ can be represented by means of a certain decomposable block eigenpair, and, vice versa, every decomposable block eigenpair $(X_1, X_2; T_1, T_2)$ corresponds to the invariant subspace with the basis matrix $[X_1\ X_2]$.

Monic and comonic block eigenpairs of a pencil $\lambda B - A$ are, evidently, particular cases of decomposable block eigenpairs.

As distinct from the case of monic and comonic block eigenpairs, when all the pairs corresponding to the same invariant subspace $\mathcal{L}$ are uniquely determined with due regard for similarity, decomposable pairs associated with the same invariant subspace generally are not uniquely determined up to similarity. However, if for two decomposable pairs $(X_1, X_2; T_1, T_2)$ and $(X_1', X_2'; T_1', T_2')$ the spectrum of $T_i$, $i = 1, 2$, coincides with the spectrum of $T_i'$ and the spectrum of $T_1$ does not intersect with the spectrum of $T_2$ then there exist nondegenerate matrices $Q_1$ and $Q_2$ such that

$$X_1' = X_1 Q_1,\ X_2' = X_2 Q_2, \qquad T_1' = Q_1^{-1} T_1 Q_1,\ T_2' = Q_2^{-1} T_2 Q_2.$$

Finally, we introduce several more definitions. An invariant subspace $\mathcal{L}$ of a regular matrix pencil $\lambda B - A$ will be called a spectral invariant subspace if the spectrum of the pencil $\lambda B_{\mathcal{L}} - A_{\mathcal{L}}$ with the multiplicities taken into account does not intersect with its complement to the spectrum of $\lambda B - A$ with its multiplicities. In other words, $\mathcal{L}$ is a spectral invariant subspace of a regular pencil $\lambda B - A$ if and only if $\mathcal{L}$ is a root subspace of $\lambda B - A$ or the sum of several root subspaces. We remind that the root subspace of a regular pencil $\lambda B - A$ corresponding to an eigenvalue $\lambda_0$ designates the linear span of all Jordan chains of $\lambda B - A$ corresponding to the eigenvalue $\lambda_0$. In view of the above observations we shall say that a spectral invariant subspace $\mathcal{L}$ corresponds to the part of the spectrum of $\lambda B - A$ which is equal to the spectrum of $\lambda B_{\mathcal{L}} - A_{\mathcal{L}}$. If the spectrum of $\lambda B - A$ is the union of two disjoint sets $\Lambda_1$ and $\Lambda_2$, then, obviously, there exists a unique spectral invariant subspace $\mathcal{L}_1$ of the pencil $\lambda B - A$ corresponding to $\Lambda_1$. Respectively, there exists a spectral invariant subspace $\mathcal{L}_2$ corresponding to $\Lambda_2$, with $\mathbf{C}^N$ being equal to the direct sum of the spectral invariant subspaces $\mathcal{L}_1$, $\mathcal{L}_2$ of the pencil $\lambda B - A$. For instance, the finite and the infinite spectra of $\lambda B - A$ can be used as $\Lambda_1$ and $\Lambda_2$.

We shall say also that $(X; T)$ is a block eigenpair of a matrix $A$ when $(X; T)$ is a monic block eigenpair for the matrix pencil $\lambda I - A$.

*Notes and references*

A proof of Theorem 1, which is really a suitable application of the classical theorem on the Jordan canonical form of matrices, can be found in [13, 39]. Theorem 2 and methods to compute the Schur form are discussed in [34]. It should be noted that in the literature on numerical linear algebra the invariant subspaces for regular linear pencils are often called the deflating subspaces [40, 41, 10, 27].

Block eigenpairs are intensively used in [18].

We do not discuss at all here the continuity properties of invariant subspaces of $\lambda B - A$ with respect to perturbations of the elements of the matrices $A$, $B$. These are extremely important for computational mathematics. General qualitative results on this subject are presented in [19]; some quantitative estimates of the Lipschitz continuity for spectral invariant subspaces can be found in [10, 39, 30].

## 2. Linear matrix equations

Let us consider the linear matrix equation of a general form with respect to $X$:

$$A_1XB_1 + A_2XB_2 + \cdots + A_nXB_n = C, \tag{5}$$

where $A_i$, $B_i$ are of size $M \times M$, $N \times N$, respectively. The way used to study (5) is to represent the matrices $X$ and $C$ as vectors in $C^{MN}$, with the linear operator in the left-hand side of (5) being an $MN \times MN$ matrix.

Denote the columns of the matrix $X$ by $x_1, x_2, \ldots, x_N \in \mathbf{C}^M$ and introduce the operation $\mathrm{vec}(X) = [x_1^T\, x_2^T\, \ldots\, x_N^T]^T$ to represent a matrix $X$ in column vector form. Then $\mathrm{vec}(AXB) = (B^T \otimes A)\mathrm{vec}(X)$, where the Kronecker product $M_1 \otimes M_2$ is by definition the following block matrix:

$$M_1 \otimes M_2 = \begin{bmatrix} (M_1)_{11}M_2 & (M_1)_{12}M_2 & \ldots & (M_1)_{1l}M_2 \\ (M_1)_{21}M_2 & (M_1)_{22}M_2 & \ldots & (M_1)_{2l}M_2 \\ \ldots & \ldots & \ldots & \ldots \\ (M_1)_{k1}M_2 & (M_1)_{k2}M_2 & \ldots & (M_1)_{kl}M_2 \end{bmatrix}.$$

Here $k \times l$ is the size of $M_1$. Thus,

$$\mathrm{vec}\left(\sum_{i=1}^{n} A_iXB_i\right) = \sum_{i=1}^{n} (B_i^T \otimes A_i)\,\mathrm{vec}(X),$$

i.e. equation (5) is equivalent to the system of linear equations in the usual form:

$$\left(\sum_{i=1}^{n} B_i^T \otimes A_i\right) \mathrm{vec}(X) = \mathrm{vec}(C). \tag{6}$$

The properties of the operator of equation (5) are entirely defined by properties of the matrix

$$\sum_{i=1}^{n} B_i^T \otimes A_i,$$

which are, in general, "hidden" at first glance.

Let us consider the problem in case $n = 2$, that is the equation

$$A_1XB_1 - A_2XB_2 = C. \tag{7}$$

For convenience in the formulations of the subsequent results we put the minus sign in (7). It is possible to simplify the structure of the matrix $B_1^T \otimes A_1 - B_2^T \otimes A_2$ by taking advantage of the triangular canonical forms of the pencils $A_1 - \lambda A_2$ and $\lambda B_1 - B_2$. We suppose that these pencils are regular and use the Weierstrass canonical form (Theorem 1). Similar results can be also obtained with the aid of the Schur form from Theorem 2.

So, using

$$A_1 - \lambda A_2 = V_A(T_{A_1} - \lambda T_{A_2})U_A^{-1}, \quad \lambda B_1 - B_2 = V_B(\lambda T_{B_1} - T_{B_2})U_B^{-1} \quad (8)$$

with nondegenerate $V_A$, $V_B$, $U_A$, $U_B$ and triangular $T_{A_1}$, $T_{A_2}$, $T_{B_1}$, $T_{B_2}$, (7) can be rewritten as

$$V_A T_{A_1} U_A^{-1} X V_B T_{B_1} U_B^{-1} - V_A T_{A_2} U_A^{-1} X V_B T_{B_2} U_B^{-1} = C.$$

Therefore,

$$T_{A_1} \widetilde{X} T_{B_1} - T_{A_2} \widetilde{X} T_{B_2} = \widetilde{C}, \quad (9)$$

where $\widetilde{X} = U_A^{-1} X V_B$, $\widetilde{C} = V_A^{-1} C U_B$. Now

$$\left[T_{B_1}^T \otimes T_{A_1} - T_{B_2}^T \otimes T_{A_2}\right] \operatorname{vec}(\widetilde{X}) = \operatorname{vec}(\widetilde{C}),$$

$$\operatorname{vec}(\widetilde{X}) = \left[V_B^T \otimes U_A^{-1}\right] \operatorname{vec}(X),$$

$$\operatorname{vec}(\widetilde{C}) = \left[U_B^T \otimes V_A^{-1}\right] \operatorname{vec}(C).$$

As a result, we have

$$\begin{aligned} &B_1^T \otimes A_1 - B_2^T \otimes A_2 \\ &\quad = \left[U_B^T \otimes V_A^{-1}\right]^{-1} \left[T_{B_1}^T \otimes T_{A_1} - T_{B_2}^T \otimes T_{A_2}\right] \left[V_B^T \otimes U_A^{-1}\right] \end{aligned}$$

with nondegenerate matrices $U_B^T \otimes V_A^{-1}$, $V_B^T \otimes U_A^{-1}$. If, for instance,

$$V_A = U_A, \qquad V_B = U_B,$$

then $U_B^T \otimes V_A^{-1} = V_B^T \otimes U_A^{-1}$, and the matrices $B_1^T \otimes A_1 - B_2^T \otimes A_2$ and $T_{B_1}^T \otimes T_{A_1} - T_{B_2}^T \otimes T_{A_2}$ are similar.

In any case, the solvability of equation (7) is equivalent to the solvability of equation (9), i.e. (in general) to nondegeneration of the matrix $T_{B_1}^T \otimes T_{A_1} - T_{B_2}^T \otimes T_{A_2}$. Without loss of generality we assume that the matrices $T_{A_1}$ and $T_{A_2}$ are upper triangular and that the matrices $T_{B_1}$, $T_{B_2}$ are lower triangular. The matrix $T_{B_1}^T \otimes T_{A_1} - T_{B_2}^T \otimes T_{A_2}$ is then upper triangular.

Using the triangularity of the matrices involved we are able to analyze conditions for the matrix $T_{B_1}^T \otimes T_{A_1} - T_{B_2}^T \otimes T_{A_2}$ to be nondegenerate. Since $T_{A_1} - \lambda T_{A_2}$ and $\lambda T_{B_1} - T_{B_2}$ are in Weierstrass canonical form, any pair of equally placed diagonal elements of the matrices $T_{A_1}$ and $T_{A_2}$ is of the form $(\lambda_A, 1)$ for a finite eigenvalue of the pencil $A_1 - \lambda A_2$ and $(1, 0)$ for an infinite one. Analogously, a pair of equally placed diagonal elements of the matrices $T_{B_1}$, $T_{B_2}$ is of the form $(1, \lambda_B)$ for a finite eigenvalue of the pencil $\lambda B_1 - B_2$ and $(0, 1)$ for an infinite one. As a result, we obtain the following table of all possible cases:

| eigenvalue of $T_{B_1}^T \otimes T_{A_1} - T_{B_2}^T \otimes T_{A_2}$ | type of an eigenvalue of $A_1 - \lambda A_2$ | type of an eigenvalue of $\lambda B_1 - B_2$ |
|---|---|---|
| $\lambda_A - \lambda_B$ | finite $\lambda_A$ | finite $\lambda_B$ |
| 1 | finite | infinite |
| 1 | infinite | finite |
| 0 | infinite | infinite |

Thus, the matrix $B_1^T \otimes A_1 - B_2^T \otimes A_2$ is nondegenerate if and only if the set of all eigenvalues of the pencil $A_1 - \lambda A_2$ does not intersect with the set of all eigenvalues of the pencil $\lambda B_1 - B_2$.

A particular case of equation (7) is the Sylvester equation frequently arising in matrix analysis:

$$AX - XB = C. \tag{10}$$

In order to obtain (10) one has to put $A_1 = A$, $A_2 = I$, $B_1 = I$, $B_2 = B$ in (7). Hence, $V_A = U_A$, $V_B = U_B$, and the Kronecker product matrices $I \otimes A - B^T \otimes I$ and $I \otimes T_A - T_B^T \otimes I$ are similar. The eigenvalues of the matrix $I \otimes A - B^T \otimes I$ are equal to all possible differences of the form $\lambda_A - \lambda_B$, where $\lambda_A$, $\lambda_B$ are eigenvalues of the matrices $A$ and $B$ respectively. Therefore, (10) is solvable if and only if the spectra of $A$ and $B$ are disjoint.

There is an integral formula for the unique solution of (10). Indeed, let $\gamma$ be a rectifiable finite contour in the complex plane containing the eigenvalues of the matrix $A$ inside and the eigenvalues of $B$ outside, then

$$X = \frac{1}{2\pi \mathrm{i}} \int_\gamma (\lambda I - A)^{-1} C (\lambda I - B)^{-1} \, \mathrm{d}\lambda$$

is the unique solution of equation (10).

### 2.1. *The Lyapunov equations*

We pay special attention to a particular case of equation (1), which is the so-called matrix Lyapunov equation $X$:

$$A^* X + XA = -C. \tag{11}$$

Here $A$, $X$ and $C$ are $N \times N$ matrices. The previous arguments yield the following necessary and sufficient condition for the unique solvability of equation (11): $\lambda_1 + \bar{\lambda}_2 \neq 0$ for any eigenvalues $\lambda_1$ and $\lambda_2$ of the matrix $A$. This condition obviously holds if all eigenvalues of the matrix $A$ lie in the open left half-plane, $\operatorname{Re}\lambda_j < 0$; in this case, $A$ is said to be a stable matrix.

THEOREM 3 (Lyapunov). 1) *If the matrix $A$ is stable then equation* (11) *has a unique solution for every right-hand side $C$, and the solution can be represented as*

$$X = \int_0^\infty \mathrm{e}^{tA^*} C \mathrm{e}^{tA}\, \mathrm{d}t.$$

*Furthermore, if $C = C^* > 0$, then the solution $X$ of equation* (11) *satisfies $X = X^* > 0$.*

2) *If equation* (11) *holds for some matrix $X = X^* > 0$ and $C = C^* > 0$, then the matrix $A$ is stable.*

A more general result is contained in the following

THEOREM 4 (Ostrowsky–Schneider). *The matrix $A$ has no pure imaginary eigenvalues if and only if* (11) *has a solution for some Hermitian positive definite matrix $C$. If $X$ is a Hermitian solution of* (11) *for some Hermitian positive definite matrix $C$ then the number of eigenvalues of the matrix $A$ with negative (positive) real part equals the number of positive (negative) eigenvalues of the matrix $X$.*

There is another important generalization of Lyapunov's results. Consider the system of matrix equations with respect to a pair of matrices $P$ and $X$:

$$\begin{cases} P^2 - P = 0, \\ AP - PA = 0, \\ XP - P^*X = 0, \\ A^*X + XA = -P^*CP + (I - P^*)C(I - P). \end{cases} \tag{12}$$

THEOREM 5 (Godunov–Bulgakov). 1) *If the matrix $A$ has no pure imaginary eigenvalues, then* (12) *has a unique solution $(P, X)$ for every matrix $C$. Moreover, if $C = C^* > 0$, then $X = X^* > 0$.*

2) *If the system* (12) *is satisfied for some $P$, $X$ and $C$, where $X = X^* > 0$ and $C = C^* > 0$, then the matrix $A$ has no pure imaginary eigenvalues and $P$ is the spectral projector onto the invariant subspace of the matrix $A$ corresponding to the eigenvalues in the left half-plane.*

We recall that the spectral projector $P$ onto the invariant subspace of a matrix $A$ corresponding to an isolated group of spectrum points is the linear operator

$$P = \frac{1}{2\pi \mathrm{i}} \int_\gamma (\lambda I - A)^{-1}\, \mathrm{d}\lambda, \tag{13}$$

where the integration is carried out along a rectifiable closed contour $\gamma$ enclosing the isolated group of spectrum points and leaving the rest of the spectrum outside $\gamma$.

In the discrete case the Lyapunov equation takes the form

$$A^*XA - X = -C. \tag{14}$$

This is a particular case of equation (7) with $A_1 = A^*$, $B_1 = A$, $A_2 = B_2 = I$. The eigenvalues of the matrix $A^T \otimes A^* - I \otimes I$ are all possible expressions $\lambda_i\bar{\lambda}_j - 1$, where the $\lambda_i$, $\lambda_j$ are eigenvalues of the matrix $A$. Hence equation (14) has a solution (for arbitrary $C$) if and only if $\lambda_i\bar{\lambda}_j - 1 \neq 0$ for any eigenvalues $\lambda_i$, $\lambda_j$ of $A$. This condition obviously holds if all eigenvalues of the matrix $A$ lie in the open unit disk $|\lambda_i| < 1$. A matrix $A$ is said to be discrete stable if all its spectrum lies in the open unit disk.

In the discrete case there are also analogs of the theorems by Lyapunov and Ostrowsky–Schneider. Indeed, one has to substitute in the formulations for the continuous case: stable by discrete stable, the integral

$$\int_0^\infty \mathrm{e}^{tA^*} C \mathrm{e}^{tA}\, \mathrm{d}t$$

by

$$\sum_{k=0}^{\infty} (A^*)^k C A^k,$$

the imaginary axis by the unit circle, negative (positive) real parts of eigenvalues by the condition to lie inside (outside) the open unit disk. Generalizations of the Godunov–Bulgakov theorem to the discrete case are rather nontrivial.

*Notes and references*

Linear matrix equations of the form (5) and the Kronecker products of matrices are considered in most textbooks on matrix analysis [3, 21, 25].

The canonical Schur form is always used in practice instead of the canonical Weierstrass form for numerical solution of equation (7) and its particular cases. This is accounted for by the instability of the computation of the Weierstrass form. The method of solution of (7) with the aid of the Schur form is called the Bartels–Stewart method [1]. Some additional developments of the idea of [1] are discussed in [20].

A description of all solutions of the equation $AX = XB$ can be found in [13, 25].

Together with the necessity to solve just the equation (5) or (7), the need to study and solve more general systems of linear matrix equations often appears. For example, when studying perturbations of spectral invariant subspaces of regular matrix pencils or singular subspaces of matrices, the generalized Sylvester equation

$$\begin{cases} A_1X - YB_1 = C_1, \\ A_2X - YB_2 = C_2 \end{cases}$$

is widely used [10, 39]. The technique of the Kronecker products is also useful for the generalized Sylvester equation: indeed,

$$\begin{bmatrix} I \otimes A_1 & -B_1^T \otimes I \\ I \otimes A_2 & -B_2^T \otimes I \end{bmatrix} \begin{bmatrix} \mathrm{vec}(X) \\ \mathrm{vec}(Y) \end{bmatrix} = \begin{bmatrix} \mathrm{vec}(C_1) \\ \mathrm{vec}(C_2) \end{bmatrix}.$$

The continuous Lyapunov equation (11) naturally arises in the study of stability in the Lyapunov sense of a system of ordinary differential equations

$$\frac{\mathrm{d}x}{\mathrm{d}t} = Ax$$

by the Lyapunov method of quadratic functions. The idea of this method consists in the selection of a self-adjoint positive definite matrix $X$ such that the quadratic form $(Xx, x)$ would decay along the solutions $x(t)$ of the system

$$\frac{\mathrm{d}x}{\mathrm{d}t} = Ax.$$

Since

$$\frac{\mathrm{d}}{\mathrm{d}t}(Xx, x) = \big([A^*X + XA]x, x\big),$$

the self-adjoint matrix $C = -[A^*X + XA]$ is necessarily positive definite.

When studying the Lyapunov stability of solutions of a finite difference equation $x_n = Ax_n$, the discrete Lyapunov equation (14) arises. Indeed, since

$$(Xx_{n+1}, x_{n+1}) - (x_n, x_n) = \big([A^*XA - I]x_n, x_n\big),$$

the matrix $C = -[A^*XA - I]$ has to be positive definite.

A thorough study of the Lyapunov equation is found in [13, 25]. The Ostrowsky–Schneider theorem, published first in [36], is discussed in detail in [25].

The generalized Lyapunov equation (12) appeared for the first time in [14], then in [6] the ultimate form (12) was derived. Discrete generalized Lyapunov equations were derived and thoroughly studied in [30].

Special attention should be paid to one of the main directions in modern matrix analysis: the study of the conditioning of matrix equations. By conditioning in numerical linear algebra we mean a quantitative characterization of a problem that reflects the degree of continuity of its solution to perturbations (mainly, infinitely small) of the problem data (coefficients of equations and right-hand sides). For a discussion of the notion of conditioning see, e.g., [42, 39]. For equations of the form (5) the condition number is commonly defined as the condition number of the matrix $M = \sum_{i=1}^{n} B_i^T \otimes A_i$:

$$\mathrm{cond}\, M = \|M\|\,\|M^{-1}\| \tag{15}$$

with an appropriate matrix norm. However, the condition number (15) is not always easy for computations, and from time to time a great many attempts were undertaken to find other condition numbers which are easier for computations; for the Lyapunov equations see [5, 27].

While developing the approaches from [14, 6], it became possible to obtain condition numbers for the generalized Lyapunov equations (12) and its discrete analogs [30].

## 3. Nonlinear matrix equations

It is worthwhile mentioning first a linearization method coupled with a perturbation theory. This approach is rather general but only permits local, in a certain sense, results. To demonstrate this, consider the equation

$$\mathcal{L}(X) + f(X) = C, \tag{16}$$

which inherits properties of the quadratic equation $AX - XB - XDX = C$. The linear operator $\mathcal{L}$: $C^{MN} \to C^{MN}$ is supposed to be invertible. The continuous mapping $f$ satisfies the following estimates:

a) $\|f(X)\| \leqslant K\|X\|$ with some constant $K$;

b) $\|f(X) - f(Y)\| \leqslant 2K \max\{\|X\|, \|Y\|\}\|X - Y\|$ with the same constant $K$.

Any consistent matrix norm will do as the norm $\|\cdot\|$.

THEOREM 6 (G.W. Stewart). *If $k = K\|C\|\,\|\mathcal{L}^{-1}\|^2 < 1/4$, then the matrix sequence*

$$X_0 = 0, \qquad X_{k+1} = \mathcal{L}^{-1}\big(C - f(X_k)\big), \quad k = 0, 1, \ldots,$$

*converges to the unique solution $X$ of equation* (16) *in the ball*

$$\|X\| \leqslant \frac{1 - \sqrt{1 - 4k}}{2k}\|C\|\,\|\mathcal{L}^{-1}\| < 2\|C\|\,\|\mathcal{L}^{-1}\|. \tag{17}$$

*Furthermore, if equation* (16) *has a solution that does not belong to the ball* (17), *then such a solution must be found outside the ball $\|X\| \geqslant (1 + \sqrt{1 - 4k})/(2\|\mathcal{L}^{-1}\|K)$.*

Now we start to present the method of invariant subspaces for companion linear matrix pencils which allows one in many cases to obtain global results. This method is applicable to matrix equations with a certain structure, which, fortunately, include several important matrix equations.

Consider the matrix equation

$$\begin{pmatrix} A_1 & A_2 \\ A_3 & A_4 \end{pmatrix} \begin{pmatrix} I \\ X \end{pmatrix} = \begin{pmatrix} B_1 & B_2 \\ B_3 & B_4 \end{pmatrix} \begin{pmatrix} I \\ X \end{pmatrix} \Lambda \tag{18}$$

with matrices $A_i$, $B_j$, $X$, $\Lambda$ of appropriate size. This equality means that the pencil

$$\lambda \begin{pmatrix} B_1 & B_2 \\ B_3 & B_4 \end{pmatrix} - \begin{pmatrix} A_1 & A_2 \\ A_3 & A_4 \end{pmatrix}$$

has an invariant subspace with basis matrix

$$\begin{pmatrix} I \\ X \end{pmatrix}.$$

Suppose, for instance, that $B_1 + B_2X$ is an invertible matrix, then, obviously, $\Lambda = (B_1 + B_2X)^{-1}(A_1 + A_2X)$ and

$$A_3 + A_4X = (B_3 + B_4X)(B_1 + B_2X)^{-1}(A_1 + A_2X). \tag{19}$$

Conversely, if equation (19) is satisfied, then equality (18) holds with

$$\Lambda = (B_1 + B_2X)^{-1}(A_1 + A_2X).$$

Thus, solving equation (19) is reduced to finding invariant subspaces of the matrix pencil

$$\lambda \begin{pmatrix} B_1 & B_2 \\ B_3 & B_4 \end{pmatrix} - \begin{pmatrix} A_1 & A_2 \\ A_3 & A_4 \end{pmatrix},$$

which possess basis matrices of the form

$$\begin{pmatrix} I \\ X \end{pmatrix}$$

for some matrix $X$. All such matrices $X$ are solutions of equation (19).

Let

$$\begin{pmatrix} B_1 & B_2 \\ B_3 & B_4 \end{pmatrix} = I$$

in (19), then (19) turns into a quadratic equation (sometimes called the Riccati equation; but we reserve the name of Riccati for equations of a more special kind):

$$A_3 + A_4X - XA_1 - XA_2X = 0. \tag{20}$$

As was shown above, $X$ is a solution of (20) if and only if the matrix

$$\begin{pmatrix} A_1 & A_2 \\ A_3 & A_4 \end{pmatrix}$$

has an invariant subspace with a basis matrix of the form

$$\begin{pmatrix} I \\ X \end{pmatrix}.$$

### 3.1. *The Riccati equation*

In applications structured nonlinear matrix equations often arise. One of the most important such equations, the Riccati algebraic matrix equation, is widely used in the theory of optimal control.

#### 3.1.1. *Continuous case*

Let the state of a linear system be governed by the differential equation

$$\frac{\mathrm{d}}{\mathrm{d}t}x(t) = Ax(t) + Bu(t), \tag{21}$$

where $A$ is an $N \times N$ matrix, $B$ is an $N \times M$ matrix. The cost of a control is defined by the functional

$$\Phi(u) = \int_0^\infty \big[(Qx, x) + (Ru, u)\big]\,\mathrm{d}t, \tag{22}$$

where $R$ and $Q$ are self-adjoint matrices, $R$ is positive definite, and $Q$ is non-negative definite.

The matrix Riccati equation

$$XBR^{-1}B^*X - A^*X - XA - Q = 0 \tag{23}$$

is associated to the system (21), (22).

DEFINITION.

1) A pair of matrices $(A, B)$ is called stabilizable if

$$\operatorname{rank}\big([\lambda I - A \;\; B]\big) = N$$

for all eigenvalues $\lambda$ of $A$ with $\operatorname{Re}\lambda \geqslant 0$.

2) A pair of matrices $(C, A)$ is called detectable if

$$\operatorname{rank}\left(\begin{bmatrix} C \\ \lambda I - A \end{bmatrix}\right) = N$$

for all eigenvalues $\lambda$ of $A$ with $\operatorname{Re}\lambda \geqslant 0$.

THEOREM 7. *If the matrix pair $(A, B)$ is stabilizable and the pair $(C, A)$ is detectable, where $C$ is any matrix satisfying the equality $C^*C = Q$, then equation (23) has a unique self-adjoint solution $X_0$ and the optimal control $u^*(t)$ is given by the formula $u^* = -R^{-1}B^*X_0x$. In addition, the closed loop matrix $\Lambda = A - BR^{-1}B^*X_0$ is stable, and the minimal value of the cost equals $\Phi_{\min} = (X_0x(0), x(0))$.*

We also associate to equation (23) the block matrix of order $2N$

$$H = \begin{pmatrix} A & -BR^{-1}B^* \\ -Q & -A^* \end{pmatrix}. \tag{24}$$

Since

$$\begin{aligned} &H \begin{bmatrix} I \\ X \end{bmatrix} - \begin{bmatrix} I \\ X \end{bmatrix} T \\ &\quad = \begin{bmatrix} A - BR^{-1}B^*X - T \\ XBR^{-1}B^*X - A^*X - XA - Q + X(A - BR^{-1}B^*X - T) \end{bmatrix}, \end{aligned}$$

solving (23) is equivalent to finding an invariant subspace of the matrix $H$ with a block eigenpair of the form

$$\left( \begin{bmatrix} I_N \\ X \end{bmatrix}; T \right).$$

The matrix $H$ is Hamiltonian, i.e. $J^{-1}H^*J = -H$ with

$$J = \begin{bmatrix} 0 & I_N \\ -I_N & 0 \end{bmatrix}.$$

This implies that together with every eigenvalue $\lambda$ the matrix $H$ has an eigenvalue $-\bar{\lambda}$ of the same multiplicity. The conditions of Theorem 7 imposed on the matrices $A$, $B$, $Q$ guarantee that $H$ has no pure imaginary eigenvalues. Moreover, the matrix $H$ has a unique invariant subspace with the block eigenpair

$$\left( \begin{bmatrix} I_N \\ X \end{bmatrix}, \Lambda \right),$$

where $\Lambda$ is a stable matrix and $X$ is a solution of equation (23). Thus, solving (23) is reduced to finding a basis matrix of the invariant subspace of the Hamiltonian matrix $H$ corresponding to the eigenvalues in the open left half-plane. If, for instance, this basis matrix is equal to

$$\begin{bmatrix} X_1 \\ X_2 \end{bmatrix},$$

then, obviously, the matrix $X_1$ is invertible and $X = X_2X_1^{-1}$.

There exists another variant of the reduction of equation (23) to the problem of finding invariant subspaces of regular matrix pencils. Namely, consider the pencil

$$\lambda\widetilde{H}_2 - \widetilde{H}_1 = \lambda \begin{bmatrix} I & 0 & 0 \\ 0 & I & 0 \\ 0 & 0 & 0 \end{bmatrix} - \begin{bmatrix} A & 0 & B \\ -Q & -A^* & 0 \\ 0 & B^* & R \end{bmatrix} \tag{25}$$

and note the following identity

$$\begin{bmatrix} A & 0 & B \\ -Q & -A^* & 0 \\ 0 & B^* & R \end{bmatrix} \begin{bmatrix} I \\ X \\ Z \end{bmatrix} - \begin{bmatrix} I & 0 & 0 \\ 0 & I & 0 \\ 0 & 0 & 0 \end{bmatrix} \begin{bmatrix} I \\ X \\ Z \end{bmatrix} T$$
$$= \begin{bmatrix} A - BR^{-1}B^*X - T + B(R^{-1}B^*X + Z) \\ -Q - A^*X - XA + XBR^{-1}B^*X + X(A - BR^{-1}B^*X - T) \\ R(R^{-1}B^*X + Z) \end{bmatrix}.$$

Under the conditions of Theorem 7 the pencil $\lambda\widetilde{H}_2 - \widetilde{H}_1$ has $M$ infinite eigenvalues, $N$ finite eigenvalues in the open left half-plane and $N$ finite eigenvalues in the open right half-plane. In fact,

$$\begin{bmatrix} I & 0 & -BR^{-1} \\ 0 & I & 0 \\ 0 & 0 & I \end{bmatrix} (\lambda\widetilde{H}_2 - \widetilde{H}_1) = \lambda \begin{bmatrix} I & 0 & 0 \\ 0 & I & 0 \\ 0 & 0 & 0 \end{bmatrix} - \begin{bmatrix} A & -BR^{-1}B^* & 0 \\ -Q & -A^* & 0 \\ 0 & B^* & R \end{bmatrix},$$

and, therefore, the set of all eigenvalues of the pencil $\lambda\widetilde{H}_2 - \widetilde{H}_1$ coincides with the union of the spectra of the pencils $\lambda I - H$ and $\lambda 0 - R$. As a result, finding a solution $X = X^*$ of equation (23) is reduced to finding the unique invariant subspace of the pencil $\lambda\widetilde{H}_2 - \widetilde{H}_1$ with a monic block eigenpair of the form

$$\left( \begin{bmatrix} I \\ X \\ Z \end{bmatrix} ; \Lambda \right),$$

where $\Lambda$ is stable.

### 3.1.2. *Discrete case*

Let the state of a discrete linear system be governed by the finite difference equation

$$x(k+1) = Ax(k) + Bu(k), \tag{26}$$

where $A$ is an $N \times N$ matrix and $B$ is an $N \times M$ matrix. The cost of control for (26) is given by the functional

$$\Phi(u) = \sum_{k=0}^{\infty} \left[ \left(Qx(k), x(k)\right) + \left(Ru(k), u(k)\right) \right], \tag{27}$$

where $R = R^*$ is positive definite and $Q = Q^*$ is non-negative definite.

The discrete algebraic matrix Riccati equation

$$A^*XA - X - A^*XB(R + B^*XB)^{-1}B^*XA + Q = 0 \tag{28}$$

corresponds to (26), (27).

DEFINITION.

1) A pair of matrices $(A, B)$ is discrete stabilizable if

$$\operatorname{rank}\left([\lambda I - A \; B]\right) = N$$

for all eigenvalues $\lambda$ of $A$ with $|\lambda| \geqslant 1$.

2) A pair of matrices $(C, A)$ is discrete detectable if

$$\operatorname{rank}\left(\begin{bmatrix} C \\ \lambda I - A \end{bmatrix}\right) = N$$

for all eigenvalues $\lambda$ of $A$ with $|\lambda| \geqslant 1$.

THEOREM 8. *If the matrix pair $(A, B)$ is discrete stabilizable and the pair $(C, A)$ is discrete detectable, where $C$ is any matrix satisfying $C^*C = Q$, then there exists a unique self-adjoint solution $X_0$ of equation* (28). *Additionally, the optimal control is expressed by the formula*

$$u^*(k) = -(R + B^*X_0B)^{-1}B^*X_0Ax(k),$$

*the closed loop matrix $\Lambda = A - B(R + B^*X_0B)^{-1}B^*X_0A$ is discrete stable, and the minimal value of $\Phi(u)$ is equal to $(X_0x(0), x(0))$.*

Equation (28) is associated with the regular matrix pencil

$$\lambda S_2 - S_1 = \lambda \begin{pmatrix} I & BR^{-1}B^* \\ 0 & -A^* \end{pmatrix} - \begin{pmatrix} A & 0 \\ Q & -I \end{pmatrix}, \tag{29}$$

so that the identity

$$\begin{pmatrix} A & 0 \\ Q & -I \end{pmatrix} \begin{bmatrix} I \\ X \end{bmatrix} - \begin{pmatrix} I & BR^{-1}B^* \\ 0 & -A^* \end{pmatrix} \begin{bmatrix} I \\ X \end{bmatrix} T$$

$$= \begin{bmatrix} (I + BR^{-1}B^*X)[A - B(R + B^*XB)^{-1}B^*XA - T] \\ F(X) - A^*X[A - B(R + B^*XB)^{-1}B^*XA - T] \end{bmatrix}$$

holds, where $F(X)$ stands for $Q - X + A^*XA - A^*XB(R + B^*XB)^{-1}B^*XA$.

The regular $2N \times 2N$ matrix pencil $\lambda B - A$ will be called symplectic if $AJ^{-1}A^* = BJ^{-1}B^*$ with

$$J = \begin{bmatrix} 0 & I_N \\ -I_N & 0 \end{bmatrix}.$$

Obviously, the pencil $\lambda S_2 - S_1$ is symplectic. By virtue of this property, if the pencil has an eigenvalue $\lambda$, then $1/\bar{\lambda}$ is also an eigenvalue of this pencil with taking into account multiplicities, i.e. the spectrum of a symplectic matrix pencil is located symmetrically with respect to the unit circle.

Summarizing, finding a solution $X = X^*$ of equation (28) under the conditions of Theorem 8 is reduced to finding the unique invariant subspace of the pencil (29) with a monic block eigenpair of the form

$$\left( \begin{bmatrix} I_N \\ X \end{bmatrix} ; \Lambda \right),$$

where $\Lambda$ is discrete stable.

Finally, let us form the matrix pencil

$$\lambda \widetilde{S}_2 - \widetilde{S}_1 = \lambda \begin{bmatrix} I & 0 & 0 \\ 0 & -A^* & 0 \\ 0 & -B^* & 0 \end{bmatrix} - \begin{bmatrix} A & 0 & B \\ Q & -I & 0 \\ 0 & 0 & R \end{bmatrix} \tag{30}$$

for which the identity

$$\begin{bmatrix} A & 0 & B \\ Q & -I & 0 \\ 0 & 0 & R \end{bmatrix} \begin{bmatrix} I \\ X \\ Z \end{bmatrix} - \begin{bmatrix} I & 0 & 0 \\ 0 & -A^* & 0 \\ 0 & -B^* & 0 \end{bmatrix} \begin{bmatrix} I \\ X \\ Z \end{bmatrix} T$$
$$= \begin{bmatrix} [A - B(R + B^*XB)^{-1}B^*XA - T] + B[(R + B^*XB)^{-1}B^*XA + Z] \\ F(X) - A^*X[A - B(R + B^*XB)^{-1}B^*XA - T] \\ -B^*X[A - B(R + B^*XB)^{-1}B^*XA - T] + R[(R + B^*XB)^{-1}B^*XA + Z] \end{bmatrix}$$

holds. By virtue of the identity

$$\begin{bmatrix} I & 0 & -BR^{-1} \\ 0 & I & 0 \\ 0 & 0 & I \end{bmatrix} (\lambda \widetilde{S}_2 - \widetilde{S}_1) = \lambda \begin{bmatrix} I & BR^{-1}B^* & 0 \\ 0 & -A^* & 0 \\ 0 & -B^* & 0 \end{bmatrix} - \begin{bmatrix} A & 0 & 0 \\ Q & -I & 0 \\ 0 & 0 & R \end{bmatrix},$$

the spectrum of $\lambda\widetilde{S}_2 - \widetilde{S}_1$ is the union of the spectra of $\lambda S_2 - S_1$ and $\lambda 0 - R$ taking into account the multiplicities. Therefore, in order to find a solution $X = X^*$ of equation (28) satisfying Theorem 8 one has to find the unique monic block eigenpair

$$\left(\begin{bmatrix} I \\ X \\ Z \end{bmatrix} ;\ \Lambda\right)$$

with a discrete stable $\Lambda$.

### 3.2. *Polynomial matrix equations*

A polynomial matrix equation for an $N \times N$ matrix $X$ looks as follows

$$A_0 + A_1 X + \cdots + A_n X^n = 0 \tag{31}$$

with $N \times N$ matrix coefficients $A_i$, $i = 0, 1, \ldots, n$. Let us associate to (31) the companion regular linear matrix pencil

$$\lambda C_2 - C_1 = \lambda \begin{bmatrix} I & & & & 0 \\ & I & & & \\ & & \ddots & & \\ & & & I & \\ 0 & & & & A_n \end{bmatrix} - \begin{bmatrix} 0 & I & & & \\ & 0 & I & & \\ & & \ddots & \ddots & \\ & & & 0 & I \\ -A_0 & -A_1 & \ldots & -A_{n-2} & -A_{n-1} \end{bmatrix}. \tag{32}$$

Since, obviously,

$$C_1 \begin{bmatrix} I \\ X \\ \vdots \\ X^{n-1} \end{bmatrix} - C_2 \begin{bmatrix} I \\ X \\ \vdots \\ X^{n-1} \end{bmatrix} T = \begin{bmatrix} X - T \\ X(X - T) \\ \vdots \\ -A_0 - A_1 X - \ldots - A_n X^n + A_n X^{n-1}(X - T) \end{bmatrix},$$

solving equation (31) is equivalent to finding invariant subspaces of the pencil (32) with monic block eigenpairs of the form

$$\left( \begin{bmatrix} U_1 \\ U_2 \\ \vdots \\ U_n \end{bmatrix}, \Lambda \right)$$

with an invertible matrix $U_1$. In this case the matrix $X = U_2 U_1^{-1} = U_1 \Lambda U_1^{-1}$ will be a solution of equation (31).

To conclude, we mention the following simple result.

THEOREM 9. *A matrix $X$ is a solution of* (31) *if and only if the matrix polynomial $A(\lambda) = A_0 + A_1\lambda + \cdots + A_n\lambda^n$ has the right divisor $\lambda I - X$, i.e. $A(\lambda) = A_2(\lambda)(I\lambda - X)$ for an appropriate matrix polynomial $A_2(\lambda)$.*

*Notes and references*

G.W. Stuart's theorem is proven in, e.g., [39]. Some applications of this theorem are in [10, 39, 31].

A vast bibliography has been devoted to iterative methods of solution of nonlinear equations whose particular case is (16). A detailed treatment of such methods is found, e.g., in [35].

The quadratic equation (20) has been studied in many papers. A thorough treatment of the theory of this equation, including such questions as continuity and analyticity of its solution with respect to the coefficients $A_i$, $i = 1, 2, 3, 4$, is contained in [19].

The algebraic Riccati equations are probably the most deeply studied structured matrix equations after the Lyapunov equations. Among the monographs entirely devoted to the theory of the Riccati equations are [38, 33]. Of the articles on this subject we mention [26, 37, 40, 32, 27, 15, 17]. Theorems about the Riccati equation are proven, e.g., in [43]. A vast bibliography about the Riccati equations is found in [38].

The use of the pencils (25), (30) is stimulated by the possibility of avoiding the inversion of the matrix $R$, when forming the pencil. Details of the numerical implementation of such an approach are discussed in [40].

Polynomial equations, as Theorem 9 (proved in [13]) shows, give rise to a particular case of the problem of factorization of matrix polynomials. A nice exposition of the theory of polynomial matrix equations is in [25].

## 4. Factorization of matrix polynomials

### 4.1. *Spectral characteristics of matrix polynomials*

We consider only regular matrix polynomials $A(\lambda) = A_0 + A_1\lambda + \cdots + A_n\lambda^n$, i.e. polynomials such that $\det A(\lambda) \not\equiv 0$. If $A_n = I$, then such a polynomial $A(\lambda)$ is called monic. If $A_0 = I$ then $A(\lambda)$ is called comonic. In general, we admit the possibility of $A_n$ being equal to the zero matrix, but, usually, $A_n \neq 0$, and in this latter case $n$ is referred to as the degree of the polynomial $A(\lambda)$, $n = \deg A(\lambda)$.

The roots of the scalar polynomial $\det A(\lambda)$ are called finite eigenvalues of $A(\lambda)$. A polynomial $A(\lambda)$ is said to have infinite eigenvalue if the dual polynomial $\tilde{A}(\mu) = \mu^n A(1/\mu)$ has zero eigenvalue.

Finite elementary divisors of the $\lambda$-matrix $A(\lambda)$ are supplemented by infinite elementary divisors which are defined as elementary divisors of the form $\mu^n$ of the dual polynomial $\tilde{A}(\mu)$. Thus, the multiplicities of eigenvalues of the polynomial $A(\lambda)$ are accounted for by the elementary divisors of $A(\lambda)$ in a proper way.

A set of vectors $x_0, x_1, \ldots, x_{l-1} \in C^N$, $x_0 \neq 0$, is a right Jordan chain of $A(\lambda)$ corresponding to a finite eigenvalue $\lambda_0$ if

$$A(\lambda_0)x_i + \frac{1}{1!}A^{(1)}(\lambda_0)x_{i-1} + \cdots + \frac{1}{i!}A^{(i)}(\lambda_0)x_0 = 0, \quad i = 0, \ldots, l-1. \tag{33}$$

Here $A^{(p)}(\lambda)$ stands for the $p$-th derivative of $A(\lambda)$ with respect to $\lambda$. Similarly, $x_0, x_1, \ldots, x_{l-1} \in C^N$, $x_0 \neq 0$, is a right Jordan chain of $A(\lambda)$ corresponding to the infinite eigenvalue if

$$\tilde{A}(0)x_i + \frac{1}{1!}\tilde{A}^{(1)}(0)x_{i-1} + \cdots + \frac{1}{i!}\tilde{A}^{(i)}(0)x_0 = 0, \quad i = 0, \ldots, l-1. \tag{34}$$

Left Jordan chains of the polynomial $A(\lambda)$ are defined analogously. Henceforward we usually omit the word "right" before "Jordan chains".

One of the main tools for the study of spectral properties of matrix polynomials is the companion linear matrix pencil

$$C_A(\lambda) = \begin{bmatrix} 0 & -I & & \\ & \ddots & \ddots & \\ & & 0 & -I \\ A_0 & \ldots & A_{n-2} & A_{n-1} \end{bmatrix} + \begin{bmatrix} I & & & \\ & \ddots & & \\ & & I & \\ & & & A_n \end{bmatrix} \lambda, \tag{35}$$

which is equivalent to the following $\lambda$-matrix

$$\begin{bmatrix} I & & & & \\ & I & & & \\ & & \ddots & & \\ & & & I & \\ & & & & A(\lambda) \end{bmatrix} = \begin{bmatrix} I & & & & \\ & I & & & \\ & & \ddots & & \\ & & & I & \\ A_1 & A_2 & \ldots & A_{n-1} + A_n\lambda & I \end{bmatrix}$$

$$\times \begin{bmatrix} I & & & \\ \lambda I & I & & \\ & \ddots & \ddots & \\ & & \lambda I & I \end{bmatrix} C_A(\lambda) \begin{bmatrix} 0 & & & I \\ -I & 0 & & \lambda I \\ & \ddots & \ddots & \vdots \\ & & -I & \lambda^{n-1} I \end{bmatrix}.$$

Due to this equivalence the set of elementary divisors of $A(\lambda)$ coincides with the set of those of $C_A(\lambda)$.

A decisive reason to introduce the companion linear matrix polynomial $C_A(\lambda)$ is the structure of block eigenpairs for $C_A(\lambda)$, particularly, that of block Jordan pairs.

THEOREM 10. *A decomposable block eigenpair of dimension $d$ for the pencil $C_A(\lambda)$ has the following structure:*

$$\left( \begin{bmatrix} X_1 \\ X_1 T_1 \\ \vdots \\ X_1 T_1^{n-1} \end{bmatrix}, \begin{bmatrix} X_2 T_2^{n-1} \\ X_2 T_2^{n-2} \\ \vdots \\ X_2 \end{bmatrix}; T_1, T_2 \right),$$

*where*

$$A_0 X_1 + A_1 X_1 T_1 + \cdots + A_n X_1 T_1^n = 0$$

*and*

$$A_0 X_2 T_2^n + A_1 X_2 T_2^{n-1} + \cdots + A_n X_2 = 0.$$

*Here $X_1$, $X_2$, $T_1$ and $T_2$ are matrices of sizes $N \times d_1$, $N \times d_2$, $d_1 \times d_1$, $d_2 \times d_2$, respectively, where $d_2 = d - d_1$,*

The proof of this theorem is obvious.

DEFINITION. A pair $(X_1, X_2; T_1, T_2)$ consisting of an $N \times d_1$ matrix $X_1$, a $N \times d_2$ matrix $X_2$, a $d_1 \times d_1$ matrix $T_1$, a $d_2 \times d_2$ matrix $T_2$ is a decomposable block eigenpair of dimension $d$ for the matrix polynomial $A(\lambda)$ if

$$\operatorname{rank} \left( \begin{bmatrix} X_1 & X_2 T_2^{n-1} \\ \vdots & \vdots \\ X_1 T_1^{n-1} & X_2 \end{bmatrix} \right) = d = d_1 + d_2$$

and

$$A_0 X_1 + A_1 X_1 T_1 + \cdots + A_n X_1 T_1^n = 0,$$

$$A_0 X_2 T_2^n + A_1 X_2 T_2^{n-1} + \cdots + A_n X_2 = 0.$$

If $d_2 = 0$, then such a pair is called monic and it is called comonic when $d_1 = 0$. Such a (co)monic eigenpair is denoted by $(X;T)$, where $X = X_1$, $T = T_1$ for the former case and $X = X_2$, $T = T_2$ for the latter one.

By Theorem 10, block eigenpairs of $A(\lambda)$ are nothing other than generators of the block eigenpairs for $C_A(\lambda)$.

The decomposable eigenpairs $(X_1, X_2; T_1, T_2)$ and $(X_1', X_2'; T_1', T_2')$ for a polynomial $A(\lambda)$ are said to be similar if there exist $Q_1$ and $Q_2$ such that $X_1' = X_1 Q_1$, $T_1' = Q_1^{-1} T_1 Q_1$, $X_2' = X_2 Q_2$, $T_2' = Q_2^{-1} T_2 Q_2$. It is easy to show that similar eigenpairs define the same invariant subspace of the companion linear pencil.

If the matrices $J_1$ and $J_2$ in a decomposable block eigenpair $(X_1, X_2; J_1, J_2)$ are Jordan, then such a pair is called a block Jordan pair for the polynomial $A(\lambda)$.

Let

$$J_1 = \begin{bmatrix} J_{11} & & \\ & \ddots & \\ & & J_{1k_1} \end{bmatrix},$$

where the $J_{1j}$ are Jordan blocks, then, partitioning $X_1 = [X_{11} \ldots X_{1k_1}]$, we obtain that the columns of $X_{1j}$ form a Jordan chain of $A(\lambda)$ corresponding to the eigenvalue of the block $J_{1j}$. The matrix $X_2$ has an analogous structure corresponding to the Jordan blocks in $J_2$.

Similar to the case of linear matrix pencils one can define block eigenpairs for a matrix polynomial $A(\lambda)$ which are associated with isolated parts of the spectrum of $A(\lambda)$, for instance the finite spectrum, i.e. with all finite eigenvalues of $A(\lambda)$. Indeed, if the spectrum of $A(\lambda)$ is the union of two disjoint sets $\Lambda_1$ and $\Lambda_2$ with the multiplicities taken into account, then a decomposable block eigenpair $(X_1, X_2; T_1, T_2)$ is associated with $\Lambda_1$ when the spectrum of the pencil

$$\lambda \begin{pmatrix} I & 0 \\ 0 & T_2 \end{pmatrix} - \begin{pmatrix} T_1 & 0 \\ 0 & I \end{pmatrix}$$

including the multiplicities coincides with $\Lambda_1$.

### 4.2. *The factorization problem*

DEFINITION. A matrix polynomial $A_1(\lambda) = A_{10} + \cdots + A_{1n_1}\lambda^{n_1}$ of order $n_1$ is called a right divisor of a polynomial $A(\lambda)$ of order $n$ if there exists a matrix polynomial $A_2(\lambda) = A_{20} + \cdots + A_{2n_2}\lambda^{n_2}$ of order $n_2$, $n_2 \geqslant n - n_1$ such that $A(\lambda) = A_2(\lambda)A_1(\lambda)$.

THEOREM 11. *Let $(X_F, T_F)$ be a monic block eigenpair of a matrix polynomial $A_1(\lambda)$ corresponding to the finite spectrum of $A_1(\lambda)$. The matrix polynomial $A_1(\lambda)$ is a right divisor of the polynomial $A(\lambda)$ if and only if*

$$A_0X_F + A_1X_FT_F + \cdots + A_nX_FT_F^n = 0. \tag{36}$$

PROOF. Since

$$A(\lambda) = \sum_{j=0}^{n_2-1} A_{2j}(A_{10} + \cdots + A_{1n_1}\lambda^{n_1})\lambda^j,$$

then

$$A_0X_F + A_1X_FT_F + \cdots + A_nX_FT_F^n \\ = \sum_{j=0}^{n_2-1} A_{2j}(A_{10}X_F + \cdots + A_{1n_1}X_FT_F^{n_1})T_F^j.$$

The inverse statement is more nontrivial. First of all, we carry out, if necessary, a change of the variable $\lambda$ by a transformation $\lambda = \lambda' + \alpha$, in order that the coefficient $A'_{10} = A'_1(0)$ of the polynomial $A'_1(\lambda') = A_1(\lambda' + \alpha)$ is a nondegenerate matrix. Such a transformation conserves the property of being a divisor: $A(\lambda) = A_2(\lambda)A_1(\lambda)$ is equivalent to $A'(\lambda') = A'_2(\lambda')A'_1(\lambda')$. At the same time the pair $(X_F, T_F - \alpha I)$ is a monic eigenpair for the polynomial $A'_1(\lambda')$ corresponding to the finite eigenvalues, and the equality $A_0X_F + \cdots + A_nX_FT_F^n = 0$ holds iff the equality $A'_0X_F + \cdots + A'_nX_F(T_F - \alpha I)^n = 0$ holds. Therefore, without loss of generality, we can suppose in what follows that $A_1(0)$ is a nondegenerate matrix.

By virtue of the nondegeneration of $A_1(0)$ the matrix $T_F$ is also nondegenerate. Hence the pair $(X_F, T_F^{-1})$ is a monic block eigenpair for the dual polynomial $\tilde{A}_1(\mu) = A_{10}\mu^{n_1} + \cdots + A_{1n_1}$, corresponding to all nonzero eigenvalues. Let $(X_\infty; T_\infty)$ be a monic block eigenpair for the polynomial $\tilde{A}_1(\mu)$ corresponding to the zero eigenvalue, then the matrix $T_\infty$ is, evidently, nilpotent, i.e. $T_\infty^\nu = 0$ for some positive integer $\nu$.

Let us divide the polynomial $\mu^\nu\tilde{A}(\mu)$ by $\tilde{A}_1(\mu)$ with remainder: $\mu^\nu\tilde{A}(\mu) = \tilde{A}_2(\mu)\tilde{A}_1(\mu) + \widetilde{R}(\mu)$. Here $\tilde{A}(\mu) = A_0\mu^n + \cdots + A_n$ is the dual to the polynomial $A(\lambda)$ and the degree of the polynomial $\widetilde{R}(\mu) = R_0\mu^{n_3} + \cdots + R_{n_3}$ is $n_3$, $n_3 < n_1$. It follows from the conditions of Theorem 11 that $A_0X_FT_F^{-n} + \cdots + A_nX_F = 0$. Therefore, $A_0X_F(T_F^{-1})^{n+\nu} + \cdots + A_nX_F(T_F^{-1})^\nu = 0$. Since $T_\infty^\nu = 0$,

$$A_0X_\infty T_\infty^{n+\nu} + \cdots + A_nX_\infty(T_\infty)^\nu = 0.$$

As a result, we obtain from $\widetilde{R}(\mu) = \mu^\nu\tilde{A}(\mu) - \tilde{A}_2(\mu)\tilde{A}_1(\mu)$ that

$$R_0X_F(T_F^{-1})^{n_3} + \cdots + R_{n_3}X_F = 0, \qquad R_0X_\infty(T_\infty)^{n_3} + \cdots + R_{n_3}X_\infty = 0.$$

It remains now to observe that these two equalities can be rewritten in a block form as

$$[R_0 \quad R_1 \quad \dots \quad R_{n_3}] \begin{bmatrix} X_F T_F^{-n_3} & X_\infty T_\infty^{n_3} \\ \vdots & \vdots \\ X_F T_F^{-1} & X_\infty T_\infty \\ X_F & X_\infty \end{bmatrix} = 0,$$

and that the matrix

$$\begin{bmatrix} X_F T_F^{-n_1} & X_\infty T_\infty^{n_1} \\ \vdots & \vdots \\ X_F T_F^{-1} & X_\infty T_\infty \\ X_F & X_\infty \end{bmatrix}$$

is nondegenerate. The nondegeneration of this matrix is a consequence of the fact that the pair

$$\left( [X_F\ X_\infty];\ \begin{pmatrix} T_F^{-1} & 0 \\ 0 & T_\infty \end{pmatrix} \right)$$

is a monic block eigenpair of dimension $n_1 N$ for the polynomial $\tilde{A}_1(\mu)$. As a result, $R_0 = R_1 = \cdots = R_{n_3} = 0$ and $\widetilde{R}(\mu) \equiv 0$.

From the identity $\mu^\nu \tilde{A}(\mu) = \tilde{A}_2(\mu)\tilde{A}_2(\mu)$ it follows that the polynomial $A_2(\lambda)$, being dual to $\tilde{A}_2(\mu)$, is the quotient of division of $A(\lambda)$ by $A_1(\lambda)$ with no remainder. The degree of the polynomial $A_2(\lambda)$ equals to $n_2 = n + \nu - n_1$. □

Using this theorem we consider the problem of the description of all right divisors for a given matrix polynomial $A(\lambda) = A_0 + \cdots + A_n\lambda^n$. Let us choose an invariant subspace of the companion pencil $C_A(\lambda)$ with a monic block eigenpair $(X; T)$ of dimension $d$ such that

$$\bigcap_{k=0}^{\infty} \operatorname{Ker}\left(XT^k\right) = 0. \tag{37}$$

Denote by $n_1$ the minimal integer for which

$$\bigcap_{k=0}^{n_1-1} \operatorname{Ker}\left(XT^k\right) = 0,$$

then obviously,

$$\operatorname{rank}\begin{bmatrix} X \\ XT \\ \vdots \\ XT^{n_1-1} \end{bmatrix} = d \leqslant n_1 N. \tag{38}$$

According to Theorem 11 the following auxiliary problem arises: given a pair of matrices $X$ of size $N \times d$ and $T$ of size $d \times d$ satisfying (38), one needs to construct a matrix polynomial $A_1(\lambda)$ of degree $n_1$, for which $(X;T)$ will be a monic block eigenpair corresponding to the finite spectrum of $A_1(\lambda)$. It turns out that this auxiliary problem is always solvable. Furthermore, all its solutions are described by the formula $A_1(\lambda) = Q(\lambda)\breve{A}_1(\lambda)$, where $\breve{A}_1(\lambda)$ is a particular solution and $Q(\lambda)$ is any unimodular $\lambda$-matrix, i.e. $\det Q(\lambda) \equiv \text{const} \neq 0$.

By Theorem 11 the polynomial $A_1(\lambda)$, which is one of solutions to the auxiliary problem, is a right divisor of $A(\lambda)$.

The most important is the case of $d = Nn_1$ in (38). Then a particular solution $\breve{A}_1(\lambda)$ of the auxiliary problem can be chosen as a monic polynomial $B_0 + B_1\lambda + \cdots + I\lambda^{n_1}$ with coefficients $B_i$, $i = 0, 1, \ldots, n_1 - 1$, calculated by the formula

$$[B_0 \quad B_1 \quad \ldots \quad B_{n_1-1}] = -XT^{n_1}\begin{bmatrix} X \\ XT \\ \vdots \\ XT^{n_1-1} \end{bmatrix}^{-1}, \tag{39}$$

which is derived from the identity

$$\begin{bmatrix} 0 & -I & & \\ & \ddots & \ddots & \\ & & 0 & -I \\ B_0 & \ldots & B_{n_1-2} & B_{n_1-1} \end{bmatrix}\begin{bmatrix} X \\ XT \\ \vdots \\ XT^{n_1-1} \end{bmatrix} + \begin{bmatrix} X \\ XT \\ \vdots \\ XT^{n_1-1} \end{bmatrix} T = 0.$$

Thus, when $d = Nn_1$, it is possible to find a unique monic right divisor of $A(\lambda)$ with monic block eigenpair $(X;T)$ of dimension $Nn_1$.

This case also admits the following interesting geometric characterization. Namely, there is a bijective correspondence between monic right divisors of the matrix polynomial $A(\lambda)$ and invariant subspaces of the companion linear pencil $C_A(\lambda)$ which satisfy two properties:

a) the dimension $d$ of the invariant subspace is a multiple of $N$, i.e. $d = n_1 N$ for some integer $n_1$;

b) if

$$\begin{bmatrix} X_1 \\ X_2 \end{bmatrix}$$

is any basis matrix of the invariant subspace, where $X_1$ is a $d \times d$ matrix, then $X_1$ is nondegenerate.

The following uniqueness theorem is a consequence of this bijective correspondence.

THEOREM 12 (Uniqueness of a monic divisor). *Let $A(\lambda) = A_2(\lambda)A_1(\lambda) = A_2'(\lambda)A_1'(\lambda)$, where $A_1$ and $A_1'$ are monic. Suppose that the spectrum of $A_1(\lambda)$ does not intersect with the spectrum of $A_2(\lambda)$, the spectrum of $A_1(\lambda)$ coincides with the spectrum of $A_1'(\lambda)$, and the spectrum of $A_2(\lambda)$ coincides with the spectrum of $A_2'(\lambda)$. Then $A_1(\lambda) = A_1'(\lambda)$.*

Now we formulate a theorem about factorization of a monic $N \times N$ matrix polynomial $A(\lambda)$ of degree $n$ into linear monic divisors.

THEOREM 13. *Let $(X;T)$ be a monic block eigenpair of dimension $nN$ for a monic polynomial $A(\lambda)$, where $T$ can be diagonalized by similarity transformations; that is, all elementary divisors of $T$ are linear as well as elementary divisors of the polynomial $A(\lambda)$ itself. Then there exist matrices $T_1, T_2, \ldots, T_n$ such that*

$$A(\lambda) = (I\lambda - T_n) \cdots (I\lambda - T_2)(I\lambda - T_1).$$

The proof of Theorem 13 is based on the fact that out of the matrix

$$\begin{bmatrix} X \\ XT \\ \vdots \\ XT^{n-1} \end{bmatrix},$$

where $T$ is diagonal, one can pick out a nondegenerate submatrix

$$\begin{bmatrix} \widetilde{X} \\ \widetilde{X}\widetilde{T} \\ \vdots \\ \widetilde{X}\widetilde{T}^{n-2} \end{bmatrix},$$

where $\widetilde{X}$ is an $N \times [N(n-1)]$ submatrix of $X$ and $\widetilde{T}$ is a submatrix of $T$, thus defining a left monic divisor of degree $n-1$ for $A(\lambda)$.

Finally, we give a sketch of the solution to the auxiliary problem. We recall that we are given a matrix pair $(X, T)$ consisting of an $N \times d$ matrix $X$ and $d \times d$ matrix $T$ and satisfying the following condition:

$$\operatorname{rank} \begin{bmatrix} X \\ XT \\ \vdots \\ XT^{l-1} \end{bmatrix} = d < lN \quad \text{for some } l.$$

First, one needs to construct a decomposable block eigenpair $(X, X_\infty; T, T_\infty)$ of dimension $lN$ with a nilpotent matrix $T_\infty$. To this end, we consider a construction where $T_\infty$ is of the shape of a Jordan matrix with zero eigenvalues.

It is not very hard to understand that it is sufficient to construct a sequence of matrices $Y_{l-1}, \ldots, Y_0$ of full column rank such that $Y_j = [Y_{j+1}\, Z_j]$, $j = 1, 2, \ldots, l$, and satisfying the following properties: the sum of the linear span of the columns of the matrix

$$\Xi_j = \begin{bmatrix} Y_{l-1} & & & 0 \\ & Y_{l-2} & & \\ & & \ddots & \\ 0 & & & Y_{l-j} \end{bmatrix}$$

and the linear span of the columns of the matrix

$$\Delta_j = \begin{bmatrix} X \\ XT \\ \vdots \\ XT^{j-1} \end{bmatrix}$$

is equal to all of the space $\mathbf{C}^{jN}$, and the intersection of these two linear spans equals the null space $\{0\}$. Such a sequence $Y_j$ is constructed recursively. For $j = 1$ the columns of the matrix $X$ must simply be supplemented by a minimal set of vectors, which are columns of $Z_{l-1} = Y_{l-1}$, to a vector system linearly spanning $\mathbf{C}^N$.

Let the matrices $Y_{l-i}$ have been constructed for all $i = 1, 2, \ldots, j$, $1 \leqslant j \leqslant l$. It is somewhat nontrivial to observe that the intersection of the linear span of the columns of the matrix

$$\begin{bmatrix} \Xi_j & 0 \\ 0 & Y_{l-j} \end{bmatrix}$$

with the linear span of the columns of the matrix $\Delta_{j+1}$ is equal to the null space $\{0\}$. Let us augment the direct sum of these two linear spans by a minimal set of vectors, which are columns of the matrix $\widetilde{Z}_{l-j-1}$, to a vector system linearly generating $\mathbf{C}^{N(j+1)}$. Due to the completeness of the columns of the matrix $[\Delta_j \ \Xi_j]$ in $\mathbf{C}^{jN}$ the matrix $\widetilde{Z}_{l-j-1}$ can be taken to be of the following form

$$\begin{bmatrix} 0 \\ \vdots \\ 0 \\ Z_{l-j-1} \end{bmatrix}.$$

Having concluded the recursive procedure we obtain the following nondegenerate square matrix

$$[\Delta_l \quad \Xi_l] = \begin{bmatrix} X & Y_{l-1} & & 0 \\ XT & & Y_{l-2} & \\ \vdots & & & \ddots \\ XT^{l-1} & 0 & & Y_0 \end{bmatrix}.$$

After a suitable permutation of the columns of the matrix $\Xi_l$ one can define matrices $X_\infty$ and $T_\infty$ such that after a column permutation the matrix $\Xi_l$ looks like

$$\begin{bmatrix} X_\infty T_\infty^{l-1} \\ \vdots \\ X_\infty \end{bmatrix}.$$

The second step in the proof of the auxiliary proposition is to recover the matrix polynomial $A(\lambda) = A_0 + A_1\lambda + \cdots + A_l\lambda^l$ from the matrix pair $(X, X_\infty; T, T_\infty)$ so that this pair will be a decomposable block eigenpair for $A(\lambda)$. Getting ahead of ourselves, we remark that this problem is always solvable; furthermore, if $A(\lambda)$ and $A'(\lambda)$ are any two solutions of the problem, then there exists such a nondegenerate matrix $Q$ that $A'(\lambda) = QA(\lambda)$.

So, to solve the question at hand one has to find matrices $A_0$, $A_1, \ldots, A_l$, $W$ satisfying the equation

$$\begin{bmatrix} \lambda I & -I & & & \\ & \lambda I & -I & & \\ & & \ddots & \ddots & \\ & & & \lambda I & -I \\ A_0 & A_1 & \ldots & A_{l-2} & A_{l-1} + \lambda A_l \end{bmatrix} S = \begin{bmatrix} M \\ W \end{bmatrix} \begin{bmatrix} \lambda I - T & 0 \\ 0 & \lambda T_\infty - I \end{bmatrix}, \tag{40}$$

where

$$S = \begin{bmatrix} X & X_\infty T_\infty^{l-1} \\ \vdots & \vdots \\ XT^{l-1} & X_\infty \end{bmatrix}, \qquad M = \begin{bmatrix} X & X_\infty T_\infty^{l-2} \\ XT & X_\infty T_\infty^{l-3} \\ \vdots & \vdots \\ XT^{l-2} & X_\infty \end{bmatrix},$$

and the matrix

$$\begin{bmatrix} M \\ W \end{bmatrix}$$

is nondegenerate. There results the equivalent matrix system

$$\begin{cases} [A_0 \quad A_1 \quad \dots \quad A_{l-1}]S = W \begin{pmatrix} -T & 0 \\ 0 & -I \end{pmatrix}, \\ [0 \quad 0 \quad \dots \quad 0 \quad A_l]S = W \begin{pmatrix} I & 0 \\ 0 & T_\infty \end{pmatrix}. \end{cases}$$

Having solved the latter equation with respect to $A_l$ and $W$, the matrices $A_0$, $A_1, \dots, A_{l-1}$ are uniquely determined from the former one because $S$ is nondegenerate.

The matrix $M$ is of full rank as $S$ is nondegenerate. Let us choose a matrix $V$ such that

$$\begin{bmatrix} M \\ V \end{bmatrix}$$

is nondegenerate. Define the matrices $Z$ and $A_l$ from the equality

$$(Z \quad A_l) = V \begin{pmatrix} I & 0 \\ 0 & T_\infty \end{pmatrix} S^{-1}.$$

As the matrix $W$ we take $W = V - ZM$. Then with the aid of the obvious identity

$$M \begin{pmatrix} I & 0 \\ 0 & T_\infty \end{pmatrix} = (I \quad 0)S,$$

the following chain of equalities is deduced:

$$W \begin{pmatrix} I & 0 \\ 0 & T_\infty \end{pmatrix} = (V - ZM) \begin{pmatrix} I & 0 \\ 0 & T_\infty \end{pmatrix} = V \begin{pmatrix} I & 0 \\ 0 & T_\infty \end{pmatrix} - ZM \begin{pmatrix} I & 0 \\ 0 & T_\infty \end{pmatrix}$$
$$= (Z \quad A_l)S - (Z \quad 0)S = (0 \quad \dots \quad 0 \quad A_l).$$

Since

$$\begin{bmatrix} M \\ W \end{bmatrix} = \begin{bmatrix} I & 0 \\ -Z & I \end{bmatrix} \begin{bmatrix} M \\ V \end{bmatrix},$$

the matrix

$$\begin{bmatrix} M \\ W \end{bmatrix}$$

is nondegenerate.

Further, let $A(\lambda)$ satisfy (40) for some $W$. Assume that $M = (L\ 0)U$ with nondegenerate $L$ and $U$, and, therefore, $W = (W_1\ W_2)U$. By virtue of the nondegeneration of

$$\begin{bmatrix} M \\ W \end{bmatrix},$$

the $N \times N$ matrix $W_2$ is also nondegenerate. The identity

$$M \begin{pmatrix} I & 0 \\ 0 & T_\infty \end{pmatrix} = (I \quad 0)S$$

implies that

$$U \begin{pmatrix} I & 0 \\ 0 & T_\infty \end{pmatrix} S^{-1} = \begin{pmatrix} L^{-1} & 0 \\ Z_1 & Z_2 \end{pmatrix}$$

for suitable $Z_1$, $Z_2$. Therefore,

$$\begin{aligned}(0 \quad A_l) = W \begin{pmatrix} I & 0 \\ 0 & T_\infty \end{pmatrix} S^{-1} &= (W_1 \quad W_2)U \begin{pmatrix} I & 0 \\ 0 & T_\infty \end{pmatrix} S^{-1} \\ &= (W_1 L^{-1} + W_2 Z_1 \quad W_2 Z_2).\end{aligned}$$

From this we have $W_1 = -W_2 Z_1 L$, $W = W_2(-Z_1 L \quad I)U$. As a result,

$$[A_0\, A_1\, \ldots\, A_{l-1}] = W_2(-Z_1 L \quad I)U \begin{pmatrix} -T & 0 \\ 0 & -I \end{pmatrix} S^{-1}, \quad A_l = W_2 Z_2.$$

If $A'(\lambda)$ is another matrix polynomial with the same decomposable block eigenpair, and, consequently, with the same $Z_1$, $Z_2$, $L$, $U$, then

$$[A'_0\, A'_1\, \ldots\, A'_{l-1}] = W'_2(-Z_1 L \quad I)U \begin{pmatrix} -T & 0 \\ 0 & -I \end{pmatrix} S^{-1},$$

$A'_l = W'_2 Z_2$. Hence $A'(\lambda) = QA(\lambda)$ with $Q = W'_2 W_2^{-1}$.

### 4.3. *Factorization of self-adjoint matrix polynomials*

Let $A(\lambda) = I\lambda^{2n} + A_{2n-1}\lambda^{2n-1} + \cdots + A_0$ be a monic matrix polynomial, whose coefficients are self-adjoint matrices, i.e. $A_j^* = A_j$, $j = 0,1,\ldots,2n-1$. We also assume that $A(\lambda)$ has no eigenvalues on the real axis. Since the coefficients of $A(\lambda)$ are self-adjoint, the spectrum of the polynomial $A(\lambda)$ is symmetric with respect to the real axis.

THEOREM 14. *A self-adjoint monic matrix polynomial $A(\lambda)$ which has no eigenvalues on the real axis factorizes into the product $A(\lambda) = L^*(\lambda)L(\lambda)$, where the polynomial*

$$L(\lambda) = I\lambda^n + \cdots + L_0$$

*has all its spectrum above the real axis and*

$$L^*(\lambda) = I\lambda^n + \cdots + L_0^*.$$

PROOF. All arguments are based on the fact that $GC = C^*G$, where

$$C = \begin{bmatrix} 0 & -I & & \\ & \ddots & \ddots & \\ & & 0 & -I \\ A_0 & \ldots & A_{2n-2} & A_{2n-1} \end{bmatrix}, \qquad G = \begin{bmatrix} A_1 & \ldots & A_{2n-1} & I \\ A_2 & \ldots & I & \\ \vdots & \iddots & & \\ I & & & 0 \end{bmatrix}. \tag{41}$$

Let

$$C = U \begin{pmatrix} J & 0 \\ 0 & J^* \end{pmatrix} U^{-1}$$

be a Jordan decomposition of the matrix $C$, with $J$ containing all eigenvalues above the real axis. Write

$$U^*GU = \begin{pmatrix} P & Q \\ Q^* & R \end{pmatrix},$$

where $P = P^*$ and $R = R^*$ are matrices of size $Nn \times Nn$, and obtain from the identity $GC = C^*G$ the following system of matrix equations:

$$\begin{cases} PJ - J^*P = 0, \\ RJ^* - JR = 0, \\ QJ^* - J^*Q = 0. \end{cases}$$

Since the spectra of the matrices $J$ and $J^*$ do not intersect one another, it follows from the first two matrix equations that $P = R = 0$. By virtue of the nondegeneration of the matrix $G$, the matrix $Q$ is nondegenerate. Introduce the matrix

$$V = U \begin{pmatrix} I & 0 \\ 0 & Q \end{pmatrix}^{-1},$$

then

$$C = V \begin{pmatrix} J & 0 \\ 0 & QJ^*Q^{-1} \end{pmatrix} V^{-1}, \qquad G = V^{-*} \begin{pmatrix} 0 & I \\ I & 0 \end{pmatrix} V^{-1}. \tag{42}$$

Let $V = [V_1 \quad V_2]$ be the block partition consistent with the blocks in (42), then $CV_1 = V_1 J$, $V_1^* G V_1 = 0$. It is necessary for the existence of a right monic divisor having its spectrum coinciding with the spectrum of $J$ that the square matrix $V_{11}$ in the representation

$$V_1 = \begin{bmatrix} V_{11} \\ V_{21} \end{bmatrix}$$

be nondegenerate. Nondegeneration of $V_{11}$ follows from the identity $V_1^* G V_1 = 0$ and the structure of the matrices $G$ and $V_1$. Indeed, assume the opposite. Let $x \in \operatorname{Ker} V_{11}$, $x \neq 0$. Since

$$V_1 = \begin{bmatrix} X \\ \vdots \\ XJ^{2n-1} \end{bmatrix},$$

$$V_1 x = y = \begin{bmatrix} y_0 \\ \vdots \\ y_{2n-1} \end{bmatrix},$$

where $y_i = XJ^i x$; moreover, $y_0 = y_1 = \cdots = y_{n-1} = 0$. Writing

$$G = \begin{pmatrix} G_{11} & G_{12} \\ G_{21} & 0 \end{pmatrix}$$

we obtain the equality $V_{11}^* G_{11} V_{11} + V_{11}^* G_{12} V_{21} + V_{21}^* G_{21} V_{11} = 0$, and

$$0 = V_{11}^* G_{12} V_{21} x = [X^* \quad J^* X^* \quad \dots \quad (J^*)^{n-1} X^*] G_{12} \begin{bmatrix} y_n \\ \vdots \\ y_{2n-1} \end{bmatrix}.$$

Now calculate the scalar product of the latter identity with $Jx$ to obtain

$$0 = (0 \quad \dots \quad 0 \quad y_n^*) G_{12} \begin{bmatrix} y_n \\ \vdots \\ y_{2n-1} \end{bmatrix} = y_n^* y_n,$$

which implies $y_n = 0$. Multiplying then by $J^3 x$ we obtain $y_{n+1}^* y_{n+1} = 0$. Proceeding similarly we find the equality $y = 0$. This contradicts the condition that rank $(V_1) = nN$.

Finally, let $L(\lambda)$ be a right monic divisor of $A(\lambda)$. From the equality $A(\lambda) = L_1(\lambda) L(\lambda)$ we deduce the equality $A(\lambda) = A^*(\lambda) = L^*(\lambda) L_1^*(\lambda)$. By the uniqueness theorem $L_1^*(\lambda) = L(\lambda)$, therefore, $A(\lambda) = L^*(\lambda) L(\lambda)$. □

There is another important result on self-adjoint matrix polynomials. Given a trigonometric self-adjoint matrix polynomial

$$A(\phi) = \sum_{k=-n}^{n} A_k e^{ik\phi},$$

where $A_{-k} = A_k^*$, and $A(\phi)$ is nondegenerate for any real $\phi$. Then there exists a unique trigonometric polynomial $B(\phi) = B_0 + B_1 e^{i\phi} + \dots + B_n e^{in\phi}$ such that $A(\phi) = [B(\phi)]^* B(\phi)$ and $B(\lambda) = B_0 + B_{21}\lambda + \dots + B_n \lambda^n$ has all its eigenvalues inside the unit circle.

*Notes and references*

The main source of references for the theory of matrix polynomials is [18]. Other relevant publications are [28, 23, 24, 29, 25, 19].

More general results about factorization of self-adjoint matrix polynomials can also be found in [18] and references therein. A proof of the theorem on factorization of trigonometric self-adjoint matrix polynomials is given, e.g., in [29].

The questions about continuity and analyticity of monic divisors are discussed in detail in [19].

A stable algorithm for numerical factorization of matrix polynomials is proposed in [30].

Additionally, in [18] there are results on the least common multiple and the greatest common divisor of matrix polynomials.

## References


[1] R.H. Bartels and G.W. Stewart, *Algorithm 432: the solution of the matrix equation $AX - XB = C$*, Comm. Assoc. Comp. Math. **8** (1972), 820–826.

[2] M.A. Beitia and I. Zaballa, *Factorization of the matrix polynomial $A(\lambda) = A_0\lambda^t + A_1\lambda^{t-1} + \cdots + A_{t-1}\lambda + A_t$*, Linear Algebra Appl. **121** (1989), 423–432.

[3] R. Bellman, *Introduction to Matrix Analysis*, McGraw-Hill, New York (1960).

[4] D.S. Bernstein, *Some open problems in matrix theory arising in linear systems and control*, Linear Algebra Appl. **162–164** (1992), 409–432.

[5] A.Ya. Bulgakov, *An efficiently calculable parameter for the stability property of a system of linear differential equations with constant coefficients*, Siberian Math. J. **21**(3) (1980), 339–347.

[6] A.Ya. Bulgakov, *Generalization of the matrix Lyapunov equation*, Siberian Math. J. **30**(4) (1989), 30–39.

[7] F.M. Caller, *On polynomial matrix spectral factorization by symmetric extraction*, IEEE Trans. Automat. Control. **AC-30**(5) (1985), 453–464.

[8] S. Campbell and J. Daughtry, *The stable solutions of quadratic matrix equations*, Proc. Amer. Math. Soc. **74**(1) (1979), 19–23.

[9] W.A. Coppel, *Matrix quadratic equations*, Bull. Austral. Math. Soc. **10** (1974), 377–401.

[10] J.W. Demmel and B. Kågström, *Computing stable eigendecompositions of matrix pencils*, Linear Algebra Appl. **88/89** (1987), 139–186.

[11] J. Feinstein and Y. Bar-Ness, *The solution of the matrix polynomial equation $A(s)X(s) + B(s)Y(s) = C(s)$*, IEEE Trans. Automat. Control. **AC-29**(1) (1984), 75–77.

[12] H. Flanders and H.K. Wimmer, *On the matrix equations $AX - XB = C$ and $AX - YB = C$*, SIAM J. Appl. Math. **32** (1977), 707–710.

[13] F.R. Gantmacher, *The Theory of Matrices*, Moscow, Nauka (1966). (In Russian; English translation of an earlier edition in vols I and II, Chelsea, New York, 1959.)

[14] S.K. Godunov, *The problem of dichotomy of the spectrum of a matrix*, Siberian Math. J. **27**(5) (1986), 649–660.

[15] S.K. Godunov, *Norms of solutions to the Lourie–Riccati matrix equations as criteria of the quality of stabilizability and detectability*, Siberian Adv. Math. **2**(3) (1992), 135–157.

[16] S.K. Godunov, *Verification of boundedness for the powers of symplectic matrices with the help of averaging*, Siberian Math. J. **33**(6) (1992), 939–949.

[17] S.K. Godunov, *An estimate for Green's matrix of a Hamiltonian system in the optimal control problem*, Siberian Math. J. **34**(4) (1993), 70–80.

[18] I. Gohberg, P. Lancaster and L. Rodman, *Matrix Polynomials*, Academic Press, New York (1982).

[19] I. Gohberg, P. Lancaster and L. Rodman, *Invariant Subspaces of Matrices with Applications*, John Wiley, New York (1986).

[20] G.H. Golub, S. Nash and C. Van Loan, *Hessenberg–Schur method for the problem $AX + XB = C$*, IEEE Trans. Automat. Control **AC-24**(1979), 909–913.

[21] H.D. Ikramov, *Numerical Solution of Matrix Equations*, Moscow, Nauka (1984) (in Russian).

[22] V. Ionesku and M. Weiss, *On computing the stabilizing solution of the discrete-time Riccati equation*, Linear Algebra Appl. **174** (1992), 229–238.

[23] V.A. Jakubovič, *Factorization of symmetric matrix polynomials*, Soviet Math. Dokl. **11** (1970), 1261–1264.

[24] P.S. Kazimirskii, *Solution of the problem of extraction of a regular divisor of a matrix polynomial*, Ukranian Math. J. **32**(4) (1980), 483–498.

[25] P. Lancaster and M. Tismenetsky, *The Theory of Matrices with Applications*, Academic Press, New York (1985).

[26] A.J. Laub, *A Schur method for solving algebraic Riccati equations*, IEEE Trans. Automat. Control. **24**(6) (1979), 913–921.

[27] A.J. Laub, *Invariant subspace methods for the numerical solution of Riccati equations*, The Riccati Equations, S. Bittani, A. Laub and J. Willems, eds, Springer, Berlin (1991).

[28] Ya.B. Lopatinskii, *Factorization of a polynomial matrix*, Nauch. Zap. L'vov Politekh. Inst., Ser. Fiz.-Mat. **38** (1956), 3–9 (in Russian).

[29] A.N. Malyshev, *Factorization of matrix polynomials*, Siberian Math. J. **23**(3) (1982), 136–146.

[30] A.N. Malyshev, *Guaranteed accuracy in spectral problems of linear algebra*, Siberian Adv. Math. **2**(1) (1992), 144–197; **2**(2) (1992), 153–204.

[31] A.N. Malyshev, *On iterative refinement for the spectral decomposition of symmetric matrices*, East-West J. Num. Math. **1**(1) (1993), 27–50.

[32] V. Mehrmann, *Existence, uniqueness, and stability of solutions to singular linear quadratic optimal control problems*, Linear Algebra Appl. **121** (1989), 291–331.

[33] V. Mehrmann, *The Autonomous Linear Quadratic Control Problem: Theory and Numerical Solution*, Lecture Notes in Control and Information Sciences vol. 163, Springer, Heidelberg (1991).

[34] C. Moler and G.W. Stewart, *An algorithm for generalized matrix eigenvalue problems*, SIAM J. Numer. Anal. **10** (1973), 241–256.

[35] J.M. Ortega and W.C. Rheinboldt, *Iterative Solution of Nonlinear Equations in Several Variables*, Academic Press, New York (1970).

[36] A. Ostrowski and H. Schneider, *Some theorems on the inertia of general matrices*, J. Math. Anal. Appl. **4** (1962), 72–84.

[37] T. Pappas, A.J. Laub and N.R. Sandell, *On the numerical solution of the discrete-time algebraic Riccati equation*, IEEE Trans. Automat. Control. **AC-25**(8) (1980), 631–641.

[38] S. Bittani, A. Laub and J. Willems, eds, *The Riccati Equation*, Springer, Berlin (1991).

[39] G.W. Stewart and Ji-guang Sun, *Matrix Perturbation Theory*, Academic Press, San Diego, CA (1990).

[40] P. Van Dooren, *A generalized eigenvalue approach for solving Riccati equations*, SIAM J. Sci. Statist. Comp. **2** (1981), 121–135.

[41] P. Van Dooren, *Reducing subspaces: definitions, properties and algorithms*, Matrix Pencils, B. Kågström and A. Ruhe, eds, SLNM 973, Springer, Berlin (1983).

[42] J.H. Wilkinson, *The Algebraic Eigenvalue Problem*, Clarendon Press, Oxford, England (1965).

[43] W.M. Wonham, *Linear Multivariable Control: A Geometric Approach*, Springer, Berlin (1979).

# Matrix Functions

L. Rodman*

*Department of Mathematics, College of William and Mary, Williamsburg, VA 23187-8795, USA*
*e-mail: lxrodm@facstaff.wm.edu*

*Contents*



*Partially supported by NSF Grant DMS-9000839 and by the NSF International Cooperation Grant with the Netherlands. This work was done while the author visited the Vrije Universiteit, Amsterdam.

HANDBOOK OF ALGEBRA, VOL. 1
Edited by M. Hazewinkel

## 1. Introduction

In this chapter we take the point of view according to which matrices are considered as changing quantities (rather than given and constant). We consider a matrix as an independent variable and study functions of that matrix; on the other hand we also can consider the set of matrices as the target space of a function, in which case a matrix valued function appears. And, of course, one can combine both approaches and study matrix valued functions of a matrix argument. We encompass all these situations by using the term "matricial functions".

The need for matricial functions and their theory is apparent in many applications in mathematics, sciences and engineering. The first such instance appears in the study of systems of first order linear differential equations with constant coefficients

$$\frac{\mathrm{d}x}{\mathrm{d}t} = Ax(t),$$

where $A$ is an $n \times n$ matrix. The solution is given in terms of the initial value $x(0)$ by the matrix exponential $x(t) = \exp\{tA\}x(0)$. The theory of vibrating systems (mechanical or electrical) with a finite number of degrees of freedom involves matrix polynomials $\lambda^2 A_2 + \lambda A_1 + A_0$, where $A_2,\ A_1,\ A_0$ are $n \times n$ matrices with certain symmetry properties (for example, positive definite). In engineering, the transfer function of a linear time invariant multivariable control system is a matrix valued rational function. In numerical analysis, one is often interested in the behaviour of various quantities derived from a matrix (such as eigenvalues, singular values, eigenvectors, invariant subspaces etc.) if the matrix is subject to small perturbations; in other words, the matrix is considered as a variable quantity.

Driven by these and many other applications, as well as a simple mathematical interest, the recent 30 years or so witnessed an explosion of research works on matricial functions, scattered in mathematical, physical and engineering literature. In particular, several books devoted solely to various aspects of matricial functions appeared recently. It is clear therefore that the material selected for this chapter has to be severely restricted and represents only a small fraction of the material available in the literature. For one thing, we shall put aside all applications, and focus on the theoretical aspects of matricial functions. It is hoped, however, that practitioners interested mostly in applications will be able to relate at least some of the material to their needs. Even so, many interesting and important theoretical developments are excluded from this chapter as well; in particular, the perturbation theory of matrices and their derived quantities is not considered here (the reader is referred to the texts Stewart and Sun (1990), Kato (1982), Bhatia (1987) for this theory). A reasonably extensive bibliography (again, far from being complete), should compensate somewhat for these exclusions.

Only finite matrices with entries in a fixed field $F$ will be considered; since analytical properties (such as continuity, differentiability etc.) of matricial functions will be important here, the field often must be topological in this chapter. Simplicity of exposition and uses in applications make it natural to choose either the real field $F = \mathbb{R}$ or the complex field $F = \mathbb{C}$.

Some notation which will be frequently used: $I$ (or $I_n$) stands for the $n \times n$ identity matrix. The linear space of all $m \times n$ matrices with entries in a field $F$ will be denoted $M_{m\times n}(F)$. For an $m \times n$ matrix $A$, we write

$$\operatorname{Ker} A = \{x \in F^n\colon\ Ax = 0\}, \qquad \operatorname{Range} A = \{Ax\colon\ x \in F^n\}.$$

$A^T$ (resp. $A^*$) stands for the transpose (resp. conjugate transpose) of a matrix $A$.

## 2. Functions of matrices

### 2.1. *Basic definitions and properties*

The simplest functions of an $n \times n$ matrix $A$ are polynomials: for

$$f(\lambda) = \sum_{j=0}^{m} a_j \lambda^j, \quad a_j \in F,$$

define

$$f(A) = \sum_{j=0}^{m} q_j A^j = a_0 I + a_1 A + \cdots + a_m A^m. \tag{2.1}$$

The definition makes good sense for matrices over an arbitrary (commutative) field $F$, or, more generally, for matrices over a unital ring (provided the coefficients $a_j$ belong to the center of that ring).

Several basic properties of polynomials of matrices are summarized in the following theorem. We denote by $p_A(\lambda) = \det(\lambda I - A)$ the characteristic polynomial of $A \in M_{n\times n}(F)$. The *spectrum*, or the set of *eigenvalues* of $A \in M_{n\times n}(F)$ will be denoted $\sigma(A)$; thus,

$$\sigma(A) = \{\lambda \in F_0 \mid p_A(\lambda) = 0\},$$

where $F_0$ is a fixed algebraic closure of the field $F$.

THEOREM 2.1. *Let $F$ be a field, and let $A$ be a fixed $n \times n$ matrix over $F$.*

(a) *The map $f \mapsto f(A)$ defined by* (2.1) *is an algebra homomorphism from the algebra $F[\lambda]$ of polynomials in one variable with coefficients in $F$ into the algebra of all $n \times n$ matrices over $F$.*

(b) $f(\sigma(A)) = \sigma(f(A))$ *for every $f \in F[\lambda]$.*

(c) $f(T^{-1}AT) = T^{-1}f(A)T$ *for any nonsingular $T \in M_{n\times n}(F_0)$, where $F_0$ is a fixed algebraic closure of $F$.*

(d) $p_A(A) = 0$ *(Cayley–Hamilton theorem).*

PROOF. The parts (a) and (c) are easily verified. For part (b), recall a basic result in linear algebra that for any $A \in M_{n\times n}(F)$ there is an invertible $T \in M_{n\times n}(F_0)$ such that $T^{-1}AT$ is in the Jordan normal form:

$$T^{-1}AT = J_{m_1}(\lambda_1) \oplus \cdots \oplus J_{m_k}(\lambda_k), \tag{2.2}$$

where $\lambda_1, \dots, \lambda_k$ are (possibly with repetitions) all the elements in $\sigma(A)$, and where $J_m(\lambda_0)$ stands for the $m \times m$ upper triangular Jordan block with eigenvalue $\lambda_0$:

$$J_m(\lambda_0) = \begin{pmatrix} \lambda_0 & 1 & 0 & \dots & 0 \\ 0 & \lambda_0 & 1 & \dots & 0 \\ \vdots & \vdots & & \ddots & \vdots \\ & & & & 1 \\ 0 & 0 & & \dots & \lambda_0 \end{pmatrix}. \tag{2.3}$$

Rewriting the polynomial $f(\lambda)$ in the form

$$f(\lambda) = \sum_{j=0}^{m} \frac{f^{(j)}(\lambda_0)}{j!} (\lambda - \lambda_0)^j,$$

it easily follows by the definition of $f(T^{-1}AT)$ that

$$f(T^{-1}AT) = f\big(J_{m_1}(\lambda_1)\big) \oplus \cdots \oplus f\big(J_{m_k}(\lambda_k)\big), \tag{2.4}$$

where

$$f(J_{m_j}(\lambda_j)) = \begin{pmatrix} \alpha_{j0} & \alpha_{j1} & \dots & \alpha_{jm_j-1} \\ 0 & \alpha_{j0} & \dots & \alpha_{jm_j-2} \\ \vdots & \vdots & \ddots & \vdots \\ 0 & 0 & \dots & \alpha_{j0} \end{pmatrix}, \quad \alpha_{jp} = \frac{f^{(p)}(\lambda_j)}{p!}. \tag{2.5}$$

In particular, $\sigma(f(T^{-1}AT)) = \{\alpha_{10}, \alpha_{20}, \dots, \alpha_{k0}\} = \{f(\lambda_1), \dots, f(\lambda_k)\}$. But in view of (c), $\sigma(f(T^{-1}AT)) = \sigma(f(A))$, and (b) follows.

Finally, to prove (d), let $B(\lambda)$ be the algebraic adjoint of $\lambda I - A$; in other words, the $(i,k)$ element of $B(\lambda)$ is equal to $(-1)^{i+k}$ {determinant of the $(n-1)\times(n-1)$ matrix obtained from $\lambda I - A$ by crossing out the $k$-th row and $i$-th column}. The properties of the determinant ensure that

$$(\lambda I - A)B(\lambda) = B(\lambda)(\lambda I - A) = p_A(\lambda)I. \tag{2.6}$$

The right-hand side of (2.6) is a polynomial with matrix coefficients which is divisible by $\lambda I - A$. The Bezout theorem (which is applicable in this situation as one can easily check by using long division of polynomials with matrix coefficients) gives $\big[p_A(\lambda)I\big]_{\lambda=A} = 0$, i.e. $p_A(A) = 0$. □

Observe that the parts (a), (c) and (d) of Theorem 2.1 are valid also for polynomials of matrices over unital commutative rings (see, e.g., Brewer et al. (1986)).

Let us remark that the Jordan normal form (2.1) is a particular case of a more general and very useful *Kronecker normal form* of linear matrix functions $\lambda A+B$, under the group of transformations $\lambda A + B \mapsto P(\lambda A + B)Q$. Here $A, B \in M_{m\times n}(F)$, $P \in M_{m\times m}(F)$, $Q \in M_{m\times m}(F)$ and $P$ and $Q$ are invertible. (The field $F$ is assumed to be algebraically closed.) We refer the reader to Gantmacher (1959), Gohberg et al. (1982b) for a full description of the Kronecker normal form.

*From now on until the end of Subsection* 1.2 *we assume* $F = \mathbb{C}$.

We now extend the definition of functions of matrices to more general classes of functions (beyond polynomials). The possibilities for such extensions are suggested by the formulas (2.4), (2.5). Let $A \in M_{n\times n}(\mathbb{C})$, and let $\mathcal{A}(A)$ be the class of complex valued functions which are defined and analytic in a neighborhood of $\sigma(A)$. If

$$T^{-1}AT = J_{m_1}(\lambda_1) \oplus \cdots \oplus J_{m_k}(\lambda_k) \tag{2.7}$$

is the Jordan form of $A$, then for $f \in \mathcal{A}(A)$ define

$$f(A) = T\big(f\big(J_{m_1}(\lambda_1)\big) \oplus \cdots \oplus f\big(J_{m_k}(\lambda_k)\big)\big)T^{-1}, \tag{2.8}$$

where $f(J_{m_1}(\lambda_1))$ is given by (2.5). This definition is correct, i.e. does not depend on the choice of the nonsingular matrix $T$ that reduces $A$ to its Jordan form. Indeed, we have

$$f(A) = \frac{1}{2\pi}\int_\Gamma f(\lambda)(\lambda I - A)^{-1}\,\mathrm{d}\lambda, \tag{2.9}$$

where the contour $\Gamma$ consists of a small circle around each eigenvalue of $A$; to verify (2.7), use the reduction of $A$ to its Jordan form and the easily verified formula

$$\big(\lambda I - J_k(\lambda_0)\big)^{-1} = \begin{pmatrix} (\lambda-\lambda_0)^{-1} & (\lambda-\lambda_0)^{-2} & \cdots & (\lambda-\lambda_0)^{-k} \\ 0 & (\lambda-\lambda_0)^{-1} & \cdots & (\lambda-\lambda_0)^{-k+1} \\ \vdots & \vdots & & \vdots \\ 0 & 0 & \cdots & (\lambda-\lambda_0)^{-1} \end{pmatrix}.$$

Thus, in view of (2.9), $f(A)$ depends on $A$ and $f(\lambda)$ only. The properties (a), (b), (c) of Theorem 2.1 remain valid for $f \in \mathcal{A}(A)$.

As it follows from (2.8), we have $f(A) = g(A)$ for every $f, g \in \mathcal{A}(A)$ such that

$$f^{(m)}(\lambda_j) = g^{(m)}(\lambda_j); \quad m = 0, \ldots, r_j - 1;\ j = 1, \ldots, p, \tag{2.10}$$

where $\lambda_1, \ldots, \lambda_p$ are all the distinct eigenvalues of $A$ and $r_j$ is the maximal size of the Jordan blocks corresponding to $\lambda_j$ $(j = 1, \ldots, p)$. This observation allows us to write

$f(A)$ as a linear combination of certain polynomials of $A$, as follows. Let

$$q(\lambda) = \prod_{j=1}^{p} (\lambda - \lambda_j)^{r_j}$$

be the minimal polynomial of $A$, with distinct roots $\lambda_1, \ldots, \lambda_p$. For $j = 1, \ldots, p$ and $k = 0, \ldots, r_j - 1$ denote by $\varphi_{jk}(\lambda)$ the polynomial of minimal degree such that

$$\varphi_{jk}^{(k)}(\lambda_j) = 1; \quad \varphi_{jk}^{(\ell)}(\lambda_j) = 0 \quad \text{for } \ell = 0, \ldots, r_j - 1 \text{ and } \ell \neq k;$$
$$\varphi_{jk}^{(\ell)}(\lambda_s) = 0; \quad \text{for } \ell = 0, \ldots, r_s - 1 \text{ and } s = 1, \ldots, p; \ s \neq j.$$

Clearly, such $\varphi_{jk}(\lambda)$ exists; in fact, $\varphi_{jk}$ has the form

$$\varphi_{jk}(\lambda) = \psi_j(\lambda) \prod_{s \neq j} (\lambda - \lambda_s)^{r_s}$$

for some polynomial $\psi_j(\lambda)$ of degree less that $r_j$. Let

$$Z_{jk} = \varphi_{jk}(A).$$

The matrices $Z_{jk}$ are called *components* of $A$. Being polynomials in $A$, the matrices $Z_{jk}$ commute with every matrix that commutes with $A$. One can show that $Z_{jk}$ ($j = 1, \ldots, p$; $k = 0, \ldots, r_j - 1$) are linearly independent. Now given $f \in \mathcal{A}(A)$, let $g(\lambda)$ be the polynomial defined by

$$g(\lambda) = \sum_{j=1}^{p} \sum_{k=0}^{r_j - 1} g^{(k)}(\lambda_j) \varphi_{jk}(\lambda).$$

Because of the construction of $\varphi_{jk}(\lambda)$, the equalities (2.10) hold, and we have

$$f(A) = \sum_{j=1}^{p} \sum_{k=0}^{r_j - 1} f^{(k)}(\lambda_j) Z_{jk}. \tag{2.11}$$

This formula is convenient if many functions of the same $A$ are to be studied, as, for example, is the case when $f(\lambda)$ depends on parameters.

Besides the formulas (2.9) and (2.11), for many important functions a useful power series representation is available. Thus, let $f(\lambda)$ be an analytic function given by a power series

$$f(\lambda) = \sum_{j=0}^{\infty} f_j (\lambda - \lambda_0)^j$$

which converges in a disc $D = \{|\lambda - \lambda_0| < r\}$. Then for any matrix $A \in M_{n\times n}(\mathbb{C})$ all eigenvalues of which are in the disc $D$ we have $f \in \mathcal{A}(A)$ and therefore $f(A)$ is defined. It turns out that in fact

$$f(A) = \sum_{j=0}^{\infty} f_j(A - \lambda_0 I)^j \tag{2.12}$$

and the matrix series in the right hand side is absolutely convergent. One can verify (2.12) by reduction of $A$ to the Jordan form and by using the formula (2.5). For example, we have the power series

$$\mathrm{e}^A = \sum_{m=0}^{\infty} \frac{A^n}{m!}; \qquad \sin A = \sum_{m=0}^{\infty} (-1)^m \frac{A^{2m+1}}{(2m+1)!}$$

valid for every $n \times n$ matrix $A$. The algebraic relations for scalar functions continue to be valid when the variable is a matrix; for example

$$(\sin A)^2 + (\cos A)^2 = I$$

for any $n \times n$ matrix $A$.

For many applications the class $\mathcal{A}(A)$ is not sufficiently wide, and one would like to extend the definition of $f(A)$ to a wider class of functions. To do this in a coherent fashion, we have to restrict the class of matrices. Let $\Omega$ be an open interval of a straight line in the complex plane, and denote by $C^p(\Omega)$ the class of $p$ times differentiable complex valued functions on $\Omega$ (differentiability is understood in the sense of $\Omega$:

$$f'(t_0) = \lim_{\substack{t\to t_0 \\ t\in\Omega}} \frac{f(t) - f(t_0)}{t - t_0}, \qquad t_0 \in \Omega).$$

Then for any $A \in M_{n\times n}(\mathbb{C})$ with eigenvalues in $\Omega$, and any $f \in C^{p-1}(\Omega)$, where $p$ is the biggest size of a Jordan block in the Jordan form of $A$, we can define $f(A)$ by the same formulas (2.8), (2.5), where the Jordan form of $A$ is given by (2.7). Again, the basic functorial properties (Theorem 2.1(a), (b),(c)) are valid for the class $C^{p-1}(\Omega)$. The formula (2.11) is valid also, which proves, in particular, that $f(A)$ is correctly defined (i.e. is independent of the choice of the order of Jordan blocks in the Jordan form of $A$, and of the choice of the similarity transformation that reduces $A$ to its Jordan form).

*Literature guide.* The material of Section 2.1 is fairly standard and various parts of it can be found in many texts (see, e.g., Bellman (1970), Wedderburn (1964), Pease (1965), Gohberg et al. (1986a)). More or less complete and detailed expositions of this and related material are given in Gantmacher (1959), Lancaster and Tismenetsky (1985), Horn and Johnson (1991). A thorough exposition of old and new results concerning solutions $X$ of the equation $f(X) = A$, where $f(\lambda)$ is a given analytic function and $A \in M_{n\times n}(F)$, $F = \mathbb{R}$ or $F = \mathbb{C}$, is a given matrix, is found in Evard and Uhlig (1992).

**2.2.** *Formulas for the derivative of a function of matrices*

Let $A(t)$ be an $n \times n$ matrix depending on a real parameter $t \in (a, b)$. In this subsection we will give formulas for the derivative of the composite function $f(A(t))$, where $f(\lambda)$ belongs to a suitable class of functions.

In the next theorem it will be assumed that $A(t)$ is continuously differentiable.

THEOREM 2.2. (a) *Assume that, for a fixed $t_0 \in (a, b)$, $f(\lambda)$ is an analytic function in an open set containing the eigenvalues of $A(t_0)$. Then*

$$\frac{\mathrm{d}}{\mathrm{d}t} f(A(t)) = \frac{1}{2\pi} \int_{\Gamma} f(\lambda)\big(\lambda I - A(t)\big)^{-1} A'(t) \big(\lambda I - A(t)\big)^{-1} \, \mathrm{d}\lambda, \tag{2.13}$$

*for all $t$ sufficiently close to $t_0$, where $\Gamma$ is a simple closed rectifiable curve that encloses all the eigenvalues of $A(t_0)$.*

(b) *Assume that all eigenvalues of $A(t)$ lie in an open-ended interval $\Omega \subset \mathbb{C}$, and assume that $f(\lambda)$ is a complex valued continuously differentiable function of $\lambda \in \Omega$. Assume, in addition, that $A(t)$ is diagonalizable for all $t \in (a, b)$. Then*

$$\frac{\mathrm{d}}{\mathrm{d}t} f(A(t)) = \sum_{j,k=1}^{s} \frac{f(\lambda_j) - f(\lambda_k)}{\lambda_j - \lambda_k} P_j A'(t) P_k, \tag{2.14}$$

*where $\lambda_1 = \lambda_1(t), \ldots, \lambda_s = \lambda_s(t)$ are all the distinct eigenvalues of $A(t)$ and*

$$P_j = \frac{1}{2\pi} \int_{|\lambda - \lambda_j| = \varepsilon} (\lambda I - A)^{-1} \, \mathrm{d}\lambda \quad (\varepsilon > 0 \textit{ sufficiently small}),$$

*is the Riesz projector corresponding to the eigenvalue $\lambda_j$.*

The quotient $(\lambda_j - \lambda_k)^{-1}(f(\lambda_j) - f(\lambda_k))$ in (2.14) is interpreted as $f'(\lambda_j)$ if $j = k$. We emphasize that under the hypothesis of Theorem 2.2(b) the multiplicities of the eigenvalues $\lambda_j(t)$, as well as their number $s$, may depend on $t$.

Formula (2.13) is a rather simple consequence of (2.9). Formula (2.14) is a special case of a general formula for $\frac{\mathrm{d}}{\mathrm{d}t} f(A(t))$ obtained in Daleckii (1965) (without the diagonalizability assumption). An analogous formula for hermitian operators was obtained in Daleckii and Krein (1965). Formulas for the second derivative of $f(A(t))$ are given in Chapter 6 of Horn and Johnson (1991). The book Rogers (1980) contains formulas for the derivative of scalar or matrix functions of a matrix variable, and several useful applications, for example, the derivative of the generalized inverse, and the derivatives of elementary symmetric functions.

**2.3.** *Entrywise functions of matrices*

In this subsection we adopt a completely different approach to define a function of a matrix. We assume here $F = \mathbb{R}$, as this is the case studied mostly (if not exclusively) in the literature. Given a function $f\colon \mathbb{R} \to \mathbb{R}$, we define for every $m \times n$ real matrix $A = [a_{ij}]_{i=1,j=1}^{m,n}$

$$f(A) = \left[f(a_{ij})\right]_{i=1,j=1}^{m,n}. \tag{2.15}$$

With this definition, a functorial property analogous to Theorem 2.1(a) holds with respect to entrywise multiplication (also called *Hadamard multiplication*) of matrices:

$$[a_{ij}]_{i=1,j=1}^{m,n} \circ [b_{ij}]_{i=1,j=1}^{m,n} = [a_{ij}b_{ij}]_{i=1,j=1}^{m,n}.$$

We present here several results concerning entrywise functions of matrices which are, in a sense, typical of problems that have been studied for such functions.

THEOREM 2.3. *Let $A \in M_{n\times n}(\mathbb{R})$ be positive semidefinite with non-negative entries ($n \geqslant 2$), and let $f(x) = x^{\alpha}$. If $\alpha \geqslant n-2$, then $f(A)$ defined by* (2.15) *is positive semidefinite. If $0 < \alpha < n-2$ and $\alpha$ is not a positive integer, then for some positive semidefinite $A_0 \in M_{n\times n}(\mathbb{R})$ with non-negative entries the matrix $f(A_0)$ is not semidefinite.*

Theorem 2.3 was proved in FitzGerald and Horn (1977), (see also Section 6.3 in Horn and Johnson (1991)).

Observe that (under the hypotheses of Theorem 2.3) $f(A)$ is positive semidefinite for every integer $\alpha \geqslant 0$. This follows from the general and very important result (due to Schur (1891)):

THEOREM 2.4. *The entrywise product of two positive (semi)definite matrices is again positive (semi)definite.*

A proof can be given by using the spectral theorem for a positive semidefinite $n \times n$ matrix $A$:

$$A = \sum_{j=1}^{n} \lambda_j P_j,$$

where $\lambda_j \geqslant 0$ and $P_1, \ldots, P_n$ are one-dimensional orthogonal projectors which are orthogonal to each other. For a detailed proof see, e.g., Section 5.2 in Horn and Johnson (1991) or Section 7.5 in Horn and Johnson (1985).

Another useful result concerns the entrywise exponential. An $n \times n$ hermitian matrix $A$ is called *conditionally positive semidefinite* if $x^*Ax \geqslant 0$ for every

$$x = (x_1, x_2, \ldots, x_n)^{\top} \in \mathbb{C}^n$$

such that

$$\sum_{j=1}^{n} x_j = 0.$$

THEOREM 2.5. *Let $A = [a_{ij}]_{i,j=1}^n$ be a hermitian matrix. Then the entrywise exponential $[\exp(ta_{ij})]_{i,j=1}^n$ is positive semidefinite for all $t > 0$ if and only if $A$ is conditionally positive semidefinite.*

PROOF. The proof of the "if" part is found in Section 6.3 of Horn and Johnson (1991). For the "only if" part observe that the positive semidefiniteness of $[\exp(ta_{ij})]_{i,j=1}^n$ implies the conditional positive semidefiniteness of $t^{-1}[\exp(ta_{ij})]_{i,j=1}^n$. It remains to pass to the limit when $t \to 0$. See also Section 1 in Parthasarathy and Schmidt (1972). □

We conclude this subsection with a result concerning spectral radii of entrywise functions of matrices. Denote by $\rho(A)$ the spectral radius (i.e. the maximal modulus of eigenvalues) of an $n \times n$ matrix $A$, and denote $f(A)$ by (2.10), i.e. entrywise.

THEOREM 2.6. *A function $f$: $\{x \in \mathbb{R}\colon x \geqslant 0\} \to \{x \in \mathbb{R}\colon x \geqslant 0\}$ satisfies the inequality*

$$\rho(f(A)) \leqslant f(\rho(A))$$

*for any $n \times n$ matrix $A$ with real non-negative entries, and for any size $n$, if and only if the following two conditions are satisfied:*

(i) $f(a) + f(b) \leqslant f(a+b)$ *for all* $a, b \geqslant 0$,
(ii) $(f(a)f(b))^{1/2} \leqslant f((ab)^{1/2})$ *for all* $a, b \geqslant 0$.

Theorem 2.6, as well as its generalization to functions of several real variables, and a characterization of functions $f$ satisfying the opposite inequality $\rho(f(A)) \geqslant f(\rho(A))$, are proved in Elsner et al. (1990).

*Literature guide.* For additional information concerning entrywise powers of matrices, with applications to infinitely divisible matrices, see Horn (1967, 1969). Various applications of Theorem 2.5 and related properties of conditionally positive semidefinite matrices are found in Bapat (1988), Parthasarathy and Schmidt (1972) (probability theory), Donoghue (1974) and Micchelli (1986) (two dimensional data fitting). An in-depth discussion of Theorem 2.4 and related results is given in Horn (1990).

## **2.4.** *Monotone matrix functions*

Here we consider functions of hermitian matrices. An $n \times n$ hermitian matrix $A$ is diagonalizable with real eigenvalues; moreover, there exists a unitary matrix $U$ such that

$$U^*AU = U^{-1}AU = \begin{pmatrix} \lambda_1 & 0 & \dots & 0 \\ 0 & \lambda_2 & \dots & 0 \\ \vdots & \vdots & \ddots & \vdots \\ 0 & 0 & \dots & \lambda_n \end{pmatrix},$$

where $\lambda_1, \ldots, \lambda_n$ are (not necessarily distinct) eigenvalues of $A$. These properties allow us to define $f(A)$ for any complex valued function $f(\lambda)$ whose domain of definition contains $\sigma(A)$, by a formula analogous to (2.3):

$$f(A) = U \begin{pmatrix} f(\lambda_1) & 0 & \ldots & 0 \\ 0 & f(\lambda_2) & \ldots & 0 \\ \vdots & \vdots & \ddots & \vdots \\ 0 & 0 & \ldots & f(\lambda_n) \end{pmatrix} U^{-1}.$$

We assume in this section that $f(\lambda)$ is real valued (this guarantees that $f(A)$ is hermitian as well) and is defined on a real interval $(a, b)$ $(-\infty \leqslant a < b \leqslant \infty)$.

There is a natural partial order (sometimes called *Loewner partial order*) on the set $H_n$ of all $n \times n$ hermitian matrices. Namely, for $A, B \in H_n$ we define $A \leqslant B$ (or $B \geqslant A$) to mean that $B - A$ is positive semidefinite. A real function $f(\lambda)$, $\lambda \in (a, b)$, is called a *monotone matrix function* on $H_n$ with respect to $(a, b)$ if $A \leqslant B$, where $A, B \in H_n$ and $\sigma(A) \cup \sigma(B)$ is contained in $(a, b)$, implies that $f(A) \leqslant f(B)$. Some important examples of monotone matrix functions on $H_n$ (for all $n = 1, 2, \ldots$) are:

1) $f(\lambda) = -\lambda^{-1}$ with respect to $(0, \infty)$ as well as with respect to $(-\infty, 0)$;

2) $f(\lambda) = \sqrt{\lambda}$ with respect to $(0, \infty)$;

3) $f(\lambda) = \log \lambda$ with respect to $(0, \infty)$, where the branch of the logarithm is chosen so that $f(\lambda)$ is real valued for real positive $\lambda$.

Functions that are matrix monotone on $H_n$ for all $n$ can be characterized as follows:

THEOREM 2.7. *The following statements are equivalent for a real valued function $f(\lambda)$, $\lambda \in (a, b)$.*

(i) *$f(\lambda)$ is a monotone matrix function on $H_n$ with respect to $(a, b)$, for all $n = 1, 2, \ldots$;*

(ii) *$f(\lambda)$ is analytic on $(a, b)$, admits analytic continuation to the open upper halfplane and the open lower halfplane, and (unless $f(\lambda)$ is constant) $f(\lambda_0)$ has positive imaginary part for every $\lambda_0$ in the open upper halfplane;*

(iii) *$f(\lambda)$ admits an integral representation*

$$(\lambda) = \alpha\lambda + \beta + \int_{-\infty}^{\infty} \left[(t - \lambda)^{-1} - t(t^2 + 1)^{-1}\right] \mathrm{d}\mu(t), \tag{2.16}$$

*where $\lambda \geqslant 0$, $\beta$ real, and $\mu(t)$ is a positive Borel measure on the real t-axis which has no mass on $(a, b)$ and such that*

$$\int_{-\infty}^{\infty} (t^2 + 1)\, \mathrm{d}\mu(t) < \infty.$$

The equivalence (i) $\Leftrightarrow$ (ii) is known as Loewner's theorem (Loewner (1934)). The functions $f(\lambda)$ which are analytic in the open upper halfplane and map this halfplane into itself are called *Pick functions*. The formula (2.16) is a well-known integral representation of Pick functions, taking into account the additional analytic continuation properties stated in (ii).

We now restrict the matrix monotonicity property to a fixed $H_n$:

THEOREM 2.8. *Let $f(\lambda)$ be a continuously differentiable real valued function on $(a,b)$. Then $f(\lambda)$ is matrix monotone on $H_n$ with respect to $(a,b)$ if and only if for all $\lambda_1,\dots,\lambda_n \in (a,b)$ the matrix*

$$\left[(\lambda_i-\lambda_j)^{-1}\big(f(\lambda_i)-f(\lambda_j)\big)\right]_{i,j=1}^{n}$$

*is positive semidefinite. (If $\lambda_i = \lambda_j$, the expression $(\lambda_i-\lambda_j)^{-1}(f(\lambda_i)-f(\lambda_j))$ is interpreted as $f'(\lambda_i)$.)*

Theorem 2.8 is again due to Loewner (1934) (in fact, every matrix monotone function on $H_n$ is $(2n-3)$ times continuously differentiable; thus, the differentiability hypothesis in Theorem 2.8 is superfluous if $n \geqslant 2$). A relatively easy proof of Theorem 2.8 based on formula (2.14) and on Schur's theorem 2.4 is found in Section 6.6 in Horn and Johnson (1991).

*Literature guide.* The book Donoghue (1974) contains a full proof of Loewner's theorems, as well as several important related results and subsequent developments. For several other criteria (besides Theorem 2.7) for matrix monotonicity on $H_n$ see Section 6.6 in Horn and Johnson (1991) and Bendat and Sherman (1955). Additional sources containing information on monotone matrix functions include Davis (1963), Horn (1990). A real function $f(x)$, $x \in \mathbb{R}$, is called *matrix convex* if

$$f\big((1-\lambda)A+\lambda B\big) \leqslant (1-\lambda)f(A)+\lambda f(B)$$

for every pair of $n \times n$ hermitian matrices $A$ and $B$ and every $\lambda \in [0,1]$. This class of functions is closely related to matrix monotone functions (see Krauss (1936), Davis (1963), Bendat and Sherman (1955), Section 16E in Marshall and Olkin (1979) and Section 6.6 in Horn and Johnson (1991) for the basic results on matrix convex functions).

## 3. Matrices dependent on parameters

Let $A(t)$ be an $n\times n$ complex matrix depending on parameters $t$. In applications, it is often desirable to find out what is the nature of dependence of $t$ of many important quantities associated with $A(t)$, such as eigenvalues, eigenvectors, Jordan form, triangular (Schur) form, singular values, basis in $\operatorname{Ker} A(t)$, basis in $\operatorname{Range} A(t)$ etc. Without attempting to cover, or even mention, many important results in that area, we present here some basic facts and ideas.

### 3.1. *Analytic matrix functions*

We start with the complex analytic dependence on $t$. Thus, assume that $A(t)$ (i.e. every entry of $A(t)$) is an analytic function of the complex variable $t \in \Omega$, where $\Omega$ is a domain in the complex plane. Easy examples show that the eigenvalues of $A(t)$ need not be analytic functions of $t$ (even if one allows for an arbitrary permutation of eigenvalues for each $t$). Moreover, when the eigenvalues of $A(t)$ are analytic (even constant) the Jordan form of $A(t)$ need not be analytic:

EXAMPLE 3.1. The Jordan form of

$$A(t) = \begin{pmatrix} 0 & 1 & 0 \\ 0 & 0 & 0 \\ t & 0 & 0 \end{pmatrix}, \quad t \in \mathbb{C},$$

is

$$\begin{pmatrix} 0 & 1 & 0 \\ 0 & 0 & 1 \\ 0 & 0 & 0 \end{pmatrix} \text{ if } t \neq 0 \quad \text{and} \quad \begin{pmatrix} 0 & 1 & 0 \\ 0 & 0 & 0 \\ 0 & 0 & 0 \end{pmatrix} \text{ if } t = 0.$$

The set of points $t_0 \in \Omega$ at which the continuity of the Jordan form of $A(t)$ breaks down is at most countable with limit points (if any) on the boundary of $\Omega$. Denote this set $\Omega_0$.

THEOREM 3.1 (see Baumgartel (1985)). *The eigenvalues (suitably ordered)* $\lambda_1(t), \ldots, \lambda_n(t)$ *of* $A(t)$ *are given in a neighborhood* $U(t_0)$ *of every* $t_0 \in \Omega$ *by the fractional power series (p is a positive integer)*

$$\lambda_j(t) = \sum_{k=0}^{\infty} \alpha_{jk}(t - t_0)^{k/p}; \quad \alpha_{jk} \in \mathbb{C}. \tag{3.1}$$

*A basis* $x_1(t), \ldots, x_n(t)$ *in* $\mathbb{C}^n$ *consisting of chains of eigenvectors and generalized eigenvectors (Jordan chains) of* $A(t)$ *exists which is given by fractional Laurent series*

$$x_j(t) = \sum_{k=-q}^{\infty} x_{jk}(t - t_0)^{k/p}; \quad x_{jk} \in \mathbb{C}^n$$

*for* $t \in U(t_0)\backslash\{t_0\}$ *(here* $q \geqslant 0$ *is an integer). Moreover, if* $t_0 \notin \Omega_0$*, then* $p = 1$*, and* $q = 0$*, i.e.* $\lambda_j(t)$ *and* $x_j(t)$ *are in fact analytic at* $t_0$*.*

A generalization of Theorem 3.1 to matrices depending analytically on several complex parameters is obtained in Baumgartel (1974).

Behaviour of eigenvalues and their multiplicities of analytic matrix functions under small analytic perturbations was studied for hermitian valued function in Gohberg et al. (1985, 1986b) (see also Gohberg et al. (1983) and references there), and for general functions in Najman (1986), Langer and Najman (1989) (where the results and proofs are given using Newton diagrams).

Since the Jordan form of $A(t)$ is not necessarily analytic on $\Omega$, a natural question arises: find a simplest possible form of $A(t)$ (for every $t \in \Omega$) which is guaranteed to be analytic on $t$, at least in a neighborhood of every $t \in \Omega$. The answer is given by the following result.

THEOREM 3.2 (Arnold (1971)). *Let $t_0 \in \Omega$, and assume that $A(t_0)$ is the Jordan form*

$$A(t_0) = J_{m_{11}}(\lambda_1) \oplus \cdots \oplus J_{m_{1r_1}}(\lambda_1) \oplus \cdots \oplus J_{m_{s1}}(\lambda_s) \oplus \cdots \oplus J_{m_{sr_s}}(\lambda_s),$$

*where $\lambda_1, \ldots, \lambda_s$ are the distinct eigenvalues of $A(t_0)$, and $m_{i1} \geqslant \cdots \geqslant m_{ir_i}$ ($i = 1, \ldots, s$). ($J_m(\lambda_0)$ is defined by (2.2)). Then there exists an invertible matrix $S(t)$ depending analytically on $t$ in a neighborhood of $t_0$ such that $S(t)A(t)S(t)^{-1}$ has the form*

$$S(t)A(t)S(t)^{-1} = K_1 \oplus \cdots \oplus K_s. \tag{3.2}$$

*Here $K_j$ has the same size as $J_{m_{j1}}(\lambda_1) \oplus \cdots \oplus J_{m_{jr_j}}(\lambda_j)$ and its structure is given by*

$$K_j = \left[K_j^{(pq)}\right]_{p,q=1}^{r_j}$$

*where $K_j^{(pq)}$ is a $m_{jp} \times m_{jq}$ matrix having all entries zero with the possible exception of the bottom $\min(m_{jp}, m_{jq})$ entries in the first column.*

Theorem 3.2 holds verbatim for matrix functions that are analytic functions of several complex variables.

A refinement of (3.2) is given in Kashchenko (1988) under additional hypotheses on $A(t)$. Other special cases are studied in Tovbis (1992).

In Krein and Tovbis (1990), Tovbis (1992) various forms are described that can be obtained from $A(t)$ by applying similarities $S(t)A(t)S(t)^{-1}$ or transformations of the form

$$A(t) \to S(t)^{-1}A(t)S(z) - z^r S(t)^{-1}S'(t),$$

where the invertible matrix $S(t)$ is assumed to depend meromorphically on $t$, or have expansion in fractional powers, in a neighborhood of a given point. The study of these forms is motivated by transformations of systems of linear differential equations.

Another approach for studying analytic matrix functions concerns characterizations of various types of similarities between such functions. We say that $n \times n$ matrix functions $A(\lambda)$ and $B(\lambda)$ analytic at $\lambda_0$ are *pointwise similar* if $A(\lambda)$ is similar to $B(\lambda)$ for every $\lambda$ sufficiently close to $\lambda_0$; $A(\lambda)$ and $B(\lambda)$ are called *analytically similar* if $A(\lambda) = T(\lambda)^{-1}B(\lambda)T(\lambda)$ for some matrix function $T(\lambda)$ which is analytic and invertible at $\lambda_0$. The pointwise similarity does not always imply analytic similarity. An important problem (that arises in the study of singular ordinary differential equations) is to find conditions on $A(\lambda)$ that guarantee the equivalence of pointwise and analytic similarity. This problem was studied in Wasow (1962), Friedland (1980a, 1980b). Some of the results of these papers have been interpreted and generalized in the framework of matrices over commutative rings Guralnick (1981).

### 3.2. *Real analytic matrix functions*

For some important classes of matrix functions (e.g., hermitian matrix valued) it is natural to consider dependence on real rather than complex parameter. Here the main result concerns analytic behaviour of eigenvalues and eigenvectors, and (as a consequence) of a triangular form:

THEOREM 3.3. *Let $A(t)$ be an $n \times n$ matrix which is an analytic function of a real parameter $t \in (a,b)$, $-\infty \leqslant a < b \leqslant \infty$. Assume that all the eigenvalues of $A(t)$, for all $t \in (a,b)$, lie on a differentiable (i.e. having tangent at each point) curve $\Gamma \subset \mathbb{C}$. Then there exists an analytic (on $t \in (a,b)$) $n \times n$ matrix function $U(t)$ such that*

$$U(t)^* U(t) = I$$

*(i.e. $U(t)$ is unitary valued) and*

$$U(t)^* A(t) U(t) = \left[x_{ij}(t)\right]_{i,j=1}^{n} \tag{3.3}$$

*is triangular: $x_{ij}(t) \equiv 0$ for $i > j$.*

PROOF. Let $t_0 \in (a,b)$. By Theorem 3.1 the eigenvalues $\lambda_j(t)$ of $A(t)$ are given by fractional power series (3.1) in a neighborhood of $t_0$. Assume that $p \geqslant 2$ and that at least one of the coefficients $\alpha_{jk}$, $k$ not a multiple of $p$, is nonzero. Let $k_0$ be the smallest integer, not a multiple of $p$, such that $\alpha_{jk_0} \neq 0$. Then letting $t \to T_0$ firstly for $t > t_0$ and secondly for $t < t_0$ we obtain

$$\lim \left[\lambda_j(t) - \lambda_j(t_0)\right](t - t_0)^{-k_0/p} = \alpha_{jk_0},$$
$$\lim \left[\lambda_j(t) - \lambda_j(t_0)\right](t - t_0)^{-k_0/p} = (-1)^{k_0/p} \alpha_{jk_0}.$$

Clearly, the numbers $\alpha_{jk_0}$ and $(-1)^{k_0/p}\alpha_{jk_0}$ have arguments that either coincide with the tangential direction to $\Gamma$ at $\lambda_j(t_0)$, or are opposite to this direction. Hence $(-1)^{k_0/p}$ must be real, a contradiction. We have proved that the eigenvalues (suitably ordered) of $A(t)$ are analytic in a neighborhood of $t_0$. By analytic continuation, the eigenvalues $\lambda_1(t), \ldots, \lambda_n(t)$ of $A(t)$ are analytic on $t \in (a,b)$. Now we use a result (see, e.g., Theorem 5.6.1 in Gohberg et al. (1978c)) according to which an analytic (on $t \in (a,b)$) column vector valued function $y_1(t)$ can be found such that $y_1(t) \neq 0$ and $(A(t) - \lambda_1(t)I)y_1(t) = 0$ for all $t \in (a,b)$. In other words, $y_1(t)$ is an eigenvector of $A(t)$ corresponding to $\lambda_1(t)$. By the same result, there exists an analytic basis $y_2(t), \ldots, y_n(t)$ in $\operatorname{Ker}(y_1(t))^*$ (at this point we use the fact that $t$ is a real variable and therefore $(y_1(t))^*$ is analytic on $t$ as well). Performing the Gram–Schmidt orthonormalization on $y_1(t), \ldots, y_n(t)$ (which does not spoil the analyticity) we obtain a unitary analytic (on $t \in (a,b)$) matrix function $U_1(t)$ whose first column is a scalar multiple of $y_1(t)$. Clearly,

$$U_1(t)^* A(t) U_1(t) = \begin{pmatrix} \lambda_1(t) & * \\ 0 & A_1(t) \end{pmatrix}$$

for some $(n-1) \times (n-1)$ analytic matrix function $A_1(t)$, and so on, until the proof is completed. □

The most important particular cases of Theorem 3.3 are when $\Gamma$ is a straight line (e.g., the real axis) or a unit circle, or when $A(t)$ satisfies additional hypotheses (e.g., being hermitian valued, or unitary valued) that make the hypothesis on the location of eigenvalues of $A(t)$ satisfied automatically. For hermitian valued $A(t)$ the result of Theorem 3.3 goes back to Rellich (1937, 1953); see also Porsching (1968), Kato (1966, 1982), Gohberg et al. (1978c), Gingold and Hsieh (1992).

Note that Theorem 3.3 is false if $A(t)$ depends analytically on more than one real variable. The following well-known example illustrating this fact can be found, for example, in Section II.5.7 of Kato (1966):

EXAMPLE 3.2. Let

$$A(t_1, t_2) = \begin{pmatrix} t_1 & t_2 \\ t_2 & -t_1 \end{pmatrix}.$$

$A(t_1, t_2)$ is obviously analytic and hermitian as function of $(t_1, t_2) \in \mathbb{R}^2$. However, the eigenvalues $\pm(t_1^2 + t_2^2)^{1/2}$ are not analytic at $t_1 = t_2 = 0$.

### 3.3. *Matrices with entries in a function algebra*

Let $K \subset \mathbb{R}^n$ be a connected compact set, and let $X$ be an algebra (over $\mathbb{C}$) of continuous complex valued functions on $K$ with the following properties:

(i) $X$ admits partitions of unity: for every relatively open finite covering $\{V_j\}_{j=1}^r$ of $K$ there exist non-negative functions $\varphi_1(t), \ldots, \varphi_r(t)$ in $X$ such that

$$\sum_{j=1}^{r} \varphi_j(t) = 1$$

and $\varphi_j(t) = 0$ for $t \in K \backslash V_j$.

(ii) if $f(t) \in X$ and $f(t) \neq 0$ for all $t \in K$, then $(f(t))^{-1} \in X$.

Denote by $M_{m\times p}(X)$ the set of $m \times p$ matrices with entries in $X$.

A typical question one is interested in when studying matrices with entries in $X$ is whether a certain quantity associated with a matrix can be expressed in terms of the algebra $X$. We state here one result in this spirit concerning the kernel and the range of a matrix.

THEOREM 3.4 (Gochberg and Leiterer (1976)). *Let $A(t) \in M_{m\times p}(X)$, $t \in K$, and assume that the dimension of* Ker $A(t)$ *(and therefore also the dimension of* Range $A(t)$ *is independent of $t \in K$. If $K$ is contractible, or if $n \leqslant 2$, then there exist a basis $x_1(t), \ldots, x_s(t)$ in* Ker $A(t)$ *and a basis $y_1(t), \ldots, y_q(t)$ in* Range $A(t)$*) such that $x_j \in M_{p\times 1}(X)$ (for $k = 1, \ldots, s$) and $y_k \in M_{m\times 1}(X)$ (for $k = 1, \ldots, q$).*

A example is given in Evard (1990) of a $2 \times 2$ hermitian valued matrix function $A(t)$ which is infinitely differentiable for parameter $t \in K$, where $K$ is the unit sphere in $\mathbb{R}^3$, and such that rank $A(t) = 1$ for all $t \in K$; nevertheless, there is no continuous vector function $x(t)$ which is a basis in $\operatorname{Ker} A(t)$ for all $t \in K$.

*Literature guide.* The theory of matrices with entries in function algebras is only at the beginning of its systematic development. Gochberg and Leiterer (1976) is an early paper on this subject, and some later work in this direction includes Gingold (1979), Evard (1990), Evard and Gracia (1990). In particular, the following result is proved in Evard and Gracia (1990): Let $A(t)$ and $B(t)$, $t \in \Omega$ be two $n \times n$ matrix functions of the $C^p$ class, where $\Omega \subseteq \mathbb{R}^q$ is an open set $C^p$-diffeomorphic to $\mathbb{R}^q$, such that $A(t)$ and $B(t)$ have constant Jordan structure for all $t \in \Omega$, and for every fixed $t_0 \in \Omega$ the matrices $A(t_0)$ and $B(t_0)$ are similar (here $p$ is a non-negative integer or $\infty$). Then there is a $C^p$ class similarity between $A(t)$ and $B(t)$. A related result (known as Doležal's theorem, Doležal (1964)) states that the kernel of a $C^p$-class matrix function $A(t)$, $t \in [0, \infty)$ with constant rank can be transformed to a constant subspace by means of a $C^p$-class invertible matrix function. This theorem and its generalizations are well-known and widely used in control systems: Silverman and Bucy (1970), Weiss and Falb (1969). A far reaching generalization of this result was obtained in Guralnick (1991) for matrices over certain commutative reduced rings $R$. Without setting up the precise framework for such a generalization, we just mention that the key property of the ring $R$ needed here is that every finitely generated projective $R$-module is free. Many results concerning matrices whose entries are continuous or analytic functions are exposed in Gohberg et al. (1986a). A completely different problem – positive semidefinite completions of partial matrices – was treated in Johnson and Rodman (1988) from the point of view of matrices over function rings.

Analytic properties of singular values and singular value decompositions of analytic matrix valued functions are studied in Boyd and Balakrishnan (1992), Bunse and Gerstner et al. (1992), Boyd and De Moor (1990). Derivatives (sensitivities) of eigenvalues and eigenvectors of matrix functions depending analytically on several real or complex variables are studied in Sun (1990), Andrew et al. (1992); and see Burke and Overton (1991) for analogous questions concerning the maximum real part and the maximum modulus of eigenvalues. We note also the paper Overton (1992) (and references therein), devoted to the problem of minimization of the maximum eigenvalue of $A(x)$ subject to linear constraints and bounds on $x \in \mathbb{R}^q$; here $A(x)$ is a real symmetric matrix function of $x$ which is continuously differentiable.

## 4. Matrix polynomials

Let $F$ be a field, $F[\lambda]$ the algebra of polynomials in one variable $\lambda$ with coefficients in $F$. Matrices with entries in $F[\lambda]$ are called *matrix polynomials*, or *polynomial matrices*.

## 4.1. *The Smith form*

We start with the Smith canonical form which plays an important role in the analysis of matrix polynomials.

THEOREM 4.1. *Let $A(\lambda) \in M_{m\times n}(F[\lambda])$ be a matrix polynomial. Then $A(\lambda)$ admits the representation*

$$A(\lambda) = E_1(\lambda)D(\lambda)E_2(\lambda), \tag{4.1}$$

*where*

$$D(\lambda) = \begin{pmatrix} d_1(\lambda) & 0 & \dots & \dots & 0 \\ 0 & d_2(\lambda) & \dots & \dots & 0 \\ \vdots & \ddots & \vdots & \vdots & \vdots \\ & & d_r(\lambda) & \dots & \\ & & & 0 & \\ & & & & \ddots \\ 0 & 0 & \dots & \dots & 0 \end{pmatrix} \tag{4.2}$$

*is a diagonal matrix polynomial with monic (i.e. having leading coefficient 1) scalar polynomials $d_j(\lambda) \in F[\lambda]$ such that $d_i(\lambda)$ is divisible by $d_{i-1}(\lambda)$ $(i = 2, \dots, r)$; $E_1(\lambda) \in M_{m\times m}(F[\lambda])$ and $E_2(\lambda) \in M_{n\times n}(F[\lambda])$ have constant (i.e. independent of $\lambda$) nonzero determinants. Moreover, the polynomials $d_1(\lambda), d_2(\lambda), \dots, d_r(\lambda)$ as well as their number $r$, are uniquely determined by $A(\lambda)$.*

The proof of Theorem 4.1 is accomplished by applying elementary row and column operations (see, e.g., Thrall and Tornheim (1957), MacDuffee (1946), Gantmacher (1959) or Gohberg et al. (1982a) for the proof). The book Newman (1972) contains a detailed exposition of the Smith form (4.1) as well as of other related forms.

The polynomials $d_i(\lambda)$ are called the *invariant polynomials* of $A(\lambda)$. They can be determined by $A(\lambda)$ as follows: Let $r \times r$ be the maximal size of a square submatrix in $A(\lambda)$ with not identically zero determinant, and for $i = 1, \dots, r$, let $D_i(\lambda)$ be the monic greatest common divisor of all $i \times i$ minors (= determinants of $i \times i$ submatrices) of $A(\lambda)$. Then

$$d_i(\lambda) = D_i(\lambda)/D_{i-1}(\lambda), \quad i = 1, \dots, r,$$

where we put $D_0(\lambda) \equiv 1$.

Two matrix polynomials $A(\lambda), B(\lambda) \in M_{m\times n}(F[\lambda])$ are called *equivalent* if $A(\lambda) = E_1(\lambda)B(\lambda)E_2(\lambda)$ for some $E_1(\lambda) \in M_{m\times m}(F[\lambda])$ and $E_2(\lambda) \in M_{n\times n}(F[\lambda])$ with constant nonzero determinants. Theorem 4.1 can be recast in the following alternative form: *$A(\lambda), B(\lambda) \in M_{m\times n}(F[\lambda])$ are equivalent if and only if they have the same invariant polynomials.*

The Smith form (4.1) has been studied in the more general framework of matrices over rings. We call a commutative unital ring $R$ without divisors of zero a *Smith domain* if every matrix $A$ over $R$ has the Smith form: i.e. a representation $A = E_1 D E_2$, where $E_1$ and $E_2$ are invertible matrices (over $R$) and $D = \mathrm{diag}(d_1, \dots, d_r, 0, \dots, 0)$, $d_j \in R$ is diagonal where $d_i$ is divisible by $d_{i-1}$ $(i = 2, \dots, r)$. It is known that every principal ideal domain is a Smith domain (see Section III.8 of Jacobson (1964)); on the other hand, every Smith domain is a *Bezout domain*, i.e. every finitely generated ideal is principal. The ring $\mathbb{Z}[\lambda]$ is an example of a Smith domain which is not a PID. These and other properties of Smith domains and related rings are found in Brewer et al. (1986), Den Boer (1981); see Kaplansky (1949) for results concerning rings admitting the Smith form (in the general framework of not necessarily commutative rings possibly having divisors of zero). The question whether every Bezout domain is a Smith domain seems to be still open.

Since $F[\lambda]$ is a principal ideal domain, the ring $M_{n\times n}(F[\lambda])$ enjoys the divisibility properties common to matrix rings over principal ideal domains (or, more generally, Bezout domains). Namely, every pair of $n \times n$ matrix polynomials $A(\lambda)$ and $B(\lambda)$ has a *greatest common right divisor* $D(\lambda)$ (which also belongs to $M_{n\times n}(F[\lambda])$), and moreover $D(\lambda)$ can be expressed in the form

$$D(\lambda) = X(\lambda)A(\lambda) + Y(\lambda)B(\lambda)$$

for some $X, Y \in M_{n\times n}(F[\lambda])$. Also, every pair of $n \times n$ matrix polynomials $A(\lambda)$ and $B(\lambda)$ which are not divisors of zero in $M_{n\times n}(F[\lambda])$ have a *least common left multiple* $C(\lambda) \in M_{n\times n}(F[\lambda])$; moreover, $C(\lambda)$ is unique up to left factor with constant nonzero determinant. Proof of these facts can be found, for example, in MacDuffee (1946). In the next section, we will study divisibility and factorization of matrix polynomials from a different (geometric) point of view based on invariant subspaces.

### 4.2. *Factorization of matrix polynomials*

One of the main problems in the theory of matrix polynomials is the problem of factorization:

$$A(\lambda) = B(\lambda)C(\lambda), \tag{4.3}$$

where $A(\lambda)$, $B(\lambda)$ and $C(\lambda)$ are $n \times n$ matrix polynomials. In the sequel we consider factorization of matrix polynomials which are *monic*, i.e. with the leading coefficient $I$.

In contrast with scalar polynomials, even when $F$ is algebraically closed, not every monic matrix polynomial admits a factorization (4.3) into product if monic matrix polynomials of smaller degrees:

EXAMPLE 4.1.

$$A(\lambda) = \begin{pmatrix} \lambda^2 & 1 \\ 0 & \lambda^2 \end{pmatrix}$$

has no factorization (4.3), where $B(\lambda)$ and $C(\lambda)$ are monic nonconstant polynomials. This follows easily from the nonexistence (over any field containing $F$) of a square root of the matrix

$$\begin{pmatrix} 0 & 1 \\ 0 & 0 \end{pmatrix}.$$

Let

$$A(\lambda) = \lambda^m I + \sum_{j=0}^{m-1} \lambda^j A_j, \quad A_j \in M_{n\times n}(F). \tag{4.4}$$

The $mn \times nm$ matrix

$$C_A = \begin{pmatrix} 0 & I & 0 & \dots & 0 \\ 0 & 0 & I & \dots & 0 \\ & & & & \vdots \\ \vdots & \vdots & \vdots & & I \\ -A_0 & -A_1 & -A_2 & \dots & -A_{m-1} \end{pmatrix} \tag{4.5}$$

is called the *companion matrix* associated with $A(\lambda)$. It turns out that the factorization of $A(\lambda)$ can be described in terms of certain $C_A$-invariant subspaces. A subspace $M \subseteq F^{nm}$ is called *$C_A$-invariant* if $C_A x \in M$ for every $x \in M$, where $C_A$ is considered in the natural way as a linear transformation $F^{nm} \to F^{nm}$.

For $A(\lambda) \in M_{m\times n}(F[\lambda])$ with the invariant polynomials $d_1(\lambda), \dots, d_r(\lambda)$, the roots of $d_j(\lambda)$ (in some algebraic closure of $F$) will be called the *zeros* of $A(\lambda)$; the multiplicities of $\lambda_0$ as a root of $d_1(\lambda), \dots, d_r(\lambda)$ are called the *partial multiplicities* of $\lambda_0$ as a zero of $A(\lambda)$.

THEOREM 4.2. *Let be given monic matrix polynomial* $A(\lambda)$ (4.4) *with its companion matrix* (4.6). *Then the factorizations* (4.3) *with monic matrix polynomials* $B(\lambda)$ *and* $C(\lambda)$, *where* $B(\lambda)$ *has degree* $k$ *and* $C(\lambda)$ *has degree* $m-k$, *are in one-to-one correspondence with* $C_A$*-invariant subspaces* $M$ *such that* $M$ *is a direct complement to the subspace*

$$\{x \in F^{mn}\colon \textit{ the first } (m-k)n \textit{ components of } x \textit{ are zeros}\}. \tag{4.6}$$

*Moreover, the zeros of* $C(\lambda)$ *coincide with the eigenvalues of the restriction* $C_A \mid M$, *and the partial multiplicities of a zero* $\lambda_0$ *of* $C(\lambda)$ *coincide with the multiplicities of* $\lambda_0$ *as an eigenvalue of* $C_A \mid M$. *Given the subspace* $M$ *as above, the corresponding matrix polynomials* $B(\lambda)$ *and* $C(\lambda)$ *are given by the formulas*

$$C(\lambda) = \lambda^{m-k} I - [I\ 0\ \dots\ 0](C_A \mid M)^{m-k}[V_1 + V_2\lambda + \dots + V_{m-k}\lambda^{m-k-1}],$$

*where*

$$[V_1\ V_2\ \dots\ V_{m-k}] = \big(\big(I_{(m-k)n} \quad 0\big) \mid M\big)^{-1}; \tag{4.7}$$

$$B(\lambda) = \lambda^k I - \left(Z_1 + Z_2\lambda + \cdots + Z_k\lambda^{k-1}\right) PC_A^k PY,$$

*where*

$$Y = \begin{pmatrix} 0 \\ \vdots \\ 0 \\ I \end{pmatrix},$$

*$P$ is the projector on the subspace* (4.6) *along $M$* (*understood as a linear transformation from $F^{mn}$ onto* (4.6)), *and*

$$\begin{pmatrix} Z_1 \\ Z_2 \\ \vdots \\ Z_k \end{pmatrix} = \left[PY, PC_A PY, \ldots, PC_A^{k-1} PY\right]^{-1}. \tag{4.8}$$

Observe that the existence of inverses in (4.7) and (4.8) is guaranteed by the condition that $M$ is a direct complement to (4.6).

The proof on Theorem 4.2 is given in Gohberg et al. (1978c) (see also Chapter 3 in Gohberg et al. (1982b)) for the case $F = \mathbb{C}$; the general case is proved in the same way.

Of special interest are right divisors of $A(\lambda)$ of the form $\lambda I - Z$, $Z \in M_{n\times n}(F)$. This happens if and only if $Z$ is a *right solvent* of $A(\lambda)$, i.e.

$$Z^m + \sum_{j=0}^{m-1} A_j Z^j = 0.$$

Right solvents of matrix polynomials are studied in Markus and Mereutsa (1973), Gohberg et al. (1978a), Maroulas (1985) using generalized Vandermonde matrices. In this connection we note a result proved in Krupnik (1991) according to which $A(\lambda)$ admits factorization

$$\prod_{j=1}^{m} (\lambda I - Z_j), \quad Z_j \in M_{n\times n}(F),$$

provided all elementary divisors of $A(\lambda)$ are either linear or quadratic ($F$ is assumed algebraically closed here).

Theorem 4.2 allows one to reduce factorization problems to invariant subspace problems. This approach is especially useful when $F$ is algebraically closed, or when $A(\lambda)$ has a special structure. We shall illustrate this approach for an important class of matrix polynomials (over $\mathbb{C}$) with hermitian coefficients.

THEOREM 4.3. *Let* $A(\lambda) \in M_{n\times n}(\mathbb{C}[\lambda])$ *be a matrix polynomial given by* (4.4) *and assume that* $A_j$ $(j = 0, \ldots, m-1)$ *are hermitian matrices. Then* $A(\lambda)$ *admits factorizations*

$$A(\lambda) = B(\lambda)C(\lambda) \tag{4.9}$$

*where* $C(\lambda)$ *is a monic matrix matrix polynomial of degree* $\left[\frac{m+1}{2}\right]$ *such that all zeros of* $C(\lambda)$ *have nonpositive imaginary part, and all zeros of* $B(\lambda)$ *have non-negative imaginary part.*

PROOF. We provide only an outline of the proof, and refer the reader to Section II.3.2 in Gohberg et al. (1983) for the full proof.

Let $C_A$ be the companion matrix of $A(\lambda)$, and let

$$H_A = \begin{pmatrix} A_1 & A_2 & & \ldots & I \\ A_2 & & & & 0 \\ \vdots & & & \vdots & \vdots \\ & I & & & 0 \\ I & 0 & \ldots & 0 & 0 \end{pmatrix} \in M_{mn\times mn}(\mathbb{C}). \tag{4.10}$$

Clearly, $H_A$ is invertible and hermitian. A straightforward calculation shows that

$$H_A C_A = C_A^* H_A. \tag{4.11}$$

This equality can be interpreted in the following way: Introduce the *indefinite scalar product* $[\cdot,\cdot]$ on $\mathbb{C}^{mn}$ by

$$[x, y] = y^* H_A x, \quad x, y \in \mathbb{C}^{mn}.$$

The equality (4.11) means that $C_A$ is selfadjoint with respect to $[\cdot,\cdot]$:

$$[C_A x, y] = [x, C_A y], \quad x, y \in \mathbb{C}^{mn}.$$

Now the theory of linear transformations that are selfadjoint in an indefinite scalar product (see, e.g., Gohberg et al. (1983)) guarantees existence of an $\left[\frac{m+1}{2}\right]n$-dimensional $C_A$-invariant subspace $M \subset \mathbb{C}^{mn}$ with the additional properties that

$$[x, x] \geqslant 0 \quad \text{for all } x \in M,$$

and that all eigenvalues of the restriction $C_A \mid M$ lie in the closed lower halfplane. Moreover, it turns out that every such $M$ is a direct complement to the subspace (4.6), where $k = \left[\frac{m}{2}\right]$. It remains to apply Theorem 4.2. □

Under the hypotheses of Theorem 4.3, $A(\lambda)$ also admits factorizations (4.9) with all zeros of $C(\lambda)$ having non-negative imaginary part.

Of special interest are matrix polynomials which are positive semidefinite on the real line:

THEOREM 4.4. *The following statements are equivalent for a matrix polynomial* (4.4) (*over* $\mathbb{C}$)*:*

(i) $A(\lambda)$ *is positive semidefinite for every real* $\lambda$*;*

(ii) $A(\lambda)$ *admits factorization of the form*

$$A(\lambda) = \left(M(\overline{\lambda})\right)^* M(\lambda),$$

*where* $M(\lambda)$ *is an* $n \times n$ *matrix polynomial;*

(iii) *the degree* $m$ *of* $A(\lambda)$ *is even, and, letting* $C_A$ *and* $H_A$ *be matrices defined by* (4.5) *and* (4.10) *respectively, there exists a* $C_A$*-invariant* $\frac{mn}{2}$*-dimensional subspace* $M \subset \mathbb{C}^{mn}$ *which is* $H_A$*-neutral:* $y^* H_A x = 0$ *for all* $x, y \in M$*.*

(iv) *all partial multiplicities of real zeros* (*if any*) *of* $A(\lambda)$ *are even.*

The proof of Theorem 4.4 is found in Section II.3.2 of Gohberg et al. (1983).

*Literature guide.* Much of the development of the theory of matrix polynomials (with real or complex coefficients) was motivated by applications in mechanical and electrical systems with finite number of degrees of freedom (see, e.g., Whittaker (1952), Frazer et al. (1955) and especially Lancaster (1966)). Other important applications of matrix polynomials are found in modern control theory (see the books Barnett (1983), Kailath (1980), Rosenbrock (1970)). The spectral analysis of matrix polynomials (leading, in particular, to Theorem 4.2) has been initiated in Gohberg et al. (1978c, 1978d); a comprehensive exposition of this theory (including nonmonic matrix polynomials) is given in Gohberg et al. (1982b); see also Lancaster and Tismenetsky (1985), Gohberg et al. (1986a). Perturbation theory for divisors of monic matrix polynomials was developed in Gohberg et al. (1979) in the context of both continuous and analytic perturbations. For the theory of common multiples and divisors of matrix polynomials from the spectral analysis point of view, see Gohberg et al. (1978a, 1978b, 1981, 1982a), also the book Gohberg et al. (1982b). The book Kazimirskii (1981) contains the factorization theory of matrix polynomials over a general field, from the algebraic point of view.

Matrix polynomials with hermitian coefficients, as well as with other symmetries, are of special interest because of numerous applications, especially in vibrating systems and linear control systems: Lancaster (1966), Coppel (1972), Gohberg et al. (1983). The comprehensive spectral theory of hermitian matrix polynomials was developed starting with Gohberg et al. (1980), also Gohberg et al. (1982c, 1982d) (it should be noted, however that the spectral theory for certain classes of operator polynomials with selfadjoint coefficients was developed before, see Krein and Langer (1978), Langer (1976)). The book Gohberg et al. (1983) contains an exposition of this theory, as well as many applications; see also the review papers Lancaster (1982), Rodman (1987).

Matrix polynomials of second degree with hermitian coefficients are especially important in applications. Besides the above references, we mention here Gohberg et al. (1986b) and Lancaster and Maroulas (1988), where the behaviour of zeros of such polynomials is studied under analytic perturbations of the linear term and under feedback, respectively, and Lancaster and Maroulas (1987), where the problems concerning the determination of such polynomials from the knowledge of their spectral properties are studied.

We conclude with a brief mention of some other aspects of the theory of matrix polynomials. The volume Kagström and Ruhe (1983) is devoted mainly to the computational aspects of matrix polynomials; among numerous papers on this subject we mention only Belyi et al. (1989), Khazanov and Kublanovskaya (1988), Van Dooren and Dewilde (1983) (see also references in those papers), where algorithms are given for computing the zero structure of rectangular matrix polynomials. Orthogonal matrix polynomials have been studied in Delsarte et al. (1978), Fuhrmann (1987) and in the volume Gohberg (1988a), (among others); in Gohberg and Lerer (1988) the spectral analysis of matrix polynomials, and in particular connections with coprime and Wiener–Hopf factorizations, play a prominent role.

## 4.3. *Bezoutian of matrix polynomials*

Let

$$a(\lambda) = \sum_{i=0}^{\ell} a_i \lambda^i, \qquad b(\lambda) = \sum_{i=0}^{m} b_i \lambda^i \quad (m \leqslant \ell)$$

be scalar polynomials with coefficients in the field $F$ (we assume $a_\ell \neq 0$, $b_m \neq 0$). The concepts of the resultant and Bezoutian matrices of $a(\lambda)$ and $b(\lambda)$ are classical. The resultant is the $(\ell + m) \times (\ell + m)$ matrix

$$\operatorname{Res}(a, b) = \begin{pmatrix} a_0 & a_1 & \ldots & a_\ell & 0 & \ldots & 0 \\ 0 & a_0 & a_1 & \ldots & a_\ell & \ldots & 0 \\ \vdots & \vdots & & & & & \vdots \\ 0 & 0 & & \ldots & a_0 & \ldots & a_\ell \\ b_0 & b_1 & \ldots & b_m & 0 & \ldots & 0 \\ 0 & b_0 & \ldots & & b_m & \ldots & 0 \\ \vdots & \vdots & & & & & \vdots \\ 0 & 0 & \ldots & & b_0 & \ldots & b_m \end{pmatrix}, \tag{4.12}$$

and the Bezoutian is the $\ell \times \ell$ matrix $\operatorname{Bez}(a, b) = [x_{ij}]_{i,j=0}^{\ell-1}$ defined by

$$\sum_{i,j=0}^{\ell-1} x_{ij} \lambda^i \mu^j = (\lambda - \mu)^{-1} \big[a(\lambda) b(\mu) - a(\mu) b(\lambda)\big]. \tag{4.13}$$

The fundamental property of these matrices is that the dimension of $\operatorname{Ker} \operatorname{Res}(a, b)$, as well as the dimension of $\operatorname{Ker} \operatorname{Bez}(a, b)$ is equal to the degree of the greatest common divisor of $a(\lambda)$ and $b(\lambda)$. This property has been used, in particular, to prove various root separation and inertia results for scalar polynomials (with real or complex coefficients).

Recently, the concepts of resultant and Bezoutian matrices and their fundamental property have been extended to matrix polynomials. We will focus here on the Bezoutian matrix.

Let $L_1(\lambda)$ and $L(\lambda)$ be two $n\times n$ matrix polynomials with $\det L_1(\lambda) \not\equiv 0$, $\det L(\lambda) \not\equiv 0$. Since $L_1$ and $L$ generally do not commute, we cannot use the same definition as in the scalar case based on (4.13). However, there exists a common left multiple of $L_1$ and $L$, i.e. $n \times n$ matrix polynomials $M_1(\lambda)$ and $M(\lambda)$ exist such that $\det M_1(\lambda) \not\equiv 0$, $\det M(\lambda) \not\equiv 0$ and the equality

$$M_1(\lambda)L_1(\lambda) = M(\lambda)L(\lambda) \tag{4.14}$$

holds. The Bezoutian associated with the equality (4.14) is defined as the block matrix

$$B = [\Gamma_{ij}]_{i=0,j=0}^{m-1,\ell-1} \tag{4.15}$$

where the block entries $\Gamma_{ij}$ are given by

$$\sum_{i=0}^{m-1}\sum_{j=0}^{\ell-1} \Gamma_{ij}\lambda^i\mu^j = (\lambda-\mu)^{-1}\big[M_1(\lambda)L_1(\mu) - M(\lambda)L(\mu)\big],$$

and where $\ell$ (respectively, $m$) is the maximal degree of $L$ and $L_1$ (respectively, of $M$ and $M_1$).

The fundamental property of the matrix Bezoutian can be stated as follows:

THEOREM 4.5. *Assume that one of the matrix polynomials $L(\lambda)$ and $L_1(\lambda)$ has invertible leading coefficient, and one of them has invertible constant term. Further assume that* (4.14) *holds. Then*

$$\dim \operatorname{Ker} B = \operatorname{degree}\big(\det L_0(\lambda)\big), \tag{4.16}$$

*where $L_0(\lambda)$ is the greatest common divisor of $L(\lambda)$ and $L_1(\lambda)$.*

This result as well as its analogue for the case when the hypotheses on the invertibility of coefficients is omitted (in this case the equality (4.16) should be modified), was proved in Lerer and Tismenetsky (1982). (It was assumed there $F = \mathbb{C}$, the generalization for any field is immediate.) Furthermore, a description of $\operatorname{Ker} B$ in terms of the zeros of $L_0(\lambda)$ and the corresponding eigenvectors and generalized eigenvectors is given also in Lerer and Tismenetsky (1982).

*Literature guide.* For the theory and applications of resultant and Bezoutian matrices for scalar polynomials see, e.g., the books Uspensky (1978), Lancaster and Tismenetsky (1985) and review papers Krein and Naimark (1981), Helmke and Fuhrmann (1989). The definition of the Bezoutian for matrix polynomials based on (4.14) was introduced in Anderson and Jury (1976), Bitmead et al. (1978), inspired by some problems in linear control systems. The theory of Bezoutians for matrix polynomials and its applications and connections to inertia and root separation of matrix polynomials and to various matrix equations have been developed in a series of papers Lerer and Tismenetsky (1982, 1984, 1988), Lerer (1989), Lerer et al. (1991) (see also references in these papers). Other applications of the Bezoutian are found in Barnett (1972), Wimmer (1988) (factorization of matrices) and in Lerer and Tismenetsky (1986), Gohberg and Shalom (1990) (inversion of structured matrices; the idea of this application goes back to Lander (1974)). See Kailath and Sayed (1996), and the extensive bibliography therein, for applications of Bezoutians and related matrix functions in developing fast computational algorithms for structured matrices. Another concept of Bezoutian for matrix polynomials $L_1(\lambda)$ and $L(\lambda)$ based on the equality

$$\sum_{ij} \Gamma_{ij}\lambda^i\mu^j = (\lambda-\mu)^{-1}\left[L_1(\lambda)\otimes L(\mu) - L_1(\mu)\otimes L(\lambda)\right]$$

was also studied in the literature: Bitmead et al. (1978), Barnett and Lancaster (1980), Heinig (1979), and see Wimmer (1989) for the concept of Bezoutian based on pairs of coprime matrix polynomials. Very recently, the notion of the Bezoutian, and its key properties and applications have been extended to rational matrix functions in Lerer and Rodman (1996).

For results concerning generalization of the resultant matrix to the case of matrix polynomials see Barnett (1969), Gohberg and Heinig (1975), Gohberg and Lerer (1976), Gohberg et al. (1982a), Lerer and Tismenetsky (1982). In Helton and Rodman (1987) the resultant matrices have been studied from the abstract point of view of matrices over rings.

## 5. Rational matrices

In this section we study $r \times n$ matrices $W(\lambda)$ whose elements are rational functions over a fixed field $F$. Thus,

$$W(\lambda) = \left[p_{ij}(\lambda)/q_{ij}(\lambda)\right]_{i=1,j=1}^{r,n} \tag{5.1}$$

where $p_{ij}(\lambda) \in F[\lambda]$, $q_{ij}(\lambda) \in F[\lambda]$ and $q_{ij}$ are not identically zero. The matrices of the form (5.1) are called *rational matrices* (over $F$).

Rational matrices appear in linear systems theory as follows (in this paragraph we assume $F = \mathbb{C}$): Consider a system of linear differential equations

$$\begin{cases} \dfrac{\mathrm{d}x}{\mathrm{d}t} = Ax(t) + Bu(t); \qquad x(0) = 0;\ t \geqslant 0, \\ y(t) = Cx(t) + Du(t); \end{cases} \tag{5.2}$$

where $A \in M_{m\times m}(\mathbb{C})$, $B \in M_{m\times n}(\mathbb{C})$, $C \in M_{r\times m}(\mathbb{C})$, $D \in M_{r\times n}(\mathbb{C})$ are constant (i.e. independent of $t$), $u(t)$ is an $n$-dimensional vector function that is at our disposal and is referred to as the *input* (or *control*), and $y(t)$ is the *output*. Taking the Laplace transform

$$Z(\lambda) = \int_0^\infty \mathrm{e}^{-\lambda s} z(s)\,\mathrm{d}s$$

and denoting by the capital Roman letter the Laplace transform designated by the corresponding small letter, the system (5.2) becomes

$$\begin{aligned} \lambda X(\lambda) &= AX(\lambda) + BU(\lambda), \\ Y(\lambda) &= CX(\lambda) + DU(\lambda), \end{aligned}$$

which can be solved for $Y(\lambda)$ in terms of $U(\lambda)$:

$$Y(\lambda) = \left[D + C(\lambda I - A)^{-1}B\right]U(\lambda). \tag{5.3}$$

The matrix $W(\lambda) = D+C(\lambda I-A)^{-1}B$, called the *transfer function* of (5.2), is obviously rational. Thus, the input-output map of a linear time invariant system of differential equations is given (after Laplace transforms) in terms of a rational matrix. This fact, and an analogous fact concerning systems of difference equations, explains the crucial role the theory of rational matrices is playing in modern linear systems theory.

Let $W(\lambda)$ be a rational $r \times n$ matrix over the field $F$. A representation of the form

$$W(\lambda) = D + C(\lambda I - A)^{-1}B, \tag{5.4}$$

where $D \in M_{r\times n}(F)$, $C \in M_{r\times m}(F)$, $A \in M_{m\times m}(F)$, $D \in M_{m\times n}(F)$, is called a *realization* of $W(\lambda)$ (cf. formula (5.3)).

THEOREM 5.1. *$W(\lambda)$ admits a realization if and only if $W(\lambda)$ is finite at infinity, i.e.*

$$\operatorname{degree} p_{ij}(\lambda) \leqslant \operatorname{degree} q_{ij}(\lambda) \tag{5.5}$$

*for every pair of indices $i, j$ $(1 \leqslant i \leqslant r,\ 1 \leqslant j \leqslant n)$ such that $p_{ij}(\lambda) \not\equiv 0$.*

PROOF. If $W(\lambda)$ has a realization (5.4), then the representation

$$(\lambda I - A)^{-1} = \frac{\operatorname{Adj}(\lambda I - A)}{\det(\lambda I - A)}, \tag{5.6}$$

where $\operatorname{Adj}(\lambda I - A)$ is the algebraic adjoint of $\lambda I - A$, shows that (5.5) holds. Conversely, assume that $W(\lambda)$ is finite at infinity. Let $p(\lambda)$ be a monic scalar polynomial such that $p(\lambda)W(\lambda)$ is a (matrix) polynomial. Denoting $H(\lambda) = p(\lambda)(W(\lambda) - W(\infty))$, $L(\lambda) = p(\lambda)I_n$, we have

$$W(\lambda) = W(\infty) + C(\lambda I - A)^{-1}B, \tag{5.7}$$

where

$$B = \begin{pmatrix} 0 \\ \vdots \\ 0 \\ I \end{pmatrix}, \quad A = \begin{pmatrix} 0 & I & \dots & 0 \\ \vdots & \vdots & & \vdots \\ 0 & 0 & \dots & I \\ -L_0 & -L_1 & \dots & L_{\ell-1} \end{pmatrix}, \quad C = [H_0, H_1, \dots, H_{\ell-1}],$$

and where the matrices $H_j$ and $L_j$ are the coefficients of $H(\lambda)$ and $L(\lambda)$:

$$H(\lambda) = \sum_{j=0}^{\ell-1} \lambda^j H_j, \qquad L(\lambda) = \lambda^\ell J + \sum_{j=0}^{\ell-1} \lambda^j L_j.$$

A full proof of (5.7) is found in Bart et al. (1979) and in Gohberg et al. (1986a). □

A realization (5.4) is far from being unique. For example, one can replace $A, B$ and $C$ in (5.4) by $S^{-1}AS, S^{-1}B$ and $CS$, where $S$ is an invertible matrix (this transformation is called *similarity*. There is an important class of minimal realizations (to be defined below), which enjoy many useful properties, and in particular, any two minimal realizations are similar. A realization (5.4) is called *minimal* if the size $m$ of the matrix $A$ is minimal among all realizations of $W(\lambda)$. The basic properties of minimal realizations are summarized in the following theorem.

THEOREM 5.2. (a) *A realization* (5.4) *is minimal if and only if*

$$\operatorname{rank}\left[B, AB, \dots, A^{p-1}B\right] = \operatorname{rank}\begin{pmatrix} C \\ CA \\ \vdots \\ CA^{p-1} \end{pmatrix} = m$$

*for sufficiently large integers* $p$.

(b) *If* (5.4) *is a* (*not necessarily minimal*) *realization of* $W(\lambda)$, *then, after a suitable similarity transformation,* $A, B$ *and* $C$ *have the form*

$$B = \begin{pmatrix} * \\ B_1 \\ 0 \end{pmatrix}, \quad C = \begin{pmatrix} 0 & C_1 & * \end{pmatrix}, \quad A = \begin{pmatrix} * & * & * \\ 0 & A_1 & * \\ 0 & 0 & * \end{pmatrix}, \tag{5.8}$$

*where*

$$W(\lambda) = D + C_1(\lambda I - A_1)^{-1}B_1$$

*is a minimal realization* (*the stars in* (5.8) *denote block entries of no immediate interest*).

(c) *Let*

$$W(\lambda) = D + C_j(\lambda I - A_j)^{-1}B_j \quad (j = 1, 2) \tag{5.9}$$

*be two minimal realizations of $W(\lambda)$. Then there exists a unique invertible matrix $S$ such that*

$$A_1 = S^{-1}A_2S, \quad B_1 = S^{-1}B_2, \quad C_1 = C_2S. \tag{5.10}$$

*The matrix $S$ is given by*

$$S = \begin{pmatrix} C_2 \\ C_2A_2 \\ \vdots \\ C_2A_2^{p-1} \end{pmatrix}^{-L} \begin{pmatrix} C_1 \\ C_1A_1 \\ \vdots \\ C_1A_1^{p-1} \end{pmatrix}$$
$$= [B_2, A_2B_2, \dots, A_2^{p-1}B_2] \cdot [B_1, A_1B_1, \dots, A_1^{p-1}B_1]^{-R} \tag{5.11}$$

*where the subscripts "$-L$" and "$-R$" denote left inverse and right inverse, respectively, and the integer $p$ is large enough so that the existence of the one-sided inverses is guaranteed (by the part* (a)*).*

PROOF. We prove the part (c) only (see, e.g., Section 7.1 in Gohberg et al. (1986a) for a complete proof). Using formula (5.6), we develop $(\lambda I - A_j)^{-1}$ into formal power series

$$(\lambda I - A_j)^{-1} = \sum_{k=1}^{\infty} \lambda^{-k} A_{jk} \quad (j = 1, 2).$$

Write

$$I = (\lambda I - A_j) \sum_{k=1}^{\infty} \lambda^{-k} A_{jk}$$

and compare coefficients; it follows that $A_{jk} = A_j^{k-1}$. Now (5.9) takes the form

$$C_1A_1^kB_1 = C_2A_2^kB_2, \quad k = 0, 1, \dots.$$

For $j = 1, 2$, let

$$\Omega_j = \begin{pmatrix} C_j \\ \vdots \\ C_jA_j^{p-1} \end{pmatrix}; \quad \Delta_j = [B_j, A_jB_j, \dots, A_j^{p-1}B_j].$$

We have $\Omega_1\Delta_1 = \Omega_2\Delta_2$. Premultiplying by $\Omega_2^{-L}$ and postmultiplying by $\Delta_2$, we verify the second equality in (5.11). Now define $S$ as in (5.11). The formulas

$$(\Omega_1^{-L}\Omega_2)S = I, \qquad S(\Delta_1\Delta_2^{-R}) = I \tag{5.12}$$

hold; therefore, $S$ is invertible. Furthermore,

$$\Omega_2 A_2 \Delta_2 = \Omega_2 \Delta_2 \Delta_1^{-R} A_1 \Delta_1,$$

which (in view of (5.12)) implies $A_2 S = S A_1$. The other two equalities in (5.10) can be verified directly. Finally, if $S$ satisfies (5.10), then

$$S\big[B_1, A_1 B_1, \dots, A_1^{p-1} B_1\big] = \big[B_2, A_2 B_2, \dots, A_2^{p-1} B_2\big]. \tag{5.13}$$

By the part (a),

$$\operatorname{rank}\big[B_1, A_1 B_1, \dots, A_1^{p-1} B_1\big] = \{\text{the size of } A_1\} = \{\text{the size of } A_2\},$$

and therefore the matrix $S$ satisfying (5.13) is unique. □

We pass to factorization of rational matrices. Here, the concept of a McMillan degree will be fundamental. Let $W(\lambda)$ be an $r \times n$ rational matrix (not necessarily finite at infinity), and write

$$W(\lambda) = P(\lambda) + W_0(\lambda),$$

where $P(\lambda)$ is a polynomial, and $W_0(\lambda)$ is finite at infinity. By Theorem 5.1 there exists minimal realizations

$$P(\lambda^{-1}) = D_1 + C_1(\lambda I - A_1)^{-1} B_1,$$
$$W_0(\lambda) = D_2 + C_2(\lambda I - A_2)^{-1} B_2,$$

where $A_1$ (resp. $A_2$) is $m_1 \times m_1$ (resp. $m_2 \times m_2$). The sum $m_1 + m_2$ is called the *McMillan degree* of $W(\lambda)$ and will be denoted $\delta(W)$. A factorization $W(\lambda) = W_1(\lambda) W_2(\lambda)$ of rational matrices is called *minimal* if $\delta(W) = \delta(W_1) + \delta(W_2)$. Informally, it means that there is no pole-zero cancellation between the factors $W_1$ and $W_2$, and represents a natural extension of factorization of matrix polynomials to the class of rational matrices.

It turns out that minimal factorizations can be described in terms of certain subspace decompositions. For simplicity, we present here such description in case $W(\lambda)$ takes value $I$ at infinity; then $W(\lambda)$ admits a minimal realization

$$W(\lambda) = I + C(\lambda I - A)^{-1} B,$$

where $A$ is $m \times m$. Let $A^\times = A - BC$. We say that a direct sum decomposition

$$F^m = \mathcal{L} \dotplus \mathcal{N} \tag{5.14}$$

is a *supporting decomposition* for $W(\lambda)$ if the subspace $\mathcal{L}$ is $A$-invariant, and the subspace $\mathcal{N}$ is $A^\times$-invariant.

THEOREM 5.3. *Let* (5.14) *be supporting decomposition for* $W(\lambda)$. *Then* $W(\lambda)$ *admits a minimal factorization*

$$W(\lambda) = \big[I + C\pi_{\mathcal{L}}(\lambda I - A)^{-1}\pi_{\mathcal{L}} B\big]\big[I + C\pi_{\mathcal{N}}(\lambda I - A)^{-1}\pi_{\mathcal{N}} B\big]$$

$$= \left[I + C(\lambda I - A)^{-1}\pi_{\mathcal{L}}B\right]\left[I + C\pi_{\mathcal{N}}(\lambda I - A)^{-1}B\right] \tag{5.15}$$

*where $\pi_{\mathcal{L}}$ is the projector on $\mathcal{L}$ along $\mathcal{N}$, and $\pi_{\mathcal{N}} = I - \pi_{\mathcal{L}}$. Conversely, for every minimal factorization $W(\lambda) = W_1(\lambda)W_2(\lambda)$ where the factors are rational matrices with value $I$ at infinity there exists a unique supporting decomposition $F^m = \mathcal{L} \dot{+} \mathcal{N}$ such that*

$$W_1(\lambda) = I + C\pi_{\mathcal{L}}(\lambda I - A)^{-1}\pi_{\mathcal{L}}B, \qquad W_2 = I + C\pi_{\mathcal{N}}(\lambda I - A)^{-1}\pi_{\mathcal{N}}B.$$

Note that the second equality in (5.15) follows from the relations $\pi_{\mathcal{L}}A\pi_{\mathcal{L}} = A\pi_{\mathcal{L}}$ and $\pi_{\mathcal{N}}A\pi_{\mathcal{N}} = \pi_{\mathcal{N}}A$, which express the $A$-invariance of $\mathcal{L}$.

Theorem 5.3 (for the case $F = \mathbb{C}$) is proved in Bart et al. (1979); see Section 7.3 of Gohberg et al. (1986a) for another variant of this result that involves three factors. The proofs given in these books for the case $F = \mathbb{C}$ are applicable verbatim to any field $F$. Other relevant references are Bart et al. (1980) and Gohberg et al. (1984). The importance of minimal factorizations is widely recognized and used in the modern theory of linear systems; we refer to Van Dooren and Dewilde (1981), Vanderwalle and Dewilde (1978), where minimal factorizations are studied from this point of view.

*Literature guide.* Realization theory is a major tool in modern control systems theory, and is developed and used in many texts on control systems (Brockett (1970), Barnett and Cameron (1985), Kailath (1980), Kalman et al. (1969), Rosenbrock (1970), Anderson and Vongpanitlerd (1973) is a representative sample). The theory of rational matrices, in particular, problems concerning various types of factorization and interpolation, and the applications of this theory (notably in $H_\infty$-control) has been extensively developed during the last twenty years or so. This development is based on the realization representation of rational matrices. The theory and its applications are to be found in the books Bart et al. (1979), Gohberg et al. (1983, 1986a), Ball et al. (1990a), several collections of papers Gohberg (1988a, 1988b, 1990), Gohberg and Kaashoek (1986), and see also the special issues Ball et al. (1990b), Fuhrmann et al. (1989) where many papers on this subject appear. Besides these volumes, we will mention here only few selected topics and references.

Rational matrices which enjoy certain symmetries (such as having hermitian values on the imaginary axis, having unitary values on the unit circle, or having real coefficients, etc.) play an important role in applications (see, e.g., Anderson and Vongpanitlerd (1973)), and therefore attracted considerable attention in the engineering literature. Efimov and Potapov (1973) is an early work on the factorization theory for a certain class of symmetric matrix functions (motivated by applications in circuit theory). From the standpoint of realization representations, the factorization and interpolation problems of rational matrices with various symmetries have been studied in Ran (1982), Fuhrmann (1983), Genin et al. (1983), Alpay and Gohberg (1988), Alpay et al. (1990, 1992); see also the books Anderson and Vongpanitlerd (1973), Ball et al. (1990), Gohberg et al. (1983).

The geometric approach to the minimal factorizations (Theorem 5.3) was extended in Ball et al. (1987), where they have been described in terms of local pole and zero structure of the rational matrices. Cascade decompositions of linear systems of the type

(5.2) correspond to factorizations in a linear fractional form of the corresponding transfer functions. In Helton and Ball (1982), such minimal linear fractional factorizations have been characterized in terms of generalized invariant subspaces (see also Gohberg and Rubinstein (1986)). In another direction, the result of Theorem 5.3 has been extended to more general classes of rational matrices (not necessarily of square size and having invertible value at infinity), see Cohen (1983), Van Dooren (1984).

Canonical factorization (defined below) is a very important special case of minimal factorization (when applied to a rational matrix function). Let $\Gamma$ be a simple closed rectifiable contour in $\mathbb{C}\cup\{\infty\}$ dividing the set $(\mathbb{C}\cup\{\infty\})\backslash\Gamma$ into two disjoint open sets $\Gamma_+$ and $\Gamma_-$. A matrix function $W(\lambda)$ is said to admit canonical factorization if it can be represented in the form $W(\lambda) = W_-(\lambda)W_+(\lambda)$, where $W_\pm(\lambda)$ is analytic in $\Gamma_\pm$, and is continuous and takes invertible values on $\Gamma_\pm \cup \Gamma$. Canonical factorizations, as well as the more general Wiener–Hopf factorizations $W(\lambda) = W_-(\lambda)D(\lambda)W_+(\lambda)$, where $D(\lambda)$ is a diagonal rational matrix function with poles and zeros allowed only in two preselected points $\lambda_\pm \in \Gamma_\pm$, are studied in numerous books and papers, of which we mention here only Bart et al. (1979), Clancey and Gohberg (1981), Gohberg and Kaashoek (1986), Litvinchuk and Spitkovskii (1987).

For a description of poles and zeros (including multiplicities) of a rational matrix in module theoretic terms see the expository paper Wyman et al. (1991) and references therein.

For the purpose of reference, the papers "to appear" are arbitrarily assigned year (1996). The letters (a), (b) etc., are used to distinguish references having the same year and the same authors (or the same first author if the number of authors is three or more).

## References

Alpay, D., J.A. Ball, I. Gohberg and L. Rodman (1990). *Realizations and factorizations of rational matrix functions with symmetries*, Operator Theory: Advances and Applications vol. 47, I. Gohberg, ed., Birkhäuser, Basel, 1–60.

Alpay, D., J.A. Ball, I. Gohberg and L. Rodman (1992). *State space theory of automorphisms of rational matrix functions*, Integral Equations Operator Theory **15**, 349–377.

Alpay, D. and I. Gohberg (1988). *Unitary rational matrix functions*, Operator Theory: Advances and Applications vol. 33, I. Gohberg, ed., Birkhäuser, Basel, 175–222.

Anderson, B.D.O. and E.I. Jury (1976). *Generalized Bezoutian and Sylvester matrices in multivariable linear control*, IEEE Trans. Automat. Control **AC-21**, 551–556.

Anderson, B.D.O. and S. Vongpanitlerd (1973). *Network Analysis and Synthesis*, Prentice-Hall, Englewood Cliffs.

Andrew, A.L., K.-W.E. Chu and P. Lancaster (1993). *Derivatives of eigenvalues and eigenvectors of matrix functions*, SIAM J. Matrix Anal. Appl. **14**, 903–926.

Arnold, V.I. (1971). *On matrices depending on parameters*, Russian Math. Surv. **26** (2), 29–43.

Ball, J.A., I. Gohberg and L. Rodman (1987). *Minimal factorization of meromorphic matrix functions in terms of local data*, Integral Equations Operator Theory **10**, 309–348.

Ball, J.A., I. Gohberg and L. Rodman (1990a). *Interpolation of Rational Matrix Functions*, Operator Theory vol. 45, Birkhäuser, Basel.

Ball, J.A., L. Rodman and P. Van Dooren, eds (1990b). *Matrix Valued Functions (Special Issue)*, Linear Algebra Appl. **137/138**.

Bapat, R.B. (1988). *Multinomial probabilities, permanents and a conjecture of Karlin and Rinott*, Proc. Amer. Math. Soc. **102**, 467–472.

Barnett, S. (1983). *Polynomials and Linear Control Systems*, Dekker, New York.

Barnett, S. (1969). *Regular polynomial matrices having relatively prime determinants*, Proc. Cambridge Philos. Soc. **65**, 585–590.

Barnett, S. (1972). *A note on the Bezoutian matrix*, SIAM J. Appl. Math. **22**, 84–86.

Barnett, S. and R.G. Cameron (1985). *Introduction to Mathematical Control Theory*, 2nd ed., Clarendon Press, Oxford.

Barnett, S. and P. Lancaster (1980). *Some properties of the Bezoutian for polynomial matrices*, Linear and Multilinear Algebra **9**, 99–111.

Bart, H., I. Gohberg and M.A. Kaashoek (1979). *Minimal Factorization of Matrix and Operator Functions*, Operator Theory vol. 1, Birkhäuser, Basel.

Bart, H., I. Gohberg, M.A. Kaashoek and P. Van Dooren (1980). *Factorizations of transfer functions*, SIAM J. Control Optim. **18**, 675–696.

Baumgartel, H. (1985). *Analytic Perturbation Theory for Matrices and Operators*, Birkhäuser, Basel.

Baumgartel, H. (1974). *Theory of analytic perturbations of linear operators depending on several complex variables*, Mat. Issled. **9**, 17–39 (in Russian).

Bellman, R. (1970). *Introduction to Matrix Analysis*, McGraw-Hill, New York, etc.

Belyi, V.A., V.B. Khasanov and V.N. Kublanovskaya (1989). *Spectral problems for matrix pencils. Methods and algorithms, III*, Soviet J. Numer. Anal. Math. Modelling **4**, 19–51.

Bendat, J. and S. Sherman (1955). *Monotone and convex operator functions*, Trans. Amer. Math. Soc. **79**, 58–71.

Bhatia, R. (1987). *Perturbation Bounds for Matrix Eigenvalues*, Pittman Research Notes in Mathematics, Longmann Scientific & Technical, Harlow, Essex.

Bitmead, R.R., S.Y. Kung, B.D.O. Anderson and T. Kailath (1978). *Greatest common divisors via generalized Sylvester and Bezout matrices*, IEEE Trans. Automat. Control **AC-23**, 1043–1047.

Boyd, S. and V. Balakrishnan (1990). *A regularity result for the singular values of a transfer matrix and a quadratically convergent algorithm for computing its $L_\infty$-norm*, Systems Control Lett. **15**, 1–7.

Boyd, S. and B.L.R. De Moor (1990). *Analytic properties of singular values and vectors*, Katholieke Universiteit Leuven, Dept. of Electrical Engineering, B-3030 Leuven (Heverlee) Belgium, Technical Report.

Brewer, J.W., J.W. Bunce and F.C. Van Vleck (1986). *Linear Systems over Commutative Rings*, Dekker, New York and Basel.

Brockett, R. (1970). *Finite Dimensional Linear Systems*, Wiley, New York.

Bunse-Gerstner, A., R. Byers, V. Mehrmann and N.K. Nichols (1991). *Numerical computation of an analytic singular value decomposition of a matrix valued function*, Numer. Math. **60**, 1–39.

Burke, J.V. and M.L. Overton (1991). *Differential properties of eigenvalues*, NYU Computer Science Department Technical Report 579.

Clancey, K. and I. Gohberg (1981). *Factorization of Matrix Functions and Singular Integral Operators*, Operator Theory vol. 3, Birkhäuser, Basel.

Cohen, N. (1983). *On minimal factorizations of rational matrix functions*, Integral Equations and Operator Theory **6**, 647–671.

Coppel, W.A. (1972). *Linear Systems*, Notes on Pure Mathematics vol. 6, The Australian National University, Canberra.

Daleckii, Ju.L. (1965). *Differentiation of non-hermitian matrix functions depending on a parameter*, Amer. Math. Soc. Transl. vol. 47, Providence, RI, 73–87 (Russian original: Izv. Vysš. Učebn. Zaved. Mat. (1962), 52–64).

Daleckii, Ju.L. and S.G. Krein (1956). *Integration and differentiation of functions of hermitian operators and applications to the theory of perturbations*, Amer. Math. Soc. Transl. vol. 47, Providence, RI (1965), 1–30 (Russian original: Voronež. Gos. Univ. Trudy Sem. Functional. Anal. **1**, 81–105).

Davis, C. (1963). *Notions generalizing convexity for functions defined on spaces of matrices*, Convexity, V. Klee, ed., Proc. of Symposia in Pure Math. VII, Amer. Math. Soc., Providence, RI, 187–203.

Delsarte, Ph., Y. Genin and Y. Kamp (1978). *Orthogonal polynomial matrices on the unit circle*, IEEE Trans. Circuits Systems **25**, 145–160.

Den Boer, H. (1981). *Block diagonalization of matrix functions*, Ph.D. Thesis, Vrije Universiteit, Amsterdam.

Doležal, V. (1964). *The existence of a continuous basis of a certain linear subspace of $E_r$ which depends on a parameter*, Časopis Pěst. Mat. **89**, 466–469.

Donoghue, W.F., Jr. (1974). *Monotone Matrix Functions and Analytic Continuation*, Springer, Berlin.

Efimov, A. and V. Potapov (1973). *J-expanding matrix functions and their role in the analytical theory of electric circuits*, Russian Math. Surv. **28**, 69–140.

Elsner, L., D. Hershkowitz and A. Pincus (1990). *Functional inequalities for spectral radii of non-negative matrices*, Linear Algebra Appl. **129**, 103–130.

Evard, J.C. (1990). *On the existence of bases of class $C^p$ of the kernel and the image of a matrix function*, Linear Algebra Appl. **135**, 33–67.

Evard, J.C. and J.M. Gracia (1990). *On similarities of class $C^p$ and applications to matrix differential equations*, Linear Algebra Appl. **137/138**, 363–386.

Evard, J.C. and F. Uhlig (1992). *On the matrix equation $f(X) = A$*, Linear Algebra Appl. **162–164**, 447–519.

FitzGerald, C.H. and R.A. Horn (1977). *On fractional Hadamard powers of positive definite matrices*, J. Math. Anal. Appl. **61**, 633–642.

Frazer, R.A., W.J. Duncan and A.R. Collar (1955). *Elementary Matrices*, Cambridge Univ. Press, Cambridge.

Friedland, S. (1980a). *Analytic similarity of matrices*, Lectures in App. Math. **18**, 43–85.

Friedland, S. (1980b). *On pointwise and analytic similarity of matrices*, Israel J. Math. **35**, 89–108.

Fuhrmann, P.A. (1983). *On symmetric rational transfer functions*, Linear Algebra Appl. **50**, 167–250.

Fuhrmann, P.A. (1987). *Orthogonal matrix polynomials and system theory*, Rend. Sem. Math. Univ. Politec. Torino, Fasc. Spec. Control Theory, 68–124.

Fuhrmann, P.A., H. Kimura and J.C. Willems, eds (1989). *Linear Systems and Control* (*Special Issue*), Linear Algebra Appl. **122/123/124**.

Gantmacher, F.R. (1959). *The Theory of Matrices*, 2 volumes, Chelsea, New York.

Genin, Y., P. Van Dooren, T. Kailath, J.M. Delosme and M. Morf (1983). *On $\Sigma$-lossless transfer functions and related questions*, Linear Algebra Appl. **50**, 251–275.

Gingold, H. (1979). *On continuous triangularization of matrix functions*, SIAM J. Math. Anal. **10**, 709–720.

Gingold, H. and P.F. Hsieh (1992). *Globally analytic triangularization of a matrix function*, Linear Algebra Appl. **169**, 75–101.

Gochberg, I.Z. and J. Leiterer (1976). *Über Algebren stetiger Operatorfunctionen*, Studia Mat. **57**, 1–26.

Gohberg, I., ed. (1988a). *Topics on Interpolation Theory of Rational Matrix Valued Functions*, Operator Theory vol. 33, Birkhäuser, Basel.

Gohberg, I., ed. (1988b). *Orthogonal Matrix Valued Polynomials and Applications*, Operator Theory vol. 34, Birkhäuser, Basel.

Gohberg, I., ed. (1990). *Extension and Interpolation of Linear Operators and Matrix Functions*, Operator Theory vol. 47, Birkhäuser, Basel.

Gohberg, I. and G. Heinig (1975). *The resultant matrix and its generalizations, I. The resultant operator for matrix polynomials*, Acta Sci. Math. **37**, 41–61 (in Russian).

Gohberg, I. and M.A. Kaashoek, eds (1986). *Constructive Methods of Wiener–Hopf Factorization*, Operator Theory vol. 21, Birkhäuser, Basel.

Gohberg, I., M.A. Kaashoek, L. Lerer and L. Rodman (1981). *Common multiples and common divisors of matrix polynomials, I. Spectral method*, Indiana Univ. J. Math. **30**, 321–356.

Gohberg, I., M.A. Kaashoek, L. Lerer and L. Rodman (1982a). *Common multiples and common divisors of matrix polynomials, II. The Vandermonde and the resultant matrix*, Linear and Multilinear Algebra **12**, 159–203.

Gohberg, I., M.A. Kaashoek, L. Lerer and L. Rodman (1984). *Minimal divisors of rational matrix functions with prescribed zero and pole structure*, Operator Theory: Advances and Applications vol. 12, H. Dym and I. Gohberg, eds, Birkhäuser, Basel, 241–275.

Gohberg, I., M.A. Kaashoek and L. Rodman (1978a). *Spectral analysis of families of operator polynomials and a generalized Vandermonde matrix, I. The finite dimensional case*, Topics in Functional Analysis, I. Gohberg and M. Kac, eds, Academic Press, New York, 91–128.

Gohberg, I., M.A. Kaashoek and F. van Schagen (1978b). *Common multiples of operator polynomials with analytic coefficients*, Manuscripta Math. **25**, 279–314.

Gohberg, I., P. Lancaster and L. Rodman (1978c). *Spectral analysis of matrix polynomials I. Canonical forms and divisors*, Linear Algebra Appl. **20**, 1–44.

Gohberg, I., P. Lancaster and L. Rodman (1978d). *Spectral analysis of matrix polynomials, II. The resolvent form and spectral divisors*, Linear Algebra Appl. **21**, 65–88.

Gohberg, I., P. Lancaster and L. Rodman (1979). *Perturbation theory for divisors of operator polynomials*, SIAM J. Math. Anal. **10**, 1161–1183.

Gohberg, I., P. Lancaster and L. Rodman (1980). *Spectral analysis of selfadjoint matrix polynomials*, Ann. Math. **112**, 33–71.

Gohberg, I., P. Lancaster and L. Rodman (1982b). *Matrix Polynomials*, Academic Press, New York, etc.

Gohberg, I., P. Lancaster and L. Rodman (1982c). *Perturbations of H-selfadjoint matrices, with applications to differential equations*, Integral Equations Operator Theory **5**, 718–757.

Gohberg, I., P. Lancaster and L. Rodman (1982d). *Factorization of selfadjoint matrix polynomials with constant signature*, Linear and Multilinear Algebra **11**, 209–226.

Gohberg, I., P. Lancaster and L. Rodman (1983). *Matrices and Indefinite Scalar Products*, Operator Theory vol. 8, Birkhäuser, Basel.

Gohberg, I., P. Lancaster and L. Rodman (1985). *Perturbations of analytic hermitian matrix functions*, Appl. Anal. **20**, 23–48.

Gohberg, I., P. Lancaster and L. Rodman (1986a). *Invariant Subspaces of Matrices with Applications*, Wiley, New York.

Gohberg, I., P. Lancaster and L. Rodman (1986b). *Quadratic matrix polynomials with a parameter*, Adv. Appl. Math. **7**, 253–281.

Gohberg, I. and L. Lerer (1976). *Resultants of matrix polynomials*, Bull. Amer. Math. Soc. **82**, 465–467.

Gohberg, I. and L. Lerer (1988). *Matrix generalizations of M.G. Krein theorems on orthogonal polynomials*, Operator Theory: Advances and Applications vol. 34, I. Gohberg, ed., Birkhäuser, Basel, 137–202.

Gohberg, I. and S. Rubinstein (1986). *Stability of minimal fractional decompositions of rational matrix functions*, Operator Theory: Advances and Applications vol. 18, I. Gohberg, ed., Birkhäuser, Basel, 249–270.

Gohberg, I. and T. Shalom (1990). *On Bezoutians of nonsquare matrix polynomials and inversion of matrices with nonsquare blocks*, Linear Algebra Appl. **137/138**, 249–323.

Guralnick, R.M. (1981). *Similarity of matrices over local rings*, Linear Algebra Appl. **41**, 161–174.

Guralnick, R.M. (1991). *Similarity of matrices over commutative rings*, Linear Algebra Appl. **157**, 55–68.

Heinig, G. (1979). *Bezoutiante, Resultante und Spektralverteilungsprobleme für Operatorpolynome*, Math. Nachr. **91**, 23–43.

Helmke, U. and P.A. Fuhrmann (1989). *Bezoutians*, Linear Algebra Appl. **122/123/124**, 1039–1097.

Helton, J.W. and J.A. Ball (1982). *The cascade decomposition of a given system is the linear fractional decompositions of its transfer function*, Integral Equations and Operator Theory **5**, 341–385.

Helton, J.W. and L. Rodman (1987). *Vandermonde and resultant matrices: an abstract approach*, Math. Systems Theory **20**, 169–192; *Correction*, **21**, (1988), 61.

Horn, R.A. (1967). *On infinitely divisible matrices, kernels and functions*, Z. Wahrscheinlichkeitsth. **8**, 219–230.

Horn, R.A. (1969). *The theory of infinitely divisible matrices and kernels*, Trans. Amer. Math. Soc. **136**, 269–286.

Horn, R.A. (1990). *The Hadamard product*, Matrix Theory and Applications, C.R. Johnson, ed., Proc. of Symposia in Applied Math. vol. 40, Amer. Math. Soc., Providence, RI, 87–169.

Horn, R.A. and C.R. Johnson (1985). *Matrix Analysis*, Cambridge Univ. Press, Cambridge.

Horn, R.A. and C.R. Johnson (1991). *Topics in Matrix Analysis*, Cambridge Univ. Press, Cambridge.

Jacobson, N. (1964). *Lectures in Abstract Algebra*, Van Nostrand, Princeton, NJ.

Johnson, C.R. and L. Rodman (1988). *Chordal inheritance principles and positive definite completions of partial matrices over function rings*, Operator Theory: Advances and Applications vol. 35, I. Gohberg, J.W. Helton and L. Rodman, eds, Birkhäuser, Basel, 107–127.

Kagström, B. and A. Ruhe, eds (1983). *Matrix Pencils*, SLNM 973, Springer, Berlin.

Kailath, T. (1980). *Linear Systems*, Prentice-Hall, Englewood Cliffs.

Kailath, T. and A.H. Sayed (1996). *Displacement structure: theory and applications*, SIAM Review, to appear.

Kalman, R.E., P.L. Falb and M.A. Arbib (1969). *Topics in Mathematical System Theory*, McGraw-Hill, New York.

Kaplansky, I. (1949). *Elementary divisors and modules*, Trans. Amer. Math. Soc. **66**, 464–491.

Kashchenko, S.A. (1988). *On miniversal deformations of matrices*, Uspekhi Mat. Nauk **43**, 201–202 (in Russian).

Kato, T. (1966). *Perturbation Theory for Linear Operators*, Springer, Berlin.

Kato, T. (1982). *Introduction to Perturbation Theory for Linear Operators*, Springer, Berlin.

Kazimirskii, P.S. (1981). *Factorization of Matrix Polynomials*, Kiev, Naukova Dumka (in Ukrainian).

Khazanov, V.B. and V.N. Kublanovskaya (1988). *Spectral problems for matrix pencils. Methods and algorithms, I, II*, Soviet J. Numer. Anal. Math. Modelling **3**, 337–371; 467–485.

Krauss, F. (1936). *Über konvexe Matrixfunktionen*, Math. Z. **41**, 18–42.

Krein, M.G. and H. Langer (1978). *On some mathematical principles in the linear theory of damped oscillations of continua, I, II*, Integral Equations Operator Theory **1**, 364–399; 539–566 (English translation; Russian original: 1965).

Krein, M.G. and M.A. Naimark (1981). *The method of symmetric and hermitian forms in the theory of the separation of the roots of algebraic equation*, Linear and Multilinear Algebra **10**, 265–308 (English translation; Russian original: 1936).

Krein, S.G. and A.I. Tovbis (1990). *Linear singular differential equations in finite dimensional and Banach spaces*, Leningrad Math. J. **2**, 931–985.

Krupnik, I. (1991). *Decomposition of a matrix polynomial into a product of linear pencils*, Math. Notes.

Lancaster, P. (1966). *Lambda Matrices and Vibrating Systems*, Pergamon Press, Oxford.

Lancaster, P. (1982). *A review of some recent results concerning factorization of matrix and operator valued functions*, Nonlinear Analysis and Applications, S.P. Singh and J.H. Burry, eds, Dekker, New York and Basel, 117–139.

Lancaster, P. and J. Maroulas (1987). *Inverse eigenvalue problems for damped vibrating systems*, J. Math. Anal. Appl. **123**, 238–261.

Lancaster, P. and J. Maroulas (1988). *Selective perturbations of spectral properties of vibrating systems using feedback*, Linear Algebra Appl. **98**, 309–330.

Lancaster, P. and M. Tismenetsky (1985). *The Theory of Matrices with Applications*, 2nd ed., Academic Press, New York.

Lander, F. (1974). *Bezoutian and inversion of Hankel and Toeplitz matrices*, Mat. Issled. **9**, 69–87 (in Russian).

Langer, H. (1976). *Factorization of operator pencils*, Acta Sci. Math. (Szeged) **38**, 83–96.

Langer, H. and B. Najman. (1989). *Remarks on the perturbation of analytic matrix functions, II*, Integral Equations Operator Theory **12**, 392–407.

Lerer, L. (1989). *The matrix quadratic equation and factorization of matrix polynomials*, Operator Theory: Advances and Applications vol. 40, H. Dym, S. Goldberg, M.A. Kaashoek and P. Lancaster, eds, Birkhäuser, Basel, 279–325.

Lerer, L. and L. Rodman (1996). *Bezoutians for rational matrix functions*, to appear.

Lerer, L., L. Rodman and M. Tismenetsky (1991). *Inertia theorems for matrix polynomials*, Linear and Multilinear Algebra **30**, 157–182.

Lerer, L. and M. Tismenetsky (1982). *The eigenvalue separation problem for matrix polynomials*, Integral Equations Operator Theory **5**, 386–445.

Lerer, L. and M. Tismenetsky (1984). *Bezoutian for several matrix polynomials and matrix equations*, Technical Report 88.145, IBM–Israel Scientific Center, Haifa, November.

Lerer, L. and M. Tismenetsky (1986). *Generalized Bezoutians and the inversion problem for block matrices, I. General scheme*, Integral Equations Operator Theory **9**, 790–819.

Lerer, L. and M. Tismenetsky (1988). *Generalized Bezoutian and matrix equations*, Linear Algebra Appl. **99**, 123–160.

Litvinchuk, G.S. and I.M. Spitkovskii (1987). *Factorization of Measurable Matrix Functions*, Operator Theory vol. 25, Birkhäuser, Basel.

Löwner, K. (1934). *Über monotone Matrix functionen*, Math. Z. **38**, 177–216.

MacDuffee, C.C. (1946). *The Theory of Matrices*, Chelsea, New York.

Markus, A.S. and I. Mereutsa (1973). *On the complete $n$-tuple of roots of the operator equation corresponding to a polynomial operator bundle*, Izv. Akad. Nauk SSSR **37**, 1105–1128.

Maroulas, J. (1985). *Factorization of matrix polynomials with multiple roots*, Linear Algebra Appl. **69**, 9–32.

Marshall, A.W. and I. Olkin (1979). *Inequalities: Theory of Majorization and Its Applications*, Academic Press, New York.

Micchelli, C.A. (1986). *Interpolation of scattered data: distance matrices and conditionally positive definite functions*, Constr. Approx. **2**, 11–22.

Najman, B. (1986). *Remarks on the perturbation of analytic matrix functions*, Integral Equations Operator Theory **9**, 593–599.

Newman, M. (1972). *Integral Matrices*, Academic Press, New York and London.

Overton, M.L. (1992). *Large-scale optimization of eigenvalues*, SIAM J. Control Optim., to appear.

Parthasarathy, K.R. and K. Schmidt (1972). *Positive Definite Kernels, Continuous Tensor Products and Central Limit Theorems of Probability Theory*, Springer, Berlin.

Pease, M.C., III (1965). *Methods of Matrix Algebra*, Academic Press, New York and London.

Porsching, T.A. (1968). *Analytic eigenvalues and eigenvectors*, Duke Math. J. **35**, 363–367.

Ran, A.C.M. (1982). *Minimal factorization of selfadjoint rational matrix functions*, Integral Equations Operator Theory **5**, 850–869.

Rellich, F. (1937). *Störungstheorie der Spektralzerlegung*, Math. Ann. **113**, 600–619.

Rellich, F. (1953). *Perturbation Theory of Eigenvalue Problems*, Lecture Notes, New York University.

Rodman, L. (1987). *Hermitian matrix polynomials*, Current Trends in Matrix Theory, F. Uhlig and R. Grone, eds, Elsevier, Amsterdam, 305–329.

Rogers, G.S. (1980). *Matrix Derivatives*, Dekker.

Rosenbrock, H.H. (1970). *State Space and Multivariate Theory*, Nelson, London.

Schur, I. (1911). *Bemerkungen zür Theorie der beschränkten Bilinearformen*, J. Reine Angew. Math. **140**, 1–29.

Silverman, L.M. and R.S. Bucy (1970). *Generalizations of a theorem of Dolezal*, Math. Systems Theory **4**, 334–339.

Stewart, G.W. and J.-g. Sun (1990). *Matrix Perturbation Theory*, Academic Press, Boston etc.

Sun, J.-g. (1990). *Multiple eigenvalue sensitivity analysis*, Linear Algebra Appl. **137/138**, 183–211.

Thrall, R.M. and L. Tornheim (1957). *Vector Spaces and Matrices*, Wiley, New York.

Tovbis, A.I. (1992). *Normal forms of holomorphic matrix-valued functions and corresponding forms for singular differential operators*, Linear Algebra Appl. **162–164**, 389–407.

Uspensky, J.V. (1978). *Theory of Equations*, McGraw-Hill, New York.

Van Dooren, P. (1984). *Factorization of a rational matrix: the singular case*, Integral Equations Operator Theory **7**, 704–741.

Van Dooren, P. and P. Dewilde (1983). *The eigenstructure of an arbitrary polynomial matrix: computational aspects*, Linear Algebra Appl. **50**, 545–579.

Van Dooren, P.M. and P. Dewilde (1981). *Minimal cascade factorization of real and complex rational transfer matrices*, IEEE Trans. Circuits Systems, **CAS-28**, 390–400.

Vandewalle, J. and P. Dewilde (1978). *A local I/O structure theory for multivariable systems and its application to minimal cascade realization*, IEEE Trans. Circuits Systems, **CAS-25**, 279–289.

Wasow, W. (1962). *On holomorphically similar matrices*, J. Math. Anal. Appl. **4**, 202–206.

Wedderburn, J.H.M. (1964). *Lectures on Matrices*, Dover (first published: Colloquium Publications vol. SVII, Amer. Math. Soc., 1934).

Weiss, L. and P.L. Falb (1969). *Dolezal's theorem, linear algebra with continuously parametrized elements, and time varying systems*, Math. Systems Theory **3**, 67–75.

Whittaker, E.T. (1952). *Analytical Dynamics*, Cambridge Univ. Press, Cambridge.

Wimmer, H.K. (1988). *A factorization of the generalized Bezoutian of polynomial matrices*, Linear and Multilinear Algebra **23**, 255–261.

Wimmer, H.K. (1989). *Bezoutians of polynomial matrices and their generalized inverses*, Linear Algebra Appl. **122/123/124**, 475–481.

Wyman, B.F., M.K. Sain, G. Conte and A.M. Perdon (1991). *Poles and zeros of matrices of rational functions*, Linear Algebra Appl. **157**, 113–139.

# Section 1B
# Linear (In)dependence

# Matroids

Joseph P.S. Kung

*Department of Mathematics, University of North Texas, Denton, TX 76203, USA*

*Contents*



HANDBOOK OF ALGEBRA, VOL. 1
Edited by M. Hazewinkel

## 1. Dependence

Just as group axioms formalize the intuitive notion of symmetry, matroid axioms formalize the notion of dependence. Steinitz in 1910 wrote down the defining properties of a matroid in a recognizable form. His paper [280] was edited and reprinted as a book in 1930 and through R. Baer (see [6]), it probably exerted a strong influence. Nakasawa [235] and Whitney [328] postulated axiom systems for matroids in 1935. While Nakasawa's paper fell into obscurity, Whitney's paper founded a new subject in mathematics.

In this chapter, we will survey matroid theory with an algebraist's eye. We begin with the basic axiom systems in Section 2. In Section 3, we take a geometric approach and introduce exchange closures and geometric lattices. Two fundamental constructions – minor and orthogonal duality – are the topics of Section 4. Sections 5 and 6 are concerned with examples. Many important families of matroids are minor-closed. These families are the subject of Section 6. Matroid theory is related to classical invariant theory through basis exchange properties: this connection is explained in Section 7. Another connection, described in Section 8, is with synthetic geometry and "geometric algebra". In Section 9, we describe three commonly used categories, weak maps or specializations, strong maps, and comaps. Finally, in Section 10, we survey enumerative results; these results form a major area of algebraic combinatorics. Applications of matroids to combinatorics, graph theory, and optimization will be mentioned briefly. Some books about matroids are [37, 53, 77, 186, 248, 299, 311, 321–323]. Some general survey articles are [52, 66, 76, 118, 253, 297, 340].

## 2. Cryptomorphisms

One of the distinctive features of matroids is that they have many equivalent axiomatizations, or, in Birkhoff's terminology [15], p. 154, *Cryptomorphisms*. In [328], Whitney gave axiom systems for matroids in terms of rank, independent sets, bases, and circuits.

**2.1.** *Rank.* A matroid $M$ on the set $S$ is specified by a rank function $r$ from $2^S$, the collection of all subsets of $S$, to the non-negative integers $\mathbb{N}$ satisfying the following axioms:

($R_1$) $r(\varnothing) = 0$.

($R_2$) If $a \in S$ and $A \subseteq S$, then $r(A) \leqslant r(A \cup \{a\}) \leqslant r(A) + 1$.

($R_3$) Submodularity. If $A, B \subseteq S$, then $r(A \cup B) + r(A \cap B) \leqslant r(A) + r(B)$.

The *rank* $r(M)$ of a matroid $M$on $S$ is the rank $r(S)$ of its set of elements. By definition, $r(M)$ is finite.

**2.2.** *Independent sets.* A *matroid* $M$ on the set $S$ is specified by a collection $\mathcal{I}$ of finite subsets of $S$ called *independent sets* satisfying the following axioms:

($I_1$) $\varnothing \in \mathcal{I}$.

($I_2$) If $J \subseteq I$ and $I \in \mathcal{I}$, then $J \in \mathcal{I}$.

($I_3$) *Independent set augmentation.* If $I$ and $J$ are independent and $|I| < |J|$, then there exists an element $a$ in $J$ but not in $I$ such that $I \cup \{a\}$ is independent.

**2.3.** *Bases.* A *matroid* $M$ on the set $S$ is specified by a collection $\mathcal{B}$ of finite subsets of $S$ called bases satisfying the following axioms:

($B_1$) If $B_1$ and $B_2$ are bases, then $B_1 \not\subset B_2$.

($B_2$) *Basis replacement.* If $B_1$ and $B_2$ are bases and $a$ is any element in $B_1$, then there exists an element $a'$ in $B_2$ such that $(B_1 \backslash \{a\}) \cup \{a'\}$ is a basis.

**2.4.** *Circuits.* A *matroid* $M$ on the set $S$ is specified by a collection $\mathcal{C}$ of finite nonempty subsets of $S$ called *circuits* satisfying the following axioms:

($C_1$) If $C_1$ and $C_2$ are circuits, then $C_1 \not\subset C_2$.

($C_2$) *Circuit elimination.* If $C_1$ and $C_2$ are circuits and $a \in C_1 \cap C_2$, then there exists a circuit $C_3$ contained in $(C_1 \cup C_2) \backslash \{a\}$.

By rephrasing proofs from linear algebra, it is not hard to prove that these axiom systems are equivalent. Start with the independent set axioms as the standard axiomatization and use the following translations: $r(A) =$ size of a maximal independent set contained in $A$, bases are maximal independent sets, and circuits are minimal dependent sets.

One of Whitney's examples in [328] is the *linear matroid* $M_T$ on the set of columns of a matrix $T$ with independent sets the linearly independent sets. Thus, he considered matroids to be combinatorial variants of matrices and therefore named them "matroids". Despite attempts to rename matroids – "independence structures" and "combinatorial pregeometries" have been suggested – they will perhaps always be called "matroids".

The other example in [328] arises from graphs. Let $\Gamma$ be a graph on a finite vertex set $V$ and edge set $S$. The *cycle or polygon matroid* $M(\Gamma)$ of $\Gamma$ is the matroid on $S$ with circuits the cycles of $\Gamma$. Cycle matroids suggest the following terminology. A *loop* is an element $a$ such that $r(\{a\}) = 0$, or, equivalently, $\{a\}$ is a circuit. In $M_T$, the loops are the columns all of whose entries are zero. Two elements $a$ and $b$ which are not loops are said to be *parallel* if $r(\{a, b\}) = 1$. In $M(\Gamma)$, two edges are parallel whenever they have the same endpoints. In $M_T$, two nonzero columns are parallel if and only if they are scalar multiples of each other. A (*combinatorial*) *geometry* or *simple matroid* is a matroid containing no loops or parallel elements.

Another natural way to axiomatize matroids is use the exchange and transitivity properties of dependence relations [139, 280, 235, 305].

**2.5.** *Dependence relation.* A *matroid* on the set $S$ is specified by a relation in $S \times 2^S$, *$a$ is dependent* on $A$, satisfying the following axioms:

($D_1$) If $a \in A$, then $a$ is dependent on $A$.

($D_2$) *Exchange.* If $a$ is dependent on $A \cup \{b\}$ but $a$ is not dependent on $A$, then $b$ is dependent on $A \cup \{a\}$.

($D_3$) *Transitivity.* If $a$ is dependent on $A$ and every element in $A$ is dependent on $B$, then $a$ is dependent on $B$.

($D_4$) *Finite rank.* If $a$ is dependent on $A$, then there exists a finite subset $A_0 \subseteq A$ such that $a$ is dependent on $A_0$.

A radically different axiomatization for finite matroids was discovered by Edmonds [103]. Let $S$ be a finite set, $\mathcal{I}$ a collection of subsets of $S$ containing $\varnothing$, and $w$: $S \to \mathbb{R}^+$ a positive real-valued "weight" function on $S$. If $E \subseteq S$, the weight $w(E)$ is defined to be the sum $\sum_{a \in E} w(a)$ of the weights of its elements. The *greedy algorithm* attempts to find a subset $I$ in $\mathcal{I}$ of maximum weight in the following way: Start with $I = \varnothing$. Suppose that $I$ has been chosen. Among all the elements in $S \backslash I$, choose an element $a$ such that $I \cup \{a\} \in \mathcal{I}$ and $w(a)$ is maximum. Replace $I$ by $I \cup \{a\}$. Continue until $I$ is a maximal subset in $\mathcal{I}$.

**2.6.** *Greedy algorithm.* A *matroid* $M$ on the finite set $S$ is specified by a collection $\mathcal{I}$ of subsets of $S$ called *independent sets* satisfying ($I_1$), ($I_2$), and

(Gr) The greedy algorithm outputs a subset in $\mathcal{I}$ of maximum weight for *every* weight function $w$: $S \to \mathbb{R}^+$.

This axiomatization is one of the reasons why matroids and, more generally, submodular functions are important in combinatorial optimization. Key papers in this area are [102, 103]; [313, 106] are useful surveys. Greedoids are matroid-like structures motivated originally by how certain maximum-weight sets are constructed using the greedy algorithm [170, 171]. A *greedoid* on the set $S$ is specified by a collection $\mathcal{I}$ of subsets of $S$ satisfying ($I_1$), ($I_3$) and the following weaker version of ($I_2$): If $J$ is a nonempty subset in $\mathcal{I}$, then there exists an element $e \in J$ such that $J \backslash \{e\} \in \mathcal{I}$. Many dependence structures in algebra which are not matroids are greedoids. Much work has been done on greedoids; see [24] and [172] for an introduction to the subject.

Other axiomatizations of dependence have been studied. A very small sample can be found in [51, 62, 99, 100, 123, 243]. Model- and recursion-theoretic aspects of dependence structures not having the finite rank property are discussed in the survey [7].

## 3. Geometric lattices and exchange closures

A geometric way of defining matroids is to abstract the properties of taking linear span [217]. A *closure* (*operator*) on a partially ordered set $P$ is a function $x \mapsto \bar{x}$ defined from $P$ to itself satisfying: $x \leqslant \bar{x}$, $\bar{x} = \bar{\bar{x}}$, and $x \leqslant y \Rightarrow \bar{x} \leqslant \bar{y}$.

**3.1.** *Exchange closure.* A *matroid* $M$ on the set $S$ is specified by a closure $A \mapsto \bar{A}$ on $2^S$ (partially ordered by containment) satisfying the following axioms:

($CL_1$) *MacLane–Steinitz exchange property.* Let $a$ and $b$ be elements not in $\bar{A}$. Then $a \in \overline{A \cup \{b\}}$ implies that $b \in \overline{A \cup \{a\}}$.

($CL_2$) *Finite rank.* For every subset $A \subseteq S$, there exists a finite subset $A_0$ such that $\bar{A} = \bar{A}_0$.

A subset $X \subseteq S$ is a *flat* or *closed set* if $\overline{X} = X$. The flats of a matroid $M$ form a lattice $L(M)$ called the *lattice of flats* of $M$ under the partial order of set containment. The meet and joint in $L(M)$ are given by: $X \vee Y = \overline{X \cup Y}$ and $X \wedge Y = X \cap Y$. The minimum $\widehat{0}$ of $L(M)$ is $\bar{\varnothing}$ and the maximum $\widehat{1}$ is $S$. A flat $Y$ *covers* a flat $X$ if $Y > X$ and there is no flat $Z$ such that $Y > Z > X$; equivalently, $Y$ covers $X$ if $Y = \overline{X \cup \{a\}}$ for some element $a$.

A *point* or *atom* is a flat covering $\widehat{0}$. A *copoint* is a flat covered by the maximum flat $S$. A *bond* or *cocircuit* is the set-theoretic complement of a copoint. A subset is *spanning* if its closure is the entire set $S$. A matroid can be completely described in many ways: by its rank function, its independent sets, its circuits, its copoints, its bonds, its spanning sets, etc. The algorithmic complexity of converting from one description to another is studied in [141].

Lattices of flats satisfy two important properties:

($L_1$) *Semimodularity*. If $X$ and $Y$ cover $X \wedge Y$, then $X \vee Y$ covers $X$ and $Y$.

($L_2$) *Atomicity*. Every flat is a join of points.

A chain $X_0 < X_1 < X_2 < \cdots < X_r$ of flats is *saturated* if for every $i$, $X_{i+1}$ covers $X_i$. It follows from semimodularity that if $Y$ and $X$ are flats, then every saturated chain $Y = X_0 < X_1 < X_2 < \cdots < X_l = X$ from $Y$ to $X$ has the same length $l$. The rank $r(X)$ of a flat $X$ in the matroid $M$ equals the length of a saturated chain from $\widehat{0}$ to $X$. Because $r(S)$ is finite, $L(M)$ satisfies the additional property:

($L_3$) *Finite rank*. Every saturated chain from $\widehat{0}$ to $\widehat{1}$ is finite.

A lattice satisfying ($L_1$), ($L_2$), and ($L_3$) is said to be *geometric*.

THEOREM 3.1. *The lattice $L(M)$ of flats of a matroid $M$ is a geometric lattice. Conversely, a geometric lattice $L$ defines a geometry $G$ on the set $S$ of points in $L$ such that $L$ and $L(G)$ are isomorphic lattices.*

The geometry $G$ is defined by the closure relation: for a set $A$ of points,

$$\bar{A} = \left\{ b\colon\ b \text{ is a point and } b \leqslant \bigvee_{a \in A} a \right\}.$$

If $M$ is a matroid, then the geometry $G$ defined on the points of $L(M)$ is called the *simplification* of $M$; $G$ is the unique geometry (up to isomorphism) such that $L(M) \cong L(G)$; $G$ can be obtained from $M$ by removing all the loops and all but one element from each class of parallel elements.

The lattice-theoretic approach to matroids was initiated by Birkhoff [13, 15]; [80–82, 169, 336, 338] are some early papers using this approach. Topological geometric lattices are studied in [337, 132, 133]. See also [140]. Work has also been done on nonatomic lattices satisfying semimodularity or analogous properties. See [105, 122, 257, 281, 304].

## 4. Minors, direct sums, and orthogonal duals

Let $M$ be a matroid on the set $S$ with rank function $r_M$. If $T \subseteq S$, the *restriction* $M|T$ of $M$ to $T$ is the matroid on $T$ with rank function: $r_{M|T}(A) = r_M(A)$ for $A \subseteq T$. Three other ways of describing this situation are: (a) $M|T$ is a *submatroid* of $M$, (b) $M$ is an *extension* of $M|T$ by the elements in $S \backslash T$, and (c) $M|T$ equals the *deletion* $M \backslash (S \backslash T)$.

Now let $U \subseteq S$. The *contraction* $M/U$ of $M$ by $U$ is the matroid on $S \backslash U$ with rank function:

$$r_{M/U}(A) = r_M(A \cup U) - r_M(U) \quad \text{for } A \subseteq S \backslash U.$$

Its lattice $L(M/U)$ of flats is isomorphic to the upper interval

$$[\overline{U}, \widehat{1}] = \{Z \in L(M)\colon\ Z \geqslant \overline{U}\}.$$

If $e$ is an edge with two distinct endpoints in the graph $\Gamma$, then $M(\Gamma)/\{e\}$ is isomorphic to $M(\Gamma/\{e\})$, where $\Gamma/\{e\}$ is the graph obtained from $\Gamma$ by deleting the edge $e$ and identifying the endpoints of $e$. If $a$ is a nonzero column of a matrix $T$, then $M_T/\{a\}$ is the linear matroid of the matrix obtained as follows: (a) Reduce $T$ by row operations so that all but one of the entries in $a$, say the entry at row $u$, are zero, and, (b) Delete row $u$ and column $a$.

Contraction and deletion commute, in the sense that if $T$ and $U$ are disjoint subsets of $S$, then $(M/U)\backslash T$ and $(M\backslash T)/U$ are the same matroid on $S\backslash(T \cup U)$. A *minor* of $M$ is a matroid obtainable from $M$ by a sequence of contractions and deletions [295].

If $M$ and $N$ are matroids on disjoint sets $S$ and $T$ with rank functions $r_M$ and $r_N$, the *direct sum* $M \oplus N$ is the matroid on $S \cup T$ with rank function $r_{M\oplus N}$ given by: for $A \subseteq S \cup T$,

$$r_{M\oplus N}(A) = r_M(A \cap S) + r_N(A \cap T).$$

If $M = M_1 \oplus M_2$, then the lattice $L(M)$ is the direct product $L(M_1) \times L(M_2)$ [83]. A subset $A \subseteq S$ is a *separator* of a matroid $M$ on $S$ if $M = M|A \oplus M|(S\backslash A)$. An element $a$ is an *isthmus* if $r(\{a\}) = 1$ and $\{a\}$ is a separator. A matroid $M$ on $S$ is *connected* if $M$ does not have any separators apart from $\varnothing$ and $S$, or, equivalently, given any two distinct elements $a$ and $b$ in $S$, there exists a circuit of $M$ containing $a$ and $b$. Motivated from graph theory, notions of $k$-connectivity have been defined. See [298].

Let $M$ be a matroid on a *finite* set $S$ with rank function $r$. Then, the formula

$$r^{\perp}(A) = |A| + r(S\backslash A) - r(S) \quad \text{for } A \subseteq S$$

defines the rank function $r^{\perp}$ of a matroid $M^{\perp}$ on $S$ called the (*orthogonal*) *dual* of $M$. The dual $M_T^{\perp}$ of the linear matroid $M_T$ of a matrix $T$ is the linear matroid of any matrix whose rows span the orthogonal complement of the row space of $T$. Duality can also be defined using other cryptomorphisms ([328]). Three such definitions are:

(1) $I$ is an independent set of $M \Leftrightarrow S\backslash I$ is a spanning set of $M^{\perp}$.

(2) $B$ is a basis of $M \Leftrightarrow S\backslash B$ is a basis of $M^{\perp}$.

(3) $C$ is a circuit of $M \Leftrightarrow C$ is a bond of $M^{\perp}$.

Minty [231] has given a self-dual axiomatization of matroids in terms of circuits and bonds. Duality is involutory, i.e. $(M^{\perp})^{\perp} = M$, and interchanges deletion and contraction, i.e.

$$(M/T)^{\perp} = (M^{\perp})\backslash T \quad \text{and} \quad (M\backslash T)^{\perp} = (M^{\perp})/T.$$

These two properties characterize duality [179, 27].

The *cocycle* or *bond matroid* $M^{\perp}(\Gamma)$ of $\Gamma$ is the dual of the cycle matroid $M(\Gamma)$. Whitney [325, 327] characterized planarity of graphs using duality.

THEOREM 4.1. *A finite graph $\Gamma$ can be drawn on the plane if and only if its bond matroid $M^{\perp}(\Gamma)$ is isomorphic to the cycle matroid of a graph $\Delta$.*

The graph $\Delta$ is the dual graph formed on the regions of a planar drawing of $\Gamma$.

A *clutter* $\mathcal{C}$ on the set $S$ is a collection of subsets of $S$ such that if $C_1$ and $C_2$ are in $\mathcal{C}$, then $C_1 \not\subset C_2$. The collection of circuits of a matroid is a clutter. Some papers on matroids as clutters are [68, 203, 250, 265, 266].

## 5. Some examples

**5.1.** *Projective geometries and modular lattices.* Matroid theory differs from projective geometry in that the intersection of two flats may not have the rank predicted by linear algebra. A pair $X$ and $Y$ of elements in a lattice is a *modular pair* – symbolically, $(X,Y)M$ – if for every element $Z \leqslant X$, $(X \wedge Y) \vee Z = X \wedge (Y \vee Z)$. Symmetry of the relation of being a modular pair [that is, $(X,Y)M$ implies $(Y,X)M$] is equivalent to semimodularity. See [338, 222]. In a geometric lattice, $(X,Y)M$ if and only if

$$r(X \vee Y) + r(X \wedge Y) = r(X) + r(Y).$$

A flat $X$ which forms a modular pair with every flat is said to be *modular.* A *modular* lattice is one in which every element is modular.

The lattice $L(n,\mathbb{F})$ of subspaces of the $n$-dimensional vector space $\mathbb{F}^n$ over a skew field $\mathbb{F}$ is a modular geometric lattice. The matroid defined on the points of $L(n,\mathbb{F})$ is the *projective geometry* $\mathrm{PG}(n-1,\mathbb{F})$. Birkhoff [14] showed that a *geometry has a modular lattice of flats if and only if it is a direct sum of projective geometries and points.* See also [58].

**5.2.** *Linear matroids.* A *representation of* a matroid $M$ on $S$ over the skew field $\mathbb{F}$ is a function $\rho$ defined from $S$ to an $\mathbb{F}$-vector space $V$ such that for all $I \subseteq S$, $I$ is independent in $M$ if and only if $|\rho(I)| = |I|$ and $\rho(I)$ is linearly independent. A matroid is said to be ($\mathbb{F}$-)*linear* if it has a representation (over $\mathbb{F}$).

**5.3.** *Algebraic matroids.* Let $\mathbb{K}$ be an extension field of the field $\mathbb{F}$. A set $\{x_1, x_2, \ldots, x_n\}$ of elements is *algebraically dependent* over $\mathbb{F}$ if there exists a nonzero polynomial with coefficients in $\mathbb{F}$ such that $p(x_1, x_2, \ldots, x_n) = 0$. If $\mathbb{K}$ has finite transcendence degree over $\mathbb{F}$, then any subset $S \subseteq \mathbb{K}$ has a matroid structure given by algebraic dependence. Such matroids are said to be *algebraic.* Much work has been done on finding conditions on a matroid to be algebraic [150, 208, 209, 211]. The *algebraic closure geometry* $G(\mathbb{K}/\mathbb{F})$ is the simplification of the algebraic matroid on $\mathbb{K}$. These matroids are analogues of projective geometries. MacLane [217–219] used them to find invariants of field extensions. More recent work can be found in [22, 93, 104, 146, 147, 207]. Other papers on algebraic matroids are [120, 121, 204, 206, 210–213, 308].

**5.4.** *Transversal matroids.* Let $R \subseteq S \times T$ be a relation between $S$ and $T = \{1,2,\ldots,m\}$. A *partial transversal* in $R$ is a subset $I \subseteq S$ for which there exists an injection $f$: $I \to T$ such that the ordered pairs $(a, f(a))$ are in $R$. The *transversal matroid* $T(R)$ *of the relation* $R$ is the matroid on $S$ with independent sets the partial transversals. The matroid $T(R)$ can be represented over every "sufficiently large" field

$\mathbb{F}$ [111, 233]. Let $\{X_{(a,i)}\colon (a,i) \in R\}$ be a set of elements algebraically independent over the prime field of $\mathbb{F}$. Then the function $S \to \mathbb{F}^m$, $a \mapsto (a_1, a_2, \ldots, a_m)$, where $a_i = X_{(a,i)}$ if $(a,i) \in R$ and 0 otherwise is a representation of $T(R)$ over $\mathbb{F}$. Matroids play a central rôle in transversal theory. For example, a unifying result in transversal theory is Rado's extension of the marriage theorem [254]:

*Let $M$ be a matroid on $S$ with rank function $r$ and let $R \subseteq S \times T$ be a relation. Then there exists an independent transversal $I$ of size $m$ if and only if for all subsets $J \subseteq \{1, 2, \ldots, m\}$,*

$$r\Big(\bigcup_{j \in J} R(j)\Big) \geqslant |J|.$$

*Here, $R(j)$ is the subset of elements in $S$ related to $j \in T$.*

See [35, 36, 214, 232, 233] for surveys.

**5.5.** *Arrangements of hyperplanes.* A *hyperplane* $H$ in $\mathbb{F}^n$ is a subspace of codimension 1; equivalently, $H$ is the kernel of a nonzero linear functional. An *arrangement of hyperplanes* $\mathcal{A}$ is a finite collection of hyperplanes. The (*intersection*) *lattice* $L(\mathcal{A})$ of $\mathcal{A}$ is the lattice formed by all intersections of hyperplanes in $\mathcal{A}$ under reverse set-inclusion. $L(\mathcal{A})$ is a geometric lattice. Its associated geometry is the linear geometry on the hyperplanes in $\mathcal{A}$ considered as linear functionals. Many topological invariants of arrangements (such as the singular cohomology ring of the complement of a complex arrangement [240]) depend only on the lattice of intersection. See [63, 239, 241, 343] for surveys.

**5.6.** *Simplicial matroids.* Simplicial matroids are generalizations of cycle matroids of graphs where the "edges" are $k$-element subsets of vertices. Dependence in simplicial matroids is defined using a boundary operator. Two key papers are [78, 70].

## 6. Minor-closed classes

A *minor-closed class* $\mathcal{C}$ is a collection of matroids satisfying the conditions:
(Min$_1$) If $M \in \mathcal{C}$ and the lattice $L(N)$ of flats is isomorphic to $L(M)$, then $N \in \mathcal{C}$.
(Min$_2$) If $N$ is a minor of $M$ and $M \in \mathcal{C}$, then $N \in \mathcal{C}$.

Examples of minor-closed classes are the class $\mathcal{L}(\mathbb{F})$ of $\mathbb{F}$-linear matroids and the class $\mathcal{G}$ of cycle matroids of graphs. The class $\mathcal{L}(\mathrm{GF}(q))$, where $\mathrm{GF}(q)$ is the finite field of order $q$, is usually denoted by $\mathcal{L}(q)$. A matroid is *regular* if it can be represented over any field. The class $\mathcal{R}$ of regular matroids is minor-closed and equals the intersection $\bigcap \mathcal{L}(\mathbb{F})$. Two papers on matroid classes are [273, 303].

**6.1.** *Forbidden minors.* Let $\{N_\alpha\}$ be a collection of matroids. Then the class $\mathcal{EX}(N_\alpha)$ of matroids not having any of the matroids $N_\alpha$ as minors is a minor-closed class. Conversely, taking $\{N_\alpha\}$ to be all the matroids not in $\mathcal{C}$, every minor-closed class is of this form. A matroid $N$ is a *forbidden minor* for $\mathcal{C}$ if $N$ is not in $\mathcal{C}$ but every proper minor of $N$ is in $\mathcal{C}$. The classical forbidden-minor theorem is Kuratowski's theorem for planar graphs [194]. In matroid terminology, this states: *A graph $\Gamma$ can be drawn on the plane if and only if its cycle matroid does not contain as minors the cycle matroids $M(K_5)$ and $M(K_{3,3})$.* (The complete graph $K_5$ is the graph on $\{v_1, v_2, v_3, v_4, v_5\}$ and all 10 edges between distinct vertices; the graph $K_{3,3}$ is the graph on $\{v_1, v_2, v_3, u_1, u_2, u_3\}$ with all 9 edges between $v_i$ and $u_j$.)

A major research area is to determine whether the set of forbidden minors for the classes $\mathcal{L}(q)$ is finite. The answer is known for $\mathcal{L}(2)$, the class of *binary matroids* [295], and $\mathcal{L}(3)$, the class of *ternary matroids* [17, 161, 165, 267, 288]. The *uniform matroid* $U_{r,s}$ is the rank-$r$ matroid consisting of $s$ points "in general position": more specifically, it is the matroid with $s$ elements in which all the $r$-element subsets are bases.

THEOREM 6.1. *$\mathcal{L}(2)$ has one forbidden minor, the 4-point line $U_{2,4}$.*

The *Fano plane* $F_7$ is the binary projective plane $\mathrm{PG}(2,2)$.

THEOREM 6.2. *$\mathcal{L}(3)$ has four forbidden minors: the 5-point line $U_{2,5}$, its dual $U_{3,5}$, the Fano plane $F_7$, and its dual $F_7^{\perp}$.*

The forbidden minors are also known for the classes $\mathcal{R}$ and $\mathcal{G}$. Using a homotopy theorem on a graph formed from the copoints, Tutte [295] found the forbidden minors for $\mathcal{R}$. See [16, 117, 267] for shorter proofs or extensions. Some papers on regular matroids or generalizations are [5, 142, 187, 188, 193, 201, 202, 335].

THEOREM 6.3. *$\mathcal{R}$ has three forbidden minors: $U_{2,4}, F_7$, and $F_7^{\perp}$.*

It follows from (6.3) that $\mathcal{R} = \mathcal{L}(2) \cap \mathcal{L}(3)$. Direct proofs can be found in [42, 267]. The next theorem [296] is an extension of Kuratowski's theorem. See also [268, 306].

THEOREM 6.4. *$\mathcal{G}$ has five forbidden minors: $U_{2,4}$, $F_7$, $F_7^{\perp}$, and the cocycle matroids of the Kuratowski graphs, $M^{\perp}(K_5)$ and $M^{\perp}(K_{3,3})$.*

The set of isomorphism classes of matroids is partially ordered by the relation of being a minor. Robertson and Seymour [259] have shown that graphic matroids under the minor-order is a *well-quasi-order*, that is, it is a partial order with no infinite strictly descending chains or antichains (i.e. sets of mutually incomparable elements). This result shows that every minor-closed class of graphic matroids has finitely many forbidden minors.

**6.2.** *Gain-graphic matroids.* Let $e_1, e_2, \ldots, e_n$ be a basis in the projective geometry $\mathrm{PG}(n-1, \mathbb{F})$. The rank-$n$ Dowling geometry $Q_n(\mathbb{F}^{\times})$ over the multiplicative group $\mathbb{F}^{\times}$ is the matroid consisting of the points $e_1, e_2, \ldots, e_n$, and $e_i - \alpha e_j$ in $\mathrm{PG}(n-1, \mathbb{F})$, for all pairs $i \neq j$ and all $\alpha \in \mathbb{F}^{\times}$. Because the points in $Q_n(\mathbb{F}^{\times})$ are linear combinations of at most two basis elements, the dependencies can be specified combinatorially without using the additive structure of $\mathbb{F}$. Thus, one can define a Dowling geometry $Q_n(A)$ for any group $A$ [86, 87]. See also [10, 29–31, 164, 193]. A matroid $M$ is said to be *gain-graphic (with gains in the group $A$)* if $M$ is a restriction of $Q_n(A)$ for some $n$. The class $\mathcal{Z}(A)$ of all gain-graphic matroids over a group $A$ is a minor-closed class. Some papers in this area are [343, 345, 346].

A variety $\mathcal{V}$ is a class closed under minors and direct sums satisfying: for every non-negative integer $n$, there exists a unique geometry $T_n$ such that the simplification of every rank-$n$ matroid in $\mathcal{V}$ is a submatroid of $T_n$. The geometries $T_n$ can be regarded as "ambient spaces" for $\mathcal{V}$.

CLASSIFICATION OF VARIETIES 6.5 ([163]). *There are five families of varieties of finite matroids: three degenerate varieties constructed from lines, $\mathcal{L}(q)$, and $\mathcal{Z}(A)$.*

Two papers on varieties are [184, 134].

**6.3.** *Regular matroids and decomposition theory.* Examples of regular matroids are cycle and bond matroids of graphs. Seymour [269] proved a decomposition theorem for regular matroids: *every regular matroid can be put together by taking* 1-, 2- *or* 3-*sums* (*roughly speaking, gluing two matroids together at an empty set, a point, or a line*) *of graphic matroids, cographic matroids, and copies of a* 10-*element rank*-5 *matroid* $R_{10}$. This result is the first of many decomposition theorems [245–247, 270, 272, 291, 292].

Seymour's theorem implies that cycle matroids of the complete graph $K_{n+1}$ are the rank-$n$ regular matroids having the maximum number $\binom{n+1}{2}$ of points. This was proved earlier in Heller's paper [142]. This paper initiated *extremal matroid theory* which is concerned with determining the maximum number $h(n)$ of points in a rank-$n$ matroid in a given class $\mathcal{C}$ of matroids. This area is surveyed in [189].

## 7. Basis exchange, matroid partitions, and determinantal identities

The best known application of the basis replacement axiom is to prove the following elementary result.

THEOREM 7.1. *Bases of a matroid have the same size.*

To prove that two bases $B$ and $B'$ have the same size, one starts with $B$, and, using the basis replacement axiom, constructs a sequence of bases $B = B_1, B_2, \dots, B_k = B'$ such that $B_i$ and $B_{i+1}$ differ in one element. The *basis graph* of a matroid $M$ is the graph with vertex set the set $\mathcal{B}$ of bases of $M$ with two bases $B$ and $B'$ joined by an edge if and only if $B$ and $B'$ differ in exactly one element. See [224] for results (including a homotopy theorem for paths) on basis graphs.

A deep result which can be proved by using basis exchanges is the following theorem due to Edmonds [148, 101].

MATROID PARTITION THEOREM 7.2. *Let $M_1, M_2, \dots, M_m$ be matroids with rank function $r_i$ on the finite set $S$. Then there exists a partition $S_1 \cup \cdots \cup S_m = S$ such that $S_i$ is independent in $M_i$ if and only if for every subset $A \subseteq S$,*

$$\sum_{i=1}^{m} r_i(A) \geqslant |A|.$$

The *basis monomial ring* of a matroid $M$ on the set $S$ is the subring of the ring $k[X_a]$ of polynomials in $|S|$ indeterminates $X_a, a \in S$, over a field $k$ generated by the monomials $\{\prod_{a \in B} X_a$: $B$ is a basis of $M\}$. White [319] has used Theorem 7.2 to show that basis monomial rings are Cohen–Macauley. Applying Theorem 7.2 to the matroids $M$ and $N^{\perp}$, one obtains the matroid intersection theorem [102]:

*Let M and N be matroids with rank function $r_M$ and $r_N$ on the same finite set S. Then the maximum size of a subset I independent in both M and N equals*

$$\min\{r_M(A) + r_N(B)\colon\ A \cup B = S\}.$$

The basis replacement axiom can be regarded as an abstraction of Laplace's expansion for determinants. Let $V$ be a vector space of dimension $n$. If $x_1, x_2, \ldots, x_n$ are $n$ vectors in $V$, their *bracket* is defined by:

$$[x_1, x_2, \ldots, x_n] = \det(x_{ij})_{1 \leqslant i,j \leqslant n},$$

where $x_i = (x_{ij})$ relative to a chosen coordinate system. Brackets satisfy *Laplace's expansion*:

$$[x_1, \ldots, x_n][y_1, \ldots, y_n] = \sum_{i=1}^{n} [y_i, x_2, \ldots, x_n][y_1, \ldots, y_{i-1}, x_1, y_{i+1}, \ldots, y_n].$$

If $\{x_1, \ldots, x_n\}$ and $\{y_1, \ldots, y_n\}$ are bases, then the left hand side of Laplace's expansion is nonzero. This implies that for some $i$, the $i$-th term on the right hand side is nonzero, that is, both $\{y_i, x_2, \ldots, x_n\}$ and $\{y_1, \ldots, y_{i-1}, x_1, y_{i+1}, \ldots, y_n\}$ are bases. This is the *(symmetric) basis exchange property*, a strong form of the basis replacement axiom which holds in all matroids. *Which determinantal identities translate combinatorially into exchange properties holding in all matroids?* There are two results in this direction: the multiple exchange property [84, 125, 227, 341] (which allows subsets to be exchanged) and the alternating exchange property [127, 174] (based on the fact that an alternating multilinear form is zero on any dependent set of vectors). See [182] for a survey.

The fundamental theorems of classical projective invariant theory [314, 324] say that (1) Brackets generate the relative invariants of the general linear group GL($V$), and (2) Every identity amongst brackets can be derived algebraically from Laplace's expansion. Thus, structures similar to matroids can be defined for other classical group actions [263]. For GL($V$) acting on both vectors and dual vectors in $V$, one abstracts the properties of nonsingularity of submatrices of a matrix to obtain a *bimatroid* or *linking system* ([173, 264]; see also [64, 145, 234]). Many matrix properties can be carried over to bimatroids; for example, using the Cauchy–Binet identity for determinants and matroid intersection, one can define an analogue of matrix multiplication. Two structures have been proposed for the orthogonal group: symmetric bimatroids [173] and metroids (or metric matroids) [33, 91, 92]. Both abstract nonsingularity properties of Gramians. For the symplectic group, one has Pfaffian structures [173]. Another approach, using greedoids, can be found in [115, 112, 113].

Over an ordered field, one can also take into account the sign of the bracket.

**7.1.** *Orientation.* An *rank-$n$ oriented matroid* $M$ on the set $S$ is given by a *sign function* $\varphi$ defined from $n$-tuples of elements in $S$ to $\{-, 0, +\}$ (with the usual multiplication) satisfying the following axioms:

($\mathrm{Or}_0$) $\varphi$ is not identically zero.

($\mathrm{Or}_1$) *Alternation.* For any permutation $\sigma$,

$$\varphi[x_1, x_2, \ldots, x_n] = \operatorname{sgn}(\sigma)\varphi[x_{\sigma(1)}, x_{\sigma(2)}, \ldots, x_{\sigma(n)}].$$

($\mathrm{Or}_2$) *Signed basis exchange.* If $\varphi[x_1, x_2, \ldots, x_n]\varphi[y_1, y_2, \ldots, y_n] = -$, then there exists $i$ such that

$$\varphi[y_i, x_2, \ldots, x_n]\varphi[y_1, y_2, \ldots, y_{i-1}, x_1, y_{i+1}, \ldots, x_n] = -.$$

This axiomatization was discovered by Gutierrez Novoa [135] in 1965. Oriented matroids were rediscovered in the 1970's [110, 28, 196, 198]. See also [98]. Oriented matroids are used in linear programming [26] and the theory of combinatorial differential manifolds [116]. An application of algebraic geometry to oriented matroids can be found in [2]. See [25] for a comprehensive account of oriented matroids.

## 8. Geometric algebra and linear representability

**8.1.** *Geometric algebra.* The problem of representability was first considered by Whitney in [328]. He showed that the Fano plane $F_7$ is representable over a field $\mathbb{F}$ if and only if $\mathbb{F}$ has characteristic 2. Using a method of von Staudt [279] which converts addition and multiplication into geometric configurations, MacLane [216] showed that any algebraic equation can be coded by a configuration. Hence given an algebraic number $\alpha$, there exists a matroid representable only over fields containing $\alpha$.

An easy way to obtain matroids not representable over any field is to code the equations $m = 0$, but $k \neq 0$ for $1 \leqslant k < m$, where $m$ is a positive composite integer [124, 187, 256]. Another method is to start with a "suitable" theorem of projective geometry, convert it into a geometric configuration, and modify the configuration so that it remains a matroid but no longer satisfies the theorem. For example, by declaring the three points on one of the lines in the Desargues configuration independent, one obtains a nonrepresentable matroid. The informal method, called "relaxing a circuit", has yield many useful examples. Some papers in this area are [149, 199, 220, 119].

**8.2.** *Characteristic sets.* The *characteristic set* $\chi(M)$ of a matroid $M$ is the set of characteristic of fields over which $M$ has a representation. Thus, $\chi(M) \subseteq \mathcal{P} \cup \{0\}$, where $\mathcal{P}$ is the set of primes. Using algebraic number theory, Rado [255] proved that if $M$ is finite and $0 \in \chi(M)$, then $\chi(M)$ contains all sufficiently large primes. On the other hand, the compactness theorem in logic implies that for a finite matroid $M$, $\chi(M)$ is infinite implies $0 \in \chi(M)$ [301, 200, 308]. Kahn [160] showed that every finite subset of primes is the characteristic set of some finite matroid. Related results can be found in [47, 120, 121, 204, 206].

**8.3.** *Bracket rings and abstract coordinates.* Suppose $M$ is a rank-$n$ matroid on the set $S$. Let $R$ be the polynomial ring over $\mathbb{Z}$ whose variables are symbolic brackets $[x_1, x_2, \ldots, x_n]$, where $x_1, x_2, \ldots, x_n$ ranges over all $n$-tuples of distinct elements in $S$. The bracket ring $B_M$ is the quotient of $R$ by the relations: (a) $[x_1, x_2, \ldots, x_n] = \operatorname{sgn}(\sigma)[x_{\sigma(1)}, x_{\sigma(2)}, \ldots, x_{\sigma(n)}]$ for any permutation $\sigma$, (b) $[x_1, x_2, \ldots, x_n] = 0$ if $\{x_1, x_2, \ldots, x_n\}$ is dependent in $M$, and (c) Laplace's expansion. A representation $\rho$: $S \to \mathbb{F}^n$ of $M$ defines a homomorphism $\eta$: $B_M \to \mathbb{F}$ such that $\eta([x_1, x_2, \ldots, x_n]) \neq 0$ if and only if $\{x_1, x_2, \ldots, x_n\}$ is a basis in $M$, and conversely [318]. There are two other rings with similar universal properties [300, 108]. These rings allow methods of computational algebra (such as straightening and Gröbner bases) to be applied to the problem of characterizing linear matroids [59–61, 258]. (See also [320, 284].) Any purely matroid-theoretic characterization must be complicated, since it is known that (a) there does not exist a finite set of first-order axioms for the theory of linear matroids ([302]; see also [7]), (b) the number of forbidden minors for the class of linear matroids is infinite (the earliest reference is [216]), and (c) deciding linearity is highly complex algorithmically [159, 260, 289]. Brackets are abstract coordinates; other ways to introduce abstract coordinates can be found in [90, 94–98, 166, 307, 315–317]. Other papers on embeddings and representations are [41, 48, 54, 82, 162, 167, 168, 252, 290, 309].

## 9. Categories of matroids

**9.1.** *Weak maps.* Weak maps formalize the idea of *special position.* To allow the analogue of mapping a nonzero vector to the zero vector, we need to add a loop to every matroid. This is done as follows: Let $M$ be a matroid on the set $S$. Then $M_o$ is the matroid $M \oplus \{o\}$ on the disjoint union $S \cup \{o\}$, where $\{o\}$ is the matroid of rank zero on the set $\{o\}$. Suppose that $M$ is a matroid on the set $S$ and $N$ is a matroid on $T$. A *weak map* $\tau$ from $M$ to $N$ is a function $\tau$: $S \cup \{o\} \to T \cup \{o\}$ mapping $o$ to $o$ and satisfying the following condition: For every subset $A \subseteq S$, $r_{N_o}(\tau(A)) \leqslant r_M(A)$ [143]. When $S = T$ and $\tau$ is the identity function, we say that $N$ is a *specialization* or *weak map image* of $M$ and write $M \to N$. In particular, $M \to N$ if and only if every $N$-independent set is also $M$-independent. The classical example of a specialization is obtained by imposing extra algebraic relations on the coordinates of a set of vectors. More precisely, let $M$ be a matroid on a set $S$ of elements in a module $U$ over the integral domain $R$ under $R$-linear dependence and let $P$ be a prime ideal in $R$. Then the matroid obtained by regarding $S$ as a subset of $U \otimes_R R/P$ is a specialization of $M$. Specializations define a partial order, called the *weak order*, on the set of all matroids on a given set $S$ by: $M \geqslant N$ whenever $M \to N$.

The *weak cut* of the specialization $M \to N$ is defined to be the collection of sets independent in $M$ but dependent in $N$. Weak cuts have been characterized. This characterization can be used to construct all the minimal weak cuts containing a given collection of independent sets. See [176, 191, 238]. A specialization $M \to N$ is *simple* if $M$ covers

$N$ in the weak order. Lucas [215] determined the structure of simple specializations of binary matroids.

THEOREM 9.1. *Let $M$ be a binary matroid on the set $S$ and let $N$ be a simple specialization of $M$ having the same rank as $M$. Then there exists a subset $F \subseteq S$ such that $N = M/F \oplus M|F$.*

Other results about weak maps can be found in [215, 237, 284, 316]. Weak maps can be defined for oriented matroids [116] and are used in combinatorial calculations in differential geometry.

**9.2.** *Strong maps.* Another category is obtained by abstracting the properties of linear transformations. A *strong map* $\sigma$ from $M$ to $N$ is a function $\sigma$: $S \cup \{o\} \to T \cup \{o\}$ mapping $o$ to $o$ and satisfying the condition: the inverse image of any closed set of $N_o$ is closed in $M_o$. The two basic examples of strong maps are *injections* and *contractions*. Let $M$ be a matroid on $S$ and $T \subseteq S$. The injection $T \cup \{o\} \hookrightarrow S \cup \{o\}$ is a strong map from the restriction $M|T$ to $M$. The function $\sigma$: $S \cup \{o\} \to (S \backslash T) \cup \{o\}$ defined by $\sigma(a)$ equals $o$ if $a \in T$ and $a$ otherwise is a strong map from $M$ to the contraction $M/T$.

THE FACTORIZATION THEOREM FOR STRONG MAPS 9.2. *Every strong map can be factored into an injection followed by a contraction.*

This result [144] implies that injections and contractions generate all strong maps, and hence, strong maps form the smallest category with minors as subobjects.

When $S = T$ and $\tau$ is the identity function, we say that $N$ is a *quotient* of $M$. Quotients can be represented by bimatroid products [173]. If $r(N) \geqslant r(M) - 1$, then $N$ is said to be an *elementary* quotient if $M$. Elementary quotients have been intensively studied. Their weak cuts are called *modular cuts* and there are natural one-to-one correspondences between elementary quotients, extensions by a single element, and modular cuts [71, 144]. A useful result proved using strong maps is the *scum theorem* [144]:

*Let $N$ be a simple minor of $M$. Then there exists an upper interval $[U, \widehat{1}]$ in $L(M)$ such that $N$ is a submatroid of the simplification of $M/U$.*

Two surveys are [49] on constructions and [183] on strong maps. Some recent papers are [48, 67, 137, 221, 225, 226, 236, 293, 331, 334]. The automorphism group of a matroid is studied in [32, 138, 228–230, 311].

**9.3.** *Comaps.* The third category has geometric lattices as objects. Let $K$ and $L$ be geometric lattices. A (*normalized*) *comap* is a function $\gamma$: $K \to L$ satisfying the following conditions:

($\mathrm{Cm}_0$) $\gamma(\widehat{0}) = \widehat{0}$.

($\mathrm{Cm}_1$) If $X$ covers $Y$, then $\gamma(X)$ covers or equals $\gamma(Y)$.

($\mathrm{Cm}_2$) If $X$ and $Y$ form a modular pair in $K$, then $\gamma(X \wedge Y) = \gamma(X) \wedge \gamma(Y)$.

Injections of submatroids are comaps. The *retraction* of $K$ to a modular flat $Z$ in $K$ is the function $\rho$: $K \to [\widehat{0}, Z]$, $X \mapsto X \wedge Z$. Injections and retractions generate all comaps; indeed, *any normalized comap can be factored into an injection followed by a retraction to a modular flat* [178]. The proof uses Crapo's construction [73] "joining" two geometric lattices along a comap.

## 10. Enumeration

**10.1.** *Möbius functions, characteristic polynomials, and Whitney numbers.* The *Möbius function* $\mu$: $P \times P \to \mathbb{Z}$ of a finite partially ordered set $P$ is defined recursively by:

$$\mu(x,y) = 0 \quad \text{if } x \not\leqslant y, \quad \mu(x,x) = 1$$

and

$$\sum_{z:\ x \leqslant z \leqslant y} \mu(x,z) = 0 \quad \text{for } x < y.$$

Möbius functions are used to invert summations. Let $f, g$: $P \to R$, where $R$ is a commutative ring with identity. Then

$$g(x) = \sum_{y:\ y \leqslant x} f(y) \Leftrightarrow f(x) = \sum_{y:\ y \leqslant x} g(y)\mu(y,x).$$

Some papers on Möbius functions are [9, 126, 262, 275].

ROTA'S THEOREM 10.1 ([262]). *Let $X$ be a flat in a finite geometric lattice. Then $(-1)^{r(X)}\mu(\widehat{0}, X) > 0$.*

The *characteristic polynomial* $\chi(L;\lambda)$ of a finite rank-$n$ geometric lattice $L$ is the polynomial in the indeterminate $\lambda$ defined by:

$$\chi(L;\lambda) = \sum_{X \in L} \mu(\widehat{0}, X)\lambda^{n-r(X)} = \sum_{m=0}^{n} (-1)^m w_m \lambda^{n-m}.$$

The coefficients $w_m$ are called *Whitney numbers of the first kind.* By Theorem 10.1, $w_m$ is positive. The Whitney numbers $w_m$ have many combinatorial and homological interpretations. For example, they count the number of $m$-simplices in the "broken-circuit complex" [8, 11, 12, 45, 55, 56, 23, 158, 326, 339] and they are the dimensions of the $m$-graded part of a quotient of the exterior algebra on the points [107, 240, 287]. (See also [261].) They also count acyclic orientations and similar objects [69, 130]. In addition, $w_0(L) = |\mu(\widehat{0}, \widehat{1})|$, the *Möbius invariant* of $L$, also has homological interpretations [4, 18, 19, 109, 114, 262]. See [21] for a survey.

Stanley [277] showed that *if $X$ is a modular flat of $L$, then $\chi([\widehat{0}, X];\lambda)$ divides $\chi(L;\lambda)$.* In particular, if $L$ is *supersolvable*, that is, $L$ contains a saturated chain of modular flats, then all the roots of $\chi(L;\lambda)$ are positive integers [278]. Some papers in this area are

[40, 45, 55, 56, 23, 128, 136, 157, 158, 164, 190, 243, 244, 276, 283, 285, 286, 309, 348–350].

For the lattice $L$ of flats of the cycle matroid of a graph $\Gamma$ with $k$ connected components, $c^k\chi(L;c)$ equals the number of ways of assigning $c$ colors to the vertices of $\Gamma$ so that no two adjacent vertices are assigned the same color [262]. The *critical problem of Crapo and Rota* [77] is a geometric variant of the graph coloring problem. Let $S$ be a set of nonzero vectors in the finite vector space $[\mathrm{GF}(q)]^n$. A $c$-tuple $(L_1, L_2, \ldots, L_c)$ of linear functionals *distinguishes* $S$ if for all vectors $a \in S$, there exists $i$ such that $L_i(a) \neq 0$. The *critical exponent* of $S$ is the minimum number $c$ such that there exists a $c$-tuple of linear functionals distinguishing $S$.

THEOREM 10.2. *Let $S$ be a spanning set of nonzero vectors in $[\mathrm{GF}(q)]^n$ and let L be the lattice of flats of the linear matroid on $S$. Then $\chi(L;q^c)$ equals the number of c-tuples of linear functionals distinguishing $S$.*

Finding critical exponents includes the fundamental problem of linear coding theory (to determine given $n$ and $t$ the maximum dimension $k$ of a code in $[\mathrm{GF}(q)]^n$ having minimum weight greater than $t$) [85] and finding nowhere-zero flows on graphs [151]. Other papers on critical exponents are [34, 50, 152, 177, 185, 192, 223, 242, 310, 312, 329, 330, 332, 333].

The *Whitney number $W_k$ of the second kind* of a finite geometric lattice $L$ is the number of rank-$k$ flats in $L$. There are many quite difficult conjectures about the Whitney numbers of both kinds, chief among them is *the logarithmic unimodality conjecture*: $w_{k-1}w_{k+1} \leqslant w_k^2$ and $W_{k-1}W_{k+1} \leqslant W_k^2$. See [282, 271]. For the Whitney numbers $W_n$, Dowling and Wilson has obtained the following inequalities [89]: In a geometric lattice of rank $n$,

$$W_0 + W_1 + \cdots + W_k \leqslant W_{n-k} + W_{n-k+1} + \cdots + W_{n-1} + W_n$$

for all $k \leqslant n/2$. See [1, 131, 181] for surveys of this and related areas. Other papers on Whitney numbers are [3, 88, 44, 175, 180, 190, 344].

**10.2.** *Tutte invariants.* A function $f$ defined from the class of finite matroids to a ring $R$ is said to be a *Tutte or Tutte–Grothendieck invariant* if it satisfies the following conditions:

($\mathrm{TG}_0$) If $M_1$ is isomorphic to $M_2$, then $f(M_1) = f(M_2)$.

($\mathrm{TG}_1$) *Direct-sum rule.* $f(M_1 \oplus M_2) = f(M_1)f(M_2)$.

($\mathrm{TG}_2$) *Deletion-contraction rule.* For every point $e$ that is neither a loop nor an isthmus, $f(M) = f(M\backslash e) + f(M/e)$.

The *Tutte polynomial* $t(M;x,y)$ of a matroid $M$ on $S$ with rank function $r$ is the polynomial in the variables $x$ and $y$ defined by

$$t(M;x,y) = \sum_{A \subseteq S} (x-1)^{r(M)-r(A)}(y-1)^{|A|-r(A)}.$$

THEOREM 10.3. *Every Tutte invariant is an evaluation of the Tutte polynomial.*

Theorem 10.3 is due to Brylawski [38, 39]; it is implicit in [75, 294]. (See also [274].) Theorem 10.3 can be proved by defining a Grothendieck ring. This is the method used in Tutte's 1947 paper [294] which is perhaps the earliest paper in $K$-theory.

Many numerical invariants are Tutte invariants. These include the number of bases of a matroid and the characteristic polynomial $\chi(L(M); \lambda)$ of the lattice of flats. The number of regions and the number of bounded regions in the complement of an arrangement $\mathcal{A}$ of hyperplanes are Tutte invariants [43, 342]. The weight enumerator of a linear code is related to a Tutte invariant of the linear matroid of its generator matrix [129]. Finally, the Tutte polynomial is related to polynomials in knot theory [153, 156, 205, 347]. See [46] for a survey. Related papers are [65, 72, 74, 154, 155, 177, 195, 197, 244, 251, 347].

## References


[1] M. Aigner, *Whitney numbers*, Combinatorial Geometries, N.L. White, ed., Cambridge Univ. Press, Cambridge (1987), 139–160.

[2] N. Alon, *The number of polytopes, configurations and real matroids*, Mathematika **33** (1986), 62–71.

[3] K. Baclawski, *Whitney numbers of geometric lattices*, Adv. Math. **16** (1975), 125–138.

[4] K. Baclawski, *Cohen–Macauley connectivity and geometric lattices*, European J. Combin. **3** (1982), 293–305.

[5] K. Baclawski and N.L. White, *Higher order independence in matroids*, J. London Math. Soc. (2) **19** (1979), 193–202.

[6] R. Baer, *A unified theory of projective spaces and finite abelian groups*, Trans. Amer. Math. Soc. **52** (1942), 283–343.

[7] J.T. Baldwin, *Recursion theory and abstract dependence*, Ann. Pure Appl. Logic **26** (1984), 215–243.

[8] M.O. Ball and S.J. Provan, *Bounds on the reliability polynomial for shellable independence structures*, SIAM J. Algebraic Discrete Methods **3** (1982), 166–181.

[9] M. Barnabei, A. Brini and G.-C. Rota, *The theory of Möbius functions*, Russian Math. Surveys **41** (3) (1986), 135–188.

[10] M.K. Bennett, K.P. Bogart and J.E. Bonin, *The geometry of Dowling lattices*, Adv. Math. **103** (1994), 131–161.

[11] L.J. Billera and S.J. Provan, *Decompositions of simplicial complexes related to diameters of convex polyhedra*, Math. Oper. Res. **5** (1980), 576–594.

[12] L.J. Billera and S.J. Provan, *Leontief substitution systems and matroid complexes*, Math. Oper. Res. **7** (1982), 81–87.

[13] G. Birkhoff, *Abstract linear dependence in lattices*, Amer. J. Math. **57** (1935), 800–804.

[14] G. Birkhoff, *Combinatorial relations in projective geometries*, Ann. Math. (2) **36** (1935), 743–748.

[15] G. Birkhoff, *Lattice Theory*, 3rd edn., Amer. Math. Soc., Providence, RI (1967).

[16] R.E. Bixby, *A strengthened form of Tutte's characterization of regular matroids*, J. Combin. Theory Ser. B **20** (1976), 216–221.

[17] R.E. Bixby, *On Reid's characterization of the matroids representable over $GF(3)$*, J. Combin. Theory Ser. B **26** (1979), 174–204.

[18] A. Björner, *Some matroid inequalities*, Discrete Math. **31** (1980), 101–103.

[19] A. Björner, *On the homology of geometric lattices*, Algebra Universalis **14** (1982), 107–128.

[20] A. Björner, *On matroids, groups and exchange languages*, Matroid Theory (Szeged, 1982), Colloq. Math. Soc. János Bolyai vol. 40, North-Holland, Amsterdam (1985), 25–60.

[21] A. Björner, *Homology and shellability of matroids and geometric lattices*, Matroid Applications, N.L. White, ed., Cambridge Univ. Press, Cambridge (1992), 226–283.

[22] A. Björner and L. Lovász, *Pseudomodular lattices and continuous matroids*, Acta Sci. Math. (Szeged) **51** (1987), 295–308.

[23] A. Björner and G.M. Ziegler, *Broken circuit complexes: factorizations and generalizations*, J. Combin. Theory Ser. B **51** (1991), 96–126.
[24] A. Björner and G.M. Ziegler, *Introduction to greedoids*, Matroid Applications, N.L. White, ed., Cambridge Univ. Press, Cambridge (1992), 284–357.
[25] A. Björner, M. Las Vergnas, B. Sturmfels, N.L. White and G.M. Ziegler, *Oriented Matroids*, Cambridge Univ. Press, Cambridge (1992).
[26] R.G. Bland, *A combinatorial abstraction of linear programming*, J. Combin. Theory Ser. B **23** (1977), 33–57.
[27] R.G. Bland and B.L. Dietrich, *An abstract duality*, Discrete Math. **70** (1988), 203–208.
[28] R.G. Bland and M. Las Vergnas, *Orientability of matroids*, J. Combin. Theory Ser. B **24** (1978), 94–123.
[29] K.P. Bogart and J.E. Bonin, *A geometric characterization of Dowling lattices*, J. Combin. Theory Ser. A **56** (1991), 195–202.
[30] J.E. Bonin, *Modular elements of higher-weight Dowling lattices*, Discrete Math. **119** (1993), 3–11.
[31] J.E. Bonin, *Automorphism groups of higher-weight Dowling geometries*, J. Combin. Theory Ser. B, to appear.
[32] J.E. Bonin and J.P.S. Kung, *Every group is the automorphism group of a planar geometry*, Geom. Dedicata **50** (1994), 243–246.
[33] A. Bouchet, A.W.M. Dress and T.F. Havel, *$\Delta$-matroids and metroids*, Adv. Math. **62** (1992), 136–142.
[34] A. Brini, *Some remarks on the critical problem*, Matroid Theory and Its Applications, Liguori, Naples (1982), 111–124.
[35] R.A. Brualdi, *Introduction to matching theory*, Combinatorial Geometries, N.L. White, ed., Cambridge Univ. Press, Cambridge (1987), 53–71.
[36] R.A. Brualdi, *Transversal matroids*, Combinatorial Geometries, N.L. White, ed., Cambridge Univ. Press, Cambridge (1987), 72–97.
[37] V. Bryant and H. Perfect, *Independence Theory in Combinatorics*, Chapman and Hall, London and New York (1980).
[38] T.H. Brylawski, *The Tutte–Grothendieck ring*, Algebra Universalis **2** (1972), 375–388.
[39] T.H. Brylawski, *A decomposition for combinatorial geometries*, Trans. Amer. Math. Soc. **171** (1972), 235–282.
[40] T.H. Brylawski, *Modular constructions for combinatorial geometries*, Trans. Amer. Math. Soc. **203** (1975), 1–44.
[41] T.H. Brylawski, *An affine representation for transversal geometries*, Stud. Appl. Math. **54** (1975), 143–160.
[42] T.H. Brylawski, *A note on Tutte's unimodular representation theorem*, Proc. Amer. Math. Soc. **52** (1975), 499–502.
[43] T.H. Brylawski, *A combinatorial perspective on the Radon convexity theorem*, Geom. Dedicata **5** (1976), 459–446.
[44] T.H. Brylawski, *Connected matroids with the smallest Whitney numbers*, Discrete Math. **18** (1977), 243–252.
[45] T.H. Brylawski, *The broken-circuit complex*, Trans. Amer. Math. Soc. **234** (1977), 417–443.
[46] T.H. Brylawski, *The Tutte polynomial, Part I: General theory*, Matroid Theory and Its Applications, Liguori, Naples (1982), 125–275.
[47] T.H. Brylawski, *Finite prime-field characteristic sets for planar configurations*, Linear Algebra Appl. **46** (1982), 155–176.
[48] T. Brylawski, *Coordinatizing the Dilworth truncation*, Matroid Theory (Szeged, 1982), Colloq. Math. Soc. János Bolyai vol. 40, North-Holland, Amsterdam (1985), 61–95.
[49] T. Brylawski, *Constructions*, Theory of Matroids, N.L. White, ed., Cambridge Univ. Press, Cambridge (1986), 127–223.
[50] T. Brylawski, *Blocking sets and the Möbius function*, Symposia Mathematica vol. 28 (Rome, 1983), Academic Press, London and New York (1986), 231–249.
[51] T.H. Brylawski and E. Dieter, *Exchange systems*, Discrete Math. **69** (1988), 123–151.
[52] T. Brylawski and D.G. Kelly, *Matroids and combinatorial geometries*, Combinatorics, G.-C. Rota, ed., Math. Assoc. Amer., Washington, DC (1978), 179–217.

[53] T. Brylawski and D.G. Kelly, *Matroids and Combinatorial Geometries*, Univ. North Carolina, Chapel Hill, NC (1980).

[54] T.H. Brylawski and D. Lucas, *Uniquely representable combinatorial geometries*, Colloquio Internazionale sulle Teorie Combinatorie (Roma, 1973), Tomo I, Accad. Naz. Lincei, Rome (1976), 83–104.

[55] T. Brylawski and J.G. Oxley, *Several identities for the characteristic polynomial of a combinatorial geometry*, Discrete Math. **31** (1980), 161–170.

[56] T. Brylawski and J.G. Oxley, *The broken-circuit complex: its structure and factorization*, European J. Combin. **2** (1981), 107–121.

[57] T. Brylawski and J.G. Oxley, *The Tutte polynomial and its applications*, Matroid Applications, N.L. White, ed., Cambridge Univ. press, Cambridge (1992), 123–225.

[58] S. Buechler, *"Geometrical" stability theory*, Logic Colloquium '85, North-Holland, Amsterdam (1987), 53–66.

[59] J. Bukowski and J. Richter, *On the finding of final polynomials*, European J. Math. **11** (1990), 21–34.

[60] J. Bukowski and B. Sturmfels, *On the coordinatization of oriented matroids*, Discrete Comput. Geom. **1** (1986), 293–306.

[61] J. Bukowski and B. Sturmfels, *Computational Synthetic Geometry*, SLNM 1355, Springer, Berlin (1989).

[62] P.J. Cameron and M. Deza, *On permutation geometries*, J. London Math. Soc. (2) **20** (1979), 373–386.

[63] P. Cartier, *Les arrangements d'hyperplanes: Un chapitre de géométric combinatoire*, Séminaire Bourbaki, 1980/81, SLNM 901, Springer, Berlin (1981), 1–22.

[64] S. Chaiken, *A matroid abstraction of the Bott–Duffin constrained inverse*, SIAM J. Algebra Discrete Methods **4** (1983), 467–475.

[65] S. Chaiken, *The Tutte polynomial for ported matroids*, J. Combin. Theory Ser. B **46** (1989), 96–117.

[66] A.L.C. Cheung and H.H. Crapo, *A combinatorial perspective on algebraic geometry*, Adv. Math. **20** (1976), 388–414.

[67] A.L.C. Cheung and H.H. Crapo, *On relative position in extensions of combinatorial geometries*, J. Combin. Theory Ser. B **44** (1988), 201–229.

[68] R. Cordovil, K. Fukuda and M.L. Moreira, *Clutters and matroids*, Discrete Math. **89** (1991), 161–171.

[69] R. Cordovil, M. Las Vergnas and A. Mandel, *Euler relations, Möbius functions and matroid identities*, Geom. Dedicata **12** (1982), 147–162.

[70] R. Cordovil and B. Lindström, *Simplicial matroids*, Combinatorial Geometries, N.L. White, ed., Cambridge Univ. Press, Cambridge (1987), 98–113.

[71] H.H. Crapo, *Single-element extensions of matroids*, J. Res. Nat. Bur. Standards Sect. B **69B** (1965), 55–65.

[72] H.H. Crapo, *A higher invariant for matroids*, J. Combin. Theory **2** (1967), 406–417.

[73] H.H. Crapo, *The joining of exchange geometries*, J. Math. Mech. **17** (1967/68), 837–852.

[74] H.H. Crapo, *Möbius inversion in lattices*, Arch. Math. (Basel) **19** (1968), 595–607.

[75] H.H. Crapo, *The Tutte polynomial*, Aequationes Math. **3** (1969), 211–229.

[76] H.H. Crapo, *The combinatorial theory of structures*, Matroid Theory (Szeged, 1982), Colloq. Math. Soc. János Bolyai vol. 40, North-Holland, Amsterdam (1985), 107–213.

[77] H.H. Crapo and G.-C. Rota, *On the Foundations of Combinatorial Theory: Combinatorial Geometries*, M.I.T. Press, Cambridge, MA (1970).

[78] H.H. Crapo and G.-C. Rota, *Simplicial geometries*, Combinatorics (Proc. Sympos. Pure Math. vol. 19), Amer. Math. Soc., Providence, RI (1971), 71–75.

[79] B.L. Dietrich, *Matroids and antimatroids – A survey*, Discrete Math. **78** (1988), 223–237.

[80] R.P. Dilworth, *The arithmetic theory of Birkhoff lattices*, Duke Math. J. **8** (1941), 286–299.

[81] R.P. Dilworth, *Ideals in Birkhoff lattices*, Trans. Amer. Math. Soc. **49** (1941), 325–353.

[82] R.P. Dilworth, *Dependence relations in a semi-modular lattice*, Duke Math. J. **11** (1944), 575–587.

[83] R.P. Dilworth, *The structure of relatively complemented lattices*, Ann. Math. (2) **51** (1950), 348–359.

[84] J. Donald and M. Tobey, *A generalised exchange theorem for matroid bases*, Bull. Austral. Math. Soc. **43** (1991), 177–180.

[85] T.A. Dowling, *Codes, packing and the critical problem*, Atti del Convegno di Geometria Combinatoria e sue Applicazioni, Instituto di Matematica, Univ. di Perugia, Perugia (1971), 209–224.

[86] T.A. Dowling, *A q-analog of the partition lattice*, A Survey of Combinatorial Theory, J.N. Srivastava, ed., North-Holland, Amsterdam (1973), 101–115.

[87] T.A. Dowling, *A class of geometric lattices based on finite groups*, J. Combin. Theory Ser. B **14** (1973), 61–86; *erratum*, ibid. **15** (1973), 211.

[88] T.A. Dowling and R.M. Wilson, *The slimmest geometric lattices*, Trans. Amer. Math. Soc. **196** (1974), 203–215.

[89] T.A. Dowling and R.M. Wilson, *Whitney number inequalities for geometric lattices*, Proc. Amer. Math. Soc. **47** (1975), 504–512.

[90] A.W.M. Dress, *Duality theory for finite and infinite matroids with coefficients*, Adv. Math. **59** (1986), 97–123.

[91] A.W.M. Dress, *On matroids which have precisely one basis in common*, European J. Combin. **9** (1988), 149–151.

[92] A.W.M. Dress and T.F. Havel, *Some combinatorial properties of discriminants in metric vector spaces*, Adv. Math. **62** (1986), 285–312.

[93] A.W.M. Dress and L. Lovász, *On some combinatorial properties of algebraic matroids*, Combinatorica **7** (1987), 39–48.

[94] A.W.M. Dress and W. Wenzel, *Geometric algebra for combinatorial geometries*, Adv. Math. **77** (1989), 1–36.

[95] A.W.M. Dress and W. Wenzel, *On combinatorial and projective geometry*, Geom. Dedicata **34** (1990), 161–197.

[96] A.W.M. Dress and W. Wenzel, *Grassmann–Plücker relations and matroids with coefficients*, Adv. Math. **86** (1991), 68–110.

[97] A.W.M. Dress and W. Wenzel, *Perfect matroids*, Adv. Math. **91** (1992), 158–208.

[98] A.W.M. Dress and W. Wenzel, *Valuated matroids*, Adv. Math. **93** (1992), 214–250.

[99] F.D.J. Dunstan, A.W. Ingleton and D.J.A. Welsh, *Supermatroids*, Combinatorics (Proc. Conf. Combinatorial Math., Math. Inst., Oxford, 1972), Inst. Math. Appl., Southend-on-Sea (1972), 72–122.

[100] P.H. Edelman and R.E. Jamison, *Theory of convex geometries*, Geom. Dedicata **19** (1985), 249–270.

[101] J. Edmonds, *Minimal partition of a matroid into independent sets*, J. Res. Nat. Bur. Standards Sect. B **69B** (1965), 67–77.

[102] J. Edmonds, *Submodular functions, matroids, and certain polyhedra*, Combinatorial Structures and Their Applications, Gordon and Breach, New York (1970), 69–97.

[103] J. Edmonds, *Matroids and the greedy algorithm*, Math. Programming **1** (1971), 127–136.

[104] M. Evans and E. Hrushovski, *Projective planes in algebraically closed fields*, Proc. London Math. Soc. **62** (1991), 1–24.

[105] U. Faigle, *Geometries on partially ordered sets*, J. Combin. Theory Ser. B **28** (1980), 26–51.

[106] U. Faigle, *Matroids in combinatorial optimization*, Combinatorial Geometries, N.L. White, ed., Cambridge Univ. Press, Cambridge (1987), 161–210.

[107] M. Falk, *On the algebra associated with a geometric lattice*, Adv. Math. **80** (1990), 152–163.

[108] N.E. Fenton, *Matroid representations – an algebraic treatment*, Quart. J. Math. Oxford (2) **35** (1984), 263–280.

[109] J. Folkman, *The homology groups of a lattice*, J. Math. Mech. **15** (1966), 631–636.

[110] J. Folkman and J. Lawrence, *Oriented matroids*, J. Combin. Theory Ser. B **25** (1978), 199–236.

[111] G. Frobenius, *Über zerlegbare Determinanten*, Sitzber. Preuss. Akad. Wiss. (1917), 274–277.

[112] I.M. Gelfand and V.V. Serganova, *On the general definition of a matroid and a greedoid*, Dokl. Akad. Nauk SSSR **292** (1987), 15–20 (in Russian).

[113] I.M. Gelfand and V.V. Serganova, *Combinatorial geometries and torus strata on homogeneous compact manifolds*, Russian Math. Surveys **42**(2) (1987), 133–168.

[114] I.M. Gelfand and A.V. Zelevinskii, *Algebraic and combinatorial aspects of the general theory of hypergeometric functions*, Functional Anal. Appl. **20** (1986), 183–197.

[115] I.M. Gelfand, R.M. Goresky, R.D. MacPherson and V.V. Serganova, *Combinatorial geometries, convex polyhedra and Schubert cells*, Adv. Math. **63** (1987), 301–316.

[116] I.M. Gelfand and R.D. MacPherson, *A combinatorial formula for the Pontryagin classes*, Bull. Amer. Math. Soc. (N.S.) **26** (1992), 304–309.

[117] A.M.H. Gerards, *A short proof of Tutte's characterization of totally unimodular matrices*, Linear Algebra Appl. **114/115** (1989), 207–212.
[118] K. Glazek, *Some old and new problems in the independence theory*, Colloq. Math. **42** (1979), 127–189.
[119] G. Gordon, *Matroids over $F_p$ which are rational excluded minors*, Discrete Math. **52** (1984), 51–65.
[120] G. Gordon, *Constructing prime-field planar configurations*, Proc. Amer. Math. Soc. **91** (1984), 492–502.
[121] G. Gordon, *Algebraic characteristic sets of matroids*, J. Combin. Theory Ser. B **44** (1988), 64–74.
[122] K.M. Gragg and J.P.S. Kung, *Consistent dually semimodular lattices*, J. Combin. Theory Ser. A. **60** (1992), 246–263; *Erratum*, ibid. **71** (1995), 173.
[123] W.H. Graves, *A categorical approach to combinatorial geometry*, J. Combin. Theory Ser. A **11** (1971), 222–232.
[124] C. Greene, *Lectures in combinatorial geometries*, Notes taken by D. Kennedy from the National Science Foundation Seminar in Combinatorial Theory, Bowdoin College, Maine, unpublished (1971).
[125] C. Greene, *A multiple exchange property for bases*, Proc. Amer. Math. Soc. **39** (1973), 45–50.
[126] C. Greene, *On the Möbius algebra of a partially ordered set*, Adv. Math. **10** (1973), 177–187.
[127] C. Greene, *Another exchange property for bases*, Proc. Amer. Math. Soc. **46** (1974), 155–156.
[128] C. Greene, *An inequality for the Möbius function of a geometric lattice*, Stud. Appl. Math. **54** (1975), 71–74.
[129] C. Greene, *Weight enumeration and the geometry of linear codes*, Stud. Appl. Math. **55** (1976), 119–128.
[130] C. Greene and T. Zaslavsky, *On the interpretation of Whitney numbers through arrangements of hyperplanes, zonotopes, non-Radon partitions, and orientations of graphs*, Trans. Amer. Math. Soc. **280** (1983), 97–126.
[131] J.R. Griggs, *The Sperner property in geometric and partition lattices*, The Dilworth Theorems, K.P. Bogart, R. Freese and J.P.S. Kung, eds, Birkhäuser, Boston (1990), 298–304.
[132] H. Groh, *Geometric lattices with topology*, J. Combin. Theory Ser. A **42** (1986), 111–125.
[133] H. Groh, *Embedding geometric lattices with topology*, J. Combin. Theory Ser. A **42** (1986), 126–136.
[134] H. Groh, *Varieties of topological geometries*, Trans. Amer. Math. Soc., to appear.
[135] L. Gutierrez Novoa, *On $n$-ordered sets and order completeness*, Pacific J. Math. **15** (1965), 1337–1345.
[136] M.D. Halsey, *Line-closed combinatorial geometries*, Discrete Math. **65** (1987), 245–248.
[137] M.D. Halsey, *Extending a combinatorial geometry by adding a unique line*, J. Combin. Theory Ser. B **46** (1989), 118–120.
[138] F. Harary, M.J. Piff and D.J.A. Welsh, *On the automorphism group of a matroid*, Discrete Math. **2** (1972), 163–171.
[139] O. Haupt, G. Nöbeling and C. Pauc, *Über Abhängigkeitsräume*, J. Reine Angew. Math. **181** (1940), 193–217.
[140] O. Haupt, G. Nöbeling and C. Pauc, *Sekanten und Paratingenten in topologischen Abhängigkeitsräumen*, J. Reine Angew. Math. **182** (1940), 105–121.
[141] D. Hausmann and B. Korte, *Algorithmic versus axiomatic definitions of matroids*, Math. Programming Stud. No. 14 (1981), 99–111.
[142] I. Heller, *On linear systems with integral valued solutions*, Pacific J. Math. **7** (1957), 1351–1364.
[143] D.A. Higgs, *Maps of geometries*, J. London Math. Soc. **41** (1966), 612–618.
[144] D.A. Higgs, *Strong maps of geometries*, J. Combin. Theory **5** (1968), 185–191.
[145] S. Hocquenghem, *Tabloïdes*, J. Combin. Theory Ser. B **26** (1979), 233–250.
[146] K.L. Holland, *On finding fields from their algebraic closure geometries*, Proc. Amer. Math. Soc. **116** (1992), 1135–1142.
[147] K.L. Holland, *Projective geometries of algebraically closed fields of characteristic zero*, Ann. Pure Appl. Logic **60** (1993), 237–260.
[148] A. Horn, *A characterization of unions of linearly independent sets*, J. London Math. Soc. **30** (1955), 494–496.
[149] A.W. Ingleton, *Representations of matroids*, Combinatorial Mathematics and Its Applications, D.J.A. Welsh, ed., Academic Press, London and New York (1971), 149–169.
[150] A.W. Ingleton and R. Main, *Non-algebraic matroids exist*, J. London Math. Soc. (2) **7** (1975), 144–146.
[151] F. Jaeger, *Flows and generalized coloring theorems in graphs*, J. Combin. Theory Ser. B **26** (1979), 205–216.

[152] F. Jaeger, *A constructive approach to the critical problem for matroids*, European J. Combin. **2** (1981), 137–144.
[153] F. Jaeger, *On Tutte polynomials and link polynomials*, Proc. Amer. Math. Soc. **103** (1988), 647–654.
[154] F. Jaeger, *On Tutte polynomials and bicycle dimensions of ternary matroids*, Proc. Amer. Math. Soc. **107** (1989), 17–25.
[155] F. Jaeger, *On Tutte polynomials of matroids representable over* $GF(q)$, European J. Combin. **10** (1989), 247–255.
[156] F. Jaeger, D.L. Vertigan and D.J.A. Welsh, *On the computational complexity of the Jones and Tutte polynomials*, Math. Proc. Cambridge Philos. Soc. **108** (1990), 35–53.
[157] M. Jambu and H. Terao, *Free arrangements of hyperplanes and supersolvable lattices*, Adv. Math. **52** (1984), 248–258.
[158] M. Jambu and H. Terao, *Arrangements of hyperplanes and broken circuits*, Singularities, R. Randell, ed., Amer. Math. Soc., Providence, RI (1989).
[159] P.M. Jensen and B. Korte, *Complexity of matroid property algorithms*, SIAM J. Comput. **11** (1982), 184–190.
[160] J. Kahn, *Characteristic sets of matroids*, J. London Math. Soc. (2) **26** (1982), 207–217.
[161] J. Kahn, *A geometric approach to forbidden minors for* $GF(3)$, J. Combin. Theory Ser. A **37** (1984), 1–12.
[162] J. Kahn, *On the uniqueness of matroid representations over* $GF(4)$, Bull. London Math. Soc. **20** (1988), 5–10.
[163] J. Kahn and J.P.S. Kung, *Varieties of combinatorial geometries*, Trans. Amer. Math. Soc. **271** (1982), 485–499.
[164] J. Kahn and J.P.S. Kung, *A classification of modularly complemented geometric lattices*, European J. Combin. **7** (1986), 243–248.
[165] J. Kahn and P.D. Seymour, *On forbidden minors for* $GF(3)$, Proc. Amer. Math. Soc. **103** (1988), 437–440.
[166] F.B. Kalhoff, *The Tutte group of projective planes*, Geom. Dedicata **43** (1992), 225–231.
[167] W.M. Kantor, *Dimension and embedding theorems for geometric lattices*, J. Combin. Theory Ser. A **17** (1974), 173–195.
[168] W.M. Kantor, *Envelopes of geometric lattices*, J. Combin. Theory Ser. A **18** (1975), 12–27.
[169] F. Klein-Barmen, *Birkhoffsche und harmonische Verbände*, Math. Z. **42** (1937), 58–81.
[170] B. Korte and L. Lovász, *Greedoids – a structural framework for the greedy algorithm*, Progress in Combinatorial Optimization, W.R. Pulleyblank, ed., Academic Press, New York and London (1984), 221–243.
[171] B. Korte and L. Lovász, *Poset, matroids and greedoids*, Matroid Theory (Szeged, 1982), Colloq. Math. Soc. János Bolyai vol. 40, L. Lovász and A. Recski, eds, North-Holland, Amsterdam (1985), 239–265.
[172] B. Korte, L. Lovász and R. Schrader, *Greedoids*, Springer, Berlin (1991).
[173] J.P.S. Kung, *Bimatroids and invariants*, Adv. Math. **30** (1978), 238–249.
[174] J.P.S. Kung, *Alternating basis exchanges in matroids*, Proc. Amer. Math. Soc. **71** (1978), 355–358.
[175] J.P.S. Kung, *The Radon transforms of a combinatorial geometry, I*, J. Combin. Theory Ser. A **26** (1979), 97–102; *II. Partition lattices*, Adv. Math. **101** (1993), 114–132.
[176] J.P.S. Kung, *On specialization of matroids*, Stud. Appl. Math. **62** (1980), 183–187.
[177] J.P.S. Kung, *The Rédei function of a relation*, J. Combin. Theory Ser A **29** (1980), 287–296.
[178] J.P.S. Kung, *A factorization theorem for comaps of geometric lattices*, J. Combin. Theory Ser. B **34** (1983), 40–47.
[179] J.P.S. Kung, *A characterization of orthogonal duality in matroid theory*, Geom. Dedicata **15** (1983), 69–72.
[180] J.P.S. Kung, *Matchings and Radon transforms in lattices, I. Consistent lattices*, Order **2** (1985), 105–112; *II. Concordant sets*, Math. Proc. Cambridge Philos. Soc. **101** (1987), 221–231.
[181] J.P.S. Kung, *Radon transforms in combinatorics and lattice theory*, Combinatorics and Ordered Sets, I. Rival, ed., Amer. Math. Soc., Providence, RI (1986), 33–74.
[182] J.P.S. Kung, *Basis-exchange properties*, Theory of Matroids, N.L. White, ed., Cambridge Univ. Press, Cambridge (1986), 62–75.

[183] J.P.S. Kung, *Strong maps*, Theory of Matroids, N.L. White, ed., Cambridge Univ. Press, Cambridge (1986), 224–253.

[184] J.P.S. Kung, *Numerically regular hereditary classes of combinatorial geometries*, Geom. Dedicata **21** (1986), 85–105.

[185] J.P.S. Kung, *Growth rates and critical exponents of minor-closed classes of binary geometries*, Trans. Amer. Math. Soc. **293** (1986), 837–857.

[186] J.P.S. Kung (ed.), *A Source Book in Matroid Theory*, Birkhäuser, Boston and Basel (1986).

[187] J.P.S. Kung, *The long-line graph of a combinatorial geometry, II. Geometries representable over two fields of different characteristics,* J. Combin. Theory Ser. B **50** (1990), 41–53.

[188] J.P.S. Kung, *Combinatorial geometries representable over $GF(3)$ and $GF(q)$, I. The number of points*, Discrete Comput. Geom. **5** (1990), 84–95.

[189] J.P.S. Kung, *Extremal matroid theory*, Graph Structure Theory, N. Robertson and P.D. Seymour, eds, Amer. Math. Soc., Providence, RI (1993), 21–61.

[190] J.P.S. Kung, *Flags and Whitney numbers of matroids*, J. Combin. Theory Ser. B **59** (1993), 85–88.

[191] J.P.S. Kung and H.Q. Nguyen, *Weak maps,* Theory of Matroids, N.L. White, ed., Cambridge Univ. Press, Cambridge (1986), 254–271.

[192] J.P.S. Kung, M.R. Murty and G.-C. Rota, *On the Rédei zeta function*, J. Number Theory **12** (1980), 421–436.

[193] J.P.S. Kung and J.G. Oxley, *Combinatorial geometries representable over $GF(3)$ and $GF(q)$, II. Dowling geometries*, Graphs Combin. **4** (1988), 323–332.

[194] C. Kuratowski, *Sur les problèmes des courbes gauches en topologie*, Fund. Math. **15** (1930), 271–283.

[195] M. Las Vergnas, *Acylic and totally cyclic orientations of combinatorial geometries,* Discrete Math. **20** (1977/78), 51–61.

[196] M. Las Vergnas, *Bases in oriented matroids*, J. Combin. Theory Ser. B **25** (1978), 283–289.

[197] M. Las Vergnas, *On the Tutte polynomial of a morphism of matroids*, Ann. Discrete Math. **8** (1980), 7–20; *II. Activities of orientations*, Progress in Graph Theory (Waterloo, Ont. 1982), Academic Press, Toronto, Ont. (1984), 367–380.

[198] J. Lawrence, *Oriented matroids and multiply ordered sets*, Linear Algebra Appl. **48** (1982), 1–12.

[199] T. Lazarson, *The representation problem for independence functions*, J. London Math. Soc. **33** (1958), 21–25.

[200] I. Leader, *A short proof of a theorem of Vámos on matroid representation*, Discrete Math. **75** (1989), 315–317.

[201] J. Lee, *Subspaces with well-scaled frames*, Linear Algebra Appl. **114/115** (1989), 21–56.

[202] J. Lee, *The incidence structure of subspaces with well-scaled frames*, J. Combin. Theory Ser. B **35** (1991), 265–287.

[203] A. Lehman, *Matroids and ports (Abstract)*, Notices Amer. Math. Soc. **12** (1965), 342.

[204] M. Lemos, *A extension of Lindström's result about characteristic sets of matroids*, Discrete Math. **68** (1988), 85–101.

[205] W.B.R. Lickorish, *Polynomials for links*, Bull. London Math. Soc. **20** (1988), 558–588.

[206] B. Lindström, *A class of algebraic matroids with simple characteristic set,* Proc. Amer. Math. Soc. **95** (1985), 147–151.

[207] B. Lindström, *On harmonic conjugates in full algebraic combinatorial geometries*, European J. Combin. **7** (1986), 259–262.

[208] B. Lindström, *A class of nonalgebraic matroids of rank three*, Geom. Dedicata **23** (1987), 255–258.

[209] B. Lindström, *A generalization of the Ingleton–Main lemma and a class of nonalgebraic matroids*, Combinatorica **8** (1988), 87–90.

[210] B. Lindström, *On p-polynomial representations of projective geometries in algebraic combinatorial geometries*, Math. Scand. **63** (1988), 36–42.

[211] B. Lindström, *Matroids, algebraic and non-algebraic,* Algebraic Extremal and Metric Combinatorics, 1986 (Montreal, PQ, 1986), Cambridge Univ. Press, Cambridge (1988), 166–174.

[212] B. Lindström, *Matroids algebraic over $F(t)$ are algebraic over $F$,* Combinatorica **9** (1989), 107–109.

[213] B. Lindström, *p-independence implies pseudomodularity*, European J. Combin. **11** (1990), 489–490.

[214] L. Lovász and M.D. Plummer, *Matching Theory,* North-Holland, Amsterdam (1986).

[215] D. Lucas, *Weak maps of combinatorial geometries*, Trans. Amer. Math. Soc. **206** (1975), 247–279.

[216] S. MacLane, *Some interpretation of abstract linear dependence in terms of projective geometry*, Amer. J. Math. **58** (1936), 236–240.

[217] S. MacLane, *A lattice formulation for transcendence degree and p-bases*, Duke Math. J. **4** (1938), 455–468.

[218] S. MacLane, *Modular fields*, Amer. Math. Monthly **47** (1940), 259–274.

[219] S. MacLane, *Topology and logic as a source of algebra*, Bull. Amer. Math. Soc. **82** (1976), 1–40.

[220] J.H. Mason, *Matroids as the study of geometric configurations*, Higher Combinatorics, M. Aigner, ed., Reidel, Dordrecht (1977), 133–176.

[221] J.H. Mason, *Gluing matroids together: a study of Dilworth truncations and matroid analogues of exterior and symmetric powers*, Algebraic Methods in Graph Theory (Szeged, 1978), Colloq. Math. Soc. János Bolyai vol. 25, North-Holland, Amsterdam (1981), 519–561.

[222] F. Maeda and S. Maeda, *Theory of Symmetric Lattices*, Springer, Berlin (1970).

[223] K.R. Matthews, *An example from power residues of the critical problem of Crapo and Rota*, J. Number Theory **9** (1977), 203–208.

[224] S.B. Maurer, *Matroid basis graphs, I*, J. Combin. Theory Ser. B **14** (1973), 216–240; *II*, J. Combin. Theory Ser. B **15** (1973), 121–145.

[225] F. Mazzocca, *On a characterization of Dilworth truncation of combinatorial geometries*, J. Geometry **20** (1983), 63–73.

[226] F. Mazzocca, *Extensions of combinatorial geometries by the addition of a unique line*, J. Combin. Theory Ser. A **37** (1984), 32–45.

[227] C.J.H. McDiarmid, *An exchange theorem for independence structures*, Proc. Amer. Math. Soc. **47** (1975), 513–514.

[228] E. Mendelsohn, *Every group is the collineation group of some projective plane*, J. Geometry **2** (1972), 97–106.

[229] E. Mendelsohn, *Pathological projective planes: associated affine planes*, J. Geometry **4** (1974), 161–165.

[230] E. Mendelsohn, *On the groups of automorphisms of Steiner triple and quadruple systems*, Proceedings of a Conference on Algebraic Aspects of Combinatorics, Utilitas Math., Winnipeg (1975), 255–264.

[231] G.J. Minty, *On the axiomatic foundations of the theories of directed linear graphs, electrical networks and network-programming*, J. Math. Mech. **15** (1966), 485–520.

[232] L. Mirsky, *Transversal Theory*, Academic Press, New York and London (1971).

[233] L. Mirsky and H. Perfect, *Applications of the notion of independence to problems of combinatorial analysis*, J. Combin. Theory **2** (1967), 327–357.

[234] K. Murota, *Eigensets and power products of a bimatroid*, Adv. Math. **80** (1990), 78–91.

[235] T. Nakasawa, *Zur Axiomatik der linearen Abhängigheit, I*, Sci. Rep. Toyko Bunrika Daigaku **2** (1935), 235–255; *II*, Sci. Rep. Toyko Bunrika Daigaku **3** (1936), 123–136; *III*, Sci. Rep. Toyko Bunrika Daigaku **3** (1936), 45–69.

[236] H.Q. Nguyen, *Functors of the category of combinatorial geometries and strong maps*, Discrete Math. **20** (1977/78), 143–158.

[237] H.Q. Nguyen, *Projections and weak maps in combinatorial geometries*, Discrete Math. **24** (1978), 281–289.

[238] H.Q. Nguyen, *Weak cuts of combinatorial geometries*, Trans. Amer. Math. Soc. **250** (1979), 247–262.

[239] P. Orlik, *Introduction to Arrangements*, Amer. Math. Soc., Providence, RI (1989).

[240] P. Orlik and L. Solomon, *Combinatorics and topology of complements of hyperplanes*, Invent. Math. **56** (1980), 167–189.

[241] P. Orlik and H. Terao, *Arrangements of Hyperplanes*, Springer, Berlin (1992).

[242] J.G. Oxley, *Colouring, packing and the critical problem*, Quart. J. Math. Oxford (2) **29** (1978), 11–22.

[243] J.G. Oxley, *Infinite matroids*, Proc. London Math. Soc. (3) **37** (1978), 259–272.

[244] J.G. Oxley, *On a matroid identity*, Discrete Math. **44** (1983), 55–60.

[245] J.G. Oxley, *A characterization of the ternary matroids with no $M(K_4)$-minor*, J. Combin. Theory Ser. B **42** (1987), 212–249.

[246] J.G. Oxley, *The binary matroids with no 4-wheel minor*, Trans. Amer. Math. Soc. **301** (1987), 63–75.

[247] J.G. Oxley, *The regular matroids with no 5-wheel minor*, J. Combin. Theory Ser. B **46** (1989), 292–305.

[248] J.G. Oxley, *Matroid Theory*, Oxford Univ. Press, Oxford (1992).

[249] J.G. Oxley, *Infinite matroids*, Matroid Applications, N.L. White, ed., Cambridge Univ. Press, Cambridge (1992), 73–90.

[250] J.G. Oxley and D.J.A. Welsh, *The Tutte polynomial and percolation*, Graph Theory and Related Topics (Proc. Conf. Univ. Waterloo, 1977), Academic Press, New York (1979), 329–339.

[251] J.G. Oxley and G.P. Whittle, *A characterisation of Tutte invariants of 2-polymatroids*, J. Combin. Theory Ser. B **59** (1993), 210–244.

[252] N. Percsy, *Locally embeddable geometries*, Arch. Math. (Basel) **37** (1981), 184–192.

[253] H. Perfect, *Independence theory and matroids*, Math. Gaz. **65** (1981), 103–111.

[254] R. Rado, *A theorem on independence relations*, Quart. J. Math. Oxford **13** (1942), 83–89.

[255] R. Rado, *Note on independence functions*, Proc. London Math. Soc. (3) **7** (1957), 300–320.

[256] R. Reid, *Obstructions to representations of combinatorial geometries*, Appendix in *Matroids and Combinatorial Geometries*, by T. Brylawski and D.G. Kelly, Univ. North Carolina Press, Chapel Hill, NC (1980).

[257] K. Reuter, *The Kurosh–Ore exchange property*, Acta Math. Hungar. **53** (1989), 119–127.

[258] J. Richter, *Kombinatorische Realisierbarkeitskriterian für orientierte Matroide*, Mitt. Math. Sem. Geissen No. 194 (1989).

[259] N. Robertson and P.D. Seymour, *Graph minors, VIII. A Kuratowski theorem for general surfaces*, J. Combin. Theory Ser. B **48** (1990), 255–288.

[260] G.C. Robinson and D.J.A. Welsh, *The computational complexity of matroid properties*, Math. Proc. Cambridge Philos. Soc. **87** (1980), 29–45.

[261] L.L. Rose and H. Terao, *Hilbert polynomials and geometric lattices*, Adv. Math. **84** (1990), 209–225.

[262] G.-C. Rota, *On the foundations of combinatorial theory, I. Theory of Möbius functions*, Z. Wahrsh. Verw. Gebiete **2** (1964), 340–368.

[263] G.-C. Rota, *Combinatorial theory and invariant theory*, Notes taken by L. Guibas from the National Science Foundation Seminar in Combinatorial Theory, Bowdoin College, Maine, unpublished (1971).

[264] A. Schrijver, *Matroids and linking systems*, J. Combin. Theory Ser. B **26** (1979), 343–369.

[265] P.D. Seymour, *The forbidden minors of binary clutters*, J. London Math. Soc. (2) **12** (1976), 356–360.

[266] P.D. Seymour, *A forbidden minor characterization of matroid ports*, Quart. J. Math. Oxford (2) **27** (1976), 407–413.

[267] P.D. Seymour, *Matroid representation over $GF(3)$*, J. Combin. Theory Ser. B **26** (1979), 159–173.

[268] P.D. Seymour, *On Tutte's characterization of graphic matroids*, Ann. Discrete Math. **8** (1980), 83–90.

[269] P.D. Seymour, *Decomposition of regular matroids*, J. Combin. Theory Ser. B **28** (1980), 305–359.

[270] P.D. Seymour, *Some applications of matroid decomposition,* Algebraic Methods in Graph Theory (Szeged, 1978), Colloq. Math. Soc. János Bolyai vol. 25, North-Holland, Amsterdam (1981), 713–726.

[271] P.D. Seymour, *On the points-lines-planes conjectures*, J. Combin. Theory Ser. B **33** (1982), 17–26.

[272] P.D. Seymour, *Applications of regular matroid decomposition*, Matroid Theory (Szeged, 1982), Colloq. Math. Soc. János Bolyai vol. 40, North-Holland, Amsterdam (1985), 345–357.

[273] J.A. Sims, *A complete class of matroids*, Quart. J. Math. Oxford (2) **28** (1977), 449–451.

[274] C.A.B. Smith, *On Tutte's dichromatic polynomial*, Ann. Discrete Math. **3** (1978), 247–257.

[275] L. Solomon, *The Burnside algebra of a finite group*, J. Combin. Theory **2** (1967), 603–615.

[276] L. Solomon and H. Terao, *A formula for the characteristic polynomial of an arrangement*, Adv. Math. **64** (1987), 305–325.

[277] R.P. Stanley, *Modular elements of geometric lattices*, Algebra Universalis **1** (1971), 214–217.

[278] R.P. Stanley, *Supersolvable lattices*, Algebra Universalis **2** (1972), 197–217.

[279] G.K.C. von Staudt, *Beiträge zur Geometrie der Lage*, Nuremberg (1856).

[280] E. Steinitz, *Algebraische Theorie der Körper*, J. Reine Angew. Math. **137** (1910), 167–309; Reprinted as a book, de Gruyter, Berlin (1930).

[281] M. Stern, *Semimodular Lattices*, B.G. Teubner, Stuttgart and Leipzig (1991).

[282] J.R. Stonesifer, *Logarithmic concavity for edge lattices of graphs*, J. Combin. Theory Ser. A **18** (1975), 36–46.

[283] J.R. Stonesifer, *Modularly complemented geometric lattices*, Discrete Math. **32** (1980), 85–88.

[284] B. Sturmfels, *On the matroid stratification of Grassmann varieties, specialization of coordinates, and a problem of N. White*, Adv. Math. **75** (1989), 202–211.

[285] H. Terao, *Arrangements of hyperplanes and their freeness*, J. Faculty Sci., Univ. Tokyo, Sci. IA **27** (1980), 293–312; *II*, J. Faculty Sci., Univ. Tokyo, Sci. IA **27** (1980), 313–320.

[286] H. Terao, *Modular elements of lattices and topological fibration*, Adv. Math. **62** (1986), 135–154.

[287] H. Terao, *Factorization of the Orlik–Solomon algebra*, Adv. Math. **92** (1992), 45–53.

[288] K. Truemper, *Alpha-balanced graphs and matrices and $GF(3)$-representability of matroids*, J. Combin. Theory Ser. B **32** (1982), 112–139.

[289] K. Truemper, *On the efficiency of representability tests for matroids*, European J. Combin. **3** (1982), 275–291.

[290] K. Truemper, *Partial matroid representations*, European J. Combin. **5** (1984), 377–394.

[291] K. Truemper, *A decomposition theory for matroids, I. General results*, J. Combin. Theory Ser. B **39** (1985), 43–76; *II. Minimal violation matroids*, J. Combin. Theory Ser. B **39** (1985), 282–297; *III. Decomposition conditions*, J. Combin. Theory Ser. B **41** (1986), 275–305; *IV. Decompositions of graphs*, J. Combin. Theory Ser. B **39** (1988), 259–292; *V. Testing of matrix total unimodularity*, J. Combin. Theory Ser. B. **49** (1990), 241–281; *VI. Almost regular matroids*, J. Combin. Theory Ser. B **55** (1992), 253–301; *VII. Analysis of minimal violation matrices*, J. Combin. Theory Ser. B **55** (1992), 302–335.

[292] K. Truemper, *Matroid Decomposition*, Academic Press, New York (1992).

[293] J. Tůma, *Dilworth truncations and modular cuts*, Matroid Theory (Szeged, 1982), Colloq. Math. Soc. János Bolyai vol. 40, North-Holland, Amsterdam (1985), 383–400.

[294] W.T. Tutte, *A ring in graph theory*, Proc. Cambridge Philos. Soc. **43** (1947), 26–40.

[295] W.T. Tutte, *A homotopy theorem for matroids, I*, Trans. Amer. Math. Soc. **88** (1958), 144–160; *II, ibid.* **88** (1958), 161–174.

[296] W.T. Tutte, *Matroids and graphs*, Trans. Amer. Math. Soc. **90** (1959), 527–552.

[297] W.T. Tutte, *Lectures on matroids*, J. Res. Nat. Bur. Standards Sect. B **69B** (1965), 1–47.

[298] W.T. Tutte, *Connectivity in matroids*, Canad. J. Math. **18** (1966), 1301–1324.

[299] W.T. Tutte, *Introduction to the Theory of Matroids*, American Elsevier, New York (1971).

[300] P. Vámos, *A necessary and sufficient condition for a matroids to be linear*, Möbius Algebras (Proc. Conf., Univ. Waterloo, Waterloo, Ontario, 1971), Univ. of Waterloo, Waterloo, Ontario (1971), 162–169.

[301] P. Vámos, *Linearity of matroids over division rings* (*Notes by G. Roulet*), Möbius Algebras (Proc. Conf., Univ. Waterloo, Waterloo, Ontario, 1971), Univ. of Waterloo, Waterloo, Ontario (1971), 170–174.

[302] P. Vámos, *The missing axiom of matroid theory is lost forever*, J. London Math. Soc. (2) **18** (1978), 403–409.

[303] D.L. Vertigan, *Minor classes*, Graph Structure Theory, N. Robertson and P.D. Seymour, eds, Amer. Math. Soc., Providence, RI (1993), 495–509.

[304] M. Wachs and J.W. Walker, *Geometric semilattices*, Order **2** (1986), 367–385.

[305] B.L. van der Waerden, *Moderne Algebra*, 2 Aufl., Springer, Berlin (1937).

[306] D.K. Wagner, *On theorems of Whitney and Tutte*, Discrete Math. **57** (1985), 147–154.

[307] M. Wagowski, *The Tutte group of a weakly orientable matroid*, Linear Algebra Appl. **117** (1989), 21–24.

[308] S.S. Wagstaff, *Infinite matroids*, Trans. Amer. Math. Soc. **175** (1973), 141–153.

[309] T. Wanner and G. Ziegler, *Supersolvable and modularly complemented matroid extensions*, European J. Combin. **12** (1991), 341–360.

[310] P.N. Walton and D.J.A. Welsh, *On the chromatic number of binary matroids*, Mathematika **27** (1980), 1–9.

[311] D.J.A. Welsh, *Matroid Theory*, Academic Press, London and New York (1976).

[312] D.J.A. Welsh, *Colourings, flows, and projective geometry*, Nieuw Arch. Wisk. (3) **28** (1980), 159–176.

[313] D.J.A. Welsh, *Matroids and combinatorial optimisation*, Matroid Theory and Its Applications, Liguori, Naples (1982), 323–416.

[314] H. Weyl, *The classical groups*, 2nd edn., Princeton Univ. Press, Princeton, NJ (1946).

[315] W. Wenzel, *A group-theoretic interpretation of Tutte's homotopy theory*, Adv. Math. **77** (1989), 37–75.

[316] W. Wenzel, *Algebraic relations between matroids connected by weak homomorphisms*, J. Combin. Theory Ser. A **54** (1990), 214–224.

[317] W. Wenzel, *Projective equivalence of matroids with coefficients*, J. Combin. Theory Ser. A **57** (1991), 15–45.

[318] N.L. White, *The bracket ring of a combinatorial geometry, I*, Trans. Amer. Math. Soc. **202** (1975), 79–95; *II. Unimodular geometries*, Trans. Amer. Math. Soc. **214** (1975), 233–248.

[319] N.L. White, *The basis monomial ring of a matroid*, Adv. Math. **24** (1977), 292–297.
[320] N.L. White, *The transcendence degree of a coordinatization of a combinatorial geometry*, J. Combin. Theory Ser. B **29** (1980), 168–175.
[321] N.L. White (ed.), *Theory of Matroids*, Cambridge Univ. Press. Cambridge (1986).
[322] N.L. White (ed.), *Combinatorial Geometries*, Cambridge Univ. Press, Cambridge (1987).
[323] N.L. White (ed.), *Matroid Applications*, Cambridge Univ. Press, Cambridge (1992).
[324] W. Whiteley, *Logic and invariant theory, I. Invariant theory of projective properties*, Trans. Amer. Math. Soc. **177** (1973), 121–39; *III. Axioms systems and basic syzygies*, J. London Math. Soc. (2) **15** (1977), 1–15; *IV. Invariants and syzygies in combinatorial geometry*, J. Combin. Theory Ser. B **26** (1979), 251–267.
[325] H. Whitney, *Non-separable and planar graphs*. Trans. Amer. Math. Soc. **34** (1932), 339–362.
[326] H. Whitney, *The coloring of graphs*, Ann. Math. (2) **33** (1932), 688–718.
[327] H. Whitney, *Planar graphs*, Fund. Math. **21** (1933), 73–84.
[328] H. Whitney, *On the abstract properties of linear dependence*, Amer. J. Math. **57** (1935), 509–533.
[329] G.P. Whittle, *On the critical exponent of transversal matroids*, J. Combin. Theory Ser. B **37** (1984), 94–95.
[330] G.P. Whittle, *Quotients of tangential k-blocks*, Proc. Amer. Math. Soc. **102** (1988), 1088–1098.
[331] G.P. Whittle, *A generalization of the matroid lift construction*, Trans. Amer. Math. Soc. **316** (1989), 141–159.
[332] G.P. Whittle, *q-lifts of tangential k-blocks*, J. London Math. Soc. (2) **39** (1989), 9–15.
[333] G.P. Whittle, *Dowling group geometries and the critical problem,* J. Combin. Theory Ser. B **47** (1989), 80–92.
[334] G.P. Whittle, *Quotients of Dilworth truncations*, J. Combin. Theory Ser. B **49** (1990), 78–86.
[335] G.P. Whittle, *A characterization of the matroids representable over* GF(3) *and the rationals*, J. Combin. Theory Ser. B, to appear.
[336] L.R. Wilcox, *Modularity in the theory of lattices*, Ann. Math. (2) **40** (1939), 490–505.
[337] L.R. Wilcox, *A topology for semi-modular lattices*, Duke Math. J. **8** (1941), 273–285.
[338] L.R. Wilcox, *Modularity in Birkhoff lattices*, Bull. Amer. Math. Soc. **50** (1944), 135–138.
[339] H.S. Wilf, *Which polynomials are chromatic?*, Colloquio Internazionale sulle Teorie Combinatorie (Roma, 1973), Tomo I, Accad. Naz. Lincei, Rome (1976), 247–257.
[340] R.J. Wilson, *An introduction to matroid theory,* Amer. Math. Monthly **80** (1973), 500–525.
[341] D.R. Woodall, *An exchange theorem for bases of matroids*, J. Combin. Theory Ser. B **16** (1974), 227–228.
[342] T. Zaslavsky, *Facing up to arrangements: Face-count formulas for partition of space by hyperplanes*, Mem. Amer. Math. Soc. No. 154, Amer. Math. Soc., Providence, RI (1975).
[343] T. Zaslavsky, *The geometry of root systems and signed graphs*, Amer. Math. Monthly **88** (1981), 88–105.
[344] T. Zaslavsky, *The slimmest arrangements of hyperplanes: I. Geometric lattices and projective arrangements*, Geom. Dedicata **14** (1983), 243–259; *II. Basepointed geometric lattices and Euclidean arrangements*, Mathematika **28** (1981), 169–190.
[345] T. Zaslavsky, *Signed graphs*, Discrete Appl. Math. **4** (1982), 47–74; *erratum*, ibid. **5** (1983), 248.
[346] T. Zaslavsky, *Biased graphs, I. Bias, balance, and gains*, J. Combin. Theory Ser. B **47** (1989), 32–52; *II. The three matroids*, ibid. **51** (1991), 46–72; *III. Chromatic and dichromatic invariants*, J. Combin. Theory Ser. B **64** (1995), 17–88; *IV. Geometric realizations*, to appear.
[347] T. Zaslavsky, *Strong Tutte functions of matroids and graphs*, Trans. Amer. Math. Soc. **334** (1992), 317–347.
[348] G.M. Ziegler, *Combinatorial construction of logarithmic differential forms*, Adv. Math. **76** (1989), 116–154.
[349] G.M. Ziegler, *Matroid representations and free arrangements*, Trans. Amer. Math. Soc. **320** (1990), 525–541.
[350] G.M. Ziegler, *Binary supersolvable matroids and modular constructions*, Proc. Amer. Math. Soc. **113** (1991), 817–829.

# Section 1D
# Fields, Galois Theory, and Algebraic Number Theory

# Higher Derivation Galois Theory of Inseparable Field Extensions

James K. Deveney

*Department of Math. Sci., Virginia Commonwealth Univ., 1015 West Main Street, Box 2014, Richmond, Virginia 23284-2014, USA*
*e-mail: jdeveney@cabell.vcu.edu*

John N. Mordeson

*Department of Mathematics, Creighton University, Omaha, Nebraska 68178, USA*
*e-mail: mordes@blvejay.creighton.edu*

*Contents*



HANDBOOK OF ALGEBRA, VOL. 1
Edited by M. Hazewinkel

**Introduction**

Let $L/\mathrm{K}$ denote a field extension of characteristic $p \neq 0$. If $L$ is inseparable algebraic over $K$, then there will not be sufficient automorphisms to construct a complete correspondence between subgroups of $\mathrm{Aut}(L/K)$ and the intermediate fields. Indeed, if $L$ is purely inseparable over $K$, the group $\mathrm{Aut}(L/K)$ will be trivial. This problem was the motivation for developing derivations and higher derivations, but as we shall see these maps not only provided information on the correspondence problem, but have led to an understanding of the structure of inseparable field extensions, both algebraic and transcendental.

A *derivation* $d$ on $L$ is an additive map of $L$ into $L$ such that $d(ab) = d(a)b + ad(b)$. The constants of a set of derivations will be a subfield of $L$ containing $L^p$. It had been known that $\mathrm{Der}_K(L)$, the space of derivations on $L$ trivial on $K$, had field of constants $K(L^p)$ and moreover that any intermediate field of $L/K(L^p)$ would be the field of constants of a subspace. The problem of determining when a subspace was the space of all derivations over its field of constants was first solved by Jacobson [55] for the finite dimensional case and Gerstenhaber [38] for the infinite dimensional case. A key ingredient for the higher exponent theory is the notion of a higher derivation due to Hasse and Schmidt [45]. A rank $t$ higher derivation on $L$ is a sequence $d = \{d_i \mid 0 \leqslant i < t+1\}$ of additive maps of $K$ into $K$ such that

$$d_r(ab) = \sum \{d_i(a)d_j(b) \mid i + j = r\}$$

and $d_0$ is the identity map. For purely inseparable Galois theory, higher derivations of finite rank are used. Weisfeld [101] characterized the fields of constants of groups of finite rank $t$ higher derivations as those $F$ over which $L$ has a subbasis $C$, i.e. $C = \{x_\alpha\}$ and $L$ is the tensor product over $F$ of the simple extensions $F(x_\alpha)$. One of the most useful ingredients in the theory was provided by Sweedler [92]. He established that $L$ having a subbasis over $F$ was equivalent to $L^{p^n}$ and $F$ being linearly disjoint[1] over their intersection for all $n$. This has proven versatile for two main reasons. Firstly, the definition applies to arbitrary field extensions of characteristic $p > 0$, ones satisfying the condition now being called modular. Secondly, Waterhouse [99] established the principle that linear disjointness is preserved by intersections. These results have been the keys to determining the structure of inseparable extensions which we shall discuss shortly.

Let $L$ be a finitely generated purely inseparable modular extension of $K$. Then, as noted, any intermediate field over which $L$ is modular will be the field of constants of a group of higher derivations and the remaining obstruction to establishing a Galois type correspondence was to determine when a subgroup was actually a full group. This was established by Gerstenhaber and Zaromp [40] and Heerema and Deveney [53] by determining certain canonical generating sets. There is also a theory for special intermediate fields and this is discussed in section two of the paper.

[1] Two subextensions $L/K$, $\Pi/K$ of a containing extension $N/K$ are linearly disjoint over $K$ if there is a basis of $L$ over $K$ that is still independent over $\Pi$ (or vice versa). Or, equivalently if the natural map $L \otimes_K \Pi \to L\Pi$ of the tensor product to the compositum is an isomorphism.

Heerema [49] developed a theory which incorporated both the classical Galois theory and the purely inseparable Galois theory in a single framework. The groups are essentially groups of higher derivations where the first map of a higher derivation is allowed to be an automorphism and not just the identity map. The fields of constants $F$ of such a group is characterized by the conditions that $L$ is a normal modular extension of $F$ of bounded exponent. Mordeson [67] has developed a theory relating invariance of subgroups to the structure of intermediate fields. The concept of linear disjointness and its intersection preservation is also applied to extend the theory of distinguished subfields to this setting.

The basic properties of infinite rank higher derivations, especially the iterative ones, were developed by Zerla [103]. The fields of constants of these higher derivations are the subfields $F$ of $L$ over which $L$ is regular (separable and algebraically closed) and

$$\bigcap F\big(L^{p^n}\big) = F.$$

As such, they do not properly belong to the study of inseparable Galois theory. However, Heerema [51] was able to combine both finite rank and infinite rank higher derivations in a single group, the group of pencils, by using a direct limit technique. The fields of constants in this theory are the fields $F$ which are separably algebraically closed in $L$ and over which $L$ is modular and of finite inseparability exponent. The characterization of the full subgroups is once again in terms of certain canonical generating sets.

Aside from the intrinsic value of having a Galois type correspondence, the information obtained on the structure of the fields involved is also important. The Galois theories of higher derivations and the concepts developed along the way have given a nice picture of the structure of inseparable field extensions. As an illustration, let $L$ be a finite dimensional extension of $K$. If $L$ is modular over $K$, then $L = J \otimes_K D$ where $D$ is separable over $K$ and $J$ is a tensor product of simple purely inseparable extensions of $K$. In general, one uses Sweedler's characterization of modularity and Waterhouse's results to find a unique minimal intermediate field $Q^*$ over which $L$ is modular. If $Q^*$ is separable over $K$, then $Q^*$ is the unique minimal field over which $L$ splits as above.

## 1.

Throughout this section $L/K$ denotes an arbitrary field extension of characteristic $p > 0$. If $L/K$ is not separable, then $L/K$ is called *inseparable*.

DEFINITION 1.1. If $\exists$ a non-negative integer $e$ such that $K(L^{p^e})/K$ is separable, then the smallest such non-negative integer is called the *inseparability exponent* of $L/K$ and is denoted by $\operatorname{inex}(L/K)$.

DEFINITION 1.2. If

$$\min\big\{[L:S] \mid S \text{ is a maximal separable intermediate field of } L/K\big\} < \infty,$$

then this number is called the *inseparability order* of $L/K$ and is denoted by $\operatorname{inor}(L/K)$.

Maximal separable intermediate fields of $L/K$ exist. If $L/K$ is finitely generated, then $\operatorname{inor}(L/K)$ exists as does $\operatorname{inex}(L/K)$.

DEFINITION 1.3. If $L/K$ has a maximal separable intermediate field $D$ such that $L \subseteq K^{p^{-\infty}}(D)$, then $D$ is called *distinguished.*

It is shown in [13] that not every field extension has a distinguished maximal separable intermediate field.

THEOREM 1.4 ([35]). *Suppose that $L/K$ has finite inseparability exponent $e$. Then $L/K$ has distinguished maximal separable intermediate fields $D$. If $D$ is a distinguished maximal separable intermediate field of $L/K$, then $K(L^{p^e}) = K(D^{p^e})$.*

PROOF. From a relative $p$-basis[2] $X$ of $L/K$ select a subset $Y$ such that $Y^{p^e}$ is a relative $p$-basis of $K(L^{p^e})/K$. Since the latter extension is separable, $Y$ is algebraically independent over $K$ and since $K(L^{p^e})/K(Y^{p^e})$ is separable so is $K(L^{p^e})(Y)/K(Y)$. Then $D = K(L^{p^e})(Y)$ is a distinguished maximal separable intermediate field of $L/K$. □

Although every field extension $L/K$ has a maximal separable intermediate field $S$, not every such $S$ need be distinguished [35]. Necessary and sufficient conditions for this to be the case can be found in [27, 33]. In [52, 31, 34], conditions are determined for the maximal separable intermediate fields to be of bounded codegree. For $L/K$ such that $\operatorname{inor}(L/K) < \infty$, intermediate fields $L'$ of $L/K$ with the property that $\operatorname{inor}(L'/K) = \operatorname{inor}(L/K)$ are characterized in [24]. Other properties of distinguished maximal separable intermediate fields are determined in [28, 30, 32, 50].

THEOREM 1.5 ([62]). *Suppose that $L/K$ is finitely generated. Then for every distinguished maximal separable intermediate field $D$ of $L/K$,* $\operatorname{inor}(L/K) = [L : D]$.

PROOF. Let $S$ be a maximal separable intermediate field of $L/K$. Let $r = \operatorname{inex}(L/S)$. Let $D$ be a distinguished maximal separable intermediate field of $L/K$. Since $r \geqslant e$ where $e = \operatorname{inex}(L/K)$, $K(S^{p^r}) \subseteq K(D^{p^r})$. Thus

$$[L : S]\big[S : K\big(D^{p^r}\big)\big] = [L : D]\big[D : K\big(D^{p^r}\big)\big].$$

Now with $t = \text{trans.deg.}(L/K)$,

$$\big[D : K\big(D^{p^r}\big)\big] = p^{rt} = \big[S : K\big(S^{p^r}\big)\big] \geqslant \big[S : K\big(D^{p^r}\big)\big].$$

Thus $[L : S] \geqslant [L : D]$.

[2] A relatively $p$-independent subset $B$ of $L/K$ is a subset $B$ of $L$ such that for all proper subsets $B'$ of $B$, $K(L^p, B') \subsetneqq K(L^p, B)$. A relative $p$-basis for $L/K$ is a relatively $p$-independent subset $B$ such that $L = K(L^p, B)$.

DEFINITION 1.6. $L/K$ is said to *split* if and only if $L = J \otimes_K D$, i.e. $L$ is the field composite $JD$ and $J$ and $D$ are linearly disjoint over $K$, where $J$ and $D$ are intermediate fields of $L/K$ such that $J/K$ is purely inseparable and $D/K$ is separable.

DEFINITION 1.7. $L/K$ is said to be *modular* if and only if $K$ and $L^{p^i}$ are linearly disjoint over $K \cap L^{p^i}$ for $i = 1, 2, \ldots$ .

As we will see in the next section, finite modular purely inseparable field extensions play a role like that of finite separable extensions which are their own splitting field.

THEOREM 1.8 ([68]). *$L/K$ splits if and only if $L/K$ has a distinguished maximal separable intermediate field and $L/J$ is separable where $J$ is the maximal purely inseparable intermediate field of $L/K$.*

PROOF. If $L/K$ has a distinguished maximal separable intermediate field $D$ and $L/J$ is separable, then

$$L(K^{p^{-\infty}}) = K^{p^{-\infty}} \otimes_J L$$

and so

$$\begin{aligned} L \subseteq K^{p^{-\infty}} \otimes_K D &= K^{p^{-\infty}} \otimes_J (J \otimes_K D) \\ &\subseteq K^{p^{-\infty}} \otimes_J L \subseteq K^{p^{-\infty}} \otimes_J (J \otimes_K D). \end{aligned}$$

Thus $L = J \otimes_K D$. □

In [13] an example is given showing that there exist $L/K$ such that $L/J$ is separable, but $L/K$ does not split.

COROLLARY 1.9 ([63]). *If $L/J$ is separable and $J/K$ is of bounded exponent, then $L/K$ splits.*

PROOF. The result here follows from Theorems 1.4 and 1.8. □

COROLLARY 1.10. *Suppose that $L/K$ is modular. Then $L/K$ splits if and only if $L/K$ has a distinguished maximal separable intermediate field.*

PROOF. Since $L/K$ is modular it follows that $L/(K^{p^{-i}} \cap L)$ is modular for $i = 1, 2, \ldots$ . Thus $L/J$ is modular and so separable. □

THEOREM 1.11 ([54]). *If $L/J$ has separating transcendence basis*[3] *where $J$ is the maximal purely inseparable intermediate field of $L/K$, then $L/K$ splits.*

PROOF. Let $X$ be a separating transcendence basis of $L/J$ and let $S$ be the maximal separable intermediate field of $L/K(X)$. Then $L/J(S)$ is separable algebraic and purely inseparable. □

[3] A separating transcendency basis for $L/J$ is a transcendency basis $B$ such that $L/J(B)$ is separable.

THEOREM 1.12 ([54]). *$L/K$ is modular if and only if $L/J$ is separable and $J/K$ is modular where $J$ is the maximal purely inseparable intermediate field of $L/K$.*

PROOF. $J \cap L^{p^i} = J^{p^i}$, $i = 1, 2, \ldots$. The result follows from definitions and [56], Lemma, p. 162. □

THEOREM 1.13 ([68, 97]). *Suppose that $L/K$ has a distinguished maximal separable intermediate field. Then there exists a unique minimal intermediate field $J^*$ of $K^{p^{-\infty}}/J$ where $J$ is the maximal purely inseparable intermediate field of $L/K$ such that $L(J^*)/K$ splits. $J^*$ has the following properties:*

1. *$J^*$ is the unique minimal purely inseparable field extension of $J$ such that for every distinguished maximal separable intermediate field $D$ of $L/K$, $L \subseteq J^* \otimes_K D$.*
2. *$J^*$ is the unique minimal purely inseparable field extension of $J$ such that $L(J^*)/J^*$ is separable.*
3. *If $L/K$ has finite inseparability exponent $e$, then $J^*/K$ has exponent $e$.*
4. *If $\operatorname{inor}(L/K) < \infty$, then $[J^* : K] < \infty$.*

PROOF. $L(K^{p^{-\infty}}) = K^{p^{-\infty}} \otimes_K D$ where $D$ is a distinguished maximal separable intermediate field of $L/K$. Thus

$$\mathcal{J} = \{J' \mid J' \text{ is an intermediate field of } K^{p^{-\infty}}/J \text{ such that } L(J')/K \text{ splits}\}$$

is not empty. It follows that

$$J^* = \bigcap \{J' \mid J' \in \mathcal{J}\},$$

i.e. $J^*$ is the unique minimal purely inseparable field extension of $J$ such that $L(J^*)/K$ splits. If $\operatorname{inex}(L/K) = e$, then $J^*/K$ has exponent $e$ since $D \subseteq L \subseteq D(J^*) \subseteq D(K^{p^{-e}})$. If $[L : D] < \infty$, then $\exists$ a finite subset $X \subseteq J^*$ such that $L \subseteq D(K(X))$. By the minimality of $J^*$, $J^* = K(X)$. □

THEOREM 1.14 ([53, 99]). *Let $K$ and $F_t$ be subfields of some common field and suppose that $K$ is linearly disjoint from each $F_t$. Then $K$ is linearly disjoint from $\bigcap F_t$.*

PROOF. Suppose

$$\exists x_1, \ldots, x_n \in \bigcap F_t$$

linearly independent over $K \cap F$, but linearly dependent over $K$ where $F = \bigcap F_t$. We assume $n$ is minimal. Now

$$\sum k_i x_i = 0$$

with $k_i \in K$ not zero. We may take $k_1 = 1$. With $k_1 = 1$, the $k_i$ are unique. Since the $x_i$ are each in $F_t$, they are linearly dependent over each $K \cap F_t$. Since the $k_i$ are unique, the $k_i \in K \cap F_t$ and so the $k_i \in K \cap F$, a contradiction. □

THEOREM 1.15 ([92, 68]). *In any field containing* $L(K^{p^{-\infty}})$ $\exists$ *a unique minimal field extension* $L^m/L$ *such that* $L^m/K$ *is modular.* $L^m/L$ *is necessarily purely inseparable.*

PROOF. $L(K^{p^{-\infty}})/K$ is modular since $L(K^{p^{-\infty}})/K^{p^{-\infty}}$ is separable and $K^{p^{-\infty}}/K$ is purely inseparable and modular. Thus

$$\mathcal{L} = \{L' \mid L' \text{ is an intermediate field of } L(K^{p^{-\infty}})/L \text{ and } L'/K \text{ is modular}\}$$

is not empty.

$$L^m = \bigcap\{L' \mid L' \in \mathcal{L}\}$$

is the desired field. □

We call $L^m$ the *modular closure* of $L/K$.

THEOREM 1.16 ([68]). *Suppose that* $L/K$ *has a distinguished maximal separable intermediate field. Let* $L^m$ *be the modular closure of* $L/K$. *Then every distinguished maximal separable intermediate field* $D$ *of* $L/K$ *is one of* $L^m/K$ *and* $L^m = J^{*m} \otimes_K D$ *where* $J^{*m}$ *is the modular closure of* $J^*/K$. *Furthermore,*

1) *if* $L/K$ *has finite inseparability exponent* $e$, *then* $L^m/K$ *has inseparability exponent* $e$;

2) *if* $\operatorname{inor}(L/K) < \infty$, *then* $\operatorname{inor}(L^m/K) < \infty$.

PROOF. $D \subseteq L \subseteq L^m \subseteq K^{p^{-\infty}} \otimes_K D$ and so $D$ is a distinguished maximal separable intermediate field of $L^m/K$. Since $L^m/K$ is modular and has a distinguished maximal separable intermediate field $D$, $L^m/K$ splits, say $L^m = J' \otimes_K D$ where $J'/K$ is purely inseparable and modular. It follows that $J^{*m} = J'$. □

THEOREM 1.17 ([18, 53]). $\exists$ *unique minimal intermediate fields* $H^*$, $C^*$, *and* $Q^*$ *of* $L/K$ *such that* $L/H^*$ *is regular,* $L/C^*$ *is separable, and* $L/Q^*$ *is modular. These intermediate fields satisfy the properties* $H^* \supseteq C^* \supseteq Q^*$, $H^* = \overline{C^*} = \overline{Q^*}$ *(the algebraic closure of* $C^*$, $Q^*$ *in* $L$, *respectively),* $C^*/Q^*$ *is purely inseparable modular, and* $H^* = S \otimes_{Q^*} C^*$ *where* $S$ *is the maximal separable intermediate field of* $H^*/Q^*$.

PROOF. Let $\mathcal{H} = \{H \mid H \text{ is an intermediate field of } L/K \text{ such that } L/H \text{ is regular}\}$. Now $L \in \mathcal{H}$ and $L^p$ and $H$ are linearly disjoint over $H^p$ for all $H \in \mathcal{H}$. An application of Theorem 1.14 yields $L^p$ and

$$\bigcap\{H \mid H \in \mathcal{H}\}$$

are linearly disjoint over

$$\left(\bigcap\{H \mid H \in \mathcal{H}\}\right)^p.$$

It follows that

$$H^* = \bigcap\{H \mid H \in \mathcal{H}\}.$$

The existence of $C^*$ and $Q^*$ follow in a similar manner. □

DEFINITION 1.18. 1. If $L/K(M)$ is separable algebraic for every relative $p$-basis $M$ of $L/K$, then $L/K$ is called *relatively separated.*

2. If $L = K(M)$ for every relative $p$-basis $M$ of $L/K$, then $L/K$ is called *reliable.*

The characterization of relatively separated and reliable field extensions can be found in [57].

THEOREM 1.19 ([18]). *If $L^*/K$ is relatively separated, then $C^*$ has the following properties:*

1. *$C^*$ is a maximal intermediate field which is reliable over $K$;*
2. *$C^*$ is the only intermediate field of $L/K$ such that $L/C^*$ is separable and $C^*/K$ is reliable.*

THEOREM 1.20 ([54, 18]). *Suppose that $C^*/K$ is reliable. Then $L/Q^*$ has finite inseparability exponent, $C^*/Q^*$ is purely inseparable modular with bounded exponent, and $L = F \otimes_S (S \otimes_{Q^*} C^*)$ where $S$ is the maximal separable intermediate field of $H^*/Q^*$ and $F$ is an intermediate field of $L/S$ which is regular over $S$ and separable over $Q^*$.*

The representation of $L/K$ in Theorems 1.17 and 1.20 display the intermediate fields $H^*$, $C^*$, and $Q^*$ which are related to the Galois theories discussed below.

THEOREM 1.21 ([18]). *Suppose that $L/K$ is algebraic and let $S$ denote the maximal separable intermediate field of $L/K$. Then $L/S$ is modular if and only if $Q^*$ is the maximal separable intermediate field of $C^*/K$. If $L/S$ is modular, then $Q^*$ is the unique minimal intermediate field over which $L$ splits.*

PROOF. $L/Q^*$ splits since $L/Q^*$ is modular and algebraic. Suppose that $L/S$ is modular and $L/Q$ splits where $Q$ is an intermediate field of $L/K$, say $L = C \otimes_Q S'$ where $C$ and $S'$ are intermediate fields of $L/Q$ such that $C/Q$ is purely inseparable and $S'/Q$ is separable algebraic. Let $Q'$ be the maximal separable intermediate field of $Q/K$. It follows that $L = C \otimes_{Q'} S''$ where $S''$ is an intermediate field of $S/Q'$ such that $S''/K$ is separable algebraic. Since $L/S''$ is purely inseparable and $S'' \subseteq S$, $S'' = S$. Since $L/S$ is modular, $Q^* \subseteq S \cap C^* \subseteq S \cap C = Q'$. Thus $Q^* \subseteq Q$. □

Results concerning the transitivity of modularity can be found in [54]. A discussion of modularly perfect fields, i.e. fields which have only modular extensions can be found in [63, 25].

## 2.

DEFINITION 2.1. A *derivation* on a field $L$ is an additive map $d$: $L \to L$ such that $d(ab) = d(a)b + ad(b)$.

An element $c$ such that $d(c) = 0$ is called a constant of $d$. Since $d(x^n) = nx^{n-1}d(x)$, elements of $L^p$ are always constants. If $S$ is a set of derivations on $L$, the set of constants of $S$ is $\{x \mid d(x) = 0$ for all $d$ in $S\}$. It is straightforward that the set of constants of $S$ is a subfield of $L$ which contains $L^p$.

THEOREM 2.2 ([47]). *Let $L/K$ be a field extension and let $B$ be a relative $p$ basis of $L/K$. Let $\delta$: $B \to L$ be an arbitrary map from $B$ to $L$. Then there is one and only one derivation $d$ of $L$ over $K$ such that $d(x) = \delta(x)$ for every $x \in B$.*

Let $L$ be purely inseparable exponent one over a subfield $K$ and suppose $B$ is a finite relative $p$-basis for $L$ over $K$. If $x$ is an element of $L$ not in $K$, then $x$ is part of a relative $p$-basis $B'$ of $L$ over $K$ and hence by the last theorem is not a constant for some derivation of $L$ over $K$. Thus $K$ is the field of constants of $\mathcal{D}(L/K)$, the set of all derivations of $L$ over $K$. However, $K$ could also be the field of constants of a smaller set of derivations. Thus to establish a Galois type correspondence it is necessary to determine when a set of derivations is as large as possible.

DEFINITION 2.3. A set of derivations $\mathcal{D}$ is called a *restricted p-Lie algebra* if
1) $\mathcal{D}$ is closed under addition;
2) $\mathcal{D}$ is closed under left multiplication by elements of $L$;
3) $\mathcal{D}$ is closed under $p$th powers;
4) $\mathcal{D}$ is closed under Lie commutation, $[d_1 d_2] = d_1 d_2 - d_2 d_1$.

THEOREM 2.4 ([55]). *Let $L$ be a field of characteristic $p \neq 0$ and let $\mathcal{D}$ be a restricted Lie algebra of derivations on $L$ which is of finite dimension $m$ as a vector space over $L$. If $K$ is the field of constants of $\mathcal{D}$, then $L$ is purely inseparable of exponent $\leqslant 1$ over $K$ and $[L : K] = p^m$. If $d$ is any derivation of $L$ over $K$, then $d \in \mathcal{D}$.*

Gerstenhaber [38] has generalized this Galois theory to the infinite case and showed that with the natural Krull-topology on $Der L$, there is a bijective correspondence between closed restricted Lie algebras of derivations and subfields of L containing $L^p$. Ojanguren and Sridharan [74] show that a subspace which is closed under $p$th powers is automatically closed under Lie product.

DEFINITION 2.5. A *rank $t$ higher derivation* on a field $L$ is a sequence

$$d = \{d_i \mid 0 \leqslant i < t + 1\}$$

of additive maps of $L$ into $L$ such that

$$d_r(ab) = \sum \{d_i(a)d_j(b) \mid i + j = r\}$$

and $d_0$ is the identity map.

Let $x$ be an indeterminate and form $L[x]/(x^{t+1})$. There is a 1–1 correspondence between finite rank $t$ higher derivations $d$ of $L$ and algebra homomorphisms $\rho$: $L \to L[x]/(x^{t+1})$ such that $\rho(a) - a$ has zero constant term. If $\rho$ is given by

$$\rho(a_0) = a + a_1 x + \cdots + a_t x^t,$$

then $d$ is specified by $d_i(a) = a_i$. For infinite rank $d$ one uses homomorphisms as above from $L$ to $L[[x]]$.

If $S$ is a set of rank $t$ higher derivations, the field of constants of $S$ is

$$\{a \in L \mid d_i(a) = 0,\ i > 0,\ (d_i) \in S\}.$$

THEOREM 2.6. 1. ([47]). *Let $B$ be a $p$-basis for $L$ and let $f$: $Z \times B \to L$ be an arbitrary function. There is a unique $(d_i)$ such that for each $b \in B$ and $i \in Z$, $d_i(b) = f(i, b)$.*
2. ([101]). *$d_{ip}(a^p) = (d_i(a))^p$ and if $p$ and $j$ are relatively prime, then $d_j(a^p) = 0$.*

THEOREM 2.7 ([92]). *Let $L$ be a purely inseparable extension of $K$ of finite exponent. The following are equivalent.*
1. *$L$ is the tensor product of simple extensions of $K$.*
2. *$K$ is the field of constants of a set of higher derivations on $L$.*
3. *$L^{p^n}$ and $K$ are linearly disjoint over $L^{p^n} \cap K$ for all positive $n$.*

PROOF. (1) implies (2). Since the field of constants of a set of higher derivations is the intersection of the respective fields of constants, it suffices to show if $L = K(x)$ with

$$x^{p^n} = K$$

then $K$ is the field of constants of a higher derivation on $L$. Let

$$\{x^{p^n}\} \cup B$$

be a $p$-basis of $K$ and define a higher derivation $(d_i)$ on $L$ where $d_1(x) = 1$, $d_1(b) = 0$ for all $b \in B$, $d_i(y) = 0$, $y \in \{x\} \cup B$, $i \neq 1$. The rank $p^{n-1}$ higher derivation $d = \{d_i \mid 0 \leqslant i < p^{n-1} + 1\}$ will have field of constants $K$.

(2) implies (3). If they are not linearly disjoint, we can find a minimal length relation of the form $0 = x_1 + a_2 x_2 + \cdots + a_t x_t$ where $\{x_i\} \subset K$ is independent over $L^{p^n} \cap K$ and $a_i \in L^{p^n}$ with $t \geqslant 2$. Since $\{x_i\}$ is independent over $L^{p^n} \cap K$, we can assume $a_2 \notin L^{p^n} \cap K$. Thus there is a higher derivation $(d)$ with $d_m(a_2) \neq 0, m \geqslant 1$. Theorem 2.6 shows $L^{p^n}$ is invariant under $(d)$ and applying $d_m$ to the relation we get one of shorter length.

(3) implies (1). If $n \geqslant 1$, the linear disjointness condition implies that if $S$ is a $p$-basis for $K^{p^{-n}} \cap L$ over $K^{p^{-n+1}} \cap L$, then $S^p$ is $p$-independent in $K^{p^{-n+1}} \cap L$ over $K^{p^{-n+2}} \cap L$. This condition allows one to construct a set of elements which will be the generators for the single factors in the tensor product. □

[101] provides the following example which shows not every purely inseparable extension is modular. Let $Z_p$ be the prime field, and $x$, $y$, $z$ be indeterminates. Let $K = Z_p(x^p, y^p, z^{p^2})$ and $L = K(z, xz + y)$. $L$ has exponent 2 over $K$ and is not modular.

If modularity for purely inseparable extensions is to correspond to normality for separable extensions, then there should be a minimal extension of a purely inseparable extension which is modular. This is the content of the next result due to Sweedler.

THEOREM 2.8 ([92]). *Let $L$ be a purely inseparable extension of $K$ of exponent $n$. There is a unique minimal field extension $M$ of $L$ which is modular over $K$. $M$ is purely inseparable of exponent $n$ over $K$. If $[L : K] < \infty$, $[M : K] < \infty$.*

PROOF. $K^{p^{-n}}$ can be seen to be modular over $K$ by using the linear disjointness condition of Theorem 2.7. $K^{p^{-n}} \supseteq L$ and the intersection of all subfields of $K^{p^{-n}}$ which contain $L$ and are modular over $K$ will be $M$. If $[L : K]$ is finite, then a set of generators of $L$ over $K$ will involve only a finite number of tensor product generators of $K^{p^{-n}}$ over $K$, and $L$ will be contained in a finite dimensional modular extension of $K$.

The maps which give the Galois correspondence when the exponent is greater than one are the higher derivations. The set $H^t(L)$ of all rank $t$ higher derivations of $L$ is a group with respect to the composition $d \circ e = f$ where $f_j = \sum\{d_m e_n \mid m + n = j\}$ [48]. The first nonzero map of positive subscript is a derivation. If $d = (d_i)$ is a higher derivation of rank $t$, the $s$-section of $d$ is the higher derivation $e = \{d_i \mid i = 0, 1, \ldots, s\}$, $1 \leqslant s < t$. □

DEFINITION 2.9. The index $i(d)$ for a nonzero higher derivation is either 1 or, if $d$ has the property $d_q \neq 0$ and $d_j = 0$ if $q \nmid j$, then $i(d) = q$. A $d \in H^\infty(L)$ is *iterative* of index $q$ if

$$\binom{i}{j} d_{qi} = d_{qi} d_{q(i-j)}$$

for all $i$ and $j \leqslant i$, whereas $d_m = 0$ if $q \nmid m$. A finite rank $t$ higher derivation is iterative if it is the $t$-th section of an infinite higher derivation. If $d$ has index $q$, and $a \in L$, then $ad = e$ where $e_{qi} = a^i d_{qi}$ and $e_j = 0$ if $q \nmid j$.

A complete description of iterative higher derivations has been given by Zerla [103]. However, it should be noted that his finite rank iterative higher derivations are only required to satisfy the combinatorial identity and not be sections of infinite ones. This extra requirement is needed to control the last map in a finite rank iterative higher derivation. A set $F = \{d^\alpha \mid \alpha \in A\}$ of higher derivations is abelian if $d_i^\alpha d_j^\beta = d_j^\beta d_i^\alpha$ for all $\alpha, \beta \in A$. A set of nonzero higher derivations is independent if the set of first nonzero maps of $F$ with positive subscript is independent over $L$.

Before beginning the higher derivation Galois theory, we give a reformulation of the exponent one theory which follows the intended approach.

THEOREM 2.10 ([40, 53]). *Let $F = \{\rho_1, \ldots, \rho_n\}$ be derivations on $L$. The following are equivalent.*

1. *$F$ is abelian, independent, and $\rho_i^p = 0$, $1 \leqslant i \leqslant n$.*
2. *$L = K(x_1, \ldots, x_n)$ where $K$ is the field of constants of $F$ and $\rho_i(x_j) = \delta_{ij}$. The set $\{x_1, \ldots, x_n\}$ is a relative $p$-basis of $L/K$.*

PROOF. (1) implies (2). The idea is to induct on $n$. For a single derivation $\rho$, choose $x$ with $\rho(x) \neq 0$. Since $\rho^p = 0$, there is an $n < p$ such that $\rho^n(x) \neq 0$ and $\rho^{n+1}(x) = 0$. Then $\rho(\rho^{n-2}(x)/\rho^{n-1}(x) = 1$, so there is a $y$ with $\rho(y) = 1$. If $K$ is the field of constants of $\rho$, then $L = K(y)$. For if $x$ is a nonconstant, let $\rho^r$ be the least power of $\rho$ which does not map $x$ to 0. Then

$$\rho^r\left(x - \frac{\rho^r(x)}{r!}\, y^r\right) = 0.$$

Continuing this approximation process will express $x$ as a linear combination of $\{1, y, \ldots, y^{p-1}\}$ with $\rho$-constant coefficients.

Inductively, one can find $x_1, \ldots, x_{n-1}$ a relative $p$-basis of $L$ over $K_1$, the field of constants of $\rho_1, \ldots, \rho_{n-1}$. By commutativity, $\rho_n(K_1) \subset K_1$ and hence by induction $K_1 = K(x_n)$ with $\rho_n(x_n) = 1$ and $\rho_i(x_n) = 0$ for $i \neq n$. Also by commutativity, $\rho_n(x_i) \in K_n$. By a similar approximation process as above, one can subtract an element $z_i$ of $K_n$ from each $x_i$ to force $\rho_n(x_i - z_i) = 0$. The reverse implication is straightforward. □

DEFINITION 2.11. A relative $p$-basis for $L$ over $K$ as in Theorem 2.10 will be called a *dual $p$-base* with respect to $\{\rho_1, \ldots, \rho_n\}$.

The group generated over $L$ by a subset $F$ of higher derivations (or derivations) is the subgroup generated by $\{ad \mid a \in L,\ d \in F\}$.

In view of Theorem 2.10, the exponent one Galois theory could be restated as: a finite-dimensional subspace of $Der(L)$ is Galois if and only if it is generated by a set $\{\rho_1, \ldots, \rho_n\}$ of commuting independent derivations such that $\rho_i^p = 0$, $1 \leqslant i \leqslant n$.

DEFINITION 2.12. For $d \neq 0$ in $H^t(L)$ with first nonzero map $d_r$,

$$p(d) = \min\{s \mid p^s r > r\}.$$

An iterative $d$ of rank $t$ is *normal* if for some $j > 0$, $i(d)$ is $[t/p^j] + 1$, where $[t/p^j]$ is the greatest integer less than one equal to $t/p^j$.

THEOREM 2.13 ([53]). *Let $F = \{d^{(1)}, \ldots, d^{(n)}\}$ be an abelian set of independent iterative derivations of finite rank $t$ and let $K$ be the field of constants of $F$. Then*

$$[L : K] = p^{p(d^{(1)}) + \cdots + p(d^{(n)})}.$$

PROOF. The proof is by induction and the consequence of Theorem 2.6 that for any $d \neq 0$ in $H^t(L)$, $p(d)$ is the exponent of $L$ over the field of constants of $d$. For an iterative higher derivation the first nonzero map is a derivation $d$ with $d^p = 0$. Thus if $K_1$ is the field of constants of $d$, $L = K_1(x)$ and $[L : K] = p$. Since the higher derivation is abelian, $K_1$ is an invariant field and if the first nonzero map has subscript $r$, the first nonzero map of the higher derivation restricted to $K_1$ has subscript $pr$. □

DEFINITION 2.14. Let $F = \{d^{(1)}, \dots, d^{(n)}\}$ be a set of rank $t$ higher derivations on $L$. $\{x_1, \dots, x_n\}$ is a *dual basis* for $F$ if

1) $L = K(x_1, \dots, x_n)$, $K$ the field of constants of $F$,

2) $d^{(i)}_{r_i}(x_i) = 1$, where $d^{(i)}_{r_i}$ is the first nonzero map of $d^{(i)}$ and all other maps in $F$ with nonzero subscript take $x_i$ to zero.

THEOREM 2.15 ([53]). *Let $F = \{d^{(1)}, \dots, d^{(n)}\}$ be a subset of $H^t(L)$. The following are equivalent.*

1. *$F$ is an abelian set of independent iterative higher derivations.*
2. *$F$ has a dual basis $\{x_1, \dots, x_n\}$. If $\{x_1, \dots, x_n\}$ is a dual basis, then*

$$L = K(x_1) \otimes \cdots \otimes K(x_n),$$

*$K_i = K(x_1, \dots, \hat{x}_i, \dots, x_n)$ is the field of constants of $d^{(i)}$.*

PROOF. The proof proceeds by induction. For a single abelian iterative higher derivation $d = \{d_0, \dots, d_t\}$ with first nonzero map $d_r$, $d_r$ is a derivation with $d_r^p = 0$. A lengthy approximation process [103], Theorem 2, p. 411, is used to determine a dual basis for a single higher derivation. Commutativity of the maps allows the induction to proceed.

Given $d \in H^t(L)$ of index $q$, $v(d) = e \in H^t(L)$ where $e_{(q+1)i} = d_{qi}$ for $(q+1)i \leqslant t$ and $e_j = 0$ if $q + 1 \nmid j$. The $v$ closure $\bar{v}(F)$ of a set $F$ in $H^t(L)$ is $F \cup \{v^i(d) \mid d \in F,\ i \geqslant 1\}$. A subgroup $G$ of $H^t(L)$ with field of constants $K$, $[L : K] < \infty$, will be called Galois if $G$ is the group of all higher derivations in $H^t(L)$ which contain $K$ in their fields of constants. □

THEOREM 2.16. *A subgroup $G$ of $H^t(L)$ is Galois if and only if $G$ is generated over $L$ by $\bar{v}(F)$ where $F$ is a finite normal independent iterative subset of $H^t(L)$.*

PROOF. Let $G$ have field of constants $K$. By Theorem 2.7,

$$L = K(x_1) \otimes K(x_2) \otimes \cdots \otimes K(x_n).$$

Let $F$ be a normal set of higher derivations with dual basis $\{x_1, \dots, x_n\}$. Normality insures that the first nonzero map of $d^{(i)}$ has lowest possible positive subscript for a map which does not map $x_i$ to zero. The idea of the proof is to show that all higher derivations can be obtained from $F$ by using the $v$ operation, scalar multiplication, and the group operation. If $d$ is an arbitrary higher derivation on $L$ over $K$, the first nonzero map of $d$ is a derivation and as such is uniquely determined by where it maps $\{x_1, \dots, x_n\}$. By using the $v$ operation, scalar multiplication and product of elements in $F$ one creates a higher derivation $e$ which has the same first nonzero map as $d$. The process is then applied to $de^{-1}$ which has first nonzero map of higher subscript. □

*Intermediate theory*

If the higher derivation theory were to exactly parallel the classical Galois theory of automorphisms, the distinguished intermediate fields $F$ should be those which are invariant under all higher derivations of $L$ over $K$ and such that all higher derivations on $F$ over $K$ could be extended to $L$.

THEOREM 2.17 ([11]). *Let $F$ be a Galois subfield of a Galois extension $L/K$. Then $F$ is invariant under $H^t_K(L)$ of and only if $F = K(L^{p^r})$ for some $r$.*

PROOF. Theorem 2.6 shows $K(L^{p^r})$ is an invariant subfield. To show the converse, assume $F \subseteq K(L^{p^r}$ but $\nsubseteq K(L^{p^{r+1}})$ (otherwise $F = K$ and let $x \in L \setminus K(L^{p^{r+1}})$. Using the generating set $F$ one constructs a higher derivation $(d)$ of index $s$ such that $d_{sp^r}(x) \neq 0$. For any $a \in L$, $(ad)$ has $sp^r$ map $a^{p^r} d_{sp^r}$. $F$ being invariant forces $a^{p^r} \in F$, i.e. $F = K(L^{p^r})$. □

THEOREM 2.18 ([11]). *Let $F$ be a Galois subfield of a Galois extension $L/K$ and assume $F$ is modular over $K$. Then every rank $t$ higher derivation on $F/K$ extends to $L$ if and only if $L = F \otimes_K J$ for some modular extension $J$ of $K$.*

DEFINITION 2.19. Let $F$ be a Galois subfield of $L$ containing $K$. $F$ is *distinguished* if and only if there exists a standard generating set for $H^t_K(L)$ which leaves $F$ invariant.

DEFINITION 2.20. $L$ is an *equiexponential modular extension* of $K$ if and only if $L$ has a subbasis over $K$ all of whose elements have the same exponent over $K$.

THEOREM 2.21. *Assume $L$ is an equiexponential modular extension of $K$. If $L$ is modular over an intermediate field $F$, then $F$ is also modular over $K$.*

PROOF. For a modular extension a subbasis is a relative $p$-basis of minimal total exponent and hence for an equiexponential modular extension any relative $p$-basis is a subbasis. Let $\{x_1, \dots, x_r\}$ be a subbasis of $L$ over $F$ where $x_i$ has exponent $r_i$. Let $\{y_1, \dots, y_t\} \subset F$ be such that $\{x_1, \dots, x_r, y_1, \dots, y_t\}$ is a relative $p$-basis, hence subbasis, of $L$ over $K$. A dimension argument shows

$$\{x^{p_1^{r_1}}, \dots, x^{p_r^{r_r}}, y_1, \dots, y_t\}$$

is a subbasis of $F$ over $K$. □

THEOREM 2.22 ([11]). *An intermediate field $F$ is distinguished if and only if $L$ has a subbasis $T_1 \cup \cdots \cup T_n$ over $K$, the elements of $T_i$ being of exponent $i$ over $K$, and*

$$F = F \cap K(T_1) \otimes \cdots \otimes F \cap K(T_n)$$

*and $K(T_i)$ is modular over $F \cap K(T_1)$ for all $i$.*

**3.**

In this section we exhibit an automorphism group invariant field correspondence which incorporates both the Krull infinite Galois theory [56], p. 147, and the purely inseparable theory of the second section. The invariant subfields $K$ of $L$ are those for which $L/K$ is algebraic, normal, modular and the purely inseparable part has finite exponent. The associated automorphism groups are subgroups of the automorphism group of the local ring described below. They can also be described as groups of rank $p^e$ higher derivations with the modification that $d_0$ is an automorphism of $L$ rather than restricting $d_0$ to be the identity map. Let $A$ denote the group of all automorphisms $\alpha$ of the local ring

$$L[\bar{x}] = L[x]/x^{p^{e+1}}L[x]$$

such that $\alpha(\bar{x}) = \bar{x}$ where $x$ is an indeterminate over $L$, $e$ is a non-negative integer, $x^{p^{e+1}}L[x]$ is the ideal in $L[x]$ generated by $x^{p^{e+1}}$, and $\bar{x}$ is the coset

$$x + x^{p^{e+1}}L[x].$$

We use the following notation: For a subgroup $G$ of $A$,

$$G_L = \{\alpha \in G \mid \alpha(L) \subseteq L\},$$

$$G_0 = \{\alpha \in G \mid \alpha(c) - c \in \bar{x}L[\bar{x}] \; \forall c \in L\},$$

and

$$L^G = \{c \in L \mid \alpha(c) = c \; \forall \alpha \in G\}.$$

For $K$ a subfield of $L$,

$$G^K = \{\alpha \in G \mid \alpha(c) = c \; \forall c \in K\}.$$

For $f(\bar{x})$ in $L[\bar{x}]$, let $\zeta(f(\bar{x})) = f(0)$. Then, for $\alpha \in A$, $\alpha^c (= \zeta\alpha|_L)$ is an automorphism of $L$. For $\beta$ an automorphism of $L$, $\beta^e$ will denote its unique extension to $A$. The map $\alpha \mapsto \alpha^{ce}$ is a homomorphism of $A$ onto $A_L$. With a subgroup $G$ of $A$ we associate the groups $G^c = \{\alpha^c \mid \alpha \in G\}$ and $G^{ce} = \{\alpha^{ce} \mid \alpha \in G\}$. Recall that $d = \{d_i \mid 0 \leqslant i \leqslant p^e\}$ denotes a rank $p^e$ higher derivation of $L$ into $L$. Let $\mathcal{H}$ denote the group of all rank $p^e$ higher derivations on $L$.

PROPOSITION 3.1. *The map* $\delta$: $\mathcal{H} \mapsto A_0$ *given by*

$$\delta(d)|_L = \sum \{\bar{x}^i d_i \mid i = 0, 1, \ldots, p^e\}$$

*and* $\delta(d)(\bar{x}) = \bar{x}$ *is an isomorphism of* $\mathcal{H}$ *with* $A_0$.

PROOF. For $\alpha \in A_0$ and $c \in L$,

$$\alpha(c) = \sum c_i \bar{x}^i$$

and $c_0 = c$. For $i = 0, \dots, p^e$, let $d_i(c) = c_i$. Then $d = \{d_i \mid 0 \leqslant i \leqslant p^e\} \in \mathcal{H}$ and $\delta(d) = \alpha$. This and the fact that $\sum \bar{x}^i d_i$ is an isomorphism for $d$ in $\mathcal{H}$ can be found in [45]. Also

$$\begin{aligned}\delta(d)\delta(d')|_L &= \sum \{\bar{x}^i d_i(\bar{x}^j d'_j) \mid 0 \leqslant i, j \leqslant p^e\} \\ &= \sum \{\bar{x}^i (d \circ d')_i \mid 0 \leqslant i \leqslant p^e\} = \delta(d \circ d')|_L.\end{aligned}$$

For $\mathcal{K}$ a subgroup of $\mathcal{H}$, let

$$L^{\mathcal{K}} = \{c \in L \mid d_i(c) = 0,\ i > 0,\ \forall d \in \mathcal{K}\}.$$

For $K$ a subfield of $L$, let

$$\mathcal{H}^K = \{d \in \mathcal{H} \mid d_i(c) = 0,\ i > 0,\ \forall c \in K\}.$$

PROPOSITION 3.2. *For $\mathcal{K}$ a subgroup of $\mathcal{H}$, $L^{\mathcal{K}} = L^{\delta\mathcal{K}}$, and for $K$ a subfield of $L$, $\delta(\mathcal{H}^K) = A_0^K$.*

Let $L^\alpha = L^G$ where $G$ is the group generated by $\alpha$ in $G$.

PROPOSITION 3.3. *$L^{p^{e+1}} \subseteq L^\alpha$ for $\alpha \in A_0$.*

PROPOSITION 3.4. *Each $\alpha \in A$ has a unique representation as a product $\beta\gamma$, $\beta \in A_L$, $\gamma \in A_0$. In fact, $\beta = \alpha^{ce}$ and thus $\gamma = (\alpha^{ce})^{-1}\alpha$.*

PROPOSITION 3.5. *$L^\alpha = L^{\alpha^{ce}} \cap L^{(\alpha^{ce})^{-1}\alpha}$.*

COROLLARY 3.6. *For $G$ a subgroup of $A$, let $H$ be the group generated by $G^{ce}$ and $G$. Then $L^H = L^G = L^{H_L} \cap L^{H_0}$.*

PROPOSITION 3.7. *If $L/K$ is normal and $L^{p^e} \subseteq S$ for some non-negative integer $e$, then $L = S \otimes_K J$ where $J^{p^e} \subseteq K$.*

PROOF. The proof follows from Corollary 1.9 and [56], Theorem 13, p. 52. □

PROPOSITION 3.8. *Let $K$ be a subfield of $L$ such that $L/K$ is algebraic.*
1. *$\sigma(L^{G_0}) = L^{G_0}$ for $\sigma \in G^c$.*
2. *$L$ is a normal extension of $L^G$.*
3. *$L^{G_1} = J$ where $G_1$ is the group of extensions to $A$ of the automorphism group of $S/K$ and $J$ is the maximal purely inseparable intermediate field of $L/K$.*

PROOF. 1. Suppose for some $\sigma \in G^c$ and $c \in L^{G_0}$ that $\sigma(c) = b \notin L^{G_0}$. Choose $\alpha \in G_0$ for which $\alpha(b) \neq b$. Then $\beta = (\sigma^e)^{-1}\alpha\sigma^e \in G_0$ while $\beta(c) \neq c$, a contradiction. Thus $\sigma(L^{G_0}) \subseteq L^{G_0}$ and $\sigma^{-1}(L^{G_0}) \subseteq L^{G_0}$. Hence $\sigma(L^{G_0}) = L^{G_0}$.

2. By (1), the restriction to $L^{G_0}$ of $\alpha$ in $G^c$ is an automorphism. Let $H$ be the group of all such automorphisms on $L^{G_0}$. Since $L^G = L^{G_L} \cap L^{G_0}$ and $G_L = G^{ce}$, the subfield of $L^{G_0}$ invariant under $H$ is $L^G$. Thus $L^{G_0}/L^G$ is normal separable. By Proposition 3.3, $L^{p^{e+1}} \subseteq L^{G_0}$ from which we conclude that $L^{G_0}$ is the separable closure of $L^G$ in $L$. Let $c \in L$ and let $f(x)$ be its minimal polynomial over $L^G$. Then $f(x) = g(x^{p^r})$ where $r$ is the exponent of inseparability of $c$ over $L^G$ and so $g(x)$ is separable over $L^G$. Since $g(x)$ has $c^{p^r}$ in $L^{G_0}$ as a root, $g(x)$ splits over $L^{G_0}$. It thus follows that $f(x)$ splits over $K$.

3. Clearly $J \subseteq L^{G_1}$. If $c \in L^{G_1}$, then since $K(L^{p^{e+1}}) \subseteq S$, $c^{p^{e+1}} \in S \cap L^{G_1} = K$. Hence $c \in J$ and so $L^{G_1} = J$.

LEMMA 3.9. *Suppose $L/K$ is an algebraic field extension such that $L = J \otimes_K S$ where $S$ is the maximal separable intermediate field of $L/K$ and $J$ is the maximal purely inseparable intermediate field of $L/K$. Then the following conditions are equivalent:*

1. *$L/K$ is modular;*
2. *$L/S$ is modular;*
3. *$J/K$ is modular.*

PROOF. One first shows that $(K \cap L^{p^i})(S^{p^i}) = S \cap L^{p^i}$, $i = 1, 2, \ldots$. Then an application of [56], Lemma, p. 162, yields the equivalence of (1) and (2). The equivalence of (1) and (3) follows from [72], Lemma 1.61 (c), p. 56.

We now give the Galois correspondence.

THEOREM 3.10 ([49]). *Let $K$ be a subfield of $L$ such that $L/K$ is algebraic. The following four conditions are equivalent.*

1. *$K = L^G$ for a subgroup $G$ of $A$.*
2. *$L$ is a normal modular extension of $K$ such that $K(L^{p^{e+1}})/K$ is separable.*
3. *There are intermediate fields $S$ and $J$ such that $J^{p^{e+1}} \subseteq K$, $J/K$ is modular, $S/K$ is normal separable, and $L = S \otimes_K J$.*
4. *There are intermediate fields $S$ and $J$ such that $S/K$ is normal separable, $J$ is the tensor product over $K$ of simple purely inseparable extensions of $K$ having degree $\leqslant p^{e+1}$ and $L = S \otimes_K J$.*

*If $L$ satisfies one of (1)–(4) and $G = A^K$, then $S = L^{G_0}$ and $J = L^{G_L}$ where $S$ and $J$ are given by (3) or (4).*

PROOF. (1) implies (2): $L/K$ is normal by Proposition 3.8 (2). The field of constants of the group $\delta^{-1}(G_0)$ of higher derivations is $L^{G_0}$ and thus $L/L^{G_0}$ is modular by Theorem 2.7. An application of Proposition 3.7 and Lemma 3.9 yield the modularity of $L/K$.

(2) implies (3): The result here follows from Proposition 3.7 and Lemma 3.9.

(3) implies (4): The result here follows from Theorem 2.7 and Lemma 3.9.

(4) implies (1): Let $\mathcal{H}_1$ represent the group of all rank $p^e$ higher derivations of $J$ into $J$ which are trivial on $K$. Each $d \in \mathcal{H}_1$ has a unique extension to $L$ since $L/J$ is separable algebraic [47], Theorem 3. Then

$$\mathcal{H} = \{d \mid d \text{ is an extension to } L \text{ of an element of } \mathcal{H}_1\}$$

is a group of rank $p^e$ higher derivations on $L$ with the property $S \subseteq L^{\mathcal{H}}$. Let $G$ be the subgroup of $A$ generated by $G_1$ and $\delta\mathcal{H}$. By Corollary 3.6,

$$L^G = L^{G_1} \cap L^{G_0} = J \cap L^{\mathcal{H}} = K.$$

In establishing that (1) implies (2) it was shown that $L^{G_0} = S$. Proposition 3.8 (3) yields $L^{G_L} = J$.

DEFINITION 3.11. A subgroup $G$ of $A$ is *Galois* if $G = A^K$ for a subfield $K$ of $L$ such that $L/K$ is algebraic.

DEFINITION 3.12. A subfield $K$ of $L$ is *Galois* if
1) $L/K$ is algebraic, and
2) $K = L^G$ for a subgroup $G$ of $A$.

Theorem 3.10 identifies those subfields of $L$ which are Galois. The Krull infinite Galois theory asserts that a subgroup $G$ of $A_L$ is Galois if and only if $G^c$ is compact in the finite topology [56], Example 5, p. 151. For $[L : K] < \infty$, those subgroups of $\mathcal{H}$ having the form $\mathcal{H}^K$ and hence, via $\delta$, those subgroups of $A_0$ which are Galois have been characterized in [7–9].

THEOREM 3.13 ([49]). *A subgroup $G$ of $A$ is Galois if and only if*
1) $G^{ce} \subseteq G$, *and*
2) $G^{ce}$ *and* $G_0$ *are Galois.*

PROOF. Suppose that $G$ is Galois. Then $G \supseteq A^S$ and $G \supseteq A^J$ where $L = J \otimes_{L^G} S$. For $\alpha \in A^S$, $\alpha^c$ is an automorphism on $L$ which is the identity on $L^{p^{e+1}} \subseteq S$ and hence the identity on $L$. Thus $\alpha \in A_0$ or $A^S \subseteq G_0$. By Theorem 3.10, $G_0 \subseteq A^S$. Let $\alpha \in A^J$ and $\beta = (\alpha^{ce})^{-1}\alpha$. Then $L/L^\beta$ is separable algebraic since $J \subseteq L^\beta$ by Proposition 3.5. Since $L^\beta$ is the field of constants of a finite higher derivation, $L = L^\beta$, $\alpha \in A_L$ or $A^J \subseteq A_L \cap G = G_L$. Hence $G^{ce} = G_L = A^J$ and $G^{ce}$ is Galois. Conversely, suppose that $G$ satisfies (1) and (2). Using Proposition 3.8 (1) and the fact that $L^{p^{e+1}} \subseteq L^{G_0}$, it follows that $L/L^G$ is algebraic. Let $K = L^G$ and $H = A^K$. Then $G \subseteq H$, $G^c \subseteq H^c$, and $G_0 \subseteq H_0$. By Theorem 3.10, $L = J \otimes_K S$, $J = L^{G_L} = L^{H_L}$ and $S = L^{G_0} = L^{H_0}$. But $L^{G_L}$ and $L^{H_L}$ are the fields of invariants of $G^c$ and $H^c$, respectively, and since $G^c$ is Galois, $G^c \supseteq H^c$. Hence $G^c = H^c$. Similarly, $G_0 = H_0$ and $G = G^{ce}G_0 = H^{ce}H_0 = H$. □

DEFINITION 3.14. Given subgroups $H_1$ and $H_2$ of a group $H$, we say $H_1$ is $H_2$ *invariant* if for $\alpha \in H_2$, $\alpha^{-1}H_1\alpha \subseteq H_1$.

Let $H_1$ be a subgroup of $A_L$ and $H_2$ a subgroup of $A_0$. Then $H_1$ and $H_2$ are compatible in the sense that there is a group $G$ in $A$ for which $G_L = H_1$ and $G_0 = H_2$ if and only if $H_2H_1$ is such a group, and since $H_2$ must be an invariant subgroup of $G$, $H_2$ and $H_1$ will be compatible if and only if $H_2$ is $H_1$ invariant.

Let $\mathcal{G}$ be the set of groups of automorphisms on $L$ and $\mathcal{D}$ the set of groups of rank $p^e$ higher derivations on $L$.

DEFINITION 3.15. A pair $(H, D)$ in $\mathcal{G} \times \mathcal{D}$ is *compatible* if there is a subgroup $G$ of $A$ such that $G^c = H$ and $G_0 = \delta(D)$. A pair $(H, D)$ is *Galois* if it is compatible and $H^e\delta(D)$ is Galois.

Given $(H, D)$ in $\mathcal{G} \times \mathcal{D}$, $D$ is invariant under $H$ if given $\sigma \in H$ and $d = \{d_i\} \in D$, then $\sigma^{-1}d\sigma = \{\sigma^{-1}d_i\sigma\} \in D$.

PROPOSITION 3.16. *A pair $(H, D)$ in $\mathcal{G} \times \mathcal{D}$ is compatible if and only if $D$ is $H$ invariant. A compatible pair $(H, D)$ is Galois if and only if $H^e$ and $\delta(D)$ are Galois.*

We now consider the subgroup subfield correspondence. Let $H \subseteq G$ be Galois subgroups of $A$. We consider the consequences for $L^H/L^G$ of invariance of $H_0$ in $G_L$ and of $H_L$ in $G_L$. The objective is the identification of conditions on $H$ relative to $G$ equivalent to $L^H/L^G$ being Galois.

THEOREM 3.17 ([49]). *Let $G$ be a Galois subgroup of $A$. Then $G_L$ is $G_0$ invariant if and only if $G_L$ is $G$ or $\{1\}$.*

PROOF. If $G_L$ is $G_0$ invariant, then $G_L$ is invariant in $G$. Hence since $G_L \cap G_0 = \{1\}$, $G$ is the direct product of $G_L$ and $G_0$. Thus for $d \in \delta^{-1}(G_0)$ and $\alpha \in G_L^c$, $\alpha d_i = d_i\alpha$, $i = 1, \ldots, p^e$. Assume that $G_L$ and $G_0$ are nontrivial. By Theorem 3.10 (4), $J/K$, where $K = L^G$, has a subbasis $B$ and $C = \{b^{p^i} \mid b \in B,\ i \text{ is the exponent of } b \text{ over } K\}$ is $p$-independent in $K$. Extend $C$ to a $p$-basis $C \cup C_1$ of $S$. Then $B \cup C_1$ is a $p$-basis of $L$. By Theorem 2.6, a higher derivation $d$ is determined by its action on a $p$-basis and this action may be arbitrarily prescribed for each $d$. We defined $d$ by the requirement $d_i(c) = 0$ for $c \in B \cup C_1$ and $i < p^e$. For $c_1 \in B$, we let $d_{p^e}(c_1) = s \in S$, $s \notin J$, and let $d_{p^e}$ map every other element of $B \cup C_1$ to 0. Clearly $d$ is trivial on $B \cup C_1$ and hence $\delta(d) \in G_0$ since $G_0$ is Galois. However, $\alpha(c) \neq c$ for some $c$ in $G_L$ and $\alpha d_{p^e}(c_1) = \alpha(c) \neq c = d_{p^e}\alpha(c_1)$. Thus if $G_L$ is $G_0$ invariant, either $G_0$ or $G_L$ must be the trivial group.

THEOREM 3.18 ([49]). *Let $H \subseteq G$ be Galois subgroups of $A$. Let $L = J \otimes_{L^G} S$ as in Theorem 3.10. Then $L^H = J_1 \otimes_{L^G} S_1$ with $S_1 \subseteq S$ and $J_1 \subseteq J$ if and only if $H_0$ is $G_L$ invariant. Moreover, if $L^H = J_1 \otimes_{L^G} S_1$, then $L^{H_0} = J_1 \otimes_{L^G} S$ and $L^{H_L} = J \otimes_{L^G} S_1$.*

PROOF. Suppose that $L^H = J_1 \otimes_{L^G} S_1$. Then $L^{H_0} \supseteq J_1 \otimes S$ and $L^{H_L} \supseteq J \otimes S_1$. But

$$L = L^{H_0} \otimes_{L^H} L^{H_L} = (J_1 \otimes S) \otimes_{L^H} (J \otimes S_1).$$

Hence $L^{H_0} = J_1 \otimes S$ and $L^{H_L} = J \otimes S_1$. If $\alpha \in G_L$, then $\alpha(L) \subseteq L$ and $\alpha|_J$ is the identity. Hence $\alpha(L^{H_0}) = L^{H_0}$ from which it follows that if $d \in \delta^{-1}(H_0)$, then

$\alpha^{-1}d\alpha \in \delta^{-1}(H_0)$ or $H_0$ is $G_L$ invariant. Conversely, assume that $H_0$ is $G_L$ invariant. By Theorem 3.13, $G_L$ and $H_0$ are Galois. Since $(G_L H_0)^{ce} = (G_L)^{ce}(H_0)^{ce} = G_L$ and $(G_L H_0)_0 = H_0$, $G_L H_0$ is also Galois. It follows that $L^{H_0} = J_1 \otimes_{L^G} S$ where $J_1 = L^{H_0} \cap L^{G_L}$ and $L^{H_L} = J \otimes_{L^G} S_1$ where $S_1 = L^{H_L} \cap L^{G_0}$, [49], Lemma 3, p. 199. Now

$$L^H = L^{H_L} \cap L^{H_0} = (J \otimes S_1) \otimes (J_1 \otimes S) = J_1 \otimes S_1.$$

□

As a Corollary to Theorem 3.18, it is shown in [49] that if $H$ is an invariant subgroup of $G$, then $L^H/L^G$ is normal, but not conversely. Let $H \subseteq G$ be Galois subgroups of $A$. We determine a necessary and sufficient condition for $H$ to be $G$ invariant.

LEMMA 3.19. *Let $E/K$ be a field extension and $F$ an intermediate field. If $E/K$ and $E/F$ are modular, then $E/K(F \cap E^{p^j})$ is modular for $j = 0, 1, \ldots$.*

PROOF. Let $j$ be a fixed non-negative integer. Suppose $i$ is an integer such that $i \geqslant j$. Then

$$F \cap E^{p^i} = F \cap E^{p^j} \cap E^{p^i} \subseteq K\big(F \cap E^{p^j}\big) \cap E^{p^i} \subseteq F \cap E^{p^i}.$$

Thus

$$K(F \cap E^{p^j}) \cap E^{p^i} = F \cap E^{p^i}.$$

Since also $F \supseteq K(F \cap E^{p^j})$ and $E/F$ are modular, $K(F \cap E^{p^j})$ and $E^{p^i}$ are linearly disjoint over $F \cap E^{p^i}$. Now suppose $i < j$. That $K(F \cap E^{p^j})$ and $E^{p^i}$ are linearly disjoint over $(K \cap E^{p^i})(F \cap E^{p^j})$ follows from the modularity $E/K$ and use of [56], Lemma, p. 162. □

LEMMA 3.20. *Let $J/K$ be a purely inseparable field extension of bounded exponent $e$ and let $F$ be an intermediate field of $J/K$. If $J/K$ and $F/K$ are modular and if for every subbasis $B$ of $J/K$ every $b \in B$ has the same exponent over $F$ that it has over $K$, then $F = K$.*

PROOF. We first prove the result when $F/K$ has exponent $\leqslant 1$. There does not exist $c \in (F \cap J^{p^i}) - K(K^{p^{-1}} \cap J^{p^{i+1}})$ else it follows that $c^{p^{-i}}$ is in a subbasis of $J/K$ [72], Proposition 1.55 (c), p. 49, and has exponent $i+1$ over $K$ and exponent $i$ over $F$. Since $J/K$ is modular, $K$ and $F \cap J^{p^i}$ are linearly disjoint over $K \cap J^{p^i}$, $i = 0, 1, \ldots$. Also since $J/F$ is modular, $F$ and $K(J^{p^{i+1}})$ are linearly disjoint over $K(F \cap J^{p^{i+1}})$, $i = 0, 1, \ldots$, by [72], Lemma 1.60 (c), p. 55. It follows that

$$K\big(F \cap J^{p^{i+1}}\big) \subseteq K\big(F \cap J^{p^i}\big) \subseteq K\big(J^{p^{i+1}}\big)$$

and so $K(F \cap J^{p^i}) = K(F \cap J^{p^{i+1}})$ for $i = 0, 1, \ldots, e$. Thus

$$F = K(F \cap J^p) = \cdots = K(F \cap J^{p^e}) = K$$

Now suppose $F$ has arbitrary exponent ($\leqslant e$) over $K$ and $F \supset K$. Clearly every subbasis of $J/K$ has the same property concerning exponents over any intermediate field of $F/K$. There exists a non-negative integer $i$ such that $K(F \cap J^{p^i}) \not\subseteq K$ and $K(F \cap J^{p^{i+1}}) \subseteq K$. Then $K(F \cap J^{p^i})/K$ has exponent 1 and $J/K(F \cap J^{p^i})$ is modular by Lemma 3.19. By the exponent 1 argument, $K(F \cap J^{p^i}) = K$, a contradiction. Thus $F = K$.

THEOREM 3.21 ([67]). *Let $H \subseteq G$ be Galois subgroups of $A$ and let $S$ denote the maximal separable intermediate field of $L/L^G$. Then $H$ is $G$ invariant if and only if either $L^H \subseteq S$ and $H_L$ is $G_L$ invariant, or $L^H \supseteq S$, $L^H/L^G$ splits, and $H_0$ is $G_0$ invariant.*

PROOF. Suppose that $L/L^G$ is inseparable, but not purely inseparable. Let $J$ denote the maximal purely inseparable intermediate field of $L/L^G$. Assume $H$ is $G$ invariant. Then $L^H/L^G$ is normal by [49], Corollary 4.4, p. 200, and so $L^H/L^G$ splits. Also $H_0$ is $G_0$ invariant and $H_L$ is $G_L$ invariant. Suppose $L^H \not\subseteq S$ and $L^H \not\supseteq S$. Since $L^H \not\subseteq S$, $L^H \cap J \supset L^G$. Since $H$ is Galois, $L^H J$ is modular over $L^H$ by Theorem 3.10. Thus $J/(L^H \cap J)$ is modular by Lemma 3.9. By Lemma 3.20, there exists a subbasis $M$ of $J/L^G$ and an element $m$ of $M$ such that $m$ has exponent $n$ over $L^G$ and exponent $t$ over $L^H \cap J$ with $n > t$. There exists a subset $X$ of $L^G$ such that $X \cup M$ is a $p$-basis of $J$. Since $L/J$ is separable algebraic, $X \cup M$ is a $p$-basis of $L$. Set $B = X \cup M$ and $C = \{b^{p^i} \mid b \in B,\ i \text{ is the exponent of } b \text{ over } L^G\}$. Then $C$ is a $p$-basis of $L^G$ by [72], Proposition 1.22, p. 14. Since $L^H \not\supseteq S$, $S \supset L^H \cap S$. Let $s \in S - L^H \cap S$. Let $q$ be an integer such that $p^{e-n} < q \leqslant p^{e-n+1}$. Then $\exists d = \{d_i \mid i = 0, 1, \ldots, p^e\}$ in $\mathcal{H}$ such that $d_i(m) = 0$, $i = 1, \ldots, q-1$, $d_q(m) = s$, and $d_i(b) = 0 (i = 1, \ldots, p^e)$ for all $b \in B - \{m\}$. For all $c \in C - \{m^{p^n}\}$, $d_i(c) = 0$ for $i = 1, \ldots, p^e$. Now $d_i(m^{p^n}) = (d_j(m))^{p^n}$ if $i = jp^n$ for some j and $d_i(m^{p^n}) = 0$ otherwise. Consider those $i$ such that $i = jp^n$. Then $1 \leqslant j \leqslant p^{e-n} < q$ whence $d_i(m^{p^n}) = 0$. Thus $d \in \mathcal{H}^{L^G}$. Since $s \notin L^H$, there exists $h_1 \in H_L$ such that $h_1(s) = s' \in S$ with $s' \neq s$. Now $p^{e-n+t} < qp^t < p^{e-n+t+1} \leqslant p^e$ and so $d_{qp^t}$ is defined. Also $m^{p^t} \in L^H \cap J$, $m^{p^t} \notin L^G$, and $d_{qp^t}(m^{p^t}) = (d_q(m))^{p^t} = s^{p^t}$. For any integer $i$ such that $1 \leqslant i < qp^t$, we have that $d_i(m^{p^t}) = (d_j(m))^{p^t}$ if $i = jp^t$ for some $j$ and $d_i(m^{p^t}) = 0$ otherwise. For those $i$ such that $i = jp^t$, $jp^t < qp^t$ so $j < q$. Thus $d_i(m^{p^t}) = 0$ when $1 \leqslant i < qp^t$. One can show $h_1 g_0(m^{p^t}) \neq g_0(m^{p^t})$ where $g_0 = \delta(d)$ and thus $H$ is not $G$ invariant, a contradiction. Thus either $L^H \subseteq S$ or $L^H \supseteq S$. Conversely, suppose $L^H \subseteq S$ and $H_L$ is $G_L$ invariant. One uses Proposition 3.4 to show $H$ is $G$ invariant. Now suppose $L^H \supseteq S$, $L^H/L^G$ splits, and $H_0$ is $G_0$ invariant. Since $L^H/L^G$ splits, we have that $H_0$ is $G_L$ invariant by Theorem 3.18. From this and the fact that $H_0$ is $G_0$ invariant, it follows that $H_0 = H$ is $G$ invariant. □

[20], Theorem, p. 277, can be applied to Theorem 3.21 to give a necessary and sufficient condition concerning group invariance.

Let $H \subseteq G$ be Galois subgroups of $A$. Let $G_H$ denote the group of all automorphisms $g_H$ for

$$L^H[\bar{x}] = L^H[x]/x^{p^{e+1}} L^H[x]$$

such that $g_H(\bar{x}) = \bar{x}$ and $g_H$ is the identity on $L^G$. The proof of the next result follows along lines similar to that of classical Galois theory.

THEOREM 3.22 ([67]). *Let $H \subseteq G$ be Galois subgroups of $A$. If $H$ is $G$ invariant and $H_0 = G_0$, then $G/H \simeq G_H$.*

PROPOSITION 3.23. *Let $H \subseteq G$ be Galois subgroups of $A$. If $H$ is $G$ invariant and $H_0 = H$, then $G_L \simeq (G_H)_{L^H}$.*

Let $G' = \{g \in G \mid g(L^H[\bar{x}]) = L^H[\bar{x}]\}$ where $H \subseteq G$ are Galois subgroups of $A$. Then $G'$ is a subgroup of $G$ and $H \subseteq G'$.

PROPOSITION 3.24. *Let $H \subseteq G$ be Galois subgroups of $A$ such that $H$ is $G$ invariant and $H_0 = H$. If $L = L^H \otimes_S J'$ for some intermediate field $J'$ of $L/S$ such that $L^H/S$ and $J'/S$ are modular, then $G'/H \simeq G_H$.*

LEMMA 3.25. *Let $F/K$ be an inseparable but not purely inseparable, algebraic field extension such that $F = S \otimes_K J$ where $S$ is the maximal separable intermediate field of $F/K$ and $J$ is the maximal purely inseparable intermediate field of $F/K$ and $J/K$ has a subbasis. Then there exists an intermediate field of $F/K$ over which $F$ is modular and which is an exceptional* [86] *and reliable extension of $K$ if and only if $(K^{p^{-1}} \cap J)/K$ is not simple.*

THEOREM 3.26 ([67]). *Suppose $K$ is a Galois subfield of $L$. Then the following conditions are equivalent.*

1. *Every Galois intermediate field of $L/K$ splits over $K$.*
2. *Every intermediate field of $L/K$ splits over $K$.*
3. *Every intermediate field of $L/K$ is Galois and splits over $K$.*
4. *Every intermediate field of $L/K$ is Galois, splits over $K$, and is modular over $K$.*
5. *$L/S$ is simple where $S$ is the maximal separable intermediate field of $L/K$.*

COROLLARY 3.27. *Suppose $G$ is a Galois subgroup of $A$. Then $L/S$ is simple where $S$ is the maximal separable intermediate field of $L/L^G$ if and only if for every subgroup $H$ of $G$ which is Galois, $H_0$ is $G_L$ invariant.*

The description of a necessary and sufficient condition for every intermediate field of $L/K$ to be Galois where $K$ is a Galois subfield of $L$ can be found in [67], Section 4.

We now give a new characterization of the distinguished subfields for the purely inseparable case in terms of linear disjointness properties to incorporate the purely inseparable intermediate theory as a special case of the inseparable theory developed here.

Throughout the remainder of this section, $K$ will be a given Galois subfield of $L$ with Galois group $G$. Let $L^{G_0} = S$, $L^{G_L} = J$, and $\delta(\mathcal{H}^S) = G_0$. We assume $[L : S] < \infty$ in order to apply the Galois theory in Section 2. In particular, $S$ is normal over $K$ and is the maximal separable intermediate field of $L/K$, and $J$ is a finite dimensional purely inseparable modular extension of $K$.

LEMMA 3.28. *Assume $L/K$ is purely inseparable modular of exponent $e$. Let*

$$T_e \cup T_{e-1} \cup \cdots \cup T_1$$

*be a subbasis for $L/K$ where the elements of $T_i$ are of exponent $i$ over $K$. Let $\{b_1, \ldots, b_r\} \subseteq L$ be such that $\{b_1^{p^i}, \ldots, b_r^{p^i}\}$ is relatively $p$-independent in $K^{p^{-s}} \cap L$ over $(K^{p^{-s+1}} \cap L)(L^{p^{i+1}} \cap K^{p^{-s}})$. Then there exist $T'_{s+i} \supseteq \{b_1, \ldots, b_r\}$ such that*

$$T_e \cup \cdots \cup T_{s+i+1} \cup T'_{s+i} \cup \cdots \cup T_1$$

*is also a subbasis for $L/K$.*

PROOF. $T_e$ is a relative $p$-basis for $L/(K^{p^{-e+1}} \cap L)$. Since $T_e \cup \cdots \cup T_1$ is assumed to be a subbasis for $L/K$, we can proceed to the stage of constructing a relative $p$-basis for $K^{p^{-(i+s)}} \cap L$ over $K^{p^{-(i+s)+1}} \cap L$. Since $L/K$ is modular,

$$T_e^{p^{e-(i+s)}} \cup \cdots \cup T_{i+s+1}^p$$

is $p$-independent [92], Theorem 1, p. 403, and in fact is a relative $p$-basis for

$$\left(K^{p^{-(i+s)+1}} \cap L\right)\left(L^p \cap K^{p^{-(i+s)}}\right)$$

over $K^{p^{-(i+s)+1}} \cap L$. The set $\{b_1, \ldots, b_r\}$ is in $K^{p^{-(i+s)}} \cap L$ since

$$\{b_1^{p^i}, \ldots, b_r^{p^i}\} \subseteq K^{p^{-s}} \cap L.$$

Moreover, it is $p$-independent over

$$\left(K^{p^{-(i+s)+1}} \cap L\right)\left(L^p \cap K^{p^{-(i+s)}}\right).$$

Thus $\{b_1, \ldots, b_r\}$ can be completed to a relative $p$-basis $T'_{s+1}$ for $K^{p^{-(i+s)}} \cap L$ over

$$\left(K^{p^{-(i+s)+1}} \cap L\right)\left(L^p \cap K^{p^{-(i+s)}}\right).$$

Thus $T_e \cup \cdots \cup T'_{i+s}$ is part of a subbasis for $L/K$. In constructing a relative $p$-basis for $K^{p^{-h}} \cap L$ over $K^{p^{-h+1}} \cap L$ where $h < i + s$,

$$T_e^{p^{e-h}} \cup \cdots \cup T_{i+s}'^{p^{(i+s)-h}} \cup \cdots \cup T_{h+1}^p$$

will be a relative $p$-basis for $(K^{p^{-h+1}} \cap L)(L^p \cap K^{p^{-h}})$ over $K^{p^{-h+1}} \cap L$ and hence can be completed to a relative $p$-basis with $T_h$. □

LEMMA 3.29. *Assume $L \supseteq M \supseteq K$ where $L$ is purely inseparable modular of exponent $e$ over $K$. If*

*1) $K^{p^{-r}} \cap L$ and $M$ are linearly disjoint over $K^{p^{-r}} \cap M$ for all $r$, and*

*2) $(K^{p^{-r}} \cap L)(L^{p^{i+1}} \cap K^{p^{-r-1}})$ and $(K^{p^{-r}} \cap L)(L^{p^i} \cap K^{p^{-r-1}} \cap M)$ are linearly disjoint over $(K^{p^{-r}} \cap L)(L^{p^{i+1}} \cap K^{p^{-r-1}} \cap M)$ for all $i$ and $r$, then any relative $p$-basis for $(K^{p^{-r}} \cap M)(L^{p^i} \cap K^{p^{-r-1}} \cap M)$ over $(K^{p^{-r}} \cap M)(L^{p^{i+1}} \cap K^{p^{-r-1}} \cap M)$ remains $p$-independent over $(K^{p^{-r}} \cap L)(L^{p^{i+1}} \cap K^{p^{-r-1}})$.*

PROOF. The proof here uses [56], Lemma, p. 162. □

THEOREM 3.30 ([23]). *Assume $L \supseteq M \supseteq K$ where $L$ is purely inseparable modular of exponent $e$ over $K$. Then there is a subbasis $B$ of $L$ over $K$ and a subset $B'$ of $B$ such that $C = \{b^{p^r} \mid b \in B',\ r$ is the exponent of $b$ over $M\}$ is a subbasis of $M$ over $K$ if and only if*

*1) $K^{p^{-r}} \cap L$ and $M$ are linearly disjoint over $K^{p^{-r}} \cap M$ for all $r$;*

*2) $(K^{p^{-r}} \cap L)(L^{p^{i+1}} \cap K^{p^{-r-1}})$ and $(K^{p^{-r}} \cap L)(L^{p^i} \cap K^{p^{-r-1}} \cap M)$ are linearly disjoint over $(K^{p^{-r}} \cap L)(L^{p^{i+1}} \cap K^{p^{-r-1}} \cap M)$ for all $r$, $i$.*

PROOF. The idea of the proof is to simultaneously construct subbases for $L/K$ and $M/K$ with the desired property. Assume conditions (1) and (2) hold. Then $M$ is modular over $K$, [99], Proposition 1.4, p. 41. Let $A_e$ be a relative $p$-basis for $M$ over $K^{p^{-e+1}} \cap M$. By (1), $A_e$ remains $p$-independent over $K^{p^{-e+1}} \cap L$ and hence can be completed to a relative $p$-basis for $K^{p^{-e}} \cap L = L$ over $K^{p^{-e+1}} \cap L$ with $B_{e1}$. We now construct a relative $p$-basis for $K^{p^{-e+1}} \cap M$ over $K^{p^{-e+2}} \cap M$. $A_e^p$ is $p$-independent in

$$(K^{p^{-e+2}} \cap M)(L^p \cap K^{p^{-e+1}} \cap M)/(K^{p^{-e+2}} \cap M)$$

since $L/K$ is modular. Using Lemma 3.29, there exists $C_{e1} \subseteq L$ such that $A_e^p \cup C_{e1}^p$ is $p$-independent in $(K^{p^{-e+2}} \cap L)(L^p \cap K^{p^{-e+1}})$ over

$$K^{p^{-e+2}} \cap L = (K^{p^{-e+2}} \cap M)(L^{p^2} \cap K^{p^{-e+1}} \cap M)$$

and hence by Lemma 3.28, $A_e \cup B_{e1}$ can be replaced with $A_e \cup C_{e1} \cup B_{e2}$. Let $A_{e-1}$ be a relative $p$-basis for $K^{p^{-e+1}} \cap M$ over

$$(K^{p^{-e+2}} \cap M)(L^p \cap K^{p^{-e+1}} \cap M).$$

By Lemma 3.29, $A_{e-1}$ is $p$-independent in $K^{p^{-e+1}} \cap L$ over $(K^{p^{-e+2}} \cap L)(L^p \cap K^{p^{-e+1}})$, and hence $A_e^p \cup C_{e1}^p \cup B_{e2}^p \cup A_{e-1}$ is $p$-independent over $K^{p^{-e+2}} \cap L$ and can be completed to a relative $p$-basis for $K^{p^{-e+1}} \cap L$ over $K^{p^{-e+2}} \cap L$ with $B_{e-1,1}$. Thus we now have $T_e = A_e \cup C_{e1} \cup B_{e2}$, $T_{e-1} = A_{e-1} \cup B_{e-1,1}$ as part of a subbasis for $L/K$ and $T'_e = A_e$, $T'_{e-1} = C_{e1}^p \cup A_{e-1}$ as part of a subbasis for $M/K$. We assume that after the completion of the $(i-1)$ stage, we have constructed partial subbases

$$T_r = A_r \cup C_{r1} \cup \cdots \cup C_{r,i-e+r-2} \cup B_{r,i-e+r-1}$$

and

$$T'_r = A_r \cup C^{p^{e-r}}_{e,e-r} \cup \cdots \cup C^{p}_{r-1,1},$$

$e \geqslant r \geqslant e-i+2$. We now construct a relative $p$-basis for $K^{p^{-e+i-1}} \cap M$ over $K^{p^{-e+i}} \cap M$. This is done in $i$ steps via the intermediate fields

$$(K^{p^{-e+i}} \cap M)(L^{p^i} \cap K^{p^{-e+i-1}} \cap M)/(K^{p^{-e+i}} \cap M)(L^{p^{j+1}} \cap K^{p^{-e+i-1}} \cap M), \quad i-1 \geqslant j \geqslant 0,$$

and is done in descending order of $j$. Since $L/K$ is of bounded exponent $e$, the desired subbases are constructed in a finite number of steps.

DEFINITION 3.31. Let $G = A^K$ and let $B = \{x_{11}, \ldots, x_{1j_1}, \ldots, x_{n1}, \ldots, x_{nj_n}\}$ be a subbasis for $L/S$ where $x_{ij}$ has exponent $i$ over $S$. Let

$$B^J = \{d^{ij} \mid 1 \leqslant n,\ 1 \leqslant j \leqslant j_i\}$$

be the set of rank $p^e$ higher derivations defined on $L$ by $d^{ij}_u(x_{rs}) = \delta_{(i,j),(r,s)}$ if $u = p^{e-i}+1$ and 0 otherwise, and $\delta_{(i,j),(r,s)} = 1$ if $i = r$, $j = s$, and is 0 otherwise. Let $H^J = \delta(B^J)$. Then $G_L H^J = \{\sigma\delta(d^{ij}) \mid \sigma \in G_L,\ \delta(d^{ij}) \in H^J\}$ is called a *standard generating set* for $G$ with respect to $B$. An intermediate field $F$ is *distinguished* if and only if $F[\bar{x}]$ is invariant under some standard generating set.

The linear disjointness conditions of Theorem 3.30 yield a characterization of the distinguished subfields. We now derive a characterization of the distinguished subfields for the inseparable Galois theory.

THEOREM 3.32 ([23]). *Let $K$ be a Galois subfield of $L$ such that $[L : S] < \infty$. Let $G = A^K$ and let $F$ be a Galois intermediate field of $L/K$. Then the following conditions are equivalent.*
1. *$F/K$ is normal and $F \cap J$ is homogeneous in $J/K$.*
2. *$F[\bar{x}]$ is invariant under a standard generating set for $G$.*
3. *$F/K$ is normal and $SF$ is homogeneous in $L/S$.*

PROOF. (1) implies (2): Let

$$B = \{x_{11}, \ldots, x_{1j_1}, \ldots, x_{n1}, \ldots, x_{nj_n}\}$$

be a subbasis of $J/K$ such that for $k_1 \leqslant j_1, \ldots, k_n \leqslant j_n$,

$$C = \{x_{11}^{p^{e_{11}}}, \ldots, x_{1k_1}^{p^{e_{1k_1}}}, \ldots, x_{n1}^{p^{e_{n1}}}, \ldots, x_{nk_n}^{p^{e_{nk_n}}}\}$$

is a subbasis of $(F \cap J)/K$. Let $G_L H^J$ be a standard generating set for $G$ with respect to $B$ and $\sigma f^{ij} \in G_L H^J$ where $f^{ij} = \delta(d^{ij})$. Since $C$ generates $F$ over $S \cap F$, it suffices to show $\sigma f^{ij}(x_{rs}^{p^{e_{rs}}}) \in F[\bar{x}]$ and $\sigma f^{ij}(s) \in F$ where $s \in S \cap F$. Clearly $\sigma f^{ij}(s) \in S \cap F$ since $f^{ij}$ is the identity on $S$ and $(S \cap F)/K$ is necessarily normal. It follows that $\sigma f^{ij}(x_{rs}^{p^{e_{rs}}}) \in (F \cap J)[\bar{x}]$ since $\sigma$ is the identity on $J[\bar{x}]$.

(2) implies (3): Let $G_L H^J$ be the standard generating set. Since the identity map is in $G_L$, $F[\bar{x}]$ is invariant under $H^J$. Thus $F[\bar{x}]$ is invariant under $G_L$. Since also $L$ is invariant under $G_L$, $F[\bar{x}] \cap L = F$ is invariant under $G_L$ and $F/K$ is normal. Since $F[\bar{x}]$ is invariant under $H^J$ and every element of $H^J$ is the identity on $S$, $SF[\bar{x}]$ is invariant under $H^J$. Thus $SF$ is invariant under $\delta^{-1}(H^J)$ which is a standard generating set for $L/S$. Thus $SF$ is homogeneous in $L/S$.

(3) implies (1): We show that $F \cap J = M$ satisfies conditions (1) and (2) of Theorem 3.30. Condition (1) follows using the fact that $SF$ is homogeneous in $L/S$. To show condition (2), we have $(S^{p^{-r}} \cap L)(L^{p^{i+1}} \cap S^{p^{-r-1}})$ and $(S^{p^{-r}} \cap L)(L^{p^i} \cap S^{p^{-r-1}} \cap SF)$ are linearly disjoint over $(S^{p^{-r}} \cap L)(L^{p^{i+1}} \cap S^{p^{-r-1}} \cap SF)$ since $SF$ is homogeneous in $L/S$. Since $S^{p^{-r}} \cap L = S(K^{p^{-r}} \cap J)$ and $S = S^{p^{i+1}} \otimes_{K^{p^{i+1}}} K$, it follows that

$$(S^{p^{-r}} \cap L)(L^{p^{i+1}} \cap S^{p^{-r-1}} \cap SF) = S(K^{p^{-r}} \cap J)(J^{p^{i+1}} \cap K^{p^{-r-1}} \cap M).$$

Similarly,

$$(S^{p^{-r}} \cap L)(L^{p^{i+1}} \cap S^{p^{-r-1}}) = S(K^{p^{-r}} \cap J)(J^{p^{i+1}} \cap K^{p^{-r-1}})$$

and

$$(S^{p^{-r}} \cap L)(L^{p^i} \cap S^{p^{-r-1}} \cap F) = S(K^{p^{-r}} \cap J)(J^{p^i} \cap K^{p^{-r-1}} \cap M).$$

It follows that $(K^{p^{-r}} \cap J)(J^{p^{i+1}} \cap K^{p^{-r-1}})$ and $(K^{p^{-r}} \cap J)(J^{p^i} \cap K^{p^{-r-1}} \cap M)$ are linearly disjoint over $(K^{p^{-r}} \cap J)(J^{p^{i+1}} \cap K^{p^{-r-1}} \cap M)$. Thus $M = F \cap J$ satisfies (1) and (2) of Theorem 3.30. Hence $F \cap J$ is homogeneous in $J/K$. □

COROLLARY 3.33. *Let $K$ be a Galois subfield of $L$. Let $G = A^K$, $S = L^{G_0}$, and $J = L^{G_L}$. Let $F$ be a Galois intermediate field. Then $F$ is distinguished if and only if $F = S_1 \otimes_K J_1$ where $S_1$ is normal over $K$ and there exists a subbasis $\{x_1, \ldots, x_n\}$ for $J$ over $K$ such that $\{x_1^{p^{n_1}}, \ldots, x_s^{p^{n_s}}\}$ is a subbasis of $J_1$ over $K$, $s \leqslant n$.*

## 4.

THEOREM 4.1 ([53]). *Let $H$ be the field of constants of a set of infinite rank higher derivations on $L$. Then $L$ is regular over $H$.*

PROOF. We show $L^p$ and $H$ are linearly disjoint over $H^p$. Let $\{z_1, \ldots, z_n\} \subset H$ be independent over $H^p$ and assume we can find a relation of the form $z_1 + a_2^p z_2 + \cdots + a_s^p z_s$ of minimal length with $a_i \in L$, $a_2 \notin H$. If $d_j$ is a map of some higher derivation with $d_j(a_2) \neq 0$, applying $d_{jp}$ to the relation yields a shorter one. Thus $L$ is separable over

$H$. An element of $L$ separable algebraic over $H$ can be expressed as an arbitrarily high $p^n$-th power, and hence is mapped to zero by any map in a higher derivation. □

The dimension of an infinite rank higher derivation is the transcendence degree of $L$ over its field of constants.

The following result is the infinite analogue of Theorem 2.10.

THEOREM 4.2 ([53]). *Let $P = \{d^{(1)}, \dots, d^{(n)}\}$ be an abelian set of one-dimensional higher derivations on $L$ over $M$ and let their field of constants be $H$. Then*

1) $\mathrm{tr.d}(L/H) \leqslant n$;
2) *if $F$ is independent, then* $\mathrm{tr.d}(L/H) = n$.

PROOF. For (1) we use induction. Let $H_1$ be the field of constants of $d^{(1)}$ and $H_{n-1}$ the field of constants of $\{d^{(2)}, \dots, d^{(n)}\}$. By induction the transcendence degree of $L$ over $H_1$ is one and over $H_{n-1}$ is at most $n-1$ and $H = H_1 \cap H_{n-1}$. The abelian condition is used to show $H_1$ and $H_{n-1}$ are linearly disjoint, and hence free. Let $H_{1,r}$ be the field of constants of the first $p^r$ maps of $d^{(1)}$ and let $H_{n-1,r} = H_{1,r} \cap H_{n-1}$. Then $H_{n-1}$ is purely inseparable over $H_{n-1,r}$ and since $\{d^{(2)}, \dots, d^{(n)}\}$ restricted to $H_{1,r}$ have field of constants $H_{n-1,r}$, $H_{1,r}$ is a regular extension of $H_{n-1,r}$. Thus $H_{1,r}$ and $H_{n-1}$ are linearly disjoint for all $r$ and hence $\bigcap H_{1,r} = H_1$ and $H_{n-1}$ are linearly disjoint. (2) follows since the dimension of the space of derivations of $L$ over $H$ is $n$. □

We assume $L$ is finitely generated over $M$.

The fields of constants in this section will be the fields of constants of sets of higher derivations of finite or infinite rank. If $S$ is such a set and $S_n$ denotes the set of $n$-th sections of $S$, the field of constants of $S$ will be the intersection of the fields of constants of the $S_n$. Since $L$ is modular over $L^{S_n}$, $L$ is modular over $L^S$. Since $L/L^S$ is finitely generated and modular, $L = H \otimes_{L^S} M$ where $L/H$ is regular, $H/L_S$ is purely inseparable modular of finite exponent.

DEFINITION 4.3. A set $\{x_1, \dots, x_n\} \subset L$ will be called a *tensor basis* of $L/M$ if

$$L = M\big(L^{p^t}\big)(x_1) \otimes \cdots \otimes M\big(L^{p^t}(x_n)\big)$$

for all $t \geqslant 1$, tensor product being over $M(L^{p^t})$.

THEOREM 4.4 ([49]). *If $M$ is separably algebraically closed in $L$ then $M$ is a field of constants if and only if $L/M$ has a tensor basis. The tensor basis of $L/M$ are the sets $S \cup T$ where $S$ is a tensor basis (subbasis) for $\overline{M}/M$ and $T$ is a separating transcendence basis of $L/\bar{M}$.*

PROOF. The idea of the proof is that $L$ is modular over $H$ and as such splits as a tensor product $M \otimes_H R$ where $M$ is purely inseparable with a subbasis over $H$ and $R$ has a separating transcendence basis over $H$. □

To establish a Galois type correspondence it is necessary to make a group generated by $H^n(L/M)$ and $H^\infty(L/M)$. For simplicity, [49] uses only the $H^{p^n}(L/M)$.

Define the maps $V_{m,n} : H^{p^n}(L/M) \to H^{p^m}(L/M)$ where $m > n$ by $V_{m,n}(d) = (f)$; $f_{p^{m-n}i} = d_i$ for $1 \leqslant i \leqslant p^n$, and $f_j = 0$ for $1 \leqslant j \leqslant p^m$ and $j \nmid p^{m-n}$. Then

1) $V_{m,n}(fg) = V_{m,n}(f)V_{m,n}(g)$;
2) $V_{m,n}$ is injective;
3) $V_{r,m}V_{m,n} = V_{r,n}$.

Thus $\{H^{p^n}(L/M),\ V_{m,n} \mid n \geqslant 0,\ m \geqslant 1\}$ is a directed set of groups. Let $\overline{H}(L/M)$ be the direct limit of this system. Let $\bar{d}$ be the element of $\overline{H}(L/M)$ containing $d$. $\bar{d}$ is called the pencil of $d$. A $d$ in $H^{p^n}(L/M)$ is noncontractible if $d$ does not have the form $V_{m,n}(f)$ for some $n$ and $f$. The rank of the unique noncontractible $d$ in any pencil is the rank of the pencil.

A higher derivation $d$ of rank $n$ is the $n$-th section of $f$ if rank $f \geqslant n$ and $d_i = f_i$, $1 \leqslant i \leqslant n$. $\bar{d}$ is a section of $\bar{f}$ if each $d \in \bar{d}$ is a section of some $f \in \bar{f}$. The extended rank of $\bar{d} = \sup\{\text{rank}\bar{f} \mid \bar{d} \text{ is a section of } \bar{f}\}$. Let $\overline{H}^{\infty}(L/M) = \{\bar{d} \in \overline{H}(L/M) \mid \text{extended rank } \bar{d} = \infty\}$.

THEOREM 4.5 ([49]). *If $\overline{H}(L/M)$ is nontrivial, $\overline{H}^{\infty}(L/M) = \overline{H}(L/H)$, where as usual $H$ is the unique minimal intermediate field of $L/M$ over which $L$ is regular.*

PROOF. If $x$ is purely inseparable over $M$ of exponent $m$, then

$$0 = d_{jp^m}\left(x^{p^m}\right) = \left(d_j(x)\right)^{p^m}$$

for any map $d_{jp^m}$ of a higher derivation. Thus if $d_j$ is a section of $f \in H^{p^m}(L/M)$ for arbitrary $m > 0$, $d_j(x) = 0$. Thus $\overline{H}^{\infty}(L/M) \subset \overline{H}(L/H)$. The other containment is straightforward. □

The fields of constants of our subgroups will be the subfields $M$ over which $L$ has a tensor basis. Thus it remains to determine when a subgroup will be a full group. The characterization will be similar to that in Section 2. However, in this section iterative higher derivations are required to have index 1 or to be normal, i.e. $d_1 \neq 0$. A higher derivation $d$ on $L$ with constant field $M$ is one dimensional if $L$ has a tensor basis of 1 element. The proofs of the next two results are similar to those of Section 2.

PROPOSITION 4.6. *Let $d$ be an iterative higher derivation on $L$ with constant field $M$. The following are equivalent.*

1. *$d$ is one dimensional.*
2. *The constant field of the $p^n$-th section of $d$ is $M(L^{p^{n+1}})$ for all $n \geqslant 0$.*
3. *The constant field of the $p^n$-th section of $d$ is $M(L^{p^{n+1}})$ for some $n \geqslant 0$.*

If $A$ is a finite set of normal one dimensional higher derivations, let $A_t = \{d \mid d \in A$ and rank $d \leqslant p^t$ or $d$ is the $p^t$-th section of some $f$ in $A\}$.

THEOREM 4.7 ([49]). *Let $A$ be a finite set of one dimensional higher derivations on $L$. The following are equivalent.*

1. *$A$ is an abelian independent set of iterative higher derivations.*

2. $A_t$ *has a dual basis* $A_t^*$ *(see Section* 2) *for all* $t > 0$. *If* $A$ *satisfies* (1) *or* (2), *then* $A_t^*$ *is a tensor basis of* $L$ *over the constant field of* $A_t$.

DEFINITION 4.8. A finite set of higher derivations satisfying the conditions of Theorem 4.7 is called a *standard set of generators.*

Obtaining a set of generators of a Galois group from a standard set of generators requires the following constructions and definitions. If $d$ has rank $t$ and $q > 0$ is an integer then $v_q(d) = f$ where rank $f = qt$, $f_{qi} = d_i$ and $f_i = 0$ if $q \nmid j$. If $p^m \leqslant q^t$ then $d_{(q,m)}$ is the $p^m$-th section of $v_q(d)$. Given $a$ in $L$ and a higher derivation $d$, then $ad = \{a^i d_i\}$ is a higher derivation. Given a set $D$ of higher derivations on $L$, let $(LD)_m$ be the group of rank $p^m$ higher derivations generated by the set of all $(ad)_{(q,m)}$ for $a \in L$, $d \in D$ and any allowable $q$. $(L\overline{D})$ represents the group of pencils generated by $\{d \mid d \in (LD)_m \text{ for some } m \geqslant 0\}$.

THEOREM 4.9 ([49]). *Let* $L/K$ *be finitely generated. A subgroup* $\overline{H}$ *of* $\overline{H}(L/K)$ *has the form* $\overline{H}(L/M)$ *if and only if* $\overline{H} = (L\overline{A})$ *where* $A$ *is a standard set of generators constant on* $M$. *Let* $\mathcal{F} = \{M \mid L \supset M \supset K \text{ with } L/M \text{ Galois}\}$ *and* $\mathcal{G} = \{(L\overline{A}) \mid A$ *a standard set of generators on* $L$ *with constant field* $L_{\mathcal{A}} \supset K\}$. *The map* $\sigma$: $\mathcal{F} \to \mathcal{G}$ *where* $\sigma(M) = \overline{H}(L/M)$ *and* $\tau$: $\mathcal{G} \to \mathcal{F}$ *where* $\tau((L\overline{A})) = L_{\mathcal{A}}$ *are inverse bijections.*

PROOF. The proof is an approximation process similar to that of Theorem 2.16, with however many more technical obstacles to be overcome. □

Heerema developed the theory in the situation where $L$ is finitely generated over $M$. He also proved the following results on the intermediate theory in this situation.

THEOREM 4.10 ([49]). *Let* $L/M$ *be Galois. An intermediate field* $H$ *is invariant under* $H(L/M)$, *the set of all higher derivations on* $L/M$ *if and only if* $H = M(L^{p^r})$ *for some* $r \geqslant 0$.

THEOREM 4.11 ([49]). *Given* $L/H$ *and* $H/M$ *Galois, every higher derivation on* $H/M$ *into* $L$ *extends to a higher derivation on* $L$ *if and only if there is a finite purely inseparable modular extension* $T$ *of* $M$ *in* $K$ *such that* $H(T) = H \otimes_M T$ *and* $L/H(T)$ *is regular.*

It is natural to attempt to extend the Galois theory of pencils of higher derivations to the situation where $L/M$ is not finitely generated.

PROPOSITION 4.12 ([21]). *Let* $K$ *be a subfield of* $L$. $K$ *is the field of constants of a set (and hence a group) of pencils on* $L$ *if and only if* $L/K$ *is modular and* $\bigcap_n K(L^{p^n}) = K$.

COROLLARY 4.13 ([21]). *Let* $K$ *be any subfield of* $L$. *The field of constants of the group of all pencils on* $L$ *over* $K$ *is* $\bigcap Q^*(L^{p^n})$ *where* $Q^*$ *is the unique minimal intermediate field over which* $L$ *is modular.*

The following result gives the most general situation where a complete theory could be developed.

THEOREM 4.14 ([21]). *Suppose $L/K$ is modular. Then every intermediate field $F$ such that $L/F$ is modular and $F$ is separably algebraically closed in $L$ is the field of constants of a group of pencils on $L$ if and only if $K(L^{p^e})$ has a finite separating transcendence basis over $K$ for some non-negative e.*

PROOF. The essence of the proof is to construct two examples. The first is to show if $L$ is purely inseparable modular over $K$ with a subbasis of unbounded exponent, then there is a proper intermediate field $F$ with $L$ modular and relatively perfect, $(L = F(L^p))$, over $F$. Let $B = \bigcup B_i$ with the elements of $B_i$ of exponent $i$ over $K$. Let $x_{ij} \in B$ be such that $x_{ij}$ has exponent $i_j$ over $K$, $i_j \leqslant i_{j+1}$, $1 \leqslant j < \infty$. Then

$$F = K(B \setminus \{x_{ij}, x_{i_1} - x_{i_2}^{p^{i_2 - i_1}}, \dots\}$$

is the desired proper intermediate field since [72], p. 20, shows the intermediate fields are chained and infinite in number. □

The second is to show that if $L$ is regular over $K$ with an infinite separating transcendence basis, then there is a proper intermediate field $F$ with $L$ regular and relatively perfect over $F$. We can assume $L$ has $\{x_i \mid 1 \leqslant i < \infty\}$ as a separating transcendence basis. The $\{x_i\}$ is a relative $p$-basis of $L$ over $K$. But $\{x_1 x_2^p, x_2 x_3^p, \dots\}$ is also a relative $p$-basis. Thus $L$ is separable over $K(\{x_1 x_2^p, x_2 x_3^p, \dots\})$. But $\{x_1 x_2^p, x_2 x_3^p, \dots\}$ is algebraically independent over $K$, and hence $L$ is regular and relatively perfect over the algebraic closure of $K(\{x_1 x_2^p, x_2 x_3^p, \dots\})$ in $L$.

Let $K \subset F$ be Galois subfields of $L$. [21] and [22] examine the question of when the group of pencils of $L/F$ will be a normal subgroup of the group of pencils of $L/K$. If the characteristic of $L$ is not 2, [22] shows that a necessary and sufficient condition is that $F = K(L^{p^n})$ for some $n$.

Let $K$ be a Galois subfield of $L$ and $\overline{K}$ the algebraic closure of $K$ in $L$. It is always true that $\overline{K}$ is Galois over $K$. In the situation where $K(L^{p^e})$ has a finite separating transcendence basis over $K$, it is also true that $L/\overline{K}$ is Galois and $L/K$ splits as a tensor product of a purely inseparable modular extension and a regular extension. However, in a general setting this is no longer true. $L$ need not be Galois over $\overline{K}$ [22] and even if it is, $L/K$ need not split as a tensor product [22].

## References


[1] M.F. Becker and S. MacLane, *The minimum number of generators for inseparable algebraic extensions*, Bull. Amer. Math. Soc. **46** (1940), 182–186.

[2] F.P. Callahan, *An identity satisfied by derivations of a purely inseparable extension*, Amer. Math. Monthly **80** (1973), 40–42.

[3] S. Chase, *On inseparable Galois theory*, Bull. Amer. Math. Soc. **77** (1971), 413–417.

[4] S.U. Chase and A. Rosenberg, *A theorem of Harrison, Kummer theory, and Galois algebras*, Nagoya J. Math. **27** (1966), 665–685.

[5] S.U. Chase and M.E. Sweedler, *Hopf algebras and Galois theory*, SLNM 97, Springer, Berlin (1969).

[6] P.M. Cohn, *On the decomposition of a field as a tensor product*, Glasgow Math. J. **20** (1979), 141–145.

[7] R.L. Davis, *A Galois theory for a class of purely inseparable field extensions*, Dissertation, Florida State University, Tallahassee, Florida, USA (1969).

[8] R.L. Davis, *A Galois theory for a class of purely inseparable exponent two field extensions*, Bull. Amer. Math. Soc. **75** (1969), 1001–1004.

[9] R.L. Davis, *Higher derivations and field extensions*, Trans. Amer. Math. Soc. **180** (1973), 47–52.

[10] J.K. Deveney, *Fields of constants of infinite higher derivations*, Proc. Amer. Math. Soc. **41** (1973), 394–398.

[11] J.K. Deveney, *An intermediate theory for a purely inseparable Galois theory*, Trans. Amer. Math. Soc. **198** (1974), 287–295.

[12] J.K. Deveney, *Pure subfields of purely inseparable field extensions*, Canad. J. Math. **28** (1976), 1162–1166.

[13] J.K. Deveney, *A counter-example concerning inseparable field extensions*, Proc. Amer. Math. Soc. **55** (1976), 33–34.

[14] J.K. Deveney, *Generalized primitive elements for transcendental field extensions*, Pacific J. Math. **68** (1977), 41–45.

[15] J.K. Deveney, *Induced subspaces of derivations*, Arch. Math. **35** (1981), 528–532.

[16] J.K. Deveney, *$\omega_0$-generated field extensions*, Arch. Math. **47** (1986), 410–412.

[17] J.K. Deveney and N. Heerema, *Higher derivations and distinguished subfields*, Canad. J. Math. **40** (1988), 131–141.

[18] J.K. Deveney and J.N. Mordeson, *Subfields and invariants of inseparable field extensions*, Canad. J. Math. **29** (1977), 1304–1311.

[19] J.K. Deveney and J.N. Mordeson, *Invariants of reliable field extensions*, Arch. Math. **29** (1977), 141–147.

[20] J.K. Deveney and J.N. Mordeson, *Invariant subgroups of groups of higher derivations*, Proc. Amer. Math. Soc. **68** (1978), 277–280.

[21] J.K. Deveney and J.N. Mordeson, *On Galois theory using pencils of higher derivations*, Proc. Amer. Math. Soc. **72** (1978), 233–238.

[22] J.K. Deveney and J.N. Mordeson, *Pencils of higher derivations of arbitrary field extensions*, Proc. Amer. Math. Soc. **74** (1979), 205–211.

[23] J.K. Deveney and J.N. Mordeson, *Invariance in inseparable Galois theory*, Rocky Mountain J. Math. **9** (1979), 395–403.

[24] J.K. Deveney and J.N. Mordeson, *The order of inseparability of fields*, Canad. J. Math. **31** (1979), 655–662.

[25] J.K. Deveney and J.N. Mordeson, *Splitting and modularly perfect fields*, Pacific J. Math. **83** (1979), 45–54.

[26] J.K. Deveney and J.N. Mordeson, *Inseparable extensions and primary abelian groups*, Arch. Math. **33** (1979), 538–545.

[27] J.K. Deveney and J.N. Mordeson, *Distinguished subfields*, Trans. Amer. Math. Soc. **260** (1980), 185–193.

[28] J.K. Deveney and J.N. Mordeson, *Distinguished subfields of intermediate fields*, Canad. J. Math. **33** (1981), 1085–1096.

[29] J.K. Deveney and J.N. Mordeson, *Calculating invariants of inseparable field extensions*, Proc. Amer. Math. Soc. **81** (1981), 373–376.

[30] J.K. Deveney and J.N. Mordeson, *Transcendence bases for field extensions*, J. Math. Soc. Japan **34** (4) (1982), 703–707.

[31] J.K. Deveney and J.N. Mordeson, *Maximal separable subfields of bounded codegree*, Proc. Amer. Math. Soc. **88** (1) (1983), 16–20.

[32] J.K. Deveney and J.N. Mordeson, *Subfields containing distinguished subfields*, Arch. Math. **40** (1983), 509–515.

[33] J.K. Deveney and J.N. Mordeson, *Uniqueness of subfields*, Canad. Math. Bull. **29** (2) (1986), 191–196.

[34] J.K. Deveney and J.N. Mordeson, *Modularity and separability of function fields*, Chinese J. Math. **14** (1986), 125–131.

[35] J. Dieudonné, *Sur les extensions transcendantes separables*, Summa Brazil Math. **2** (1) (1947), 1–20.

[36] B.I. Eke, *Special generating sets of purely inseparable fields of unbounded exponent*, Pacific J. Math. **128** (1987), 73–79.

[37] M. Gerstenhaber, *On the Galois theory of inseparable extensions*, Bull. Amer. Math. Soc. **70** (1964), 561–566.

[38] M. Gerstenhaber, *On infinite inseparable extensions of exponent one*, Bull. Amer. Math. Soc. **71** (1965), 878–881.

[39] M. Gerstenhaber, *On modular field extensions*, J. Algebra **10** (1968), 478–484.

[40] M. Gerstenhaber and A. Zaromp, *On the Galois theory of purely inseparable field extensions*, Bull. Amer. Math. Soc. **76** (1970), 1011–1014.

[41] R. Gilmer and W. Heinzer, *On the existence of exceptional field extensions*, Bull. Amer. Math. Soc. **74** (1968), 545–547.

[42] G.F. Haddix, J.N. Mordeson and B. Vinograde, *Lattice isomorphisms of purely inseparable extensions*, Math. Z. **111** (1969), 169–174.

[43] G.F. Haddix, J.N. Mordeson and B. Vinograde, *On purely inseparable extensions of unbounded exponent*, Canad. J. Math. **21** (1969), 1526–1532.

[44] G.F. Haddix, J.N. Mordeson and B. Vinograde, *On the two main intermediate field towers of purely inseparable extensions*, Math. Z. **119** (1971), 21–27.

[45] H. Hasse and R. Schmidt, *Noch eine Begrundung der Theorie der höheren Differentialquotienten in einem algebraischen Functionenkörper einem Unbestimmten*, J. Reine Angew. Math. **17** (1937), 215–237.

[46] N. Heerema, *Derivations and embeddings of a field in its power series ring*, Proc. Amer. Math. Soc. **11** (1960), 188–194.

[47] N. Heerema, *Derivations and embeddings of a field in its power series ring, II*, Michigan Math. J. **8** (1961), 129–134.

[48] N. Heerema, *Convergent higher derivations on local rings*, Trans. Amer. Math. Soc. **132** (1968), 31–44.

[49] N. Heerema, *A Galois theory for inseparable field extensions*, Trans. Amer. Math. Soc. **154** (1971), 193–200.

[50] N. Heerema, *p-th powers of distinguished subfields*, Proc. Amer. Math. Soc. **55** (1976), 287–292.

[51] N. Heerema, *Higher derivation Galois theory of fields*, Trans. Amer. Math. Soc. **265** (1981), 169–179.

[52] N. Heerema, *Maximal separable intermediate fields of large codegree*, Proc. Amer. Math. Soc. **82** (1982), 351–354.

[53] N. Heerema and J.K. Deveney, *Galois theory for fields $K/k$ finitely generated*, Trans. Amer. Math. Soc. **189** (1974), 263–274.

[54] N. Heerema and D. Tucker, *Modular field extensions*, Proc. Amer. Math. Soc. **53** (1975), 1–6.

[55] N. Jacobson, *Galois theory of purely inseparable fields of exponent one*, Amer. J. Math. **66** (1944), 645–648.

[56] N. Jacobson, *Lectures in Abstract Algebra, Vol. III: Theory of Fields and Galois Theory*, Van Nostrand, Princeton, NJ (1964).

[57] G. Karpilovsky, *Topics in Field Theory*, Mathematics Studies 155, North-Holland, Amsterdam (1989).

[58] L.A. Kime, *Purely inseparable modular extensions of unbounded exponent*, Trans. Amer. Math. Soc. **176** (1973), 335–349.

[59] L.A. Kime, *Allowable diagrams for purely inseparable field extensions*, Proc. Amer. Math. Soc. **41** (1973), 389–393.

[60] A. Kinohara, *On the derivations and the relative differents in commutative fields*, J. Sci. Hiroshima Univ. **16** (1953), 441–456.

[61] K. Kosaki and H. Yanagihara, *On purely inseparable extensions of algebraic function fields*, J. Sci. Hiroshima Univ. Ser. A I **34** (1970), 69–72.

[62] H. Kraft, *Inseparable Körperweiterungen*, Comment. Math. Helv. **45** (1970), 110–118.

[63] H. Kreimer and N. Heerema, *Modularity vs. separability for field extensions*, Canad. J. Math. **27** (1975), 1176–1182.

[64] J. Lipman, *Balanced field extensions*, Amer. Math. Monthly **73** (1966), 373–374.

[65] S. MacLane, *Modular fields, I. Separating transcendency bases*, Duke Math. J. **5** (1939), 372–393.

[66] M. Miyanishi, *A remark on an iterative infinite higher derivation*, J. Math. Kyoto Univ. **8** (3) (1968), 411–415.

[67] J.N. Mordeson, *On a Galois theory for inseparable field extensions*, Trans. Amer. Math. Soc. **214** (1975), 337–347.

[68] J.N. Mordeson, *Splitting of field extensions*, Arch. Math. **26** (1975), 606–610.

[69] J.N. Mordeson, *Modular extensions and abelian groups*, Arch. Math. **36** (1981), 13–20.

[70] J.N. Mordeson, *Algebraically compact field extensions*, Arch. Math. **38** (1982), 175–183.

[71] J.N. Mordeson and B. Vinograde, *Generators and tensor factors of purely inseparable fields*, Math. Z. **107** (1968), 326–334.

[72] J.N. Mordeson and B. Vinograde, *Structure of Arbitrary Purely Inseparable Extension Fields*, SLNM 173, Springer, Berlin (1970).

[73] M. Norris and W.Y. Velez, *A characterization of the splitting of inseparable algebraic field extensions*, Amer. Math. Monthly **85** (1978), 338–341.

[74] M. Ojanguren and R. Sridharan, *A note on purely inseparable extensions*, Comment. Math. Helv. **44** (1969), 457–561.

[75] U. Orbanz, *Tensorprodukte und höhere Derivationen*, Arch. Math. **24** (1973), 513–520.

[76] G. Paregis, *Bemerkung über Tensorprodukte von Körpern*, Arch. Math. **31** (1970), 479–482.

[77] F. Pauer, *Invariante Zwischenkörper endlicher Körperweiterungen*, Manuscripta Math. **29** (1979), 147–157.

[78] G. Pickert, *Inseparable Körperweiterungen*, Math. Z. **52** (1949), 81–135.

[79] G. Pickert, *Eine Normalform für endliche rein-inseparable Körperweiterungen*, Math. Z. **53** (1950), 133–135.

[80] G. Pickert, *Zwischenkörperverbande endlicher inseparabler Erweiterungen*, Math. Z. **55** (1952), 355–363.

[81] P. Ponomarenko, *The Galois theory of infinite purely inseparable extensions*, Bull. Amer. Math. Soc. **71** (1965), 876–877.

[82] A. Popescu, *Galois type correspondence for non-separable normal extensions of fields*, Proc. Japan Acad. Ser. A Math. Sci. **62** (1986), 213–215.

[83] A. Popescu, *Galois correspondence for algebraic extensions of fields*, Stud. Cerc. Mat. **39** (1987), 187–227.

[84] E.L. Popescu and N. Popescu, *On a class of intermediate subfields*, Stud. Cerc. Mat. **39** (1987), 156–162.

[85] R. Rasala, *Inseparable splitting theory*, Trans. Amer. Math. Soc. **162** (1971), 411–448.

[86] J. Reid, *A note on inseparability*, Michigan Math. J. **13** (1966), 219–223.

[87] P. Rygg, *On minimal sets of generators of purely inseparable field extensions*, Proc. Amer. Math. Soc. **14** (1963), 742–745.

[88] F.K. Schmidt and S. MacLane, *The generation of inseparable fields*, Proc. Nat. Acad. Sci. **27** (1941), 583–587.

[89] S. Shatz, *Galois theory*, Proc. Battelle Conference on Categorical Algebra, SLNM 86, Springer, Berlin (1969).

[90] E. Steinitz, *Algebraische Theorie der Körper*, de Gruyter, Berlin (1930).

[91] S. Suzuki, *Some types of derivations and their applications to field theory*, J. Math. Kyoto Univ. **21** (2) (1981), 375–382.

[92] M.E. Sweedler, *Structure of inseparable extensions*, Ann. Math. (2) **87** (1968), 401–410.

[93] M.E. Sweedler, *Correction to: Structure of inseparable extensions*, Ann. Math. (2) **89** (1969), 206–207.

[94] M.E. Sweedler, *The Hopf algebra of an algebra applied to field theory*, J. Algebra **8** (1968), 262–276.

[95] M. Takeuchi, *A characterization of the Galois subalgebras* $H_k(H/F)$, J. Algebra **42** (1976), 315–362.

[96] O. Teichmüller, *p-Algebren*, Deutsche Mathematik vol. 1 (1936), 362–388.

[97] D. Tucker, *Finitely generated inseparable field extensions*, Notices Amer. Math. Soc. **21** (1974), A-607.

[98] K. Uchida, *On fields that satisfy* $[K : K^p] = p$, Sugaku **24** (1972), 313–316.

[99] W. Waterhouse, *The structure of inseparable field extensions*, Trans. Amer. Math. Soc. **211** (1975), 39–56.

[100] A. Weil, *Foundations of Algebraic Geometry*, Amer. Math. Soc. Colloq. Publ. vol. 29, Amer. Math. Soc., Providence, RI (1946).

[101] M. Weisfeld, *Purely inseparable extensions and higher derivations*, Trans. Amer. Math. Soc. **116** (1965), 435–449.

[102] D. Winter, *Structure of Fields*, Graduate Texts in Mathematics vol. 16, Springer, Berlin (1974).

[103] F. Zerla, *Iterative higher derivations in fields of prime characteristic*, Michigan Math. J. **15** (1968), 402–415.

# Complete Discrete Valuation Fields. Abelian Local Class Field Theories

I.B. Fesenko

*Department of Mathematics, University of Nottingham, NG7 2RG Nottingham, United Kingdom*
*e-mail: i.fesenko@maths.nott.ac.uk*

*Contents*



HANDBOOK OF ALGEBRA, VOL. 1
Edited by M. Hazewinkel

Local behaviour of a 1-dimensional scheme $X$ near a "nice" point $x$ is described by the local ring $\mathcal{O}_{X,x}$ whose completion is a complete discrete valuation ring with residue field $k(x)$. When the latter is finite this ring is the ring of integers of a local field.

The first local fields in characteristic zero – the $p$-adic fields $\mathbb{Q}_p$ and their finite extensions for a prime $p$ were introduced by Hensel in a series of papers beginning from 1897. These fields possess some properties similar to those of formal power series fields $\mathbb{F}_q((X))$ over a finite field $\mathbb{F}_q$, $q = p^f$, $f \geqslant 1$. Though there are essential distinctions, common features are crucial. Numerous profound theories were borned as an attempt to translate an existing theory from positive characteristic to characteristic zero and conversely.

In general, the class of complete discrete valuation fields seems to be the next in importance and comparatively simple after the class of finite fields. It is closely connected with global fields – algebraic number and rational function fields. The famous Hasse local-global principle solves global problems by appealing to local ones.

Local class field theory is one of the highest tops of classical algebraic number theory. It establishes a 1–1 correspondence between abelian extensions of a complete discrete valuation field $F$ whose residue field is finite and subgroups in the multiplicative group $F^*$. Historically this theory appeared as a consequence of the global one in the 1930's in the work of Hasse. Later F.K. Schmidt and Chevalley found an independent of global exposition. Postwar period of the theory may be characterized as comprising the incorporation of cohomological methods. A modern statement of class field theory employs calculations in cohomology groups (see, e.g., [Se2]).

One can now observe a new period in evolution of class field theory. The first work in this direction was a paper of Dwork [Dw]. He pointed out a way to compute values of the reciprocity map. This trend was continued by Hazewinkel [Haz1, Haz2], who gave a noncohomological exposition of the theory. A still more simple construction for local and global fields was proposed by Neukirch [N3, N4]. The Hazewinkel and Neukirch constructions were generalized for the case of arbitrary residue field of positive characteristic (perfect [Fe5] and imperfect [Fe7]). As corollaries utmost generalizations of classical results follow.

Proper objects which describe local behaviour of an $n$-dimensional scheme near a closed point are so-called $n$-discrete valuation fields studied by Parshin and Kato in the middle of the 1970-s. They developed two independent approaches to higher local class field theory. Abelian extensions of complete $n$-discrete valuation fields are described by closed subgroups in topological $K$-groups. Later Koya found by using Lichtenbaum complex a 2-dimensional formation classes approach. An easy and explicit higher local class field theories is yielded if one extends the Hazewinkel and Neukirch constructions [Fe3, Fe4, Fe9].

We discuss in this review only main topics connected with local fields. For proofs and details see [FV]. For more details about higher local fields see [Fe7]. The bibliography gives references to some topics uncovered in this review.

I am grateful to many mathematicians for their remarks on a preliminary version of this work [Fe8] published in 1992.

## 1. Discrete valuation fields

### 1.1. *Definitions and examples*

**1.1.1.** Let $\Gamma$ be an additively written totally ordered group. Denote $\Gamma' = \Gamma \cup \{+\infty\}$, where $+\infty$ is a formal element with properties: $a \leqslant +\infty$, $+\infty \leqslant +\infty$, $a+(+\infty) = +\infty$, $(+\infty) + (+\infty) = +\infty$. Let $F$ be a field. A map $v\colon\ F \to \Gamma'$ with the properties:

$$v(\alpha) = +\infty \Leftrightarrow \alpha = 0,$$
$$v(\alpha\beta) = v(\alpha) + v(\beta),$$
$$v(\alpha+\beta) \geqslant \min\big(v(\alpha), v(\beta)\big)$$

is called a *valuation* on $F$, $F$ is called a *valuation field*. The map $v$ induces a homomorphism of $F^*$ to $\Gamma$. If $v(F^*) = \{0\}$ then $v$ is called a trivial valuation.

**1.1.2.** For any valuation $v$ one can define the *ring of integers*

$$\mathfrak{O}_v = \{\alpha \in F^*\colon\ v(\alpha) \geqslant 0\}$$

and its ideal

$$\mathfrak{M}_v = \{\alpha \in F^*\colon\ v(\alpha) > 0\}.$$

Then $\mathfrak{M}_v$ is the unique *maximal ideal* of $\mathfrak{O}_v$ and the field $\overline{F}_v = \mathfrak{O}_v/\mathfrak{M}_v$ is called the *residue field* of $F$ with respect to $v$. The image of an element $\alpha \in \mathfrak{O}_v$ in $\overline{F}_v$ is denoted by $\overline{\alpha}$ and is called the residue of $\alpha$ in $\overline{F}_v$. The set $U_v = \mathfrak{O}_v - \mathfrak{M}_v$ forms the multiplicative group of invertible elements of $\mathfrak{O}_v$ and is called the *group of units*.

**1.1.3.** Let $(m_1,\dots,m_n) < (m'_1,\dots,m'_n)$ if $m_i < m'_i$ for the least index $i$ with $m_i \neq m'_i$. Then the group

$$(\mathbb{Z})^n = \underbrace{\mathbb{Z} \oplus \cdots \oplus \mathbb{Z}}_{n \text{ times}}$$

is ordered lexicographically. A valuation $v$ is called $n$-discrete if the group $\Gamma$ is isomorphic as an ordered abelian group with $(\mathbb{Z})^n$ for some $n \geqslant 1$. The classical case is $n = 1$, then $v$ is called 1-discrete or discrete. It is convenient to assume that the map $v\colon\ F^* \to (\mathbb{Z})^n$ is surjective.

**1.1.4.** *Examples.* 1. For an integer $n$ put $v_p(n) = k$, where $k$ is the highest integer such that $p^k$ divides $n$. For rational $m/n \neq 0$ with integer $m, n$ put $v_p(m/n) = v_p(m) - v_p(n)$. Then $v_p$ is well defined and the map $v_p\colon\ \mathbb{Q}^* \to \mathbb{Z}$ is 1-discrete valuation which is called $p$-adic. The ring of integers is

$$\mathfrak{O}_{v_p} = \{m/n\colon\ m, n \in \mathbb{Z},\ (n,p) = 1\}$$

and the residue field is $\mathbb{F}_p$. The well-known Ostrowsky's theorem asserts that any metric on $\mathbb{Q}$ is equivalent to the usual absolute value $|\ |$ or to a metric $|\alpha|_p = p^{-v_p(\alpha)}$ induced by the $p$-adic valuation for some prime $p$.

2. Let $F$ be a field of rational functions over a coefficient field $K$, $F = K(X)$. Then there is the 1-discrete valuation $v_{1/X}$ on $F$:

$$v_{1/X}\big(p(X)/q(X)\big) = \deg p(X) - \deg q(X) \quad \text{for } p(X), q(X) \in K[X].$$

The ring of integers with respect to $v_{1/X}$ is $K[X]$ and the residue field is isomorphic with $K$. For each monic irreducible polynomial $p(X)$ of positive degree over $K$ there is the 1-discrete valuation $v_{p(X)}$ on $F$ which is an analog of the $p$-adic valuation:

$$v_{p(X)}\big(f(X)\big) = k,$$

where $k$ is the highest integer such that $p(X)^k$ divides $f(X)$. The residue field with respect to $v_{p(X)}$ is the field $K[X]/p(X)K[X]$ which is a finite extension of $K$. It is obtained by adjoining a root of $p(X)$. Any discrete valuation on $F$ which is trivial on $K$ coincides with some $v_{p(X)}$ or $v_{1/X}$.

3. Let $v_i\colon F^* \to (\mathbb{Z})^{n_i}$, $1 \leqslant i \leqslant k$ be $n_i$-discrete valuations. Then

$$v = (v_1, \dots, v_k)\colon F^* \to (\mathbb{Z})^n$$

is $n$-discrete valuation with $n = n_1 + \dots + n_k$.

4. Let $v = (v^{(n)}, \dots, v^{(1)})\colon F^* \to (\mathbb{Z})^n$ be an $n$-discrete valuation. Then $F$ is 1-discrete with respect to the first component $v^{(n)}$ of $v$ and the residue field $F_{n-1} = \overline{F}_{v^{(n)}}$ is an $(n-1)$-discrete valuation field with respect to the induced valuation from $v^{(n-1)}, \dots, v^{(1)}$. Continuing in this way we get a tower of discrete valuation fields $F = F_n, F_{n-1}, \dots, F_1$ such that $F_i$ is the residue field of $F_{i+1}$ with respect to a 1-discrete valuation and the residue field $F_0$ of $F_1$ coincides with $\overline{F}_v$.

5. Let $F$ be a field with a valuation $v$.

a) For a polynomial $f(X) = \alpha_m X^m + \dots + \alpha_M X^M \in F[X]$ with $\alpha_m \neq 0$, $m \leqslant M$ put

$$v^*\big(f(X)\big) = \big(m, v(\alpha_m)\big) \in \mathbb{Z} \times v\big(F^*\big).$$

One can naturally extend $v^*$ to $F(X)$. Ordering the group $\mathbb{Z} \times v(F^*)$ lexicographically we get a valuation $v^*$ on $F(X)$ with residue field isomorphic with $\overline{F}_v$.

b) For a polynomial $f(X) = \alpha_m X^m + \dots + \alpha_M X^M \in F[X]$ with $\alpha_m \neq 0$, $m \leqslant M$ put

$$w\big(f(X)\big) = \min_{m \leqslant i \leqslant M} v(\alpha_i).$$

The residue field of the extension $w$ to $F(X)$ is $\overline{F}_v(X)$.

c) For a polynomial $f(X) = \alpha_m X^m + \dots + \alpha_M X^M \in F[X]$ with $\alpha_m \neq 0$, $m \leqslant M$ put

$$v_*(f(X)) = \min_{m \leqslant i \leqslant M} (v(\alpha_i), i) \in v(F^*) \times \mathbb{Z},$$

where $v(F^*) \times \mathbb{Z}$ is ordered lexicographically. Then the map $v_*$ can be extended to $F(X)$ and the residue field with respect to $v_*$ is isomorphic with $\overline{F}_v$.

6. A valuation $v$ on $F$ is said to be $p$-valuation of rank $d$ for a prime $p$ if $\operatorname{char}(F) = 0$, $\operatorname{char}(\overline{F}) = p > 0$ and $\mathfrak{O}_v/p\mathfrak{O}_v$ is of order $p^d$. $F$ is said to be a formally $p$-adic field if it admits a nontrivial $p$-valuation. See [PR, Po].

**1.1.5.** *Prime elements.* Let $F$ be a field and $v$ be an $n$-discrete valuation. Let $v(\pi_n, \dots, \pi_1) = (1, \dots, 1) \in (\mathbb{Z})^n$. Then the elements $\pi_n, \dots, \pi_1$ are called local parameters of $F$ with respect to $v$. The maximal ideal $\mathfrak{M}_v$ coincides with the ideal generated by $\pi_n, \dots, \pi_1$. For $n = 1$ such an element $\pi = \pi_1$ is called a *prime (uniformizing) element* of $F$. The ring of integers $\mathfrak{O}_v$ is a principal ideal ring only for $n = 1$ and in this case any proper ideal of $\mathfrak{O}_v$ can be written as $\pi^m \mathfrak{O}_v$, $m > 0$.

Any element $\alpha \in F^*$ can be uniquely written as

$$\pi_n^{a_n} \cdots \pi_1^{a_1} \varepsilon \quad \text{with } a_i \in \mathbb{Z},\ \varepsilon \in U_v,$$

and we get a noncanonical decomposition $F^* \simeq (\mathbb{Z})^n \times U_v$.

## 1.2. *Completion*

**1.2.1.** Let $F$ be discrete valuation field with respect to $v$. A sequence $(\alpha_m)_{m \geqslant 0}$ of elements in $F$ is called Cauchy if for any integer $c$ there exists $m$ such that $v(\alpha_k - \alpha_l) \geqslant c$ for $k, l \geqslant m$. Then there exists $\lim v(\alpha_m) \in \mathbb{Z} \cup \{+\infty\}$. The set of all Cauchy sequences forms a ring $R$ with respect to componentwise addition and multiplication. The set of all Cauchy sequences $(\alpha_m)_{m \geqslant 0}$ such that $\lim v(\alpha_m) = +\infty$ forms a maximal ideal $I$. The field $\widehat{F}_v = \widehat{F} = R/I$ admits the discrete valuation $\widehat{v}$: $\widehat{v}((\alpha_m)) = \lim v(\alpha_m)$. This field $\widehat{F}_v$ is called the *completion* of $F$ with respect to $v$. $F$ can be identified with a dense subfield in $\widehat{F}_v$ under the map: $\alpha \to (\alpha)_{m \geqslant 0} \in \widehat{F}_v$. The ring of integers $\mathfrak{O}_v$ is dense in $\mathfrak{O}_{\widehat{v}}$, the residue field $\overline{F}_v$ coincides with the residue field of $\widehat{F}_v$ with respect to $\widehat{v}$.

A field $F$ is called complete if any Cauchy sequence $(\alpha_m)_{m \geqslant 0}$ is convergent, i.e. there exists $\alpha = \lim \alpha_m \in F$ with respect to $v$. The completion of $F$ can be treated as the minimal up to an isomorphism over $F$ complete field which contains $F$.

An $n$-discrete valuation field $F$ for $n > 1$ is called complete if it is complete with respect to the first component $v^{(n)}$ of $v$ and the residue field $\overline{F}_{v^{(n)}}$ is complete. The completion of $F$ is the minimal (up to an isomorphism over $F$) complete $n$-discrete valuation field.

**1.2.2.** *Examples.* 1. (See Example 1 in 1.1.4.) The completion of $\mathbb{Q}$ with respect to $p$-adic valuation $v_p$ is denoted by $\mathbb{Q}_p$ and is called the *p-adic field*. Note that the completion of $\mathbb{Q}$ with respect to $|\ |$ is $\mathbb{R}$. Imbedding $\mathbb{Q}$ in $\mathbb{Q}_p$ for all primes $p$ and in $\mathbb{R}$ permits the solving of many problems. The famous Hasse local-global principle for a variety $V$ over $\mathbb{Q}$ declares that the existence of nontrivial $\mathbb{R}$- and $\mathbb{Q}_p$-points in $V$ for all prime $p$ is equivalent to the existence of a nontrivial $\mathbb{Q}$-point in $V$. In general this principle doesn't hold but there are important instances where it works. For example, this principle holds for conics defined by an equation $\sum a_{ij}X_iX_j = 0$. Note that from the point of view of model theory the complex number field $\mathbb{C}$ is locally equivalent for any prime $p$ with the algebraic closure $\mathbb{Q}_p^{\mathrm{alg}}$ of $\mathbb{Q}_p$, see [Roq2].

The ring of integers of $\mathbb{Q}_p$ is denoted by $\mathbb{Z}_p$ and is called the ring of $p$-adic integers.

2. (See Example 2 in 1.1.4.) The completion of $K(X)$ with respect to $v_X$ is the formal power series field $K((X))$ of all formal power series

$$\sum_{-\infty}^{+\infty} \alpha_n X_n,$$

$\alpha_n = 0$ for $n < n_0$.

3. (See Example 5 in 1.1.4.) Let $F$ be a field with an $n$-discrete valuation $v$, and let $\widehat{F}_v$ be its completion. Then the field $\widehat{F}_v((X))$ and the field $\widehat{F}_v\{\{X\}\}$ of all formal power series

$$\sum_{-\infty}^{+\infty} \alpha_m X_m,$$

$\alpha_m \in F^*$ such that $\inf\{\widehat{v}(\alpha_m)\} > -\infty$ and $\widehat{v}(\alpha_m) \to +\infty$ when $m \to -\infty$ are complete $(n+1)$-discrete with respect to $v^*$, $v_*$. The field $\widehat{F}_v(X)$ is an $n$-discrete complete valuation field with respect to $w$.

**1.2.3.** The completion $\widehat{F}_v$ of $F$ with respect to a 1-discrete valuation $v$ coincides with the completion of $F$ with respect to the $\mathfrak{M}_v$-adic topology (i.e. regarding $\mathfrak{M}_v^m$, $m \geqslant 0$ as a basis of neighborhoods of 0). In this case the ring of integers $\mathfrak{O}_{\widehat{v}}$ is isomorphic algebraically and topologically to $\varprojlim \mathfrak{O}_v/\pi^m\mathfrak{O}_v$ with the discrete topology on $\mathfrak{O}_v/\pi^m\mathfrak{O}_v$.

**1.2.4.** Let $F$ be a $n$-discrete valuation field with the residue field $\overline{F}_v$. Let $r\colon \overline{F}_v \to \mathfrak{O}_v$, $r(0) = 0$ be a map such that its composition with the residue map $\mathfrak{O}_v \to \overline{F}_v$ is the identity map. The set $R = r(\overline{F}_v)$ is called a *set of representatives* (of $\overline{F}_v$ in $F$). If $n = 1$ and $F$ is complete then there is a map

$$\oplus R \to F, \quad (a_i)_{i \geqslant i_0} \to \sum a_i \pi^i,$$

where $\pi$ is prime in $F$. This map is a topological bijection with respect to the discrete topology on $R$. In general, one can introduce, following Parshin [Pa4], a topology on a complete $n$-discrete valuation field $F$, which takes into consideration the topologies on the residue fields $F_{n-1}, \ldots, F_1$. Assume that $\operatorname{char}(\overline{F}_{v^{(n)}}) \neq 0$. This topology is the strongest one in which any element $\alpha \in F$ is uniquely expressed as a convergent sum

$$\alpha = \sum_{i_n} \sum_{i_{n-1} \geqslant I_{n-1}(i_n)} \cdots \sum_{i_1 \geqslant I_1(i_n,\ldots,i_2)} \theta_{i_n,\ldots,i_1} \pi_n^{i_n} \cdots \pi_1^{i_1},$$

where $\pi_n, \ldots, \pi_1$ are local parameters, $\theta_{i_n,\ldots,i_1} \in R$, $(i_n, \ldots, i_1) \geqslant (a_n, \ldots, a_1) \in (\mathbb{Z})^n$.

The multiplicative group $F^* \simeq (\mathbb{Z})^n \times U_v$ is equipped with the product of the discrete topology on $(\mathbb{Z})^n$ and the induced one from $F$ on $U_v$. For $n \geqslant 2$ $F^*$ is not a topological group with respect to this topology but the multiplication is sequentially continuous.

*From now on we confine our attention to discrete* (1-*discrete*) *valuation fields until Sections* 6 *and* 7.

### 1.3. *The group of units*

Let $F$ be a discrete valuation field, $U = U_v$, $\mathfrak{O} = \mathfrak{O}_v$, $\mathfrak{M} = \mathfrak{M}_v$, $\overline{F} = \overline{F}_v$. Put $U_i = 1 + \mathfrak{M}^i$. $U_1$ is called the *group of principal units*.

**1.3.1.** Fix a prime element $\pi$ and introduce maps $\lambda_0$: $U \to \overline{F}^*$, $\lambda_i$: $U_i \to \overline{F}$ by the formulas $\lambda_0(\alpha) = \overline{\alpha}$, $\lambda_i(1 + \pi^i \beta) = \overline{\beta}$ for $\alpha \in U$, $\beta \in \mathfrak{O}$. They induce isomorphisms

$$\lambda_0\colon\ U/U_1 \simeq \overline{F}^*, \qquad \lambda_i\colon\ U_i/U_{i+1} \simeq \overline{F}.$$

Therefore, the group $U_1$ is uniquely $l$-divisible for $(l, \operatorname{char}(F)) = 1$.

**1.3.2.** We are interested in a description of the raising to $p$-th power, $\uparrow p$, with respect to the filtration $U_i$. If $\operatorname{char}(F) = p$ then $(1+\alpha)^p = 1 + \alpha^p$ and therefore the following diagram

$$\begin{array}{ccc} U_i/U_{i+1} & \xrightarrow{\uparrow p} & U_{pi}/U_{pi+1} \\ \downarrow{\scriptstyle \lambda_i} & & \downarrow{\scriptstyle \lambda_{pi}} \\ \overline{F} & \xrightarrow{\uparrow p} & \overline{F} \end{array}$$

is commutative.

If $\operatorname{char}(F) = 0$ then $(1+\alpha)^p = 1 + \alpha^p + p\alpha + \cdots$, $\alpha \in \mathfrak{M}$, where dots denote terms of higher order than the preceding. Put $e = v(p)$, $e_1 = e/(p-1)$. Let $p - \eta\pi^e \in \mathfrak{M}^{e+1}$ for a suitable $\eta \in \mathfrak{O}$. Then the following diagrams are commutative:

$$\text{for } i < e_1 \qquad \begin{array}{ccc} U_i/U_{i+1} & \xrightarrow{\uparrow p} & U_{pi}/U_{pi+1} \\ \downarrow{\scriptstyle \lambda_i} & & \downarrow{\scriptstyle \lambda_{pi}} \\ \overline{F} & \xrightarrow{\uparrow p} & \overline{F} \end{array},$$

$$\text{for } i = e_1 \qquad \begin{array}{ccc} U_{e_1}/U_{e_1+1} & \xrightarrow{\uparrow p} & U_{pe_1}/U_{pe_1+1} \\ \lambda_{e_i}\downarrow & & \downarrow\lambda_{pe_i} \\ \overline{F} & \xrightarrow{\overline{\theta}\mapsto\overline{\theta}^p+\overline{\eta}\overline{\theta}} & \overline{F} \end{array},$$

$$\text{for } i > e_1 \qquad \begin{array}{ccc} U_i/U_{i+1} & \xrightarrow{\uparrow p} & U_{i+e}/U_{i+e+1} \\ \lambda_i\downarrow & & \downarrow\lambda_{i+e} \\ \overline{F} & \xrightarrow{\overline{\theta}\mapsto\overline{\eta}\overline{\theta}} & \overline{F} \end{array}.$$

**1.3.3.** Assume that $F$ is complete with respect to its discrete valuation. Then it follows from 1.3.1 that any element $\alpha \in U_1$ can be uniquely expressed as a convergent product

$$\alpha = \prod_{i\geqslant 1}(1+\theta_i\pi^i), \quad \theta_i \in R.$$

If $\operatorname{char}(\overline{F}) = p > 0$ then the group of principal units $U_1$ can be seen as a multiplicative $\mathbb{Z}_p$-module: for $a = \lim a_n \in \mathbb{Z}_p$, $a_n \in \mathbb{Z}$ put $\varepsilon^a = \lim \varepsilon^{a_n}$. The commutative diagrams of 1.3.2 imply that $U_1$ is a free $\mathbb{Z}_p$-module of infinite rank when $\operatorname{char}(F) = p$ and is a $\mathbb{Z}_p$-module of rank $n$ (resp. $n+1$ with one relation) if $F$ is a $p$-adic field of degree $n$ over $\mathbb{Q}_p$ and a primitive $p$-th root of unity doesn't belong (resp. belongs) to $F$.

**1.3.4.** Assume that $F$ is complete with respect to a discrete valuation and $\operatorname{char}(\overline{F}) = p > 0$. A representative $\alpha \in \mathfrak{O}$ of $\overline{\alpha} \in \overline{F}$ is called *multiplicative* if

$$\alpha \in \bigcap_{m\geqslant 1} \mathfrak{O}^{p^m}.$$

The set of nonzero multiplicative representatives forms a group which is isomorphically mapped onto the maximal perfect subfield

$$\overline{F}_0 = \bigcap \overline{F}^{p^m}$$

in $\overline{F}$. If the residue field $\overline{F}$ is perfect then any element $\overline{\alpha} \in \overline{F}$ has a multiplicative representative $\alpha \in \mathcal{R}$ in $F$. The corresponding map $r$: $\overline{\alpha} \to \alpha$ is called the *Teichmüller map*. It induces an isomorphism of groups $\overline{F}^* \simeq r(F^*) = \mathcal{R}^*$ and an isomorphism of fields $\overline{F} \simeq \mathcal{R}$ when $\operatorname{char}(F) = p$. The group $U$ is canonically decomposed as the product $\mathcal{R}^* \times U_1$.

## 1.4. *The Witt ring*

Let $\alpha = \sum\theta_i\pi^i$, $\beta = \sum\eta_i\pi^i$ be expansions with $\theta_i, \eta_i \in \mathcal{R}$. Then a description of coefficients of $\alpha+\beta$, $\alpha\beta$ naturally leads to the notion of the Witt vectors (see [Se2], Chapter 2, §6).

**1.4.1.** Let $S$ be an arbitrary commutative ring with unity. For $(a_i)_{i\geqslant 0}$, $a_i \in S$, put $(a^{(i)}) = (w_0(a_0), w_1(a_0, a_1), \dots)$, where

$$w_n(X_0, \dots, X_n) = \sum_{i=0}^{n} p^i X_i^{p^{n-i}}.$$

The map $(a_i) \to (a^{(i)})$ is a bijection of $(S)_0^{+\infty}$ with $(S)_0^{+\infty}$ if $p$ is invertible in $S$. In this case one can transfer the ring structure from $(a^{(i)}) \in (S)_0^{+\infty}$ under the componentwise addition and multiplication to $(a_i) \in (S)_0^{+\infty}$. Then for $(a_i), (b_i) \in (S)_0^{+\infty}$

$$(a_i) * (b_i) = \big(\omega_0^{(*)}(a_0, b_0), \omega_1^{(*)}(a_0, a_1, b_0, b_1), \dots\big), \quad * = + \text{ or } * = \times.$$

Here $\omega_i^{(*)}$ is the image of the polynomial $\nu_i^{(*)} \in \mathbb{Z}[X_0, X_1, \dots, Y_0, Y_1, \dots]$ under the canonical homomorphism $\mathbb{Z} \to S$, where $\nu_i^{(*)}$ are defined by the equations

$$\begin{aligned} &w_n(X_0, \dots, X_n) * w_n(Y_0, \dots, Y_n) \\ &\quad = w_n\big(\nu_0^{(*)}, \dots, \nu_n^{(*)}(X_0, \dots, X_n; Y_0, \dots, Y_n)\big) \end{aligned} \qquad (*)$$

The sequences $(a_i)_{i\geqslant 0}$ are called Witt vectors and the $a^{(i)}$ for $i \geqslant 0$ are called the ghost components of $(a_i)_{i\geqslant 0}$. The set of Witt vectors is then a commutative ring. This is still the case if $p$ is not necessarily invertible. Indeed one shows without much trouble that the polynomials $\nu_i^{(*)}$, $* = +, \times$ defined by $(*)$ above have their coefficients in $\mathbb{Z}$. Thus, the set of Witt vectors is a commutative ring with unity $(1, 0, \dots)$. This ring is called the ring of Witt vectors $W(S)$ of $S$. For ramified Witt vectors see [Haz4] and [FV], Chapter I, Section 7.

**1.4.2.** Assume that $p = 0$ in $S$. Then one can define maps

$r_0\colon S \to W(S)$,

$V\colon W(S) \to W(S)$ (the "Verschiebung" map),

$\mathbf{F}\colon W(S) \to W(S)$ (the "Frobenius" map)

by the formulas

$$\begin{aligned} &r_0(a) = (a, 0, 0, \dots), \quad V(a_0, a_1, \dots) = (0, a_0, a_1, \dots), \\ &\mathbf{F}(a_0, a_1, \dots) = \big(a_0^p, a_1^p, \dots\big). \end{aligned}$$

Then $\mathbf{F}$ is a ring homomorphism and $V\mathbf{F}(\alpha) = \mathbf{F}V(\alpha) = p\alpha$. The ring $W_n(S) = W(S)/V^n W(S)$ consists of the Witt vectors $(a_0, \dots, a_{n-1})$ of length $n$.

**1.4.3.** Assume that $S = K$ is a perfect field of characteristic $p$. For a Witt vector $\alpha = (a_0, a_1, \dots) \in W(K)$ put

$$v(\alpha) = \min\{i\colon\ \alpha \in V^i W(K),\ \alpha \notin V^{i+1} W(K)\}, \quad v(0) = +\infty.$$

Then the field of fractions $F_0$ of $W(K)$ is a complete discrete valuation field of characteristic 0 with respect to the extension of $v$. The element $p$ is a prime element of $F_0$ and its residue field is isomorphic with $K$. The set of multiplicative representatives $\mathcal{R}$ coincides with $r_0(K)$. In particular, $W(\mathbb{F}_p) = \mathbb{Z}_p$.

## 2. Extensions of discrete valuation fields

### 2.1. *The Hensel Lemma*

Let $F$ be a valuation field with ring of integers $\mathfrak{O}$, maximal ideal $\mathfrak{M}$ and residue field $\overline{F}$. For a polynomial $f(X) = \alpha_n X^n + \dots + \alpha_0 \in \mathfrak{O}[X]$ we shall denote by $\overline{f}(X) \in \overline{F}[X]$ the polynomial $\overline{\alpha}_n X^n + \dots + \overline{\alpha}_0$. We shall write

$$f(X) \equiv g(X) \pmod{\mathfrak{M}^m}$$

if $f(X) - g(X) \in \mathfrak{M}^m \mathfrak{O}[X]$.

**2.1.1.** PROPOSITION. *Let $F$ be a complete discrete valuation field. Let $f(X)$, $g_0(X)$, $h_0(X)$ be polynomials over $\mathfrak{O}$ and let $f(X)$, $g_0(X)$ be monic polynomials. Let the resultant $R(g_0(X), h_0(X)) \notin \mathfrak{M}^{s+1}$ and*

$$f(X) \equiv g_0(X) h_0(X) \pmod{\mathfrak{M}^{2s+1}}$$

*for an integer $s \geqslant 0$. Then there is a polynomial $h(X)$ and a monic polynomial $g(X)$ over $\mathfrak{O}$ such that $f(X) = g(X)h(X)$ and*

$$g(X) \equiv g_0(X) \pmod{\mathfrak{M}^{s+1}}, \quad h(X) \equiv h_0(X) \pmod{\mathfrak{M}^{s+1}},$$

$\deg g(X) = \deg g_0(X)$.

The proof is carried out by constructing polynomials $g_i(X)$, $h_i(X)$ over $\mathfrak{O}$ with the properties: the $g_i(X)$ are monic polynomials, $\deg g_i(X) = \deg g_0(X)$,

$$g_i(X) \equiv g_{i-1}(X) \pmod{\mathfrak{M}^{i+s}},$$

$$h_i(X) \equiv h_{i-1}(X) \pmod{\mathfrak{M}^{i+s}},$$

$$f(X) \equiv g_i(X) h_i(X) \pmod{\mathfrak{M}^{i+2s+1}}$$

and proceeding by induction on $i$. Then $g(X) = \lim g_i(X)$, $h(X) = \lim h_i(X)$ are the desired polynomials.

**2.1.2.** COROLLARY (Hensel Lemma). *Let $F$ be as in* 2.1.1 *and $\overline{F}$ be the residue field of $F$. Let $f(X)$, $g_0(X)$, $h_0(X)$ be monic polynomials over $\mathfrak{O}$ and $\overline{f}(X) = \overline{g}_0(X)\overline{h}_0(X)$. Let $\overline{g}_0(X)$, $\overline{h}_0(X)$ be relatively prime in $\overline{F}[X]$. Then there are monic polynomials $g(X)$, $h(X)$ with coefficients in $\mathfrak{O}$ such that $f(X) = g(X)h(X)$ and $\overline{g} = \overline{g}_0$, $\overline{h} = \overline{h}_0$.*

Valuation fields satisfying the assertion of this corollary are called *Henselian*.

**2.1.3.** COROLLARY. *Let $F$ be as in* 2.1.1 *and $f(X)$ be a monic polynomial over $\mathfrak{O}$. Let $f(\alpha_0) \in \mathfrak{M}^{2s+1}$, $f'(\alpha_0) \notin \mathfrak{M}^{s+1}$ for some $\alpha_0 \in \mathfrak{O}$, $s \geqslant 0$. Then there is $\alpha \in \mathfrak{O}$ such that $\alpha - \alpha_0 \in \mathfrak{M}^{s+1}$ and $f(\alpha) = 0$.*

Other characterizations of Henselian fields can be found in [Bou, Ra].

## 2.2. *Extensions*

**2.2.1.** Let $F$ be a field and $L$ be an extension of $F$ with a valuation $w$: $L^* \to \Gamma$. Then $w$ induces a valuation $w_0$: $F^* \to \Gamma$ on $F$. In this context the extension $L/F$ is said to be an extension of valuation fields. The group $w_0(F^*)$ is a totally ordered subgroup of $w(L^*)$ and the index of $w(F^*)$ in $w(L^*)$ is called the *ramification index* $e(L/F, w)$. The ring of integers $\mathfrak{O}_{w_0}$ is a subring in $\mathfrak{O}_w$ and $\mathfrak{M}_{w_0}$ coincides with $\mathfrak{M}_w \cap \mathfrak{O}_{w_0}$. Hence the residue field $\overline{F}_{w_0}$ can be regarded as a subfield of the residue field $\overline{L}_w$. The residue of an element $\alpha \in \mathfrak{O}_{w_0}$ in $\overline{F}_{w_0}$ can be identified with the image of $\alpha \in \mathfrak{O}_w$ in $\overline{L}_w$. The degree of the extension $\overline{L}_w/\overline{F}_{w_0}$ is called the *residue degree* $f(L/F, w)$. This immediately implies that for $F \subset M \subset L$ and the induced valuation $w_0$ on $M$ from $w$ on $L$

$$e(L/F, w) = e(L/M, w)e(M/F, w_0),$$
$$f(L/F, w) = f(L/M, w)f(M/F, w_0).$$

If $L/F$ is a finite extension and $w_0$ is discrete for a valuation $w$ on $L$ then $w$ is discrete.

In what follows we shall consider discrete valuations.

**2.2.2.** Let $F$ and $L$ be fields with discrete valuations $v$, $w$, $F \subset L$. The valuation $w$ is said to be an *extension* of $v$ if the topology defined by $w_0$ is equivalent with the topology defined by $v$. In this case we write $w|v$ and use the notations $e(w|v)$, $f(w|v)$ instead of $e(L/F, w)$, $f(L/F, w)$. Then $e(w|v) = |\mathbb{Z}\colon w(F^*)|$ and if $\pi_v$, $\pi_w$ are prime elements with respect to $v$, $w$ then $\pi_v = \pi_w^{e(w|v)}\varepsilon$ with $\varepsilon \in U_w$.

If $L$ is a finite extension of $F$ then $e(w|v)f(w|v) \leqslant |L : F|$.

For instance if $L$ is a finite extension of $F$ in the completion $\widehat{F}_v$ then $e(w|v) = f(w|v) = 1$. Therefore, in general the inequality is not an equality. However, if $L$ is a finite extension of a complete discrete valuation field $F$ then $L$ is complete and

$$e(w|v)f(w|v) = |L : F|.$$

Moreover, if $\theta_1, \dots, \theta_f$ are elements of $\mathfrak{O}_w$ of which the residues form a basis of $\overline{L}_w$ over $\overline{F}_v$ and $\pi_w$ is prime in $L$ then $\mathfrak{O}_w = \mathfrak{O}_v[\{\theta_i \pi_w^j\}]$, $L = F(\{\theta_i \pi_w^j\})$ with $1 \leqslant i \leqslant f(w|v)$, $0 \leqslant j \leqslant e(w|v) - 1$.

**2.2.3.** Complete discrete valuation fields also possess the following property: there is exactly one extension $w$ of the discrete valuation $v$ of $F$ to a finite extension $L$ of $F$. It is defined by the formula $w = (1/f) v \circ N_{L/F}$ with $f = f(w|v)$, where $N_{L/F}$ is the norm map from $L$ to $F$, see [CF, Bou].

A general case now can be deduced from this one.

**2.2.4.** PROPOSITION. *Let $F$ be a field with the discrete valuation $v$, $\widehat{F}$ the completion of $F$ with respect to $v$. Let $L = F(\alpha)$ be a finite extension of $F$ and $f(X)$ the monic irreducible polynomial of $\alpha$ over $F$. Let*

$$f(X) = \prod_{i=1}^{k} g_i(X)^{e_i}$$

*be the decomposition of the polynomial $f(X)$ into irreducible monic factors in $\widehat{F}[X]$. Let $\alpha_i$ be a root of the polynomial $g_i(X)$ $(\alpha_1 = \alpha)$ and $L_i = \widehat{F}(\alpha_i)$. Let $\widehat{w}_i$ be the unique extension of $\widehat{v}$ to $L_i$. Then $L$ is embedded as the dense subfield in the complete discrete valuation field $L_i$ under $F \hookrightarrow \widehat{F}$, $\alpha \mapsto \alpha_i$. The restriction $w_i$ of $\widehat{w}_i$ to $L$ is a discrete valuation on $L$ which extends $v$. The valuations $w_i$ are distinct and any extension of $v$ on $L$ coincides with some $w_i$ for $1 \leqslant i \leqslant k$.*

Thus, this assertion establishes a connection between extensions of a discrete valuation and the decomposition of the irreducible polynomial over the completed field.

In particular, there is a unique extension of a discrete valuation $v$ of $F$ on $L$ for purely inseparable extension $L/F$. Indeed, in this case $L = F(\alpha)$ and $f(X)$ decomposes as $(X - \alpha)^{p^m}$ in the fixed algebraic extension of $F$, therefore $k = 1$.

Now we are able to describe extensions of discrete valuations on Henselian fields.

**2.2.5.** PROPOSITION. *The following conditions are equivalent:*

1) *$F$ is a Henselian field with respect to a discrete valuation $v$.*

2) *The discrete valuation extends uniquely to a finite algebraic extension $L$ of $F$.*

3) *If $L$ is a finite separable extension of $F$ of degree $n$ then $n = e(w|v) f(w|v)$, where $w$ is the extension of $v$ on $L$.*

4) *$F$ is separably closed in $\widehat{F}$.*

A proof follows from 2.2.4. The separable closure of a discrete valuation field $F$ in $\widehat{F}$ is called the *henselization* $F^h$ of $F$, it is the minimal Henselian field which contains $F$. For instance, the elements in $\mathbb{Q}_p$ algebraic over $\mathbb{Q}$ form a Henselian countable field, but $\mathbb{Q}_p$ is uncountable.

**2.2.6.** COROLLARY. *Let $F$ be a Henselian discrete valuation field and $L$ an algebraic extension of $F$. Then there is a unique valuation $w$: $L^* \to \mathbb{Q}$ (not necessarily discrete)* *such that the restriction* $w|_F$ *coincides with the discrete valuation* $v$ *on* $F$. *Moreover,* $w$ *is Henselian.*

**2.2.7.** COROLLARY. *Let* $F$ *be a Henselian discrete valuation field and* $L/F$ *a finite separable extension. Let* $w$ *be a discrete valuation on* $L$ *and* $\sigma\colon L \to F^{\mathrm{alg}}$ *be an imbedding of* $L$ *in a fixed algebraic closure* $F^{\mathrm{alg}}$ *over* $F$. *Then* $w \circ \sigma^{-1}$ *is a discrete valuation on* $\sigma L$ *and* $\mathfrak{M}_{\sigma L} = \sigma \mathfrak{M}_L$, $\mathfrak{O}_{\sigma L} = \sigma \mathfrak{O}_L$.

## 2.3. *Unramified and ramified extensions*

Let $F$ be a Henselian discrete valuation field and $L$ be an algebraic extension over $F$. If the unique extension $w$ of the valuation $v$ on $F$ is discrete on $L$ then we shall write $e(L|F)$, $f(L|F)$ instead of $e(w|v)$, $f(w|v)$. We shall write $\mathfrak{O}$ or $\mathfrak{O}_F$, $\mathfrak{M}$ or $\mathfrak{M}_F$, $U$ or $U_F$, $\pi$ or $\pi_F$, and $\overline{F}$ instead of $\mathfrak{O}_v$, $\mathfrak{M}_v$, $U_v$, $\pi_v$, and $\overline{F}_v$.

**2.3.1.** A finite extension $L$ of $F$ is called *unramified* if $\overline{L}/\overline{F}$ is separable of the same degree as $L/F$. A finite extension $L/F$ is called *totally ramified* if $f(L|F) = 1$. A finite extension $L/F$ is called *tamely unramified* if $\overline{L}/\overline{F}$ is separable and if $p = \operatorname{char}(\overline{F}) > 0$ then $(p, e(L|F)) = 1$, $e(L|F) < \infty$.

Then it follows from 2.2.2, 2.2.3 that $f(L|F) = |L : F|$ when $L/F$ is unramified and $e(L|F) \leqslant |L : F|$ if $L/F$ is totally ramified.

**2.3.2.** We first treat the case of unramified extensions. The next assertion follows from the Hensel Lemma.

PROPOSITION. 1) *Let* $L/F$ *be an unramified extension and* $\overline{L} = \overline{F}(\theta)$ *for some* $\theta \in \overline{L}$. *Let* $\alpha \in \mathfrak{O}_L$ *be such that* $\overline{\alpha} = \theta$. *Then* $L = F(\alpha)$ *is separable over* $F$ *and* $\mathfrak{O}_L = \mathfrak{O}_F[\alpha]$. $\theta$ *is a simple root of the irreducible over* $\overline{F}$ *polynomial* $\overline{f}(X)$, *where* $f(X) \in \mathfrak{O}_F[X]$ *is the monic irreducible polynomial of* $\alpha$ *over* $F$.

2) *Let* $g(X)$ *be a monic separable polynomial over* $\overline{F}$ *and* $f(X)$ *a monic polynomial over* $\mathfrak{O}_F$, $\overline{f}(X) = g(X)$. *If* $\alpha$ *is a root of* $f(X)$ *in* $F^{\mathrm{alg}}$ *then the extension* $L/F$ *for* $L = F(\alpha)$ *is unramified and* $\overline{L} = \overline{F}(\theta)$ *for a root* $\theta = \overline{\alpha}$ *of the polynomial* $g(X)$.

COROLLARY. 1) *If* $M/F$, $L/M$ *are unramified then* $L/F$ *is unramified.*

2) *If* $L_1/F$, $L_2/F$ *are unramified then* $L_1L_2/F$ *is unramified.*

**2.3.3.** An algebraic extension $L$ of a Henselian discrete valuation field $F$ is called unramified if $L/F$, $\overline{L}/\overline{F}$ are separable extensions and $e(L|F) = 1$. The compositum of all finite unramified extensions of $F$ in a fixed algebraic closure $F^{\mathrm{alg}}$ is unramified and this field is a Henselian discrete valuation field (not complete in general). This field is called the *maximal unramified extension* $F^{\mathrm{ur}}$ of $F$. For instance, $\mathbb{Q}_p^{\mathrm{ur}}$ is obtained from $\mathbb{Q}_p$ by adjoining of all roots of unity of order relatively prime to $p$.

**2.3.4.** By using Corollary 2 in 2.2.7 and the Hensel Lemma we deduce

PROPOSITION. 1) *Let $L/F$ be unramified and $\overline{L}/\overline{F}$ be Galois. Then $L/F$ is Galois.*

2) *Let $L/F$ be unramified Galois. Then $\overline{L}/\overline{F}$ is Galois. Let for an automorphism $\sigma \in \mathrm{Gal}(L/F)$ the automorphism $\overline{\sigma}$ in $\mathrm{Gal}(\overline{L}/\overline{F})$ satisfy the relation $\overline{\sigma}(\overline{\alpha}) = \overline{\sigma(\alpha)}$ for $\alpha \in \mathfrak{O}_L$. Then the map $\sigma \mapsto \overline{\sigma}$ induces an isomorphism of $\mathrm{Gal}(L/F)$ onto $\mathrm{Gal}(\overline{L}/\overline{F})$.*

COROLLARY. *The residue field of $F^{\mathrm{ur}}$ coincides with the separable closure $\overline{F}^{\mathrm{sep}}$ of $\overline{F}$ and*

$$\mathrm{Gal}\left(F^{\mathrm{ur}}/F\right) \simeq \mathrm{Gal}\left(\overline{F}^{\mathrm{sep}}/\overline{F}\right).$$

*If $L$ is an algebraic extension of $F$ and $L$ is a discrete valuation field then $L^{\mathrm{ur}} = LF^{\mathrm{ur}}$ and $L_0 = L \cap F^{\mathrm{ur}}$ is the maximal unramified subextension of $F$ in $L$.*

**2.3.5.** Now we consider tamely ramified extensions.

PROPOSITION. 1) *Let $L$ be a finite separable tamely ramified extension of a Henselian discrete valuation field $F$ and $L_0/F$ be the maximal unramified subextension in $L/F$. Then $L = L_0(\pi)$ and $\mathfrak{O}_L = \mathfrak{O}_{L_0}[\pi]$ with a prime element $\pi$ in $L$ satisfying an equation $X^e - \pi_0 = 0$ for a proper prime element $\pi_0$ in $L_0$, where $e = e(L|F)$.*

2) *Let $L/F$ be a finite unramified extension and $L = L_0(\alpha)$ with $\alpha^e = \beta \in L_0$, $(p, e) = 1$ if $p = \mathrm{char}(F) > 0$. Then $L/F$ is separable tamely ramified.*

The proof follows from writing $\pi_1 = \pi_L^e \varepsilon$ for prime elements $\pi_L$ in $L$, $\pi_1$ in $L_0$ and $\varepsilon \in U_L$ and the $e$-divisibility of the group of principal units.

The field $L_0$ is called the *inertia* subfield of the extension $L/F$.

COROLLARY. 1) *If $M/F$, $L/M$ are separable tamely ramified then $L/F$ is also tamely ramified.*

2) *If $L_1/F$, $L_2/F$ are separable tamely ramified then so is $L_1L_2/F$.*

**2.3.6.** The last and most complicated case concerns totally ramified extensions.

Let $F$ be a Henselian discrete valuation field. A polynomial

$$f(X) = X^n + \alpha_{n-1}X^{n-1} + \cdots + \alpha_0$$

over $\mathfrak{O}$ is called an *Eisenstein polynomial* if $\alpha_0, \ldots, \alpha_{n-1} \in \mathfrak{M}$, $\alpha_0 \notin \mathfrak{M}^2$.

PROPOSITION. 1) *An Eisenstein polynomial $f(X)$ is irreducible over $F$. If $\alpha$ is a root of $f(X)$ then $F(\alpha)/F$ is a totally ramified extension of degree $n$ and $\alpha$ is a prime element in $F(\alpha)$, $\mathfrak{O}_{F(\alpha)} = \mathfrak{O}_F[\alpha]$.*

2) *Let $L/F$ be a separable totally ramified extension of degree $n$ and $\pi$ be a prime element in $L$. Then $\pi$ is a root of an Eisenstein polynomial over $F$ of degree $n$.*

Note that properties analogous to those in Corollary 2.3.2, 2.3.5 don't hold for totally ramified extensions.

**2.4.** *Galois extensions and ramification groups*

Let $F$ be a Henselian discrete valuation field.

**2.4.1.** Let $L$ be a finite Galois extension of $F$, $G = \mathrm{Gal}(L/F)$. Put for an integer $i \geqslant -1$

$$G_i = \{\sigma \in G\colon\ \sigma\alpha - \alpha \in \mathfrak{M}_L^{i+1} \text{ for all } \alpha \in \mathfrak{O}_L\}.$$

Then $G_{-1} = G$ and $G_i$ is a normal subgroup of $G$. If $\overline{L}/\overline{F}$ is separable then the subgroup $G_0$ corresponds to the field $L_0$ which was defined in 2.3.5 and is called the *inertia* subgroup of $G$. In this case the group $G_1$ corresponds to the maximal tamely ramified extension of $F$ in $L$.

The definitions imply that $G_i = \{\sigma \in G\colon\ \sigma\pi - \pi \in \mathfrak{M}_L^{i+1}\}$ for a prime $\pi$ in $L$, $G_i = \{1\}$ for sufficiently large $i$.

**2.4.2.** Let $L$ be a finite Galois extension of $F$, $\overline{L}$ separable over $\overline{F}$. Let $\pi$ be prime in $L$. Introduce maps

$$\psi_0\colon\ G_0 \to \overline{L}^*, \qquad \psi_i\colon\ G_i \to \overline{L} \quad (i > 0)$$

by the formulas $\psi_i(\sigma) = \lambda_i(\sigma\pi/\pi)$, where the maps $\lambda_i$ were defined in 1.3.1. Then the $\psi_i$, $i > 0$, induce injective homomorphisms $G_0/G_1 \to \overline{L}^*$, $G_i/G_{i+1} \to \overline{L}$ for $i > 0$. By the structure of the groups $\overline{L}^*$, $\overline{L}$ this implies that the group $G_0/G_1$ is cyclic of order relatively prime with $\mathrm{char}(\overline{F})$ if $\mathrm{char}(\overline{F}) > 0$. If $\mathrm{char}(\overline{F}) = 0$ then $G_1 = 1$ and $G_0$ is cyclic. If $\mathrm{char}(\overline{F}) = p > 0$ then $G_i/G_{i+1}$ are abelian $p$-groups and $G_1$ is the maximal $p$-subgroup of $G_0$. Therefore, $G_0$ is a solvable group and $G$ is solvable if $\mathrm{Gal}(\overline{L}/\overline{F})$ is solvable.

**2.4.3.** For further properties of ramification groups see [Se2], Chapter 4, [Sen1, Sen2]. There exists a metatheorem which claims that an assertion about properties of ramification groups of totally ramified extensions which holds for a perfect residue field is true for a finite residue field as well, see [Lau1]. A case of an imperfect residue field is treated in [Lo, Hy, Kat5].

**2.5.** *Structure theorems for complete fields*

Let $F$ be a discrete valuation field. If $\mathrm{char}(F) = p > 0$ then $p = 0$ in $\overline{F}$ and $\mathrm{char}(\overline{F}) = p$. Therefore, there are the equal-characteristic cases $\mathrm{char}(F) = \mathrm{char}(\overline{F}) = 0$ or $\mathrm{char}(F) = \mathrm{char}(\overline{F}) = p > 0$ and the unequal-characteristic case $\mathrm{char}(F) = 0$, $\mathrm{char}(\overline{F}) = p$. For proofs of the following assertions see [Coh].

Now let $F$ be a complete discrete valuation field.

**2.5.1.** The simplest case is $\mathrm{char}(F) = \mathrm{char}(\overline{F}) = 0$. In this case there exists a (not unique in general) field in $\mathfrak{O}_F$ which isomorphically mapped onto $\overline{F}$. This field is a maximal one which is contained in $\mathfrak{O}_F$ and its existence is verified by using the Hensel Lemma. Therefore, the field $F$ is isomorphic algebraically and topologically to the field $\overline{F}((X))$, where $X$ corresponds to a prime element $\pi$ in $F$.

**2.5.2.** The next case is $\operatorname{char}(F) = \operatorname{char}(\overline{F}) = p$. If $\overline{F}$ is perfect then the set of multiplicative representatives as was noticed in 1.3.3 is a field in $\mathfrak{O}_F$ which is mapped isomorphically onto $\overline{F}$. This field is the unique one which has this property. By using the notion of a $p$-basis the existence of such a field can also be proved for $\overline{F}$ not perfect (in this case there are many such fields). Therefore, in this case the field $F$ is isomorphic and homeomorphic with the field of formal power series $\overline{F}((X))$.

**2.5.3.** The most complicated case is $\operatorname{char}(F) = 0$, $\operatorname{char}(\overline{F}) = p$. In this case $e(F) = v(p)$ is called the *absolute index* of ramification of $F$.

The preceding assertions show that in the equal-characteristic cases for an arbitrary field $K$ there exists a complete discrete valuation field $F$, whose residue field is isomorphic to $K$. The same assertion holds for the unequal-characteristic case: if $K$ is a field of characteristic $p$ then there is a complete discrete valuation field $F$ of characteristic 0 with prime element $p$ and residue field $K$. If $K$ is perfect then one can take the quotient field of $W(K)$ by using 1.4.3. If $K$ is imperfect, let

$$K' = \bigcup_{n \geqslant 0} K^{1/p^n}$$

be its extension. Then $K'$ is perfect and one can take the subring $S$ in $W(K')$ generated by the multiplicative representatives of $K$. Then the quotient field of $S$ is complete with prime $p$ and its residue field is $K$.

**2.5.4.** Now let $F$, $L$ be complete discrete valuation fields of characteristic 0 with the residue field $\overline{F}$ of characteristic $p$ and $\overline{F} = \overline{L}$. Let $p$ be prime in $F$. Then there is a homomorphism $\varphi$: $F \to L$ such that $v_L \circ \varphi = e(L) v_F$ and $\overline{\varphi(\alpha)} = \overline{\alpha}$. We deduce that $L$ can be regarded as a totally ramified extension of degree $e$ over $F$. In particular, if $\overline{L}$ is perfect then $L$ can be regarded as a finite totally ramified extension of the quotient field of $W(\overline{L})$. If $p$ is prime also in $L$ then $\varphi$ is an isomorphism. For more details see [FV], Chapter II, Section 5.

## 3. The norm

From now on we treat complete discrete valuation fields.

### 3.1. *Cyclic extensions of prime degree*

To describe the action of the norm map $N_{L/F}$ with respect to the filtration of 1.3 there are four cases to consider: $L/F$ is unramified, $L/F$ is tamely and totally ramified,

$L/F$ is totally ramified of degree $p = \mathrm{char}(\overline{F}) > 0$, $\overline{L}/\overline{F}$ is inseparable of degree $p = \mathrm{char}(\overline{F}) > 0$ and $e(L|F) = 1$. We confine our attention to the first three cases.

**3.1.1.** For the proposition to follow it is convenient to use the next assertion: If $L/F$ is a separable finite extension and $\gamma \in \mathfrak{O}_L$ then

$$N_{L/F}(1+\gamma) = 1 + N_{L/F}(\gamma) + Tr_{L/F}(\gamma) + Tr_{L/F}(\delta)$$

for some $\delta \in \mathfrak{O}_L$ with $v_L(\delta) \geqslant 2v_L(\gamma)$, where $N_{L/F}$ is the norm map, $Tr_{L/F}$ is the trace map.

**3.1.2.** PROPOSITION. *Let $L/F$ be an unramified extension of degree $n$. Then a prime element $\pi_F$ in $F$ is prime in $L$. Let $U_{i,L} = 1 + \pi_F^i \mathfrak{O}_L$, $U_{i,F} = 1 + \pi_F^i \mathfrak{O}_F$ and $\lambda_{i,F}$, $\lambda_{i,L}$ for $F$ and $L$ be as in* 1.3.1. *Then the following diagrams are commutative:*

$$\begin{array}{ccc} L^* & \xrightarrow{v_L} & \mathbb{Z} \\ {\scriptstyle N_{L/F}}\downarrow & & \downarrow{\scriptstyle \times n} \\ F^* & \xrightarrow{v_F} & \mathbb{Z} \end{array} \qquad \begin{array}{ccc} U_L & \xrightarrow{\lambda_{0,L}} & \overline{L}^* \\ {\scriptstyle N_{L/F}}\downarrow & & \downarrow{\scriptstyle N_{L/F}} \\ U_F & \xrightarrow{\lambda_{0,F}} & \overline{F}^* \end{array}$$

$$\begin{array}{ccc} U_{i,L} & \xrightarrow{\lambda_{i,L}} & \overline{L}^* \\ {\scriptstyle N_{L/F}}\downarrow & & \downarrow{\scriptstyle Tr_{L/F}} \\ U_{i,F} & \xrightarrow{\lambda_{i,F}} & \overline{F} \end{array}, \qquad i \geqslant 1.$$

**3.1.3.** PROPOSITION. *Let $L/F$ be a totally and tamely ratified Galois extension of degree $n$. Then for some prime element $\pi_L$ in $L$ the element $\pi_F = \pi_L^n$ is prime in $F$. Let $U_{i,L} = 1 + \pi_L^i \mathfrak{O}_L$, $U_{i,F} = 1 + \pi_F^i \mathfrak{O}_F$ and $\lambda_{i,F}$, $\lambda_{i,L}$ for $F$ and $L$ be as in* 1.3.1. *Then the following diagrams are commutative:*

$$\begin{array}{ccc} L^* & \xrightarrow{v_L} & \mathbb{Z} \\ {\scriptstyle N_{L/F}}\downarrow & & \downarrow{\scriptstyle \mathrm{id}} \\ F^* & \xrightarrow{v_F} & \mathbb{Z} \end{array} \qquad \begin{array}{ccc} U_L & \xrightarrow{\lambda_{0,L}} & \overline{L}^* = \overline{F}^* \\ {\scriptstyle N_{L/F}}\downarrow & & \downarrow{\scriptstyle \uparrow n} \\ U_F & \xrightarrow{\lambda_{0,F}} & \overline{F}^* \end{array}$$

*if $i \geqslant 1$*

$$\begin{array}{ccc} U_{ni,L} & \xrightarrow{\lambda_{ni,L}} & \overline{L} = \overline{F} \\ {\scriptstyle N_{L/F}}\downarrow & & \downarrow{\scriptstyle \times \overline{n}} \\ U_{i,F} & \xrightarrow{\lambda_{i,F}} & \overline{F} \end{array}$$

*where* id *is the identity map, $\uparrow n$ is raising to the $n$-th power, $\times \overline{n}$ is multiplication by $\overline{n} \in \overline{F}$.*

**3.1.4.** PROPOSITION. *Let $L/F$ be a totally ratified Galois extension of degree $p = \operatorname{char}(\overline{F}) > 0$. Let $\sigma$ be a generator of $\operatorname{Gal}(L/F)$ and for a prime element $\pi_L$ in $L$*

$$\sigma(\pi_L)/\pi_L = 1 + \eta\pi_L^s \quad \text{with } \eta \in U_L,\ s \geqslant 1.$$

*Then $s$ doesn't depend on the choice of $\pi_L$. Let $\pi_F = N_{L/F}\pi_L$, then $\pi_F$ is prime in $F$. Let $U_{i,L} = 1 + \pi_L^i\mathfrak{O}_L$, $U_{i,F} = 1 + \pi_F^i\mathfrak{O}_F$ and $\lambda_{i,F}$, $\lambda_{i,L}$ for $F$ and $L$ be as in* 1.3.1. *Then the following diagrams are commutative:*

$$\begin{array}{ccc} L^* & \xrightarrow{v_L} & \mathbb{Z} \\ {\scriptstyle N_{L/F}}\downarrow & & \downarrow{\scriptstyle \mathrm{id}} \\ F^* & \xrightarrow{v_F} & \mathbb{Z} \end{array} \qquad \begin{array}{ccc} U_L & \xrightarrow{\lambda_{0,L}} & \overline{L}^* \\ {\scriptstyle N_{L/F}}\downarrow & & \downarrow{\scriptstyle N_{\uparrow p}} \\ U_F & \xrightarrow{\lambda_{0,F}} & \overline{F}^* \end{array}$$

*if* $1 \leqslant i < s$

$$\begin{array}{ccc} U_{i,L} & \xrightarrow{\lambda_{i,L}} & \overline{L} = \overline{F} \\ {\scriptstyle N_{L/F}}\downarrow & & \downarrow{\scriptstyle \uparrow p} \\ U_{i,F} & \xrightarrow{\lambda_{i,F}} & \overline{F} \end{array} \qquad \begin{array}{ccc} U_{s,L} & \xrightarrow{\lambda_{s,L}} & \overline{L} = \overline{F} \\ {\scriptstyle N_{L/F}}\downarrow & & \downarrow{\scriptstyle \xi} \\ U_{s,F} & \xrightarrow{\lambda_{s,F}} & \overline{F}^* \end{array}$$

*if* $i > 0$

$$\begin{array}{ccc} U_{s+pi,L} & \xrightarrow{\lambda_{s+pi,L}} & \overline{L} = \overline{F} \\ {\scriptstyle N_{L/F}}\downarrow & & \downarrow{\scriptstyle \times(-\overline{\eta}^{p-1})} \\ U_{s+i,F} & \xrightarrow{\lambda_{s+i,F}} & \overline{F} \end{array}$$

*where* $\xi(\overline{\theta}) = \overline{\theta}^p - \overline{\theta}\overline{\eta}^{p-1}$.

In particular, $U_{s+1,F} \subseteq N_{L/F}U_{s+1,L}$.

## 3.2. *The Hasse–Herbrand function*

We now assume that $F$ is a complete discrete valuation field whose residue field is perfect.

**3.2.1.** Let the residue field $\overline{F}$ of $F$ be infinite. Let $L/F$ be a finite Galois extension, $N = N_{L/F}$. The commutative diagrams of 3.1 and the solvability of $G_0$ in 2.4.2 imply that there exists the unique function $h = h_{L/F}\colon \mathbb{N} \to \mathbb{N}$ such that $h(0) = 0$ and

$$N\left(U_{h(i),L}\right) \subseteq U_{i,F}, \quad N\left(U_{h(i),L}\right) \not\subseteq U_{i+1,F}, \quad N\left(U_{h(i)+1,L}\right) \subseteq U_{i+1,F}.$$

Then $h_{L/F} = h_{L/L_0}$. For the case of finite residue fields we put $h_{L/F} = h_{\widehat{L^{ur}}/\widehat{F^{ur}}}$, where $\widehat{F^{ur}}$ is the completion of the maximal unramified extension $F^{ur}$ of $F$. If $M$ is a subextension in $L/F$ then $h_{L/F} = h_{L/M} \circ h_{M/F}$. Consequently, for a finite separable extension $L/F$ we put $h_{L/F} = h_{E/L}^{-1} \circ h_{E/F}$ for a finite Galois extension $E/F$, $L \subseteq E$. Then $h_{L/F}$ is well-defined.

**3.2.2.** It is more convenient to extend the Hasse–Herbrand function to be defined not only for natural numbers. For real $a \geqslant 0$ one sets $h(a) = a$, $h(a) = |L : F|a$,

$$h(a) = \begin{cases} a, & a \leqslant s, \\ s(1-p) + pa, & a \geqslant s, \end{cases}$$

for $L/F$ unramified, Galois totally ramified of degree prime to $p = \mathrm{char}(\overline{F})$ if $\mathrm{char}(F) > 0$, Galois totally ramified of degree $p = \mathrm{char}(\overline{F}) > 0$, respectively. Then

$$h_{L/F}\colon \mathbb{R}_{\geqslant 0} \to \mathbb{R}_{\geqslant 0}$$

is determined by employing these building block functions for any Galois or separable extension. The function $h_{L/F}$ is a well-defined, piecewise linear, continuous and increasing [FV], Chapter III.

**3.2.3.** Let $L/F$ be a finite Galois extension, $G = \mathrm{Gal}(L/F)$, $h = h_{L/F}$. Let $G_a$ for a real $a \geqslant 0$ denote the ramification group $G_m$, where $m$ is the smallest natural $\geqslant a$. Let $h'_l$, $h'_r$ be the left and the right derivatives of $h$. Then

$$h'_l(a) = |G_0 : G_{h(a)}|, \quad h'_r(a) = |G_0 : G_{h(a)}| \text{ if } h(a) \notin \mathbb{N},$$
$$h'_r(a) = |G_0 : G_{h(a)+1}| \text{ if } h(a) \in \mathbb{N}.$$

**3.2.4.** The traditional notation for $h_{L/F}$ is $\psi_{L/F}$. We call it Hasse–Herbrand, since Hasse introduced it in this form and Herbrand was the first who studied it (in a different form). This is the inverse function to $\varphi_{L/F}$ which plays a central role in expositions of ramification theory, see [Kaw1, Kaw2, Se3, Sen1, Sen2, CF, Lau1, Lau2, Lau3, Mar1, Mar2, Mau1, Mau2, Mau3, Mau4, Mau5, Win1, Win2]. Introduce an upper numbering of the ramification groups by setting $G^a = G_{h(a)}$, $a \geqslant 0$. Then for a normal subgroup $H$ in $G$ one can deduce by using the properties of $h$ that $(G/H)^a = G^a H/H$ for $a \geqslant 0$. For an infinite Galois extension $L/F$ with group $G$ the upper numbering is defined as $G^a = \varprojlim G(M/F)^a$, where $M/F$ runs over all finite subextensions in $L/F$.

### 3.3. *The norm and ramification groups*

Let $F$ be as in 3.2.

**3.3.1.** Let $L/F$ be a finite Galois totally ramified extension, $G = \mathrm{Gal}(L/F)$, $h = h_{L/F}$. Then for any integer $i \geqslant 0$ the sequence

$$1 \to G_{h(i)}/G_{h(i)+1} \xrightarrow{\psi_{h(i)}} U_{h(i),L}/U_{h(i)+1,L} \xrightarrow{N_i} U_{i,F}/U_{i+1,F}$$

is exact, where $\psi_{h(i)}$ is induced by the homomorphisms of 2.4.2, $N_i$ is induced by $N_{L/F}$.

**3.3.2.** Abelian extensions have some additional properties.

THEOREM (Hasse–Arf). *Let $L/F$ be a finite abelian extension with group $G = \mathrm{Gal}(L/F)$. Then $G_j = G_{j+1}$ for $j \in \mathbb{N}$ such that $j \notin h_{L/F}(\mathbb{N})$.*

For an assertion converse to the Hasse–Arf theorem see [Fe6].

### 3.4. *The Fontaine–Wintenberger fields of norms*

Let $F$ be as in 3.2.

**3.4.1.** Let $L$ be a separable extension of $F$ with a finite residue field extension $\overline{L}/\overline{F}$. Let $L$ be the union of an increasing directed family of subfields $L_i$, $i \geqslant 0$, which are finite extensions of $F$. The extension $L/F$ is said to be *arithmetically profinite* if the composition $\cdots \circ h_{L_i/L_{i-1}} \circ \cdots$ can be defined. In other words, taking into consideration 3.2.3 for any real $c > 0$ there is an integer $j$ such that

$$h_{L_i/L_j}(a) = a \qquad \text{for } a < h_{L_j/F}(c),\ i > j.$$

We put $h_{L/F} = \cdots \circ h_{L_i/L_{i-1}} \circ \cdots$. Then the function $h_{L/F}$ doesn't depend on the choice of $L_i$ and is piecewise linear, continuous, increasing. If $M/F$ is a subextension in $L/F$ then $M/F$ is arithmetically profinite. If, in addition, $M/F$ is finite then $h_{L/F} = h_{L/M} \circ h_{M/F}$. An extension $L/F$ is arithmetically profinite if and only if $G(F^{\mathrm{sep}}/L)G(F^{\mathrm{sep}}/F)^a$ is of finite index in $G(F^{\mathrm{sep}}/F)$ for any $a \geqslant 0$.

An important example of infinite arithmetically profinite extensions is a Galois extension $L/F$ with a finite residue field extension whose Galois group $\mathrm{Gal}(L/F)$ is a $p$-adic Lie group, see [Sen2, Win1].

A Galois totally ramified extension $L$ of a local field $F$ with finite residue field is arithmetically profinite if and only if $G(L/F)$ has a discrete set of breaks with respect to the upper numbering.

**3.4.2.** Let $L$ be an infinite arithmetically profinite extension of $F$ and $L_i$ be an increasing directed family of subfields which are finite extensions of $F$, $L = \bigcup L_i$. Let

$$N(L|F)^* = \varprojlim L_i^*$$

be the projective limit of the multiplicative groups with respect to norm homomorphisms $N_{L_i/L_j}$, $i \geqslant j$. Put $N(L|F) = N(L|F)^* \cup \{0\}$. Then $N(L|F)^*$ doesn't depend on the choice of $L_i$. Let $A = (\alpha_{L_i})_i$, $B = (\beta_{L_i})_i$ be elements of $N(L|F)$. Then the sequence $N_{L_j/L_i}(\alpha_{L_j} + \beta_{L_j})$, $j \to \infty$ is convergent in $L_i$. Let $\gamma_{L_i}$ be its limit. Then put $C = (\gamma_{L_i})_i = A + B$. The set $N(L|F)$ possesses the structure of a field under the multiplication and addition thus defined.

For $A = (\alpha_{L_i})_i$ put $v(A) = v(\alpha_{L_0})$, where $L_0$ is the maximal unramified subextension in $L/F$. Then the map $v$ is a discrete valuation and $N(L|F)$ is complete of characteristic $p$. There is an isomorphism of $\overline{L}$ onto a subfield in $N(L|F)$ which is mapped isomorphically onto the residue field of $N(L|F)$.

**3.4.3.** If $M/F$ is a finite subextension in $L/F$ then $N(L|F) = N(L|M)$. On the other hand, if $E/L$ is a finite separable extension then $N(L|F)$ can be identified with a subfield of $N(E|F)$ and $N(E|F)/N(L|F)$ is an extension of complete discrete valuation fields.

**3.4.4.** For an arbitrary separable extension $E/L$ denote by $N(E, L|F)$ the injective limit of $N(E'|F)$ for a finite separable subextension $E'/L$ in $E/L$. If $E/L$ is finite then $N(E, L|F) = N(E|F)$. If $E/L$ is Galois extension then $\mathrm{Gal}(E/L)$ is isomorphic with the Galois group of $N(E, L|F)$ over $N(L|F)$. Moreover, the group $\mathrm{Gal}(F^{\mathrm{sep}}/L)$ is isomorphic with the Galois group of $N(L|F)^{\mathrm{sep}}$ over $N(L|F)$.

Further properties of fields of norms can be found in [Win1, Win2, Win3, Lau4, Ke]. For some connections between complete discrete valuation fields of characteristic 0 and $p$ see [Del]. The objects that have been discussed are closely related with the theory of $p$-adic representations and $p$-adic periods, see [Win3, Fo1, Fo2, FI].

## 4. Local class field theory

We describe here abelian extensions of some classes of discrete valuation fields. In 4.1–4.5 we assume that $F$ is a complete discrete valuation field with finite residue field.

### 4.1. *Complete discrete valuation fields with a finite residue field*

For a finite field $\mathbb{F}_q$ its absolute Galois group $\mathrm{Gal}(\mathbb{F}_q^{\mathrm{sep}}/\mathbb{F}_q)$ is isomorphic with $\widehat{\mathbb{Z}}$ and topologically generated by the automorphism $\overline{\varphi}$: $\mathbb{F}_q^{\mathrm{sep}} \to \mathbb{F}_q^{\mathrm{sep}}$, $\overline{\varphi}(\theta) = \theta^q$.

**4.1.1.** Let $\overline{F} = \mathbb{F}_q$ for $q = p^f$, $p = \mathrm{char}(\overline{F})$. $f$ is called the *absolute inertia degree* of $F$. It follows from 2.5 that either $\mathrm{char}(F) = 0$ or $\mathrm{char}(F) = p$. In the first case $e = v(p) > 0$ and the restriction of $v$ to $\mathbb{Q}$ is equivalent to $p$-adic valuation by 1.1.4. Then $F$ can be regarded as containing the field $\mathbb{Q}_p$ and $F/\mathbb{Q}_p$ is a finite extension of degree $n = ef$. Such a field is called a $\mathfrak{p}$-adic field. Fields of the second class are called local function fields, they are isomorphic with $\mathbb{F}_q((X))$. Complete discrete valuation fields with perfect residue fields are often called *local.*

**4.1.2.** The ring of integers $\mathfrak{O}$ of $F$ and the unit group $U$ are compact with respect to the valuation topology, $F$ is locally compact. The commutative diagrams of 1.3.2 imply that subgroups of finite index $n$ in $F^*$ are open if $\mathrm{char}(F) = 0$ or if $\mathrm{char}(F) = p$, $(n, p) = 1$.

Thus, topological properties of $\mathfrak{p}$-adic fields are determined by their algebraic structure. This is not the case for local function fields.

**4.1.3.** One can deduce from 2.3.2 that there exists a uniquely determined unramified extension $L$ of $F$ of degree $n \geqslant 1$: $L = F(\mu_{q^n-1})$, where $\mu_{q^n-1}$ is the group of all $(q^n - 1)$-th roots of unity in $F^{\mathrm{sep}}$. The extension $L/F$ is cyclic and by 2.3.4 and the previous remark the maximal unramified extension $F^{\mathrm{ur}}$ of $F$ is Galois with the group isomorphic to $\widehat{\mathbb{Z}}$ and topologically generated by an automorphism $\varphi_F$ such that

$$\varphi_F(\alpha) \equiv \alpha^q \pmod{\mathfrak{M}_{F^{\mathrm{ur}}}} \quad \text{for } \alpha \in \mathfrak{O}_{F^{\mathrm{ur}}}.$$

The automorphism $\varphi_F$ is called the *Frobenius* automorphism of $F$.

**4.1.4.** By using 1.3.2 one can deduce that if $\mathrm{char}(F) = p$ then any element $\alpha \in U_1$ can be uniquely expressed as a convergent product

$$\alpha = \prod_{\substack{(i,p)=1 \\ i \geqslant 1}} \prod_{j \in J} (1 + \theta_{ij}\pi^i)^{a_{ij}}$$

with the index-set $J$ enumerating $f$ elements in $\mathfrak{O}_F$ of which the residues form a basis of $\overline{F}$ over $\mathbb{F}_p$, the elements $\theta_{ij}$ belonging to this set, $\pi$ is a prime element in $F$, $a_{ij} \in \mathbb{Z}_p$.

Denote the polynomial $X^p - X$ by $\wp(X)$. Note that the subgroup $\wp(\overline{F})$ is of index $p$ in $\overline{F}$. If $\mathrm{char}(F) = 0$ then any element $\alpha \in U_1$ can be expressed as a convergent product

$$\alpha = \prod_{i \in I} \prod_{j \in J} (1 + \theta_{ij}\pi^i)^{a_{ij}} \omega_*^a$$

with $I = \{1 \leqslant i < pe/(p-1),\ (i,p) = 1\}$, the index-set $J$ and $\theta_{ij}$ being as above, $a_{ij} \in \mathbb{Z}_p$. If there is no primitive $p$-th root of unity in $F$ then $\omega_* = 1$, $a = 0$ and the writing is unique. If there is a primitive $p$-th root of unity in $F$ then $\omega_* = 1 + \theta_* \pi^{pe/(p-1)}$ such that $\omega_* \notin F^{*p}$, $a \in \mathbb{Z}_p$. In this case the expression above isn't unique.

**4.1.5.** The commutative diagrams of 3.1 imply that the norm group $N_{L/F}L^*$ is of index $l = |L : F|$ in $F^*$ for a cyclic extension $L/F$ of degree $l$.

## 4.2. *The Neukirch construction of the reciprocity map*

**4.2.1.** Let $L/F$ be a finite Galois extension. Denote by $\phi(L|F)$ the set of those automorphisms $\widetilde{\sigma}$ in $\mathrm{Gal}(L^{\mathrm{ur}}/F)$ for which $\widetilde{\sigma}|_{F^{\mathrm{ur}}}$ is a positive integer power of $\varphi_F$. Then the set $\phi(L|F)$ is closed with respect to multiplication, but $1 \notin \phi(L|F)$. The map $\phi(L|F) \to \mathrm{Gal}(L/F)$: $\widetilde{\sigma} \to \widetilde{\sigma}|_L$ is surjective. The fixed field $\Sigma$ of $\widetilde{\sigma} \in \phi(L|F)$ is of finite degree over $F$ and $\widetilde{\sigma}$ is the Frobenius automorphism of $\Sigma$, $\Sigma^{\mathrm{ur}} = L^{\mathrm{ur}}$.

**4.2.2.** Let $L/F$ be a finite Galois extension. Introduce the map

$$\widetilde{\Upsilon}_{L/F}\colon \phi(L|F) \to F^*/N_{L/F}L^*$$

by the formula

$$\widetilde{\Upsilon}_{L/F}(\widetilde{\sigma}) \equiv N_{\Sigma/F}\pi_\Sigma \pmod{N_{L/F}L^*},$$

where $\Sigma$ is the fixed field of $\widetilde{\sigma} \in \phi(L|F)$, $\pi_\Sigma$ is prime in $\Sigma$. Then the map $\widetilde{\Upsilon}_{L/F}$ is well defined.

The next assertion is of great importance in this exposition: let $\widetilde{\sigma}_1, \widetilde{\sigma}_2 \in \phi(L|F)$ and $\widetilde{\sigma}_3 = \widetilde{\sigma}_2\widetilde{\sigma}_1 \in \phi(L|F)$ then

$$N_{\Sigma_3/F}\pi_3 \equiv N_{\Sigma_1/F}\pi_1 N_{\Sigma_2/F}\pi_2 \pmod{N_{L/F}L^*},$$

where $\pi_i$ is a prime element in the fixed field $\Sigma_i$ of $\widetilde{\sigma}_i$. This assertion is verified by technical but not complicated computations, see [N3]. The congruence can be proved easier if the Hazewinkel construction of the reciprocity map (4.3) comes into play. It shows that the map $\widetilde{\Upsilon}_{L/F}$ induces a homomorphism

$$\Upsilon_{L/F}\colon \mathrm{Gal}(L/F) \to F^*/N_{L/F}L^*,$$

where $\Upsilon_{L/F}(\sigma) = \widetilde{\Upsilon}_{L/F}(\widetilde{\sigma})$ and $\widetilde{\sigma}$ be any element of $\phi(L|F)$ such that $\widetilde{\sigma}|_L = \sigma$.

**4.2.3.** The homomorphism $\Upsilon_{L/F}$ has natural properties. If $L/F$ is an unramified finite extension then $\Upsilon_{L/F}$ is an isomorphism and $\Upsilon_{L/F}(\varphi_F|_L) \equiv \pi_F \pmod{N_{L/F}L^*}$ for a prime element $\pi_F$ in $F$.

If $M/F$ is a finite separable extension and $L/M$ a finite Galois extension,

$$\sigma \in \mathrm{Gal}\big(F^{\mathrm{sep}}/F\big)$$

then the diagram

$$\begin{array}{ccc} \mathrm{Gal}(L/M) & \xrightarrow{\ \Upsilon_{L/M}\ } & M^*/N_{L/M}L^* \\ \downarrow{\scriptstyle \sigma^*} & & \downarrow{\scriptstyle \sigma} \\ \mathrm{Gal}(\sigma L/\sigma M) & \xrightarrow{\ \Upsilon_{\sigma L/\sigma M}\ } & (\sigma M)^*/N_{\sigma L/\sigma M}(\sigma L)^* \end{array}$$

is commutative, where $\sigma^*(\tau) = \sigma\tau\sigma^{-1}|_{\sigma L}$ for $\tau \in \mathrm{Gal}(L/M)$.

If $M/F$, $E/L$ are finite separable extensions and $L/F$, $E/M$ are finite Galois extensions then the diagram

$$\begin{array}{ccc} \mathrm{Gal}(E/M) & \xrightarrow{\Upsilon_{E/M}} & M^*/N_{E/M}E^* \\ \downarrow & & \downarrow N^*_{M/F} \\ \mathrm{Gal}(L/F) & \xrightarrow{\Upsilon_{L/F}} & F^*/N_{L/F}L^* \end{array}$$

is commutative, where the left vertical homomorphism is the restriction $\sigma|_L$ of $\sigma \in \mathrm{Gal}(E/M)$ and the right vertical homomorphism is induced by the norm map $N_{M/F}$.

As the image of $\Upsilon_{L/F}$ is abelian, one can define a homomorphism

$$\Upsilon_{L/F}\colon \mathrm{Gal}(L/F)^{\mathrm{ab}} \to F^*/N_{L/F}L^*,$$

where $\mathrm{Gal}(L/F)^{\mathrm{ab}}$ is the maximal abelian quotient of $\mathrm{Gal}(L/F)$.

If $L/F$ is a finite Galois extension and $M/F$ a subextension in $L/F$ then the diagram

$$\begin{array}{ccc} \mathrm{Gal}(L/F)^{\mathrm{ab}} & \xrightarrow{\Upsilon_{L/F}} & F^*/N_{L/F}L^* \\ \mathrm{Ver}\downarrow & & \downarrow \\ \mathrm{Gal}(L/M)^{\mathrm{ab}} & \xrightarrow{\Upsilon_{L/M}} & M^*/N_{L/M}L^* \end{array}$$

is commutative, where the right vertical homomorphism is induced by the imbedding $F \hookrightarrow M$ and Ver is the transfer map (Verlagerung) for finite groups.

**4.2.4.** It is easy to verify by using 4.1.5 that $\Upsilon_{L/F}$ is an isomorphism for a cyclic extension $L/F$. By induction on degree one can show that $\Upsilon_{L/F}$ is an isomorphism for an abelian extension $L/F$. Thus, the Neukirch map

$$\Upsilon_{L/F}\colon \mathrm{Gal}(L/F)^{\mathrm{ab}} \to F^*/N_{L/F}L^*$$

is an isomorphism.

**4.2.5.** The inverse to $\Upsilon_{L/F}$ homomorphism induces a surjective homomorphism

$$(\ ,L/F)\colon\ F^* \to \mathrm{Gal}(L/F)^{\mathrm{ab}}.$$

Denote by $F^{\mathrm{ab}}$ the maximal abelian extension of $F$ in $F^{\mathrm{sep}}$. Passing to the projective limit via 4.2.3 we get a well defined homomorphism

$$\Psi_F\colon\ F^* \to \mathrm{Gal}(F^{\mathrm{ab}}/F),$$

which is called the *reciprocity map*. Its image is dense in $\mathrm{Gal}(F^{\mathrm{ab}}/F)$ and its kernel coincides with the intersection of all norm subgroups $N_{L/F}L^*$ in $F^*$ for finite Galois extensions $L/F$.

If $L/F$ is a finite Galois extension then $\Psi_F(\alpha)$ for $\alpha \in F^*$ acts trivially on $L \cap F^{\mathrm{ab}}$ if and only if $\alpha \in N_{L/F}L^*$.

For $\alpha \in F^*$

$$\Psi_F(\alpha)|_{F^{\mathrm{ur}}} = \varphi_F^{v_F(\alpha)}.$$

The reciprocity map possesses natural functorial properties analogous to those in 4.2.3.

### 4.3. *The Hazewinkel construction of the reciprocity map*

Let, for simplicity, $L/F$ be a cyclic totally ramified extension. Let $\varepsilon \in U_F$. By using the surjectivity of the norm map $N\colon U_{\widehat{L^{\mathrm{ur}}}} \to U_{\widehat{F^{\mathrm{ur}}}}$, where $\widehat{L^{\mathrm{ur}}}$, $\widehat{F^{\mathrm{ur}}}$ are the completions of $L^{\mathrm{ur}}$, $F^{\mathrm{ur}}$, it can be verified that there exists an element $\beta \in U_{\widehat{L^{\mathrm{ur}}}}$ such that $N\beta = \varepsilon$. Let $\varphi$ be a continuous extension of the Frobenius automorphism $\varphi_L$ on $\widehat{L}^{\mathrm{ur}}$. Then $N(\varphi(\beta)/\beta) = 1$. By the Hilbert 90 theorem there exists an element $\alpha \in \widehat{L^{\mathrm{ur}}}$ such that $\sigma(\alpha)/\alpha = \varphi(\beta)/\beta$, where $\sigma$ is a generator of $\mathrm{Gal}(L/F)$. Moreover, if $\pi$ is prime in $L$ then $\sigma(\alpha)/\alpha$ can be written as $(\tau(\pi)/\pi)\,(\sigma(\varepsilon)/\varepsilon)$ for some $\tau \in \mathrm{Gal}(L/F)$, $\varepsilon \in U_{\widehat{L^{\mathrm{ur}}}}$. Then the map $\varepsilon \to \tau$ induces the homomorphism

$$U_F/N_{L/F}U_L \to \mathrm{Gal}(L/F)$$

which is an isomorphism and inverse to $\Upsilon_{L/F}$, see [Haz1, Haz2, Iw5].

### 4.4. *Cohomological approach*

Another construction of the reciprocity map follows also from considerations of the Brauer group $\mathrm{Br}(F)$. A Theorem of Hasse asserts that $\mathrm{Br}(F) \simeq \mathbb{Q}/\mathbb{Z}$. There is a pairing for $\mathrm{char}(F) = 0$:

$$\mathrm{Hom}\left(\mathrm{Gal}(F^{\mathrm{sep}}/F), \mathbb{Q}/\mathbb{Z}\right) \times F^* \to H^1(F, \mathbb{Q}/\mathbb{Z}) \times \varinjlim H^1(F, \mu_n) \to \varinjlim H^2(F, \mu_n) \simeq \mathbb{Q}/\mathbb{Z}$$

where the injective limit is taken with respect to the natural maps $\mu_n \to \mu_{nm}$, $m \geqslant 1$ (there is also a pairing for $\mathrm{char}(F) = p$, see 7.2.4). This pairing induces a homomorphism

$$F^* \to \mathrm{Hom}\left(\mathrm{Hom}\left(\mathrm{Gal}(F^{\mathrm{sep}}/F), \mathbb{Q}/\mathbb{Z}\right), \mathbb{Q}/\mathbb{Z}\right) = \mathrm{Gal}(F^{\mathrm{ab}}/F),$$

which coincides with the reciprocity map. For details see [Se3, Se4, CF].

### 4.5. *Existence theorem*

This theorem makes the description of abelian extensions more precise: there is a one-to-one correspondence between open subgroups of finite index in $F^*$ and the norm subgroups $N_{L/F}L^*$ of finite abelian extensions $L/F$. If $L_1$, $L_2$ are finite abelian extensions over $F$ then $L_1 \subseteq L_2$ if and only if $N_{L_2/F}{L_2}^* \subseteq N_{L_1/F}{L_1}^*$. If $L_3 = L_1L_2$, $L_4 = L_1 \cap L_2$ then

$$N_{L_3/F}{L_3}^* = N_{L_1/F}{L_1}^* \cap N_{L_2/F}{L_2}^*, \qquad N_{L_4/F}{L_4}^* = N_{L_1/F}{L_1}^* N_{L_2/F}{L_2}^*.$$

The proof employs the fact that any open subgroup of prime index in $F^*$ is a norm group $N_{L/F}L^*$ for a suitable cyclic extension $L/F$, see below 5.1.2, 5.3.2, 5.4.1.

Existence Theorem implies that the reciprocity map $\Psi_F$ is injective and continuous.

### 4.6. *Generalizations*

**4.6.1.** Existence Theorem can be extended to the case of abelian (not necessarily finite) extensions of $F$. For an abelian extension $L/F$ put

$$N_{L/F}L^* = \bigcap_M N_{M/F}M^*,$$

where $M$ runs over all finite subextensions in $L/F$. In particular, $\Psi_F$ maps the group $U_{i,F}$ isomorphically onto the ramification group $\mathrm{Gal}(F^{\mathrm{ab}}/F)^i$, where the upper numbering was defined in 3.2.4. See also 5.4.1 below.

**4.6.2.** The same theory can be established for a complete discrete valuation field $F$ whose residue field is *quasi-finite*, i.e.

$$\mathrm{Gal}\left(\overline{F}^{\mathrm{sep}}/\overline{F}\right) \simeq \widehat{\mathbb{Z}},$$

see [Mo1, Mo2, Mo3, Wh1, Wh2, Wh3, Wh4]. A distinction is that there are no canonical generators of $\mathrm{Gal}(\overline{F}^{\mathrm{sep}}/\overline{F})$ as in the case of a finite residue field and an open subgroup of finite index in $F^*$ isn't in general a norm subgroup and one has introduce a notion of a normic subgroup, see [Wh1].

**4.6.3.** If $\mathcal{F}$ is an infinite separable extension of $F$ with finite residue extension then put $\mathcal{F}^{\times} = \varprojlim M^*$, where $M$ runs all finite subextensions of $F$ in $\mathcal{F}$ and the projective limit is taken with respect to the norm maps. For a finite separable extension $\mathcal{L}/\mathcal{F}$ one can define the norm map $\mathcal{N}_{\mathcal{L}/\mathcal{F}}\colon \mathcal{L}^{\times} \to \mathcal{F}^{\times}$. There is an isomorphism

$$\Upsilon_{\mathcal{L}/\mathcal{F}}\colon \mathrm{Gal}(\mathcal{L}/\mathcal{F})^{\mathrm{ab}} \to \mathcal{F}^{\times}/\mathcal{N}_{\mathcal{L}/\mathcal{F}}\mathcal{L}^{\times}.$$

For more details see [Sch, Kaw2, N3], Chapter 2, §5. This isomorphism is compatible with the construction of fields of norms in 3.4.

**4.6.4.** The same theory can be established for Henselian discrete valuation fields with a quasi-finite residue field (existence theorem is different!), see [FV], Chapter V.

**4.6.5.** Let $F$ be a complete discrete valuation field with an algebraically closed residue field. Serre's geometric class field theory describes abelian extensions of $F$ in terms of the fundamental subgroup $\pi_1(U_F)$ regarding $U_F$ as a pro-algebraic group, see [Se2, Haz1].

**4.6.6.** Generalizations for the case of a perfect residue field can be found in [Haz1] (via Serre's theory). Another approach is described in [Fe5]. Let $F$ be a local field with perfect residue field $\overline{F}$ of characteristic $p$. Denote by $\widetilde{F}$ the maximal abelian unramified $p$-extension of $F$. Then for a finite abelian totally ramified extension $L/F$ the group $\mathrm{Hom}_{\mathrm{cont}}(G(\widetilde{F}/F), G(L/F))$ of continuous homomorphisms from the profinite group $G(\widetilde{F}/F)$ (we assume $\widetilde{F} \neq F$) to the discrete finite group $G(L/F)$ is canonically isomorphic to the quotient group $U_{1,F}/N_{L/F}U_{1,L}$ [Fe5].

**4.6.7.** For the case of imperfect residue field see [Fe7] and Section 7 below.

**4.6.8.** Let $K$ be a local field with finite residue field and let $L$ be a Galois totally ramified extension of $K$. Let $F$ be a formal group over $\mathfrak{O}_K$ which is isomorphic to $\mathbb{G}_m^d$ over the maximal unramified extension. Let $N_{(F),L/K}$ be the formal norm from $F(\mathfrak{M}_L)$ to $F(\mathfrak{M}_K)$ (see Subsection 5.4.1 below). Then, according to Mazur [Maz] there is a canonical isomorphism of the group $F(\mathfrak{M}_K)/N_{(F),L/K}F(\mathfrak{M}_L)$ onto the group

$$\left(G(L/K)^{\mathrm{ab}}\right)^d/(E-M)\left(G(L/K)^{\mathrm{ab}}\right)^d,$$

where $M \in GL_d(\mathbb{Z}_p)$ is a twisted matrix of $F$. Its construction (see also [LR]) is a generalization of the Hazewinkel homomorphism.

**4.6.9.** Using the theory of fields of norms one can derive (Koch, de Shalit) overcoming technical difficulties the so-called metabelian local class field theory which describes a maximal abelian extension of the maximal abelian extension of $F$.

## 5. Pairings on the multiplicative group

We assume that $F$ is a complete discrete valuation field with a finite residue field.

### 5.1. *The Hilbert symbol*

**5.1.1.** Let the group $\mu_n$ of all $n$-th roots of unity in $F^{\mathrm{sep}}$ be contained in $F$ and $n$ be relatively prime with $p$ if $\mathrm{char}(F) = p > 0$. The *Hilbert norm residue symbol* $(\,,\,)_n$: $F^* \times F^* \to \mu_n$ is defined by the formula

$$(\alpha, \beta)_n = \gamma^{-1}\Psi_F(\alpha)(\gamma), \quad \text{where } \gamma^n = \beta,\ \gamma \in F^{\mathrm{sep}}.$$

PROPOSITION. *The Hilbert symbol is well defined. It possesses the following properties:*

1) $(\,,\,)_n$ *is bilinear;*
2) $(1-\alpha,\alpha)_n = 1$ *for* $\alpha \in F^*$, $\alpha \neq 1$ *(Steinberg property);*
3) $(-\alpha,\alpha)_n = 1$ *for* $\alpha \in F^*$;
4) $(\alpha,\beta)_n = (\beta,\alpha)_n^{-1}$;
5) $(\alpha,\beta)_n = 1$ *if and only if* $\alpha \in N_{F(\sqrt[n]{\beta})/F}F(\sqrt[n]{\beta})^*$ *and if and only if* $\beta \in N_{F(\sqrt[n]{\alpha})/F}F(\sqrt[n]{\alpha})^*$;
6) $(\alpha,\beta)_n = 1$ *for all* $\beta \in F^*$ *if and only if* $\alpha \in F^{*n}$;
7) $(\alpha,\beta)_{nm}^m = (\alpha,\beta)_n$ *for* $m \geqslant 1$;
8) $(\alpha,\beta)_{n,L} = (N_{L/F}\alpha,\beta)_{n,F}$ *for* $\alpha \in L^*$, $\beta \in F^*$;
9) $(\sigma\alpha,\sigma\beta)_{n,\sigma L} = \sigma(\alpha,\beta)_{n,L}$, *where* $L$ *is a finite separable extension of* $F$, $\sigma \in \mathrm{Gal}(F^{\mathrm{sep}}/F)$, $\mu_n \subset L^*$.

Thus, the Hilbert symbol induces a nondegenerate pairing

$$(\,,\,)_n\colon\ F^*/F^{*n} \times F^*/F^{*n} \to \mu_n.$$

**5.1.2.** Let $\mu_n \subset F^*$ and $n$ be as in 5.1.1. The theory of Kummer extensions (see [La1], Chapter 8) asserts that abelian extensions $L/F$ of exponent $n$ are in one-to-one correspondence with subgroups $B_L \subseteq F^*$ with $F^{*n} \subseteq B_L$:

$$L = F(\sqrt[n]{B_L}) = F(\gamma_i; \gamma_i^n \in B_L)$$

and the group $B_L/F^{*n}$ has the same structure as $\mathrm{Gal}(L/F)$.

Now let $A$ be a subgroup in $F^*$ such that $F^{*n} \subseteq A$. Denote by $B = A^\perp$ its orthogonal supplement with respect to the Hilbert symbol. Then $A = N_{L/F}L^*$, where $L = F(\sqrt[n]{B})$. Conversely, if $B$ is a subgroup in $F^*$ such that $F^{*n} \subseteq B$ then its orthogonal supplement $A = B^\perp$ coincides with $N_{L/F}L^*$ for $L = F(\sqrt[n]{B})$. It follows that any subgroup of a prime index $l$ in $F^*$, $l \neq \mathrm{char}(F)$ if $\mathrm{char}(F) > 0$, is a norm subgroup.

**5.1.3.** Hilbert's 9th Problem is to find explicit formulas for the global norm residue symbol. In the case under consideration this means to discover a formula for the Hilbert symbol $(\alpha,\beta)_n$ in terms of elements $\alpha$, $\beta$ of the field $F$.

There is a simple answer to this question when $n$ is relatively prime with $\mathrm{char}(\overline{F})$ Then $(\alpha,\beta)_n = t(\alpha,\beta)^{(q-1)/n}$, where $t\colon F^* \times F^* \to \mu_{q-1}$ is the tame symbol defined by the formula

$$t(\alpha,\beta) = pr\big(\beta^{v_F(\alpha)}\alpha^{-v_F(\beta)}(-1)^{v_F(\alpha)v_F(\beta)}\big)$$

with the projection $pr\colon U_F \to \mu_{q-1}$ induced by the decomposition $U_F \simeq \mu_{q-1} \times U_{1,F}$ as in 1.3.3.

## 5.2. *Explicit formulas for the Hilbert $p^n$-th symbol*

In his celebrated work [Sha2] Shafarevich proposed an explicit formula for the Hilbert $p$-th symbol in terms of his basis of the group of principal units. His idea was then to apply this formula pairing for an independent construction of local class field theory. At the end of the 70's Vostokov obtained explicit formulas for the Hilbert $p^n$-th symbol, $p > 2$.

Let $F$ be a $\mathfrak{p}$-adic field, $\zeta_{p^n}$ a primitive $p^n$-th root of unity which is contained in $F$, $n \geqslant 1$. Let $\mathfrak{O}_0$ be the ring of integers of the field $F_0 = F \cap \mathbb{Q}_p^{\mathrm{ur}}$. Let $\pi$ be a fixed prime element in $F$.

For an element $\alpha \in F^*$ let $\psi(X) \in 1 + X\mathfrak{O}_0[[X]]$ be such that $\pi^m \theta \psi(\pi) = \alpha$, where $m \in \mathbb{Z}$, $\theta \in \mu_{q-1}$. Put $\alpha(X) = X^m \theta \psi(X)$.

Put

$$l_X\big(\alpha(X)\big) = \left(1 - \frac{\Delta_X}{p}\right) \log \psi(X),$$

where

$$\log(1+X) = \sum_{i \geqslant 1} (-1)^{i-1} X^i / i, \qquad \Delta_X\Big(\sum a_i X^i\Big) = \sum \varphi(a_i) X^{pi},$$

$a_i \in \mathfrak{O}_0$, and $\varphi$ is the Frobenius automorphism of $\mathbb{Q}_p$.

For $\alpha, \beta \in F^*$ put

$$\begin{aligned} \Phi_{\alpha,\beta}(X) = {} & l_X\big(\alpha(X)\big) l_X\big(\beta(X)\big)' - l_X\big(\alpha(X)\big)\beta(X)'/\beta(X) \\ & + l_X\big(\beta(X)\big)\alpha(X)'/\alpha(X). \end{aligned}$$

Let $z(X) \in 1 + X\mathfrak{O}_0[[X]]$ be such that $z(\pi) = \zeta_{p^n}$. Put $s(X) = z(X)^{p^n} - 1$.

Let $p > 2$. Employing Shafarevich's canonical basis of the group of principal units, Vostokov ([V1, V2]) established the following explicit formula for the $p^n$-th Hilbert symbol:

$$(\alpha, \beta)_{p^n} = \zeta_{p^n}^{Tr\, res\, \Phi_{\alpha,\beta}(X)/s(X)},$$

where

$$res\Big(\sum \alpha_i X^i\Big) = \alpha_{-1}, \qquad Tr = Tr_{F_0/\mathbb{Q}_p}.$$

For $p = 2$ the formulas are more complicated, see [VF, Fe1, Fe2]. Details are in [FV], Chapter VII.

Among various applications of the explicit formulas there is an exposition of the correspondence between Kummer's extensions of $F$ and open subgroups in $F^*$ that is independent of class field theory, see [FV], Chapter VII. Independently, approximately the same formula was obtained by Brückner [Bru1, Bru2] by using different methods.

This formula can be generalized for complete discrete valuation fields with quasi-finite residue field. For other formulas for the Hilbert symbol in general and special cases see [AH1, AH2, Kn, Iw3, Henn1, Henn2, Sen3].

**5.3.** *Pairings using the Witt vectors*

Let $F$ be a local function field with $\overline{F} = \mathbb{F}_q$. We shall consider an analog of the Hilbert symbol for such a field.

**5.3.1.** Define a map

$$( \, , \, ]\colon \; F^* \times F \to \mathbb{F}_p$$

by the formula $(\alpha, \beta] = \Psi_F(\alpha)(\gamma) - \gamma$, where $\wp(\gamma) = \gamma^p - \gamma = \beta$, $\gamma \in F^{\mathrm{sep}}$, see 4.1.4.

PROPOSITION. *This map is well defined and has the following properties:*
1) $(\alpha_1\alpha_2, \beta] = (\alpha_1, \beta] + (\alpha_2, \beta]$, $(\alpha, \beta_1 + \beta_2] = (\alpha, \beta_1] + (\alpha, \beta_2]$;
2) $(-\alpha, \alpha] = 0$ *for* $\alpha \in F^*$;
3) $(\alpha, \beta] = 0$ *if and only if* $\alpha \in N_{F(\gamma)/F}F(\gamma)^*$, *where* $\wp(\gamma) = \beta$;
4) $(\alpha, \beta] = 0$ *for all* $\alpha \in F^*$ *if and only if* $\beta \in \wp(F)$;
5) $(\alpha, \beta] = 0$ *for all* $\beta \in F$ *if and only if* $\alpha \in F^{*p}$.

Thus, this map determines the nondegenerate pairing

$$F^*/F^{*p} \times F/\wp(F) \to \mathbb{F}_p.$$

**5.3.2.** Any open subgroup $A$ in $F^*$ of index $p$ coincides with $N_{L/F}L^*$, where $L = F(\gamma\colon \wp(\gamma) \in B)$ and $B = A^{\perp}$ is the orthogonal supplement of $A$ with respect to $( \, , \, ]$. This assertion is applied for the proof of the Existence Theorem in 4.5.

**5.3.3.** There is a formula for the pairing $( \, , \, ]$:

$$(\alpha, \beta] = Tr_{\mathbb{F}_q/\mathbb{F}_p} res\left(\beta(X)\alpha(X)'/\alpha(X)\right),$$

where $\alpha(X) \in \mathbb{F}_q((X))$, $\beta(X) \in \mathbb{F}_q((X))$ such that $\alpha(\pi) = \alpha$, $\beta(\pi) = \beta$, $\pi$ prime in $F$. Compare this formula with 5.2.

**5.3.4.** The pairing $( \, , \, ]$ can be generalized using the ring of Witt vectors:

$$( \, , \, ]_n\colon \; F^* \times W_n(F) \to W_n(\mathbb{F}_p) \simeq \mathbb{Z}/p^n\mathbb{Z},$$

(see 1.4.2) by the formula $(\alpha, y]_n = \Psi_F(\alpha)(z) - z$, where $z \in W_n(F^{\mathrm{sep}})$, $\wp(z) = y$.

By using this pairing one can construct the reciprocity map independently, see [Sek1, Sek2].

**5.4.** *Pairings using formal groups*

**5.4.1.** Let $K$ be a $\mathfrak{p}$-adic field, $\overline{K} = \mathbb{F}_q$, $\pi$ be prime in $K$.

Denote by $\mathcal{F}_\pi$ the set of formal power series $f(X) \in X\mathfrak{O}_K[[X]]$ such that $f(X) = \pi X + X^2 g(X)$ with $g(X) \in \mathfrak{O}_K[[X]]$ and $f(X) = X^q + \pi h(X)$ with $h(X) \in \mathfrak{O}_K[[X]]$. Then there exists a unique formal power series $F(X,Y) \in \mathfrak{O}_K[[X,Y]]$ (*Lubin–Tate formal group*) such that $F(f(X), f(Y)) = f(F(X,Y))$ and

$$F(X,0) = F(0,X) = 0, \quad F(X, F(Y,Z)) = F(F(X,Y), Z),$$
$$F(X,Y) = F(Y,X).$$

In particular, if $\pi = p$ then there is the multiplicative formal group $F_m(X,Y) = X + Y + XY$, which corresponds to multiplication.

Denote by $\operatorname{End}_{\mathfrak{O}_K}(F)$ the set

$$\{g(X) \in \mathfrak{O}_K[[X]]\colon\ F(g(X), g(Y)) = g(F(X,Y))\}.$$

There is a ring homomorphism $\mathfrak{O}_K \to \operatorname{End}_{\mathfrak{O}_K}(F)\colon\ \alpha \to [\alpha]_F$ such that

$$[\alpha]_F(X) = \alpha X + \cdots,$$

$f = [\pi]_F$, see [CF], Chapter 6, [Iw5], §7.3, [N4], Chapter 3.

Let $L$ be an algebraic extension of $K$. One can define on the set $\mathfrak{M}_L$ a structure of $\mathfrak{O}_K$-module $F(\mathfrak{M}_L)$:

$$\alpha + \beta = F(\alpha, \beta), \quad a\alpha = [a]_F(\alpha), \quad a \in \mathfrak{O}_K,\ \alpha, \beta \in \mathfrak{M}_L.$$

Denote by $\kappa_n$ the group of $\pi^n$-division points $\{\alpha \in \mathfrak{M}_{K^{\mathrm{sep}}}\colon\ [\pi^n]_F(\alpha) = 0\}$.

Then the field $L_n = K(\kappa_n)$ is a totally ramified abelian extension of degree $q^{n-1}(q-1)$ over $K$ and corresponds to the subgroup $(\pi) \times U_{n,K}$ in $K^*$. $\operatorname{Gal}(L_n/K)$ is isomorphic with $U_K/U_{n,K}$. Put

$$K_\pi = \bigcup_{n \geqslant 1} L_n.$$

Then the field $K_\pi$ corresponds to the subgroup generated by $\pi$ and

$$\operatorname{Gal}(K^{\mathrm{ab}}/K) \simeq \operatorname{Gal}(K^{\mathrm{ur}}/K) \times \operatorname{Gal}(K_\pi/K),$$
$$\Psi_K(\pi^a \varepsilon)(\zeta) = [\varepsilon^{-1}]_F(\zeta) \quad \text{for } \zeta \in \bigcup_{n \geqslant 1} \kappa_n,\ a \in \mathbb{Z},\ \varepsilon \in U_K.$$

In particular, putting $K = \mathbb{Q}_p$, $F = F_m$, we deduce the local Kronecker–Weber theorem: $\mathbb{Q}_p^{\mathrm{ab}}$ is generated by all roots of unity.

**5.4.2.** Let $\pi_0$ be prime in $K$, $F(X,Y)$ be a formal Lubin–Tate group for $f(X) \in \mathcal{F}_{\pi_0}$. Let $L/K$ be a finite extension such that $\kappa_n \subset L$. Define the generalized Hilbert pairing

$$(\ ,\ )_{F,n}\colon\ L^* \times F(\mathfrak{M}_L) \to \kappa_n$$

by the formula

$$(\alpha, \beta)_{F,n} = F\big(\Psi_F(\alpha)(\gamma), [-1]_F(\gamma)\big),$$

where $\gamma \in F(\mathfrak{M}_{K^{\mathrm{sep}}})$ with $[\pi_0^n](\gamma) = \beta$.

Explicit formulas for $(\ ,\ )_{F,n}$ and applications can be found in [V2, V3, VF, Fe2, CW1, Wil, Kol, Col1, Col2, Col3, dSh1, dSh2, Sue]. This pairing can be generalized to the case of Honda formal groups with corresponding explicit formulas, see [BeV].

## 6. The Milnor $K$-groups of local fields

### 6.1. *The Milnor $K$-groups*

**6.1.1.** Let $F$ be a field. The $n$-th Milnor $K$-group of a field $F$ is defined as

$$K_n(F) = (F^* \otimes \cdots \otimes F^*)/I_n,$$

where $I_n$ is the subgroup generated by the elements $\alpha_1 \otimes \cdots \otimes \alpha_n$ with $\alpha_i + \alpha_j = 1$ for some $i \neq j$. Put $K_0(F) = \mathbb{Z}$. The image of $\alpha_1 \otimes \cdots \otimes \alpha_n$ in $K_n(F)$ is denoted by $\{\alpha_1, \dots, \alpha_n\}$. There is a natural map $K_n(F) \times K_m(F) \to K_{n+m}(F)$.

An imbedding of fields $F \hookrightarrow L$ induces a map $j_{F/L}\colon K_n(F) \to K_n(L)$. The norm map $N_{L/F}\colon L^* \to F^*$ for a finite extension $L/F$ induces a norm map

$$N_{L/F}\colon K_n(L) \to K_n(F)$$

with the following properties: $N_{L/F}$ acts on $K_0(L) = \mathbb{Z}$ as multiplication by $|L:F|$, on $K_1(L) = L^*$ as the norm map; $N_{L/F} \circ j_{F/L} = |L:F|$; if $L/F$ is Galois then

$$j_{F/L} \circ N_{L/F} = \sum_{\sigma_i \in \mathrm{Gal}(L/F)} \sigma_i,$$

where $\sigma_i\colon K_n(L) \to K_n(L)$ is induced by $\sigma_i \in \mathrm{Gal}(L/F)$.

**6.1.2.** If $F$ is a discrete valuation field, $v$ its valuation, $\overline{F}_v$ its residue field then there is a homomorphism

$$\partial_\pi\colon K_n(F) \to K_n(\overline{F}_v) \oplus K_{n-1}(\overline{F}_v)$$

defined by the formula

$$\partial_\pi\left(\{\varepsilon_1, \dots, \varepsilon_n\} + \{\pi, \eta_1, \dots, \eta_{n-1}\}\right) = \left(\{\overline{\varepsilon}_1, \dots, \overline{\varepsilon}_n\}, \{\overline{\eta}_1, \dots, \overline{\eta}_{n-1}\}\right),$$

where $\pi$ is prime in $F$, $\varepsilon_i, \eta_i \in U_v$. The second component $\partial_v$ of $\partial_\pi$ doesn't depend on the choice of $\pi$. If $F$ is a complete discrete valuation field with finite residue field then $\partial_v\{\alpha, \beta\} = \overline{t(\alpha, \beta)}$, where $t$ is the tame symbol defined in 5.1.3.

### 6.2. *The Milnor $K$-groups of a complete discrete valuation field*

Let $F$ be a complete discrete valuation field, $\overline{F} = \mathbb{F}_q$, $q = p^f$. A new role of the Hilbert symbol consists in its application for a description of the Milnor $K$-groups.

**6.2.1.** The properties in 5.1.1 imply that the Hilbert symbol $( \, , \, )_n$ induces a surjective homomorphism $H_n: K_2(F) \to \mu_n$.

PROPOSITION (C. Moore). *Let $m$ be the cardinality of the torsion group in $F^*$. Then $H_m$ induces an exact splitting sequence*

$$0 \to mK_2(F) \to K_2(F) \to \mu_m \to 1.$$

*The group $mK_2(F)$ is divisible.*

**6.2.2.** Let a primitive $l$-th root of unity $\zeta_l$ be contained in $F$. A general conjecture of Tate for arbitrary field $F$ asserts that if $lx = 0$ for $x \in K_n(F)$ then $x = \{\zeta_l\}y$ for some $y \in K_{n-1}(F)$. It was proved for $n = 2$ by Suslin ([Sus3]). For a field $F$ such as under consideration here this assertion for $l$ relatively prime with $p$ was elementarily verified by Carroll ([Car]) and for $l = p$ was deduced by Tate from a similar result for global fields ([T6]). Employing this assertion Merkurjev proved that $mK_2(F)$ is a uniquely divisible uncountable group ([Me]).

Sivitskii showed that $K_n(F)$ for $n \geqslant 3$ is a uniquely divisible uncountable group ([Si]). For details and proofs see [FV], Chapter IX.

### 6.3. *The Milnor $K$-groups of a complete $n$-discrete valuation field*

**6.3.1.** Let $F$ be a complete $n$-discrete valuation field with a finite residue field (see 1.2.1). Let $\tau$ be the strongest topology on $K_m(F)$ for which the map

$$\underbrace{F^* \times \dots \times F^*}_{m \text{ times}} \to K_m(F)$$

is sequentially continuous with respect to the topology on $F^*$ defined in 1.2.4 and $x_i + y_i \to x + y$, $-x_i \to -x$ in $K_m(F)$ if $x_i \to x$, $y_i \to y$. Let $\Lambda_m(F)$ be the intersection of all neighborhoods of 0 in $K_m(F)$. Parshin introduced the topological $K$-groups as

$$K_m^{\text{top}}(F) = K_m(F)/\Lambda_m(F)$$

for fields of characteristic $p$. The same definition is valid for $\text{char}(F) = 0$, $\text{char}(F_{n-1}) = p$. In the general case

$$\Lambda_m(F) = \bigcap_{l \geqslant 1} lK_m(F)$$

is the maximal divisible subgroup of $K_m(F)$, see [Fe9].

Then $K_0^{\text{top}}(F) = K_0(F)$, $K_1^{\text{top}}(F) = K_1(F)$, $K_{n+1}^{\text{top}}(F) \simeq \mu_F$, where $\mu_F$ is the torsion group of $F^*$, $K_m^{\text{top}}(F) = 0$ if $m \geqslant n+2$.

**6.3.2.** Let $v$ be $n$-discrete valuation on $F$. For elements $\alpha, \beta \in \mathfrak{M}_v$ the following equality holds:

$$\{1-\alpha, 1-\beta\} = -\{1+\alpha\beta(1-\alpha)^{-1}, \alpha\} - \{1+\alpha\beta(1-\alpha)^{-1}, 1-\beta\}.$$

Then the definition of $K_m^{\text{top}}(F)$ and 1.2.4 imply that this group is topologically generated by the elements $\{1+\theta\pi_n^{i_n}\cdots\pi_1^{i_1}, \pi_{j_1}, \dots, \pi_{j_{m-1}}\}$, where $\pi_n, \dots, \pi_1$ are local parameters, $\theta \in R$, $R$ is a set of representatives of $F_0 = \mathbb{F}_q$ in $F$, $1 \leqslant j_1, \dots, j_{m-1} \leqslant n$.

**6.3.3.** If $\operatorname{char}(F) = 0$ then the conjecture of 6.2.2 holds for $K_m^{\text{top}}(F)$, see [Fe3, Fe4]. If $\operatorname{char}(F) = p$ then there is no nontrivial $p$-torsion in $K_m^{\text{top}}(F)$ and a full description of these groups can be obtained by generalizing the pairings of 5.3, see [Pa4].

**6.3.4.** One can define surjective homomorphisms

$$w_F\colon K_n^{\text{top}}(F) \to K_{n-1}^{\text{top}}(F_{n-1}) \to \cdots \to K_0(F_0) \simeq \mathbb{Z},$$

induced by $\partial_v\colon K_m(F) \to K_{m-1}(\overline{F}_v)$, see 6.1.2.

## 7. Higher local class field theory

### 7.1. *Origins*

Let $k$ be a finite field. Then there is an injective homomorphism

$$K_0(k) = \mathbb{Z} \to \operatorname{Gal}(k^{\text{ab}}/k) \simeq \widehat{\mathbb{Z}},$$

where $k^{\text{ab}} = k^{\text{sep}}$ is the maximal abelian extension of $k$.

Let $K$ be a 1-dimensional complete discrete valuation field. Then there is an injective homomorphism (the reciprocity map)

$$K_1(K) = K^* \to \operatorname{Gal}(K^{\text{ab}}/K),$$

and the image of $K_1^{\text{top}}(K)$ is dense in $\operatorname{Gal}(K^{\text{ab}}/K)$.

We shall show that for a complete $n$-discrete valuation field there is a homomorphism

$$K_n^{\text{top}}(F) \to \operatorname{Gal}(F^{\text{ab}}/F),$$

which is injective and such that the image is dense in $\operatorname{Gal}(F^{\text{ab}}/F)$.

The $K$-theoretic generalization of class field theory (not only local but global too) was first studied by Parshin ([Pa1, Pa2, Pa3, Pa4, Pa5]). A cohomological approach to such a theory was proposed by Kato ([Kat1, Kat2, Kat3, KtS]). For another construction of the reciprocity map via an extension of the Neukirch map, see [Fe3, Fe4, Fe9].

Note that the residue field of a complete $n$-discrete valuation field $F$ when regarding it as 1-discrete is imperfect if $n > 1$, $\mathrm{char}(F_{n-1}) = p > 0$. So higher local class field theory may imply a description of abelian extensions of a complete discrete valuation field with arbitrary residue field. For a class field theory of such fields without $K$-groups see [Fe7].

## 7.2. *The reciprocity map*

Let $F$ be a complete $n$-discrete valuation field with the residue field $\mathbb{F}_q$, $q = p^f$.

**7.2.1.** For any finite extension $L$ of $F$ there is a unique extension of the $n$-discrete valuation to $L$. A separable extension $L/F$ is called *purely unramified* if its degree coincides with those of $\overline{L}/\overline{F}$. There is an analog of the assertion of 2.3.2 for purely unramified extensions. The compositum of all finite purely unramified extensions of $F$ in a fixed separable closure is denoted by $F^{\mathrm{pur}}$. Then

$$F^{\mathrm{pur}} = \bigcup_{(l,p)=1} F(\zeta_l),$$

where $\zeta_l$ is a primitive $l$-th root of unity. A generator $\varphi_F$ of $\mathrm{Gal}(F^{\mathrm{pur}}/F)$ which is mapped on the generator $\varphi$ of $\mathrm{Gal}(\mathbb{F}_q^{\mathrm{sep}}/\mathbb{F}_q)$ is called the Frobenius automorphism of $F$.

**7.2.2.** Let $L/F$ be a finite Galois extension, $\sigma \in \mathrm{Gal}(L/F)$. Let $\widetilde{\sigma}$ be an element of $\mathrm{Gal}(L^{\mathrm{pur}}/F)$ such that $\widetilde{\sigma}|_L = \sigma$ and $\widetilde{\sigma}|_{F^{\mathrm{pur}}}$ is a positive integer power of the Frobenius automorphism $\varphi_F$. Let $\Sigma$ be the fixed field of $\widetilde{\sigma}$ and $\pi_\Sigma \in K_n^{\mathrm{top}}(\Sigma)$ be a "prime" element of $K_n^{\mathrm{top}}(\Sigma)$, i.e. $w_\Sigma(\pi_\Sigma) = 1$ (see 6.3.3).

Then the map

$$\sigma \mapsto N_{\Sigma/F}\pi_\Sigma \pmod{N_{L/F}K_n^{\mathrm{top}}(L)}$$

is well defined and is a homomorphism, where $N_{L/F}$ for topological $K$-groups is induced by the norm map for the Milnor $K$-groups. Moreover, this map determines an isomorphism

$$\mathrm{Gal}(L/F)^{\mathrm{ab}} \simeq K_n^{\mathrm{top}}(F)/N_{L/F}K_n^{\mathrm{top}}(L).$$

This isomorphism possesses natural functorial properties analogous to 4.2.3. The inverse homomorphism induces a reciprocity map

$$\Psi_F\colon\ K_n^{\mathrm{top}}(F) \to \mathrm{Gal}(F^{\mathrm{ab}}/F).$$

It is injective and continuous. This shows that $\Lambda_n(F)$ is exactly the kernel of the homomorphism $K_n(F) \to \mathrm{Gal}(F^{\mathrm{ab}}/F)$.

In particular, the diagram

$$\begin{array}{ccc} K_n^{\mathrm{top}}(F) & \xrightarrow{\Psi_F} & \mathrm{Gal}(F^{\mathrm{ab}}/F) \\ \downarrow & & \downarrow \\ K_{n-1}^{\mathrm{top}}(\overline{F}) & \xrightarrow{\Psi_{\overline{F}}} & \mathrm{Gal}(\overline{F}^{\mathrm{ab}}/\overline{F}) \end{array}$$

is commutative, where the left vertical homomorphism is induced by $\partial_{v^{(n)}}$, see 6.1.2, and $v^{(n)}$ is the first component of $v$.

Existence Theorem for the fields under consideration asserts that any open subgroup of finite index in $K_n^{\mathrm{top}}(F)$ is a norm subgroup $N_{L/F}K_n^{\mathrm{top}}(L)$ for a suitable abelian extension $L/F$, see [Fe3, Fe4].

**7.2.3.** Parshin constructed the reciprocity map for the fields of positive characteristic [Pa4, Pa5] especially elegant via his generalization of Artin–Schneider–Witt pairings and Kawada–Satake's theory [KwS].

**7.2.4.** Another construction of the reciprocity map follows from cohomological considerations due to Kato ([Kat1, Kat2, Kat3, Kat4]).

If $\mathrm{char}(F) = 0$ put

$$H^m(F) = \varinjlim H^m\big(F, \mu_n^{\otimes(m-1)}\big),$$

where $\mu_n$ is the group of all $n$-th roots of unity in $F^{\mathrm{sep}}$, $\mu_n^{\otimes(m-1)}$ is the $(m-1)$-th tensor power over $\mathbb{Z}/n\mathbb{Z}$, $n \geqslant 1$, and the homomorphisms of the injective system are induced by the canonical injections $\mu_n^{\otimes(m-1)} \to \mu_d^{\otimes(m-1)}$ when $n$ divides $d$. If $\mathrm{char}(F) = p > 0$ put

$$H^m(F) = \varinjlim H^m\big(F, \mu_n^{\otimes(m-1)}\big) \oplus \varprojlim H^m_{p^d}(F),$$

where $n$ runs over all positive integers prime to $p$, $d$ runs over all positive integers. Here

$$H^m_{p^d}(F) = W_d(F) \otimes \underbrace{(F^* \otimes \cdots \otimes F^*)}_{m-1 \text{ times}}/J,$$

where $J$ is the subgroup generated by the elements $(\mathbf{F}y - y) \otimes \beta_1 \otimes \cdots \otimes \beta_{m-1}$, where $y \in W_d(F)$, $\beta_i \in F^*$, $\mathbf{F}$ as in 1.4.2; $y_i(\beta_1) \otimes \beta_1 \otimes \cdots \otimes \beta_{m-1}$, where

$$y_i(\beta_1) = (\underbrace{0, \ldots, 0}_{i \text{ times}}, \beta_1, 0, \ldots, 0) \in W_d(F), \quad 0 \leqslant i < d;$$

$y \otimes \beta_1 \otimes \cdots \otimes \beta_{m-1}$, where $\beta_i = \beta_j$ for some $i \neq j$.

For any field $F$ the group $H^1(F)$ is isomorphic to the group of all continuous homomorphisms $\mathrm{Gal}(F^{\mathrm{ab}}/F) \to \mathbb{Q}/\mathbb{Z}$ and $H^2(F)$ is isomorphic to $\mathrm{Br}(F)$.

If $F$ is a complete $n$-discrete valuation field with a finite residue field then the canonical homomorphism $H^{n+1}(F) \simeq \mathbb{Q}/\mathbb{Z}$ is an analog of 4.5. Then by using the canonical pairing

$$H^1(F) \times K_n(F) \to H^1(F) \times H^n(F) \to H^{n+1}(F) \simeq \mathbb{Q}/\mathbb{Z}$$

one obtains a homomorphism $K_n(F) \to \mathrm{Gal}(F^{\mathrm{ab}}/F)$, which coincides with the reciprocity map up to the projection $K_n(F) \to K_n^{\mathrm{top}}(F)$.

**7.2.5.** The Kato theory can be treated as a generalization of Tate's approach in classical class field theory. Koya found a generalization of class formations to higher class field theory using bounded complexes of Galois modules and their modified hypercohomology groups [Koy1, 2]. For a 2-dimensional field a shifted Lichtenbaum complex satisfies generalized axioms of formation classes, and thus 2-dimensional class field theory follows.

**7.2.6.** For a description of abelian totally ramified $p$-extensions of an $n$-dimensional complete field with arbitrary perfect residue field see [Fe9].

## 8. Absolute Galois group of a local field

Let $F$ be a complete discrete valuation field with residue field $\mathbb{F}_q$.

### 8.1. *The maximal tamely ramified extension*

Let $F^{\mathrm{sep}}$ be a fixed separable closure of $F$ and $G_F = \mathrm{Gal}(F^{\mathrm{sep}}/F)$. Let $F^{\mathrm{tr}}$ be the maximal tamely ramified extension of $F$ in $F^{\mathrm{sep}}$. Then

$$F^{\mathrm{tr}} = \bigcup_{(l,p)=1} F^{\mathrm{ur}}(\sqrt[l]{\pi}),$$

where $\pi$ is a prime element in $F$.

Let $n_1 < n_2 < \cdots$ be a sequence of natural numbers such that $n_{i+1}$ is divisible by $n_i$ and for any natural $m$ there exists an index $i$ for which $n_i$ is divisible by $m$. Put $l_i = q^{n_i} - 1$. Choose primitive $l_i$-th roots of unity $\zeta_{l_i}$ and roots $\sqrt[l_i]{\pi}$ such that $\zeta_{l_j}^{l_j l_i^{-1}} = \zeta_{l_i}$, $\sqrt[l_j]{\pi}^{l_j l_i^{-1}} = \sqrt[l_i]{\pi}$ for $j > i$. Take $\sigma \in \mathrm{Gal}(F^{\mathrm{tr}}/F)$ such that $\sigma(\sqrt[l_i]{\pi}) = \sqrt[l_i]{\pi}$, $\sigma(\zeta_{l_i}) = \zeta_{l_i}^q$ and $\tau \in \mathrm{Gal}(F^{\mathrm{tr}}/F)$ such that $\tau(\sqrt[l_i]{\pi}) = \zeta_{l_i} \sqrt[l_i]{\pi}$, $\tau(\zeta_{l_i}) = \zeta_{l_i}$. Then $\sigma|_{F^{\mathrm{ur}}}$ coincides with the Frobenius automorphism of $F$ and $\sigma\tau\sigma^{-1} = \tau^q$. The theorem of Hasse–Iwasawa (see [Has12, Iw1]) asserts that $G_{\mathrm{tr}} = \mathrm{Gal}(F^{\mathrm{tr}}/F)$ is topologically generated by $\sigma$ and $\tau$ with a relation $\sigma\tau\sigma^{-1} = \tau^q$.

**8.2.** *Absolute Galois group*

**8.2.1.** Now let $I$ be an index-set and $F_I$ be the free profinite group with a basis $z_i$, $i \in I$. Let $F_I * G_{\text{tr}}$ be the free profinite product of $F_I$ and $G_{\text{tr}}$, see [N2, BNW]. Let $H$ be the normal closed subgroup of $F_I * G_{\text{tr}}$ generated by $(z_i)_{i\in I}$ and $K$ be the normal closed subgroup of $H$ such that the quotient group $H/K$ is the maximal $p$-factor group of $H$. Put $F(I, G_{\text{tr}}) = (F_{n+1} * G)/K$ and denote by $x_i$ the image of $z_i$ in $F(I, G_{\text{tr}})$. The group $F(I, G_{\text{tr}})$ has topological generators $\sigma$, $\tau$, $x_i$, $i \in I$ with a relation $\sigma\tau\sigma^{-1} = \tau^q$.

**8.2.2.** Assume first that $\operatorname{char}(F) = p$ (the function field case). Then Koch's Theorem (see [Ko3]) asserts that the group $G_F$ is topologically isomorphic with $F(\mathbb{N}, G_{\text{tr}})$. Note that in this case $U_{1,F}$ is a free $\mathbb{Z}_p$-module of rank = cardinality of $\mathbb{N}$, see 1.3.3.

**8.2.3.** Assume next that $\operatorname{char}(F) = 0$ and there is no a nontrivial $p$-torsion in $F^*$. Shafarevich's Theorem ([Sha2, JW]) implies that the group $G_F$ is topologically isomorphic to $F(n, G_{\text{tr}})$, where $n = |F : \mathbb{Q}_p|$. See also [Se4, Mik1, Mar2] for a case of a perfect residue field. Note that in this case $U_{1,F}$ is a free $\mathbb{Z}_p$-module of rank $n$, see 1.3.3.

**8.2.4.** Assume finally that $\operatorname{char}(F) = 0$ and $r \geqslant 1$ is the maximal integer such that $\mu_{p^r} \subset F^*$. This is the most complicated case. Let $\chi_0$ be the homomorphism of $G_{\text{tr}}$ onto $(\mathbb{Z}/p^r\mathbb{Z})^*$ such that $\rho(\zeta_{p^r}) = \zeta_{p^r}^{\chi_0(\rho)}$ for $\rho \in G_{\text{tr}}$, where $\zeta_{p^r}$ is a primitive $p^r$-th root of unity. Let $\chi$: $G_{\text{tr}} \to \mathbb{Z}_p^*$ be a lifting of $\chi_0$. Let $l$ be prime, $\{p_1, p_2, \dots\}$ be a set of all primes $\neq l$. For $m \geqslant 1$ there exist integers $a_m$, $b_m$ such that $1 = a_m l^m + b_m p_1^m p_2^m \cdots p_m^m$. Put

$$\pi_l = \lim b_m p_1^m p_2^m \cdots p_m^m \in \widehat{\mathbb{Z}}.$$

For an element $\rho \in G_{\text{tr}}$ and $\xi \in F(I, G_{\text{tr}})$ put

$$(\xi, \rho) = \left(\xi^{\chi(1)} \rho \xi^{\chi(\rho)} \rho \cdots \xi^{\chi(\rho^{p-2})} \rho\right)^{\pi_p/(p-1)},$$

$$\{\xi, \rho\} = \left(\xi^{\chi(1)} \rho^2 \xi^{\chi(\rho)} \rho^2 \cdots \xi^{\chi(\rho^{p-2})} \rho^2\right)^{\pi_p/(p-1)}$$

If $n = |F : \mathbb{Q}_p|$ is even, put

$$\lambda = \sigma x_0^{-1} \sigma^{-1} (x_0, \tau)^{\chi(\sigma)^{-1}} x_1^{p^r} x_1 x_2 x_1^{-1} x_2^{-1} x_3 x_4 x_3^{-1} x_4^{-1} \cdots x_{n-1} x_n x_{n-1}^{-1} x_n^{-1}.$$

If $n = |F : \mathbb{Q}_p|$ is odd, let $a$, $b$ be integers such that $-\chi_0(\sigma\tau^a)$ is a square $\bmod p$ and $-\chi_0(\sigma\tau^b)$ isn't. Put

$$\begin{aligned}\lambda_1 = {} & \tau_2^{p+1} x_1 \tau_2^{-(p+1)} \sigma_2 \tau_2^a \{x_1, \tau_2^{p+1}\} \tau_2^{-a+b} \{\{x_1, \tau_2^{p+1}\}, \sigma_2 \tau_2^a\} \\ & \times \tau_2^{-b} \sigma_2^{-1} \tau_2^{(p+1)/2} \{\{x_1, \tau_2^{p+1}\}, \sigma_2 \tau_2^a\} \tau_2^{-(p+1)/2},\end{aligned}$$

where $\sigma_2 = \sigma^{\pi_2}$, $\tau_2 = \tau^{\pi_2}$. Put

$$\lambda = \sigma x_0^{-1}\sigma^{-1}(x_0,\tau)^{\chi(\sigma)^{-1}} x_1^{p^r+1}\lambda_1 x_1^{-1}\lambda_1^{-1} x_2 x_3 x_2^{-1} x_3^{-1}\cdots x_{n-1}x_n x_{n-1}^{-1} x_n^{-1}.$$

For $n+1$ we choose the index set $I = \{0,\dots,n\}$.

Jakovlev's theorem and Jannsen–Wingberg's theorem (see [Ja1, Ja2, Ja3, Ja4, Ja5], Koch [Ko1, Ko2, Ko4, Ko5], [Jan, Wig, JW], Demushkin [Dem1, Dem2], Labute [Lab]) assert that for $p > 2$ the group $G_F$ is topologically isomorphic to $F(n+1, G_{\mathrm{tr}})/(\lambda)$, where $(\lambda)$ is the closed normal subgroup of $F(n+1, G_{\mathrm{tr}})$ generated by $\lambda$. Note that $U_{1,F}$ is a $\mathbb{Z}_p$-module of rank $n+1$ with one relation, see 1.3.3.

**8.2.5.** For the case $p=2$, $\sqrt{-1}\in F$ see [Di, Ze]. See also [Gor, JR2] and [Mik2, Kom] for a brief discussion of proofs. Jarden and Ritter ([JR1, Rit]) showed that two absolute Galois groups $G_F$ and $G_L$ for $\mathfrak{p}$-adic fields $F$ and $L$ are topologically isomorphic if and only if $|F:\mathbb{Q}_p| = |L:\mathbb{Q}_p|$ and $F\cap \mathbb{Q}_p{}^{\mathrm{ab}} = L\cap \mathbb{Q}_p{}^{\mathrm{ab}}$ (for $p>2$ or $p=2$, $\sqrt{-1}\in F, L$).

*Comments on the bibliography*

Textbooks on local fields and local class field theory: [AT, Has11, Se3, Wel, N4, Iw5, Iw6, Mi, CF, Cas, Ko4, Sch, Na, Ch1, FV].

Textbooks on related subjects: formal groups [Fr, Haz3], cyclotomic fields [Wa, La3], algebraic number theory [Ar, Wes, IR, BSh, La2, Ko6], valuation theory [E, Rib], formally $p$-adic fields [PR], non-Archimedean analysis [BGR, Kob1, Kob2], $K$-theory [Bas, Mil1, Mil2].

Galois groups of local fields: [Sha1, Iw1, Iw2, Iw4, Ko1, Ko2, Ko3, Ko5, Dem1, Dem2, Lab, Ja1, Ja2, Ja3, Ja4, Ja5, Jan, JW, Wig, Di, Mar2, Mik1, Mik2, Rit, MSh, Sek1, Gor, Ze, Kom].

Ramification theory: [Her, Kaw1, Lau1, Lau2, Lau3, Lau4, Lau5, Mar1, Mau2, Mau3, Mau4, Mau5, Mik5, Mik6, Sa, Sen1, Sen2, ST, Tam, Fe6, Hy, Kat5, Lo].

$p$-periods and Galois representations: [Fo2, FI, T1, FM].

Fields of norms and related subjects: [Fo1, FW, Win1, Win2, Win3, Del, Lau4, Ke].

Symbols and explicit formulas: [AH1, AH2, Has1, Has2, Has3, Has4, Has5, Has6, Has7, Has8, Has9, Has10, Sha2, Kn, Bru1, Bru2, Rot, V1, V2, V3, V4, V5, V6, V7, Fe1, Fe2, Iw3, Col1, Col3, Wil, CW1, dSh1, dSh2, Henn1, Henn2, Sen3, Shi, Sue, Kol, Kuz].

Milnor $K$-groups of local fields: [BT, Mil2, T5, Car, Me, MS, Si, Bog].

Explicit constructions of the reciprocity map: [Ya1, Dw, Haz1, Haz2, N3].

Local class field theories: [Mik4, Kur, Fe5, Fe7, LR].

Higher local class field theory: [Pa1, Pa2, Pa3, Pa4, Pa5, Kat1, Kat2, Kat3, Kat4, Kat5, Ko1, Ko2, Fe3, Fe4, Fe9, FVZ].

Diophantine problems over local fields via logic: [AxK, Er1, Er2].

## References


[Am] S. Amano, *Eisenstein equations of degree $p$ in a $\mathfrak{p}$-adic field*, J. Fac. Sci. Univ. Tokyo, Sect. 1A **18** (1971), 1–22.

[Ar] E. Artin, *Algebraic Numbers and Algebraic Functions*, Gordon and Breach, New York (1967).

[AH1] E. Artin and H. Hasse, *Über den zweiten Ergänzungssatz zum Reziprozitätsgesetz der $l$-ten Potenzreste im Körper $k_\zeta$ der $l$-ten Einheitswurzeln und Oberkörpern von $k_\zeta$*, J. Reine Angew. Math. **154** (1925), 143–148.

[AH2] E. Artin and H. Hasse, *Die beiden Ergänzungssatz zum Reziprozitätsgesetz der $l^n$-ten Potenzreste im Körper der $l^n$-ten Einheitswurzeln*, Hamb. Abh. **6** (1928), 146–162.

[ASch] E. Artin and O. Schreier, *Eine Kennzeichnung der reell abgeschlossen Körper*, Hamb. Abh. **5** (1927), 225–231.

[AT] E. Artin and J. Tate, *Class Field Theory*, Harvard (1961).

[AxK] J. Ax and S. Kochen, *Diophantine problems over local fields, I*, Amer. J. Math. **87** (1965), 605–630; *II*, Amer. J. Math. **87** (1965), 631–648; *III*, Ann. Math. **83** (1966), 437–456.

[Bah] G. Bachman, *Introduction to $p$-adic Numbers and Valuation Theory*, Academic Press, New York and London (1964).

[Bas] H. Bass, *Algebraic $K$-theory*, W.A. Benjamin, New York (1968).

[Bog] R.A. Bogomolov, *Two theorems on divisibility and torsion in the Milnor $K$-groups*, Mat. Sb. **130** (1986), 404–412 (in Russian; Engl. transl. in Math. USSR Sbornik).

[Bor1] Z.I. Borevich, *On the multiplicative group of cyclic $p$-extensions of a local field*, Proc. Steklov Inst. Math. **80** (1965), 15–30.

[Bor2] Z.I. Borevich, *Groups of principal units of $p$-extension of a local field*, Soviet Math. Dokl. **8** (1967), 359–361.

[Bou] N. Bourbaki, *Algébre Commutative*, Hermann, Paris (1965).

[Br] R. Brauer, *Über die Konstruktion der Schiefkörper, die von endlichem Rang in bezug auf ein gegebenes Zentrum sind*, J. Reine Angew. Math. **168** (1932), 44–64.

[Bru1] H. Brückner, *Eine explizite Formel zum Reziprozitätsgesetz für Primzahlexponent $p$*, Alg. Zahlentheorie, Bibiliogr. Inst., Mannheim (1967).

[Bru2] H. Brückner, *Hilbertsymbole zum Exponenten $p^n$ und Pfaffische Formen*, Prepubl., Hamburg (1979).

[Bu] D.J. Burns, *Factorisability and the arithmetic of wildly ramified Galois extensions*, Sém. Théor. Nombres Bordeax (2) **1** (1989), 59–65.

[BeV] D.G. Benois and S.V. Vostokov, *Norm pairing in formal groups and Galois representations*, Algebra i Analiz **2** (6) (1990), 69–97 (in Russian; Engl. transl. in Leningrad Math. J.).

[BoV] Z.I. Borevich and S.V. Vostokov, *The ring of integers of a local field as Galois module in the case of a prime degree extension*, Zap. Nauch. Sem. Leningrad. Otdel. Mat. Inst. **31** (1973), 24–37 (in Russian; English transl. in J. Soviet Math. **6** (1976)).

[BSh] Z.I. Borevich and I.R. Shafarevich, *Number Theory*, New York (1966).

[BSk] Z.I. Borevich and A.I. Skopin, *Extensions of a local field with normal basis for principal units*, Proc. Steklov Inst. Math. **80** (1965), 48–55.

[BT] H. Bass and J. Tate, *The Milnor ring of a global field*, SLNM 342, Springer, Berlin (1973), 349–446.

[BGR] S. Bosch, U. Günter and R. Remmert, *Non-Archimedean Analysis*, Springer, Berlin (1984).

[BNW] E. Binz, J. Neukirch and G.H. Wenzel, *A subgroup theorem for free products of pro-finite groups*, J. Algebra **19** (1971), 104–109.

[Car] J.E. Carroll, *On the torsion in $K_2$ of local fields*, SLNM 342 (1973), 464–473.

[Cas] J.W.C. Cassels, *Local Fields*, Cambridge Univ. Press, London (1986).

[Ch1] C. Chevalley, *Sur la théorie du corps de classes dans les corps finis et les corps locaux*, J. Fac. Sci. Tokio Imp. Univ. **2** (1933), 363–476.

[Ch2] C. Chevalley, *Class Field Theory*, Nagoya Univ., Nagoya (1954).

[Coh] I.S. Cohen, *On the structure and ideal theory of complete local rings*, Trans. Amer. Math. Soc. **59** (1946), 54–106.

[Col1] R. Coleman, *Division values in local fields*, Invent. Math. **53** (1979), 91–116.

[Col2] R. Coleman, *The dilogarithm and the norm residue symbol*, Bull. Soc. Math. France **109** (1981), 373–402.

[Col3] R. Coleman, *Arithmetic of Lubin–Tate division towers*, Duke Math. J. **48** (1981), 449–466.

[CF] J.W.C. Cassels and A. Fröhlich, (eds), *Algebraic Number Theory*, Thompsen Book Comp. Inc., Washington, D.C. (1967).

[CW1] J. Coates and A. Wiles, *Explicit reciprocity laws*, Astérisque **41–42** (1977), 7–17.

[CW2] J. Coates and A. Wiles, *On the conjecture of Birch and Swinnerton–Dyer*, Invent. Math. **39** (1977), 223–251.

[Del] P. Deligne, *Les corps locaux des caracteristique p, limites de corps locaux de caracteristique* 0, Representations des Groupes Réductifs sur un Corps Local, Hermann, Paris (1984).

[Dem1] S.P. Demushkin, *The group of the maximal p-extension of a local field*, Dokl. Akad. Nauk SSSR **128** (1959), 657–660 (in Russian).

[Dem2] S.P. Demushkin, *On the maximal p-extension of a local field*, Izv. Akad. Nauk SSSR, Ser. Mat. **25** (1961), 329–346 (in Russian).

[Di] V. Diekert, *Über die absolute Galoisgruppe dyadischer Zahlkörper*, J. Reine Angew. Math. **350** (1984), 152–172.

[Dw] B. Dwork, *Norm residue symbol in local number fields*, Abh. Math. Sem. Univ. Hamburg **22** (1958), 180–190.

[dSh1] E. de Shalit, *The explicit reciprocity laws in Lubin–Tate relative groups*, Duke Math. J. **53** (1986), 163–186.

[dSh2] E. de Shalit, *Iwasawa Theory of Elliptic Curves with Complex Multiplication*, Academic Press, New York (1987).

[DG] H. Demasure and P. Gabriel, *Groupes Algebriques, vol. 1*, North-Holland, Amsterdam (1970).

[DS] R.K. Dennis and M.R. Stein, *$K_2$ of discrete valuation rings*, Adv. Math. **18** (1975), 182–238.

[E] O. Endler, *Valuation Theory*, Springer, Berlin (1972).

[Ep] H. Epp, *Eliminating wild ramification*, Invent. Math. **19** (1973), 235–249.

[Er1] Yu.L. Ershov, *On the elementary theory of maximal normed fields*, Soviet Math. Dokl. **6** (1965), 1390–1393.

[Er2] Yu.L. Ershov, *On elementary theories of local fields*, Algebra i Logika **4** (2) (1965), 5–30 (in Russian).

[Er3] Yu.L. Ershov, *On the elementary theory of maximal valued fields, I*, Algebra i Logika **4** (3) (1965), 31–70; *II*, Algebra i Logika **5** (1) (1966), 5–40; *III*, Algebra i Logika **6** (3) (1967), 31–38 (in Russian).

[Fe1] I.B. Fesenko, *The generalized Hilbert symbol in 2-adic case*, Vestn. Leningrad Univ., no. 22 (1985), 112–114 (in Russian; English transl. in Vestnik Leningrad Univ. Math. **18** (1985)).

[Fe2] I.B. Fesenko, *Explicit constructions in local class field theory*, Thesis, Leningrad Univ. (1987) (in Russian).

[Fe3] I.B. Fesenko, *Class field theory of multidimensional local fields of characteristic* 0 *with the residue field of positive characteristic*, Algebra i Analiz **3** (3) (1991), 165–196 (in Russian; Engl. transl. in Leningrad Math. J.).

[Fe4] I.B. Fesenko, *On class field theory of multidimensional local fields of positive characteristic*, Adv. Soviet Math. **4** (1991).

[Fe5] I.B. Fesenko, *Local class field theory: perfect residue field case*, Izv. Russ. Akad. Nauk Ser. Mat. **57** (4) (1993), 72–91 (in Russian; English transl. in Russ. Acad. Sci. Izv. Math. **43** (1994), 65–81).

[Fe6] I.B. Fesenko, *Hasse–Arf property and abelian extensions*, Math. Nachr. **174** (1995).

[Fe7] I.B. Fesenko, *Abelian extensions of complete discrete valuation fields*, to appear in Paris Number Theory Seminar, Cambridge Univ. Press, Cambridge (1996).

[Fe8] I.B. Fesenko, *Local fields. Local class field theory. Higher local class field theory via algebraic K-theory*, Algebra i Analiz **4** (3) (1992) (in Russian; English transl. in St. Petersburg Math. J. **4**(3) (1993), 403–438).

[Fe9] I.B. Fesenko, *Abelian local p-class field theory*, Math. Ann. **301** (1995), 561–586.

[Fo1] J.-M. Fontaine, *Corps de series formelles et extensions galoisiennes des corps locaux*, Sém. Théor. Nombres Grenoble (1971–72), 28–38.

[Fo2] J.-M. Fontaine, *Groupes p-divisibles sur les corps locaux*, Astérisque **47–48** (1977).

[Fr] A. Fröhlich, *Formal Groups*, SLNM 74, Springer, Berlin (1968).

[FI] J.-M. Fontaine and L. Illusie, *p-adic periods: a survey*, Prepubl. Univ. Paris-Sud (1990).

[FM] J.-M. Fontaine and W. Messing, *p-adic periods and p-etale cohomology*, Contemp. Math. vol. 67 (1987), 179–207.

[FV] I.B. Fesenko and S.V. Vostokov, *Local Fields and Their Extensions: A Constructive Approach*, Transl. of Math. Monogr. vol. 121 (1993).

[FVZ] I.B. Fesenko, S.V. Vostokov and Zhukov, *On multidimensional local fields. Methods and constructions*, Algebra i Analiz **2**(3) (1990), 91–118 (in Russian; Engl. transl. in Leningrad Math. J.).

[FW] J.-M. Fontaine and J.-P. Wintenberger, *Le "corps des normes" de certaines extensions algebriques de corps locaux*, C. R. Acad. Sci. Paris **288** (1979), 367–370.

[Gi] D. Gilbarg, *The structure of the group of* $\mathfrak{p}$*-adic* 1*-units*, Duke Math. J. **9** (1942), 262–271.

[Gol] L.J. Goldstein, *Analytic Number Theory*, Prentice-Hall, New Jersey (1971).

[Gor] N.L. Gordeev, *An infinity of relations in the Galois group of the maximal p-extension of a local field with a boundered ramification*, Izv. Akad. Nauk SSSR Ser. Mat. **45** (1981), 592–607 (in Russian; Engl. transl. in Math. USSR Izv.).

[Gr] D.R. Grayson, *On the* $K$*-theory of fields*, Contemp. Math. **83** (1989), 31–55.

[GR] H. Grauert and R. Remmert, *Über die Methode der diskret bewerteten Ringe in der nicht archimedischen Analysis*, Invent. Math. **2** (1966), 87–133.

[Has1] H. Hasse, *Über die Normenreste eines relativ-zyklischen Körpers vom Primzahlgrad* $l$ *nach Einem Primteiler* $\mathfrak{l}$ *von* $l$, Math. Ann. **90** (1923), 262–278.

[Has2] H. Hasse, *Das allgemeine Reziprozitätsgesetz und seine Ergänzungssätze in beliebigen algebraischen Zahlkörpern für gewisse, nicht-primäre Zahlen*, J. Reine Angew. Math. **153** (1924), 192–207.

[Has3] H. Hasse, *Direkter Beweis des Zerleguns- und Vertauschungssatzes für das Hilbertische Normenrestesymbol in einem algebraischen Zahlkörper im Falle eines Primteiler* $\mathfrak{l}$ *des Relativgrades* $l$, J. Reine Angew. Math. **154** (1925), 20–35.

[Has4] H. Hasse, *Über das allgemeine Reziprozitätsgesetz der* $l$*-ten Potenzreste im Körper* $k_\zeta$ *der* $l$*-ten Einheitswurzeln und in Oberkörpern von* $k_\zeta$, J. Reine Angew. Math. **154** (1925), 96–109.

[Has5] H. Hasse, *Das allgemeine Reziprozitätsgesetz der* $l$*-ten Potenzreste für beliebege, zu* $l$ *prime Zahlen in gewissen Oberkörpern des Körpers der* $l$*-ten Einheitswurzeln*, J. Reine Angew. Math. **154** (1925), 199–214.

[Has6] H. Hasse, *Zum expliziten Reziprozitätsgesetz*, Hamb. Abh. **7** (1929), 52–63.

[Has7] H. Hasse, *Normenresttheorie galoischer Zahlkörpern mit Anwendungen auf Führer und Discriminante abelscher Zahlkörper*, J. Fac. Sci. Tokio Imp. Univ. Ser. Math. **2** (1934), 477–498.

[Has8] H. Hasse, *Die Gruppe der* $p^n$*-primären Zahlen für einen Primteiler* $\mathfrak{p}$ *von* $p$, J. Reine Angew. Math. **176** (1936), 174–183.

[Has9] H. Hasse, *Zur Arbeit von I.R. Šafarevič über das allgemeine Reziprozitätsgesetz*, Math. Nachr. **5** (1951), 302–327.

[Has10] H. Hasse, *Der* $2^n$*-te Potenzcharacter von* 2 *im Körper der* $2^n$*-ten Einheitswurzeln*, Rend. Circ. Mat. Palermo **7** (1958), 185–243.

[Has11] H. Hasse, *Zum expliziten Reziprozitätsgesetz*, Arch. Math. **13** (1961), 479–485.

[Has12] H. Hasse, *Zahlentheorie*, Akademie-Verlag, Berlin (1949).

[Haz1] M. Hazewinkel, *Corps de classes local*, [DG].

[Haz2] M. Hazewinkel, *Local class field theory is easy*, Adv. Math. **18** (1975), 148–181.

[Haz3] M. Hazewinkel, *Formal Groups and Application*, Academic Press, New York (1978).

[Haz4] M. Hazewinkel, *Twisted Lubin–Tate formal group laws, ramified Witt vectors and (ramified) Artin–Hasse exponentials*, Trans. Amer. Math. Soc. **259** (1980), 47–63.

[Henn1] G. Henniart, *Lois de reciprocité explicites*, Sém. Théor. Nombres, Paris 1979–80, Birkhäuser, Boston (1981), 135–149.

[Henn2] G. Henniart, *Sur les lois de reciprocité explicites, I*, J. Reine Angew. Math. **329** (1981), 177–203.

[Her] J. Herbrand, *Sur la théorie des groups de décomposition, d'inertie et de ramification*, J. Math. Pures Appl. **10** (1931), 481–498.

[Ho] G. Hochshild, *Local class field theory*, Ann. Math. **51** (1950), 331–347.

[Hon] T. Honda, *On the theory of commutative formal groups*, J. Math. Soc. Japan **22** (1970), 213–243.

[Hy] O. Hyodo, *Wild ramification in the imperfect residue case*, Adv. Stud. Pure Math. **12** (1987), 287–314.

[HSch] H. Hasse and F.K. Schmidt, *Die Struktur discret bewerteten Körper*, J. Reine Angew. Math. **170** (1934), 4–63.

[Iw1] K. Iwasawa, *On Galois groups of local fields*, Trans. Amer. Math. Soc. **80** (1955), 448–469.

[Iw2] K. Iwasawa, *On local cyclotomic fields*, J. Math. Soc. Japan **12** (1960), 16–21.

[Iw3] K. Iwasawa, *On explicit formulas for the norm residue symbol*, J. Math. Soc. Japan **20** (1968), 151–165.

[Iw4] K. Iwasawa, *On* $\mathbb{Z}_l$*-extensions of algebraic number fields*, Ann. Math. **98** (1973), 246–327.

[Iw5] K. Iwasawa, *Local Class Field Theory*, Iwanami-Shoten, Tokyo (1980) (in Japanese; Russian transl. by Mir, Moscow, 1983).

[Iw6] K. Iwasawa, *Local Class Field Theory*, Oxford Univ. Press & Clarendon Press, New York (1986).

[IR] K. Ireland and M. Rosen, *A Classical Introduction to Modern Number Theory*, Springer, Berlin (1982).

[Ja1] A.V. Jakovlev, *The Galois group of the algebraic closure of a local field*, Math. USSR Izv. **2** (1968), 1231–1269.

[Ja2] A.V. Jakovlev, *Remarks on my paper "The Galois group of the algebraic closure of a local field"*, Math. USSR Izv. **12** (1978), 205–206.

[Ja3] A.V. Jakovlev, *Symplectic spaces with operators over commutative rings*, Vestn. Leningrad Univ. Math., no. 3 (1976), 339–346.

[Ja4] A.V. Jakovlev, *An abstract characterization of the Galois group of the algebraic closure of a local field*, Zap. Nauch. Sem. Leningrad. Otdel. Mat. Inst. **75** (1978), 179–199 (in Russian).

[Ja5] A.V. Jakovlev, *Structure of the multiplicative group of a simply ramified extension of a local field of odd degree*, Math. USSR Sb. **35** (1979), 581–591.

[Jan] U. Jannsen, *Über Galoisgruppen lokaler Körper*, Invent. Math. **70** (1982), 53–69.

[JR1] M. Jarden and J. Ritter, *On the characterization of local fields by their absolute Galois groups*, J. Number Theory **11** (1979), 1–13.

[JR2] M. Jarden and J. Ritter, *Normal automorphisms of absolute Galois group of* $\mathfrak{p}$*-adic fields*, Duke Math. J. **47** (1980), 47–56.

[JW] U. Jannsen and K. Wingberg, *Die struktur der absoluten Galoisgruppe* $\mathfrak{p}$*-adischer Zahlkörpers*, Invent. Math. **70** (1982), 71–98.

[Kah] B. Kahn, *L'anneau de Milnor d'un corps local*, Thése, Univ. Bordeaux (1983).

[Kat1] K. Kato, *A generalization of local class field theory by using* $K$*-groups, I*, J. Fac. Sci. Tokio, Sect. 1A **26** (1979), 303–376; *II*, J. Fac. Sci. Tokio, Sect. 1A **27** (1980), 603–683; *III*, J. Fac. Sci. Tokio, Sect. 1A **29** (1982), 31–43.

[Kat2] K. Kato, *The existense theorem for higher local class field theory*, Preprint IHES (1980).

[Kat3] K. Kato, *Galois cohomology of complete discrete valuation fields*, SLNM 967, Springer, Berlin (1982), 215–238.

[Kat4] K. Kato, *Swan conductors with differential values*, Adv. Stud. Pure Math. **12** (1987), 315–342.

[Kat5] K. Kato, *Swan conductors for characters of degree one in the imperfect residue field case*, Contemp. Math. vol. 83 (1989), 101–131.

[Kaw1] Y. Kawada, *On the ramification theory of infinite algebraic extensions*, Ann. Math. **58** (1953), 24–47.

[Kaw2] J. Kawada, *Class formations*, Duke Math. J. **22** (1955), 165–178; *IV*, J. Math. Soc. Japan **9** (1957), 395–405; *V*, J. Math. Soc. Japan **12** (1960), 34–64.

[Ke] K. Keating, *Galois extensions associated to deformations of formal A-modules*, J. Fac. Sci. Tokio, Sect. 1A **37** (1990), 151–170.

[Kn] M. Kneser, *Zum expliziten Reziprozitätsgesetz von Šafarevič*, Math. Nachr. **6** (1951), 89–96.

[Ko1] H. Koch, *Über Darstellungsträume und die Struktur der multiplikativen Gruppe eines* $p$*-adischen Zahlkörpers*, Math. Nachr. **26** (1963), 67–100.

[Ko2] H. Koch, *Über Galoissche Gruppen von* $\mathfrak{p}$*-adischen Zahlkörpern*, Math. Nachr. **29** (1965), 77–111.

[Ko3] H. Koch, *Über Galoissche Gruppen der algebraischen Abschließung eines Potenzreihenkörpers mit endlichem Konstantenkörper*, Math. Nachr. **35** (1967), 323–327.

[Ko4] H. Koch, *Galoische Theorie der* $p$*-Erweiterungen*, Deutscher Verlag Wissenschaften, Berlin (1970).

[Ko5] H. Koch, *The Galois group of a* $p$*-closed extension of a local field*, Soviet Math. Dokl. **19** (1978), 10–13.

[Ko6] H. Koch, *Algebraic Number Theory*, Itogi Nauki i Techniki. Sovr. Probl. Mat. 62, VINITI, Moscow (1990) (in Russian).

[Kob1] N. Koblitz, *p-adic Analysis: A Short Course on Recent Works*, London Math. Soc., Lect. Note Ser. 46, London (1980).

[Kob2] N. Koblitz, *p-adic Numbers, p-adic Analysis and Zeta-Functions*, 2nd ed., Springer, Berlin (1984).

[Kol] V.A. Kolyvagin, *Formal groups and norm residue symbol*, Math. USSR Izv. **15** (1980), 289–348.

[Kom] K. Komatsu, *On the absolute Galois group of local fields, II*, Adv. Stud. Pure Math. **2** (1983), 63–68.

[Koy1] Y. Koya, *A generalization of class formation by using hypercohomology*, Inv. Math. **101** (1990), 705–715.

[Koy2] Y. Koya, *A generalization of Tate–Nakayama theorem by using hypercohomology*, Proc. Japan Acad., Ser. A **69**(3) (1993), 53–57.

[Kr1] M. Krasner, *Sur la représentation exponentielle dans les corps relativement galoisiens de nombres* $\mathfrak{p}$*-adiques*, Acta Arithm. **3** (1939), 133–173.

[Kr2] M. Krasner, *Rapport sur le prolongement analytique dans les corps valués complets par la méthode des éléments analytiques quasi-connexes*, Bull. Soc. Math. France **39–40** (1974), 131–254.

[Kub] T. Kubota, *Geometry of numbers and class field theory* Japan J. Math. **63** (1987), 237–257.

[Kud] A. Kudo, *On Iwasawa's explicit formulas for the norm residue symbol*, Mem. Fac. Sci. Kyuchu Univ., Ser. A **26** (1972), 139–148.

[Kur] M. Kurihara, *Abelian extensions of an absolutely unramified local field with general residue field*, Invent. Math. **93** (1988), 451–480.

[Kuz] L.V. Kuzmin, *New explicit formulas for the norm residue symbol and their applications*, Izv. Akad. Nauk SSSR, Ser. Mat. **54** (1990), 1196–1228 (in Russian; Engl. transl. in Math. USSR Izv.).

[KnS] K. Kanesaka and K. Sekiguchi, *Representation of Witt vectors by formal power series and its applications*, Tokyo J. Math. **2** (1979), 349–370.

[KtS] K. Kato and S. Saito, *Two dimensional class field theory*, Adv. Stud. Pure Math. **2** (1983), 103–152.

[KwS] Y. Kawada and I. Satake, *Class formations, II*, J. Fac. Sci. Tokio, Sect. 1A **7** (1956), 353–389.

[La1] S. Lang, *Algebra*, Addison-Wesley, Reading, MA (1965).

[La2] S. Lang, *Algebraic Number Theory*, 2nd print., Springer, Berlin (1986).

[La3] S. Lang, *Cyclotomic Fields*, 2nd ed., Springer, Berlin (1986).

[Lab] J.-P. Labute, *Classification of Demushkin groups*, Canad. J. Math. **19** (1967), 106–132.

[Lau1] F. Laubie, *Groupes de ramification et corps résiduels*, Bull. Sci. Math. (2) **105** (1981), 309–320.

[Lau2] F. Laubie, *Sur la ramification des extensions de Lie*, Compositio Math. **55** (1985), 253–262.

[Lau3] F. Laubie, *Sur la ramification des extensions infinies des corps locaux*, Sém. Théor. Nombres, Paris 1985–86, Birkhäuser, Boston (1987).

[Lau4] F. Laubie, *Extensions de Lie et groups d'automorphismes de corps locaux*, Compositio Math. **67** (1988), 165–190.

[Lau5] F. Laubie, *La ramification des extensions galoisiennes est déterminée par les discriminants de certaines sous-extensions*, Acta Arithm. **65** (1993), 283–291.

[Le] H.W. Leopoldt, *Zur Approximation des p-adischen Logarithmus*, Abh. Math. Sem. Univ. Hamburg **25** (1961), 77–81.

[Lo] V.G. Lomadze, *On the ramification theory of two dimensional local fields*, Mat. Sb. **151** (1979), 378–394 (in Russian; Engl. transl. in Math. USSR Sb.).

[LR] I. Lubin and M.I. Rosen, *The norm map for ordinary abelian varieties*, J. Algebra **52** (1978), 236–240.

[LT] J. Lubin and J. Tate, *Formal complex multiplication in local fields*, Ann. Math. **81** (1965), 380–387.

[Mah] K. Mahler, *Introduction to p-adic Numbers and Their Functions*, Cambridge Univ. Press, Cambridge (1973).

[Man] Ju.I. Manin, *Cyclotomic fields and modular curves*, Usp. Mat. Nauk **26** (1971), 7–78 (in Russian).

[Mar1] M.A. Marshall, *Ramification groups of abelian local field extensions*, Canad. J. Math. **23** (1971), 184–203.

[Mar2] M.A. Marshall, *The maximal p-extension of a local field*, Canad. J. Math. **23** (1971), 398–402.

[Mau1] E. Maus, *Arithmetisch disjunkte Körper*, J. Reine Angew. Math. **226** (1967), 184–203.

[Mau2] E. Maus, *Die gruppentheoretische Struktur der Verzweigungsgruppenreihen*, J. Reine Angew. Math. **230** (1968), 1–28.

[Mau3] E. Maus, *On the jumps in the series of ramification groups*, Bull. Soc. Math. France **25** (1971), 127–133.

[Mau4] E. Maus, *Über die Verteilung der Grundverzweigungszahlen von wild verzweigen Erzweiterungen p-adischer Zahlkörper*, J. Reine Angew. Math. **257** (1972), 47–79.

[Mau5] E. Maus, *Relationen in Verzweigungsgruppen*, J. Reine Angew. Math. **258** (1973), 23–50.

[Maz] B. Mazur, *Rational points of abelian varieties with values in towers of number fields*, Invent. Math. **18** (1972), 183–266.

[McC] W.G. Mc Callum, *Tate duality and wild ramification*, Math. Ann. **288** (1990), 553–558.

[McL] S. MacLane, *Subfields and automorphism groups of p-adic fields*, Ann. Math. **40** (1939), 423–442.

[Me] A.S. Merkurjev, *On the torsion in $K_2$ of local fields*, Ann. Math. **118** (1983), 375–381.

[Mi] I.S. Milne, *Arithmetic Duality Theorems*, Academic Press, New York (1986).

[Mik1] H. Miki, *On $\mathbb{Z}_p$-extensions of complete p-adic power series fields and function fields*, J. Fac. Sci. Tokio, Sect. 1A **21** (1974), 377–393.

[Mik2] H. Miki, *On some Galois cohomology groups of a local field and its application to the maximal p-extension*, J. Math. Soc. Japan **28** (1976), 114–122.

[Mik3] H. Miki, *On the absolute Galois group of a local fields, I*, Adv. Stud. Pure Math. **2** (1983), 55–61.

[Mik4] H. Miki, *On unramified abelian extensions of a complete field under a discrete valuation with arbitrary residue field of characteristic $p \neq 0$ and its application to wildly ramified $\mathbb{Z}_p$-extensions*, J. Math. Soc. Japan **29** (1977), 363–371.

[Mik5] H. Miki, *A note on Maus' theorem on ramification groups*, Tôhoku Math. J. (2) **29** (1977), 61–68.

[Mik6] H. Miki, *On the ramification numbers of cyclic p-extensions over local fields*, J. Reine Angew. Math. **328** (1981), 99–115.

[Mil1] J. Milnor, *Introducion to Algebraic K-theory*, Princeton Univ. Press and Univ. Tokyo Press, Princeton (1971).

[Mil2] J. Milnor, *Algebraic K-theory and quadratic forms*, Invent. Math. **9** (1970), 318–344.

[Miy] K. Miyake, *A fundamental theorem on p-extensions of algebraic number fields*, Japan J. Math., N.S. **16**, 307–316.

[Mo1] M. Moriya, *Einige Eigenschaften der endlicshen separablen algebraischen Erzweiterungen über perfekten Körpern*, Proc. Imper. Acad. Tokyo **7** (1941), 405–410.

[Mo2] M. Moriya, *Die Theorie des Klassenkörper im Kleinen über diskret perfekten Körpern, I*, Proc. Imper. Acad. Tokyo **18** (1942), 39–44; *II*, Proc. Imper. Acad. Tokyo **18** (1942), 452–459.

[Mo3] M. Moriya, *Zur theorie der Klassenkörper im Kleinen*, J. Math. Soc. Japan **3** (1951), 195–203.

[Moo] C.C. Moore, *Group extensions of p-adic and adelic linear groups*, Inst. Hautes Études Sci. Publ. Math. **35** (1968), 157–221.

[MS] A.S. Merkurjev and A.A. Suslin, *K-cohomology of Severi–Brauer varieties and the norm residue homomorphism*, Math. USSR Izv. **21** (1983), 307–340.

[MSh] O.V. Melnikov and A.A. Sharomet, *The Galois group of multidimensional local field of positive characteristic*, Mat. Sb. **180** (1989), 1132–1147 (in Russian; Engl. transl. in Math. USSR Sb.).

[MW] R. Mackenzie and G. Whaples, *Artin–Schreier equations in characteristic zero*, Amer. J. Math. **78** (1956), 473–485.

[N1] J. Neukirch, *Kennzeichning der $\mathfrak{p}$-adischen and der endlichen Zählkörper*, Invent. Math. **6** (1969), 296–314.

[N2] J. Neukirch, *Freie Produkte pro-endlicher Gruppen und ihre Kohomologie*, Arch. Math. **12** (1971), 337–357.

[N3] J. Neukirch, *Neubegründung der Klassenkörpertheorie*, Math. Z. **186** (1984), 557–574.

[N4] J. Neukirch, *Class Field Theory*, Springer, Berlin (1986).

[Na] M. Nagata, *Local Rings*, Interscience, New York (1962).

[Pa1] A.N. Parshin, *Class fields and algebraic K-theory*, Usp. Mat. Nauk **30** (1975), 253–254 (in Russian; Engl. transl. in Russian Math. Surveys).

[Pa2] A.N. Parshin, *On the arithmetic of two dimensional schemes, I. Repartitions and residues*, Math. USSR Izv. **10** (1976), 695–747.

[Pa3] A.N. Parshin, *Abelian coverings of arithmetic schemes*, Dokl. Akad. Nauk SSSR **243**(4) (1978), 855–858 (in Russian; Engl. transl. in Soviet Math. Dokl.).

[Pa4] A.N. Parshin, *Local class field theory*, Trudy Mat. Inst. Akad. Nauk SSSR **165** (1985), 143–170 (in Russian; Engl. transl. in Proc. Steklov Inst. Math.).

[Pa5] A.N. Parshin, *Galois cohomology and Brauer group of local fields*, Trudy Mat. Inst. Akad. Nauk SSSR **183** (1990), 159–169 (in Russian; Engl. transl. in Proc. Steklov Inst. Math.).

[Po] F. Pop, *Galoissche Kennzeichnung p-adisch abgeschlossener Körper*, J. Reine Angew. Math. **392** (1988), 145–175.

[PR] A. Prestel and P. Roquette, *Formally p-adic Fields*, Springer, Berlin (1984).

[Ra] M. Raynaud, *Anneaux Locaux Henseliens*, Springer, Berlin (1970).

[Rib] P. Ribenboim, *Theorie des Valuations*, 2nd ed., Les Presses Univ. Montreal, Montreal (1968).

[Rit] J. Ritter, *p-adic fields having the same type of algebraic extensions*, Math. Ann. **238** (1978), 281–288.

[Roq1] P. Roquette, *Abspaltung des Radikals in vollständingen lokalen Ringen*, Abh. Math. Sem. Univ. Hamburg **23** (1959), 75–113.

[Roq2] P. Roquette, *Some tendencies in contemporary algebra*, Perspectives in Math., Anniv. Oberwolfach 1984, Basel (1984) 393–404.

[Ros] M. Rosen, *An elementary proof of the local Kronecker–Weber theorem*, Trans. Amer. Math. Soc. **265** (1981), 599–605.

[Rot] H. Rothgiesser, *Zum Reziprotitätsgesetz für $p^n$*, Hamb. Abh. **11** (1934).

[RT] S. Rossett and J. Tate, *A reciprocity law for $K_2$-traces*, Comment. Math. Helv. **52** (1983), 38–47.

[Sa] I. Satake, *On a generalization of Hilbert's theory of ramifications*, Sci. Papers Coll. Gen. Ed. Univ. Tokyo **2** (1952), 25–39.

[Sch] O. Schilling, *The Theory of Valuations*, Amer. Math. Soc., New York (1950).

[Schi] A. Schinzel, *The number of zeros of polynomials in valuation rings of complete discretely valued fields*, Fund. Math. **124** (1984), 41–97.

[Se1] J.-P. Serre, *Groupes Algébriques et Corps de Classes*, Hermann, Paris (1959).

[Se2] J.-P. Serre, *Sur les corps locaux á corps residuel algébriqument clos*, Bull. Soc. Math. France **89** (1961), 105–154.

[Se3] J.-P. Serre, *Corps Locaux*, 2nd ed., Hermann, Paris (1968).

[Se4] J.-P. Serre, *Cohomologie Galoissiene*, 4th ed., Springer, Berlin (1973).

[Se5] J.-P. Serre, *A Course in Arithmetic*, Springer, Berlin (1978).

[Sek1] K. Sekiguchi, *Class field theory of p-extensions over a formal power series field with a p-quasi-finite coefficient field*, Tokyo J. Math. **6** (1983), 167–190.

[Sek2] K. Sekiguchi, *The Lubin–Tate theory for formal power series fields with finite coefficient fields*, J. Number Theory **18** (1984), 360–370.

[Sen1] S. Sen, *On automorphisms of local fields*, Ann. Math. **90** (1969), 33–46.

[Sen2] S. Sen, *Ramification in p-adic Lie extensions*, Invent. Math. **17** (1972), 44–50.

[Sen3] S. Sen, *On explicit reciprocity laws, I*, J. Reine Angew. Math. **313** (1980), 1–26; *II*, J. Reine Angew. Math. **323** (1981), 68–87.

[Sha1] I.R. Shafarevich, *On p-extensions*, Amer. Math. Soc. Transl. Ser. 2 vol. 4 (1956), 59–72 (original Russian paper in Mat. Sb. **20** (**62**) (1947), 351–363).

[Sha2] I.R. Shafarevich, *A general reciprocity law*, Amer. Math. Soc. Transl. Ser. 2 vol. 4 (1956), 73–106 (original Russian paper in Mat. Sb. **26** (**68**) (1950), 113–146).

[Shi] K. Shiratani, *Note on the Kummer–Hilbert reciprocity law*, J. Math. Soc. Japan **12** (1960), 412–421.

[Si] I.Ya. Sivitskii, *Torsion in Milnor's $K$-groups of a local field*, Math. USSR Sb. **54** (1985), 561–567.

[Sue] Y. Sueyoshi, *The explicit reciprocity law in Lubin–Tate relative groups*, Acta Arithm. **55** (1990), 291–299.

[Sus1] A.A. Suslin, *Homology of $GL_n$, characteristic classes and Milnor $K$-theory*, SLNM 1046, Springer, Berlin (1984), 357–375.

[Sus2] A.A. Suslin, *On the $K$-theory of local fields*, J. Pure Appl. Algebra **34** (1984), 301–318.

[Sus3] A.A. Suslin, *Torsion in $K_2$ of fields*, $K$-theory **1** (1987), 5–29.

[ST] S. Sen and J. Tate, *Ramification groups of local fields*, J. Ind. Math. Soc. **27** (1963), 197–202.

[T1] J. Tate, *$WC$-groups over p-adic fields*, Sem. Bourbaki **156** (1957/58).

[T2] J. Tate, *p-divisible groups* Proc. Conf. Local Fields, Springer, Berlin (1967), 158–183.

[T3] J. Tate, *Symbols in arithmetic*, Acta Congr. Int. Math. Nice (1970), 201–211.

[T4] J. Tate, *Rigid analytic spaces*, Invent. Math. **36** (1976), 257–289.

[T5] J. Tate, *Relations between $K_2$ and Galois cohomology*, Invent. Math. **36** (1976), 257–274.

[T6] J. Tate, *On the torsion in $K_2$ of fields*, Alg. Number Theory, Intern. Symp. Kyoto (1977), 243–261.

[Tam] T. Tamagawa, *On the theory of ramification groups and conductors*, Japan J. Math. **21** (1951), 197–215.

[Te] O. Teichmüller, *Diskret bewertete perfekte Körper mit unvollkommen Restklassenkörper*, J. Reine Angew. Math. **176** (1936), 141–152.

[V1] S.V. Vostokov, *Explicit form of the law of reciprocity*, Math. USSR Izv. **13** (1979), 557–588.

[V2] S.V. Vostokov, *A norm pairing in formal modules*, Math. USSR Izv. **15** (1980), 25–51.

[V3] S.V. Vostokov, *Symbols on formal groups*, Math. USSR Izv. **19** (1982), 261–284.

[V4] S.V. Vostokov, *The Hilbert symbol for Lubin–Tate formal groups, I*, Zap. Nauch. Sem. Leningrad. Otdel. Mat. Inst. **114** (1982), 77–95 (in Russian).

[V5] S.V. Vostokov, *Explicit construction in class field theory for a multidimensional local field*, Math. USSR Izv. **26** (1986), 263–287.

[V6] S.V. Vostokov, *Lutz filtration as Galois module in tamely ramified extensions*, Zap. Nauch. Sem. Leningrad. Otdel. Mat. Inst. **160** (1987), 182–192 (in Russian).

[V7] S.V. Vostokov, *A note on cyclotomic units*, Vestn. Leningrad Univ., no. 1 (1988), 14–17 (in Russian).

[VF] S.V. Vostokov and I.B. Fesenko, *The Hilbert symbol for Lubin–Tate formal groups, II*, Zap. Nauch. Sem. Leningrad. Otdel. Mat. Inst. **132** (1983), 85–96 (in Russian).

[Wa] L.C. Washington, *Introduction to Cyclotomic Fields*, Springer, Berlin (1982).

[Wd] A.R. Wadsworth, *p-Henselian fields: $K$-theory, Galois cohomology and graded Witt rings*, Pacific J. Math. **105** (1983), 473–496.

[Wel] A. Weil, *Basic Number Theory*, 3rd ed., Springer, Berlin (1974).

[Wes] E. Weiss, *Algebraic Number Theory*, McGraw-Hill Book, New York (1963).

[Wh1] G. Whaples, *Generalized local class field theory, I*, Duke Math. J. **19** (1952), 505–517; *II*, Duke Math. J. **21** (1954), 247–256; *III*, Duke Math. J. **21** (1954), 575–581; *IV*, Duke Math. J. **21** (1954), 583–586.

[Wh2] G. Whaples, *Additive polynomials*, Duke Math. J. **21** (1954), 55–66.

[Wh3] G. Whaples, *Galois cohomology of additive polynomials and $n$th power mapping of fields*, Duke Math. J. **24** (1957), 143–150.

[Wh4] G. Whaples, *The generality of local class field theory* (*Generalized local class field theory, V*), Proc. Amer. Math. Soc. **8** (1957), 137–140.

[Wig] K. Wingberg, *Der Eindentigkeitssatz für Demuškinformationen*, Invent. Math. **70** (1982), 99–113.

[Win1] J.-P. Wintenberger, *Extensions de Lie et groupes d'automorphismes des corps locaux de caracteristique p*, C. R. Acad. Sci. Paris **288** (1979), 477–479.

[Win2] J.-P. Wintenberger, *Extensions abéliennes et groupes d'automorphismes der corps locaux*, C. R. Acad. Sci. Paris **290** (1980), 201–204.

[Win3] J.-P. Wintenberger, *Le corps des normes de certaines extensions infinies des corps locaux, applications*, Ann. Sci. École Norm. Sup. **16** (1983), 59–89.

[Wil] A. Wiles, *Higher explicit reciprocity laws*, Ann. Math. **107** (1978), 235–254.

[Wit1] E. Witt, *Der Existenzsatz für abelsche Functionenkörper*, J. Reine Angew. Math. **173** (1935), 43–51.

[Wit2] E. Witt, *Zyklische Körper und Algebren der Characteristik $p$ vom grade $p^n$*, J. Reine Angew. Math. **176** (1936), 126–140.

[Wit3] E. Witt, *Schiefkörper über diskret bewerteten Körpern*, J. Reine Angew. Math. **176** (1936), 153–156.

[Ya1] K. Yamamoto, *Isomorphism theorem in the local class field theory*, Mem. Fac. Sci. Kyuchu Univ. **12** (1958), 67–103.

[Ya2] K. Yamamoto, *On the Kummer–Hilbert reciprocity law*, Mem. Fac. Sci. Kyuchu Univ. **13** (1959), 85–95.

[Ze] I.G. Zel'venskiĭ, *On the algebraic closure of a local field for $p = 2$*, Izv. Akad. Nauk SSSR, Ser. Mat. **40** (1976), 3–25 (in Russian; Engl. transl. in Math. USSR Izv.).

# Infinite Galois Theory

Moshe Jarden

*School of Mathematical Sciences, Raymond and Beverly Sackler Faculty of Exact Sciences,
Tel Aviv University, Ramat Aviv, Tel Aviv 69978, Israel
e-mail: jarden@math.tau.ac.il*

*Contents*



HANDBOOK OF ALGEBRA, VOL. 1
Edited by M. Hazewinkel

## Introduction

The main problem of Galois theory is to find out whether or not each finite group occurs as a Galois group over the field $\mathbb{Q}$ of rational numbers. The solution of this one hundred years old problem is still out of reach. Yet one hopes for an affirmative solution. This hope is based on a long list of finite groups which have been realized over $\mathbb{Q}$. Cyclotomic extensions supply all finite abelian groups as Galois groups over $\mathbb{Q}$. The Hilbert irreducibility theorem combined with the Riemann existence theorem gives many nonabelian simple groups and quasi simple groups.

To go beyond this list, one has to solve 'embedding problems'. Here one starts with a finite Galois extension $L/\mathbb{Q}$ and an epimorphism $\alpha: G \to \mathcal{G}(L/\mathbb{Q})$ with $G$ finite, and one looks for a Galois extension $N$ of $Q$ which contains $L$ and for an isomorphism $\gamma: \mathcal{G}(N/\mathbb{Q}) \to G$ such that $\alpha \circ \gamma = \mathrm{res}_L$. Not every embedding problem over $\mathbb{Q}$ is solvable. So, in order to realize $G$ one has to find another Galois extension $L'/\mathbb{Q}$ with the same Galois group as $L/\mathbb{Q}$ such that the corresponding embedding problem has a solution.

This method has led Scholz, Reichardt and Shafarevich to realize each finite $l$-group ($l$ is a prime) and eventually each finite solvable group over $\mathbb{Q}$.

Solving embedding problems with a nonabelian kernel is in some cases simpler. If a finite nonabelian group $C$ can be realized with some extra conditions (GAR-realization), then each embedding problem as above with $\mathrm{Ker}(\alpha) \cong C^r$ is solvable. For example, all $A_n$ with $n \geqslant 5$ and $n \neq 6$ and all sporadic groups with the possible exception of $M_{24}$ have GAR-realization over $\mathbb{Q}$.

One therefore faces the possibility to continue solving embedding problems infinitely many times. In this way one arrives at infinite Galois extensions $N$ of $\mathbb{Q}$ and eventually at the algebraic closure $\widetilde{\mathbb{Q}}$ of $\mathbb{Q}$. We call $\mathcal{G}(\widetilde{\mathbb{Q}}/\mathbb{Q})$ the *absolute Galois group* of $\mathbb{Q}$ and denote it by $G(\mathbb{Q})$. This group is the inverse limit of all Galois groups of finite Galois extensions $L/\mathbb{Q}$. It is a *profinite group*. As such it is compact, Hausdorff, and totally disconnected. In particular, $G(\mathbb{Q})$ carries a natural unique Haar measure. Since not each finite embedding problem over $\mathbb{Q}$ is solvable, $G(\mathbb{Q})$ is not a free profinite group $\widehat{F}_\omega$ on countably many generators (Iwasawa). The main problem of Galois theory becomes therefore a partial problem of the more general problem about the structure of $G(\mathbb{Q})$ as a profinite group. Namely, are all finite groups quotients of $G(\mathbb{Q})$?

We are very far from understanding $G(\mathbb{Q})$. Nevertheless, we know quite a bit about it:

(1a) The only elements of finite order of $G(\mathbb{Q})$ are involutions. They are conjugate to each other. The closed subgroup generated by the involutions is isomorphic to the free product of groups of order 2 over the Cantor set.

(1b) Each open subgroup of $G(\mathbb{Q})$ (i.e. each absolute Galois group of a number field) which contains no involutions has cohomological dimension 2.

(1c) The only closed abelian subgroups are procyclic (i.e. generated as profinite groups by one element).

(1d) Almost all $e$-tuples $(\sigma_1, \ldots, \sigma_e)$ of $G(\mathbb{Q})$ generate a free profinite group of rank $e$. Moreover, the closed normal subgroup generated by almost all $(\sigma_1, \ldots, \sigma_e)$ is $\widehat{F}_\omega$, which is the free profinite group on countably many generators.

(1e) The maximal abelian quotient of $G(\mathbb{Q})$ (i.e. $\mathcal{G}(\mathbb{Q}_{\text{ab}}/\mathbb{Q})$) is isomorphic to the direct product $\prod \mathbb{Z}_l^\times$, where $l$ ranges over all primes $l$, and where $\mathbb{Z}_l$ is the ring of $l$-adic integers.

(1f) There is a short exact sequence

$$1 \longrightarrow \widehat{F}_\omega \longrightarrow G(\mathbb{Q}) \longrightarrow \prod_{n=2}^{\infty} S_n \longrightarrow 1.$$

(1g) $G(\mathbb{Q})$ has no closed normal nontrivial prosolvable subgroup. In particular, its Frattini group and its center are trivial.

(1h) Every isomorphism between open subgroups of $G(\mathbb{Q})$ is induced by an inner automorphism. In particular every automorphism of $G(\mathbb{Q})$ is inner. So, every closed normal subgroup of $G(\mathbb{Q})$ is characteristic.

Infinite Galois theory extends the question about the structure of $G(\mathbb{Q})$ to a question about the structure of absolute Galois groups of other distinguished fields. In some cases we have the full answer:

(2a) $G(R) \cong \mathbb{Z}/2\mathbb{Z}$ if $R$ is real closed;

(2b) $G(K) \cong \widehat{\mathbb{Z}}$ if $K$ is a finite field or if $K \cong C((t))$ with $C$ algebraically closed of characteristic 0;

(2c) For each prime $p$, $G(\mathbb{Q}_p)$ is generated by 4 elements. If $p \neq 2$, generating relations between them are explicitly given;

(2d) If $C$ is an algebraically closed field, then $G(C(t))$ is the free profinite group of rank card$(C)$;

(2e) $G(R(t))$ is real free;

(2f) Let $S$ be a finite set of rational primes and possibly $\infty$. Denote the maximal Galois extension of $\mathbb{Q}$ in which each $p \in S$ totally splits by $\mathbb{Q}_{\text{tot},S}$. If $S$ consists of one finite prime $p$, we write $\mathbb{Q}_{tp}$ instead of $\mathbb{Q}_{\text{tot},S}$. If $S = \{\infty\}$, we also write $\mathbb{Q}_{\text{tr}}$ for $\mathbb{Q}_{\text{tot},S}$. Then $G(\mathbb{Q}_{\text{tot},S})$ is the free product of the groups $G(\mathbb{Q}_{\text{tp}})$, $p \in S$, and each $G(\mathbb{Q}_{\text{tp}})$ is a free product of isomorphic copies of $G(\mathbb{Q}_p)$, $p \in S$ (and where $\mathbb{Q}_\infty = \mathbb{R}$).

We have a partial knowledge about few other absolute Galois groups. They should be next in line to be studied.

(3a) The maximal prosolvable quotient of $G(\mathbb{Q}_{\text{ab}})$ is the free prosolvable group on countably many generators. Shafarevich's conjecture says that $G(\mathbb{Q}_{\text{ab}}) \cong \widehat{F}_\omega$.

(3b) Each finite group occurs as a Galois group over $\mathbb{Q}_p(t)$ but the cohomological dimension of $G(\mathbb{Q}_p(t))$ is 3.

(3c) Again, each finite group occurs as a Galois group over $\mathbb{C}(t_1, t_2)$ and the cohomological dimension of $\mathbb{C}(t_1, t_2)$ is 2.

(3d) The same goes for $\mathbb{C}((t_1, t_2))$.

(3e) The field $\mathbb{F}_p((t))$ plays the analog to $\mathbb{Q}_p$ in characteristic $p$. Its absolute Galois group is prosolvable, and of infinite rank.

The theory of finite groups partially emerged out of Galois theory and has become a subject of research in its own right. The theory of profinite groups is an outcome of infinite Galois theory. As for finite groups, each profinite group occurs as a Galois group of some Galois extension. The inverse problem of infinite Galois theory is to characterize those profinite groups which occur as absolute Galois groups of fields.

There exist several partial results in this direction. They play off projectivity of groups against pseudo finiteness of fields:

(4a) A profinite group $G$ is isomorphic to the absolute Galois group of a PAC (resp. PRC, P$p$C) field if and only if $G$ is projective (resp. real projective, $p$-adically projective). In particular, every free profinite group is projective and therefore occurs as the absolute Galois group of a PAC field.

(4b) A profinite group $G$ of at most countable rank is isomorphic to the absolute Galois group of a PAC (resp. PRC, P$p$C) field which is algebraic over $\mathbb{Q}$ if and only if $G$ is projective (resp. real projective, $p$-adically projective).

A good knowledge of the absolute Galois group of a field or of a family of fields is a vital ingredient in the study of their model theory. For example, the Riemann hypothesis for function fields of one variable over finite fields (= Weil's theorem) combined with $G(\mathbb{F}_q) \cong \widehat{\mathbb{Z}}$ are the basic facts in the decidability of the theory of finite fields. Likewise, the indecidability of the theory of PAC fields is based on (4a).

The purpose of this survey is to expand the above mentioned points to the story of infinite Galois theory as it stands when these lines are written. We have put the main emphasis on the absolute Galois group of fields. Therefore, we have not covered interesting results on relative Galois groups, like Wingberg's work on Galois extensions of number fields and $\Gamma$-extensions of number fields or the Galois groups of maximal pro-2 extensions and their connection to quadratic forms.

*Acknowledgement.* The author is indebted to Ido Efrat, Wulf-Dieter Geyer, Dan Haran, and Aharon Razon for thorough reading and constructive criticism. He also thanks Helmut Völklein and Michael Fried for useful remarks.

## 1. Infinite Galois theory

Consider a *Galois extension* $N$ of a field $K$. This is the splitting field of a set of separable polynomials in $K[X]$ over $K$. Let $G = \mathcal{G}(N/K)$ be the group of all automorphisms of $N$ that fix each element of $K$. This is the *Galois group* of $N/K$. For each subgroup $H$ of $G$ let

$$N(H) = \{x \in N \mid \sigma x = x \text{ for each } \sigma \in H\}$$

be the *fixed field* of $H$ in $N$. Unlike in the case where $N/K$ is a finite extension, there need not exist an intermediate field $M$ between $K$ and $N$ such that $\mathcal{G}(N/M) = H$ [Rib], p. 3. Krull restored the Galois correspondence between subgroups and intermediate fields by introducing a topology to $G$. A basis for the neighborhoods of 1 in this topology are all the subgroups $\mathcal{G}(N/L)$, where $L$ ranges over all finite Galois extensions of $K$ which are contained in $N$. Under this *Krull topology* $G$ is a Hausdorff, totally disconnected, compact group [Rib], p. 7. It turns out that the fundamental theorems of Galois theory of finite extensions remain unchanged if we replace each occurrence of 'subgroup' by 'closed subgroup':

**1.1.** THEOREM ([FrJ], Proposition 1.8). *Let $N$ be a Galois extension of a field $K$. Then the map $M \mapsto \mathcal{G}(N/M)$ is a bijection from the family of fields lying between $K$ and $N$ onto the family of closed subgroups of $\mathcal{G}(N/K)$. The inverse map is given by $H \mapsto N(H)$.*

As in finite Galois theory [La2], pp. 192–199, Theorem 1.1 gives the following rules for the Galois correspondence:

(1a) $M_1 \subseteq M_2$ if and only if $\mathcal{G}(N/M_2) \leqslant \mathcal{G}(N/M_1)$;

(1b) $H_1 \leqslant H_2$ if and only if $N(H_2) \subseteq N(H_1)$;

(1c) $N(H_1) \cap N(H_2) = N(\langle H_1, H_2 \rangle)$, where $\langle H_1, H_2 \rangle$ is the closed subgroup of $G$ generated by the closed subgroups $H_1$ and $H_2$;

(1d) $N(H_1 \cap H_2) = N(H_1)N(H_2)$;

(1e) $\mathcal{G}(N/M_1 \cap M_2) = \langle \mathcal{G}(N/M_1), \mathcal{G}(N/M_2) \rangle$;

(1f) $\mathcal{G}(N/M_1M_2) = \mathcal{G}(N/M_1) \cap \mathcal{G}(N/M_2)$;

(1g) $N(\sigma H \sigma^{-1}) = \sigma N(H)$;

(1h) $\mathcal{G}(N/\sigma M) = \sigma \mathcal{G}(N/M)\sigma^{-1}$, for each $\sigma \in G$;

(1i) A closed subgroup $H$ of $G$ is normal if and only if $L = N(H)$ is a Galois extension of $K$;

(1j) If $M$ is a Galois extension of $K$ and $M \subseteq N$, then the map

$$\text{res}\colon \mathcal{G}(N/K) \to \mathcal{G}(M/K)$$

that assigns to each $\sigma \in \mathcal{G}(N/K)$ its restriction to $M$ is a continuous open epimorphism with kernel $\mathcal{G}(N/M)$ and we have $\mathcal{G}(M/K) \cong \mathcal{G}(N/K)/\mathcal{G}(N/M)$;

(1k) If $E$ is any extension of $K$, then res: $\mathcal{G}(NE/E) \to \mathcal{G}(N/N \cap E)$ is an isomorphism; and

(1l) If in (1k), $E$ is also a Galois extension of $K$, then the map $\sigma \mapsto (\text{res}_N \sigma, \text{res}_E \sigma)$ is an isomorphism

$$\mathcal{G}(NE/N \cap E) \cong \mathcal{G}(N/N \cap E) \times \mathcal{G}(E/N \cap E),$$

where the right hand side is equipped with the product topology, and

$$\mathcal{G}(NE/K) \cong \{(\sigma, \tau) \in \mathcal{G}(N/K) \times \mathcal{G}(E/K) \mid \text{res}_{N \cap E} \sigma = \text{res}_{N \cap E} \tau\},$$

that is, $\mathcal{G}(NE/K)$ is the *fiber product* $\mathcal{G}(N/K) \times_{\mathcal{G}(N \cap E/K)} \mathcal{G}(E/K)$.

This correspondence holds in particular in the case where $N$ is the separable closure $K_s$ of $K$. We denote $\mathcal{G}(K_s/K)$ by $G(K)$ and call it the *absolute Galois group* of $K$. We also denote the algebraic closure of $K$ by $\widetilde{K}$ and the maximal purely inseparable extension of $K$ by $K_{\text{ins}}$. If char$(K)$ is $p$, then

$$K_{\text{ins}} = \{a^{1/p^n} \mid a \in K,\ n = 0, 1, 2, \ldots\}.$$

It is a perfect field and res: $G(K_{\text{ins}}) \to G(K)$ is an isomorphism. So, when studying absolute Galois groups of a field we may assume that it is perfect.

## 2. Profinite groups

Profinite groups are intimately connected to general Galois theory in the same way that finite groups are linked to Galois theory of finite extensions. One considers a set $I$ with a partial order such that for each $i, j \in I$ there exists $k \in I$ with $i, j \leqslant k$. An *inverse system* of finite groups over $(I, \leqslant)$ is a system $\langle G_i, \pi_{ji} \rangle_{i,j \in I}$ where $G_i$ is a finite group and $\pi_{ji}\colon G_j \to G_i$ is a homomorphism whenever $j \geqslant i$. These objects satisfy the rules $\pi_{ii} = \text{Identity}_{G_i}$ and $\pi_{ki} = \pi_{ji} \circ \pi_{kj}$ if $i \leqslant j \leqslant k$. The *inverse limit* of this system is the subgroup $G = \varprojlim G_i$ of the direct product (equipped with the product topology) $\prod_{i \in I} G_i$ consisting of all elements $g = (g_i)_{i \in I}$ such that $\pi_{ji} g_j = g_i$ if $j \geqslant i$. This is a *profinite group*. The group $G$ is closed in $\prod_{i \in I} G_i$ and is therefore compact. It is also Hausdorff and totally disconnected [FrJ], Lemma 1.2. More precisely, the closed subgroups of $G$ of a finite index (= open subgroups) form a basis for the open neighborhoods of 1 in $G$.

The case where $I$ consists of one element shows that each finite group is also a profinite group. The simplest infinite profinite group is the group

$$\mathbb{Z}_p = \varprojlim \mathbb{Z}/p^i\mathbb{Z}$$

of $p$-adic integers. The direct product of all $\mathbb{Z}_p$ is the *Prüfer group*

$$\widehat{\mathbb{Z}} = \varprojlim \mathbb{Z}/n\mathbb{Z}$$

[FrJ], Lemma 1.12. Here we order the set of positive integers $\mathbb{N}$ by divisibility. If $m \mid n$, then we take the map $\mathbb{Z}/n\mathbb{Z} \to \mathbb{Z}/m\mathbb{Z}$ as the natural homomorphism.

The diagonal embedding embeds $\mathbb{Z}$ as a dense subgroup of $\widehat{\mathbb{Z}}$. Thus $\widehat{\mathbb{Z}}$ is the closed subgroup generated by 1. Moreover, for each profinite group $G$ and each element $g \in G$, the map $1 \mapsto g$ uniquely extends to a homomorphism of $\widehat{\mathbb{Z}}$ into $G$. Thus, $\widehat{\mathbb{Z}}$ is the free profinite group generated by one element. Here and in general for profinite groups, whenever we use the term 'homomorphism' we mean 'continuous homomorphism'.

In general, a profinite group which is generated by one element is *procyclic*. It is the direct product $\prod Z_p$, where $p$ ranges over all primes and each $Z_p$ is either $\mathbb{Z}/p^n\mathbb{Z}$ for some $n \geqslant 0$ or $\mathbb{Z}_p$.

However, the most prominent example for a profinite group is the Galois group of a Galois extension $N/K$. Indeed, we order the finite sub-Galois extensions $L/K$ of $N/K$ by inclusion. If $L \subseteq L'$, then we take $\text{res}_L\colon \mathcal{G}(L'/K) \to \mathcal{G}(L/K)$ as the corresponding homomorphism. We find that $\mathcal{G}(N/K) \cong \varprojlim \mathcal{G}(L/K)$, as topological groups.

Conversely, generalizing a construction of Emil Artin, Waterhouse constructed for each profinite group $G$ a Galois extension $N/K$ with $\mathcal{G}(N/K) \cong G$ [FrJ], Corollary 1.11.

The inverse problem of (finite) Galois theory is to determine which finite groups occur as Galois groups over the field $\mathbb{Q}$ of rational numbers, and more generally over other distinguished fields $K$. In terms of infinite Galois theory this problem can be rephrased as "which finite groups are quotients of $G(K)$?"

Infinite Galois theory deals with two basic problems:

1. Given a distinguished field $K$, describe $G(K)$ in group theoretic terms.

2. Give a necessary and sufficient group theoretic conditions on a profinite group $G$ to be isomorphic to the absolute Galois group of some field $K$.

Both problems are very far from being settled. However, there are already quite a few interesting results that shed light on both problems. This article surveys some of them.

## 3. Separably closed fields, real closed fields, and finite fields

There are three classes of fields with absolute Galois groups which are easy to describe.

**3.1.** *Separably closed fields.* A field $K$ is *separably closed* if every irreducible separable polynomial has a root in it. If $\mathrm{char}(K) = 0$, then separably closed and algebraically closed are the same. If $\mathrm{char}(K) = p$, then these notions may differ. For example, the separable closure of $\mathbb{F}_p(t)$ ($t$ is transcendental over $\mathbb{F}_p$) is different from its algebraic closure. The fundamental theorem of algebra says that the field $\mathbb{C}$ is algebraically closed. This theorem has been proved in many ways, e.g., in the theory of analytic functions as a consequence of Cauchy's integral formula [Car], p. 80, or by Galois theory, as a consequence of Sylow theorems [La2], p. 202. Finally, $K$ is separably closed if and only if $G(K)$ is trivial.

**3.2.** *Real closed fields.* A field $K$ is *formally real* if $-1$ is not a sum of squares in $K$. Alternatively, $K$ admits an ordering [La2], p. 274. For example, $\mathbb{Q}$ and $\mathbb{Q}(t)$ are formally real but $\mathbb{Q}(\sqrt{-1})$, $\mathbb{C}$ and $\mathbb{F}_p$ are not. If $K$ is formally real, then $\mathrm{char}(K) = 0$.

We say that a field $K$ is *real closed* if it is formally real but no proper algebraic extension of $K$ is formally real. Then $K$ admits a unique ordering. For example, $\mathbb{R}$ and $\widetilde{\mathbb{Q}} \cap \mathbb{R}$ are real closed fields. If $K$ is a real closed field and $K_0$ is a subfield which is algebraically closed in $K$, then $K_0$ is also real closed [La2], p. 280.

The theory of Artin and Schreier says that $K$ is a real closed field if and only if $G(K)$ is of order 2, i.e. $G(K) \cong \mathbb{Z}/2\mathbb{Z}$. Moreover, if $K$ is an arbitrary field such that $[K_s : K] < \infty$, then $K$ is either separably closed or real closed [La2], pp. 223 and 224.

The latter theorem gives the first necessary condition on a profinite group $G$ to be isomorphic to the absolute Galois group of a field $K$: The only elements of $G$ of finite order are *involutions* (i.e. elements of order 2). Moreover, if $\mathrm{char}(K) \neq 0$, then $G(K)$ is torsion free.

**3.3.** *Finite fields.* So, we are forced now to consider fields with infinite absolute Galois group. The easiest to handle among them are the finite fields. Recall, that if $K$ is a finite field, then it has $q$ elements, where $q$ is a power of $p = \mathrm{char}(K)$. Moreover, $K$ is the splitting field over $\mathbb{F}_p = \mathbb{Z}/p\mathbb{Z}$ of the polynomial $X^q - X$. In particular, there is, up to an isomorphism, a unique field with $q$ elements. We denote it by $\mathbb{F}_q$.

For each $n$ the field $\mathbb{F}_q$ has a unique extension $\mathbb{F}_{q^n}$ of degree $n$. This extension is *cyclic* (i.e. Galois with a cyclic Galois group). The map $x \mapsto x^q$ is a canonical generator of $\mathcal{G}(\mathbb{F}_{q^n}/\mathbb{F}_q)$. It is the *Frobenius automorphism* and we denote it by $\mathrm{Frob}(\mathbb{F}_{q^n}/\mathbb{F}_q)$ [La2], p. 185. It follows that

$$G(\mathbb{F}_q) \cong \varprojlim \mathcal{G}(\mathbb{F}_{q^n}/\mathbb{F}_q) \cong \varprojlim \mathbb{Z}/n\mathbb{Z} = \widehat{\mathbb{Z}}.$$

The inverse limit of the relative Frobenius automorphisms is the *absolute Frobenius automorphism* $\mathrm{Frob}_q$. It is a (topological) generator of $G(\mathbb{F}_q)$.

**3.4.** *Quasifinite fields.* Unlike in the case of separably closed fields and real closed fields, the absolute Galois group of finite fields does not characterize this class of fields. For example, the compositum $K$ of all $\mathbb{F}_{q^l}$ with $l$ prime is an infinite field with $G(K) \cong \widehat{\mathbb{Z}}$. More interesting, by a theorem of Puiseux, if $C$ is an algebraically closed field of characteristic 0, then the absolute Galois group of the field $C((t))$ of formal power series over $K$ is isomorphic to $\widehat{\mathbb{Z}}$ [Se1], p. 199. In Section 12 we explain that 'almost all' $\sigma \in G(\mathbb{Q})$ generate a subgroup of $G(\mathbb{Q})$ which is isomorphic to $\widehat{\mathbb{Z}}$. Each perfect field with an absolute Galois group isomorphic to $\widehat{\mathbb{Z}}$ is *quasifinite*.

**3.5.** *Model theory of algebraically closed fields.* The simple structure of the absolute Galois groups of the three classes that we have described here has a favorable impact on their elementary theories. Here we assume that the reader is familiar with the basic notions and results of Model theory and ultraproducts, say as presented in [FrJ], Chapters 6 and 7. We consider the first order language, $\mathcal{L}(\mathrm{ring})$, of the theory of rings. Given a basic field $K$, we also add a constant symbol for each element of $K$ to $\mathcal{L}(\mathrm{ring})$ and denote the resulting language by $\mathcal{L}(\mathrm{ring}, K)$. The *elementary theory* of a class $\mathcal{F}$ of fields (resp. that contain $K$) is the set of all sentences in $\mathcal{L}(\mathrm{ring})$ (resp. $\mathcal{L}(\mathrm{ring}, K)$) that are true in each $F \in \mathcal{F}$.

It turns out that the elementary theory of algebraically closed fields (resp. of fixed characteristic) is decidable. Moreover, the division algorithm for polynomials leads to a primitive recursive elimination of quantifiers procedure for these theories [FrJ], Section 8.2. Thus, there is an effective procedure that determines whether a given sentence of $\mathcal{L}(\mathrm{ring})$ is true in all algebraically closed fields (resp. of a given characteristic). It follows that this theory is *model complete*, that is, if $F \subseteq F'$ are algebraically closed fields, then $F'$ is an elementary extension of $F$.

**3.6.** *Model theory of real closed fields.* Similarly, the theory of real closed fields is decidable and model complete. Moreover, if one adds a binary symbol for the ordering relation to $\mathcal{L}(\mathrm{ring})$, then, by a theorem of Tarsky, the theory even has an elimination of quantifiers [Pr1], p. 48, or [Coh], Section 1. As an application one proves that an absolutely irreducible variety $V$ which is defined over a real closed field $R$ has a simple $R$-rational point if and only if its function field over $R$ is formally real [Pr1], p. 59, or [La2], p. 282.

**3.7.** *Pseudo finite fields.* Let $\mathcal{C}$ be either the class of algebraically closed fields of a fixed characteristic or the class of real closed fields. Then $\mathcal{C}$ has a *complete* theory. That is, all the fields in $\mathcal{C}$ satisfy exactly the same sentences (i.e. they are *elementarily equivalent*). This is obviously not the case for the class of finite fields. Moreover, there exist infinite models of the theory of finite fields. They are called *pseudo finite fields*. For example, each nonprincipal ultraproduct of finite fields is pseudo finite.

Ax [Ax1], p. 262, proves that a field $K$ is pseudo finite if and only if it satisfies:

(1a) Each nonempty absolutely irreducible variety defined over $K$ has a $K$-rational point;

(1b) $G(K) \cong \widehat{\mathbb{Z}}$; and

(1c) $K$ is perfect.

These conditions are then used to establish a (recursive) decision procedure for the theory of finite fields, for the theory of pseudo finite fields, for the theory of statements true in all but finitely many fields $\mathbb{F}_p$, and for some more related theories [Ax1], Section 11.

A field which satisfies Condition (1a) is said to be *pseudo algebraically closed* (abbreviated *PAC*). In Section 12 we put these decidability results in the more general context of decidability and undecidability results for PAC fields.

## 4. More about profinite groups

Several concepts and results of the theory of finite groups can be carried over to profinite groups by 'taking limits'. Among those are the Sylow theorems, the Frattini subgroup, and cohomology.

**4.1.** *Pro-p groups.* A profinite group $G$ is a *pro-p group* if each of its finite quotients is a $p$-group. If

$$1 \longrightarrow A \longrightarrow B \longrightarrow C \longrightarrow 1 \tag{1}$$

is a short exact sequence of finite groups, then $B$ is a $p$-group if and only if $A$ and $C$ are. The same is true for pro-$p$ groups. Each profinite group $G$ has a *p-Sylow group* $G_p$. By definition, $G_p$ is a closed subgroup of $G$ which is pro-$p$ and which is maximal with this property. Every pro-$p$ subgroup of $G$ is contained in a $p$-Sylow subgroup and every two $p$-Sylow groups of $G$ are conjugate. Finally, an epimorphism of $G$ onto a profinite group $H$ maps $G_p$ onto a $p$-Sylow group of $H$ [FrJ], Section 20.10.

**4.2.** *Full families.* In general, if $\mathcal{C}$ is a family of finite groups, then a *pro-$\mathcal{C}$ group* is a profinite group all of its finite quotients belong to $\mathcal{C}$. If in (1), $B$ belongs to $\mathcal{C}$ if and only if $A$ and $C$ belong to $\mathcal{C}$, then this is the case for pro-$\mathcal{C}$ groups. We then say that $\mathcal{C}$ is *full*. For example, the family abelian groups or of all finite groups, the family of $p$-groups, and the family of solvable groups are full but the family of nilpotent groups is not full.

**4.3.** *The Frattini group.* The intersection of all closed maximal proper subgroups of a profinite group $G$ is a closed characteristic subgroup of $G$ called the *Frattini group* of $G$ and denoted by $\Phi(G)$. As for finite groups, $\Phi(G)$ is the set of all nongenerators of $G$. That is $\Phi(G)$ is the set of all $g \in G$ with the following property: for each subset $S$ of $G$, the relation $\langle g, S\rangle = G$ implies $\langle S\rangle = G$. Here $\langle S\rangle$ is the closed subgroup of $G$ generated by $S$ [FrJ], Section 20.1.

For example, if $G$ is a pro-$p$ group, then $\Phi(G)$ is the intersection of all open subgroups of index $p$. Thus $\Phi(G) = G^p[G,G]$ is the closed subgroup generated by all $p$-powers and the commutators of $G$ [FrJ], Lemma 20.36. In general $\Phi(G)$ is the direct product of its $p$-Sylow groups. This is equivalent to saying that $\Phi(G)$ is an inverse limit of finite nilpotent groups. In other words, $\Phi(G)$ is *pronilpotent*. Likewise, a *prosolvable group* is an inverse limit of finite solvable groups.

**4.4.** *Finitely generated profinite groups.* Obviously, every finite group has a finite set of generators. We say that a profinite group $G$ is *finitely generated* if it has elements $x_1, \ldots, x_e$ such that $G = \langle x_1, \ldots, x_e \rangle$. The minimal $e$ with this property is the *rank* of $G$. If $G$ is a finitely generated profinite group, then for each $n$, $G$ has only finitely many open subgroups of index at most $n$ [FrJ], Lemma 15.1. The intersection of all these subgroups is a characteristic open subgroup $G_n$ of $G$, and the intersection of all $G_n$ is 1. The former property implies, like for finite sets or for vector spaces of a finite dimension, that if $\alpha\colon G \to G$ is an epimorphism of finitely generated profinite groups, then $\alpha$ is an automorphism [FrJ], Proposition 15.3.

For each finite group $G$ and a field $K$, the statement '$G$ occurs as a Galois group over $K$' is equivalent to the truth in $K$ of a sentence in $\mathcal{L}(\text{ring})$ [FrJ], proof of Proposition 18.12. Hence, for each $e$, the statement 'the finite Galois groups over $K$ have at most $e$ generators' is equivalent to the truth in $K$ of a conjunction of a sequence of sentences of $\mathcal{L}(\text{ring})$. It follows that any ultraproduct of fields with absolute Galois groups of rank at most $e$ also has a Galois group of rank at most $e$.

**4.5.** *Rank of a profinite group.* A theorem of Douady, says that every profinite group $G$ has a system of generators $X$ which *converges to* 1. That is, for each open normal subgroup $H$ of $G$, all but finitely many elements of $X$ belong to $H$ [FrJ], Proposition 15.11. If $G$ is not finitely generated, then the cardinality of $X$ is equal to the cardinality of the set of all open normal subgroups of $G$. This is then the *rank* of $G$. In particular, if $\text{rank}(G) \leqslant \aleph_0$, then $G$ has a descending sequence of open normal subgroups whose intersection is 1. Also, if $G$ is a pro-$p$ group, then $\Phi(G)$ is the intersection of all open subgroups of index $p$, and $G/\Phi(G)$ is a vector space over $\mathbb{F}_p$ whose dimension is equal to the rank of $G$ [FrJ], Lemma 20.36.

**4.6.** *Free profinite group.* Given a set $X$, one constructs the free discrete group $F$ on $X$. Then one considers the inverse limit $\widehat{F} = \varprojlim F/N$, where $N$ ranges over all normal subgroups of $F$ of finite index which contain almost all elements of $X$. This is the *free profinite group* with basis $X$. The group $F$ naturally embeds in $\widehat{F}$ such that $X$ becomes a set of generators which converges to 1. The pair $(\widehat{F}, X)$ has a universal property in the category of profinite groups similar to the one that $(F, X)$ has in the category of discrete groups. Each map $\alpha$ of $X$ into a profinite group $G$ such that $\alpha(X)$ converges to 1 uniquely extends to a homomorphism of $F$ into $G$. In particular, $\widehat{F}$ is determined by the cardinality of $X$ up to an isomorphism. So, for each cardinal number $m$ we denote the free profinite group of rank $m$ by $\widehat{F}_m$. In particular, $\widehat{F}_\omega$ is the free profinite group on countably many generators.

**4.7.** *Embedding problems.* The universal property of $(\widehat{F}, X)$ is responsible for the solvability of 'embedding problems' for $\widehat{F}$. In general, an *embedding problem* for a profinite group $G$ is a pair

$$(\varphi\colon G \to A,\ \alpha\colon B \to A), \tag{2}$$

where $\varphi$ and $\alpha$ are epimorphisms of profinite groups. If $\varphi$ is only a homomorphism, we call (2) a *weak embedding problem*. The *kernel* of the embedding problem is the kernel of $\alpha$. If $B$ is finite, then the embedding problem is *finite*. A *weak solution* to (2) is a homomorphism $\gamma\colon G \to B$ such that $\alpha \circ \gamma = \varphi$. We say that $\gamma$ is a *solution* if it is surjective (then necessarily $\varphi$ is surjective).

**4.8.** *Characterization of $\widehat{F}_\omega$ by embedding problems.* Each embedding problem $(\varphi\colon \widehat{F} \to A,\ \alpha\colon B \to A)$ for $\widehat{F}$ in which $\operatorname{rank}(B) \leqslant \operatorname{rank}(\widehat{F})$ has a solution. The proof of this result uses an argument of Iwasawa if $\operatorname{rank}(F) = \infty$ [FrJ], Lemma 24.14, and a Lemma of Gaschütz if $\operatorname{rank}(F)$ is finite [FrJ], Proposition 15.31. Iwasawa used this argument to characterize $\widehat{F}_\omega$ as a profinite group of rank $\aleph_0$ for which every finite embedding problem is solvable [FrJ], Corollary 24.2.

**4.9.** *Free pro-$\mathcal{C}$ groups.* Let $\mathcal{C}$ be a full family of finite groups. If we put the extra condition on $N$ (in §4.6) in the construction of $\widehat{F}$ that $F/N \in \mathcal{C}$, then the resulting inverse limit is the *free pro-$\mathcal{C}$ group* on $X$. We denote it by $\widehat{F}_X(\mathcal{C})$ or also by $\widehat{F}_m(\mathcal{C})$ if $X$ is of cardinality $m$. The notation and results of the preceding paragraph hold if we restrict them to the category of pro-$\mathcal{C}$ groups. If $\mathcal{C}$ is the family of all $p$-groups (resp. solvable groups), then we also write $\widehat{F}_X(p)$ and $\widehat{F}_m(p)$ (resp. $\widehat{F}_X(\text{solv})$ and $\widehat{F}_m(\text{solv})$) for $\widehat{F}_X(\mathcal{C})$ and $\widehat{F}_m(\mathcal{C})$, respectively.

**4.10.** *Index and order.* Like for finite groups one may speak about an 'index' and an 'order' for profinite groups. Let $M$ be a closed subgroup of a profinite group $G$. Then $(G : M)$ is defined as $\prod l^{\lambda(l)}$, where $l$ ranges over all primes and for each $l$, $\lambda(l)$ is the maximal power of $l$ which divides the index $(G : H)$ of an open subgroup $H$ of $G$ which contains $M$. If these powers are not bounded, we put $\lambda(l) = \infty$. Note that $\lambda(l)$ may be different from 0 for infinitely many $l$'s. So, $(G : M)$ is a *super natural number*. The index is multiplicative: $(G : N) = (G : M)(M : N)$ if $N \leqslant M \leqslant G$. The *order* of $G$ is defined as $\#G = (G : 1)$. For example, the order of an infinite pro-$p$ group is $p^\infty$. Finally, one translates indices of profinite groups to degrees of infinite algebraic field extensions. If $L/K$ is an algebraic extension, then $[L : K] = (G(K) : G(L))$.

## 5. Cohomology of profinite groups

The action of profinite groups and in particular Galois groups on discrete abelian groups and the cohomology groups attached to this action capture valuable information about them. In this section we briefly survey the main concepts and results of the cohomology of profinite groups which enter into the study of Galois groups.

**5.1.** *Cohomology groups.* Let $G$ be a profinite group. A *$G$-module* is a discrete abelian group $A$ (usually additive) on which $G$ acts continuously (usually from the left). Continuous functions from $G^n$ to $A$ are called (nonhomogeneous) *$n$-cochains* [Rib], p. 95. They form an additive abelian group $C^n(G,A)$. For each $n$ there is a homomorphism $\partial_{n+1}\colon C^n(G,A) \to C^{n+1}(G,A)$, known as the *coboundary operator*:

$$(\partial_{n+1}f)(\sigma_1,\sigma_2,\ldots,\sigma_n) = \sigma_1 f(\sigma_2,\sigma_3,\ldots,\sigma_{n+1}) + \sum_{i=1}^{n}(-1)^i f(\sigma_1,\sigma_2,\ldots,\sigma_i\sigma_{i+1},\ldots,\sigma_{n+1}) + (-1)^{n+1} f(\sigma_1,\sigma_2,\ldots,\sigma_n).$$

It satisfies the rule: $\partial_{n+2}\circ\partial_{n+1} = 0$. One considers the subgroup

$$B^n(G,A) = \mathrm{Ker}(\partial_{n+1})$$

of *$n$-coboundaries* and the subgroup

$$Z^n(G,Z) = \partial_n\big(C^{n-1}(G,A)\big)$$

of *$n$-cocycles*. The *$n$-th cohomology group* of the $G$-module $A$ is the quotient $H^n(G,A) = Z^n(G,A)/B^n(G,A)$.

**5.2.** *Low dimensions.* In low dimensions, these groups have useful interpretation. For $n=0$ we have $H^0(G,A) = A^G = \{a \in A \mid \sigma a = a \text{ for all } \sigma \in G\}$. In particular, if the action of $G$ on $A$ is trivial (i.e. $\sigma a = a$ for all $\sigma \in G$ and $a \in A$), then $H^0(G,A) = A$. For $n = 1$, the cocycles are *crossed homomorphisms*. That is functions $f\colon G \to A$ such that $f(\sigma\tau) = f(\sigma) + \sigma f(\tau)$. A *1-coboundary* is a map $\sigma \mapsto \sigma a - a$ for some fixed $a \in A$. In particular, if the action of $G$ on $A$ is trivial, then $H^1(G,A) = \mathrm{Hom}(G,A)$. Finally, for $n = 2$, there is a natural bijection between the elements of $H^2(G,A)$ and equivalence classes of short exact sequences

$$0 \longrightarrow A \longrightarrow \widetilde{G} \longrightarrow G \longrightarrow 1, \tag{1}$$

where the action of $\widetilde{G}$ on $A$ through conjugation induces the given action of $G$ on $A$. Under this bijection, the zero element of $H^2(G,A)$ corresponds to a split exact sequence. In particular, if $H^2(G,A) = 0$, then each short exact sequence (1) with the prescribed action of $G$ on $A$ splits.

**5.3.** *Functorial properties.* The $n$-cohomology group $H^n(G,A)$ is a *contravariant functor* in $G$ and *covariant functor* in $A$. That is, to each homomorphism $f\colon G \to H$ of profinite groups there corresponds a homomorphism $f^*\colon H^n(H,A) \to H^n(G,A)$ which satisfies the rules $\mathrm{id}^* = \mathrm{id}$ and $(f\circ g)^* = g^*\circ f^*$. Also, to each homomorphism $f\colon A \to B$ of $G$-modules there corresponds a homomorphism $f_*\colon H^n(G,A) \to H^n(G,B)$ such that $\mathrm{id}_* = \mathrm{id}$ and $(f\circ g)_* = f_*\circ g_*$. Accordingly, we can present the cohomology groups of a profinite group as a direct limit of cohomology groups of finite groups:

$$H^n(G,A) = \varinjlim H^n\big(G/N, A^N\big)$$

with $N$ ranging over all open normal subgroups. Similarly, for direct limits of $G$-modules we have the rule: $H^n(G, \varinjlim A_i) = \varinjlim H^n(G, A_i)$ [Rib], p. 109.

**5.4.** *Exact sequences.* Each short exact sequence of $G$-modules

$$0 \longrightarrow A \longrightarrow B \longrightarrow C \longrightarrow 0$$

naturally induces a long exact sequence

$$\cdots \longrightarrow H^n(G,A) \longrightarrow H^n(G,B) \longrightarrow H^n(G,C) \xrightarrow{\delta} H^{n+1}(G,A) \longrightarrow \cdots, \qquad n \geqslant 0.$$

The map $\delta$ is called the *connecting homomorphism.*

On the other hand, if $N$ is a closed normal subgroup of $G$ and $A$ is a $G$-module, then $A^N$ is a $\overline{G} = G/N$-module. If in addition $H^i(N,A) = 0$ for all $0 < i < n$, then we have the 5-term exact sequence of Hochschild and Serre [Rib], p. 177:

$$0 \longrightarrow H^n(\overline{G}, A^N) \xrightarrow{\text{inf}} H^n(G,A) \xrightarrow{\text{res}} H^n(N,A)^{\overline{G}} \xrightarrow{\text{tr}} H^{n+1}(\overline{G}, A^N) \xrightarrow{\text{inf}} H^{n+1}(G,A).$$

Here inf (*inflation*) is the homomorphism that corresponds to the canonical map $G \to \overline{G}$, res (*restriction*) is the homomorphism that corresponds to the inclusion map $N \to G$, tr is a special map called *transgression*, and see [Rib], p. 173, for the action of $\overline{G}$ on $H^n(N,A)$. One derives this sequence from a spectral sequence whose initial elements are the groups $E_2^{p,q} = H^p(\overline{G}, H^q(N,A))$ and which converges to $H^n(G,A)$ (see also [Sht], Section II4).

**5.5.** *Cup products.* Another operation that connects cohomology groups of different dimensions is the *cup product*. Given $G$-modules $A$ and $B$ there is for each $m$ and $n$ a natural homomorphism $a \otimes b \mapsto a \cup b$:

$$H^m(G,A) \otimes_{\mathbb{Z}} H^n(G,B) \longrightarrow H^{m+n}(G, A \otimes_{\mathbb{Z}G} B)$$

that for $m = n = 0$ is the identity map, and as a functor of both $A$ and $B$ commutes with the connecting homomorphism. Moreover, the cup product is associative, it commutes with restriction and inflation, and satisfies $a \cup b = (-1)^{mn} b \cup a$ [Rib], pp. 178–195.

**5.6.** *Induced modules.* A lemma of Shapiro connects the cohomology groups of a profinite group $G$ and those of a closed subgroup $H$ of $G$. Each $H$-module $B$ *induces* a $G$-module

$$A = \text{Ind}_H^G B = \{f\colon G \to B \mid f(\eta\sigma) = \eta f(\sigma) \text{ for all } \eta \in H \text{ and } \sigma \in G\}.$$

The action of $G$ on $A$ is given by $(\tau f)(\sigma) = f(\sigma\tau)$. The lemma of Shapiro then states that $H^n(G,A) \cong H^n(H,B)$ for all $n$. In particular, for $H = 1$, we have $H^n(G, \text{Ind}_1^G B) = 0$ for all abelian groups $B$ [Rib], p. 146.

**5.7.** *Cohomological dimension.* An important invariant that cohomology theory supplies is the *cohomological dimension* of a profinite group $G$. Given a prime $p$, $\mathrm{cd}_p(G)$ is the least positive integer $n$ such that $H^q(G, A) = 0$ for all $q > n$ and all finite $G$-modules $A$ of a $p$-power order. (Note that the definition in [Rib], p. 196, asks $A$ to range over all torsion $G$-modules, but the proof of (iii) $\Rightarrow$ (ii) on page 201 of [Rib] shows that it suffices to consider only finite $G$-modules.) Then $\mathrm{cd}(G)$ is the supremum over all $\mathrm{cd}_p(G)$.

Several rules help to compute the cohomological dimension of a profinite group [Rib], Chapter 4:

(2a) $\mathrm{cd}_p(G) = \mathrm{cd}_p(G_p)$, where $G_p$ is a $p$-Sylow group of $G$;

(2b) $\mathrm{cd}_p(G) = 0$ if and only if $G_p = 1$;

(2c) if $G$ has an element of order $p$, then $\mathrm{cd}_p(G) = \infty$; in particular, $\mathrm{cd}(G) = \infty$ if $G$ is a nontrivial finite group;

(2d) $H \leqslant G$ implies that $\mathrm{cd}_p(H) \leqslant \mathrm{cd}_p(G)$;

(2e) equality holds in (2d) if $p \nmid (G : H)$;

(2f) if $H$ is an open subgroup of $G$ and $G$ has no element of order $p$, then $\mathrm{cd}_p(H) = \mathrm{cd}_p(G)$ [Se2], Theorem, or [Ha1], Theorem A.

**5.8.** *Projective groups.* The interpretation of the second cohomology group as a collection of equivalence classes of short exact sequences allows us to draw an important consequence from the assumption $\mathrm{cd}(G) \leqslant 1$. In this case $H^2(G, C) = 0$ for each finite $G$-module $C$. Hence, each short exact sequence

$$1 \longrightarrow C \longrightarrow \widehat{G} \longrightarrow G \longrightarrow 1$$

splits. Suppose now that

$$(\varphi\colon G \to A,\ \alpha\colon B \to A) \tag{2}$$

is a finite embedding problem with an abelian kernel $C$. Then the fiber product $\widetilde{G} = B \times_A G$ gives rise to a commutative diagram of short exact sequences:

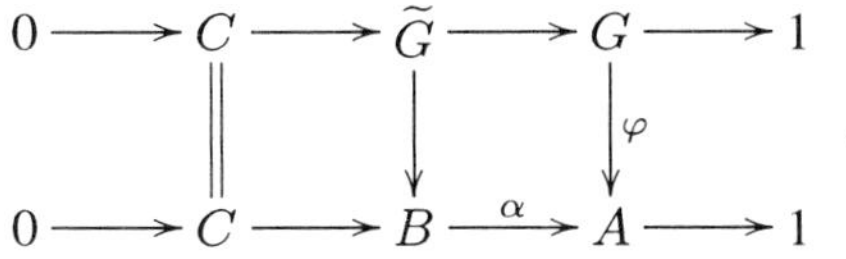


The splitting of the upper exact sequence gives a homomorphism $\gamma\colon G \to B$ such that $\alpha \circ \gamma = \varphi$. Thus, (2) is weakly solvable. If this happens for each finite embedding problem of $G$, we say that $G$ is *projective*. The above argument can be reversed to prove that, conversely, if $G$ is projective, then $\mathrm{cd}(G) \leqslant 1$. It turns out that if $G$ is projective, then each weak embedding problem for $G$ (i.e. one in which $A$, $B$, and $C$ are arbitrary profinite groups) is weakly solvable [FrJ], Lemma 20.8. In particular, each epimorphism $\pi\colon \widetilde{G} \to G$ has a *section*, i.e. a homomorphism $\theta\colon G \to \widetilde{G}$ such that $\pi \circ \theta = \mathrm{id}$.

If only $\mathrm{cd}_p(G) \leqslant 1$, then each weak embedding problem for $G$ with a pro-$p$ kernel is weakly solvable [Rib], p. 211, Proposition 3.1. In particular, every epimorphism $\pi: \widetilde{G} \to G$ with a pro-$p$ kernel has a section.

In Section 4, we pointed out that each finite embedding problem for a free profinite group $F$ is solvable. Hence, $F$ is projective. By (2d), each closed subgroup $G$ of $F$ is projective. Conversely, each profinite group $G$ is a quotient of some free profinite group $F$ (Douady [FrJ], Corollary 15.20). In particular, if $G$ is projective, the conclusion of the preceding paragraph implies that $G$ is isomorphic to a closed subgroup of $F$. This gives us the third characterization of projective groups.

The fourth characterization of projective groups comes from a theorem of Tate: Every projective pro-$p$ group is free pro-$p$ [FrJ], Proposition 20.37. This, together with (2a) and (2b), implies that a profinite group $G$ is projective if and only if its $p$-Sylow groups are free pro-$p$ for all primes $p$.

Note that the intersection of all open normal subgroups $H$ of a profinite group $G$ such that $G/H$ is a $p$-group is a closed normal subgroup $N$ and $G/N$ is the *maximal* pro-$p$ quotient of $G$. That is, each epimorphism of $G$ onto a pro-$p$ group factors through $G/N$. If $\mathrm{cd}_p(G) \leqslant 1$, then $G/N$ is a free pro-$p$ group [Rib], Corollary 3.2.

**5.9.** *Cohomology of pro-p groups.* Cohomology is most useful to analyze pro-$p$ groups. If $G$ is a pro-$p$ group, then $\mathrm{cd}(G)$ is the minimal number $n$ such that $H^{n+1}(G, \mathbb{Z}/p\mathbb{Z}) = 0$, where $G$ acts trivially on $\mathbb{Z}/p\mathbb{Z}$. In general, each of the groups $H^n(G, \mathbb{Z}/p\mathbb{Z})$ is annihilated by $p$ and can therefore be considered as a vector space over $\mathbb{F}_p$. We have $H^1(G, \mathbb{Z}/p\mathbb{Z}) = \mathrm{Hom}(G, \mathbb{Z}/p\mathbb{Z}) \cong G/\Phi(G)$ and $\dim_{\mathbb{F}_p} H^1(G, \mathbb{Z}/p\mathbb{Z}) = \mathrm{rank}(G)$. Also, $\dim_{\mathbb{F}_p} H^2(G, \mathbb{Z}/p\mathbb{Z})$ is the *relation rank* of $G$. Thus, if $e = \mathrm{rank}(G)$ and $k =$ relation rank$(G)$ are finite, then $\widehat{F}_e(p)$ has $k$ elements $r_1, \ldots, r_k$ such that $G \cong \widehat{F}_e(p)/R$, where $R$ is the smallest closed normal subgroup of $\widehat{F}_e(p)$ that contains $r_1, \ldots, r_k$. This is a *presentation* of $G$ by $e$ generators and $k$ relations $r_i = 1$.

## 6. Galois cohomology

Galois cohomology is the theory that applies cohomological methods to Galois groups and their action on various modules which come up in a natural way in field theory.

**6.1.** *The additive group of a field.* Denote the additive group of a field by $K^+$ and its multiplicative group by $K^\times$. The normal basis theorem for finite Galois extensions and Shapiro's lemma (Section 5.6) imply that

$$H^n(\mathcal{G}(L/K), L^+) = 0 \tag{1}$$

for an arbitrary Galois extension $L/K$ and each $n \geqslant 1$ [Rib], p. 246. In particular for $\mathrm{char}(K) = p$, we may use the long exact sequence that corresponds to the short one

$$0 \longrightarrow \mathbb{Z}/p\mathbb{Z} \longrightarrow K_s \xrightarrow{\wp} K_s \longrightarrow 0$$

with $\wp(x) = x^p - x$, to conclude that $H^n(G(K), \mathbb{Z}/p\mathbb{Z}) = 0$ for all $n \geqslant 2$. Thus $\mathrm{cd}_p(G) \leqslant 1$. Now denote the maximal pro-$p$-extension of $K$ by $K^{(p)}$ and observe that $\mathcal{G}(K^{(p)}/K)$ is the maximal pro-$p$ quotient of $G(K)$. It follows that $\mathcal{G}(K^{(p)}/K)$ is a free pro-$p$ group (last paragraph of §5.8). Moreover, the first part of the above long exact sequence shows that $\mathrm{Hom}(G(K), \mathbb{Z}/p\mathbb{Z}) = K^+/\wp(K^+)$. So, the rank of $\mathcal{G}(K^{(p)}/K)$ is the dimension $d$ of $K^+/\wp(K^+)$ over $\mathbb{F}_p$. In particular, every finite $p$-group of rank at most $d$ occurs as a Galois group over $K$. This is a theorem of Witt [Wit], p. 237.

**6.2.** *The multiplicative group of a field.* The multiplicative counterpart of (1) is known as Hilbert's theorem 90 [Rib], p. 246: For each Galois extension $L/K$

$$H^1(\mathcal{G}(L/K), L^\times) = 1.$$

(One uses 1 instead of 0, because $L^\times$ is a multiplicative module.) If $L/K$ is a finite cyclic extension with generator $\sigma$, one obtains as a consequence that the norm of an element $a \in L$ is 1 if and only if there exists $b \in L$ such that $a = \sigma b/b$. Also, if $n$ is prime to $\mathrm{char}(K)$ and the group $\mu_n$ of $n$-th roots of 1 is contained in $K$, we may consider the short exact sequence

$$1 \longrightarrow \mu_n \longrightarrow K_s^\times \stackrel{n}{\longrightarrow} K_s^\times \longrightarrow 1,$$

where $n$ is the map $x \mapsto x^n$. The beginning of the corresponding long exact sequence gives Kummer's correspondence: $K^\times/(K^\times)^n \cong \mathrm{Hom}(G(K), \mu_n)$.

For arbitrary Galois extensions $N \supseteq L \supseteq K$ we may write the following special case of the Hochschild–Serre exact sequence:

$$1 \longrightarrow H^2(\mathcal{G}(L/K), L^\times) \stackrel{\mathrm{inf}}{\longrightarrow} H^2(\mathcal{G}(N/K), N^\times) \stackrel{\mathrm{res}}{\longrightarrow} H^2(\mathcal{G}(N/L), N^\times)^{\mathcal{G}(N/L)}$$
$$\stackrel{\mathrm{tr}}{\longrightarrow} H^3(\mathcal{G}(L/K, L^\times)) \stackrel{\mathrm{inf}}{\longrightarrow} H^3(\mathcal{G}(N/K), N^\times).$$

In particular the first inflation map is injective.

The *Brauer group* of a field $K$ is the group $\mathrm{Br}(K)$ of all equivalence classes of finite dimensional central simple $K$-algebras. Here two such algebras $A$ and $A'$ are said to be *equivalent* if there exist division rings $D$ and $D'$ over $K$ and positive integers $n$ and $n'$ such that $A \cong M_{n\times n}(D)$, $A' \cong M_{n'\times n'}(D')$, and $D \cong_K D'$. The product of the equivalence classes of two such algebras $A$ and $B$ is represented by $A \otimes_K B$. There is a canonical isomorphism $\mathrm{Br}(K) \cong H^2(G(K), K_s^\times)$ [Se1], X5, p. 165, or [Rib], pp. 250–252. The latter group is the direct limit of the *relative Brauer groups* $\mathrm{Br}(L/K) = H^2(\mathcal{G}(L/K), L^\times)$, where $L$ ranges over all finite Galois extensions of $K$.

If $L/K$ is a finite cyclic extension of degree $n$ and $\sigma$ is a generator of $\mathcal{G}(L/K)$, then the following sequence is exact:

$$1 \longrightarrow K^\times \longrightarrow L^\times \stackrel{1-\sigma}{\longrightarrow} L^\times \stackrel{N_{L/K}}{\longrightarrow} K^\times \stackrel{\alpha}{\longrightarrow} \mathrm{Br}(K) \stackrel{\mathrm{res}}{\longrightarrow} \mathrm{Br}(L).$$

Here $x^{1-\sigma} = x/x^\sigma$, $N_{L/K}$ is the norm map, and $\alpha$ associates with each $a \in K^\times$ the factor system $c$ defined by $c(\sigma^i, \sigma^j) = 1$ if $i + j < n$ and $c(\sigma^i, \sigma^j) = a$ if $i + j \geqslant n$ [Deu], p. 64.

If $\mathrm{Br}(L) = 0$ for each finite separable extension of $K$, then $\mathrm{cd}(G(K)) \leqslant 1$. Conversely, if $K$ is perfect and $\mathrm{cd}(G(K)) \leqslant 1$, then $\mathrm{Br}(L) = 0$ for each finite extension $L$ of $K$ [Rib], p. 263. It follows that in this case, the norm map $N_{L/K}\colon L^\times \to K^\times$ is surjective for each finite Galois extension $L/K$.

**6.3.** *Cohomological dimension.* The rule (2) of Section 5.7 for arbitrary profinite groups applies also to absolute Galois groups. In particular, since elements of order $p$ appear in $G(K)$ only if $p = 2$ and $K$ is formally real, Condition (2f) of Section 5 improves to the following one:

(2a) If $L/K$ is a finite extension, then $\mathrm{cd}_p(K) = \mathrm{cd}_p(L)$, unless $p = 2$, $K$ is a formally real field but $L$ is not.

In addition we have:

(2b) If $t$ is transcendental over a field $K$ and $p \neq \mathrm{char}(K)$, then $\mathrm{cd}_p(G(K(t)) = 1 + \mathrm{cd}_p(G(K))$ [Rib], p. 271, and [Ax2], p. 1221.

(2c) Let $(K, v)$ be a Henselian valued field with value group $\Gamma$ and residue field $\overline{K}$. If $p \neq \mathrm{char}(\overline{K})$, then $\mathrm{cd}_p(G(K)) = \dim_{\mathbb{F}_p}(\Gamma/p\Gamma) + \mathrm{cd}_p(G(\overline{K}))$. In particular, if $v$ is discrete, then $\mathrm{cd}_p(G(K)) = 1 + \mathrm{cd}_p(G(\overline{K}))$ [Me2], Theorem 3.

Recall that a valued field $(K, v)$ is *Henselian* if it satisfies the lemma of Hensel and Rychlik: Let $O$ be the valuation ring of $(K, v)$. If $f \in O[X]$ and $a \in O$ satisfy $v(f(a)) > 2v(f'(a))$, then there exists a unique $x \in O$ such that $f(x) = 0$ and $v(x - a) > v(f'(a))$. Equivalently, $v$ has a unique extension to each finite extension of $K$ [CaF], p. 56, or [Ja1], Proposition 11.1. For example, $\mathbb{Q}_p$ and the field $K_0((t))$ of formal power series over an arbitrary field $K_0$ are Henselian.

If $K$ is separably closed, then $G(K) = 1$ and therefore $\mathrm{cd}_p(G(K)) = 0$ for each $p$. By (2b), $\mathrm{cd}_p(G(K(t))) = 1$ for $p \neq \mathrm{char}(K)$. Hence $\mathrm{cd}(G(F)) = 1$ also for all algebraic extensions of $K(t)$, except those which are separably closed.

If $K$ is a finite field, then $G(K) \cong \widehat{\mathbb{Z}}$ is free and therefore projective. By (2b), $\mathrm{cd}_p(G(K(t))) = 2$ for all primes $p \neq \mathrm{char}(K)$. By (2c), $\mathrm{cd}_p(G(K((t)))) = 2$, and $\mathrm{cd}_p(G(\mathbb{Q}_p)) = 2$. It follows from (2a), that if $F$ is a finite extension of any of these three fields, then $\mathrm{cd}_p(F) = 2$.

Finally, we explain in §8.2 that if $K$ is a number field, then $\mathrm{cd}_p(K) = 2$, unless $p = 2$ and $K$ is formally real. In the latter case $\mathrm{cd}_2(K) = \infty$.

**6.4.** *Connection to Milnor's algebraic K-theory. Milnor's $n$-th $K$-group* of a field $F$ is the quotient

$$K_n^{\mathrm{M}}(F) = (F^\times \otimes \cdots \otimes F^\times)/I,$$

with $n$ factors $F^\times$ and where $I$ is the additive subgroup generated by all elements $x_1 \otimes \cdots \otimes x_n$, with $x_i + x_j = 1$ for some $1 \leqslant i < j \leqslant n$ (one writes Milnor's groups additively). Milnor's conjecture states that if $F$ contains a primitive root of 1 of order $p$, then

$$K_n^{\mathrm{M}}(F)/pK_n^{\mathrm{M}}(F) \cong H^n(G(F), \mathbb{Z}/p\mathbb{Z}).$$

For $n = 1$, this is the classical isomorphism $F^\times/(F^\times)^p \cong H^1(G(F), \mathbb{Z}/p\mathbb{Z})$ (§6.2). Denoting the element of $H^1(G(F), \mathbb{Z}/p\mathbb{Z})$ which corresponds to a coset $x \cdot (F^\times)^p$ with $x \in F^\times$ by $(x)$, the conjectured isomorphism for arbitrary $n$ maps

$$x_1 \otimes \cdots \otimes x_n + I \bmod p$$

to the cup product $(x_1) \cup \cdots \cup (x_n)$. So, if $S_n(F)$ denotes the subgroup of

$$H^1\big(G(F), \mathbb{Z}/p\mathbb{Z}\big) \otimes \cdots \otimes H^1\big(G(F), \mathbb{Z}/p\mathbb{Z}\big)$$

generated by all $(x_1) \otimes \cdots \otimes (x_n)$ with $x_i \in F^\times$ and $x_i + x_j = 1$ for some $1 \leqslant i < j \leqslant n$, then the following short sequence should be exact:

$$\begin{aligned} 0 \longrightarrow S_n(F) \longrightarrow H^1(G(F), \mathbb{Z}/p\mathbb{Z}) \otimes \cdots \otimes H^1(G(F), \mathbb{Z}/p\mathbb{Z}) \\ \xrightarrow{\cup} H^n(G(F), \mathbb{Z}/p\mathbb{Z}) \longrightarrow 0. \end{aligned}$$

Merkurjev and Suslin [MeS] prove Milnor's conjecture for $n = 2$. For $n = 3$, partial results have been obtained by Merkurjev and Suslin and independently by Rost.

## 7. The field of $p$-adic numbers

A study of a 'global question' often starts with the study of 'local questions' associated with it. In particular, a good understanding of $G(\mathbb{Q}_p)$ leads to information about $G(\mathbb{Q})$. The former group is much simpler than the latter. Nevertheless, the structure of $G(\mathbb{Q}_p)$ is complicated enough to be the subject of numerous pieces of research.

**7.1.** *The field $\mathbb{Q}_p$.* The field $\mathbb{Q}_p$ is the completion of $\mathbb{Q}$ with respect to the $p$-adic valuation $v_p$:

$$v_p\left(\frac{a}{b}\,p^n\right) = n$$

if $a, b, n$ are integers and $p \nmid a, b$. In particular, its value group is $\mathbb{Z}$ (so the valuation is *discrete*), its valuation ring is $\mathbb{Z}_p$, its unique maximal ideal is $p\mathbb{Z}_p$, and its residue field is $\mathbb{F}_p$. The completeness of $\mathbb{Q}_p$ implies that $\mathbb{Q}_p$ is Henselian. The finiteness of the residue field is responsible for the compactness of $\mathbb{Z}_p$ and hence for the local compactness of $\mathbb{Q}_p$ under the $p$-adic topology [CaF], p. 50. Consequently, $\mathbb{Q}_p$ has for each $n$ only finitely many extensions of degree at most $n$ [La3], p. 54.

**7.2.** *Henselian fields.* Let $(K, v)$ be a Henselian field with residue field $\overline{K}$ of characteristic $p$ (which may be 0). Each finite extension $L$ of $K$ satisfies the formula $[L : K] = def$, where $f = [\bar{L} : \overline{K}]$ is the *residue degree*, $e = (v(L^\times) : v(K^\times))$ is the *ramification index*, and $d$ is a $p$-power, called the *defect* of $L/K$. This formula is due to Ostrowski [Rbn], p. 236. If $d = 1$ for each $L$, we say that $K$ is *defectless*.

Denote the unique extension of $v$ to $K_s$ also by $v$ and let $O_s$ be the corresponding valuation ring. Define the *inertia group* of $G(K)$ as

$$I(K) = \{\sigma \in G(K) \mid v(\sigma x - x) > 0 \text{ for all } x \in O_s\}.$$

It is a closed normal subgroup of $G(K)$ and we denote its fixed field in $K_s$ by $K_{\text{ur}}$. For each $\sigma \in \mathcal{G}(K_{\text{ur}}/K)$ define $\bar{\sigma} \in G(\overline{K})$ by $\bar{\sigma}\bar{a} = \overline{\sigma a}$ for each $a \in K_{\text{ur}}$ with $v(a) \geqslant 0$. Here $\bar{a}$ is the image of $a$ under the residue map. The map $\sigma \mapsto \bar{\sigma}$ is an isomorphism $\mathcal{G}(K_{\text{ur}}/K) \cong G(\overline{K})$ [End], Theorem 19.13. In particular, $K_{\text{ur}}$ is the compositum of all finite extensions $L$ of $K$ for which $[L : K] = f$ is the residue degree. If $K$ is defectless, $K_{\text{ur}}$ is the *maximal unramified* extension of $K$. That is, $K_{\text{ur}}$ is the compositum of all finite extensions $L$ of $K$ for which $\bar{L}/\overline{K}$ is separable and $e = 1$.

The *ramification group* of $G(K)$ is

$$R(K) = \left\{\sigma \in G(K) \;\middle|\; v\left(\frac{\sigma x}{x} - 1\right) > 0 \text{ for all } x \in K_s^\times\right\}.$$

It is a closed normal subgroup of $G(K)$ which is contained in $I(K)$. We denote its fixed field in $K_s$ by $K_{\text{tr}}$. If $K$ is defectless, then $K_{\text{tr}}$ is the maximal *tamely ramified* extension of $K$. That is, $K_{\text{tr}}$ is the compositum of all finite extensions $L$ of $K$ for which $\bar{L}/\overline{K}$ is separable and $p \nmid e$. Combined with the preceding paragraph, we have the following exact sequence:

$$1 \longrightarrow \mathcal{G}(K_{\text{tr}}/K_{\text{ur}}) \longrightarrow \mathcal{G}(K_{\text{tr}}/K) \longrightarrow G(\overline{K}) \longrightarrow 1. \tag{1}$$

Let $\Gamma = v(K^\times) = v(K_{\text{ur}}^\times)$ and let $\Delta = v(K_{\text{tr}}^\times)$. For each

$$\sigma \in \mathcal{G}(K_{\text{tr}}/K_{\text{ur}}) = I(K)/R(K)$$

we define a homomorphism $h_\sigma\colon \Delta/\Gamma \to \overline{K}_s^\times$ by

$$h_\sigma(v(x) + \Gamma) = \overline{\sigma x/x}, \quad x \in K_{\text{tr}}^\times.$$

Then the map $\sigma \to h_\sigma$ gives an isomorphism $\mathcal{G}(K_{\text{tr}}/K_{\text{ur}}) \cong \text{Hom}(\Delta/\Gamma, \overline{K}_s^\times)$ [End], Theorem 20.12, or [ZaS], p. 75, (18). Moreover,

$$\text{Hom}(\Delta/\Gamma, \overline{K}_s^\times) \cong \prod_{l \neq p} \mathbb{Z}_l^{\delta_l},$$

where $l$ ranges over all primes $\neq p$ and $\delta_l = \dim_{\mathbb{F}_l} \Gamma/l\Gamma$. Thus [Me2], Theorem 1,

$$\mathcal{G}(K_{\text{tr}}/K_{\text{ur}}) \cong \prod_{l \neq p} \mathbb{Z}_l^{\delta_l}.$$

In particular, $\mathcal{G}(K_{\text{tr}}/K_{\text{ur}})$ is an abelian group of order prime to $p$. Moreover, the exact sequence (1) splits. To describe the action of $G(\overline{K})$ on $\mathcal{G}(K_{\text{tr}}/K_{\text{ur}})$ let $\mu(\overline{K}_s)$ be the group of roots of unity of $\overline{K}_s$ and let $\chi$: $G(\overline{K}) \to \mathrm{Aut}(\mu(\overline{K}_s))$ be the *cyclotomic character*: $\zeta^\sigma = \zeta^{\chi(\sigma)}$. Then the action of each $\tau \in G(\overline{K})$ (viewed also as an element of $\mathcal{G}(K_{\mathrm{tr}}/K)$) on $\mathcal{G}(K_{\mathrm{tr}}/K_{\mathrm{ur}})$ is given by the following formula:

$$\tau\sigma\tau^{-1} = \sigma^{\chi(\tau)}, \quad \sigma \in \mathcal{G}(K_{\mathrm{tr}}/K_{\mathrm{ur}}).$$

If $p = 0$, then $K_{\mathrm{tr}} = K_s$. Suppose therefore that $p > 0$. Then group $G(K_{\mathrm{tr}})$ is a pro-$p$ group [ZaS], p. 77, Theorem 24. Since the order of $\mathcal{G}(K_{\mathrm{tr}}/K_{\mathrm{ur}})$ is prime to $p$, the Schur–Zassenhaus theorem [FrJ], Lemma 20.45, implies that the short exact sequence

$$1 \longrightarrow G(K_{\mathrm{tr}}) \longrightarrow G(K_{\mathrm{ur}}) \longrightarrow \mathcal{G}(K_{\mathrm{tr}}/K_{\mathrm{ur}}) \longrightarrow 1$$

splits.

For the same reason the $p$-Sylow subgroup of $\mathcal{G}(K_{\mathrm{tr}}/K)$ are isomorphic to those of $\mathcal{G}(K_{\mathrm{ur}}/K)$, hence to those of $G(\overline{K})$. It follows that they are pro-$p$ free (§6.1). Conclude that $\mathrm{cd}_p(\mathcal{G}(K_{\mathrm{tr}}/K)) \leqslant 1$ (fourth paragraph of §5.8). Since $G(K_{\mathrm{tr}})$ is a pro-$p$ group the second paragraph of §5.8, implies now that the short exact sequence

$$1 \longrightarrow G(K_{\mathrm{tr}}) \longrightarrow G(K) \longrightarrow \mathcal{G}(K_{\mathrm{tr}}/K) \longrightarrow 1$$

splits [KPR], Theorem 2.2. It follows that also the short exact sequence

$$1 \longrightarrow G(K_{\mathrm{ur}}) \longrightarrow G(K) \longrightarrow G(\overline{K}) \longrightarrow 1$$

splits.

**7.3.** *Finite extensions of* $\mathbb{Q}_p$. We specialize now the results of §7.2 to a finite extension $K$ of $\mathbb{Q}_p$. It is defectless [CaF], p. 19, and $\overline{K} = \mathbb{F}_q$ where $q$ is a power of $p$. Hence $\mathcal{G}(K_{\mathrm{ur}}/K) \cong \widehat{\mathbb{Z}}$. Moreover, the Frobenius automorphism $\mathrm{Frob}_q$ of $G(\mathbb{F}_q)$ lifts to a generator $\mathrm{Frob}(K_{\mathrm{ur}}/K)$ of $\mathcal{G}(K_{\mathrm{ur}}/K)$. It is uniquely determined by the condition $v_p(\mathrm{Frob}(K_{\mathrm{ur}}/K)a - a^q) > 0$, for all $a \in K_{\mathrm{ur}}$ with $v(a) \geqslant 0$.

Since the valuation of $K$ is discrete, $\mathcal{G}(K_{\mathrm{tr}}/K_{\mathrm{ur}})$ is also procyclic. More precisely, it is isomorphic to $\prod_{l \neq p} \mathbb{Z}_l$ [CaF], p. 31. The group $\mathcal{G}(K_{\mathrm{tr}}/K)$ is generated by two elements $\sigma$, $\tau$, where $\sigma$ is a lifting of $\mathrm{Frob}(K_{\mathrm{ur}}/K)$ (as such $\langle\sigma\rangle \cong \widehat{\mathbb{Z}}$), $\tau$ generates $\mathcal{G}(K_{\mathrm{tr}}/K_{\mathrm{ur}})$ and

$$\sigma\tau\sigma^{-1} = \tau^q. \tag{2}$$

Indeed, $\mathcal{G}(K_{\mathrm{tr}}/K)$ is the free profinite group with two generators and the above relation. That is, if $\bar\tau, \bar\sigma$ are elements of a profinite group $G$ and $\bar\sigma\bar\tau\bar\sigma^{-1} = \bar\tau^q$, then the map $(\sigma, \tau) \mapsto (\bar\sigma, \bar\tau)$ extends to a homomorphism $\mathcal{G}(K_{\mathrm{tr}}/K) \to G$. Relation (2) is known as the *Hasse–Iwasawa relation*.

By (5b) below, $\mathrm{cd}(G(K_{\mathrm{ur}})) = 1$. It follows that $G(K_{\mathrm{tr}})$ is a free pro-$p$ group (Section 4.9). Its rank is $\aleph_0$. Since each of the three factors $\mathcal{G}(K_{\mathrm{ur}}/K)$, $\mathcal{G}(K_{\mathrm{tr}}/K_{\mathrm{ur}})$, and $G(K_{\mathrm{tr}})$ of $G(K)$ is prosolvable, so is $G(K)$.

**7.4.** *$G(K)$ is finitely generated.* Consider now the maximal abelian extension $K_{\mathrm{tr,ab}}$ of $K_{\mathrm{tr}}$. Iwasawa proves that $\mathcal{G}(K_{\mathrm{tr,ab}}/K)$ is generated by $n+3$ elements, where $n = [K:\mathbb{Q}_p]$ [Iw3], Theorem 3 and use of local class field theory. Moreover, $G(K_{\mathrm{tr,ab}})$ is contained in the Frattini subgroup of $G(K_{\mathrm{tr}})$ and therefore also of $G(K)$. Hence, $G(K)$ itself is generated by $n+3$ elements [JaR], Introduction.

Jannsen [Jan], Satz 3.6, goes one step further and presents $G(K)$ as a quotient of a semidirect product $\widehat{F}_{n+1}(p) \rtimes \mathcal{G}(K_{\mathrm{tr}}/K)$ by a subgroup $N$ which is the closed normal subgroup generated by one element. Thus, in addition to the Hasse–Iwasawa relation of $\mathcal{G}(K_{\mathrm{tr}}/K)$, the generators of $G(K)$ satisfy only one additional profinite relation.

**7.5.** *Explicit presentation of $G(K)$ by generators and relations.* Jannsen and Wingberg [JaW] improve earlier results of Jakovlev and Koch and give the exact structure of $G(K)$ by generators and relations in the case where $p \neq 2$. This depends on several invariants of $K$. The first two are $n = [K:\mathbb{Q}_p]$ and $q = |\overline{K}|$. Then one proves that the group of roots of unity of a $p$-power order in $K_{\mathrm{tr}}$ is finite. So, it is cyclic and generated by an element $\zeta$ of order $p^s$. The Iwasawa generators $\sigma$ and $\tau$ act on $\zeta$ and define two positive integers $g$ and $h$ modulo $p^s$:

$$\zeta^\sigma = \zeta^g, \qquad \zeta^\tau = \zeta^h.$$

Also, let $\pi$ be the element of $\widehat{\mathbb{Z}}$ with $l$-coordinate 0 for each prime $l \neq p$ and with $p$-coordinate 1. In particular $\pi$ is divisible (in the ring $\widehat{\mathbb{Z}}$) by $p-1$. Then $G(K)$ is the free profinite group on the generators $\sigma, \tau, x_0, \ldots, x_n$ with the following conditions and relations:

(3a) The closed normal subgroup which is generated by $x_0, \ldots, x_n$ is a free pro-$p$ group;

(3b) The elements $\sigma$ and $\tau$ satisfy $\sigma\tau\sigma^{-1} = \tau^q$;

(3c) If $n$ is even, then $x_0^\sigma = f(x_0,\tau)^g x_1^{p^s}(x_1,x_2)(x_3,x_4)\cdots(x_{n-1},x_n)$;

(3c′) If $n$ is odd, then

$$x_0^\sigma = f(x_0,\tau)^g x_1^{p^s}(x_1,y_1)(x_2,x_3)(x_4,x_5)\cdots(x_{n-1},x_n),$$

where,

$$(x,y) = xyx^{-1}y^{-1}, \quad f(x_0,\tau) = \left(x_0^{h^{p-1}}\tau x_0^{h^{p-2}}\tau\cdots x_0^h\tau\right)^{\frac{\pi}{p-1}},$$

and $y_1$ is an element in $\langle x_1,\sigma,\tau\rangle$ which is given in [JaW], p. 74.

Diekert [Die] completes the work of Jannsen and Wingberg in the case $p=2$ and $s>1$ (that is, $K(\sqrt{-1})/K$ is unramified). He proves that $G(K)$ is generated by $n+3$ generators with relations (3a), (3b) and (3c) as above. Note that in this case $n=[K:\mathbb{Q}_2]$ is even (argue with the index of ramification). The structure of $G(K)$ if $\sqrt{-1}\notin K$ and in particular if $K=\mathbb{Q}_2$ is not known yet.

**7.6.** *Characterization of finite extensions of $\mathbb{Q}_p$ by their absolute Galois groups.* The description of the absolute Galois group of finite extensions of $\mathbb{Q}_p$ by generators and relations leads to characterizations of these fields by their absolute Galois groups [JaR] and [Rit]:

Let $K$ and $L$ be finite extensions of $\mathbb{Q}_p$. Suppose that $\sqrt{-1} \in K$ if $p = 2$ (actually we could assume that $K(\sqrt{-1})/K$ is unramified). Then $G(K) \cong G(L)$ if and only if $[K : \mathbb{Q}_p] = [L : \mathbb{Q}_p]$ and $K \cap \mathbb{Q}_{p,\mathrm{ab}} = L \cap \mathbb{Q}_{p,\mathrm{ab}}$.

**7.7.** *Demushkin groups.* A forerunner to the results of 7.5 and an important ingredient in their proof is the determination of the structure of the maximal pro-$p$-quotient of $G(K)$ by Demushkin and Labute. In other words, we let $K^{(p)}$ be the maximal pro-$p$ extension of $K$ and study $G = \mathcal{G}(K^{(p)}/K)$. Denote a primitive root of unity of order $n$ by $\zeta_n$. If $\zeta_p \notin K$, then $G \cong \widehat{F}_{n+1}(p)$ [Sh1] and [Se3], 4.1. In particular, if $K = \mathbb{Q}_p$ and $p \neq 2$, then $G \cong \widehat{F}_2(p)$.

If $\zeta_p \in K$, then $G$ is isomorphic to a *Demushkin group* of rank $n+2$. This means that

(4a) $\dim_{\mathbb{F}_p}(H^1(G, \mathbb{Z}/p\mathbb{Z})) = n + 2$;

(4b) $\dim_{\mathbb{F}_p}(H^2(G, \mathbb{Z}/p\mathbb{Z})) = 1$ (thus $H^2(G, \mathbb{Z}/p\mathbb{Z}) \cong \mathbb{Z}/p\mathbb{Z}$);

(4c) the cup product $\cup$: $H^1(G, \mathbb{Z}/p\mathbb{Z}) \times H^1(G, \mathbb{Z}/p\mathbb{Z}) \to \mathbb{Z}/p\mathbb{Z}$ is a nondegenerate bilinear alternating form.

It follows that $\mathrm{rank}(G) = n+2$, and $G$ is generated by elements $x_1, x_2, \ldots, x_{n+2}$ with a single relation. Moreover, $\mathrm{cd}(G) = 2$ [Se3], 9.1.

In order to write down this relation we consider the maximal power $p^s$ of $p$ such that $K$ contains $\zeta_{p^s}$. Then the maximal abelian quotient of $G$ has the form $(\mathbb{Z}/p^s\mathbb{Z}) \oplus (\mathbb{Z}/p\mathbb{Z})^{n+1}$.

If $p^s \neq 2$, then the relation is:

$$x_1^{p^s}[x_1, x_2] \cdots [x_{n+1}, x_{n+2}] = 1,$$

where $[x, y] = x^{-1}y^{-1}xy$. If $p^s = 2$ and $n$ is odd, then

$$x_1^2 x_2^4 [x_2, x_3] \cdots [x_{n+1}, x_{n+2}] = 1$$

[Se3], p. 4, or [Lb1], Theorems 7 and 8.

In the case where $p^s = 2$ and $n$ is even, Labute considers $L = K(\zeta_4, \zeta_8, \zeta_{16}, \ldots)$. Then $\mathcal{G}(L/K)$ is isomorphic to the group $U_2$ of invertible elements of $\mathbb{Z}_2$. Moreover, $L \subseteq K^{(2)}$ and so restriction gives rise to the cyclotomic character $\chi$: $G \to U_2$. For each $\sigma \in G$, $\chi(\sigma)$ is the element $a \in U_2$ such that $\zeta^\sigma = \zeta^a$ for each root of unity of 2-power order. Consider the image $A$ of $\chi$ in $U_2$. There are two cases [Lb1], Theorem 9:

*Case* 1: $A = (1 + 2^f \mathbb{Z}_2)$ with $f \geqslant 2$. In this case the single relation is:

$$x_1^{2+2^f}[x_2, x_3] \cdots [x_{n+1}, x_{n+2}] = 1.$$

*Case* 2: $A = \{\pm 1\} \times (1 + 2^f \mathbb{Z}_2)$ with $f \geqslant 2$. Then the single relation of $G$ has the form:

$$x_1^2 [x_1, x_2] x_3^{2^f} [x_3, x_4] \cdots [x_{n+1}, x_{n+2}] = 1.$$

In particular, if $K = \mathbb{Q}_2$, then $G$ is generated by 3 elements $x, y, z$ with the single relation

$$x^2 y^4 [y, z] = 1.$$

If we replace the condition (4a) on $G$ by "rank$(G) = \aleph_0$", we get a *Demushkin group of rank* $\aleph_0$. Labute [Lb2], Theorem 5, proves that the $p$-Sylow group of $G(K)$ is a Demushkin group of rank $\aleph_0$ and determines the single profinite relation that defines that group [Lb2], Corollaries 2 and 3.

Mináč and Ware [MW1] and [MW2] determine all Demushkin groups $G$ of a countable rank which appear as the maximal pro-$p$ Galois group over of some field $F$. The analogous problem for Demushkin groups of finite rank is still open.

**7.8.** *Local class field theory.* A central tool in the proof of the above statements on the absolute Galois group of $K$ and its maximal $p$-quotient is the *reciprocity law* of local class field theory. For each finite abelian extension $L$ of $K$ there is an exact sequence

$$1 \longrightarrow N_{L/K} L^\times \longrightarrow K^\times \xrightarrow{\psi_{L/K}} \mathcal{G}(L/K) \longrightarrow 1,$$

where $N_{L/K}$ is the norm map [Ne1], p. 42, or [CaF], p. 140. The *reciprocity map*[1] $\psi_{L/K}$ behaves 'well' under extensions of $L$ and therefore gives rise to a continuous homomorphism $\psi_K\colon K^\times \to \mathcal{G}(K_{\mathrm{ab}}/K)$, which is injective. Note that $K_{\mathrm{ur}} \subseteq K_{\mathrm{ab}}$. If $\pi$ is a prime element of $K$, then $\psi_K(\pi)$ is a lifting of $\mathrm{Frob}(K_{\mathrm{ur}}/K)$. Let $O_K$ be the valuation ring of $K$, let $U_K$ be its group of units, and let $U_{K,1} = 1 + \pi O_K$ be the group of 1-units of $K$. Then $\psi_K(U_K)$ is the inertia group of $\mathcal{G}(K_{\mathrm{ab}}/K)$, while $\psi_K(U_{K,1})$ is its ramification group [CaF], pp. 142–145.

The reciprocity map has good functorial properties. If $L$ is a finite extension of $K$, then $K_{\mathrm{ab}} L \subseteq L_{\mathrm{ab}}$ and we have a commutative diagram:

$$\begin{array}{ccc} L^\times & \xrightarrow{\psi_L} & \mathcal{G}(L_{\mathrm{ab}}/L) \\ \downarrow{\scriptstyle N_{L/K}} & & \downarrow{\scriptstyle \mathrm{res}} \\ K^\times & \xrightarrow{\psi_K} & \mathcal{G}(K_{\mathrm{ab}}/K) \end{array}$$

If $\sigma \in G(K)$, $K^\sigma = K'$, and $x \in K^\times$, then $\psi_{K'}(x^\sigma) = \psi_K(x)^\sigma$ [Ne1], p. 26.

**7.9.** *The center of* $G(K)$. The latter property may be used to extend the power of local class field theory beyond abelian extension. As an illustration we repeat here an argument of Ikeda [Ike], proof of Lemma 2.1.8, which proves that the center of $G(K)$ is trivial.

Let $\sigma$ be an element of the center of $G(K)$ and let $x \in \widetilde{K}^\times$. Find a finite Galois extension $L$ of $K$ which contains $x$. Then $L_{\mathrm{ab}}$ is a Galois extension of $K$ and $\psi_L(x^\sigma) = \psi_L(x)^\sigma = \psi_L(x)$. Hence $x^\sigma = x$. Conclude that $\sigma = 1$.

[1] Note that Neukirch [Ne1], p. 22, uses the term 'reciprocity map' for the 'inverse' map $r\colon \mathcal{G}(L/K) \to K^\times/N_{L/K}L^\times$ to $\psi_{L/K}$.

**7.10.** *Cohomological dimension.* The cohomological dimension of an arbitrary algebraic extension $L$ of $\mathbb{Q}_p$ obeys the following rules [Rib], p. 291:

(5a) $\mathrm{cd}_l(G(L)) = 0$ if and only if $l \nmid [\widetilde{\mathbb{Q}}_p : L]$;

(5b) $\mathrm{cd}_l(G(L)) = 1$ if and only if $l \mid [\widetilde{\mathbb{Q}}_p : L]$ and $l^\infty \mid [L : \mathbb{Q}_p]$;

(5c) $\mathrm{cd}_l(G(L)) = 2$ if and only if $l^\infty \nmid [L : \mathbb{Q}_p]$.

They are used to prove that $\Gamma = G(K)$ is in some sense determined by a finite 'big' quotient [Ja2], Theorem 7.4:

(6) $\Gamma$ has a finite quotient $\overline{\Gamma}$ such that if a closed subgroup $H$ of $\Gamma$ is a quotient of $\Gamma$ and $\overline{\Gamma}$ is a quotient of $H$, then $H \cong \Gamma$.

**7.11.** *The field* $\mathbb{F}_q((t))$. Let $q$ be a power of a prime number $p$. Then $K = \mathbb{F}_q((t))$ is the field of all formal power series in $t$ with coefficients in $\mathbb{F}_q$. It is the completion of $\mathbb{F}_q(t)$ with respect to the zero of $t$ and shares many properties with $\mathbb{Q}_p$.

For example, let $K_{\mathrm{tr}}$ be the maximal tamely ramified extension of $K$. Then $\mathcal{G}(K_{\mathrm{tr}}/K)$ is the free profinite group with the generators $\sigma, \tau$ and a unique defining relation (2). The ramification group $G(K_{\mathrm{tr}})$ is isomorphic to $\widehat{F}_\omega(p)$. The restriction map $G(K) \to \mathcal{G}(K_{\mathrm{tr}}/K)$ splits and therefore $G(K) \cong \widehat{F}_\omega(p) \rtimes \mathcal{G}(K_{\mathrm{tr}}/K)$. Moreover, the action of $\mathcal{G}(K_{\mathrm{tr}}/K)$ on $G(K_{\mathrm{tr}})$ is free in the following sense: $G(K_{\mathrm{tr}})$ contains a closed subgroup $E$ such that (a) $E \cong \widehat{F}_\omega(p)$, (b) $G(K_{\mathrm{tr}})$ is the closed normal subgroup of $G(K)$ generated by $E$, (c) if $G = F \rtimes \mathcal{G}(K_{\mathrm{tr}}/K)$ is a semidirect product and $F$ is a pro-$p$ group, then each homomorphism $E \to F$ uniquely extends to a homomorphism $G(K) \to G$ whose restriction to $\mathcal{G}(K_{\mathrm{tr}}/K)$ is the identity map [Koc], Satz 3.

**7.12.** *Arithmetically profinite extensions of a local field.* A valued field $K$ is *local* if it is locally compact under the topology which is determined by the valuation. If $\mathrm{char}(K) = 0$, then $K$ is a finite extension of $\mathbb{Q}_p$. If $\mathrm{char}(K) = p$, then $K \cong \mathbb{F}_q((t))$, for some power $q$ of $p$ [Bou], p. 433.

Let $K$ be a global field with a residue field $\overline{K}$ of characteristic $p$. Using the closed subgroups of $G(K)$ with the "upper numeration", Wintenberger [Win], p. 62, distinguishes among all algebraic separable extensions of $K$ those which are *strictly arithmetically profinite* (*SAPF*). We do not repeat here the definition of APF extensions but rather quote some of its properties:

Let $M$ and $N$ be separable algebraic extensions of $K$ such that $M \subseteq N$.

(6a) If $M/K$ is finite, then it is SAPF [Win], 1.2.2.

(6b) If $M/K$ is finite, then $N/K$ is SAPF if and only if $N/M$ is SAPF [Win], 1.2.3(i).

(6c) If $N/M$ is finite, then $N/K$ is SAPF if and only if $M/K$ is SAPF [Win], 1.2.3(ii).

(6d) If $N/K$ is SAPF, then so is $M/K$.

(6e) If $N/K$ is a Galois extension, $\overline{N}$ is a finite field, and $\mathcal{G}(N/K)$ is a $p$-adic Lie group,[2] then $N/K$ is SAPF [Win], 1.1.2.

(6f) If $L/K$ is a SAPF extension, then the maximal tamely ramified subextension of $L/K$ has a finite degree [Win], 2.1.2. In particular, $\bar{L}$ is a finite field and the value group of $L$ is isomorphic to $\mathbb{Z}$.

---

[2] For a definition of a $p$-adic Lie group see, e.g., [DMS], Definition 9.17.

(6g) For each infinite SAPF extension $L/K$ there is a local field $X_K(L)$ of characteristic $p$ [Win], Theorem 2.1.3(ii), with residue field isomorphic to $\bar{L}$ and with absolute Galois group isomorphic to $G(L)$ [Win], Corollary 3.2.3. Thus $G(L) \cong G(\bar{L}((t)))$.

**7.13.** *Infinite extensions of $\mathbb{Q}_p$ with isomorphic Galois groups.* Consider the field $N = \mathbb{Q}_p(\zeta_p, \zeta_{p^2}, \zeta_{p^3}, \ldots)$, where $\zeta_{p^i}$ is the $p^i$-th root of unity. Then $N/K$ is a totally ramified infinite Galois extension and $\mathcal{G}(N/K) \cong \mathbb{Z}_p^\times$ [Se1], Chapter IV, Proposition 17. In particular $\mathcal{G}(N/\mathbb{Q}_p)$ is a $p$-adic Lie group. By (6e), $N/\mathbb{Q}_p$ is a SAPF extension. By (6g), $G(N) \cong G(\mathbb{F}_p((t)))$. Also, $\mathbb{Z}_p^\times = \mathbb{Z}_p \times A$, where $A \cong \mathbb{Z}/2\mathbb{Z}$ if $p = 2$ and $A \cong \mathbb{Z}/(p-1)\mathbb{Z}$ if $p > 2$. Let $M$ be the fixed field of $A$ in $N$. By (6d) (or by (6e)), $M/\mathbb{Q}_p$ is also SAPF and therefore $G(M) \cong G(\mathbb{F}_p((t))) \cong G(N)$. Since $M$ is a proper subextension of $N/\mathbb{Q}_p$, it is not isomorphic to $N$ over $\mathbb{Q}_p$ [FrJ], end of proof of Lemma 18.19.

**7.14.** CONJECTURE. *For every infinite algebraic extension $M$ of $\mathbb{Q}_p$ which is not $\widetilde{\mathbb{Q}}_p$ there exists another algebraic extension $M'$ of $\mathbb{Q}_p$ such that $G(M) \cong G(M')$ but $M \not\cong_{\mathbb{Q}_p} M'$.*

## 8. Number fields

Our knowledge of the absolute Galois group of a number field $K$ is a consequence of the arithmetic of $K$ and of the Hilbert irreducibility theorem, which $K$ satisfies. In this section we survey the consequences of the arithmetic and defer the discussion of Hilbert irreducibility theorem to Section 11.

**8.1.** *Primes.* A *prime* $\mathfrak{p}$ of $K$ is either an equivalence class of valuations of $K$ or of *archimedean absolute values*. The latter correspond to the embeddings of $K$ in $\mathbb{C}$. We denote the completion of $K$ at $\mathfrak{p}$ by $K_\mathfrak{p}$, embed $\widetilde{K}$ in $\widetilde{K}_\mathfrak{p}$, and let $K_{\mathfrak{p},\mathrm{alg}} = \widetilde{K} \cap K_\mathfrak{p}$. If $\mathfrak{p}$ is nonarchimedean, and lies over a rational prime $p$, then $K_\mathfrak{p}$ is a finite extension of $\mathbb{Q}_p$ and $K_{\mathfrak{p},\mathrm{alg}}$ is the *Henselian closure* of $K$ with respect to $\mathfrak{p}$. If $\mathfrak{p}$ is archimedean, then $K_{\mathfrak{p},\mathrm{alg}}$ is either a *real closure* of $K$ or the algebraic closure of $K$. In all cases $K_{\mathfrak{p},\mathrm{alg}}$ is determined by $\mathfrak{p}$ up to a $K$-isomorphism. Also, $\widetilde{K} K_\mathfrak{p} = \widetilde{K}_\mathfrak{p}$ (by Krasner's lemma if $\mathfrak{p}$ is nonarchimedean). Hence res: $G(K_\mathfrak{p}) \to G(K_{\mathfrak{p},\mathrm{alg}})$ is an isomorphism. So, we may and we will identify $G(K_\mathfrak{p})$ with a closed subgroup of $G(K)$.

**8.2.** *Global class field theory.* Local class field theory teaches us that the group $K_\mathfrak{p}^\times$ controls the abelian extensions of $K_\mathfrak{p}$. To control the abelian extensions of the number field $K$, global class field theory combines all groups $K_\mathfrak{p}^\times$ to the group of *ideles* of $K$:

$$I_K = \left\{\alpha \in \prod K_\mathfrak{p}^\times \mid \alpha_\mathfrak{p} \in U_\mathfrak{p} \text{ for all but finitely many } \mathfrak{p}\right\}.$$

Here $U_\mathfrak{p}$ is the group of units of $K_\mathfrak{p}$. The multiplicative group $K^\times$ embeds diagonally in $I_K$ and $C_K = I_K/K^\times$ is the group of *idele classes* of $K$. There is a natural topology on $C_K$ which makes it a locally compact group. The quotient map $I_K \to C_K$ is injective on each $K_\mathfrak{p}^\times$. We can therefore view $K_\mathfrak{p}^\times$ as a closed subgroup of $C_K$. To each finite abelian extension $L/K$ global class field theory associates an *Artin map* $\psi_{L/K}\colon C_K \to \mathcal{G}(L/K)$ which is surjective and with kernel $N_{L/K}C_L$ [CaF], p. 172. The good functorial properties of the Artin map allow us to take inverse limits on all $\psi_{L/K}$ and to obtain an Artin map $\psi_K\colon C_K \to \mathcal{G}(K_{\mathrm{ab}}/K)$. This map is surjective and its kernel is the connected component of 1. The restriction of $\psi_{L/K}$ to $K_{\mathfrak{p}}^{\times}$ maps it into the decomposition group of $\mathfrak{p}$ in $K_{\mathrm{ab}}$, that is into $\mathcal{G}(K_{\mathrm{ab}}/K_{\mathrm{ab}} \cap K_{\mathfrak{p}})$. It coincides then with the local Artin map.

One approach to class field theory is via Galois cohomology [CaF] (the other one is through analytic methods [Gol]). The cohomology of number fields is partly associated to the cohomology of its completions through the *local global principle* for the Brauer groups: The map

$$H^2(G(\widetilde{K}), \widetilde{K}^{\times}) \xrightarrow{\mathrm{res}} \prod H^2(G(K_{\mathfrak{p}}), \widetilde{K}_{\mathfrak{p}}^{\times}) \tag{1}$$

where $\mathfrak{p}$ ranges over all primes $\mathfrak{p}$ of $K$ and res is the product of all local restriction maps is injective [Ne4], p. 244. This extends also to the case where $K$ is an arbitrary algebraic extension $E$ of $\mathbb{Q}$ [Rib], p. 296. Similarly, the map

$$H^2(G(E), \mathbb{Z}/p\mathbb{Z}) \xrightarrow{\mathrm{res}} \prod H^2(G(E_{\mathfrak{p}}), \mathbb{Z}/p\mathbb{Z}) \tag{2}$$

is injective for each algebraic extension $E$ of $\mathbb{Q}$ and each prime $p$ (see [Se4], p. 12, for the case $E = \mathbb{Q}$ and [GJ2], proof of Lemma 4.3, for the general case). As a consequence, one gets the following rules for the cohomological dimension of $G(E)$ [Rib], p. 302:

(3a) $\mathrm{cd}_p(G(E)) = \infty$ if and only if $p = 2$ and $E$ embeds into $\mathbb{R}$;

(3b) Assume that either $p \neq 2$ or $E$ does not embed into $\mathbb{R}$. Then

(i) $\mathrm{cd}_p(G(E)) = 0$ if and only if $p \nmid [\widetilde{\mathbb{Q}} : E]$;

(ii) $\mathrm{cd}_p(G(E)) = 1$ if and only if $p \mid [\widetilde{\mathbb{Q}} : E]$ and $p^{\infty} \mid [E_{\mathfrak{p}} : \mathbb{Q}_p]$ for every extension $\mathfrak{p}$ of $p$ to $E$;

(iii) $\mathrm{cd}_p(G(E)) = 2$ if and only if $p \mid [\widetilde{\mathbb{Q}} : E]$ and there exists an extension $\mathfrak{p}$ of $p$ to $E$ such that $p^{\infty} \nmid [E_{\mathfrak{p}} : \mathbb{Q}_p]$.

In particular the cohomological dimension of each number field which does not embed into $\mathbb{R}$ is 2. Also, $\mathrm{cd}_p(G(\mathbb{Q}_{\mathrm{ab}})) = 1$ for each prime $p$. Thus $G(\mathbb{Q}_{\mathrm{ab}})$ is a projective group.

**8.3.** *Closed abelian subgroups.* Geyer uses the information about the cohomological dimension of closed subgroups of $G(\mathbb{Q})$ to prove that each closed abelian subgroup of $G(\mathbb{Q})$ is procyclic [Rib], p. 306.

**8.4.** *$\mathbb{Z}_l$-extensions.* Class field theory becomes concrete in the case $K = \mathbb{Q}$. The Kronecker–Weber theorem states that the maximal abelian extension $\mathbb{Q}_{\mathrm{ab}}$ of $\mathbb{Q}$ is obtained by adjoining all roots of unity to $\mathbb{Q}$ [Ne1], p. 46. Consequently

$$\mathcal{G}(\mathbb{Q}_{\mathrm{ab}}/\mathbb{Q}) \cong \widehat{\mathbb{Z}}^{\times} \cong \prod_l \mathbb{Z}_l^{\times}.$$

In particular, for each prime $l$, $\mathbb{Q}$ has a unique Galois extension $N$ such that $\mathcal{G}(N/\mathbb{Q}) \cong \mathbb{Z}_l$ (we call $N$ a *$\mathbb{Z}_l$-extension* of $\mathbb{Q}$). Iwasawa [Iw2], Theorem 2, considers the compositum

$N$ of all $\mathbb{Z}_l$-extensions of an arbitrary number field $K$. He proves that $\mathcal{G}(N/K)$ is a free $\mathbb{Z}_l$-module whose rank $a(K)$ satisfies $r_2 + 1 \leqslant a(K) \leqslant [K : \mathbb{Q}]$, where $r_2$ is the number of nonconjugate imaginary embeddings of $K$ into $\mathbb{C}$.

**8.5.** *Characterization of a number field by its absolute Galois group.* The real closures of $\mathbb{Q}$ and the $p$-adic closures of $\mathbb{Q}$ are characterized by their absolute Galois groups. If $E$ is an algebraic extension of $\mathbb{Q}$ and $G(E)$ is isomorphic to $G(\mathbb{R})$, then $E$ is isomorphic over $\mathbb{Q}$ to $\mathbb{R}_{\text{alg}} = \widetilde{\mathbb{Q}} \cap \mathbb{R}$ (because $\mathbb{Q}$ has only one ordering). Similarly, Neukirch [Ne2] proves that if $G(E) \cong G(\mathbb{Q}_p)$, then $E \cong \mathbb{Q}_{p,\text{alg}}$. He also proves that if $K$ and $L$ are Galois extensions of $\mathbb{Q}$ such that $G(K) \cong G(L)$, then $K = L$. It follows that every open normal subgroup and hence every closed normal subgroup of $G(\mathbb{Q})$ is characteristic. Uchida [Uc1] and Iwasawa [Iw1] independently generalize Neukirch's result: If $K$ and $L$ are number fields and $G(K) \cong G(L)$, then $K$ and $L$ are conjugate over $\mathbb{Q}$. Moreover, if $\sigma: G(K) \to G(L)$ is an isomorphism, then $\sigma$ can be extended to an inner automorphism of $G(\mathbb{Q})$. In particular, every automorphism of $G(\mathbb{Q})$ is inner. Since $G(\mathbb{Q})$ has a trivial center (a result of F.K. Schmidt; see also Section 12), this means that $G(\mathbb{Q})$ is a *complete group*. The latter result is also proved by Ikeda [Ike].

**8.6.** *Realization of finite solvable groups.* Each embedding problem for $G(K)$ induces by restriction an embedding problem for $G(K_\mathfrak{p})$ for each prime $\mathfrak{p}$ of $K$. The local global principle (2) for the groups $H^2(*, \mathbb{Z}/l\mathbb{Z})$ implies that a weak embedding problem for $G(K)$ with a finite abelian kernel has a weak solution if and only if it has locally a weak solution [GJ2], Lemma 4.3. Scholz used this principle to realize each finite $l$-group with $l \neq 2$ over $\mathbb{Q}$ [Se4], Chapter 2. Shafarevich extends this result to arbitrary number field $K$ and also to $l = 2$ [Sh2], p. 96, and finally proves that each split embedding problem for $G(K)$ with a nilpotent kernel is solvable [Sh2], p. 205. As a consequence he is able to realize each finite solvable group over $K$ [Sh2], p. 180.

## 9. $p$-adically closed fields

The field $\mathbb{Q}_p$ of $p$-adic numbers behaves in many respects like the field $\mathbb{R}$ of real numbers. The '$p$-adic' theory is an analog of the 'real theory', which is however more complicated.

**9.1.** *Definitions.* To define 'formally $p$-adic field' one replaces $-1$ and squaring in the definition of 'formally real' (§3.2), respectively, by $p$ and the *Kochen operator*:

$$\gamma(z) = \frac{1}{p}\frac{z^p - z}{(z^p - z)^2 - 1}.$$

A field $K$ is then *formally $p$-adic* if $p^{-1}$ does not belong to the ring $\mathbb{Z}[\gamma(K)]$. Alternatively, $K$ is formally $p$-adic if it admits a *$p$-adic valuation* $v$. That is, $v(p)$ is the least positive element of $v(K^\times)$ and $\overline{K}_v \cong \mathbb{F}_p$. For example, $\mathbb{Q}$ and $\mathbb{Q}_p$ are formally $p$-adic fields. A formally $p$-adic field $K$ is *$p$-adically closed* if it has no proper algebraic formally $p$-adic extensions. In this case $K$ has a unique $p$-adic valuation $v$ and $(K, v)$ is Henselian.

**9.2.** *Model theory.* A natural language for valued fields is obtained from $\mathcal{L}$(ring) by adding a unary predicate for the valuation ring. We denote this language by $\mathcal{L}$(valued ring). As in the real case, the elementary theory of $p$-adically closed fields is model complete in this language. Moreover all $p$-adically closed fields $K$ are elementarily equivalent in $\mathcal{L}$(valued ring) to $\mathbb{Q}_p$ (Ax, Kochen and Ershov [PrR], Theorem 5.1). In particular, since $G(\mathbb{Q}_p)$ is finitely generated, $G(K) \cong G(\mathbb{Q}_p)$ (Section 4.4). Moreover, if $K_0$ is algebraically closed in $K$, then $K_0$ is also $p$-adically closed and res: $G(K) \to G(K_0)$ is an isomorphism.

Again, as for real closed fields, an absolutely irreducible variety $V$ which is defined over a $p$-adically closed field $K$ has a simple $K$-rational point if and only if its function field over $K$ is formally $p$-adic [PrR], Theorem 7.2.

**9.3.** *$p$-adic closure.* Zorn's lemma implies that each $p$-adically valued field $(K, v)$ has a $p$-adic closure $(\overline{K}, \bar{v})$. That is, $(\overline{K}, \bar{v})$ is a $p$-adically closed field and $\overline{K}/K$ is an algebraic extension. For example, $\mathbb{Q}_{p,\mathrm{alg}}$ is the unique (up to isomorphism) $p$-adic closure of $\mathbb{Q}$. Unlike in the real case, two $p$-adic closures of $(K, v)$ need not be isomorphic. Macintyre's isomorphism theorem says that if $(E, v)$ and $(F, v)$ are $p$-adic closures of $(K, v)$, then $(E, v) \cong_K (F, v)$ if and only if $K \cap E^n = K \cap F^n$ for all positive integers $n$ (here $E^n = \{x^n \mid x \in E\}$) [PrR], Corollary 3.11. As a result, the theory of $p$-adically closed fields has an elimination of quantifiers in an extension of $\mathcal{L}$(valued ring) which contains an $n$-ary predicate symbol $P_n$ for each positive integer $n$ (Macintyre [PrR], Theorem 5.6). For a $p$-adically closed field $(K, v)$ one interprets $P_n$ as the set of all $n$-powers of elements of $K$.

As for real closed fields, $G(\mathbb{Q}_p)$ characterizes $\mathbb{Q}_p$ up to an elementary equivalence. In other words, if $K$ is a field such that

$$G(K) \cong G(\mathbb{Q}_p), \tag{1}$$

then $K$ is $p$-adically closed and is therefore elementarily equivalent to $\mathbb{Q}_p$ (§9.1).

We have already mentioned (§8.5) that Neukirch proved this theorem for algebraic extensions of $\mathbb{Q}$. Pop [Po3], Theorem E9, proves that if in addition to (1), $K$ is Henselian, then $K$ is $p$-adically closed. Efrat [Ef5], Proof of Theorem A, (for $p \neq 2$) and Koenigsmann [Koe], Proposition 4.4, (for arbitrary $p$) prove that indeed, (1) implies that $K$ is Henselian. Hence, (1) implies that $p$-adically closed.

The concept of '$p$-adically closed field' has been extended by Prestel and Roquette [PrR] to take care of finite extensions of $\mathbb{Q}_p$. A valuation $v$ of a field $K$ of characteristic 0 is *generalized $p$-adic*[3] of *rank $d$* if its valuation ring $O$ satisfies $\dim_{\mathbb{F}_p} O/pO = d$. We say that $K$ is *generalized $p$-adically closed* if it admits a generalized $p$-adic valuation but no proper algebraic extension of $K$ admits a generalized $p$-adic valuation with the same rank. For example, every finite extension $K$ of $\mathbb{Q}_p$ is generalized $p$-adically closed with rank $[K : \mathbb{Q}_p]$ (This follows from the formula $d = ef$ [PrR], p. 15.)

Most of the results for $p$-adically closed fields generalize to generalized $p$-adically closed fields. For example, if $K$ is a finite extension of $\mathbb{Q}_p$ and $L$ is another field

[3] Prestel and Roquette use the term '$p$-adic'.

with $G(L) \cong G(K)$, then $L$ is a generalized $p$-adically closed field of rank $[K : \mathbb{Q}_p]$. However, there exist pairs $(E, F)$ of finite extensions of $\mathbb{Q}_p$ such that $G(E) \cong G(F)$ but $E \not\cong F$; hence $E$ and $F$ are not isomorphic [JaR], p. 10, and therefore not elementarily equivalent. (One has to use here [FrJ], Lemma 18.19, Krasner's Lemma, and a theorem of F.K. Schmidt.)

By the Chebotarev density theorem and a group theoretic argument [FrJ], Lemma 12.4, the intersection of $\mathbb{Q}_{p,\mathrm{alg}}$, where $p$ ranges over all primes and $\mathbb{Q}_{p,\mathrm{alg}}$ is a $p$-adic closure of $\mathbb{Q}$ (one for each $p$), is $\mathbb{Q}$. Suppose that $K$ is a field with $G(K) \cong G(\mathbb{Q})$. Then $\mathrm{char}(K) = 0$ (because $\mathrm{cd}_p(G(K)) \leqslant 1$ if $\mathrm{char}(K) = p$ while $\mathrm{cd}_p(G(\mathbb{Q})) \geqslant 2$). By the theorem of Neukirch, Pop, Efrat and Koenigsmann, $\widetilde{\mathbb{Q}} \cap K = \mathbb{Q}$.

CONJECTURE. *Let $K$ be a field of characteristic 0. Suppose that $G(L) \cong G(K)$ implies that $L$ is elementarily equivalent to $K$. Then $K$ is an algebraically closed field, a real closed field, or a finite abelian extension of $\mathbb{Q}_p$.*

The assumption that $\mathrm{char}(K) = 0$ is necessary. Indeed, Efrat [Ef5], Proposition 4.7, proves that for every field $K$ of positive characteristic there exists a field $L$ of characteristic 0 such that $G(L) \cong G(K)$. Of course, $K$ and $L$ are not elementarily equivalent.

Similarly, for each field $K$ there exists an algebraic extension $L$ of $K((t))$ such that $L/K$ is regular and $G(L) \cong G(K)$ [GJ1], Proposition 4.1. In particular $L$ is Henselian. It follows that $L$ is non-Hilbertian (§11.5).

A theorem of Prestel [Pr3], p. 200, gives another evidence to Conjecture: Let $K$ be an algebraic extension of $\mathbb{Q}$. Suppose that '$G(L) \cong G(K)$ and $L\widetilde{\mathbb{Q}} = \tilde{L}$' implies that $L$ is elementarily equivalent to $K$. Then $K$ is isomorphic to $\widetilde{\mathbb{Q}}$, or to $\widetilde{\mathbb{Q}} \cap \mathbb{R}$, or to an algebraic extension of $\mathbb{Q}_{p,\mathrm{alg}}$ for some prime number $p$.

## 10. Function fields of one variable

A *function field of one variable over a field $K$* (which we shorten in this section to just a *function field over $K$*) is a finitely generated regular extension $F$ of $K$ of transcendence degree 1. The elements of $K$ are referred to as *constants*.

**10.1.** *General field of constants.* The arithmetic of $F$ is due in the first place to the set of discrete valuations which are trivial on $K$. A *prime* $\mathfrak{p}$ of $F/K$ is an equivalence class of such valuations. The completion $F_\mathfrak{p}$ of $F$ with respect to $\mathfrak{p}$ is isomorphic to the field of formal power series $L((u))$, where $L = \overline{F}_\mathfrak{p}$ is the residue field and $u$ is a prime element of $F$ with respect to $\mathfrak{p}$. It is a discrete Henselian valued field, which is defectless [CaF], p. 19.

Here $F_{\mathfrak{p},\mathrm{ur}} = K_s F_\mathfrak{p}$. By 7.2,

$$\mathrm{res}\colon \mathcal{G}(F_{\mathfrak{p},\mathrm{ur}}/F_\mathfrak{p}) \to G(L)$$

is an isomorphism. If $\mathrm{char}(K) = 0$, then $F_{\mathfrak{p},\mathrm{tr}}$ is the algebraic closure of $F_{\mathfrak{p}}$ and $\mathcal{G}(F_{\mathfrak{p},\mathrm{tr}}/F_{\mathfrak{p},\mathrm{ur}}) \cong \widehat{\mathbb{Z}}$. If $\mathrm{char}(K) = p > 0$, then

$$\mathcal{G}(F_{\mathfrak{p},\mathrm{tr}}/F_{\mathfrak{p},\mathrm{ur}}) \cong \prod_{l \neq p} \mathbb{Z}_l$$

and the restriction map res: $G(F_{\mathfrak{p}}) \to \mathcal{G}(F_{\mathfrak{p},\mathrm{tr}}/F_{\mathfrak{p}})$ has a section (§7.2).

In the latter case one checks that the set $\{u^{-i} \mid p \nmid i\}$ is linearly disjoint over $F_p$ modulo the additive group $\{x^p - x \mid x \in K_s F_{\mathfrak{p}}\}$. By 6.1 the maximal pro-$p$ quotients of $G(F_{\mathfrak{p}})$ and of $G(F_{\mathfrak{p},\mathrm{ur}})$ are free pro-$p$ groups of infinite rank. Since, by definition, $p \nmid [F_{\mathfrak{p},\mathrm{tr}} : F_{\mathfrak{p},\mathrm{ur}}]$, the maximal pro-$p$ quotient of $G(F_{\mathfrak{p},\mathrm{ur}})$ is a quotient of $G(F_{\mathfrak{p},\mathrm{tr}})$. Hence, $\mathrm{rank}(G(F_{p,\mathrm{tr}})) = \infty$.

**10.2.** *Finite field of constants.* We now assume that $K$ is a finite field. Class field theory works for $F$ in the same way as for number fields [CaF], 162–203. In particular the idele class group of $F$ controls the abelian extensions of $F$. One application of class field theory follows Scholz and Reichardt and realizes every finite $l$-group over $F$, if $l \neq p$ and if $\zeta_l \notin F$ [GJ2]. One can probably follow Shafarevich and realize each $l$-group over $F$ also in the case where $\zeta_l \in F$.

**10.3.** *Function fields with isomorphic absolute Galois groups.* One of the distinctions between number fields and function fields over finite fields is that the latter have no smallest subfield. Thus, $K(t) \cong K(\sqrt{t})$ although these fields do not have the same degree over a common field, as is the case by number fields. Nevertheless, Uchida [Uc2] proves that if $F$ and $F'$ are function fields of one variable over finite fields (of possibly different characteristic) and $\Phi$: $G(F) \to G(F')$ is an isomorphism, then there is a unique isomorphisms of fields $\varphi$: $F_s \to F'_s$ such that $\Phi(\sigma) = \varphi\sigma\varphi^{-1}$ for each $\sigma \in G(F)$. In particular $\varphi(F) = F'$, and so every automorphism of $G(F)$ is inner.

Pop [Po1] proves the same result for function fields of one variable over number fields. He falls short however in proving the conjecture that if $F$ and $F'$ are fields which are finitely generated over their prime fields and $G(F) \cong G(F')$, then $F \cong F'$. Instead he adds in [Po2], Theorem 2, a certain structure to $G(F)$ and proves that if $G(F)$ and $G(F')$ have isomorphic structures, then $F \cong F'$.

**10.4.** *The absolute Galois group of* $\mathbb{C}(t)$. Algebraic topology teaches us that the fundamental group of a sphere punctured in $r$ points is generated by $r$ elements $\sigma_1, \ldots, \sigma_r$ with the single relation $\sigma_1 \cdots \sigma_r = 1$. The theory of Riemann surfaces and in particular *Riemann existence theorem* translates this result to a theorem about Galois groups over $\mathbb{C}(t)$:

Let $F$ be a finite Galois extension of $\mathbb{C}(t)$. Let $\mathfrak{p}_1, \ldots, \mathfrak{p}_r$ be the prime divisors of $\mathbb{C}(t)$ which are ramified in $F$. Then there exist generators $\sigma_1, \ldots, \sigma_r$ of $\mathcal{G}(F/\mathbb{C}(t))$ with $\sigma_1 \cdots \sigma_r = 1$ such that $\sigma_i$ generates an inertia group over $\mathfrak{p}_i$, $i = 1, \ldots, r$. Conversely, if $G$ is a finite group generated by $\sigma_1, \ldots, \sigma_r$ with $\sigma_1 \cdots \sigma_r = 1$, then $\mathbb{C}(t)$ has a finite Galois extension $F$ over $\mathbb{C}(t)$ which is unramified outside $S = \{\mathfrak{p}_1, \ldots, \mathfrak{p}_r\}$ such that $\sigma_i$ generates an inertia group over $\mathfrak{p}_i$, $i = 1, \ldots, r$ [Ma1], p. 31, Satz 1.

It is not difficult to replace $\mathbb{C}$ in this theorem by an arbitrary algebraically closed field $C$ of characteristic 0. One then works with profinite groups to prove that the Galois group of the maximal extension $C(t)_S$ of $C(t)$ which is unramified outside $S$ is generated by $r$ elements $\sigma_1, \ldots, \sigma_r$ with a single relation $\sigma_1 \cdots \sigma_r = 1$ [Vo2], Theorem 2.12. In particular $\mathcal{G}(C(t)_S/C(t))$ is isomorphic to the free profinite groups on $r-1$ generators. Finally, one lets $S$ range over all finite sets of prime divisors of $C(t)$ and proves that $G(C(t))$ is isomorphic to the free profinite group of rank equal to the cardinality $m$ of $C$ [Rib], pp. 70–80, or [Dou]. In particular, each finite group occurs as a Galois group over $C(t)$. Since open subgroups of $\widehat{F}_m$ are isomorphic to $\widehat{F}_m$ [JaL], p. 217, we have $G(F) \cong \widehat{F}_m$ for each finite algebraic extension $F$ of $C(t)$.

One may also start directly from a finite extension $F$ of $\mathbb{C}(t)$. Let $g$ be the genus of $F$ and let $S = \{\mathfrak{p}_1, \ldots, \mathfrak{p}_r\}$ be a set of $r$ prime divisors of $F$. Denote the maximal extension of $F$ which is unramified outside $S$ by $F_S$. Then $F_S/F$ is a Galois extension and $\mathcal{G}(F_S/F)$ is the group generated by $f + 2g$ generators $\sigma_1, \ldots, \sigma_r, \tau_1, \tau_1', \ldots, \tau_g, \tau_g'$ with the single relation

$$\sigma_1 \cdots \sigma_r [\tau_1, \tau_1'] \cdots [\tau_g, \tau_g'] = 1.$$

Moreover, for each $i$ between 1 and $r$ there exists a prime divisor $\mathfrak{P}_i$ of $F_S$ lying over $\mathfrak{p}_i$ such that $\sigma_i$ generates the inertia group of $\mathfrak{P}_i$ over $F$ [Ma1], p. 31, Satz 1 and p. 34, Satz 2.

**10.5.** *The absolute Galois group of* $\mathbb{R}(t)$. Krull and Neukirch [KrN] consider a finite set $S$ of prime divisors of $\mathbb{R}(t)$ and the maximal Galois extension $F_S$ of $\mathbb{R}(t)$ which is unramified outside $S$. They present $\mathcal{G}(F_S/\mathbb{R}(t))$ by generators and relations. Schuppar [Sch] replaces $\mathbb{R}$ in this result by an arbitrary real closed field $R$. If one lets $S$ range over all finite sets of prime divisors of $R(t)$, one presents $G(R(t))$ as a *real free* profinite group. More precisely, the set of involutions of $G(R(t))$ contains a closed subset $X$ which bijectively corresponds to the space of orderings of $R(t)$ and there exists a closed subset $Y$ of $G(R(t))$ which is disjoint to $X$, contains 1 and bijectively corresponds to the set $\{a + b\sqrt{-1} \mid a, b \in R \text{ and } b > 0\}$, such that the following holds [HJ1]:

Every continuous map $\varphi$ from $X \cup Y$ into a profinite group $G$ such that $\varphi(x)^2 = 1$ for each $x \in X$ and $\varphi(1) = 1$, uniquely extends to a homomorphism of $G(R(t))$ into $G$. The set $X \cup Y$ is said to be a *basis* for $G(R(t))$. One also says that $G$ is the *free product* of the groups in $\{\langle x \rangle \mid x \in X\} \cup \{\langle y \rangle \mid y \in Y\}$ [Ha2], p. 274.

**10.6.** *Realization of finite groups over* $\widetilde{\mathbb{F}}_p(t)$. Let now $C$ be an algebraically closed field of positive characteristic $p$. Grothendieck [Grt], XIII, Corollary 2.12, proves the analog of the Riemann existence theorem (10.4) for $C$ instead of $\mathbb{C}$ in the case where $p \nmid [F : C(t)]$. Raynaud [Ray] (for $r = 1$) and Harbater [Hr2] (for $r \geqslant 1$) prove a conjecture of Abhyankar [Abh]: Let $S = \{\mathfrak{p}_1, \ldots, \mathfrak{p}_r\}$ be a set of prime divisors of $C(t)$. Consider a finite group $G$ and denote the subgroup generated by all $p$-Sylow subgroups of $G$ by $p(G)$. Suppose that $\operatorname{rank}(G/p(G)) \leqslant r - 1$. Then $C(t)$ has a Galois extension $F$ with $\mathcal{G}(F/C(t)) \cong G$ which is unramified outside $S$. In particular, each finite group occurs as a Galois group over $C(t)$. Nevertheless, unlike in characteristic 0, the structure of $\mathcal{G}(C(t)_S/C(t))$ (in the notation of 10.4) is still unknown.

**10.7.** *The absolute Galois group of* $\widetilde{\mathbb{F}}_p(t)$. By 6.3 and in the notation of 10.6, $G(C(t))$ is projective. Since $C(t)$ is Hilbertian (by 11.3), this implies, at least if $C$ is countable, that the maximal prosolvable quotient of $G(C(t))$ is free. In other words, $\mathcal{G}(C(t)_{\text{solv}}/C(t)) \cong \widehat{F}_\omega(\text{solv})$ (by 11.12).

Actually, we know now much more. Harbater [Hr3], Theorem 4.1, uses formal patching to prove that if $C$ is an algebraically closed field (of an arbitrary characteristic) of cardinality $m$, then each finite embedding problem for $C(t)$ has $m$ solutions. This implies that $G(C(t))$ is the free profinite group of rank $m$ [Hr3], Theorem 4.4. In particular $G(\widetilde{\mathbb{F}}_p(t)) \cong \widehat{F}_\omega$.

Pop [Po5] proves the latter result by methods of rigid analytic geometry. The main tool in his proof is a certain strengthening of his $\frac{1}{2}$ Riemann existence theorem which we present below.

Haran and Völklein [HaV] give a third proof to the isomorphism $G(C(t)) \cong \widehat{F}_\omega$, where $C$ is a countable algebraically closed field. In addition to algebraic arguments, only basic properties of convergent power series with coefficients in $C((t))$ are used in the proof.

**10.8.** *Function fields over Henselian fields.* Pop's '$\frac{1}{2}$ Riemann existence theorem' considers a Henselian field $K$ with respect to a nontrivial valuation $v$ of rank 1. That is, $v(K^\times)$ is isomorphic to a subgroup of $\mathbb{R}$. Let $S$ be a finite set of prime divisors of $K(t)$, none of them is a pole of $t$. Denote the set of all extensions of $S$ to $\widetilde{K}(t)$ by $\widetilde{S}$. Suppose that the set of residues of the primes in $\widetilde{S}$ at $t$ can be ordered in pairs $(x_i, y_i)$, $i = 1, \ldots, n$ such that $v(x_i - y_i) > v(x_i - x_j)$ for all $i \neq j$. Let $\overline{K}$ be the residue field of $K$ at $v$. If $\text{char}(K) = 0$ and $\text{char}(\overline{K}) = p > 0$, define $e_i'$ to be the maximal integer satisfying

$$v(x_i - y_i) > \left(e_i' + \frac{1}{p-1}\right)v(p) + v(x_i - x_j) \quad \text{for all } i \neq j,$$

and let $e_i = \max\{0, e_i'\}$. Then $K(t)$ has a Galois extension $N$ which is ramified at most in $S$ and contains $K_s$. The Galois group $\mathcal{G}(N/K_s(t))$ is the free profinite group generated by elements $\sigma_{x_1}, \tau_{y_1}, \ldots, \sigma_{x_n}, \tau_{y_n}$ subjected to the relations $\sigma_{x_i}\tau_{y_i} = 1$, $i = 1, \ldots, n$, and to the condition

$$\langle \tau_{y_i} \rangle = \begin{cases} \widehat{\mathbb{Z}} & \text{if } \text{char}(K) = \text{char}(\overline{K}), \\ \mathbb{Z}/p^{e_i}\mathbb{Z} \times \prod_{l \neq p} \mathbb{Z}_l & \text{if } \text{char}(K) = 0 \text{ and } \text{char}(\overline{K}) = p > 0. \end{cases}$$

The element $\sigma_{x_i}$ (resp. $\tau_{y_i}$) belongs to an inertia group corresponding to $x_i$ (resp. $y_i$), $i = 1, \ldots, n$.

Moreover, $\mathcal{G}(N/K(t))$ is isomorphic to the semidirect product $\mathcal{G}(N/K_s(t)) \rtimes G(K)$ and the action of $G(K)$ on $\mathcal{G}(N/K_s(t)$ is defined by

$$(\sigma_{x_i})^\beta = (\sigma_{\sigma^{-1}x_i})^{\chi(\beta^{-1})}, \quad (\tau_{y_i})^\beta = (\tau_{\sigma^{-1}y_i})^{\chi(\beta^{-1})}, \quad \beta \in G(K).$$

Here $\chi\colon G(K) \to \widehat{\mathbb{Z}}^\times$ is the homomorphism defined by the action of the elements of $G(K)$ on the roots of unity.

The theorem describes a group which has approximately half of the rank of $\mathcal{G}(\widetilde{K}(t)_{\widetilde{S}}/\widetilde{K}(t))$ (where $\widetilde{K}(t)_{\widetilde{S}}$ is the maximal Galois extension of $\widetilde{K}(t)$ which is ramified at most at $\widetilde{S}$). Also, $\widetilde{S}$ is not an arbitrary finite subset of $\widetilde{K}$. So, it is not the full analog of Riemann existence theorem in characteristic 0.

Nevertheless, this theorem is strong enough to deduce that each Hilbertian PAC field $F$ is $\omega$-free (§12.9) and to describe the absolute Galois group of totally $p$-adic numbers as a free profinite product of copies of $G(\mathbb{Q}_p)$ (§13.12).

## 11. Hilbertian fields

Elementary Galois theory teaches us that the Galois group of the *general polynomial of degree* $n$,

$$f(\mathbf{T}, X) = X^n + T_1X^{n-1} + \cdots + T_n$$

is the symmetric group $S_n$. To explain this statement consider a polynomial $g$ of degree $n$ with coefficients in a field $F$ and assume that it has $n$ distinct roots $x_1, \ldots, x_n$. Then $\widehat{F} = F(x_1, \ldots, x_n)$ is a Galois extension of $F$ and $\mathcal{G}(\widehat{F}/F)$ permutes $x_1, \ldots, x_n$. This gives a *permutation representation* of $\mathcal{G}(\widehat{F}/F)$ into $S_n$. We denote the image of $\mathcal{G}(\widehat{F}/F)$ in $S_n$ by $\mathcal{G}(g, F)$. This is the *Galois group of $g$ over $F$*. The opening statement of this section then means that if $K$ is an arbitrary field, then $\mathcal{G}(f(\mathbf{T}, X), K(\mathbf{T})) \cong S_n$.

**11.1.** *Hilbert irreducibility theorem.* Hilbert [Hil] proved in 1892 that it is possible to specialize $\mathbf{T}$ to an $n$-tuple $\mathbf{a} \in \mathbb{Q}^n$ such that $\mathcal{G}(f(\mathbf{a}, X), \mathbb{Q}) \cong S_n$. By this he realized $S_n$ over $\mathbb{Q}$. More generally, he proved that given a polynomial $f \in \mathbb{Q}[\mathbf{T}, X]$ with distinct roots it is possible to specialize $\mathbf{T}$ to an $n$-tuple $\mathbf{a} \in \mathbb{Q}^n$ such that

$$\mathcal{G}(f(\mathbf{a}, X), \mathbb{Q}) \cong \mathcal{G}(f(\mathbf{T}, X), \mathbb{Q}(\mathbf{T}))$$

as permutation groups. This is one form of what we now call the *Hilbert irreducibility theorem*. It turns out that this theorem alone is responsible for much of the structure of $G(\mathbb{Q})$.

Hilbert himself and then others found that the same theorem holds for many other fields. Consequently, all of them share common features of their absolute Galois groups. They were therefore given the name 'Hilbertian fields'.

**11.2.** *Separable Hilbert sets.* Actually, the notion which is responsible for the structure of the Galois group is 'separably Hilbertian field'. To make a precise definition let $K$ be a field and consider separable irreducible polynomials $f_1, \dots, f_m \in K(T_1, \dots, T_r)[X]$ and a nonzero polynomial $g \in K[T_1, \dots, T_r]$. Denote the set of all $\mathbf{a} \in K^r$ such that $f_i(\mathbf{a}, X)$ is a separable irreducible polynomial in $K[X]$, $i = 1, \dots, m$, and $g(\mathbf{a}) \neq 0$ by $H_K(f_1, \dots, f_m; g)$ and call it a *separable Hilbert subset* of $K^r$ (or just *separable Hilbert set* of $K$). The field $K$ is *separably Hilbertian* if each of its separable Hilbert sets is nonempty.

If one omits the condition on the $f_i$ above to be separable, one obtains *Hilbert sets* of $K$. Then $K$ is *Hilbertian* if each of its Hilbert sets are nonempty. It turns out that $K$ is Hilbertian if and only if $K$ is separably Hilbertian and imperfect [FrJ], Proposition 11.16. Also, denote the maximal purely inseparable extension of $K$ by $K_{\text{ins}}$. A simple observation shows that if $K$ is Hilbertian, than $K_{\text{ins}}$ is separably Hilbertian. Since res: $G(K_{\text{ins}}) \to G(K)$ is an isomorphism, one may always assume for the study of the absolute Galois group that $K$ is perfect.

**11.3.** *Examples of Hilbertian fields.* The following fields are Hilbertian: $\mathbb{Q}$ (Hilbert [FrJ], Corollary 12.8, or [La1], p. 148), $K_0(T)$ [FrJ], Theorem 12.9, and $K_0((T_1, \dots, T_r))$ for $r \geqslant 2$ (Weissauer [FrJ], Example 14.3) for an arbitrary field $K_0$. If $L$ is a finite extension of a separably Hilbertian field, then each separable Hilbert subset of $L^r$ contains a separable Hilbert subset of $K^r$ [FrJ], Corollary 11.7. The same is true if $L$ is a Galois extension of $K$ and $\mathcal{G}(L/K)$ is finitely generated [FrJ], Proposition 15.5. In particular $L$ is separably Hilbertian. If $N$ is a Galois extension of a separably Hilbertian field and $M$ is a finite proper extension of $N$, then $M$ is separably Hilbertian (Weissauer [FrJ], Corollary 12.15). If $L$ is an abelian extension of a separably Hilbertian field, then $L$ is separably Hilbertian (Kuyk [FrJ], Theorem 15.6). The compositum of two Galois extensions of a separably Hilbertian field neither of which contains the other is Hilbertian [HJ3]. If $L$ is an algebraic extension of a separably Hilbertian field $K$ whose degree is divisible by at least two primes and $L$ is contained in a pronilpotent extension $N$ of $K$, then $L$ is separably Hilbertian [Uc3], Theorem 3.

**11.4.** *Regular realization of finite groups.* Hilbert himself proves in [Hil] that if $f \in K[\mathbf{T}, X]$ is a separable polynomial, then the set of all $\mathbf{a} \in K^r$ such that

$$\mathcal{G}(f(\mathbf{a}, X), K) \cong \mathcal{G}(f(\mathbf{T}, X), K(T))$$

contains a Hilbert set of $K$ [FrJ], Lemma 12.12. Thus, if $K$ is a separably Hilbertian field, each finite group which occurs over $K(\mathbf{T})$ as a Galois group also occurs over $K$ as a Galois group. More interesting is the case where $f(\mathbf{T}, X)$ is an absolutely irreducible polynomial which is *Galois* in $X$. The latter means that the splitting field $\widehat{F}$ of $f(\mathbf{T}, X)$ over $K(\mathbf{T})$ is already generated by each single root of $f(\mathbf{T}, X)$. Hence

$$G = \mathcal{G}(f(\mathbf{T}, X), K(\mathbf{T})) \cong \mathcal{G}(f(\mathbf{T}, X), L(\mathbf{T}))$$

for each algebraic extension $L$ of $K$. Equivalently, $\widehat{F}$ is a *regular extension* of $K$, i.e. $\widehat{F}$ is linearly disjoint from $\widetilde{K}$ over $K$.[4] We then say that $f$ is *stable with respect to* $X$ and that $G$ is *regular over* $K$. (One may also say that $G$ has a *$K$-regular realization over* $K(\mathbf{T})$.) In this case $K$ has a linearly disjoint sequence of Galois extensions $L_1, L_2, L_3, \ldots$ such that $\mathcal{G}(L_i/K) \cong G$ [FrJ], Lemma 15.8. This rule applies in particular to $S_n$, to $A_n$ [Se4], Section 5.5, (at least if $p \nmid n(n-1)$ where $p = \mathrm{char}(K)$) and to each finite abelian group [FrJ], Lemma 24.46. Note also, that if $G$ is regular over a field $K$, then $G$ is also regular over every extension of $K$.

**11.5.** *On the absolute Galois group of a Hilbertian field.* In particular $G(K)$ has an infinite rank and $G(K)$ is not prosolvable. It follows that $K_s$, the maximal pro-$p$ extension $K^{(p)}$, and the maximal prosolvable extension $K_{\mathrm{solv}}$ of $K$ are not Hilbertian. Hence, by 11.3, none of these fields is the compositum of two Galois extensions of $K$ neither of which contains the other. Moreover, Weissauer's theorem implies that $G(K)$ has no normal prosolvable closed subgroup. In particular, the Frattini subgroup of $G(K)$ is trivial, i.e. the compositum of all minimal separable algebraic extensions of $K$ is $K_s$. Here a proper algebraic extension of $K$ is said to be *minimal* if there exists no field $K'$ such that $K \subset K' \subset L$. Also, the center of $G(K)$ is trivial [FrJ], p. 186, Theorem 15.10.

Note also, that no Henselian field can be separably Hilbertian [FrJ], p. 181, Exercise 8. In particular local fields and fields of formal power series of one variable are not Hilbertian.

**11.6.** *Embedding problems.* Let $K$ be a field and let $t_1, \ldots, t_n$ be independent indeterminates. Let $L$ be a finite Galois extension of $K$. Then each epimorphism $\alpha\colon H \to \mathcal{G}(L/K)$ gives rise to two embedding problems:

$$\big(\mathrm{res}\colon G(K) \to \mathcal{G}(L/K),\ 1 \longrightarrow C \longrightarrow H \stackrel{\alpha}{\longrightarrow} \mathcal{G}(L/K) \longrightarrow 1\big) \tag{1a}$$

$$\big(\mathrm{res}\colon G\big(K(\mathbf{t})\big) \to \mathcal{G}(L/K),\ 1 \longrightarrow C \longrightarrow H \stackrel{\alpha}{\longrightarrow} \mathcal{G}(L/K) \longrightarrow 1\big) \tag{1b}$$

We call (1a) an *embedding problem* for $K$. We call (1b) a *constant field extension embedding problem* for $K(\mathbf{t})$. If $\gamma$ is a solution of (1a) and $M$ is the fixed field in $K_s$ of $\mathrm{Ker}(\gamma)$, then $M$ is a Galois extension of $K$ which contains $L$ and $\gamma$ induces an isomorphism $\bar{\gamma}\colon \mathcal{G}(M/K) \to H$ such that $\alpha \circ \bar{\gamma} = \mathrm{res}$. We call $M$ a *solution field* for (1a). Note that the map res: $\mathcal{G}(L(\mathbf{t})/K(\mathbf{t})) \to \mathcal{G}(L/K)$ is an isomorphism. A *solution field* for (1b) is therefore a Galois extension $F$ of $K(\mathbf{t})$ which contain $L(\mathbf{t})$. We say that it is *regular* if $F/L$ is a regular extension.

If $K$ is Hilbertian, and (1b) has a solution $F$, then so does (1a). If in addition the solution $F$ is regular, then it can be specialized to an infinite sequence $M_1, M_2, M_3, \ldots$ of solutions of (1a) which are linearly disjoint over $L$.

**11.7.** *Abelian kernels.* Suppose now that the embedding problems (1) split and their kernel $C$ is abelian. Then (1b) has a regular solution [FrJ], Lemma 24.46, and therefore

[4] Some authors (e.g., Serre [Se4], Section 4.1) use the expression '$\widehat{F}$ is regular over $K(\mathbf{T})$' to mean that $\widehat{F}$ is regular over $K$.

(1a) has a solution. In particular, every finite Abelian group appears as a Galois group over $K$.

As a consequence consider an automorphism $\Phi$ of $G(K)$ which maps each $\sigma \in G(K)$ to a conjugate of $\sigma$. That is, $\Phi$ is *locally inner*. The proof of [Uc1], Lemma 3, then shows that $\Phi$ is an inner automorphism.

Nonsplit finite embedding problems with an abelian kernel for a Hilbertian field $K$ are not always solvable. For example, $\mathbb{Q}$ has no Galois group $L$ that contains $\mathbb{Q}(\sqrt{-1})$ with $\mathcal{G}(L/\mathbb{Q}) \cong \mathbb{Z}/4\mathbb{Z}$. In general, for an arbitrary field $K$ of characteristic $\neq 2$, a quadratic extension $K(\sqrt{a})$ can be embedded in a cyclic extension of degree 4 if and only if $a$ is a sum of two squares in $K$ [Se4], Theorem 1.2.4.

**11.8.** *Wreath products.* Let $G$ and $C$ be finite groups. Consider the direct product

$$C^G = \prod_{\sigma \in G} C^\sigma$$

of $|G|$ copies of $C$. Let $G$ acts on $C^G$ by $(c^\sigma)^\tau = c^{\sigma\tau}$. Then, the semidirect product $C^G \rtimes G$ is known as the *wreath product* $H = C \operatorname{wr} G$ of $C$ and $G$. In the notation of (11.6) (with $C$ replaced by $C^G$) let $G = \mathcal{G}(L/K)$ and $\alpha$: $H \to G$ be the projection on the second factor. Assume that $C$ is regular over $K$. Then (1b) has a regular solution (e.g., [Ku2], p. 114, [Ma1], p. 228, or [HJ2], the proof of Part B of Lemma 2.1). In particular, if $K$ is Hilbertian, then (1a) has a solution.

Kuyk [Ku1], Theorem 3, uses this construction to prove that each embedding problem (1a) over a Hilbertian field $K$ can be solved after a certain 'shift'. That is, there exists a finite extension $K'$ of $K$ which is linearly disjoint from $L$ over $K$ such that the embedding problem (res: $G(K') \to \mathcal{G}(LK'/K')$, $\alpha$: $H \to \mathcal{G}(LK'/K')$) is solvable (see also [Ja3], Theorem 15.1). He then applies it to prove that each profinite group $G$ of rank at most $\aleph_0$ occurs as a Galois group of a Galois extension $F/E$ for some separable algebraic extension $E$ of $K$ [Ku1], Theorem 4.

**11.9.** *GAR realizations.* If the kernel $C = \operatorname{Ker}(\alpha)$ of the embedding problem (1b) is regular over $K$ with some additional properties, then (1b) has a regular solution. We follow Matzat [Ma1] and Völklein [Vo1] and consider a finite group $C$ with a trivial center and a field $K$. We say that $C$ is $GA_r$ (resp. $GAR_r$, $GAT_r$) over $K$ (we leave out the subscript $r$ unless we wish to specify it) if there exist algebraically independent elements $t_1, \ldots, t_r$ over $K$ such that Condition (GA) (resp. (GA) and (R), (GA) and (T)) below is satisfied:

(GA) $K(\mathbf{t})$ has a subfield $E$ and an extension $F$ such that $F/K$ is regular, $F/E$ is Galois and there exists an isomorphism of $\mathcal{G}(F/E)$ onto $\operatorname{Aut}(C)$ which maps $\mathcal{G}(F/K(\mathbf{t}))$ onto $\operatorname{Inn}(C) \cong C$.

(R) If a finite extension $E'$ of $E$ satisfies $K_sE' = K_s(\mathbf{t})$, then $E'$ is purely transcendental over $E' \cap K_s$.[5]

(T) The $K$ vector space spanned by $t_1, \ldots, t_r$ is invariant under $\mathcal{G}(K(\mathbf{t})/E)$.

[5] The present formulation of Condition (R) is taken from a new book of Matzat and differs from the one that appears on p. 234 of [Ma1].

Matzat [Ma1], p. 235, proves that if a nontrivial finite group $C$ is $\mathrm{GAR}_r$ over a field $K$, then each constant field extension embedding problem (1b) for $K(\mathbf{t})$ has a regular solution. In general it is difficult to prove that $C$ is GAR over $K$. However, one can apply induction on the order of $C$ and assume that it is a minimal normal subgroup of $H$. Then

$$C \cong \prod_{i=1}^{r} C_i$$

where $C_i$ are isomorphic finite simple groups. Matzat [Ma1], p. 243, proves then that if $C_1$ is $\mathrm{GAR}_1$ over $K$ and $C_1$ is nonabelian simple, then $C$ is $\mathrm{GAR}_r$ over $K$ and embedding problem (1b) has a regular solution.

**11.10.** *Free pro-$\mathcal{D}$ groups.* Consider now a family $\mathcal{D}_0$ of finite simple groups and let $\mathcal{D}$ be the family of all finite groups whose composition factors are in $\mathcal{D}_0$. Construct the *free pro-$\mathcal{D}$ group* $\widehat{F}_\omega(\mathcal{D}) = \varprojlim F_\omega/N$, where $F_\omega$ is the free discrete group on $\aleph_0$ generators and $N$ ranges over all normal subgroups of $F_\omega$ with $F_\omega/N \in \mathcal{D}$. As in the case of $\widehat{F}_\omega$, a pro-$\mathcal{D}$ group $F$ of at most a countable rank is isomorphic to $\widehat{F}_\omega(\mathcal{D})$ if and only if each finite embedding problem for $F$ with kernel in $\mathcal{D}$ is solvable (use [Me1], Lemma 2.2).

Suppose now that $K$ is a Hilbertian field and take $\mathcal{D}_0$ to be the family $\mathrm{GAR}(K)$ of all finite nonabelian simple groups which are GAR over $K$. It follows from Matzat's theorem, that $\widehat{F}_\omega(\mathcal{D})$ occurs as a Galois group over $K$.

**11.11.** *Examples of GAR-realizations.* Matzat [Ma2], Satz 11.4(b), lists all finite simple groups which were known by 1990 to be GAR over $\mathbb{Q}$. Among them are: $A_n$ for $n \neq 6$, several other one parameter families of classical simple groups of Lie-type like $\mathrm{PSL}_2(\mathbb{F}_p)$ for $p \not\equiv \pm 1 \mod 24$, $\mathrm{PSL}_{2n+1}(p)$ for $p \not\equiv -1 \mod 12$ and all sporadic simple groups except $\mathrm{M}_{23}$.

Völklein [Vo1] proves that if a finite group $C$ is GAT over a field $K$, then $C$ is also GAR over $K$. He notes that if $C$ is GAT over $K$, then it is GAT, hence also GAR, over each extension of $K$. He then gives a two 2-parameter families of simple groups which are GAT over $\mathbb{Q}$. These are all the groups $\mathrm{PSL}_n(4^s)$ and $\mathrm{PU}_n(4^s)$, where $s$ is odd and $n \geqslant \max\{4, 4^{s-1}\}$ is an even integer such that $\gcd(n, 4^s - 1) = 1$ [Vo1], Section 2.4, Corollary and Section 2.5, Corollary. These are the first GAR-realizations over $\mathbb{Q}$ of nonabelian simple groups with an arbitrarily large outer automorphism group. Völklein also proves that if $q$ is an odd prime power and $n \geqslant q$, then the 'almost simple groups' $\mathrm{PGL}(q)$ and $\mathrm{PU}_n(q)$ are GAT over $\mathbb{Q}$. Here a group $G$ is *almost simple* if $G$ lies between a simple group and its automorphism group.

**11.12.** *Projective absolute Galois groups.* Embedding problems for $K$ become much easier if $G(K)$ is projective. Indeed, each finite embedding problem for $G(K)$ can then be reduced to a split finite embedding problem with the same kernel [Ma1], p. 231. If in addition, $K$ is Hilbertian, this leads to the solution of each finite embedding problem for $G(K)$ with abelian kernel. Thus, if in addition, $K$ is countable, then $G(K_{\text{solv}}/K) \cong \widehat{F}_\omega(\text{solv})$ [FrJ], Theorem 24.50.

Even finite embedding problems with nonabelian kernel profit from the assumption '$G(K)$ is projective'. Thus, if a finite nonabelian group $C$ is $\text{GA}_1$ over $K$, then it is also GAR [Ma1], p. 238 (at least if $\text{char}(K) \neq 2$). It follows that if this were the case for each such $C$ and $K$ is Hilbertian and countable, then $G(K) \cong \widehat{F}_\omega$. The latter consequence is a conjecture of Fried and Völklein [FV2], p. 470. However, not every finite simple group $S$ is $\text{GA}_1$ over $K$, because $\text{Out}(S)$ does not always embed into $\text{PGL}_2(K)$. So, one must think of an alternative way to prove the conjecture of Fried and Völklein.

**11.13.** *The field* $\mathbb{Q}_{\text{ab}}$. The most prominent case of a countable Hilbertian field with projective absolute Galois group is that of $\mathbb{Q}_{\text{ab}}$ ([FrJ], Theorem 15.6, and §8.2). The result $G(\mathbb{Q}_{\text{ab,solv}}/\mathbb{Q}_{\text{ab}}) \cong \widehat{F}_\omega(\text{solv})$ is due to Iwasawa [Iw3], Theorem 6, and the conjecture that $G(\mathbb{Q}_{\text{ab}}) \cong \widehat{F}_\omega$ is due to Shafarevich [Bel]. Many more finite nonabelian simple groups belong to $\text{GAR}(\mathbb{Q}_{\text{ab}})$ than to $\text{GAR}(\mathbb{Q})$. They include all $A_n$, $\text{PSL}_2(p)$, other one parameter families [Ma2], Satz 11.3, all sporadic simple groups, and two 2-parameter families: $\text{PSL}_n(4^s)$ and $\text{PU}_n(4^s)$ mentioned in 11.11.

In the next section we discuss an important case in which the conjecture of Fried and Völklein is true.

## 12. PAC fields

The existence of a $K$-rational point on each absolutely irreducible algebraic variety has a decisive influence on the absolute Galois group of a field $K$ and on its model theory. This section describes the main consequences of this assumption.

**12.1.** DEFINITION. A field $K$ is *pseudo algebraically closed* (abbreviated *PAC*) if one of the following equivalent conditions is satisfied:

(1a) Each nonempty absolutely irreducible variety $V$ defined over $K$ has a $K$-rational point.

(1b) For each function field $F$ of several variables which is regular over $K$ there exists a place $\varphi\colon F \to K \cup \{\infty\}$ which fixes each element of $K$.

(1c) For each absolutely irreducible polynomial $f \in K[X,Y]$ there exist $a, b \in K$ such that $f(a,b) = 0$ (Frey and Geyer [FrJ], Lemma 10.3).

**12.2.** *First examples.* In particular every separably closed field is PAC. The first non-trivial examples for PAC fields come from Weil's theorem on rational points of absolutely irreducible varieties over finite fields. It implies that each infinite algebraic extension of a finite field and each nonprincipal ultraproduct of distinct finite fields is PAC (Ershov [FrJ], Corollary 10.5, and Ax [FrJ], Corollary 10.6). The method of Descent due to Weil implies that each algebraic extension of a PAC field is PAC (Ax and Roquette [FrJ], Corollary 10.7).

**12.3.** *The fields* $K_s(\sigma)$. Each separably Hilbertian field $K$, in particular $\mathbb{Q}$, has a host of algebraic extensions which are PAC. The first examples for such fields have a probabilistic nature. Each profinite group $G$ and in particular $G(K)$ has a unique Haar measure $\mu$ such that $\mu(G) = 1$. We denote the fixed field in $K_s$ of $\sigma_1, \ldots, \sigma_e \in G(K)$ by $K_s(\boldsymbol{\sigma})$. If $K$ is countable and we take $\sigma_1, \ldots, \sigma_e$ at random (that is we leave out a set of measure 0), then $K_s(\boldsymbol{\sigma})$ is PAC [FrJ], Theorem 16.18. Moreover, $G(K_s(\boldsymbol{\sigma})) = \langle \sigma_1, \ldots, \sigma_e \rangle \cong \widehat{F}_e$.

The latter result holds without the restriction on $K$ to be countable [FrJ], Theorem 16.13, while the former one may become false [FrJ], Corollary 16.37. Also, if $\sigma_1, \ldots, \sigma_e$ are not chosen at random, then the above conclusion may be false. For example, if $\sigma$ is the restriction to $\widetilde{\mathbb{Q}}$ of the complex conjugation, then $\widetilde{\mathbb{Q}}(\sigma)$ is a real closure of $\mathbb{Q}$. As such, it is not PAC [FrJ], Theorem 10.17.

**12.4.** *The fields $\widetilde{K}[\boldsymbol{\sigma}]$.* Choose again $\sigma_1, \ldots, \sigma_e \in G(K)$ at random. Let $K_s[\boldsymbol{\sigma}]$ be the maximal Galois extension of $K$ which is contained in $K_s(\boldsymbol{\sigma})$ and let $\widetilde{K}[\boldsymbol{\sigma}] = K_s[\boldsymbol{\sigma}]_{\mathrm{ins}}$. Then $\widetilde{K}[\boldsymbol{\sigma}]$ is PAC and $G(\widetilde{K}[\boldsymbol{\sigma}]) \cong \widehat{F}_\omega$ [Ja4], Theorem 2.7. Hence $\widetilde{K}[\boldsymbol{\sigma}]$ is separably Hilbertian (§12.9).

**12.5.** *Stable fields.* The basic fact that allows Galois extensions of separably Hilbertian fields to be PAC is the 'stability of fields'. A field $K$ is *stable* if each finitely generated regular extension $F$ of $K$ of transcendence degree $r$ has elements $t_1, \ldots, t_r$ which are algebraically independent over $K$ such that $F/K(\mathbf{t})$ is separable and its Galois closure $\widehat{F}$ is regular over $K$. It is known that every infinite perfect field is stable [GJ3], Corollary I. Also, every PAC field is stable [FrJ], Theorem 16.41. Finally, Neumann's Thesis [Neu] proves that every field is stable.

**12.6.** *Symmetric extensions of $K$.* Suppose now that $K$ is a perfect countable separably Hilbertian field. Then $K$ has a Galois extension $N$ which is PAC and

$$\mathcal{G}(N/K) \cong \prod_{n=1}^{\infty} S_n.$$

By Weissauer's theorem (§11.3) $N$ is also separably Hilbertian (Remark 1 on p. 476 of [FV2] which proves this result for $\mathrm{char}(K) = 0$ also holds in the general case). Following this example, denote the compositum of all finite Galois extensions of $K$ with symmetric (resp. alternating) Galois group by $K_{\mathrm{symm}}$ (resp. $K_{\mathrm{alt}}$). Then $G(K_{\mathrm{symm}})$ is a closed characteristic subgroup of $G(K)$. By the extension theorem for PAC fields, $K_{\mathrm{symm}}$ is PAC. Moreover, it is Hilbertian. It is not known if $K_{\mathrm{alt}}$ is PAC [FV2], p. 176.

**12.7.** *Henselian fields are not PAC.* If $v$ is a valuation of a PAC field $K$, then the Henselian closure of $K$ with respect to $v$ is separably closed (Frey and Prestel [FrJ], Theorem 10.14). This implies that $\mathbb{Q}_{\mathrm{ab}}$ and $\mathbb{Q}_{\mathrm{nil}}$ (= the maximal pronilpotent extension of $\mathbb{Q}$) are not PAC [FrJ], Corollary 10.15. However, all Henselian closures of $\mathbb{Q}_{\mathrm{solv}}$ are algebraically closed. So, it is not known if $\mathbb{Q}_{\mathrm{solv}}$ is PAC [FrJ], Problem 10.16.

**12.8.** *Regular realizations.* Harbater [Hr1], Theorem 2.3, considers a complete local domain $R$ with a quotient field $F$ such that $R \neq F$. He proves that each finite group $G$ is regular over $F$. In particular, $G$ is regular over $\mathbb{Q}_p$ and over $K((t))$, for an arbitrary base field $K$. Haran and Völklein [HaV] supply a much simpler proof of the latter result (see also 10.7). In particular, if $K$ is PAC, then, by the Bertini–Noether theorem, $G$ is also regular over $K$ [Ja5], Theorem 2.6. Hence, if in addition $K$ is also Hilbertian, then $G$ occurs over $K$ as a Galois group. Since a nonprincipal ultraproduct of distinct finite fields is PAC, there exists $q_0 = q_0(G)$ such that if $q \geqslant q_0$ is a prime power, then $G$ is regular over $\mathbb{F}_q$ [Ja9] and [FV1], p. 784, Corollary 2. No upper bound is known on $q_0(G)$. Fried [Fri], Proposition E4, says that $q_0(G)$ is computable but gives no concrete formula for it.

**12.9.** *Hilbertian and PAC imply $\omega$-free.* The regularity of finite groups over Hilbertian PAC fields of characteristic 0 was first obtained by Fried and Völklein by complex analytic methods. Starting from the Riemann existence theorem, they construct for each finite group $G$ an absolutely irreducible variety $\mathcal{H}$ (that they call *Hurwitz space*) which is defined over $\mathbb{Q}$, such that if $K$ is a field of characteristic 0 and $\mathbf{q} \in \mathcal{H}(K)$, then $G$ is regular over $K$ [FV1], p. 772. The existence of $\mathbf{q} \in \mathcal{H}(K)$ is guaranteed if $K$ is in addition PAC.

[FV2], p. 474, improves on that and proves that each constant field extension embedding problem over a PAC field $K$ of characteristic 0 has a regular solution. If in addition, $K$ is Hilbertian, then every finite embedding problem for $K$ is solvable. Thus, by Iwasawa's theorem, if in addition $K$ is countable, then $G(K) \cong \widehat{F}_\omega$.

Pop [Po4], Theorem 1, applies his '$\frac{1}{2}$ Riemann existence theorem' (§10.8) to prove that the implication '$K$ is countable PAC and Hilbertian implies $G(K) \cong \widehat{F}_\omega$' holds in general.

The converse of this result is also true. If $K$ is a PAC field with $G(K) \cong \widehat{F}_\omega$, then $K$ is separably Hilbertian (Roquette [FrJ], Corollary 24.38).

**12.10.** *Characterization of projective groups.* The absolute Galois group of a PAC field is projective (Ax and Haran [FrJ], Theorem 10.17). Conversely, if $G$ is a projective group and $K$ is a field, then there exists an extension $F$ of $K$ which is PAC such that $G(F) \cong G$ (Lubotzky and v.d. Dries [FrJ], Corollary 20.16). Moreover, if $L/K$ is a Galois extension and $\alpha\colon G \to \mathcal{G}(L/K)$ is an epimorphism, then $F$ can be chosen together with an isomorphism $\gamma\colon G(F) \to G$ such that $\gamma \circ \alpha = \mathrm{res}$. If in addition, $\mathrm{rank}(G) \leqslant \aleph_0$, $K$ is a countable Hilbertian field and $L/K$ is finite, then $F$ can be chosen to be separable algebraic over $K$ [FrJ], Proposition 20.21. Replacing the latter $F$ by $F_{\mathrm{ins}}$ we may also assume that $F$ is algebraic over $K$ and perfect.

The characterization of projective groups as absolute Galois groups of PAC fields makes it possible to interpret the theory of finite graphs in the theory of PAC fields. Since the former one is nonrecursive (i.e. undecidable), so is the latter. Thus, there is no recursive decision procedure to determine whether a given sentence of the language of rings is true in all PAC fields (Ershov and Cherlin, v.d. Dries and Macintyre [FrJ], Section 22.10).

**12.11.** *$C_1$-fields.* A field $K$ is a *$C_1$-field* if each form of degree $d$ in more than $d$ variables over $K$ has a nontrivial $K$-zero. The absolute Galois group of $K$ is then projective [Rib], p. 269. If $K$ is a perfect PAC field and $G(K)$ is abelian (hence, procyclic), then

$K$ is $C_1$ (Ax [FrJ], Theorem 19.16). Ax' problem, whether each perfect PAC field is $C_1$ is still open. On the other hand, there are $C_1$ fields, like finite fields (a theorem of Chevalley), which are not PAC fields.

**12.12.** *The algebraic nature of the theory of PAC fields.* Let $K$ be a countable Hilbertian field. The characterization of projective groups of at most countable rank as the absolute Galois groups of perfect PAC fields which are algebraic over $K$ implies that the theory of perfect PAC fields which contain $K$ coincides with the theory of perfect PAC fields which are algebraic over $K$ [FrJ], Corollary 20.25.

**12.13.** *The elementary equivalence theorem.* Let $K$ be a Hilbertian field and let $F$ be a perfect PAC which contains $K$. Then the elementary class of $F$ is determined by the equivalence class of the map res: $G(F) \to G(K)$. Thus, if $F'$ is another perfect PAC field that contains $K$ and there exists an isomorphism $\Phi$: $G(F) \to G(F')$ and $\varphi \in G(K)$ such that $\mathrm{res}_{K_s} \circ \Phi(\sigma) = \mathrm{res}_{K_s}(\sigma)^{\varphi}$ for each $\sigma \in G(F)$, then $F$ and $F'$ are elementarily equivalent as structures of $\mathcal{L}(\mathrm{ring}, K)$ [FrJ], Theorem 18.6. Note that $\varphi$ maps $K_s \cap F$ onto $K_s \cap F'$.

**12.14.** *Frobenius fields.* There is one type of PAC fields where the isomorphism class of the absolute Galois group and the algebraic part of the fields determine the equivalence class of the field. To this end we say that a profinite group $G$ has the *embedding property* if every finite embedding problem ($\varphi$: $G \to A$, $\alpha$: $B \to A$) in which $B$ is a quotient of $G$ is solvable. We say that $F$ is a *Frobenius field* if it is PAC and $G(F)$ has the embedding property. The elementary equivalence theorem then implies that if $F$ and $F'$ are two perfect Frobenius fields which contain $K$, then $F$ and $F'$ are elementarily equivalent as structures of $\mathcal{L}(\mathrm{ring}, K)$ if and only if same finite groups occur as Galois groups over both $F$ and $F'$, and $K_s \cap F \cong K_s \cap F'$.

It turns out that the theory of all Frobenius fields (resp. of a given characteristic) is decidable. Indeed, there is a primitive recursive procedure called *Galois Stratification* which is based on explicit Galois theory and Algebraic Geometry which, in some sense, eliminates quantifiers and allows one to determine for each given sentence of $\mathcal{L}(\mathrm{ring})$ whether it holds, say, in all Frobenius fields [FrJ], Theorem 25.11.

Examples of Frobenius fields are PAC fields whose absolute Galois groups are free. Thus, for each $m$ between 1 and $\omega$, and each prime $p$, Galois Stratification gives a primitive recursive decision procedure for the class of PAC fields $F$ of characteristic $p$ such that $G(F) \cong \widehat{F}_m$ [FrJ], Theorems 25.15 and 25.17.

**12.15.** *The probability of a sentence to be true.* For a countable Hilbertian field $K$, a sentence $\theta$ of $\mathcal{L}(\mathrm{ring}, K)$ and a positive integer $e$, denote the set of all $\boldsymbol{\sigma} \in G(K)^e$ such that $\theta$ is true in $\widetilde{K}(\boldsymbol{\sigma})$ by $S(\theta)$. Denote the Haar measure of $G(K)^e$ by $\mu$. Galois stratification implies that $\mu(S(\theta))$ is a rational number and allows us to explicitly compute it if $K$ is a given global field.

In the latter case and for $e = 1$, let $A(\theta)$ be the set of all primes of $K$ such that $\theta$ is true in the residue field $\overline{K}_{\mathfrak{p}}$. The transfer theorem says that the Dirichlet density $\delta(A(\theta))$ of $A(\theta)$ is equal to $\mu(S(\theta))$. (This is one of the main results of the thesis of the author [FrJ], Theorem 18.26.) Moreover, this number is positive if and only if $A(\theta)$ is an infinite set (Ax [FrJ], Theorem 18.27).

Thus Galois Stratification gives a primitive recursive procedure for the theory of all sentences which are true in all but finitely many residue fields $\overline{K}_{\mathfrak{p}}$. This is not very far from establishing a primitive recursive procedure for the theory of all finite fields [FrJ, Theorem 26.9]. This improves earlier recursive procedures of Ax [FrJ], Corollary 18.28 and Theorem 18.29.

## 13. Pseudo closed fields

Pseudo algebraically closed fields lack any kind of arithmetic. That is, they have no orderings and all their Henselizations are separably closed [FrJ], Theorems 10.12 and 10.14. In this section we generalize the concept of PAC fields and bring arithmetic into the game. The new definitions will involve local global principles.

**13.1.** P$\mathcal{K}$C *fields.* Let $K$ be a field and let $\mathcal{K}$ be a family of algebraic extensions of $K$. We say that $K$ is *pseudo $\mathcal{K}$-closed* (and abbreviate it by P$\mathcal{K}$C) if every nonempty absolutely irreducible variety $V$ defined over $K$ with a simple $\overline{K}$-point for each $\overline{K} \in \mathcal{K}$ has a $K$-rational point.

If $\mathcal{K} \subseteq \{K_s\}$, then $K$ is PAC. If $\mathcal{K}$ is the family of all real closures of $K$, then $K$ is *PRC* (*pseudo real closed*). If $\mathcal{K}$ is the family of all $p$-adic closures of $K$ ($p$ a fixed prime), then $K$ is *PpC* (*pseudo p-adically closed*).

In each of these cases the family $\mathcal{K}$ is closed in the space of separable algebraic extensions of $K$ (which we denote by Sext($K$)). A basic open neighborhood for $E \in$ Sext($K$) in this space is determined by a finite Galois extension $N$ of $K$. It is the set of all separable algebraic extensions of $K$ whose intersection with $N$ is $E \cap N$. Thus this space is the inverse limit of the finite spaces of all intermediate fields between $K$ and $N$. In other words, it is a *profinite space* (also called *Boolean space*), and as such it is compact.

Also, each of the above families is closed under the action of $G(K)$.

**13.2.** *Relative projective groups.* Let $\mathcal{D}$ be a set of closed subgroups of a profinite group $G$. A *weak $\mathcal{D}$-embedding problem* for $G$ is a weak embedding problem ($\varphi$: $G \to A$, $\alpha$: $B \to A$), such that for each $H \in \mathcal{D}$ there exists a homomorphism $\gamma_H$: $H \to B$ such that $\alpha \circ \gamma_H = \varphi|_H$. We say that $G$ is *$\mathcal{D}$-projective* if each finite weak $\mathcal{D}$-embedding problem is weakly solvable. That is, there exists a homomorphism $\gamma$: $G \to B$ such that $\alpha \circ \gamma = \varphi$. In other words, finite weak embedding problems for $G$ are solvable if and only if they are locally solvable for each $H \in \mathcal{D}$.

The set Subg($G$) of all closed subgroups of $G$ is the inverse limit of the finite spaces Subg($G/N$), where $N$ ranges over all open normal subgroups of $G$. Thus Subg($G$) is a profinite space.

**13.3.** *Extension theorems.* Assume now that $\mathcal{D}$ is a closed subfamily of Subg($G$) and that $G$ is $\mathcal{D}$-projective. If $\mathcal{D}$ is the set of all subgroups $H$ which are isomorphic to

$G(\mathbb{R}) \cong \mathbb{Z}/2\mathbb{Z}$, we say that $G$ is *real projective* (that $\mathcal{D}$ is closed in this case means that 1 is not a limit point of involutions of $G$). If $\mathcal{D}$ is the set of all subgroups of $G$ which are isomorphic to $G(\mathbb{Q}_p)$, then $G$ is *p-adically projective.*

Every algebraic extension of a PRC field is PRC (Prestel [Pr1], Theorem 3.1). If $K$ is however a P$p$C field and $L$ is an algebraic extension of $K$, then $L$ is P$p$C if and only if for each $p$-adic closure $\overline{K}$ of $K$ we have: $L \subseteq \overline{K}$ or $\overline{K}L = \widetilde{K}$ [Ja6], Proposition 8.3. For example, out of the algebraic extensions of $\mathbb{Q}_p$ only $\mathbb{Q}_p$ and $\widetilde{\mathbb{Q}}_p$ are P$p$C.

One uses these extension theorems in the proof of the following theorems:

**13.4.** THEOREM ([HJ3], Theorem 10.4, and [HJ4], Theorems 15.1 and 15.4). *A profinite group $G$ is real* (*resp. p-adically*) *projective if and only if $G$ is isomorphic to the absolute Galois group of a PRC* (*resp.* P$p$C) *field $F$.*

**13.5.** THEOREM ([HJ1], Theorem 5.1, and [Ja6], Corollary 9.4). *Let $K$ be a formally real* (*resp. p-adic*) *countable Hilbertian field. Then, a profinite group $G$ of rank at most $\aleph_0$ is real* (*resp. p-adically*) *projective if and only if $G$ is isomorphic to the absolute Galois group of a PRC* (*resp.* P$p$C) *field $F$ which is separably algebraic over $K$.*

**13.6.** *On the proofs of Theorems* 13.4 *and* 13.5. Out of the two directions involved in these theorems, the construction of a field $F$ with a given absolute Galois group $G$ is the more difficult. It is done along the same line as for PAC fields. Starting from a formally real (resp. $p$-adic) field $K$ one constructs a regular extension $E$ which is PRC (resp. P$p$C) and a Galois extension $\widehat{E}$ such that $\mathcal{G}(\widehat{E}/E) \cong G$ and res: $G(E) \to \mathcal{G}(\widehat{E}/E)$ maps the set of involutions (resp. closed subgroups isomorphic to $G(\mathbb{Q}_p)$ ) of $G(E)$ onto the set of involutions (resp. closed subgroups isomorphic to $G(\mathbb{Q}_p)$) of $\mathcal{G}(\widehat{E}/E)$. One proves that res has a section and so one gets an algebraic extension $F$ of $E$ such that $G(F) \cong G$.

These steps become more and more complicated as one moves from PAC fields to PRC fields and from PRC fields to P$p$C fields. This difficulty is partially caused by the growing complexity of the absolute Galois groups of the local fields associated with the various pseudo closed fields. For PAC fields it is the trivial group, for PRC fields it is the group $\mathbb{Z}/2\mathbb{Z}$, and for P$p$C fields it is the group $G(\mathbb{Q}_p)$. What makes up for the infiniteness of the latter group is the fact that it is finitely generated, has a trivial center, and has a 'big' finite quotient (see (5) of Section 7).

**13.7.** *Examples of real* (*resp. p-adically*) *projective groups.* First examples of real (resp. $p$-adically) projective groups are 'free products' of several copies of $\mathbb{Z}/2\mathbb{Z}$ (resp. $G(\mathbb{Q}_p)$) and of $\widehat{\mathbb{Z}}$. In general, the *free product* of profinite groups $G_1, \dots, G_m$ is a profinite group $G = G_1 * \cdots * G_m$ which contains each $G_i$ as a closed subgroup and such that each system of homomorphisms

$$\alpha_i\colon G_i \to H, \quad i = 1, \dots, m,$$

uniquely extend to a homomorphism $\alpha\colon G \to H$. (See [FrJ], Lemma 20.18, for the existence and [HeR] for several properties of free products.) In particular, take $e$ copies $G_1, \dots, G_e$ of $\mathbb{Z}/2\mathbb{Z}$ (resp. $G(\mathbb{Q}_p)$) and define $\widehat{D}_{e,m}(\text{real})$ (resp. $\widehat{D}_{e,m}(p)$) to be the free product $G_1 * \cdots * G_e * \widehat{F}_m$. Then $\widehat{D}_{e,m}(\text{real})$ (resp. $\widehat{D}_{e,m}(p)$) is a real (resp. $p$-adically) projective group.

**13.8.** *The free product theorem.* Let $K$ be a countable Hilbertian field. For each $i$ between 1 and $e$ let $\overline{K}_i$ be either a Henselian field or a real closed field which is separable algebraic over $K$. For $\boldsymbol{\sigma} = (\sigma_1, \dots, \sigma_{e+m}) \in G(K)^{e+m}$ consider the field

$$K_{\boldsymbol{\sigma}} = \overline{K}_1^{\sigma_1} \cap \cdots \cap \overline{K}_e^{\sigma_e} \cap \widetilde{K}(\sigma_{e+1}, \dots, \sigma_{e+m}).$$

Then, for almost all $\boldsymbol{\sigma}$,

$$G(K_{\boldsymbol{\sigma}}) \cong G(\overline{K}_1) * \cdots * G(\overline{K}_e) * \widehat{F}_m.$$

([Ja1], the free product theorem, see also [Gey], Section 4.)

**13.9.** *Examples of PRC and* P$p$C *fields.* If in particular $\overline{K}_i$ is real (resp. $p$-adically) closed, $i = 1, \dots, e$, and $\boldsymbol{\sigma}$ is taken at random, then $G(K_{\boldsymbol{\sigma}})$ is isomorphic to $\widehat{D}_{e,m}(\text{real})$ (resp. $\widehat{D}_{e,m}(p)$) and $K_{\boldsymbol{\sigma}}$ is PRC (resp. P$p$C). ([HJ3], Proposition 5.6, and [EfJ], Intersection Theorem.) The latter result enters into the proof of the Characterization theorems 13.4 and 13.5.

It is an open problem whether the free product theorem holds for arbitrary separable extensions $K_1, \dots, K_e$ with finitely generated absolute Galois groups.

**13.10.** *Free products of pro-p groups.* Efrat and Haran [EfH], Lemma 2.2, prove that if pro-$p$ groups $G_1, \dots, G_m$ are isomorphic to the absolute Galois groups of fields, then their free pro-$p$ product is also isomorphic to the absolute Galois group of a field. It is not known if the latter statement holds for arbitrary profinite groups.

**13.11.** *Local primes.* A *local prime* $\mathfrak{p}$ of a field $K$ is either an equivalence class of archimedean absolute values of $K$ or an equivalence class of discrete valuations of $K$ with finite residue fields. We refer to the first type as *archimedean* and to the second as *nonarchimedean*. We denote the completion of $K$ with respect to $\mathfrak{p}$ by $K_{\mathfrak{p}}$ and let $K_{\mathfrak{p},\text{alg}} = K_s \cap K_{\mathfrak{p}}$. If $\mathfrak{p}$ is archimedean, then $K_{\mathfrak{p}}$ is either $\mathbb{R}$ or $\mathbb{C}$. In the first case $K_{\mathfrak{p},\text{alg}}$ is a real closure of $K$ with respect to the ordering of $K$ that $K_{\mathfrak{p}}$ induces. If $\mathfrak{p}$ is nonarchimedean, then $K_{\mathfrak{p}}$ is locally compact [CaF], p. 41, and $K_{\mathfrak{p},\text{alg}}$ is a Henselization of the valuation $v_{\mathfrak{p}}$ associated with $\mathfrak{p}$. In both cases $K_{\mathfrak{p},\text{alg}}$ is determined by $\mathfrak{p}$ up to $K$-isomorphism.

**13.12.** *The field of totally S-adic numbers.* Consider now a finite set $S$ of local primes of $K$. For each $\mathfrak{p} \in S$ choose a field $K_{\mathfrak{p},\text{alg}}$ as above and let

$$K_{\text{tot},S} = \bigcap_{\mathfrak{p} \in S} \bigcap_{\sigma \in G(K)} K_{\mathfrak{p},\text{alg}}^{\sigma}.$$

This is a Galois extension of $K$ which we call the field of *totally $S$-adic* elements of $K_s$. It is the maximal separable algebraic extension of $K$ in which each $\mathfrak{p} \in S$ totally splits. Pop [Po4], Theorem $\mathfrak{S}$, proves that $K_{\mathrm{tot},S}$ is PSC. That is, $K_{\mathrm{tot},S}$ is P$\mathcal{K}$C with

$$\mathcal{K} = \{K^{\sigma}_{\mathfrak{p},\mathrm{alg}} \mid \mathfrak{p} \in S,\ \sigma \in G(K)\}.$$

In particular, take $K = \mathbb{Q}$ and let $S$ consist of the unique archimedean prime of $\mathbb{Q}$. In this case, the field $\mathbb{Q}_{\mathrm{tr}} = \mathbb{Q}_{\mathrm{tot},S}$ of *totally real numbers* is PRC. Fried, Haran and Völklein [FHV], Corollary 6, prove that $G(\mathbb{Q}_{\mathrm{tr}})$ is real free (§10.5) with a basis isomorphic to the Cantor set consisting of involutions only.

By Weissauer's theorem [FrJ], Corollary 12.15, any proper finite extension $F$ of $\mathbb{Q}_{\mathrm{tr}}$ is Hilbertian. By Prestel's extension theorem it is PRC. If $F$ is not formally real, it is PAC. It follows that $G(F) \cong \widehat{F}_{\omega}$. For example, this is the case for $F = \mathbb{Q}_{\mathrm{tr}}(\sqrt{-1})$.

If $S$ consists of a unique prime $p$, then Pop's theorem asserts that the field $\mathbb{Q}_{tp}$ of *totally p-adic numbers* is P$p$C.

Pop [Po4], Theorem 3, generalizes the theorem of Fried, Haran and Völklein to arbitrary finite sets $S$. For each $\mathfrak{p} \in S$ let $X_{\mathfrak{p}}$ be the set of all extensions of $\mathfrak{p}$ to a local prime of $K_{\mathrm{tot},S}$. For each $\mathfrak{q} \in X_{\mathfrak{p}}$ let $N_{\mathfrak{q}}$ be a Henselization of $K_{\mathrm{tot},S}$ at $\mathfrak{q}$ (It is one of the fields $K^{\sigma}_{\mathfrak{p},\mathrm{alg}}$ which induces $\mathfrak{q}$ on $K_{\mathrm{tot},S}$.) It is possible to choose the various $N_{\mathfrak{q}}$ in such a way that $\{G(N_{\mathfrak{q}}) \mid \mathfrak{q} \in X_{\mathfrak{p}}\}$ is closed in $\mathrm{Subg}(G(K))$ (a consequence of [HJ4], Corollary 2.5). Theorem 3 of [Po4] then says that $G(K_{\mathrm{tot},S})$ is the *free product* of the $G(N_{\mathfrak{q}})$:

$$G(K_{\mathrm{tot},S}) = \mathop{\prod\nolimits^{*}}_{\mathfrak{p}\in S} \mathop{\prod\nolimits^{*}}_{\mathfrak{q}\in X_{\mathfrak{p}}} G(N_{\mathfrak{q}}).$$

This means that for every finite group $A$, each continuous map

$$\varphi_0\colon \bigcup_{\mathfrak{p}\in S} \bigcup_{\mathfrak{q}\in X_{\mathfrak{p}}} G(N_{\mathfrak{q}}) \to A$$

whose restriction to each $G(N_{\mathfrak{q}})$ is a homomorphism uniquely extends to a homomorphism $\varphi\colon G(K_{\mathrm{tot},S}) \to A$.

**13.13.** *References to the model theory of PSC fields.* Like for PAC fields, the knowledge of the absolute Galois group of PSC fields leads to an understanding of their model theory. We do not elaborate on this point and refer the reader to a series of articles on this subject: [Pr2, Ja7] and [Ja8] for PRC fields and [Gro, Ja6, Ef1, Ef2] and [Ef3] for P$p$C fields.

**13.14.** *Generalization of Theorems* 13.4 *and* 13.5. It seems not too difficult to generalize Theorems 13.4 and 13.5 and to treat finitely many local primes and an ordering in characteristic 0. Pop [Po6] generalizes Theorem 13.4 even further to cover the case of infinitely many local primes of characteristic 0. However, a main ingredient of the proof, that 'relative projectivity' implies 'strong relative projectivity', has yet to be cleared up.

## 14. Open problems

The first four problems on this lists are the basic problems of Galois theory. The results of this survey can be viewed as partial solutions of these problems. Problems 5–18 are specific problems of the theory. They are listed in the order of appearance of this survey.

1. Given a distinguished field $K$, list the set of finite groups which occur as Galois groups over $K$.

2. Given a distinguished field $K$, describe $G(K)$ in group theoretic terms.

3. Give necessary and sufficient group theoretic conditions on a profinite group $G$ to be isomorphic to the absolute Galois group of some field $K$.

4. Give a necessary and sufficient condition on a pro-$p$ group to be isomorphic to $\mathcal{G}(K^{(p)}/K)$ (= the maximal pro-$p$ quotient of $G(K)$) for some field $K$.

5. Is every Demushkin group of a finite rank isomorphic to the absolute Galois group of some field?

6. Present the Galois group $G(\mathbb{Q}_2)$ by generators and relations.

7. Condition (1) below on a prosolvable group $\Gamma$ is necessary for $\Gamma$ to be isomorphic to an open subgroup of $G(\mathbb{Q}_p)$ for some prime $p$ [Ja2], Theorem 7.2. Is it also sufficient?

(1a) The center of $\Gamma$ is trivial;

(1b) $\Gamma$ is finitely generated;

(1c) There exist distinct primes $l, q$ such that $\Gamma_l$ is a torsionfree nonfree pro-$l$ group and $\Gamma_q$ is a nonfree pro-$q$ group;

(1d) $\Gamma$ has a finite quotient $\overline{\Gamma}$ such that if a closed subgroup $H$ of $\Gamma$ is a quotient of $\Gamma$ and $\overline{\Gamma}$ is a quotient of $H$, then $H \cong \Gamma$.

8. Let $K$ be a field with the following property: if $G(L) \cong G(K)$, then $L$ is elementarily equivalent to $K$. Is it true that $K$ is real closed or a finite abelian extension of $\mathbb{Q}_p$, for some $p$?

9. For every infinite algebraic extension $M$ of $\mathbb{Q}_p$ which is not $\widetilde{\mathbb{Q}}_p$ there exists another algebraic extension $M'$ of $\mathbb{Q}_p$ such that $G(M) \cong G(M')$ but $M \not\equiv_{\mathbb{Q}_p} M'$.

10. Let $F$ be a function field of one variable over $\widetilde{\mathbb{F}}_p$ and let $S$ be a finite set of primes of $F$. Denote the maximal extension of $F$ which is unramified outside $S$ by $F_S$. What is the structure of $\mathcal{G}(F_S/F)$ as a profinite group?

11. Let $K$ be a field, $E$ a function field of one variable over $K$, and $S$ a finite set of primes of $K$. Prove or disprove: For each positive integer $n$, $E$ has a Galois extension $F$ of degree at least $n$ which is unramified outside $S$ and regular over $K$.

12. Let $F = \mathbb{F}_q(t)$. Describe the structure of $\mathcal{G}(F_{ab}/F)$.

13. Let $K$ be a field such that $K_{ins}$ is separably Hilbertian. Is $K$ separably Hilbertian?

14. Let $K$ be a countable separably Hilbertian field. Denote the compositum of all Galois extensions of $K$ with an alternating Galois group by $K_{alt}$. Is $K_{alt}$ PAC?

15. Is $\mathbb{Q}_{solv}$ PAC?

16. For each finite group $G$ compute a positive integer $q_0(G)$ such that $q > q_0(G)$ is a prime power, then $G$ occurs as a Galois group over $\mathbb{F}_q(t)$.

17. Let $K$ be a countable Hilbertian field and let $G_1, \ldots, G_e$ be finitely generated closed subgroups of $G(K)$. Is it true that $\langle G_1^{\sigma_1}, \ldots, G_e^{\sigma_e} \rangle \cong G_1 * \cdots * G_e$ for almost all $\sigma \in G(K)^e$?

18. Suppose that $G_1, \ldots, G_e$ are isomorphic to absolute Galois groups of fields. Is $G_1 * \cdots * G_e$ also isomorphic to an absolute Galois group of a field?

## References


[Abh] S.S. Abhyankar, *Coverings of algebraic curves,* Amer. J. Math. **79** (1957), 825–856.

[Ax1] J. Ax, *The elementary theory of finite fields,* Ann. Math. **88** (1968), 239–271.

[Ax2] J. Ax, *Proof of some conjectures on cohomological dimension,* Proc. Amer. Math. Soc. **16** (1965), 1214–1221.

[Bel] G.V. Belyi, *On extensions of the maximal cyclotomic field having a given classical Galois group,* J. Reine Angew. Math. **341** (1980), 147–158.

[Bou] N. Bourbaki, *Elements of Mathematics, Commutative Algebra, Chapters* 1–7, Springer, Berlin (1989).

[Car] H. Cartan, *Elementary Theory of Analytic Functions of One or Several Complex Variables,* Addison-Wesley, Reading, MA (1963).

[CaF] J.W.S. Cassels and A. Fröhlich, *Algebraic Number Theory,* Academic Press, London (1967).

[Coh] P.J. Cohen, *Decision procedures for real and p-adic fields,* Comm. Pure Appl. Math. **22** (1969), 131–151.

[Deu] M. Deuring, *Algebren,* Ergebnisse der Mathematik und ihrer Grenzgebiete vol. 4, Springer, Berlin (1935).

[Die] V. Diekert, *Über die absolute Galoisgruppe dyadischer Zahlkörper,* J. Reine Angew. Math. **350** (1984), 152–172.

[Dou] A. Douady, *Détermination d'un groupe de Galois,* C. R. Acad. Sci. Paris **258** (1964), 5305–5308.

[DMS] J.D. Dixon, M.P.F. du Sautoy, A. Mann and D. Segal, *Analytic pro-p Groups,* London Mathematical Society Lecture Notes Series vol. 157, Cambridge Univ. Press, Cambridge (1991).

[Ef1] I. Efrat, *The elementary theory of free pseudo p-adically closed fields of finite corank,* J. Symbolic Logic **56** (1991), 484–496.

[Ef2] I. Efrat, *On the model companion of e-fold p-adically valued fields,* Manuscripta Math. **73** (1991), 259–371.

[Ef3] I. Efrat, *Undecidability of pseudo p-adically closed fields,* Arch. Math. **58** (1992), 444–452.

[Ef4] I. Efrat, *Absolute Galois groups of p-adically maximal PpCF fields,* Forum Math. **3** (1991), 437–460.

[Ef5] I. Efrat, *A Galois-theoretic characterization of p-adically closed fields,* Israel J. Math. (91), 273–284.

[EfH] I. Efrat and D. Haran, *On Galois groups over Pythagorean and semi-real closed fields,* Israel J. Math. **85** (1994), 57–78.

[EfJ] I. Efrat and M. Jarden, *Free pseudo p-adically closed fields of finite corank,* J. Algebra **133** (1990), 132–150.

[End] O. Endler, *Valuation Theory,* Springer, Berlin (1972).

[FHV] M.D. Fried, D. Haran and H. Völklein, *The absolute Galois group of the totally real numbers,* C. R. Acad. Sci. Paris, to appear.

[Fri] M.D. Fried, *Introduction to Modular Towers,* Contemp. Math.

[FrJ] M.D. Fried and M. Jarden, *Field Arithmetic,* Ergebnisse der Mathematik (3) vol. 11, Springer, Heidelberg (1986).

[FV1] M.D. Fried and H. Völklein, *The inverse Galois problem and rational points on moduli spaces,* Math. Ann. **290** (1991), 771–800.

[FV2] M.D. Fried and H. Völklein, The embedding problem over a Hilbertian PAC-field, Ann. Math. **135** (1992), 469–481.

[FV3] M.D. Fried and H. Völklein, *The absolute Galois group of a Hilbertian prc field,* Israel J. Math. **85** (1994), 85–101.

[Gey] W.-D. Geyer, *Galois groups of intersections of local fields,* Israel J. Math. **30** (1978), 382–396.

[GJ1] W.-D. Geyer and M. Jarden, *On the normalizer of finitely generated subgroups of absolute Galois groups of uncountable Hilbertian fields of characteristic* 0, Israel J. Math. **63** (1988), 323–334.

[GJ2] W.-D. Geyer and M. Jarden, *Realization of l-groups as Galois groups over global fields, I. The method of Scholz–Reichardt,* Manuscript, Tel Aviv (1995).

[GJ3] W.-D. Geyer and M. Jarden, *On stable fields in positive characteristic,* Geom. Dedicata **29** (1989), 335–375.

[Gol] L.J. Goldstein, *Analytic Number Theory,* Prentice-Hall, Englewood Cliffs (1971).

[GPR] B. Green, F. Pop and P. Roquette, *On Rumely's local-global principle,* Jahresber. Deutsch. Math.-Verein. **97** (1995), 43–74.

[Gro] C. Grob, *Die Entscheidbarkeit der Theorie der maximalen pseudo p-adisch abgeschlossenen Körper,* Dissertation, Konstanz (1988).

[Grt] A. Grothendieck, *Revêtement Étales et Groupe Fondamental (SGA* 1), SLNM 224, Springer, Berlin (1971).

[Ha1] D. Haran, A proof of Serre's theorem, J. Indian Math. Soc. **55** (1990), 213–234.

[Ha2] D. Haran, *On closed subgroups of free products of profinite groups,* Proc. London Math. Soc. **55** (1987), 266–289.

[HJ1] D. Haran and M. Jarden, *The absolute Galois group of a pseudo real closed algebraic field,* Pacific J. Math. **123** (1986), 55–69.

[HJ2] D. Haran and M. Jarden, *Compositum of Galois extensions of Hilbertian fields,* Ann. Sci. École Norm. Sup. (4) **24** (1991), 739–748.

[HJ3] D. Haran and M. Jarden, *The absolute Galois group of a pseudo real closed field,* Ann. Scuola Norm. Sup. Pisa (4) **12** (1985), 449–489.

[HJ4] D. Haran and M. Jarden, *The absolute Galois group of a pseudo p-adically closed field,* J. Reine Angew. Math. **383** (1988), 147–206.

[HaV] D. Haran and H. Völklein, *Galois groups over complete valued fields,* Israel J. Math.

[Hr1] D. Harbater, *Galois coverings of the arithmetic line,* Number Theory, New York 1984–85, D.V. and G.V. Chudnovsky, eds, SLNM 1240, Springer, Berlin (1987), 165–195.

[Hr2] D. Harbater, *Abhyankar's conjecture on Galois groups over curves,* Invent. Math. **117** (1994), 1–25.

[Hr3] D. Harbater, *Fundamental groups and embedding problems in characteristic p,* Contemp. Math. **186** (1995), 353–369.

[Has] H. Hasse, *Number Theory,* Grundlehren der mathematischen Wissenschaften vol. 229, Springer, Berlin (1980).

[HeR] W. Herfort and L. Ribes, *Torsion elements and centralizers in free products of profinite groups,* J. Reine Angew. Math. **358** (1985), 155–161.

[Hil] D. Hilbert, *Über die Irreduzibilität ganzer rationaler Funktionen mit ganzzahligen Koeffizienten,* J. Reine Angew. Math. **110** (1892), 104–129.

[Ike] M. Ikeda, *Completeness of the absolute Galois group of the rational number field,* J. Reine Angew. Math. **291** (1977), 1–21.

[Iw1] K. Iwasawa, *Automorphisms of Galois groups over number fields,* Preprint (1976).

[Iw2] K. Iwasawa, *On* $\mathbb{Z}_l$*-extensions of algebraic number fields,* Ann. Math. **98** (1973), 246–326.

[Iw3] K. Iwasawa, *On solvable extensions of algebraic number fields,* Ann. Math. **58** (1953), 548–572.

[Jan] U. Jannsen, *Über Galoisgruppen lokaler Körper,* Invent. Math. **70** (1982), 53–69.

[JaW] U. Jannsen und K. Wingberg, *Die Struktur der absoluten Galoisgruppe* $\mathfrak{p}$*-adischer Zahlkörper,* Invent. Math. **70** (1982), 71–98.

[Ja1] M. Jarden, *Intersection of local algebraic extensions of a Hilbertian field (edited by Barlotti et al.),* NATO ASI Series C vol. 333, Kluwer, Dordrecht (1991), 343–405.

[Ja2] M. Jarden, *Prosolvable subgroups of free products of profinite groups,* Comm. Algebra.

[Ja3] M. Jarden, *Algebraic extensions of finite corank of Hilbertian fields,* Israel J. Math. **18** (1974), 279–307.

[Ja4] M. Jarden, *Large normal extensions of Hilbertian fields,* Math. Z.

[Ja5] M. Jarden, *The inverse Galois problem over formal power series fields,* Israel J. Math. **85** (1994), 263–275.

[Ja6] M. Jarden, *Algebraic realization of p-adically projective groups,* Compositio Math. **79** (1991), 21–62.

[Ja7] M. Jarden, *The elementary theory of large e-fold ordered fields,* Acta Math. **149** (1982), 239–260.

[Ja8] M. Jarden, *The algebraic nature of the elementary theory of PRC fields,* Manuscripta Math. **60** (1988), 463–475.

[Ja9] M. Jarden, *A letter to M. Fried,* February 1990.

[JaL] M. Jarden and A. Lubotzky, *Hilbertian fields and free profinite groups,* J. London Math. Soc. (2) **46** (1992), 205–227.

[JaR] M. Jarden and J. Ritter, *On the characterization of local fields by their absolute Galois groups,* J. Number Theory **11** (1979), 1–13.

[Koe] J. Koenigsmann, *From p-rigid elements to valuations (with a Galois-characterisation of p-adic fields),* Manuscript, Konstanz (1995).

[Koc] H. Koch, *Über die Galoissche Gruppe der algebraischen Abschließung eines Potenzreihnkörpers mit endlichem Konstantenkörper,* Math. Nach. **35** (1967).

[KrN] W. Krull and J. Neukirch, *Die Struktur der absoluten Galoisgruppe über dem Körper* $\mathbb{R}(t)$, Math. Ann. **193** (1971), 197–209.

[KPR] F.-V. Kuhlmann, M. Pank and P. Roquette, *Immediate and purely wild extensions of valued fields,* Manuscripta Math. **55** (1986), 39–67.

[Ku1] W. Kuyk, *Generic approach to the Galois embedding and extension problem,* J. Algebra **9** (1968), 393–407.

[Ku2] W. Kuyk, *Extensions de corps hilbertiens,* J. Algebra **14** (1970), 112–124.

[Lb1] J.P. Labute, *Classification of Demushkin groups,* Canad. J. Math. **19** (1967), 106–132.

[Lb2] J.P. Labute, *Demushkin groups of rank* $\aleph_0$, Bull. Soc. Math. France **94** (1966), 211–244.

[La1] S. Lang, *Diophantine Geometry,* Interscience Tracts in Pure and Applied Mathematics vol. 11, Interscience, New York (1962).

[La2] S. Lang, *Algebra,* Addison-Wesley, Reading, MA (1970).

[La3] S. Lang, *Algebraic Number Theory,* Addison-Wesley, Reading, MA (1970).

[Ma1] B.H. Matzat, *Konstruktive Galoistheorie,* SLNM 1284, Springer, Berlin (1987).

[Ma2] B.H. Matzat, *Der Kenntnisstand in der Konstruktiven Galoisschen Theorie,* Manuscript, Heidelberg (1990).

[Me1] O.V. Melnikov, *Normal subgroups of free profinite groups,* Math. USSR Izv. **12** (1978), 1–20.

[Me2] O.V. Mel'nikov, *The absolute Galois group of a Henselian field,* Dokl. Akad. Nauk BSSR **29** (1985), 581–583 (in Russian).

[MeS] A.S. Merkurjev and A.A. Suslin, *K-cohomology of Brauer–Severi varieties and the norm residue homomorphism,* Math. USSR Izv. **21** (1983), 307–340.

[MW1] J. Mináč and R. Ware, *Pro-2 Demuškin groups of rank* $\aleph_0$ *as Galois groups of maximal 2-extensions of fields,* Math. Ann. **292**, 337–353.

[MW2] J. Mináč and R. Ware, *Demuškin groups of rank* $\aleph_0$ *as Galois groups of automorphisms of algebraically closed fields,* Manuscripta Math. **73** (1991), 411–421.

[Ne1] J. Neukirch, *Class Field Theory,* Grundlehren der mathematischen Wissenschaften vol. 280, Springer, Berlin (1985).

[Ne2] J. Neukirch, *Kennzeichnung der p-adischen und der endlichen algebraischen Zahlkörper,* Invent. Math. **6** (1969), 296–314.

[Ne3] J. Neukirch, *Klassenkörpertheorie,* Hochschulskripten, Bibliographisches Institut, Mannheim (1969).

[Ne4] J. Neukirch, *Über die absoluten Galoisgruppen algebraischer Zahlkórper,* Astérisque **41–42** (1977), 67–79.

[Neu] K. Neumann, Dissertation, Erlangen (1995).

[Po1] F. Pop, *On the Galois theory of function fields of one variable over number fields,* J. Reine Angew. Math. **406** (1990), 200–218.

[Po2] F. Pop, *On Grothendieck's conjecture of birational anabelian geometry,* Ann. Math. **138** (1994), 145–182.

[Po3] F. Pop, *Galoissche Kennzeichnung p-adisch abgeschlossener Körper,* J. Reine Angew. Math. **392** (1988), 145–175.

[Po4] F. Pop, *Embedding problems over large fields,* Ann. Math.

[Po5] F. Pop, *The geometric case of a conjecture of Shafarevich, –* $G_{\tilde{K}(t)}$ *is profinite free,* Preprint, Heidelberg (1993).

[Po6] F. Pop, *Classically projective groups,* Preprint, Heidelberg (1993).

[Pr1] A. Prestel, *Lectures on Formally Real Fields,* SLNM 1093, Springer, Berlin (1984).

[Pr2] A. Prestel, *Pseudo real closed fields*, Set Theory and Model Theory, SLNM 872, Springer, Berlin (1981), 127–156.

[Pr3] A. Prestel, *Algebraic number fields elementarily determined by their absolute Galois group*, Israel J. Math. **73** (1991), 199–205.

[PrR] A. Prestel and P. Roquette, *Formally p-adic Fields*, SLNM 1050, Springer, Berlin (1984).

[Ray] M. Raynaud, *Revêtements de la droite affine en caractéristique $p > 0$ et conjecture d'Abhyankar*, Invent. Math.

[Rbn] P. Ribenboim, *Théorie des Valuations*, Les Presses de l'Université de Montréal, Montréal (1964).

[Rib] L. Ribes, *Introduction to Profinite Groups and Galois Cohomology*, Queen's Papers in Pure and Applied Mathematics vol. 24, Queen's University, Kingston (1970).

[Rit] J. Ritter, *$\mathfrak{P}$-adic fields having the same type of algebraic extensions*, Math. Ann. **238** (1978), 281–288.

[Sch] B. Schuppar, *Elementare Aussagen zur Arithmetik und Galoistheorie von Funktionenkörpern*, J. Reine Angew. Math. **313** (1980), 59–71.

[Se1] J.-P. Serre, *Corps Locaux*, Actualités Scientifiques et Industrielles vol. 1296, Hermann, Paris (1968).

[Se2] J.-P. Serre, *Sur la dimension cohomologique des groups profinis*, Topology **3** (1965), 413–420.

[Se3] J.-P. Serre, *Structure de certains pro-p groups (d'apres Demuskin)*, Séminaire Bourbaki **252** (1962–1963), 1–11.

[Se4] J.-P. Serre, *Topics in Galois Theory*, Jones and Barlett, Boston (1992).

[Se5] J.-P. Serre, *Cohomologie Galoisienne*, SLNM 5, Springer, Heidelberg (1965).

[Sh1] I.R. Shafarevich, *On p-extensions*, Amer. Math. Soc. Transl. Ser. 2 vol. 4 (1956), 59–72.

[Sh2] I.R. Shafarevich, *On the construction of fields with a given Galois group of order $l^a$*, Collected Mathematical Papers, Springer, Berlin (1989), 107–142.

[Sht] S. Shatz, *Profinite Groups, Arithmetic, and Geometry*, Ann. of Math. Studies vol. 67, Princeton Univ. Press, Princeton (1972).

[Uc1] K. Uchida, *Isomorphisms of Galois groups*, J. Math. Soc. Japan **28** (1976), 617–620.

[Uc2] K. Uchida, *Isomorphisms of Galois groups of algebraic function fields*, Ann. Math. **106** (1977), 589–598.

[Uc3] K. Uchida, *Separably Hilbertian fields*, Kodai Math. J. **3** (1980), 83–95.

[Vo1] H. Völklein, *Braid group action, embedding problems and the groups* $\mathrm{PGL}_n(q)$, $\mathrm{PU}_n(q^2)$, Forum Math., to appear.

[Vo2] H. Völklein, *Groups as Galois groups – an Introduction*, Manuscript, Gainesville (1995).

[Win] J.-P. Wintenberger, *Le corps des normes de certaines extensions infinies de corps locaus; applications*, Ann. Sci. École Norm. Sup. **16** (1983), 59–89.

[Wit] E. Witt, *Konstruktion von galoisschen Körpern der Charakteristik p zu vorgegebener Gruppe der Ordnung $p^f$*, J. Reine Angew. Math. **174** (1936), 237–245.

[ZaS] O. Zariski and P. Samuel, *Commutative Algebra, II*, Springer, New York (1975).

# Finite Fields and Their Applications

Rudolf Lidl

*Department of Mathematics, University of Tasmania, G.P.O. Box 252 C, Hobart, Tas. 7001, Australia*
*e-mail: rudi.lidl@admin.utas.edu.au*

Harald Niederreiter

*Institute for Information Processing, Austrian Academy of Sciences, Sonnenfelsgasse 19, A-1010 Vienna, Austria*
*e-mail: harald.niederreiter@oeaw.ac.at*

*Contents*



HANDBOOK OF ALGEBRA, VOL. 1
Edited by M. Hazewinkel

## 1. Introduction

In this brief introduction we have chosen to trace the historical development of finite fields, outlining some of the basic properties along the way and concluding with comments on the books on, or involving in a significant way, finite fields and applications. Throughout this article, $F_q$ denotes the finite field of order $q$.

The origins of finite fields reach back into the 17th and 18th centuries, with such eminent mathematicians as Pierre de Fermat, Leonhard Euler, Joseph-Louis Lagrange, and Adrien-Marie Legendre contributing to the structure theory of special finite fields, the finite prime fields $F_p$. The general theory of finite fields may be said to begin with the work of Carl Friedrich Gauss and Evariste Galois. Amongst the fundamental contributions of Gauss to this subject are arithmetic in polynomial rings $F_p[x]$, in particular the Euclidean algorithm and unique factorization. In 1830 in Férussac's Bulletin, Galois constructed finite fields in his paper "On the theory of numbers":

"If one agrees to regard as zero all quantities which in algebraic calculations are found to be multiplied by $p$, and if one tries to find, under this convention, the solution of an algebraic equation $Fx = 0$, which Mr. Gauss designates by the notation $Fx \equiv 0$, the custom is to consider integer solutions only. Having been led, by my own research, to consider incommensurable solutions, I have attained some results which I consider new."

Galois supposes $Fx$ to be irreducible mod $p$ and of degree $\nu$ and asks to solve $Fx \equiv 0$ by introducing new 'symbols', which might be just as useful as the imaginary unit $i$ in analysis. He forms $p^\nu$ expressions $a + a_1 i + \cdots + a_{\nu-1} i^{\nu-1}$, where $a$ and $a_k$ are integers mod $p$. These $p^\nu$ elements form a field, nowadays called a Galois field or finite field of order $p^\nu$. If $A$ is an element of that form, not all $a$ and $a_k$ zero, he shows that $1, A, \ldots, A^{n-1}$ are different if $n$ is the smallest positive integer for which $A^n$ is 1. Then $n$ divides $p^\nu - 1$ and $A^{p^\nu - 1} = 1$. (One proves that there exist primitive elements for which $n$ is exactly $p^\nu - 1$; all nonzero elements of the field are powers of a primitive element.) All elements of the field are roots of $x^{p^\nu} - x$, and every irreducible polynomial $Fx$ of degree $\nu$ is a divisor of this polynomial. If $\alpha$ is a root of $Fx$, then the others are $\alpha^p, \ldots, \alpha^{p^{\nu-1}}$. At the end of his article Galois reverses the situation. He starts with a field in which $x^{p^\nu} - x$ can be completely factored, then restricts himself to the subfield generated by a primitive element $i$. Every such $i$ is a root of an irreducible polynomial $Fx$, according to Galois, and no matter which irreducible of degree $\nu$ one chooses, one always obtains the same field $F_{p^\nu}$ of order $p^\nu$. Gauss comments later "... perhaps we shall have the opportunity to describe our views on this in detail", but apparently he never did. Richard Dedekind, in a paper of 1857 on higher congruences (i.e. finite fields), shows that a complete system of incongruent polynomials with respect to double modulus congruences (congruences mod $p$ and mod $M$) contains exactly $p^n$ elements. He writes such congruences as $A \equiv B$ modd. $p, M$ for $A, B, M$ in $\mathbf{Z}[x], \deg(M) = n$ in $F_p[x]$. If $M$ is irreducible mod $p$, then finite fields of order $p^n$ are constructed as residue class rings $F_p[x]/(M)$. Thus, in his 1857 paper Dedekind put the theory of finite fields on a sound basis. B.L. van der Waerden notes that "E. Galois and R. Dedekind gave modern algebra its structure, the framework is due to them".

Also in 1857 we see the first general statement and proof of the Möbius inversion formula, which can be used to show that the number of monic irreducible polynomials of degree $n$ over $F_p$ is

$$\frac{1}{n}\sum_{d|n}\mu(n/d)p^d.$$

This formula was also known to Gauss. Dedekind establishes the multiplicative form of the Möbius inversion formula and proves that $x^{p^n}-x$ equals the product of all monic irreducible polynomials over $F_p$ of degree dividing $n$. Moreover, he shows that the product of all monic irreducible polynomials of degree $n$ over $F_p$ is

$$\prod_{d|n}\left(x^{p^d}-x\right)^{\mu(n/d)},$$

which is now called Dedekind's formula. It was known by then how to construct a finite field of any prime-power order. E.H. Moore proved in 1893 that finite fields must have prime-power order and that finite fields of the same order are isomorphic.

Galois' approach via imaginary roots and Dedekind's approach via residue class rings were shown to be essentially equivalent by Kronecker. It was also known then that if $M$ is an irreducible polynomial over $F_p$, then the group of units of $F_p[x]/(M)$ is cyclic, hence the existence of primitive elements for any finite field was established. By the end of the 19th century Dickson was able to summarize in his book [34] the basic properties of finite fields. These properties are:

1. In any finite field, the number of elements is a power of a prime and this prime is the characteristic of the field.

2. If $p$ is a prime and $m$ is a positive integer, then there is a finite field of order $p^m$ which is unique up to field isomorphisms.

3. The multiplicative group $F_q^*$ of nonzero elements of $F_q$ is cyclic. Any generating element is a primitive element of $F_q$.

4. Let $q=p^m$. Then every subfield of $F_q$ has order $p^d$, where $d$ is a positive divisor of $m$. Conversely, if $d \mid m$, then there is exactly one subfield of $F_q$ of order $p^d$.

5. Any finite field $F_q$ is isomorphic to the splitting field of $x^q-x$ over $F_p$, where $q=p^m$.

6. Every element $a \in F_q$ satisfies $a^q=a$.

In Section 2 we shall make use of the trace function. Let $p$ be a prime and $q=p^m$ with $m \geqslant 1$. Let $F_{q^n}$ be the extension of degree $n$ of the finite field $F_q$. The Galois group $G$ of $F_{q^n}$ over $F_q$ is cyclic of order $n$ and generated by the Frobenius automorphism $\sigma(\alpha)=\alpha^q$ for $\alpha \in F_{q^n}$. The *trace function* of $F_{q^n}$ over $F_q$ is defined as

$$\mathrm{Tr}_{F_{q^n}/F_q}(\alpha)=\sum_{\tau\in G}\tau(\alpha)=\sum_{i=0}^{n-1}\alpha^{q^i}.$$

In Section 3 we shall need the concept of a primitive polynomial. Let $f$ be a nonzero polynomial over a finite field $F_q$. If $f(0)\neq 0$, then the least positive integer $e$ for which $f(x)$ divides $x^e-1$ is called the *order* $\mathrm{ord}(f)$ of $f$. If $f(0)=0$, then $f(x)=x^h g(x)$

with $h \geqslant 1$ and $g(0) \neq 0$; then $\mathrm{ord}(f)$ is defined to be the order of $g$. A polynomial $f \in F_q[x]$ of degree $n \geqslant 1$ is called a *primitive polynomial* over $F_q$ if it is the minimal polynomial over $F_q$ of a primitive element of $F_{q^n}$. According to [87], Theorem 3.16, $f$ is a primitive polynomial over $F_q$ if and only if $f$ is monic, $f(0) \neq 0$, and $\mathrm{ord}(f) = q^n - 1$.

As far as applications are concerned, the following sections of this article on finite fields will include a number of different topics. The big area of coding will be covered in a separate article. A number of applications could not be included because of space limitations. They include: Boolean functions in $n$ variables over finite fields, the discrete Fourier transform and spectral theory, digital signal processing and systems designs, precision measurements, radar camouflage, noise abatement, light diffusers, waveform and radiation patterns, and concert hall acoustics. We refer to Lidl and Niederreiter [87] and Schroeder [138] for some details on these topics.

We conclude with some comments on the book literature on finite fields. Journal articles of a more specialized nature will be referred to in the subsequent sections. One of the most important early books dealing with finite fields is Dickson [34]. The books Lidl and Niederreiter [87, 88] and the Russian translation of [87] represented probably the most comprehensive treatment of finite fields at the time, although several topics had to be excluded because of the vastness of the subject. The books by McEliece [93] and Small [149] emphasize specific topics in finite field theory; the former expands on linear recurrences and maximal period sequences, the latter concentrates on topics linking finite fields with number theory and algebraic geometry. See also Lüneburg [90], which emphasizes cyclotomy. A number of books contain substantial parts that address finite field theory or applications. Pohst and Zassenhaus [133] gives an introduction to constructive algebraic number theory, but it is also of interest to experimental number theorists. Lidl and Pilz [89] contains some theory and a variety of applications of finite fields, among other algebraic (discrete) structures. The book by Blake et al. [9] is devoted entirely to finite fields, with some applications in algebraic geometry and cryptology. Hirschfeld [65, 66] and Hirschfeld and Thas [67] deal with projective spaces over a finite field. Fried and Jarden [49] cover more specialized topics involving finite fields, as does Nechvatal [106] on irreducibility, primitivity, and duality. The book by Brawley and Schnibben [14] deals with generalizations to arbitrary algebraic extensions of $F_q$. A very recent book publication is the proceedings volume edited by Mullen and Shiue [100]. For open problems see that book and also Lidl and Mullen [84, 85]. Another recent book is Jungnickel [71] which emphasizes special bases and constructive aspects of finite fields.

The analogs in $F_q[x]$ of the Waring problem and the Vinogradov 3-primes problem are the main topics of Effinger and Hayes [38]. Schmidt [137] studies solutions of equations over finite fields. Several books on abstract algebra or on some topics of applied discrete mathematics contain at least brief summaries or surveys of finite field properties, e.g., Schroeder [138], van Tilborg [156], and Wallis [161]. Shparlinski [146] represents a survey of the literature on finite fields from the computational and constructive point of view.

Finally, a brief comment on computer systems with specific facilities for carrying out computations in finite fields. There are now a number of computer algebra packages available for dealing with a great variety of mathematical tasks. Large packages are, in alphabetical order, Aldes/Sac 2, Cayley, Kant, Macsyma, Magma, Maple, Mathematica,

Reduce, and Scratchpad (now Axiom). They all allow computations in finite fields with highly variable speed, efficiency, ease of use, and capabilities. There are also some smaller, purpose-built packages such as APL Classlib, Galois, Macaulay, and Simath. For references to the literature for these packages we refer to the survey article of Lidl [83].

## 2. Bases

Research in cryptology and coding theory, especially requirements for fast arithmetic in large finite fields, have motivated some of the recent advances of exhibiting properties of bases for finite fields. A careful choice of the representation of a finite field $F_q$ may assist in the algorithms for implementing arithmetic operations in $F_q$; see Beth [4], Beth and Fumy [5], and Beth and Gollmann [7]. We shall consider various bases of $F_{q^n}$ over $F_q$. The book Blake et al. [9] contains a wealth of details on bases.

There are

$$\prod_{i=0}^{n-1} (q^n - q^i)$$

ordered bases of $F_{q^n}$ over $F_q$, since this number represents the order of the general linear group $\mathrm{GL}(n, F_q)$. An ordered basis $A = \{\alpha_1, \dots, \alpha_n\}$ of $F_{q^n}$ over $F_q$ is called a *polynomial basis* if for some $\alpha \in F_{q^n}$ we have $\alpha_i = \alpha^{i-1}$, $i = 1, 2, \dots, n$. A *normal basis* is a basis $A$ where $\alpha_i = \alpha^{q^{i-1}}$, $i = 1, 2, \dots, n$, for some $\alpha \in F_{q^n}$. If $\sigma$ denotes the Frobenius automorphism $\sigma(\alpha) = \alpha^q$ for $\alpha \in F_{q^n}$, then a normal basis is a basis consisting of the orbit of a suitable $\alpha$ under $\sigma$. The element $\alpha$ is said to generate a normal basis or is called a *normal element* of $F_{q^n}$ over $F_q$. Normal bases are useful for implementing fast arithmetic in $F_{q^n}$, in particular exponentiation (see Itoh and Tsujii [70]), since computing $q$-th powers in $F_{q^n}$ is just a cyclic shift of the corresponding coordinate vectors. The additive order $\mathrm{ord}(\alpha) \in F_q[x]$ of an element $\alpha \in F_{q^n}$ is defined as the monic generator of the principal ideal $\{f \in F_q[x]\colon f(\sigma)(\alpha) = 0\}$ of $F_q[x]$. It is a divisor of $x^n - 1$. An element $\alpha \in F_{q^n}$ is normal over $F_q$ if and only if $\mathrm{ord}(\alpha) = x^n - 1$. Von zur Gathen and Giesbrecht [53] consider several aspects of normal elements. Since the existence of normal bases is a classical result, attention has shifted to normal bases of special type. Lenstra and Schoof [82] have shown that for every extension $F_{q^n}/F_q$ there exists a *primitive normal basis*, i.e. a normal basis consisting of primitive elements of $F_{q^n}$. For primes $q$ this was established earlier by Davenport [30], see also Carlitz [17]. Bshouty and Seroussi [15] give a generalization of the normal basis theorem.

Other types of bases are obtained by making use of the trace function of $F = F_{q^n}$ over $K = F_q$. If $A = \{\alpha_1, \dots, \alpha_n\}$ and $B = \{\beta_1, \dots, \beta_n\}$ are ordered bases of $F$ over $K$, then $B$ is called the *dual basis* of $A$ if and only if

$$\mathrm{Tr}_{F/K}(\alpha_i \beta_j) = \delta_{ij} \quad \text{for } 1 \leqslant i, j \leqslant n.$$

The following is easy to verify.

THEOREM 2.1. *For any given ordered basis $A = \{\alpha_1, \ldots, \alpha_n\}$ of $F_{q^n}$ over $F_q$ there exists a unique dual basis.*

It is also straightforward to show that the elements $\alpha_1, \ldots, \alpha_n$ form a basis of $F_{q^n}$ over $F_q$ if and only if the matrix

$$\begin{pmatrix} \alpha_1 & \alpha_2 & \cdots & \alpha_n \\ \alpha_1^q & \alpha_2^q & \cdots & \alpha_n^q \\ \vdots & \vdots & & \vdots \\ \alpha_1^{q^{n-1}} & \alpha_2^{q^{n-1}} & \cdots & \alpha_n^{q^{n-1}} \end{pmatrix}$$

is nonsingulár. This implies immediately that the dual basis of a normal basis is a normal basis. The dual basis of a polynomial basis

$$A = \{1, \alpha, \ldots, \alpha^{n-1}\}$$

of $F_{q^n}$ over $F_q$ is obtained as follows. Let $g \in F_q[x]$ be the minimal polynomial of $\alpha$ over $F_q$ and

$$\frac{g(x)}{x-\alpha} = \beta_0 + \beta_1 x + \cdots + \beta_{n-1}x^{n-1} \in F_{q^n}[x].$$

Then the dual basis of $A$ is $\{\beta_0\gamma^{-1}, \beta_1\gamma^{-1}, \ldots, \beta_{n-1}\gamma^{-1}\}$, where $\gamma = g'(\alpha)$.

In a normal basis $N = \{\alpha_0, \ldots, \alpha_{n-1}\}$ we assume that the elements $\alpha_i = \alpha^{q^i}$, $i = 0, 1, \ldots, n-1$, are given in that order. For any $0 \leqslant i, j \leqslant n-1$ the product $\alpha_i\alpha_j$ is a linear combination of $\alpha_0, \ldots, \alpha_{n-1}$ with coefficients in $F_q$. Multiplication of basis elements can be represented by

$$\alpha \begin{pmatrix} \alpha_0 \\ \alpha_1 \\ \vdots \\ \alpha_{n-1} \end{pmatrix} = T \begin{pmatrix} \alpha_0 \\ \alpha_1 \\ \vdots \\ \alpha_{n-1} \end{pmatrix},$$

where $T$ is an $n \times n$ matrix over $F_q$. The number of nonzero entries in $T$ is called the *complexity* $C_N$ of the normal basis $N$. A polynomial in $F_q[x]$ is called an $N$*-polynomial* if it is irreducible over $F_q$ and its roots are linearly independent over $F_q$. The elements in a normal basis are exactly the roots of an $N$-polynomial, thus $N$-polynomials describe a normal basis. Important questions are: given $n$ and $q$, construct a normal basis of $F_{q^n}$ over $F_q$, or equivalently, construct an $N$-polynomial in $F_q[x]$ of degree $n$. For practical purposes, one is interested in constructing low-complexity normal bases. We shall return to this topic later. A related question is how one can find normal elements efficiently. The algorithmic aspects of this problem area will be described elsewhere in this series of volumes. Therefore, we only refer to a few papers, such as von zur Gathen and

Giesbrecht [53], Lenstra [81], Semaev [141], and Stepanov and Shparlinski [150]. The book by Shparlinski [146] contains further detailed references.

It is shown in [87], Theorem 2.39, that $\alpha$ is a normal element of $F_{q^n}$ over $F_q$ if and only if the polynomial

$$\alpha x^{n-1} + \alpha^q x^{n-2} + \cdots + \alpha^{q^{n-1}} \in F_{q^n}[x]$$

is relatively prime to $x^n - 1$. Normal elements have been investigated further by Pincin [132], Schwarz [139], and Semaev [141]. It is of interest to construct normal bases of large finite fields, given normal bases of some smaller fields. Here the following theorem is relevant; see Blake et al. [9], Pincin [132], Séguin [140], and Semaev [141] for this result and for related ones.

THEOREM 2.2. *Let $n = vt$ with coprime positive integers $v$ and $t$ and let $\alpha \in F_{q^v}$, $\beta \in F_{q^t}$. Then $\alpha\beta \in F_{q^n}$ is a normal element of $F_{q^n}$ over $F_q$ if and only if $\alpha$ and $\beta$ are normal elements of $F_{q^v}$ and $F_{q^t}$, respectively, over $F_q$.*

An ordered basis $A$ of $F_{q^n}$ over $F_q$ is called *self-dual* (or *trace-orthonormal*) if $A$ is its own dual basis. Seroussi and Lempel [142] showed that the extension $F_{q^n}/F_q$ has a self-dual basis if and only if either $q$ is even or both $q$ and $n$ are odd. Self-dual normal bases are useful for special computational tasks. Lempel and Weinberger [79] proved that $F_{q^n}/F_q$ permits a self-dual normal basis if and only if either $n$ is odd, or $q$ is even and $n \equiv 2 \bmod 4$. A polynomial basis of $F_{q^n}$ over $F_q$ cannot be self-dual if $n \geqslant 2$, as can be seen by elementary arguments. A polynomial basis

$$A = \{1, \alpha, \ldots, \alpha^{n-1}\}$$

of $F_{q^n}$ over $F_q$ is called *weakly self-dual* if there exists an element $\gamma \in F_{q^n}$ such that $\{\gamma\beta_0, \ldots, \gamma\beta_{n-1}\}$ is a permutation of the basis $A$, where $\{\beta_0, \ldots, \beta_{n-1}\}$ is the dual of $A$. Geiselmann and Gollmann [54] show that $A$ is weakly self-dual if and only if the minimal polynomial of $\alpha$ over $F_q$ is either a trinomial with constant term $-1$ or a binomial.

Enumeration theorems for ordered bases of various types are known. The number of polynomial bases of $F_{q^n}$ over $F_q$ is clearly $n$ times the number of monic irreducible polynomials over $F_q$ of degree $n$ and so equal to

$$\sum_{d|n} \mu(n/d) q^d.$$

The number of normal elements of $F_{q^n}$ over $F_q$, and thus the number of normal bases of $F_{q^n}$ over $F_q$, is given by $\Phi_q(x^n - 1)$, where $\Phi_q(f)$ is the number of polynomials over $F_q$ of degree less than $\deg(f)$ and relatively prime to $f \in F_q[x]$ (see [87], Chapter 3). According to [87], Lemma 3.69, if $\deg(f) = n \geqslant 1$, then

$$\Phi_q(f) = q^n \prod_{j=1}^{r} \left(1 - q^{-n_j}\right),$$

where $n_1, \ldots, n_r$ are the degrees of the distinct monic irreducible factors of $f$ in $F_q[x]$. The number of normal bases of $F_{q^n}$ over $F_q$ is also equal to the number of nonsingular $n \times n$ circulant matrices over $F_q$. The number of self-dual bases of $F_{q^n}$ over $F_q$ is equal to the order of the group $\mathrm{O}(n, F_q)$ of orthogonal $n \times n$ matrices over $F_q$. This latter number is well known, namely

$$|\mathrm{O}(n, F_q)| = \begin{cases} \prod_{i=1}^{n-1} (q^i - a_i) & \text{if } q \text{ even}, \\ 2\prod_{i=1}^{n-1} (q^i - a_i) & \text{if } q \text{ and } n \text{ odd}, \\ 0 & \text{otherwise}, \end{cases}$$

where $a_i = 1$ if $i$ is even and $a_i = 0$ otherwise; see Jungnickel, Menezes and Vanstone [72]. The number of self-dual normal bases of $F_{q^n}$ over $F_q$ is equal to the order of the group of orthogonal $n \times n$ circulant matrices over $F_q$. The order of this group can be determined, but the formula is quite involved; see Beth and Geiselmann [6].

We have already introduced the complexity $C_N$ of a normal basis $N$ of $F_{q^n}$ over $F_q$. It is evident that a normal basis $N$ with low complexity $C_N$ facilitates fast arithmetic in the extension field $F_{q^n}$. Mullin, Onyszchuk, Vanstone and Wilson [104] showed that always $C_N \geqslant 2n - 1$, and they called a normal basis $N$ *optimal* if equality holds. Furthermore, they gave the following constructions of optimal normal bases.

THEOREM 2.3. *If $n + 1$ is a prime and $q$ is a primitive root modulo $n + 1$, then the $n$ primitive $(n+1)$-st roots of unity in $F_{q^n}$ are linearly independent over $F_q$ and they form an optimal normal basis of $F_{q^n}$ over $F_q$.*

THEOREM 2.4. *Suppose that $2n+1$ is a prime and that $F^*_{2n+1}$ is generated by $-1$ and $2$, and let $\gamma$ be a primitive $(2n+1)$-st root of unity in $F_{2^{2n}}$. Then $\alpha = \gamma + \gamma^{-1}$ is a normal element of $F_{2^n}$ over $F_2$ which determines an optimal normal basis of $F_{2^n}$ over $F_2$.*

Based on computer searches, it was conjectured in [104] that Theorems 2.3 and 2.4 describe essentially all optimal normal bases. Recently, Gao and Lenstra [50] verified this conjecture. In fact, they confirmed the conjecture not only for finite fields, but even for finite Galois extensions of general fields. We refer to Ash, Blake and Vanstone [2], Séguin [140], and Wassermann [164] for constructions of low-complexity normal bases of finite fields.

## 3. Irreducible and primitive polynomials and primitive elements

We saw in Section 1 that irreducible polynomials of degree $n$ over $F_q$ are important for the construction of the field $F_{q^n}$. In the constructive theory of finite fields, a problem of particular importance is that of the effective construction of irreducible polynomials over $F_q$. Thus for given $n$ and $q$, it is desired to find an effective construction of an irreducible polynomial of degree $n$ over $F_q$. A similar problem can be posed for primitive polynomials of degree $n$ over $F_q$, which is basically equivalent to the problem of effectively constructing a primitive element of the extension field $F_{q^n}$. Since the publication of the book of Lidl and Niederreiter [87] there has been a spectacular development of algorithmic and constructive aspects of the theory of finite fields; some of these results will be covered in other volumes of this Handbook series. In this section we shall emphasize these newer developments rather than go over the well-trodden territory of factorization algorithms, root-finding algorithms, or irreducibility tests that is described in the book Lidl and Niederreiter [87]. The book by Shparlinski [146] and several recent survey articles, such as Lenstra [80], Lidl [83], and Niederreiter [116, 118], refer to the 'traditional' and the constructive aspects of finite field theory.

There is presently no deterministic polynomial-time algorithm known for the explicit construction of irreducible polynomials of degree $n$ over $F_q$. Adleman and Lenstra [1] and Evdokimov [44] developed a deterministic polynomial-time algorithm under the assumption of the generalized Riemann hypothesis. If probabilistic algorithms are allowed, then the problem of constructing irreducible polynomials can be solved in polynomial time; see Lenstra [80] and Shoup [143]. The best deterministic algorithm currently available is due to Shoup [143], with running time approximately $\mathrm{O}(n^4 p^{1/2})$ for irreducible polynomials of degree $n$ over $F_p$. Shoup [143] also shows that the problem can be deterministically reduced in time bounded by a polynomial in $n$ and $\log p$ to the problem of factoring polynomials over $F_p$. For sketching Shoup's approach, let $n = p_1^{e_1} \cdots p_t^{e_t}$, where the $p_i$ are distinct primes and $e_i \geqslant 1$. For each $1 \leqslant i \leqslant t$, one has to construct an irreducible polynomial of degree $p_i^{e_i}$ over $F_p$. Thus, the critical step in Shoup's algorithm is to construct an irreducible polynomial of prime-power degree $r^e$ for any given prime $r$ and $e \geqslant 1$. The cases $r = p$ and $r \mid (p-1)$ are easy to tackle. The main task is achieved through the following result due to Shoup [143].

THEOREM 3.1. *Let $p$ be a prime, $r \neq p$ an odd prime, $m$ the multiplicative order of $p$ modulo $r$, and $a \in F_{p^m}$ an $r$-th nonresidue in $F_{p^m}$. Given a positive integer $e$, let $\beta$ be a root of $x^{r^e} - a$. Then*

$$\gamma = \sum_{i=0}^{m-1} \beta^{p^{ir^e}}$$

*has degree $r^e$ over $F_p$. Thus, the minimal polynomial of $\gamma$ over $F_p$ is an irreducible polynomial over $F_p$ of degree $r^e$.*

The final step in Shoup's algorithm requires the irreducible polynomials of degree $p_i^{e_i}$ obtained from Theorem 3.1 to be combined to get an irreducible polynomial of degree $n$ over $F_p$. The paper of Shoup [143] contains also a deterministic algorithm for constructing irreducible polynomials over an arbitrary finite field $F_q$.

A special procedure for generating irreducible polynomials over the binary field $F_2$ of arbitrarily large degrees was recently analyzed by Meyn [96]. In this procedure we start from a polynomial

$$f(x) = \sum_{j=0}^{n} a_j x^j \in F_2[x]$$

of degree $n \geqslant 2$ and form the $Q$-transform

$$f^Q(x) = x^n f\left(x + \frac{1}{x}\right),$$

which is a self-reciprocal polynomial of degree $2n$. By repeated applications of the $Q$-transform we obtain a sequence $f, f^Q, (f^Q)^Q, \ldots$ of polynomials. Meyn [96] has shown that all polynomials in this sequence are irreducible over $F_2$ if and only if $f$ is irreducible over $F_2$ and $a_1 = a_{n-1} = 1$. This raises the question for which degrees $n \geqslant 2$ there exist such $f$. Let the counting function $A(n)$ be defined as the number of irreducible $f$ over $F_2$ with $\deg(f) = n$ and $a_1 = a_{n-1} = 1$. From a result of Hayes [61] we deduce an asymptotic formula for $A(n)$, namely

$$A(n) = \frac{1}{n} 2^{n-2} + \mathrm{O}\left(\frac{1}{n} 2^{\theta n}\right) \quad \text{for some } \theta < 1.$$

This implies that $A(n) > 0$ for all sufficiently large $n$. An explicit formula for $A(n)$ was obtained by Niederreiter [115]. It follows from the formula for $A(n)$ that $A(n) > 0$ for all $n \geqslant 2$ with $n \neq 3$, whereas it is seen by inspection that $A(3) = 0$. Thus, the procedure of generating irreducible self-reciprocal polynomials over $F_2$ by iteration of the $Q$-transform can be applied exactly for all initial degrees $n \geqslant 2$ with $n \neq 3$. This procedure is also useful for so-called iterated presentations of infinite algebraic extensions of $F_2$ (compare with Brawley and Schnibben [14], Chapter 3, and Meyn [96]). Other recursive procedures for generating irreducible polynomials over finite fields were studied by Kyuregyan [75, 76] and Varshamov [159, 160].

Gao and Mullen [51] construct irreducible polynomials of arbitrarily large degrees involving the *Dickson polynomials*

$$D_n(x, a) = \sum_{j=0}^{\lfloor n/2 \rfloor} \frac{n}{n-j} \binom{n-j}{j} (-a)^j x^{n-2j} \in F_q[x],$$

where $n \geqslant 1$, $a \in F_q$, and $\lfloor u \rfloor$ denotes the greatest integer $\leqslant u$ for real $u$. We refer to the book of Lidl, Mullen and Turnwald [86] for a detailed treatment of Dickson polynomials. The construction of Gao and Mullen [51] is based on necessary and sufficient conditions for $D_n(x, a) + b$ to be irreducible over $F_q$, where $a, b \in F_q$. In addition, they show that if $n$ is neither a prime power nor of the form $2 \cdot 3^k$, $k \geqslant 1$, then there are infinitely many primes $p$ such that $D_n(x, a)$ does not permute $F_p$ for any $a \in F_p$ and there are no $a, b \in F_p$ so that $D_n(x, a) + b$ is irreducible over $F_p$. This disproves a conjecture of Chowla and Zassenhaus [21]. Gao and Mullen [51] prove also that the minimal polynomials of elements which generate the optimal normal bases in Theorem 2.4 can be derived from Dickson polynomials.

Niederreiter [122, 123] studies differential equations in the rational function field $F_q(x)$ over $F_q$ which lead to new irreducibility tests and factorization algorithms. Two types of procedures can be followed, depending on whether one wants to work with ordinary derivatives or Hasse–Teichmüller derivatives. For simplicity we describe only the first approach, and we refer to Niederreiter [123] for the second approach. The starting point is the ordinary differential equation

$$y^{(p-1)} + y^p = 0$$

of order $p-1$ in $F_q(x)$, where $p$ is the characteristic of $F_q$. The left-hand side of this differential equation is an $F_p$-linear operator on $F_q(x)$, and therefore the solutions form an $F_p$-linear subspace of $F_q(x)$. The solution space can be described explicitly: if we fix a monic nonconstant $f \in F_q[x]$, then the solutions $y$ of the form $y = h/f$ with $h \in F_q[x]$ are exactly given by

$$y = \sum_{i=1}^{m} c_i \frac{g_i'}{g_i} \quad \text{with } c_1, \ldots, c_m \in F_p,$$

where $g_1, \ldots, g_m \in F_q[x]$ are the distinct monic irreducible factors of $f$ over $F_q$ and $g_i'$ denotes the first derivative of $g_i$. If we keep $f \in F_q[x]$ with $d = \deg(f) \geqslant 1$ fixed and view $h \in F_q[x]$ as the unknown, then the differential equation is equivalent to the system of algebraic equations

$$M_p(f)\mathbf{h}^\top + (\mathbf{h}^p)^\top = \mathbf{0},$$

where $M_p(f)$ is a $d \times d$ matrix over $F_q$, $\mathbf{h} = (h_0, \ldots, h_{d-1}) \in F_q^d$ is the coefficient vector of $h$, and $\mathbf{h}^p = (h_0^p, \ldots, h_{d-1}^p) \in F_q^d$. If $q = p$, then $\mathbf{h}^p = \mathbf{h}$, and so this system of algebraic equations reduces to a system of homogeneous linear equations. If $q = p^t$ with $t \geqslant 2$, then the system of algebraic equations can be linearized by working with a normal basis $N$, preferably a low-complexity normal basis, of $F_q$ over $F_p$. One expresses the entries of $M_p(f)$ and the unknowns $h_k$, $0 \leqslant k \leqslant d-1$, as $F_p$-linear combinations of the elements of $N$ and then carries out a comparison of coefficients of the elements of $N$. In this way one arrives at the system of homogeneous linear equations

$$K_q(f, N)\mathbf{H}^\top = \mathbf{0},$$

where $K_q(f, N)$ is a $dt \times dt$ matrix over $F_p$ and $\mathbf{H} \in F_p^{dt}$ contains the unknowns $h_k^{(i)}$, $0 \leqslant k \leqslant d-1$, $0 \leqslant i \leqslant t-1$, that is, the coordinates of the $h_k$ relative to the basis $N$. Because of the equivalence of this system of linear equations with the differential equation, the matrix $K_q(f, N)$ has rank $dt - m$. This leads to the following irreducibility criterion: $f$ is irreducible over $F_q$ if and only if $\gcd(f, f') = 1$ and $K_q(f, N)$ has rank $dt - 1$. The case $t = 1$ can of course be included by formally putting $K_p(f, N) = M_p(f) + I_d$ with $I_d$ the $d \times d$ identity matrix over $F_p$.

We next turn to primitive polynomials of degree $n$ over $F_q$. The problem of the efficient construction of such polynomials, which is basically equivalent to that of efficiently constructing a primitive element of the extension field $F_{q^n}$, has been taken up only very recently. Shparlinski [145] has shown that for any prime $p$ we can find in time $\mathrm{O}(n^{\mathrm{O}(1)})$ a subset of $F_{p^n}$ of cardinality $\mathrm{O}(n^{10})$ containing at least one primitive element of $F_{p^n}$, where the implied constants may depend on $p$. The currently best result is due to Shoup [144] who replaced $\mathrm{O}(n^{10})$ by $\mathrm{O}(n^{6+\varepsilon})$.

Tables of primitive polynomials over finite fields can be found in the book Lidl and Niederreiter [87], Chapter 10. More recent work on the search for primitive polynomials was carried out by Hansen and Mullen [60] and Rybowicz [136]. Hansen and Mullen [60] tabulate for each prime power $p^n < 10^{50}$ with $p \leqslant 97$ a primitive polynomial of degree $n$ over $F_p$. They also state a conjecture on the existence of primitive polynomials with one prescribed coefficient and an analogous conjecture for irreducible polynomials. The chief theoretical result in support of the conjecture on primitive polynomials is a theorem of Cohen [24] to the effect that if $n \geqslant 2$ and $a \in F_q$ with $a \neq 0$ if $n = 2$ or if $n = 3$ and $q = 4$, then there exists a primitive polynomial over $F_q$ of degree $n$ for which the coefficient of $x^{n-1}$ is equal to $a$.

The paper Golomb [57] stimulated considerable research activity from 1984 on by posing some conjectures postulating that for sufficiently large $q$ an element of $F_q^*$ could always be expressed as the sum of two primitive elements. The context of these conjectures was the construction of Costas arrays, which are useful for radar. A *Costas array* is an $n \times n$ permutation matrix with the property that the $\binom{n}{2}$ vectors connecting two 1's of the matrix are all distinct as vectors. Golomb [57] stated that all known systematic constructions for Costas arrays involve the use of primitive elements in finite fields. Cohen and Mullen [26] provide a summary of the work related to Golomb's conjectures and resolve some unanswered questions. Cohen [25] gives an even more recent survey and shows that the first three of Golomb's conjectures can be subsumed by the following more general conjecture: for all $q$, except those in a precisely identifiable small set, and for all $a, b \in F_q^*$, there exists a primitive element $c$ of $F_q$ such that $ac + b$ is also a primitive element of $F_q$. Cohen [25] uses an alliance of careful character sum analysis with sieve methods to obtain lower bounds for the cardinality of the set of primitive elements with the desired property. We refer to that paper for details.

## 4. Permutation polynomials

Every mapping $\phi$ from $F_q$ into itself can be represented by a polynomial $g \in F_q[x]$, in the sense that $\phi(c) = g(c)$ for all $c \in F_q$. If we impose the condition $\deg(g) < q$, then $g$ is uniquely determined and given by the formula

$$g(x) = \sum_{c \in F_q} \phi(c)\bigl(1 - (x - c)^{q-1}\bigr).$$

Two polynomials over $F_q$ both represent $\phi$ if and only if they are congruent modulo $x^q - x$. Heisler [62] characterized finite fields as the only nonzero rings $R$ for which any mapping from $R$ into itself can be represented by a polynomial over $R$.

A polynomial $f \in F_q[x]$ is called a *permutation polynomial* of $F_q$ if it represents a bijection of $F_q$. In other words, $f$ is a permutation polynomial of $F_q$ if the cardinality $V(f)$ of its value set $\{f(c)\colon\ c \in F_q\}$ is equal to $q$. For small $q$ this condition can be checked directly. The classical algebraic criterion for determining whether a given polynomial is a permutation polynomial is due to Hermite [64] for finite prime fields and Dickson [33] in the general case. As usual, we denote by $p$ the characteristic of $F_q$.

THEOREM 4.1. *$f \in F_q[x]$ is a permutation polynomial of $F_q$ if and only if the following two conditions hold:*

(i) *$f$ has exactly one root in $F_q$;*

(ii) *for each integer $t$ with $1 \leqslant t \leqslant q-2$ and $t \not\equiv 0 \bmod p$, the reduction of $f(x)^t \bmod (x^q - x)$ has degree $\leqslant q-2$.*

Condition (i) may be replaced by the requirement that the reduction of

$$f(x)^{q-1} \bmod (x^q - x)$$

has degree $q-1$. It is an immediate consequence of Theorem 4.1 that there is no permutation polynomial of $F_q$ of degree $d$ whenever $d > 1$ is a divisor of $q-1$. To test whether a polynomial of degree $n$ is a permutation polynomial of $F_q$, von zur Gathen [52] designed a probabilistic polynomial-time algorithm, i.e. one with running time $\mathrm{O}((n \log q)^{\mathrm{O}(1)})$, and Shparlinski [147] developed a deterministic algorithm with running time $\mathrm{O}((nq)^{6/7}(\log nq)^{\mathrm{O}(1)})$.

Obvious examples of permutation polynomials of $F_q$ are linear polynomials. A more interesting class of examples is obtained from the Dickson polynomials $D_n(x,a)$ introduced in Section 3. For $a = 0$ we have $D_n(x,0) = x^n$, and this monomial is a permutation polynomial of $F_q$ if and only if $\gcd(n, q-1) = 1$. For $a \neq 0$, $D_n(x,a)$ is a permutation polynomial of $F_q$ if and only if $\gcd(n, q^2-1) = 1$. We note the simple principle that the set of permutation polynomials of $F_q$ is closed under composition, which can be used to generate further examples of permutation polynomials of $F_q$.

Apart from these classical examples of permutation polynomials, there are also other families of permutation polynomials that are known, but a complete classification of such polynomials seems out of reach at present. Special classes of permutation polynomials of $F_q$, such as linearized polynomials and polynomials of the form $x^r(g(x^s))^{(q-1)/s}$, are described in the book of Lidl and Niederreiter [87], Chapter 7. Some attention has also been devoted to permutation binomials $ax^n + bx^k$ with $a, b \in F_q$ and $n > k \geqslant 1$; see, e.g., Niederreiter and Robinson [125] and Turnwald [158]. More recent examples of families of permutation polynomials can be found in Cohen [23] and Tautz, Top and Verberkmoes [155]. Several classes of permutation polynomials form interesting groups under composition; we refer again to [87], Chapter 7, as well as to the recent paper of Wan and Lidl [163].

For the deeper analysis of permutation polynomials, an important connection is that between permutation polynomials and exceptional polynomials. A polynomial $f \in F_q[x]$ of degree $\geqslant 2$ is *exceptional* over $F_q$ if every irreducible factor of $(f(x) - f(y))/(x-y)$ in $F_q[x,y]$ is reducible over some algebraic extension of $F_q$.

THEOREM 4.2. *Every exceptional polynomial over $F_q$ is a permutation polynomial of $F_q$. Conversely, if $f \in F_q[x]$ is a permutation polynomial of $F_q$ with* $\deg(f) = n \geqslant 2$, *if* $\gcd(n, q) = 1$, *and if $q$ is sufficiently large relative to $n$, then $f$ is exceptional over $F_q$.*

A relatively elementary proof of the first part of Theorem 4.2 was recently given by Wan [162]. For a proof of the second part we refer to [87], Section 7.4. Methods of algebraic geometry lead to profound results such as the following criterion: if $n$ is a positive integer, if $\gcd(n, q) = 1$, and if $q$ is sufficiently large relative to $n$, then there exists a permutation polynomial of $F_q$ of degree $n$ if and only if $\gcd(n, q-1) = 1$ (see [87], Corollary 7.33). For a detailed recent treatment of exceptional polynomials we refer to Cohen [23].

The following conjecture of Carlitz on the degrees of permutation polynomials has attracted a lot of attention: if $n$ is a positive even integer and $q$ is odd and sufficiently large relative to $n$, then there are no permutation polynomials of $F_q$ of degree $n$. A number of papers treated various special cases, but in an important breakthrough the conjecture was recently proved in full generality by Fried, Guralnick and Saxl [48]. Another famous conjecture on permutation polynomials was settled much earlier by Fried [47], namely Schur's conjecture to the effect that any $f \in \mathbf{Z}[x]$ which is a permutation polynomial of $F_p$ (when considered modulo $p$) for infinitely many primes $p$ must be a composition of binomials $ax^n + b$ and Dickson polynomials; see also Turnwald [157] for some clarifying remarks on Fried's result. Based on the work of Fried, Cohen [22] verified a conjecture of Chowla and Zassenhaus [21] by proving that if $f \in F_p[x]$ with $\deg(f) = n \geqslant 2$ and a prime $p > (n^2 - 3n + 4)^2$, then $f(x) + cx$ is a permutation polynomial of $F_p$ for at most one $c \in F_p$. A more general result on permutation polynomials of the form $f(x) + cg(x)$ was shown by Cohen, Mullen and Shiue [27].

If both $f(x)$ and $f(x) + x$ are permutation polynomials of $F_q$, then $f$ is called a *complete mapping polynomial* of $F_q$. This notion was studied in detail by Niederreiter and Robinson [125] and allows interesting applications in combinatorics (see [32]) and in the theory of check digits (see [37]). More generally, we can consider for any $f \in F_q[x]$ the number $C(f)$ of elements $c \in F_q$ for which $f(x) + cx$ is a permutation polynomial of $F_q$. Since $C(f)$ depends only on the mapping properties of $f$, we can assume $\deg(f) < q$. It is trivial that $C(f) = q - 1$ whenever $\deg(f) \leqslant 1$. For $1 < \deg(f) < q$, Chou [19, 20] proved that

$$C(f) \leqslant q - 1 - \deg(f),$$

and Evans, Greene and Niederreiter [43] showed that

$$C(f) \leqslant q - \left\lceil \frac{q-1}{\deg(f) - 1} \right\rceil,$$

where $\lceil u \rceil$ is the least integer $\geqslant u$. The latter paper contains also the proof of the conjecture of Stothers which states that if $C(f) \geqslant \lfloor q/2 \rfloor$, then $f$ has the form $f(x) = ax + g(x^p)$ for some $a \in F_q$ and $g \in F_q[x]$.

The cardinality $V(f)$ of the value set of a polynomial $f \in F_q[x]$ with $\deg(f) = n \geqslant 1$ satisfies

$$\left\lceil \frac{q}{n} \right\rceil \leqslant V(f) \leqslant q,$$

where the lower bound follows from the fact that a polynomial over $F_q$ of degree $n$ has at most $n$ roots in $F_q$. We can have equality in both bounds, with equality in the upper bound corresponding of course to the case of a permutation polynomial. A discussion of polynomials with small value set can be found, e.g., in Gomez-Calderon and Madden [59]. Birch and Swinnerton-Dyer [8] proved that if $f$ is "general", in the sense that the Galois group of the equation $f(x) = y$ over $\overline{F}_q(y)$ is the symmetric group $S_n$, then

$$V(f) = q \sum_{j=1}^{n} \frac{(-1)^{j-1}}{j!} + \mathrm{O}(q^{1/2}),$$

where the implied constant depends only on $n$. Mullen [99] recently proposed a refinement of the Carlitz conjecture which can be expressed in terms of $V(f)$: if $f \in F_q[x]$ has even degree $n \geqslant 2$ and $q$ is odd with $q > n(n-2)$, then

$$V(f) \leqslant q - \left\lceil \frac{q-1}{n} \right\rceil.$$

Permutation polynomials have also been considered in other algebraic settings, for instance over residue class rings $\mathbf{Z}/N\mathbf{Z}$ (see the book of Narkiewicz [105], Chapter 2) or over more general commutative rings with identity (see the book of Lausch and Nöbauer [78], Chapter 4). An interesting theory can also be developed for permutation polynomials of matrix rings over finite fields, as the survey article by Brawley [12] demonstrates.

Another extension of the theory of permutation polynomials concerns permutation polynomials in several indeterminates. A polynomial $f \in F_q[x_1, \ldots, x_m]$ is called a *permutation polynomial* in $m$ indeterminates over $F_q$ if the equation $f(x_1, \ldots, x_m) = a$ has $q^{m-1}$ solutions in $F_q^m$ for each $a \in F_q$. More generally, a system of polynomials $f_1, \ldots, f_k \in F_q[x_1, \ldots, x_m]$ with $1 \leqslant k \leqslant m$ is *orthogonal* in $F_q$ if the system of equations

$$f_i(x_1, \ldots, x_m) = a_i \quad \text{for } 1 \leqslant i \leqslant k$$

has $q^{m-k}$ solutions in $F_q^m$ for each $(a_1, \ldots, a_k) \in F_q^k$. Every polynomial occurring in an orthogonal system is a permutation polynomial. On the other hand, a system $f_1, \ldots, f_k$ is orthogonal in $F_q$ if and only if for all nonzero $(b_1, \ldots, b_k) \in F_q^k$ the polynomial

$$\sum_{i=1}^{k} b_i f_i$$

is a permutation polynomial over $F_q$ (see [87], Corollary 7.39). Nontrivial examples of orthogonal systems, and thus of permutation polynomials in several indeterminates, can be obtained from Dickson polynomials in several indeterminates (see [87], Theorem 7.46). A simple principle for the construction of permutation polynomials in several indeterminates is the following: if $f \in F_q[x_1, \ldots, x_m]$ has the form

$$f(x_1, \ldots, x_m) = g(x_1, \ldots, x_r) + h(x_{r+1}, \ldots, x_m)$$

with $1 \leqslant r < m$ and if at least one of $g$ and $h$ is a permutation polynomial over $F_q$, then $f$ is a permutation polynomial over $F_q$. Useful applications of permutation polynomials and orthogonal systems in several indeterminates arise in the work of Mullen [98] on the construction of frequency squares for experimental designs in statistics.

Expository accounts of the theory and the applications of permutation polynomials are given in the survey articles of Lidl and Mullen [84, 85] and Mullen [99] and in the book of Lidl and Niederreiter [87], Chapter 7.

## 5. Discrete logarithms

The cyclic group $F_q^*$ is generated by a primitive element $b$ of $F_q$. Thus, for every $a \in F_q^*$ there exists a uniquely determined integer $r$ with $0 \leqslant r \leqslant q-2$ and $b^r = a$. This integer $r$ is called the *discrete logarithm* (or the *index*) of $a$ relative to $b$ and is denoted by $\mathrm{ind}_b(a)$. The computational problem of calculating $\mathrm{ind}_b(a)$ given $a$ and $b$ is called the *discrete logarithm problem*. Several cryptographic schemes are based on the presumed difficulty of the discrete logarithm problem for large $q$ (see Section 7). The discrete logarithm problem can also be formulated for an arbitrary group $G$, in the sense that if $g$ lies in the cyclic subgroup of $G$ generated by a given $h \in G$, then we are asked to find an integer $s$ such that $h^s = g$. However, we will restrict the attention to the case $G = F_q^*$.

If $q = p^t$ with $p$ being the characteristic of $F_q$, then $\mathrm{ind}_b(a)$ has a digit expansion

$$\mathrm{ind}_b(a) = \sum_{i=0}^{t-1} n_i p^i$$

in the base $p$, with integers $0 \leqslant n_i \leqslant p-1$ for $0 \leqslant i \leqslant t-1$. It suffices to know how to determine the least residue $n_0$ of $\mathrm{ind}_b(a)$ modulo $p$. For if $n_0$ has been calculated, then $\mathrm{ind}_b(a) = n_0 + mp$ for some integer $m \geqslant 0$, hence with $c = (ab^{-n_0})^{q/p}$ we get

$$c = \left(b^{mp}\right)^{q/p} = b^{mq} = b^m,$$

and so $m = \mathrm{ind}_b(c)$. Continuing in this manner, we can successively calculate $n_1, \ldots, n_{t-1}$, and so we obtain the value of $\mathrm{ind}_b(a)$. There is an explicit formula for the least residue of $\mathrm{ind}_b(a)$ modulo $p$ due to Mullen and White [102], namely

$$\mathrm{ind}_b(a) \equiv -1 + \sum_{j=1}^{q-2} \frac{a^j}{b^{-j}-1} \bmod p$$

for any $a \in F_q^*$ with $q \geqslant 3$. However, it is not yet clear whether this formula can somehow be used for the efficient computation of discrete logarithms for large $q$. For further work on explicit formulas for discrete logarithms we refer to Meletiou and Mullen [95] and Niederreiter [114].

If $q-1$ has no large prime factors, then the *Silver–Pohlig–Hellman algorithm* provides an efficient technique for solving the discrete logarithm problem in $F_q$. Recall that $r = \mathrm{ind}_b(a)$ satisfies $0 \leqslant r \leqslant q-2$, and so it suffices to determine $r$ modulo $q-1$. If $q-1 = q_1 \cdots q_k$ is the factorization of $q-1$ into pairwise coprime prime powers, then in view of the Chinese Remainder Theorem it is enough to determine $r$ modulo $q_h$ for $1 \leqslant h \leqslant k$. The latter task is solved by a procedure reminiscent of that in the previous paragraph, namely by a reduction to the problem of calculating the least residue of $r$ modulo $p_h$, where $p_h$ is the prime of which $q_h$ is a power. To determine the least residue $s_0$ of $r$ modulo $p_h$, we form

$$a^{(q-1)/p_h} = b^{(q-1)r/p_h} = c_h^r = c_h^{s_0},$$

where $c_h = b^{(q-1)/p_h}$ is a primitive $p_h$-th root of unity in $F_q$. If the distinct powers of $c_h$ have been precomputed, then $a^{(q-1)/p_h}$ uniquely determines $s_0$. For further information we refer to Lidl and Niederreiter [88], pp. 350–351, and McCurley [92], pp. 58–60.

A powerful method for computing discrete logarithms in $F_q$ is the *index-calculus algorithm*. This algorithm is of interest in the important cases where $q$ is a large prime or a large power of a small prime. The algorithm proceeds in two stages. In the first stage, we select special elements $a_1, \ldots, a_m$ of $F_q^*$ and we generate identities of the form

$$\prod_{j=1}^{m} a_j^{e_{ij}} = b^{f_i}$$

with integers $e_{ij}$ and $f_i$. These identities can be interpreted as a system of linear congruences

$$\sum_{j=1}^{m} e_{ij} \mathrm{ind}_b(a_j) \equiv f_i \bmod (q-1).$$

If sufficiently many congruences of this type have been collected, then we can expect that the system can be solved uniquely for the unknowns $\mathrm{ind}_b(a_j)$, $1 \leqslant j \leqslant m$. In the second stage of the algorithm, we calculate a desired discrete logarithm $\mathrm{ind}_b(a)$ by constructing an identity of the form

$$\prod_{j=1}^{m} a_j^{g_j} = ab^f$$

with integers $g_j$ and $f$. From this identity we obtain

$$\mathrm{ind}_b(a) \equiv \sum_{j=1}^{m} g_j \mathrm{ind}_b(a_j) - f \bmod (q-1),$$

and so $\mathrm{ind}_b(a)$ is determined. The first stage of the index-calculus algorithm is a precomputation and thus has to be carried out only once for each finite field $F_q$.

The above paragraph describes only the rough outlines of the index-calculus algorithm. The real difficulties are hidden in the details, namely how to select the special elements $a_1, \ldots, a_m$ in the first stage and how to generate the identities required in both stages. If $q$ is large, then the generation of suitable identities can be made feasible only by a probabilistic algorithm. In the case where $q$ is prime, we identify the elements of $F_q^*$ with positive integers less than $q$. Then $a_1, \ldots, a_m$ are chosen to be small primes (usually the first $m$ primes) and the desired identities are obtained by canonical factorization of integers. If $q = p^t$ with a prime $p$ and an exponent $t \geqslant 2$, then we identify the elements of $F_q^*$ with nonzero polynomials over $F_p$ of degree less than $t$. Here we take $a_1, \ldots, a_m$ to be irreducible polynomials over $F_p$ of small degree and we obtain the desired identities by canonical factorization of polynomials. Detailed expositions of the index-calculus algorithm can be found in the survey articles of McCurley [92] and Odlyzko [128] and in the book of Lidl and Niederreiter [88], Chapter 9.

For $q = 2^t$ the best available version of the index-calculus algorithm is that of Coppersmith [28] which, under certain heuristic assumptions, can be shown to be a probabilistic algorithm with a subexponential expected running time of the form $\mathrm{O}(\mathrm{e}^{ct^{1/3} \log t^{2/3}})$ with some constant $c > 0$. For the following it is convenient to put

$$L(q) = \mathrm{e}^{(\log q)^{1/2} (\log \log q)^{1/2}}$$

for any prime power $q$. In the case where $q$ is prime, the index-calculus algorithm of Coppersmith, Odlyzko and Schroeppel [29] represents a probabilistic algorithm for which a heuristic analysis yields an expected running time $\mathrm{O}(L(q)^{1+\mathrm{o}(1)})$. A probabilistic discrete logarithm algorithm that can be subjected to a rigorous complexity analysis was designed by Pomerance [134]. This algorithm uses the elliptic curve method for factoring integers as a subroutine. The algorithm of Pomerance has expected running time $\mathrm{O}(L(q)^{\sqrt{2}+\mathrm{o}(1)})$ if either $q = 2^t$ or $q$ is prime, and no unproved heuristic assumptions are needed to derive this result.

For more extensive discussions of discrete logarithms, including their cryptographic applications, we refer to Lidl and Niederreiter [88], Chapter 9, McCurley [92], Odlyzko [128], and van Oorschot [130].

## 6. Linear recurring sequences

Let $k$ be a positive integer and let $a_0, a_1, \ldots, a_{k-1}$ be fixed elements of a finite field $F_q$. A sequence $s_0, s_1, \ldots$ of elements of $F_q$ satisfying the ($k$-th-order) linear recurrence relation

$$s_{n+k} = \sum_{i=0}^{k-1} a_i s_{n+i} \quad \text{for } n = 0, 1, \ldots$$

is called a (*$k$-th-order*) *linear recurring sequence* in $F_q$. We often abbreviate the sequence $s_0, s_1, \ldots$ by $(s_n)$. The sequence $(s_n)$ is uniquely determined by the linear recurrence relation and by the initial values $s_0, s_1, \ldots, s_{k-1}$. In electrical engineering, linear recurring sequences in $F_q$ are generated by special switching circuits called "linear feedback shift registers" (compare with [88], Chapter 6), and so one speaks also of *linear feedback shift-register sequences*. Linear recurring sequences have been studied for centuries from the theoretical point of view, and in the last few decades they have become important in applied areas such as algebraic coding theory, cryptology, digital signal processing, and pseudorandom number generation.

Any linear recurring sequence $(s_n)$ in $F_q$ is *periodic*, in the sense that there exists a preperiod $n_0 \geqslant 0$ and a period $r \geqslant 1$ such that $s_{n+r} = s_n$ for all $n \geqslant n_0$. If $a_0 \neq 0$, then the sequence is purely periodic, i.e. we can take $n_0 = 0$. An easy way to prove the periodicity, and also to obtain an upper bound on the least period, is based on the consideration of the *state vectors*

$$\mathbf{s}_n = (s_n, s_{n+1}, \ldots, s_{n+k-1}) \in F_q^k \quad \text{for } n = 0, 1, \ldots .$$

If the given linear recurring sequence is such that no state vector is the zero vector, then it follows from the pigeon-hole principle that $\mathbf{s}_j = \mathbf{s}_h$ for some $h$ and $j$ with $0 \leqslant h < j \leqslant q^k - 1$. In view of the linear recurrence relation, this implies $\mathbf{s}_{n+j-h} = \mathbf{s}_n$ for all $n \geqslant h$, and so the linear recurring sequence is periodic with least period $r \leqslant j - h \leqslant q^k - 1$. On the other hand, if one of the state vectors is the zero vector, then all subsequent state vectors are zero vectors, and so the linear recurring sequence is periodic with least period $r = 1 \leqslant q^k - 1$. Thus, the least period of a $k$-th-order linear recurring sequence in $F_q$ is always at most $q^k - 1$.

State vectors are linked by the $k \times k$ matrix

$$A = \begin{pmatrix} 0 & 0 & 0 & \ldots & 0 & a_0 \\ 1 & 0 & 0 & \ldots & 0 & a_1 \\ 0 & 1 & 0 & \ldots & 0 & a_2 \\ \vdots & \vdots & \vdots & & \vdots & \vdots \\ 0 & 0 & 0 & \ldots & 1 & a_{k-1} \end{pmatrix}$$

over $F_q$ associated with the linear recurrence relation. Indeed, a straightforward induction shows that

$$\mathbf{s}_n = \mathbf{s}_0 A^n \quad \text{for } n = 0, 1, \ldots .$$

Since $A^n$ can be calculated by O(log $n$) matrix multiplications using the standard square-and-multiply technique, the above identity leads to an efficient algorithm for computing remote terms of the linear recurring sequence $(s_n)$. The currently fastest algorithm for this task is due to Fiduccia [45]. Note that the matrix $A$ is nonsingular if $a_0 \neq 0$, and so in this case the identity $\mathbf{s}_n = \mathbf{s}_0 A^n$ implies that the least period of the linear recurring sequence divides the order of $A$ in the general linear group $\mathrm{GL}(k, F_q)$.

From the linear recurrence relation for a linear recurring sequence $(s_n)$ we obtain the polynomial

$$f(x) = x^k - \sum_{i=0}^{k-1} a_i x^i \in F_q[x],$$

which is called a *characteristic polynomial* of $(s_n)$. If $A$ is the matrix above, then $f$ is also the characteristic polynomial of $A$, and on the other hand, $A$ is the companion matrix of $f$. Let $F_q^\infty$ be the sequence space over $F_q$, viewed as a vector space over $F_q$ under termwise operations, and let $T$ be the shift operator

$$T(v_n) = (v_{n+1}) \quad \text{for all } (v_n) \in F_q^\infty .$$

Then for a characteristic polynomial $f$ of $(s_n)$ we have

$$f(T)(s_n) = (0),$$

with the zero sequence on the right-hand side. The set

$$\{g \in F_q[x]\colon\ g(T)(s_n) = (0)\}$$

of annihilating polynomials is a nonzero ideal in $F_q[x]$ and is therefore generated by a uniquely determined monic polynomial over $F_q$, called the *minimal polynomial* of the linear recurring sequence $(s_n)$. It is clear that the minimal polynomial divides any characteristic polynomial of $(s_n)$. A characteristic polynomial of degree $k$ is the minimal polynomial of $(s_n)$ if and only if the corresponding state vectors $\mathbf{s}_0, \mathbf{s}_1, \ldots, \mathbf{s}_{k-1}$ are linearly independent over $F_q$.

The minimal polynomial contains all the information about the periodicity properties of a linear recurring sequence. We use the notion of the order $\mathrm{ord}(f)$ of a polynomial $f$ over $F_q$ introduced in Section 1. See [88], Chapter 3, for further information on the order of polynomials and [88], Chapter 6, for the result below.

THEOREM 6.1. *If $m \in F_q[x]$ is the minimal polynomial of the linear recurring sequence $(s_n)$ in $F_q$, then the least period of $(s_n)$ is equal to* $\mathrm{ord}(m)$ *and the least preperiod of $(s_n)$ is equal to the multiplicity of* 0 *as a root of $m$.*

A linear recurring sequence $(s_n)$ in $F_q$ whose minimal polynomial $m$ is a primitive polynomial over $F_q$ is called a *maximal period sequence* in $F_q$. If $\deg(m) = k$, then by Theorem 6.1 the maximal period sequence $(s_n)$ is purely periodic with least period $q^k - 1$. Note that by an earlier discussion, $q^k - 1$ is the largest value that can be achieved by the least period of a $k$-th-order linear recurring sequence in $F_q$. From that discussion it is also clear that the state vectors $\mathbf{s}_0, \mathbf{s}_1, \ldots, \mathbf{s}_{q^k-2}$ of a $k$-th-order maximal period sequence in $F_q$ with $k = \deg(m)$ run exactly through all nonzero vectors in $F_q^k$. This property of maximal period sequences is basic for many applications of these sequences, such as the construction of de Bruijn sequences (compare also with Section 9).

A very useful viewpoint in the theory of linear recurring sequences is that of generating functions. In the classical approach described in [88], Chapter 6, the formal power series

$$\sum_{n=0}^{\infty} s_n x^n \in F_q[[x]]$$

is associated with the sequence $s_0, s_1, \ldots$ of elements of $F_q$. However, it is often more convenient to associate with this sequence the formal Laurent series

$$\sum_{n=0}^{\infty} s_n x^{-n-1} \in F_q\big((x^{-1})\big).$$

For instance, according to Niederreiter [108] we then get the following simple characterization of linear recurring sequences with given minimal polynomial.

THEOREM 6.2. *Let $m \in F_q[x]$ be a monic polynomial. Then the sequence $s_0, s_1, \ldots$ of elements of $F_q$ is a linear recurring sequence with minimal polynomial $m$ if and only if*

$$\sum_{n=0}^{\infty} s_n x^{-n-1} = \frac{g(x)}{m(x)}$$

*with $g \in F_q[x]$ and* $\gcd(g, m) = 1$.

Thus, the formal Laurent series associated with a linear recurring sequence is a rational function, and the monic denominator of the reduced form of this rational function is the minimal polynomial of the linear recurring sequence. The approach via generating functions yields important explicit formulas for the terms of a linear recurring sequence $(s_n)$ in $F_q$ with characteristic polynomial $f \in F_q[x]$. Let $e_0$ be the multiplicity of 0 as a root of $f$, where we can have $e_0 = 0$, and let $\alpha_1, \ldots, \alpha_h$ be the distinct nonzero roots of $f$ (in its splitting field $F$ over $F_q$) with multiplicities $e_1, \ldots, e_h$, respectively. Then

$$s_n = t_n + \sum_{i=1}^{h} \sum_{j=0}^{e_i-1} \binom{n+j}{j} \beta_{ij} \alpha_i^n \quad \text{for } n = 0, 1, \ldots,$$

where all $t_n \in F_q$, $t_n = 0$ for $n \geqslant e_0$, and all $\beta_{ij} \in F$. If $f$ is irreducible over $F_q$, then this formula can be put into the much simpler form

$$s_n = \mathrm{Tr}_{F/F_q}(\theta\alpha^n) \quad \text{for } n = 0, 1, \ldots,$$

where $\alpha \in F$ is a fixed root of $f$ and $\theta \in F$ is uniquely determined. We refer to [88], Chapter 6, for the proof of these formulas.

For a monic $f \in F_q[x]$ let $S(f)$ be the kernel of the linear operator $f(T)$ on $F_q^\infty$. If $\deg(f) \geqslant 1$, then $S(f)$ consists exactly of all linear recurring sequences in $F_q$ with characteristic polynomial $f$, whereas $S(1) = \{(0)\}$. Any $S(f)$ is a linear subspace of $F_q^\infty$ of dimension $\deg(f)$. The following result (see [88], Chapter 6, for its proof) characterizes the spaces $S(f)$.

THEOREM 6.3. *A subset $E$ of $F_q^\infty$ is equal to $S(f)$ for some monic $f \in F_q[x]$ if and only if $E$ is a finite-dimensional subspace of $F_q^\infty$ which is closed under the shift operator $T$.*

The subspaces $S(f)$ of $F_q^\infty$ are linked by various identities. For instance, for any monic $f_1, \ldots, f_h \in F_q[x]$ we have

$$S(f_1) \cap \cdots \cap S(f_h) = S(\gcd(f_1, \ldots, f_h))$$

and

$$S(f_1) + \cdots + S(f_h) = S(\mathrm{lcm}(f_1, \ldots, f_h)).$$

In particular, if a monic $f \in F_q[x]$ is factored in the form $f = g_1 \cdots g_h$ with pairwise coprime and monic $g_1, \ldots, g_h \in F_q[x]$, then $S(f)$ is the direct sum

$$S(f) = S(g_1) \oplus \cdots \oplus S(g_h).$$

An important operation is that of termwise multiplication of linear recurring sequences. If $\sigma_i = (s_n^{(i)})$, $1 \leqslant i \leqslant h$, are $h$ sequences of elements of $F_q$, then their (termwise) product $\sigma_1 \cdots \sigma_h$ is the sequence $(s_n)$ with terms $s_n = s_n^{(1)} \cdots s_n^{(h)}$ for $n = 0, 1, \ldots$. For monic polynomials $f_1, \ldots, f_h \in F_q[x]$ let $S(f_1) \cdots S(f_h)$ be the subspace of $F_q^\infty$ spanned by all products $\sigma_1 \cdots \sigma_h$ with $\sigma_i \in S(f_i)$ for $1 \leqslant i \leqslant h$. Since $S(f_1) \cdots S(f_h)$ satisfies the conditions in Theorem 6.3, it follows that

$$S(f_1) \cdots S(f_h) = S(g)$$

for some monic $g \in F_q[x]$. The general problem of determining the polynomial $g$ is not easy. There is a trivial case, namely when $f_i = 1$ for some $i$, since then obviously $g = 1$. Thus we can assume that $f_1, \ldots, f_h$ are nonconstant. Let $f_1 \vee \cdots \vee f_h$ be the monic polynomial whose roots are the distinct elements of the form $\alpha_1 \cdots \alpha_h$, where each $\alpha_i$ is a root of $f_i$ in the splitting field of $f_1 \cdots f_h$ over $F_q$. Since the conjugates over $F_q$ of such a product $\alpha_1 \cdots \alpha_h$ are again elements of this form, it follows that $f_1 \vee \cdots \vee f_h$ is a polynomial over $F_q$. If each $f_i$, $1 \leqslant i \leqslant h$, has only simple roots, then we have the formula

$$S(f_1)\cdots S(f_h) = S(f_1 \vee \cdots \vee f_h).$$

The general case is considerably more complicated and is treated in Zierler and Mills [166].

If $\sigma = (s_n)$ is a sequence of elements of $F_q$ and $d$ is a positive integer, then the operation of *decimation* produces the decimated sequence $\sigma^{(d)} = (s_{nd})$. Thus, $\sigma^{(d)}$ is obtained by taking every $d$-th term of $\sigma$, starting from $s_0$. It turns out that if $\sigma$ is a linear recurring sequence in $F_q$, then so is $\sigma^{(d)}$. In fact, it follows from a general result of Niederreiter [110] that if $f \in F_q[x]$ is a characteristic polynomial of $\sigma$ and

$$f(x) = \prod_{j=1}^{k}(x - \alpha_j)$$

is the factorization of $f$ in its splitting field over $F_q$, then

$$g_d(x) = \prod_{j=1}^{k}(x - \alpha_j^d),$$

which is again a polynomial over $F_q$, is a characteristic polynomial of $\sigma^{(d)}$. Moreover, if $f$ is the minimal polynomial of $\sigma$ and $d$ is coprime to the least period of $\sigma$, then $g_d$ is the minimal polynomial of $\sigma^{(d)}$. Further information on characteristic polynomials of $\sigma^{(d)}$ can be found in Duvall and Mortick [36].

A linear recurring sequence $\sigma$ in $F_q$ which has a characteristic polynomial $f \in F_q[x]$ and satisfies $\sigma^{(q)} = \sigma$ is called a *characteristic sequence* for $f$. Characteristic sequences play an important role in the recent algorithm of Niederreiter for factoring polynomials over finite fields; see Niederreiter and Göttfert [124] for this application of characteristic sequences. In that paper, characteristic sequences are also described explicitly in terms of their generating functions: $(s_n)$ is a characteristic sequence for $f$ if and only if its generating function

$$\sum_{n=0}^{\infty} s_n x^{-n-1}$$

has the form

$$\sum_{i=1}^{m} c_i \frac{p_i'}{p_i} \quad \text{with } c_1, \ldots, c_m \in F_q,$$

where $p_1, \ldots, p_m \in F_q[x]$ are the distinct monic irreducible factors of $f$ and $p_i'$ is the first derivative of $p_i$. A consequence of this result is that the minimal polynomial of any characteristic sequence has only simple roots.

Linear recurring sequences in $F_q$ can be characterized in various ways. An obvious criterion states that a sequence of elements of $F_q$ is a linear recurring sequence if and only if it is periodic. Another criterion follows from Theorem 6.2, namely that the sequence $(s_n)$ of elements of $F_q$ is a linear recurring sequence if and only if its generating function

$$\sum_{n=0}^{\infty} s_n x^{-n-1}$$

is a rational function. Still another approach employs techniques from linear algebra. For an arbitrary sequence $(s_n)$ of elements of $F_q$ and for integers $n \geqslant 0$ and $b \geqslant 1$ define the Hankel determinant

$$D_n^{(b)} = \begin{vmatrix} s_n & s_{n+1} & \cdots & s_{n+b-1} \\ s_{n+1} & s_{n+2} & \cdots & s_{n+b} \\ \vdots & \vdots & & \vdots \\ s_{n+b-1} & s_{n+b} & \cdots & s_{n+2b-2} \end{vmatrix}.$$

Then $(s_n)$ is a linear recurring sequence in $F_q$ if and only if there exists an integer $b \geqslant 1$ such that $D_n^{(b)} = 0$ for all sufficiently large $n$. Also, $(s_n)$ is a linear recurring sequence with minimal polynomial of degree $k$ if and only if

$$D_0^{(b)} = 0 \quad \text{for all } b \geqslant k+1$$

and $k+1$ is the least positive integer for which this holds. Proofs of these results can be found in [88], Chapter 6.

If a linear recurring sequence in $F_q$ is known to have a minimal polynomial of degree $k \geqslant 1$, then the minimal polynomial is determined by the first $2k$ terms of the sequence. This is seen by writing down the linear recurrence relation for $n = 0, 1, \ldots, k-1$, thereby obtaining a system of $k$ linear equations for the unknown coefficients $a_0, a_1, \ldots, a_{k-1}$ of the minimal polynomial. The determinant of this system is $D_0^{(k)}$, which is $\neq 0$ by one of the criteria in the previous paragraph. Therefore, the system can be solved uniquely. More generally, if a linear recurring sequence in $F_q$ is known to have a minimal polynomial of degree $\leqslant k$ for some integer $k \geqslant 1$, then the minimal polynomial is determined by the first $2k$ terms of the sequence. An algorithm which, under this condition, produces the minimal polynomial from the first $2k$ terms of the sequence was developed by Berlekamp [3] and Massey [91]. This algorithm is of importance in the decoding of cyclic codes and in the analysis of keystreams used in stream ciphers (compare with Section 7 for the latter application).

Detailed expository accounts of the theory of linear recurring sequences in finite fields are given in the books of Lidl and Niederreiter [87], Chapter 8, [88], Chapter 6, and McEliece [93], Chapters 9–11. A fundamental paper in the area is that of Zierler [165]. A good source for applications of linear recurring sequences is the book of Golomb [56].

## 7. Finite fields in cryptology

Finite fields have found many applications in cryptology, the theory of data security and integrity, which is a subject of increasing importance in an era relying more and more on electronic information and communication. This is not the place to present an introduction to cryptology, for which we refer the reader to the textbook literature, such as the mathematically oriented book of van Tilborg [156]. However, to provide some background information we describe at least rudimentarily the concept of a cryptosystem, this being the basic tool for data security that transforms data in original form (= plaintext messages) into protected data in scrambled form (= ciphertexts) and vice versa.

Formally, a *cryptosystem* consists of an *enciphering scheme* $E = \{E_k\}$ and a *deciphering scheme* $D = \{D_{k'}\}$, both of which are families of injective functions parameterized by keys. Given a (plaintext) message $m$ and a key $k$, the enciphering scheme produces the ciphertext $c = E_k(m)$. The deciphering scheme recovers $m$ by using a key $k'$ and producing $D_{k'}(c) = m$. If $k = k'$, or more generally if $k$ and $k'$ are computationally equivalent in the sense that they can easily be obtained from each other, then we speak of a *symmetric cryptosystem*. In this case, both keys $k$ and $k'$ have to be kept secret from unauthorized users. Examples of symmetric cryptosystems are the well-known block cipher DES and stream ciphers. In a *public-key cryptosystem*, the encryption key is public knowledge and only the decryption key $k'$ is kept secret. The security of a public-key cryptosystem is based on the assumption that it is computationally infeasible to derive $k'$ from $k$. A standard example of a public-key cryptosystem is the RSA cryptosystem whose security rests on the difficulty of factoring large integers.

Several cryptographic schemes are based on the computational complexity of the discrete logarithm problem (compare with Section 5). A simple scheme of this type is the key-exchange system of Diffie and Hellman [35], which can be used to distribute secret keys for symmetric cryptosystems. In this system, the large finite field $F_q$ and the primitive element $b$ of $F_q$ are publicly known. If two participants A and B want to establish a common key for secret communication, they first select arbitrary integers $r$ and $s$, respectively, with $2 \leqslant r,\ s \leqslant q-2$, and then A sends $b^r$ to B, while B transmits $b^s$ to A. Now they take $b^{rs}$ as their common key, which A computes as $(b^s)^r$ and B as $(b^r)^s$. An opponent may observe $b^r$ and $b^s$ passing over the communication channel, but it seems that the only way to infer from these data the secret key $b^{rs}$ is to calculate the discrete logarithms $r$ and $s$ of $b^r$ and $b^s$, respectively, relative to $b$. If $q$ is well chosen, then the discrete logarithm problem for $F_q$ can be regarded as computationally infeasible.

A public-key cryptosystem based on the difficulty of the discrete logarithm problem was designed by El Gamal [42]. As above, the large finite field $F_q$ and the primitive element $b$ of $F_q$ are supposed to be public knowledge. The secret key of a typical participant A is an integer $r$ with $2 \leqslant r \leqslant q-2$ and the public key of A is the element $b^r$ of $F_q$. The admissible messages are nonzero elements of $F_q$. If another participant B wants to send a message $m \in F_q^*$ to A, then B selects an arbitrary integer $s$ with $2 \leqslant s \leqslant q-2$ and transmits the pair $(b^s, mb^{rs})$ to A. For decryption, A calculates $b^{rs} = (b^s)^r$ and recovers $m = (mb^{rs})(b^{rs})^{-1}$. A different type of public-key cryptosystem using a finite field $F_q$ was proposed by Chor and Rivest [18]; here $q$ is chosen such that discrete logarithms in $F_q$ can be calculated with a reasonable effort, and the security is based on the difficulty of a knapsack problem.

In view of the progress on solving the discrete logarithm problem for finite fields (see Section 5), more general problems have been proposed as the basis for cryptographic schemes. One direction of research aims to replace $F_q^*$ by seemingly more complicated groups, such as the group of rational points on an elliptic curve over $F_q$ (see Koblitz [73], Chapter 6) or the class group of an algebraic number field (see Buchmann and Williams [16]). Another idea is to view the elements $b^r$, $r = 0, 1, \ldots$, as stemming from a first-order linear recurring sequence in $F_q$ and to generalize by considering linear recurring sequences in $F_q$ of arbitrary order (compare with Section 6). The appropriate generalization of the transition from $b^r$ to $b^{rs}$ is then the operation of decimation for linear recurring sequences, and the analog of the discrete logarithm problem is the problem of calculating – given a characteristic polynomial $f$ of a linear recurring sequence $\sigma$ and a characteristic polynomial $g_d$ of a decimated sequence $\sigma^{(d)}$ – the decimation index $d$ (compare again with Section 6). A family of cryptosystems based on these principles was introduced by Niederreiter [111].

There are interesting applications of algebraic coding theory to the design of public-key cryptosystems. The theoretical basis for these applications is the fact that the decoding problem for general linear codes over $F_q$ is NP-complete. The standard example of a public-key cryptosystem based on linear codes over $F_q$ is the Goppa-code cryptosystem which is described in detail in [88], Chapter 9.

Extensive use of the theory of linear recurring sequences in $F_q$ is made in the area of stream ciphers. A *stream cipher* is a symmetric cryptosystem in which messages and ciphertexts are strings of elements of a finite field $F_q$, and encryption and decryption proceed by termwise addition, respectively subtraction, of the same secret string of elements of $F_q$. This secret string, called the *keystream*, is generated by a (possibly known) deterministic algorithm from certain secret seed data and should possess good statistical randomness properties and a high complexity, so that the keystream cannot be inferred from a small portion of its terms. Many keystream generators use linear recurring sequences in $F_q$ as building blocks; see Rueppel [135] for a survey of algorithms for keystream generation.

Since most hardware-based keystream generators produce periodic sequences, a relevant measure of complexity in this context is the *linear complexity* $L(\sigma)$, which is simply the degree of the minimal polynomial of a periodic, and thus linear recurring, sequence $\sigma$ of elements of $F_q$. In the periodic case, only sequences with a very large linear complexity are acceptable as keystreams. A fundamental problem in the analysis of keystreams is that of bounding the linear complexity of sequences obtained by algebraic operations on periodic sequences. The following result contains the basic information.

THEOREM 7.1. *If $\sigma + \tau$ and $\sigma\tau$ denote, respectively, the termwise sum and the termwise product of the periodic sequences $\sigma$ and $\tau$ of elements of $F_q$, then*

$$L(\sigma + \tau) \leqslant L(\sigma) + L(\tau) \quad \text{and} \quad L(\sigma\tau) \leqslant L(\sigma)L(\tau).$$

The first inequality follows immediately from the identity

$$S(f_1) + S(f_2) = S\big(\mathrm{lcm}(f_1, f_2)\big)$$

stated in Section 6, and the second inequality can be found in Herlestam [63]. Refinements of Theorem 7.1 are discussed in Rueppel [135].

A more subtle complexity analysis, which also allows the treatment of nonperiodic sequences, is based on the following notions. If $N$ is a positive integer and $\sigma$ is an arbitrary (infinite) sequence of elements of $F_q$ or a finite string of at least $N$ elements of $F_q$, then the *$N$-th linear complexity* $L_N(\sigma)$ is the least linear complexity of any periodic sequence whose first $N$ terms agree with those of $\sigma$. Furthermore, the *linear complexity profile* of $\sigma$ is the sequence $L_1(\sigma), L_2(\sigma), \ldots$, extended as long as $L_N(\sigma)$ is defined. Since $L_N(\sigma) \leqslant L_{N+1}(\sigma)$, the linear complexity profile is a nondecreasing sequence of non-negative integers. The linear complexity profile can be efficiently calculated by the Berlekamp–Massey algorithm mentioned in Section 6. A crucial fact for the theory of the linear complexity profile is the close connection with continued fraction expansions. For simplicity, we describe this connection only for (infinite) sequences $\sigma$ of elements $s_0, s_1, \ldots$ of $F_q$. Let

$$S = \sum_{n=0}^{\infty} s_n x^{-n-1} \in F_q\big((x^{-1})\big)$$

be the generating function of $\sigma$. Then $S$ has a unique continued fraction expansion

$$S = 1/\big(A_1 + 1/(A_2 + \cdots)\big)$$

with partial quotients $A_j \in F_q[x]$ for which $d_j = \deg(A_j) \geqslant 1$ for $j \geqslant 1$. This expansion is finite if $S$ is a rational function, i.e. if $\sigma$ is a periodic sequence, and infinite otherwise. We put $d_j = \infty$ whenever $A_j$ does not exist, and also $d_0 = 0$. Now we can state the following special case of a result of Niederreiter [108].

THEOREM 7.2. *With the above notation we have*

$$L_N(\sigma) = \sum_{i=0}^{j(N)} d_i$$

*for every $N \geqslant 1$, where $j(N) \geqslant 0$ is uniquely determined by*

$$2\sum_{i=0}^{j(N)-1} d_i + d_{j(N)} \leqslant N < 2\sum_{i=0}^{j(N)} d_i + d_{j(N)+1}.$$

Therefore, the linear complexity profile of $\sigma$ has the form

$$0, \ldots, 0, d_1, \ldots, d_1, d_1 + d_2, \ldots, d_1 + d_2, \ldots$$

with 0 occurring $d_1 - 1$ times and

$$\sum_{i=1}^{j} d_i$$

occurring $d_j + d_{j+1}$ times for all $j \geqslant 1$, where the positive integers $d_1, d_2, \ldots$ are the degrees of the partial quotients in the continued fraction expansion of the generating function of $\sigma$.

Theorem 7.2 is basic for the probabilistic theory of the linear complexity profile developed by Niederreiter [112]. This theory describes the behavior of the linear complexity profile for random sequences of elements of $F_q$ and establishes benchmarks for statistical randomness tests using the linear complexity profile. We mention a typical result, namely that – in a suitable stochastic model – for a random sequence $\sigma$ of elements of $F_q$ we have

$$L_N(\sigma) = \frac{N}{2} + \mathrm{O}(\log N) \quad \text{for all } N \geqslant 2.$$

A survey of this probabilistic theory is given in Niederreiter [119].

Algorithms for the generation of keystreams satisfying stronger complexity requirements than those connected with linear complexity have also been developed, but the verification of the desired properties is usually conditional on heuristic complexity-theoretic hypotheses. For instance, the randomness properties of the keystream generator of Blum and Micali [10] are based on the presumed difficulty of the discrete logarithm problem for finite prime fields. Surveys of such cryptographically strong keystream generators can be found in Kranakis [74] and Lagarias [77]. An algorithm for keystream generation based on polynomials over finite prime fields was recently designed by Niederreiter and Schnorr [126].

Permutation polynomials of finite fields of characteristic 2 and Boolean functions on such fields are of interest in the design of cryptographic functions for block ciphers. Desirable properties of such functions are a large deviation from linearity (which can be measured by the Hamming distance to the set of affine functions), equidistribution properties, and uncorrelatedness, among others. Recent papers on this topic, which contain also further references, include Meier and Staffelbach [94], Mitchell [97], Nyberg [127], and Pieprzyk [131].

For further information on cryptology we refer to the textbook of van Tilborg [156] and to the state-of-the-art survey articles in the book edited by Simmons [148]. Applications of finite fields to cryptology are covered in more detail in the book of Lidl and Niederreiter [88], Chapter 9, and in the review article of Niederreiter [119].

## 8. Finite fields in combinatorics

Combinatorics or combinatorial theory is a relatively young but very vigorous discipline of mathematics and many parts of it represent interesting applications of finite fields within mathematics. We shall only give some samples of these applications, such as Latin squares, block designs, and difference sets.

A square array $L = (a_{ij})$, $i, j = 1, 2, \ldots, n$, is called a *Latin square* of order $n$ if each row and each column contains every element of a set of $n$ elements exactly once. Two Latin squares $(a_{ij})$ and $(b_{ij})$ of order $n$ are said to be *orthogonal* if the $n^2$ ordered pairs $(a_{ij}, b_{ij})$ are all different. Orthogonal Latin squares were first studied by L. Euler who, in 1782 in a systematic study of designs, posed the 36 Officers Problem: "Given 6 officers from each of 6 different regiments so that a selection includes one officer from each of 6 ranks, is it possible for the officers to parade in a $6 \times 6$ formation such that each row and each column contains one member of each rank and one member of each regiment?" Euler conjectured that there is no pair of orthogonal Latin squares of order 6. This was verified by Tarry [154] in 1901. Euler conjectured also that there is no pair of orthogonal Latin squares of order $n$ for any $n \equiv 2 \bmod 4$. This was disproved by Bose and Shrikhande [11] in 1959 through the construction of a pair of orthogonal Latin squares of order 22.

It is easy to see that a Latin square of order $n$ exists for every positive integer $n$. Simply let $(a_{ij})$ be defined by $a_{ij} \equiv i + j \bmod n$, $1 \leqslant a_{ij} \leqslant n$. If a collection of Latin squares is orthogonal in pairs, we speak of *mutually orthogonal Latin squares*, or MOLS. It can be shown that there are at most $n - 1$ MOLS of order $n$. The following theorem represents an application of finite fields. Its proof is straightforward.

THEOREM 8.1. *Let $n = q = p^k$, $p$ a prime and $k$ a positive integer. Let $a_0 = 0$, $a_1, a_2, \ldots, a_{n-1}$ be the elements of $F_q$ and define $n \times n$ arrays $L_m = (a_{ij}^{(m)})$, $0 \leqslant i, j \leqslant n-1$, $m = 1, 2, \ldots, n-1$, where $a_{ij}^{(m)} = a_m a_i + a_j$. Then the $n - 1$ arrays $L_1, \ldots, L_{n-1}$ form a set of MOLS of order $n$.*

The following result shows, in particular, the existence of a pair of orthogonal Latin squares of order $n$ for any $n > 1$ with $n \not\equiv 2 \bmod 4$.

THEOREM 8.2. *Let $q_1, \ldots, q_s$ be prime powers and let $a_0^{(i)} = 0$, $a_1^{(i)}, \ldots, a_{q_i-1}^{(i)}$ be the elements of $F_{q_i}$. Define $b_k = (a_k^{(1)}, \ldots, a_k^{(s)})$ for $0 \leqslant k \leqslant r = \min_{1 \leqslant i \leqslant s}(q_i - 1)$ and let $b_{r+1}, \ldots, b_{n-1}$ with $n = q_1 \cdots q_s$ be the remaining $s$-tuples that can be formed by taking in the $i$-th coordinate an element of $F_{q_i}$. Then the arrays $L_k = (l_{ij}^{(k)})$, $0 \leqslant i, j \leqslant n - 1$, $k = 1, 2, \ldots, r$, with $l_{ij}^{(k)} = b_k b_i + b_j$ form a set of $r$ MOLS of order $n$.*

Most of the results on Latin squares are summarized in the comprehensive book by Dénes and Keedwell [31]. For a recent update see Dénes and Keedwell [32]. Latin squares can be generalized to $d$-dimensional hypercubes. For $d \geqslant 2$, a *$d$-dimensional hypercube* of order $q$ is a $q \times \cdots \times q$ array with $q^d$ points based upon $q$ distinct symbols. For $0 \leqslant i \leqslant d - 1$, such a hypercube has type $i$ if, whenever any $i$ of the coordinates are fixed, each of the $q$ symbols appears $q^{d-i-1}$ times in that subarray. A collection of $d$ hypercubes of dimension $d$ is called *$d$-orthogonal* if, when superimposed, each of the $q^d$ possible $d$-tuples appears once. A set of $t \geqslant d$ hypercubes is *$d$-orthogonal* if every subset of $d$ hypercubes is $d$-orthogonal. When $d = 2$ and $i = 0$ we obtain the notion of mutually orthogonal squares, and if $i = 1$ then they are MOLS. Golomb and Posner [58] showed the equivalence between a set of $t$ MOLS of order $q$ and a code of word length $t+2$ and minimum distance $t+1$ having $q^2$ codewords over an alphabet with $q$ symbols. Mullen and Whittle [103] generalized this result to hypercubes. Brawley and Mullen [13] give examples of some sets of MOLS of infinite order containing nested sets of MOLS of finite order by using an iterated presentation of an infinite algebraic extension of $F_p$ as in Brawley and Schnibben [14].

Latin squares have also been generalized in a different way as frequency squares. An $F(n; \lambda_1, \ldots, \lambda_m)$ *frequency square* based on $m$ symbols $a_1, \ldots, a_m$ is an $n \times n$ array such that, for each $i = 1, 2, \ldots, m$, the symbol $a_i$ occurs $\lambda_i$ times in each row and column. Hence $n = \lambda_1 + \cdots + \lambda_m$, and an $F(n; 1, \ldots, 1)$ frequency square is a Latin square. If $\lambda_1 = \cdots = \lambda_m$ we write $F(n; \lambda)$ where $n = \lambda m$. Two frequency squares $F_1(n; \lambda_1, \ldots, \lambda_m)$ based on $a_1, \ldots, a_m$ and $F_2(n; \mu_1, \ldots, \mu_l)$ based on $b_1, \ldots, b_l$ are *orthogonal* if on superposition each of the ordered pairs $(a_i, b_j)$ occurs $\lambda_i \mu_j$ times for $i = 1, \ldots, m$, $j = 1, \ldots, l$. If each of a set of $t$ mutually orthogonal frequency squares is of type $F(n; \lambda)$ with $n = \lambda m$, then $t \leqslant (n-1)^2/(m-1)$, and a set is *complete* if equality holds. Mullen [98] gives the following construction of mutually orthogonal frequency squares (MOFS). Let $i \geqslant 1$ be an integer. Label the rows of a $q^i \times q^i$ square using all $i$-tuples over $F_q$. Similarly, label the columns. Then any such square may be viewed as a function $f$: $F_q^{2i} \to F_q$, where the element $f(x_1, \ldots, x_{2i})$ is placed at the intersection of row $(x_1, \ldots, x_i)$ and column $(x_{i+1}, \ldots, x_{2i})$.

THEOREM 8.3. *The* $(q^i - 1)^2/(q-1)$ *polynomials*

$$f_{(a_1,\ldots,a_{2i})}(x_1, \ldots, x_{2i}) = a_1 x_1 + \cdots + a_{2i} x_{2i}$$

*over* $F_q$, *where*

(i) $(a_1, \ldots, a_i) \neq (0, \ldots, 0)$,

(ii) $(a_{i+1}, \ldots, a_{2i}) \neq (0, \ldots, 0)$,

(iii) *no two sets of a's are scalar multiples of each other, i.e.* $(a'_1, \ldots, a'_{2i}) \neq c(a_1, \ldots, a_{2i})$ *for any* $c \in F_q$,

*represent a complete set of MOFS of type* $F(q^i; q^{i-1})$.

This technique has been developed further by Suchower [153] to construct sets of mutually orthogonal (Youden) frequency hyperrectangles whose dimensions are prime powers by using the theory of subfield permutation polynomials and orthogonal subfield systems described in Suchower [152]. See also Mullen and Suchower [101] for a study of certain complete sets of mutually orthogonal Youden frequency hyperrectangles and their equivalence to error-correcting codes and fractional replication plans.

We now consider block designs, finite geometries, and difference sets and note a close connection between them. A *design* based on the nonempty set $V$ of points is a pair $(V, B)$, where $B = \{B_i\colon i \in I\}$ is a nonempty family of subsets of $V$, called blocks. The terminology that is normally used in this area has its origin in the applications in statistics, namely the design of experiments. The points are called varieties. Usually their number is denoted by $v$ and the number of blocks by $b$. A design in which every block is incident with the same number $k$ of varieties and every variety is incident with the same number $r$ of blocks is called a *tactical configuration*. Clearly $vr = bk$. If $v = b$ and hence $r = k$, it is called *symmetric*. A tactical configuration is called a *balanced incomplete block design* (BIBD) if $v > k \geqslant 2$ and every pair of distinct varieties is incident with the same number $\lambda$ of blocks, and we refer to it as a BIBD with parameters $(v, b, r, k, \lambda)$. Some basic relationships between the parameters of a BIBD are: $r(k-1) = \lambda(v-1)$, $\lambda < r, v \leqslant b$.

Finite fields come into play in construction methods for BIBDs. First we introduce the basic definitions of finite geometries. A *finite affine plane* is a finite set $\mathcal{P}$ of objects called points, together with a set of nonempty subsets of $\mathcal{P}$ called lines, which satisfy the axioms:

(A1) given any two distinct points, there is exactly one line that contains them both;

(A2) there is a set of four points, no three of which belong to one common line;

(A3) given any point $P$ and any line $L$ that does not contain $P$, there is exactly one line that contains $P$ and contains no point of $L$.

Any finite affine plane must contain at least four points. There is a four-point plane denoted by $AG(2,2)$, which contains exactly six lines. In a finite affine plane there is a parameter $n$ such that every line contains exactly $n$ points and every point lies on exactly $n+1$ lines. This is expressed in the notation $AG(2,n)$. If we identify points with varieties and lines with blocks, then a finite affine plane $AG(2,n)$ becomes a BIBD with parameters $(n^2, n^2+n, n+1, n, 1)$. In higher dimensions we define a *finite affine geometry* $AG(d,n)$ of dimension $d$ over the finite field $F_n$ to consist of the $n^d$ vectors of length $d$ over $F_n$, called points. If $V$ is any $k$-dimensional subspace of $AG(d,n)$ and $\mathbf{p}$ is any member of $AG(d,n)$, then $\mathbf{p}+V = \{\mathbf{p}+\mathbf{v}\colon\ \mathbf{v} \in V\}$ is called a $k$-flat. Moreover, 1-flats are called lines. It is easy to see that an $AG(d,n)$ with points as varieties and lines as blocks forms a BIBD with parameters

$$\left(n^d, \frac{n^{d-1}(n^d-1)}{n-1}, \frac{n^d-1}{n-1}, n, 1\right).$$

Further BIBDs are obtained by taking $k$-flats as blocks.

A *finite projective plane* consists of a finite set $\mathcal{P}$ of points and a set of nonempty subsets of $\mathcal{P}$ called lines, which satisfy the axioms:

(P1) given two distinct points, there is exactly one line that contains them;

(P2) given two distinct lines, there is exactly one point that lies in both;

(P3) there are four points, no three of which are collinear.

In a finite projective plane every line contains exactly $n+1$ points for some parameter $n$. Such a plane is denoted by $PG(2,n)$. It is straightforward to show that $PG(2,n)$ is a symmetric BIBD with parameters

$$v = n^2+n+1, \quad k = n+1, \quad \lambda = 1,$$

if points are identified with varieties and lines with blocks.

More generally, we define a finite projective geometry of dimension $d \geqslant 2$ over $F_n$, or $PG(d,n)$, to be a set of points which are $(d+1)$-vectors over $F_n$, where the zero vector is not allowed, and two points are considered equal if the vector of one is a scalar multiple of the vector of the other. Applying these rules to a $(t+1)$-dimensional space, we obtain a $t$-flat, where $1 \leqslant t < d$. Taking the $t$-flats of a $PG(d,n)$ as blocks and the points as varieties, we obtain a BIBD with parameters

$$v = \frac{n^{d+1}-1}{n-1}, \quad b = \prod_{i=1}^{t+1} \frac{n^{d-t+i}-1}{n^i-1}, \quad r = \prod_{i=1}^{t} \frac{n^{d-t+i}-1}{n^i-1},$$

$$k = \frac{n^{t+1} - 1}{n - 1}, \quad \lambda = \prod_{i=1}^{t-1} \frac{n^{d-t+i} - 1}{n^i - 1},$$

where the last product is interpreted to be 1 if $t = 1$. The BIBD is symmetric in case $t = d - 1$.

An *oval* of $PG(2, q)$ over $F_q$ is a set of $q + 2$ points, no three of which are collinear. An account of ovals can be found in the book by Hirschfeld [65]. Ovals exist in $PG(2, q)$ if and only if $q$ is even. They can be written as sets $\{(1, t, f(t)): t \in F_q\} \cup \{(0, 1, 0), (0, 0, 1)\}$ for $q > 2$ even, where $f$ is a permutation polynomial of $F_q$ of degree at most $q - 2$ satisfying $f(0) = 0$, $f(1) = 1$, and that for each $s \in F_q$ the polynomial $f_s(x) = (f(x + s) + f(s))/x$ is a permutation polynomial of $F_q$ with $f_s(0) = 0$. See Glynn [55] and O'Keefe and Penttila [129] for further results.

A set $D = \{d_1, \ldots, d_k\}$ of $k \geqslant 2$ distinct residues modulo $v$ is called a $(v, k, \lambda)$ *difference set* if for every $d \not\equiv 0 \bmod v$ there are exactly $\lambda$ ordered pairs $(d_i, d_j)$ with $d_i, d_j \in D$ such that $d_i - d_j \equiv d \bmod v$. Let $D$ be such a difference set. Then it is easy to show that by interpreting all residues modulo $v$ as varieties and with blocks given as

$$B_t = \{d_1 + t, \ldots, d_k + t\}, \quad t = 0, 1, \ldots, v - 1,$$

we obtain a symmetric BIBD with parameters $(v, v, k, k, \lambda)$ under the obvious incidence relation. If $D$ is a $(v, k, 1)$ difference set with $k \geqslant 3$, then these varieties and blocks satisfy the conditions of a finite projective plane $PG(2, k - 1)$. Further construction methods can be found in Lidl and Niederreiter [87], but in much more detail in Hughes and Piper [68], Street and Street [151], and Wallis [161].

Hultquist, Mullen and Niederreiter [69] use the analog $\Phi_q$ of the Euler function in the polynomial ring $F_q[x]$ to construct a class of association schemes of prime-power order. Let $f \in F_q[x]$ be monic of degree $n \geqslant 1$ and let $\Phi_q(f)$ be as in Section 2. Let $M_f$ denote the complete residue system modulo $f$ containing all the $q^n$ polynomials over $F_q$ of degree $< n$. Suppose $f = f_1^{e_1} \cdots f_r^{e_r}$, where $e_j \geqslant 1$ and the $f_j$ are distinct monic irreducible polynomials over $F_q$. Let $t_1, \ldots, t_s$ be the monic divisors of $f$ except $f$ itself, then

$$s = \prod_{j=1}^{r} (e_j + 1) - 1.$$

For $i = 1, \ldots, s$, let $A_i = \{g \in M_f\colon \gcd(g, f) = t_i\}$ of cardinality $\Phi_q(f/t_i)$. Two polynomials $g$ and $h$ in $M_f$ are said to be $i$-th associates if $g - h \in A_i$. Given a set of $v$ elements called treatments, a symmetric relation is an *association scheme* with $s$ association classes if:

(i) any two distinct treatments are $i$-th associates for a unique $i = 1, \ldots, s$;

(ii) each treatment has $n_i$ $i$-th associates, $n_i$ being independent of the treatment;

(iii) if two treatments $g$ and $h$ are $k$-th associates, then the number $p_{ij}^k$ of treatments which are $i$-th associates of $g$ and $j$-th associates of $h$ is independent of $g$ and $h$.

Hultquist, Mullen and Niederreiter [69] prove that the relation of $i$-th associates of polynomials in $M_f$ yields an association scheme with

$$s = \prod_{j=1}^{r} (e_j + 1) - 1$$

association classes and parameters $v = q^n$ and $n_i = \Phi_q(f/t_i)$ for $1 \leqslant i \leqslant s$. They also show that the number of nonisomorphic association schemes constructible in this way is given by the number of factorization patterns of polynomials over $F_q$ of degree $n$.

Because of space limitations we can only indicate some other applications of finite fields in combinatorics by referring to the literature. For example, point sets in a unit cube of $\mathbf{R}^s$ with special uniformity properties, called nets, are constructed in Niederreiter [107] by relating them to systems of vectors in finite-dimensional vector spaces over $F_q$. Finite fields are instrumental in the solution by Niederreiter [121] of Williams' problem on special experimental designs. Some books on combinatorial designs, such as Street and Street [151] and Wallis [161], describe the use of finite fields in the construction of Room squares, Hadamard matrices, and other types of designs. Applications to chemical balance weighing designs are referred to in Lidl and Pilz [89], Chapter 5. A separate article on coding theory in this Handbook of Algebra also contains relevant material.

## 9. Applications to pseudorandom numbers and quasirandom points

Finite fields are eminently useful for the design of algorithms for generating pseudorandom numbers and quasirandom points and in the analysis of the output of such algorithms. Pseudorandom numbers and quasirandom points are frequently employed in various tasks of scientific computing, such as simulation methods, computational statistics, numerical integration, and the implementation of probabilistic algorithms. A sequence of *pseudorandom numbers* is generated by a deterministic algorithm and should simulate a sequence of independent and uniformly distributed random variables on the interval $[0, 1]$. In order to be acceptable, a sequence of pseudorandom numbers must pass a variety of statistical tests for randomness. *Quasirandom points* are also generated by a deterministic algorithm, but they have to satisfy only certain equidistribution properties that are required for special applications such as numerical integration and global optimization. For further background on pseudorandom numbers and quasirandom points we refer to the book of Niederreiter [117].

A family of classical methods for the generation of pseudorandom numbers is formed by *shift-register methods*. These are based on linear recurring sequences in finite fields, and in most practical implementations on maximal period sequences (compare with Section 6). For an integer $k \geqslant 2$ and a prime $p$, let $y_0, y_1, \ldots$ be a $k$-th-order maximal period sequence in $F_p$; it is purely periodic with least period $p^k - 1$. This sequence has to be transformed into a sequence of elements of the interval $[0, 1]$ to obtain pseudorandom numbers. One transformation method is the *digital multistep method* in which we choose an integer $m$ with $2 \leqslant m \leqslant k$ and $\gcd(m, p^k - 1) = 1$ and put

$$x_n = \sum_{j=1}^{m} y_{mn+j-1} p^{-j} \quad \text{for } n = 0, 1, \ldots .$$

Thus, we obtain the numbers $x_n \in [0, 1)$ by splitting up the sequence $y_0, y_1, \ldots$ into consecutive strings of length $m$ and then interpreting each string as the $p$-ary expansion of a number in $[0, 1)$. The sequence $x_0, x_1, \ldots$ is again purely periodic with least period $p^k - 1$. Another transformation method is the *generalized feedback shift-register method* in which we choose integers $m \geqslant 2$ and $h_1, \ldots, h_m \geqslant 0$ and put

$$x_n = \sum_{j=1}^{m} y_{n+h_j} p^{-j} \quad \text{for } n = 0, 1, \ldots .$$

This sequence $x_0, x_1, \ldots$ of numbers in $[0, 1)$ is also purely periodic with least period $p^k - 1$. The fact that the state vectors of the sequence $y_0, y_1, \ldots$ run exactly through all nonzero vectors in $F_p^k$ (see Section 6) leads to almost perfect equidistribution properties of pseudorandom numbers generated by shift-register methods. Detailed discussions of shift-register methods can be found in Niederreiter [113], [117], Chapter 9.

The general family of *nonlinear congruential methods* was introduced by Eichenauer, Grothe and Lehn [39]. These methods work with a large finite prime field $F_p$ which is identified with the set $Z_p = \{0, 1, \ldots, p - 1\}$ of integers. A sequence $y_0, y_1, \ldots$ of elements of $F_p$ is generated by the recurrence relation

$$y_{n+1} = f(y_n) \quad \text{for } n = 0, 1, \ldots ,$$

where the mapping $f$ from $F_p$ into itself is chosen in such a way that the sequence $y_0, y_1, \ldots$ is purely periodic with least period $p$. Corresponding pseudorandom numbers in $[0, 1)$ are obtained by setting

$$x_n = \frac{1}{p} y_n \quad \text{for } n = 0, 1, \ldots .$$

One may also describe the $y_n$ by the uniquely determined polynomial $g \in F_p[x]$ such that $y_n = g(n)$ for all $n \in F_p$ and $\deg(g) < p$. Since $\{y_0, y_1, \ldots, y_{p-1}\} = F_p$, we have $\{g(0), g(1), \ldots, g(p - 1)\} = F_p$, and so $g$ is a permutation polynomial of $F_p$. The polynomial $g$ has to be chosen carefully to obtain pseudorandom numbers of good quality. A recent proposal of Eichenauer-Herrmann [41], namely to take $g(x) = (ax + b)^{p-2}$ with $a, b \in F_p$ and $a \neq 0$, leads to very attractive properties. The specific choice $f(x) = ax^{p-2} + b$, with $a, b \in F_p$, for the feedback function $f$ in the above recurrence relation was suggested earlier by Eichenauer and Lehn [40]. Here there arises the problem of characterizing those $a, b \in F_p$ for which the generated sequence $y_0, y_1, \ldots$ is purely periodic with least period $p$. According to a result of Flahive and Niederreiter [46], this property holds if and only if the quotient of the roots of the polynomial $x^2 - bx - a$ is an element of order $p+1$ in the group $F_{p^2}^*$. For further information on nonlinear congruential methods we refer to Niederreiter [117], Chapter 8.

Multidimensional analogs of pseudorandom numbers are *pseudorandom vectors*, which can also be generated by means of finite fields. A standard algorithm is provided by the *matrix method*. For a given dimension $d \geqslant 2$ we choose a large prime $p$ and generate a sequence $\mathbf{z}_0, \mathbf{z}_1, \ldots$ of row vectors in $F_p^d$ by starting from an initial vector $\mathbf{z}_0 \neq \mathbf{0}$ and using the recurrence relation

$$\mathbf{z}_{n+1} = \mathbf{z}_n A \quad \text{for } n = 0, 1, \ldots,$$

where $A$ is a nonsingular $d \times d$ matrix over $F_p$. Then a sequence of $d$-dimensional pseudorandom vectors is derived by identifying $F_p$ with the set $Z_p$ of integers and putting

$$\mathbf{u}_n = \frac{1}{p}\mathbf{z}_n \in [0, 1)^d \quad \text{for } n = 0, 1, \ldots.$$

The sequence $\mathbf{u}_0, \mathbf{u}_1, \ldots$ is purely periodic with least period at most $p^d - 1$. Its least period is $p^d - 1$ if and only if the characteristic polynomial of $A$ is primitive over $F_p$. A nonlinear method for pseudorandom vector generation can be based on the recurrence relation

$$\gamma_{n+1} = \alpha\gamma_n^{q-2} + \beta \quad \text{for } n = 0, 1, \ldots$$

in the finite field $F_q$ of order $q = p^d$, where $\alpha, \beta \in F_q$ are selected suitably. The vector $\mathbf{z}_n \in F_p^d$ is then obtained as the coordinate vector of $\gamma_n$ relative to a fixed basis of $F_q$ over $F_p$, and pseudorandom vectors are derived as above by setting $\mathbf{u}_n = p^{-1}\mathbf{z}_n$ for $n = 0, 1, \ldots$. An expository account of these methods for pseudorandom vector generation is given in Niederreiter [117], Chapter 10.

Quasirandom points in $[0, 1]^d$ can be constructed by several methods based on finite fields. We describe the method of Niederreiter [109] which produces quasirandom points with the currently best equidistribution properties for $d \geqslant 2$. We choose an arbitrary finite field $F_q$ and $d$ pairwise coprime polynomials $p_1, \ldots, p_d \in F_q[x]$ with $\deg(p_i) = e_i \geqslant 1$ for $1 \leqslant i \leqslant d$. For $1 \leqslant i \leqslant d$ and integers $j \geqslant 1$ and $u \geqslant 0$ we have the Laurent series expansion

$$\frac{x^u}{p_i(x)^j} = \sum_{r=w}^{\infty} a^{(i)}(j, u, r)x^{-r-1}$$

in $F_q((x^{-1}))$, where the integer $w \leqslant 0$ may depend on $i, j$, and $u$. Then we define

$$c_{jr}^{(i)} = a^{(i)}\big(Q(i, j) + 1, u(i, j), r\big) \in F_q \quad \text{for } 1 \leqslant i \leqslant d,\ j \geqslant 1,\ r \geqslant 0,$$

where $j - 1 = Q(i, j)e_i + u(i, j)$ with integers $Q(i, j)$ and $u(i, j)$ satisfying $0 \leqslant u(i, j) < e_i$. Next we choose bijections $\psi_r$, $r \geqslant 0$, from the set $Z_q = \{0, 1, \ldots, q-1\}$ of integers onto $F_q$, with $\psi_r(0) = 0$ for all sufficiently large $r$, and bijections $\eta_{ij}$ from $F_q$ onto $Z_q$ for $1 \leqslant i \leqslant d$ and $j \geqslant 1$, with $\eta_{ij}(0) = 0$ for $1 \leqslant i \leqslant d$ and all sufficiently large $j$. For $n = 0, 1, \ldots$ let

$$n = \sum_{r=0}^{\infty} a_r(n) q^r$$

with all $a_r(n) \in Z_q$ be the digit expansion of $n$ in base $q$. Then we put

$$y_{nj}^{(i)} = \eta_{ij}\left(\sum_{r=0}^{\infty} c_{jr}^{(i)} \psi_r\big(a_r(n)\big)\right) \in Z_q \quad \text{for } n \geqslant 0,\ 1 \leqslant i \leqslant d,\ j \geqslant 1,$$

where the sum over $r$ is actually a finite sum since $\psi_r(0) = 0$ and $a_r(n) = 0$ for all sufficiently large $r$. Now we set

$$x_n^{(i)} = \sum_{j=1}^{\infty} y_{nj}^{(i)} q^{-j} \quad \text{for } n \geqslant 0 \text{ and } 1 \leqslant i \leqslant d,$$

and then we obtain the quasirandom points

$$\mathbf{x}_n = \big(x_n^{(1)}, \ldots, x_n^{(d)}\big) \in [0,1]^d \quad \text{for } n \geqslant 0.$$

For fixed $d$ and $q$ this construction is optimized by letting $p_1, \ldots, p_d$ be the "first $d$" monic irreducible polynomials over $F_q$, i.e. the first $d$ terms of a sequence in which all monic irreducible polynomials over $F_q$ are listed according to nondecreasing degrees.

An in-depth discussion of this construction of quasirandom points and of other constructions using finite fields is presented in the book of Niederreiter [117]. A review of the material in this section can also be found in Niederreiter [120].

## References


[1] L.M. Adleman and H.W. Lenstra, Jr., *Finding irreducible polynomials over finite fields*, Proc. 18th ACM Symp. on Theory of Comp. (1986), 350–355.

[2] D.W. Ash, I.F. Blake and S.A. Vanstone, *Low complexity normal bases*, Discrete Appl. Math. **25** (1989), 191–210.

[3] E.R. Berlekamp, *Algebraic Coding Theory*, McGraw-Hill, New York (1968).

[4] T. Beth, *On the arithmetics of Galois fields and the like*, Algebraic Algorithms and Error-Correcting Codes (Grenoble, 1985), Lecture Notes in Computer Science vol. 229, Springer, Berlin (1986), 2–16.

[5] T. Beth and W. Fumy, *Hardware-oriented algorithms for the fast symbolic calculation of the DFT*, Electronics Letters **19** (1983), 901–902.

[6] T. Beth and W. Geiselmann, *Selbstduale Normalbasen über $GF(q)$*, Arch. Math. **55** (1990), 44–48.

[7] T. Beth and D. Gollmann, *Aspekte der technischen Realisierung von Public-Key-Verfahren*, Elektrotechnik und Informationstechnik **105** (1988), 12–18.

[8] B.J. Birch and H.P.F. Swinnerton-Dyer, *Note on a problem of Chowla*, Acta Arith. **5** (1959), 417–423.

[9] I. Blake, X.H. Gao, A. Menezes, R. Mullin, S. Vanstone and T. Yaghoobian, *Applications of Finite Fields*, Kluwer, Boston (1993).

[10] M. Blum and S. Micali, *How to generate cryptographically strong sequences of pseudo-random bits*, SIAM J. Comput. **13** (1984), 850–864.

[11] R.C. Bose and S.S. Shrikhande, *On the falsity of Euler's conjecture about the non-existence of two orthogonal Latin squares of order* $4t + 2$, Proc. Nat. Acad. Sci. USA **45** (1959), 734–737.

[12] J.V. Brawley, *Polynomial functions on matrices over a finite field*, Finite Fields, Coding Theory, and Advances in Communications and Computing, G.L. Mullen and P.J.-S. Shiue, eds, Dekker, New York (1993), 15–31.

[13] J.V. Brawley and G.L. Mullen, *Infinite latin squares containing nested sets of mutually orthogonal finite latin squares*, Publ. Math. Debrecen **39** (1991), 1–7.

[14] J.V. Brawley and G.E. Schnibben, *Infinite Algebraic Extensions of Finite Fields*, Contemp. Math. vol. 95, Amer. Math. Soc., Providence, RI (1989).

[15] N.H. Bshouty and G. Seroussi, *Generalizations of the normal basis theorem of finite fields*, SIAM J. Discrete Math. **3** (1990), 330–337.

[16] J. Buchmann and H.C. Williams, *A key-exchange system based on imaginary quadratic fields*, J. Cryptology **1** (1988), 107–118.

[17] L. Carlitz, *Primitive roots in a finite field*, Trans. Amer. Math. Soc. **73** (1952), 373–382.

[18] B. Chor and R.L. Rivest, *A knapsack-type public key cryptosystem based on arithmetic in finite fields*, IEEE Trans. Inform. Theory **34** (1988), 901–909.

[19] W.-S. Chou, *Permutation polynomials on finite fields and combinatorial applications*, Ph.D. Thesis, Pennsylvania State University (1990).

[20] W.-S. Chou, *Set-complete mappings on finite fields*, Finite Fields, Coding Theory, and Advances in Communications and Computing, G.L. Mullen and P.J.-S. Shiue, eds, Dekker, New York (1993), 33–41.

[21] S. Chowla and H. Zassenhaus, *Some conjectures concerning finite fields*, Norske Vid. Selsk. Forh. (Trondheim) **41** (1968), 34–35.

[22] S.D. Cohen, *Proof of a conjecture of Chowla and Zassenhaus on permutation polynomials*, Canad. Math. Bull. **33** (1990), 230–234.

[23] S.D. Cohen, *Exceptional polynomials and the reducibility of substitution polynomials*, Enseign. Math. (2) **36** (1990), 53–65.

[24] S.D. Cohen, *Primitive elements and polynomials with arbitrary trace*, Discrete Math. **83** (1990), 1–7.

[25] S.D. Cohen, *Primitive elements and polynomials: existence results*, Finite Fields, Coding Theory, and Advances in Communications and Computing, G.L. Mullen and P.J.-S. Shiue, eds, Dekker, New York (1993), 43–55.

[26] S.D. Cohen and G.L. Mullen, *Primitive elements in finite fields and Costas arrays*, Appl. Algebra Engrg. Comm. Comput. **2** (1991), 45–53; *Erratum*, ibid. **2** (1992), 297–299.

[27] S.D. Cohen, G.L. Mullen and P.J.-S. Shiue, *The difference between permutation polynomials over finite fields*, Proc. Amer. Math. Soc. **123** (1995), 2011–2015.

[28] D. Coppersmith, *Fast evaluation of logarithms in fields of characteristic two*, IEEE Trans. Inform. Theory **30** (1984), 587–594.

[29] D. Coppersmith, A.M. Odlyzko and R. Schroeppel, *Discrete logarithms in* $GF(p)$, Algorithmica **1** (1986), 1–15.

[30] H. Davenport, *Bases for finite fields*, J. London Math. Soc. **43** (1968), 21–39; *Addendum*, ibid. **44** (1969), 378.

[31] J. Dénes and A.D. Keedwell, *Latin Squares and Their Applications*, Academic Press, New York (1974).

[32] J. Dénes and A.D. Keedwell, *Latin Squares – New Developments in the Theory and Applications*, Ann. Discrete Math. vol. 46, North-Holland, Amsterdam (1991).

[33] L.E. Dickson, *The analytic representation of substitutions on a power of a prime number of letters with a discussion of the linear group*, Ann. Math. **11** (1897), 65–120, 161–183.

[34] L.E. Dickson, *Linear Groups with an Exposition of the Galois Field Theory*, Teubner, Leipzig (1901); Dover, New York (1958).

[35] W. Diffie and M.E. Hellman, *New directions in cryptography*, IEEE Trans. Inform. Theory **22** (1976), 644–654.

[36] P.F. Duvall and J.C. Mortick, *Decimation of periodic sequences*, SIAM J. Appl. Math. **21** (1971), 367–372.

[37] A. Ecker and G. Poch, *Check character systems*, Computing **37** (1986), 277–301.

[38] G.W. Effinger and D.R. Hayes, *Additive Number Theory of Polynomials over a Finite Field*, Oxford Univ. Press, Oxford (1991).

[39] J. Eichenauer, H. Grothe and J. Lehn, *Marsaglia's lattice test and non-linear congruential pseudo random number generators*, Metrika **35** (1988), 241–250.

[40] J. Eichenauer and J. Lehn, *A non-linear congruential pseudo random number generator*, Statist. Papers **27** (1986), 315–326.

[41] J. Eichenauer-Herrmann, *Statistical independence of a new class of inversive congruential pseudorandom numbers*, Math. Comp. **60** (1993), 375–384.

[42] T. ElGamal, *A public key cryptosystem and a signature scheme based on discrete logarithms*, IEEE Trans. Inform. Theory **31** (1985), 469–472.

[43] R.J. Evans, J. Greene and H. Niederreiter, *Linearized polynomials and permutation polynomials of finite fields*, Michigan Math. J. **39** (1992), 405–413.

[44] S.A. Evdokimov, *Factoring a solvable polynomial over a finite field and generalized Riemann hypothesis*, Zap. Nauchn. Sem. Leningrad. Otdel. Mat. Inst. Steklov. **176** (1989), 104–117 (in Russian).

[45] C.M. Fiduccia, *An efficient formula for linear recurrences*, SIAM J. Comput. **14** (1985), 106–112.

[46] M. Flahive and H. Niederreiter, *On inversive congruential generators for pseudorandom numbers*, Finite Fields, Coding Theory, and Advances in Communications and Computing, G.L. Mullen and P.J.-S. Shiue, eds, Dekker, New York (1993), 75–80.

[47] M. Fried, *On a conjecture of Schur*, Michigan Math. J. **17** (1970), 41–55.

[48] M. Fried, R. Guralnick and J. Saxl, *Schur covers and Carlitz's conjecture*, Israel J. Math. **82** (1993), 157–225.

[49] M. Fried and M. Jarden, *Field Arithmetic*, Springer, New York (1986).

[50] S. Gao and H.W. Lenstra, Jr., *Optimal normal bases*, Designs, Codes and Cryptography **2** (1992), 315–323.

[51] S. Gao and G.L. Mullen, *Dickson polynomials and irreducible polynomials over finite fields*, J. Number Theory **49** (1994), 118–132.

[52] J. von zur Gathen, *Tests for permutation polynomials*, SIAM J. Comput. **20** (1991), 591–602.

[53] J. von zur Gathen and M. Giesbrecht, *Constructing normal bases in finite fields*, J. Symbolic Comput. **10** (1990), 547–570.

[54] W. Geiselmann and D. Gollmann, *Self-dual bases in* $\mathbb{F}_{q^n}$, Designs, Codes and Cryptography **3** (1993), 333–345.

[55] D.G. Glynn, *A condition for the existence of ovals in* $PG(2,q)$, $q$ *even*, Geom. Dedicata **32** (1989), 247–252.

[56] S.W. Golomb, *Shift Register Sequences*, Aegean Park Press, Laguna Hills, CA (1982).

[57] S.W. Golomb, *Algebraic constructions for Costas arrays*, J. Combin. Theory Ser. A **37** (1984), 13–21.

[58] S.W. Golomb and E.C. Posner, *Rook domains, latin squares, affine planes, and error-distribution codes*, IEEE Trans. Inform. Theory **10** (1964), 196–208.

[59] J. Gomez-Calderon and D.J. Madden, *Polynomials with small value set over finite fields*, J. Number Theory **28** (1988), 167–188.

[60] T. Hansen and G.L. Mullen, *Primitive polynomials over finite fields*, Math. Comp. **59** (1992), 639–643, S47–S50.

[61] D.R. Hayes, *The distribution of irreducibles in* $GF[q,x]$, Trans. Amer. Math. Soc. **117** (1965), 101–127.

[62] J. Heisler, *A characterization of finite fields*, Amer. Math. Monthly **74** (1967), 537–538; *Correction*, ibid. **74** (1967), 1211.

[63] T. Herlestam, *On functions of linear shift register sequences*, Advances in Cryptology – EUROCRYPT '85, F. Pichler, ed., Lecture Notes in Computer Science vol. 219, Springer, Berlin (1986), 119–129.

[64] C. Hermite, *Sur les fonctions de sept lettres*, C. R. Acad. Sci. Paris **57** (1863), 750–757.

[65] J.W.P. Hirschfeld, *Projective Geometries over Finite Fields*, Oxford Univ. Press, Oxford (1979).

[66] J.W.P. Hirschfeld, *Finite Projective Spaces of Three Dimensions*, Oxford Univ. Press, Oxford (1985).

[67] J.W.P. Hirschfeld and J.A. Thas, *General Galois Geometries*, Oxford Univ. Press, Oxford (1991).

[68] D.R. Hughes and F.C. Piper, *Design Theory*, 2nd ed., Cambridge Univ. Press, Cambridge (1988).

[69] R.A. Hultquist, G.L. Mullen and H. Niederreiter, *Association schemes and derived PBIB designs of prime power order*, Ars Combin. **25** (1988), 65–82.

[70] T. Itoh and S. Tsujii, *A fast algorithm for computing multiplicative inverses in* $GF(2^m)$ *using normal bases*, Inform. and Comput. **78** (1988), 171–177.

[71] D. Jungnickel, *Finite Fields: Structure and Arithmetics*, Bibliographisches Institut, Mannheim (1993).

[72] D. Jungnickel, A.J. Menezes and S.A. Vanstone, *On the number of self-dual bases of* $GF(q^n)$ *over* $GF(q)$, Proc. Amer. Math. Soc. **109** (1990), 23–29.

[73] N. Koblitz, *A Course in Number Theory and Cryptography*, Springer, New York (1987).

[74] E. Kranakis, *Primality and Cryptography*, Wiley, Chichester (1986).

[75] M.K. Kyuregyan, *Polynomial expansion and the synthesis of irreducible polynomials over finite fields*, Dokl. Akad. Nauk Armyan. SSR **81** (1985), 69–73 (in Russian).

[76] M.K. Kyuregyan, *A new method for constructing irreducible polynomials over finite fields*, Dokl. Akad. Nauk Armyan. SSR **87**(2) (1988), 51–55 (in Russian).

[77] J.C. Lagarias, *Pseudorandom number generators in cryptography and number theory*, Cryptology and Computational Number Theory, C. Pomerance, ed., Proc. Symp. Applied Math. vol. 42, Amer. Math. Soc., Providence, RI (1990), 115–143.

[78] H. Lausch and W. Nöbauer, *Algebra of Polynomials,* North-Holland, Amsterdam (1973).

[79] A. Lempel and M.J. Weinberger, *Self-complementary normal bases in finite fields*, SIAM J. Discrete Math. **1** (1988), 193–198.

[80] H.W. Lenstra, Jr., *Algorithms for finite fields*, Number Theory and Cryptography, J.H. Loxton, ed., London Math. Soc. Lecture Note Series vol. 154, Cambridge Univ. Press, Cambridge (1990), 76–85.

[81] H.W. Lenstra, Jr., *Finding isomorphisms between finite fields*, Math. Comp. **56** (1991), 329–347.

[82] H.W. Lenstra, Jr. and R.J. Schoof, *Primitive normal bases for finite fields*, Math. Comp. **48** (1987), 217–231.

[83] R. Lidl, *Computational problems in the theory of finite fields*, Appl. Algebra Engrg. Comm. Comput. **2** (1991), 81–89.

[84] R. Lidl and G.L. Mullen, *When does a polynomial over a finite field permute the elements of the field?*, Amer. Math. Monthly **95** (1988), 243–246.

[85] R. Lidl and G.L. Mullen, *When does a polynomial over a finite field permute the elements of the field?, II*, Amer. Math. Monthly **100** (1993), 71–74.

[86] R. Lidl, G.L. Mullen and G. Turnwald, *Dickson Polynomials*, Pitman Monographs and Surveys in Pure and Applied Math., Longman, Harlow, Essex (1993).

[87] R. Lidl and H. Niederreiter, *Finite Fields*, Addison-Wesley, Reading, MA (1983); now distributed by Cambridge Univ. Press.

[88] R. Lidl and H. Niederreiter, *Introduction to Finite Fields and Their Applications*, rev. ed., Cambridge Univ. Press, Cambridge (1994).

[89] R. Lidl and G. Pilz, *Applied Abstract Algebra*, Springer, New York (1984).

[90] H. Lüneburg, *Galoisfelder, Kreisteilungskörper und Schieberegisterfolgen,* Bibliographisches Institut, Mannheim (1979).

[91] J.L. Massey, *Shift-register synthesis and BCH decoding*, IEEE Trans. Inform. Theory **15** (1969), 122–127.

[92] K.S. McCurley, *The discrete logarithm problem*, Cryptology and Computational Number Theory, C. Pomerance, ed., Proc. Symp. Applied Math. vol. 42, Amer. Math. Soc., Providence, RI (1990), 49–74.

[93] R.J. McEliece, *Finite Fields for Computer Scientists and Engineers*, Kluwer, Boston (1987).

[94] W. Meier and O. Staffelbach, *Nonlinearity criteria for cryptographic functions*, Advances in Cryptology – EUROCRYPT '89, J.-J. Quisquater and J. Vandewalle, eds, Lecture Notes in Computer Science vol. 434, Springer, Berlin (1990), 549–562.

[95] G. Meletiou and G.L. Mullen, *A note on discrete logarithms in finite fields*, Appl. Algebra Engrg. Comm. Comput. **3** (1992), 75–78.

[96] H. Meyn, *On the construction of irreducible self-reciprocal polynomials over finite fields*, Appl. Algebra Engrg. Comm. Comput. **1** (1990), 43–53.

[97] C. Mitchell, *Enumerating Boolean functions of cryptographic significance*, J. Cryptology **2** (1990), 155–170.

[98] G.L. Mullen, *Polynomial representation of complete sets of mutually orthogonal frequency squares of prime power order*, Discrete Math. **69** (1988), 79–84.

[99] G.L. Mullen, *Permutation polynomials over finite fields*, Finite Fields, Coding Theory, and Advances in Communications and Computing, G.L. Mullen and P.J.-S. Shiue, eds, Dekker, New York (1993), 131–151.

[100] G.L. Mullen and P.J.-S. Shiue (eds), *Finite Fields, Coding Theory, and Advances in Communications and Computing*, Dekker, New York (1993).

[101] G.L. Mullen and S.J. Suchower, *Frequency hyperrectangles, error-correcting codes, and fractional replication plans*, J. Combin. Inform. System Sci. **18** (1993), 5–18.

[102] G.L. Mullen and D. White, *A polynomial representation for logarithms in $GF(q)$*, Acta Arith. **47** (1986), 255–261.

[103] G.L. Mullen and G. Whittle, *A generalization of the Golomb–Posner code*, Preprint (1992).

[104] R.C. Mullin, I.M. Onyszchuk, S.A. Vanstone and R.M. Wilson, *Optimal normal bases in $GF(p^n)$*, Discrete Appl. Math. **22** (1988/89), 149–161.

[105] W. Narkiewicz, *Uniform Distribution of Sequences of Integers in Residue Classes*, SLNM 1087, Springer, Berlin (1984).

[106] J. Nechvatal, *Irreducibility, Primitivity, and Duality in Finite Fields*, Book manuscript (1991).

[107] H. Niederreiter, *Point sets and sequences with small discrepancy*, Monatsh. Math. **104** (1987), 273–337.

[108] H. Niederreiter, *Sequences with almost perfect linear complexity profile*, Advances in Cryptology – EUROCRYPT '87, D. Chaum and W.L. Price, eds, Lecture Notes in Computer Science vol. 304, Springer, Berlin (1988), 37–51.

[109] H. Niederreiter, *Low-discrepancy and low-dispersion sequences*, J. Number Theory **30** (1988), 51–70.

[110] H. Niederreiter, *A simple and general approach to the decimation of feedback shift-register sequences*, Problems of Control and Information Theory **17** (1988), 327–331.

[111] H. Niederreiter, *Some new cryptosystems based on feedback shift register sequences*, Math. J. Okayama Univ. **30** (1988), 121–149.

[112] H. Niederreiter, *The probabilistic theory of linear complexity*, Advances in Cryptology – EUROCRYPT '88, C.G. Günther, ed., Lecture Notes in Computer Science vol. 330, Springer, Berlin (1988), 191–209.

[113] H. Niederreiter, *Pseudorandom numbers generated from shift register sequences*, Number-Theoretic Analysis, E. Hlawka and R.F. Tichy, eds, SLNM 1452, Springer, Berlin (1990), 165–177.

[114] H. Niederreiter, *A short proof for explicit formulas for discrete logarithms in finite fields*, Appl. Algebra Engrg. Comm. Comput. **1** (1990), 55–57.

[115] H. Niederreiter, *An enumeration formula for certain irreducible polynomials with an application to the construction of irreducible polynomials over the binary field*, Appl. Algebra Engrg. Comm. Comput. **1** (1990), 119–124.

[116] H. Niederreiter, *Finite fields and their applications*, Contributions to General Algebra 7 (Vienna, 1990), Teubner, Stuttgart (1991), 251–264.

[117] H. Niederreiter, *Random Number Generation and Quasi-Monte Carlo Methods*, SIAM, Philadelphia (1992).

[118] H. Niederreiter, *Recent advances in the theory of finite fields*, Finite Fields, Coding Theory, and Advances in Communications and Computing, G.L. Mullen and P.J.-S. Shiue, eds, Dekker, New York (1993), 153–163.

[119] H. Niederreiter, *Finite fields and cryptology*, Finite Fields, Coding Theory, and Advances in Communications and Computing, G.L. Mullen and P.J.-S. Shiue, eds, Dekker, New York (1993), 359–373.

[120] H. Niederreiter, *Finite fields, pseudorandom numbers, and quasirandom points*, Finite Fields, Coding Theory, and Advances in Communications and Computing, G.L. Mullen and P.J.-S. Shiue, eds, Dekker, New York (1993), 375–394.

[121] H. Niederreiter, *Proof of Williams' conjecture on experimental designs balanced for pairs of interacting residual effects*, European J. Combin. **14** (1993), 55–58.

[122] H. Niederreiter, *A new efficient factorization algorithm for polynomials over small finite fields*, Appl. Algebra Engrg. Comm. Comput. **4** (1993), 81–87.

[123] H. Niederreiter, *Factorization of polynomials and some linear-algebra problems over finite fields*, Linear Algebra Appl. **192** (1993), 301–328.

[124] H. Niederreiter and R. Göttfert, *Factorization of polynomials over finite fields and characteristic sequences*, J. Symbolic Comput. **16** (1993), 401–412.

[125] H. Niederreiter and K.H. Robinson, *Complete mappings of finite fields*, J. Austral. Math. Soc. Ser. A **33** (1982), 197–212.

[126] H. Niederreiter and C.P. Schnorr, *Local randomness in polynomial random number and random function generators*, SIAM J. Comput. **22** (1993), 684–694.

[127] K. Nyberg, *Perfect nonlinear S-boxes*, Advances in Cryptology – EUROCRYPT '91, D.W. Davies, ed., Lecture Notes in Computer Science vol. 547, Springer, Berlin (1991), 378–386.

[128] A.M. Odlyzko, *Discrete logarithms in finite fields and their cryptographic significance*, Advances in Cryptology (Paris, 1984), Lecture Notes in Computer Science vol. 209, Springer, Berlin (1985), 224–314.

[129] C.M. O'Keefe and T. Penttila, *A new hyperoval in* $PG(2, 32)$, J. Geometry **44** (1992), 117–139.

[130] P.C. van Oorschot, *A comparison of practical public key cryptosystems based on integer factorization and discrete logarithms*, Contemporary Cryptology – The Science of Information Integrity, G.J. Simmons, ed., IEEE Press, New York (1992), 289–322.

[131] J. Pieprzyk, *Bent permutations*, Finite Fields, Coding Theory, and Advances in Communications and Computing, G.L. Mullen and P.J.-S. Shiue, eds, Dekker, New York (1993), 173–181.

[132] A. Pincin, *Bases for finite fields and a canonical decomposition for a normal basis generator*, Comm. Algebra **17** (1989), 1337–1352.

[133] M. Pohst and H. Zassenhaus, *Algorithmic Algebraic Number Theory*, Cambridge Univ. Press, Cambridge (1989).

[134] C. Pomerance, *Fast, rigorous factorization and discrete logarithm algorithms*, Discrete Algorithms and Complexity (Kyoto, 1986), Academic Press, Boston (1987), 119–143.

[135] R.A. Rueppel, *Stream ciphers*, Contemporary Cryptology – The Science of Information Integrity, G.J. Simmons, ed., IEEE Press, New York (1992), 65–134.

[136] M. Rybowicz, *Search of primitive polynomials over finite fields*, J. Pure Appl. Algebra **65** (1990), 139–151.

[137] W.M. Schmidt, *Equations over Finite Fields: An Elementary Approach*, SLNM 536, Springer, Berlin (1976).

[138] M.R. Schroeder, *Number Theory in Science and Communication*, 2nd ed., Springer, Berlin (1986).

[139] Š. Schwarz, *Construction of normal bases in cyclic extensions of a field*, Czechoslovak Math. J. **38** (1988), 291–312.

[140] G.E. Séguin, *Low complexity normal bases for* $F_{2^{mn}}$, Discrete Appl. Math. **28** (1990), 309–312.

[141] I.A. Semaev, *Construction of polynomials irreducible over a finite field with linearly independent roots*, Mat. Sb. **135** (1988), 520–532 (in Russian).

[142] G. Seroussi and A. Lempel, *Factorization of symmetric matrices and trace-orthogonal bases in finite fields*, SIAM J. Comput. **9** (1980), 758–767.

[143] V. Shoup, *New algorithms for finding irreducible polynomials over finite fields*, Math. Comp. **54** (1990), 435–447.

[144] V. Shoup, *Searching for primitive roots in finite fields*, Proc. 22nd ACM Symp. on Theory of Comp. (1990), 546–554.

[145] I.E. Shparlinski, *On primitive elements in finite fields and on elliptic curves*, Mat. Sb. **181** (1990), 1196–1206 (in Russian).

[146] I.E. Shparlinski, *Computational and Algorithmic Problems in Finite Fields*, Kluwer, Dordrecht (1992).

[147] I.E. Shparlinski, *A deterministic test for permutation polynomials*, Comput. Complexity **2** (1992), 129–132.

[148] G.J. Simmons (ed.), *Contemporary Cryptology – The Science of Information Integrity*, IEEE Press, New York (1992).

[149] C. Small, *Arithmetic of Finite Fields*, Dekker, New York (1991).

[150] S.A. Stepanov and I.E. Shparlinski, *Construction of a normal basis of a finite field*, Acta Arith. **49** (1987), 189–192 (in Russian).

[151] A.P. Street and D.J. Street, *Combinatorics of Experimental Design*, Oxford Univ. Press, Oxford (1987).

[152] S.J. Suchower, *Subfield permutation polynomials and orthogonal subfield systems in finite fields*, Acta Arith. **54** (1990), 307–315.

[153] S.J. Suchower, *Polynomial representations of complete sets of frequency hyperrectangles with prime power dimensions*, J. Combin. Theory Ser. A **62** (1993), 46–65.

[154] G. Tarry, *Le problème des* 36 *officiers*, C. R. Assoc. Française Avancement Sci. Nat. **1** (1900), 122–123; ibid. **2** (1901), 170–203.

[155] W. Tautz, J. Top and A. Verberkmoes, *Explicit hyperelliptic curves with real multiplication and permutation polynomials*, Canad. J. Math. **43** (1991), 1055–1064.

[156] H.C.A. van Tilborg, *An Introduction to Cryptology*, Kluwer, Dordrecht (1988).

[157] G. Turnwald, *On a problem concerning permutation polynomials*, Trans. Amer. Math. Soc. **302** (1987), 251–267.

[158] G. Turnwald, *Permutation polynomials of binomial type*, Contributions to General Algebra vol. 6 (To the Memory of W. Nöbauer), Teubner, Stuttgart (1988), 281–286.

[159] R.R. Varshamov, *A general method of synthesis for irreducible polynomials over Galois fields*, Dokl. Akad. Nauk SSSR **275** (1984), 1041–1044 (in Russian).

[160] R.R. Varshamov, *A method for constructing irreducible polynomials over finite fields*, Dokl. Akad. Nauk Armyan. SSR **79** (1) (1984), 26–28 (in Russian).

[161] W.D. Wallis, *Combinatorial Designs*, Dekker, New York (1988).

[162] D.Q. Wan, *A p-adic lifting lemma and its applications to permutation polynomials*, Finite Fields, Coding Theory, and Advances in Communications and Computing, G.L. Mullen and P.J.-S. Shiue, eds, Dekker, New York (1993), 209–216.

[163] D.Q. Wan and R. Lidl, *Permutation polynomials of the form* $x^r f(x^{(q-1)/d})$ *and their group structure*, Monatsh. Math. **112** (1991), 149–163.

[164] A. Wassermann, *Konstruktion von Normalbasen*, Bayreuth. Math. Schr. **31** (1990), 155–164.

[165] N. Zierler, *Linear recurring sequences*, J. Soc. Indust. Appl. Math. **7** (1959), 31–48.

[166] N. Zierler and W.H. Mills, *Products of linear recurring sequences*, J. Algebra **27** (1973), 147–157.

# Global Class-Field Theory

W. Narkiewicz

*Institute of Mathematics, Wrocław University, pl. Grunwaldzki 2/4, 50-384, Wrocław, Poland*
*e-mail: narkiew@math.uni.wroc.pl*

*Contents*



HANDBOOK OF ALGEBRA, VOL. 1
Edited by M. Hazewinkel

## 1. Introduction

**1.1.** The aim of the class-field theory is to describe all Abelian extensions of a given field $k$ and at its source lies the *Kronecker–Weber Theorem*, which solves this problem for $k = Q$, the field of rational numbers (L. Kronecker [72, 73], H. Weber [150]). The first complete proof of it has been given by D. Hilbert [57] (see also [Zber], Satz 131). An exposition of the early proofs is given in O. Neumann [104]. An elementary proof can be found in M.J. Greenberg [40] and a proof which uses local methods was given by I.R. Shafarevich [130]. This proof is exposed in [Na] and [Wa].

THEOREM 1.1 (Kronecker–Weber). *Every Abelian extension of $Q$ is contained in a cyclotomic extension $K/Q$, i.e. one has $K = Q(\zeta)$, where $\zeta$ is a root of unity.*

The problem of describing all Abelian extensions of an algebraic number field has been stated as the twelfth problem in the famous list of problems given by D. Hilbert [60] in his lecture at the Second International Congress of Mathematicians in Paris in August 1900. The work of H. Weber [151], D. Hilbert [58], P. Furtwängler [32], T. Takagi [141] and E. Artin [1] (subsumed in the report of H. Hasse [HBer]) in the first quarter of our century led to its solution. In it Abelian extensions of an algebraic number field $k$ have been associated with certain ideal class-groups related to $k$. This classical approach will be described in Section 2.

**1.2.** The analogy between the theory of algebraic numbers and the theory of algebraic functions in one variable had already been observed in 1882 (R. Dedekind and H. Weber [20]). This led later to the theory of Dedekind domains and culminated in the theory of ideals in commutative rings. In 1931 F.K. Schmidt [117] succeeded in constructing the analogue of the class-field theory for algebraic function fields in one variable over a finite field. His result gave a biunique correspondence between finite Abelian extensions $K/k$ of a given algebraic function field $k$ and certain subgroups of the divisor group of $k$.

**1.3.** Modern class-field theory begins with the invention of ideles by C. Chevalley [12] who in C. Chevalley [13] reinterpreted classical class-field theory in terms of ideles, using the theory of associative algebras. This approach led to a simultaneous proof of the class-field theory in both cases. (The book [Wei1] presents such a proof using algebras.) Later development eliminated the use of algebras in the proofs of the main results of class-field theory and replaced them by the formalism of cohomology. Expositions of this method are given in [AT, Ch, CF, Iy]. Recently a simplification in the theory was presented by J. Neukirch [103, Ne1, Ne2], whose axiomatical approach reduced the whole problematics to purely group-theoretical reasonings utilizing only rudiments of cohomology. We shall sketch the main ideas in Section 3 and the next sections will be devoted to a choice of applications of the class-field theory.

**1.4.** Expositions of class-field theory can be found in [AT, CF, Ch, Ha, Iy, Ja, La2, Ne1, Ne2, Wei1].

Generalizations of the class-field theory to the non-Abelian case (the *Langlands program*) and to other classes of fields are the subject of other chapters.

## 2. The classical approach

**2.1.** Let $k$ be an algebraic number field, and let $Z_k$ be its ring of integers. Let $H(k)$ be the *class-group* of $k$, i.e. the factor group of the group $\mathrm{Id}(k)$ of all fractional ideals of $k$ by its subgroup formed by all principal ideals, let $K^*(k)$ be the *narrow class-group* of $k$, i.e. the factor group of $\mathrm{Id}(k)$ by the subgroup of all principal ideals having a totally positive generator, and let $h(k)$, $h^*(k)$ be the cardinalities of $H(k)$ and $H^*(k)$, respectively. A finite extension $K/k$ is called *unramified* (more precisely, *unramified at finite places*) if its relative discriminant equals the unit ideal, which is equivalent to the statement that for all prime ideals $P$ of $Z_k$ the ideal $PZ_K$ of $Z_K$ is square-free. The extension $K/k$ is called *unramified at all places* provided it is unramified at finite primes and moreover if $\phi$ is an embedding of $k$ into the field of complex numbers mapping $k$ into the field $\mathbf{R}$ of reals and $\Phi$ is an extension of $\phi$ to $K$, then $\Phi(K) \subset \mathbf{R}$.

In 1898 D. Hilbert [58] formulated a series of conjectures dealing with Abelian extensions of $k$:

HILBERT'S CONJECTURES.

(i) There exists a unique maximal unramified Abelian extension $K/k$.

(ii) The Galois group of $K/k$ is isomorphic with $H^*(k)$.

(iii) The decomposition in $K$ of any prime ideal $P$ of $Z_k$ depends only on the class in $H^*(k)$ to which $P$ belongs. In particular $P$ splits completely in $K/k$ if and only if $P$ is a principal ideal generated by a totally positive number.

(iv) (*Principal ideal theorem, Hauptidealsatz*). Every fractional ideal $I$ of $k$ becomes a principal ideal in $K$, i.e. the ideal $IZ_K$ is principal.

All these statements, with the exception of (iv), were established by P. Furtwängler [32] in 1907 and the last by the same author in 1930 (P. Furtwängler [35]).

Hilbert called the field $K$ occurring in (i) the *class-field of* $k$. Today this field is called the *Hilbert class-field of* $k$. Note that nowadays this name is often used to denote the field which arises if one replaces in (i) the word "unramified" by "unramified at all places". In that case the role of $H^*(k)$ in (ii) and (iii) is taken by the group $H(k)$. For fields $k$ in which there exist units of all signatures, and in particular for totally complex fields $k$ these two notions of Hilbert class-field coincide. (Recall that a signature of a nonzero element $x \in k$ is defined as the sequence of signs of these conjugates of $x$ which correspond to embeddings of $k$ into the field of reals.)

In the case $k = Q$ it turns out that $K = Q$; in fact it has been stated by L. Kronecker [75] and proved by H. Minkowski [96] as a consequence of this *convex body theorem* that there are no nontrivial unramified extensions of $Q$. (For other proofs see J. Calloway [9], E. Landau [86], L.J. Mordell [99, 100], C. Müntz [101], I. Schur [124], C.L. Siegel [138], H. Weber and J. Wellstein [152]).

**2.2.** The main idea of the classical version of class-field theory reveals itself even in the simplest case, viz. $k = Q$, which we shall now describe.

If $K/Q$ is Abelian and, according to the Kronecker–Weber Theorem, $K$ is contained in $Q(\zeta_f)$ with $\zeta_f$ being a primitive root of unity of order $f$ and $f$ is as small as possible, then $f$ is called the *conductor* of $K$. The group $G(f)$ of invertible residue classes (mod $f$) can be identified with the Galois group of $Q(\zeta_f)$, residue $m$ (mod $f$) corresponding to that element of the Galois group which maps $\zeta_f$ to $\zeta_f^m$.

THEOREM 2.1 (Class-field theory for the rational field). *Let $H$ be the subgroup of $G(f)$ corresponding to $K$ by Galois theory. Then the following assertions hold:*

(a) *A rational prime $p$ ramifies in $K/Q$ if and only if $p$ divides $f$.*

(b) *The Galois group of $K/Q$ is isomorphic to $G(f)/H$.*

(c) *For every rational prime $p$ the decomposition in $K$ of $pZ_K$ depends only on the class in $G(f)/H$ to which $p$ (mod $f$) belongs and $p$ splits in $K/Q$ if and only if $p$ lies in $H$.*

The first and the last assertion follow from the law of decomposition of rational primes in cyclotomic fields and their subfields (which can be established in an elementary way, see, e.g., [Na], Theorem 8.1) and the second is a trivial consequence of Galois theory. We have moreover:

(d) *If a positive integer $f$ which is not congruent to* 2 (mod 4) *and a subgroup $H$ of $G(f)$ is given then there exists a unique Abelian extension $K/Q$ for which the conditions* (a)–(c) *hold.*

(The congruence restriction is needed here, because if it is not satisfied then the fields $Q(\zeta_f)$ and $Q(\zeta_{f/2})$ coincide, hence there is no Abelian extension of the rationals with conductor $f$.)

**2.3.** It is convenient to express the properties (a)–(d) in terms of characters of $G(f)$ (*Dirichlet characters*). Let $X$ be the subgroup of the dual group of $G(f)$, consisting of characters trivializing on $H$. One sees easily that if we apply the same procedure starting with an embedding of $K$ into an arbitrary cyclotomic field, not necessarily minimal, then the group $X$ will remain essentially the same, and hence in this way we can associate with every Abelian extension $K/Q$ a group $X = X(K)$ of Dirichlet characters. To make this statement precise we need the notion of the *conductor* of a character. If $m, n$ are positive integers and $m$ divides $n$, then there is a natural surjection $G(n) \to G(m)$ and thus every character of $G(m)$ has a canonical lift to a character of $G(n)$. If $\chi_1, \chi_2$ are characters of $G(m)$ and $G(n)$ and there exists $N$ and a character $\chi$ of $G(N)$ such that both $\chi_1$ and $\chi_2$ are such liftings of $\chi$ then $\chi_1$ and $\chi_2$ are called *equivalent*. One sees easily that for every equivalence class $C$ there exists a unique integer $N$ and a character $\chi$ of $G(N)$ such that all characters in $C$ are liftings of $\chi$. Such a $\chi$ is called a *primitive character* (mod $N$) and $N$ is called the *conductor* of every character in $C$. One can assume, by identifying equivalent characters, that $X(K)$ consists of primitive characters. The value $\chi(a)$ of a primitive character $\chi$ is well-defined for all integers $a$ prime to the conductor of $\chi$. We adopt the convention that in other cases $\chi(a) = 0$.

The conditions (a)–(d) can now be reformulated in terms of character groups:

THEOREM 2.2. *There is a one-to-one and inclusion-preserving correspondence $K/Q \Leftrightarrow X(K)$ between finite Abelian extensions of Q and finite groups of primitive characters X with the following properties:*

(aa) *A prime p is ramified in $K/Q$ if and only if it divides the conductor of some character in $X(K)$.*

(bb) *The groups* $\mathrm{Gal}(K/Q)$ *and* $X(K)$ *are isomorphic.*

(cc) *For every rational prime p the decomposition in K of $pZ_K$ depends only on the set*

$$\{\chi(p)\colon\ \chi \in X(K)\}.$$

*More precisely: one has*

$$pZ_K = (P_1 \cdots P_g)^e,$$

*where g is the number of $\chi \in X(K)$ with $\chi(p) = 1$, e equals the index in $X(K)$ of the group of characters vanishing at p and finally $f = [K:Q]/eg$.*

**2.4.** The main theorems of class-field theory for an arbitrary algebraic number field $k$ are parallel to (a)–(d). To state them one needs an analogue of the groups $G(f)$ in arbitrary algebraic number fields and this has been provided by H. Weber [151] (cf. R. Fueter [31]):

For any ideal $f$ of $Z_k$ let $G_f$ be the group of all fractional ideals prime to $f$ and let $G_{f,1}$ be its subgroup consisting of all principal ideals of $G_f$ having a totally positive generator congruent to unity mod $f$. If $G$ is a group lying between $G_{f,1}$ and $G_f$ then one says that $G$ *is defined* (mod $f$) and the factor group $H = H(G,f) = G_f/G$ is called an *ideal class-group* (mod $f$). Two such groups $G_1, G_2$, defined (mod $f_1$) and (mod $f_2$) respectively, are called *equivalent*, provided there is an ideal $I$ with the property that the sets of ideals in $G_1$ and $G_2$ which are prime to $I$ coincide. If $G_1$ and $G_2$ are equivalent, then the corresponding class groups are isomorphic and for a given class-group $H$ the GCD of all ideals $f$ for which $H = H(f,G)$ is called *the conductor* of $H$.

The group $G_f/G_{f,1}$ is called the *ray-class-group* (mod $f$). It forms the desired analogue of $G(f)$.

Using these notions H. Weber in the second part of [151] defined the *class-field* $K$ (at first only for quadratic $k$) associated with an ideal class-group $H(f,G)$ as an extension of $k$ such that a prime ideal $P$ of $Z_k$ splits completely in $K$ if and only if it lies in $G$, the *principal class*. He was able to establish its existence for imaginary quadratic fields $k$.

**2.5.** The decisive step was made by T. Takagi [141]. His main results can be stated in the following way:

THEOREM 2.3 (Takagi's class-field theory). *Let k be an algebraic number field.*

(A) *For every ideal class-group H there exists a unique class-field $K = K(H)$. Two such class-fields coincide if and only if the corresponding principal classes are equivalent.*

(B) *The extension $K(H)/k$ is Abelian and its Galois group is isomorphic to H.*

(C) *A prime ideal ramifies in $K(H)/k$ if and only if it divides the conductor of H.*

(D) *The degree of prime ideals of $Z_K$ lying over an unramified prime ideal $P$ of $Z_k$ equals the order of the image of $P$ in $H$. In particular $P$ splits in $K/k$ if and only if $P$ lies in the principal class.*

(E) (The Existence Theorem) *Every finite Abelian extension $K/k$ is a class-field for a suitable ideal class-group $H$.*

Thus all Abelian extensions of $k$ are described in terms of ideal class-groups $H$. The class-field associated with the ray-class-group (mod $f$) is called the *ray-class-field* (mod $f$) and one sees that every Abelian extension of $k$ is contained in a suitable ray-class-field. Thus the ray-class-fields play the same role as cyclotomic fields for Abelian extensions of the rationals.

**2.6.** Actually Takagi used another definition of the class-field which is related to norm residues:

An element $a \in k^*$, prime to $f$, is called a *norm-residue* (mod $f$) in the extension $K/k$, if in the residue class $a$ (mod $f$) there is an element of $N_{K/k}K^*$, i.e. the congruence

$$N_{K/k}(x) \equiv a \pmod f$$

is solvable. The norm-residues (mod $f$) form a subgroup $N$ of finite index in $G_f$. To obtain Takagi's definition of the class-field one has first to observe that the following inequality holds for all finite Galois extensions $K/k$:

$$[G_f : N] \leqslant [K : k]. \tag{$*$}$$

(It is called the *Second Inequality of class-field theory*. Its analogue in non-Abelian case was later established by H. Hasse and A. Scholz [48].)

The first proofs of this inequality were analytical, based on properties of the series

$$L(s, \chi) = \sum_I \chi(I) N(I)^{-s},$$

where $\chi$ is a character of a subgroup of $G_f/N$.

It has been shown by H. Weber [We] that this series converges for $\mathrm{Re}(s) \geqslant c$ (with a suitable $c = c(K) < 1$) in the case of nonprincipal $\chi$ and in the case of the principal character $\chi_0$ it converges for $\mathrm{Re}(s) > 1$ and there exists a nonzero limit

$$\lim_{s \to 1+0} (s-1) L(s, \chi_0).$$

Later it was established that $L(s, \chi_0)$ is meromorphic in the plane with a single pole at $s = 1$ (E. Hecke [52]) and for $\chi \neq \chi_0$ the function $L(s, \chi)$ is entire (E. Hecke [53], where a much more general class of Dirichlet series has been dealt with). These functions also satisfy certain functional equations, similar to that obeyed by Riemann's $\zeta$-function.

A modern proof of Hecke's result was given in 1950 in the thesis of J. Tate, published in [CF]. For expositions of his proof see [La1, Wei1, Na].

If one has equality in $(*)$, then $K$ is called the *Takagi class-field* associated with $N$. (The two definitions of the class-field can be shown to be equivalent. See, e.g., [HBer].)

The proof of (A)–(E) is first reduced by an elementary reasoning to the case of cyclic extensions of prime degree, which is then treated in a rather technical way. One of the main steps involves the proof of the *First Inequality of class-field theory* for cyclic extensions:

$$[G_f : N] \geqslant [K : k].$$

The argument, which seems now to be mainly of historic interest, is exposed with all details in [HBer].

**2.7.** In 1923 E. Artin [1] established a canonical isomorphism in (B). He utilized a trick used first by N.G. Chebotarev [11] in his proof of the density theorem, with which we shall deal later.

To formulate Artin's result (now called *Artin's Reciprocity Law*) we need certain definitions, the first being due to G. Frobenius [30]:

Let $k$ be an algebraic number field and let $K/k$ be a finite Galois extension with Galois group $G$. If $p$ is a prime ideal of $Z_k$, the ring of integers of $k$, which is unramified in $K/k$ and $P$ is a prime ideal divisor of $pZ_K$ then Galois theory implies the existence of a unique automorphism $s = s(P) \in G$ (the *Frobenius automorphism of* $P$) satisfying

$$s(x) \equiv x^{N_{K/k}(p)} \pmod{P} \quad \text{for all } x \in Z_K.$$

One sees easily that conjugate Frobenius automorphisms correspond to conjugate prime ideals, and hence in the case of Abelian $G$ the element $s$ depends only on $p$. In this case one defines the *Artin symbol* by

$$\left(\frac{K/k}{p}\right) = s.$$

THEOREM 2.4 (Artin's Reciprocity Law). *If $K/k$ is an Abelian extension with Galois group* $\mathrm{Gal}(K/k)$ *and $K$ is the class-field with respect to $H = H(f, G)$, then the map*

$$F_{K/k}\colon p \mapsto \left(\frac{K/k}{p}\right)$$

*induces an isomorphism*

$$H \cong \mathrm{Gal}(K/k).$$

For the proof one has to show that

$\alpha$) the value $F_{K/k}(p)$ depends only on the class in $H$ containing $p$, and

$\beta$) if $X, Y \in H$ and $p \in X$, $q \in Y$, $P \in XY$ are prime ideals, then $F_{K/k}(P) = F_{K/k}(p)F_{K/k}(q)$, the full theorem being an easy consequence of $\alpha$) and $\beta$). These assertions are easy to establish in the case when $k$ is contained in a cyclotomic field and in the general case their proof is based on a construction of certain auxiliary cyclotomic fields $M$. If the prime ideals $p_1$, $p_2$ are in the same class of $H$ then one constructs $M$ with the property that these ideals lie in the same class of the ideal class-group corresponding to $kM/k$. This is achieved by means of an elementary number-theoretic lemma and then a short computation leads to $\alpha$). To obtain $\beta$) a similar procedure is used.

The map $F_{K/k}$ can be extended by multiplicativity to a map (denoted again by $F_{K/k}$) defined on the set of all ideals which do not have ramified prime divisors. This map we shall use later on.

**2.8.** We now come to the last of Hilbert's conjectures, the *Principal Ideal Theorem*. It has been observed by E. Artin [2] that (v) can be reduced to a purely group-theoretical statement, which was later proved by P. Furtwängler [35]. It concerns the *transfer map* which we shall now define:

Let $G$ be a finite group and $H$ its normal subgroup. Denote by $G'$, $H'$ the commutator subgroups of $G$ and $H$ and let $C$ be a full set of representatives of $G/H$. The *transfer map* (*Verlagerung*)

$$\mathrm{Ver}_{G,H}\colon\ G/G' \to H/H'$$

is defined in the following way: if $g \in G,\ c \in C,\ gc \in c'H$ then define

$$h(g,c) = (c')^{-1}gc$$

and

$$\mathrm{Ver}(g,G') = \prod_{c\in C} h(g,c)H'.$$

The principal ideal theorem is a consequence of the triviality of the transfer map $\mathrm{Ver}_{G,G''}$ for finite groups $G$ with Abelian commutator group $G''$.

Later other proofs were provided by Z.I. Borevich [5], S. Iyanaga [63], Y. Kawada [66], W. Magnus [89], H.G. Schumann [123], K. Taketa [142], E. Witt [155, 156].

There are many generalizations and analogues of this result. See, e.g., K. Iwasawa [62], J.F. Jaulent [64], K. Miyake [97, 98] and the references on p. 203 of [Na].

## 3. The modern approach

**3.1.** To state the main theorem of class-field theory in modern form we have first to define the *adeles* and *ideles*. Ideles were introduced by C. Chevalley [12], who in [13] used them to reformulate class-field theory. Adeles occur under the name *valuation vectors* in E. Artin and G. Whaples [3] and they were named "adeles" by A. Weil [153].

We start with notion of *restricted product* of topological groups. Let $G_v$ be a sequence of locally compact groups in almost all of which a compact and open subgroup $H_v$ is selected. (We use the phrase "almost all" to mean "all with at most a finite number of exceptions".) The restricted product of the $G_v$'s with respect to the $H_v$'s is defined as the subgroup of the direct product $\prod_v G_v$ consisting of all sequences $(a_v)_v$, $(a_v \in G_v)$ with $a_v \in H_v$ for almost all $v$'s. The topology in $G$ is defined by taking for the basis of open sets all products $\prod_v O_v$ where $O_v$ is open in $G_v$ and $O_v = H_v$ holds for almost all $v$.

Let $k$ be a global field, i.e. either an algebraic number field or a field of algebraic functions in one variable over a finite field and let $V$ be a complete set of inequivalent valuations of $k$ normalized so that the *product formula*

$$\prod_{v \in V} v(x) = 1,$$

holds for all nonzero $x \in X$. (Note that as shown in E. Artin and G. Whaples [3] the existence of the product formula characterizes global fields.) Let $k_v$ be the completion of $k$ with respect to $v$ and denote the extension of $v$ to $k_v$ by the same letter $v$. In the case of nonarchimedean $v$ we put

$$R_v = \{a \in k_v\colon\ v(a_v) \leqslant 1\} \quad \text{and} \quad U_v = \{a \in k_v\colon\ v(a_v) = 1\}.$$

**3.2.** The group of adeles of $k$ is defined as the restricted product of the additive groups $k_v^+$ with respect to $R_v^+$ and defining in it multiplication coordinate-wise we get the *ring* $A_k$ *of adeles* of $k$.

The group $I_k$ of ideles of $k$ is defined similarly as the restricted product of the multiplicative groups $k_v^*$ with respect to the groups $U_v$. One sees immediately that $I_k$ equals the group of invertible elements of $A_k$; however, its topology differs from that induced by that of $A_k$.

One distinguishes two important subgroups in $I_k$: the group $U_k$ of *unit ideles* consisting of all ideles $(a_v)_v$ with $a_v \in U_v$ and the group $J_k$ of all ideles $(a_v)$ satisfying

$$\prod_{v \in V} v(a_v) = 1.$$

The embedding $k \hookrightarrow k_v$ induces an injection of $k^+$ in $A_k^+$ mapping $k^*$ into $J_k$ and one identifies $k^+$ and $k^*$ with their images. The factor groups $A_k^+/k^+$ and $C(k) = I_k/k^*$ are called *adele* and *idele class-groups*, respectively. The adele class group is compact. If $k$ is an algebraic number field and we denote by $D(K)$ the connected component of the unit element of $C(K)$, then the factor group $C(K)/D(K)$ is compact and totally disconnected. This implies in particular that every character of $C(K)/D(K)$ is of finite order. In the functional case $C(K)$ is totally disconnected.

If $K/k$ is a finite extension then the norm maps $K_v^* \to k_v^*$ induce a map

$$N_{K/k}\colon\ I_K \to I_k$$

which in turn leads to norm map $N\colon\ C(K) \to C(k)$.

We need also the definition of the Artin map for ideles. Let $a = (a_v)_v \in I_k$, and let $a_v$ be a nonunit of $k_v$ for $v \in S$ where $S$ is a finite set of valuations containing all Archimedean valuations. Then for $v \in S$ we may write

$$a_v \in \varepsilon_v \pi_v^{N_v}$$

with a unit $\varepsilon_v$ and $N_v \in Z$ and if $p_v$ denotes the prime ideal corresponding to $v$ then we define the Artin map $F_{K/k}$ in terms of the Artin map for ideals (which in the functional case is defined analogously to the number case) by putting

$$F_{K/k}\big((a_v)\big) = \prod_{v \notin S} F_{K/k}(p_v)^{N_v}.$$

**3.3.** The idelic formulation of the class-field theory gives the existence of a one-to-one correspondence (given by the norm map) between Abelian extensions of $k$ and closed subgroups of finite index of $C(k)$. More precisely we have the following assertions:

THEOREM 3.1. I. Reciprocity Law. *If $K/k$ is Abelian then there exists a surjective continuous homomorphism*

$$\psi\colon\ I_k \to \mathrm{Gal}(K/k),$$

*with kernel equal to $k^* N_{K/k}(I_K)$, hence inducing a continuous isomorphism*

$$C(k)/N_{K/k}\big(C(K)\big) \cong \mathrm{Gal}(K/k),$$

*and if $a = (a_v)_v \in I_k$ satisfies $a_v = 1$ for all $v$ which are either Archimedean or ramified in $K/k$, then*

$$\psi(a) = F_{K/k}(a),$$

*where $F_{K/k}$ denotes the Artin map for ideles.*

II. Existence Theorem. *If $N$ is a closed subgroup of finite index of $C(k)$ then there exists a unique Abelian extension $K/k$ satisfying*

$$N_{K/k}\big(C(K)\big) = N.$$

This formulation is essentially due to C. Chevalley [13], who used the convenient language of infinite extensions:

Let $K$ be a field and let $L/K$ be an infinite Galois extension. The Galois group $G$ of $L/K$ has a natural topology, the *Krull topology* in which the Galois groups of $L/K$ (where $M/K$ runs over all finite extensions contained in $L$) is taken as a fundamental set of open neighborhoods of the unit element. In this topology $G$ is compact and zero-dimensional. It has been established by W. Krull [76] that there is, as in the usual Galois theory, a biunique correspondence between fields lying between $K$ and $L$ and closed subgroups of $G$. Denote by $K^{ab}$ the maximal Abelian extension of $K$, i.e. the union (or the direct limit) of all finite Abelian extensions of $K$ contained in a fixed algebraic closure, and let $G$ be the Galois group of $K^{ab}/K$. Chevalley established in the number field case an isomorphism between the dual groups of the Galois group of $K^{ab}/K$ and $C(K)/D(K)$ which behaves in a nice way with regard to the arithmetical properties of $K$ and from this Theorem 3.1 can be deduced.

It was later shown by G. Hochschild and T. Nakayama [61, 102] that the use of group cohomology leads to an essential simplification of Chevalley's proof.

**3.4.** To show that the classical and idelic formulations of class-field theory in the number case are equivalent one establishes a topological isomorphism of the groups $C(K)/D(K)$ and $H_\infty = \operatorname{lim\,inv} H(f, G_{f,1})$. By Pontryagin duality it suffices to establish an isomorphism for the corresponding (discrete) character groups.

Every character $X$ of $C(K)/D(K)$ is of finite order and can be regarded as a character of $I_K$ trivial on $K^*D(K)$. We can write

$$X\big((x_v)_v\big) = \prod_v X_v(x_v),$$

where each $X_v$ is a character of finite order of $k_v^*$, which for almost all $v$ satisfies $X_v(U_v) = 1$. If $S_1$ denotes the set of all nonarchimedean $v$'s, for which this equality holds, $S_2$ is the set of the remaining nonarchimedean $v$'s and $p_v$ denotes the prime ideal of $Z_K$ inducing $v$, then for $v \in S_2$ the value $X_v((x_v))$ depends only on $v(x_v)$. Now for each nonarchimedean $v$ choose $\pi_v \in k_v^*$ generating $p_v$ and define

$$\chi(p_v) = \begin{cases} 0 & \text{if } v \in S_1, \\ X_v(\pi_v) & \text{if } v \in S_2. \end{cases}$$

This defines a character of the group of all fractional ideals of $K$ prime to

$$I = \prod_{v \in S_1} p_v,$$

and hence is a character of a suitable $H(f, G_{f,1})$. This leads to a homomorphism of the character group of $C(K)/D(K)$ to $\operatorname{lim\,dir} H(f, G_{f,1})$ which is the dual group of $H_\infty(K)$. To show that it is in fact an isomorphism one uses the easily established isomorphism

$$H(f, G_{f,1}) \cong I_K / I_f U_K,$$

where $I_f$ denotes the group of principal ideles $(x_v)_v$ with $x$ totally positive and congruent to unity $(\operatorname{mod} f)$. (For details see, e.g., [Na], Chapter VII.)

There are several ways of proving the assertions I and II. Usually one deduces them from the corresponding theorems of local class-field theory. Using cohomology groups this has been done in [AT] and an exposition can be found, e.g., in [CF].

**3.5.** An important step towards simplification of the class-field theory has been made by J. Neukirch [103, Ne1, Ne2], who adopted an axiomatic approach utilizing only the first few cohomology groups. One starts with a profinite group $G$ and a *degree map* deg: $G \to \widehat{Z}$, where $\widehat{Z}$ denotes the inverse (projective) limit of the cyclic groups $Z/nZ$ with the natural homomorphisms. The closed subgroups of $G$ are denoted by $G_K$ where $K$ runs over a set of indices. Abusing language a little one calls these indices $K$ fields and the index corresponding to $G$ is called the ground field and denoted by $k$. The index corresponding to the one-element group is denoted by $\tilde{k}$. One defines the intersection of a family $\{K_i\}$ of fields as the index of the group corresponding to the smallest closed subgroup of $G$ containing all groups $G_{K_i}$ and the compositum of fields from this family is defined as the index of the intersection of the corresponding groups. For every extension $K/k$ of finite degree one defines $\widetilde{K}$ as the compositum of $K$ and $\tilde{k}$.

If $G_L \subset G_K$ then the pair $(L, K)$ is called a field extension and denoted by $L/K$. The degree $[L : K]$ of such an extension is defined as the index $[G_K : G_L]$ and if this index is finite then $L/K$ is called a finite extension. If $G_L$ is a normal subgroup of $G_K$ then $L/K$ is called a normal extension and the factor-group $G_K/G_L$ is denoted by $\mathrm{Gal}(L/K)$ and called the Galois group of $L/K$.

Finally let $A$ be a multiplicative $G$-module and for every field $K$ put $A_K = A^{G_K}$, the set of elements invariant under $G_K$, and for any finite extension $L/K$ define the norm map $N_{L/K}$: $A_L \to A_K$ by

$$N_{L/K}(a) = \prod a^g,$$

where $g$ runs over a system of representatives of right cosets of $G_K$ with respect to $G_L$.

Now we may state the axioms:

Axiom I. *If the extension $L/K$ is finite and $L \subset \widetilde{K}$ (i.e. $G_{\widetilde{K}} \subset G_L$), then*

$$\#H^0\big(\mathrm{Gal}(L/K), A_L\big) = [L : K]$$

*and*

$$\#H^{-1}\big(\mathrm{Gal}(L/K), A_L\big) = 1.$$

Axiom II. *If the extension $L/K$ is finite and the group* $\mathrm{Gal}(L/K)$ *is cyclic, then*

$$\#H^0\big(\mathrm{Gal}(L/K), A_L\big) = [L : K]$$

*and*

$$\#H^{-1}\big(\mathrm{Gal}(L/K), A_L\big) = 1.$$

From these two axioms one deduces for every finite extension $L/K$ the existence of a canonical isomorphism of $A_K/N_{L/K}A_L$ and $\mathrm{Gal}(L/K)^{ab}$, the maximal abelian factor group of $\mathrm{Gal}(L/K)$.

To obtain the main theorems of the class-field theory for global fields let $k$ be either the field of rationals or the field of rational functions in one variable over $F_q$ and let $\bar{k}$ be an algebraic closure of $k$. One takes for $G$ the *absolute Galois group*, i.e. $G = \mathrm{Gal}(\bar{k}/k)$ and for the $G$-module $A$ the union of the idele class groups of all finite extensions of $k$.

## 4. Reciprocity laws

**4.1.** If $p$ is an odd prime then the unique quadratic character $(\bmod p)$, written after Legendre usually in the form $\left(\frac{x}{p}\right)$, obeys the following *Quadratic Reciprocity Law*, stated first by L. Euler in 1772 and proved in 1801 by C.F. Gauss [DA] (§135–144):

*If $p$, $q$ are distinct odd primes then*

$$\left(\frac{p}{q}\right)\left(\frac{q}{p}\right) = (-1)^{a(p,q)},$$

*where* $a(p,q) = (p-1)(q-1)/4$. (For its early history see [Wei2].)

The search for the analogue of it for higher powers formed one of the main topics of research in number theory of the XIXth century. The case of biquadratic residues has been considered by C.F. Gauss [36], who introduced complex integers, i.e. numbers of the ring $Z[i]$ to solve the problem. G. Eisenstein [28] obtained a kind of reciprocity theorem (he assumed that one of the considered moduli is a rational integer) for $p$-th powers (with $p$ prime) in the $p$-th cyclotomic fields. A simpler proof of his result can be found in [Zber] and a generalization to arbitrary exponents was given by H. Hasse [44], using Artin's Reciprocity Law.

E.E. Kummer [80, 81] dealt with the reciprocity law for $p$-th powers in the $p$-th cyclotomic field $Q(\zeta_p)$ in the case when $p$ is a *regular prime*, i.e. $p$ does not divide the class-number of $Q(\zeta_p)$. (We do not know, even today, whether there are infinitely many such primes, but it has been established by K.L. Jensen [65] in 1915 that there are infinitely many irregular primes. These numbers are related to *Fermat's Last Theorem*, which is known to be true for every regular prime exponent (E.E. Kummer [79]). It has been shown by E.E. Kummer [78] that a prime $p$ is regular if and only if it does not divide the numerator of any nonzero Bernoulli number with index $\leqslant p-3$.) The restriction on $p$ has been removed by P. Furtwängler [33] and the case of an arbitrary exponent has been settled by H. Hasse [45]. Part II of [HBer] contains all details and a complete bibliography up to 1929.

The analogue of the Quadratic Reciprocity Law in arbitrary number fields has been given by D. Hilbert [59] and E. Hecke [He, 54]. (See also L. Auslander, R. Tolimieri and S. Winograd [4].)

**4.2.** Artin's reciprocity law of the class-field theory gives a common generalization of all previously known reciprocity laws and brings several new aspects to this question.

Let $k$ be an algebraic number field containing all $n$-th roots of unity and let $a \in k^*$. Denote by $S(a)$ the set of all these nonarchimedean valuations $v$ for which $v(a) = 1$ and let $f = f(a)$ be the product of all prime ideals corresponding to $v \in S(a)$. Put $\vartheta = a^{1/n}$, $K = k(\vartheta)$ and let

$$F_{K/k}\colon G_f \to \mathrm{Gal}(K/k)$$

be the Artin map (with $G_f$ denoting the group of all fractional ideals prime to $f$). For every ideal $I \in G_f$ the number

$$\frac{F_{K/k}(I)(\vartheta)}{\vartheta}$$

is a root of unity, which we shall denote by

$$\left(\frac{a}{I}\right)_n,$$

the *n-th power-residue symbol*. One sees that this symbol is a character of $G_f$ having the property that if $P$ is a prime ideal of $Z_k$ not dividing $f(a)$, then

$$\left(\frac{a}{P}\right)_n = 1$$

holds if and only if the congruence

$$X^n \equiv a \pmod P$$

is solvable in $Z_k$, or, which in view of Hensel's Lemma means the same, $a$ is an $n$-th power in the completion of $k$, corresponding to $P$.

Artin's Reciprocity Law implies that if $I_1, I_2$ are ideals prime to $f(a)$ lying in the same ideal class and the ratio $I_1/I_2$ (which is principal) has a generator $b$ which for all $v \notin S(a)$ is an $n$-th power in $k_v$, then

$$\left(\frac{a}{I_1}\right)_n = \left(\frac{a}{I_2}\right)_n.$$

In case $n = 2$, $k = Q$ this leads after a short elementary computation to the Quadratic Reciprocity Law. One can also similarly recover the reciprocity laws of Gauss, Eisenstein and Kummer (see [HBer] for details).

**4.3.** To state the reciprocity law for $n$-th power residues we need the local *Hilbert symbol* $(a,b)_v$ occurring in the local class-field theory. Here $a, b \in k_v^*$, $k_v$, the completion of $k$ at $v$, is assumed to contain all $n$-th roots of unity and moreover the extensions $k_v(a^{1/n})/k_v$ and $k_v(b^{1/n})/k_v$ are both unramified.

The local class-field theory gives the existence of a canonical isomorphism

$$\mathrm{Gal}(K/k_v) \to k_v^*/N_{K/k}(K^*)$$

for all Abelian $K/k_v$ and its inverse leads to a homomorphism

$$k_v^* \ni x \mapsto (x, K/k_v) \in \mathrm{Gal}(K/k_v).$$

The symbol $(x, K/k_v)$ is called the *norm-residue symbol.* The local Hilbert symbol $(a,b)_v$ is now defined as that root $\zeta$ of unity for which one has

$$\left((a, k_v(b^{1/n})/k_v)\right)b^{1/n} = \zeta b^{1/n}.$$

In the case of Archimedean $v$ the Hilbert symbol is defined in the following way: Denote by $\Phi_v$ the embedding of $k$ in the complex field corresponding to $v$ and put

$$(a,b)_v = \begin{cases} -1 & \text{if } v \text{ is real and } \Phi(ab) < 0, \\ 1 & \text{otherwise.} \end{cases}$$

Now we can state:

THEOREM 4.1 (General Reciprocity Law for $n$-th powers). *Let $a, b \in k^*$ and denote by $A$, $B$ the principal ideals generated by them. If $A$, $B$ are relatively prime and prime to $nZ_k$ then one has*

$$\left(\frac{a}{B}\right)_n \left(\frac{b}{A}\right)_n^{-1} = \prod_v (a,b)_v,$$

*where $v$ runs through all valuations which either are Archimedean or satisfy $v(n) < 1$.*

The usual proof utilizes the compatibility of the reciprocal Artin maps in the local and global class-field theory. (See, e.g., [HBer, Iy].)

**4.4.** To be able to apply the reciprocity law for power residues one needs explicit formulas for the local Hilbert symbols. This belongs properly to the local class-field theory, so we restrict ourselves to few bibliographical remarks. Such formulas were given by I.R. Shafarevich [129] (see H. Hasse [46] for an exposition). Other deductions of explicit reciprocity have been given by M. Kneser [67], W.H. Mills [94, 95], H. Brückner [6], S.V. Vostokov [146, 147, 148], S.V. Vostokov and V.A. Lecko [149], S. Sen [126], E. de Shalit [132, 133], S. Helou [55]. There is a recent survey by S. Helou [56].

## 5. Density theorems

**5.1.** Let $K$ be an algebraic number field and let $\mathcal{P}$ be a set of prime ideals in its ring of integers $Z_K$. One says that $\mathcal{P}$ has *natural density* $\lambda$, provided the number of prime ideals in $\mathcal{P}$ with norms not exceeding $x$ is asymptotically equal to $\lambda x/\log x$ when $x \to \infty$. It has been shown by E. Landau [84] that the density of the set of all prime ideals, as well as of the set of all prime ideals of degree one, equals 1 (*Prime Ideal Theorem,*

*Primidealsatz*). Later E. Hecke [53] proved that the density of the set of all prime ideals lying in a given class of $H_f(K)$ exists and is the same for each class (*Hecke's Theorem on Progressions*) and E. Landau [85] extended that to the classes of $H_f^*(K)$.

A far-reaching generalization of these results is the following consequence of Artin's Reciprocity Theorem, conjectured by G. Frobenius [30] and proved by N.G. Chebotarev [11]:

THEOREM 5.1. *Let $L/K$ be a finite Galois extension of an algebraic number field, denote by $G$ its Galois group and let $A$ be an arbitrary class of conjugated elements in $G$. Let $\mathcal{P}_A$ be the set of all prime ideals $P$ of $Z_K$ for which the Frobenius automorphism $s(P)$ lies in $A$. Then the set $\mathcal{P}_A$ has a natural density, which is equal to $\#A/\#G$.*

One considers first Abelian extensions $L/K$ and in this case Artin's Reciprocity Law implies that the set $P_A$ is a union of certain cosets $(\operatorname{mod} G_{f,1})$ and hence Hecke's Theorem on Progressions (in Landau's form) may be used to obtain the assertion in this case. The final step consists in reducing the general case to the case of a cyclic extension and this is done by considering the subfield of $L/K$ corresponding to the cyclic group generated by any element of $A$ and using the proper behaviour of the Frobenius automorphism with regard to subextensions.

Chebotarev actually proved a weaker form of this theorem, since Artin's Reciprocity Law was not yet at his disposal. (In fact Artin's proof of his reciprocity depended on ideas from Chebotarev's paper.) He showed namely that the set $P_A$ has a *Dirichlet density* equal to $\#A/\#G$.

A set $\mathcal{P}$ of ideals are said to have the Dirichlet density $\lambda$, provided the function $f(s)$ defined for $\operatorname{Re}(s) > 1$ by

$$f(s) = \sum_P \frac{1}{N(P)^s}$$

satisfies

$$\lim_{x\to 1+0} f(x)/\log\frac{1}{x-1} = \lambda.$$

Note that the existence of the Dirichlet density does not imply the existence of the natural density. This implication holds, however, if the difference

$$f(s) - \lambda \log\frac{1}{s-1}$$

can be prolonged to a function regular in the closed half-plane $\operatorname{Re}(s) \geqslant 1$. (Theorem of Ikehara–Delange, see H. Delange [21].)

The original proof of Theorem 5.1 has been greatly simplified by O. Schreier [122], A. Scholz [120], M. Deuring [24] and C.R. McCluer [92]. In certain special cases there exist purely algebraic proofs (J. Wójcik [157], H.W. Lenstra Jr. and P. Stevenhagen [87]). Proofs giving an effective bound for the smallest norm of an ideal lying in $\mathcal{P}_A$ have been given by J. Lagarias, H.L. Montgomery and A. Odlyzko [82], J. Lagarias and A. Odlyzko [83] (cf. J. Oesterlé [109]) and V. Schulze [125].

Chebotarev's result immediately implies two previously known density theorems, the first due to L. Kronecker [74] and the second to G. Frobenius [30] (in both cases $K$ is an algebraic number field):

THEOREM 5.2 (Density Theorem of Kronecker). *If $L/K$ is finite of degree $n$ and $\mathcal{P}_k$ denotes the set of all prime ideals $P$ of $Z_K$ for which $PZ_L$ has exactly $k$ prime ideal divisors of the first degree, then $P_k$ has a Dirichlet density and*

$$\sum_{k=0}^{n} d_k = \sum_{k=1}^{n} k d_k = 1.$$

THEOREM 5.3 (Density Theorem of Frobenius). *If $L/K$ is Galois, $g \in \mathrm{Gal}(L/K)$ and $A$ is the union of conjugacy classes of all powers of $g$, then the set $\mathcal{P}_A$ has its Dirichlet density equal to $\#A/[L:K]$.*

**5.2.** It has been shown by J.P. Serre [127] that Chebotarev's theorem can be applied in the theory of modular forms and elliptic curves. These applications utilize the theory of $l$-adic representations (P. Deligne [22], P. Deligne and J.P. Serre [23, Se]).

The analogue of Chebotarev's density theorem holds also in the case of algebraic function fields in one variable over a finite field. If one defines the Frobenius automorphism and Artin's symbol in the same way as in the case of number fields then the following assertion holds:

THEOREM 5.4. *If $K$ is a field of algebraic functions in one variable over a finite field and $L/K$ is a finite Galois extension with Galois group $G$ and $C$ is a fixed conjugacy class in $G$ then the set of all prime ideals of $K$ unramified in $L/K$ whose Frobenius automorphism lies in $C$ has Dirichlet density which is equal to $\#C/[L:K]$.*

A special case of this theorem has been settled by H. Reichardt [113]. J.P. Serre [Se] indicated the deduction of the general case. A proof has been given by M. Fried [29]. (See also [FJ], Theorem 5.6, and F. Halter-Koch [42].)

## 6. Kronecker's "Jugendtraum" and explicit class-fields

**6.1.** The Kronecker–Weber theorem shows that every finite Abelian extension of the rationals is contained in an extension generated by the value of $e(z) = \mathrm{e}^{2\pi i z}$ at a rational point. This leads to the problem of whether a similar result holds for other base-fields and clearly one can limit the attention to ray class-fields. In the case of imaginary quadratic base-fields this question has already been considered by L. Kronecker.

Let $k$ be an imaginary quadratic field and let $I$ be a fractional ideal. Clearly $I$ can be regarded as a lattice in $\mathbf{C}$ and hence we may associate with $I$ an elliptic curve, viz. $E = \mathbf{C}/I$. Fractional ideals from different ideal classes lead to nonisomorphic curves, so we get $h = h(K)$ elliptic curves $E_1, \dots, E_h$ associated with $k$. Replacing them with isomorphic curves one may assume that each of them can be written in the form

$$Y^2 = 4X^3 - aX - b.$$

The *invariant of* $E$, $j(E)$ is defined by

$$j(E) = \frac{1728a^3}{a^3 - 27b^2},$$

and since every elliptic curve is isomorphic to one of the form $\mathbf{C}/(Z + zZ)$ with a suitable $z$ ($\mathrm{Im}(z) > 0$), we obtain in this way a function $j(z)$ defined in the upper half-plane. It can also be given explicitly by the formula

$$\begin{aligned} j(z) &= \frac{(1 + 240 \sum_{k=1}^{\infty} \sigma_3(k) q^n)^3}{q \prod_{k=1}^{\infty} (1 - q^k)^{24}} \\ &= q^{-1} + 744 + 196884q + 21493760q^2 + \cdots, \end{aligned}$$

where $q = \mathrm{e}^{2\pi i z}$ and $\sigma_3(n)$ denotes the sum of cubes of positive divisors of $n$. (Note that the coefficients 196884, 21493760, ... turned out to be closely related to the degrees of irreducible representations of the Monster. See J. Conway and S.P. Norton [19].)

Already L. Kronecker conjectured that every Abelian extension of an imaginary quadratic field $k$ is contained in a field generated by a value of $j(z)$ at a point from $k$. (*Kronecker's Youth Dream, Kroneckers Jugendtraum.*) This conjecture is closely connected with the theory of *complex multiplication*, the main result of which asserts that the Hilbert class-field $H(k)$ of $k$ is generated by each of the values $j(E_i)$, which are conjugate algebraic integers of degree equal to $h(k)$. H. Hasse [43], M. Deuring [25, 26], cf. [De, SCM, Shi, Fu, We].

The original conjecture of Kronecker turned out to be not true, but it has been shown by H. Hasse [43] that the maximal Abelian extension of $k$ can be obtained by adjoining to $H(k)$ a sequence of values at points in $k$ of a function $\tau(z)$, which is related to the Weierstrass $\wp$-function. (See, e.g., [SCM].) An important result was obtained in 1964 by K. Ramachandra [112], who for every quadratic imaginary $k$ constructed a holomorphic function, whose values at points of $k$ generate all ray class-fields of $k$.

**6.2.** The description of the Hilbert class-fields $H_k$ of an imaginary quadratic field $k$ given by values of the function $j(z)$ does not lead easily to an explicit algebraic description of its generators. However, in certain cases it is possible to utilize purely algebraic means to obtain such a description. This has been demonstrated by H. Hasse [47] in the cases $k = Q(\sqrt{-d})$ with $d = 23, 31, 47$. See also H. Cohn [14, 15, 16], H. Cohn and G. Cooke [17].

A description of Hilbert class-fields and ray class-fields for other classes of fields still forms in general an open problem, although there have been important advances in certain cases. It turned out that the generalization of complex multiplication to higher-dimensional varieties leads in certain cases to class-fields for totally complex extensions of totally real fields (the *CM-fields*). (See [ST, Sh], G. Shimura [134, 135, 136].) In [Shi] one finds applications of modular functions to the construction of class-fields for real quadratic fields.

**6.3.** The Kronecker–Weber theorem shows that the maximal Abelian extension of the rational number field is generated by the torsion points of the mappings $x \mapsto x^n$ of the multiplicative group of all algebraic numbers. (A point $a$ is a torsion point for a map $T$ provided for some $n$ one has $T^n(a) = a$, where $T^n$ denotes the $n$-th iterate of $T$.) An analogous result in the function fields case has been obtained by D.R. Hayes [49] for the maximal Abelian extension of the field $k = F_q(X)$. It turned out that it is generated (modulo certain extensions of the constant fields) by torsion points of a certain family of maps of the additive group of the algebraic closure of $k$, which has been described by L. Carlitz [10].

A description of Abelian extensions for arbitrary global function fields $k$ has been given by V.G. Drinfeld [27], who generalized the classical theory of complex multiplication by introducing *elliptic modules*, which are certain homomorphisms of a suitable subring of $k$ into the ring of additive polynomials in $k[X]$ with composition as multiplication. Using these homomorphisms one can construct all ray class-fields of $k$. (See D. Goss [39], D.R. Hayes [50].)

## 7. Class-field-tower problem

**7.1.** For a given algebraic number-field $k$ one can construct a sequence $k_1 \subset k_2 \subset \cdots$ of fields by putting $k_1 = k$ and defining $k_{i+1}$ as the Hilbert class-field of $k_i$. P. Furtwängler (see [HBer], I, p. 46) asked whether this sequence contains only finitely many distinct fields (*class-field-tower problem, Klassenkörperturmproblem*). Were it so then we could infer that every algebraic number field can be embedded in a field with class-number one and in particular the existence of infinitely many fields whose rings of integers have unique factorization would follow. Unfortunately the answer to Furtwängler's question is negative. At first A. Scholz [119] observed that the sequence $k_i$ may be arbitrarily long and in 1964 E.S. Golod and I.R. Shafarevich [38] proved that the class-field-tower may be infinite.

Let $p$ be a prime. For any algebraic number field $k$ define the *p-class-field-tower* as the sequence

$$k = k_0 \subset k_1 \subset k_2 \subset \cdots \tag{1}$$

of fields where $k_{i+1}$ is defined as the maximal unramified Abelian $p$-extension of $k_i$. If $k$ has a finite class-field-tower then for every $p$ the $p$-class-field-tower is also finite. For any finite group $G$ and prime $p$ let $\dim_p G$ be the number of $p$-primary factors in a decomposition of the maximal Abelian factor group of $G$ into cyclic summands. If now the sequence (1) is finite, $K$ denotes its last term, $G = \mathrm{Gal}(K/k)$ and $K^{ab}$ is the maximal subfield of $K$ which is Abelian over $k$, then class-field theory implies

$$\dim_p G = \dim_p \mathrm{Gal}(K^{ab}/k) = \dim_p H(k).$$

Golod and Shafarevich obtained an upper bound for this dimension in the case of a finite class-field-tower (1). We state their result in a stronger version, obtained by W. Gaschütz and E.B. Vinberg [145] (see P. Roquette [CF] for a presentation of Gaschütz's proof. Another proof has been given by H. Koch [68]):

THEOREM 7.1. *If* $[k : Q] = n$ *and* $\dim_p H(k)$ *exceeds* $2(1 + \sqrt{n+1})$ *then* $k$ *has an infinite* $p$*-class-field-tower.*

**7.2.** The proof is based on the following auxiliary group-theoretic theorem:

THEOREM 7.2. *If* $G$ *is a finite* $p$*-group and* $g = g(G)$ *is the minimal number of generators of* $G$ *then the minimal number* $r = r(G)$ *of relations among them which define* $G$ *satisfies*

$$r > g^2/4.$$

(Golod and Shafarevich had here $(g-1)^2$ in place of $g^2$.)
The exact sequences

$$1 \to \mu_K \to U_K \to U_K/\mu_K \to 1$$

and

$$1 \to U_K/\mu_K \to I_K \to H(K) \to 1$$

(where $U_K$ denotes the unit group of $K$ and $\mu_K$ the group of roots of unity contained in $K$) imply with the use of Tate's theorem (J. Tate [143]), stating that for $n > 2$ the groups $H^n(G, I_K)$ and $H^{n-2}(G, Z)$ are isomorphic, the isomorphisms

$$\widehat{H}^0(G, \mu_K) \cong H^{-1}(G, U_K/\mu_K) \cong H^{-1}(G, I_K) \cong H^{-3}(G, Z).$$

Since for every finite $p$-group $G$ one has $\dim_p H^{-3}(G, Z) = r(G) - g(G)$ and Dirichlet's unit theorem implies

$$\dim_p \widehat{H}^0(G, U_K) = \dim_p \big(U_k/N_{K/k}(U_k)\big) < n$$

thus

$$r(G) - g(G) < n$$

and Theorem 7.2 leads finally to

$$g^2/4 - g < n,$$

thus in view of $g(G) = \dim_p(G) = \dim_p H(k)$ the assertion follows.

It has been conjectured (J. Mennicke [93]) that one always has $r \geqslant g^2/2$ but it turned out that the bound in Theorem 8.2 cannot be essentially improved, since there is a sequence $G_n$ to $p$-groups for which $r(G_n)/g^2(G_n)$ tends to 1/4 (J. Wisliceny [154]). For previous results on this question see A.I. Kostrikin [71] and H. Koch [69]. The proof of Theorem 8.2 given by Golod and Shafarevich used the theory of nilpotent algebras and formed the basis of the solution of two old problems, obtained by E.S. Golod:

THEOREM 7.3 (E.S. Golod [37]). (i) *For every field $k$ there exists a finitely generated nil-algebra which is not nilpotent,*

(ii) *There exists an infinite, finitely approximated $p$-group every element of which has finite order.*

Recall that a *nil-algebra* is an associative algebra in which all elements are nilpotent and an algebra $A$ is called *nilpotent*, provided there exists $n > 0$ such that for all $x \in A$ one has $x^n = 0$. (i) answers a question of J. Levitzky [88] and (ii) settles a problem posed in 1902 by W. Burnside [8]. (It was established later by P.S. Novikov and S.I. Adjan [105] that there exists a finitely generated infinite group every element of which has bounded order. See [Ad] for an exposition of the proof.)

**7.3.** Gauss's *Theorem on Genera* (C.F. Gauss [DA], § 286–287, see, e.g., [Na], Theorem 8.8) implies that any imaginary quadratic field with at least six ramified primes and any real quadratic field with at least eight such primes must have an infinite class-field-tower. To obtain examples of such fields with higher degree one can utilize the following result of A. Brumer [7]:

THEOREM 7.4. *If $[k : Q] = n$ and there exist a rational prime $p$ and at least $t$ rational primes $q$ with the property that the ramification indices of all prime ideal divisors of $q$ in $k$ are divisible by $p$, then*

$$\dim_p H(k) \geqslant t - n^2.$$

(For improvements see P. Roquette and H. Zassenhaus [114], I. Connell and D. Sussmann [18, Na], Theorem 8.10.)

The existence of a finite class-field-tower for a field $k$ induces severe restrictions on the class-group of $k$. It has been shown in B.B. Venkov and H. Koch [144] that if $K$ is the last field in the finite $p$-class-field tower of an imaginary quadratic field $k$ (with $p$ an odd prime), then the Galois group of $K/k$ is either cyclic or has 2 generators and 2 relations.

R. Schoof [121] proved the existence of infinitely many both real and imaginary quadratic fields with two ramified primes and an infinite class-field-tower (improving in the imaginary case a result of B. Schmithals [118]), and gave examples of such fields with only one ramified prime, e.g., $Q(\sqrt{-3321607})$ and $Q(\sqrt{-39345017})$.

The paper of Schoof also contains examples of cyclotomic fields $Q(\zeta_n)$ with infinite class-field-tower, with the smallest $n$ being equal to 363 and the smallest prime $n$ being 877. It is apparently unknown whether these examples are minimal. His method utilizes a generalization of a result of H. Koch and B.B. Venkov [70].

The negative solution of the class-field-tower problem led to the solution of another old problem: denote by $M(n)$ the minimal absolute value of the discriminant of an algebraic number field of degree $n$ and put

$$D = \liminf_{n \to \infty} M(n)^{1/n}.$$

It has been conjectured that $D$ is infinite but this contradicts Theorem 7.1. Indeed, if

$$k_0 \subset k_1 \subset \cdots \subset k_n \cdots$$

is an infinite class-field-tower, $n_i$ denotes the degree of $k_i$ and $d_i$ denotes the absolute value of its discriminant, then the formula for the discriminant in a field-tower leads to $d_i = d_0^{n_i}$, because $k_i/k_0$ is unramified. This shows

$$M(n_i)^{1/n_i} \leqslant d_i^{1/n_i} = d_0 < \infty,$$

hence $D \leqslant d_0$. The proof of E.S. Golod and I.R. Shafarevich [38] leads to $D \leqslant 4404.5$, and later one got $D \leqslant 347$ (A. Brumer [7]), $D \leqslant 92.4$ (J. Martinet [90]). On the other hand one has $D \geqslant 22.38$ (A. Odlyzko [106, 107]) and even $D \geqslant 44.7$ (J.P. Serre [128]) if the General Riemann Hypothesis is true. (Cf. G. Poitou [110, 111].) These lower bounds are obtained by analytical means. This approach was initiated by H.M. Stark [139, 140]. Surveys of these methods were carried out by J. Martinet [91] and A. Odlyzko [108].

## References

### A. Books and Surveys


[Ad] S.I. Adjan, *Burnside's Problem and Identities in Groups*, Moscow (1975) (in Russian).

[AT] E. Artin and J. Tate, *Class Field Theory*, Harvard (1961).

[CF] J.W.S. Cassels and A. Fröhlich (eds), *Algebraic Number Theory*, Academic Press, New York (1967).

[Ch] C. Chevalley, *Class Field Theory*, Nagoya (1954).

[CM] B. Chandler and W. Magnus, *The History of Combinatorial Group Theory: A Case Study in the History of Ideas*, Springer, Berlin (1982).

[DA] C.F. Gauss, *Disquisitiones Arithmeticae*, Gottingae (1801).

[De] M. Deuring, *Der Klassenkörper der komplexen Multiplikation*, in Enzyklopädie der Math. Wiss., Bd. I 2, Teubner (1958).

[FJ] M.D. Fried and M. Jarden, *Field Arithmetics*, Springer, Berlin (1986).

[Fu] R. Fueter, *Vorlesungen über die singulären moduls und die komplexe Multiplikation der elliptischen Funktionen*, Leipzig (1924–1927).

[Ha] H. Hasse, *Vorlesungen über Klassenkörpertheorie*, Würzburg (1967).

[HBer] H. Hasse, *Bericht über neuere Untersuchungen und Probleme aus der Theorie der algebraischen Zahlkörpern*, Jahresber. DMV **35** (1926), 1–55; **36** (1927), 233–2311; VI Erg. Bd. 1930. Reprints: Wien (1965, 1970).

[He] E. Hecke, *Vorlesungen über die Theorie der algebraischen Zahlen*, Leipzig (1923), reprinted 1954.

[Iy] S. Iyanaga, *The Theory of Numbers*, North-Holland, Amsterdam (1975).

[Ja] G. Janusz, *Algebraic Number Fields*, Academic Press, New York (1973).

[Ko] H. Koch, *Algebraic Number Fields*, in Number Theory, II, Encyclopaedia of Mathematical Sciences vol. 62, Springer, Berlin (1992).

[Kos] A.I. Kostrikin, *Around Burnside*, Moscow (1986) (in Russian).

[La1] S. Lang, *Algebraic Numbers*, Reading (1964).

[La2] S. Lang, *Algebraic Number Theory*, Reading, London (1970).

[MH] J. Milnor and D. Husemoller, *Symmetric Bilinear Forms*, Springer, Berlin (1973).

[Na] W. Narkiewicz, *Elementary and Analytic Theory of Algebraic Numbers*, PWN (1974); 2nd ed. PWN and Springer (1990).

[Ne1] J. Neukirch, *Class Field Theory*, Springer, Berlin (1986).

[Ne2] J. Neukirch, *Algebraische Zahlentheorie*, Springer, Berlin (1992).

[SCM] A. Borel, S. Chowla, C.S. Herz, K. Iwasawa and J.P. Serre, *Seminar on Complex Multiplication*, SLNM 21, Springer, Berlin (1966).

[Se] J.P. Serre, *Abelian l-adic Representations and Elliptic Curves*, Benjamin (1968).

[Sh] G. Shimura, *Automorphic Functions and Number Theory*, SLNM 54, Springer, Berlin (1968).

[Shi] G. Shimura, *Introduction to the Arithmetic Theory of Automorphic Functions*, Iwanami Shoten and Princeton Univ. Press (1971).

[ST] G. Shimura and Y. Taniyama, *Complex Multiplication of Abelian Varieties and its Application to Number Theory*, Publ. Math. Soc. Japan vol. 6 (1961).

[Wa] L.C. Washington, *Introduction to Cyclotomic Fields*, Springer, Berlin (1982).

[We] H. Weber, *Lehrbuch der Algebra*, Braunschweig (1908).

[Wei1] A. Weil, *Basic Number Theory*, Springer, Berlin (1967).

[Wei2] A. Weil, *Number Theory*, Birkhäuser, Basel (1983).

[Zber] D. Hilbert, *Die Theorie der algebraischen Zahlkörper*, Jahresber. DMV **4** (1897), 175–546 = Ges. Abhandl. I, 63–363, Berlin (1932); reprinted by Chelsea (1965).

## *B. Papers*

[1] E. Artin, *Beweis des allgemeinen Reziprozitätsgesetzes*, Abh. Math. Sem. Hamburg **5** (1927), 353–363 = Coll. Papers, Reading (1965), 131–141.

[2] E. Artin, *Idealklassen in Oberkörpern und allgemeines Reziprozitätsgesetz*, Abh. Math. Sem. Hamburg **7** (1930), 46–51 = Coll. Papers, Reading (1965), 159–164.

[3] E. Artin and G. Whaples, *Axiomatic characterization of fields by the product formula for valuations*, Bull. Amer. Math. Soc. **51** (1945), 469–492 = E. Artin, Coll. Papers, Reading (1965), 202–225.

[4] L. Auslander, R. Tolimieri and S. Winograd, *Hecke's theorem in quadratic reciprocity, finite nilpotent groups and the Cooley–Tukey algorithm*, Adv. Math. **43** (1982), 122–172.

[5] Z.I. Borevich, *On the demonstration of the principal ideal theorem*, Vestnik Leningrad Gos. Univ. **12** (1957) (13), 5–8 (in Russian).

[6] H. Brückner, *Explizites Reziprozitätsgesetz und Anwendungen*, Vorlesungen Fachbereich Math. Univ. Essen **2** (1979).

[7] A. Brumer, *Ramification and class-field-towers of number fields*, Michigan Math. J. **12** (1965), 129–131.

[8] W. Burnside, *On an unsettled question in the theory of discontinuous groups*, Quart. J. Pure Appl. Math. **33** (1902), 230–238.

[9] J. Calloway, *On the discriminant of arbitrary algebraic number fields*, Proc. Amer. Math. Soc. **6** (1952), 482–489.

[10] L. Carlitz, *A class of polynomials*, Trans. Amer. Math. Soc. **43** (1938), 167–182.

[11] N.G. Chebotarev, *Die Bestimmung der Dichtigkeit einer Menge von Primzahlen, die zu einer gegebener Substitutionsklasse gehören*, Math. Ann. **95** (1926), 191–228.

[12] C. Chevalley, *Généralisations de la théorie du corps de classes pour les extensions infinies*, J. Math. Pures Appl. **15** (1936), 359–371.

[13] C. Chevalley, *La théorie du corps de classes*, Ann. Math. **41** (1940), 394–418.

[14] H. Cohn, *Iterated class fields and the icosahedron*, Math. Ann. **225** (1981), 107–122.

[15] H. Cohn, *The explicit Hilbert 2-cyclic class field for* $Q(\sqrt{-p})$, J. Reine Angew. Math. **321** (1981), 64–77.

[16] H. Cohn, *Some examples of Weber–Hecke ring class-field theory*, Math. Ann. **265** (1983), 83–100.

[17] H. Cohn and G. Cooke, *Parametric from of an eight class field*, Acta Arith. **30** (1976), 367–377.

[18] I. Connell and D. Sussman, *The p-dimension of class groups of number fields*, J. London Math. Soc. (2) **2** (1970), 525–529.

[19] J.H. Conway and S.P. Norton, *Monstrous moonshine*, Bull. London Math. Soc. **11** (1979), 308–339.

[20] R. Dedekind and H. Weber, *Theorie der algebraischen Funktionen einer Veränderlichen*, J. Reine Angew. Math. **92** (1882), 181–290 = R. Dedekind, Ges. Math. Werke, I, Vieweg (1930), 238–249.

[21] H. Delange, *Généralisation du théorème de Ikehara*, Ann. Sci. École Norm. Sup. (3) **71** (1954), 213–242.

[22] P. Deligne, *Formes modulaires et représentations l-adiques*, Sém. Bourbaki 1968–1969, exp. 355, SLNM 179, Springer, Berlin (1971), 139–172.

[23] P. Deligne and J.P. Serre, *Formes modulaires de poids* 1, Ann. Sci. École Norm. Sup. (4) **7** (1974), 507–530.

[24] M. Deuring, *Über den Tchebotareffschen Dichtigkeitssatz*, Math. Ann. **110** (1935), 414–415.

[25] M. Deuring, *Algebraische Begründung der komplexen Multiplikation*, Abh. Math. Sem. Hamburg **16** (1947), 32–47.

[26] M. Deuring, *Die Struktur der elliptischen Funktionenkörper und die Klassenkörper der imaginär-quadratischen Körper*, Math. Ann. **124** (1952), 393–426.

[27] V.G. Drinfeld, *Elliptic modules*, Mat. Sb. **94** (136) (1974), 596–627 (in Russian).

[28] G. Eisenstein, *Beweis der allgemeinsten Reziprozitätsgesetze zwischen reellen und complexen Zahlen*, Mber. Preuß. Akad. Wiss. Berlin (1850), 189–198 = Math. Werke, II, New York (1975), 712–721.

[29] M. Fried, *On Hilbert's irreducibility theorem*, J. Number Theory **6** (1974), 211–231.

[30] G. Frobenius, *Über die Beziehungen zwischen den Primidealen eines algebraischen Körpers und den Substitutionen seiner Gruppe*, SBer. Kgl. Preuß. Akad. Wiss. (1896), 689–703.

[31] R. Fueter, *Die Theorie der Zahlstrahlen*, J. Reine Angew. Math. **130** (1905), 197–237; **132** (1907), 255–269.

[32] P. Furtwängler, *Allgemeiner Existenzbeweis für den Klassenkörper eines beliebigen algebraischen Zahlkörpers*, Math. Ann. **63** (1907), 1–37.

[33] P. Furtwängler, *Die Reziprozitätsgesetze für Potenzreste mit Primzahlexponenten*, Math. Ann. **67** (1909), 1–31; **72** (1912), 346–386; **74** (1913), 413–429.

[34] P. Furtwängler, *Über das Verhalten der Ideale des Grundkörpers im Klassenkörper*, Monatsh. Math.-Phys. **27** (1916), 1–15.

[35] P. Furtwängler, *Beweis des Hauptidealsatzes für die Klassenkörper algebraischer Zahlkörper*, Abh. Math. Sem. Hamburg **7** (1930), 14–36.

[36] C.F. Gauss, *Theoria residuorum biquadraticorum*, Comm. Soc. Reg. Gottingae **7** (1832) = Werke, II, Göttingen (1863), 269–291.

[37] E.S. Golod, *On nil-algebras and finitely-approximable p-groups*, Izv. Akad. Nauk SSSR **28** (1964), 273–276 (in Russian).

[38] E.S. Golod and I.R. Shafarevich, *On the class-field tower*, Izv. Akad. Nauk SSSR **28** (1964), 261–272 (in Russian).

[39] D. Goss, *The algebraists upper plane*, Bull. Amer. Math. Soc. **2** (1980), 391–415.

[40] M.J. Greenberg, *An elementary proof of the Kronecker–Weber theorem*, Amer. Math. Monthly **81** (1974), 601–607; *Corr.*, ibid. **82** (1975), 803.

[41] F. Halter-Koch, *A note on ray class fields of global fields*, Nagoya Math. J. **120** (1990), 61–66.

[42] F. Halter-Koch, *Der Čebotarev'sche Dichtigkeitssatz und ein Analogon zum Dirichlet'schen Primzahlsatz für algebraische Funktionenkörper*, Manuscripta Math. **72** (1991), 205–211.

[43] H. Hasse, *Neue Begründung der komplexen Multiplikation*, J. Reine Angew. Math. **157** (1927), 115–139; **165** (1931), 64–88.

[44] H. Hasse, *Das Eisensteinsche Reziprozitätsgesetz der n-ten Potenzreste*, Math. Ann. **97** (1927), 599–623.

[45] H. Hasse, *Über das Reziprozitätsgesetz der m-ten Potenzreste*, Math. Ann. **158** (1927), 228–259.

[46] H. Hasse, *Zur Arbeit von I.R. Šafarevič über das allgemeine Reziprozitätsgesetz*, Math. Nachr. **5** (1951), 301–327.

[47] H. Hasse, *Über den Klassenkörper zum quadratischen Zahlkörper mit der Diskriminante* −47, Acta Arith. **9** (1964), 419–434.

[48] H. Hasse and A. Scholz, *Zur Klassenkörpertheorie auf Takagischer Grundlage*, Math. Z. **29** (1929), 60–69.

[49] D.R. Hayes, *Explicit class field theory for rational function fields*, Trans. Amer. Math. Soc. **189** (1974), 77–91.

[50] D.R. Hayes, *Explicit class field theory in global function fields*, Studies in Algebra and Number Theory, Academic Press (1979), 173–217.

[51] E. Hecke, *Über die Konstruktion relativ-Abelscher Zahlkörper durch Modulfunktionen von zwei Variablen*, Math. Ann. **74** (1914), 465–510.

[52] E. Hecke, *Über die Zetafunktion beliebiger algebraischer Zahlkörper*, Nachr. Ges. Wiss. Göttingen (1917), 77–89.

[53] E. Hecke, *Über die L-Funktionen und den Dirichletschen Primzahlsatz für einen beliebigen Zahlkörper*, Nachr. Ges. Wiss. Göttingen (1917), 90–95.

[54] E. Hecke, *Reziprozitätsgesetz und Gausssche Summen in quadratischen Zahlkörpern*, Nachr. Ges. Wiss. Göttingen (1919), 265–278.

[55] S. Helou, *An explicit $2^n$-th reciprocity law*, J. Reine Angew. Math. **389** (1988), 64–89.

[56] S. Helou, *Classical explicit reciprocity*, Th. des Nombres (Quebec 1987), de Gruyter (1989), 359–370.

[57] D. Hilbert, *Neuer Beweis des Kronecker'schen Fundamentalsatzes über Abelsche Zahlkörper*, Nachr. Ges. Wiss. Göttingen (1896), 29–39.

[58] D. Hilbert, *Über die Theorie der relativ-Abelschen Zahlkörper*, Nachr. Ges. Wiss. Göttingen (1898), 370–399 = Acta Math. **26** (1902), 99–132.

[59] D. Hilbert, *Über die Theorie des relativ-quadratisches Zahlkörpers*, Math. Ann. **51** (1899), 1–127.

[60] D. Hilbert, *Mathematische Probleme*, Nachr. Ges. Wiss. Göttingen (1900), 253–297.

[61] G. Hochschild and T. Nakayama, *Cohomology in class field theory*, Ann. Math. **55** (1952), 348–366.

[62] K. Iwasawa, *A note on the capitulation problem for number fields*, Proc. Japan Acad. Sci. **65** (1989), 59–61; 183–186.

[63] S. Iyanaga, *Zum Beweis des Hauptidealsatzes*, Abh. Math. Sem. Hamburg **10** (1934), 349–357.

[64] J.F. Jaulent, *L'état actuel du problème de capitulation*, Sém. Th. des Nombres Bordeaux, 1987/1988, exp. 17.

[65] K.L. Jensen, *Numbertheoretical properties of Bernoulli numbers*, Nyt Tidsskr. Mat. **26** B (1915), 73–83 (in Danish).

[66] Y. Kawada, *A remark on the principal ideal theorem*, J. Math. Soc. Japan **20** (1968), 166–169.

[67] M. Kneser, *Zum expliziten Reziprozitätsgesetz von I.R. Šafarevič*, Math. Nachr. **6** (1951), 86–96.

[68] H. Koch, *Zum Satz von Golod–Schafarewitsch*, Math. Nachr. **42** (1969), 321–333.

[69] H. Koch, *Zum Satz von Golod–Schafarewitsch*, J. Reine Angew. Math. **274/275** (1975), 240–243.

[70] H. Koch and B.B. Venkov, *Über den p-Klassenkörperturm eines imaginär quadratischen Zahlkörpers*, Astérisque **24/25** (1975), 57–67.

[71] A.I. Kostrikin, *On the construction of groups by generators and defining relations*, Izv. Akad. Nauk SSSR **29** (1965), 1119–1122 (in Russian).

[72] L. Kronecker, *Über die algebraisch auflösbaren Gleichungen*, MBer. Kgl. Preuß. Akad. Wiss. Berlin (1853), 365–374; (1856), 203–215 = Werke, IV, Leipzig–Berlin (1929), 1–11, 25–37.

[73] L. Kronecker, *Über Abelsche Gleichungen*, MBer. Kgl. Preuß. Akad. Wiss. Berlin (1877), 845–851 = Werke, IV, Leipzig–Berlin (1929), 63–71.

[74] L. Kronecker, *Über die Irreduktibilität von Gleichungen*, MBer. Kgl. Preuß. Akad. Wiss. Berlin (1880), 155–163 = Werke, II, Leipzig (1897), 85–93.

[75] L. Kronecker, *Grundzüge einer arithmetischer Theorie der algebraischen Grössen*, J. Reine Angew. Math. **92** (1882), 1–122 = Werke, II, Leipzig (1897), 237–387.

[76] W. Krull, *Galoissche Theorie der unendlichen algebraischen Erweiterungen*, Math. Ann. **100** (1928), 687–698.

[77] E.E. Kummer, *Über die Zerlegung der aus Wurzeln der Einheit gebildeten complexen Zahlen in ihre Primfactoren*, J. Reine Angew. Math. **35** (1847), 327–367 = Coll. Papers, I, Springer (1975), 211–251.

[78] E.E. Kummer, *Bestimmung der Anzahl nicht Äquivalenter Classen für die aus $\lambda$-ten Wurzeln der Einheit*

*gebildeten complexen Zahlen und die ideale Factoren derselben*, J. Reine Angew. Math. **40** (1850), 93–116 = Coll. Papers, I, Springer (1975), 299–322.

[79] E.E. Kummer, *Allgemeiner Beweis des Fermatschen Satzes, daß die Gleichung $x^\lambda + y^\lambda = z^\lambda$ durch ganze Zahlen unlösbar ist, für alle diejenige Potenz-Exponenten $\lambda$, welche ungerade Primzahlen sind und in den Zählern der ersten $\frac{1}{2}(\lambda - 3)$ Bernoulli'schen Zahlen als Factoren nicht vorkommen*, J. Reine Angew. Math. **40** (1850), 130–138 = Coll. Papers, I, Springer (1975), 336–344.

[80] E.E. Kummer, *Über die allgemeinen Reciprocitätsgesetze unter den Resten und Nichtresten der Potenzen, deren Grad eine Primzahl ist*, Math. Abh. Kgl. Akad. Wiss. Berlin (1859), 19–159 = Coll. Papers, I, Springer (1975), 699–839.

[81] E.E. Kummer, *Zwei neue Beweise der allgemeinen Reciprocitätsgesetze unter den Resten und Nichtresten der Potenzen, deren Grad eine Primzahl ist*, J. Reine Angew. Math. **100** (1887), 10–50 = Coll. Papers, I, Springer (1975), 842–882.

[82] J. Lagarias, H.L. Montgomery and A. Odlyzko, *A bound for the least prime ideal in the Chebotarev density theorem*, Invent. Math. **54** (1979), 271–296.

[83] J. Lagarias and A. Odlyzko, *Effective versions of the Chebotarev density theorem*, Algebraic Number Fields (Durham Symp.), Academic Press (1977), 409–464.

[84] E. Landau, *Neuer Beweis des Primzahlsatzes und Beweis des Primidealsatzes*, Math. Ann. **56** (1903), 645–670.

[85] E. Landau, *Über Ideale und Primideale in Idealklassen*, Math. Z. **2** (1918), 52–154.

[86] E. Landau, *Der Minkowskische Satz über die Körperdiskriminante*, Nachr. Ges. Wiss. Göttingen (1922), 80–82.

[87] H.W. Lenstra, Jr. and P. Stevenhagen, *Primes of degree one and algebraic cases of Čebotarev's theorem*, Enseign. Math. **37** (1991), 17–30.

[88] J. Levitzki, *On three problems concerning nil-rings*, Bull. Amer. Math. Soc. **51** (1945), 913–919.

[89] W. Magnus, *Über den Beweis des Hauptidealsatzes*, J. Reine Angew. Math. **170** (1934), 235–240.

[90] J. Martinet, *Tours de corps de classes et estimation de discriminants*, Invent. Math. **44** (1978), 65–73.

[91] J. Martinet, *Petits discriminants des corps de nombres*, Journées Arithmétiques (1980), Cambridge Univ. Press (1982), 151–193.

[92] C.R. McCluer, *A reduction of Čebotarev density theorem to the cyclic case*, Acta Arith. **15** (1968), 45–47.

[93] J. Mennicke, *Einige endliche Gruppen mit drei Erzeugenden und drei Relationen*, Arch. Math. **10** (1959), 409–418.

[94] W.H. Mills, *The $m$-th power residue symbol*, Amer. J. Math. **73** (1951), 59–64.

[95] W.H. Mills, *Reciprocity in algebraic number fields*, Amer. J. Math. **73** (1951), 65–77.

[96] H. Minkowski, *Über die positiven quadratischen Formen und über kettenbruchähnlichen Algorithmen*, J. Reine Angew. Math. **107** (1891), 278–297 = Ges. Abh., Leipzig–Berlin (1911), 244–260.

[97] K. Miyake, *On the general principal ideal theorem*, Proc. Japan Acad. Sci. **56** (1980), 171–174.

[98] K. Miyake, *Algebraic investigations of Hilbert's Theorem 94, the principal ideal theorem and the capitulation problem*, Exposition. Math. **7** (1989), 289–346.

[99] L.J. Mordell, *On trigonometric series involving algebraic numbers*, Proc. London Math. Soc. (2) **21** (1922), 493–496.

[100] L.J. Mordell, *On Hecke's modular functions, zeta functions, and some other analytic functions in the theory of numbers*, Proc. London Math. Soc. (2) **32** (1931), 501–556.

[101] C. Müntz, *Der Summensatz von Cauchy in beliebigen algebraischen Zahlkörpern und die Diskriminante derselben*, Math. Ann. **90** (1923), 279–291.

[102] T. Nakayama, *Idèle-class factor sets and class field theory*, Ann. Math. **55** (1952), 73–84.

[103] J. Neukirch, *Neubegründung der Klassenkörpertheorie*, Math. Z. **186** (1984), 557–574.

[104] O. Neumann, *Two proofs of the Kronecker–Weber theorem "according to Kronecker and Weber"*, J. Reine Angew. Math. **323** (1981), 105–126.

[105] P.S. Novikov and S.I. Adjan, *On infinite periodic groups*, Izv. Akad. Nauk SSSR **32** (1968), 212–244; 251–524; 709–731.

[106] A. Odlyzko, *Some analytic estimates of class numbers and discriminants*, Invent. Math. **29** (1975), 275–286.

[107] A. Odlyzko, *Lower bounds for discriminants of number fields*, Acta Arith. **29** (1976), 275–297; II, Tôhoku Math. J. **29** (1977), 209–216.

[108] A. Odlyzko, *Bounds for discriminants and related estimates for class numbers, regulators and zeros of zeta functions: a survey of recent results*, Sém. Théor. Nombres, Bordeaux **2** (1990), 119–141.

[109] J. Oesterlé, *Versions effectives du théorème de Chebotarev sous l'hypothése Riemann généralisée*, Astérisque **61** (1979), 165–167.

[110] G. Poitou, *Minoration de discriminants*, Sém. Bourbaki **28** (1975/76), SLNM 567, Springer, Berlin (1977).

[111] G. Poitou, *Sur les petits discriminants*, Sém. Delange–Pisot–Poitou **18** (1976/77), exp. 6.

[112] K. Ramachandra, *Some applications of Kronecker's limit formula*, Ann. Math. **80** (1964), 104–148.

[113] H. Reichardt, *Der Primdivisorensatz für algebraische Funktionenkörper über einem endlichen Konstantenkörper*, Math. Z. **40** (1936), 713–719.

[114] P. Roquette and H. Zassenhaus, *A class rank estimate for algebraic number fields*, J. London Math. Soc. **44** (1969), 31–38.

[115] M. Rosen, *The Hilbert class field in function fields*, Exposition. Math. **5** (1987), 365–378.

[116] M. Rosen, *S-units and S-class group in algebraic function fields*, J. Algebra **26** (1973), 98–108.

[117] F.K. Schmidt, *Die Theorie der Klassenkörper über einem Körper algebraischer Funktionen in einer Unbestimmten und mit endlichem Koeffizientenbereich*, SBer. Phys.-Med. Soz. Erlangen **62** (1931), 267–284.

[118] B. Schmithals, *Konstruktion imaginärquadratischer Körper mit unendlichen Klassenkörperturm*, Arch. Math. **34** (1980), 307–312.

[119] A. Scholz, *Zwei Bemerkungen zum Klassenkörperturmproblem*, J. Reine Angew. Math. **161** (1929), 201–207.

[120] A. Scholz, *Die Abgrenzungssätze für Kreiskörper und Klassenkörper*, SBer. Preuß. Akad. Wiss. **20/1** (1931), 417–426.

[121] R. Schoof, *Infinite class field towers of quadratic fields*, J. Reine Angew. Math. **372** (1986), 209–220.

[122] O. Schreier, *Über eine Arbeit von Herrn Tschebotareff*, Abh. Math. Sem. Hamburg **5** (1927), 1–6.

[123] H.G. Schumann, *Zum Beweis des Hauptidealsatzes*, Abh. Math. Sem. Hamburg **12** (1937), 42–47.

[124] I. Schur, *Einige Bemerkungen über die Diskriminante eines algebraischen Zahlkörpers*, J. Reine Angew. Math. **167** (1932), 264–269.

[125] V. Schulze, *Die Primteilerdichte von ganzzahligen Polynomen, III*, J. Reine Angew. Math. **273** (1975), 144–145.

[126] S. Sen, *On explicit reciprocity laws*, J. Reine Angew. Math. **313** (1980), 1–26; *II*. ibid. **323** (1981), 68–87.

[127] J.P. Serre, *Quelques applications du théorème de densité de Chebotarev*, Inst. Hautes Études Sci. Publ. Math. **54** (1981), 323–401.

[128] J.P. Serre, *Minorations de discriminants*, in J.P. Serre, Coll. Papers, III, 240–243, Springer (1986).

[129] I.R. Shafarevich, *General reciprocity law*, Mat. Sb. **26**(68) (1950), 113–146 (in Russian).

[130] I.R. Shafarevich, *A new proof of the Kronecker–Weber theorem*, Trudy Mat. Inst. Steklov. **38** (1951), 382–387 (in Russian).

[131] I.R. Shafarevich, *Extensions with given ramifications*, Inst. Hautes Études Sci. Publ. Math. **18** (1963), 71–95.

[132] E. Shalit, de, *The explicit reciprocity law in local class field theory*, Duke Math. J. **53** (1986), 163–176.

[133] E. Shalit, de, *Making class field theory explicit*, Number Theory, Proc. Conf. CMS, Providence (1987), 55–58.

[134] G. Shimura, *Moduli and fibre systems of abelian varieties*, Ann. Math. **83** (1966), 294–338.

[135] G. Shimura, *Construction of class fields and zeta functions of algebraic curves*, Ann. Math. **85** (1967), 58–159.

[136] G. Shimura, *Algebraic number fields and symplectic discontinuous groups*, Ann. Math. **86** (1967), 503–592.

[137] T. Shintani, *On certain ray class invariants for real quadratic fields*, J. Math. Soc. Japan **30** (1978), 139–167.

[138] C.L. Siegel, *Über die Diskriminanten total reeller Körper*, Nachr. Ges. Wiss. Göttingen (1922), 17–24 = Ges. Abhandl. I, Springer (1966), 157–164.

[139] H.M. Stark, *Some effective cases of the Brauer–Siegel theorem*, Invent. Math. **23** (1974), 135–152.

[140] H.M. Stark, *The analytic theory of algebraic numbers*, Bull. Amer. Math. Soc. **81** (1975), 961–972.

[141] T. Takagi, *Über eine Theorie des relativ-Abelsches Zahlkörpers*, J. Coll. Sci. Tokyo **41** (1920), 1–133 = Collected Papers, Tokyo (1973), 73–167.

[142] K. Taketa, *Neuer Beweis eines Satzes von Herrn Furtwängler über die metabelsche Gruppen*, Japan J. Math. **9** (1932), 199–218.

[143] J. Tate, *The higher-dimensional cohomological groups of class field theory*, Ann. Math. **56** (1952), 294–297.

[144] B.B. Venkov and H. Koch, *The p-tower of class fields of an imaginary quadratic field*, Zap. Nauchn. Sem. Leningrad. Otdel. Mat. Inst. Steklov. **46** (1974), 5–13 (in Russian).

[145] E.B. Vinberg, *On the theorem of infinite dimensionality of an associative algebra*, Izv. Akad. Nauk SSSR **29** (1965), 209–214 (in Russian).

[146] S.V. Vostokov, *On the reciprocity law of a field of algebraic numbers*, Trudy Mat. Inst. Steklov. **148** (1978), 77–81 (in Russian).

[147] S.V. Vostokov, *Explicit form of the reciprocity law*, Izv. Akad. Nauk SSSR **42** (1978), 1288–1321 (in Russian).

[148] S.V. Vostokov, *The second factor in the reciprocity law*, Zap. Nauchn. Sem. Leningrad. Otdel. Mat. Inst. Steklov. **75** (1978), 59–66 (in Russian).

[149] S.V. Vostokov and V.A. Lecko, *The Hilbert symbol in an extension of the field of 2-adic numbers*, Zap. Nauchn. Sem. Leningrad. Otdel. Mat. Inst. Steklov. **103** (1980), 58–61 (in Russian).

[150] H. Weber, *Theorie der Abelscher Zahlkörper*, Acta Math. **8** (1886), 193–263; **9** (1886/7), 105–130.

[151] H. Weber, *Über Zahlengruppen in algebraischen Körpern*, Math. Ann. **48** (1987), 433–473; **49** (1897), 83–100; **50** (1898), 1–26.

[152] H. Weber and J. Wellstein, *Der Minkowskische Satz über die Körperdiskriminante*, Math. Ann. **73** (1913), 275–285.

[153] A. Weil, *Sur la théorie du corps de classes*, J. Math. Soc. Japan **3** (1951), 1–3.

[154] J. Wisliceny, *Zur Darstellung der pro-p-Gruppen und Lieschen Algebren durch Erzeugende und Relationen*, Math. Nachr. **102** (1981), 57–78.

[155] E. Witt, *Bemerkungen zum Beweis des Hauptidealsatzes*, Abh. Math. Sem. Hamburg **11** (1936), 221.

[156] E. Witt, *Verlagerungen von Gruppen und Hauptidealsatz*, Proc. ICM Amsterdam, II (1954), 71–73.

[157] J. Wójcik, *A purely algebraic proof of special cases of Tschebotarev's theorem*, Acta Arith. **28** (1975), 137–145.

# Finite Fields and Error-Correcting Codes

Henk C.A. van Tilborg

*Department of Mathematics and Computing Science, Eindhoven University of Technology, Eindhoven, The Netherlands*
*e-mail: henkvt@win.tue.nl*

*Contents*



HANDBOOK OF ALGEBRA, VOL. 1
Edited by M. Hazewinkel

## 1. Introduction

One of the most appealing areas of application of abstract algebra is the theory of error-correcting codes. This theory dates back to the fourties, but at that time the importance of abstract algebra and in particular of finite fields for it was not clear at all.

Whenever information is communicated from one place to another (say be satellite) or stored (e.g., on a magnetic tape or optical disc) for later retrieval – this can be viewed as communication in time – the receiver will sometimes be faced with errors due to noise or system errors. When this information is represented in a digital way, the use of so-called error-correcting codes makes it possible to correct these errors.

What the full potential of error-correcting codes is, has been exactly determined in the fundamental work of C.E. Shannon [48]. That one is nowadays still not even close to those theoretical possibilities is cause for continuing and growing research in which finite fields play an essential role, but in which all kinds of other algebraic tools have been instrumental. The last few years for instance algebraic geometry has made quite an impact on the development of coding theory.

In the context of this chapter there will always be two parties involved in the transmission of information: the *sender* of the message(s) and the *receiver*. The medium over which the information is sent, together with its characteristics, is called the *channel*. These characteristics consist of an input alphabet $X$, an output alphabet $Y$ and a transition probability function $P$, which gives the probability $P(y \mid x)$ that a symbol $y$ in $Y$ is received given that $x$ in $X$ was transmitted. Here we shall discuss the most common case: the Binary Symmetric Channel (BSC), depicted in Fig. 1.

DEFINITION 1.1. The Binary Symmetric Channel has input and output alphabets equal to $\{0, 1\}$. The probability that a received symbol is actually equal to the transmitted symbol is given by $1 - p$, while the probability that they are not equal to each other is $p$, for some $0 \leqslant p \leqslant 1$. If the transmitted and received symbols are not equal to each other one says that an *error* has occurred.

It shall always be assumed here that $0 \leqslant p \leqslant \frac{1}{2}$. When using the BSC, one of course has to convert the information into a stream of binary data and upon arrival recover the original information. For that reason it is more convenient to represent the information sequence as a binary sequence.

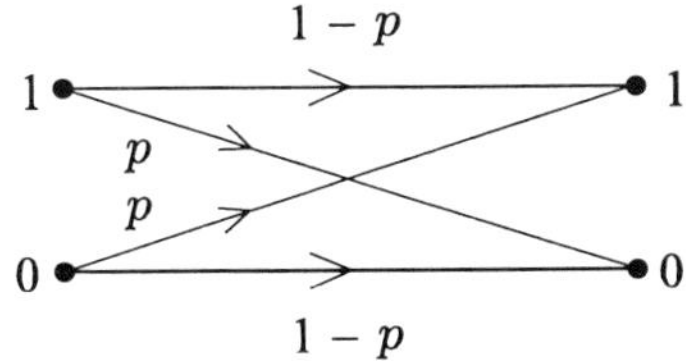


Figure 1. The binary symmetric channel.

If one wants to transmit a 1 over the BSC, with error probability $p$, one can increase the reliability of the transmission by repeating the transmission of each bit a few times, say five. The receiver can use a simple majority vote on the received sequence to decide what the most likely transmitted bit is. For instance, if $1, 1, 0, 0, 1$ is received, the most likely transmitted sequence is of course $1, 1, 1, 1, 1$.

With this system, it is still possible that the receiver makes an error, namely if three or more errors have occurred. If at most two errors occurred during the transmission the receiver will make the correct estimate of the transmitted information. It may also be clear that repeating each symbol a few times may not be the most efficient way of protecting the messages against errors. Transmitting more symbols than is strictly necessary to convey the message is called adding *redundancy* to the message. Regular languages know the same phenomenon. The redundant letters make it possible to apply error-correction techniques.

DEFINITION 1.2. An $[n, k]$ binary *encoder* is a mapping that transforms $k$-tuples $\underline{a}$ of binary (information) symbols to binary $n$-tuples $\underline{c}$ (called *codewords*). The collection of all possible codewords is called an $[n, k]$ binary *block code* $C$. The coordinates of the codewords are the input symbols for the BSC. A *decoder* maps a received $n$-tuple back to the most likely transmitted $n$-tuple $\hat{\underline{c}}$ and further back to the original $k$-tuple $\hat{\underline{a}}$ (hopefully $\underline{a} = \hat{\underline{a}}$).

The channel, encoder and decoder together with the sender and receiver form a so-called *communication system* (depicted in Fig. 2).

Note that the most likely transmitted codeword will be the codeword that differs from the received vector in the fewest number of coordinates (since $p \leqslant 0.5$). A decoder that always finds the closest codeword is called a *maximum likelihood decoder.*

The parameter $n$ is called the *length* of the code. The $2^k$ possible outcomes of the encoder are called *codewords*. Note that the encoder converts $k$-tuples to $n$-tuples, somehow adding $n - k$ redundant bits. The information *rate* $R$ of $C$ is the ratio $k/n$. It is the relative information content of each transmitted bit.

By using block codes, the sender tries to get information to the receiver in a more reliable way than without use of codes. By repeating each information symbol sufficiently many times, one can achieve this and obtain a reliability arbitrarily close to 1. However, the price that one pays is inefficient use of the channel: the rates of theses code tend to zero when their length goes to infinity! For other families of codes, the information rate does not go down to zero, but the fraction of errors per codeword that they are able to

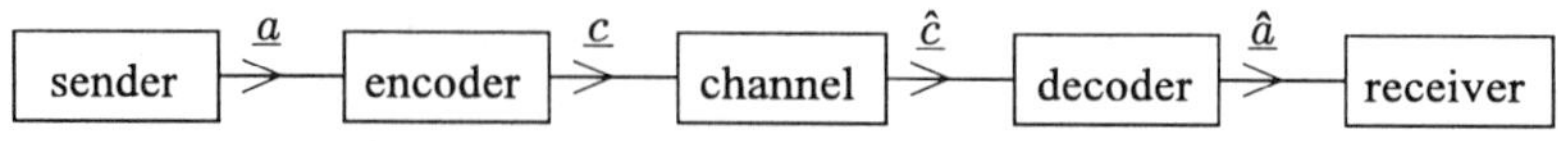


Figure 2. A communication system.

correct tends to zero, while the BSC model gives an expected number of $pn$ errors per transmitted codeword.

What Shannon was able to prove in 1948 is that, by using sufficiently long codes, information can be transmitted reliably while the rate of the codes does not tend to zero.

Let the *entropy function* $h(p)$, $0 \leqslant p \leqslant 1$, be defined by

$$h(p) = -p\log_2 p - (1-p)\log_2(1-p),$$

if $0 < p < 1$, and by 0 for $p = 0$ or 1.

Although it will not be further discussed here, Shannon's information theory makes it possible to interpret $h(p)$ as the uncertainty that the receiver of a specific binary symbol (transmitted over the BSC with error probability $p$) still has about the actually transmitted symbol. In other words, $1 - h(p)$ is the amount of information that the received symbol carries about the transmitted symbol.

THEOREM 1.3 (Shannon). *Consider the BSC with error probability $p$ and let $C = 1-h(p)$. Then, for each rate $R$ with $R < C$, an infinite sequence of $[n_l, k_l]$ codes $C_l$ exists, with $k_l = \lceil Rn_l \rceil$ (so $C_l$ has rate $> R$), such that the corresponding maximum-likelihood decoding algorithm has a probability of incorrect decoding that goes exponentially fast to 0 for $l \to \infty$.*

*For rates $R$ greater than $C$, no encodings can be made with error probabilities tending to zero.*

The quantity $C$ in Theorem 1.3 is called the *capacity* of the channel.

It is the ultimate goal of coding theory to find (families of) codes of which the rate approaches the capacity of the BSC.

A result related to the entropy functions that can be proved by standard means and that will be needed later is:

LEMMA 1.4. *Let $0 \leqslant \alpha \leqslant 0.5$. Then*

$$\sum_{i=0}^{\lfloor \alpha n \rfloor} \binom{n}{i} = 2^{(h(\alpha)+o(1))n}, \quad n \to \infty. \tag{1}$$

This chapter is organized in the following way. In Section 2 the basic concepts of block codes are explained. Projective codes (no coordinate is a scalar multiple of another coordinate) of maximal size are constructed and an important relation between a linear code and its orthogonal complement is given. In Section 3 it is shown that ideals in the residue class ring of $q$-ary polynomials modulo $x^n - 1$ define a very large class of codes. The zeros of the generator polynomial of such an ideal determine their error-correcting capability. In Section 4 generalizations of cyclic code are given by means of algebraic geometry. They lead to a powerful error-correcting algorithm and to codes that are asymptotically very interesting and may soon even be of practical value. In Section 5, a brief discussion of the available books on coding theory will be given.

## 2. Linear codes

Although the input and output alphabet of the BSC is simply the set $\{0,1\}$, this assumption would be too restrictive to set up the theory of error-correcting codes. On the other hand, assuming no structure at all about the input and output alphabet will make it difficult or impossible to construct codes and prove properties regarding information rate or error-correcting capability. For this reason, it is assumed that both the input and the output alphabets have cardinality $q$, where $q = p^a$, for some prime number $p$. In this way the letters in the alphabet can be identified with the elements of the finite field of size $q$. This field will be denoted by $GF(q)$ or $GF(p^a)$ for *Galois Field*. In many applications, $p = 2$ and $a$-tuples of binary symbols can now be identified with symbols in $GF(2^a)$.

The set $(GF(q))^n$ of $q$-ary sequences of length $n$ can now be given the additional structure of a $n$-dimensional vectorspace over $GF(q)$. This will be denoted by $V_n(q)$. If one vector, also called *word*, is transmitted while another vector is received, the number of errors made is simply the number of coordinates where the two vectors differ.

DEFINITION 2.1. The *Hamming distance* $d(\underline{x}, \underline{y})$ between $\underline{x} = (x_1, x_2, \ldots, x_n)$ and $\underline{y} = (y_1, y_2, \ldots, y_n)$ in $V_n(q)$ is given by

$$d(\underline{x}, \underline{y}) = \bigl|\{1 \leqslant i \leqslant n \mid x_i \neq y_i\}\bigr|. \tag{2}$$

It is very simple to verify that (2) defines a metric on $V_n(q)$.

Although a code can be defined as just any subset of $V_n(q)$, in this chapter only linear subspaces of $V_n(q)$ will be considered. They are called *linear codes*. If a linear code $C$ in $V_n(q)$ has dimension $k$, one simply speaks of a $q$-ary $[n,k]$ code. Its elements are called *codewords*. To maximize the error-protection, one wants codewords to have sufficiently large mutual distance.

DEFINITION 2.2. The *minimum distance* $d$ of a nontrivial code $C$ (i.e. of cardinality at least 2) is given by

$$d = \min\bigl\{d(\underline{x}, \underline{y}) \mid \underline{x}, \underline{y} \in C,\ \underline{x} \neq \underline{y}\bigr\}. \tag{3}$$

The *error-correcting capability* $e$ of $C$ is defined by

$$e = \left\lfloor \frac{d-1}{2} \right\rfloor. \tag{4}$$

The reason for the name error-correcting capability is quite obvious. If $d$ is the minimum distance of a code $C$ and if during the transmission of the codeword $\underline{c}$ over the channel at most $e$ errors have been made, the received word $\underline{r}$ will still be closer to $\underline{c}$ than to any other codeword. So a maximum likelihood decoding algorithm applied to $\underline{r}$ will result in $\underline{c}$. If the minimum distance $d$ of an $[n,k]$ code is known, one also speaks of an $[n,k,d]$ code.

A different interpretation of Definition 2.2 is that *spheres* of radius $e$ around the codewords are disjoint, where the sphere of radius $r$ around $\underline{x}$, is defined by $B_r(\underline{x}) = \{\underline{y} \in V_n(q) \mid d(\underline{y}, \underline{x}) \leqslant r\}$. Clearly

$$|B_r(\underline{x})| = \sum_{i=0}^{r} \binom{n}{i} (q-1)^i.$$

Since all spheres with radius $e$ around the $|C|$ codewords are disjoint and there are only $q^n$ distinct words in $V_n(q)$, one has:

THEOREM 2.3 (Hamming bound). *Let $C$ be a $q$-ary $[n, k]$ code that is $e$-error-correcting. Then*

$$q^k \sum_{i=0}^{e} \binom{n}{i} (q-1)^i \leqslant q^n. \tag{5}$$

For $q = 2$, it follows from a slight strengthening of (1) that the information rate $R = k/n$ of an $[n, k]$ $e$-error-correcting code $C$ satisfies $R \leqslant 1 - h(e/n)$. If equality holds in (5), the code is called *perfect*. With perfect codes the spheres with radius $e$ around the codewords partition the whole vector space $V_n(q)$. In [30] the reader can find a good survey on the results regarding the (non-)existence of perfect codes. It turns out that all linear perfect codes are already known. They are the binary [24, 12, 7] and ternary [12, 6, 5] Golay codes, the $q$-ary $[n = \frac{q^m - 1}{q - 1}, n - m, 3]$ Hamming codes and the binary repetition code of odd length, consisting of $\underline{0}$ and $\underline{1}$. The Golay and Hamming codes will be defined in the sequel. A full discussion can be found in for instance [37] but also in the remarkable one-page article [19]. Perfect codes are also of great interest to algebraists because their automorphisms groups are highly regular.

It is also possible to derive a (nonconstructive) lower bound on the size of a code.

THEOREM 2.4 (Gilbert–Varshamov bound). *There exist $q$-ary $[n, k, d]$ codes satisfying*

$$q^k \geqslant \frac{q^n}{\sum_{i=0}^{d-1} \binom{n}{i} (q-1)^i}. \tag{6}$$

PROOF. A long as the product of the cardinality of a linear code $C$ and the volume of a sphere with radius $d - 1$ is strictly less than $q^n$, a word $\underline{u}$ at distance $\geqslant d$ to $C$ exists. It follows from the linearity of $C$ that the entire linear span of $C$ and $\underline{u}$ forms a larger linear code with minimum distance still at least $d$. □

It follows from (1) that for large values of $n$, binary $[n, k, d]$ codes $C$ exist with information rate $R = k/n$ satisfying $R \geqslant 1 - h((d-1)/n)$. Despite the fact that Theorem 2.4 looks rather wasteful in its approach, it was not until [20] that a family of codes was constructed for which neither $k/n$ nor $e/n$ tended to zero. It was not until

1982 that in [53] and [54] (nonbinary) classes of codes are described that perform better than the Gilbert–Varshamov bound.

The *Hamming weight* $w(\underline{x})$ of a vector $\underline{x}$ in $V_n(q)$ is the number of nonzero coordinates in $\underline{x}$. So $w(\underline{x}) = d(\underline{x}, \underline{0})$ and $d(\underline{x}, \underline{y}) = d(\underline{x} - \underline{y}, \underline{0}) = w(\underline{x} - \underline{y})$. The linearity of a code now implies that:

THEOREM 2.5. *The minimum distance of a linear code $C$ is equal to the minimum nonzero weight in $C$.*

The *extended* code $C^{\text{ext}}$ of a code $C$ is defined by

$$C^{\text{ext}} = \left\{ \left( c_1, c_2, \ldots, c_n, -\sum_{i=1}^{n} c_i \right) \middle| \underline{c} \in C \right\}.$$

Note that sum of the coordinates of a codeword in the extended code is zero. The extended code of a binary $[n, k, 2e+1]$ code has parameters $[n+1, k, 2e+2]$.

There are two common ways of describing a $k$-dimensional linear code: one by means of $k$ independent basis vectors, the other as the null space of $n-k$ linearly independent equations.

DEFINITION 2.6. A *generator matrix* $G$ of an $[n, k, d]$ code $C$ is a $k \times n$ matrix, of which the $k$ rows form a basis of $C$.

It follows that $C = \{\underline{a}G \mid \underline{a} \in V_k(q)\}$.

DEFINITION 2.7. A *parity check matrix* $H$ of an $[n, k, d]$ code $C$ is an $(n-k) \times n$ matrix, satisfying

$$\underline{c} \in C \Leftrightarrow H\underline{c}^T = \underline{0}^T. \tag{7}$$

Let $(\underline{x}, \underline{y})$ denote the regular inner product

$$\sum_{i=1}^{n} x_i y_i$$

in $V_n(q)$. We shall say that two vectors are *orthogonal* to each other if they have inner product zero. A word of warning is in place: in $V_n(q)$ a word can be orthogonal to itself without being $\underline{0}$. For instance, in $V_7(2)$ the vector $(1, 0, 1, 0, 0, 1, 1)$ is orthogonal to itself!

DEFINITION 2.8. The *dual code* $C^\perp$ of an $[n, k, d]$ code $C$ is defined by

$$C^\perp = \left\{ \underline{x} \in V_n(q) \mid (\underline{x}, \underline{c}) = 0 \text{ for all } \underline{c} \in C \right\}. \tag{8}$$

It is quite clear that $C^\perp$ is a linear code of dimension $n-k$. Also, it is straightforward to check that $(C^\perp)^\perp = C$ and that $C^\perp$ has as its generator matrix the parity check matrix $H$ of $C$ and as its parity check matrix the generator matrix $G$ of $C$.

Since a nonzero word can be orthogonal to itself, it is possible that a nonzero vector can be in both $C$ and $C^{\perp}$. Codes that are completely contained in their dual $C^{\perp}$ are called *self-orthogonal*. If $C = C^{\perp}$, the code is called *self-dual*.

It follows from (7) that the existence of a codeword $\underline{c}$ in a code $C$ with parity check matrix $H$ implies that the columns in $H$ where $\underline{c}$ has its nonzero coordinates must be dependent and, conversely, if a set of columns in $H$ is dependent, then $C$ contains a codeword with all its nonzero coordinates confined to the positions corresponding to those columns. This proves the following theorem:

THEOREM 2.9. *The minimum distance $d$ of a linear code $C$ with parity check matrix $H$ satisfies:*

$$d = 1 + \max\{l \mid \textit{each } l \textit{ columns of } H \textit{ are linearly independent}\}.$$

EXAMPLE 2.10. The matrices

$$G = \left(\begin{array}{cccc|ccc} 1&0&0&0&1&1&0 \\ 0&1&0&0&1&0&1 \\ 0&0&1&0&0&1&1 \\ 0&0&0&1&1&1&1 \end{array}\right) \quad \text{and} \quad H = \left(\begin{array}{cccc|ccc} 1&1&0&1&1&0&0 \\ 1&0&1&1&0&1&0 \\ 0&1&1&1&0&0&1 \end{array}\right)$$

are the generator resp. parity check matrix of a binary [7, 4, 3] code. Note that since $G$ has the form $(I_4\ P)$, the matrix $(-P^T\ I_3)$ is indeed a parity check matrix. That $d = 3$ follows directly from Theorem 2.9.

If a linear code has small dimension, one may decode a received word by simply comparing it with all possible codewords and select the closest. If the dimension is very high, a different technique can be used that will be explained now.

Let the *syndrome* $s$ of a received word $\underline{r}$ be defined by $\underline{s}^T = H\underline{r}^T$. Since a linear code $C$ is a subgroup of $V_n(q)$ and a word is in the code if and only if (iff) its syndrome is $\underline{0}$, it follows that two words are in the same coset iff their syndrome is the same. To find the closest codeword to $\underline{r}$, one has to find the lowest weight error pattern $\underline{e}$ such that $\underline{r} - \underline{e}$ is in $C$, or, equivalently, one has to find the lowest weight $\underline{e}$ with the same syndrome as $\underline{r}$.

ALGORITHM 2.11 (Syndrome decoding). *Let $\underline{r}$ be the received vector.*
1. *Compute the syndrome $\underline{s}^T = H\underline{r}^T$ of the received vector $\underline{r}$.*
2. *Find the coset leader $\underline{e}$ of the coset with syndrome $\underline{s}$.*
3. *Decode $\underline{r}$ into $\underline{c} = \underline{r} - \underline{e}$.*

Often one simply makes a table of all error patterns of weight at most $e$ together with their syndrome (these are all different) and does not attempt to decode if the syndrome of the received word does not occur in this list. The complexity of decoding a received word by comparing it with all possible codewords from a binary $[n, k]$ code is $2^{Rn}$, while syndrome decoding has complexity $2^{(1-R)n}$. In [14, 16] and [52] decoding algorithms are described of complexity $2^{an}$ with $a$ smaller that $\min\{R, 1-R\}$. An open question still is how much further the constant $a$ in this exponent can be reduced.

To state a surprising result in the theory of linear codes, a new definition is needed.

DEFINITION 2.12. Let $C$ be a code. Then the *weight enumerator* $A(z)$ of $C$ is given by

$$A(z) = \sum_{i=0}^{n} A_i z^i = \sum_{\underline{c} \in C} z^{w(\underline{c})}.$$

So, $A_i$, $0 \leqslant i \leqslant n$, counts the number of codewords of weight $i$ in $C$.

For instance, the code in Example 2.10 has weight enumerator $1 + 7z^3 + 7x^4 + z^7$ and its dual code has weight enumerator $1 + 7z^4$.

In 1963, F.J. MacWilliams [36] showed that the weight enumerators of a linear code $C$ and of its dual code $C^{\perp}$ are related by a rather simple formula.

THEOREM 2.13 (MacWilliams). *Let $A(z)$ be the weight enumerator of a q-ary $[n,k]$ code $C$ and let $B(z)$ be the weight enumerator of the dual code $C^{\perp}$. Then*

$$B(z) = \frac{1}{q^k} \sum_{i=0}^{n} A_i (1-z)^i \bigl(1 + (q-1)z\bigr)^{n-i}. \tag{9}$$

The proof of Theorem 2.13 follows by evaluating

$$\sum_{\underline{c} \in C} \sum_{\underline{u} \in V_n(q)} \chi\bigl((\underline{c}, \underline{u})\bigr) z^{w(\underline{u})}$$

in two different ways. Here $\chi$ denotes a nonprincipal character of the additive group of $GF(q)$ in the field of complex numbers. Since $GF(q)$ has characteristic $p$ it follows that $\chi^p = 1$ and that

$$\sum_{\alpha \in GF(q)} \chi(\alpha) = q,$$

of $\chi$ is the principal character, and equal to 0 otherwise. For further details the reader is referred to [37].

For nonlinear codes with weight enumerator $A(z)$, the right hand side of (9) can also be evaluated. It is not clear at all what kind of interpretation can be given to the outcome $B(z)$ in this case. In particular, it would be of importance to know if the inequalities that hold for $A(z)$ (like for instance the Hamming bound on $|C| = A(1)$ in Theorem 2.3) also hold for $B(z)$.

It follows from Theorem 2.9 that for an $[n,k,d]$ code $C$ with parity check matrix $H, d$ is greater than or equal to 2 iff $H$ does not contain the all-zero column. Similarly, $d \geqslant 3$ iff $H$ does not contain two columns that are linearly dependent ($H$ generates a code that is often called a *projective code*).

In view of the above, we now know that the length of a $q$-ary $[n, k, 3]$ code is bounded above by the maximum number of pairwise linearly independent vectors in $V_r(q)$, where $r = n - k$ is the redundancy of $C$. This is the same as the number of distinct points in $PG(r-1, q)$, the projective space of dimension $r - 1$ over $GF(q)$. This number is $(q^r - 1)/(q - 1)$.

DEFINITION 2.14 (Hamming code). The $q$-ary Hamming code of length

$$n = (q^r - 1)/(q - 1)$$

and redundancy $r$ is defined by the parity check matrix, that has as columns all the projective points of $PG(r-1, q)$. It is a $[n = (q^r - 1)/(q - 1), n - r, 3]$ code.

Example 2.10 gives the parity check matrix of the binary [7, 4, 3] Hamming code. That the Hamming codes are perfect is straightforward to check.

The dual code of a Hamming code is called Simplex code. With the properties of $PG(r-1, q)$ in mind it is not so difficult to show that all the nonzero codewords in a Simplex code have weight $q^{r-1}$. From the MacWilliams relations (Theorem 2.13) It now follows that:

THEOREM 2.15. *The weight enumerator of the q-ary Hamming code of length* $n = (q^r - 1)/(q - 1)$ *is given by*

$$A(z) = \frac{1}{q^r}\Big\{\big(1 + (q-1)z\big)^n + (q^r - 1)(1 - z)^{q^{r-1}}\big(1 + (q-1)z\big)^{n - q^{r-1}}\Big\}. \tag{10}$$

Although nonlinear codes are not further discussed in this chapter, we would like to draw the attention of the reader to the following recent development.

Kerdock codes [27] and Preparata codes [42] are two families of binary nonlinear codes. Both have length $n = 2^{2m}$ with $m \geqslant 2$. They have cardinality $2^{2^m - 2m}$ resp. $2^{2m}$ and minimum distance 6 resp. $2^{m-1} - 2^{m/2-1}$. Both have a larger minimum distance than any linear code of the same length and size.

The product of the cardinalities of the Kerdock codes and the Preparata code of length $n = 2^{2m}$ is $2^n$. Further, their weight enumerators satisfy the MacWilliams relation. Despite their nonlinearity, the above suggests some kind of mutual duality. Based on intensive studies, researchers came to believe that the apparent relation between the Kerdock codes and the Preparata codes is purely coincidental.

However, in 1994 Hammons, Kumar, Calderbank, Sloane and Solé [22] observed that both codes can be described as the image under the Gray map of two mutually dual, linear codes over $\mathbb{Z}_4$ of length $2^{2m-1}$. The Gray map $\phi$ is defined by $\phi(0) = 00$, $\phi(1) = 01$, $\phi(2) = 11$, $\phi(3) = 10$. To prove these statements one needs to develop the theory of Galois Rings, just as Galois Fields will be heavily needed in the next two sections.

## 3. Cyclic codes

To reduce the complexity of the encoding and decoding algorithms significantly as well as to be able to construct more powerful codes, linear codes that are invariant under cyclic shifts were the most natural to be looked at.

DEFINITION 3.1. A linear code $C$ is called *cyclic* if for each $(c_0, c_1, c_2, \dots, c_{n-1})$ in $C$ also $(c_{n-1}, c_0, c_1, \dots, c_{n-2})$ is in $C$.

In this context it is more natural to number the coordinates from 0 to $n-1$ and to identify the words in $V_n(q)$ with $q$-ary polynomials over $GF(q)$ in the following way:

$$(c_0, c_1, c_2, \dots, c_{n-1}) \leftrightarrow c_0 + c_1 x + \dots + c_{n-1}x^{n-1}. \tag{11}$$

So, instead of writing $\underline{c}$ is in $C$, we shall often write $c(x)$ is in $C$. Notice that multiplying $c(x)$ by $x$ gives the polynomial corresponding to the cyclic shift of $\underline{c}$ if the result $xc(x)$ is reduced modulo $x^n - 1$. For this reason the polynomials associated with (code)words will be regarded as elements in the residue class ring $GF(q)[x]/(x^n - 1)$.

THEOREM 3.2. *Let $C$ be a code in $V_n(q)$. Then $C$ is a cyclic code iff (when viewed as a subset of $GF(q)[x]/(x^n - 1)$) it is an ideal.*

*There exists a unique monic polynomial $g(x)$ dividing $x^n - 1$ with the property*

$$c(x) \text{ is in } C \quad \text{iff} \quad g(x) \text{ divides } c(x). \tag{12}$$

*The polynomial $g(x)$ is called the* generator polynomial *of $C$.*

PROOF. The existence of a generator of the ideal (that $C$ is) follows from the fact that $GF(q)[x]/(x^n - 1)$ is a principal ideal ring. The only monic generator of $C$ dividing $x^n - 1$ is the nonzero (monic) polynomial of lowest degree in $C$. □

THEOREM 3.3. *Let $C$ be a cyclic code in $V_n(q)$ with generator polynomial $g(x) = g_0 + g_1 x + \dots + g_{n-k}x^{n-k}$ with $g_{n-k} \neq 0$. Then $C$ has dimension $k$ and is generated by*

$$G = \begin{pmatrix} g_0 & g_1 & \cdots & \cdots & g_{n-k} & 0 & 0 & \cdots & 0 \\ 0 & g_0 & g_1 & \cdots & \cdots & g_{n-k} & 0 & \cdots & 0 \\ 0 & 0 & g_0 & g_1 & \cdots & \cdots & g_{n-k} & \cdots & 0 \\ \vdots & & & \ddots & \ddots & & & \ddots & \vdots \\ 0 & 0 & \cdots & 0 & g_0 & g_1 & \cdots & \cdots & g_{n-k} \end{pmatrix}. \tag{13}$$

*A parity check matrix $H$ of $C$ is given by*

$$H = \begin{pmatrix} 0 & \cdots & 0 & 0 & h_k & \cdots & \cdots & h_1 & h_0 \\ 0 & \cdots & 0 & h_k & \cdots & \cdots & h_1 & h_0 & 0 \\ \vdots & & \cdot & & & \cdot & \cdot & & \vdots \\ \vdots & \cdot & & & \cdot & \cdot & & & \vdots \\ h_k & \cdots & \cdots & h_1 & h_0 & 0 & \cdots & 0 & 0 \end{pmatrix}, \tag{14}$$

*where* $h(x) = h_0 + h_1 x + \cdots + h_k x^k$ *is defined by* $g(x)h(x) = x^n - 1$ *and is called the parity check polynomial of C.*

Again the proof is very elementary and will be omitted here. Note that the dual code of a cyclic code with parity check polynomial $h(x)$ is again cyclic and is generated by the reciprocal of $h(x)$. It also follows from the above that

$$c(x) \text{ is in } C \quad \text{iff} \quad c(x)h(x) = 0. \tag{15}$$

The complete factorization of $x^{15} - 1$ in $GF(2)[x]$ is given by

$$(x+1)(x^2+x+1)(x^4+x+1)(x^4+x^3+1)(x^4+x^3+x^2+x+1).$$

Taking $g(x) = (x+1)(x^4+x+1)$ gives a [15, 10] code of which the minimum distance has not been determined yet.

To be able to say something about the minimum distance of cyclic codes, it will be necessary to consider an extension field of $GF(q)$ in which $x^n - 1$ factors completely into linear factors. This will be $GF(q^m)$ with $n$ dividing $q^m - 1$. Therefore one has to assume (as will be done from now on) that $q$ and $n$ are coprime. Let $\omega$ be a primitive element in $GF(q^m)$. Then $\alpha = \omega^{(q^m-1)/n}$ will be a primitive $n$-th root of unity in $GF(q^m)$ and

$$x^n - 1 = \prod_{i=0}^{n-1} (x - \alpha^i).$$

It follows that the generator polynomial $g(x)$ of a $q$-ary cyclic code $C$ of length $n$ factors into

$$g(x) = \prod_{i \in I} (x - \alpha^i)$$

over $GF(q^m)$, where $I$ is a subset of $\{0, 1, \ldots, n-1\}$, called the *defining set* of $C$ with respect to $\alpha$.

Let $f(x)$ be an irreducible $q$-ary polynomial dividing $x^n - 1$ and let $\alpha^i$ be a zero of $f(x)$ in $GF(q^m)$. It is well known that its conjugates $\alpha^{iq}, \alpha^{iq^2}, \ldots$ are also zeros of $f(x)$. Of course the exponents have to be reduced modulo $n$, since $\alpha^n = 1$. The set

$\{iq^j \bmod n \mid j = 0, 1, \ldots\}$ consisting of the exponents modulo $n$ of these conjugates is called the *cyclotomic coset* $C_i$ of $i$ modulo $n$.

The set of all conjugates of the zero $\alpha^i$ of an irreducible polynomial $f(x)$ gives the complete factorization of $f(x)$ into linear factors:

$$f(x) = \prod_{l \in C_i} (x - \alpha^l).$$

This polynomial $f(x)$ is called the *minimal polynomial* of $\alpha^i$ and will be denoted by $m_i(x)$.

What we have shown above is that a generator polynomial of a cyclic code is the product of some minimal polynomials and that the defining set of a cyclic code is the union of the corresponding cyclotomic cosets. A necessary and sufficient condition for this is that the defining set $I$ has the property $i \in I \Rightarrow qi \in I$, where $qi$ of course has to be reduced modulo $n$.

EXAMPLE 3.4 (*To be continued*). Let $q = 3$ and $n = 11$. To find the smallest extension field of $GF(3)$ that contains the 11-th roots of unity, one has to determine (the smallest) $m$ with $11 \mid (q^m - 1)$. One obtains $m = 5$. So

$$x^{11} - 1 = \prod_{i=0}^{10} (x - \alpha^i),$$

where $\alpha = \omega^{(3^5-1)/11}$ for some (each) primitive element $\omega$ in $GF(3^5)$.

There are three cyclotomic cosets. The first is $C_0 = 0$, giving rise to the ternary polynomial $m_0(x) = x - 1$. The other two are

$$C_1 = \{1, 3, 9, 5, 4\}$$

and

$$C_{-1} = \{2, 6, 7, 10, 8\}.$$

They correspond to the irreducible, ternary polynomials

$$\begin{aligned} m_1(x) &= (x - \alpha)(x - \alpha^3)(x - \alpha^9)(x - \alpha^5)(x - \alpha^4) \\ &= x^5 + x^4 - x^3 + x^2 - 1 \end{aligned}$$

and

$$\begin{aligned} m_{-1}(x) &= (x - \alpha^2)(x - \alpha^6)(x - \alpha^7)(x - \alpha^{10})(x - \alpha^8) \\ &= x^5 - x^3 + x^2 - x - 1 \end{aligned}$$

(or the other way around depending on the choice of $\omega$).

The code generated by $m_1(x)$ (or by $m_{-1}(x)$) has dimension $k = 6$.

In view of the above it is sufficient to give just one element of each cyclotomic coset in the defining set $I = \{i_1, i_2, \ldots, i_l\}$. One now has the following equivalent descriptions of the cyclic code in $V_n(q)$ with defining set $I$:

$$C = \{c(x) \mid m_i(x) \text{ divides } c(x) \text{ for all } i \text{ in } I\}, \tag{16}$$

$$C = \{c(x) \mid c(\alpha^i) = 0 \text{ for all } i \text{ in } I\}, \tag{17}$$

$$C = \{\underline{c} \in V_n(q) \mid H\underline{c}^T = \underline{0}^T\}, \tag{18}$$

where

$$H = \begin{pmatrix} 1 & \alpha^{i_1} & \alpha^{2i_1} & \cdots & \cdots & \alpha^{(n-1)i_1} \\ 1 & \alpha^{i_2} & \alpha^{2i_2} & \cdots & \cdots & \alpha^{(n-1)i_2} \\ \vdots & \vdots & & & & \vdots \\ 1 & \alpha^{i_l} & \alpha^{2i_l} & \cdots & \cdots & \alpha^{(n-1)i_l} \end{pmatrix}.$$

Let $\alpha$ be a primitive element in $GF(2^m)$. Clearly the $n = 2^m - 1$ elements $\alpha^i$, $0 \leqslant i < n$, are all distinct. It follows from (18) that the binary cyclic code of length $n$ with defining set $\{1\}$, which is generated by $m_1(x)$, is in fact a Hamming code.

The next (very general) class of cyclic codes will have certain guaranteed minimum distance properties. This class is named after R.C. Bose, D.K. Ray-Chaudhuri and A. Hocquenghem (the first two [9] found this class independently of the last [25]).

DEFINITION 3.5. Let $\alpha$ be a primitive $n$-th root of unity in an extension field of $GF(q)$ and let $I$ be the defining set of a $q$-ary cyclic code $C$ of length $n$.

If $I$ contains $d_{BCH} - 1$ consecutive integers (taken modulo $n$), $C$ is called a *BCH code* of *designed distance* $d_{BCH}$.

If $I$ contains $\{1, 2, \ldots, d_{BCH} - 1\}$ as a subset, the code $C$ will be called a *narrow-sense* BCH code. If $n = q^m - 1$, the BCH code $C$ is called *primitive*.

The justification of the notation $d_{BCH}$ will be given in the following theorem.

THEOREM 3.6 (BCH bound). *The minimum distance $d$ of a BCH code with designed distance $d_{BCH}$ satisfies $d \geqslant d_{BCH}$.*

PROOF. Let $I$ contain $\{i+1, i+2, \ldots, i+d_{BCH}-1\}$. Then the parity check matrix $H$ contains the following $d_{BCH} - 1$ rows:

$$H = \begin{pmatrix} 1 & \alpha^{i+1} & \alpha^{2(i+1)} & \cdots & \cdots & \alpha^{(n-1)(i+1)} \\ 1 & \alpha^{i+2} & \alpha^{2(i+2)} & \cdots & \cdots & \alpha^{(n-1)(i+2)} \\ \vdots & \vdots & & & & \vdots \\ 1 & \alpha^{i+d_{BCH}-1} & \alpha^{2(i+d_{BCH}-1)} & \cdots & \cdots & \alpha^{(n-1)(i+d_{BCH}-1)} \end{pmatrix}.$$

Now the determinant of any $(d_{BCH} - 1) \times (d_{BCH} - 1)$ submatrix of $H$ is a nonzero Vandermonde determinant. It follows from Theorem 2.9 that the BCH code has minimum distance at least equal to $d_{BCH}$. □

How to decode up to $e_{BCH} = \lfloor (d_{BCH} - 1)/2 \rfloor$ errors will be discussed in the next section. There are many cyclic codes with a minimum distance that is actually more than guaranteed by the BCH bound. This led researchers to look for techniques improving on the BCH bound. See [23, 47] and [32]. A related question of course is how to decode $e$ errors algebraically if the cyclic code is indeed $e > e_{BCH}$ error-correcting. For some results, see [17] and [10]. A third open question is the information rate and error-correcting capability of BCH codes when $n$ tends to infinity. See [3] and [4], Chapter 12.

EXAMPLE 3.4 (*Continued*). Since the cyclic code $C$ in Example 3.4 has a defining set containing $3, 4$ and 5, its minimum distance is at least 4. Let $G$ be the generator matrix of $C$ consisting of six cyclic shifts of $g(x) = m_1(x)$. Add as 12-th coordinate a $-1$ to each row. This new matrix generates a $[12, 6, \geqslant 4]$ code which is the extended code $C^{\text{ext}}$ of $C$. It is in this small example easy to check that $C^{\text{ext}}$ is a self-dual code. In particular, each word in $C^{\text{ext}}$ is orthogonal to itself. Over $GF(3)$ this means that the weight of each codeword in $C^{\text{ext}}$ is divisible by 3. Hence $C^{\text{ext}}$ is a $[12, 6, 6]$ code and thus $C$ is a $[12, 6, 5]$ code. This code is perfect and is the ternary Golay code mentioned earlier.

*Reed–Solomon codes* are defined as narrow-sense $q$-ary BCH codes of length $n = q - 1$. They have many additional properties that will not be discussed here (see [37], Chapter 10). In many applications RS codes are implemented with $q$ equal to a power of 2, say $2^a$, so that $a$ bits at a time can be regarded as one symbol in $GF(q)$.

A very special class of cyclic codes is the following.

DEFINITION 3.7. Let $n$ be a prime such that $n \equiv \pm 1 \pmod 8$ and let $QR$ and $NQR$ denote the set of quadratic residues resp. quadratic nonresidues modulo $n$ (2 is in $QR$, so $QR$ and $NQR$ are closed under multiplication by 2).

Then the binary cyclic codes of length $n$ with defining set $QR$ resp. $QR \cup \{0\}$ are both called *quadratic residue codes* (for short *QR* codes).

Let

$$q(x) = \prod_{r \in QR} (x - \alpha^r) \quad \text{and} \quad n(x) = \prod_{r \in NQR} (x - \alpha^r),$$

where $\alpha$ is a primitive $n$-th root of unity. Then $x^n - 1$ factors into $(x - 1)q(x)n(x)$. Clearly the dimension of the two QR codes is $(n + 1)/2$ resp. $(n - 1)/2$. Before giving bounds on the minimum distance a different property of QR codes will be derived. To this end, the coordinates of the QR code will be indexed by the elements in $GF(n)$ and the extra coordinate in the extended code by $\infty$. That the QR code is invariant under the cyclic shift $S$: $x \to x + 1$ is obvious. With some more work, one can also show that the extended QR code is invariant under the mapping $T$: $x \to -x^{-1}$. Together, these two coordinate permutations generate the projective special linear group $PSL(2, n)$, consisting of all mapping $x \to (ax + b)/(cx + d)$ with $ad - bc = 1$.

LEMMA 3.8. *The extended QR code is invariant under* $PSL(2, n)$.

Since $PSL(2,n)$ is a transitive group, it follows that the number of codewords of weight $2i-1$ and of weight $2i$ in the QR code are related by $2i(A_{2i-1}+A_{2i})=(n+1)A_{2i-1}$ (both expressions count the number of ones in the words of weight $2i$ in the extended code). In particular, $A_{2i-1}=0$ iff $A_{2i}=0$ and hence the minimum distance in the QR code is odd. More can be said.

THEOREM 3.9. *The minimum distance $d$ in the QR code with generator polynomial $q(x)$ is odd. Further*

1) $d^2 \geqslant n$,
2) *if* $n \equiv -1 \pmod 8$, *then* $d^2-d+1 \geqslant n$ *and* $d \equiv 3 \pmod 4$.

PROOF. Consider a codeword $c(x)$ of (odd) weight $d$ in $QR$, say

$$c(x) = x^{i_1} + x^{i_2} + \cdots + x^{i_d}.$$

Clearly $c(x)$ is divisible by $q(x)$, but not by $x-1$.

Let $u \in NQR$ and consider the coordinate permutation $\pi_u$: $i \to ui \pmod n$. Then $\pi_u$ will map $c(x)$ into a word $c'(x)$ which is divisible by $n(x)$, but not by $x-1$. Since $q(x)n(x) = (x^n-1)/(x-1)$, it follows that

$$c(x)c'(x) \equiv 1 + x + \cdots + x^{n-1} \pmod{x^n - 1}.$$

Since the left hand side has at most $d^2$ terms (cancellations may occur), the first statement follows.

If $n \equiv -1 \pmod 8$, one can take $u=-1$ above, so $n(x) = x^{(n-1)/2}q(1/x)$. This implies that $c(x)c'(x)$ has at most $d^2-d+1$ nonzero terms ($d$ terms give a 1). Moreover terms cancel four at a time: if $i_u - i_v \equiv i_{u'} - i_{v'} \pmod n$, also $i_v - i_u \equiv i_{v'} - i_{u'} \pmod n$. □

EXAMPLE 3.10. The QR code of length 23 has parameters $[23, 12, \geqslant 7]$ by the above theorem. For $d=7$, the Hamming bound holds with equality. In other words: this code is perfect, 3-error-correcting. It is the binary Golay code mentioned in Section 2. The BCH bound in this example would only give $d \geqslant 5$.

A surprising connection between QR codes and projective planes is given by the following result (see [50], Chapter 3):

LEMMA 3.11. *Let the minimum distance $d$ of a QR code of length $n$, $n \equiv -1 \pmod 8$ satisfy $d^2-d+1=n$. Then a projective plane of order $d-1$ exists.*

PROOF. Continuing with the last part of the proof of Theorem 3.9, define $n_i$ as the number of nonzero exponents in $1+x+\cdots+x^{n-1}$ that appear exactly $i$ times as $i_u - i_v \bmod n$. Clearly the $n_i$ are zero for even $i$. It follows that

$$d(d-1) = \sum_{i\ \mathrm{odd}} i n_i \quad \text{and} \quad n-1 = \sum_{i\ \mathrm{odd}} n_i.$$

So equality in $d(d-1) = n-1$ shows that $n_i = 0$ for $i \neq 1$. In other words: each nonzero $j$ modulo $n$ occurs exactly once as a difference of two exponents (the set $\{i_1, i_2, \ldots, i_d\}$ is a so-called *difference set* mod $n$).

The $n \times n$ $\{0,1\}$-circulant with top row entries 1 at the coordinates $i_u$, $1 \leqslant u \leqslant d$, now is the incidence matrix of $PG(2, d-1)$. □

Apart from determining the actual minimum distance of QR codes, a completely different (and open) problem of course is how to decode QR codes up to their error-correcting capability. For some codes this problem has been solved. For the [24, 12, 7] (Golay), [34, 16, 8] and [44, 21, 9] (extended) QR codes, the reader is referred to [15, 44], and [45], respectively.

## 4. Goppa and algebraic geometry codes

In [37], Chapter 9, Section 5, it is shown that primitive BCH codes are asymptotically bad, in the sense that either $e/n$ or $R = k/n$ tends to 0 for $n$ to infinity. To obtain better asymptotical results (and possibly also shorter codes with improved performance) a generalization is needed.

Now it is easy to check that the $q$-ary narrow sense BCH code of length $n$ with designed distance $d_{BCH}$ is equivalent under $\alpha \to \alpha^{-1}$ to the code

$$\left\{ \underline{c} \in V_n(q) \;\middle|\; \sum_{i=0}^{n-1} \frac{c_i}{x - \alpha^i} \equiv 0 \; \left(\text{mod } x^{d_{BCH}-1}\right) \right\},$$

where $\alpha$ is an $n$-th root of unity in an extension field of $GF(q)$ and where the denominator $\frac{1}{x-\alpha^i}$ simply should be interpreted as the multiplicative inverse of $x - \alpha^i$ modulo $x^{d_{BCH}-1}$.

DEFINITION 4.1. Let $L = \{\alpha_0, \alpha_1, \ldots, \alpha_{n-1}\}$ be a subset of $GF(q^m)$ of size $n$ and let $G(x)$ be a $q$-ary polynomial of degree $s$ that is not zero in any of the elements $\alpha_i$. The *Goppa code* $\Gamma(L, G)$ is defined by

$$\Gamma(L, G) = \left\{ \underline{c} \in V_n(q) \;\middle|\; \sum_{i=0}^{n-1} \frac{c_i}{x - \alpha_i} \equiv 0 \; \left(\text{mod } G(x)\right) \right\}. \tag{19}$$

Take $G(x) = x^{d_{BCH}-1}$ and $\alpha_i = \alpha^i$, $0 \leqslant i < n$, in relation (19) to see that Goppa codes contain BCH codes as a subclass. Quite clearly, Goppa codes are linear.

In the sequel, the coordinates with either be indexed by the numbers $i$, $0 \leqslant i < n$, or by the elements in $\alpha_i$, $0 \leqslant i < n-1$.

THEOREM 4.2. *The Goppa code $\Gamma(L, G)$ of length $n$ with $G(x)$ of degree $s$ has parameters $[n, k \geqslant n - ms,\ d \geqslant s+1]$.*

PROOF. i) Let $\underline{c}$ be a codeword of weight $w > 0$ and let the nonzero coordinates of $\underline{c}$ be at coordinates $\{\alpha_{i_1}, \alpha_{i_2}, \ldots, \alpha_{i_w}\}$.

Write the summation in (19) as one fraction. Then, because the denominator has no factor in common with $G(x)$, condition (19) is equivalent to

$$G(x) \text{ divides } \sum_{l=1}^{w(\underline{c})} c_{i_l} \prod_{1 \leqslant j \leqslant w,\ j \neq l} (x - \alpha_{i_j}).$$

However this numerator has degree at most $w-1$. It follows that $w - 1 \geqslant s$ and thus (by the linearity) also $d \geqslant s+1$.

ii) Writing $1/(x - \alpha_i)$, $0 \leqslant i \leqslant n-1$, as polynomial

$$G_i(x) = \sum_{j=0}^{s-1} G_{ij} x^j$$

modulo $G(x)$, condition (19) can be rewritten as

$$\sum_{i=0}^{n-1} c_i G_i(x) \equiv 0 \pmod{G(x)}$$

or, alternatively, by considering the coefficients of $x^j$, $0 \leqslant j \leqslant s-1$,

$$\sum_{i=0}^{n-1} c_i G_{ij} = 0 \quad \text{for } 0 \leqslant j \leqslant s-1.$$

This means that $\Gamma(L, G)$ can be defined by $s$ linear equations over $GF(q^m)$ and thus by $\leqslant ms$ linear equations over $GF(q)$. Hence, $\Gamma(L, G)$ has dimension at least $n - ms$. □

In some cases, much better bounds on the minimum distance can be given than the bound in Theorem 4.2.

THEOREM 4.3. *Let the defining Goppa polynomial $G(x)$ of the Goppa code $\Gamma(L, G)$ be a polynomial over $GF(2^m)$ of degree $s$ that has no multiple zeros. Then, $\Gamma(L, G)$ will have minimum distance at least $\geqslant 2s + 1$.*

PROOF. Let $\underline{c}$ be a codeword of weight $w > 0$. Note that

$$\sum_{i=0}^{n-1} c_i/(x - \alpha_i)$$

can be written as $f'(x)/f(x)$, with

$$f(x) = \prod_{i=0}^{n-1} (x - \alpha_i)^{c_i}.$$

So equation (19) is equivalent to $G(x)$ divides $f'(x)$. But, because $q = 2$, $f'(x)$ is a perfect square. Since $G(x)$ has no multiple zeros, one now has that $G(x)$ divides a polynomial of degree $(w-1)/2$ and thus $w - 1 > 2u \geqslant 2s$. □

Goppa codes (and thus also BCH codes and Reed–Solomon codes) can be decoded by an efficient decoding technique that makes use of Euclid's Algorithm. For the correct decoding of a received word $\underline{r}$, which is the sum of a codeword $\underline{c}$ and an error pattern $\underline{e}$, one needs to know two things: where the errors occurred and what their values are.

Define the set $B$ of *error locations* by $B = \{\alpha_i \mid e_i \neq 0\}$ and for each $\beta$ in $B$ the corresponding *error value* $e_\beta = e_i$, where $\beta = \alpha_i$. The *error locator polynomial* $\sigma(x)$ and the *error evaluator polynomial* $\omega(x)$ of the error vector are defined by

$$\sigma(x) = \prod_{\beta \in B} (x - \beta), \tag{20}$$

$$\omega(x) = \sum_{\beta \in B} e_\beta \prod_{\gamma \in B,\ \gamma \neq \beta} (x - \gamma). \tag{21}$$

The error locations are simply the zeros of $\sigma(x)$. The corresponding error value follows from $e_\beta = \omega(\beta)/\sigma'(\beta)$, for $\beta \in B$. So, the decoding problem reduces to finding $\sigma(x)$ and $\omega(x)$ from $L$, $G(x)$ and the syndrome $S(x)$ defined by

$$S(x) = \sum_{i=0}^{n-1} \frac{r_i}{x - \alpha_i} \pmod{G(x)}.$$

The following relation (which can be verified by simply substituting the various definitions) plays the key role in determining $\sigma(x)$ and $\omega(x)$.

$$S(x)\sigma(x) \equiv \omega(x) \pmod{G(x)}. \tag{22}$$

Determining $\sigma(x)$ and $\omega(x)$ from (22) amounts to applying the extended version of Euclid's Algorithm to the polynomials $G(x)$ and $S(x)$. This would not uniquely define $\sigma(x)$ and $\omega(x)$, except for the fact that degree $(\sigma(x)) = t = |B|$, degree$(\omega(x)) < t$, and $\gcd(\sigma(x), \omega(x)) = 1$.

ALGORITHM 4.4 (Euclid's Algorithm). *Let $a(x)$ and $b(x)$ be two $q$-ary polynomials, where* degree$(a(x)) \geqslant$ degree$(b(x))$.

*Define the sequences of polynomials $s_i(x)$, $u_i(x)$, $v_i(x)$ and $q_i(x)$, where the degrees of $s_i(x)$ are strictly decreasing, recursively as follows.*

$$s_0(x) = a(x),\ u_0(x) = 1,\ v_0(x) = 0,$$
$$s_1(x) = b(x),\ u_1(x) = 0,\ v_1(x) = 1,$$
$$i = 1.$$

*While* $s_i(x) \neq 0$ *do* **begin**

$i := i + 1$

*write* $s_{i-2}(x) = q_i(x)s_{i-1}(x) + s_i(x), \quad degree(s_i(x)) < degree(s_{i-1}(x))$.

*Define* $u_i(x)$ *and* $v_i(x)$ *by*

$u_{i-2}(x) = q_i(x)u_{i-1}(x) + u_i(x), \quad v_{i-2}(x) = q_i(x)v_{i-1}(x) + v_i(x)$.

**end**

$n = i$.

*Then*

$$\gcd\big(a(x), b(x)\big) = s_{n-1}(x) = u_{n-1}(x)a(x) + v_{n-1}(x)b(x). \tag{23}$$

A full discussion of the decoding algorithm of Goppa codes can be found in [38]. Here it shall presented without the (rather technical) proof. Note that only the $v_i$'s play a role and not the $u_i$'s.

ALGORITHM 4.5 (Decoding Goppa codes). *Let $\Gamma(L, G)$ be a Goppa code with $G(x)$ of degree $s$ and let $\underline{r} = (r_0, r_1, \ldots, r_{n-1})$ be a received vector.*

1. *Compute the syndrome*

$$S(x) \equiv \sum_{i=0}^{n-1} \frac{r_i}{x - \alpha_i} \pmod{G(x)}.$$

2. *Apply Euclid's Algorithm to $a(x) = G(x)$ and $b(x) = S(x)$ until $degree(s_i(x)) < \lfloor s/2 \rfloor$ for the first time. Let $\nu$ be the leading coefficient of $v_i(x)$. Set $\sigma(x) = v_i(x)/\nu$ and $\omega(x) = s_i(x)/\nu$.*
3. *Find the set $B = \{\beta \text{ in } GF(q^m) \mid \sigma(\beta) = 0\}$ of error locations.*
4. *Determine the error values $e_\beta = \omega(\beta)/\sigma'(\beta)$ for all $\beta$ in $B$.*
5. *Determine $\underline{e} = (e_0, e_1, \ldots, e_{n-1})$ from $e_i = e_\beta$ if $\beta$ is in $B$ and $\beta = \alpha_i$ and $e_i = 0$ otherwise.*
6. *Set $\underline{c} = \underline{r} - \underline{e}$.*

For BCH and Goppa codes it is known that in some/many cases the actual error-correcting capability $e$ is larger than the bound given in Theorem 4.2. For BCH codes it is in some cases known how to decode up to $e$ errors when $e > e_{BCH}$. For Goppa codes results of this type are unknown.

The code constructions thus far were asymptotically bad in the sense that either $d/n \to 0$ or $R = k/n \to 0$ for $n \to \infty$. The next theorem shows that with Goppa codes that situation is over now.

THEOREM 4.6. *For each $q$ there exists a sequence of $q$-ary Goppa codes meeting the Gilbert–Varshamov bound.*

The proof will not be given in full detail but boils down to the following reasoning. Take $n = q^m, s, d$ and $L = GF(q^m)$. For each word $\underline{u}$ of weight $w$, $w < d$, there are at most $\lfloor (w-1)/s \rfloor < d/s$ irreducible polynomials $g(x)$ of degree $s$ over $GF(q^m)$ with $\underline{u} \in \Gamma(L, g)$. So if $d/s$ times the volume of a sphere with radius $d-1$ is strictly less than the total number of irreducible polynomials of degree $s$ over $GF(q^m)$, one has shown that there are polynomials $G(x)$ left that define Goppa codes $\Gamma(L, G)$ with minimum distance at least equal to $d$. Using well known estimates on the number of irreducible polynomials of given degree, one can show that the above mentioned inequality is satisfied if the rate of the Goppa code meets the Gilbert–Varshamov bound (asymptotically).

It is important to notice that Theorem 4.6 still is nonconstructive in the sense that it does not say how to choose $G(x)$. Any further results in this direction would be extremely important.

How Goppa codes can be further generalized to yield codes that are asymptotically better than the Gilbert–Varshamov [53] will be the final topic of this section. Since deep results from algebraic geometry are needed, a full discussion of this important development is beyond the scope of this chapter. The reader is referred to [33–35, 39] and [55]. Here [35] is followed closely.

For the class of $q$-ary Reed–Solomon codes (with parameters $[n = q, k, d = n-k+1]$) the following yields an equivalent description:

$$\{(f(0), f(1), f(\alpha), \ldots, f(\alpha^{n-1}) \mid f \in GF(q)[x], \text{ degree}(f) < k\},$$

where $\alpha$ is a primitive $n$-th root of unity. Yet another notation will be needed. Consider $X = PG(1, F)$, the projective line over $F$, where $F$ is the algebraic closure of $GF(q)$. Its points can be described by coordinates $(x, y)$ with $x, y \in F$ and where $(x, y)$ and $(xz, yz)$, $z \neq 0$, denote the same point. Points on $X$ with coordinates in $GF(q)$ are called *rational* points. They can be represented by $P_i = (\alpha_i, 1)$, $0 \leqslant i < q$, and $Q = (1, 0)$, where the elements $\alpha_i$, $0 \leqslant i < q$, form a particular numbering of the field elements in $GF(q)$.

Let $\mathcal{R}$ be the subset of rational forms $a(x, y)/b(x, y)$ on $X$ (so both $a(x, y)$ and $b(x, y)$ are homogeneous polynomials of the same degree) that are defined on each $P_i$ and have coefficients in $GF(q)$. Now, the code above can be described by

$$\{(f(P_0), f(P_1), f(P_2), \ldots, f(P_{q-1}) \mid f = a/b \in \mathcal{R}, \text{ degree}(a) < k)\}.$$

To generalize this description of Reed–Solomon codes, $X$ will now be an irreducible, nonsingular projective curve in $PG(N, F)$ of genus $g$ (see [39] or [55] for precise definitions; here an intuitive concept of $X$ is good enough, $g$ is a number uniquely determined by $X$).

DEFINITION 4.7. A *divisor* $D$ on $X$ is a formal sum $\sum_{P \text{ on } X} n_P P$, where the coefficients $n_P$ are integers and only finitely many of them can be nonzero.

The *degree* of a divisor $D = \sum_{P \text{ on } X} n_P P$ is defined by $\sum_{P \text{ on } X} n_P$.

Let $f$ be a nonzero rational function on $X$ and let $P$ lie on $X$. Then $f$ has *order* $n$

in $P$ if $P$ is a zero (of $f$) of multiplicity $n$ and has order $-n$ in $P$ if $P$ is a pole of multiplicity $n$.

EXAMPLE 4.8. Take $q = 4$, let $GF(4)$ be generated by $\omega$, $\omega^2 + \omega + 1 = 0$, and let $F$ be the algebraic closure of $GF(4)$. The curve $X$ in $PG(2, F)$ is defined by the equation $x^3 + y^3 + z^3 = 0$.

The rational function $f = (y^2 + yz + z^2)/(x^2)$ has order 1 in $(0, \omega, 1)$ and order $-2$ in $(0, 1, 1)$.

Note that in Example 4.8 $f$ can also be written as $x/(y + z)$, since

$$y + z = (y^3 + z^3)/(y^2 + yz + z^2) = x^3/(y^2 + yz + z^2),$$

but that with that description the behavior of $f$ in $P$ cannot be determined.

For a divisor $D = \sum_{P \text{ on } X} n_P P$ let $\mathcal{L}(D)$ be the linear space of all rational functions $f$ on $X$ such that the order of $f$ in any point $P$ on $X$ is at least equal to $-n_P$. For divisors of negative degree $\mathcal{L}(D)$ only consists of 0. The Riemann–Roch theorem (a fundamental result in algebraic geometry; for a proof see [39], Theorem 2.5) implies that the dimension $l(D)$ of $\mathcal{L}(D)$ satisfies

$$l(D) \geqslant \text{degree}(D) - q + 1. \tag{24}$$

Let $X$ be an irreducible, nonsingular projective curve in $PG(N, F)$ of genus $g$, where $F$ is the algebraic closure of the finite field $GF(q)$, and let $P_i$, $1 \leqslant i \leqslant n$, and $Q$ be the rational points on $X$.

DEFINITION 4.9. In the notation of above, choose $m$ such that $2g - 2 < m < n$. Then

$$C = \{\underline{f} = (f(P_1), f(P_2), \ldots, f(P_n) \mid f \in \mathcal{L}(mQ))\},$$

is called an *algebraic geometry code* (AG code).

THEOREM 4.10. *The AG code $C$ in Definition* 4.9 *is a $q$-ary $[n, k \geqslant m-g+1,\ d \geqslant n-m]$ code.*

PROOF. $C$ is indeed a code over $GF(q)$, since the points $P_i$ are rational and $f$ has its coefficients in $GF(q)$. The linearity of $C$ is obvious.

Consider the codeword $\underline{0}$. The corresponding polynomial $f$ in $\mathcal{L}(mQ)$ is in fact in

$$\mathcal{L}\left(mQ - \sum_{i=1}^{n} P_i\right).$$

Since the divisor

$$mQ - \sum_{i=1}^{n} P_i$$

has negative degree, it follows that $f = 0$, which implies that the dimension of $C$ is the same as that of $\mathcal{L}(mQ)$. By the Riemann–Roch theorem, this dimension is at least $m - g + 1$.

Let $\underline{c}$ be a codeword of weight $d$. The corresponding polynomial $f$ in $\mathcal{L}(mQ)$ is zero in $n - d$ points $P_i$, say $P_{i_1}, P_{i_2}, \ldots, P_{i_{n-d}}$, so in fact $f$ is in

$$\mathcal{L}\left(mQ - \sum_{j=1}^{n-d} P_{i_j}\right).$$

Since the divisor

$$mQ - \sum_{i=1}^{n-d} P_{i_j}$$

has degree $m - (n - d)$, which must be at least 0 (otherwise $f$ would be 0), one may conclude that $d \geqslant n - m$. □

It is quite obvious that the particular choice of $X$ above is vital in finding good AG codes. For the construction of asymptotically good AG codes, an infinite sequence of curves $X$ is necessary. In [53] a sequence is described for $q = p^{2r}$ such that for $n \to \infty$

$$\frac{g}{n} \to \bar{\gamma} = \frac{1}{q^{1/2} - 1}.$$

For the AG codes defined by this sequence, write $\delta = d/n$, $\gamma = g/n$ and $R = k/n$ for $n \to \infty$. Then

$$\begin{aligned} R = \frac{k}{n} &\geqslant \frac{m}{n} - \frac{g}{n} \\ &\geqslant 1 - \frac{d}{n} - \frac{g}{n} = 1 - \delta - \gamma = 1 - \delta - \frac{1}{q^{1/2} - 1}. \end{aligned} \tag{25}$$

It is now a matter of simple calculus that for $q \geqslant 49$ values of $R$ satisfying inequality (25) lie above the asymptotic version of the Gilbert–Varshamov bound in Theorem 2.4 in an appropriate subinterval of $[0, (q-1)/q]$ for $\delta$. In [35] it is shown that also the dual codes of the AG codes defined above exceed the Gilbert–Varshamov bound.

That algebraic geometry codes can also be decoded efficiently, which may very well make them of practical use in the near future, was demonstrated for the first time in [26].

## 5. Further reading

For the interested reader, quite a few books are available nowadays. A (very) brief discussion of them will be given here.

[1] tries to minimize the role of algebra by presenting the concepts in coding theory in terms of gates, shift registers and elementary linear algebra; well suited for undergraduate students in electrical engineering and computer science.

[2] contains a series of papers presented at a conference on Cryptography and Coding in 1986.

[4] is a thorough introduction to coding theory with an excellent chapter on finite fields; it also includes ingenious circuits showing how to implement various functions.

[5] contains a collection of (reprints of) key papers in coding theory.

[6] is a very well written introduction to coding theory, discussing also rather new developments as combined coding and modulation; does not assume a deep background in mathematics and is, as such, well suited for electrical engineers.

[7] contains a collection of 35 "benchmark papers" on various aspects of algebraic coding theory.

[8] assumes an introductory knowledge of modern algebra; it contains a chapter on group codes for the Gaussian channel.

[11] is based on a series of lectures attended by mainly design theorists to present the (for them) relevant developments in graph theory and coding theory.

[12] is a textbook on communications in general with chapters on coding theory and also cryptography.

[13] is about digital communication but with an emphasis on coding, including soft decision decoding and convolutional codes.

[18] is intended for graduate students in electrical engineering; it discusses various channels, some coding theory and also contains a chapter on source coding.

[21] is an undergraduate textbook, well suited for students in electrical engineering.

[24] is an elementary treatment of the theory of error-correcting codes; well suited for undergraduates in mathematics.

[28] discusses error-control coding in general, so also burst-correcting codes and convolutional codes; it includes a chapter on finite geometry codes.

[29] is very suited for a graduate course for students in mathematics, but [31] is a more up to date replacement for that.

[31] is well suited for a graduate course for students in mathematics; it contains a large chapter on the nonexistence of perfect codes.

[34] is both an introduction to coding theory and algebraic geometry.

[37] The Bible of algebraic coding theory with over 1000 references.

[38] is an excellent graduate textbook on information theory and coding theory.

[39] is well suited for a course in algebraic geometry ending with a discussion of algebraic geometry codes.

[40] is the second edition of one of the earliest books on coding theory; it also covers arithmetic, burst correcting, and convolutional codes, it contains valuable tables on cyclic codes and on irreducible/primitive polynomials.

[41] puts emphasis on the connections between coding theory and design theory and can be used for undergraduate students in mathematics.

[43] is about codes for computer memories; it includes a chapter on asymmetric and unidirectional codes.

[46] explains how the basic concepts and techniques of error control are applied to digital transmission and storage systems.

[49] is a textbook discussing topics like cyclic codes, convolutional codes, burst correcting codes, but also combined coding and modulation and some error-detection methods.

[51] is a textbook for coding theory, suited for students in mathematics, computer science and electrical engineering.

[55] gives a introduction to algebraic curves, discusses algebraic geometry codes and the relevant asymptotic results.

[56] gives a good introduction to linear codes; background in abstract algebra is not assumed; it emphasizes the practical considerations of efficient encoding and decoding.

[57] is based on a course on the mathematics of communication theory for undergraduate students; it covers error-correcting codes and cryptography (including a chapter on complexity theory).

## References


[1] B. Arazi, *A Commonsense Approach to the Theory of Error Correcting Codes*, MIT Press Series in Computer Systems, Cambridge, MA (1988).

[2] H.J. Beker and F.C. Piper, *Cryptography and Coding*, Clarendon Press, Oxford (1989).

[3] E.R. Berlekamp, *The enumeration of information symbols in BCH codes*, Bell System Tech. J. **46** (1967), 1853–1859.

[4] E.R. Berlekamp, *Algebraic Coding Theory*, McGraw-Hill, New York (1968).

[5] E.R. Berlekamp, *Key Papers in the Development of Coding Theory*, IEEE Press Selected Reprint Series, New York (1974).

[6] R.E. Blahut, *Theory and Practice of Error Control Codes*, Addison-Wesley, Reading, MA (1983).

[7] I.F. Blake, *Algebraic Coding Theory*, Benchmark Papers in Electrical Engineering and Computer Science, Hutchinson & Ross, Inc., Dowden (1973).

[8] I.F. Blake and R.C. Mullin, *The Mathematical Theory of Coding*, Academic Press, New York (1975).

[9] R.C. Bose and D.K. Ray-Chaudhuri, *On a class of error correcting binary group codes*, Inform. and Control. **3** (1960), 68–79.

[10] P. Bours, J.C.M. Janssen, M. van Asperdt and H.C.A. van Tilborg, *Algebraic decoding beyond* $e_{BCH}$, IEEE Trans. Inform. Theory **36** (1990), 214–222.

[11] P.J. Cameron and J.H. van Lint, *Designs, Graphs, Codes and Their Links*, London Math. Soc. Student Texts vol. 22, Cambridge Univ. Press, Cambridge (1991).

[12] W.G. Chambers, *Basics of Communications and Coding*, Clarendon Press, Oxford (1985).

[13] G.C. Clark, Jr. and J.B. Cain, *Error-Correction Coding for Digital Communications*, Plenum, New York (1981).

[14] J.T. Coffey and R.M. Goodman, *The complexity of information set decoding*, IEEE Trans. Inform. Theory **36** (1990), 1031–1037.

[15] M. Elia, *Algebraic decoding of the* $(23, 12, 7)$ *Golay code*, IEEE Trans. Inform. Theory **33** (1987), 150–151.

[16] G.S. Evseev, *Complexity of decoding for linear codes*, Problemy Peredachi Informatisii **19** (1983), 3–8.

[17] G.-L. Feng and K.K. Tzeng, *Decoding Cyclic and BCH codes up to actual minimum distance using nonrecurrent syndrome dependence relations*, IEEE Trans. Inform. Theory **37** (1991), 1716–1723.

[18] R.G. Gallager, *Information Theory and Reliable Communication*, Wiley, New York (1968).

[19] M.J.E. Golay, *Notes on digital coding*, Proc. IEEE **37** (1949), 657.

[20] V.D. Goppa, *A new class of linear error-correcting codes*, Problems Inform. Transmission **6** (1970), 207–212.

[21] R.W. Hamming, *Coding and Information Theory*, Prentice-Hall, Englewood Cliffs, NJ (1980).

[22] A.R. Hammons, P.V. Kumar, A.R. Calderbank, N.J.A. Sloane and P. Solé, *The* $\mathbb{Z}_4$*-linearity of Kerdock, Preparata, Goethals, and related codes*, IEEE Trans. Inform. Theory **40** (1994), 301–319.

[23] C.R.P. Hartmann and K.K. Tzeng, *Generalizations of the BCH bound*, Inform. and Contr. **20** (1972), 489–498.

[24] R. Hill, *A First Course in Coding Theory*, Clarendon Press, Oxford (1986).

[25] A. Hocquenghem, *Codes correcteurs d'erreurs*, Chiffres (Paris) **2** (1959), 147–156.

[26] J. Justesen, K.J. Larsen, H.E. Jensen, A. Havemose and T. Høholdt, *Construction and decoding of a class of algebraic geometry codes*, IEEE Trans. Inform. Theory **35** (1989), 811–821.

[27] A.M. Kerdock, *A class of low-rate non-linear binary codes*, Inform. and Contr. **20** (1972), 182–187.

[28] S. Lin and D.J. Costello, Jr., *Error Control Coding; Fundamentals and Applications*, Prentice-Hall, Englewood Cliffs, NJ (1983).

[29] J.H. van Lint, *Coding Theory*, SLNM 201, Springer, Berlin (1971).

[30] J.H. van Lint, *A survey of perfect codes*, Rocky Mountain J. Math. **5** (1975), 199–224.

[31] J.H. van Lint, *Introduction to Coding Theory*, Graduate Texts in Mathematics vol. 86, Springer, New York (1982).

[32] J.H. van Lint and R.M. Wilson, *On the minimum distance of cyclic codes*, IEEE Trans. Inform. Theory **32** (1986), 23–40.

[33] J.H. van Lint and T.A. Springer, *Generalized Reed–Solomon codes from algebraic geometry*, IEEE Trans. Inform. Theory **33** (1987), 305–309.

[34] J.H. van Lint and G. van der Geer, *Introduction to Coding Theory and Algebraic Geometry*, DMV Seminar Band 12, Birkhäuser, Basel (1988).

[35] J.H. van Lint, *Algebraic Geometry Codes*, Coding Theory and Designs, Part I, D. Ray-Chauduri, ed., IMA Volumes in Math. and its Appl. vol. 20, Springer, New York (1990), 137–162.

[36] F.J. MacWilliams, *A theorem on the distribution of weights in a systematic code*, Bell Syst. Tech. J. **42** (1963), 79–94.

[37] F.J. MacWilliams and N.J.A. Sloane, *The Theory of Error-Correcting Codes*, North-Holland, Amsterdam (1977).

[38] R.J. McEliece, *The Theory of Information and Coding*, Encyclopedia of Math. and its Applications vol. 3, Addison-Wesley, Reading, MA (1977).

[39] C.J. Moreno, *Algebraic Curves over Finite Fields*, Cambridge Tracts in Mathematics 97, Cambridge Univ. Press, Cambridge (1991).

[40] W.W. Peterson and E.J. Weldon, *Error-Correcting Codes*, 2nd ed., MIT Press, Cambridge, MA (1972).

[41] V. Pless, *Introduction to the Theory of Error-Correcting Codes*, 2nd ed., Wiley, New York (1989).

[42] F.P. Preparata, *A class of optimum non-linear double-error correcting codes*, Inform. and Control. **13** (1968), 378–400.

[43] T.R.N. Rao and E. Fujiwara, *Error-Control Coding for Computer Systems*, Prentice-Hall Series in Computer Engineering, Prentice-Hall, Englewood Cliffs, NJ (1989).

[44] I.S. Reed, X. Yin and T.K. Truong, *Algebraic decoding of the* $(32, 12, 8)$ *quadratic residue code*, IEEE Trans. Inform. Theory **36** (1990), 676–880.

[45] I.S. Reed, T.K. Truong, X. Chen and X. Yin, *The algebraic decoding of the* $(41, 21, 9)$ *quadratic residue code*, IEEE Trans. Inform. Theory **38** (1992), 974–986.

[46] M.Y. Rhee, *Error-Correcting Coding Theory*, McGraw-Hill, New York (1989).

[47] C. Roos, *A new lower bound for the minimum distance of a cyclic code*, IEEE Trans. Inform. Theory **29** (1983), 330–332.

[48] C.E. Shannon, *A mathematical theory of communication*, Bell Syst. Tech. J. **27** (1948), 379–423; 623–656.

[49] P. Sweeney, *Error Control Coding: An Introduction*, Prentice-Hall, New York (1991).

[50] H.C.A. van Tilborg, *On weights in codes*, Eindhoven University of Technology Report 71-WSK-03 (1971).

[51] H.C.A. van Tilborg, *Coding Theory, a First Course*, Chartwell Bratt Studentlitteratur, Lund, Sweden (1993).

[52] H. van Tilburg, *On the McEliece public-key cryptosystem*, Advances in Cryptography: Proc. of Crypto '88, S. Goldwasser, ed., Lecture Notes in Computer Science 403, Springer, Berlin (1989), 119–131.

[53] M.A. Tsfasman, *Goppa codes that are better than the Gilbert–Varshamov bound*, Problems Inform. Transmission **18** (1982), 163–165.

[54] M.A. Tsfasman, S.G. Vlăduţ and Z. Zink, *Modular curves, Shimura curves and Goppa codes better than the Varshamov–Gilbert bound*, Math. Nachr. **109** (1982), 21–28.

[55] M.A. Tsfasman and S.G. Vlăduţ, *Algebraic-Geometric Codes*, Kluwer, Dordrecht (1991).

[56] S.A. Vanstone and P.C. van Oorschot, *An Introduction to Error Correcting Codes with Applications*, Kluwer, Boston, MA (1989).

[57] D. Welsh, *Codes and Cryptography*, Clarendon Press, Oxford (1988).

# Section 1F
# Generalizations of Fields and Related Objects

# Semirings and Semifields

Udo Hebisch

*Inst. Theor. Math., TU Bergakademie Freiberg, 09596 Freiberg, Germany*
*e-mail: hebisch@koala.mathe.tu-freiberg.de*

Hanns Joachim Weinert

*Inst. Math., Erzstraße 1, 38678 Clausthal, Germany*

*Contents*



HANDBOOK OF ALGEBRA, VOL. 1
Edited by M. Hazewinkel

## 1. Basic concepts and examples

There are many concepts of universal algebras generalizing that of a ring $(R, +, \cdot)$. Among them are those called *semirings,* which originate from rings, roughly speaking, by cancelling the assumption that $(R, +)$ has to be a group. Depending on how much other ring-like properties are also cancelled or added, various different concepts of semirings $(S, +, \cdot)$ have been considered in the literature since 1934, when the first abstract concept of this kind was introduced by Vandiver [203]. (The list of papers given in our references is by no means complete, and we thank K. Głazek for his helpful collection [59].) Nowadays, semirings with different properties have become important in Theoretical Computer Science as we will see below. According to the following definition, we use the term semiring here in a rather general meaning and assume further properties explicitly if it is necessary or advisable to smooth our presentation.

DEFINITION 1.1. a) A universal algebra $S = (S, +, \cdot)$ with a nonempty set $S$ and two binary operations, written as addition and multiplication, is called a *semiring* if $(S, +)$ and $(S, \cdot)$ are arbitrary semigroups such that $a(b + c) = ab + ac$ and $(b + c)a = ba + ca$ hold for all $a, b, c \in S$. In particular, $(S, +, \cdot)$ is called a *proper semiring* if $(S, +)$ is not a group. The meaning of an *additively* or *multiplicatively commutative semiring* is clear, and $(S, +, \cdot)$ is called *commutative* if it has both properties.

b) We denote by $|A|$ the cardinality of a set $A$ and call $|S|$ the *order* of the semiring $(S, +, \cdot)$. A semiring $(S, +, \cdot)$ is called *trivial* if $|S| = 1$ holds.

c) If $(S, +)$ [or $(S, \cdot)$] has a neutral element, it is called the *zero* $o$ [the *identity* $e$] of the semiring $(S, +, \cdot)$. Notions as *left zero* and *left identity* are used accordingly. An element $O$ of a semiring $(S, +, \cdot)$ is called *multiplicatively absorbing* if $Oa = aO = O$ holds for all $a \in S$. (If such an element exists, it is unique and satisfies $O + O = O$.) Note that the zero $o$ of a semiring $(S, +, \cdot)$ need not be multiplicatively absorbing (cf. Example 1.9 c)). Otherwise we call $o$ briefly an *absorbing zero.*

REMARK 1.2. a) Except from some remarks, *we always consider semirings as universal* $(2, 2)$*-algebras,* and the class of all semirings clearly forms a variety. Consequently, *homomorphisms* (or morphisms) *and congruences of semirings* are determined (cf. Definition 6.1). Moreover, *each subalgebra of a semiring is a subsemiring, and any direct product of semirings is again a semiring* (cf. Remark 3.13).

b) Whereas (associative) rings are exactly those semirings $(S, +, \cdot)$ for which $(S, +)$ is a commutative group, there are also various semirings with a noncommutative group $(S, +)$. They have been investigated as distributive nearrings (cf. Remark 1.11) or "additively noncommutative rings" (cf. [81, 215] and [216]).

c) Some authors, e.g., [53, 125] and [142], use the term *hemiring* for additively commutative semiring $(S, +, \cdot)$. Sometimes, e.g., in [238, 26] and [197], also *halfring* is used if $(S, +)$ is commutative and cancellative and hence $(S, +, \cdot)$ embeddable into a ring (cf. Theorem 5.7). In both cases, assumptions concerning a zero of $(S, +, \cdot)$ may be added or not. Terms of this kind are hard to translate in other languages, which is one reason not to use them here.

d) Out of the scope of this chapter are *topological semirings*, defined by the assumption that both operations are continuous with respect to a (Hausdorff) topology (cf., e.g., [33, 35, 183, 154, 96] and [27]).

DEFINITION 1.3. a) An element $a$ of a semiring $(S, +, \cdot)$ is called *additively* [*multiplicatively*] *idempotent* if it satisfies $a + a = a$ [$aa = a$]. If all elements $a \in S$ have this property, $(S, +, \cdot)$ is called an *additively* [*multiplicatively*] *idempotent semiring*.

b) Let $(S, +, \cdot)$ be a semiring with a zero $o$ [an identity $e$]. If elements $a, b \in S$ satisfy $a + b = b + a = o$ [$ab = ba = e$], we write $b = -a$ [$b = a^{-1}$] and call $-a$ the *opposite element of* $a$ [$a^{-1}$ the *inverse of* $a$].

c) For each semiring $(S, +, \cdot)$ we introduce the notation

$$S^* = \begin{cases} S \setminus \{o\} & \text{if } (S, +, \cdot) \text{ has a zero } o \\ S & \text{otherwise.} \end{cases} \tag{1.1}$$

d) A semiring $(S, +, \cdot)$ with a zero $o$ is called *zero-sum free* if $a + b = o$ implies $a = b = o$ for all $a, b \in S$, equivalently, if $(S^*, +)$ is either empty or a subsemigroup of $(S, +)$.

e) A semiring $(S, +, \cdot)$ is called *semisubtractive* if for all $a, b \in S$ such that $a \neq b$ there is some $x \in S$ satisfying $a + x = b$ or $x + a = b$ or $b + x = a$ or $x + b = a$.

REMARK 1.4. We illustrate the various elementary statements on calculations with elements in semirings (cf. [165] or [89]) by the following one. If $(S, +, \cdot)$ has an absorbing zero $o$ and $a \in S$ has an opposite element $-a \in S$, then all elements $as$ and $sa$ for $s \in S$ have an opposite, namely $(-a)s = -(as)$ and $s(-a) = -(sa)$.

DEFINITION 1.5. A nontrivial semiring $(S, +, \cdot)$ is called a *semifield* if $(S, \cdot)$ is a group or $(S^*, \cdot)$ is a subgroup of $(S, \cdot)$, the latter clearly if $(S, +, \cdot)$ has a zero $o$.

REMARK 1.6. a) Whereas a zero of a semiring may also be its identity, one easily checks (cf., e.g., [89], I.5) that a nontrivial semiring such that $(S, \cdot)$ is a group has no zero. Hence, using (1.1), the two cases of the above definition can be combined: *a nontrivial semiring* $(S, +, \cdot)$ *is a semifield iff* $(S^*, \cdot)$ *is a subgroup of* $(S, \cdot)$.

b) Since the multiplication of a semifield is not assumed to be commutative, also the term *division semiring* is used instead of semifield (cf., e.g., [30, 193] and [196]).

c) Sometimes also semirings consisting of one single element are considered as semifields.

EXAMPLE 1.7. a) Clearly, each ring is a semiring and each (not necessarily commutative) field is a semifield. As usual, we denote by $(\mathbf{Z}, +, \cdot)$ the ring of integers and by $(\mathbf{Q}, +, \cdot)$ and $(\mathbf{R}, +, \cdot)$ the fields of rational and real numbers.

b) The positive as well as the non-negative integers form, again with the usual operations, semirings $(\mathbf{N}, +, \cdot)$ and $(\mathbf{N}_0, +, \cdot)$. Likewise we denote by $(\mathbf{H}, +, \cdot)$ and $(\mathbf{P}, +, \cdot)$ the semifields of positive rational and real numbers, and by $(\mathbf{H}_0, +, \cdot)$ and $(\mathbf{P}_0, +, \cdot)$ the corresponding semifields including the number 0.

c) Let $\mathbf{m}$ be a transfinite cardinal and $\mathbf{K}$ the set of all cardinals less or equal to $\mathbf{m}$. Then, with the usual operations of cardinals, $(\mathbf{K}, +, \cdot)$ is a commutative semiring containing

$(\mathbf{N}_0, +, \cdot)$ as a subsemiring. Note that each transfinite cardinal is idempotent with respect to both operations and that $\mathbf{m}$ is additively absorbing.

EXAMPLE 1.8. a) Each distributive lattice $(L, \cup, \cap) = (L, +, \cdot)$ (and likewise $(L, \cap, \cup) = (L, +, \cdot)$) is a semiring, clearly commutative and idempotent with respect to both operations. It has a zero or an identity iff it is bounded from below or above, respectively.

b) A special case is the semiring $(\mathbf{P}(X), \cup, \cap)$ of all subsets of a set $X$. For $|X| = 1$, it is a semifield consisting of an absorbing zero $o = \varnothing$ and an identity $e = X$ satisfying $e + e = e$. This semifield is important for various applications and mostly called the *Boolean semiring* $\mathbf{B}$ instead of *Boolean semifield.*

EXAMPLE 1.9. a) Define on $\mathbf{R}$ another addition by $a \oplus b = \min\{a, b\}$ with respect to the usual total order on $\mathbf{R}$ and consider the usual addition $a+b$ as multiplication $a \odot b = a+b$. Then $(\mathbf{R}, \oplus, \odot) = (\mathbf{R}, \min, +)$ is a semifield with 0 as identity. Adjoining an absorbing zero (cf. Lemma 3.1), denoted in this case as $\infty$, the semifield $(\mathbf{R} \cup \{\infty\}, \min, +)$ is one of the most important path algebras. It is used to deal with the problem of "shortest paths" in a finite directed valuated graph (cf. Definition 10.6).

b) Likewise one obtains the semifields $(\mathbf{R}, \max, +)$ and $(\mathbf{R} \cup \{-\infty\}, \max, +)$. The latter is the path algebra corresponding to the problem of "critical paths", and also called the schedule algebra.

c) Another semifield is obtained on the set $\mathbf{P}_0$ of non-negative real numbers by $(\mathbf{P}_0, \max, \cdot)$, where $\cdot$ denotes the usual multiplication. It has 1 as identity and 0 as absorbing zero. Here the subsemiring $([0, 1], \max, \cdot)$ is a path algebra corresponding to the problem of "paths of greatest reliability". Other interesting subsemirings are $([c, \infty), \max, \cdot)$ for each $c \geqslant 1$, forming obviously a chain of subsemirings. Each of it has $c$ as its special zero, an element which is cancellable in $([c, \infty), \cdot)$ and hence far away from being absorbing. In particular, the identity 1 of $([1, \infty), \max, \cdot)$ is at the same time the zero of this semiring. (For semirings with such a "double-neutral" cf. [84].)

EXAMPLE 1.10. a) Each semigroup $(S, \cdot)$ can be considered as a semiring $(S, +, \cdot)$ together with the left absorbing addition on $S$, defined by $a + b = a$ for all $a, b \in S$.

b) Likewise, each idempotent semigroup $(S, +)$ is part of a semiring $(S, +, \cdot)$, where the multiplication can be defined by $a \cdot b = a$ for all $a, b \in S$.

c) A semiring $(S, +, \cdot)$ is called a *mono-semiring* (cf. [240] and [84]) if addition and multiplication are the same operation. Obviously, mono-semirings correspond to normal (or distributive) semigroups $(S, \cdot)$ defined by $abc = abac$ and $bca = baca$ for all $a, b, c \in S$ (cf. [116, 79] and [100]).

We mention already here that additively commutative semirings arise in a natural way from endomorphisms of a commutative semigroup $(S, +)$, and that each semiring of this kind is isomorphic to such a subsemiring of endomorphisms (cf. Result 7.5 b)).

REMARK 1.11. a) If one demands in Definition 1.1 only one of both distributive laws, one obtains the concept of a (left or right distributive) *seminearring* $(S, +, \cdot)$ and correspondingly as in Definition 1.5 that of a *seminearfield* (cf., e.g., [206, 205, 217] and [220]). Seminearrings and seminearfields share many properties with semirings and semifields, and allow a common treatment of the latter and of *nearrings* and *nearfields*, defined by the assumption that $(S,+)$ is a group (cf. [155]).

b) There are even concepts of semi(near)rings in the literature (e.g., [238, 221] and [222]) such that multiplication or addition is not assumed to be associative.

c) Note that in papers dealing with projective planes sometimes another concept of "semifield" is used, which means in fact "nonassociative division ring" $(S,+,\cdot)$, i.e. $(S^*,\cdot)$ is a loop (cf., e.g., [97]).

## 2. Cancellativity and zero-divisors

DEFINITION 2.1. a) An element $a$ of a semiring $(S,+,\cdot)$ is called *additively* [*multiplicatively*] *left cancellable* if $a+b=a+c \Rightarrow b=c$ [$ab=ac \Rightarrow b=c$] holds for all $b,c \in S$, i.e. if $a$ is left cancellable in the semigroup $(S,+)$ [$(S,\cdot)$].

b) A semiring $(S,+,\cdot)$ is called *additively* [*multiplicatively*] *left cancellative* if each $a \in S$ [$a \in S^*$, cf. (1.1)] is additively [multiplicatively] left cancellable.

c) The corresponding concepts of *right cancellativity* and (two-sided) *cancellativity* are obvious.

RESULT 2.2. a) *If each element of a semiring $(S,+,\cdot)$ is additively cancellable from at least one side and if $(S,+,\cdot)$ has a zero $o$, then $o$ is absorbing.*

b) *Let $(S,+,\cdot)$ be additively cancellative. Then $ab+cd = cd+ab$ holds for all $a,b,c,d \in S$. Hence, if there is one element in $S$ which is multiplicatively cancellable from one side, e.g., a one-sided identity, then $(S,+)$ is commutative.*

DEFINITION 2.3. Let $(S,+,\cdot)$ be a semiring with a (not necessarily absorbing) zero $o$. An element $a \in S$ is called a *left zero-divisor of* $(S,+,\cdot)$ if $ab=o$ holds for some $b \neq o$ of $S$. Moreover, $(S,+,\cdot)$ is called *zero-divisor free* if it has no left (and hence no right) zero divisors different from $o$, equivalently, if $(S^*,\cdot)$ is either empty or a subsemigroup of $(S,\cdot)$.

Note that a nonabsorbing zero $o$ need neither be a left nor a right zero-divisor, and apply the next result also to the case where $o$ is also the identity of $(S,+,\cdot)$.

RESULT 2.4. *For a semiring $(S,+,\cdot)$ with a zero, a multiplicatively left cancellable element is not a left zero-divisor. In particular, a multiplicatively left cancellative semiring is zero-divisor free. The converse statements (well known to be true for rings) do not hold.*

REMARK 2.5. Concerning the latter, there are (even semisubtractive and additively commutative) semirings with an absorbing zero which are multiplicatively right cancellative and hence zero-divisor free, but not multiplicatively left cancellative. For example, $(\mathbf{P}_0, \max, \cdot)$ with the multiplication defined by $a \cdot b = a$ for all $a,b \in \mathbf{P}$ and $a \cdot 0 = 0 \cdot a = 0$ for all $a \in \mathbf{P}_0$ is such a semiring. However, suitable assumptions make the situation more ring-like (for b) and c) below cf. [86], §4 and §3):

RESULT 2.6. a) *Let* $(S, +, \cdot)$ *be a semiring with zero which is additively cancellative and semisubtractive. Then* $a \in S$ *is not a left zero-divisor iff it is multiplicatively left cancellable. Consequently, such a semiring is multiplicatively left cancellative iff it is zero-divisor free and hence iff it is multiplicatively right cancellative.*

b) *Let* $(S, +, \cdot)$ *have a zero and zero-sums. Then* $(S, +, \cdot)$ *is multiplicatively left cancellative iff it is multiplicatively right cancellative.*

c) *Let* $(S, +, \cdot)$ *be a finite semiring with an absorbing zero which is zero-divisor free. Then* $(S, +, \cdot)$ *is either zero-sum free or a ring (and hence a commutative field).*

The following theorem characterizes multiplicatively left, right and two-sided cancellative semirings and shows that there are nine possible classes for those semirings, each of which is in fact not empty (cf. [219] and [89], I.4). Typical examples for the two-sided class corresponding to b) are the semirings $([c, \infty), \max, \cdot)$ considered in Example 1.9. c).

THEOREM 2.7. *A nontrivial semiring* $(S, +, \cdot)$ *is multiplicatively (left) cancellative iff one of the following statements holds:*

a) $(S, +, \cdot)$ *has no zero, and* $(S, \cdot)$ *is (left) cancellative.*

b) $(S, +, \cdot)$ *has a zero, and* $(S, \cdot)$ *is (left) cancellative.*

c) $(S, +, \cdot)$ *has an absorbing zero, and* $(S^*, \cdot)$ *is a (left) cancellative subsemigroup of* $(S, \cdot)$.

## 3. Elementary extensions of semirings

By an *extension* $(T, +, \cdot)$ *of a semiring* $(S, +, \cdot)$ we mean a semiring $(T, +, \cdot)$ containing $(S, +, \cdot)$ as a subsemiring. At first we state that an absorbing zero can be adjoined to any semiring, and similarly a double-absorbing element $\infty$.

LEMMA 3.1. *Let* $(S, +, \cdot)$ *be a semiring and* $z \notin S$. *Extend the operations on* $S$ *to those on* $T = S \cup \{z\}$ *by* $z + a = a + z = a$ *and* $z \cdot a = a \cdot z = z$ *for all* $a \in T$. *Then* $(T, +, \cdot)$ *is an extension of* $(S, +, \cdot)$ *with* $z$ *as absorbing zero. Clearly,* $(T, +, \cdot)$ *is zero-sum free and zero-divisor free.*

REMARK 3.2. a) Omitting trivial statements concerning commutativity of the operations and the existence of an identity we only state that $(T, +, \cdot)$ is additively cancellative iff $(S, +, \cdot)$ is and has no zero $o_S$. Moreover, $(T, +, \cdot)$ is multiplicatively (left) cancellative iff $(S, +, \cdot)$ is and satisfies a) or b) of Theorem 2.7 or $|S| = 1$.

b) Clearly, Lemma 3.1 is extremely useful, and semirings with an absorbing zero are the most important ones. However, it does not make it superfluous to investigate also other semirings (cf., e.g., [132, 133, 161, 162] and [145]), in particular those ones with a zero which is not absorbing or only absorbing from one side.

LEMMA 3.3. *Let* $(S, +, \cdot)$ *be a semiring and* $\infty \notin S$. *Extend the operations on* $S$ *to those on* $T = S \cup \{\infty\}$ *by* $\infty + a = a + \infty = \infty$ *and* $\infty \cdot a = a \cdot \infty = \infty$ *for all* $a \in T$. *Then* $(T, +, \cdot)$ *is an extension of* $(S, +, \cdot)$ *for which the element* $\infty$ *is absorbing with respect to both operations.*

For several applications it is more important to have semirings with an absorbing zero $o$ and an element $\infty$ which is double-absorbing with the exception of $o \cdot \infty = \infty \cdot o = o$. Here we state (cf. [226], §5):

LEMMA 3.4. *Let $(S, +, \cdot)$ be a semiring with an absorbing zero $o$ and $\infty \notin S$. Extend the operations on $S$ to those on $T = S \cup \{\infty\}$ by*

$$\infty + a = a + \infty = \infty \quad \textit{for all } a \in T,$$

$$\infty \cdot a = a \cdot \infty = \infty \quad \textit{for all } a \in T \setminus \{o\}, \quad \textit{and} \quad \infty \cdot o = o \cdot \infty = o.$$

*Then $(T, +, \cdot)$ is a semiring and hence an extension of $(S, +, \cdot)$ as claimed above iff $(S, +, \cdot)$ is zero-sum free and zero-divisor free.*

Next we deal with the embedding of a semiring $(S, +, \cdot)$ into one with an identity. *If $(S, +, \cdot)$ is additively not commutative, this is not always possible.* Counter-examples, necessary and sufficient conditions and corresponding constructions are given in [66, 73] and [227]. However, for each semiring with commutative addition there exist extensions with an identity. This follows from Lemma 3.1 and the following construction which, in the case of rings, is due to Dorroh [48].

LEMMA 3.5. *Let $(S, +, \cdot)$ be an additively commutative semiring with an absorbing zero $o$ and $(\mathbf{N}_0, +, \cdot)$ the semiring of non-negative integers. Define on $D = \mathbf{N}_0 \times S$ addition and multiplication by $(n, a) + (m, b) = (n + m, a + b)$ and $(n, a) \cdot (m, b) = (nm, ma + nb + ab)$, where $ma$ means $\sum_{i=1}^{m} a$ for $m \in \mathbf{N}_0$ and $a \in S$. Then $(D, +, \cdot)$ is an additively commutative semiring with $(0, o)$ as absorbing zero and $(1, o)$ as identity, and $a \mapsto (0, a)$ defines an embedding of $(S, +, \cdot)$ into $(D, +, \cdot)$.*

REMARK 3.6. Let us call this semiring $D$ the Dorroh-extension of $S$. Corresponding to the same situation for rings (cf. [207]), the Dorroh-extension itself is in general not the right one to reduce statements on a semiring $S$ without an identity to those with an identity. However, for each minimal extension $T$ of $S$ with an identity $e_T$ there is an epimorphism $\varphi$: $(D, +, \cdot) \to (T, +, \cdot)$ leaving $S$ elementwise fixed, and a suitable epimorphic image $T$ of this kind may share more properties with $S$ than $D$ does. For example, if $S$ is multiplicatively cancellative, $D$ is in general not, whereas some (uniquely determined) $T$ has this property (cf. [194]). For more details also in the additively noncommutative case and the concept of the *characteristic of an arbitrary semiring* (involved in these questions) we refer to [64, 65] and [218].

Next we turn to matrices over a semiring, defined as in the case of rings:

LEMMA 3.7. *Let $(S, +, \cdot)$ be an additively commutative semiring and $M_{n,n}(S)$ the set of all $n \times n$-matrices over $S$. Then, provided with the usual operations, $(M_{n,n}(S), +, \cdot)$ is a semiring. For $|S| \geqslant 2$ and $n \geqslant 2$, the multiplication on $M_{n,n}(S)$ is neither left nor right cancellative and, apart from very restrictive assumptions on $(S, +, \cdot)$, not commutative either.*

REMARK 3.8. a) If $S$ is nontrivial and has an absorbing zero, the same holds for $M_{n,n}(S)$, and $M_{n,n}(S)$ has zero-divisors if $n \geqslant 2$. Under these assumptions, an identity of $S$ yields the usual identity $E$ of $M_{n,n}(S)$.

b) The assumption in Lemma 3.7 that $(S,+)$ is commutative avoids all difficulties concerning the associativity of the multiplication in $M_{n,n}(S)$. Moreover, define $\big(M_{n,n}(S),+,\cdot\big)$ as above over an arbitrary semiring $(S,+,\cdot)$ and assume that $S$ has an absorbing zero and an identity. Then $\big(M_{n,n}(S),\cdot\big)$ is associative for some $n \geqslant 2$ only if $(S,+)$ is commutative.

c) A hard to prove result is due to [168]: Let $S$ be a commutative semiring with absorbing zero and identity and $A, B \in M_{n,n}(S)$. Then $AB = E$ implies $BA = E$.

Now we consider polynomial semirings and other constructions of semirings.

DEFINITION 3.9. Let $S = (S,+,\cdot)$ be an additively commutative semiring with an absorbing zero $o$ and an identity $e \neq o$. Then an element $x$ of an extension $(T,+,\cdot)$ of $(S,+,\cdot)$ is called an *indeterminate over* $S$ if it has the following properties:

i) $ax = xa$ holds for all $a \in S$ as well as $ex = x$.

ii) $\sum_{\nu=0}^{n} a_\nu x^\nu = \sum_{\nu=0}^{n} b_\nu x^\nu$ for $a_\nu, b_\nu \in S$ implies $a_\nu = b_\nu$ for all $\nu = 0,\dots,n$.

For each $x$ of this kind, the subset

$$S[x] = \left\{ \sum_{\nu=0}^{n} a_\nu x^\nu \;\middle|\; a_\nu \in S,\ n \in \mathbf{N}_0 \right\} \subseteq T$$

forms a subsemiring $(S[x],+,\cdot)$ of $(T,+,\cdot)$ and an extension of $(S,+,\cdot)$, called the *polynomial semiring over* $S$ *in the indeterminate* $x$.

THEOREM 3.10. *For each semiring $S$ as above, such a polynomial semiring $S[x]$ exists and is, up to isomorphisms leaving $S$ elementwise fixed, uniquely determined.*

REMARK 3.11. a) From ii) it follows that $\sum_{\nu=0}^{n} c_\nu x^\nu = o$ always implies $c_\nu = o$ for all $\nu = 0,\dots,n$. For semirings, however, the latter does not imply ii).

b) The proof of Theorem 3.10 as well as further properties of polynomial semirings are more or less similar to the ring case (cf. [89], II.1). The same holds for polynomial semirings $S[x_1, x_2, \dots, x_n]$ or even $S[X]$ in a set $X = \{x_i \mid i \in I\} \neq \varnothing$ of independent indeterminates $x_i$ over $S$. Each polynomial semiring $S[X]$ can also be obtained as the semigroup semiring over $S$ of the free commutative semigroup with identity generated by $X$ (cf. Example 9.3).

REMARK 3.12. As known for universal algebras, each variety $\mathcal{V}$ of semirings contains for any set $X \neq \varnothing$ a *free $\mathcal{V}$-semiring* $(F_{\mathcal{V},X},+,\cdot)$ *over* $X$. For the variety $\mathcal{V}$ of all semirings, an elementary construction of these free $\mathcal{V}$-semirings was given in [67]. For commutative semirings with an absorbing zero and an identity, considered as a variety $\mathcal{V}$ of $(2,2,0,0)$-algebras, the free $\mathcal{V}$-semirings over $X = \{x_i \mid i \in I\}$ are just the polynomial semirings $S[X]$ considered above (cf. also [15]).

REMARK 3.13. As already mentioned, the *direct product of a family* $((S_i,+,\cdot))_{i\in I}$ *of arbitrary semirings* is again a semiring $(T,+,\cdot)$. It is defined in the usual way on the Cartesian product

$$T=\prod_{i\in I} S_i$$

by pointwise operations such that the projections $\pi_i\colon\ T\to S$ become epimorphisms $\pi_i\colon\ (T,+,\cdot)\to(S_i,+,\cdot)$. However, injective homomorphisms

$$\iota_i\colon\ (S_i,+,\cdot)\to(T,+,\cdot)$$

can be defined only in the case that each $(S_i,+,\cdot)$ has an element $s_i$ which is idempotent with respect to both operations, for instance an absorbing zero $o_i$. In the latter case, $(T,+,\cdot)$ can be considered as an extension of each $(S_i,+,\cdot)$.

REMARK 3.14. Sometimes it is useful to consider an *inflation* $(T,+,\cdot)$ *of a semiring* $(S,+,\cdot)$. According to the corresponding concept for semigroups (cf. [43], Chapter 3), associate to each $a\in S$ a set $T_a$ such that all sets $T_a$ and $S$ are mutually disjoint. If $T_a\neq\varnothing$, the elements of $T_a$ are called *shadows of* $a$. Extending the operations of $S$ to

$$T=S\cup\Big(\bigcup_{a\in S}T_a\Big)$$

by $a+b'=a'+b=a'+b'=a+b$ and $a\cdot b'=a'\cdot b=a'\cdot b'=a\cdot b$ for all $a,b\in S$, $a'\in T_a$ and $b'\in T_b$, one obtains the extension $(T,+,\cdot)$.

REMARK 3.15. Other constructions of semirings and semifields start with a set of disjoint semigroups $(S_\lambda,+)$ [or semirings or rings $(S_\lambda,+,\cdot)$] for $\lambda\in\Lambda$, where $(\Lambda,+,\cdot)$ is assumed to be an additively [and multiplicatively] idempotent semiring. Depending on further assumptions, various possibilities have been investigated to extend the given operations to those on

$$T=\bigcup_{\lambda\in\Lambda}S_\lambda$$

in such a way that $(T,+,\cdot)$ is a semiring and $a_\lambda\mapsto\lambda$ for all $a_\lambda\in S_\lambda\subseteq T$ defines an epimorphism of $(T,+,\cdot)$ onto $(\Lambda,+,\cdot)$ (cf., e.g., [210, 75, 57, 170, 58, 14, 15] and [159]).

## 4. More about semifields

REMARK 4.1. Besides the field of order 2 and the Boolean semifield (cf. Example 1.8b)) there are four other nonisomorphic semifields $(S,+,\cdot)$ of order 2 which have a zero. The latter are given by the tables:

| $+$ | $o$ | $e$ |
|---|---|---|
| $o$ | $o$ | $e$ |
| $e$ | $e$ | $e$ |

| $\cdot$ | $o$ | $e$ |
|---|---|---|
| $o$ | $o$ | $e$ |
| $e$ | $e$ | $e$ |

| $\cdot$ | $o$ | $e$ |
|---|---|---|
| $o$ | $e$ | $e$ |
| $e$ | $e$ | $e$ |

| $\cdot$ | $o$ | $e$ |
|---|---|---|
| $o$ | $o$ | $o$ |
| $e$ | $e$ | $e$ |

| $\cdot$ | $o$ | $e$ |
|---|---|---|
| $o$ | $o$ | $e$ |
| $e$ | $o$ | $e$ |

For each of these semifields, the identity $e$ of the group $(S^*, \cdot) = (\{e\}, \cdot)$ is neither the identity of $(S, +, \cdot)$ nor cancellable in $(S, \cdot)$, and the zero $o$ is not absorbing. Fortunately, things become much better if one assumes $|S^*| \geqslant 2$. However, these four exceptional semifields show that the following two field-like theorems are not trivial. In fact, some proofs (cf. [208] and [221]) are even somewhat sophisticated.

THEOREM 4.2. *Let $(S, +, \cdot)$ be a semifield such that $|S^*| \geqslant 2$. Then the identity $e$ of $(S^*, \cdot)$ is the identity of $(S, +, \cdot)$, and $(S, +, \cdot)$ is multiplicatively cancellative. If $(S, +, \cdot)$ has a zero $o$, it is absorbing and $(S, +, \cdot)$ is zero-divisor free.*

THEOREM 4.3. *Let $(S, +, \cdot)$ be a semiring such that $|S^*| \geqslant 2$. Then $(S, +, \cdot)$ is a semifield iff one of the following statements holds:*

a) *$(S, +, \cdot)$ has an identity $e$, and each $a \in S^*$ is invertible in $(S, \cdot)$.*

b) *$(S, +, \cdot)$ has a left identity $e_l$ such that for each $a \in S^*$ there is some $y \in S$ satisfying $ya = e_l$.*

c) *For all $a \in S^*$ and $b \in S$ there are some $x, y \in S$ satisfying $ax = b$ and $ya = b$.*

d) *For all $a \in S^*$ and $b \in S$ there is some $x \in S$ satisfying $ax = b$, and $S$ has a right identity. (The latter can be replaced by the existence of a unique multiplicative idempotent in $S^*$.)*

THEOREM 4.4. *Let $(S, +, \cdot)$ be a semifield with a zero $o$ and $|S^*| \geqslant 2$. Then $(S, +, \cdot)$ is either a field or zero-sum free and hence $(S^*, +, \cdot)$ a subsemifield of $(S, +, \cdot)$. In the latter case, $(S, +, \cdot)$ is obtained from $(S^*, +, \cdot)$ by adjoining an absorbing zero.*

REMARK 4.5. Consequently, excluding the six semifields of order 2, each proper semifield occurs in two corresponding versions, one without a zero and one with an absorbing zero. Hence investigations on proper semifields can be done considering only those with or only those without a zero. In the latter case one investigates all semifields $(S, +, \cdot)$ such that $(S, \cdot)$ is a group. Clearly, these algebras form a variety of $(2, 2, 0, 1)$-algebras, provided that one includes the semirings of order 1 in the definition of semifields (cf. [99, 201] and [228]). The following both theorems are due to [208].

THEOREM 4.6. a) *Let $(S, +, \cdot)$ be a proper semifield with commutative addition such that $|S^*| \geqslant 2$. Then the group $(S^*, \cdot)$ is torsion free and hence $(S, +, \cdot)$ is of infinite order, whereas the semigroup $(S, +)$ is uniquely divisible.*

b) *The finite semifields with commutative addition are the Galois-fields, the Boolean semifield and the four semifields of order 2 described in Remark* 4.1.

THEOREM 4.7. *Each additively commutative and idempotent semifield $(S, +, \cdot)$ without a zero is a lattice ordered group $(S, \cdot, \leqslant)$, and conversely.*

Note in this context, that $a \leqslant b \Leftrightarrow a + b = b$ defines a partial order on $S$ such that $\sup\{a, b\} = a \vee b = a + b$ and $\inf\{a, b\} = a \wedge b = \left(a^{-1} + b^{-1}\right)^{-1}$ exist for all $a, b \in S$.

Conversely, $a+b=a\vee b$ defines an addition for $(S,\cdot,\leqslant)$ such that the distributive laws are satisfied.

REMARK 4.8. a) There are various semifields $(S,+,\cdot)$ such that $(S,+)$ is idempotent, but not commutative. For instance, the direct product $(G_1,\cdot)\times(G_2,\cdot)$ of two groups (at least one nontrivial) provides such a semifield defining $(a_1,a_2)+(b_1,b_2)=(a_1,b_2)$. In fact, *all finite semifields with noncommutative addition and without a zero* (cf. Theorem 4.4) *are obtained in this way* (cf. [208] and [210]).

b) For other detailed investigations on semifields, including objects such as algebraic or transcendental (simple) semifield extensions, we refer to [211, 118, 55, 220] and [99]. Here we only mention the following characterization of all subsemifields of an algebraic number field.

THEOREM 4.9. *Let $(K,+,\cdot)$ be an algebraic number field and $(S,+,\cdot)$ a proper subsemifield containing the zero 0 of $K$. Let $(K',+,\cdot)$ be the smallest subfield of $K$ which contains $S$. Then there exist finitely many subsemifields $T_1,\ldots,T_r$ of $K'$ which are semisubtractive, hence determined as the positive cones of all total orders on $(K',+,\cdot)$, and the $2^r-1$ intersections*

$$T_1,\ldots,T_r,\ T_1\cap T_2,\ldots,T_{r-1}\cap T_r,\ldots,T_1\cap T_2\cap\cdots\cap T_r$$

*are pairwise distinct subsemifields of $K'$ and $S$ is one of them. Moreover, $S$ is a simple semifield extension of $\mathbf{H}_0$.*

## 5. Extensions of semirings by quotients and differences

Recall that a *semigroup of right quotients (briefly a $Q_r$-semigroup) $(T,\cdot)=Q_r(S,\Sigma)$ of a semigroup $(S,\cdot)$ with respect to a subsemigroup $\Sigma$* of $(S,\cdot)$ is defined as follows: $(T,\cdot)$ contains $(S,\cdot)$ as a subsemigroup and has an identity $e_T$, each $\alpha\in\Sigma$ has an inverse $\alpha^{-1}\in T$, and the subset $\{a\alpha^{-1}\mid a\in S,\ \alpha\in\Sigma\}\subseteq T$ coincides with $T$. (Note that $e_T=e_S$ holds if $S$ has an identity $e_S$.) *Given $(S,\cdot)$ and $\Sigma$, such a semigroup $(T,\cdot)=Q_r(S,\Sigma)$ exists iff each $\alpha\in\Sigma$ is cancellable in $(S,\cdot)$ and*

$$\alpha S\cap a\Sigma\neq\varnothing \quad \text{holds for all } \alpha\in\Sigma,\ a\in S. \tag{5.1}$$

If this is the case, $(T,\cdot)$ is completely described by the rules

$$\begin{aligned} a\alpha^{-1}=b\beta^{-1} &\Leftrightarrow \alpha x=\beta\xi \text{ and } ax=b\xi \text{ for some } (x,\xi)\in S\times\Sigma\\ &\Leftrightarrow \alpha u=\beta v \text{ implies } au=bv \text{ for all } (u,v)\in S\times S, \end{aligned} \tag{5.2}$$

$$\begin{aligned} a\alpha^{-1}\cdot b\beta^{-1}=(ax)(\beta\xi)^{-1} \quad &\text{for any } (x,\xi)\in S\times\Sigma\\ &\text{satisfying } \alpha x=b\xi. \end{aligned} \tag{5.3}$$

*This yields that a $Q_r$-semigroup $(T,\cdot)$ of $(S,\cdot)$ with respect to $\Sigma$ is, up to isomorphisms leaving $S$ elementwise fixed, uniquely determined by $(S,\cdot)$ and $\Sigma$,* which justifies the notation $(T,\cdot) = Q_r(S,\Sigma)$. (For these results and most of the following ones we refer to [144, 209, 212] and [222].) Clearly, (5.1) holds trivially if $(S,\cdot)$ is commutative (or at least $\Sigma$ is in the center of $(S,\cdot)$), and in these cases (5.2) and (5.3) turn into the usual rules on fractions.

DEFINITION 5.1. An extension $(T,+,\cdot)$ of a semiring $(S,+,\cdot)$ is called a *$Q_r$-semiring $(T,+,\cdot) = Q_r(S,\Sigma)$ of $(S,+,\cdot)$ with respect to a subsemigroup $\Sigma$ of $(S,\cdot)$* if $(T,\cdot)$ is a $Q_r$-semigroup of $(S,\cdot)$ with respect to $\Sigma$.

THEOREM 5.2. *Let $(T,+,\cdot) = Q_r(S,\Sigma)$ be a $Q_r$-semiring of a semiring $(S,+,\cdot)$. Then the addition on $T$ is uniquely determined by the addition on $S$ according to*

$$a\alpha^{-1} + b\beta^{-1} = (ax + b\xi)(\beta\xi)^{-1} \quad \text{for any } (x,\xi) \in S \times \Sigma \text{ satisfying } \alpha x = \beta\xi. \tag{5.4}$$

*Conversely, let $(S,+,\cdot)$ be a semiring and $(T,\cdot) = Q_r(S,\Sigma)$ a $Q_r$-semigroup of $(S,\cdot)$, then (5.4) defines an addition on $T$ such that $(T,+,\cdot)$ is an extension of $(S,+,\cdot)$. Hence a $Q_r$-semiring $(T,+,\cdot) = Q_r(S,\Sigma)$ of $(S,+,\cdot)$ exists iff each element $\alpha \in \Sigma$ is cancellable in $(S,\cdot)$ and (5.1) holds, and $(T,+,\cdot)$ is then, up to isomorphisms leaving $S$ elementwise fixed, uniquely determined by $(S,+,\cdot)$ and $\Sigma$.*

REMARK 5.3. The following statements on $Q_r$-semirings are essentially those on $Q_r$-semigroups:

a) If $(S,+,\cdot)$ is multiplicatively left or right cancellative, the same holds for each $Q_r$-semiring $(T,+,\cdot)$ of $(S,+,\cdot)$.

b) If the $Q_r$-semiring $(T,+,\cdot) = Q_r(S,\Sigma)$ of $(S,+,\cdot)$ exists and the left-right dual of (5.1) is also true, then $(T,+,\cdot) = Q_r(S,\Sigma)$ is also a $Q_l$-semiring of $(S,+,\cdot)$. In this case we write $(T,+,\cdot) = Q(S,\Sigma)$ and call it a *$Q$-semiring of $(S,+,\cdot)$*.

c) There may be various subsemigroups $\Sigma_i$ of $(S,\cdot)$ such that $(T,+,\cdot) = Q_r(S,\Sigma_i)$ holds. In this case there is a unique greatest one among these $\Sigma_i$.

d) If $(S,+,\cdot)$ has a $Q_r$-semiring, then there is a subsemigroup $\Sigma_m$ of $(S,\cdot)$ such that $(T_m,+,\cdot) = Q_r(S,\Sigma_m)$ exists and $\Sigma \subseteq \Sigma_m$ holds for each $\Sigma$ yielding a $Q_r$-semiring $(T,+,\cdot) = Q_r(S,\Sigma)$. Clearly, $(T_m,+,\cdot)$ contains (an isomorphic copy of) each $Q_r$-semiring of $(S,+,\cdot)$. If $(S,\cdot)$ is commutative, $\Sigma_m$ consists of all cancellable elements of $(S,\cdot)$.

DEFINITION 5.4. The $Q_r$-semiring $(T_m,+,\cdot) = Q_r(S,\Sigma_m)$ of $(S,+,\cdot)$ just described is called *the maximal $Q_r$-semiring of* $(S,+,\cdot)$ and denoted by $Q_r(S)$. In the following, we also write $(T,+,\cdot) = Q_r(S,\Sigma_m) = Q_r(S)$. In particular, we call it *the $Q_r$-semifield of* $(S,+,\cdot)$ if it happens to be a semifield.

THEOREM 5.5. *Let $(S,+,\cdot)$ be a nontrivial semiring which is multiplicatively commutative. Then $(S,+,\cdot)$ is embeddable into a semifield iff it is multiplicatively cancellative. In this case, the $Q$-semifield $(T,+,\cdot) = Q(S) = Q(S,\Sigma_m)$ is, unique up to isomorphisms leaving $S$ elementwise fixed, the smallest semifield-extension of $(S,+,\cdot)$.*

Note that $\Sigma_m = S^*$ holds iff $(S, +, \cdot)$ has an absorbing zero, and $\Sigma_m = S$ in the other cases of Theorem 2.7. In particular, each semiring $(S, +, \cdot) = ([c, \infty), \max, \cdot)$ with $c$ as its zero (cf. Example 1.9 c)) has the $Q$-semifield $(\mathbf{P}, \max, \cdot) = Q(S)$ which has no zero.

THEOREM 5.6. *Let $(T, +, \cdot) = Q_r(S, \Sigma)$ be a $Q_r$-semiring of $(S, +, \cdot)$.*

a) *If $(S, +)$ is commutative, left [right] cancellative, idempotent or a group, $(T, +)$ has the same property. In particular, if $(S, +, \cdot)$ is a ring, then $(T, +, \cdot) = Q_r(S, \Sigma)$ is the ring of right quotients of $(S, +, \cdot)$ with respect to $\Sigma$ in the usual meaning.*

b) *$(T, +, \cdot)$ has a zero $o_T$ iff $(S, +, \cdot)$ has a zero $o_S$ which satisfies $o_S\alpha = o_S$ for all $\alpha \in \Sigma$, and in this case $o_T = o_S\alpha^{-1} = o_S$ holds for all $\alpha \in \Sigma$.*

We apply statements on $Q_r$-semigroups to the case that $(S, +)$ is a commutative semigroup and $\Theta$ a subsemigroup of $S$ such that each $u \in \Theta$ is cancellable in $(S, +)$. Then there exists, unique up to isomorphisms leaving $S$ elementwise fixed, a *semigroup of differences (briefly a D-semigroup) $(T, +) = D(S, \Theta)$ of $(S, +)$ with respect to $\Theta$*. It consists of all differences $a - u$ for $a \in S$ and $u \in \Theta$ and is determined by $a - u = b - v \Leftrightarrow a + v = b + u$ and $(a - u) + (b - v) = (a + b) - (u + v)$. For investigations on the corresponding concept of a *D-semiring $(T, +, \cdot) = D(S, \Theta)$ of a semiring $(S, +, \cdot)$ with respect to a subsemigroup $\Theta$ of $(S, +)$* we refer to [209] and [89]. Here we restrict ourselves to the case $\Theta = S$:

THEOREM 5.7. *A semiring $(S, +, \cdot)$ is embeddable into a ring iff $(S, +)$ is commutative and cancellative. In this case, the minimal ring-extension $(R, +, \cdot)$ of $(S, +, \cdot)$ is, uniquely up to isomorphisms leaving $S$ elementwise fixed, given by the D-semigroup $(R, +) = D(S, S)$ of $(S, +)$, established with the multiplication*

$$(a - b) \cdot (c - d) = (ac + bd) - (ad + bc).$$

DEFINITION 5.8. The ring $(R, +, \cdot)$ just described is called the *ring of differences* or the *D-ring of the semiring* $(S, +, \cdot)$ and denoted by $(R, +, \cdot) = D(S, S) = D(S)$.

THEOREM 5.9. *Let $(R, +, \cdot) = D(S)$ be the D-ring of a semiring $(S, +, \cdot)$. Then each $a \in S$ which is multiplicatively (left) cancellable in $(S, +, \cdot)$ has the same property in $(R, +, \cdot)$. However, if $(S, +, \cdot)$ is multiplicatively (left) cancellative or even a semifield, $(R, +, \cdot) = D(S)$ need neither be multiplicatively (left) cancellative nor, in the second case, a semifield and hence a field. The first property transfers from $(S, +, \cdot)$ to $(R, +, \cdot)$ iff $a \neq b$ and $c \neq d$ implies $ad + bc \neq ac + bd$ for all $a, b, c, d \in S$. A sufficient condition such that both properties transfer from $(S, +, \cdot)$ to $(R, +, \cdot)$ is that $(S, +, \cdot)$ is semisubtractive.*

EXAMPLE 5.10. To illustrate the negative statements of Theorem 5.9, we consider the residue class ring $\bar{R} = \mathbf{Q}[x]/(x^2)$ of the polynomial ring $\mathbf{Q}[x]$ and denote its elements by $a + bx$. Then $T = \{a + bx \mid a > 0\}$ is a subsemiring of $(\bar{R}, +, \cdot)$ satisfying $\bar{R} = D(T)$. Since each $a + bx \in T$ has $a^{-1} - ba^{-2}x \in T$ as its inverse, $(T, +, \cdot)$ is a semifield and hence multiplicatively cancellative, whereas $\bar{R}$ is not even multiplicatively cancellative.

REMARK 5.11. For the semiring $(\mathbf{N},+,\cdot)$, one clearly obtains $(\mathbf{H},+,\cdot) = Q(\mathbf{N})$ as its $Q$-semifield and $(\mathbf{Z},+,\cdot) = D(\mathbf{N})$ as its $D$-ring. As two second steps, the $D$-ring $D(\mathbf{H}) = D(Q(\mathbf{N}))$ turns out to be a field which contains $D(\mathbf{N}) = \mathbf{Z}$, and the $Q$-semifield $Q(\mathbf{Z}) = Q(D(\mathbf{N}))$ is a field containing $Q(\mathbf{N}) = \mathbf{H}$, two ways to obtain the field $(\mathbf{Q},+,\cdot) = D(Q(\mathbf{N})) = Q(D(\mathbf{N}))$. The first one is more appropriate in elementary school education (mankind has calculated in **N** and **H** thousands of years before inventing negative numbers), whereas the second way is preferred at universities. For this reason the fact that $D(Q(\mathbf{N})) = Q(D(\mathbf{N}))$ holds, sometimes considered to be self-evident, is of some interest for training teachers. However, $D(Q(S)) = Q(D(S))$ need not be true if one replaces $(\mathbf{N},+,\cdot)$ by a commutative semiring $(S,+,\cdot)$ for which the $Q$-semifield $(T,+,\cdot) = Q(S) = Q(S,\Sigma_m)$ and the $D$-ring $(R,+,\cdot) = D(S)$ exist (a sufficient condition for $D(Q(S)) = Q(D(S))$ is that $(S,+,\cdot)$ is semisubtractive). In general, the situation is more complicated:

$$\begin{array}{ccccc} & & Q(R) = Q(\bar{R}) = \bar{\bar{R}} & & \\ & & / & & \\ & & Q(R,\Sigma_m) = \bar{R} = D(T) & & \\ & / & & \backslash & \\ R = D(S) = D(S,S) & & & & T = Q(S) = Q(S,\Sigma_m) \\ & \backslash & & / & \\ & & S & & \end{array}$$

Since $(T,+,\cdot) = Q(S)$ is additively commutative and cancellative, the $D$-ring $(\bar{R},+,\cdot) = D(Q(S))$ exists and is also the $Q$-(semi)ring $Q(D(S),\Sigma_m)$ of $(R,+,\cdot) = D(S)$ with respect to the subsemigroup $\Sigma_m = \Sigma_m(S)$ of $(S,\cdot)$, cf. the first statement of Theorem 5.9. However, the subsemigroup $\Sigma_m(R)$ of all elements $a - b \in R$ which are cancellable in $(R,\cdot)$ may contain $\Sigma_m(S)$ properly. Even if this is the case, the maximal $Q$-ring $(\bar{\bar{R}},+,\cdot) = Q(R) = Q(R,\Sigma_m(R))$ of $(R,+,\cdot)$ may either coincide with $(\bar{R},+,\cdot)$ or contain it properly, and $\bar{R}$ or $\bar{\bar{R}}$ may be fields or not (cf. [209] and [89], II.6, also for the following examples).

EXAMPLE 5.12. a) According to Example 5.10, the ring $R = \mathbf{Z}[x]/(x^2)$ is the $D$-ring of its subsemifield $S = \{a + bx \mid a > 0,\ b \geqslant 0\}$. Then the semifield $T$ considered in Example 5.10 is the $Q$-semifield $T = Q(S)$, and $\bar{R} = D(T) = \mathbf{Q}[x]/(x^2)$ coincides with the $Q$-ring $Q(R,\Sigma_m(S))$. In this case we have $\Sigma_m(S) \subset \Sigma_m(R)$, but $\bar{\bar{R}} = Q(R) = Q(R,\Sigma_m(R))$ coincides with $\bar{R}$.

b) Let $S = \mathbf{N}_0[x]$ be a polynomial semiring. Then the $Q$-semifield $T = Q(S)$ and the $D$-ring $R = \mathbf{Z}[x] = D(S)$ exist, and $\bar{R} = D(T) = Q(R,\Sigma_m(S))$ holds, where $\Sigma_m(S)$ consists of all polynomials $f(x) \neq 0$ of $\mathbf{N}_0[x]$. However, the $Q$-field $\bar{\bar{R}} = \mathbf{Q}(x) = Q(R) = Q(\bar{R})$ contains $\bar{R}$ properly, such that $\bar{R}$ is merely a ring. Replacing $x$ by any transcendental real number $\tau$, these considerations take place within the totally ordered field $(\mathbf{R},+,\cdot,\leqslant)$.

## 6. Congruences, ideals and radicals

DEFINITION 6.1. Let $(S,+,\cdot)$ be a semiring and $(T,+,\cdot)$ a $(2,2)$-algebra.

a) According to Remark 1.2 a), a mapping $\varphi\colon S \to T$ is called a *homomorphism of* $(S,+,\cdot)$ *into* $(T,+,\cdot)$, briefly denoted by $\varphi\colon (S,+,\cdot) \to (T,+,\cdot)$, if $\varphi$ satisfies

$$\varphi(a+b) = \varphi(a) + \varphi(b) \text{ and } \varphi(a\cdot b) = \varphi(a)\cdot\varphi(b) \quad \text{for all } a,b \in S. \tag{6.1}$$

b) Let $\kappa \subseteq S \times S$ be an equivalence on $S$. Instead of $(a,a') \in \kappa$ we write $a\ \kappa\ a'$, and $\kappa$ is called a *congruence on* $(S,+,\cdot)$ if, for all $a,a',b,b' \in S$,

$$a\ \kappa\ a' \text{ and } b\ \kappa\ b' \quad \text{imply} \quad (a+b)\ \kappa\ (a'+b') \text{ and } (a\cdot b)\ \kappa\ (a'\cdot b'). \tag{6.2}$$

We denote by $S/\kappa$ *the set of all* $\kappa$*-classes* $[a]_\kappa = \{a' \in S \mid a'\ \kappa\ a\}$ *of* $S$, by $\kappa^\#\colon S \to S/\kappa$ *the natural mapping* according to $\kappa^\#(a) = [a]_\kappa$, and by $\mathcal{C}_{(S,+,\cdot)}$ the set of all congruences on $(S,+,\cdot)$.

REMARK 6.2. a) The homomorphic image $\varphi(S) \subseteq T$ of $(S,+,\cdot)$ is a subsemiring $(\varphi(S),+,\cdot)$ of $(T,+,\cdot)$ and shares all properties with $(S,+,\cdot)$ which are defined by equations. Moreover, if $(S,+,\cdot)$ has an (absorbing) zero $o$, then $\varphi(o)$ is an (absorbing) zero of $(\varphi(S),+,\cdot)$, and likewise for an identity $e$ of $(S,+,\cdot)$.

b) Note that "$(a\ \kappa\ a')$ implies $(a+b)\ \kappa\ (a'+b)$ and $(b+a)\ \kappa\ (b+a')$" is equivalent to the additive part of (6.2). Further, the congruences on $(S,+,\cdot)$ form a complete lattice $(\mathcal{C}_{(S,+,\cdot)},\cup,\cap)$, where the identical relation $\iota_S$ on $S$ is the smallest and $S\times S$ the greatest element of $\mathcal{C}_{(S,+,\cdot)}$. Also $\mathcal{C}_{(S,+,\cdot)} = \mathcal{C}_{(S,+)} \cap \mathcal{C}_{(S,\cdot)}$ holds in an obvious interpretation.

c) If $(U,+,\cdot)$ is a subsemiring of $(S,+,\cdot)$, then each congruence $\kappa$ on $(S,+,\cdot)$ induces a congruence $\kappa \cap (U\times U)$ on $(U,+,\cdot)$.

d) The semiring $(\mathbf{N}_0,+,\cdot)$ satisfies $\mathcal{C}_{(\mathbf{N}_0,+,\cdot)} = \mathcal{C}_{(\mathbf{N}_0,+)}$. Hence each congruence $\kappa \neq \iota_{\mathbf{N}_0}$ is determined by a pair $(v,g) \in \mathbf{N}_0 \times \mathbf{N}$ such that $a\ \kappa\ a'$ holds iff either $a = a'$ or $a \equiv a'$ modulo $g$ for $a,a' \geqslant v$, and conversely.

THEOREM 6.3. a) *Let* $\kappa$ *be a congruence on a semiring* $(S,+,\cdot)$. *Then*

$$[a]_\kappa + [b]_\kappa = [a+b]_\kappa \quad \textit{and} \quad [a]_\kappa \cdot [b]_\kappa = [a\cdot b]_\kappa \tag{6.3}$$

*define operations on* $S/\kappa$ *such that* $\kappa^\#\colon (S,+,\cdot) \to (S/\kappa,+,\cdot)$ *is an epimorphism and hence* $(S/\kappa,+,\cdot)$ *a semiring, the congruence class semiring of* $(S,+,\cdot)$ *by* $\kappa$.

b) *Conversely, let* $(S,+,\cdot)$ *be a semiring,* $\varphi\colon (S,+,\cdot) \to (T,+,\cdot)$ *a homomorphism, and* $(\varphi(S),+,\cdot)$ *the homomorphic image. Then* $\kappa = \kappa_\varphi$ *defined by* $a\ \kappa\ a' \Leftrightarrow \varphi(a) = \varphi(a')$ *is a congruence on* $(S,+,\cdot)$, *and there is a unique isomorphism*

$$\psi\colon (S/\kappa,+,\cdot) \to (\varphi(S),+,\cdot)$$

*such that* $\varphi = \iota \circ \psi \circ \kappa^\#$ *holds, where* $\iota$ *is the identical embedding of* $\varphi(S)$ *into* $T$.

c) *For* $\kappa_1,\kappa_2 \in \mathcal{C}_{(S,+,\cdot)}$ *there is a homomorphism*

$$\psi\colon (S/\kappa_1,+,\cdot) \to (S/\kappa_2,+,\cdot)$$

*satisfying* $\psi \circ \kappa_1^\# = \kappa_2^\#$ *iff* $\kappa_1 \subseteq \kappa_2$ *holds.*

DEFINITION 6.4. Let $(S, +, \cdot)$ be a semiring. Then a subsemigroup $L$ of $(S, +)$ is called a *left ideal of* $(S, +, \cdot)$ if $sa \in L$ holds for all $s \in S$ and $a \in L$. If $A$ is a left and a right ideal of $(S, +, \cdot)$, it is called a (two-sided) *ideal of* $(S, +, \cdot)$.

REMARK 6.5. a) Each semiring $(S, +, \cdot)$ has $S$ as an ideal, and if there is an ideal $A$ satisfying $|A| = 1$, it is unique and $A = \{O\}$ consists of the multiplicatively absorbing element $O$ of $(S, +, \cdot)$. (Note that $O = o$ holds if there is an absorbing zero $o$.) If $O$ exists, a left ideal $L$ is called *O-minimal* if $\{O\} \subset L$ holds and $\{O\} \subset L' \subseteq L$ implies $L' = L$ for each left ideal $L'$. In general, a left ideal $L$ of $S$ is called *minimal* if $L' \subseteq L$ implies $L' = L$ for each left ideal $L'$ of $S$.

b) Each ideal of $(\mathbf{N}, +, \cdot)$ or $(\mathbf{N}_0, +, \cdot)$ can be generated by a finite set $\{a_1, \dots, a_n\}$, but the number $n$ of elements needed for this purpose is not limited (cf. [12, 146] and [89], I.8).

DEFINITION 6.6. a) Let $A$ be an ideal of an additively commutative semiring $(S, +, \cdot)$. Then $\bar{A} = \{\bar{a} \in S \mid \bar{a} + a \in A$ holds for some $a \in A\}$ defines an ideal of $(S, +, \cdot)$ satisfying $A \subseteq \bar{A}$ and $\bar{\bar{A}} = \bar{A}$, called the *k-closure of A*. In particular, if $A = \bar{A}$ holds, $A$ is called a *k-ideal of* $(S, +, \cdot)$ or *k-closed.*

b) Likewise, $\hat{A} = \{\hat{a} \in S \mid \hat{a} + a + s \in A + s$ for some $a \in A$ and $s \in S\}$ defines the *h-closure of* $A$ and $A = \hat{A}$ an *h-ideal* of $(S, +, \cdot)$. Clearly, $A \subseteq \bar{A} \subseteq \hat{A}$ holds for each ideal $A$ of $(S, +, \cdot)$.

REMARK 6.7. a) For these and the following concepts cf. [204, 32, 94] and [101].

b) Note that $\bar{A}$ and $\hat{A}$ can be defined in the same way for each subsemigroup $A$ of a semigroup $(S, +)$. Hence Definition 6.6 applies also to left ideals.

c) Let $(S, +, \cdot)$ be a ring and $A$ an ideal of $(S, +, \cdot)$ considered as a semiring, i.e. in the meaning of Definition 6.4. Then $A$ is an ideal of $(S, +, \cdot)$ in the ring-theoretical meaning iff $A$ is $k$-closed.

THEOREM 6.8. *Let $(S, +, \cdot)$ be an additively commutative semiring.*

a) *Each ideal $A$ of $(S, +, \cdot)$ defines a congruence $\kappa_A$ on $(S, +, \cdot)$ by*

$$x \,\kappa_A\, x' \Leftrightarrow x + a_1 = x' + a_2 \text{ for some } a_i \in A, \tag{6.4}$$

*for which the k-closure $\bar{A}$ of $A$ is one congruence class, i.e. $\bar{A} = [a]_{\kappa_A}$ holds for any $a \in A$. This class is the absorbing zero of the semiring $(S/\kappa_A, +, \cdot)$. Moreover, $\kappa_A$ coincides with $\kappa_{\bar{A}}$, whereas $\kappa_{\bar{A}} = \kappa_{\bar{B}}$ holds iff the k-ideals $\bar{A}$ and $\bar{B}$ are equal.*

b) *Each ideal $A$ of $(S, +, \cdot)$ defines a congruence $\eta_A$ on $(S, +, \cdot)$ by*

$$x \,\eta_A\, x' \Leftrightarrow x + a_1 + s = x' + a_2 + s \text{ for some } a_i \in A,\ s \in S, \tag{6.5}$$

*for which the above statements hold with the h-closure $\hat{A}$ of $A$ instead of the k-closure $\bar{A}$. In particular, the semiring $(S/\eta_A, +, \cdot)$ is additively cancellative.*

REMARK 6.9. a) For each additively commutative semiring $(T, +, \cdot)$, the congruence $\gamma$ defined by $y \gamma y' \Leftrightarrow y + t = y' + t$ for some $t \in T$ is the smallest congruence on $(T, +, \cdot)$ such that $(T/\gamma, +, \cdot)$ is additively cancellative. Applying this to a semiring

$(T,+,\cdot) = (S/\kappa_A,+,\cdot)$, the congruence class semiring of $S/\kappa_A$ with respect to $\gamma$ is (isomorphic to) $(S/\eta_A,+,\cdot)$.

b) The congruences $\kappa_A$ and $\eta_A$ coincide iff $(S/\kappa_A,+,\cdot)$ is additively cancellative, which in turn implies $\bar{A} = \hat{A}$, but not conversely.

c) Even for a commutative semiring $(S,+,\cdot)$ with an absorbing zero and an identity, there are in general various ideals with the same $k$-closure or $h$-closure. On the other hand, there may be a lot of congruences on $(S,+,\cdot)$ which cannot be obtained by (6.4) or (6.5). For instance, all $k$-ideals of $(\mathbf{N}_0,+,\cdot)$ are given by $m\mathbf{N}_0$ for all $m \in \mathbf{N}_0$, and these are also all $h$-ideals (very few in view of Remark 6.5 b)). The corresponding congruences $\kappa_{m\mathbf{N}_0}$ are, apart from $\iota_{\mathbf{N}_0}$, just those characterized in Remark 6.2 d) by the pairs $(0,g) \in \mathbf{N}_0 \times \mathbf{N}$.

Also the following result shows that, for semirings, ideals do not supply very much knowledge about congruences:

RESULT 6.10. *Let $\varphi$: $(S,+,\cdot) \to (T,+,\cdot)$ be a surjective homomorphism of semirings and assume that $(T,+,\cdot)$ has an absorbing zero $o_T$. Then*

$$A = \varphi^{-1}(o_T) = \{a \in S \mid \varphi(a) = o_T\}$$

*is an ideal of $(S,+,\cdot)$, often called the "kernel" of $\varphi$. If $(S,+)$ is commutative, $A$ is a $k$-ideal. However, even in this case the congruence $\kappa_A$ of* (6.4) *is merely contained in the congruence $\kappa_\varphi$ belonging to $\varphi$* (cf. Theorem 6.3 b)), *and there are various cases such that $\kappa_A \subset \kappa_\varphi$ holds.*

Since subsemirings of rings are often used in applications, we state in this context (cf. [99], §6, and [89], II.7):

THEOREM 6.11. *Let $(R,+,\cdot) = D(S)$ be the difference ring of a semiring $(S,+,\cdot)$.*

a) *Each congruence $\varrho$ on $(R,+,\cdot)$ defines a congruence $\varrho' = \varrho \cap (S \times S)$ on $(S,+,\cdot)$ such that $(S/\varrho',+,\cdot)$ is again additively cancellative and $(R/\varrho,+,\cdot)$ is isomorphic to the difference ring $D(S/\varrho')$ of $(S/\varrho',+,\cdot)$.*

b) *Each congruence $\kappa$ on $(S,+,\cdot)$ generates a congruence $\bar{\kappa}$ on $(R,+,\cdot)$ by*

$$r\bar{\kappa}r' \Leftrightarrow r - r' = s - s' \text{ for some } s, s' \in S \text{ satisfying } s\ \kappa\ s',$$

*where $B = \{s - s' \mid s, s' \in S \text{ and } s\ \kappa\ s'\}$ is the corresponding ring ideal $B = [o]_{\bar{\kappa}}$ of $(R,+,\cdot)$.*

c) *Using these notations, we have $(\overline{\varrho'}) = \varrho$ but merely $(\bar{\kappa})' \supseteq \kappa$, and $(\bar{\kappa})' = \kappa$ holds iff $(S/\kappa,+,\cdot)$ is additively cancellative. Hence* a) *defines an isomorphism of $(\mathcal{C}_{(R,+,\cdot)},\cup,\cap)$ onto the lattice of those congruences $\kappa$ on $(S,+,\cdot)$ for which $(S/\kappa,+,\cdot)$ is additively cancellative.*

REMARK 6.12. a) There are various papers investigating semiring ideals, in particular those called, e.g., principal, maximal, ($O$-)minimal, (completely) prime, primary, irreducible etc., some of them assuming different chain conditions or considering "ideal free" or "congruence free" semirings (cf., e.g., [5–7, 11, 49, 53, 99, 103, 104, 108, 125, 141–143, 184, 188, 197, 200, 224, 235] and [231]). Also the concepts of Green's relations (cf., e.g., [68, 69] and [171]) and of quasi-ideals (cf. [110, 223] and [232]) have been transferred to semirings.

b) We also mention various investigations on semirings $(S, +, \cdot)$ (or on ideals of them) for which $(S, +)$ or $(S, \cdot)$ or both are assumed to have particular semigroup-theoretical properties as to be regular, inverse, orthodox, completely ($O$-)simple, a union of groups etc. (cf., e.g., [10, 15, 76–78, 113, 114, 156, 239, 240] and [242]).

c) As a peculiarity we emphasize that semifields, which are clearly ideal free, may have various congruences. Moreover, for each semifield $(S, +, \cdot)$ without a zero, a certain set $\mathcal{K}_{(S,+,\cdot)}$ of normal subgroups of $(S, \cdot)$ can be characterized in $(S, +, \cdot)$ such that each congruence $\kappa \in \mathcal{C}_{(S,+,\cdot)}$ corresponds uniquely to some $K \in \mathcal{K}_{(S,+,\cdot)}$, and conversely. Hence each homomorphism of a semifield really has a "kernel" as it is true for groups or rings (cf. [98, 99] and [228]).

REMARK 6.13. a) Let $(S, +, \cdot)$ be an additively commutative semiring with an absorbing zero. The first radical for such a semiring, called the *Jacobson radical* $J(S)$, was introduced in [28] as the sum of all "right semiregular" (right, left or two-sided) ideals of $(S, +, \cdot)$. Another characterization of $J(S)$ by

$$\bigcap_{i \in I} (o\colon M_i)$$

for all "irreducible representation $S$-semimodules" $M_i$ was given in [101]. Replacing "right semiregular" above by "right quasiregular", one obtains the *semiradical* $\sigma(S)$. It was introduced in [32] using the Jacobson radical of the $D$-ring of $(S/\gamma, +, \cdot)$ (for $\gamma$ as defined in Remark 6.9 a)). It was claimed in [32] that the inclusion $J(S) \subseteq \sigma(S)$ may be proper; in fact, $J(S) = \sigma(S)$ holds as was shown in [102]. In this context we also refer to [34, 126, 213, 38] and [131].

b) Various other radicals for (certain classes of) semirings have been investigated, in particular those corresponding to the nilradical, the Levitzki radical or the Brown–McCoy radical in ring theory (cf., e.g., [11, 16, 109, 127, 129, 147, 185, 213, 214] and [236]). Moreover, there are also some papers dealing with a Kurosh–Amitsur radical theory for suitable classes of semirings or semifields (cf. [92, 147, 148, 228–230] and [237]).

## 7. Structural results on semiring semimodules and semirings

DEFINITION 7.1. a) In the context of the following considerations, an arbitrary semigroup $(S, +)$ is mostly called a *semimodule*.

b) Assume that $(S, +)$ has a zero $o$ and let $\{(S_i, +) \mid i \in I\}$ be a set of subsemimodules of $(S, +)$ satisfying $o \in S_i$ for all $i \in I$. Then $(S, +)$ is called the *direct sum of the subsemimodules* $(S_i, +)$ if $a_i + a_j = a_j + a_i$ holds for all $i \neq j$, $a_i \in S_i$ and $a_j \in S_j$, and if each $a \in S$ has a unique presentation

$$a = \sum_{i \in I} a_i \quad \text{where almost all } a_i \in S_i \text{ equal } o, \tag{7.1}$$

in the obvious interpretation of (7.1) as a "formally infinite sum". For $I = \{1, \dots, n\}$ we write $S = S_1 \oplus \cdots \oplus S_n$. (Finite direct sums of this kind have been called "strong direct sums", e.g., in [31, 193, 196] and [232].)

c) A semimodule $(S, +)$ is called *subcommutative* if

$$a + b + c + d = a + c + b + d \quad \text{holds for all } a, b, c, d \in S. \tag{7.2}$$

d) Let $(S, +)$ be a semimodule, $(\Omega, +, \cdot)$ a semiring and $\Omega \times S \to S$ a mapping which assigns to each $(\alpha, a) \in \Omega \times S$ an element $\alpha a \in S$. Then $(S, +)$ is called a *semimodule with $\Omega$ as (left) operator domain,* or a *(left) $\Omega$-semimodule* $({}_\Omega S, +)$, if

$$\alpha(a + b) = \alpha a + \alpha b, \quad (\alpha + \beta)a = \alpha a + \beta a \quad \text{and} \quad (\alpha\beta)a = \alpha(\beta a) \tag{7.3}$$

hold for all $\alpha, \beta \in \Omega$ and $a, b \in S$. In particular, $({}_\Omega S, +)$ is called a *unitary $\Omega$-semimodule* if $(\Omega, +, \cdot)$ has an identity $\varepsilon$ and $\varepsilon a = a$ holds for all $a \in S$. Finally, if $(S, +)$ has a zero $o$ and $(\Omega, +, \cdot)$ a zero $\omega$, it is convenient to assume that

$$\omega a = o \quad \text{and} \quad \alpha o = o \quad \text{hold for all } a \in S \text{ and } \alpha \in \Omega. \tag{7.4}$$

e) Let $({}_\Omega S, +)$ and $({}_\Omega T, +)$ be $\Omega$-semimodules. Then the meaning of an *$\Omega$-subsemimodule* of $({}_\Omega S, +)$ is clear, and a homomorphism $\varphi$: $(S, +) \to (T, +)$ is called an *$\Omega$-homomorphism* if $\varphi(\alpha a) = \alpha\varphi(a)$ holds for all $\alpha \in \Omega$ and $a \in S$. In particular, $({}_\Omega S, +)$ and $({}_\Omega T, +)$ are called *operator-isomorphic* or *$\Omega$-isomorphic* if there is an $\Omega$-isomorphism $\varphi : (S, +) \to (T, +)$.

REMARK 7.2. a) Examples of $\Omega$-semimodules are abundant. In particular, each semimodule $(S, +)$ is an $\mathbf{N}$-semimodule for $(\mathbf{N}, +, \cdot)$ where $na$ is defined by $\sum_{\nu=1}^{n} a$. Moreover, each semiring $(S, +, \cdot)$ can by considered as an $S$-semimodule $({}_S S, +)$ defining $sa$ by the multiplication in $(S, +, \cdot)$. In this case the $S$-subsemimodules are just the left ideals of $(S, +, \cdot)$. Also the semimodule $(M_{n,n}(S), +)$ of each matrix semiring is an $S$-semimodule in an obvious way, and another example is described in Remark 7.4.

b) There are various investigations on semiring-semimodules in the literature, dealing with their structure or using them to investigate semirings (cf., e.g., [238, 193, 101, 194, 195, 157, 74, 111, 159] and [91]). Special objects of this kind are $P$-semimodules of a ring $(R, +, \cdot)$ with identity, used to investigate representations of $(R, +, \cdot)$ by rings of continuous functions (cf. [80, 51, 52, 21, 39] and [22]). The semirings $(P, +, \cdot)$ occurring in this context are certain subsemirings of $(R, +, \cdot)$, called "primes" or "preprimes", and have similar properties as positive cones of $(R, +, \cdot)$ (cf. Definition 8.1).

THEOREM 7.3. *Let $(S, +)$ be a subcommutative (or even commutative) semimodule and* $\mathrm{End}(S, +) = \mathrm{End}(S)$ *the set of all endomorphisms $\varphi$: $(S, +) \to (S, +)$. Then, for all $\varphi, \psi \in \mathrm{End}(S)$, the mapping $\varphi + \psi$ defined by*

$$(\varphi + \psi)(a) = \varphi(a) + \psi(a) \quad \textit{for all } a \in S \tag{7.5}$$

*is, due to* (7.2), *again an endomorphism of $(S, +)$, and the same holds for $(\psi \circ \varphi)(a) = \psi(\varphi(a))$. In this way one obtains an additively subcommutative (commutative) semiring* $(\mathrm{End}(S), +, \circ)$ *with the identical mapping $\iota_S$ as identity, called the endomorphism semiring of $(S, +)$. If $(S, +)$ has a zero $o$, the mapping $\zeta$ defined by $\zeta(a) = o$ for all $a \in S$*

*is a left absorbing zero of* $(\mathrm{End}(S), +, \circ)$, *whereas* $\varphi \circ \zeta = \zeta$ *holds for all* $\varphi \in \mathrm{End}(S)$ *satisfying* $\varphi(o) = o$.

REMARK 7.4. a) Considering the endomorphisms of $(S, +)$ by $\varphi(a) = \varphi a$ as (left) operators, one obtains the *endomorphism semiring* $(\mathrm{End}(S), +, \circ)$ *as an operator domain of* $(S, +)$ and hence $({}_{\mathrm{End}(S)}S, +)$ as a unitary $\mathrm{End}(S)$-semimodule. If $(S, +)$ has a zero $o$, only $\zeta a = o$ of (7.4) holds in general. If one restricts $\mathrm{End}(S)$ to those endomorphisms $\varphi$ which keep $o$ fixed, also $\varphi o = o$ is satisfied.

b) Conversely, if $({}_{\Omega}S, +)$ is any $\Omega$-semimodule, $a \mapsto \alpha a$ for each $\alpha \in \Omega$ and all $a \in S$ defines an endomorphism of $(S, +)$.

THEOREM 7.5. a) *Let* $(S, +, \cdot)$ *be an additively subcommutative semiring and* $(\mathrm{End}(S), +, \circ)$ *the endomorphism semiring of* $(S, +)$. *Then each* $s \in S$ *defines by* $\varphi_s(a) = sa$ *for all* $a \in S$ *an endomorphism* $\varphi_s$ *of* $(S, +)$, *and the mapping* $\Phi$: $S \to \mathrm{End}(S)$ *defined by* $\Phi(s) = \varphi_s$ *is a homomorphism of* $(S, +, \cdot)$ *into* $(\mathrm{End}(S), +, \circ)$. *The latter is injective iff for all* $s, t \in S$ *the following holds:* $sa = ta$ *for all* $a \in S$ *implies* $s = t$.

b) *Together with Lemma* 3.1 *and Lemma* 3.5 *this yields that each additively commutative semiring is isomorphic to a subsemiring of an endomorphism semiring of a suitable semimodule.*

We close this section with the following celebrated structural statements on semirings. In the form presented here, they are essentially due to [193] and [196]. However, basic versions of Theorem 7.8 and Theorem 7.9 (with more assumptions to obtain b) of Theorem 7.8 and c) of Theorem 7.9, in particular that $(S, +, \cdot)$ has no nilpotent left or right ideals except $\{o\}$) have already been published in [31]]. We also emphasize that $(S, +, \cdot)$ has to be additively commutative in Theorem 7.8 and Theorem 7.9 as a consequence of Remark 3.8 b). This assumption is not made in [193] and [196], and not mentioned explicitly in [31].

RESULT 7.6. *Let* $L$ *be a left ideal of a semiring* $(S, +, \cdot)$ *which is o-minimal if* $(S, +, \cdot)$ *has an absorbing zero* $o$ *and minimal otherwise. If* $L$ *contains an idempotent* $e = e^2$ *such that* $e \in S^*$ *holds in the first case, then* $eL \subseteq L$ *is a semifield with* $e$ *as identity (where* $eL$ *may consist of one single element in the second case).*

RESULT 7.7. *Let* $(S, +, \cdot)$ *be a semiring with* $o$ *as absorbing zero and a right identity* $e$, *and assume*

$$S = L_1 \oplus \cdots \oplus L_r$$

*for left ideals* $L_i \neq \{o\}$ *of* $S$. *Then the elements* $e_i \in L_i$ *occurring in the presentation* $e = e_1 + \cdots + e_r$ *satisfy* $e_i e_j = o$ *for* $i \neq j$ *and* $e_i e_i = e_i \neq o$ *as well as* $L_i = Se_i$ *for all* $i, j \in \{1, \ldots, r\}$.

THEOREM 7.8. *For an additively commutative semiring* $(S, +, \cdot)$ *with an absorbing zero* $o$ *the following statements are equivalent:*

a) $(S, +, \cdot)$ *has an identity and is the direct sum* $S = L_1 \oplus \cdots \oplus L_r$ *of o-minimal left ideals* $L_j$ *of* $S$.

b) *$(S,+,\cdot)$ is the direct sum of a finite number of ideals $A_i \neq \{o\}$ of $S$, where each $(A_i,+,\cdot)$ is isomorphic to a matrix semiring $(M_{n_i,n_i}(T_i),+,\cdot)$ over a semifield $(T_i,+,\cdot)$ for some $n_i \geqslant 1$.*

THEOREM 7.9. *For an additively commutative semiring $(S,+,\cdot)$ with an absorbing zero $o$ the following statements are equivalent:*

a) *$(S,+,\cdot)$ has an identity and no ideals except $\{o\}$ and $S$, and it is the direct sum $S = L_1 \oplus \cdots \oplus L_r$ of $o$-minimal left ideals $L_j$ of $S$.*

b) *$(S,+,\cdot)$ has an identity and is the direct sum $S = L_1 \oplus \cdots \oplus L_r$ of $o$-minimal left ideals $L_j$ of $S$, which are, considered as $S$-subsemimodules of $({}_S S,+)$, two by two operator-isomorphic.*

c) *$(S,+,\cdot)$ is isomorphic to a matrix semiring $(M_{n,n}(T),+,\cdot)$ over a semifield $(T,+,\cdot)$.*

For further investigations in this direction, in particular those which include quasi-ideals of semirings, we refer to [232, 143, 49] and [199].

## 8. Partially ordered semirings

Despite the existence of more general investigations in the literature, we restrict ourselves here to semirings with commutative addition. All concepts and results in this section are essentially due to [56, 25, 211, 212, 77] and [225]. We also refer to [89], Chapter III, for a detailed presentation with all proofs.

DEFINITION 8.1. a) Let $(S,+)$ be a commutative semigroup and $(S,\leqslant)$ a partially ordered (briefly p. o.) set. Then $(S,+,\leqslant)$ is called a *p. o. semigroup* if $a<b$ implies $a+c \leqslant b+c$ for all $a,b,c \in S$. By $P = \{p \in S \mid a+p \geqslant a$ for all $a \in S\}$ we define the *positive cone* $P$ and by $N = \{n \in S \mid a+n \leqslant a$ for all $a \in S\}$ the *negative cone* $N$ of $(S,+,\leqslant)$, both of which may be empty. In particular, $(S,+,\leqslant)$ is called a *totally ordered (t. o.) semigroup* if $(S,\leqslant)$ is a t. o. set. Moreover, a p. o. semigroup $(S,+,\leqslant)$ is called *positively* [*negatively*] *p. o.* if $P = S$ [$N = S$] is satisfied.

b) A semiring $(S,+,\cdot)$ is called a *p. o. semiring* $(S,+,\cdot,\leqslant)$ if $(S,+,\leqslant)$ is a p. o. semigroup and if it satisfies the (*multiplicative*) *monotony law*

$$a<b \quad \text{implies} \quad ac \leqslant bc \quad \text{and} \quad ca \leqslant cb$$
$$\text{for all } a,b \in S \text{ and all } c \in P, \tag{8.1}$$

where $P$ is the positive cone as defined above, t.o. semirings are defined correspondingly.

c) Sometimes we will consider the strict version of (8.1) defined by

$$a<b \quad \text{implies} \quad ac<bc \quad \text{and} \quad ca<cb$$
$$\text{for all } a,b \in S \text{ and all } c \in P \cap S^*. \tag{8.2}$$

d) By $M = \{m \in S \mid a<b \Rightarrow am \leqslant bm$ and $ma \leqslant mb$ for all $a,b \in S\}$ and by $W = \{w \in S \mid a<b \Rightarrow aw \geqslant bw$ and $wa \geqslant wb$ for all $a,b \in S\}$ we define the *monotony domain* $M$ and the *anti-monotony domain* $W$ of $(S,\cdot,\leqslant)$.

REMARK 8.2. a) Note that (8.1) is equivalent to $P \subseteq M$, and that the anti-monotony law $N \subseteq W$ (well-known to be a consequence of $P \subseteq M$ for p. o. rings) is not assumed and need not be true for a p. o. semiring $(S, +, \cdot, \leqslant)$.

b) A concept also called "partially ordered semiring" in some papers which demands (8.1) for all $c \in S$ is not meaningful, since it excludes any p. o. ring.

c) Clearly, $N \cap P$ is either empty or consists of the zero $o$ of $(S, +, \cdot, \leqslant)$. If $o$ exists, one has $N = \{n \in S \mid n \leqslant o\}$ and $P = \{p \in S \mid o \leqslant p\}$, and $P$ is a p. o. subsemiring of $(S, +, \cdot, \leqslant)$ if $o$ is absorbing.

THEOREM 8.3. a) *Let $X$ be a subsemigroup of a commutative semigroup $(S, +)$. Then*

$$a \leqslant_X b \Leftrightarrow a = b \text{ or } a + x = b \text{ for some } x \in X \tag{8.3}$$

*defines a relation on $S$ such that $(S, +, \leqslant_X)$ is a p. o. semigroup iff*

$$a + x + y = a \quad \text{implies} \quad a + x = a \text{ for all } a \in S \text{ and } x, y \in X. \tag{8.4}$$

*If this is the case, $X$ (and also $X \cup \{o\}$ if $o$ exists) may be properly contained in the positive cone $P$ of $(S, +, \leqslant_X)$, but $\leqslant_X$ and $\leqslant_P$ coincide.*

b) *Applying* a) *to a subsemiring $X$ of an additively commutative semiring $(S, +, \cdot)$, one obtains a p. o. semiring $(S, +, \cdot, \leqslant_X)$ provided that $P \subseteq M$ holds. Sufficient conditions for the latter are:*

i) *$X$ is an ideal of $(S, +, \cdot)$, hence in particular $X = S$.*
ii) *$(S, +, \cdot)$ has a zero satisfying $oX$, $Xo \subseteq X$. (This yields $X \cup \{o\} = P$.)*
iii) *$(S, +, \cdot)$ has an absorbing zero. (This yields $N \subseteq W$.)*

REMARK 8.4. a) If $X = S$ satisfies (8.4), $\leqslant_S$ is called the *difference order* on $(S, +, \cdot)$. In this case, $\leqslant_S$ is a total order iff $(S, +)$ is semisubtractive.

b) Let $(S, +, \cdot)$ be an additively cancellative semiring with a (by Result 2.2 a) absorbing) zero $o$. Then a subsemiring $X$ satisfies (8.4) iff either $o \notin X$ holds or $(X, +, \cdot)$ is zero-sum free. If this is the case, $(S, +, \cdot, \leqslant_X)$ is a p. o. semiring by iii) and satisfies $X \cup \{o\} = P$ and $N \subseteq W$. However, there are even commutative t. o. semirings $(S, +, \cdot, \leqslant)$ of this kind, for which $\leqslant$ can not be obtained by (8.3). For instance, the t. o. subsemiring $S = \{0\} \cup \{s \in \mathbf{R} \mid s \geqslant 1\}$ of $(\mathbf{R}, +, \cdot, \leqslant)$ satisfies $P = S$, but $\leqslant_S$ is properly contained in $\leqslant$.

c) Let $(X, +, \cdot)$ be an additively idempotent subsemiring of $(S, +, \cdot)$. Then $X$ satisfies (8.4), and $(S, +, \cdot, \leqslant_X)$ is a p. o. semiring according to Theorem 8.3 b) for instance if $X$ is an ideal of $(S, +, \cdot)$.

d) If $(S, +)$ itself is idempotent, then $a \leqslant_S b$ and $a + b = b$ are equivalent, and $\leqslant_S$ is the partial order turning a commutative idempotent semigroup $(S, +)$ into a semilattice such that $a + b = a \vee b$ holds. A p. o. semiring $(S, +, \cdot, \leqslant_S)$ of this kind is a semilattice ordered semigroup $(S, \cdot, \leqslant_S)$ (which usually includes $a(b \vee c) = ab \vee ac$ and $(b \vee c)a = ba \vee ca$, a stronger assumption than $M = S$ for $(S, \cdot, \leqslant_S)$), and conversely.

Remark 8.4 b) generalizes the well-known fact that for a ring $(R, +, \cdot)$ each relation $\leqslant$ such that $(R, +, \cdot, \leqslant)$ is a p. o. ring is uniquely determined by its positive cone $P$ according to $a \leqslant b \Leftrightarrow b - a \in P$. In this context we note:

RESULT 8.5. *Let $P$ be a subset of a ring $(R, +, \cdot)$. Then there exists a relation $\leqslant$ such that $P$ is the positive cone of the p. o. ring $(R, +, \cdot, \leqslant)$ iff $(P, +, \cdot)$ is a subsemiring of $(R, +, \cdot)$ which contains the zero $o$ of $R$ and is zero-sum free, where the latter is equivalent to $P \cap -P = \{o\}$. Moreover, $(R, +, \cdot, \leqslant)$ is a t. o. ring iff $(P, +, \cdot)$ is semisubtractive, and $(R, +, \cdot, \leqslant)$ satisfies* (8.2) *iff $(P, +, \cdot)$ is zero-divisor free.*

Finally, we give some typical results on the extension of partial orders, where the main part of Theorem 8.6 a) are statements on semigroups $(S, \cdot)$ which are also t. o. sets $(S, \leqslant)$ and their $Q_r$-semigroups $(T, \cdot) = Q_r(S, \Sigma)$:

THEOREM 8.6. a) *Let $(S, +, \cdot, \leqslant)$ be a t. o. semiring and $(T, +, \cdot) = Q_r(S, \Sigma)$ a $Q_r$-semiring of $S$ for which $\Sigma$ can be chosen such that $\Sigma \subseteq M(S)$ holds. Then the total order $\leqslant$ on $S$ can be extended to a partial order $\leqslant_T$ on $T$ satisfying $M(S) \leqslant M(T)$ iff $\xi c \in M(S)$ implies $c \in M(S)$ for all $\xi \in \Sigma$ and $c \in S$. If the latter condition holds, there is exactly one extension $\leqslant_T$ of this kind, determined by*

$$a\alpha^{-1} \leqslant_T b\beta^{-1} \Leftrightarrow \alpha x = \beta x \text{ and } ax \leqslant b\xi \text{ for some } (x, \xi) \in S \times \Sigma,$$

*and $\leqslant_T$ is in fact a total order on $T$. Moreover, $(T, +, \cdot, \leqslant_T)$ is a t. o. semiring.*

b) *If one additionally assumes that $\Sigma$ is in the centre of $(S, \cdot)$ (in particular, if $(S, \cdot)$ is commutative), the condition $\xi c \in M(S) \Rightarrow c \in M(S)$ is always satisfied and the total order $\leqslant_T$ is given by $a\alpha^{-1} \leqslant_T b\beta^{-1} \Leftrightarrow a\beta \leqslant b\alpha$.*

c) *If $(S, +, \cdot, \leqslant)$ is merely a p. o. semiring, the corresponding statements concerning the extension of $\leqslant$ to a (unique minimal) partial order $\leqslant_T$ on $(T, +, \cdot) = Q_r(S, \Sigma)$ as above need further assumptions even if $(S, \cdot)$ is commutative* (cf. [90] and [89], III.3).

THEOREM 8.7. *Let $(S, +, \cdot, \leqslant)$ be a p. o. semiring and $(R, +, \cdot) = D(S)$ its difference ring. Then the partial order $\leqslant$ on $S$ can be extended to a partial order $\leqslant_R$ on $R$ such that $(R, +, \cdot, \leqslant_R)$ is a p. o. ring iff, for all $a, b, c, d \in S$,*

$$a < b \text{ and } c < d \Rightarrow ad + bc + s \leqslant ac + bd + s \text{ for some } s \in S. \tag{8.5}$$

*The minimal extension $\leqslant_R$ of this kind is uniquely determined by*

$$a - u \leqslant_R b - v \Leftrightarrow a + v + s \leqslant b + u + s$$

*for some $s \in S$. Moreover, $(R, +, \cdot, \leqslant_R)$ satisfies* (8.2) *iff the same holds for $(S, +, \cdot, \leqslant)$ and the conclusion of* (8.5) *holds with $<$ instead of $\leqslant$. Finally, if $\leqslant$ is a total order, the same holds for $\leqslant_R$ (but not conversely); in that case it is superfluous to add some $s \in S$ in* (8.5) *and in the definition of $\leqslant_R$.*

EXAMPLE 8.8. a) It is indispensable to add some $s \in S$ in (8.5) and in the definition of $\leqslant_R$. An example is the p. o. subsemiring $(\mathbf{H}_0, +, \cdot, \leqslant')$ of $(\mathbf{Q}, +, \cdot, \leqslant)$, the latter with the usual total order $\leqslant$, whereas $a <' b$ is defined by $0.25 \leqslant a < b$ (for details see [89], Beispiel III.4.10).

b) The subsemiring $S = \{a + bx \mid a > 0,\ b \geqslant 0\}$ of the ring $R = \mathbf{Z}[x]/(x^2)$ considered in Example 5.12 a) is a positively t. o. semiring $(S, +, \cdot, \leqslant)$ if one defines

$a + bx \leqslant c + dx$ by $a < c$ or $a = c$ and $b \leqslant d$. Since $S$ is multiplicatively cancellative, (8.2) holds, and one checks straightforwardly that (8.5) is satisfied. The extension $\leqslant_R$ of $\leqslant$ defined in Theorem 8.7 is given by $a + bx \leqslant_R c + dx$ iff $a < c$ or $a = c$ and $b \leqslant d$ for all $a, b, c, d \in \mathbf{Z}$, and $(R, +, \cdot, \leqslant_R)$ is a t. o. ring which clearly does not satisfy (8.2). On the other hand, $S \cup \{0\}$ is also a positive cone of $(R, +, \cdot)$, and the partial order $\leqslant$ on $R$ mentioned before Result 8.5 yields a p. o. semiring $(R, +, \cdot, \leqslant)$ according to $a + bx \leqslant c + dx$ iff $a \leqslant c$ and $b \leqslant d$ for all $a, b, c, d \in \mathbf{Z}$.

## 9. Generalized semigroup semirings and formal languages

In the last 20 years semirings of formal power series have become an important algebraic tool for investigations in the theory of formal languages and automata (cf. Example 9.4), and in combinatorics (cf., e.g., [41]).

DEFINITION 9.1. a) Let $(S, +, \cdot)$ be a nontrivial additively commutative semiring with $\omega$ as absorbing zero, $U \neq \varnothing$ a set, and $S\langle\langle U\rangle\rangle$ the set of all mappings $f\colon U \to S$. Define $f + g$ and $\alpha f$ for $f, g \in S\langle\langle U\rangle\rangle$ and $\alpha \in S$ by

$$(f + g)(u) = f(u) + g(u) \quad \text{and} \quad (\alpha f)(u) = \alpha f(u) \quad \text{for all } u \in U.$$

Then $({}_S S\langle\langle U\rangle\rangle, +)$ is an $S$-semimodule, in fact the direct product of $|U|$ copies of $({}_S S, +)$. The mapping $o$ defined by $o(u) = \omega$ for all $u \in U$ is the zero of $S\langle\langle U\rangle\rangle$ and satisfies (7.4). As usual in this context, we write the elements of $S\langle\langle U\rangle\rangle$ in a formal way as (possibly infinite) sums

$$f = \sum_{u \in U} (f, u)u \quad \text{with } f(u) = (f, u), \tag{9.1}$$

and call $\operatorname{supp}(f) = \{u \in U \mid (f, u) \neq \omega\}$ the *support of* $f$. The set $S\langle U\rangle$ of all $f \in S\langle\langle U\rangle\rangle$ with finite support is an $S$-subsemimodule of $({}_S S\langle\langle U\rangle\rangle, +)$.

b) Now let $(U, \cdot)$ be a semigroup satisfying the *finite factorization property*, which means that each $w \in U$ has only a finite number of factorizations $w = u \cdot v$ for $u, v \in U$. Then

$$f \cdot g = \sum_{w \in U} \Bigl( \sum_{u \cdot v = w} (f, u)(g, v) \Bigr) w \tag{9.2}$$

defines a multiplication on $S\langle\langle U\rangle\rangle$ such that $(S\langle\langle U\rangle\rangle, +, \cdot)$ is a semiring with $o$ as absorbing zero, called the *generalized semigroup semiring of* $(U, \cdot)$ *over* $S$.

c) For each semigroup $(U, \cdot)$ there exists the *semigroup semiring* $(S\langle U\rangle, +, \cdot)$, defined on $(S\langle U\rangle, +)$ by (9.2). Clearly, if $(U, \cdot)$ has the finite factorization property, $(S\langle U\rangle, +, \cdot)$ is a subsemiring of the generalized semigroup semiring $(S\langle\langle U\rangle\rangle, +, \cdot)$. However, there may be interesting subsemirings $(T, +, \cdot)$ of $(S\langle\langle U\rangle\rangle, +, \cdot)$ containing $S\langle U\rangle$ properly, cf. [88].

REMARK 9.2. Note that so far $U$ is not a subset of $S\langle\langle U\rangle\rangle$. However, if $(S, +, \cdot)$ has a (right) identity $\varepsilon$, then each $u \in U$ can be identified with the mapping $f_u$ defined by $f_u(u) = \varepsilon$ and $f_u(v) = \omega$ for all $v \neq u$ in $U$. If this is done and if $(U, \cdot)$ has an identity $e$, also $(S, +, \cdot)$ can be considered as a subsemiring of $(S\langle U\rangle, +, \cdot)$, identifying each $\alpha \in S$ with $\alpha e \in S\langle U\rangle$.

In the following we use the term *monoid* for a semigroup with identity.

EXAMPLE 9.3. The free monoid over a set $X \neq \varnothing$, in this context usually denoted by $X^*$, has the finite factorization property. The same holds for any *partial commutative free monoid over* $X$ and for the *commutative free monoid over* $X$, defined by assuming $x_i x_j = x_j x_i$ for certain pairs or for all elements $x_i, x_j \in X$ (cf. [41]). Let us denote these semigroups by $\mathcal{PC}X^*$ or $\mathcal{C}X^*$, respectively. Then, for every such monoid and each semiring $S$ as above, the generalized semigroup semiring over $S$ exists. In particular, if $S$ has an identity (cf. Remark 9.2), $(S\langle\langle X^*\rangle\rangle, +, \cdot)$ is called the *semiring of formal power series over* $S$, and $(S\langle \mathcal{C}X^*\rangle, +, \cdot)$ is just the polynomial semiring $(S[X], +, \cdot)$ according to Remark 3.11 b).

EXAMPLE 9.4. In the theory of formal languages, any (finite) set $X \neq \varnothing$ is called an *alphabet*, and every subset $L \subseteq X^*$ a *formal language*. If for $L_1, L_2 \subseteq X^*$ the addition is defined by $L_1 \cup L_2$ and the multiplication by $L_1 \cdot L_2 = \{w_1 w_2 \mid w_i \in L_i\}$, it is straightforward to check that $(\mathbf{P}(X^*), \cup, \cdot)$ is a semiring, the *semiring of formal languages over the alphabet* $X$. Now define for each $L \subseteq X^*$ a mapping $\chi_L$ from $X^*$ into the Boolean semifield (cf. Example 1.8 b)) by $\chi_L(w) = e$ if $w \in L$ and $\chi_L(w) = o$ otherwise. Then the mapping $L \mapsto \chi_L$ shows that the semirings $(\mathbf{P}(X^*), \cup, \cdot)$ and $(\mathbf{B}\langle\langle X^*\rangle\rangle, +, \cdot)$ are isomorphic. This isomorphism was the starting point for many investigations on formal languages and automata theory by algebraic methods, cf. [178, 124, 24] and the literature cited there.

For detailed investigations in the context of Example 9.3 we refer to [88]. Here we only state the following theorem, for which a weaker version is due to [50].

THEOREM 9.5. *For each subsemiring* $(T, +, \cdot)$ *of* $(S\langle\langle \mathcal{PC}X^*\rangle\rangle, +, \cdot)$ *containing*

$$(S\langle \mathcal{PC}X^*\rangle, +, \cdot)$$

*as considered in Example* 9.3 (*including the cases* $\mathcal{PC}X^* = X^*$ *and* $\mathcal{PC}X^* = \mathcal{C}X^*$) *the following statements hold:*

a) $(T, +, \cdot)$ *is zero-divisor free iff* $(S, +, \cdot)$ *is zero-divisor free.*

b) $(T, +, \cdot)$ *is multiplicatively* (*left*) *cancellative iff* $(S, +, \cdot)$ *has the same property and is additively cancellative.*

REMARK 9.6. For more general considerations it is convenient to write the elements of the semimodule $A = S\langle U\rangle$ in the form

$$a = \sum_{u \in U} \alpha_u u \quad \text{and} \quad b = \sum_{v \in U} \beta_v v$$

for $\alpha_u = a(u)$ and $\beta_v = b(v)$. Then each mapping $U \times U \to A$, defined by

$$u \cdot v = \sum_{w \in U} \gamma_{u,v}^{w} w$$

with "structure constants" $\gamma_{u,v}^{w} \in S$, defines a multiplication on $A$ by

$$a \cdot b = \sum_{w \in U} \left( \sum_{u,v \in U} \alpha_u \beta_v \gamma_{u,v}^{w} \right) w. \tag{9.3}$$

In this way one obtains the concept of an *S-semialgebra* $({}_S A, +, \cdot)$, which is a semiring iff certain associativity conditions are satisfied. Proceeding in the same way with $A = S\langle\langle U \rangle\rangle$ to obtain *generalized S-semialgebras*, one has to make (9.3) meaningful. One possibility to do this is to choose almost all $\gamma_{u,v}^{w} = \omega$ for each $w \in U$. This was done above assuming the finite factorization property for the semigroup $(U, \cdot)$ in Definition 9.1 b). Otherwise, one has to assume, according to the next section, that the infinite sums $\sum_{u,v \in U} \alpha_u \beta_v \gamma_{u,v}^{w}$ occurring in (9.3) are defined in $(S, +, \cdot)$. For more details concerning these concepts we refer to [226] and [89], V.2 and V.3.

## 10. Semirings with infinite sums

Stimulated by [54], several concepts of semirings and other algebraic systems have been considered in which (some or even arbitrary) infinite sums exist, cf., e.g., [233, 40, 95, 139, 174, 120, 226, 85] and [119]. These concepts have been defined by various sets of axioms, where some of those definitions are equivalent and some not. For a comprehensive investigation also of related concepts we refer to [87] and [89], IV. Here we only sketch some of these concepts which are suitable for various applications.

DEFINITION 10.1. a) Let $(A, +)$ be a commutative semimodule with a zero $o$, and

$$\sum : \mathcal{S} \to A$$

a mapping, where $\mathcal{S}$ denotes a class of families $(a_i)_{i \in I}$ with $a_i \in A$ for arbitrary index sets $I$. Each family $(a_i)_{i \in I}$ in $\mathcal{S}$ will be called *summable* with $\sum (a_i)_{i \in I} = \sum_{i \in I} a_i$ as its *sum*. Then $(A, +, \sum, \mathcal{S})$, or briefly $(A, +, \sum)$, is called a $\sum$*-semimodule* over $(A, +)$ if the following three axioms hold:

$(F)$ Each finite family $(a_i)_{i \in I}$ is summable and $\sum_{i \in I} a_i$ equals the usual sum in $(A, +)$, i.e. $a_1 + \cdots + a_n$ for $I = \{1, \ldots, n\}$, including $n = 1$ and $\sum_{i \in \varnothing} a_i = o$.

$(GP)$ If $(a_i)_{i \in I}$ is summable and

$$I = \bigcup_{j \in J} I_j \tag{10.1}$$

is any *generalized partition* of $I$ (which means $I_j \cap I_{j'} = \varnothing$ for all $j \neq j'$, but allows $I_j = \varnothing$ for some $j \in J$), then all sums on the right hand side of

$$\sum_{i\in I} a_i = \sum_{j\in J}\Big(\sum_{i\in I_j} a_i\Big) \tag{10.2}$$

do also exist and the noted equality holds.

$(GP'_F)$ Let $(a_i)_{i\in I}$ be a family with $a_i \in A$ and (10.1) any generalized partition of $I$ where $J$ is finite. Then, if all sums on the right hand side of (10.2) exist, $(a_i)_{i\in I}$ is also summable and (10.2) is satisfied.

b) A $\sum$-semimodule $(A, +, \sum)$ is called *countably complete* if every family $(a_i)_{i\in I}$ with a countable index set $I$ is summable, and *complete*, if every family is summable.

REMARK 10.2. a) Let $(a_i)_{i\in I}$ be summable in $(A, +, \Sigma)$ and $(b_k)_{k\in K}$ obtained by $b_k = a_{\varphi(k)}$ for a bijection $\varphi\colon K \to I$. Then the above axioms imply that also $(b_k)_{k\in K}$ is summable and that both families have the same sum.

b) If $(a_i)_{i\in I}$ for $a_i = o$ and some (infinite) $I$ is summable, $\sum_{i\in I} o = o$ holds.

c) A $\sum$-semimodule $(A, +, \sum)$ may satisfy the stronger axiom $(GP')$, obtained from $(GP'_F)$ by cancelling the restriction on $J$. If this is the case and one infinite sum as considered in b) exists, then $(A, +)$ is zero-sum free.

d) Note that $(GP')$ follows from $(GP)$ if $(A, +, \sum)$ is complete, and also if one only considers countably infinite sums and $(A, +, \sum)$ is countably complete.

EXAMPLE 10.3. a) Each commutative semimodule $(A, +)$ with a zero $o$ can be considered as a $\sum$-semimodule $(A, +, \sum^o)$ in a natural way: Let $\mathcal{S}^o$ consist of all families $(a_i)_{i\in I}$ such that $I' = \{i \in I \mid a_i \neq o\}$ is finite and define $\sum^o\colon \mathcal{S}^o \to A$ by

$$\sum_{i\in I}^{o} a_i = \sum_{i\in I'} a_i$$

including $\sum_{i\in I}^o o = o$ for $I' = \varnothing$. Then $(A, +, \sum^o)$ is a $\sum$-semimodule, called the *$\sum$-semimodule of formal infinite sums of* $(A, +)$. Note that $\mathcal{S}^o \subseteq \mathcal{S}$ holds for any $\sum$-semimodule $(A, +, \sum) = (A, +, \sum, \mathcal{S})$ over $(A, +)$, and that $\sum$ and $\sum^o$ coincide on $\mathcal{S}^0$.

b) Consider the semimodule $(S\langle\langle U\rangle\rangle, +)$ of Definition 9.1 for a semiring $(S, +, \cdot)$ and let $(S, +, \sum)$ be any $\sum$-semimodule over $(S, +)$. Then the infinite sums in $S$ define infinite sums in $S\langle\langle U\rangle\rangle$ in a natural way (cf. [226], §6, or [89], Satz V.1.14) such that $(S\langle\langle U\rangle\rangle, +, \sum)$ is also a $\sum$-semimodule. Moreover, the infinite sums occurring in the formal notation (9.1) are then defined in each $\sum$-semimodule $(S\langle\langle U\rangle\rangle, +, \sum)$ over $(S\langle\langle U\rangle\rangle, +)$ obtained from $(S, +, \sum)$ in this way. Clearly, it is enough to start with the $\sum$-semimodule $(S, +, \sum^o)$ of formal infinite sums of $(S, +)$ in order to turn (9.1) into well defined infinite sums.

DEFINITION 10.4. a) An additively commutative semiring $(A, +, \cdot)$ with an absorbing zero $o$ is called a *$\sum$-semiring* $(A, +, \sum, \cdot)$ if $(A, +, \sum)$ is a $\sum$-semimodule and the following axiom is satisfied:

$(D)$ If $(a_i)_{i\in I}$ and $(b_j)_{j\in J}$ are summable families, then $(a_ib_j)_{(i,j)\in I\times J}$ is also summable and

$$\Big(\sum_{i\in I} a_i\Big)\Big(\sum_{j\in J} b_j\Big) = \sum_{(i,j)\in I\times J} a_ib_j$$

holds.

b) For every $\sum$-semiring $(A,+,\sum,\cdot)$ with an identity a partial unary operation $^*$ on $A$ is defined by

$$a^* = \sum_{i\in\mathbf{N}_0} a^i \quad \text{iff the family } (a^i)_{i\in\mathbf{N}_0} \text{ is summable.}$$

We call $(A,+,\sum,\cdot)$ *$^*$-complete* if $a^*$ exists for each $a\in A$. This is clearly the case if $(A,+,\sum,\cdot)$ is countably complete.

REMARK 10.5. a) Axiom $(D)$ implies the one-sided axioms $(D_l)$ and $(D_r)$, defined by $|I| = 1$ or $|J| = 1$ in $(D)$, respectively. The converse holds if $(A,+,\sum,\cdot)$ satisfies $(GP')$ (cf. Remark 10.2 c) and d)), but not in general.

b) Every semiring $(A,+,\cdot)$ as above can be considered as a $\sum$-semiring $(A,+,\sum^o,\cdot)$ with respect to Example 10.3 a), and we will do this in the following if no other infinite sums are defined on $(A,+,\cdot)$.

c) A $\sum$-semiring $(A,+,\sum,\cdot)$ is called *countably idempotent*, if each family $(a_i)_{i\in\mathbf{N}}$ with $a_i = a$ for all $i\in\mathbf{N}$ is summable and

$$\sum_{i\in\mathbf{N}} a_i = a$$

holds. Clearly, this implies that $(A,+,\cdot)$ is additively idempotent, but not conversely. If $(A,+,\cdot)$ has an identity, then an equivalent condition is that $e^* = e$ holds.

d) In Theoretical Computer Science countably complete $\sum$-semirings $(A,+,\sum,\cdot)$ with identity which are countably idempotent were called *closed semirings* (cf. [119] for this and related concepts as ($^*$-continuous) Kleene algebras, $S$-algebras, and $R$-algebras). This name originates from the following example. Let $\mathcal{R}(X) = \mathbf{P}(X\times X)$ denote the set of all binary relations $\varrho$ on an arbitrary set $X$. If addition is defined as union $\cup$ and multiplication as composition $\circ$ of relations, then it is obvious that $(\mathcal{R}(X),\cup,\bigcup,\circ)$ is a complete $\sum$-semiring such that $\varrho^*$ is just the reflexive and transitive closure for any $\varrho\in\mathcal{R}(X)$. Note that each semiring $(\mathbf{P}(X^*),\cup,\cdot)$ of formal languages (cf. Example 9.4) is also a closed semiring.

e) Let $(A,+,\sum,\cdot)$ be a $\sum$-semiring and $(M_{n,n}(A),+,\cdot)$ a matrix semiring over $A$. Then $\sum$ can be transferred to each family $(M_k)_{k\in K}$ of matrices $M_k = (m_{i,j,k}) \in M_{n,n}(A)$ for which

$$\sum_{k\in K} m_{i,j,k} = m_{i,j}$$

exists in $(A, +, \sum)$ for all $i, j \in \{1, \ldots, n\}$ by defining

$$\sum_{k \in K} M_k = (m_{i,j}).$$

In this way one obtains a $\sum$-semiring $(M_{n,n}(A), +, \sum, \cdot)$. Clearly, if $(A, +, \sum, \cdot)$ is countably complete, complete, countably idempotent or satisfies $(GP')$, the same holds for any $(M_{n,n}(A), +, \sum, \cdot)$. The same can be proved for *-completeness provided that $(GP')$ is satisfied.

One reason for developing a theory of (countably complete) $\sum$-semirings was to formulate and prove the correctness of algorithms which solve the so-called algebraic path problem which we are going to describe now.

DEFINITION 10.6. a) We call a semiring $(A, +, \cdot)$ with absorbing zero $o$ and identity $e \neq o$ a *path algebra* if $(A, +)$ is commutative (but not necessarily idempotent). With respect to Remark 10.5 b) we will always assume that it is a $\sum$-semiring.

b) By a *finite directed graph* $G = (N, E)$ we mean a finite set $N = \{1, \ldots, n\}$ of *nodes* and a set $E \subseteq N \times N$ of directed *edges*. If $(A, +, \sum, \cdot)$ is a path algebra, then a *valuation of* $G = (N, E)$ is given by a matrix $M \in M_{n,n}(A)$ satisfying $m_{i,j} = o$ if $(i, j) \notin E$ and $m_{i,j} \neq o$ otherwise. For $e_\nu = (i, j) \in E$ we call $w(e_\nu) = m_{i,j}$ the *value of* $e_\nu$. Moreover, for every path $p = (e_1, \ldots, e_r)$ in $G$ we define its value by the product $w(p) = w(e_1) \cdots w(e_r)$. Let $P_{i,j}$ be the set of all paths in $G$ from node $i$ to node $j$, including for each $i \in N$ the path $\{i\}$ from $i$ to $i$ of order 0 with $w(\{i\}) = e$ for technical reasons. Then the *algebraic path problem* is to decide whether there exist elements $d_{i,j} \in A$ such that

$$d_{i,j} = \sum_{p \in P_{i,j}} w(p)$$

holds in $(A, +, \sum, \cdot)$, and to compute $D = (d_{i,j}) \in M_{n,n}(A)$ if possible.

REMARK 10.7. If $(A, +, \cdot)$ is additively idempotent, then it determines a partial order on $A$ (cf. Remark 8.4 d)) and in this case

$$d_{i,j} = \sum_{p \in P_{i,j}} w(p)$$

is a lower or upper bound of all values $w(p)$ for paths $p$ from node $i$ to node $j$. For some concrete examples cf. Example 1.9.

THEOREM 10.8. *Under the same assumptions as in Definition* 10.6 *the following statements hold in the* $\sum$*-semiring* $(M_{n,n}(A), +, \sum, \cdot)$:

a) *If* $D$ *exists, then* $M^*$ *exists, too.*

b) *If* $(A, +, \sum, \cdot)$ *satisfies* $(GP')$, *then the converse of* a) *is true.*

c) *In both cases* a) *and* b), *one has* $D = M^*$.

In the literature, e.g., in [2, 40, 243] and [174], several algorithms were presented to solve the algebraic path problem. A detailed investigation using the concepts sketched here shows the following (cf. [87] and [89], IV.6). The proof of the correctness for each of these algorithms needs at least that the path algebras under consideration are $\sum$-semirings $(A, +, \sum, \cdot)$ satisfying $(GP')$. For some algorithms even more assumptions on $(A, +, \sum, \cdot)$ are indispensable, e.g., that $(A, +, \sum, \cdot)$ has to be countably idempotent. Sometimes not all of these assumptions are mentioned explicitly.

## References


[1] S.K. Abdali and B.D. Saunders, *Transitive closure and related semiring properties via eliminants*, Theoret. Comput. Sci. **40** (1985), 257–274.

[2] A.V. Aho, J.E. Hopcroft and J.D. Ullman, *The Design and Analysis of Computer Algorithms*, Addison-Wesley (1974).

[3] J. Ahsan, *Fully idempotent semirings*, Proc. Japan Acad. **69** (1993), 185–188.

[4] F.E. Alarcón and D.D. Anderson, *Commutative semirings and their lattices of ideals*, Houston J. Math. **20** (1994), 571–590.

[5] P.J. Allen, *A fundamental theorem of homomorphisms for semirings*, Proc. Amer. Math. Soc. **21** (1969), 412–416.

[6] P.J. Allen, *Cohen's theorem for a class of Noetherian semirings*, Publ. Math. Debrecen **17** (1970), 169–171.

[7] P.J. Allen, *An extension of the Hilbert basis theorem to semirings*, Publ. Math. Debrecen **22** (1975), 31–34.

[8] P.J. Allen and L. Dale, *Ideal theory in polynomial semirings*, Publ. Math. Debrecen **23** (1976), 183–190.

[9] P.J. Allen and L. Dale, *An extension of the Hilbert basis theorem*, Publ. Math. Debrecen **27** (1980), 31–34.

[10] E. Allevi, *A class of* $(+)$*-inverse semirings*, Istit. Lombardo Accad. Sci. Lett. Rend. A **119** (1985), 89–107.

[11] A. Almeida Costa, *Sur la théorie générale des demi-anneaux*, Publ. Math. Debrecen **10** (1963), 14–29.

[12] A. Almeida Costa and M.L. Noronha Galvao, *Sur le demi-anneau des nombres naturels*, Centro Estudos Mat. Porto Publ. **45** (1965), 1–5.

[13] A. Almeida Costa, *Cours d'Algébre Générale*, Vol. III, Gulbenkian, Lisboa (1975).

[14] H.-J. Bandelt and M. Petrich, *Subdirect products of rings and distributive lattices*, Proc. Edinburgh Math. Soc. **25** (1982), 155–171.

[15] H.-J. Bandelt, *Free objects in the variety generated by rings and distributive lattices*, SLNM 998, Springer, Berlin (1983), 255–260.

[16] E. Barbut, *On nil semirings with ascending chain conditions*, Fund. Math. **58** (1970), 261–264.

[17] L. Beasley, Chi-Kwong Li and S. Pierce, *Miscellaneous preserver problems*, Linear and Multilinear Algebra **33** (1992), 109–119.

[18] L. Beasley and N.J. Pullman, *Linear operators strongly preserving idempotent matrices over semirings*, Linear Algebra Appl. **160** (1992), 217–229.

[19] L. Beasley and Sang-Gu Lee, *Linear operators strongly preserving* $r$*-potent matrices over semirings*, Linear Algebra Appl. **162–164** (1992), 589–599.

[20] L. Beasley and N.J. Pullman, *Polynomials which permute matrices over commutative antinegative semirings*, Linear Algebra Appl. **165** (1992), 167–172.

[21] E. Becker, *Partial orders on a field and valuation rings*, Comm. Algebra **7** (1979), 1933–1976.

[22] E. Becker and N. Schwartz, *Zum Darstellungssatz von Kadison–Dubois*, Arch. Math. **40** (1983), 421–428.

[23] D.B. Benson, *Bialgebras: Some foundations for distributed and concurrent computation*, Fund. Inform. **12** (1988), 427–486.

[24] J. Berstel and C. Reutenauer, *Rational Series and Their Languages*, Springer, Berlin (1988).

[25] G. Birkhoff, *Lattice Theory*, Amer. Math. Soc., Providence, RI (1967).

[26] M.N. Bleicher and S. Bourne, *On the embeddability of partially ordered halfrings*, J. Math. Mech. **14** (1965), 109–116.

[27] M. Botero de Meza and H.J. Weinert, *Erweiterung topologischer Halbringe durch Quotienten- und Differenzenbildung*, Jahresber. Deutsch. Math.-Verein. **73** (1971), 60–85.

[28] S. Bourne, *The Jacobson radical of a semiring*, Proc. Nat. Acad. Sci. USA **37** (1951), 163–170.

[29] S. Bourne, *On the homomorphism theorem for semirings*, Proc. Nat. Acad. Sci. USA **38** (1952), 118–119.

[30] S. Bourne, *On multiplicative idempotents of a potent semiring*, Proc. Nat. Acad. Sci. USA **42** (1956), 632–638.

[31] S. Bourne and H. Zassenhaus, *On a Wedderburn–Artin structure theory of a potent semiring*, Proc. Nat. Acad. Sci. USA **43** (1957), 613–615.

[32] S. Bourne and H. Zassenhaus, *On the semiradical of a semiring*, Proc. Nat. Acad. Sci. USA **44** (1958), 907–914.

[33] S. Bourne, *On compact semirings*, Proc. Japan Acad. **35** (1959), 332–334.

[34] S. Bourne, *On the radical of a positive semiring*, Proc. Nat. Acad. Sci. USA **45** (1959), 519.

[35] S. Bourne, *On locally compact halfrings*, Proc. Japan Acad. **36** (1960), 192–195.

[36] S. Bourne, *On locally compact positive halffields*, Math. Ann. **146** (1962), 423–426.

[37] S. Bourne, *On positive Banach halfalgebras without identity*, Stud. Math. **22** (1963), 247–249.

[38] S. Bourne and H. Zassenhaus, *Certain characterizations of the semiradical of a semiring*, Comm. Algebra **3** (1975), 525–530.

[39] L. Bröcker, *Positivbereiche in kommutativen Ringen*, Abh. Math. Sem. Univ. Hamburg **52** (1982), 170–178.

[40] B. Carré, *Graphs and Networks*, Clarendon Press, Oxford (1979).

[41] P. Cartier and D. Foata, *Problemes combinatoires de commutation et rearrangements*, SLNM 85, Springer, Berlin (1969), 1–88.

[42] V.V. Chermnykh, *Representation of positive semirings by sections*, Russian Math. Surveys **47** (1992), 180–182.

[43] A.H. Clifford and G.B. Preston, *The Algebraic Theory of Semigroups*, Vol. I, Amer. Math. Soc., Providence, RI (1961).

[44] A.H. Clifford and G.B. Preston, *The Algebraic Theory of Semigroups*, Vol. II, Amer. Math. Soc., Providence, RI (1967).

[45] T.C. Craven, *Orderings on semirings*, Semigroup Forum **43** (1991), 45–52.

[46] L. Dale, *The structure of ideals in a polynomial semiring in several variables*, Kyungpook Math. J. **26** (1986), 129–135.

[47] C. Dönges, *On quasi-ideals of semirings*, Internat. J. Math. Math. Sci. **17** (1994), 47–58.

[48] I.L. Dorroh, *Concerning adjunctions to algebra*, Bull. Amer. Math. Soc. **38** (1932), 85–88.

[49] R.E. Dover and H.E. Stone, *On semisubtractive halfrings*, Bull. Austral. Math. Soc. **12** (1975), 371–378.

[50] G. Duchamp and J.Y. Thibon, *Theoremes de transfert pour les polynomes partiellement commutatifs*, Theoret. Comput. Sci. **57** (1988), 239–249.

[51] D.W. Dubois, *A note on David Harrison's theory of preprimes*, Pacific J. Math. **21** (1967), 15–19.

[52] D.W. Dubois, *Second note on David Harrison's theory of preprimes*, Pacific J. Math. **24** (1968), 57–68.

[53] B.J. Dulin and J. Mosher, *The Dedekind property for semirings*, J. Austral. Math. Soc. **14** (1972), 82–90.

[54] S. Eilenberg, *Automata, Languages, and Machines*, Vol. A, Academic Press, New York (1974).

[55] R. Eilhauer, *Zur Theorie der Halbkörper, I*, Acta Math. Acad. Sci. Hungar. **19** (1968), 23–45.

[56] L. Fuchs, *Partially Ordered Algebraic Systems*, Pergamon Press, Oxford 1963; German edition: Vandenhoek & Ruprecht (1966).

[57] J.L. Galbiati and M.L. Veronesi, *On Boolean semirings*, Istit. Lombardo Accad. Sci. Lett. Rend. A **114** (1980), 73–88.

[58] E. Giraldes, *Decomposition of a semiring into division semirings*, Portugal. Math. **39** (1980), 349–356.

[59] K. Głazek, *A short guide through the literature on semirings*, Math. Inst. Univ. Wroclaw, Poland (1985).

[60] J.S. Golan, *Linear Topologies on a Ring*, Longman, Harlow (1987).

[61] J.S. Golan, *Semirings for the ring theorist*, Rev. Roumaine Math. Pures Appl. **35** (1990), 531–540.

[62] J.S. Golan, *The Theory of Semirings with Applications in Mathematics and Theoretical Computer Science*, Pitman Monographs and Surveys in Pure and Applied Mathematics 54, Longman (1992).

[63] D.A. Gregory and N.J. Pullman, *Semiring rank: Boolean rank and non-negative factorizations*, J. Combin. Inform. System. Sci. **8** (1983), 223–233.
[64] R.D. Griepentrog and H.J. Weinert, *Embedding semirings in semirings with identity*, Colloq. Math. Soc. János Bolyai vol. 20, Algebraic Theory of Semigroups, North-Holland, Amsterdam (1979), 225–245.
[65] R.D. Griepentrog and H.J. Weinert, *Correction and remarks to our paper "Embedding semirings in semirings with identity"*, Colloq. Math. Soc. János Bolyai vol. 39, Semigroups, Szeged, Amsterdam (1981), 491–493.
[66] M.P. Grillet, *Embedding of a semiring into a semiring with identity*, Acta Math. Acad. Sci. Hungar. **20** (1969), 121–128.
[67] M.P. Grillet, *Free semirings over a set*, J. Natur. Sci. Math. **9** (1969), 285–291.
[68] M.P. Grillet, *Green's relations in a semiring*, Portugal. Math. **29** (1970), 181–195.
[69] M.P. Grillet, *A semiring whose Green's relations do not commute*, Acta Sci. Math. **31** (1970), 161–166.
[70] M.P. Grillet, *Subdivision rings of a semiring*, Fund. Math. **67** (1970), 67–74.
[71] M.P. Grillet, *Examples of semirings of endomorphisms of semigroups*, J. Austral. Math. Soc. **11** (1970), 345–349.
[72] M.P. Grillet and P.A. Grillet, *Completely* 0*-simple semirings*, Trans. Amer. Math. Soc. **155** (1971), 19–33.
[73] M.P. Grillet, *On semirings which are embeddable into a semiring with identity*, Acta Math. Acad. Sci. Hungar. **22** (1971), 305–307.
[74] M.P. Grillet, *Semisimple A-semigroups and semirings*, Fund. Math. **76** (1972), 109–116.
[75] M.P. Grillet and P.A. Grillet, *Building semirings*, Estudos de Matematica, Lisboa (1974), 71–75.
[76] M.P. Grillet, *Semirings with a completely simple additive semigroup*, J. Austral. Math. Soc. **20** (1975), 257–267.
[77] D. Haftendorn, *Additiv kommutative und idempotente Halbringe mit Faktorbedingung, I*, Publ. Math. Debrecen **25** (1978), 107–116.
[78] D. Haftendorn, *Additiv kommutative und idempotente Halbringe mit Faktorbedingung, II*, Publ. Math. Debrecen **26** (1979), 5–12.
[79] J. Hanumanthachari, K. Venu Raju and H.J. Weinert, *Some results on partially ordered semirings and semigroups*, Algebra and Order, Proc. First Int. Symp. Ordered Algebraic Structures Luminy-Marseilles 1984, Heldermann, Berlin (1986), 313–322.
[80] D.K. Harrison, *Finite and Infinite Primes for Rings and Fields*, Amer. Math. Soc., Providence, RI (1966).
[81] H.E. Heatherly, *Distributive near-rings*, Quart. J. Math. Oxford (2) **24** (1973), 63–70.
[82] U. Hebisch, *On special $Q_r$-filters of semigroups, semirings and rings*, Semigroup Forum **30** (1984), 195–210.
[83] U. Hebisch and H.J. Weinert, *On euclidean semirings*, Kyungpook Math. J. **27** (1987), 61–88.
[84] U. Hebisch and L.C.A. van Leeuwen, *On additively or multiplicatively idempotent semirings and partial orders*, SLNM 1320, Springer, Berlin (1988), 154–161.
[85] U. Hebisch, *The Kleene theorem in countably complete semirings*, Bayreuth. Math. Schr. **31** (1990), 55–66.
[86] U. Hebisch and H.J. Weinert, *Semirings without zero divisors*, Math. Pannon. **1** (1990), 73–94.
[87] U. Hebisch, *Eine algebraische Theorie unendlicher Summen mit Anwendungen auf Halbgruppen und Halbringe*, Bayreut. Math. Schr. **40** (1992), 21–152.
[88] U. Hebisch and H.J. Weinert, *Generalized semigroup semirings which are zero divisor free or multiplicatively left cancellative*, Theoret. Comput. Sci. **92** (1992), 269–289.
[89] U. Hebisch and H.J. Weinert, *Halbringe-Algebraische Theorie und Anwendungen in der Informatik*, Teubner, Stuttgart (1993) (English edition in preparation).
[90] U. Hebisch, *Partial orders in semigroups and semirings of right quotients*, Semigroup Forum **49** (1994), 165–174.
[91] U. Hebisch and H.J. Weinert, *On the rank of semimodules over semirings*, Collect. Math., to appear.
[92] U. Hebisch and H.J. Weinert, *Radical Theory for Semirings*, to appear.
[93] M. Henriksen, *The $a^{n(a)} = a$ theorem for semirings*, Math. Japon. **5** (1958), 21–24.
[94] M. Henriksen, *Ideals in semirings with commutative addition*, Notices Amer. Math. Soc. **6** (1958), 321.
[95] D. Higgs, *Axiomatic infinite sums – an algebraic approach to integration theory*, Contemp. Math. vol. 2 (1980), 205–212.

[96] K.H. Hofmann, *Über lokalkompakte positive Halbkörper*, Math. Ann. **151** (1963), 262–271.
[97] D.R. Hughes and F.C. Piper, *Design Theory*, Cambridge Univ. Press, Cambridge (1985).
[98] H. Hutchins, *Division semirings with* $1+1=1$, Semigroup Forum **22** (1981), 181–188.
[99] H.C. Hutchins and H.J. Weinert, *Homomorphisms and kernels of semifields*, Period. Math. Hungar. **21** (1990), 113–152.
[100] S.A. Huq, *Distributivity in semigroups*, Math. Japon. **33** (1988), 535–541.
[101] K. Iizuka, *On the Jacobson radical of a semiring*, Tôhoku Math. J. **11** (1959), 409–421.
[102] K. Iizuka and I. Nakahara, *A note on the semiradical of a semiring*, Kumatoto J. Sci., Ser. A **4** (1959), 1–3.
[103] K. Iseki, *Ideal theory of semirings*, Proc. Japan Acad. **32** (1956), 554–559.
[104] K. Iseki and Y. Miyanaga, *Notes on topological spaces, III*, Proc. Japan Acad. **32** (1956), 325–328.
[105] K. Iseki and Y. Miyanaga, *Notes on topological spaces, IV*, Proc. Japan Acad. **32** (1956), 392–395.
[106] K. Iseki, *Notes on topological spaces, V*, Proc. Japan Acad. **32** (1956), 426–429.
[107] K. Iseki and Y. Miyanaga, *On a radical in a semiring*, Proc. Japan Acad. **32** (1956), 562–563.
[108] K. Iseki, *Ideals in semirings*, Proc. Japan Acad. **34** (1958), 29–31.
[109] K. Iseki, *On ideals in semirings*, Proc. Japan Acad. **34** (1958), 507–509.
[110] K. Iseki, *Quasiideals in semirings without zero*, Proc. Japan Acad. **34** (1958), 79–81.
[111] J. Jezek and T. Kepka, *Simple semimodules over commutative semirings*, Acta Sci. Math. **46** (1983), 17–27.
[112] G. Karner, *On limits in complete semirings*, Semigroup Forum **45** (1992), 148–165.
[113] P.H. Karvellas, *Inversive semirings*, J. Austral. Math. Soc. **18** (1974), 277–288.
[114] P.H. Karvellas, *Extension of a semigroup embedding theorem to semirings*, Canad. Math. Bull. **18** (1975), 297–298.
[115] A. Kaya and M. Satyanarayana, *Semirings satisfying properties of distributive type*, Proc. Amer. Math. Soc. **82** (1981), 341–346.
[116] T. Kepka, *Varieties of left distributive semigroups*, Acta Univ. Carolin. Math. Phys. **25** (1984), 3–18.
[117] S.C. Kleene, *Representation of events in nerve nets and finite automata*, Automata Studies, C.E. Shannon and J. McCarthy, eds, Princeton Univ. Press, Princeton (1956), 3–42.
[118] H. Koch, *Über Halbkörper, die in algebraischen Zahlkörpern enthalten sind*, Acta Math. Acad. Sci. Hungar. **15** (1964), 439–444.
[119] D. Kozen, *On Kleene algebras and closed semirings*, Lecture Notes in Comput. Sci. vol. 452 (1990), 26–47.
[120] D. Krob, *Monoides et semi-anneaux complets*, Semigroup Forum **36** (1987), 323–339.
[121] D. Krob, *Monoides et semi-anneaux continus*, Semigroup Forum **37** (1988), 59–78.
[122] D. Krob, *Models of a $K$-rational identity system*, J. Comput. Syst. Sci. **45** (1992), 396–434.
[123] W. Kuich, *The Kleene and the Parikh theorem in complete semirings*, Lecture Notes in Comput. Sci. vol. 267 (1987), 212–225.
[124] W. Kuich and A. Salomaa, *Semirings, Automata, Languages*, Springer, Berlin (1986).
[125] D.R. LaTorre, *On $h$-ideals and $k$-ideals in hemirings*, Publ. Math. Debrecen **12** (1965), 219–226.
[126] D.R. LaTorre, *A note on the Jacobson radical of a hemiring*, Publ. Math. Debrecen **14** (1967), 9–13.
[127] D.R. LaTorre, *The Brown–McCoy radical of a hemiring*, Publ. Math. Debrecen **14** (1967), 15–28.
[128] D.R. LaTorre, *A note on quotient semirings*, Proc. Amer. Math. Soc. **24** (1970), 463–465.
[129] H. Lee, *A short proof of Barbut's theorem*, J. Korean Math. Soc. **11** (1974), 141–142.
[130] S.A. Lesin and S.N. Samborskii, *Spectra of compact endomorphisms*, Adv. Soviet Math. **13** (1992), 103–118.
[131] L.D. Li, *On the structure of hemirings*, Simon Stevin **58** (1984), 91–113.
[132] Y.-F. Lin and J.S. Ratti, *The graphs of semirings*, J. Algebra **14** (1970), 73–82.
[133] Y.-F. Lin and J.S. Ratti, *Connectivity of the graphs of semirings: lifting and product*, Proc. Amer. Math. Soc. **24** (1970), 411–414.
[134] H. Lugowski, *Über die Vervollständigung geordneter Halbringe*, Publ. Math. Debrecen **9** (1962), 213–222.
[135] H. Lugowski, *Die Charakterisierung gewisser geordneter Halbmoduln mit Hilfe der Erweiterungstheorie*, Publ. Math. Debrecen **13** (1966), 237–248.

[136] H. Lugowski, *Über die Struktur gewisser geordneter Halbringe*, Math. Nachr. **51** (1971), 311–325.
[137] H. Lugowski, *Vollständige JIV-Halbringe mit negativen Elementen*, Math. Nachr. **61** (1974), 37–46.
[138] B. Mahr, *Iteration and summability in semirings*, Ann. Discrete Math. **19** (1984), 229–256.
[139] E.G. Manes and D.B. Benson, *The inverse semigroup of a sum-ordered semiring*, Semigroup Forum **31** (1985), 129–152.
[140] S.S. Mitchell and P. Sinutoke, *The theory of semifields*, Kyungpook Math. J. **22** (1982), 325–348.
[141] S.S. Mitchell and P.B. Fengolio, *Congruence-free commutative semirings*, Semigroup Forum **37** (1988), 79–91.
[142] J.R. Mosher, *Generalized quotients of hemirings*, Compositio Math. **22** (1970), 275–281.
[143] J.R. Mosher, *Semirings with descending chain condition and without nilpotent elements*, Compositio Math. **23** (1971), 79–85.
[144] K. Murata, *On the quotient semi-group of a non-commutative semi-group*, Osaka Math. J. **2** (1950), 1–5.
[145] A. Nakassis, *The diameter of the graph of a semiring*, Proc. Amer. Math. Soc. **60** (1976), 353–359.
[146] M.L. Noronha-Galvao, *Ideals in the semiring $N$*, Portugal. Math. **37** (1978), 113–117.
[147] D.M. Olson and T.L. Jenkins, *Radical theory for hemirings*, J. Natur. Sci. Math. **23** (1983), 23–32.
[148] D.M. Olson and A.C. Nance, *A note on radicals for hemirings*, Quaestiones Math. **12** (1989), 307–314.
[149] D.M. Olson, G.A.P. Heyman and H.J. LeRoux, *Weakly special classes of hemirings*, Quaestiones Math. **15** (1992), 119–126.
[150] D.M. Olson, H.J. LeRoux and G.A.P. Heyman, *Three special radicals for hemirings*, Quaestiones Math. **17** (1994), 205–215.
[151] F. Pastijn and A. Romanowska, *Idempotent distributive semirings, I*, Acta Sci. Math. **44** (1982), 239–253.
[152] F. Pastijn, *Idempotent distributive semirings, II*, Semigroup Forum **26** (1983), 151–166.
[153] K.R. Pearson, *Interval semirings on $R_1$ with ordinary multiplication*, J. Austral. Math. Soc. **6** (1966), 273–288.
[154] K.R. Pearson, *Certain topological semirings in $R_1$*, J. Austral. Math. Soc. **8** (1968), 171–182.
[155] G. Pilz, *Near-Rings*, North-Holland, Amsterdam (1977).
[156] B. Piochi, *Congruences on inversive hemirings*, Studia Sci. Math. Hungar. **23** (1988), 251–255.
[157] F. Poyatos, *Descomposiciones irreducibles en suma directa interna de ciertas estructuras algebraicas*, Rev. Mat. Hisp.-Amer. **27** (1967), 151–170.
[158] F. Poyatos, *The Jordan–Hölder theorem for semirings*, Rev. Mat.-Hisp. Amer. **40** (1980), 49–65.
[159] F. Poyatos, *Archimedean decompositions of left $S$-semimodules and semirings*, Studia Sci. Math. Hungar. **20** (1985), 323–324.
[160] V. Raju and J. Hanumanthachari, *The additive semigroup structure of semirings*, Math. Sem. Notes Kobe Univ. **11** (1983), 381–386.
[161] J.S. Ratti and Y.-F. Lin, *The diameters of the graphs of semirings*, J. Austral. Math. Soc. **11** (1970), 433–440.
[162] J.S. Ratti and Y.-F. Lin, *The graphs of semirings, II*, Proc. Amer. Math. Soc. **30** (1971), 473–478.
[163] J.S. Ratti and Y.F. Lin, *On anti-commutative semirings*, Internat. J. Math. Math. Sci. **12** (1989), 205–207.
[164] L. Rédei, *Die Verallgemeinerung der Schreierschen Erweiterungstheorie*, Acta Sci. Math. **13** (1952), 252–273.
[165] L. Rédei, *Algebra*, Geest & Portig (1959); English edition: Akadémiai Kiadó (1967).
[166] W.H. Reynolds, *Embedding a partially ordered ring in a division algebra*, Trans. Amer. Math. Soc. **158** (1971), 293–300.
[167] W.H. Reynolds, *A note on embedding a partially ordered ring in a division algebra*, Proc. Amer. Math. Soc. **37** (1973), 37–41.
[168] C. Reutenauer and H. Straubing, *Inversion of matrices over a commutative semiring*, J. Algebra **88** (1984), 350–360.
[169] G. Rodriguez, *Bande di semianelli monoidali*, Boll. Un. Mat. Ital. **14** (1977), 569–591.
[170] G. Rodriguez, *Decomposition of a semiring in a semilattice of semirings*, Boll. Un. Mat. Ital. Suppl. (1980), 53–67.
[171] G. Rodriguez, *Green equivalence in distributive semirings*, Atti Accad. Sci. Lett. Arti Palermo Ser. (5) **2** (1981/82), 181–193.

[172] A. Romanowska, *Free idempotent distributive semirings with a semilattice reduct*, Math. Japon. **27** (1982), 467–481.

[173] A. Romanowska, *Idempotent distributive semirings with a semilattice reduct*, Math. Japon. **27** (1982), 483–493.

[174] G. Rote, *A systolic array algorithm for the algebraic path problem (shortest paths; matrix inversion)*, Computing **34** (1985), 191–219.

[175] D.E. Rutherford, *The Cayley–Hamilton theorem for semi-rings*, Proc. Roy. Soc. Edinburgh **66** (1963), 211–215.

[176] T. Saito, *Ordered idempotent semigroups*, J. Math. Soc. Japan **4** (1962), 150–169.

[177] T. Saito, *The orderability of idempotent semigroups*, Semigroup Forum **7** (1974), 264–285.

[178] A. Salomaa and M. Soittola, *Automata-Theoretic Aspects of Formal Power Series*, Springer, Berlin (1978).

[179] M. Satyanarayana, *On the additive semigroup structure of semirings*, Semigroup Forum **23** (1981), 7–14.

[180] M. Satyanarayana, *On the additive semigroup of ordered semirings*, Semigroup Forum **31** (1985), 193–199.

[181] M. Satyanarayana, *Archimedean property of ordered semirings*, Semigroup Forum **33** (1986), 57–63.

[182] M. Satyanarayana, J. Hanumanthachari and D. Umamaheswarareddy, *On the additive structure of a class of ordered semirings*, Semigroup Forum **33** (1986), 251–255.

[183] J. Selden, *A note on compact semirings*, Proc. Amer. Math. Soc. **15** (1964), 882–886.

[184] M.K. Sen and M.R. Adhikari, *On maximal k-ideals of semirings*, Proc. Amer. Math. Soc. **118** (1993), 699–703.

[185] T. Shaheen and S.M. Yusuf, *Radicals of additively inversive hemirings*, Stud. Sci. Math. Hungar. **14** (1979), 303–309.

[186] J. Sichler and V. Trnkova, *Automorphisms of semirings and of their reducts*, Period. Math. Hung. **24** (1992), 167–177.

[187] H. Simmons, *The semiring of topologizing filters of a ring*, Israel J. Math. **61** (1988), 271–284.

[188] W. Slowikowski and W. Zawadowski, *A generalization of maximal ideals method of Stone and Gelfand*, Fund. Math. **42** (1955), 216–231.

[189] D.A. Smith, *On semigroups, semirings, and rings of quotients*, J. Sci. Hiroshima Univ. Ser. A **30** (1966), 123–130.

[190] F.A. Smith, *A structure theory for a class of lattice ordered semirings*, Fund. Math. **59** (1966), 49–64.

[191] F.A. Smith, *A subdirect decomposition of additively idempotent semirings*, J. Natur. Sci. Math. **7** (1967), 253–257.

[192] F.A. Smith, *$\ell$-semirings*, J. Natur. Sci. Math. **8** (1968), 95–98.

[193] O. Steinfeld, *Über die Struktursätze der Semiringe*, Acta Math. Acad. Sci. Hungar. **10** (1959), 149–155.

[194] O. Steinfeld, *Über Semiringe mit multiplikativer Kürzungsregel*, Acta Sci. Math. **24** (1963), 190–195.

[195] O. Steinfeld, *Über die Operatorendomorphismen gewisser Operatorhalbgruppen,* Acta Math. Acad. Sci. Hungar. **15** (1964), 123–131.

[196] O. Steinfeld and R. Wiegandt, *Über die Verallgemeinerung und Analoga der Wedderburn–Artinschen und Noetherschen Struktursätze*, Math. Nachr. **34** (1967), 143–156.

[197] H.E. Stone, *Ideals in halfrings*, Proc. Amer. Math. Soc. **33** (1972), 8–14.

[198] H.E. Stone, *Semirings with non-commutative addition*, Kyungpook Math. J. **13** (1973), 141–151.

[199] H.E. Stone, *Matrix representation of simple halfrings*, Trans. Amer. Math. Soc. **233** (1977), 339–353.

[200] H. Subramanian, *Von Neumann regularity in semirings*, Math. Nachr. **45** (1970), 73–79.

[201] T. Tamura, *Notes on semirings whose multiplicative semigroups are groups*, Semigroup Theory and Its Related Fields, Proc. 5th Symp. on Semigroups, Sakado Japan 1981 (1981), 56–66.

[202] J. Thibon, *Integrité des algèbres de séries formelles sur un alphabet partiellement commutatif*, Theoret. Comput. Sci. **41** (1985), 109–112.

[203] H.S. Vandiver, *Note on a simple type of algebra in which the cancellation law of addition does not hold*, Bull. Amer. Math. Soc. **40** (1934), 916–920.

[204] H.S. Vandiver and M.V. Weaver, *A development of associative algebra and an algebraic theory of numbers, IV*, Math. Mag. **30** (1956).

[205] V.G. van Horn and B. van Rootselaar, *Fundamental notions in the theory of seminear-rings*, Compositio Math. **18** (1966), 65–78.

[206] B. van Rootselaar, *Algebraische Kennzeichnung freier Wortarithmetiken*, Compositio Math. **15** (1963), 156–186.

[207] H.J. Weinert, *Über die Einbettung von Ringen in Oberringe mit Einselement*, Acta Sci. Math. Szeged **22** (1961), 91–105.

[208] H.J. Weinert, *Über Halbringe und Halbkörper, I*, Acta Math. Acad. Sci. Hungar. **13** (1962), 365–378.

[209] H.J. Weinert, *Über Halbringe und Halbkörper, II*, Acta Math. Acad. Sci. Hungar. **14** (1963), 209–227.

[210] H.J. Weinert, *Ein Struktursatz für idempotente Halbkörper*, Acta Math. Acad. Sci. Hungar. **15** (1964), 289–295.

[211] H.J. Weinert, *Über Halbringe und Halbkörper, III*, Acta Math. Acad. Sci. Hungar. **15** (1964), 177–194.

[212] H.J. Weinert, *On the extension of partial orders on semigroups of right quotients*, Trans. Amer. Math. Soc. **142** (1969), 345–353.

[213] H.J. Weinert, *Zur Theorie Levitzkischer Radikale in Halbringen*, Math. Z. **128** (1972), 325–341.

[214] H.J. Weinert, *Halbringe mit aufsteigender Kettenbildung für Annullatorideale*, J. Reine Angew. Math. **274/275** (1975), 417–423.

[215] H.J. Weinert, *Ringe mit nichtkommutativer Addition, I*, Jahresber. Deutsch. Math.-Verein. **77** (1975), 10–27.

[216] H.J. Weinert, *Ringe mit nichtkommutativer Addition, II*, Acta Math. Acad. Sci. Hungar. **26** (1975), 295–310.

[217] H.J. Weinert, *Related representation theorems for rings, semirings, nearrings and seminearrings by partial transformations and partial endomorphisms*, Proc. Edinburgh Math. Soc. **20** (1976/77), 307–315.

[218] H.J. Weinert, *A concept of characteristics for semigroups and semirings*, Acta Sci. Math. **41** (1979), 445–456.

[219] H.J. Weinert, *Multiplicative cancellativity of semirings and semigroups*, Acta Math. Acad. Sci. Hungar. **35** (1980), 335–338.

[220] H.J. Weinert, *Zur Theorie der Halbfastkörper*, Studia Sci. Math. Hungar. **16** (1981), 201–218.

[221] H.J. Weinert, *Seminearrings, seminearfields and their semigroup-theoretical background*, Semigroup Forum **24** (1982), 231–254.

[222] H.J. Weinert, *Extensions of seminearrings by semigroups of right quotients*, SLNM 998, 412–486, Springer, Berlin (1983).

[223] H.J. Weinert, *Über Quasiideale in Halbringen*, Contribution to General Algebra 2, Proc. of the Klagenfurt Conference 1982, Hölder-Pichler-Temsky, Wien; B.G. Teubner, Stuttgart (1983), 375–394.

[224] H.J. Weinert, *On 0-simple semirings, semigroup semirings, and two kinds of division semirings*, Semigroup Forum **28** (1984), 313–333.

[225] H.J. Weinert, *Partially ordered semirings and semigroups*, Algebra and Order, Proc. First Int. Symp. Ordered Algebraic Structures Luminy-Marseilles 1984, Heldermann, Berlin (1986), 265–292.

[226] H.J. Weinert, *Generalized semialgebras over semirings*, SLNM 1320, Springer, Berlin (1988), 380–416.

[227] H.J. Weinert and R.D. Griepentrog, *Embedding semirings by translational hulls*, Semigroup Forum **14** (1977), 235–246.

[228] H.J. Weinert and R.Wiegandt, *A Kurosh–Amitsur radical theory for proper semifields*, Comm. Algebra **20** (1992), 2419–2458.

[229] H.J. Weinert and R. Wiegandt, *Complementary radical classes of proper semifields*, Colloq. Math. Soc. János Bolyai vol. 61 (1991), 297–310.

[230] H.J. Weinert and R. Wiegandt, *On the structure of semifields and lattice-ordered groups*, Period. Math. Hungar., to appear.

[231] H.J. Weinert, M.K. Sen and M.R. Adhikari, *One-sided k-ideals and h-ideals on semirings*, Math. Pannon., to appear.

[232] R. Wiegandt, *Über die Struktursätze der Halbringe*, Ann. Univ. Sci. Budapest. Eötvös Sect. Math. **5** (1962), 51–68.

[233] A. Wongseelashote, *Semirings and path spaces*, Discrete Math. **26** (1979), 55–78.

[234] M. Yoeli, *A note on a generalization of boolean matrix theory*, Amer. Math. Monthly **68** (1961), 552–557.

[235] S.M. Yusuf, *Ideals in additively inversive semirings*, J. Natur. Sci. Math. **5** (1965), 45–56.

[236] S.M. Yusuf, *The classical radical of an additively inversive semirings*, J. Natur. Sci. Math. **5** (1965), 57–69.

[237] S.M. Yusuf and M. Shabir, *Radical classes and semisimple classes for hemirings,* Stud. Sci. Math. Hungar. **23** (1988), 231–235.

[238] H. Zassenhaus, *The Theory of Groups*, Chelsea, New York (1949).

[239] J. Zeleznikow, *Orthodox semirings and rings*, J. Austral. Math. Soc. A **30** (1980), 50–54.

[240] J. Zeleznikow, *Regular semirings*, Semigroup Forum **23** (1981), 119–136.

[241] J. Zeleznikow, *The natural partial order on semirings*, SLNM 848, Springer, Berlin (1981), 255–261.

[242] J. Zeleznikow, *On regular ring-semigroups and semirings*, Comment. Math. Univ. Carolin. **25** (1984), 129–139.

[243] U. Zimmermann, *Linear and Combinatorial Optimization in Ordered Algebraic Structures*, Annals of Discrete Mathematics vol. 10, North-Holland, Amsterdam (1981).

# Near-rings and Near-fields

Günter F. Pilz

*Institut für Mathematik, Johannes-Kepler-Universität, Linz, Austria*
*e-mail: guenter.pilz@jk.uni-linz.ac.at*

*Contents*



HANDBOOK OF ALGEBRA, VOL. 1
Edited by M. Hazewinkel

## Abstract

Near-rings are generalized rings: commutativity of addition is not assumed, and – more important – only one distributive law is required. The most famous example is the collection of all maps from an additive group into itself w.r.t. function addition and composition. Compared with a standard class of rings, endomorphism rings of abelian groups, one sees that rings describe "linear" maps on groups, while near-rings handle the general nonlinear case.

The theory of near-rings is now a sophisticated theory which has found numerous applications in various areas. In this article, we concentrate on a deep structure theorem for near-rings, the density theorem for primitive near-rings. This result is applied to interpolation theory, group theory and polynomials. Connections between other parts of near-rings (especially near-fields) and geometry come up at several places. Efficient block designs and codes can be constructed from finite near-rings. Finally we mention connections to ring theory, computer science, graph theory and other parts of the "outside world".

Only some easier proofs are given. For the others, references will be given; if these results are available in book form, we cite this book and not the original paper. Up to now, three books on near-rings and one on near-fields have appeared (see [1–4] in the list of references).

$\mathbb{N}$, $\mathbb{Z}$, $\mathbb{R}$ denote the sets if natural numbers, integers and reals, respectively, and $\mathbb{Z}_n$ denotes the set of residue classes modulo $n$.


## 1. Introduction to near-rings

Endomorphisms $h$ on groups fulfill the law $h(x+y) = h(x) + h(y)$, a property which is very easy to check in most situations. But suppose now you have a collection $M = \{f, g, h, \ldots\}$ of maps on a group at hand. Can you find out *within* this system, if the functions are "linear" (= endomorphisms), without referring to the arguments $x, y$? A good first guess would be that these functions are linear iff $(*)$ holds:

$$m_1 \circ (m_2 + m_3) = m_1 \circ m_2 + m_1 \circ m_3 \quad \text{for all } m_1, m_2, m_3 \in M. \qquad (*)$$

(Recall that the other distributive law $(m_1 + m_2) \circ m_3 = m_1 \circ m_3 + m_2 \circ m_3$ is always true, just by the definition of function addition!)

In fact, $(*)$ comes pretty close to linearity, but there are "a few" situations where also nonendomorphisms fulfill $(*)$. The easiest example for that is the collection $M = \{k\,\mathrm{id} \mid k \in \mathbb{Z}\}$ of all multiples of the identity function id on a nonabelian group $G$. Recall that $2\,\mathrm{id}$ is never an endomorphism if $G$ is not abelian. But

$$(k\,\mathrm{id}) \circ (s\,\mathrm{id} + t\,\mathrm{id}) = (ks + kt)\,\mathrm{id} = (k\,\mathrm{id}) \circ (s\,\mathrm{id}) + (k\,\mathrm{id}) \circ (t\,\mathrm{id})$$

holds for all $k, s, t \in \mathbb{Z}$. The system $\{k\,\mathrm{id} \mid k \in \mathbb{Z}\}$ is also closed w.r.t. addition and composition. Hence it is a (not very typical) example of a near-ring:

DEFINITION 1.1. A *near-ring* is a set $N$ together with two binary operations "+" and "$\circ$" such that

(i) $(N, +)$ is a group (not necessarily abelian),
(ii) $(f+g) \circ h = f \circ h + g \circ h$ for all $f, g, h \in N$,
(iii) $(f \circ g) \circ h = f \circ (g \circ h)$ for all $f, g, h \in N$.

Hence there are two axioms missing for near-rings compared with rings: addition is not necessarily abelian and only one distributive law is assumed. It should be clear how to define concepts like subnear-rings and near-ring homomorphisms. More precisely, we have defined *right near-rings*. Using the law (ii)$'$: $f \circ (g+h) = f \circ g + f \circ h$ would yield *left near-rings*.

EXAMPLE 1.2. Near-rings are abundant. Let $(G, +)$ be a group (not necessarily abelian), $T$ a topological group, $V$ a vector space, and $R$ a commutative ring with identity. With respect to addition $+$ and composition $\circ$, the following sets are near-rings:

$M(G) := \{f \mid f\colon G \to G\}$
$M_0(G) := \{f \in M(G) \mid f(0) = 0\}$
$M_c(G) := \{f \in M(G) \mid f \text{ is constant}\}$
$M_{\text{cont}}(T) := \{f \in M(T) \mid f \text{ is continuous}\}$
$M_{\text{aff}}(V) := \{f \in M(V) \mid f \text{ is affine } (= \text{the sum of a linear and a constant map})\}$
$P(R) := \{f \in M(R) \mid f \text{ is a polynomial function}\}$
$R[x]$ (as $(R[x], +, \circ)$!)

Of course, every ring is a near-ring. If we define $*$ on any group $(G, +)$ by $a * b := a$, we get a near-ring $(G, +, *)$. Hence every group can be turned into a near-ring.

More examples will follow. We now show that every near-ring can be considered as a subnear-ring of some $M(G)$:

THEOREM 1.3. *For every near-ring $N$ there is some group $G$ with $N \subsetneq M(G)$.*

PROOF. Let $G$ be any group containing $(N, +)$ properly. For $n \in N$ and $g \in G$ let $\phi_n(g)$ be $= n \circ g$ if $g \in N$ and $= n$ otherwise. Then the map $\Phi\colon N \to M(G)$ sending $n$ to $\phi_n$ is easily seen to be a monomorphism. □

Since $M(G)$ contains an identity (the identity function id), we get from 1.3 instantly

COROLLARY 1.4. *Every near-ring can be embedded in a near-ring with identity.*

Note that 1.3 and 1.4 have their analogues in ring theory. Since every ring can be embedded in the endomorphism ring of some abelian group, we might view a ring as a "system of linear maps", while near-rings "consist of arbitrary mappings on groups". This reveals the basic difference between rings and near-rings, and this point will become even more apparent in the sequel. Unless indicated otherwise, we will, from now on, write "products" $f \circ g$ simply as juxtaposition $fg$.

While 1.3 is basically shown as in ring theory, the proof of 1.4 drastically differs from the one for rings (this is typical for some parts of near-ring theory). Also, it is not always possible (as it is for rings) to embed a near-ring as an ideal in one with identity ([15]).

DEFINITION 1.5. A near-ring $N = (N, +, \circ)$ is a *near-field* if the set $N^*$ of nonzero elements of $N$ forms a group w.r.t. $\circ$.

Historically, the first near-rings considered were actually these near-fields which we will study in more detail in Section 2. The early 30's saw the first "proper" near-ring considerations. If $G$ is a nonabelian group and $h_1, h_2$ are endomorphisms of $G$ then

$$(h_1 + h_2)(g_1 + g_2) = h_1(g_1) + h_1(g_2) + h_2(g_1) + h_2(g_2),$$

which is in general not equal to

$$h_1(g_1) + h_2(g_1) + h_1(g_2) + h_2(g_2) = (h_1 + h_2)(g_1) + (h_1 + h_2)(g_2).$$

Hence the sum of two endomorphisms is rarely an endomorphism again. Fitting [24] studied cases in which the sum of two automorphisms is again an automorphism. We'll return to this question later on.

In 1938, H. Wielandt [77] initiated a structure theory (semisimplicity) of near-rings (which he called "Stamm" = tribe). Much work was done by him in unpublished manuscripts. In a dissertation under the guidance of E. Artin, D.W. Blackett [16] studied simple and semisimple near-rings around 1950. A dozen years later, G. Betsch [14] defined and studied the first and still most important type of a Jacobson-type radical for near-rings. Among several nice applications of near-rings, G. Ferrero and J.R. Clay discovered the "down-to-earth-application" of constructing Balanced Incomplete Block Designs from planar near-rings in the early 70's (see Section 6.1).

For near-rings we always have the law $0x = 0$; this can be shown as for rings. The lack of the second distributive law, however, does not allow us to show that $x0 = 0$ in general. An easy counterexample is $M(G)$: not every function maps 0 into 0.

DEFINITION 1.6. For a near-ring $N$, we call
$N_0 := \{n \in N \mid n \circ 0 = 0\}$ the *zero-symmetric part* of $N$, and
$N_c := \{n \in N \mid n \circ 0 = n\}$ the *constant part* of $N$.
$N$ is called *zero-symmetric* (*constant*) if $N = N_0$ ($N = N_c$, respectively).

In fact, if $n0 = n$ holds then we get $nm = (n0)m = n(0m) = n0 = n$ for each $m \in N$; this justifies the name "constant". If we apply 1.6 to $N = M(G)$ we get $(M(G))_0 = M_0(G)$ and $(M(G))_c = M_c(G)$. $N_0$ and $N_c$ are easily seen to be subnear-rings of $N$. Moreover, $(N_0, +)$ is normal in $(N, +)$. The zero-symmetric near-ring $N_0$ is "closer" to the class of rings (sometimes it actually will be a ring – see, e.g., $M_{\mathrm{aff}}(V)$), while $N_c$ has a trivial multiplication $nm = n$, hence $N_c$ is only interesting as a group.

PROPOSITION 1.7. *For every near-ring $N$, we have $N = N_0 + N_c$ and $N_0 \cap N_c = \{0\}$.*

$n = (n - n \circ 0) + n \circ 0$ does the decomposition job in 1.7. Hence $(N, +)$ is a semidirect sum of $(N_0, +)$ and $(N_c, +)$.

DEFINITION 1.8. For a near-ring $N$,

$$N_d := \{n \in N \mid n(m+m') = nm + nm' \text{ for all } m, m' \in N\}$$

is the *distributive part* of $N$; each $n \in N_d$ is called *distributive*. $N$ is called *distributively generated* (d.g.) if $N_d$ generates $(N, +)$, and *distributive* if $N = N_d$. If $N_d = N_0$ and if $(N, +)$ is abelian then $N$ is called an *affine near-ring*.

Clearly $N_d \subseteq N_0$. If $(N, +)$ is abelian then $N_d$ is a ring. If $N = M_{\mathrm{aff}}(V)$ (see 1.2) then $N_0 = N_d = \mathrm{Hom}(V, V), N_c = M_c(V)$, so $M_{\mathrm{aff}}(V)$ is an affine near-ring. Important examples of d.g. near-rings will follow in Section 3. Distributive near-rings are the place where near-rings meet semirings (see the article "Semirings and Semifields" by U. Hebisch and H.J. Weinert in this volume of the *Handbook of Algebra*). The corresponding diagram is given by

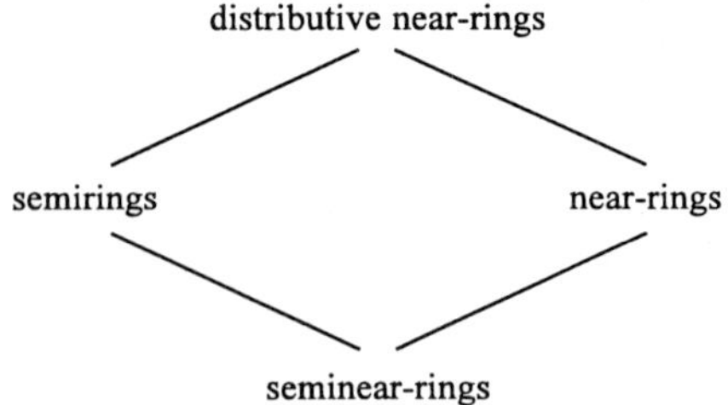


For seminear-rings see 6.7.1. We now take a brief look at ideals.

DEFINITION 1.9. Let $N$ be a near-ring and I a normal subgroup of $(N, +)$. $I$ is called a *right ideal* iff $i \circ n \in I$ for all $i \in I$, $n \in N$, *left ideal* if $n \circ (i + m) - n \circ m \in I$ for all $i \in I$, $n, m \in N$, and an *ideal* if $I$ is a left and a right ideal. $N$ is *simple* if has it no ideals besides $\{0\}$ and $N$.

Observe that the condition $n(i + m) - nm \in I$ can only be reduced to $ni \in I$ if $n \in N_d$. We now examine the case $N = M_0(G)$. For proofs, see [31] or [3].

THEOREM 1.10.
(i) *All minimal left ideals of $M_0(G)$ are given by*

$$L_g := \{f \in M_0(G) \mid f(g') = 0 \textit{ for all } g' \neq g\} \quad (g \in G,\ g \neq 0).$$

(ii) *All maximal left ideals of $M_0(G)$ are given by*

$$M_g := \{f \in M_0(G) \mid f(g) = 0\} \quad (g \in G,\ g \neq 0).$$

(iii) *$M_0(G)$ is simple.*
(iv) *$M(G)$ is simple unless $|G| = 2$.*

THEOREM 1.11. *The following are equivalent for* $M_0(G)$*:*

(i) *All left ideals are given by* $L_H := \{f \in M_0(G) \mid f|_H = 0\}$ *with* $H \subseteq G$.

(ii) $M_0(G)$ *is the direct sum of all* $L_g$ *of* 1.10(i).

(iii) $M_0(G)$ *fulfills the descending chain condition on left ideals.*

(iv) $M_0(G)$ *fulfills the ascending chain condition on left ideals.*

(v) $M_0(G)$ *is finite.*

(vi) $G$ *is finite.*

In near-ring theory, left ideals are much more important than right ideals. This is due to the fact that left ideals are precisely the kernels of "$N$-near-module" homomorphisms (cf. Chapter 3), while nothing similar applies to right ideals.

We have seen (1.3) that near-rings "describe mappings on groups". More generally, we might consider systems of mappings on semigroups (e.g., $M(\mathbb{N})$), vector spaces, lattices, and so on. Proceeding in this way, we can start with a universal algebra $(A, \Omega)$, form the collection $A^A$ of all mappings $A \to A$, define the operations of $\Omega$ pointwisely to these functions, and add composition $\circ$ as additional operation. We then get a "richer" structure $M(A) := (A^A, \Omega \cup \{\circ\})$. Examples:

| $A$ | $M(A)$ |
|---|---|
| set | semigroup |
| semigroup | seminear-ring |
| group | near-ring |
| module | $(-)$ |
| vector space | near-algebra |
| ring | composition ring |
| near-ring | $(-)$ |
| $\Omega$-group | $\Omega$-composition group |
| lattice | tri-operational lattice algebra |

$(-)$ means: not even a name was given to these structures. For instance, $(\mathbb{R}^\mathbb{R}, +, \cdot, \circ)$ is a nice example of a composition ring: $(\mathbb{R}^\mathbb{R}, +, \cdot)$ is a ring, $(\mathbb{R}^\mathbb{R}, +, \circ)$ a near-ring, and $(f \cdot g) \circ h = (f \circ h) \cdot (g \circ h)$ holds for all $f, g, h \in \mathbb{R}^\mathbb{R}$. Much remains to be done in these areas.

## 2. Near-fields

The whole thing started with a special class of near-rings: with near-fields. At the beginning of this century, L.E. Dickson constructed in [22] the first example of "proper" near-fields by "distorting" the multiplication in a field. More precisely, starting from a field $(F, +, \cdot)$ and a *coupling map* $\Phi$: $F^* \to \mathrm{Aut}(F, +)$, $f \to \phi_f$, with $\phi_f \circ \phi_g = \phi_{\phi_f(g)\cdot f}$, Dickson defined $f \cdot_\Phi g := \phi_g(f) \cdot g$ (and $f \cdot_\Phi 0 := 0$). Then $F^\Phi := (F, +, \cdot_\Phi)$, called the *$\Phi$-derivation* of $F$, is a near-field, and in general not a field.

DEFINITION 2.1. A near-field $N$ is a *Dickson near-field* if there is some field $F$ with $N = F^{\Phi}$ (for a suitable $\Phi$).

From the numerous deep results about near-fields we mention

THEOREM 2.2 (B.H. Neumann [62], Karzel [39], et al.). *If $N$ is a near-field then $(N, +)$ is abelian.*

THEOREM 2.3 (Zemmer [81]). *If $N$ is a near-field then $N_d$ is a field and* $\operatorname{char} N = \operatorname{char} N_d$ *(which is either* 0 *or a prime).*

THEOREM 2.4 (Zassenhaus [79], see also [4]). *A finite near-field is either a Dickson near-field, or it belongs to* 8 *exceptional near-fields of order* $2$, $5^2$, $7^2$, $11^2$ *(two near-fields),* $23^2$, $29^2$ *or* $59^2$.

The exceptional near-fields of orders $p^2$ in 2.4 can easily be described by means of $2 \times 2$-matrices. The exceptional near-field of order 2 is $(\mathbb{Z}_2, +, \circ)$ with $x \circ y := x$ for $x, y \in \mathbb{Z}_2$. Hence all finite near-fields can considered to be "known". The smallest "interesting" example is $(\mathrm{GF}(9), +, \circ)$ with $x \circ y := xy$ if $y$ is a square and $x \circ y = x^3 y$ otherwise. Its multiplicative group is the quaternion group.

The constant near-field $(\mathbb{Z}_2, +, \circ)$ mentioned above is, by the way, the only near-field $N$ which is not zero-symmetric. In fact, let $c$ be a nonzero constant element of $N_c^*$. If $n, n'$ are in $N$ then $c \circ n = c = c \circ n'$; multiplication by $c^{-1}$ yields $n = n'$. Hence $N = \{0, 1\} \cong (\mathbb{Z}_2, +, \circ)$.

For many years, it was an open problem if there exist infinite non-Dicksonian near-fields. A new (but complicated) construction method enabled H. Zassenhaus to show

THEOREM 2.5 (Zassenhaus [80]). *There exist infinite non-Dicksonian near-fields (for every prime characteristic).*

From 1907 on, Veblen, Wedderburn and many successors used near-fields $N$ to coordinatize geometric planes $\mathcal{G}$, so that the points of $\mathcal{G}$ are just the elements of $N \times N$ and the lines of $\mathcal{G}$ are given by all $\{(x, xa + b) \mid x \in N\}$ and $\{(c, x) \mid x \in N\}$ for $a \in N^*$ and $b, c \in N$. Given any geometric plane, it is certainly a big achievement to find a suitable "domain" that coordinatizes it. Descartes did that for the "real plane", using the field $\mathbb{R}$. For more general types of planes, more general "domains" that $\mathbb{R}$ are needed. Two results in this direction are given in 2.7 below. For this, we need another concept. In geometry, we usually want two lines $y = xa + b$ and $y = xc + d$ with $a \neq c$ to intersect in precisely one point. Since $xa + b = xc + d$ can be transformed into $xa = xc + (d - b)$, we find it natural to give the following

DEFINITION 2.6. A near-field $N$ is *planar* (or *projective*) if all equations $xa = xb + c$ $(a, b, c \in N,\ a \neq b)$ have a unique solution.

It can be shown (see, e.g., Zemmer [81]) that all finite near-fields are planar. But there do exist (infinite) nonplanar near-fields [4].

THEOREM 2.7. *Let $\mathcal{G}$ be an affine plane, coordinatized by $N$.*

(i) *$\mathcal{G}$ is a translation plane iff $N$ is a "Veblen–Wedderburn system" (i.e. a "multiplicatively nonassociative near-field").*

(ii) $\mathcal{G}$ *is a translation plane of the "Lenz–Barlotti-type IV.a.2" iff* $N$ *is a planar near-field.*

The proof of (i) is, e.g., in M. Hall's book [30], while (ii) can be found (along with many related results) in Dembowski [21].

Zassenhaus also initiated the study of the role near-rings play in group theory. A group $G$ of permutations on the set $X$ is called *sharply k-transitive* if for all $x_1, \ldots, x_k$ of pairwise different elements of $X$ and all pairwise different $y_1, \ldots, y_k$ in $X$ there is exactly one $g \in G$ with $g(x_i) = y_i$ $(1 \leqslant i \leqslant k)$.

Sharply 1-transitive permutation groups are just the regular ones. The sharply $k$-transitive groups for $k \geqslant 4$ were basically already known to C. Jordan (1872): they are finite and isomorphic either to a symmetric group of degree $n \geqslant 4$ or to alternating group of degree $n \geqslant 6$ or to one of the Mathieu groups of degree 11 or 12. Hence it remained to determine the sharply 2- and 3-transitive groups. For proofs and more details see, e.g., [44].

THEOREM 2.8. *If G is a sharply 2-transitive groups then there is a "near-domain" (i.e. roughly spoken, an "additively nonassociative near-field") N such that G is isomorphic to the group of all transformations* $x \to x \circ a + b$ $(a, b \in N,\ a \neq 0)$.

Since all finite near-domains are known to be near-fields, 2.4 allows us to say that all finite sharply 2-transitive groups are "known". The sharply 3-transitive groups were shown to consist of "fractional affine transformations"

$$x \mapsto \frac{x \circ a + b}{x \circ c + d}, \quad a, b, c, d \in N,\ a \circ d \neq b \circ c,$$

where $N$ is a "Karzel–Tits-field" (i.e. a certain near-domain). This finally concluded the description of all sharply transitive groups.

The transformations $x \mapsto x \circ a + b$ in a near-field also form important examples of Frobenius groups. Recall that a group $\Gamma$ of permutations on a set $X$ is a *Frobenius group* if each $\gamma \in \Gamma$, $\gamma \neq \mathrm{id}$, has at most one fixed point, and if the set of all fixed-point-free $\gamma$, together with id, forms a transitive proper normal subgroup $K_\Gamma$ of $\Gamma$. This $K_\Gamma$ is called the *Frobenius-kernel* of $\Gamma$. If $\Gamma$ is finite, $K_\Gamma$ is known to be characteristic and nilpotent. $\Gamma$ is always a semidirect product of $K_\Gamma$ and some "Frobenius complement".

THEOREM 2.9 (André [11]). *If N is a planar near-field with more than two elements then* $\Gamma := \{x \to x \circ a + b \mid a, b \in N,\ a \neq 0\}$ *is a Frobenius group on N, and its Frobenius kernel is given by* $K_\Gamma = \{x \to x + b \mid b \in N\}$.

Many important classes of Frobenius groups arise from planar near-fields in this manner.

Another connection to geometry was developed by H. Karzel. It concerns projective spaces in which the collineations (= automorphisms) have a simple algebraic description:

DEFINITION 2.10. $(P, \mathcal{L})$, is a *projective incidence group* if $(P, \mathcal{L})$ is a projective space, $(P, \cdot)$ a group, and each $q \to p \cdot q$ is a collineation of $(P, \mathcal{L})$.

THEOREM 2.11 (Karzel [39]). *Let $N$ be a near-field and $F$ a subfield of $N$ such that $F^*$ is normal in $(N^*, \cdot)$ and $n \circ (f + f') = n \circ f + n \circ f'$ for all $n \in N, f, f' \in F$. If $S$ denotes the set of all subspaces of ${}_FN$ of dimension 2 then $(N^*/F^*, S, \cdot)$ is a Desarguesian projective incidence group.*

*Conversely, every Desarguesian projective incidence group arises in this way: the near-field $N$ is "essentially uniquely" determined by the incidence group.*

For more on the recent developments in the theory of near-fields see the surveys [12] and [42] and the book [4] by Wähling.

## 3. The structure of near-rings

As in ring theory, one can learn a lot about near-rings if one studies how they behave on their "offsprings", i.e. on their "near-modules":

DEFINITION 3.1. Let $N$ be a near-ring and $(G, +)$ a group such that for all $n \in N$ and $g \in G$ a "product" $ng \in G$ is defined. Then $G$ is called an *$N$-group* or *$N$-near-module* if $(n_1 + n_2)g = n_1g + n_2g$ and $(n_1n_2)g = n_1(n_2g)$ hold for all $n_1, n_2 \in N$ and $g \in G$.

EXAMPLE 3.2. Every (ring-) $N$-module is an $N$-group, of course. Every group $(\Gamma, +)$ is a $\mathbb{Z}$-group and an $M(\Gamma)$-group in the natural sense. Hence every group can be considered as an $N$-group in several ways.

DEFINITION 3.3. An $N$-group $G$ is called *irreducible* if $NG \neq \{0\}$ and if there is no nontrivial subgroup $H$ of $G$ with $N_0H \subseteq H$. Kernels of $N$-group-homomorphisms (which are defined as expected) are called *$N$-ideals* of $G$. If $G$ only has the trivial $N$-ideals then $G$ is called *$N$-simple*.

If one transfers ideas from ring to near-ring theory, sometimes equivalent concepts for rings become inequivalent for near-rings. Hence there are different possible definitions for "irreducible". Our definition coincides with the concept of "$N$-groups of type 2" in the literature on near-rings.

We have to study the action of $N$ upon its $N$-groups $G$. Especially disgusting will be those $n \in N$ which "kill" everything in $G$. Also, we shall need the "$N$-automorphisms" of $G$:

DEFINITION 3.4. Let $G$ be an $N$-group.

(i) $A(G) := \{n \in N \mid ng = 0 \text{ for all } g \in G\}$ is called the *annihilator* of $G$.

(ii) $\Phi(G) := \{\phi \in \mathrm{Aut}(G, +) \mid \phi(ng) = n\phi(g) \text{ for all } n \in N_0 \text{ and } g \in G\}$.

For every $N$-group $G$, $A(G)$ is an ideal of $N$. Those elements in $N$ might be considered to be the "‘‘worst" ones which "liquidate" all "small" $N$-groups. We collect them in the "radical":

DEFINITION 3.5. The *radical* $J(N)$ of the near-ring $N$ is the intersection of all $A(G)$ where $G$ is an irreducible $N$-group. $N$ is called *semisimple* if $J(N) = \{0\}$. $N$ is *primitive* if $N$ has an irreducible $N$-group $G$ with $A(G) = \{0\}$ (we then say that $N$ is *primitive on* $G$). An ideal $I$ of $N$ is called a *primitive ideal* if $N/I$ is a primitive near-ring.

As an intersection of ideals, $J(N)$ is itself an ideal of $N$. Clearly, every primitive near-ring is semisimple, since a primitive near-ring has $A(G) = \{0\}$ for some $G$. Similar to the remark after 3.3, we have defined what is known as "2-primitivity", "2-semisimplicity" and the "2-radical, $J_2$".

If one defines subdirect products similarly to ring theory (or according to universal algebra) one can prove as in ring theory:

THEOREM 3.6.

(i) *For every near-ring* $N, N/J(N)$ *is semisimple.*

(ii) *A near-ring is semisimple iff it is isomorphic to a subdirect product of primitive near-rings.*

So far, the theory runs along the same lines as ring theory. Theorem 3.6 tells us what one has to split off from an arbitrary near-ring in order to get a semisimple one. These near-rings, in turn, are in some way known if the primitive near-rings are known. The complete determination 3.8 of all primitive near-rings will be the central result of this chapter. We need some preparations:

DEFINITION 3.7. Let $N \subseteq M \subseteq M(G)$ and $k \in \mathbb{N}$. The set $N$ is called *$k$-fold transitive* w.r.t. $M$ if for all $g_1, \dots, g_k \in G$ and all $m \in M$ there is some $n \in N$ with $n(g_i) = m(g_i)$ $(1 \leqslant i \leqslant k)$. $N$ is *dense* in $M$ if $N$ is $k$-fold transitive w.r.t. $M$ for all $k \in \mathbb{N}$.

If one introduces the "finite topology" in $M$ by taking all

$$S(m, g) := \{m' \in M \mid m'(g) = m(g)\}$$

as a subbase then the density concept of 3.7 is just the topological concept of density.

In order to describe primitive near-rings completely, we need a new class of near-rings.

DEFINITION 3.8. Let $S$ be a semigroup of endomorphisms of the group $(G, +)$. Then

$$M_S(G) := \{f \in M(G) \mid f(0) = 0 \text{ and } f \circ s = s \circ f \text{ for all } s \in S\}$$

is called the *centralizer near-ring* on $G$ w.r.t. $S$.

If $S = \{\mathrm{id}\}$ then $M_S(G) = M_0(G)$. If $R$ is a ring, $G$ an $R$-module and

$$S = \{g \to \lambda g \mid \lambda \in R\}$$

then $M_S(G) := M_R(G)$ consists of all "homogeneous functions" $f$ (which fulfill $f(\lambda g) = \lambda f(g)$ for all $\lambda \in R$ and $g \in G$). This type of near-ring will be explored in Section 6.4. We are now able to state our fundamental structure theorem.

DENSITY THEOREM 3.9. *Let the near-ring $N$ with identity be primitive on $G$.*

Case I: *$N_0$ is a ring. Then $G$ is a vector space over the skew-field $F := \mathrm{Hom}_N(G,G)$ and $N$ is dense in $\mathrm{Hom}_F(G,G)$ (if $N = N_0$) or dense in $M_{\mathrm{aff}}(G)$ (if $N \neq N_0$).*

Case II: *$N_0$ is not a ring. Then $\Phi(G)$ is a fixed-point-free group of automorphisms of $G$, and if $S$ denotes $\Phi(G) \cup \{0\}$, $N$ is dense in $M_S(G)$ (if $N = N_0$) or, otherwise, dense in $M_S(G) + M_c(G)$.*

The proof of 3.9 can be found in [3]. The first alternative in case I is just Jacobson's Density theorem for rings. 3.9 does not give a new proof of Jacobson's result; case I has to be split off by a careful investigation of the lattice of left ideals of $N$. In both cases, either $N_c = \{0\}$ or else $N_c = M_c(G)$.

COROLLARY 3.10. *Let $N$ be primitive on $G$, containing an identity and with $\Phi(G) = \{\mathrm{id}\}$. Then $N$ is dense in one of the following near-rings:*

Case I (*$N_0$ a ring*): *$N$ is dense in $\mathrm{Hom}_F(G,G)$, or in $M_{\mathrm{aff}}(G)$* ("linear case").

Case II (*$N_0$ a "nonring"*): *$N$ is dense in $M_0(G)$ (if $N = N_0$) or in $M(G)$ (if $N \neq N_0$)* ("nonlinear case").

If $N$ fulfills the descending chain condition on left ideals then "density" is the same as "equality". Let us take a look now at some applications of the Density theorem.

Given a near-ring $N$ of mappings on a group $G$, it might be interesting to ask for $k \in \mathbb{N}$, for all given $x_1, \ldots, x_k \in G$ (distinct) and $y_1, \ldots, y_k \in G$ when is there some $n \in N$ with $n(x_i) = y_i$ for $1 \leqslant i \leqslant k$. Obviously $N$ fulfills this "*$k$-interpolation property*" iff $N$ is a $k$-fold transitive subnear-ring of $M(G)$ (see 3.7). If $N$ is 3-fold transitive on $G$ then $N_0$ is 2-fold transitive on $G^* = G\backslash\{0\}$. It is then clear that $G$ cannot have a proper subgroup $H$ with $N_0 H \subseteq H$. Since $A(G) = \{0\}$, $N_0$ is primitive on $G$. It is easy to see that $N$ is primitive on $G$ with $\Phi(G) = \{\mathrm{id}\}$. Hence Corollary 3.10 shows the nontrivial part of

THEOREM 3.11. *Let $N \neq N_0$ be a subnear-ring of $M(G)$ containing* id *and suppose that $N_0$ is not a ring. Then $N$ is 3-fold transitive on $G$ iff $N$ is $k$-fold transitive on $G$ for all $k \in N$ (i.e. $N$ is dense in $M(G)$).*

Hence it is a consequence of the Density Theorem that "if $N$ interpolates at 3 places then at all (finitely many) places".

As remarked in Section 2, $\mathrm{End}\, G$ is in general not a ring if $G$ is not abelian, because the sum of two endomorphisms is usually not an endomorphism anymore. Hence it is a good idea to look at the additive closure.

DEFINITION 3.12. Let $(G,+)$ be a group. Let $E(G)$, $A(G)$ and $I(G)$ be the additive closures in $M(G)$ of $\mathrm{End}\, G, \mathrm{Aut}\, G$ and $\mathrm{Inn}\, G$, respectively.

It is easy to show the following statements.

THEOREM 3.13. *For each group $G, E(G), A(G)$ and $I(G)$ are distributively (by $\mathrm{End}\, G$, $\mathrm{Aut}\, G, \mathrm{Inn}\, G$, respectively) generated near-rings. So they are zero-symmetric. The $E(G)$-,*

*$A(G)$-, and $I(G)$-subgroups of $G$ are precisely the fully invariant, the characteristic, and the normal subgroups, respectively.*

Hence $I(G)$ is primitive on $G$ iff $G$ is simple. If $\Phi(G) \neq \mathrm{id}$ and $G$ is finite, then $\Phi(G)$ contains a fixed-point-free automorphism of $G$ of prime order. By a well-known theorem of Thompson, $G$ is then nilpotent and (since $G$ is simple) nonabelian; so $E(G) = \operatorname{End} G$ is a ring in this case. Similar considerations apply to $A(G)$ and $E(G)$; so we have shown the basic parts of

THEOREM 3.14. *Let $G$ be a finite group, $|G| > 2$.*
(i) *$E(G) = M_0(G) \Leftrightarrow G$ is nonabelian and characteristically simple.*
(ii) *$A(G) = M_0(G) \Leftrightarrow G$ is nonabelian and invariantly simple.*
(iii) *$I(G) = M_0(G) \Leftrightarrow G$ is nonabelian and simple.*

In a "near-ring-free" language we can restate, e.g., 3.14 (iii) as

COROLLARY 3.15. *Let $G$ be a group with more than 2 elements. Every map from $G$ to $G$ is the sum of inner automorphisms and a constant map $\Leftrightarrow$ $G$ is finite, simple and nonabelian.*

If $R$ is a commutative ring with identity then the near-ring $P(R)$ of polynomial functions on $R$ is easily seen to be primitive on $(R,+)$ iff $R$ is a finite field. Then $\Phi(R) = \{\mathrm{id}\}$ and the Density Theorem gives a new proof (without using Lagrange's interpolation formula) of

THEOREM 3.16. *Let $R$ be a commutative ring with identity. Every map $R \to R$ is a polynomial function $\Leftrightarrow$ $R$ is a finite field.*

But that brings us right away into the next section.

## 4. Polynomials

If we look at a polynomial $p = a_0 + a_1x + a_2x^2 + \cdots + a_nx^n$ over a commutative ring $R$ with identity, we can view $p$ as a "typical element which can be generated by $R \cup \{x\}$ using the operations and laws of commutative rings with identity". So the "shape" of $p$ depends not only on $R$, but also the class of algebras (commutative rings with identity, in our case) from which $R$ is taken. We will give a very general definition. Recall that a variety of algebras is a class $\boldsymbol{V}$ of algebras of the same type which is closed w.r.t. homomorphic images, arbitrary direct products and subalgebras. Equivalently, $\boldsymbol{V}$ is a variety iff $\boldsymbol{V}$ can be "defined by identities". For instance, the classes of rings or of commutative rings with identity are varieties, while the class of fields is not a variety. For varieties and polynomials see [46].

DEFINITION 4.1. Let $A$ be an algebra of a variety $\boldsymbol{V}$. Then the polynomial algebra $A_{\boldsymbol{V}}[x]$ over $A$ in $\boldsymbol{V}$ is defined as the free union of $A$ with the free algebra over $\{x\}$ in $\boldsymbol{V}$.

Hence $A_{\boldsymbol{V}}[x] \in \boldsymbol{V}$. It is clear that $A_{\boldsymbol{V}}[x]$ is generated by $A \cup \{x\}$.

EXAMPLE 4.2. (i) Polynomials over commutative rings with identity have the usual form indicated above.

(ii) Polynomials of $A_{\boldsymbol{R}}[x]$, $\boldsymbol{R}$ = variety of all rings, are of the form

$$a_0 + a_1 x + x a_2 + a_3 x a_4 + zx + a_5 x^2 + \cdots \quad (a_i \in A \in \boldsymbol{R},\ z \in \mathbb{Z}).$$

(iii) Let $\boldsymbol{G}$ be the variety of all groups. If $A \in \boldsymbol{G}$ then

$$A_{\boldsymbol{G}}[x] = \{a_0 + z_1 x + a_1 + z_2 x + \cdots + z_n x + a_n \mid a_i \in A,\ z_i \in \mathbb{Z}\}.$$

(iv) If $\boldsymbol{A}$ = variety of abelian groups then, for $A \in \boldsymbol{A}$,

$$A_{\boldsymbol{A}}[x] = \{a + zx \mid a \in A,\ z \in \mathbb{Z}\}.$$

(v) In the variety $\boldsymbol{M}(R)$ of unitary $R$-modules ($R$ = ring with identity) we get

$$A_{\boldsymbol{M}(R)}[x] = \{a + rx \mid a \in A,\ r \in R\}.$$

The following is easy to see and ties us to near-ring theory. Recall that an $\Omega$-group (in the sense of Higgins [33]) is a group $(G,+)$, together with possibly other operations $\omega_i$ $(i \in I)$ such that $\omega_i(0,0,\ldots,0) = 0$ for all $i \in I$.

THEOREM 4.3. *If $\boldsymbol{V}$ is a variety of $\Omega$-groups then $A_{\boldsymbol{V}}[x]$ is a near-ring w.r.t. addition $+$ and composition $\circ$.*

For each $p \in A_{\boldsymbol{V}}[x]$ we can associate the *induced polynomial function* $\bar{p}$: $A \to A$ in the obvious way. Let $P(A)$ be the set of all $\bar{p}$ $(p \in A_{\boldsymbol{V}}[x])$. Observe that polynomials depend on $\boldsymbol{V}$, while polynomial functions do not.

THEOREM 4.4. *For all $A$ in a variety $\boldsymbol{V}$ of $\Omega$-groups, $(P(A),+,\circ)$ is a near-ring and the correspondence $h$: $A_{\boldsymbol{V}}[x] \to P(A)$, $p \mapsto \bar{p}$ is an epimorphism. $P(A)$ is generated by the constant functions on $A$ and the identity function. Each $\bar{p} \in P(A)$ is compatible in the sense that for each congruence $\theta$ in $A$, we have $a\theta b \Rightarrow \bar{p}(a)\theta\bar{p}(b)$.*

The proof is easy: since $h$ can be considered as homomorphism from $A_{\boldsymbol{V}}[x]$ to $M(A) = A^A$, $\operatorname{Im} h = P(A)$ is a near-ring. Since for $a \in A$, $\bar{a}$ is the constant function with value $a$, and $\bar{x} = \mathrm{id}$, $P(A)$ is generated by the constant maps and id. Since constants and id are compatible, so are all polynomials.

We examine the varieties of commutative rings with identity and of groups more closely. We omit $\boldsymbol{V}$ in $A_{\boldsymbol{V}}[x]$, since the meaning of $\boldsymbol{V}$ will be clear. Of course, within the first variety the special case of fields deserves special interest. Proofs can be found in [3].

THEOREM 4.5 (Clay–Straus). *If $F$ is an infinite field then $F[x]$ is simple. If $F$ is a finite field with char $F \neq 2$ then all ideals of $(F[x], +, \circ)$ are precisely all principal ideals $(p)$ of $(F[x], +, \cdot)$ with $p \mid p \circ q$ for all $q \in F[x]$.*

*Polynomials $p$ with $p \mid p \circ q$ for all $q \in F[x]$ ($F$ a finite field) are precisely the lcm's of polynomials of the type $(x^{q^n} - x)^m$ $(n, m \in \mathbb{N})$.*

The case of char $F = 2$ is considerably more complicated, see [3].

THEOREM 4.6 (Nöbauer). *Let $R$ be a commutative ring with identity. $P(R)$ is simple $\Leftrightarrow$ $R$ is a field with $|R| \geqslant 3$.*

Now we turn to the variety $\boldsymbol{G}$. If $p = a_0 + z_1x + \cdots + z_nx + a_n$ is zero-symmetric then $\bar{p}(0) = 0$, hence $\sum a_i = 0$, and conversely. In this case, we are able to write

$$\begin{aligned} p &= a_0 + z_1x + a_1 + \cdots + z_nx + a_n \\ &= (a_0 + z_1x - a_0) + \big((a_0 + a_1) + z_2x - (a_0 + a_1)\big) + \cdots - \sum a_i \\ &= z_1(a_0 + x - a_0) + z_2\big((a_0 + a_1) + x - (a_0 + a_1)\big) + \cdots \\ &\quad + z_n(a' + x - a') \end{aligned}$$

with $a' = a_0 + a_1 + \cdots + a_{n-1}$. Hence we get for each $G \in \boldsymbol{G}$:

THEOREM 4.7. $P_0(G) = I(G)$.

THEOREM 4.8 (Lausch and Nöbauer, see [46]). $P(G) = M(G) = G^G \Leftrightarrow G \cong \mathbb{Z}_2$ *or $G$ is finite, simple, and nonabelian.*

This brings us back to the question of which functions are polynomial functions.

DEFINITION 4.9. For $n \in \mathbb{N}$, let $\mathrm{L}_n\mathrm{P}(A)$ be the collection of all maps $A \to A$ which can be interpolated by polynomial functions on any set of $\leqslant n$ places in $A$.

$$\mathrm{LP}(A) := \bigcap_{n \in \mathbb{N}} \mathrm{L}_n\mathrm{P}(A)$$

is the set of *local polynomial functions*. Let $C(A)$ be the set of all compatible functions (see 4.4).

Obviously, $\mathrm{L}_2\mathrm{P}(A) \subseteq C(A)$. The converse can also be seen; the remaining assertions in 4.10 are easy:

THEOREM 4.10. *If $A$ is an $\Omega$-group then*

$$\begin{aligned} P(A) &\leqslant \mathrm{LP}(A) \leqslant \cdots \leqslant \mathrm{L}_n\mathrm{P}(A) \leqslant \cdots \leqslant \mathrm{L}_3\mathrm{P}(A) \\ &\leqslant \mathrm{L}_2\mathrm{P}(A) = C(A) \leqslant \mathrm{L}_1\mathrm{P}(A) = M(A) \end{aligned}$$

*is a chain of near-rings (w.r.t. $+$ and $\circ$).*

DEFINITION 4.11. An algebra $A$ is called *locally polynomially complete* if $\mathrm{LP}(A) = M(A)$ and *polynomially complete* if $P(A) = M(A)$.

So a finite field is polynomially complete while an infinite field is "only" locally polynomially complete. Obviously an $\Omega$-group $G$ is locally polynomially complete iff $P(G)$ is dense in $M(G)$. Hence we might expect some help from the Density Theorem if $P(G)$ is primitive on $G$. Detailed considerations show that $P(G)$ is primitive on $G$ if $G$ is a simple $\Omega$-group. We then get

THEOREM 4.12. *Let $G$ be a simple $\Omega$-group. If $P_0(G)$ is a ring then $G$ is a vector space and $P(G)$ is dense in $M_{\mathrm{aff}}(G)$. If $P_0(G)$ is not a ring then $G$ is locally polynomially complete.*

For which simple $\Omega$-groups can $P_0(G)$ be a ring (w.r.t. $+$ and $\circ$)? We collect some results. Much more on this and on related subjects can be found in the substantial paper [73] by S.D. Scott.

THEOREM 4.13 ([69]).
(i) *If $G$ is a simple group then $P_0(G)$ is a ring iff $G$ is abelian.*
(ii) *If $G$ is a simple ring then $P_0(G)$ is a ring iff $G \cong \mathbb{Z}_2$.*
(iii) *If $G$ is an $R$-(ring-)module then $P_0(G)$ is always a ring.*

Hence the local completeness of fields can also be derived from 4.12 and 4.13.

For general algebras, $\mathrm{L_5P}(A) = M(A)$ implies $\mathrm{LP}(A) = M(A)$, an almost total collapse in the chain of 4.10. If $A$ is an $\Omega$-group, however, it can be shown (see [3]) that $\mathrm{L_3P}(A) = M(A)$ implies that $P(A)$ is primitive on $(A, +)$ with $\Phi(A) = \{\mathrm{id}\}$. Hence $P(A)$ is dense in $M(A)$, which shows

THEOREM 4.14. *If $A$ is an $\Omega$-group such that* $\mathrm{L_3P}(A) = M(A)$ *then* $\mathrm{LP}(A) = M(A)$, *and $A$ is locally polynomially complete.*

Polynomial functions are also useful to describe generated ideals in $\Omega$-groups. Take the case of "plain" groups $G$, for instance. We have seen before that $P_0(G)$ consists of the maps

$$g \mapsto \sum_i (h_i + z_i g - h_i).$$

Hence the normal subgroup ($=$ ideal) generated by $g \in G$ is given by $\{p(g) \mid p \in P_0(G)\}$. This turns out to be true in general:

THEOREM 4.15 ([69]). *If $G$ is an $\Omega$-group and $g \in G$, $S \subseteq G$, then the ideals $(g)$, $(S)$ generated by $g, S$, respectively, are given by*

$$(g) = \{p(g) \mid p \in P_0(G)\},$$

$$(S) = \sum_{g \in S} (g).$$

This is, for example, useful to get results about (sub)direct decompositions of $\Omega$-groups (see, e.g., [64]). From 4.2 (ii) one instantly derives the well-known result that the principal ideal (a) in a ring $R$ is given by $Ra + aR + RaR + \mathbb{Z}a$ (the items with $x^2, \ldots$ are redundant!).

Polynomial near-rings are also involved in the theory of algebraic equations over $\Omega$-groups. Intuitively we think of an algebraic equation as $p = 0$, where $p$ is a polynomial. Now "$p = 0$" is not an identity (except if $p$ *is* the zero polynomial), but a "command" to find zeros of $p$. Since "$p = 0$" is completely determined by $p$ we can more safely speak of "the equation $p$". We confine our attention to equations in a single variable.

DEFINITION 4.16. Let $A$ be an algebra in a variety $\boldsymbol{V}$ of $\Omega$-groups. A *system of equations* over $A$ in $\boldsymbol{V}$ is a family $(p_i)_{i \in I}$ of polynomials in $A_{\boldsymbol{V}}[x]$. If $B \geqslant A$, $B \in \boldsymbol{V}$ then $b \in B$ is called a *solution* of $(p_i)$ if $\bar{p}_i(b) = 0$ for all $i \in I$. The system $(p_i)$ is *solvable* if it has a solution in some suitable extension of $A$ in V.

If $b$ is a solution of $p_1$ and $p_2$, then $b$ is also a solution of $p_1 + p_2$. More generally, every solution of $(p_i)_{i \in I}$ is also a solution of all $p \in \langle p_i \rangle_{i \in I}$ = the ideal generated by $\{p_i \mid i \in I\}$. Hence solving (algebraic) equations is equivalent to finding zeros of ideals. For a fairly complete treatment on equations see [46].

There are two ties between equations and near-rings. The first is that from the last lines it is clear that one has to generate ideals. Theorem 4.15 is "responsible" for this process. The second connection is in the area of equations over groups. It turns out that important classes of ideals (= classes of equations) of $(A_{\boldsymbol{V}}[x], +)$ turn out to be also ideals of the near-ring $(A_{\boldsymbol{V}}[x], +, \circ)$, see [19], and the latter ideals are known in many cases. We have yet just scratched the surface of this interesting interplay; much remains to be done.

## 5. Matrix near-rings

Matrix rings play a central role in ring theory. In some sense, matrix rings are "the stuff rings are made of". Hence it is natural to ask if a similar situation applies to near-ring theory. The answer is: no and yes. No, because Heatherly showed in [32] that if one starts with a near-ring $N$ with identity and forms matrices as in the ring case, matrix multiplication is associative iff $N$ is a ring. Yes, because of the following lines. For quite a while it was unclear how to define matrix near-rings "correctly".

It was mainly Andries Van der Walt who came up with a "good" definition in the mid-80's. Observe that if $A^r_{i,j}$ is the matrix with $r$ at the $(i, j)$-position and zeroes elsewhere then $A^r_{i,r}(x_1, \ldots, x_n)^t = (0, \ldots, 0, rx_j, 0, \ldots, 0)^t$ with $rx_j$ at the $i$-th position, and every matrix is the sum of matrices of the $A^r_{i,j}$-type.

CONVENTION. All near-rings in this section are zero-symmetric and have an identity.

DEFINITION 5.1. Let $R$ be a near-ring (zero-symmetric, with identity), $n \in \mathbb{N}$, $r \in R$ and $1 \leqslant i, j \leqslant n$. Then $f^r_{i,j}$ denotes the map from $R^n$ to $R^n$, mapping $(x_1, \ldots, x_n)$ to

$(0, \ldots, 0, rx_j, 0, \ldots, 0)$ with $rx_j$ at the $i$-th position. The subnear-ring $M_n(R)$ generated by $F_n(R) := \{f^r_{i,j} \mid r \in R,\ 1 \leqslant i, j \leqslant n\}$ of $M_0(R)$ is called the $n \times n$-*matrix-near-ring* over $R$; its elements are called $(n \times n)$-*matrices* over $R$.

Observe that a matrix in $M_n(R)$ might have different representations by elements of $F_n(R)$, in striking contrast to the ring-theoretical situation [76]. The following is easy to see ([59]):

PROPOSITION 5.2. *Let the notation be as in* 5.1.
a) $f^r_{i,j}$ *is distributive* $\Leftrightarrow$ $r$ *is distributive.*
b) $M_n(R)$ *is distributively generated* $\Leftrightarrow$ $R$ *is distributively generated.*
c) $M_n(R)$ *is a ring* $\Leftrightarrow$ $R$ *is a ring (in this case,* $M_n(R)$ *is isomorphic to the "usual"* $n \times n$-*matrix ring over* $R$).

We now turn to ideals of $M_n(R)$. The nice situation in rings (all ideals of $M_n(R)$ are of the type $M_n(I)$ with $I \trianglelefteq R$) breaks down a bit.

DEFINITION 5.3. a) Let $I$ be an ideal of $R$. Then

$$I^* := (I^n\colon R^n) = \{f \in M_n(R) \mid f(R^n) \subseteq I^n\},$$

and let $I^+$ be the ideal generated by all $f^r_{i,j}$ with $r \in I$ and $1 \leqslant i, j \leqslant n$.
b) Let $J$ be an ideal of $M_n(R)$. Then let $J_*$ be the set of all components of $j(x_1, \ldots, x_n)$ with $j \in J$ and $(x_1, \ldots, x_n) \in R^n$.

PROPOSITION 5.4 ([76]). *Let the notation be as in* 5.3.
a) $I^*$ *and* $I^+$ *are ideals of* $M_n(R)$, *and* $J_*$ *is an ideal of* $R$.
b) $I^+ \subseteq I^*$, *and* $I^+ \subset I^*$ *might be the case (but not if* $R$ *is a ring).*
c) $(J_*)^+ \subseteq J \subseteq (J_*)^*$.
d) $M_n(R/I) \cong M_n(R)/I^*$.

Due to 5.4 c), $J_*$ is sometimes called the *enclosing ideal* of $J$. The following results on the structure of $M_n(R)$ (see [76, 77]) are also very satisfactory:

THEOREM 5.5. *Let* $R$ *be a near-ring and* $n \in \mathbb{N}$.
a) $M_n(R)$ *is simple* $\Leftrightarrow$ $R$ *is simple.*
b) $M_n(R)$ *is semisimple* $\Leftrightarrow$ $R$ *is semisimple.*
c) $M_n(R)$ *is primitive* $\Leftrightarrow$ $R$ *is primitive.*
d) $M_n(R)$ *is subdirectly irreducible* $\Leftrightarrow$ $R$ *is subdirectly irreducible.*
e) $J(M_n(R)) = J(R)^*$.
f) $J$ *is a primitive ideal of* $M_n(R) \Leftrightarrow J = I^*$ *for a primitive ideal* $I$ *of* $R$.

Other features don't come up that nicely. In ring theory one knows that, for a division ring $D$, $M_n(D)$ has minimal condition, and every primitive ring with minimal condition is isomorphic to some $M_n(D)$. In contrast to that, Meyer has shown ([60]) that if $R$ is an infinite near-field, but not a field, then $M_2(F)$ does not fulfill the minimal condition on left ideals. And it is not true that every primitive near-ring with minimal condition must be isomorphic to some $M_n(R)$. But partial results are possible (note that, by 3.9, primitive near-rings are dense in near-rings of the type $M_S(G)$):

THEOREM 5.6 ([76]). *Let $R$ be a nonring, primitive on $G$. If $S := \mathrm{End}_R G$ has only finitely many orbits on $G^n$ then $M_n(R) \cong M_S(G^n)$.*

COROLLARY 5.7. *Let $R$ be a finite near-field which is not a field. Then $M_n(R) \cong M_R(R^n)$.*

Finally, we mention a remarkable result which connects the $R$-subgroups $G$ with the $M_n(R)$-groups

THEOREM 5.8. *Suppose that the $R$-group $G$ has the property that for each $g_1, g_2 \in G$ there is some $g \in G$ with $\{g_1, g_2\} \subseteq Rg$. Then*

a) *Each $M_n(R)$-ideal of $G^n$ is of the form $H^n$ for some $R$-ideal $H$ of $G$.*
b) *$G$ is $R$-simple iff $G^n$ is $M_n(R)$-simple.*
c) $\mathrm{End}_R(G) \cong \mathrm{End}_{M_n(R)}(G^n)$.

For more on matrix near-rings see the nice survey article [58].

## 6. What else near-rings can do for you

In this last section we take brief looks at several areas inside and outside of mathematics which have connections to near-rings or even applications of near-rings to these fields. We start with a down-to-earth application to the designs of statistical experiments developed by G. Ferrero and J.R. Clay.

### 6.1. *Near-rings and experimental designs*

Certain near-rings give rise to interesting contributions to combinatorics. With a look back to 2.6, we define planar near-rings:

DEFINITION 6.1.1. In a near-ring $N$, we let, for $a, b \in N$, $a \equiv b$ iff $na = nb$ for all $n \in N$. Let $N^\# := \{n \in N \mid n \not\equiv 0\}$. $N$ *is planar* if $|N/{\equiv}| \geqslant 3$ and if all equations

$$xa = xb + c \quad (a, b, c \in N,\ a \not\equiv b)$$

have a unique solution.

A planar near-ring is zero-symmetric, since for all $n \in N$, $n0$ and $0$ are both solutions of $xa = x0 + 0$ (for some $a \in N^\#$), hence identical. See, e.g., [1] for the proof that if $\equiv$ is the identity relation in a planar near-ring then $N$ is a planar near-field in the sense of 2.6. In [1] it is also shown that planar near-rings and Frobenius groups are "basically the same".

There are good construction methods for obtaining planar near-rings. We exhibit the following one, due to J.R. Clay, which is both easy and most useful.

THEOREM 6.1.2. *Let $F$ be a field of order $p^n$, where $p$ is a prime and let $t$ be a nontrivial divisor of $p^n - 1$, so $st = p^n - 1$ for some $s$. Choose a generator $g$ of the multiplicative group of $F$. Define $g^a \bullet_t g^b := g^{a+b-[a]_s}$, where $[a]_s$ denotes the residue class of $a$ modulo $s$. Then $N = (F, +, \bullet_t)$ is a planar near-ring with $N^{\#} = N\backslash\{0\}$.*

We will use the following example in the sequel.

EXAMPLE 6.1.3. Let $F$ be the field $\mathbb{Z}_7$. Then $p^n - 1 = 6$, we choose $t = 3$ and get $s = 2$. This yields the planar near-ring $(\mathbb{Z}_7, +, \bullet_3)$:

| $+$ | 0 | 1 | 2 | 3 | 4 | 5 | 6 |
|---|---|---|---|---|---|---|---|
| 0 | 0 | 1 | 2 | 3 | 4 | 5 | 6 |
| 1 | 1 | 2 | 3 | 4 | 5 | 6 | 0 |
| 2 | 2 | 3 | 4 | 5 | 6 | 0 | 1 |
| 3 | 3 | 4 | 5 | 6 | 0 | 1 | 2 |
| 4 | 4 | 5 | 6 | 0 | 1 | 2 | 3 |
| 5 | 5 | 6 | 0 | 1 | 2 | 3 | 4 |
| 6 | 6 | 0 | 1 | 2 | 3 | 4 | 5 |

| $\bullet_3$ | 0 | 1 | 2 | 3 | 4 | 5 | 6 |
|---|---|---|---|---|---|---|---|
| 0 | 0 | 0 | 0 | 0 | 0 | 0 | 0 |
| 1 | 0 | 1 | 2 | 1 | 4 | 4 | 2 |
| 2 | 0 | 2 | 4 | 2 | 1 | 1 | 4 |
| 3 | 0 | 3 | 6 | 3 | 5 | 5 | 6 |
| 4 | 0 | 4 | 1 | 4 | 2 | 2 | 1 |
| 5 | 0 | 5 | 3 | 5 | 6 | 6 | 3 |
| 6 | 0 | 6 | 5 | 6 | 3 | 3 | 5 |

DEFINITION 6.1.4. A *balanced block design* with parameters $(v, b, r, k, \lambda) \in \mathbb{N}^5$ is a set $P$ of *points* together with a set $\mathbf{B}$ of subsets (called *blocks*) of $P$ such that

(i) $|P| = v$.
(ii) $|\mathbf{B}| = b$.
(iii) Each $p \in P$ appears in exactly $r$ blocks of $\mathbf{B}$.
(iv) Each $B \in \mathbf{B}$ has cardinality $k$.
(v) Each pair of different points belongs to precisely $\lambda$ blocks.

If $b = \binom{v}{k}$, the design is "*complete*" and rather uninteresting. Balanced incomplete block designs are abbreviated by "*BIB-designs*".

BIB-designs are, apart from combinatorics, widely used in the design of statistical experiments. Several designs can be constructed from planar near-rings; an excellent account on that can be found in [1]. Most of them turn out to be of a remarkable high "efficiency". We concentrate on the "easiest" construction.

THEOREM 6.1.5. *Let $N$ be a finite planar near-ring and*

$$\mathrm{B} = \{aN^* + b \mid a, b \in N,\ a \neq 0\}.$$

*Then $(N, \mathbf{B})$ is a BIB-design with parameters*

$$\left(v, \frac{v(v-1)}{k-1}, v-1, k, k-1\right),$$

*where $b = |N|$ and $k$ is the cardinality of each $aN^*$ with $a \neq 0$.*

So if **B** is constructed as in 6.1.2, we get the parameters $(p^n, p^n s, p^n - 1, t, t - 1)$.

EXAMPLE 6.1.6. The blocks of 6.1.5 are then given by all $a \bullet_3 F^* + b$ $(a \neq 0)$; we write juxtaposition instead of $\bullet_3$; observe that $2F^* = 4F^* = 1F^*$ and $5F^* = 6F^* = 3F^*$.

| | |
|---|---|
| $1F^* + 0 = \{1, 2, 4\} =: B_1$ | $3F^* + 0 = \{3, 5, 6\} =: B_8$ |
| $1F^* + 1 = \{2, 3, 5\} =: B_2$ | $3F^* + 1 = \{4, 6, 0\} =: B_9$ |
| $1F^* + 2 = \{3, 4, 6\} =: B_3$ | $3F^* + 2 = \{5, 0, 1\} =: B_{10}$ |
| $1F^* + 3 = \{4, 5, 0\} =: B_4$ | $3F^* + 3 = \{6, 1, 2\} =: B_{11}$ |
| $1F^* + 4 = \{5, 6, 1\} =: B_5$ | $3F^* + 4 = \{0, 2, 3\} =: B_{12}$ |
| $1F^* + 5 = \{6, 0, 2\} =: B_6$ | $3F^* + 4 = \{1, 3, 4\} =: B_{13}$ |
| $1F^* + 6 = \{0, 1, 3\} =: B_7$ | $3F^* + 4 = \{2, 4, 5\} =: B_{14}$. |

This is a design with parameters $(v, b, r, k, \lambda) = (7, 14, 6, 3, 2)$.

Designs constructed by planar near-rings as in 6.1.5 were recently actually used for agricultural experiments in Scotland. The following example refers to 6.1.6.

EXAMPLE 6.1.7. Suppose we want to test combinations of 6 out of 14 given fertilizers $D_1, \ldots, D_{14}$, with each fertilizer used on exactly 3 of 7 experimental fields. Then we can use the design just constructed in 6.1.6. We divide our field into 7 smaller test-fields with numbers $0, 1, 2, \ldots, 6$, and apply the fertilizer $D_1$ $(1 \leqslant i \leqslant 14)$ on each field in the block $B_i$:

| field 0 | field 1 | field 2 | field 3 | field 4 | field 5 | field 6 |
|---|---|---|---|---|---|---|
| | $D_1$ | $D_1$ | | $D_1$ | | |
| | | $D_2$ | $D_2$ | | $D_2$ | |
| | | | $D_3$ | $D_3$ | | $D_3$ |
| $D_4$ | | | | $D_4$ | $D_4$ | |
| | $D_5$ | | | | $D_5$ | $D_5$ |
| $D_6$ | | $D_6$ | | | | $D_6$ |
| $D_7$ | $D_7$ | | $D_7$ | | | |
| | | | $D_8$ | | $D_8$ | $D_8$ |
| $D_9$ | | | | $D_9$ | | $D_9$ |
| $D_{10}$ | $D_{10}$ | | | | $D_{10}$ | |
| | $D_{11}$ | $D_{11}$ | | | | $D_{11}$ |
| $D_{12}$ | | $D_{12}$ | $D_{12}$ | | | |
| | $D_{13}$ | | $D_{13}$ | $D_{13}$ | | |
| | | $D_{14}$ | | $D_{14}$ | $D_{14}$ | |

Then each field is supplied with 6 fertilizers, each fertilizer is applied to 3 fields, and each pair of different fields has precisely $\lambda = 2$ fertilizers in common. Try to write down such a design without any theory!

### 6.2. *Efficient codes from near-rings*

Several ways have been discovered in which near-rings produce efficient error-correcting codes. We list methods which yield nonlinear and linear codes.

The main goal of coding theory is the following. Given some alphabet $A$, a *message* over $A$ is a word $a_1a_2\dots a_k$ of length $k$. If this is transmitted over a "long" channel, errors might occur at the receiver's end. In order to detect and correct these messages, they will be encoded (= prolonged) to $a_1a_2\dots a_ka_{k+1}\dots a_n$ before transmission. The *test symbols* $a_{k+1},\dots,a_n$ are to be computed in some way from $a_1,\dots,a_k$ so that, for each other message $b_1b_2\dots b_n$, the resulting *codewords* $a_1\dots a_n$ and $b_1\dots b_n$ are "very distinct"; in this way, a "small" number of errors can be detected and even corrected. The *(Hamming) distance* $d(a_1\dots a_n, b_1\dots b_n)$ of two codewords is the number of places $i$ in which $a_i$ and $b_i$ differ. The *(Hamming) weight* $\mathrm{wt}(a_1\dots a_n)$ of $a_1\dots a_n$ is the number of places $i$ where $a_i \neq 0$ (if $0 \in A$). A *code* $C$ of length $n$ is a set of codewords of length $n$; its *minimal distance* $d_{\min}(C)$ is the minimal distance between two different codewords. If $d = d_{\min}(C)$ then up to $d-1$ errors can be detected and up to $\lfloor\frac{d-1}{2}\rfloor$ errors can even be corrected. Of course, one wants $n-k$ to be small and $d$ to be large. These are contradicting goals, and one seeks "optimal compromises". If $A$ is a field and $C$ a subspace of $A^n$ then $C$ is called a *linear code*. If $A = \mathbb{Z}_2$ then $C$ is called a *binary code*. All our codes will be binary. For more on codes see, e.g., [52].

DEFINITION 6.2.1. let $(P = \{p_1,\dots,p_v\},\ \mathbf{B} = \{B_1,\dots,B_b\})$ be a BIB-design with parameters $(v,b,r,k,\lambda)$ and let $M_{\mathbf{B}} = (m_{ij})$ be its $v\times b$-incidence matrix where $m_{ij} = 1$ iff $p_i \in B_j$ and 0 otherwise. The rows and the columns can both be viewed as a binary code, called the *row code* $C_{\mathrm{row}\,\mathbf{B}}$ (*column code* $C_{\mathrm{col}\,\mathbf{B}}$, respectively) of $\mathbf{B}$.

EXAMPLE 6.2.2. The design in 6.1.6 yields

$$M_{\mathbf{B}} = \begin{pmatrix} 0&0&0&1&0&1&1&0&1&1&0&1&0&0\\ 1&0&0&0&1&0&1&0&0&1&1&0&1&0\\ 1&1&0&0&0&1&0&0&0&0&1&1&0&1\\ 0&1&1&0&0&0&1&1&0&0&0&1&1&0\\ 1&0&1&1&0&0&0&0&1&0&0&0&1&1\\ 0&1&0&1&1&0&0&1&0&1&0&0&0&1\\ 0&0&1&0&1&1&0&1&1&0&1&0&0&0 \end{pmatrix}.$$

The row code has 7 codewords of length 14, each of weight 6; the column code is also an equal-weight-code, consisting of 14 codewords of length 7, each having weight 3.

PROPOSITION 6.2.3 ([27]). *Let the notation be as in* 6.2.1.

a) $C_{\mathrm{row}\,\mathbf{B}}$ *has* $\nu$ *codewords of length* $b$, *equal weight* $r$ *and minimal distance* $2(r-\lambda)$.

b) $C_{\mathrm{col}\,\mathbf{B}}$ *has* $b$ *codewords of length* $\nu$, *equal weight* $k$ *and minimal distance* $(k-\mu)$, *where*

$$m = \max_{i\neq j} |B_i \cap B_j|.$$

c) *Neither $C_{\text{row}\,\mathbf{B}}$ nor $C_{\text{col}\,\mathbf{B}}$ can be linear.*

It is easy to see that if $\lambda = 1$ then also $\mu = 1$. Planar near-rings with $\mu \leqslant 2$ (important for coding theory!) are called *circular* (see [1, 13, 20]).

THEOREM 6.2.4 ([27]). *If $p$ is prime with $p \equiv 1 \pmod 6$ and $\mathbf{B}$ is constructed as in* 6.1.2 *then $(P, B)$ is circular iff $p \neq \{7, 13, 19\}$.*

DEFINITION 6.2.5. Let $A(n, d, w)$ be the maximal number of codewords in a code of length $n$, equal weight $w$ and minimal distance $\geqslant d$. A code $C$ is called *maximal* if $|C| = A(n, d, w)$.

The determination of $A(n, d, w)$ and maximal codes is a discrete sphere packing problem: If $C$ is maximal then the spheres around all codewords with radius $d_{\min}(C)$ do not intersect and are maximal in number with this property. Only few formulas for $A(n, d, w)$ are actually known (see [57] and [17]).

THEOREM 6.2.6 ([71]). *Let $(P, \mathbf{B})$ be a BIB-design. Then $C_{\text{row}\,\mathbf{B}}$ and (if $\lambda = 1$) also $C_{\text{col}\,\mathbf{B}}$ are maximal.*

COROLLARY 6.2.7. *If $p^n - 1 = st$ then*

$$A\big(p^n s, 2(p^n - t), p^n - 1\big) = p^n \quad \textit{and} \quad A\big(p^n, 2(t-1), t\big) = p^n s.$$

See [25] for a decoding algorithm for these types of codes. Now we take a completely different approach; this time we use polynomials. We identify $b_0 + b_1 x + \cdots + b_n x^k$ with $(b_0, b_1, \ldots, b_k)$ and $b_0 b_1 \ldots b_k$.

DEFINITION 6.2.8. Let $f = x + x^2 + \cdots + x^m \in \mathbb{Z}_2[x]$. Let a message $a_1 a_2 \ldots a_z$ be encoded as $a_1 f + a_2 f \circ x^2 + \cdots + a_z f \circ x^z$, and let $C(m, z)$ be the resulting code.

THEOREM 6.2.9 ([66]). *$C(m, z)$ is a linear binary code of length $mz$, dimension $z$ and $\pi(m) + 2 \leqslant d_{\min}(C(m, z)) \leqslant m$, where $\pi(m)$ denotes the number of primes $\leqslant m$.*

So far, no example is known where $d_{\min}(C(m, z)) < m$.

See [26] for a different method to use polynomial near-rings in coding theory.

### **6.3.** *Near-rings, group partitions, and translation planes*

For the sake of completeness, we compile some well-known concepts of geometry. A *dilatation* of an incidence structure is an automorphism ($=$ *collineation*) which maps lines onto parallel lines. A *translation* is a dilatation which is fixed-point-free or the identity. An affine plane in which the translations form a group which is transitive on the points is called a *translation plane*.

For the following considerations, we need a generalization due to C.J. Maxson [47].

DEFINITION 6.3.1. The triple $(P, \mathcal{L}, //)$ consisting of a set $P$ of *points*, a set $\mathcal{L}$ of subsets of $P$ (the *lines*) and an equivalence relation $//$ on $\mathcal{L}$ (*parallelism*) is called a *generalized translation structure* if

(i) Every two points are contained in at least one line $\in \mathcal{L}$.

(ii) There are at least two different lines, and each line has at least two points.

(iii) Euclid's parallelism axiom holds.

(iv) There exists an injective map $\Phi$ from $P$ into the set of collineations of $(P, \mathcal{L}, //)$ such that $\Phi(P)$ is a group which acts transitively on $P$.

DEFINITION 6.3.2. Let $(P, \mathcal{L}, //)$ be a generalized translation structure. If (i) in 6.3.1 is replaced by (i)′: Every two different points are contained in precisely one line of $\mathcal{L}$, then $(P, \mathcal{L}, //)$ is called a *translation structure* (André). If, moreover, all lines are equipotent, $(P, \mathcal{L}, //)$ is called a *Sperner space* (or *weakly affine space*).

We now tie these things with group theory.

DEFINITION 6.3.3. A family $\mathcal{F} = (G_i)_{i \in I}$ of proper subgroups of a group $G = (G, +)$ is called a *cover* of $G$ if

$$\bigcup_{i \in I} G_i = G.$$

If no $G_i$ is contained in another $G_j$, $\mathcal{F}$ is called a *geometric cover*. If moreover

$$G_i \cap G_j = \{0\} \quad \text{for } i \neq j, \mathcal{F}$$

is called a *fibration* of *partition* of $G$. A fibration with $|G_i| = |G_j|$ for all $i, j \in I$ is called an *equal* one. If $G_i + G_j = G$ for all $i \neq j$, $\mathcal{F}$ is a *congruence fibration*.

For instance, if $G$ is the vector space $F^2$, $F$ a field, the collection of all lines through $(0,0)$ forms an equal congruence partition, while the set of all planes through $(0,0,0)$ is "only" a geometric cover of $G = F^3$. This examples are typical, since "fibrations and covers come from geometry":

THEOREM 6.3.4 (Andre [10]). (i) *If $(P, \mathcal{L})$ is a translation plane then choose an arbitrary point $P$ and declare it as the zero point $o$. If the translations $t, t'$ are the ones which take $o$ to $p, p'$, respectively, then $t' \circ t$ maps $o$ to a point which we denote by $p + p'$. Note that $p, p'$ determine $t, t'$ uniquely. In this way, $(P, +)$ becomes a group. The lines $G_i$ through $o$ then form an equal congruence fibration of $G$.*

(ii) *Conversely, if $\mathcal{F} = (G_i)_{i \in I}$ is an equal congruence fibration of a group $G$ then one gets a translation plane whose points are the elements of $G$, and the lines are are the cosets of the $G_i$'s. Furthermore, $(x + G_i)//(y + G_j) \Leftrightarrow i = j$.*

The concepts in 6.3.1 and 6.3.2 allow us to generalize 6.3.4, using the same type of constructions. We get a "dictionary" between geometry and group theory. For proofs see [10, 38, 47] and [75]:

| | | |
|---|---|---|
| Generalized translation structure | $\leftrightarrow$ | geometric cover |
| Translation structure | $\leftrightarrow$ | fibration |
| Sperner space | $\leftrightarrow$ | equal fibration |
| Translation plane | $\leftrightarrow$ | equal congruence fibration |
| Desarguesian translation plane | $\leftrightarrow$ | vector space with all 1-dimensional subspaces $G_i$ as fibration |

Now near-rings enter the scene. The *kernel* $E(P, \mathcal{L}, //)$ of a generalized translation structure is the set of all endomorphisms $\sigma$ of $(P, \mathcal{L}, //)$ with $\sigma(L)//L$ for all $L \in \mathcal{L}$. If $(P, \mathcal{L}, //)$ is a translation plane, $E(P, \mathcal{L}, //)$ is precisely the set (even group!) of all dilatations; so a lot of geometric information is stored in the kernel.

If we make the transition to the group theoretic side,

$$E(P, \mathcal{L}, //) = \{\sigma \in \operatorname{End} G \mid \sigma(G_i) \subseteq G_i \text{ for all } i\} =: \operatorname{End}(G, \mathcal{F}),$$

as one can see ([47]; notations are as in 6.3.4). Information about the kernel can thus be obtained both from geometry and group theory. But things now lead to rings and near-rings, and algebra pays back some information to geometry. We collect some basic results in this area. For proofs see the quoted papers. We use the group-theoretic version of 6.3.4 (and the following lines).

THEOREM 6.3.5. *Let $\mathcal{F}$ be a geometric cover of the abelian group $G$.*

(i) *The kernel* $\operatorname{End}(G, \mathcal{F})$ *is a ring.*

(ii) *If $G$ is elementary abelian and* $\operatorname{End}(G, \mathcal{F})$ *is semisimple then* $\bigcap \mathcal{F} = \{0\}$ ([41]).

(iii) *Every finite semisimple ring with identity is isomorphic to a suitable* $\operatorname{End}(G, \mathcal{F})$ *with $G$ abelian, $\mathcal{F}$ geometric* ([41]).

(iv) *If $\mathcal{F}$ is a fibration then* $\operatorname{End}(G, \mathcal{F})$ *is an integral domain* ([10]).

(v) *If $(G, \mathcal{F})$ is a finite fibred group then* $\operatorname{End}(G, \mathcal{F})$ *is a finite field* ([10]).

In the last case of 6.3.5, the fact that the kernel is a finite field allows geometers to use this field for coordinatizing the corresponding translation structure (see [21]). If, however, $G$ in 6.3.5 is not abelian, $\operatorname{End}(G, \mathcal{F})$ is usually not a ring any more. From 3.12 we know what we have to do now.

DEFINITION 6.3.6. Let $(G, \mathcal{F})$ be a covered group.

$$E(G, \mathcal{F}) := \left\{\sum \pm h_i \mid h_i \in \operatorname{End}(G, \mathcal{F})\right\}$$

is the *extended kernel* of $(G, \mathcal{F})$.

Observe that $E(G, \mathcal{F}) = \operatorname{End}(G, \mathcal{F})$ if $G$ is abelian. Some remarkable results are:

THEOREM 6.3.7. *Let $(G, +)$ be a finite group with a fibration $\mathcal{F}$.*

(i) *If with each $G_i \in \mathcal{F}$ all conjugates $g + G_i - g$ $(g \in G)$ are again in $\mathcal{F}$ then either $E(G, \mathcal{F}) = \{0, \mathrm{id}\}$ or $G$ is abelian* ([40]).

(ii) *$E(G,\mathcal{F})$ is always a ring. Either* $\mathrm{End}(G,\mathcal{F}) = \{0,\mathrm{id}\}$ *(then* $E(G,\mathcal{F}) = \{0,\mathrm{id},2\mathrm{id},\dots\} \cong \mathbb{Z}_n$ *for some* $n \in \mathbb{N}$*) or else* $E(G,\mathcal{F})$ *is a (finite) field. In the latter case, if* $p = \mathrm{char}\, E(G,\mathcal{F})$*,* $pG = 0$ *and* $G$ *is nilpotent of class* $\leqslant 2$ ([51]).

(iii) *If* $\mathcal{F}$ *is an equal fibration and* $\mathrm{End}(G,\mathcal{F}) \neq \{0,\mathrm{id}\}$ *then* $G$ *is a finite vector space* ([40]).

See [51] for an example of a nonabelian fibred group $G$ (nilpotent of class 2) such that $E(G,F)$ is a field, but not a prime field. Observe that in the second (= interesting) case of 6.3.7 (ii), the elements $e_1, e_2, e_3 \in E(G,F)$ fulfill $e_1 \circ (e_2 + e_3) = e_1 \circ e_2 + e_1 \circ e_3$, although they are *not* endomorphisms in general! Cf. Section 1. Much work remains to be done to explore the geometric significance of the appearance of fields as extended kernels of nonabelian groups.

If we turn to $E(G,\mathcal{F})$ for arbitrary covers $\mathcal{F}$ (instead of fibrations) we do get "proper" near-rings in general, see [41] and [53]; we mention one result from [53]:

THEOREM 6.3.8. *Let* $(G,\mathcal{F})$ *be a finite covered group. Then* $E(G,\mathcal{F})$ *is a field iff the lattice* $\mathrm{Lat}(G,\mathcal{F})$ *of all subgroups* $H$ *of* $G$ *with* $f(H) \subseteq H$ *for all* $f \in E(G,\mathcal{F})$ *contains a fibration and* $\exp(G)$ *is a prime.*

From a combinatorical view of incidence structures, or from a group-theoretic view of covered groups, it is also interesting to study those endomorphisms which map lines to possibly nonparallel lines, or "cells" $\in \mathcal{F}$ to possibly other cells. Again, we use the group-theoretic language.

DEFINITION 6.3.9. Let $(G,\mathcal{F})$ be a covered group.

$$\mathrm{Mix}(G,\mathcal{F}) := \{h \in \mathrm{End}\, G \mid \forall i \in I\ \exists\, j \in I\colon\ h(G_i) \subseteq G_j\};$$

$$M(G,\mathcal{F}) := \Bigl\{\sum \pm h_i \mid h_i \in \mathrm{Mix}(G,\mathcal{F})\Bigr\}.$$

Again, $M(G,\mathcal{F})$ is a near-ring, but in general not a ring if $G$ is not abelian. Even in the abelian case, $M(G,\mathcal{F})$ is usually bigger than $\mathrm{Mix}(G,\mathcal{F})$. For finite fibred groups, there almost always exists some $m \in M(G,\mathcal{F})$ which really mixes the fibers in $\mathcal{F}$:

THEOREM 6.3.10 ([52] and [43]). *If* $(G,\mathcal{F})$ *is a finite fibred group then always* $E(G,\mathcal{F}) < M(G,\mathcal{F})$*, unless* $G$ *is of the type* $\mathrm{PSL}(2,p^n)$.

This does not remain true if $\mathcal{F}$ is allowed to be a cover. C.J. Maxson even found a cover of $(\mathbb{Z}_2)^{12}$ with $E(G,\mathcal{F}) = M(G,\mathcal{F}) = \{0,\mathrm{id}\}$.

### **6.4.** *Homogeneous maps on modules*

This topic ties together centralizer near-rings (3.8) and covered groups (6.3.3). If $R$ is a ring with identity and ${}_RG$ a unitary $R$-module then it is natural to consider the endomorphism $f_r$: $x \to rx$ on $(G,+)$. If $S := \{f_r \mid r \in R\}$, the corresponding centralizer near-ring is given by

$$M_S(G) = \{f\colon G \to G \mid f(rg) = rf(g)\ \forall r \in R\ \forall g \in G\}.$$

Its elements are called the *homogeneous maps* on ${}_RG$. If $f \in M_S(G)$ then it is clear that its restriction to any cyclic submodule $H$ of $G$ acts as an endomorphism on $H$. Hence each $f \in M_S(G)$ might be called a "piecewise endomorphism" on $G$. The cyclic submodules form a cover of $(G,+)$. If the maximal cyclic submodules also cover $G$ then they automatically form a geometric cover (6.3.3). Why not take other covers $\mathcal{H}$ of $G$?

DEFINITION 6.4.1. Let ${}_RG$ be a unitary $R$-module and $\mathcal{H}$ a cover of $G$. Then the elements of $\mathrm{PE}_R(G,\mathcal{H}) := \{f \in M_R(G) \mid f/H$ can be extended to some element of $\mathrm{End}_R(G)$ for each $h \in \mathcal{H}\}$ are called *piecewise endomorphisms* on $G$ w.r.t. $\mathcal{H}$.

There are two prominent covers (if they are covers at all): $\mathcal{C} = \{H \mid H$ is a maximal cyclic submodule of ${}_RG\}$ and $\mathcal{M} = \{H \mid H$ is a maximal submodule of ${}_RG\}$. In both cases, $\mathrm{PE}_R(G,\ldots)$ is a near-ring and

$$\mathrm{End}_R G \leqslant \mathrm{PE}_R(G,\mathcal{M}) \leqslant \mathrm{PE}_R(G,\mathcal{C}) \leqslant M_R(G).$$

Equalities in this chain seem to be interesting for ring and near-ring theory. For instance, [48] contains an example for $\mathrm{End}_R G = \mathrm{PE}_R(G,\mathcal{M}) < \mathrm{PE}_R(G,\mathcal{C}) = M_R(G)$.

DEFINITION 6.4.2. An $R$-module ${}_RG$ is an *mc-module* if both $\mathcal{M}$ and $\mathcal{C}$ are covers and $\mathcal{M} \neq \mathcal{C}$.

THEOREM 6.4.3 ([48]). *Let ${}_RG$ be a semisimple mc-module. Then* $\mathrm{End}_R G = \mathrm{PE}_R(G,\mathcal{M})$ *and* $\mathrm{PE}_R(G,\mathcal{C}) = M_R(G)$.

The next two results – which can also be found in [48] – investigate $\mathrm{PE}_R(G,\ldots)$ for special types of rings.

THEOREM 6.4.4. *Let $R$ be a commutative ring with identity. The following are equivalent:*
a) *$R$ is a finite direct sum of fields.*
b) *$R$ is noetherian and* $\mathrm{End}_R G = \mathrm{PE}_R(G,\mathcal{M})$ *for each mc-module ${}_RG$.*
c) *$R$ is noetherian and* $\mathrm{PE}_R(G,\mathcal{M})$ *is a ring for each mc-module ${}_RG$.*

THEOREM 6.4.5. *Let $R$ be a PID. Then ${}_RG$ is a finitely generated module iff*

$$\mathrm{PE}_R(G,\mathcal{C}) = M_R(G).$$

We now turn to a closer look at $M_R(G)$. Questions which come up naturally include: When is $M_R(G)$ a ring? When is $M_R(G) = \mathrm{End}_R G$, i.e. when is every homogeneous map already an endomorphism? What is the structure of $M_R(G)$? Some results in this area include

THEOREM 6.4.6 ([50]). *Let $R$ be local and ${}_RG$ finitely generated. Then $M_R(G)_d = \mathrm{End}_R G$.*

Let $\mathcal{R}$ be the class of all rings with identity such that for all unitary $R$-modules $G, M_R(G)$ is a ring.

THEOREM 6.4.7 ([29]). *For a ring $R$ with identity, $R \in \mathcal{R}$ iff $M_R(G) = \mathrm{End}_R G$ for all unitary modules ${}_RG$.*

THEOREM 6.4.8 ([29]). *If $R$ is a direct product of $n_\alpha \times n_\alpha$-matrix rings $R_\alpha$ then $R \in \mathcal{R}$ iff each $n_\alpha \geqslant 2$.*

In particular, a matrix ring $M_n(R)$ is in $\mathcal{R}$ iff $n \geqslant 2$. For some rings, 6.4.8 is close to a complete characterization:

THEOREM 6.4.9 ([29]). *Let $R$ be a semiperfect ring. Then $R \in \mathcal{R}$ iff $R/J(R)$ is a direct product of $n_\alpha \times n_\alpha$-matrix rings over division rings such that all $n_\alpha \geqslant 2$.*

Definitely not in $\mathcal{R}$ are rings which have a homomorphic image which is either commutative or integral ([29]).

THEOREM 6.4.10 ([28]). *Let $R$ be a PID and ${}_RG$ be finitely generated. Then $M_R(G)$ is semisimple iff $G$ is free or elementary torsion.*

Of special interest (and well studied) is the case $G = R^2$.

THEOREM 6.4.11 ([56]). *Let $R \times S$ be the (ring-theoretic) direct product of the rings $R$ and $S$ with identity. Then $M_{R\times S}((R \times S)^2) \cong M_R(R^2) \times M_S(S^2)$.*

More on $M_R(R^2)$ can be found in [54] and [55].

### 6.5. *Near-rings and automata*

We now link near-rings with (semi-)automata.

DEFINITION 6.5.1. A *semiautomaton* is a triple $\mathcal{S} = (Q, A, F)$ where $Q$ and $A$ are sets (called the *state* and *input set*) and $F$ is a function from $Q \times A$ into $Q$ (called the *state-transition function*). If $Q$ is a group (we always write it additively) we call $\mathcal{S}$ a *group-semiautomaton* and abbreviate this by *GSA*.

For any semiautomaton $(Q, A, F)$ we get a collection of mappings $f_a\colon Q \to Q$, one for each $a \in A$, which are given by

$$f_a(q) := F(q, a).$$

If the input $a_1 \in A$ is followed by the input $a_2$, the semiautomaton "moves" from the state $q \in Q$ first into $f_{a_1}(q)$ and then into $f_{a_2}(f_{a_1}(q))$.

If we extend (as usual) $A$ to the free monoid $A^*$ over $A$ (consisting of all finite sequences of elements of $A$, including the empty sequence $\Lambda$), we get

$$f_{a_1 a_2} = f_{a_2} \circ f_{a_1},$$

i.e. the map a $\mapsto f_a$ is an anti-monomorphism from $A^*$ into the transformation monoid over $Q$ with $f_\Lambda = \mathrm{id}_Q$. In the case of GSA's, we are also able to study the superposition $f_{a_1} + f_{a_2}$ (defined pointwisely) of two "simultaneous" inputs $a_1, a_2 \in A$. Hence it is natural to consider $\{f_a \mid a \in A\} \cup \{f_\Lambda\}$ and all of its sums and products (= composition of maps). This yields a subnear-ring of $M(Q)$:

DEFINITION 6.5.2. If $\mathcal{S}$ is as in 6.5.1, $N(\mathcal{S})$, the subnear-ring of $M(Q)$ generated by all $f_a$ $(a \in A)$ and id, is called the *syntactic near-ring* of $\mathcal{S}$.

$N(\mathcal{S})$ is always a near-ring with identity. If $Q$ is finite (in particular, if $\mathcal{S}$ is finite), so is $N(\mathcal{S})$. Even for linear (semi-)automata, $N(\mathcal{S})$ is almost never a ring:

EXAMPLE 6.5.3. Linear semiautomata. Then $Q, A$ are vector spaces and $F$ is supposed to be linear (on the product space $Q \times A$). Because of

$$f_a(q) = F(q, a) = F\big((q, 0) + (0, a)\big) = F(q, 0) + F(0, a),$$

we get $f_a = f_0 + \bar{a}$, where $f_0$: $q \to F(q, 0)$ is linear and $\bar{a}$ is constant with value $F(0, a) = f_a(0)$. Hence $f_a \in M_{\text{aff}}(Q)$, whence $N(\mathcal{S}) \leqslant M_{\text{aff}}(Q)$ for linear semiautomata $\mathcal{S}$. Only if each $F(0, a) = 0$, i.e. if no input can change the zero state, $N(\mathcal{S})$ is a ring.

The situation of 6.5.3 remains basically unchanged if $Q, A$ are just abelian groups and $F$: $Q \times A \to Q$ is a homomorphism in the first component (second component = 0). In this case one can show:

PROPOSITION 6.5.4. *If $\mathcal{S}$ is as just described,*

$$N(\mathcal{S}) = \Big\{ \sum \pm f\alpha_i \mid \alpha_i \in A^* \Big\}.$$

The proofs of 6.5.4 and the following two results can be found in [3], Chapter 9i.

THEOREM 6.5.5. *For every near-ring $N$ with identity there is a suitable GSA $\mathcal{S}$ such that $N$ is isomorphic to $N(\mathcal{S})$.*

THEOREM 6.5.6. *For a near-ring $N$ with identity there is a linear semiautomaton $\mathcal{S}$ with $N \cong N(\mathcal{S})$ iff $(N, +)$ is abelian and there is some $n \in N_d$ such that $N_0$ is generated by $\{1, d\}$.*

Obviously $Q$ is a faithful $N := N(\mathcal{S})$-group in the natural way. We say $q_1 \in Q$ is *reachable* from $q_2 \in Q$ if $q_1 \in Nq_2$. The semiautomaton $\mathcal{S}$ is *reachable* or *connected* if every state is reachable from every other state. The following is easy to see, but exhibits the role nearrings play in automata theory.

PROPOSITION 6.5.7. *Let $\mathcal{S}$ be a GSA and $N := N(\mathcal{S})$. $\mathcal{S}$ is reachable* $\Leftrightarrow$ $N0 = Q$ *and for each* $q \in Q$, *either* $Nq = Q$ *or* $Nq = \{0\}$ *holds.*

$N$-groups with the property of 6.5.7 are called "strictly monogenic" in the near-ring literature. We see from 6.5.7 that for a reachable GSA $\mathcal{S}$, $N$ must "move" 0 to each other state. One gets more insight if one considers what $N_0$ has to say:

DEFINITION 6.5.8. A GSA $\mathcal{S}$ is *strictly connected* if for each $q, q' \in Q$ with $q \neq 0$ there is some $n \in N_0(\mathcal{S})$ with $n(q) = q'$.

Again, we can employ our Density Theorem 3.9. We only list the "nonlinear" case (the "hard one").

THEOREM 6.5.9 ([67]). *Let $\mathcal{S}$ be a finite strictly connected GSA such that for $N = N(\mathcal{S})$, $N_0$ is not a ring. Let*

$$C(\mathcal{S}) := \{h \in \operatorname{End} Q \mid hf_\alpha = f_\alpha h \text{ for all } f_\alpha \in N_0\}$$

*and* $c := |C(\mathcal{S})|$. *Then* $N_0 = M_{C(\mathcal{S})}(Q)$ *is primitive on* $Q$ *and*

$$|N| = |Q|^{\frac{|Q|-1}{c-1}} \quad \text{if } N = N_0,$$

$$|N| = |Q|^{\frac{|Q|+c-2}{c-1}} \quad \text{if } N \neq N_0.$$

More information can be found, e.g., in [34] and [35], where among other things the "radical" of $N(\mathcal{S})$ is used to describe "how reduced" and "how reachable" $\mathcal{S}$ is.

## 6.6. *Near-rings and dynamical systems*

DEFINITION 6.6.1. A (discrete, dynamical, time-invariant) *system* $\Sigma$ is a quintuple $\Sigma = (Q, A, B, F, G)$, where $Q$ is a set (of *states*), $A$ a set (of *inputs*), $B$ a set (of *outputs*), $F$ a function $Q \times A \to Q$ (the *state transition function*) and $G$ a function $Q \times A \to B$ (the *output function*).

The description of $\Sigma$ in Definition 6.6.1 is usually the "local description" of $\Sigma$. In order to obtain the "global" description, we do not consider a single input, but a series of input signals $a_i$, which enter the system "at time $i \in \mathbb{Z}$". Hence we'll consider input sequences $(a_i)_{i \in \mathbb{Z}}$. It is generally assumed that the inputs don't come in "since eternity"; so we assume that there exists an index $k \in \mathbb{Z}$ such that $a_i = 0$ for all $i < k$. Sequences of this type are usually called "formal Laurent series":

DEFINITION 6.6.2. For any set $X$ containing 0, let $L(X)$ be the set of all sequences $(x_i)_{i\in\mathbb{Z}}$, for which there is some $k \in \mathbb{Z}$ such that $x_i = 0$ for all $i < k$. The elements of $L(X)$ are called *(formal) Laurent series* of elements of $X$.

In this context, the interpretation is as follows. At a certain "time" $k \in \mathbb{Z}$ (hence $k$ can be negative), the system is in state $q_k$ when the first input $a_k$ arrives. The system produces an output $b_k = G(q_k, a_k)$ and changes its state $q_k$ into $q_{k+1} = F(q_k, a_k)$. Then $a_{k+1}$ arrives, and so on.

$\Sigma$, as in Definition 6.6.1, is again called *linear* if $Q, A, B$ are vector spaces over some field $K$ and $F, G$ are linear maps on the product spaces $Q \times A$. In this case, $F$ and $G$ can be decomposed into linear functions $\alpha\colon Q \to Q$, $\beta\colon A \to Q$, $\gamma\colon Q \to B$ and $\delta\colon A \to B$ such that $F(q,a) = \alpha(q) + \beta(a)$, $G(q,a) = \gamma(q) + \delta(a)$ hold for all $(q,a) \in Q \times A$. If the vector spaces in question are finite dimensional, $\alpha, \beta, \gamma, \delta$ are usually represented by matrices.

It is not true, however, that these decompositions are only possible for linear systems. We are going to introduce "separable" systems now. They are much more general than linear ones, allow highly nonlinear transition and output functions, but we can do with these systems most things we can do with linear ones.

DEFINITION 6.6.3. $\Sigma$ (as in Definition 6.6.1) is called *separable* if $Q, A, B$ are groups (written additively, but not necessarily abelian) and if there are maps $\alpha\colon Q \to Q$, $\gamma\colon Q \to B$, and homomorphisms $\beta\colon A \to Q$, $\delta\colon A \to B$ such that

$$F(q,a) = \alpha(q) + \beta(a), \quad G(q,a) = \gamma(q) + \delta(a)$$

hold for all $q \in Q$, $a \in A$. We then denote $\Sigma$ by $(Q, A, B, \alpha, \beta, \gamma, \delta)$ or simply by $(\alpha, \beta, \gamma, \delta)$.

Clearly, each linear system is separable. Separable systems fit into the classes of nonlinear systems described by [18].

The map $\alpha\colon Q \to Q$ can be extended to a map (again denoted by $\alpha$) $\alpha\colon L(Q) \to L(Q)$ by

$$\alpha(q_k, q_{k+1}, \dots) = \bigl(\alpha(q_k), \alpha(q_{k+1}), \dots\bigr).$$

The same can be done for $\beta, \gamma$ and $\delta$. Let $s$ be the shift-to-the-left operator on $L(Q)$, given by

$$s(q_k, q_{k+1}, \dots) = (q'_{k-1}, q'_k, \dots),$$

where $q'_{k-1} = q_k$, $q'_k = q_{k+1}, \dots$ . It can be shown (see [68]) that for $\Sigma$ as in 6.6.3, the map $-\alpha + s$ from $L(Q)$ into $L(Q)$ is always bijective. [68] also contains the proofs of the following results.

THEOREM 6.6.4. *Let $\Sigma$ be separable as in* 6.6.3 *and* $q_k = 0$. *Then an input sequence* $(a_k, a_{k+1}, \dots)$ *produces an output sequence* $(b_k, b_{k+1}, \dots)$ *with*

$$(a_k, b_{k+1}, \dots) = \bigl[\gamma \circ (-\alpha + s)^{-1} \circ \beta + \delta\bigr](a_k, a_{k+1}, \dots).$$

The map in brackets hence gives a way to get the outputs directly from the inputs to a "new" system ($q_k = 0$) without the need to compute the various states in between. This is an essential feature for many concrete situations: it is often very hard or even impossible to conduct measurements of the internal states of systems. For instance, in the case of a propulsion engine of an airplane it is virtually impossible to measure all relevant data of the engine at any time in order to install feedback features (see below) to stabilize the engine.

DEFINITION 6.6.5. In a separable system $\Sigma = (\alpha, \beta, \gamma, \delta)$, the map

$$f_\Sigma := \gamma \circ (-\alpha + s)^{-1} \circ \beta + \delta$$

from $L(A)$ into $L(B)$ is called the *transfer function* of $\Sigma$.

$f_\Sigma$ describes, in a far reaching sense, $\Sigma$ itself and sometimes one identifies $f_\Sigma$ and $\Sigma$. If the series connection $\Sigma_1 \leftrightarrow \Sigma_2$ and the parallel connection $\Sigma_1 \updownarrow \Sigma_2$ are defined as usual, we get

THEOREM 6.6.6. *If $\Sigma_1$ and $\Sigma$, are separable, the same applies to $\Sigma_1 \leftrightarrow \Sigma_2$ and (if the output groups are abelian) also to $\Sigma_1 \updownarrow \Sigma_2$, and we then get*

$$f_{\Sigma_1 \updownarrow \Sigma_2} = f_{\Sigma_1} + f_{\Sigma_2},$$
$$f_{\Sigma_1 \leftrightarrow \Sigma_2} = f_{\Sigma_2} \circ f_{\Sigma_1}.$$

Hence, if we have input group $A =$ output group $B$, we get

COROLLARY 6.6.7. *If $S(A)$ is the set of all separable systems with finite abelian input-group $=$ output group $A$ then $(S(A), \updownarrow, \leftrightarrow)$ is a near-ring.*

In this case (and in contrast to the situation in the last subsection on automata) the systems themselves form a near-ring. For instance, we can then say

COROLLARY 6.6.8. *If $(\Sigma_i)_{i \in I}$ is a collection of separable systems with finite, abelian input-group $=$ output group $= A$, then the set of systems which can be constructed from the $\Sigma_i$'s by means of series and/or parallel connections is precisely the subnear-ring $N$ of $S(A)$ generated by $\{\Sigma_i \mid i \in I\}$. If we identify $\Sigma$ with $f_\Sigma$, $N$ is the subnear-ring of $M(L(Q))$ generated by $\{f_{\Sigma_i} \mid i \in I\}$.*

Many more topics in systems theory can then be transcribed to near-ring theory. For example, questions of invertibility of systems (with delay) transfer to some extent to questions of "von Neumann regularity" of near-rings of systems. Feedbacks and (again) reachability questions can also be handled with near-rings. For this and more see [68].

### 6.7. *Seminear-rings and rooted trees*

We restrict our considerations to finite rooted trees, although the infinite case does not create essential problems.

Given two such trees, say

$T_1$ $T_2$

we can compose them in the following ways: Addition: Give $T_1$ and $T_2$ a new common root:

$T_1 + T_2$:

Multiplication: Connect $T_2$ to each final node of $T_1$:

$T_1T_2$

It is easy to see that $(T_1 + T_2)T_3 = T_1T_3 + T_2T_3$ holds for all $T_1, T_2, T_3$. We now compare $T_2T_2 + T_2T_2$ and $T_2(T_2 + T_2)$ in our example

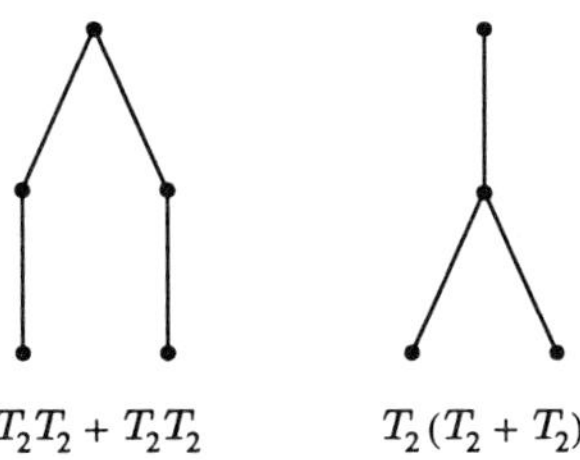

$T_2T_2 + T_2T_2$ $T_2(T_2 + T_2)$

One clearly sees the difference, so the other distributive law does not hold. We arrive at a new structure (see again the article "Semirings and Semifields" by U. Hebisch and H.J. Weinert in this volume of the *Handbook of Algebra*):

DEFINITION 6.7.1. $(S, +, \cdot)$ is a *seminear-ring* if $(S, +)$ and $(S, \cdot)$ are semigroups and $(s_1 + s_2)s_3 = s_1s_3 + s_2s_3$ holds for all $s_1, s_2, s_3 \in S$.

A prominent example is $(\mathbf{N}^{\mathbf{N}}, +, \circ)$. We get

THEOREM 6.7.2. *Rooted trees form a seminear-ring w.r.t. addition and multiplication (as defined above).*

The same can be done for valued trees and similar objects. The following interpretation was pointed out by J.D.P. Meldrum.

Consider a nondeterministic machine which can take "actions", one at a time. These are represented by nodes and lines going downward. Sometimes the machine has the "choice" of several actions; this is represented by splits in the trees. Hence a rooted tree can be though of a "computer program". In this interpretation one would identity

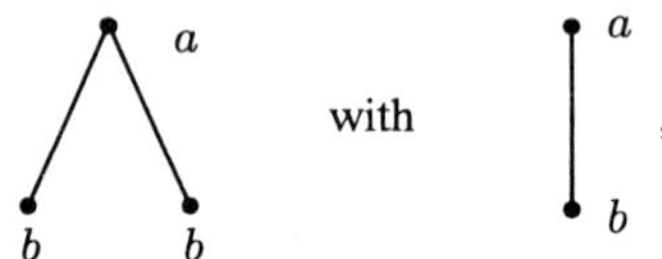


since the choice between $b$ and $b$ means that the machine has to do $b$. So we get $b + b = b$ for all $b$, and the corresponding seminear-ring has idempotent addition. People in Computer Science (e.g., [61]) also add a "deadlock" (= neutral element for addition), and so on.

Other interpretations include decision trees, travel plans, distributions of offsprings (an idea of D.W. Blackett) and so on.

This ends our trip to the wonderful world of one-sided distributive systems. If you want to keep in touch, please send your address to the author, and you will get a copy of the "Near-Ring Newsletter" about twice a year.

## References

### Books

[1] J.R. Clay, *Nearrings: Geneses and Applications*, Oxford Univ. Press, Oxford (1992).
[2] J.D.P. Meldrum, *Near-Rings and Their Links with Groups*, Pitman, London (1985).
[3] G.F. Pilz, *Near-Rings*, 2nd ed., North-Holland, Amsterdam (1983).
[4] H. Wähling, *Theorie der Fastkörper*, Thales-Verlag, Essen (1987).

### Articles

[10] J. André, *Über Parallelstrukturen, II. Translationsstrukturen*, Math. Z. **76** (1961), 155–163.
[11] J. André, *Lineare Algebra über Fastkörpern*, Math. Z. **136** (1974), 295–313.
[12] J. André, *On finite non-commutative spaces over certain nearrings*, in [102], 5–14.
[13] R. Baer, *Partitionen endlicher Gruppen*, Math. Z. **75** (1961), 333–372.
[14] G. Betsch, *Ein Radikal für Fastringe*, Math. Z. **78** (1962), 86–90.
[15] G. Betsch, *Embedding of a near-ring into a near-ring with identity*, in [100], 37–40.
[16] D.W. Blackett, *Simple and semi-simple near-rings*, Diss. Princeton Univ. (1950).
[17] A.E. Brouwer, J.B. Shearer, N.J.A. Sloane and W.D. Smith, *A new table of constant weight codes*, IEEE Trans. Inform. Theory **36** (1990), 1334–1380.

[18] J.L. Casti, *Non-linear System Theory*, Academic Press, New York (1985).
[19] J.R. Clay, *Generating balanced incomplete block designs from planar nearrings*, J. Algebra **22** (1972), 319–331.
[20] J.R. Clay, *Circular block designs from planar near-rings*, Ann. Discrete Math. **37** (1988), 95–106.
[21] P. Dembowski, *Finite Geometries*, Springer, Berlin (1968).
[22] L.E. Dickson, *Definitions of a group and a field by independent postulates*, Trans. Amer. Math. Soc. **6** (1905), 198–204.
[23] G. Ferrero, *Stems planari e BIB-disegni*, Riv. Mat. Univ. Parma (2) **11** (1970), 79–96.
[24] H. Fitting, *Die Theorie der Automorphismenringe abelscher Gruppen und ihr Analogon bei nicht kommutativen Gruppen*, Math. Ann. **107** (1932), 514–524.
[25] P. Fuchs, *A decoding method for planar near-ring codes*, Riv. Mat. Univ. Parma (2) **17** (1991), 325–331.
[26] P. Fuchs, *On the construction of codes using composition*, in [102], 71–80.
[27] P. Fuchs, G. Hofer and G. Pilz, *Codes from planar near-rings*, IEEE Trans. Inform. Theory **36** (1990), 647–651.
[28] P. Fuchs and C.J. Maxson, *Centralizer near-rings determined by PID-modules*, Arch. Math. **56** (1991), 140–147.
[29] P. Fuchs, C.J. Maxson and G. Pilz, *On rings for which homogeneous maps are linear*, Proc. Amer. Math. Soc. **112** (1991), 1–7.
[30] M. Hall, *The Theory of Groups*, McMillan, New York (1959).
[31] H.E. Heatherly, *One-sided ideals in near-rings of transformations*, J. Austral. Math. Soc. **13** (1972), 171–179.
[32] H.E. Heatherly, *Matrix near-rings*, J. London Math. Soc. (2) **7** (1973), 355–356.
[33] P.J. Higgins, *Groups with multiple operators*, Proc. London Math. Soc. (3) **6** (1956), 366–416.
[34] G. Hofer, *Ideals and reachability in machines*, in [100], 123–131.
[35] W.M.L. Holcombe, *A radical for linear sequential machines*, Proc. Roy. Irish Acad. **84A** (1984), 27–35.
[36] H. Hule and G. Pilz, *Algebraische Gleichungssysteme über universellen Algebren*, Inst. Ber. 306, Univ. Linz, Jan. 1986.
[37] H. Hule and G. Pilz, *Equations over abelian groups*, Contributions to General Algebra vol. 5, Teubner, Stuttgart (1987), 197–212.
[38] I.M. Isaacs, *Equally partitioned groups*, Pacific J. Math. **49** (1973), 109–116.
[39] H. Karzel, *Bericht über projektive Inzidenzgruppen*, Jahresber. Deutsch. Math.-Verrein. **67** (1965), 58–92.
[40] H. Karzel and C.J. Maxson, *Kinematic spaces with dilatations*, J. Geometry **22** (1984), 196–201.
[41] H. Karzel, C.J. Maxson and G. Pilz, *Kernels of covered groups*, Results Math. **9** (1986), 70–81.
[42] H. Karzel and M.J. Thomsen, *Near-rings, generalizations, near-rings with regular elements and applications, a report*, in [102], 91–110.
[43] O.H. Kegel, *Review of the paper* [52] (see below), Math. Reviews **90a** 20056 (1990).
[44] W.E. Kerby, *On Infinite Sharply Multiply Transitive Groups*, Vandenhoeck & Ruprecht, Göttingen (1974).
[45] J. Krempa and D. Niewieczerzał, *On homogeneous mappings of modules*, in [102], 123–136.
[46] H. Lausch and W. Nöbauer, *Algebra of Polynomials*, North-Holland, Amsterdam (1973).
[47] C.J. Maxson, *Near-rings associated with generalized translation structures*, J. Geometry **24** (1985), 175–193.
[48] C.J. Maxson, *Piecewise endomorphisms of PID-modules*, Results Math. **18** (1990), 125–132.
[49] C.J. Maxson, *Near-rings of piecewise endomorphisms*, in [102], 177–188.
[50] C.J. Maxson, *Homogenous functions of modules over local rings, II*, Submitted.
[51] C.J. Maxson and G. Pilz, *Near-rings determined by fibered groups*, Arch. Math. **44** (1985), 311–318.
[52] C.J. Maxson and G. Pilz, *Endomorphisms of fibered groups*, Proc. Edinburgh Math. Soc. **32** (1989), 127–129.
[53] C.J. Maxson and G. Pilz, *Kernels of covered groups, II*, Results Math. **16** (1989), 140–154.
[54] C.J. Maxson and A.P. Van der Walt, *Piecewise endomorphisms of ring modules*, Quaestiones Math. **14** (1991), 419–431.
[55] C.J. Maxson and A.P. Van der Walt, *Homogeneous maps as piecewise endomorphisms*, Comm. Algebra **20** (1992), 2755–2776.

[56] C.J. Maxson and L. Van Wyk, *The lattice of ideals of $M_R(R^2)$, R a commutative PIR*, J. Austral. Math. Soc. **52** (1992), 368–382.

[57] F.J. McWilliams and N.J.A. Sloane, *The Theory of Error-Correcting Codes, I, II*, North-Holland, Amsterdam (1977).

[58] J.D.P. Meldrum, *Matrix near-rings*, in [102], 189–204.

[59] J.D.P. Meldrum and A.P. Van der Walt, *Matrix near-rings*, Arch. Math. **47** (1986), 312–319.

[60] J.H. Meyer, *Matrix near-rings*, Doct. Diss., Univ. Stellenbosch, RSA (1986).

[61] R. Milner, *A calculus of communicating systems*, Lecture Notes in Comput. Sci. vol. 92, Springer, Berlin (1983).

[62] B.H. Neumann, *On the commutativity of addition*, J. London Math. Soc. **15** (1940), 203–208.

[63] G. Pilz, *Quasi-anelli: teoria ed applicazioni*, Rend. Sem. Mat. Fis. Milano **48** (1978), 79–86.

[64] G. Pilz, *Near-rings of compatible functions*, Proc. Edinburgh Math. Soc. **23** (1980), 87–95.

[65] G. Pilz, *Near-rings: What they are and what they are good for*, Contemp. Math. vol. 9 (1982), 97–119.

[66] G. Pilz, *Algebra – ein Reiseführer durch die schönsten Gebiete*, Trauner-Verlag, Linz (1984).

[67] G. Pilz, *Strictly connected group automata*, Proc. Roy. Irish Acad. **86A** (1986), 115–118.

[68] G. Pilz, *Near-rings and non-linear dynamical systems*, in [100], 211–232.

[69] G. Pilz, *What near-rings can do for you*, Contributions to General Algebra vol. 5, Teubner, Stuttgart (1987), 11–29.

[70] G. Pilz, *Near-rings*, 5 lectures, 2° Sem. Algebra non Commutativa, Siena (1987), 1–35. (Remark: Larger parts of this article were taken from that paper.)

[71] G. Pilz, *Codes, block designs, Frobenius groups, and near-rings*, Proc. Conf. Combinatorics '90, Gaeta (1990).

[72] G. Pilz, *On polynomial near-ring codes*, in [102], 233–238.

[73] S.D. Scott, *Tame Theory*, Amo Publishing, Auckland Univ. (1983).

[74] S.D. Scott, *Linear $\Omega$-groups, polynomial maps*, in [102], 239–294.

[75] W. Seier, *Kollineationen von Translationsstrukturen*, J. Geometry **12** (1971), 183–195.

[76] A.P. Van der Walt, *Primitivity in matrix near-rings*, Quaestiones Math. **9** (1986), 459–469.

[77] A.P. Van Der Walt, *On two-sided ideals in matrix near-rings*, in [100], 267–272.

[78] H. Wielandt, *Über Bereiche aus Gruppenabbildungen*, Dt. Mathematik **3** (1938), 9–10.

[79] H. Zassenhaus, *Über endliche Fastkörper*, Abh. Math. Sem. Univ. Hamburg **11** (1936), 187–220.

[80] H. Zassenhaus, *On Frobenius groups, II: Universal completion of near-fields of finite degrees over a field of reference*, Results Math. **11** (1987), 317–358.

[81] J.L. Zemmer, *Near-fields, planar and non-planar*, Math. Student **31** (1964), 145–150.

[82] J.L. Zemmer, *The additive group of an infinite near-field is abelian*, J. London Math. Soc. **44** (1969), 65–67.


## *Published Proceedings*


[100] G. Betsch (ed.), *Near-Rings and Near-Fields* (*Tübingen,* 1985), North-Holland, Amsterdam (1987).

[101] G. Betsch, G. Pilz and H. Wefelscheid (eds), *Near-Rings and Near-Fields* (*Oberwolfach,* 1989), Thales-Verlag, Essen (1993).

[102] G. Pilz (ed.), *Contributions to General Algebra vol.* 8 (*Linz,* 1991), Hölder-Pichler-Tempsky, Wien and Teubner, Stuttgart (1992).

[103] Y. Fong, W.E. Ke, G. Mason and G. Pilz (eds), *Near-Rings and Near-Fields* (*Fredericton, Canada,* 1993), Kluwer, Amsterdam (1995).

# Section 2A
# Category Theory

# Topos Theory

S. MacLane

*Department of Mathematics, University of Chicago, 5734 University av., Chicago IL 60637, USA*

I. Moerdijk

*Universiteit Utrecht, Fac. Wiskunde & Informatica, Postbus 80010, 35 08 TA Utrecht, The Netherlands*
*e-mail: moerdijk@math.ruu.nl*

*Contents*



HANDBOOK OF ALGEBRA, VOL. 1
Edited by M. Hazewinkel

This chapter will introduce topos theory, which arose from two separate explorations; first, Lawvere's proposal to replace the membership axioms for set theory by axioms on the composition of functions; and second, the Grothendieck initiatives in algebraic geometry. Indeed, Grothendieck and his collaborators [SGA4] needed to use cohomological ideas for algebraic varieties, not just for topological spaces, and in this context they introduced the notion of a topos as a generalized topological space. In topology the cohomology groups of a space, originally defined for a "constant" coefficient group, soon required coefficients which varied; these were codified by Leray and Cartan as sheaves on the space $X$; they were sheaves of abelian groups or of modules, but they could be described in terms of the more general (and simpler) sheaves of sets on the space. The category $Sh(X)$ of all such sheaves of sets on $X$ is an example of a topos. For algebraic geometry Grothendieck needed more general "tolopogies", defined not by open sets but by more general "coverings", and the sheaves of sets defined by such coverings provided the general notion of a "Grothendieck topos" (and the slogan: a topos is what a topologist needs to study, [SGA4], p. 301).

For many purposes, a space $X$ can be replaced by the corresponding topos $Sh(X)$. For example, continuous mappings $X \to Y$ between spaces correspond exactly to "geometric" morphisms $Sh(X) \to Sh(Y)$ of topoi. The generalization, from spaces $X$ and their sheaves to Grothendieck topoi, has proved extremely useful in algebraic geometry, as for example in P. Deligne's 1974 solution of the famous Weil conjectures about the solutions of Diophantine equations. These geometric ideas will be introduced in Section 6 (cohomologies for a topos) and Section 7 (the general fundamental group).

But a topos is not only a generalized space; it can also be viewed as a generalized "universe" of sets – as indeed the sheaves on a one-point space form the classical category of sets. This viewpoint appeared when Lawvere and Tierney developed an axiomatic treatment of sheaf theory without explicit reference to the specific Grothendieck topology used to define these sheaves. They also discovered that the process of turning a "presheaf" into a sheaf was implicitly involved in the notion of forcing used by Cohen in his proof of the independence of the Continuum Hypothesis – and by Scott and Solovay in the corresponding "Boolean valued" models of set theory. This led to the discovery of the more general "elementary" topoi described by first order (elementary) axioms. Any such topos is a universe in which one can do mathematics, classical except for the restriction that the logic in such a topos is in general "constructive" or "intuitionistic".

This chapter will begin with the Lawvere–Tierney definition of an elementary topos, followed by the definition of Grothendieck topologies and their sheaves. The third section of the chapter then describes the mappings between topoi, called geometric morphisms. Since topologies can be described in terms of open sets, with little mention of points, it is possible to discuss "pointless" spaces, or locales, and the corresponding topoi of sheaves (Section 4). The next section discusses representations of topoi in terms of such pointless spaces, including the theorem of Freyd showing how every topos can be embedded, in a suitably nice way, in the category of equivariant sheaves on some locale. In Section 6 the sheaf cohomology groups of an arbitrary topos are introduced. Certain basic spectral sequences relate these groups to Čech cohomology and to Verdier's cohomology of "hypercovers". These can be employed to define certain pro-groups, which are the "étale" homotopy groups of a topos, matching the Grothendieck fundamental group, to be discussed in Section 7. The final section describes the intimate relation between a topos and its "logic" – with Heyting (not Boolean) algebras of subobjects and with quantifiers described as adjoints – plus the "typed" language associated with a topos.

Our presentation is necessarily brief, and the interested reader is urged to consult further references. The material in Sections 1–4 and 8 is treated in detail in our recent book [MM], which also contains an extensive bibliography. Homotopy and cohomology of topoi are discussed extensively in [SGA1] (for the fundamental group), [SGA4] (vol. 2) and [AM]. For basic notions the reader may consult [CWM] for category theory, [M] for homological algebra, and [Ha] for algebraic geometry.

## 1. Elementary topoi

In any category $\mathcal{E}$ there are the notions of finite limits and colimits, see [CWM]. Specifically, $\mathcal{E}$ has all finite limits iff it has a terminal object 1 (or $1_{\mathcal{E}}$) and pullbacks $X \times_Z Y$ for any pair of arrows $X \to Z \leftarrow Y$. For a fixed object $X$ in such a category $\mathcal{E}$, taking the product with $X$ defines a functor $X \times (-)$: $\mathcal{E} \to \mathcal{E}$. The object $X$ is said to be *exponentiable* of this functor has a right adjoint, denoted $(-)^X$. This means that for any two objects $Y, Z$ in $\mathcal{E}$, there is a bijective correspondence between arrows $X \times Z \to Y$ and arrows $Z \to Y^X$, natural in $Y$ and $Z$. A *cartesian closed category* is a category which finite limits, in which each object is exponentiable.

A *subobject classifier* in a category $\mathcal{E}$ is an object $\Omega$, equipped with an arrow $t$: $1 \to \Omega$ from the terminal object 1, so that for any monomorphism $U \to X$ in $\mathcal{E}$ there is a unique arrow $c_U$: $X \to \Omega$ which makes the square

$$\begin{array}{ccc} U & \longrightarrow & 1 \\ \downarrow & & \downarrow t \\ X & \xrightarrow{c_U} & \Omega \end{array}$$

into a pullback. (One thinks of $\Omega$ as an object of "truth-values", of $t$ as "true", and of $c_U$ as a "characteristic function" for $U$.) Thus there is a natural bijection

$$Sub(X) \xrightarrow{\sim} \mathcal{E}(X, \Omega),$$

between subobjects of $X$ and arrows $X \to \Omega$.

**1.1.** DEFINITION. An *elementary topos* is a cartesian closed category with a subobject classifier.

**1.2.** REMARKS. (i) In an elementary topos $\mathcal{E}$, the exponential $\Omega^X$ is a "powerset" object for $X$, and is also denoted $PX$. Arrows $Y \to PX$ in $\mathcal{E}$ correspond naturally to subobjects of $X \times Y$.

(ii) It is a consequence of the axioms that *finite colimits* exist in $\mathcal{E}$, [P].

**1.3.** EXAMPLES. Here is a short list of easiest examples of elementary topoi.

(a) *Sets*: this is the category of sets and functions. In this category, the exponential $Y^X$ is the set of all functions $X \to Y$, while the subobject classifier $\Omega$ is $\{0,1\}$, with $t\colon 1 \to \{0,1\}$ the function from the one-element set with value 1. For a subobject $U \rightarrowtail X$ in *Sets*, $c_U\colon X \to \Omega$ is the usual characteristic function.

(b) *G-Sets*: this is the category of sets with a (right) action by a fixed group $G$. The arrows in this category are functions which preserve the action ("equivariant functions"). For two sets with $G$-action $X$ and $Y$, the exponential $Y^X$ is the set of all functions $\alpha\colon X \to Y$, equipped with the $G$-action defined as

$$(\alpha \cdot g)(x) = \alpha(x \cdot g^{-1}) \cdot g.$$

In this topos, the subobject classifier $\Omega$ is again the set $\{0,1\}$, with trivial $G$-action ($0 \cdot g = 0$ and $1 \cdot g = 1$ for all $g \in G$).

(c) $Sh(X)$: this is the topos of all sheaves (of sets) on a topological space $X$. Recall that a sheaf $E$ on $X$ is given by a set $E(U)$ for each open set $U \subseteq X$, and a restriction operation $E(U) \to E(V)$, denoted $e \rightarrowtail e \mid V$, for each inclusion $V \subseteq U$ between open sets, such that the following two properties hold:

(i) (functoriality) For inclusions $W \subseteq V \subseteq U$ and any $e \in E(U)$,

$$(e|V)|W = e|W,$$

and $e|U = e$.

(ii) (amalgamation) For any open covering $U = \bigcup_{i\in I} U_i$, and any family $e_i \in E(U_i)$ (for $i \in I$) of elements which are compatible in the sense that for any two $i, j \in I$:

$$e_i|U_i \cap U_j = e_j|U_i \cap U_j,$$

there is a unique $e \in E(U)$ with $e|U_i = e_i$.

For example, for any continuous map $\pi\colon P \to X$ there is the sheaf $\Gamma_P$ of sections of $P$, defined for each open $U \subseteq X$ by

$$\Gamma_P(U) = \{s\colon U \to P \mid s \text{ is continuous, and } \pi s(x) = x \text{ for all } x \in U\}.$$

It can be shown that, up to isomorphism, every sheaf on $X$ is of the form $\Gamma_P$ where $\pi\colon P \to X$ is a local homeomorphism (an "étale space over $X$"); up to isomorphism, this étale space is uniquely determined by the sheaf. (Recall that a continuous map $\pi\colon P \to X$ is a local homeomorphism if for every point $y \in P$ there are open neighborhoods $U$ of $y$ and $V$ of $\pi(y)$ so that $\pi$ restricts to a homeomorphism $\pi\colon U \xrightarrow{\sim} V$.)

The sheaves $E$ on $X$ form a category, where an arrow $\alpha\colon E \to E'$ between sheaves is defined as a natural transformation, i.e. a family of functions $\alpha_U\colon E(U) \to E'(U)$ (for all open $U \subseteq X$) which commute with the restriction operations of $E$ and $E'$. This category $Sh(X)$ is a topos. The correspondence between sheaves and étale spaces over $X$ is an equivalence of categories, and in practice one often identifies sheaves and étale spaces.

More examples of topoi will be given below.

**1.4.** *Slice topoi.* An important fact is that for any topos $\mathcal{E}$ and any object $X$ in $\mathcal{E}$, the comma (or "slice") category $\mathcal{E}/X$ is again a topos. Moreover, for any arrow $f\colon X \to Y$ in $\mathcal{E}$ the pullback functor

$$f^*\colon \mathcal{E}/Y \to \mathcal{E}/X$$

preserves the topos structure (finite limits, exponentials, subobject classifier) and has a left adjoint $\sum_f$ as well as a right adjoint $\prod_f$. (The functor $\sum_f$ is simply given by composition with $f$, but $\prod_f$ is harder to construct.)

**1.5.** *Constructions of elementary topoi.* There are many standard constructions of new topoi from old ones, described in detail in any good book on the subject. We mention the following: For two topoi $\mathcal{E}$ and $\mathcal{F}$, the product category $\mathcal{E} \times \mathcal{F}$ is again a topos. If $j\colon \Omega \to \Omega$ is a so-called Lawvere–Tierney topology in $\mathcal{E}$ then the category $\mathcal{E}_j$ of $j$-sheaves is again a topos. If $\mathbb{G} = (G, \varepsilon, \delta)$ is a left exact comonad on $\mathcal{E}$ then the category $\mathcal{E}_{\mathbb{G}}$ of $\mathbb{G}$-coalgebras is a topos. If $C$ is an "internal" category object in $\mathcal{E}$, then the category $\mathcal{E}^C$ of diagrams on $C$ in $\mathcal{E}$ is a topos. If $\mathcal{E}$ is a topos and $\mathbb{F}$ is a filter of subobjects of 1 in $\mathcal{E}$, then the "filterquotient" $\mathcal{E}/\mathbb{F}$ (which is essentially the directed limit $\varinjlim_{U \in \mathbb{F}} \mathcal{E}/U$) is again a topos.

## 2. Grothendieck topoi

The notion of a topos as originally introduced by Grothendieck is more restrictive than that of an elementary topos. Grothendieck's notion is based on the definition of a site: a category $\mathbb{C}$ with a generalized notion of "covers", sufficient to define "sheaves" on $\mathbb{C}$.

**2.1.** *Grothendieck sites.* Let $\mathbb{C}$ be a small category. A *Grothendieck topology* on $\mathbb{C}$ is an operation $J$ which assigns to each object $C$ a collection $J(C)$ of families of arrows in $\mathbb{C}$ with codomain $C$, called *covering families*, such that the following three conditions are satisfied:

(i) For any isomorphism $f\colon D \xrightarrow{\sim} C$ in $\mathbb{C}$, the one-element family $\{f\colon D \to C\}$ belongs to $J(C)$.

(ii) (Transitivity condition) Given a covering family $\{f_i\colon C_i \to C\}_{i \in A}$ in $J(C)$, and for each index $i \in A$ another covering family $\{g_{ij}\colon D_{ij} \to C_i\}_{j \in B_i}$ in $J(C_i)$, the family of all composites $\{f_i \circ g_{ij}\colon D_{ij} \to C\}_{i,j}$ belongs to $J(C)$.

(iii) (Stability condition) Given a covering family $\{f_i\colon C_i \to C\}_i$ and an arrow $g\colon D \to C$ in $\mathbb{C}$, there exists a covering family $\{h_j\colon D_j \to D\}_j$, such that each $gh_j$ factors through some $f_i$ (i.e. for each index $j$ there is an index $i$ and an arrow $k$ so that $g \circ h_j = f_i \circ k$).

A *site* is a category $\mathbb{C}$ equipped with a Grothendieck topology $J$.

**2.2.** Remark. In many examples of sites the category $\mathbb{C}$ has pullbacks, and the stability condition is satisfied in the following stronger form: Given $\{f_i\colon C_i \to C\}_i \in J(C)$ and $g\colon D \to C$ as in (iii), the family of pullbacks $\{D \times_C C_i \to D\}_i$ belongs to $J(D)$.

**2.3.** EXAMPLES.

(i) (Topology) Let $X$ be a topological space. Define a category $\mathcal{O}(X)$ whose objects are the open subsets $U \subseteq X$, and with exactly one arrow $U \to V$ in case $U \subseteq V$. Define a Grothendieck topology $J$ on $\mathcal{O}(X)$ by

$$\{U_i \to U\} \in J(U) \quad \text{iff} \quad U = \bigcup U_i.$$

In other words, a family is covering in the sense of $J$ iff it is covering in the usual sense.

(ii) (Algebraic geometry) Let $X$ be a scheme (over a fixed groundfield $k$), and consider the category $Et/X$, with all étale morphisms $f\colon Y \to X$ as objects and all commuting triangles as arrows. One obtains a Grothendieck topology on $Et/X$ by defining a family

$$\begin{array}{ccc} Y_i & \xrightarrow{g_i} & Y \\ & {\scriptstyle f_i}\searrow \quad \swarrow{\scriptstyle f} & \\ & X & \end{array}, \quad i \in I, \tag{1}$$

to be covering iff

$$Y = \bigcup_{i \in I} g_i(Y_i)$$

(as sets). The site thus defined is called the (small) *étale site* over the scheme $X$ and is denoted $X_{\text{ét}}$. (There is also a "big" étale site, [SGA4].)

**2.4.** *Sheaves.* The central notion of sheaf on a topological space can now be generalized to any site. Let $(\mathbb{C}, J)$ be a site, and call a functor $P\colon \mathbb{C}^{\mathrm{op}} \to \mathit{Sets}$ a *presheaf* on $\mathbb{C}$. For an arrow $f\colon C \to D$ and an element $x \in P(D)$, one also denotes $P(f)(x)$ by $x|f$. Let $\{f_i\colon C_i \to C\}_{i \in I}$ be a covering family for the Grothendieck topology $J$. A family $x_i \in P(C_i)$, $i \in I$, is called a *compatible family* of elements of $P$ if for any commutative diagram in $\mathbb{C}$ of the form

$$\begin{array}{ccc} D & \xrightarrow{h} & C_i \\ {\scriptstyle k}\downarrow & & \downarrow{\scriptstyle f_i} \\ C_j & \xrightarrow{f_j} & C \end{array}$$

the identity $x_i|h = x_j|k$ holds. An *amalgamation* for such a family $\{x_i\}$ is an $x \in P(C)$ so that $x|f_i = x_i$, for each arrow $f_i$ in the covering family. The presheaf $P$ is said to be a *sheaf* (for the topology $J$) when, for each covering family, each compatible family of elements of $P$ has a unique amalgamation. With arrows between sheaves the natural transformations, these sheaves form a category, denoted

$$Sh(\mathbb{C}, J).$$

This is the category of sheaves (of sets) on the site $(\mathbb{C}, J)$. In a similar fashion one can define sheaves of (abelian) groups, rings, modules, etc.

**2.5.** DEFINITION. A category $\mathcal{E}$ is called a *Grothendieck topos* if there exist a site $(\mathbb{C}, J)$ and an equivalence of categories

$$\mathcal{E} \cong Sh(\mathbb{C}, J).$$

For a topological space $X$ and the associated site $\mathcal{O}(X)$ described in 2.3(i), the category of sheaves is the category $Sh(X)$ already introduced in 1. For the small étale site $X_{\text{ét}}$ associated to a scheme $X$ as in 2.3(ii), the corresponding topos $Sh(X_{\text{ét}})$ is called the (small) *étale topos* associated to the scheme $X$. One often simply writes $Sh(X)$ for $Sh(X_{\text{ét}})$, and refers to sheaves on the étale site $X_{\text{ét}}$ simply as sheaves on $X$.

**2.6.** REMARKS. (i) For a given category $\mathcal{E}$, the "Giraud Theorem" gives conditions for $\mathcal{E}$ to be a Grothendieck topos without any reference to sites (see [MM], p. 575).

(ii) Every Grothendieck topos is an elementary topos, but the converse is not true: An elementary topos $\mathcal{E}$ is a Grothendieck topos precisely when $\mathcal{E}$ has coproducts indexed by arbitrary sets, as well as a (small) set of generators ([MM], p. 591).

(iii) Given a Grothendieck topos $\mathcal{E}$, there are many different sites for which there is an equivalence as in Definition 2.5. The "Comparison Lemma" ([MM], p. 588) gives conditions under which two sites $(\mathbb{C}, J)$ and $(\mathbb{C}', J')$ give rise to the same topos (or, more precisely, to an equivalence of categories $Sh(\mathbb{C}, J) \cong Sh(\mathbb{C}', J')$).

**2.7.** EXAMPLES. We list some more basic examples, in addition to the sheaves on a topological space already mentioned below 2.5.

(i) For each small category $\mathbb{C}$, there is the *trivial topology* $T$ where the only covering families are the one-element families $\{f\colon D \xrightarrow{\sim} C\}$ where $f$ is an isomorphism. For this topology $T$, every presheaf is a sheaf, and $Sh(\mathbb{C}, T)$ is the functor category $Sets^{\mathbb{C}^{\text{op}}}$ of all presheaves.

(ii) (Boolean sheaves) If $\mathbb{B} = (B, 0, 1, \vee, \wedge)$ is a complete Boolean algebra ([H]), there is a corresponding site: the objects are the elements $b \in \mathbb{B}$; there is exactly one arrow $b \to b'$ iff $b \leqslant b'$, while a family $\{b_i \to b\}$ is covering iff $b = \vee b_i$ in $\mathbb{B}$. This site is again denoted by $\mathbb{B}$, and its topos of sheaves is $Sh(\mathbb{B})$. This topos closely resembles the category of sets, and is related to Boolean-valued models of set theory.

(iii) Let $G$ be a topological group. A *continuous $G$-set* is a set $S$ equipped with a right action $S \times G \to S$ by $G$, which is continuous when we give $S$ the discrete topology. The category of all continuous $G$-sets and equivariant functions is a topos denoted $\mathcal{B}G$. (A site for this topos has as objects the cosets $G/U$ where $U$ is an open subgroup of $G$.)

(iv) More generally, let $X$ be a topological space equipped with a continuous action by the topological group $G$. As above, we may identify sheaves on $X$ with étale spaces over $X$, i.e. local homeomorphisms $p\colon E \to X$. An equivariant sheaf is such an étale space $p\colon E \to X$ equipped with a continuous action $E \times G \to E$ which makes $p$ into an equivariant map. The category of all such equivariant sheaves is a Grothendieck topos (see [MM], p. 594), and is denoted $Sh_G(X)$.

## 3. Geometric morphisms

These morphisms extend the usual notion of continuous maps between topological spaces (3.1 below). Let $\mathcal{E}$ and $\mathcal{F}$ be two topoi. A *geometric morphism* $f\colon \mathcal{F} \to \mathcal{E}$ from $\mathcal{F}$ to $\mathcal{E}$ is a pair of adjoint functors (called the inverse and direct image functors)

$$f^* : \mathcal{E} \rightleftarrows \mathcal{F} : f_*$$

where $f^*$ is left adjoint to $f_*$, with the additional property that $f^*$ is left exact (i.e. preserves finite limits). One also refers to such geometric morphisms simply as morphisms, or maps, from $\mathcal{F}$ to $\mathcal{E}$.

For each topos $\mathcal{E}$ there is an "identity-morphism" $\mathcal{E} \to \mathcal{E}$, given by the identity functor on $\mathcal{E}$ which is adjoint to itself. And given two morphisms $g\colon \mathcal{G} \to \mathcal{F}$ and $f\colon \mathcal{F} \to \mathcal{E}$, one can construct a composite geometric morphism $f \circ g\colon \mathcal{G} \to \mathcal{E}$, simply by composing the adjoints: $(f \circ g)^* = g^* \circ f^*$ while $(f \circ g)_* = f_* \circ g_*$. Furthermore, for two topoi $\mathcal{E}$ and $\mathcal{F}$, all the geometric morphisms from $\mathcal{F}$ to $\mathcal{E}$ form a category

$$\mathrm{Hom}(\mathcal{F}, \mathcal{E}),$$

with arrows $\alpha\colon f \to g$ in this category the natural transformations $f^* \to g^*$. In this way one obtains a so-called 2-dimensional category, with topoi as objects, geometric morphisms as arrows, and such natural transformations as "2-cells".

This 2-categorical structure is always (at least) implicitly present in topos theory. For example, in practice two topoi $\mathcal{E}$ and $\mathcal{F}$ are identified when they are *equivalent* as categories (they are hardly ever isomorphic). This is already apparant in the definition of a Grothendieck topos, which "is" a category of sheaves on a site. Similar remarks apply, for example, to the construction of pullbacks of topoi.

We next give some easy examples of geometric morphisms, using our stock of first examples of topoi from Sections 1 and 2.

**3.1.** EXAMPLES. (i) Let $f\colon X \to Y$ be a map between topological spaces. Then each étale space $P$ over $Y$ pulls back to such a space $f^*(P)$ over $X$, while the map $f^{-1}\colon \mathcal{O}(Y) \to \mathcal{O}(X)$ converts each sheaf $E$ on $X$ to a sheaf $f_*(E) := E \circ f^{-1}$ on $Y$. Thus $f$ induces the so-called inverse and direct image functors on sheaves:

$$f^* : Sh(Y) \leftrightarrows Sh(X) : f_*,$$

and $f^*$ is left exact (see almost any book on sheaf theory). These two functors define a geometric morphism of topoi, again denoted $f\colon Sh(X) \to Sh(Y)$. Under mild separation conditions on $Y$ (e.g., $Y$ Hausdorff), every geometric morphism of topoi $Sh(X) \to Sh(Y)$ comes from a unique map of spaces in this way.

(ii) A homomorphism $\varphi\colon G \to H$ of groups induces three functors between the category of $G$-sets and that of $H$-sets:

$$(G\text{-}Sets) \underset{\varphi_*}{\overset{\varphi_!}{\underset{\longleftarrow}{\overset{\longrightarrow}{\underset{\longrightarrow}{\varphi^*}}}}} (H\text{-}Sets),$$

with $\varphi_!$ left adjoint to $\varphi^*$, and $\varphi^*$ in turn left adjoint to $\varphi_*$. Indeed, for an $H$-Set $Y$, one defines $\varphi^*(Y)$ to be the same set $Y$ with $G$-action induced by $\varphi\colon G \to H$. In particular, $H$ itself is a right $G$-set in this way, and one defines for a $G$-set $X$,

$$\varphi_*(X) = \mathrm{Hom}_G(H, X), \quad \varphi_!(X) = X \otimes_G H.$$

Here $\mathrm{Hom}_G(H, X)$ is the set of $G$-equivariant functions $H \to X$, while $X \otimes_G H$ is the cartesian product $X \times H$ factored out by the equivalence relation $(x \cdot g, h) \sim (x, \varphi(g) \cdot h)$, much as for the tensor product of modules. Since this functor $\varphi_!$ is left adjoint to $\varphi^*$, the latter functor $\varphi^*$ must be left exact. Therefore the pair $\varphi^*$, $\varphi_*$ defines a geometric morphism (again denoted) $\varphi\colon (G\text{-}\mathit{Sets}) \to (H\text{-}\mathit{Sets})$. It can be shown that, up to isomorphism between geometric morphisms, every map $(G\text{-}\mathit{Sets}) \to (H\text{-}\mathit{Sets})$ is of this form.

(iii) If $\pi\colon P \to X$ is étale over the topological space $X$, a "global cross section" of $P$ is just a map $s\colon X \to P$ with $\pi \circ s = \mathrm{id}$; or, a map from the terminal object $(\mathrm{id}\colon X \to X)$ to $(\pi\colon P \to X)$ in the category of étale spaces over $X$. More generally, consider for a topos $\mathcal{E}$ the global sections functor:

$$\Gamma\colon \mathcal{E} \to \mathit{Sets},$$

defined by

$$\Gamma E = \mathcal{E}(1, E) = \mathrm{Hom}_{\mathcal{E}}(1, E),$$

the set of all arrows from the terminal object 1 to $E$. If $\mathcal{E}$ has small sums (e.g., if $\mathcal{E}$ is a Grothendieck topos), then $\Gamma$ has a left adjoint, the "constant sheaf functor",

$$\Delta\colon \mathit{Sets} \to \mathcal{E}, \quad \Delta(S) := \sum_{s \in S} 1.$$

This left adjoint $\Delta$ is left exact. So the pair $(\Delta, \Gamma)$ defines a morphism $\mathcal{E} \to \mathit{Sets}$. (It is not difficult to see that, up to isomorphism, there can be at most one such geometric morphism $\mathcal{E} \to \mathit{Sets}$.)

(iv) Over a topological space, each presheaf can be made into an "associated" sheaf. More generally, let $(\mathbb{C}, J)$ be a site, with associated topos $\mathit{Sh}(\mathbb{C}, J)$, and write $i\colon \mathit{Sh}(\mathbb{C}, J) \to \mathit{Sets}^{\mathbb{C}^{\mathrm{op}}}$ for the inclusion functor of sheaves into presheaves. One can prove that this functor has a left exact left adjoint, the so-called *associated sheaf functor* $\mathbf{a}$: $\mathit{Sets}^{\mathbb{C}^{\mathrm{op}}} \to \mathit{Sh}(\mathbb{C}, J)$. The pair $(\mathbf{a}, i)$ defines a geometric morphism $\mathit{Sh}(\mathbb{C}, J) \to \mathit{Sets}^{\mathbb{C}^{\mathrm{op}}}$.

Geometric morphisms between Grothendieck topoi are often constructed using flat (or filtering) functors. To explain this, let $\mathcal{E}$ and $\mathcal{F}$ be Grothendieck topoi, and let us fix an explicit site $(\mathbb{C}, J)$ for $\mathcal{E}$. (For convenience we will assume an actual equality $\mathcal{E} = \mathit{Sh}(\mathbb{C}, J)$.) Any functor

$$A\colon \mathbb{C} \to \mathcal{F}$$

can be canonically extended to a functor

$$f^*_{(A)}\colon\ Sets^{\mathbb{C}^{op}} \to \mathcal{F}$$

by "Kan extension". (Indeed, a presheaf $E$ on $\mathbb{C}$ can be viewed as a right $\mathbb{C}$-module – a set equipped with an action by $\mathbb{C}$ from the right –, while the covariant functor $A$ can be viewed as a left $\mathbb{C}$-module in $\mathcal{F}$; then $f^*_{(A)}(E)$ is simply defined as the tensor product $E \otimes_{\mathbb{C}} A$. See [MM].) Just as for modules, this "tensor product" has a hom-functor as right adjoint. Specifically, $A$ induces a functor, right adjoint to $f^*_{(A)}$,

$$f_{(A)*}\colon\ \mathcal{F} \to Sets^{\mathbb{C}^{op}},$$

sending an object $F \in \mathcal{F}$ to the presheaf $f_{(A)*}(F)$ on $\mathbb{C}$ defined by

$$f_{(A)*}(F)(C) = \mathrm{Hom}_{\mathcal{F}}\big(A(C), F\big).$$

By definition, the functor $A\colon\ \mathbb{C} \to \mathcal{F}$ is said to be *flat* if the associated tensor product functor $f^*_{(A)} = (- \otimes_{\mathbb{C}} A)$ is left exact. Furthermore, the functor $A$ is said to be *continuous* if for every covering family $\{C_i \to C\}$ in the site $(\mathbb{C}, J)$, the induced map

$$\sum A(C_i) \to A(C)$$

is an epimorphism in the topos $\mathcal{F}$. This ensures that the presheaf $f_{(A)*}(F)$ is actually a sheaf. Thus when $A$ is flat and continuous, the tensor-product functor $f^*_{(A)}$ restricted to sheaves, and the functor $f_{(A)*}$, together yield a geometric morphism $f_{(A)}\colon\ \mathcal{F} \to Sh(\mathbb{C}, J)$. Every geometric morphism is of this form:

**3.2.** PROPOSITION. *The operation $A \mapsto f_{(A)}$ induces an equivalence of categories, between the category of flat and continuous functors $A\colon\ \mathbb{C} \to \mathcal{F}$, and that of geometric morphisms $\mathcal{F} \to Sh(\mathbb{C}, J)$.*

**3.3.** REMARK. The condition for a functor $A\colon\ \mathbb{C} \to \mathcal{F}$ to be flat can be made more explicit. In case the category $\mathbb{C}$ has finite limits, $A$ is flat iff it preserves finite limits. In general, $A$ is flat iff it is "filtering", as defined in [MM].

**3.4.** EXAMPLES. We mention a "mixed" case of the examples (i) and (ii) in 3.1. Let $X$ be a topological space and let $G$ be a group, viewed as a category with just one object. A functor $A\colon\ G \to Sh(X)$ is the same thing as a sheaf (again denoted) $A$ on $X$ equipped with a left $G$-action. When we view the sheaf $A$ as the sheaf of sections of an étale space $p\colon\ E_A \to X$ (as in Example 2.7(iv)), then $G$ acts on the fibers $p^{-1}(x)$. The functor $A\colon\ G \to Sh(X)$ is flat iff this action by $G$ is free and transitive on each fiber. In other words, $E_A \to X$ is a principal $G$-bundle, or a covering projection with group $G$. Thus by Proposition 3.2, mappings of topoi $Sh(X) \to (G\text{-}Sets)$ correspond to principal $G$-bundles over $X$. One says: the topos $(G\text{-}Sets)$ *classifies* principal $G$-bundles; or, $(G\text{-}Sets)$ is a *classifying topos* for principal $G$-bundles.

Similarly, it can be shown that there exists a classifying topos for rings. This is a topos $\mathcal{B}$, with a ring object $R$ in $\mathcal{B}$, such that for any other Grothendieck topos $\mathcal{E}$, ring objects in $\mathcal{E}$ correspond to morphisms $\mathcal{E} \to \mathcal{B}$.

Many topoi can be viewed as classifying topoi for particular "structures" in this way. And conversely, if a structure can be defined by so-called geometric axioms, then one can prove that there is a classifying topos for this structure. This leads to an extensive theory of classifying topoi, in which Proposition 3.2 plays a central rôle.

**3.5.** *Morphisms of sites.* Consider two sites $(\mathbb{C}, J)$ and $(\mathbb{D}, K)$ for which both $\mathbb{C}$ and $\mathbb{D}$ have finite limits. A morphism of sites $\alpha\colon (\mathbb{C}, J) \to (\mathbb{D}, K)$ is a functor $\alpha\colon \mathbb{C} \to \mathbb{D}$ which preserves both finite limits and covers. (The latter means: for every covering family $\{C_i \to C\}$ in $(\mathbb{C}, J)$, there exists a covering family $\{D_j \to \alpha(C)\}$ in $(\mathbb{D}, K)$ which refines the family $\{\alpha(C_i) \to \alpha(C)\}$, in the sense that each $D_j \to \alpha(C)$ factors through some $\alpha(C_i) \to \alpha(C)$.) Such a morphism of sites yields a geometric morphism between the topoi of sheaves: first form the composite

$$A = \mathbf{a} \circ y \circ \alpha\colon\ \mathbb{C} \xrightarrow{\alpha} \mathbb{D} \xrightarrow{y} \mathrm{Sets}^{\mathbb{D}^{\mathrm{op}}} \xrightarrow{\mathbf{a}} Sh(\mathbb{D}, K).$$

Here $\mathbf{a}$ is the associated sheaf functor (cf. 3.1(iv)), and $y$ is the Yoneda embedding which sends each object $D \in \mathbb{D}$ to the representable presheaf $\mathbb{D}(-, D)$. The conditions above on $\alpha$ ensure that $A$ is left exact and continuous, hence (cf. 3.3) flat and continuous. Thus $A$ induces a geometric morphism $f = f_{(A)}$, as in Proposition 3.2:

$$f\colon\ Sh(\mathbb{D}, K) \to Sh(\mathbb{C}, J).$$

The direct image functor $f_*$ can be described simply in terms of composition with $\alpha$: For a sheaf $F$ on $(\mathbb{D}, K)$ and an object $C \in \mathbb{C}$,

$$f_*(F)(C) = F(\alpha C).$$

**3.6.** EXAMPLE. In algebraic geometry, each morphism between schemes $f\colon X \to Y$ induces by pullback a morphism of sites $f^{\#}\colon X_{\text{ét}} \to X_{\text{ét}}$, and hence a geometric morphism between (small) étale topoi, (again denoted) $f\colon Sh(X) \to Sh(Y)$.

**3.7.** *Points.* Motivated by the correspondence (3.1(i)) between continuous mappings $X \to Y$ between topological spaces and morphisms of topoi $Sh(X) \to Sh(Y)$, together with the observation that $Sets$ is the category $Sh(1)$ of sheaves on the one-point space, one defines a *point* of a topos $\mathcal{E}$ to be a geometric morphism $p\colon Sets \to \mathcal{E}$. A topos $\mathcal{E}$ is said to have *enough points* if all the inverse image functors $p^*\colon \mathcal{E} \to Sets$ of points $p$ are collectively faithful; or in other words, if for any two distinct parallel arrows $f, g\colon A \to B$ in $\mathcal{E}$ there exists a point $p\colon Sets \to \mathcal{E}$ so that $p^*(f)$ and $p^*(g)\colon p^*(A) \to p^*(B)$ are still distinct.

For a topos $\mathcal{E}$, having enough points is a useful property, because it implies that any statement expressible in terms of colimits and finite limits and true in $Sets$ will be true in $\mathcal{E}$. (For general topoi, Barr's Theorem (5.3 below) provides a similar useful result.)

Call a site $(\mathbb{C}, J)$ of *finite type* if $\mathbb{C}$ has pullbacks and every covering family in $J$ is finite. "*Deligne's Theorem*" states that any topos $Sh(\mathbb{C}, J)$ of sheaves on such a site of finite type has enough points.

**3.8.** *Constructions of topoi.* The 2-dimensional category of Grothendieck topoi and geometric morphisms has very good closure properties. For example, for two geometric morphisms $\mathcal{F} \to \mathcal{E}$ and $\mathcal{G} \to \mathcal{E}$ the pullback $\mathcal{F} \times_{\mathcal{E}} \mathcal{G}$ always exists (in some sense appropriate for 2-categories), see [D] and Section 8 below, and many of its basic properties have been studied. The same is true for limits of filtered inverse systems (see [M1]). Colimits of diagrams of Grothendieck topoi also exist, in a very general 2-categorical sense, and can easily be constructed explicitly. (A good exposition can be found in [MP], p. 108.) For example, for two maps $f, g$: $\mathcal{E} \rightrightarrows \mathcal{F}$, the "lax-coequalizer" is the universal solution for a map of topoi $q$: $\mathcal{F} \to \mathcal{G}$ together with a 2-cell $q \circ f \Rightarrow q \circ q$ (i.e. a natural transformation $f^*q^* \to g^*q^*$). Such a universal $\mathcal{G}$ exists, and is simply constructed as the category of pairs $(F, u)$ where $F \in \mathcal{F}$ and $u$: $f^*(F) \to g^*(F)$. This category is indeed a topos. Furthermore, for two Grothendieck topoi $\mathcal{E}$ and $\mathcal{F}$ one can also construct their exponential $\mathcal{F}^{\mathcal{E}}$, provided $\mathcal{E}$ is "locally compact" in a suitable sense; see [JJ].

## 4. Locales

Locales are like topological spaces, but without points. They play a central role in topos theory, partly because topoi need not have "enough" points (cf. 3.7). For example, in the next section we shall present various covering theorems stating that for every topos $\mathcal{E}$ there exists a "space" $X$ and a morphism $Sh(X) \to \mathcal{E}$ with good properties, but in general one should allow this space to be a locale.

The formal definition starts from the properties of the open set lattice $\mathcal{O}(T)$ of a topological space $T$. Define a *frame* to be a complete lattice $\mathcal{A}$, in which binary meets distribute over arbitrary joins, as in the identity

$$U \wedge \bigvee_{i \in I} V_i = \bigvee_{i \in I} U \wedge V_i, \tag{1}$$

for any $U \in \mathcal{A}$ and any collection $\{V_i\colon\ i \in I\}$ of elements of $\mathcal{A}$. Define a morphism of frames $\Phi$: $\mathcal{A} \to \mathcal{B}$ to be a map which preserves finite meets and arbitrary joints:

$$\Phi(1_{\mathcal{A}}) = 1_{\mathcal{B}}, \quad \Phi(U \wedge V) = \Phi(U) \wedge \Phi(V), \quad \Phi\Big(\bigvee U_i\Big) = \bigvee \Phi(U_i).$$

Here $1_{\mathcal{A}}$ denotes the largest element of $\mathcal{A}$ (the empty meet). This defines a category (Frames).

For example, for a topological space $T$ the lattice $\mathcal{O}(T)$ of open sets is a frame. And for a continuous mapping $f$: $T \to S$, the inverse image mapping $f^{-1}$: $\mathcal{O}(S) \to \mathcal{O}(T)$ is a morphism of frames.

Motivated by this change of direction between $f$ and $f^{-1}$, the category of locales is defined as the opposite (formal dual) of that of frames:

$$\text{(Locales)} = \text{(Frames)}^{\text{op}}.$$

In particular, the two categories have the same objects. However, to avoid possible confusion about whether we are considering a given object as a locale or as a frame, and to emphasize the similarity with topological spaces, we shall denote locales by $X, Y, \ldots$ and their corresponding frames by $\mathcal{O}(X)$, $\mathcal{O}(Y), \ldots$. Similarly, a morphism of locales will be denoted $f\colon X \to Y$, while the corresponding morphism of frames will be denoted $f^{-1}\colon \mathcal{O}(Y) \to \mathcal{O}(X)$. Thus, the two expressions

$$f\colon X \to Y \quad \text{and} \quad f^{-1}\colon \mathcal{O}(Y) \to \mathcal{O}(X)$$

denote the same data, but on the left in the category of locales and on the right in that of frames.

For many properties of topological spaces and mappings there are analogous properties for locales (but there are also surprising differences), and there is now an extensive literature on locales. The reader may consult [SS], or the recent survey paper [J] with its extensive bibliography.

Now recall that the definition of a sheaf on a topological space (1) only made use of the open set lattice of the space. Thus one can define sheaves on a locale in exactly the same way. More explicitly, for a locale $X$ the corresponding frame $\mathcal{O}(X)$ can be viewed as a site: its objects are the elements $U \in \mathcal{O}(X)$, there is exactly one arrow $U \to V$ if $U \leqslant V$, and a family $\{U_i \to U\}$ is covering iff $U = \bigvee U_i$ (The distributivity law (1) ensures that the stability axiom for Grothendieck topologies holds.) The topos of sheaves on this site will be denoted

$$Sh(X).$$

As for topological spaces, every sheaf on a locale $X$ can be represented as the sheaf of sections of a local homeomorphism between locales $E \to X$. This gives an equivalence of categories between $Sh(X)$ and the category of such $E \to X$.

The construction of the topos $Sh(X)$ of sheaves on a locale $X$ is functorial. Indeed, a map of locales $f\colon X \to Y$ gives a morphism of sites (cf. 3.5) $f^{-1}\colon \mathcal{O}(Y) \to \mathcal{O}(X)$, hence a geometric morphism $Sh(X) \to Sh(Y)$.

## 5. Some representation theorems

Recall from Section 2 that for a topological space $X$ equipped with a continuous action $X \times G \to X$ by a topological group $G$, one can construct a topos $Sh_G(X)$ of $G$-equivariant sheaves. Exactly the same construction can be given for a locale $X$ equipped with a continuous action by a localic group $G$ (a group object in the category of locales).

**5.1.** THEOREM (Freyd [F2]). *For every Grothendieck topos $\mathcal{E}$ there exists a locale $X$ equipped with a continuous group action by a localic group $G$, and an atomic connected map $p\colon Sh_G(X) \twoheadrightarrow \mathcal{E}$.*

We explain some of the terms: a geometric morphism $p\colon \mathcal{F} \to \mathcal{E}$ is said to be surjective of $p^*\colon \mathcal{E} \to \mathcal{F}$ is a faithful functor, and connected if $p^*$ is also full. (Intuitively, this means that $p$ has connected fibers.) It is called atomic if $p^*$ preserves exponentials as well as the subobject classifier (i.e. $p^*(B^A) \cong p^*(B)^{p^*(A)}$ for any two objects $A, B \in \mathcal{E}$, and $p^*(\Omega_{\mathcal{E}}) \cong \Omega_{\mathcal{F}}$). Since any inverse image functor $p^*$ automatically preserves finite limits and arbitrary colimits, Freyd's theorem states that for every Grothendieck topos $\mathcal{E}$ there exists an embedding $p^*$ of $\mathcal{E}$ into a category of equivariant sheaves on a locale, such that the embedding preserves "all" the topos structure.

It is possible to improve on this theorem, and get an actual *equivalence* of topoi $p\colon Sh_G(X) \xrightarrow{\sim} \mathcal{E}$, if one allows $G$ to be a localic groupoid rather than a group. To be more explicit, first recall that a *groupoid* is a category in which every arrow is an isomorphism. Similarly, a *topological groupoid* is a groupoid in the category of topological spaces. It is given by a space $X$ of objects, a space $G$ of arrows, source and target maps

$$s, t\colon G \rightrightarrows X$$

and a composition map $m\colon G \times_X G \to G$ denoted $m(g, h) = g \circ h$, a map $i\colon X \to G$ assigning to each point $x \in X$ the identity arrow $i(x)$, and a map $r\colon G \to G$ assigning to each point $g \in G$ its inverse $r(g) = g^{-1}$. These maps are all required to be continuous and to satisfy the usual identities. Now let $E$ be a sheaf on $X$, represented as (the sheaf of sections of) a local homeomorphism $p\colon E \to X$. An action by $G$ on $E$ is given by a continuous map on the pullback $E \times_X G$ along $t\colon G \to X$,

$$\alpha\colon E \times_X G \to E, \quad \text{denoted } \alpha(e, g) = e \cdot g.$$

Thus, this map is defined for every pair $(e, g)$ such that $p(e) = t(g)$, and satisfies the usual identities for an action

$$p(e \cdot g) = s(g), \quad (e \cdot g) \cdot h = e \cdot (g \circ h), \quad e \cdot i(x) = e \tag{1}$$

(for any $e \in E$, $x \in X$ and $g, h \in G$ for which these expressions are defined). A sheaf on $X$ with such an action is again called an *equivariant sheaf*. The category $Sh_G(X)$ for all such equivariant sheaves, and action-preserving morphisms, is a Grothendieck topos. (It is called the classifying topos of the groupoid $G \rightrightarrows X$.)

These definitions never make essential use of the points of the topological spaces $G$, $X$ and $E$. Indeed, the equations (1) can also be expressed by commutative diagrams. Therefore one can define, in exactly the same way, the notion of a localic groupoid $G \rightrightarrows X$, as a groupoid in the category of locales, given by locales $X$ and $G$ together with appropriate morphisms of locales ($s$, $t$, $m$, $i$ and $r$). Similarly, one can construct for such a localic groupoid $G \rightrightarrows X$ a topos $Sh_G(X)$ of equivariant sheaves. Surprisingly, every topos is of this form:

**5.2.** THEOREM (Joyal and Tierney [JT]). *For every Grothendieck topos $\mathcal{E}$ there exists a localic groupoid $G \rightrightarrows X$ and an equivalence of topoi $Sh_G(X) \cong \mathcal{E}$, between $\mathcal{E}$ and the topos of $G$-equivariant sheaves.*

The functor $Sh_G(X) \to Sh(X)$, defined as "forget the action", is the inverse image functor of a geometric morphism

$$q\colon\ Sh(X) \to Sh_G(X), \tag{2}$$

surjective because this functor is faithful. Thus from Theorem 5.2 (or from 5.1) it follows that for every topos $\mathcal{E}$ there exists a surjective geometric morphism of the form $Sh(X) \to \mathcal{E}$. Using the fact that the frame $\mathcal{O}(X)$ can be suitably embedded into a complete Boolean algebra $\mathbb{B}$, one obtains "Barr's Theorem".

**5.3.** THEOREM (Barr [B]). *For every Grothendieck topos $\mathcal{E}$ there exists a surjective morphism $r\colon\ Sh(\mathbb{B}) \to \mathcal{E}$ from the topos of sheaves on a complete Boolean algebra to $\mathcal{E}$.*

This result was of course originally proved without use of Theorem 5.2. Barr's theorem is extremely useful in practice: Since a topos of the form $Sh(\mathbb{B})$ is very much like the topos of sets (cf. 2.7(ii)), and since $r^*\colon\ \mathcal{E} \to Sh(\mathbb{B})$ preserves colimits and finite limits, one concludes that any property which can be expressed in terms of such colimits and finite limits, true for sets, is true in any Grothendieck topos.

Various further refinements of Theorem 5.2 are possible. For example, given representations $Sh_G(X) \cong \mathcal{E}$ and $Sh_{G'}(X') \cong \mathcal{E}'$ as in Theorem 5.2, one can describe geometric morphisms $\mathcal{E} \to \mathcal{E}'$ in terms of the localic groupoids $G$ and $G'$. One thus obtains the result [M2] that the category of Grothendieck topoi can be obtained as a category of fractions (in the sense of [GZ]) from that of localic groupoids.

The representation Theorem 5.2 is further improved in [JM2, JM2], by showing that it suffices to consider localic groupoids $G \rightrightarrows X$ of a very special form, namely those where $G$ is a groupoid of homotopy classes of paths in $X$ (much as in the fundamental groupoid of a topological space, or a subgroupoid thereof). Furthermore, for this path-groupoid, the geometric morphism $q\colon\ Sh(X) \to Sh_G(X) \cong \mathcal{E}$ of (2) induces isomorphisms in homotopy and cohomology. Thus although topoi originally arose for more general cohomology theories than the cohomology of topological spaces, it suffices (in theory!) to consider only cohomology of locales. We should add that there has so far been little study of the locales so arising, and their possible applications.

## 6. Cohomology

In topology, one often uses cohomology groups of a space $X$ with coefficients in a varying abelian group $A$. This variation may consist in an action of the fundamental group $\pi_1(X, x_0)$ on $A$, and $A$ is then said to be a "twisted" system of coefficients. Such a system may also be viewed as a locally constant sheaf on $X$ (see Section 7). More generally, cohomology groups with coefficients in any abelian sheaf $A$ can be defined and used, as discussed extensively in [Go]. Generalizing such sheaf cohomology groups, Grothendieck and his collaborators (M. Artin, J.-L. Verdier and others) introduced cohomology groups for an arbitrary topos. The generality and flexibility of this framework was successfully applied in the solution of the Weil conjectures about the number of zeros of polynomial equations with integer coefficients modulo a prime number, by using the so-called étale cohomology groups of schemes. These are the cohomology groups of the étale topos associated to the scheme. They fit in well with Grothendieck's earlier theory of the fundamental group of a scheme (Section 7 below), e.g., by the Hurewicz formula (1) in Section 7.

In this section we will introduce these sheaf cohomology groups for an arbitrary topos, and present some of their basic properties. In particular, we will introduce the more explicit Čech cohomology groups associated to a cover of a topos, and explain Cartan's criterion providing conditions for when these Čech cohomology groups agree with the sheaf cohomology groups. Verdier's theory of hypercovers provides a generalized Čech cohomology which *always* agrees with sheaf cohomology.

Let $\mathcal{E}$ be a Grothendieck topos, and write $\mathsf{Ab}(\mathcal{E})$ for the category of abelian groups in $\mathcal{E}$. Thus, if $\mathcal{E}$ is the topos $Sh(\mathbb{C}, J)$ of sheaves on a site $(\mathbb{C}, J)$, then an object $A \in \mathsf{Ab}(\mathcal{E})$ is sheaf $A$ on $\mathbb{C}$ such that each $A(C)$ has the structure of an abelian group, and each arrow $\alpha\colon C \to D$ in $\mathbb{C}$ induces a group homomorphism $A(D) \to A(C)$. This category $\mathsf{Ab}(\mathcal{E})$ is an abelian category with enough injectives. Now write $\mathsf{Ab} = \mathsf{Ab}(Sets)$ for the category of abelian groups. The global sections functor $\Gamma\colon \mathcal{E} \to Sets$ (see 3.1(iii)) sends abelian group objects to abelian groups, so induces a functor (again denoted) $\Gamma\colon \mathsf{Ab}(\mathcal{E}) \to \mathsf{Ab}$, which is left exact and preserves injectives. Thus one can introduce the cohomology groups $H^n(\mathcal{E}, A)$ by using the standard resolutions of homological algebra:

**6.1.** DEFINITION. For any group object $A$ in $\mathcal{E}$, the cohomology groups $H^n(\mathcal{E}, A)$ are defined to be the right derived functors of $\Gamma$, as

$$H^n(\mathcal{E}, A) := R^n\Gamma(A).$$

Thus $H^n(\mathcal{E}, A)$ is the $n$-th cohomology of the abelian cochain complex

$$\Gamma I^0 \to \Gamma I^1 \to \Gamma I^2 \to \cdots$$

obtained from an injective resolution $0 \to A \to I^0 \to I^1 \to \cdots$ in $\mathsf{Ab}(\mathcal{E})$. The construction of these groups $H^n(\mathcal{E}, A)$ is functorial, contravariant in $\mathcal{E}$ and covariant in $A$, as usual. For an object $X \in \mathcal{E}$, one also considers the right derived functors of the functor $\mathcal{E}(X, -)$ which sends an abelian group $A$ to the group of arrows $X \to A$ in $\mathcal{E}$ ("sections over $X$"). These groups are denoted $H^n(\mathcal{E}, X; A)$. For such an object $X \in \mathcal{E}$, the product functor $X^*\colon \mathcal{E} \to \mathcal{E}/X$ (sending $Y$ to $Y \times X \to X$) induces a functor $X^*\colon \mathsf{Ab}(\mathcal{E}) \to \mathsf{Ab}(\mathcal{E}/X)$ which is exact and preserves injectives. Thus for $A \in \mathsf{Ab}(\mathcal{E})$ there is a canonical isomorphism

$$H^n(\mathcal{E}, X; A) \cong H^n\big(\mathcal{E}/X, X^*(A)\big).$$

The latter group will also be denoted simply $H^n(\mathcal{E}/X, A)$.

**6.2.** Examples.

(i) (Cohomology of spaces) For a topological space $X$, the functor $\Gamma$: $\mathsf{Ab}Sh(X) \to \mathsf{Ab}$ sends an abelian sheaf $A$ to $A(X)$. The cohomology groups $H^n(Sh(X), A)$ are the usual sheaf cohomology groups of $X$ with coefficients in $A$, used extensively in topology, cf. [Go, I].

(ii) (Cohomology of groups) Let $G$ be a group, with its associated topos ($G$-$Sets$). An object $A$ of $\mathsf{Ab}(G\text{-}Sets)$ is an abelian group $A$ equipped with an action by $G$, or equivalently, a $\mathbb{Z}[G]$-module. In this case $\Gamma(A)$ is the subgroup $A^G$ if fixed points. The cohomology $H^n(G\text{-}Sets, A)$ is the usual Eilenberg–MacLane cohomology of the group with coefficients in $A$, [M].

(iii) (Algebraic geometry) Let $X$ be a scheme, with associated étale topos $Sh(X)$. The étale cohomology groups $H^n(X_{\text{ét}}, A)$ of the scheme $X$ are by definition the cohomology groups $H^n(Sh(X), A)$ of the (small) étale topos. (In fact, the big étale topos has the same cohomology; [SGA4], 2, p. 353.)

(iv) (Cohomology of categories) Let $\mathbb{C}$ be a small category, with presheaf topos $Sets^{\mathbb{C}^{\text{op}}}$. An object $A$ of $\mathsf{Ab}(Sets^{\mathbb{C}^{\text{op}}})$ is simply a functor $A$: $\mathbb{C}^{\text{op}} \to \mathsf{Ab}$. The cohomology groups $H^n(Sets^{\mathbb{C}^{\text{op}}}, A)$ are the cohomology groups of the category $\mathbb{C}$, discussed, e.g., in [Q2], p. 91, [R].

In the rest of this section, we will simply outline some very basic properties of topos cohomology. For further study the reader may consult, among others, [SGA4] (vol. 2), [Mi, AM, B, T].

**6.3.** *Leray spectral sequence.* For any geometric morphism $f$: $\mathcal{F} \to \mathcal{E}$ between Grothendieck topoi, the direct image functor $f_*$: $\mathcal{F} \to \mathcal{E}$ also defines a functor $f_*$: $\mathsf{Ab}(\mathcal{F}) \to \mathsf{Ab}(\mathcal{E})$. Grothendieck described a spectral sequence for the composite of two functors. For the composite $\Gamma \circ f_*$ this gives the *Leray* spectral sequence of $f$:

$$E_2^{p,q} = H^p(\mathcal{E}, R^q f_*(A)) \Rightarrow H^{p+q}(\mathcal{F}, A).$$

For example, if $f$ comes from a continuous map between topological spaces $f$: $X \to Y$ (cf. 3.1(i)) then this is the usual Leray spectral sequence for sheaf cohomology [Go], p. 201–202. For the geometric morphism $\varphi$: ($G$-$Sets$) $\to$ ($H$-$Sets$) induced by a group homomorphism as in 3.1(ii), the Leray spectral sequence is precisely the Lyndon–Hochschild–Serre spectral sequence in group cohomology [M]. The Leray spectral sequence in étale cohomology is discussed in [SGA4] (2), [Mi].

**6.4.** *The basic spectral sequence associated to a simplicial object.* Let $\mathcal{E}$ be a Grothendieck topos. A simplicial object $X. = \{X_p\}_{p\geqslant 0}$ in $\mathcal{E}$ gives rise to an augmented chain complex in $\mathsf{Ab}(\mathcal{E})$,

$$0 \leftarrow \mathbf{Z} \leftarrow \mathbf{Z}\cdot X_0 \overset{\partial}{\leftarrow} \mathbf{Z}\cdot X_1 \overset{\partial}{\leftarrow} \cdots . \tag{6.4.1}$$

Here $\mathbf{Z}\cdot(-)$: $\mathcal{E} \to \mathsf{Ab}(\mathcal{E})$ is the free abelian group functor (left adjoint to the forgetful functor), and $\partial$ is defined as usual by alternating sums of the boundary operations

$d_i$: $X_{n+1} \to X_n$. The simplicial object $X.$ is said to be *locally acyclic* if this complex (6.4.1) is exact. (If $\mathcal{E}$ has enough points, $X.$ is locally acyclic iff for each point $p$: *Sets* $\to \mathcal{E}$ the simplicial set $p^*(X.)$ is acyclic, i.e. $\tilde{H}_n(p^*(X.), \mathbf{Z}) = 0$ for each $n \geqslant 0$.) For any such locally acyclic simplicial object $X.$ there is a spectral sequence

$$E_2^{p,q} = E_2^{p,q}(X.) = H^p H^q(\mathcal{E}/X., A) \Rightarrow H^{p+q}(\mathcal{E}, A). \tag{6.4.2}$$

We will discuss two kinds of locally acyclic objects, Čech covers and the more general hypercovers.

**6.5.** *Čech cohomology.* Any epimorphism $U \twoheadrightarrow 1$ in a Grothendieck topos $\mathcal{E}$ gives rise to a locally acyclic simplicial object $U. = \{U_p\}_p$, with $U_p = U \times \cdots \times U$ ($p+1$ times, the factors numbered $0, \ldots, p$) and $d_i$: $U_p \to U_{p-1}$ the projection omitting the $i$-th coordinate, for $i = 0, \ldots, p$. The $E_2^{p,0}$-term of (6.4.2) is the cohomology of the complex

$$\mathrm{Hom}_{\mathcal{E}}(U, A) \to \mathrm{Hom}_{\mathcal{E}}(U \times U, A) \to \cdots \tag{6.5.1}$$

and is called the Čech cohomology of $\mathcal{E}$ for the cover $U$, denoted $\check{H}^*(U, A)$. For two such epimorphisms $U \to 1$ and $V \to 1$, any map ("refinement") $\alpha$: $V \to U$ induces a map of cochain complexes $\alpha^*$: $\mathrm{Hom}_{\mathcal{E}}(U., A) \to \mathrm{Hom}_{\mathcal{E}}(V., A)$. Up to homotopy, this map is independent of $\alpha$, so induces a well-defined homomorphism $\check{H}^*(U, A) \to \check{H}^*(V, A)$. By definition, the Čech cohomology $\check{H}^*(\mathcal{E}, A)$ is the direct limit over all these $U \twoheadrightarrow 1$:

$$\check{H}^p(\mathcal{E}, A) := \varinjlim_U \check{H}^p(U, A) = \varinjlim_U E_2^{p,0}(U.). \tag{6.5.2}$$

(This direct limit is well-defined, because the system is directed, and a small set of epis $U \twoheadrightarrow 1$ is cofinal, cf. (6.6.2) below.) The direct limit of (6.5.2) thus yields a spectral sequence

$$E_2^{p,q} = \varinjlim_U H^p H^q(\mathcal{E}/U., A) \Rightarrow H^{p+q}(\mathcal{E}, A). \tag{6.5.3}$$

This sequence has the special property that $E_2^{0,q} = 0$ for $q > 0$. For $q = 0$, there is a canonical edge homomorphism

$$\varepsilon\colon\ E_2^{p,0} = \check{H}^p(\mathcal{E}, A) \to H^p(\mathcal{E}, A), \tag{6.5.4}$$

which is an isomorphism for $p = 0, 1$ and a monomorphism for $p = 2$.

**6.6.** *Čech cohomology and sites.* Let $(\mathbb{C}, J)$ be a site for $\mathcal{E}$. We assume that $\mathbb{C}$ has finite limits, and that every representable presheaf $\mathbb{C}(-, C)$ is already a sheaf (in this case the topology $J$ is called "subcanonical"). Then, for each object $C \in \mathbb{C}$, the comma category $\mathbb{C}/C$, with the evident topology inherited from $J$, is a site for the topos $\mathcal{E}/\mathbb{C}(-, C)$. Let

$A\colon \mathbb{C}^{\mathrm{op}} \to \mathsf{Ab}$ be an object of $\mathsf{Ab}(\mathcal{E})$. For any covering family $\mathcal{U} = \{C_i \to 1\}_{i\in I}$ of the terminal object 1 in $\mathbb{C}$, one obtains a complex $C^{\cdot}(\mathcal{U}, A)$ defined by

$$C^p(\mathcal{U}, A) = \prod_{(i_0,\dots,i_p)} A(C_{i_0} \times \cdots \times C_{i_p}). \tag{6.6.1}$$

This is the complex (6.5.1) where

$$U = \sum_{i\in I} \mathbb{C}(-, C_i).$$

These covers coming from the site are cofinal in the system (6.5.2), so

$$\check{H}^p(\mathcal{E}, A) = \varinjlim_{\mathcal{U}} H^p\big(C^{\cdot}(\mathcal{U}, A)\big), \tag{6.6.2}$$

where $\mathcal{U}$ ranges over the covers of 1 in the site $\mathbb{C}$, ordered by refinement ($\mathcal{V} = \{D_s \to 1\}_{s\in S}$ refines $\{C_i \to 1\}_{i\in I}$ if for each $s \in S$ there is an $i \in I$ and a map $D_s \to C_i$). For a specific cover $\mathcal{U}$, the $E_2^{p,q}$-term of the spectral sequence (6.4.2) takes the form

$$H^p\Big(\prod_{(i_0,\dots,i_p)} H^q(\mathcal{E}/C_{i_0} \times \cdots \times C_{i_p}, A)\Big). \tag{6.6.3}$$

Suppose now that there is a class of objects $\mathcal{B}$ in $\mathbb{C}$, closed under products, such that every object $C \in \mathbb{C}$ is covered by objects from $\mathcal{B}$. Then in the direct limit (6.6.2) it suffices to consider covering families $\mathcal{U}$ consisting of objects $B \in \mathcal{B}$. Thus, if for each $B \in \mathcal{B}$ one has $H^q(\mathcal{E}/B, A) = 0$ $(q > 0)$, the spectral sequence (6.4.2) with $E_2$-term (6.6.3) collapses, and the homomorphism (6.5.4) is an isomorphism

$$\check{H}^p(\mathcal{E}, A) \xrightarrow{\sim} H^p(\mathcal{E}, A) \quad (p \geqslant 0); \tag{6.6.4}$$

i.e. Čech cohomology coincides with ordinary cohomology. If in addition $\mathcal{B}$ is closed under pullbacks, a slightly more careful inspection of the basic spectral sequence will in fact show that, to obtain an isomorphism (6.6.4), it is enough to assume that $\check{H}^q(\mathcal{E}/B, A) = 0$ for $q > 0$ (rather than $H^q(\mathcal{E}/B, A) = 0$) for every $B \in \mathcal{B}$. This condition for the isomorphism between Čech cohomology and topos cohomology is called *Cartan's criterion* [Go], p. 227. This criterion is often useful. For example, any manifold $M$ has a basis of contractible open sets, so the Čech and sheaf cohomology groups of $M$ coincide, for any locally constant sheaf $A$ of abelian coefficients. Another example is provided by the sheaves $W(F)$ on the small étale site of a scheme $X$ which are induced from quasi-coherent sheaves $F$ for the Zariski-topology on $X$; cf. [SGA4] (2), p. 355.

**6.7.** *Hypercovers.* (Verdier [SGA4], Artin and Mazur [AM], Brown [Br].) It is possible to obtain an isomorphism of the form (6.6.4) without any conditions on the site $\mathbb{C}$, if one allows more general covers $U.$ in the direct limit (6.5.2) (or (6.6.2)). Recall from [Q1] that a map $f\colon X. \to Y.$ between simplicial sets is a trivial fibration if any square of the form

$$\begin{array}{ccc} \dot{\Delta}[n] & \longrightarrow & X \\ \downarrow & \nearrow & \downarrow \\ \Delta[n] & \longrightarrow & Y \end{array}$$

has a diagonal filling (as indicated by the dotted arrow). Here $\Delta[n]$ is the standard $n$-simplex and $\dot{\Delta}[n]$ is its boundary. In other words, $f\colon X. \to Y.$ is a trivial fibration if the map

$$X_n = \mathrm{Hom}(\Delta[n], X.) \to \mathrm{Hom}(\dot{\Delta}[n], X.) \times_{\mathrm{Hom}(\dot{\Delta}[n],Y.)} \mathrm{Hom}(\Delta[n], Y.) \tag{6.7.1}$$

is surjective. If $Y. = 1$, this is the familiar requirement that $X.$ is a *contractible Kan complex*. Call a map $f\colon X. \to Y.$ between simplicial objects in $\mathcal{E}$ a *local trivial fibration* if the corresponding map (6.7.1) is an epimorphism in $\mathcal{E}$. (If $\mathcal{E}$ has enough points, this is the case iff for every point $p\colon$ *Sets* $\to \mathcal{E}$ the map $p^*(f)$ is a trivial fibration of simplicial sets.) A simplicial object $X.$ is called a *hypercover* of $\mathcal{E}$ if $X. \to 1$ is a local trivial fibration. For example, for any object $U \twoheadrightarrow 1$ in $\mathcal{E}$ the simplicial object $U.$ described above is a hypercover. Denote by $HR(\mathcal{E})$ the category of hypercovers of $\mathcal{E}$ and homotopy classes of maps. Clearly every hypercover is locally acyclic, and gives rise to a basic spectral sequence (6.4.2). As for Čech cohomology, one can form a "Verdier cohomology" direct limit over all hypercovers

$$\begin{aligned} \check{H}^p_{\text{Verdier}}(\mathcal{E}, A) &:= \varinjlim_{X.\in HR(\mathcal{E})} H^p(\mathrm{Hom}_{\mathcal{E}}(X., A)) \\ &= \varinjlim_{X.} H^p H^o(\mathcal{E}/X., A). \end{aligned}$$

The direct limit over all hypercovers of the spectral sequence (6.4.2) gives another spectral sequence

$$E_2^{p,q} = \varinjlim H^p H^q(\mathcal{E}/X., A) \Rightarrow H^{p+q}(\mathcal{E}, A),$$

with edge homomorphism

$$\varepsilon\colon \check{H}^p_{\text{Verdier}}(\mathcal{E}, A) \to H^p(\mathcal{E}, A). \tag{6.7.2}$$

This spectral sequence collapses (since $\varinjlim_{X.} H^q(\mathcal{E}/X., A) = 0$ for $q > 0$), and $\varepsilon$ is an isomorphism. (One can also prove that (6.7.2) is an isomorphism along the lines of [Bn], p. 427, by using that the local trivial fibrations in a topos form part of a "category of fibrant objects" in the sense of [Br], see [Ja].)

## 7. The fundamental group

A map $p$: $E \to X$ between topological spaces is said to be a *covering space* of $X$ if there exists a covering of $X$ by open sets $U$ with the property that $p^{-1}(U)$ is (homeomorphic to) a disjoint sum $\sum_{s\in S} V_s$ of open sets such that $p$ restricts to a homeomorphism $p$: $V_s \xrightarrow{\sim} U$ for each $V_s$. If the space $X$ is connected and locally simply connected, these covering spaces are "classified" by the fundamental group $\pi_1(X, x_0)$ where $x_0$ is any base point. Specifically, the functor which sends a covering space $p$: $E \to X$ to the set $p^{-1}(x_0)$, equipped with the action of $\pi_1(X, x_0)$ defined by "pathlifting", is an equivalence of categories.

These covering spaces can be described within the category of sheaves on $X$, when we identify sheaves with local homeomorphisms (étale spaces). Indeed, a local homeomorphism $p$: $E \to X$ is a covering space iff there is a surjective local homeomorphism $U \twoheadrightarrow X$ for which

$$E \times_X U \cong \sum_{s\in S} U$$

over $U$; in other words, the sheaf $E \to X$ becomes "constant" when pulled back to a sheaf $E \times_X U \to U$ over $U$.

This description applies to any topos. Specifically, let $\mathcal{E}$ be a Grothendieck topos, with a "base point" $p$: *Sets* $\to \mathcal{E}$. This topos is said to be *connected* if its terminal object cannot be decomposed as the sum of smaller objects (equivalently, the unique geometric morphism $\mathcal{E} \to$ *Sets* is connected, cf. just below Theorem 5.1). An object $E$ of $\mathcal{E}$ is said to be *locally constant* if there exists a set $S$, an epi $U \twoheadrightarrow 1$ in $\mathcal{E}$, and an isomorphism

$$E \times U \xrightarrow{\sim} \sum_{s\in S} U$$

over $U$. (Thus, viewing topoi as generalized spaces, $\mathcal{E}/E \to \mathcal{E}$ is a "covering space" of $\mathcal{E}$.) Such a locally constant object is said to be *finite* if $S$ is a finite set. In that case the set $p^*(E)$ is finite as well since $p^*(E) \cong S$. Let $FLC(\mathcal{E})$ be the full subcategory of $\mathcal{E}$ consisting of such finite locally constant objects. It can be shown that there exists a profinite topological group $G$, unique up to isomorphism, such that $FLC(\mathcal{E})$ is equivalent to the category of finite continuous $G$-sets. The construction of $G$ makes use of the point $p$. This group $G$ is the profinite fundamental group of the topos $\mathcal{E}$ with base point $p$, and denoted $\pi_1(\mathcal{E}, p)$ (or $\pi_1^{pf}(\mathcal{E}, p)$ for emphasis).

The explicit construction of this profinite group $G$ proceeds in an indirect way, using Grothendieck's categorical Galois theory [SGA1], Exp. V. This theory gives an axiomatic characterization of categories of continuous $G$-sets for a profinite group $G$, as follows.

**7.1.** DEFINITION. A *Galois category* is a category $\mathcal{G}$ equipped with a functor $F$: $\mathcal{G} \to$ (finite sets), satisfying the following conditions:

(i) $\mathcal{G}$ has finite limits, finite sums, and for every object $X \in \mathcal{G}$ and any finite group $H$ of automorphisms of $X$ the quotient $X/H$ exists in $\mathcal{G}$. Furthermore, every morphism

$f\colon X \to Y$ in $\mathcal{G}$ can be factored as an effective epi $X \twoheadrightarrow f(X)$ followed by a mono $f(X) \rightarrowtail X$. Finally, every mono $T \rightarrowtail X$ has a complement $T' \rightarrowtail X$ (i.e. $T+T' \xrightarrow{\sim} X$).

(ii) The functor $F$ preserves all the structure mentioned in (i): finite limits and sums, quotients by such finite groups, epi-mono factorizations and complements.

(iii) For every arrow $f$ in $\mathcal{G}$, if $F(f)$ is an isomorphism of finite sets then $f$ is an isomorphism in $\mathcal{G}$.

If $G$ is a profinite topological group, than the category $\mathcal{C}(G)$ of finite continuous $G$-sets, with the "underlying set" functor $U\colon \mathcal{C}(G) \to$ (finite sets), is a Galois category. Conversely, Grothendieck's theorem states that for any Galois category $(\mathcal{G}, F)$ there is, up to isomorphism, a unique profinite group $G$ and an equivalence of categories $\mathcal{C}(G) \cong \mathcal{G}$, so that $F$ corresponds to $U$ under this equivalence (up to natural isomorphism). For a connected Grothendieck topos $\mathcal{E}$ with a point $p$, the inverse image functor $p^*$ restricts to a functor $p^*\colon FLC(\mathcal{E}) \to$ (finite sets), and it is not difficult to verify that $(FLC(\mathcal{E}), p^*)$ is a Galois category. Thus Grothendieck's theorem for such categories gives the profinite fundamental group $\pi_1(\mathcal{E}, p)$ as described above.

**7.2.** EXAMPLES.

(i) Consider a connected locally simply connected topological space $X$ with base point $x_0$, and its associated topos $Sh(X)$. The point $x_0$ gives a point of this topos, $\bar{x}_0\colon Sets \to Sh(X)$. An object $E$ of $Sh(X)$ is locally constant iff, when viewed as an étale space $E \to X$, it is a covering projection. As mentioned at the beginning of this section, this category of covering projections is equivalent to the category of sets with an action by the usual fundamental group $\pi_1(X, x_0)$, constructed using paths. It follows that the Grothendieck profinite fundamental group $\pi_1(Sh(X), \bar{x}_0)$ is the profinite completion [GR] of $\pi_1(X, x_0)$.

(ii) (algebraic geometry) Consider a connected scheme $X$, with associated (small) étale topos $Sh(X)$. A geometric point $x_0$ of $X$ will again give a point $\bar{x}_0$ of the topos $Sh(X)$, and one can form the profinite fundamental group $\pi_1(Sh(X), \bar{x})_0)$. In this case it follows by descent theory [SGA1], VIII.7, that every finite locally constant object of $Sh(X)$ is actually representable by a finite étale cover $X' \to X$ of schemes. Thus $\pi_1(Sh(X), \bar{x}_0)$ is the usual fundamental group $\pi_1(X, x_0)$ of the scheme $X$ ([SGA1], Exp. V).

It is also possible to classify arbitrary covering spaces (not just finite ones) of a topos $\mathcal{E}$, provided $\mathcal{E}$ is *locally connected.* To define this, first call an object $E$ of $\mathcal{E}$ *connected* if $E$ cannot be decomposed as a sum $E \cong E_1 + E_2$, except in the trivial ways where $E_1 = 0$ or $E_2 = 0$. The topos $\mathcal{E}$ is said to be locally connected if every object $E$ of $\mathcal{E}$ can be decomposed as a sum of connected objects, say

$$E = \sum_{i \in I} E_i.$$

This decomposition is essentially unique, and its index set $I$ is the set of connected components of $E$, denoted $\pi_0(E)$. In this way one obtains a functor $\pi_0\colon \mathcal{E} \to Sets$. (This functor is left adjoint to the functor $\Delta$ of 3.1(iii).) For example, for a topological space $X$ the topos $Sh(X)$ is locally connected whenever $X$ is locally connected; and for a scheme $X$, the étale topos $Sh(X)$ is locally connected when $X$ is locally Noetherian.

For a locally connected topos $\mathcal{E}$, the full subcategory $SLC(\mathcal{E})$, consisting of sums of locally constant objects of $\mathcal{E}$, is again a Grothendieck topos. Furthermore, the inclusion functor $SLC(\mathcal{E}) \hookrightarrow \mathcal{E}$ is the inverse image part of a geometric morphism $\mathcal{E} \twoheadrightarrow SLC(\mathcal{E})$. In particular, a point $p$ of $\mathcal{E}$ gives by composition a point $\tilde{p}$ of $SLC(\mathcal{E})$. An infinite version of Grothendieck's Galois theory ([M3], Proposition 3.2) shows that $SLC(\mathcal{E})$ is equivalent to a topos of the form $\mathcal{B}G$ (= continuous $G$-sets) where $G$ is a prodiscrete localic group; or equivalently, a (strict) progroup. This group $G$ is essentially unique, and the equivalence identifies $\tilde{p}^*$: $SLC(\mathcal{E}) \to Sets$ with the underlying set functor $U$: $\mathcal{B}G \to Sets$. One denotes $G$ by $\pi_1(\mathcal{E}, p)$.

This "enlarged" (when compared to the profinite one) fundamental group $\pi_1(\mathcal{E}, p)$ shares many of the usual properties of fundamental groups of spaces. For example, for any abelian group $A$ there is a canonical isomorphism

$$H^1(\mathcal{E}, \Delta(A)) \cong \mathrm{Hom}(\pi_1(\mathcal{E}, p), A), \tag{1}$$

analogous to the Hurewicz theorem in topology which states that the first homology group is the abelianization of the fundamental group.

Using Verdier's hypercovers, one can also define higher homotopy groups of a connected locally connected topos $\mathcal{E}$ with base point $p$. These higher homotopy groups are pro-groups, called the *étale homotopy groups* of $(\mathcal{E}, p)$, and denoted $\pi_n(\mathcal{E}, p)$ (or $\pi_n^{\mathrm{et}}(\mathcal{E}, p)$). For $n = 1$, this agrees with the enlarged fundamental group $\pi_1(\mathcal{E}, p)$ just described. Their construction can be outlined as follows: For any hypercover $X.$ of $\mathcal{E}$, the connected components functor $\pi_0$: $\mathcal{E} \to Sets$ gives a connected simplicial set $\pi_0(X.)$. A base-point of such a hypercover ("over" the point $p$ of $\mathcal{E}$) is by definition a vertex $x_0$ of the simplicial set $p^*(X.)$. The canonical map $p^*(X.) \to p^*\Delta\pi_0(X.) \cong \pi_0(X.)$ will then give a base-point $\tilde{x}_0$ of the simplicial set $\pi_0(X.)$, and one obtains homotopy groups $\pi_n(\pi_0(X.), \tilde{x}_0)$. The étale homotopy groups are defined as the pro-groups ("formal" inverse limits)

$$\varprojlim_{(X, x_0)} \pi_n(\pi_0(X.), \tilde{x}_0),$$

indexed by all the pointed hypercovers and homotopy classes of maps between them (or rather, some small cofinal system of such, just as for Čech cohomology). These groups are described in detail in [AM].

## 8. Topoi and logic

The starting point for the relation to mathematical logic is the following observation. Let $\mathcal{E}$ be an elementary topos, and for each object $X$ in $\mathcal{E}$ let Sub $(X)$ denote the poset of subobjects of $X$.

**8.1.** PROPOSITION.

(i) *For each object $X$ the poset* Sub$(X)$ *is a Heyting algebra.*

(ii) *For each arrow* $f\colon X \to Y$ *in* $\mathcal{E}$ *the pullback functor* $f^{-1}\colon \mathrm{Sub}(Y) \to \mathrm{Sub}(X)$ *has both a left and a right adjoint, denoted* $\exists_f, \forall_f\colon \mathrm{Sub}(X) \to \mathrm{Sub}(Y)$.

(iii) *For each pullback square*

$$\begin{array}{ccc} Y \times_X Z & \xrightarrow{\pi_2} & Z \\ {\scriptstyle \pi_1}\downarrow & & \downarrow{\scriptstyle g} \\ Y & \xrightarrow{f} & X \end{array}$$

*the identities* $g^{-1} \circ \forall_f = \forall_{\pi_2} \circ \pi_1^{-1}$ *and* $g^{-1} \circ \exists_f = \exists_{\pi_2} \circ \pi_1^{-1}$ *hold.*

In the topos of sets, $\mathrm{Sub}(X)$ is the Boolean algebra of all subsets of $X$, and for a function $f\colon X \to Y$ the adjoints $\exists_f$ and $\forall_f$ are defined in terms of the existential and universal quantifiers: for any subset $U \subseteq X$, one has

$$\exists_f(U) = \{y \in Y \mid \exists x \in f^{-1}(y)\colon\ x \in U\}$$

and

$$\forall_f(U) = \{y \in Y \mid \forall x \in f^{-1}(y)\colon\ x \in U\}.$$

Part (i) of this proposition states that the poset $\mathrm{Sub}(X)$ has a largest element $1_X$ and a smallest one $0_X$, as well as operations of infimum, supremum and implication, denoted for $U, V \in \mathrm{Sub}(X)$ by

$$U \wedge V, \quad U \vee V, \quad U \Rightarrow V.$$

Furthermore, one can define a negation $\neg U$ as $\neg U = (U \Rightarrow 0_X)$. The poset $\mathrm{Sub}(X)$ is a Heyting algebra because these operations satisfy the laws of the intuitionistic propositional calculus.

The topos $\mathcal{E}$ is said to be *Boolean* if for every object $X$ the Heyting algebra $\mathrm{Sub}(X)$ is a Boolean algebra. This means that the laws of the ordinary ("classical") propositional logic hold. Thus a topos $\mathcal{E}$ is Boolean if every subobject has a complement. An arbitrary topos $\mathcal{E}$ always contains a "largest" (in some sense) Boolean subtopos $\mathcal{E}_{\neg\neg}$ (constructed as sheaves for the Lawvere–Tierney topology given by "double negation"; cf. 1.5).

It follows from the proposition above that one can interpret formulas of predicate logic in any topos. More specifically, one can associate to each topos $\mathcal{E}$ a language $L(\mathcal{E})$ of "typed" predicate logic. The types of this language are the objects of $\mathcal{E}$. Furthermore, if $X_1, \ldots, X_n, Y$ are objects in $\mathcal{E}$ then any arrow $f\colon X_1 \times \cdots \times X_n \to Y$ is a function symbol of the language (taking $n$ arguments of types $X_1, \ldots, X_n$ respectively to a value of type $Y$), and similarly every subobject $R \in \mathrm{Sub}(X_1 \times \cdots \times X_n)$ in $\mathcal{E}$ is a relation symbol of the language (taking $n$ arguments of types $X_1, \ldots, X_n$). For any formula $\varphi(x_1, \ldots, x_n)$ of this language, with free variables $x_i$ of types $X_i \in \mathcal{E}$, one can then build up an object (the "value" of $\varphi$),

$$\{(x_1, \ldots, x_n) \mid \varphi(x_1, \ldots, x_n)\} \in \mathrm{Sub}(X_1 \times \cdots \times X_n), \tag{1}$$

by induction on the construction of $\varphi$, using Proposition 8.1. For example, typical inductive clauses in the definition of this object (1) read

$$\{\vec{x} \mid \varphi(\vec{x}) \wedge \psi(\vec{x})\} = \{\vec{x} \mid \varphi(\vec{x})\} \wedge \{\vec{x} \mid \psi(\vec{x})\},$$

$$\{\vec{x} \mid \forall x_{n+1} \varphi(\vec{x}, x_{n+1})\} = \forall_\pi(\{(\vec{x}, x_{n+1}) \mid \varphi(\vec{x}, x_{n+1})\});$$

here $\vec{x}$ stands for the sequence of variables $x_1, \dots, x_n$, while $\forall_\pi$ is the "quantification" along the projection arrow $\pi$: $X_1 \times \cdots \times X_{n+1} \to X_1 \times \cdots \times X_n$ of $\mathcal{E}$. This valuation (1) obeys all the rules of intuitionistic predicate logic. Moreover, exponentials $Y^X$ and power objects $P(X) = \Omega^X$ give a corresponding structure on the types of this language $L(\mathcal{E})$, making it a "higher order" language. In this way, one obtains in fact a suitable correspondence between elementary topoi on the one hand, and intuitionistic theories in such higher order languages on the other. This correspondence and some of its applications are exposed in detail in [LS].

Thus, any topos can be viewed as some universe of sets which obeys the rules of intuitionistic logic. This can be exploited in two directions.

On the one hand, one can use topos theory to prove results about logical systems. These will in general be systems of intuitionistic logic, unless the topoi involved are Boolean. (But remember that any topos can be "Booleanized".) In order to model interesting logical theories in a topos $\mathcal{E}$, one generally assumes that $\mathcal{E}$ has a *natural numbers object* (n.n.o.). Such an n.n.o. is a universal object $N$ equipped with arrows $z$: $1_\mathcal{E} \to N$ (zero) and $s$: $N \to N$ (successor). Universality in this case means that for any other object $X$ in $\mathcal{E}$, with given arrows $a$: $1 \to X$ and $t$: $X \to X$, there is a unique arrow $f$: $N \to X$ so that $f \circ z = a$ and $f \circ s = t \circ f$. This property is essentially equivalent to the Peano axioms for $N$. In the topos of sets, the usual set of natural numbers is an n.n.o., and the unique arrow $f$ is defined by "recursion". Any Grothendieck topos has an n.n.o.

For example, Cohen's famous proof of the independence of the Continuum Hypothesis has a sheaf theoretic interpretation, due to Tierney [Ti]. Say that a monomorphism $A \rightarrowtail B$ in a topos $\mathcal{E}$ is strict of there is no nonzero object $U$ in $\mathcal{E}$ for which there exists an epimorphism $U \times B \twoheadrightarrow U \times A$ over $U$. In the "internal" logic of $\mathcal{E}$ this expresses that $A$ is a subset of $B$ with cardinality strictly smaller than that of $B$.

**8.2.** THEOREM. *There exists a Boolean (Grothendieck) topos $\mathcal{E}$, with an n.n.o. $N$, in which there are strict monomorphisms $N \rightarrowtail A \rightarrowtail P(N)$.*

From this theorem one can derive the independence of the Continuum Hypothesis from the usual axioms of Zermelo Fraenkel set theory, by imitating the construction of the cumulative hierarchy of sets

$$V = \bigcup_\alpha V_\alpha$$

inside $\mathcal{E}$; see [F].

In a similar vein, P. Freyd [F1] gave a very elegant topos-theoretic proof of the independence of the Axiom of Choice.

If any topos $\mathcal{E}$ with n.n.o., one can interpret the usual construction of the set of real numbers in terms of Dedekind cuts in the language of that topos, and construct an object $R_{\mathcal{E}}$ of real numbers in $\mathcal{E}$. The intuitionistic aspect of the "logic of topoi" is illustrated very strikingly by the fact that in many naturally arising topoi, the statement that all functions are continuous ("Brouwer's theorem") holds. This is described in detail in [MM].

Most of the proposed models for intuitionistic logic can be seen as special cases of the interpretation of logic in topoi. For example, Kripke models [Kr] describe truth in the topos $Sets^{\mathbb{P}^{op}}$ of presheaves on a poset $\mathbb{P}$, while Beth models [Be] describe truth in the topos of sheaves on the Cantor space (or Baire space $\mathbb{N}^{\mathbb{N}}$). Kleene's recursive realizability semantics [K] can also be viewed as the description of truth in a topos, the so-called effective topos [H].

In the other of the two directions mentioned above, one can construct objects in a topos and prove properties about them, just as if these objects were sets, *provided* these constructions and proofs are intuitionistically valid (that is, all constructions must be explicit, and use of the axiom of choice and the excluded middle is prohibited). This is particularly effective when one studies geometric morphisms and pullbacks of Grothendieck topoi ("change-of-base"). Thus, if $f: \mathcal{F} \to \mathcal{E}$ is a geometric morphism between Grothendieck topoi, one may view $\mathcal{E}$ as a "universe of sets" and construct a site $(\mathbb{C}, J)$ *inside* this universe, so that $\mathcal{F}$ is (equivalent to) the topos of "internal" sheaves on $(\mathbb{C}, J)$ constructed inside $\mathcal{E}$, denoted $Sh_{\mathcal{E}}(\mathbb{C}, J)$. If $p$: $\mathcal{E}' \to \mathcal{E}$ is another geometric morphism, then the pullback $\mathcal{E}' \times_{\mathcal{E}} \mathcal{F}$ can be constructed by first using the inverse image functor $p^*$: $\mathcal{E} \to \mathcal{E}'$ to obtain a site $p^*(\mathbb{C}, J)$ in $\mathcal{E}'$, and then constructing internal sheaves in $\mathcal{E}'$; thus

$$Sh_{\mathcal{E}'}\big(p^*(\mathbb{C}, J)\big) = \mathcal{E}' \times_{\mathcal{E}} \mathcal{F},$$

up to equivalence of topoi over $\mathcal{E}'$. This combination of exploiting the internal logic and change-of-base provides a powerful technique, exploited, e.g., in the references [JT, JM1, JM2, M1, M2] already mentioned in Section 5.

## References


[SGA4] M. Artin, A. Grothendieck and J.L. Verdier, *Théorie de Topos et Cohomologie Etale des Schémas*, SLNM 269 and 270, Springer, Berlin (1972).

[AM] M. Artin and B. Mazur, *Etale Homotopy*, SLNM 100, Springer, Berlin (1969).

[B] M. Barr, *Toposes without points*, J. Pure Appl. Algebra **5** (1974), 265–280.

[Be] E.W. Beth, *Semantical construction of intuitionistic logic*, Kon. Ned. Ak. Wet., Afd. Let. Med., Nwe Serie 19/11 (1956), 357–388.

[Br] L. Breen, *Bitorseurs et cohomologie non-abélienne*, The Grothendieck Festschrift I, Birkhäuser (1990), 401–476.

[Bn] K.S. Brown, *Abstract homotopy and sheaf cohomology*, Trans. Amer. Math. Soc. **186** (1973), 419–458.

[D] R. Diaconescu, *Change of base for toposes with generators*, J. Pure Appl. Algebra **6** (1975), 191–218.

[F] M.P. Fourman, *Sheaf models for set theory*, J. Pure Appl. Algebra **19** (1980), 91–101.

[F1] P.J. Freyd, *The axiom of choice*, J. Pure Appl. Algebra **19** (1980), 103–125.

[F2] P.J. Freyd, *All topoi are localic, or: why permutation models prevail*, J. Pure Appl. Algebra **46** (1987), 49–58.

[GZ] P. Gabriel and M. Zisman, *Calculus of Fractions and Homotopy Theory*, Springer, Berlin (1967).

[Go] R. Godement, *Topologie Algébrique et Théorie des Faisceaux*, Hermann, Paris (1958).

[SGA1] A. Grothendieck, *Revêtements Etales et Groupe Fondamental*, SLNM 224, Springer, Berlin (1971).

[GR] K. Gruenberg, *Profinite groups*, Algebraic Number Theory, J. Cassels and A. Fröhlich, eds, Academic Press, New York (1967).

[H] P. Halmos, *Boolean Algebras*, Springer, Berlin (1974).

[Ha] R. Hartshorne, *Algebraic Geometry*, Springer, Berlin (1977).

[H1] J.M.E. Hyland, *The effective topos*, The L.E.J. Brouwer Centenary Symposium, North-Holland, Amsterdam (1982), 165–216.

[I] B. Iversen, *Sheaf Cohomology*, Springer, Berlin (1986).

[Ja] J.F. Jardine, *Simplicial objects in a Grothendieck topos*, Contemp. Math. **55** (1) (1986), 153–239.

[JJ] P.T. Johnstone and A. Joyal, *Continuous categories and exponentiable toposes*, J. Pure Appl. Algebra **25** (1982), 255–296.

[J] P.T. Johnstone, *The art of pointless thinking: a student's guide to the category of locales*, Category Theory at Work, Herrlich and Porst, eds, Heldermann Verlag (1991).

[SS] P.T. Johnstone, *Stone Spaces*, Cambridge Univ. Press, Cambridge (1982).

[JT] A. Joyal and M. Tierney, *An extension of the Galois theory of Grothendieck*, Mem. Amer. Math. Soc. **309** (1984).

[JM1] A. Joyal and I. Moerdijk, *Toposes as homotopy groupoids*, Adv. Math. **80** (1990), 22–38.

[JM2] A. Joyal and I. Moerdijk, *Toposes are cohomologically equivalent to spaces*, Amer. J. Math. **112** (1990), 87–96.

[K] S.C. Kleene, *On the interpretation of intuitionistic number theory*, J. Symb. Logic **10** (1945), 109–124.

[Kr] S.A. Kripke, *Semantical analysis of intuitionistic logic*, Formal Systems and Recursive Functions, J. Crossley and M. Dummett, eds, North-Holland, Amsterdam (1965), 92–130.

[LS] J. Lambek and P. Scott, *Introduction to Higher Order Categorical Logic*, Cambridge Univ. Press, Cambridge (1986).

[L1] F.W. Lawvere, *Quantifiers as sheaves*, Proc. Int. Congress on Math., Gauthier-Villars, Paris (1971), 1506–1511.

[L2] F.W. Lawvere, *An elementary theory of the category of sets*, Proc. Nat. Acad. Sci. USA **52** (1964), 1506–1511.

[MM] S. MacLane and I. Moerdijk, *Sheaves in Geometry and Logic*, Springer, Berlin (1992).

[CWM] S. MacLane, *Categories for the Working Mathematician*, Springer, Berlin (1971).

[M] S. MacLane, *Homology*, Grundlehren der math. Wissenschaften 114, Springer, Berlin (1963).

[MP] M. Makkai and R. Pare, *Accessible Categories: The Foundations of Categorical Model Theory*, Contemp. Math. vol. 104 (1989).

[Mi] J.S. Milne, *Etale Cohomology*, Princeton Univ. Press, Princeton (1980).

[M1] I. Moerdijk, *Continuous fibrations and inverse limits of topoi*, Compositio Math. **58** (1986), 45–72.

[M2] I. Moerdijk, *The classifying topos of a continuous groupoid, I*, Trans. Amer. Math. Soc. **310** (1988), 629–668.

[M3] I. Moerdijk, *Prodiscrete groups and Galois toposes*, Indag. Math. **51** (1989), 219–234.

[P] R. Paré, *Colimits in topoi*, Bull. Amer. Math. Soc. **80** (1974), 556–561.

[Q1] D. Quillen, *Homotopy Algebra*, SLNM 43, Springer, Berlin (1967).

[Q2] D. Quillen, *Higher order $K$-theory, I*, SLNM 341, Springer, Berlin (1973), 85–147.

[R] J. Roos, *Sur les foncteurs dérivés de lim. Applications*, C. R. Acad. Sci. Paris **252** (1961), 3702–3704.

[T] R.W. Thomason, *Algebraic $K$-theory and étale cohomology*, Ann. Sci. École Norm. Sup. **18** (1985), 437–552.

[Ti] M. Tierney, *Sheaf theory and continuum hypothesis*, SLNM 274, Springer, Berlin (1972), 13–42.

# Categorical Structures

Ross Street

*School of Mathematics, Physics, Computing and Electronics, Macquarie University, New South Wales 2109, Australia*
*e-mail: street@macadam.mpce.mq.edu.au*

*Contents*



HANDBOOK OF ALGEBRA, VOL. 1
Edited by M. Hazewinkel

## Introduction

Category theory is a young subject yet has, by now, contributed its share of substantial theorems to the vast body of mathematics. In certain areas, I consider that it has also managed to revolutionize thinking. Examples of such areas, and the innovative categorical concepts, are:
- *homological algebra:* abelian category [F, Sch, Gt];
- *universal algebra:* triple (= monad), sketch [ML2, Sch, BW];
- *algebraic geometry:* scheme, topos [SGA, Sch, Gd, Jt, MLM];
- *set theory:* elementary topos [Jt, BW, MLM];
- *enumerative combinatorics:* Joyal species [Joy].

These matters are well covered by the indicated accessible literature; therefore, it is not the purpose of this article to repeat them. I shall be concerned more with categories as vital mathematical structures (as emphasized by Ehresmann [Eh1, Eh2] and Lawvere [L]), rather than with traditional category *theory*.

In topology texts, we read that the spaces were designed to carry continuity to a useful conceptual level. Yet, categories are *two* steps away from naturality, the concept they were designed to formalize. The intermediate notion, functor, is the expected kind of morphism between categories. From the very study of the established practice of routinely specifying morphisms along with each mathematical structure, we were presented, in the 1940's, with an extra dimension: morphisms between morphisms. We were naturally led by naturality to objects, arrow *and* 2-cells. Topology had its analogue: homotopies.

The reader will be assumed to have familiarity with categories, functors and natural transformations. My starting point is the introduction of 2-cells. I consider a category further equipped with 2-cells, but with no compositions apart from the composition of arrows already existing in the category; this is called a *derivation scheme*. With such a simple structure, this paper explores some fundamental interconnections involving:
- rewrite systems;
- free higher-order categories;
- cubes and simplexes;
- string diagrams, Penrose tensor notation, and braids;
- the $d$-simplex equations arising in the study of exactly soluble models in statistical mechanics and quantum field theory;
- homotopy theory;
- coherence in category theory.

*Convention.* The composite of arrows $\alpha\colon a \to b$, $\beta\colon b \to c$ in a category $A$ will be written in the algebraic order $\alpha \circ \beta\colon a \to c$. The other order may be regarded as "evaluation", so that parentheses $\beta(\alpha)\colon a \to c$ will be used.

## 1. Graphs, and 2-graphs

Recall that a *(directed) graph* $G$ consists of two sets $G_0$, $G_1$ and an ordered pair of functions $s, t\colon G_1 \to G_0$. Elements of $G_0$ are called *objects*, *vertices*, *or* 0-*cells*. Elements of $G_1$ are called *arrows, edges, or* 1*-cells*. Call $s(\gamma)$ the *source* of the arrow $\gamma$, call $t(\gamma)$ its *target* and denote this by $\gamma\colon\ s(\gamma) \to t(\gamma)$. For objects $a$, $b$ of $G$, we write $G(a, b)$ for the set of arrows $\gamma\colon\ a \to b$. There is a category **Grph** whose objects are graphs; the arrows $f\colon\ G \to H$, called *graph morphisms*, are pairs of functions $f_0\colon\ G_0 \to H_0$, $f_1\colon\ G_1 \to H_1$ such that, if $\gamma\colon\ a \to b$ in $G$, then $f_1(\gamma)\colon\ f_0(a) \to f_0(b)$ in $H$.

The *opposite* of a graph $G$ is the graph $G^{\mathrm{op}}$ obtained from $G$ by interchanging the functions $s$, $t$.

Each category $A$ has an underlying graph (since a category has a set $A_0$ of objects and a set $A_1$ of arrows) which we also denote by $A$. The free category on (or generated by) a graph $G$ is *the category* **F**$G$ *of paths in* $G$, described as follows. The objects of **F**$G$ are the objects of $G$. A *path* from $a_0$ to $a_n$ of length $n \geqslant 0$ is a $(2n+1)$-plet $(a_0, \gamma_1, a_1, \gamma_2, \ldots, \gamma_n, a_n)$:

$$a_0 \xrightarrow{\gamma_1} a_1 \xrightarrow{\gamma_2} a_2 \xrightarrow{\gamma_3} \cdots \xrightarrow{\gamma_n} a_n$$

where $\gamma_m\colon\ a_{m-1} \to a_m$ in $G$ for $0 < m \leqslant n$. An arrow $\alpha\colon\ a \to b$ in **F**$G$ is a path from $a$ to $b$ of any length $\ell(\alpha) \geqslant 0$. Composition of paths is given by

$$\begin{aligned}(a_0, \gamma_1, a_1, \ldots, \gamma_n, a_n) &\circ (b_0, \delta_1, b_1, \ldots, \delta_n, b_n)\\ &= (a_0, \gamma_1, a_1, \ldots, \gamma_n, a_n, \delta_1, b_1, \ldots, \delta_n, b_n)\end{aligned}$$

for $a_n = b_0$. So $\ell(\alpha \circ \beta) = \ell(\alpha) + \ell(\beta)$. It is convenient to identify the edge $\gamma\colon\ a \to b$ of $G$ with the path $(a, \gamma, b)\colon\ a \to b$, and to denote the path $(a)\colon\ a \to a$ of length 0 by $1_a\colon\ a \to a$ (as we do for identity arrows in any category). For $n > 0$, we then have

$$(a_0, \gamma_1, a_1, \gamma_2, \ldots, \gamma_n, a_n) = \gamma_1 \circ \gamma_2 \circ \cdots \circ \gamma_n.$$

A category is called *free* when it is isomorphic to a category **F**$G$ of paths in some graph $G$. For example, the category **N** which has one object 0, natural numbers $n\colon\ 0 \to 0$ as arrows, and addition as composition, is free. Each free category $A$ has a length functor

$$\ell\colon\ A \to \mathbf{N};$$

the generating graph has the same objects as $A$, but only the arrows of length 1. The generating graph for **N** is a terminal object in the category **Grph**.

Let **2** denote the free category on the graph with two objects 0, 1, and one arrow $0 \to 1$. Let **3** denote the free category on the graph with three objects 0, 1, 2, and two arrows $0 \to 1 \to 2$. Let $\partial_i\colon \mathbf{2} \to \mathbf{3}$, $i = 0, 1, 2$, denote the functor which is injective on objects and does not have $i = 0, 1, 2$ in the image.

A functor $f\colon\ A \to X$ is said to be *ulf* (for "unique lifting of factorizations") when each commutative square of functors

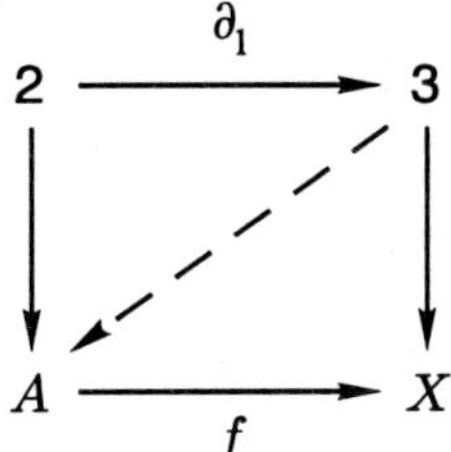


has a unique filler, as shown by the dashed functor, making the two triangles commute. A category $A$ is free if and only if there exists an ulf functor $\ell\colon A \to \mathbf{N}$.

For any category $A$, there is a functor **comp**: $\mathbf{F}A \to A$ given by "composing the paths":

$$\mathbf{comp}(\xi) = \gamma_1 \circ \cdots \circ \gamma_n \quad \text{for } \xi = (a_0, \gamma_1, a_1, \ldots, \gamma_n, a_n).$$

In fact, the category structure on the graph $A$ is encapsulated by the graph morphism **comp**: $\mathbf{F}A \to A$; the precise statement is that the underlying functor from the category **Cat** of categories to **Grph** is monadic (or "tripleable").

The *chaotic graph* $X_{\mathrm{ch}}$ on a set $X$ has source and target given by the first and second projections $X \times X \to X$. There is a unique category structure on $X_{\mathrm{ch}}$ so it is also called the *chaotic category* on $X$. The *discrete graph* $X_{\mathrm{d}}$ on the set $X$ has source and target both given by the unique function $\varnothing \to X$. The *discrete category* on $X$ is the free category $\mathbf{F}X_{\mathrm{d}}$ on $X_{\mathrm{d}}$; its source and target are both the identity function $1_X\colon X \to X$ of $X$.

Let $\pi_0 G$ denote the set of *connected components* of $G$; it is obtained from $G_0$ by identifying objects which have an arrow between them. Clearly $\pi_0 G = \pi_0 \mathbf{F}G$.

A 2-*graph* $G$ consists of three sets $G_0$, $G_1$, $G_2$ and four functions $s, t\colon G_1 \to G_0$, $s_1, t_1\colon G_2 \to G_1$ such that $s_1 \circ s = t_1 \circ s$ and $s_2 \circ t = t_1 \circ t$. The last two functions are denoted by $s, t\colon G_2 \to G_0$. Terminology for the graph $s, t\colon G_1 \to G_0$ is used for the 2-graph. Also, the elements $u$ of $G_2$ are called 2-*cells*; when $\gamma, \delta\colon a \to b$ and $s_1(u) = \gamma$, $t_1(u) = \delta$, we write either

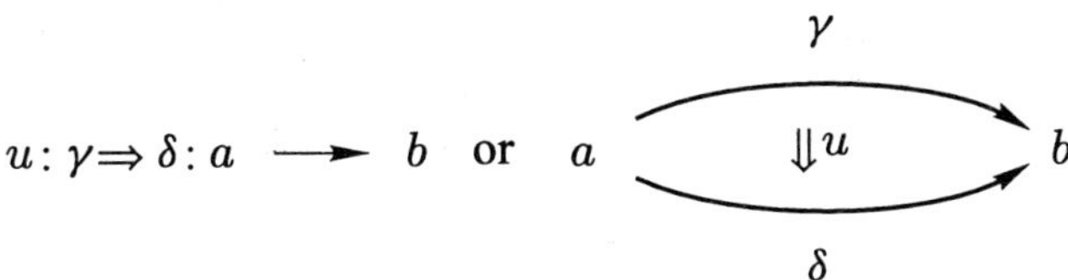


Write $G(a, b)$ for the graph whose objects are arrows $\gamma\colon a \to b$, and whose arrows are 2-cells $u\colon \gamma \Rightarrow \delta\colon a \to b$. The graph $s_1, t_1\colon G_2 \to G_1$ is the disjoint union of the graphs $G(a, b)$, $a, b \in G_0$. There is a category **2-Grph** of 2-graphs whose arrows $f\colon G \to H$, called 2-*graph morphisms*, are triplets of functions $f_i\colon G_i \to H_i$, $i = 0, 1, 2$, such that $(f_0, f_1)$, $(f_1, f_2)$ are graph morphisms.

The *opposite* $G^{\mathrm{op}}$ of a 2-graph is obtained by interchanging $s, t$: $G_1 \to G_0$. The *conjugate* $G^{\mathrm{co}}$ of $G$ is obtained by interchanging $s_1, t_1$: $G_2 \to G_1$. There is also $G^{\mathrm{coop}}$.

## 2. Derivation schemes, sesquicategories, and 2-categories

This section reviews concepts, selected from [S2] and [ES], which underpin 2-dimensional categories.

A *derivation scheme* $D$ consists of a 2-graph $D$ together with a category $\langle D\rangle$ whose underlying graph is $s, t$: $D_1 \to D_0$. We shall often provide the data for a derivation scheme $D$ in a diagram

$$s_1, t_1\colon\ M \to A$$

where $A$ is the category $\langle D\rangle$ and $M$ is the set $D_2$. There is a category **DS** of derivation schemes whose arrows $f$: $D \to E$, called *derivation scheme morphisms*, are 2-graph morphisms for which $(f_0, f_1)$: $\langle D\rangle \to \langle E\rangle$ is a functor.

Each 2-cell $u$: $\gamma \Rightarrow \delta$: $a \to b$ in a derivation scheme $D$ can be thought of as a *rewrite rule* which labels the directed replacement of $\gamma$ by $\delta$. An *application* of the rule $u$ is the replacement of any arrow of the form $\alpha \circ \gamma \circ \beta$ by $\alpha \circ \delta \circ \beta$. We label this application by the symbol $\alpha u \beta$: $\alpha \circ \gamma \circ \beta \Rightarrow \alpha \circ \delta \circ \beta$, and call it the *whiskering* of $u$ by $\alpha, \beta$ as suggested by the following diagram.

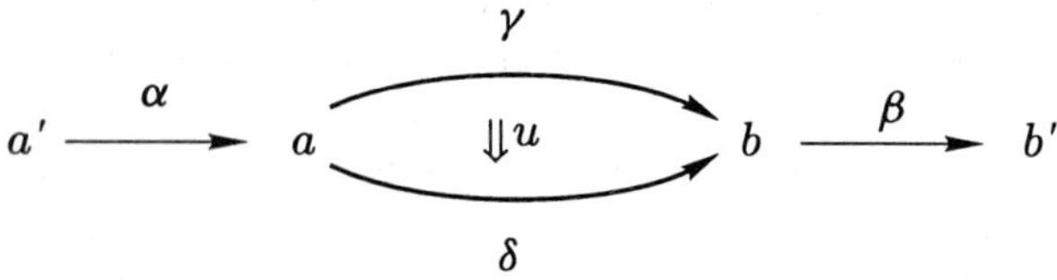


It is harmless to identify $u$ with its whiskering by identities. This gives a derivation scheme $\mathbf{w}D$ with the same category $\langle D\rangle$ and with the whiskered 2-cells; so $\mathbf{w}D$ contains $D$. A *derivation* in $D$ is a finite sequence of applications of rules; more precisely, it is a path in the graph $s_1, t_1$: $(\mathbf{w}D)_2 \to D_1$. We obtain another derivation scheme $\mathbf{d}D$ with the same category $\langle D\rangle$ and with derivations as 2-cells. We write $(\mathbf{d}D)(a, b)$ for the path category of the graph $(\mathbf{w}D)(a, b)$. In fact, $\mathbf{d}D$ is more richly structured than a mere derivation scheme, it is an example of a "sesquicategory".

A *sesquicategory* $S$ consists of a derivation scheme $S$ and a functor

$$S(-,-)\colon\ \langle S\rangle^{\mathrm{op}} \times \langle S\rangle \to \mathbf{Cat}$$

whose composite with the functor $obj$: $\mathbf{Cat} \to \mathbf{Set}$ is the homfunctor of the category $\langle S\rangle$, and whose value at an object $(a, b) \in \langle S\rangle^{\mathrm{op}} \times \langle S\rangle$ is a category with underlying graph $S(a, b)$. We now write $S(a, b)$ for the category and not just the graph; the composition of $S(a, b)$ is called *vertical composition* and denoted by $\bullet$. For each pair of arrows $\alpha$: $a' \to$

$a$, $\beta$: $b \to b'$, a functor $S(\alpha, \beta)$: $S(a, b) \to S(a', b')$ has its value at $u$: $\gamma \Rightarrow \delta$: $a \to b$ denoted by

$$\alpha \circ u \circ \beta\colon\ \alpha \circ \gamma \circ \beta \Rightarrow \alpha \circ \delta \circ \beta\colon\ a' \to b'$$

where $\circ$ between 1-cells is composition in the category $\langle S \rangle$. Let $\langle\!\langle S \rangle\!\rangle$ denote the category whose underlying graph is $s_1, t_1$: $S_2 \to S_1$ and whose composition is vertical composition $\bullet$.

There is a category **Sqc** of sesquicategories; the arrows, called *sesquifunctors*, are 2-graph morphisms which preserve all the compositions and identities.

Each sesquicategory $S$ gives rise to a category $\mathbf{q}S$, called the *quotient category* of $S$. The objects are the objects of $S$. The set of arrows is the set of components of the category $\langle\!\langle S \rangle\!\rangle$. Composition is induced by that of $\langle S \rangle$ (this uses the compatibility of $\langle S \rangle$ composition with existence of 2-cells).

A *2-category* $K$ [Eh1, Eh2] is a sesquicategory $K$ such that, for all $u$: $\gamma \Rightarrow \gamma'$: $a \to b$, $v$: $\delta \Rightarrow \delta'$: $b \to c$, the following equation holds:

$$(u \circ \delta) \bullet (\gamma' \circ v) = (\gamma \circ v) \bullet (u \circ \delta').$$

The 2-cell given by either side of the last equation is denoted by

$$u \circ v\colon\ \gamma \circ \delta \Rightarrow \gamma' \circ \delta'\colon\ a \to c$$

(and called the *horizontal composite* of the 2-cells $u, v$).
(HC)

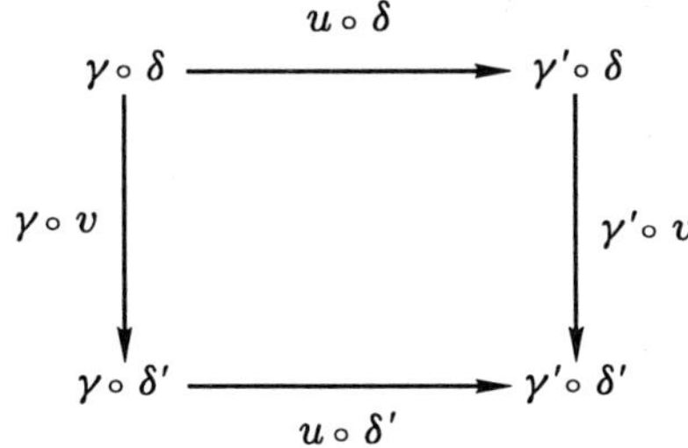


It follows that the *middle-four-interchange law* holds: that is, for each diagram

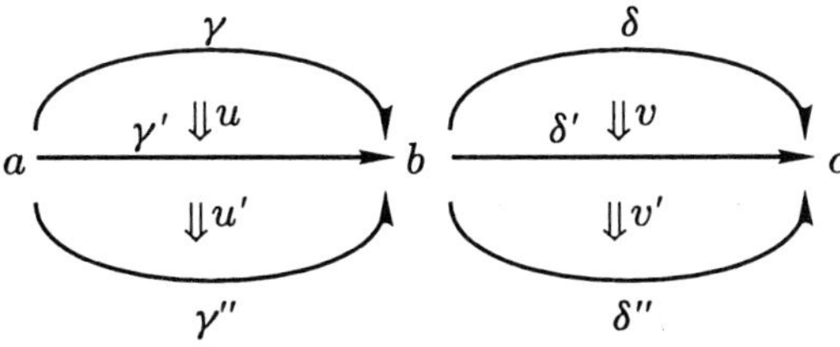

in $K$, there is an equality

$$(u \bullet u') \circ (v \bullet v') = (u \circ v) \bullet (u' \circ v').$$

So, horizontal composition $- \circ -\colon K(a,b) \times K(b,c) \to K(a,c)$ is a functor. There is a category **2-Cat** of 2-categories; the arrows, now called *2-functors*, are sesquifunctors.

The basic example of a 2-category is **Cat**: its objects are categories (subject to some size restriction, if the reader feels this is needed), arrows are functors, and 2-cells are natural transformations [Gt], Appendix. Just as one considers additive categories, which are categories whose homsets are *enriched* in the monoidal category of abelian groups, we can describe 2-categories as categories whose homsets are enriched in **Cat** (with cartesian product as tensor product); see [EK] for precise definitions. Some connection between 2-categories and homotopy theory can be found in [GZ]. The connection between 2-categories and derivations in rewrite systems was made in [Bns].

Each sesquicategory $S$ yields a 2-category $\mathbf{f}S$ by forcing commutativity in the squares (HC). This can be described by constructing a new derivation scheme $E$ which will provide rewrite rules for arrows in $S$. Take $\langle E\rangle = \langle\langle S\rangle\rangle$. Take $E_2$ to be the subset of $S_2 \times S_2$ consisting of those pairs $(u,v)$ of nonidentity 2-cells with $t(u) = s(v)$, and where $s_1, t_1\colon E_2 \to \langle S\rangle_1$ take $(u,v)$ to the lower, upper paths around the above square (HC).

$$(u,v)\frac{(\gamma \circ v) \bullet (u \circ \delta')}{(u \circ \delta) \bullet (\gamma' \circ v)}.$$

Then form the quotient category $\mathbf{qd}E$ of the sesquicategory $\mathbf{d}E$. The objects of $\mathbf{qd}E$ are the arrows of $S$. Our 2-category $\mathbf{f}S$ is given by $\langle \mathbf{f}S\rangle = \langle S\rangle$ and $\langle\langle \mathbf{f}S\rangle\rangle = \mathbf{qd}E$. There is a canonical sesquifunctor $S \to \mathbf{f}S$, and composition with it establishes a bijection between 2-functors $\mathbf{f}S \to K$ and sesquifunctors $S \to K$, for all 2-categories $K$.

For any derivation scheme $D$, we can apply the construction of the last paragraph to the sesquicategory $S = \mathbf{d}D$ where the 2-cells are derivations in $D$ and so have length. It is possible then to replace the derivation scheme $E$ by the sub-derivation-scheme $\uparrow D$ of $E$ whose 2-cells $(u,v)$ are restricted to those with $u, v$ both derivations of length 1. We call $\uparrow D$ the *lift* of $D$.

For any derivation scheme $D$, we obtain a 2-category $\mathbf{fd}D$. Two derivations in $D$ are called *equivalent* when they are identified by the canonical sesquifunctor $\mathbf{d}D \to \mathbf{fd}D$; this means there is an undirected sequence of applications of the rules of $\uparrow D$ taking one derivation to the other.

## 3. Pasting, computads, and free 2-categories

Repeated horizontal and vertical composition in a 2-category $K$ determine a more general operation called *pasting*. For example, consider the following diagram in $K$.

(P)

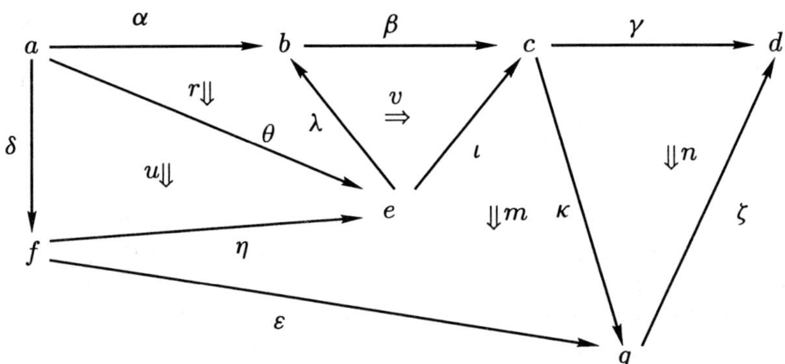


A 2-cell in a region means that its source and target are given by the composites of the indicated paths: for example, we have $v : \lambda \circ \beta \Rightarrow \iota$, $m : \eta \circ \iota \circ \kappa \Rightarrow \varepsilon$, and $r : \alpha \Rightarrow \theta \circ \lambda$. (Care is needed in placing the double arrow in each region so that it is clear which path is intended to be the source and which the target. If the arrow for $r$ had pointed from left to right instead of downward, the result would be meaningless.) The 2-cells of the diagram (P) can be whiskered in such a way as to obtain a path from $\alpha \circ \beta \circ \gamma$ to $\delta \circ \varepsilon \circ \zeta$ of length 5 in the underlying graph of the category $K(a, d)$; for example,

$$\alpha \circ \beta \circ \gamma \xrightarrow{r \circ \beta \circ \gamma} \theta \circ \lambda \circ \beta \circ \gamma \xrightarrow{\theta \circ v \circ \gamma} \theta \circ \iota \circ \gamma \xrightarrow{u \circ \iota \circ \gamma}$$
$$\delta \circ \eta \circ \iota \circ \gamma \xrightarrow{\delta \circ \eta \circ \iota \circ n} \delta \circ \eta \circ \iota \circ \kappa \circ \zeta \xrightarrow{\delta \circ m \circ \zeta} \delta \circ \varepsilon \circ \zeta.$$

Another such path is

$$\alpha \circ \beta \circ \gamma \xrightarrow{\alpha \circ \beta \circ n} \alpha \circ \beta \circ \kappa \circ \zeta \xrightarrow{r \circ \beta \circ \kappa \circ \zeta} \theta \circ \lambda \circ \beta \circ \kappa \circ \zeta \xrightarrow{u \circ \lambda \circ \beta \circ \kappa \circ \zeta}$$
$$\delta \circ \eta \circ \lambda \circ \beta \circ \kappa \circ \zeta \xrightarrow{\delta \circ \eta \circ v \circ \kappa \circ \zeta} \delta \circ \eta \circ \iota \circ \kappa \circ \zeta \xrightarrow{\delta \circ m \circ \zeta} \delta \circ \varepsilon \circ \zeta.$$

We leave it as an exercise for the reader to check that these paths have the same composite in the category $K(a, d)$. Diagrams such as (P) are called *pasting diagrams*, and the 2-cell

$$(r \circ \beta \circ \gamma) \bullet (\theta \circ v \circ \gamma) \bullet (u \circ \iota \circ \gamma) \bullet (\delta \circ \eta \circ \iota \circ n) \bullet (\delta \circ m \circ \zeta) \colon \alpha \circ \beta \circ \gamma$$
$$\Rightarrow \delta \circ \varepsilon \circ \zeta \colon a \to d$$

is called the *pasting composite* of the diagram. Notice that, if we reversed the direction of the 2-cell $r$ (say) in (P), we would no longer have a pasting diagram since no path in $K(a, d)$ could be made from it by whiskering the 2-cells.

A *computad* $C$ consists of a graph $s, t \colon C_1 \to C_0$, denoted by $C^{\#}$, together with a derivation scheme $s_1, t_1 \colon C_2 \to \mathbf{F}C^{\#}$. The elements $u$ of $C_2$ can be pictured as diagrams

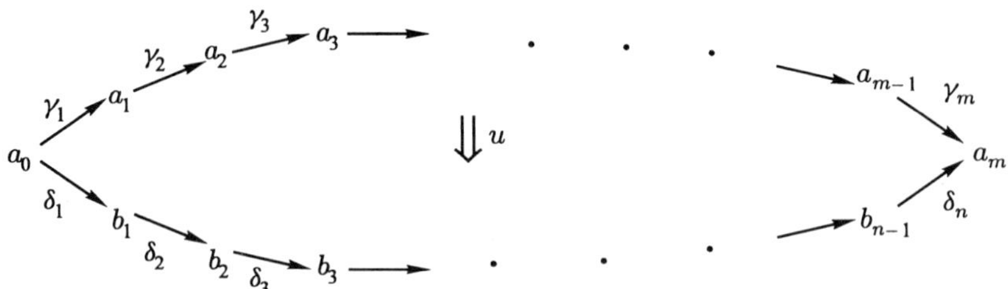


where the upper path is $s_1(u)$ and the lower is $t_1(u)$. A *computad morphism* $f\colon C \to C'$ is a triplet of functions $f_i\colon C_i \to C_i'$, $i = 0, 1, 2$, for which there is a morphism $(f_0, f_1', f_2)$ of the derivation schemes such that $f_1'$ agrees with $f_1$ on arrows of length 1. This gives a category **Cptd** of computads. Having given this precise definition, we can regard a computad as a derivation scheme $C$ with $\langle C \rangle$ a free category, so long as we take care to remember that the computad morphisms preserve the length of 1-cells.

Each 2-category $K$ has an underlying computad $C = \mathbf{U}K$ with $C^{\#}$ the underlying graph of the category $\langle K \rangle$, with

$$C_2 = \big\{(\xi, u, \eta) \mid \xi, \eta \text{ are paths in } C^{\#} \text{ and } u\colon \mathbf{comp}(\xi) \Rightarrow \mathbf{comp}(\eta) \text{ in } K\big\},$$

and with $s_1, t_1\colon C_2 \to \mathbf{F}\langle K \rangle$ taking $(\xi, u, \eta)$ to $\xi, \eta$, respectively.

The *free 2-category* $\mathbf{F}C$ on the computad $C$ is $\mathbf{fd}C$. There is an obvious inclusion computad morphism $i\colon C \to \mathbf{UF}C$. For each 2-category $K$ and each computad morphism $f\colon C \to \mathbf{U}K$, there exists a unique 2-functor $g\colon \mathbf{F}C \to K$ such that the following triangle commutes.

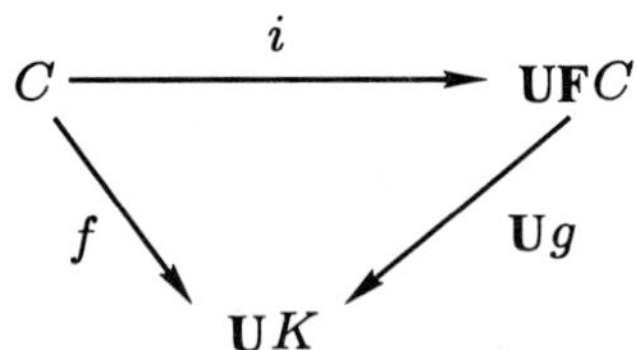


This means that the functor $\mathbf{U}$ has a left adjoint $\mathbf{F}$. Taking $C = \mathbf{U}K$, we obtain a 2-functor **past**: $\mathbf{FU}K \to K$, called the *pasting operation* for the 2-category $K$. A 2-category structure on a computad $C$ can be characterized in terms of an abstract pasting operation $\mathbf{UF}C \to C$. More precisely, the functor $\mathbf{U}$: **2-Cat** $\to$ **Cptd** is monadic.

This pasting operation will now be related to our previous discussion of the diagram (P). Suppose now that (P) is made up from data of a computad $C$. For example, there are 2-cells

$$v\colon (e, \lambda, b, \beta, c) \Rightarrow (e, \iota, c), \quad m\colon (f, \eta, e, \iota, c, \kappa, g) \Rightarrow (f, \varepsilon, g),$$

and

$$r\colon (a, \alpha, b) \Rightarrow (a, \theta, e, \lambda, b).$$

Whiskering the five 2-cells in the derivation scheme $D$ of $C$, we obtain 2-cells

$$r \circ (b, \beta, c, \gamma, d), \quad (a, \theta, e) \circ v \circ (c, \gamma, d), \quad u \circ (e, \iota, c, \gamma, d),$$

$$(a, \delta, f, \eta, e, \iota, c) \circ n, \quad (a, \delta, f) \circ m \circ (g, \zeta, d)$$

from $(a, \alpha, b, \beta, c, \gamma, d)$ to $(a, \delta, f, \varepsilon, g, \zeta, d)$ in $\mathbf{w}D$; they form a path in the graph $(\mathbf{w}D)(a, d)$. The connected component (with respect to (HC)) of this path gives a 2-cell

$$(a, \alpha, b, \beta, c, \gamma, d) \Rightarrow (a, \delta, f, \varepsilon, g, \zeta, d)\colon\ a \to d$$

in $\mathbf{F}C$. This is, of course, none other than the pasting composite of the diagram (P) in the 2-category $\mathbf{F}C$. Conversely, any other representative of this 2-cell in $\mathbf{F}C$ by a path in $(\mathbf{w}D)(a, d)$ leads us back to a planar diagram equivalent to (P). So a pasting diagram in $C$ seems to provide a geometrically invariant way of depicting a 2-cell of $\mathbf{F}C$. For a 2-category $K$, the pasting operation **past**: $\mathbf{FU}K \to K$ assigns the pasting composite to the pasting diagram.

In general, however, when there are 2-cells which have source or target paths of length 0 in the computad $C$, the faithful geometric representation of 2-cells of $\mathbf{F}C$ by pasting diagrams breaks down. The reason is that the following three geometrically inequivalent pasting diagrams all represent the same 2-cell when

$$t_1(u) = s_1(v) = 1_a\colon\ a \to a.$$

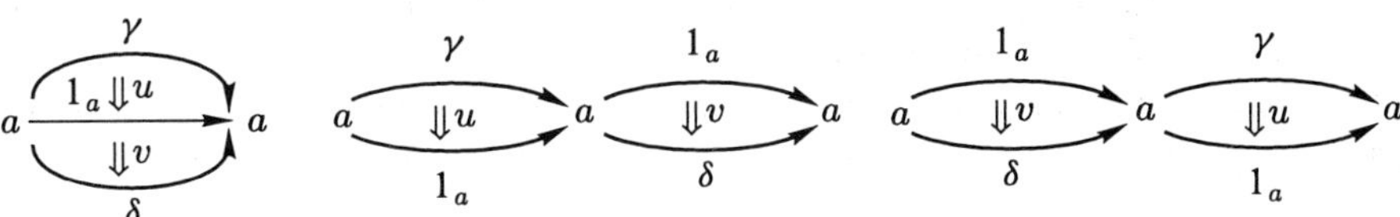

We shall see below that this problem can be overcome by using the string diagrams which are planar dual to pasting diagrams.

## 4. Strings, and the terminal computad

Consider the planar dual of the pasting diagram (P) at the beginning of Section 3. Each 2-cell $r, u, v, m, n$ becomes a node labeled by the same symbol; each arrow $\alpha, \beta, \ldots$ becomes an edge, called a *string*. A string is attached to a node when the original arrow formed part of the boundary of the region containing the 2-cell. Moreover, we require that the strings *progress* down the page from nodes that were source 2-cells towards nodes that were target 2-cells. The resultant graph, embedded in the plane, is called *a string diagram*.

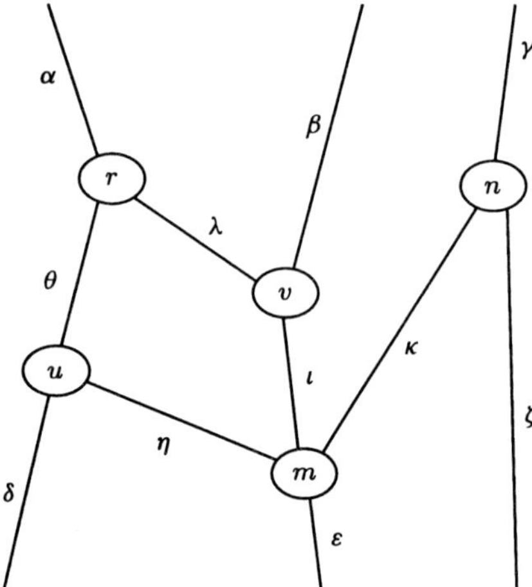


The *value* of this string diagram is the 2-cell $\alpha \circ \beta \circ \gamma \Rightarrow \delta \circ \varepsilon \circ \zeta$ obtained by breaking the diagram into horizontal layers with nodes at different levels in different layers. Reading from left to right, we obtain a horizontal composite of 2-cells from each layer; each node contributes its 2-cell, and each nodeless string contributes the identity 2-cell of its arrow. This gives the value of each layer. Then the values of the layers are composed vertically, reading down the page. For our example, we obtain:

$$(r \circ \beta \circ n) \bullet (\theta \circ v \circ \kappa \circ \zeta) \bullet (u \circ \iota \circ \kappa \circ \zeta) \bullet (\delta \circ m \circ \zeta).$$

The reader should enjoy checking that this agrees with the pasting composite of the pasting diagram (P) using the axioms for a 2-category.

The above string diagram can be deformed in the plane (as below) so as to preserve the strings' progression downward. The value remains the same [JS2].

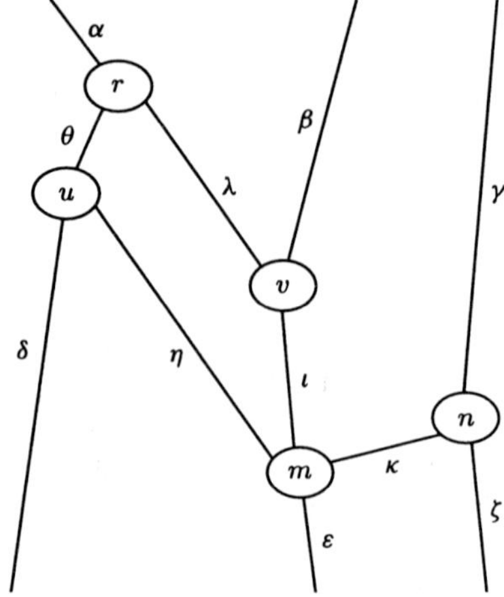

The value of this deformed string diagram is

$$(r \circ \beta \circ \gamma) \bullet (u \circ \lambda \circ \beta \circ \gamma) \bullet (\delta \circ \eta \circ v \circ \gamma) \bullet (\delta \circ \eta \circ \iota \circ n) \bullet (\delta \circ m \circ \zeta),$$

which is also equal to the pasting composite of (P).

Moreover, the string representation deals with the problem involving identities described at the end of Section 3. For suppose we have 2-cells $u$, $v$ with $t_1(u) = s_1(v) = 1_a\colon\ a \to a$. Corresponding to the pasting diagram

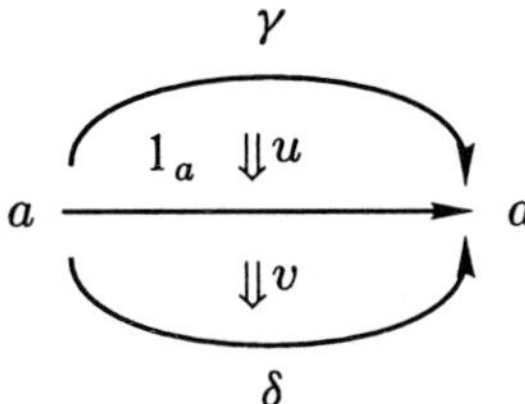


we have the string diagram

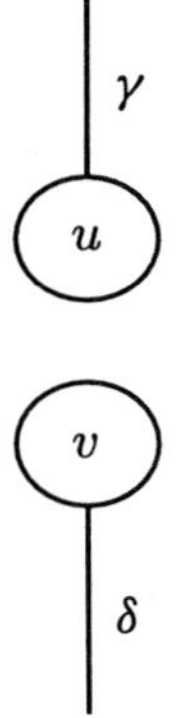


whose value is $u \bullet v$ and which can be deformed to

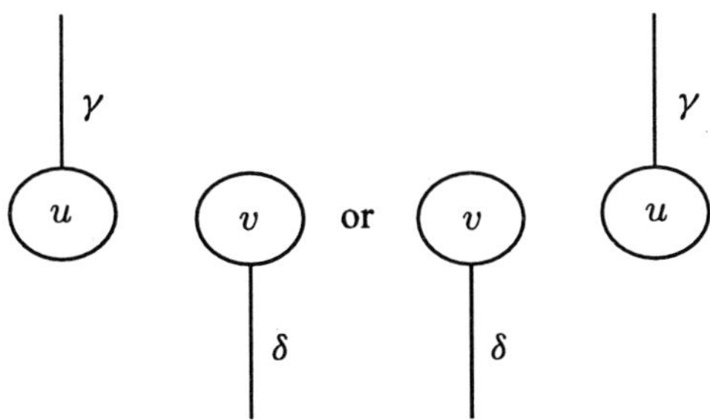

which have the values $u \circ v$ and $v \circ u$, respectively. This suggests that the geometry of the string diagram provides a faithful representation of 2-cells in free 2-categories, which is indeed the case [JS2] as we shall explain in more detail below.

Just as it is of particular interest to consider the free category $\mathsf{N}$ on the terminal graph, it is also worth considering the free 2-category $\mathsf{M}$ on the terminal computad. Recall that $\mathsf{N}$ is a one-object category, and so is really just a monoid. Similarly, $\mathsf{M}$ is a one-object 2-category, and so is really just a strict monoidal category (that is, a monoid in the category **Cat** of categories and functors).

The terminal computad $C_t$ is the terminal object in the category **Cptd**. The graph $C_t^{\#}$ is the terminal graph. So $\mathbf{F}C_t^{\#} = \mathsf{N}$. There must be exactly one 2-cell for each possible source and target path; so the derivation scheme of $C_t$ is the chaotic graph on the set $\{0, 1, 2, \ldots\}$ of natural numbers. We write the 2-cells of $C_t^{\#}$ as $m/n$: $m \Rightarrow n$.

The derivation scheme $\mathbf{w}C_t$ has 2-cells of the form $(1, m/n, r)$: $1+m+r \Rightarrow 1+n+r$ obtained by whiskering $m/n$ on the left by 1 and on the right by $r$. Thus the free 2-category $\mathsf{M}$ on $C_t$ is obtained by taking paths of these 2-cells and identifying subject to condition (HC) of Section 2.

This gives the following direct description of $\mathsf{M}$ as a strict monoidal category. Consider the graph $\mathsf{W}$ whose vertices are natural numbers and whose edges $(l, m/n, r)$: $a \to b$ consist of natural numbers $l, m, n, r$ with $l+m+r = a$ and $l+n+r = b$. Then consider the path category $\mathbf{F}\mathsf{W}$. We introduce the following "rewrite rule" on arrows of $\mathbf{F}\mathsf{W}$ of length 2:

$$\frac{(l, m/n, r) \circ (l', m'/n', r')}{(l', m'/n', r' - n + m) \circ (l - m' + n', m/n, r)} \quad \text{for } l' + m' \leqslant l,$$

and, to exclude the case where the top and bottom are equal, we ask that not all of $l = l'$, $m = n = m' = n' = 0$ hold. This rule is a directed form of the condition (HC) as with the 2-cells of the lift $\uparrow C_t$. An application of this rewrite rule is the replacement of a path $\pi \circ \sigma \circ \pi'$ by $\pi \circ \tau \circ \pi'$ where $\sigma$ is the top path and $\tau$ is the bottom path of the rule. To obtain $\mathsf{M}$, identify arrows of $\mathbf{F}\mathsf{W}$ when one arrow can be obtained from the other by a finite sequence of undirected applications of the rewrite rules. For objects $c, d$ of $\mathbf{F}\mathsf{W}$, we have functors

$$c + -, - + d\colon \mathbf{F}\mathsf{W} \to \mathbf{F}\mathsf{W}$$

taking $(l, m/n, r)$: $a \to b$ to

$$(c + l, m/n, r)\colon c + a \to c + b, \ (l, m/n, r + d)\colon a + d \to b + d,$$

respectively. The identification of arrows in $\mathbf{F}\mathsf{W}$ was introduced precisely so that these functors would induce partial functors for a functor

$$\mathsf{M} \times \mathsf{M} \to \mathsf{M}$$

which provides the tensor product for $\mathsf{M}$; it is given on objects by addition of natural numbers.

We briefly consider the question of whether the (directed) rewrite rule above can be used to find "normal representatives" in **FW** for arrows in **M**. Notice that we do have "confluence" for the rewrite rules in the sense that, starting with a path in **W** of length 3 for which two rewrite rules can be applied, we can begin by applying either rule, yet continue applying rules to obtain a common result. For, suppose we have both $l'+m' \leqslant l$ and $l''+m'' \leqslant l'$. Then $l''+m'' \leqslant l' \leqslant l-m' \leqslant l-m'+n'$; so we have the following derivation.

$$\cfrac{\cfrac{(l,m/n,r)\circ(l',m'/n',r')\circ(l'',m''/n'',r)}{(l',m'/n',r'-n+m)\circ(l-m'+n',m/n,r)\circ(l'',m''/n'',r)}}{\cfrac{(l',m'/n',r'-n+m)\circ(l'',m''/n'',r''-n+m)\circ(l-m'+n'-m''+n'',m/n,r)}{(l'',m''/n'',r''-n+m-n'+m')\circ(l'-m''+n'',m'/n',r'-n+m)\circ(l-m'+n'-m''+n'',m/n,r)}}$$

Also, $(l'-m''+n'')+m' = (l'+m')-m''+n'' \leqslant l-m''+n''$; so we have the following derivation.

$$\cfrac{\cfrac{(l,m/n,r)\circ(l',m'/n',r')\circ(l'',m''/n'',r)}{(l,m/n,r)\circ(l'',m''/n'',r''-n'+m')\circ(l'-m''+n'',m'/n',r')}}{\cfrac{(l'',m''/n'',r''-n'+m'-n+m)\circ(l-m''+n'',m/n,r)\circ(l'-m''+n'',m'/n',r)}{(l'',m''/n'',r''-n'+m'-n+m)\circ(l'-m''+n'',m'/n',r'-n+m)\circ(l-m''+n''-m'+n',m/n,r)}}$$

Notice that the derivations both lead to the same bottom line, yielding the desired confluence.

A path in **W** is called *reduced* when the rewrite rules cannot be applied to it. So a path $(l,m/n,r)\circ(l',m'/n',r')$ of length 2 is reduced when either $l < l'+m'$, or $m=n=m'=n'=0$ and $l=l'$. An arbitrary path is reduced if and only if every path of length 2 through which it factors is reduced. Notice that, if $l'+m' \leqslant l$, then the path

$$(l',m'/n',r'-n+m)\circ(l-m'+n',m/n,r')$$

is reduced if $n'+m>0$; so in this case, for paths of length 2, a reduced path is obtained in one application of a rewrite rule. For the case $n'+m=0$, notice the derivation of length 2:

$$\cfrac{\cfrac{(m',0/n,0)\circ(0,m'/0,n)}{(0,m'/0,0)\circ(0,0/n,0)}}{(0,0/n,m')\circ(n,m'/0,0)}.$$

This is why we need the second sentence of the following result.

PROPOSITION 4.1 [ES]. *Let $\pi$: $a \to b$ be a path of length $k$ in the graph* **W**. *Suppose $\pi$ does not contain both an edge $(l,m/n,r)$ with $m=0$ and an edge $(l',m'/n',r')$ with $n'=0$. Then all derivations with source $\pi$, using the above rewrite rules, have length $\leqslant k(k-1)/2$. Moreover, $\pi$ is equivalent to a unique reduced path.*

REMARK. Without the second sentence of the Proposition 4.1, the upper bound $k(k-1)/2$ must be increased (as shown by the above derivation of length 2 with $k=2$). David Benson has advised me that $k(k-1)$ is an upper bound in the general case, and that this follows from his paper [Bns]. The complication is related to the one discussed at the end of Section 3, which reminds us to look at a string model for **M**.

A *plane graph* $\Gamma$ is a compact topological subspace of $\mathbb{R}^2$ with a distinguished set $\Gamma_0$ of points whose complement $\Gamma - \Gamma_0$ in $\Gamma$ is homeomorphic to a finite union of disjoint open intervals. The elements of $\Gamma_0$ are called *vertices* and the connected components of $\Gamma - \Gamma_0$ are called *edges*. We say that $(x, y)$ is *above* $(x', y')$ in $\mathbb{R}^2$ when $y' \leqslant y$; *below* means the reverse. The plane graph $\Gamma$ is called *progressive* when aboveness is a total (linear) order on each edge. Progressive plane graphs are directed graphs: the source and target of an edge are the vertices in the closure of the edge; the source is above the target.

A *progressive plane graph with boundary* consists of a progressive plane graph $\Gamma$ with a distinguished set $\mathbf{i}\Gamma$ of vertices such that each vertex in $\partial\Gamma = \Gamma_0 - \mathbf{i}\Gamma$ is in the closure of precisely one edge, and $\mathbf{i}\Gamma$ is an interval in the aboveness order on $\Gamma_0$ (that is, if $p, q, r$ are vertices with $p$ above $q$ and $q$ above $r$, then $p, r \in \mathbf{i}\Gamma$ implies $q \in \mathbf{i}\Gamma$). Notice that $\partial\Gamma$ is the disjoint union of the subset $\mathbf{s}\Gamma$ of those vertices which are sources and the subset $\mathbf{t}\Gamma$ of those vertices which are targets. For example, in the progressive plane graph depicted below, the white nodes provide an acceptable set $\mathbf{i}\Gamma$; so the black nodes constitute $\partial\Gamma$, the cardinality of $\mathbf{s}\Gamma$ is two, and the cardinality of $\mathbf{t}\Gamma$ is eight.

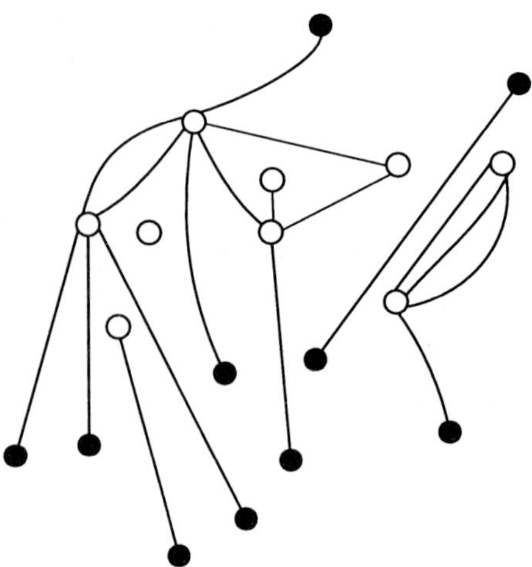

Of course, the size of the nodes is exaggerated for visibility. It is customary to omit the boundary (black) nodes from the picture, leaving loose the single edge having it in the closure.

Suppose $\Gamma$, $\Gamma'$ are progressive plane graphs with boundary. We say that $\Gamma$ is a *deformation* of $\Gamma'$ when there exists a homeomorphism $h$: $\mathbb{R}^2 \to \mathbb{R}^2$ such that $h(\Gamma) = \Gamma'$, $h(\partial\Gamma) = \partial\Gamma'$, and $h$ preserves the aboveness order on edges.

Now we give the geometric model of the strict monoidal category **M**. The objects are natural numbers. An arrow $[\Gamma]\colon m \to n$ is a deformation class of progressive plane graphs with boundary such that the cardinalities of $\mathbf{s}\Gamma$, $\mathbf{t}\Gamma$ are $m, n$, respectively. We define the composite $[\Gamma] \circ [\Lambda]\colon m \to p$ of arrows $[\Gamma]\colon m \to n$, $[\Lambda]\colon n \to p$ by choosing representatives $\Gamma$, $\Lambda$ such that

$$\mathbf{s}\Gamma = \mathbf{t}\Lambda = \{(k,0)\colon k = 1,2,\ldots,n\},$$

with $\Gamma - \mathbf{t}\Gamma$ contained in the upper half plane and $\Delta - \mathbf{s}\Delta$ in the lower half plane; then put

$$[\Gamma] \circ [\Lambda] = [\Gamma \cup \Lambda]$$

where $(\Gamma \cup \Lambda)_0 = (\Gamma_0 \cup \Lambda_0) - \mathbf{t}\Gamma$ and $\partial(\Gamma \cup \Lambda) = (\partial\Gamma \cup \partial\Lambda) - \mathbf{t}\Gamma$. We define the tensor product

$$[\Gamma] \otimes [\Gamma']\colon m + m' \to n + n'$$

of arrows $[\Gamma]\colon m \to n$, $[\Gamma']\colon m' \to n'$ by choosing the representatives $\Gamma, \Gamma'$ to be contained in the left, right half plane (respectively); then put

$$[\Gamma] \otimes [\Lambda] = [\Gamma \cup \Lambda]$$

where $(\Gamma \cup \Lambda)_0 = \Gamma_0 \cup \Lambda_0$ and $\partial(\Gamma \cup \Lambda) = \partial\Gamma \cup \partial\Lambda$. There is a graph morphism $\mathbf{W} \to \mathbf{M}$ which is the identity on objects and takes the edge $(l, m/n, r)$ in $\mathbf{W}$ to the deformation class of the following graph.

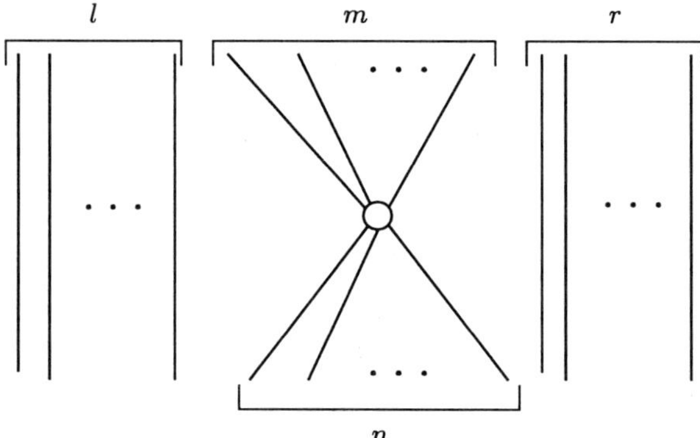


This graph morphism extends to a functor $\mathbf{FW} \to \mathbf{M}$ which is the universal functor out of $\mathbf{FW}$ identifying the rewrite rules for paths in $\mathbf{W}$ [JS2].

REMARK. When returning to the view of **M** as a 2-category, its single object will be denoted by 0, and horizontal, vertical composition will be denoted by $\circ, \bullet$ as usual in a 2-category, rather than by $\otimes, \circ$ with their usual meaning in a monoidal category.

## 5. Length 2-functors, and presentations of 2-categories

Each free 2-category $A$ has a *length 2-functor* $\ell\colon A \to \mathbf{M}$ induced by the unique computad morphism between the generating computads; recall that the generating computad of **M** is terminal. We now attempt to characterize free 2-categories in terms of the length 2-functor.

Each 2-functor $\ell\colon A \to \mathbf{M}$ determines a computad $\ell^{-1}(C_t)$ which is the subcomputad of $\mathbf{U}A$ with the same objects, the arrows $\gamma$ with $\ell(\gamma) = 1$, and the 2-cells $u\colon \alpha \Rightarrow \beta$ with $\ell(u)$ represented by the edge $(0, \ell(\alpha)/\ell(\beta), 0)$ of **W**. If $A$ is free and $\ell$ is its length 2-functor then $A$ is free on the computad $\ell^{-1}(C_t)$.

Let $\mathbf{2}_2$ and $\mathbf{3}_2$ denote the free 2-categories on the computads depicted by

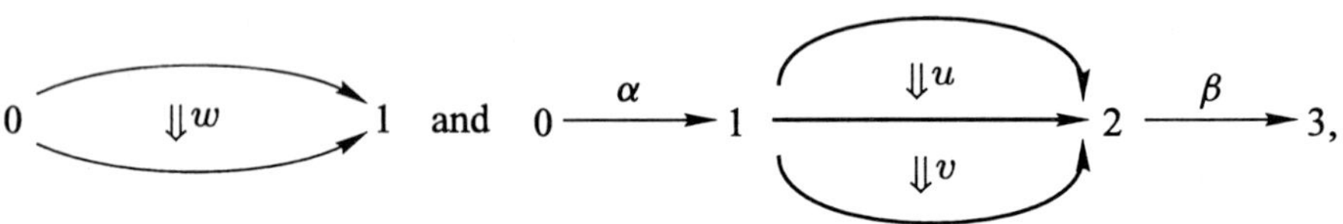


respectively, and let $\partial\colon \mathbf{2}_2 \to \mathbf{3}_2$ be the 2-functor which takes $w$ to $\alpha \circ (u \bullet v) \circ \beta$. A 2-functor $f\colon A \to X$ is said to be *ulf* when each commutative square

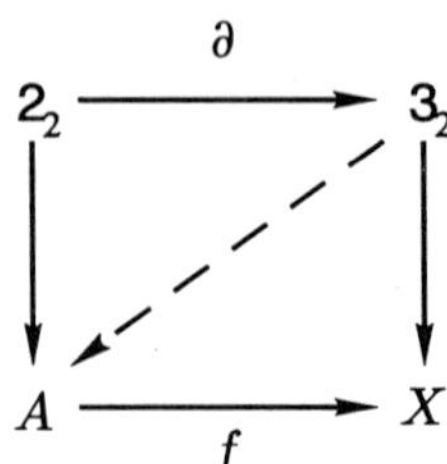


can be uniquely filled by a 2-functor as indicated by the dashed arrow.

A computad (or derivation scheme) is called *tight* when there are no 2-cells $u\colon \alpha \Rightarrow \beta\colon a \to a$ with $\alpha$ the identity of $a$. (From the rewrite view of $C$, this is a mild requirement, since the possibility of rewriting nothing as something is seldom desirable as it leads to infinite derivations.) Let $\mathsf{M}'$ denote the free 2-category on the sub-computad of the computad $C_t$ consisting of those 2-cells $m/n$: $m \Rightarrow n$ with $m \neq 0$.

PROPOSITION 5.1. *A 2-category $A$ is free on a tight computad if and only if there exists an ulf 2-functor*

$$\ell\colon A \to \mathsf{M}'.$$

To characterize general free 2-categories, we take the string viewpoint. Let $\Gamma$ denote a progressive plane graph with boundary, and let $D$ be any computad. A *valuation* $\nu\colon \Gamma \to D$ of $\Gamma$ in $D$ consists of a pair of functions

$$\nu_0\colon \Gamma_1 \to D_1 \quad \text{and} \quad \nu_1\colon \mathbf{i}\Gamma \to D_2$$

such that, for each $x \in \mathbf{i}\Gamma$, one has

$$\nu_1(x)\colon \nu_0(e_1) \circ \nu_0(e_2) \circ \cdots \circ \nu_0(e_m) \Rightarrow \nu_0(f_1) \circ \nu_0(f_2) \circ \cdots \circ \nu_0(f_n)$$

where $e_1, \ldots, e_m$ are the edges with target $x$ ordered from left to right in the plane, and $f_1, \ldots, f_m$ are the edges with source $x$ also ordered from left to right. A *string diagram* in $D$ is a pair $(\Gamma, \nu)$ consisting of a progressive plane graph $\Gamma$ with boundary and a valuation $\nu\colon \Gamma \to D$. If $(\Gamma, \nu)$ is a string diagram in $D$ and $\Gamma'$ is a deformation of $\Gamma$ then there is an obvious way to obtain a valuation $\nu'$ on $\Gamma'$; in this case, $(\Gamma', \nu')$ is called a *deformation of the string diagram* $(\Gamma, \nu)$. Write $[\Gamma, \nu]$ for the deformation class of $(\Gamma, \nu)$.

PROPOSITION 5.2. *A 2-category $A$ is free on some computad if and only if there exists a 2-functor $\ell\colon A \to \mathsf{M}$ such that, for each string diagram $(\Gamma, \nu)$ in the computad $\ell^{-1}(C_t)$, there exists a unique 2-cell $u$ in $A$ with $\ell(u) = [\Gamma, \nu]$.*

Suppose $A$ in any 2-category. By a *valuation* $\nu\colon \Gamma \to A$ and a *string diagram* in $A$, we mean a valuation and a string diagram in the computad $\mathbf{U}A$. Suppose $(\Gamma, \nu)$ is any string diagram in $A$. By Proposition 5.2, there exists a unique 2-cell $u$ in $\mathbf{FU}A$ with $\ell(u) = [\Gamma, \nu]$. The *value* $\nu(\Gamma)$ of the string diagram $(\Gamma, \nu)$ in $A$ is the value of $u$ under the 2-functor **past**: $\mathbf{FU}A \to A$; that is,

$$\nu(\Gamma) = \mathbf{past}(u).$$

Now that we have some understanding of free 2-categories, we can contemplate presentations of 2-categories. A *presentation* of a 2-category consists of a computad $C$ and a relation $R$ on the set $(\mathbf{F}C)_2$ of 2-cells of the free 2-category $\mathbf{F}C$. (The elements $(\alpha, \beta)$ of $R$ are often written as equations $\alpha = \beta$.) One obtains a 2-category $A$ (unique up to isomorphism) by constructing the universal 2-functor $\mathbf{F}C \to A$ which identifies $R$-related 2-cells; then $(C, R)$ is called a *presentation of* $A$. Of course, $C$ can be identified with a subcomputad of $\mathbf{U}A$.

EXAMPLE 1. *Monads*. Consider the computad $C$ with one object $a$, one arrow $\tau$: $a \to a$, and two 2-cells $e$: $1_a \Rightarrow \tau$, $m$: $\tau \circ \tau \Rightarrow \tau$. While this computad is not tight, its conjugate $C^{co}$ is tight, so all views of $\mathbf{F}C$ are available. Consider the relations $R$ given as follows.

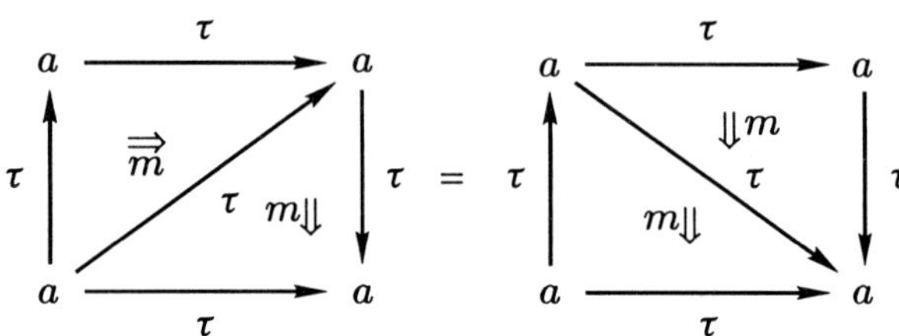

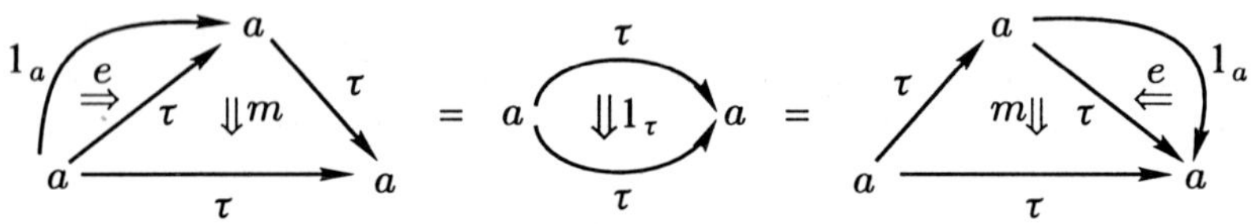

These relations can also be drawn using string diagrams, as follows.

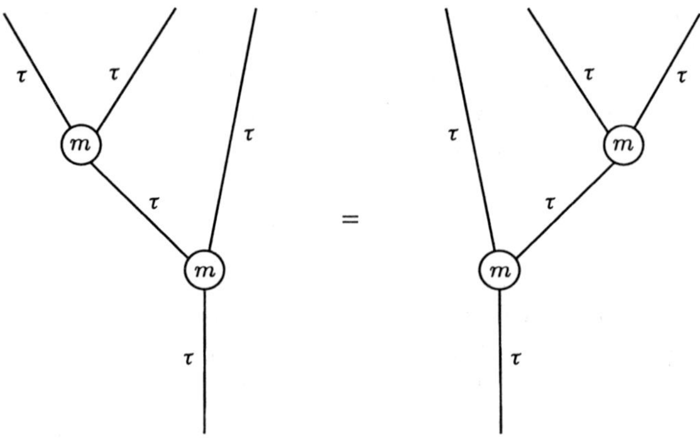

Let **Mnd** denote the 2-category with one object $a$, and with homcategory $\mathbf{Mnd}(a, a)$ equal to the category of finite ordinals and order-preserving functions; horizontal composition is ordinal sum. The $(C, R)$ provides a presentation for the 2-category **Mnd** via the interpretation of $\tau$ as the ordinal 1, and $e$: $0 \to 1$, $m$: $2 \to 1$ as the unique functions. To give a 2-functor $\mathbf{Mnd} \to K$ into a 2-category $K$ is to give a *monad* in $K$.

EXAMPLE 2. *Distributive laws between monads and comonads.* As a natural example of a computad $C$ which is neither tight nor has a tight conjugate, we take one object $a$, two arrows $\tau, \gamma$: $a \to a$, and five 2-cells $e$: $1_a \Rightarrow \tau$, $m$: $\tau \circ \tau \Rightarrow \tau$, $k$: $\gamma \Rightarrow 1_a$, $d$: $\gamma \Rightarrow \gamma \circ \gamma$, $r$: $\gamma \circ \tau \Rightarrow \tau \circ \gamma$. It will make the relations we are about to consider look more geometrically appealing if, in the string diagrams, we depict the 2-cell $r$ as a cross-over of string $\gamma$ over string $\tau$, rather than as a node. Let $R$ consist of the relations for $e$: $1_a \Rightarrow \tau$, $m$: $\tau \circ \tau \Rightarrow \tau$ as in Example 1, the relations given by inverting the string diagrams for $e$, $m$ and replacing $e$, $m$ by $k$, $d$, and the following four extra relations.

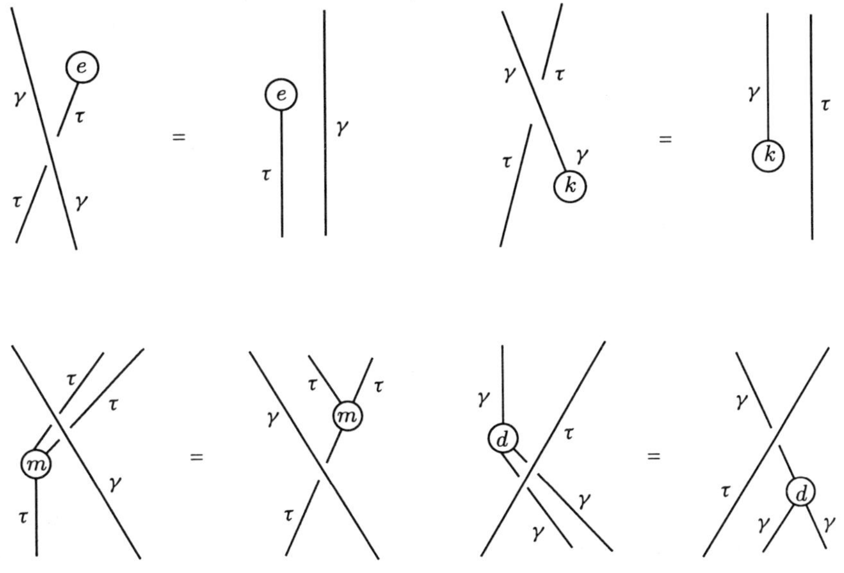

For a 2-category $A$, the computad morphisms $C \to \mathbf{U}A$, which identify $R$-related 2-cells, are in bijection with objects $a$ of $A$ equipped with a monad $\tau$, comonad $\gamma$, and a distributive law $r$ between them [Bc, S1, BW].

## 6. Cubes, and Gray's tensor product of 2-categories

By way of application of the above ideas, we now consider structures arising from consideration of cubes of all dimensions. What could be more basic than rewriting a single given symbol, say "minus", by another, say "plus"? We begin with a computad which, in a sense, is a combinatorial version of the *interval*, so we denote it by $\mathbf{I}$. The graph $\mathbf{I}^{\#}$ has one vertex (which shall remain nameless), and two edges denoted by $-$ and $+$. Paths in this graph are words $\alpha$ in the symbols $-$ and $+$; such words of length $n$ are in bijection with the $2^n$ vertices of the $n$-cube. There is only one 2-cell in $\mathbf{I}$ which we denote by $0 : - \Rightarrow +$.

An application of the rewrite rule $0 : - \Rightarrow +$ to a word $\alpha$ of length $n$ can be identified with an edge of the $n$-cube; it is a word $u$ of length $n$ in the symbols $-, 0, +$ with precisely one 0 occurring. The position of the 0 in $u$ is a position in $\alpha$ where there is a symbol $-$ and the target of $u$ is obtained from $\alpha$ by changing this $-$ to a $+$. Derivations in the derivation scheme $\mathbf{I}$ are paths around the edges of the cube. So the 2-cells of the one object sesquicategory $\mathbf{dI}$ can be regarded as paths around some $n$-cube. Write $\mathbf{I}[n, 1]$ for the subderivation scheme of $\mathbf{dI}$ consisting of the words $\alpha$ in the symbols $-$, $+$ of length $n$.

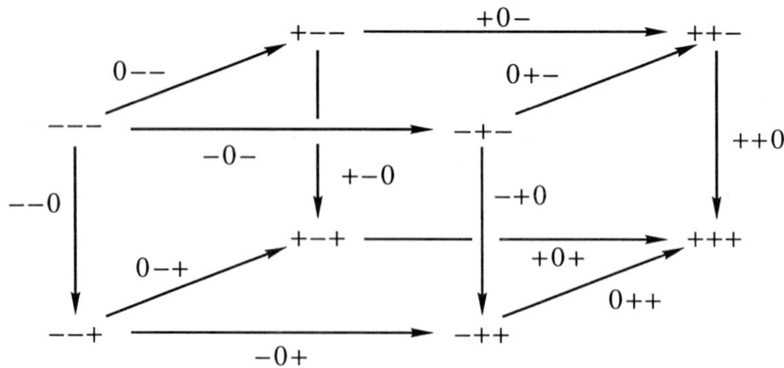


The 3-cube $\mathbf{I}[3, 1]$.

For words $\alpha, \beta \in \mathbf{I}[n, 1]$, write $\alpha \leqslant \beta$ when $\alpha, \beta$ have the same length and $\alpha$ has the symbol $-$ in every position that $\beta$ does. Clearly there exists a derivation $\alpha \to \beta$ if and only if $\alpha \leqslant \beta$. Moreover, any two derivations with the same source and target are equivalent. It follows that the homcategory of the free 2-category $\mathbf{fdI}$ on $\mathbf{dI}$ is a partially ordered set: it is a strict monoidal category whose tensor product is juxtaposition of words. If we take the full subcategory of this homcategory consisting of the words $\alpha$ of length $n$, we obtain a category $\mathbf{Cub}[n, 1]$, called the *$n$-cube with commutative 2-faces*.

However, we may not wish the 2-faces to commute. In other words, we may not wish to identify equivalent derivations. Let us examine the derivations in more detail. Suppose $\alpha, \beta$ are words of the same length $n$ in the symbols $-, +$ and suppose $\alpha \leqslant \beta$. Write $\alpha \backslash \beta$ for the set of positions where $\alpha$ has $-$ and $\beta$ has $+$. A derivation $u$ of $\mathbf{I}$ from $\alpha$ to $\beta$ can be identified with a listing $u = u_1 u_2 \ldots u_k$ of the elements of $\alpha \backslash \beta$ (each application determines an element of $\alpha \backslash \beta$ which is the position of the symbol 0 and the order is that forced by composibility of the applications making up the derivation). With this notation it must be realized that the source and target of $u$: $\alpha \to \beta$ must be specified in order to fully determine the derivation. Put

$$\mathcal{V}(u) = \{(u_i, u_j) \colon\ i < j \text{ and } u_i < u_j\}.$$

Notice that, for derivations $u$: $\alpha \to \beta$, $v$: $\beta \to \gamma$, there is a partition of $\mathcal{V}(uv)$ as

$$\mathcal{V}(uv) = \mathcal{V}(u) + \{(u_i, v_j) \colon\ u_i < v_j\} + \mathcal{V}(v).$$

We shall now describe the lift derivation scheme $\uparrow \mathbf{I}$. The objects are words $\alpha$ in the symbols $-, +$. The arrows are derivations $u$: $\alpha \to \beta$ of $\mathbf{I}$. The 2-cells are oriented 2-faces of an $n$-cube which can be depicted in pure $-, 0, +$ notation as

$$\begin{array}{ccc}
\alpha - \beta - \gamma & \xrightarrow{\ \alpha\, 0\, \beta - \gamma\ } & \alpha + \beta - \gamma \\
{\scriptstyle \alpha - \beta\, 0\, \gamma}\downarrow & \overset{\alpha\, 0\, \beta\, 0\, \gamma}{\Rightarrow} & \downarrow{\scriptstyle \alpha + \beta\, 0\, \gamma} \\
\alpha - \beta + \gamma & \xrightarrow[\ \alpha\, 0\, \beta + \gamma\ ]{} & \alpha + \beta + \gamma
\end{array}$$

or in "position of 0" notation as

$$\begin{array}{ccc}
\alpha - \beta - \gamma & \xrightarrow{\ u\ } & \alpha + \beta - \gamma \\
{\scriptstyle v}\downarrow & \overset{(u,\ v)}{\Rightarrow} & \downarrow{\scriptstyle v} \\
\alpha - \beta + \gamma & \xrightarrow[\ u\ ]{} & \alpha + \beta + \gamma
\end{array}$$

where $u = \ell(\alpha) + 1,\ v = u + \ell(\beta) + 1$. Notice in the last square that

$$\mathcal{V}(vu) = \varnothing \subset \{(u,v)\} = \mathcal{V}(uv).$$

Write $\mathbf{I}[n,2]$ for the sub-derivation scheme of $\uparrow \mathbf{I}$ obtained by taking only the objects $\alpha$ of length $n$.

The *commuting* 3*-face relations* are the following relations on 2-cells in the free 2-category $\mathbf{F} \uparrow \mathbf{I}$: for each object $\alpha$ of $\uparrow \mathbf{I}$ with the symbol – in positions $u < v < w$,

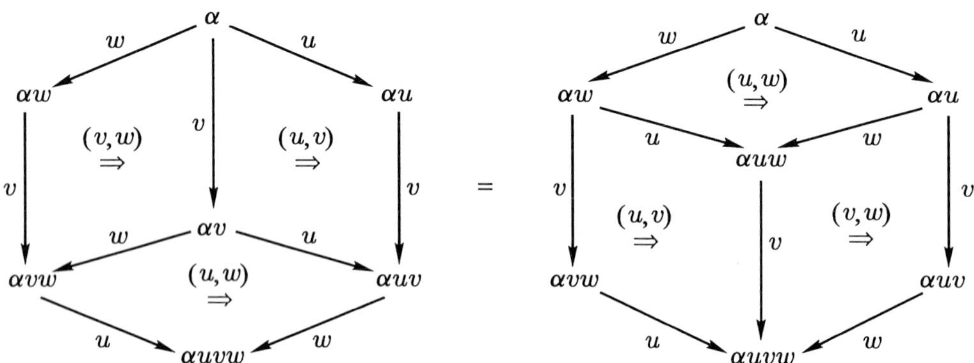


where $\alpha u$ denotes the result of changing $-$ to $+$ in position $u$ of $\alpha$.

There is a 2-category $\mathbf{Cub}[n,2]$ defined as follows. The objects are words $\alpha$ of length $n$ in the symbols $-, +$. For $\alpha \leqslant \beta$, the homcategory $\mathbf{Cub}[n,2](\alpha,\beta)$ is the ordered set of listings $u = u_1 u_2 \ldots u_k$ of the elements of $\alpha \backslash \beta$ where $u \leqslant u'$ if and only if $\mathcal{V}u \subseteq \mathcal{V}u'$. Otherwise, $\mathbf{Cub}[n,2](\alpha,\beta) = \varnothing$. Horizontal composition

$$\mathbf{Cub}[n,2](\alpha,\beta) \times \mathbf{Cub}[n,2](\beta,\gamma) \to \mathbf{Cub}[n,2](\alpha,\gamma)$$

is concatenation of listings which is order preserving (by the formula for $\mathcal{V}(uv)$).

PROPOSITION 6.1. *A presentation of the* 2*-category* $\mathbf{Cub}[n,2]$ *is provided by the computad* $\mathbf{I}[n,2]$ *subject to the commuting* 3*-face relations.*

Consequently, the 2-category $\mathbf{Cub}[n,2]$ is called the *$n$-cube with commutative* 3*-faces*. This 2-category was given in terms of generators and relations by Gray [Gy2] who used the positive part of the braid groups to show its homcategories were ordered (strong Bruhat order of the symmetric groups). To make a connection here with positive braids notice that the string diagrams for the commuting 3-face relations are as follows provided we depict the nodes as crossovers. (More will be said on this in later sections.)

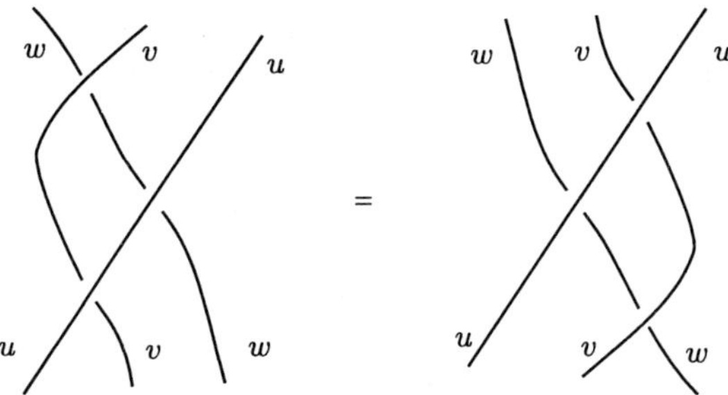


The cube 2-categories arose in Gray's work in order to prove that his *tensor product of* 2-*categories* was a monoidal structure on the category **2-Cat**. (That is, that the tensor product is associative up to isomorphisms which satisfy certain axioms.) This tensor product

$$\otimes\colon \mathbf{2\text{-}Cat} \times \mathbf{2\text{-}Cat} \to \mathbf{2\text{-}Cat}$$

is not the product in the category **2-Cat**. One way to construct it is to first define it on the cube 2-categories by putting

$$\mathbf{Cub}[m, 2] \otimes \mathbf{Cub}[n, 2] = \mathbf{Cub}[m + n, 2].$$

Then we need to observe:

PROPOSITION 6.2. *The full subcategory of* **2-Cat** *consisting of the* 2-*categories* $\mathbf{Cub}[n, 2]$, $n = 0, 1, 2, \ldots$, *is dense.* (*In fact,* $\mathbf{Cub}[3, 2]$ *alone suffices.*)

This means that every 2-category $A$ is a canonical colimit

$$A \cong \operatorname{colim}_i \mathbf{Cub}[m_i, 2]$$

of cube 2-categories. Since we wish the functors $A \otimes -$, $- \otimes B$: **2-Cat** $\to$ **2-Cat** to preserve colimits, we are forced to the formula

$$A \otimes B \cong \operatorname{colim}_{i,j} \mathbf{Cub}[m_i + n_j, 2].$$

The fact that this approach leads to a biclosed monoidal structure on **2-Cat** follows from a general result of Day [Da2, Da3] on Kan extending tensor products along dense functors. Moreover, the **2-Cat**-valued homs, which provide right adjoints for $A \otimes -$ and $- \otimes B$, are easily described as "funny 2-functor 2-categories".

Before describing these, it is worth looking at the situation with the category **Cat**. Any category **V** with products becomes a monoidal category by taking the product as the tensor product; this is called the *cartesian monoidal structure* on **V**. Call **V** *cartesian closed* when, for all objects $B, C$ of **V**, there exists an *exponential object* $[B, C]$

characterized up to isomorphism by the existence of a natural bijection between arrows $A \times B \to C$ and arrows $A \to [B, C]$. In the case $\mathbf{V} = \mathbf{Cat}$, of course, $[B, C]$ is the *functor category* whose objects are functors from $A$ to $B$ and whose arrows are natural transformations. Categories with homs enriched in the cartesian closed category **Cat** are precisely 2-categories.

However, for categories $B$, $C$, there is also the *funny functor category* $\{B, C\}$ whose objects are functors $f\colon B \to C$, and whose arrows $\theta\colon f \to g$ are families of arrows $\theta_b\colon f(b) \to g(b)$ in $C$ indexed by the objects $b \in B$ (no naturality requirement!). There is a *funny tensor product* $A \otimes B$ of categories $A$, $B$ such that functors $h\colon A \otimes B \to C$ are in natural bijection with functors $k\colon A \to \{B, C\}$. In fact, a category with homs enriched in the monoidal category **Cat** with the funny tensor product is more general than a 2-category; it is precisely a sesquicategory. (The funny tensor product was used recently [BG1, BG2] in studying Petri nets.)

The category **2-Cat** is cartesian closed. For 2-categories $B, C$, the exponential 2-category $[B, C]$ has 2-functors as objects, 2-natural transformations as arrows, and *modifications* as 2-cells. A category with homs enriched in **2-Cat**, with the cartesian structure, is called a 3-*category*.

For 2-categories $B$, $C$, the *funny* 2-*functor* 2-*category* $\{B, C\}$ has 2-functors $f\colon B \to C$ as objects, transformations $\theta\colon f \to g$ as arrows, and modifications as 2-cells (this terminology will be discussed in Section 9 in the context of bicategories). There is a natural bijection between 2-functors $h\colon A \otimes B \to C$ (where $\otimes$ is Gray's tensor product of 2-categories) and 2-functors $k\colon A \to \{B, C\}$. So $\{B, -\}$ is a right adjoint for $- \otimes B$. A right adjoint for $A \otimes -$ is obtained using the canonical isomorphism

$$(A \otimes B)^{\mathrm{co}} \cong B^{\mathrm{co}} \otimes A^{\mathrm{co}}$$

which can be seen for cubes and extended by taking colimits.

A category with homs enriched in **2-Cat**, with Gray's tensor product, we call a *Gray-category*: roughly speaking, this is a sesquicategory $X$ with each homcategory $X(x, y)$ equipped with a 2-category structure, whose 2-cells are called 3-*cells* of $X$, such that the squares (HC) have 3-cells in them, subject to appropriate axioms. Gray-categories are more general than 3-categories. In unpublished work of A. Joyal and M. Tierney, suitable algebraic models for homotopy 3-types are found to be Gray-categories in which all 1-cells, 2-cells and 3-cells are invertible.

For more details on the Gray tensor product, the interested reader should consult [Gy1, Gy2]; and, for "strong" Gray-categories, see [GPS].

## 7. Higher dimensions and parity complexes

Returning to cubes, we consider the case where the 3-faces do not commute. We consider the derivation scheme $\mathbf{I}[n, 3]$ given by

$$\begin{aligned} \mathbf{s}_1, \mathbf{t}_1\colon \{&\text{words } \theta \text{ of length } n \text{ in the symbols } -, 0, + \\ &\text{with precisely three 0's}\} \to \langle\!\langle \mathbf{FI}[n, 2] \rangle\!\rangle \end{aligned}$$

where $\mathbf{s}_1(\theta)$ = left hand side of the commuting 3-face condition, and $\mathbf{t}_1(\theta)$ = right hand side of commuting 3-face condition, for $\theta$ with 0's in the positions $u, v, w$. From the rewrite viewpoint, the words $\theta$ give a directed distinction between confluence checks beginning with the word obtained from $\theta$ by replacing the three 0's by $-$'s. Recall that the category $\langle\langle \mathbf{FI}[n, 2] \rangle\rangle$ has the arrows of $\mathbf{FI}[n, 2]$ as objects and has the 2-cells as arrows. While $\mathbf{FI}[n, 2]$ is a free 2-category and $\langle \mathbf{FI}[n, 2] \rangle$ is a free category, the category $\langle\langle \mathbf{FI}[n, 2] \rangle\rangle$ is not free. So $\mathbf{I}[n, 3]$ is *not* a computad. It is really a "3-computad".

A 3-*computad* $E$ (where we rename graphs as "1-computads", and computads as "2-computads") is a computad $E^{\#}$ together with a derivation scheme

$$\mathbf{s}_2, \mathbf{t}_2\colon\ E_3 \to \langle\langle \mathbf{F}E^{\#} \rangle\rangle;$$

elements of $E_3$ are called 3-*cells of* $E$. A 3-*computad morphism* $E \to E'$ is a computad morphism $E^{\#} \to E'^{\#}$ together with a derivation scheme morphism for which the functor $\langle\langle \mathbf{F}E^{\#} \rangle\rangle \to \langle\langle \mathbf{F}E'^{\#} \rangle\rangle$ is induced by $E^{\#} \to E'^{\#}$. Each 3-computad $E$ determines a *free* 3-*category* $\mathbf{F}E$. A *presentation of a* 3-*category* is a 3-computad together with a set of relations between 3-cells in $\mathbf{F}E$.

Here is an example of a 3-computad with one 3-cell called 0 0 0.

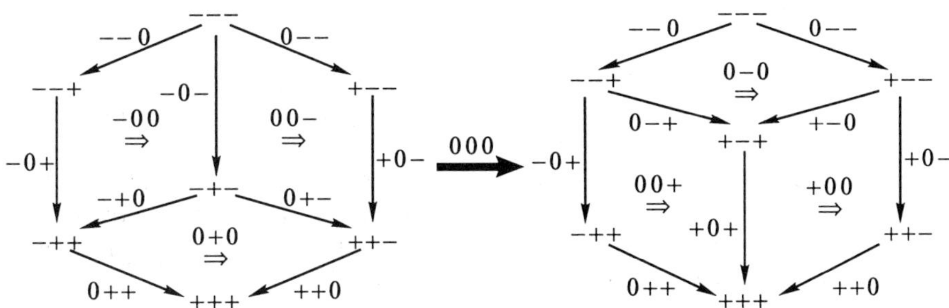


Each 3-cell $\theta \in E_3$ of a 3-computad $E$ determines two 2-cells $\mathbf{s}_2(\theta)$, $\mathbf{t}_2(\theta)$ in the free 2-category $\mathbf{F}E^{\#}$. These 2-cells can be represented by string diagrams in the computad $E^{\#}$. Write $\theta^-$ for the *set* of 2-cells of $E^{\#}$ which label the nodes of a string diagram for $\mathbf{s}_2(\theta)$, and write $\theta^+$ for the *set* of 2-cells of $E^{\#}$ which label the nodes of a string diagram for $\mathbf{t}_2(\theta)$. (These sets are independent of the choices of string diagrams in the deformation classes.) So we have two functions

$$(-)^-, (-)^+\colon\ E_3 \to \mathbf{P}(E_2)$$

where $\mathbf{P}(E_2)$ is the power set of the set of 2-cells of $E^{\#}$. In considering only the labels on nodes of a string diagram, we are, in general, disregarding quite a lot of information about the string diagram. Hence, it is a perhaps surprising consequence of the work of [S5, A1, J, S6, ASn, Pw1, Pw2] that we have:

PROPOSITION 7.1. *For 3-computads $E$ arising from many convex polytopes such as $\mathbf{I}[n,3]$ arising from cubes, the functions $\mathbf{s}_2, \mathbf{t}_2\colon E_3 \to \langle\!\langle \mathbf{F}E^{\#} \rangle\!\rangle$ are uniquely determined by the functions $(-)^-, (-)^+\colon E_3 \to \mathbf{P}E_2$.*

At the lower dimension, the corresponding result is easily understood. For, suppose $C$ is a (2-)computad. Then, for each 2-cell $u \in C_2$, we have paths $\mathbf{s}_1(u)$, $\mathbf{t}_1(u)$, and we can write $u^-$, $u^+$ for the *sets* of 1-cells of the graph $C^{\#}$ which occur in the respective paths. Provided the graph $C^{\#}$ has no circuits, the only other information we need to reconstruct the paths from the set is the order. However, the order is forced by knowledge of the functions $\mathbf{s}_0, \mathbf{t}_0\colon C_1 \to C_0$. So the 2-dimensional version of Proposition 7.1 is true. To be consistent at even the lowest dimension, we can define $\alpha^- = \{a\}$, $\alpha^+ = \{b\}$ for each 1-cell $\alpha\colon a \to b$ of $C$.

In this way, each $n$-computad $E$ leads to a graded set $E_k$, $0 \leqslant k \leqslant n$, together with functions $(-)^-, (-)^+\colon E_k \to \mathbf{P}(E_{k-1})$, $0 < k \leqslant n$. This is the basic structure involved in the higher-dimensional combinatorial notion of circuit-free graph called *parity complex* [S6, S8]; however, a parity complex is to satisfy some axioms which are not true of all such structures underlying $n$-computads. The axiom which somewhat reflects the source-target equations in a computad is, for all cells $x$ of dimension $\geqslant 2$, the equality of sets

$$x^{--} \cup x^{++} = x^{-+} \cup x^{+-},$$

where the unions are disjoint, and, for example, $S^-$ is the union of the sets $x^-$, $x \in S$, for any $S \subset E_k$. The main result of [S6] is the construction of the free $n$-category on an $n$-computad which is uniquely determined by the parity complex.

Following Aitchison's ideas [A2] for cubes and simplexes, we note that it is possible to use string-like diagrams to keep track of facial relations in consecutive dimensions of a parity complex. Specifically, suppose we have disjoint finite sets $M$, $X$ and functions $(-)^-, (-)^+\colon M \to \mathbf{P}(X)$ such that, for all $m \neq n$ in $M$,

$$(m^- \cap n^-) \cup (m^+ \cap n^+) = \varnothing.$$

Put

$$\partial M = \{(-,x)\colon\ x \in X,\ x \notin M^+\} \cup \{(+,x)\colon\ x \in X,\ x \notin M^-\}.$$

Then there is a graph $s, t\colon X \to M \cup \partial M$ given by

$$x \in s(x)^- \cap t(x)^+ \quad \text{for } x \in M^- \cap M^+, \quad s(x) = (-,x) \quad \text{for } x \notin M^+,$$

and

$$t(x) = (+,x) \quad \text{for } x \notin M^-.$$

There is no reason why such a graph should be planar; however, we do draw it in the plane, with edges directed down, sometimes crossing at non-nodes, with each *inner node*

$m \in M$ labeled by $m$, with each *outer node* in $\partial M$ left undistinguished, and with each edge $x \in X$ labeled by $x$.

Returning to cubes, we look at the 3-computad $\mathbf{I}[4,3]$. The set $\mathbf{I}[4,3]_k$ of $k$-cells contains the words of length 4 in the symbols $-, 0, +$ where the symbol 0 occurs precisely $k$ times. In particular,

$$\mathbf{I}[4,3]_3 = \{-000, 0-00, 00-0, 000-, +000, 0+00, 00+0, 000+\}$$

and the parity complex structure is recorded by the string-like diagrams of [A2] as shown below.

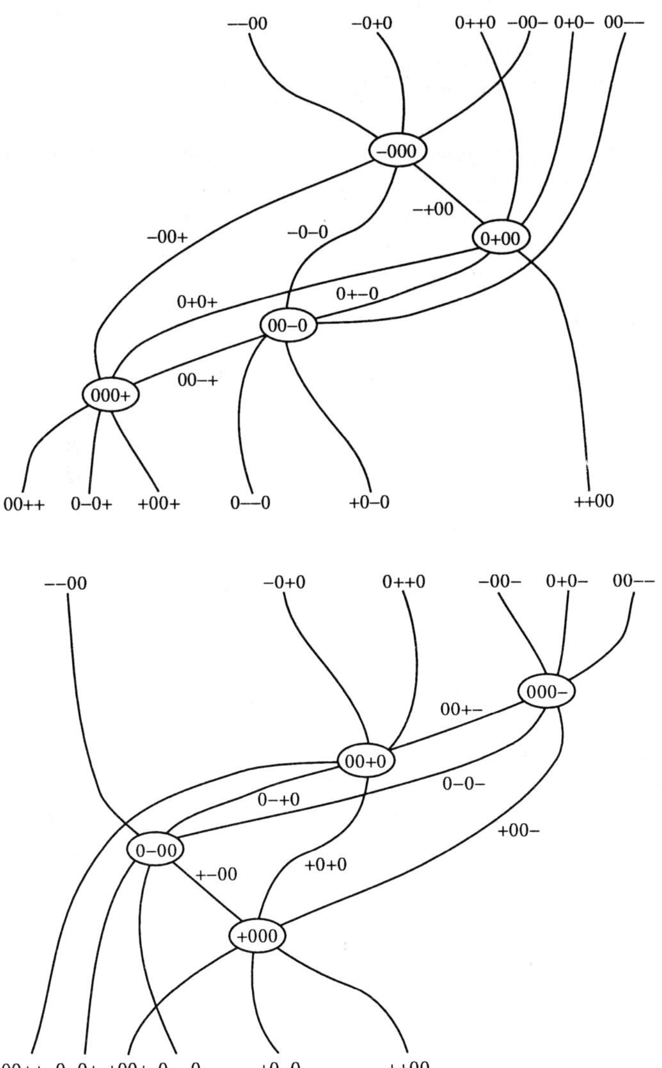

By Proposition 7.1, each of these string-like diagrams represents a 2-cell in the 2-category $\mathbf{FI}[4,2]$. The *commuting* 4-*face relation* is the equality between these two 2-cells. The 3-computad $\mathbf{I}[4,3]$ together with the commuting 4-face relation provides a presentation of a 3-category $\mathbf{Cub}[4,3]$.

There is an explicit description of the free $m$-category on an $m$-dimensional parity complex in [S6]. In particular, there is a combinatorial model for $\mathbf{FI}[n,m]$. Except in the case $m = 2$, as described above, I do not know of a combinatorial model for the $n$-cube $\mathbf{Cub}[n,m]$ with commuting $(m+1)$-faces. Of course, we do have a presentation of the $m$-category $\mathbf{Cub}[n,m]$ (as the $m$-computad $\mathbf{I}[n,m]$ and the commuting $m$-face relations), and this suffices for many purposes.

## 8. The Yang–Baxter and Zamolodchikov equations

In this section we study the connection between categories and the so-called "Bazhanov–Stroganov $d$-simplex equations" which have arisen in statistical and quantum mechanics. We discuss here only the algebraic generic forms of these equations as found in [MN1], [MN2] where other references are provided and some of the physical significance is explained (also see [Dr1, T, JS3, JS4, Dr2, Z]):

$d = 1$ *Matrix commutativity*

$$A_i^k B_k^j = B_i^k A_k^i,$$

$d = 2$ *Yang–Baxter equation*

$$A_{i_1 i_2}^{k_1 k_2} B_{k_1 i_3}^{j_1 k_3} C_{k_2 k_3}^{j_2 j_3} = C_{i_2 i_3}^{k_2 k_3} B_{i_1 k_3}^{k_1 j_3} A_{k_1 k_2}^{j_1 j_2},$$

$d = 3$ *Zamolodchikov equation*

$$A_{i_1 i_2 i_3}^{k_1 k_2 k_3} B_{k_1 i_4 i_5}^{j_1 k_4 k_5} C_{k_2 k_4 i_6}^{j_2 j_4 k_6} D_{k_3 k_5 k_6}^{j_3 j_5 j_6} = D_{i_3 i_5 i_6}^{k_3 k_5 k_6} C_{i_2 i_4 k_6}^{k_2 k_4 j_6} B_{i_1 k_4 k_5}^{k_1 j_4 j_5} A_{k_1 k_2 k_3}^{j_1 j_2 j_3}.$$

In these equations, observe that the subscript on a given subscript is the same as the subscript on the superscript directly above it. Also, superscripts are all $j$'s and $k$'s while subscripts are all $k$'s and $i$'s; in each case, there is a string of one letter followed by a string of the other. So the information in the equations can be recorded schematically as follows:

$$d = 1: \qquad (*1)(1*) = (1*)(*1),$$

$$d = 2: \qquad (*12)(1*3)(23*) = (23*)(1*3)(*12),$$

$$d = 3: \qquad (*123)(1*45)(24*6)(356*) = (356*)(24*6)(1*45)(*123).$$

The symbol $*$ indicates where the letter change-over occurs. The pattern here is made clear by recording the bracketed terms on each side as rows of a matrix; this gives the formal matrix identities:

$$\begin{bmatrix} * & 1 \\ 1 & * \end{bmatrix} = \begin{bmatrix} 1 & * \\ * & 1 \end{bmatrix}$$

$$\begin{bmatrix} * & 1 & 2 \\ 1 & * & 3 \\ 2 & 3 & * \end{bmatrix} = \begin{bmatrix} 2 & 3 & * \\ 1 & * & 3 \\ * & 1 & 2 \end{bmatrix}$$

$$\begin{bmatrix} * & 1 & 2 & 3 \\ 1 & * & 4 & 5 \\ 2 & 3 & * & 6 \\ 4 & 5 & 6 & * \end{bmatrix} = \begin{bmatrix} 4 & 5 & 6 & * \\ 2 & 3 & * & 6 \\ 1 & * & 4 & 5 \\ * & 1 & 2 & 3 \end{bmatrix}$$

So the Bazhanov–Stroganov 4-simplex equation can be reconstructed from the formal matrix identity:

$$\begin{bmatrix} * & 1 & 2 & 3 & 4 \\ 1 & * & 5 & 6 & 7 \\ 2 & 5 & * & 8 & 9 \\ 3 & 6 & 8 & * & 10 \\ 4 & 7 & 9 & 10 & * \end{bmatrix} = \begin{bmatrix} 4 & 7 & 9 & 10 & * \\ 3 & 6 & 8 & * & 10 \\ 2 & 5 & * & 8 & 9 \\ 1 & * & 5 & 6 & 7 \\ * & 1 & 2 & 3 & 4 \end{bmatrix}$$

In fact, for building up these equations dimension by dimension and for dealing with the nonsquare matrices corresponding to the other entries in Aitchison's Pascal triangle of string-like diagrams, it is more convenient to renumber the strings so that the 4-simplex equation, in matrix form, becomes:

$$\begin{bmatrix} * & 1 & 2 & 4 & 7 \\ 1 & * & 3 & 5 & 8 \\ 2 & 3 & * & 6 & 9 \\ 4 & 5 & 6 & * & 10 \\ 7 & 8 & 9 & 10 & * \end{bmatrix} = \begin{bmatrix} 7 & 8 & 9 & 10 & * \\ 4 & 5 & 6 & * & 10 \\ 2 & 3 & * & 6 & 9 \\ 1 & * & 3 & 5 & 8 \\ * & 1 & 2 & 4 & 7 \end{bmatrix}$$

These formal matrices are also related to the numerical matrices for which the vanishing determinant condition [MN2] gives the dependence of the $d(d+1)/2$ parameters in the parameterized version of the $d$-simplex equation.

Referring to the $A, B, C, \ldots$ form of the equations, Ian Aitchisom observed (1990) that the Penrose diagrams (in the sense of [PR]) for these tensor equations occurred in his "Pascal's triangle" of string-like diagrams [A2] associated with the oriented $d$-cubes (*not* the $d$-simplexes). For $d = 2$, this reflects the well-known connections between the Yang–Baxter equation, the Coxeter relations for the symmetric groups, and paths around the edges of a cube. It should be recalled here that the ordering of the strings into, and out of, nodes is ignored (as usual with parity complexes and with Penrose notation).

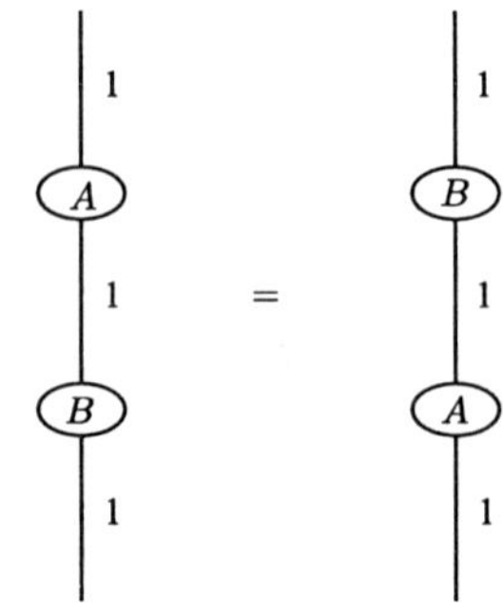


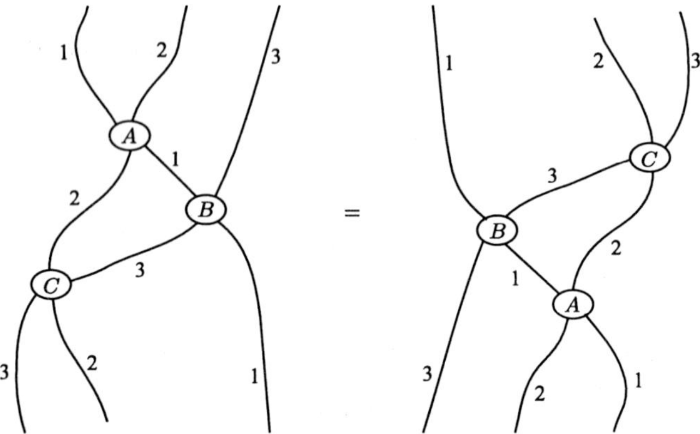


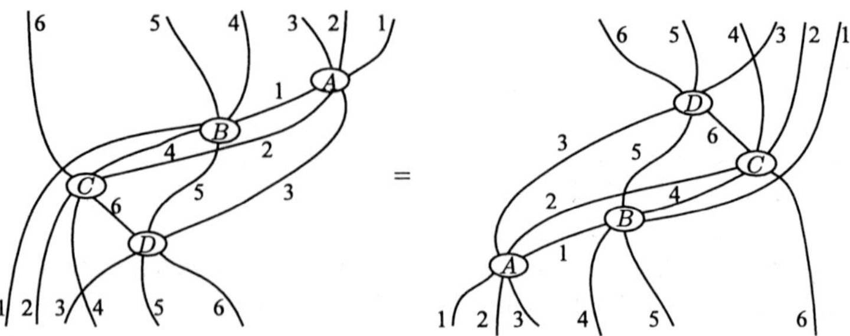


Comparison with the string-like diagrams of Section 7 shows that the $d$-simplex equation is allied with the commuting $(d+1)$-cube.

It is possible to interpret the $d$-simplex equation as a morphism from a categorical structure constructed from geometry to a categorical structure of the same kind constructed from algebra. In particular, consider the Yang–Baxter equation ($d = 2$).

On the geometric side, recall that we have the derivation scheme $\mathbf{I}[n, 2]$ which involves the 2-dimensional faces of the $n$-cube; this gives the free 2-category $\mathbf{FI}[n, 2]$.

On the algebraic side, we would like to consider a 2-category $\Sigma\mathbf{Vect}_{\mathbf{k}}$ whose only object is a field $\mathbf{k}$ whose arrows $V\colon \mathbf{k} \to \mathbf{k}$ are finite-dimensional vector spaces over $\mathbf{k}$ and whose 2-cells $t\colon V \Rightarrow W\colon \mathbf{k} \to \mathbf{k}$ are linear functions $t\colon V \to W$. However, we want the composition of arrows to be tensor product of vector spaces which is not strictly associative. This really provides an example of a "bicategory" which is the subject of the next section. For our present purposes, this problem can be avoided by using matrices instead of linear functions. More precisely, let $\Sigma\mathbf{Mat}_{\mathbf{k}}$ denote the 2-category with one object $\mathbf{k}$ whose arrows are natural numbers and whose 2-cells $A\colon m \Rightarrow n\colon \mathbf{k} \to \mathbf{k}$ are $m \times n$ matrices $A = (a_{ij})$; the vertical composition is usual multiplication of matrices, while the horizontal composite of $A\colon m \Rightarrow n$, $B\colon r \Rightarrow s$ is their Kronecker product $A \otimes B = (a_{ij}b_{pq})\colon mr \Rightarrow ns$.

Now suppose $R\colon mm \Rightarrow mm$ is a 2-cell in $\Sigma\mathbf{Mat}_{\mathbf{k}}$. We can extend this to a 2-functor

$$R^\wedge\colon\ \mathbf{FI}[n, 2] \to \Sigma\mathbf{Mat}_{\mathbf{k}}$$

determined by the following assignment.

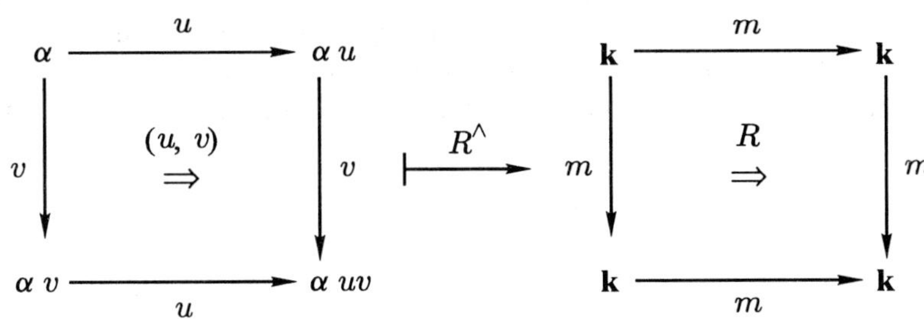


The matrix $R$ is called a *Yang–Baxter matrix* when it is invertible and the 2-functor $R^\wedge$ identifies the commuting 3-face relations for some (and hence all) $n \geqslant 3$. It should be clear now how such a matrix $R$ provides a solution to the Yang–Baxter equation. There is an induced 2-functor $R^\wedge\colon \mathbf{Cub}[n, 2] \to \Sigma\mathbf{Mat}_{\mathbf{k}}$.

Now consider applying the same ideas to the Zamolodchikov equation. On the geometric side there is no problem since we have the free 3-category $\mathbf{FI}[n, 3]$. A small difficulty arises on the algebraic side when we try to push the category of vector spaces up another dimension. This time we would like to consider a 3-category $\Sigma^2\mathbf{Vect}_{\mathbf{k}}$ whose only object is a field $\mathbf{k}$ whose only arrow is the identity of $\mathbf{k}$ whose 2-cells $V$ are finite-dimensional vector spaces over $\mathbf{k}$ and whose 3-cells $t\colon V \to M$ are linear functions. This time two of the compositions are to be tensor product with the third taken to be composition of linear functions, as before. The problem of nonstrictness of associativity of tensor product can be avoided as before by using matrices, however, now we also require the middle-four-interchange law:

$$(U \otimes V) \otimes (W \otimes X) = (U \otimes W) \otimes (V \otimes X)$$

which of course does not strictly hold; there is only a canonical isomorphism in place of the equality. This problem cannot be avoided. In fact, $\Sigma^2\mathbf{Vect}_\mathbf{k}$ is an example of a *tricategory* in the sense of [GPS]. Using matrices, we obtain a Gray-category $\Sigma^2\mathbf{Mat}_\mathbf{k}$. (It is shown in [GPS] that, more generally, every tricategory is "triequivalent" to a Gray-category.) As we mentioned at the end of Section 6, every 3-category is a Gray-category. It is therefore meaningful to consider Gray-functors from a 3-category to a Gray-category. In particular, each $m^3 \times m^3$ matrix $R$ induces such a Gray-functor

$$R^\wedge\colon\ \mathbf{Fl}[n,3] \to \Sigma^2\mathbf{Mat}_\mathbf{k}$$

we call $R$ a *Zamolodchikov matrix* when it is invertible and $R^\wedge$ identifies the commuting 4-face relations for some (and hence all) $n \geqslant 4$. Such a matrix $R$ provides a solution to the Zamolodchikov equation.

Higher dimensions offer no new problems. For the $d$-simplex equation, there is an appropriate structure $\Sigma^{d-1}\mathbf{Mat}_\mathbf{k}$ with precisely one $i$-cell for each $i \leqslant d-2$, whose $(d-1)$-cells are natural numbers, whose $d$-cells are matrices, whose first $d-1$ compositions are Kronecker product (among which the middle-four-interchange law holds only up to a coherent invertible $d$-cell), and whose remaining composition is usual matrix product (which strictly satisfies the middle-four-interchange law with each earlier composition). A *d-simplex matrix* is an invertible $m^d \times m^d$ matrix $R$ which induces a structure-preserving morphism

$$R^\wedge\colon\ \mathbf{Cub}[d+1,d] \to \Sigma^{d-1}\mathbf{Mat}_\mathbf{k}.$$

## 9. Bicategories

Bicategories (and the appropriate 3-graph with bicategories as 0-cells) were first defined by Bénabou [Bn1, Bn2].

A *bicategory* **B** is a 2-graph equipped with the following extra structure:

(Ba) for each pair of objects $a, b$, a category structure on the graph $\mathbf{B}(a,b)$ with composition called *vertical* and denoted by $\bullet$ ("invertibility" for 2-cells will mean with respect to this composition);

(Bb) for each triple of objects $a, b, c$ a functor

$$\circ\colon\ \mathbf{B}(a,b) \times \mathbf{B}(b,c) \to \mathbf{B}(a,c),$$

called the *horizontal composition* and written between the arguments;

(Bc) for each object $a$, an arrow $1_a\colon\ a \to a$ called the *identity* for $a$;

(Bd) invertible 2-cells

$$a_{\alpha,\beta,\gamma}\colon\ \alpha\circ(\beta\circ\gamma)\Rightarrow(\alpha\circ\beta)\circ\gamma\colon\ a\to d,$$

called *associativity constraints*, which are natural in $\alpha,\beta,\gamma\in\mathbf{B}(a,b)\times\mathbf{B}(b,c)\times\mathbf{B}(c,d)$;

(Be) invertible 2-cells

$$l_\alpha\colon\ 1_a\circ\alpha\Rightarrow\alpha\colon\ a\to b,\quad r_\alpha\colon\ \alpha\circ 1_b\Rightarrow\alpha\colon\ a\to b,$$

called *identity constraints*, which are natural in $\alpha\in\mathbf{B}(a,b)$;

subject to the following commutativity conditions:

(B1) *pentagon for associativity constraints*

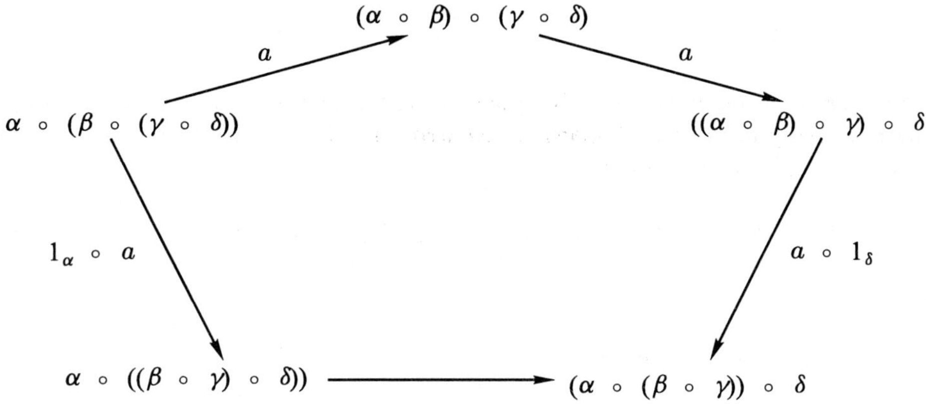


(B2) *triangle for identity constraints*

$$\alpha\circ(1_b\circ\beta)\xrightarrow{\ a\ }(\alpha\circ 1_b)\circ\beta$$

$$1_\alpha\circ l\colon\ \alpha\circ(1_b\circ\beta)\to\alpha\circ\beta,\qquad r\circ 1_\beta\colon\ (\alpha\circ 1_b)\circ\beta\to\alpha\circ\beta$$

EXAMPLE 1. Let $A$ be a category equipped with a choice of pushouts for each pair of arrows with common source. There is a bicategory $\operatorname{Cospn} A$ defined as follows. The objects are the objects of $A$. For $a,b\in A$, the category $(\operatorname{Cospn} A)(a,b)$ has as objects triples $(\rho_0,t,\rho_1)$, called *cospans from $a$ to $b$*, consisting of an object $r$ and arrows $\rho_0\colon\ a\to r$, $\rho_1\colon\ b\to r$ of $A$; an arrow $\phi\colon\ (\rho_0,r,\rho_1)\to(\sigma_0,s,\sigma_1)$ of cospans is an arrow $\phi\colon\ r\to s$ in $A$ such that

$$\rho_0\circ\phi=\sigma_0,\quad \rho_1\circ\phi=\sigma_1.$$

Given $(\rho_0, r, \rho_1) \in (\mathrm{Cospn}\, A)(a, b)$ and $(\sigma_0, s, \sigma_1) \in (\mathrm{Cospn}\, A)(b, c)$, define

$$(\rho_0, r, \rho_1) \circ (\sigma_0, s, \sigma_1) = (\rho_0 \circ \omega_0, p, \sigma_1 \circ \omega_1)$$

where the square

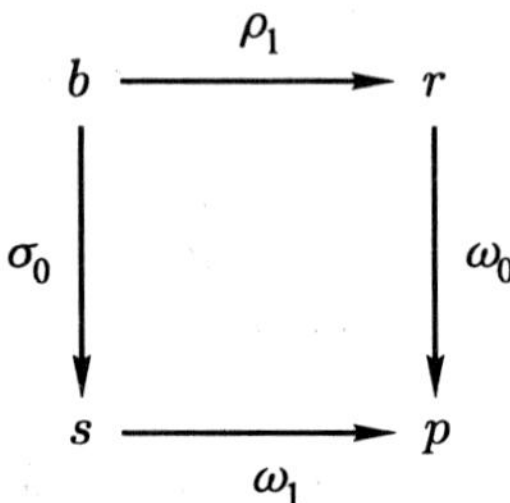


is the selected pushout of $\rho_0, \sigma_1$. Using the universal property of pushout, we can extend this functorially to arrows of spans as required for (Bb). The identity cospan for a is $1_a = (1_a, a, 1_a)$. Given cospans

$$(\rho_0, r, \rho_1) \in (\mathrm{Cospn}\, A)(a, b), (\sigma_0, s, \sigma_1) \in (\mathrm{Cospn}\, A)(b, c)$$

and

$$(\tau_0, t, \tau_1) \in (\mathrm{Cospn}\, A)(c, d),$$

we can form the diagram

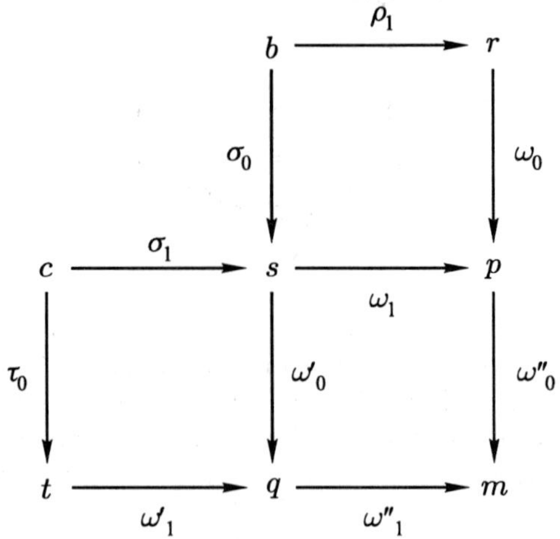

of selected pushouts. Each of the cospans

$$(\rho_0, r, \rho_1) \circ ((\sigma_0, s, \sigma_1) \circ (\tau_0, t, \tau_1)), \quad ((\rho_0, r, \rho_1) \circ (\sigma_0, s, \sigma_1)) \circ (\tau_0, t, \tau_1)$$

is canonically isomorphic to $(\rho_0 \circ \omega_0 \circ \omega_0'', m, \rho_1 \circ \omega_1 \circ \omega_1'')$, and so, to each other, yielding the associativity constraints. There are also canonical isomorphisms

$$1_a \circ (\rho_0, r, \rho_1) \cong (\rho_0, r, \rho_1) \cong (\rho_0, r, \rho_1) \circ 1_b$$

yielding the identity constraints. To check commutativity of (B1), (B2), it suffices to check after composition with the coprojections $\omega_0, \omega_1$ into the appropriate pushouts, and we recommended this as an exercise.

EXAMPLE 2. A monoidal (= "tensor") category **V** (in the sense of [EK]) can be defined to be a bicategory **B** with one object. More precisely, if **V** is a monoidal category then a bicategory with only one object $a$ is defined by $\mathbf{B}(a, a) = \mathbf{V}$; the tensor product of **V** is the horizontal composition $\circ$ of **B**. While, if a is any object of a bicategory **B** then $\mathbf{B}(a, a)$ becomes a monoidal category. For many purposes it is convenient to distinguish **V** from the one-object **B**; the notation $\Sigma\mathbf{V}$ for **B** is not bad.

EXAMPLE 3. A bicategory in which all the constraints are identities in a 2-category (Section 2). As each category $A$ can be regarded as a 2-category for which each category $A(a, b)$ is discrete, we can also regard categories as special bicategories.

EXAMPLE 4. There is a bicategory **Prof** which stands in relation to the 2-category **Cat** much as the category of sets and relations stands in relation to the category **Set** of sets and functions. The objects of **Prof** are categories. An arrow $M\colon A \to B$ is a *profunctor* (also called "distributor" [Bn2], "bimodule" [L], or just "module" [S3]); that is, a functor $M\colon A^{\mathrm{op}} \times B \to \mathbf{Set}$. The 2-cells $M \Rightarrow N$ are natural transformations between the functors. Composition of profunctors $M\colon A \to B$, $N\colon A \to B$ is given by the coend formula (see [ML1, ML2] for the history of "ends"):

$$(M \circ N)(a, c) = \int^b M(a, b) \times N(b, c).$$

Suppose $\mathbf{B}, \mathbf{X}$ are bicategories. A *lax functor* (also called "morphism of bicategories")

$$T\colon \mathbf{B} \to \mathbf{X}$$

is a 2-graph morphism which is functorial on vertical composition and is equipped with the following extra structure:

(LFa) for each object $a$ of **B**, a 2-cell $i_a\colon 1_{T(a)} \Rightarrow T(1_a)$ of **X**;

(LFb) 2-cells $m_{\alpha,\beta}\colon T(\alpha) \circ T(\beta) \Rightarrow T(\alpha \circ \beta)$ which are natural in $(\alpha, \beta) \in \mathbf{B}(a, b) \times \mathbf{B}(b, c)$; subject to the following commutativity conditions:

(LF1)

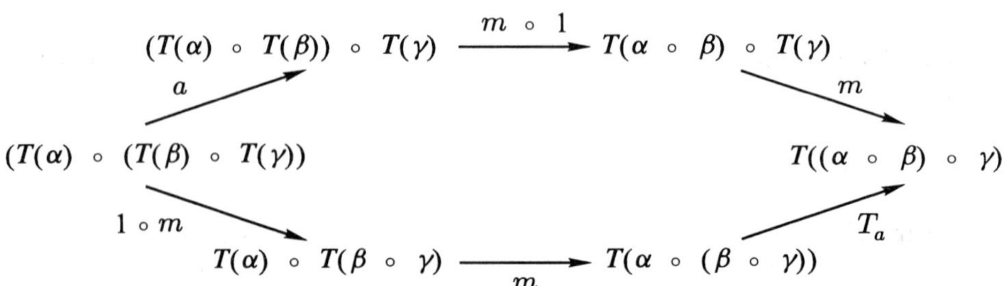


(LF2)

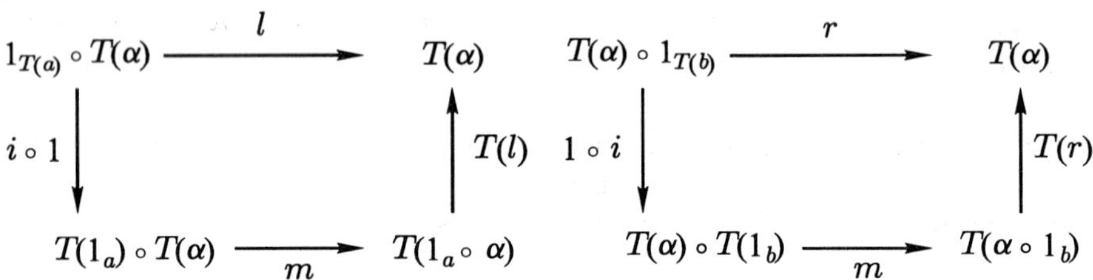


EXAMPLE 5. Suppose $F\colon A \to X$ is a functor between categories with selected pushouts. Then there is a lax functor $T = \text{Cospn}(F)\colon \text{Cospn}\, A \to \text{Cospn}\, X$ described as follows. Let $T$ take a general 2-cell $\phi\colon (\rho_0, r, \rho_1) \Rightarrow (\sigma_0, s, \sigma_1)\colon a \to b$ in Cospn $A$ to the 2-cell

$$F(\phi)\colon \big(F(\rho_0), F(r), F(\rho_1)\big) \Rightarrow \big(F(\sigma_0), F(s), F(\sigma_1)\big)\colon F(a) \to F(b)$$

in Cospn $X$. The 2-cells of (LFa) are identities. The universal property of pushouts in $X$ yields a canonical comparison arrow from the pushout of $F(\rho_0)$, $F(\sigma_1)$ to $F(p)$ (in the notation of Example 1). This gives the data for (LFb). The axioms (LF1), (LF2) are easily verified.

EXAMPLE 6. A monoidal functor $F\colon \mathbf{V} \to \mathbf{W}$ (in the sense of [EK]) amounts precisely to a lax functor $T\colon \Sigma\mathbf{V} \to \Sigma\mathbf{W}$ (see Example 2).

EXAMPLE 7. For bicategories $\mathbf{B}, \mathbf{X}$, a lax functor $T\colon \mathbf{B} \to \mathbf{X}$ is called a *pseudo functor* (also called "homomorphism" in [Bn1, Bn2]) when all the 2-cells

$$m_{\alpha,\beta}\colon T(\alpha) \circ T(\beta) \Rightarrow T(\alpha \circ \beta)$$

and $i_a$: $1_{T(a)} \Rightarrow T(1_a)$ are invertible. When these 2-cells are all identities, $T$ is called a *2-functor*; when **B**, **X** are both 2-categories (see Example 3) this agrees with the terminology in Section 2. It is perhaps of interest that, for any category $C$, pseudo functors $T$: $C^{\mathrm{op}} \to \mathbf{Cat}$ are equivalent, via the "Grothendieck construction", to functors $P$: $E \to C$ which are *fibrations*; in particular, when $C$ is a group (regarded as a category with one object and all arrows invertible), such a $T$ is a Schreier *factor system* as occur in group cohomology (for example, see [Gd]). A lax functor $T$: $\mathbf{B} \to \mathbf{X}$ is called *normalized* when all the 2-cells $i_a$: $1_{T(a)} \Rightarrow T(a_1)$ are identities. Jean Bénabou [Bn2] has shown how to construct, from *every* functor (not just fibrations!) $P$: $E \to C$, a normalized lax functor $T$: $C^{\mathrm{op}} \to \mathbf{Prof}$ (see Example 4); the Grothendieck construction generalizes to reverse this construction.

EXAMPLE 8. Let **1** denote the one-object discrete category. A lax functor $T$: $\mathbf{1} \to \mathbf{B}$ amounts to a monad in **B** (also see Section 5).

EXAMPLE 9. Lax functors can be composed in a fairly obvious way (which we leave to the reader) yielding a category **Bicat** whose objects are bicategories and whose arrows are lax functors.

EXAMPLE 10. Each object $k$ of a bicategory **B** determines a pseudo functor

$$H_k = \mathbf{B}(k, -)\colon\ \mathbf{B} \to \mathbf{Cat}$$

called the *pseudo functor represented by k*. The category $H_k(a)$ is $\mathbf{B}(k,a)$. The functor $H_k(\alpha)$: $\mathbf{B}(k,a) \to \mathbf{B}(k,b)$ is given by composing on the right with $\alpha$: $a \to b$. For each 2-cell $u$: $\alpha \Rightarrow \beta$, the natural transformation $H_k(u)$: $H_k(\alpha) \Rightarrow H_k(\beta)$ has component $H_k(u)_\varepsilon = \varepsilon \circ u$: $\varepsilon \circ \alpha \Rightarrow \varepsilon \circ \beta$ at $\varepsilon \in \mathbf{B}(k,a)$. The natural transformation

$$i_a\colon\ 1_{\mathbf{B}(k,a)} \Rightarrow - \circ 1_a$$

is provided by the inverse of the identity constraint $r$. The natural transformation

$$m_{\alpha,\beta}\colon\ H_k(\alpha) \circ H_k(\beta) \Rightarrow H_k(\alpha \circ \beta)$$

has component $(\varepsilon \circ \alpha) \circ \beta \to \varepsilon \circ (\alpha \circ \beta)$ at $\varepsilon \in \mathbf{B}(k,a)$ given by the inverse of the associativity constraint $a$. Axiom (LF1) is a pentagon since **Cat** is a 2-category and it amounts to axiom (B1) for **B**. We leave (LF2) as an exercise.

Suppose $S, T$: $\mathbf{B} \to \mathbf{X}$ are lax functors. A *transformation* $\theta$: $S \Rightarrow T$ consists of the following data:

(Ta) for each object $a$ of **B**, an arrow $\theta_a$: $S(a) \to T(a)$ of **X**;

(Tb) 2-cells $\theta_\alpha$: $S(\alpha) \circ \theta_b \Rightarrow \theta_a \circ T(\alpha)$ which are natural in $\alpha \in \mathbf{B}(a,b)$;

such that the following commutativity conditions hold:

(T1)

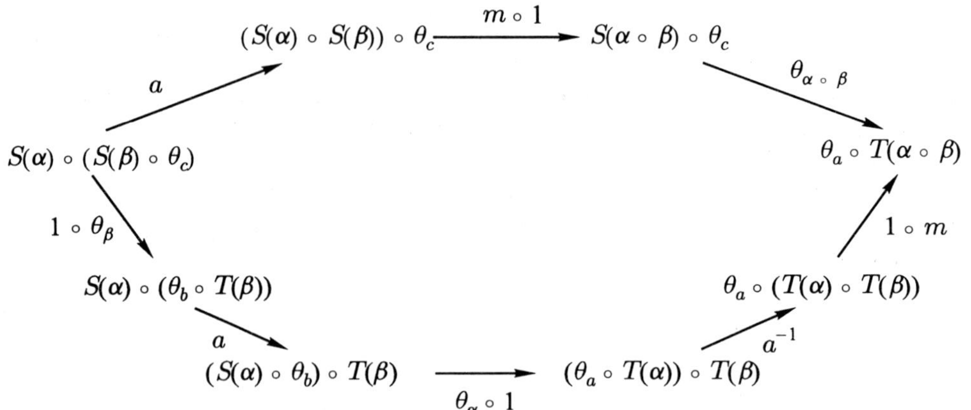


(T2)

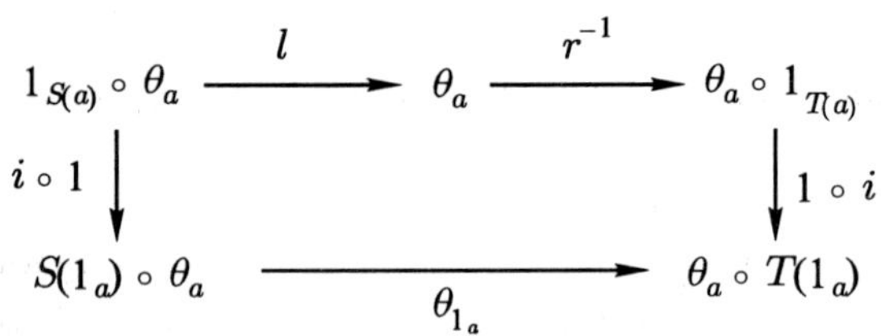


A transformation $\theta$: $S \Rightarrow T$ is called *strong* when each of the 2-cells $\theta_\alpha$: $S(\alpha) \circ \theta_b \Rightarrow \theta_a \circ T(\alpha)$ is invertible.

EXAMPLE 11. Suppose $\kappa$: $h \to k$ is an arrow of a bicategory $\mathbf{B}$. There is a strong transformation $\theta = H_\kappa$: $H_k \Rightarrow H_h$ whose component $\theta_a$: $\mathbf{B}(k,a) \to \mathbf{B}(h,a)$ is the functor given by composition on the left with $\kappa$, and whose natural isomorphism $\theta_\alpha$: $H_k(\alpha) \circ \theta_b \Rightarrow \theta_a \circ H_h(\alpha)$ has component at $\xi$: $k \to a$ given by the associativity constraint $a$: $\kappa \circ (\xi \circ \alpha) \to (\kappa \circ \xi) \circ \alpha$.

Suppose $\theta, \phi$: $S \Rightarrow T$: $\mathbf{B} \to \mathbf{X}$ are transformations. A *modification* $m$: $\theta \to \phi$ is a family of 2-cells

$$S(a) \underset{\phi_a}{\overset{\theta_a}{\rightrightarrows}} T(a) \qquad \Downarrow m_a$$

subject to the following commutativity condition:
(M)

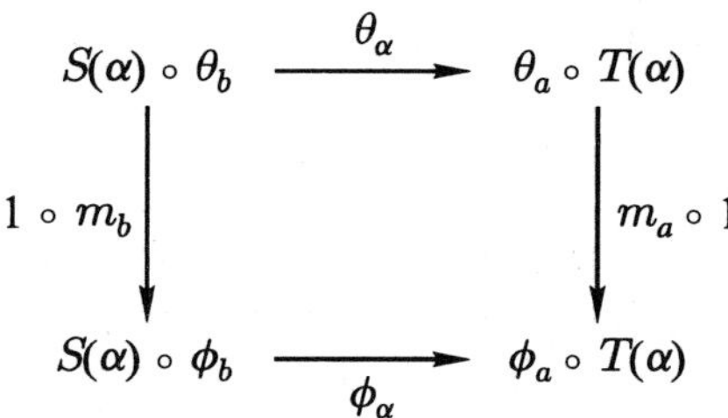

EXAMPLE 12. Each 2-cell $w\colon \kappa \Rightarrow \lambda\colon k \to h$ in a bicategory **B** yields a modification

$$H_w\colon H_\kappa \to H_\lambda\colon H_k \Rightarrow H_h\colon \mathbf{B} \to \mathbf{Cat}$$

whose component at $a \in \mathbf{B}$ is the natural transformation given by horizontal composition on the left with the 2-cell $w$.

Modifications $m\colon \theta \to \phi$, $n\colon \phi \to \psi$ can be composed to yield a modification $m \bullet n\colon \theta \to \psi$ using pointwise vertical composition in **X**. Transformations $\theta\colon S \Rightarrow T$, $\theta'\colon T \Rightarrow U$ can be composed to yield a transformation $\theta \circ \theta'\colon S \Rightarrow U$ by putting

$$(\theta \circ \theta')_a = \theta_a \circ \theta'_a$$

and

$$(\theta \circ \theta')_\alpha = \begin{pmatrix} S(\alpha) \circ (\theta_b \circ \theta'_b) \xrightarrow{a} (S(\alpha) \circ \theta_b) \circ \theta'_b \xrightarrow{\theta_\alpha \circ 1} (\theta_a \circ T(\alpha)) \circ \theta'_b \\ \xrightarrow{a^{-1}} \theta_a \circ (T(\alpha) \circ \theta'_b) \xrightarrow{1 \circ \theta'_\alpha} \theta_a \circ (\theta'_a \circ U(\alpha)) \xrightarrow{a} (\theta_a \circ \theta'_a) \circ U(\alpha) \end{pmatrix};$$

this composition is not strictly associative, but the associativity and identity constraints of **X** yield associativity and identity constraints here. This describes a bicategory $\mathbf{Lax}(\mathbf{B}, \mathbf{X})$ whose objects are lax functors, whose arrows are transformations, and whose 2-cells are modifications. Write $\mathbf{Psd}(\mathbf{B}, \mathbf{X})$ for the subbicategory of $\mathbf{Lax}(\mathbf{B}, \mathbf{X})$ consisting of the pseudo functors $T\colon \mathbf{B} \to \mathbf{X}$, the strong transformations between these, and the modifications between these. Notice that $\mathbf{Lax}(\mathbf{B}, \mathbf{X})$ and $\mathbf{Psd}(\mathbf{B}, \mathbf{X})$ are 2-categories if **X** is a 2-category (there is no need for **B** to be).

EXERCISE. Show that a lax functor $\mathbf{1} \to \mathbf{Lax}(\mathbf{1}, \mathbf{X})$ amounts to a pair of monads on the same object of **X** together with a distributive law between the monads (see Section 5).

For each bicategory **B**, there is a pseudo functor

$$\mathcal{Y}\colon \mathbf{B} \to \mathbf{Psd}(\mathbf{B}, \mathbf{Cat})^{\mathrm{op}}$$

the letter "$Y$" is for Yoneda since this is a generalization of the Yoneda embedding of categories. The value of $\mathcal{Y}$ at a 2-cell $w\colon \kappa \Rightarrow \lambda\colon k \to h$ in $\mathbf{B}$ is the displayed modification in Exercise 11. The data (LFa), (LFb) for $\mathcal{Y}$ are supplied by the identity and associativity constraints of $\mathbf{B}$.

For any pseudo functor $T\colon \mathbf{B} \to \mathbf{Cat}$, we shall describe a strong transformation

$$\mathbf{e}\colon \mathbf{Psd}(\mathbf{B},\mathbf{Cat})(\mathcal{Y},T) \Rightarrow T\colon \mathbf{B} \to \mathbf{Cat}.$$

For each $k \in \mathbf{B}$, the functor $\mathbf{e}_k\colon \mathbf{Psd}(\mathbf{B},\mathbf{Cat})(H_k,T) \to T(k)$ takes an arrow $m\colon \theta \to \phi$ in the category $\mathbf{Psd}(\mathbf{B},\mathbf{Cat})(H_k,T)$ to the arrow $m_k(1_k)\colon \theta_k(1_k) \to \phi_k(1_k)$ in the category $T(k)$. For each $\kappa\colon k \to h$ in $\mathbf{B}$, the natural isomorphism

$$\mathbf{e}_\kappa\colon H_\kappa \circ \mathbf{e}_h \Rightarrow \mathbf{e}_k \circ T(\kappa)\colon \mathbf{Psd}(\mathbf{B},\mathbf{Cat})(H_k,T) \to T(h)$$

whose component at the object $\theta$ of $\mathbf{Psd}(\mathbf{B},\mathbf{Cat})(H_k,T)$ is the isomorphism

$$\theta_\kappa\colon H_k(\kappa) \circ \theta_h \Rightarrow \theta_k \circ T(\kappa).$$

PROPOSITION 9.1 (Bicategorical Yoneda Lemma [S3]). *For each object $k$ of the bicategory $\mathbf{B}$ and each pseudo functor $T\colon \mathbf{B} \to \mathbf{Cat}$, the functor*

$$\mathbf{e}_k\colon \mathbf{Psd}(\mathbf{B},\mathbf{Cat})(H_k,T) \to T(k)$$

*is an equivalence of categories.*

An arrow $\alpha\colon a \to b$ in a bicategory $\mathbf{B}$ is called an *equivalence* when there exist an arrow $\beta\colon b \to a$ and invertible 2-cells $\alpha \circ \beta \Rightarrow 1_a$, $1_b \Rightarrow \beta \circ \alpha$; write $\alpha\colon a \xrightarrow{\sim} b$. For example, using the axiom of choice, one can see that an arrow $f\colon A \to B$ in $\mathbf{Cat}$ is an equivalence if and only if the functor $f\colon A \to B$ is full, faithful and each object $b$ of $B$ is isomorphic to an object of the form $f(a)$ for some $a \in A$. As another example, an arrow $\theta$ in $\mathbf{Psd}(\mathbf{B},\mathbf{X})$ is an equivalence if and only if each arrow $\theta_a$ is an equivalence in $\mathbf{X}$.

Hence, the bicategorical Yoneda lemma states that $\mathbf{e}$ is an equivalence in the bicategory $\mathbf{Psd}(\mathbf{B},\mathbf{Cat})$. Notice that $\mathcal{Y}$ and hence $\mathbf{Psd}(\mathbf{B},\mathbf{Cat})(\mathcal{Y},T)$ are 2-functors if $\mathbf{B}$ is a 2-category, so we obtain the following result which is an example of a "coherence theorem".

COROLLARY 9.2. *If $\mathbf{B}$ is a 2-category then every pseudo functor $T\colon \mathbf{B} \to \mathbf{Cat}$ is equivalent, in the 2-category $\mathbf{Psd}(\mathbf{B},\mathbf{Cat})$, to a 2-functor.*

A lax functor $T\colon \mathbf{B} \to \mathbf{X}$ is called a *biequivalence* when it is a pseudo functor, each of the functors $T\colon \mathbf{B}(a,b) \to \mathbf{X}(T(a),T(b))$ is an equivalence, and, for each object $x$ of $\mathbf{X}$, there exists an object $a$ of $\mathbf{B}$ and an equivalence $T(a) \xrightarrow{\sim} x$ in $\mathbf{X}$. Using the axiom of choice, we can see that $T\colon \mathbf{B} \to \mathbf{X}$ is a biequivalence if and only if there exists a lax functor $S\colon \mathbf{X} \to \mathbf{B}$ and equivalences $T \circ S \xrightarrow{\sim} 1_{\mathbf{B}}$, $1_{\mathbf{X}} \xrightarrow{\sim} S \circ T$ in the bicategories $\mathbf{Lax}(\mathbf{B},\mathbf{B})$, $\mathbf{Lax}(\mathbf{X},\mathbf{X})$, respectively.

The following proof is due to R. Gordon and A.J. Power and was made public at the 1991 Summer Category Theory Conference in Montréal.

PROPOSITION 9.3 [MP]. *For every bicategory* **B**, *there exists a 2-category* **K** *with a biequivalence* $\mathbf{B} \to \mathbf{K}$.

PROOF. It follows from the bicategorical Yoneda lemma that the functors

$$\mathcal{Y}\colon\ \mathbf{B}(a,b) \to \mathbf{Psd}(\mathbf{B},\mathbf{Cat})^{\mathrm{op}}(H_a, H_b)$$

are equivalences. So we can take **K** to be the sub-2-category of $\mathbf{Psd}(\mathbf{B},\mathbf{Cat})^{\mathrm{op}}$ obtained by restricting to those objects of the form $H_a$. Then $\mathcal{Y}$ gives the desired biequivalence. □

A direct proof, based on the above recall (Example 2), that every monoidal category is monoidally equivalent to a strict monoidal category, can be found in [JS5]. The result [GPS] for the next dimension is that every tricategory is "triequivalent" to a Gray-category (not in general to a 3-category). These references also explain how to extract from this result the coherence theorems in the more familiar form "all diagrams commute".

## 10. Nerves

The purpose of forming the nerve of a categorical structure is to create an object which contains all the information of the structure and yet is in a form more able to be compared with familiar geometric structures. There is a notion of cubical nerve, but we shall deal with the more usual simplicial nerve. In preparation for this, we need to modify our discussion of cubes to extract simplexes. For each natural number $r$, consider the word $\alpha_{r,n}$ of length $n$ in the symbols $-, +$ which begins with $r$ minuses and ends with $n-r$ pluses.

$$\alpha_{r,n} = \underbrace{- - \cdots -}_{r}\underbrace{+ + \cdots +}_{n-r}$$

Let $\mathsf{Smp}[n,m]$ denote the sub-$m$-category of $\mathsf{Cub}[n,m]$ obtained by taking only the objects $\alpha_{r,n}$. The $m$-category $\mathsf{Smp}[n,m]$ is the *$n$-simplex with commuting* $(m+1)$-*faces*. (There is an analogue of Proposition 4.1.) In particular, $\mathsf{Smp}[n,1]$ is a linearly ordered set with $n+1$ elements; it is more usual to use the ordered set

$$[n] = \{0, 1, \ldots, n\}.$$

Also, we have the 2-categories (using "position" notation):

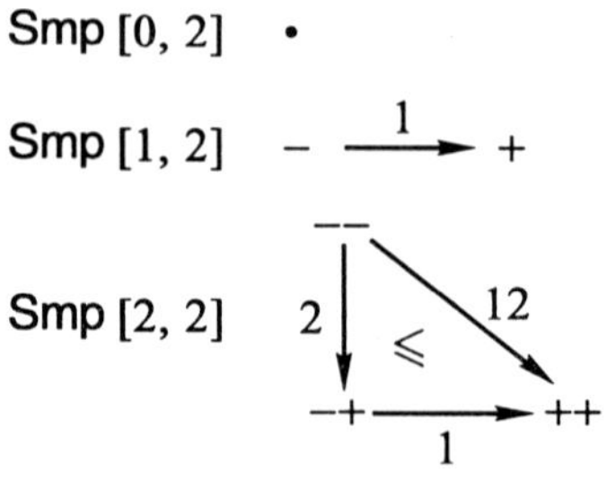

Smp [0, 2] •

Smp [1, 2] − $\xrightarrow{1}$ +

Smp [2, 2] (−− $\xrightarrow{2}$ −+, −− $\xrightarrow{12}$ ++, −+ $\xrightarrow{1}$ ++, ⩽)

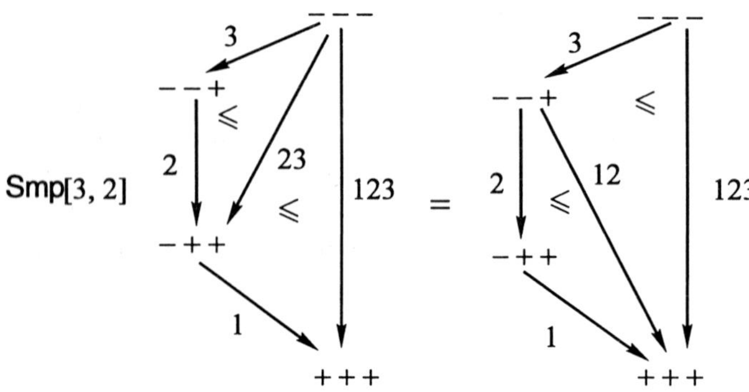


Recall that $\langle\mathbf{Cat}\rangle$ denotes the category of (small) categories and functors. The category $\boldsymbol{\Delta}$ of finite nonempty ordinals and order-preserving functions is the full subcategory $\boldsymbol{\Delta}$ of $\langle\mathbf{Cat}\rangle$ consisting of the categories $[n]$. A *simplicial set* is a functor $S\colon\ \boldsymbol{\Delta}^{\mathrm{op}} \to \mathbf{Set}$; its value at $[n]$ is denoted by $S_n$. The *nerve* $N(A)$ of a category $A$ is the simplicial set obtained by restricting the representable functor

$$\langle\mathbf{Cat}\rangle(-, A)\colon\ \langle\mathbf{Cat}\rangle^{\mathrm{op}} \to \mathbf{Set} \text{ to } \boldsymbol{\Delta}^{\mathrm{op}};$$

so

$$N(A)_n = \langle\mathbf{Cat}\rangle\big([n], A\big).$$

This construction is obviously functorial in $A \in \langle\mathbf{Cat}\rangle$, so we obtain nerve as a functor

$$N\colon\ \langle\mathbf{Cat}\rangle \to [\boldsymbol{\Delta}^{\mathrm{op}}, \mathbf{Set}]$$

into the category $[\boldsymbol{\Delta}^{\mathrm{op}}, \mathbf{Set}]$ of simplicial sets. It is easily seen that this functor is full, faithful, and has a left adjoint which preserves finite products. The simplicial sets $S$

which are isomorphic to nerves of categories can be characterized as those functors $S\colon \boldsymbol{\Delta}^{\mathrm{op}} \to \mathbf{Set}$ which preserve pullbacks; but they can also be characterized as those $S$ for which each *admissible horn* has a unique filler (see [S4, S5, S7] for this terminology).

There is a canonical 2-functor $\mathsf{Smp}[n,2] \to \mathsf{Smp}[n,1]$ which is the identity function on objects and identifies the 2-cells. Each functor $f\colon \mathsf{Smp}[n,1] \to \mathsf{Smp}[n',1]$ has a lifting to a 2-functor $f'\colon \mathsf{Smp}[n,2] \to \mathsf{Smp}[n',2]$ uniquely determined by the condition that each arrow $f'(r\colon \alpha_{r,n} \to \alpha_{r+1,n})$ is given by the natural ordering of $f(\alpha_{r,n})\backslash f(\alpha_{r+1,n})$. This gives a functor

$$j\colon \boldsymbol{\Delta} \to \langle\mathbf{2\text{-}Cat}\rangle, \quad [n] \mapsto \mathsf{Smp}[n,2], \ f \mapsto f'.$$

The *nerve* $N(K)$ of a 2-category $K$ is the simplicial set obtained by composing the functor $j^{\mathrm{op}}\colon \boldsymbol{\Delta}^{\mathrm{op}} \to \langle\mathbf{2\text{-}Cat}\rangle^{\mathrm{op}}$ with the representable functor

$$\langle\mathbf{2\text{-}Cat}\rangle(-,K)\colon \langle\mathbf{2\text{-}Cat}\rangle^{\mathrm{op}} \to \mathbf{Set}.$$

So, an element of $N(K)$ of dimension $n$ is a 2-functor $x\colon \mathsf{Smp}[n,2] \to K$; we think of this as an $n$-simplex in $K$ with commuting 3-faces. We obtain a nerve functor

$$N\colon \langle\mathbf{2\text{-}Cat}\rangle \to [\boldsymbol{\Delta}^{\mathrm{op}}, \mathbf{Set}]$$

with a left adjoint; but this time the functor is not full. We need to take account of more structure on the simplicial set $N(K)$, namely, those elements of dimension 2 which are *commutative* triangles. It is possible [S4] to characterize (up to isomorphism) nerves of 2-categories as simplicial sets, with some distinguished elements (called "hollow" or "thin"), satisfying some axioms the main one of which states that each admissible horn should have a unique thin filler.

There is also a notion of nerve for a bicategory [DS] which has not received much attention. Let $\mathbf{Bicat_{norm}}$ denote the category whose objects are bicategories and whose arrows are normalized lax functors. As every category is a bicategory, we can regard $\boldsymbol{\Delta}$ as a subcategory of $\mathbf{Bicat_{norm}}$. For each bicategory $\mathbf{B}$, the composite of the inclusion of $\boldsymbol{\Delta}^{\mathrm{op}}$ in $\mathbf{Bicat_{norm}}(-,\mathbf{B})$ with the representable

$$\mathbf{Bicat_{norm}}(-,\mathbf{B})\colon \mathbf{Bicat}^{\mathrm{op}}_{\mathbf{norm}} \to \mathbf{Set}$$

is defined to be the *nerve* $N(\mathbf{B})$ of $\mathbf{B}$; so

$$N(\mathbf{B})_n = \mathbf{Bicat_{norm}}\big([n],\mathbf{B}\big).$$

EXERCISE. For a 2-category $K$, the nerve of $K$ as a 2-category is isomorphic to the nerve of $K$ as a bicategory.

EXERCISE. Biequivalent bicategories have homotopically equivalent nerves. (See [GZ] for homotopy for simplicial sets.)

The nerve of an $m$-category was made precise in [S5], and other approaches appear in [A1, JW, ASn]. Essentially each proceeds as above after giving a precise description of **Smp**$[n, m]$. Verity [V] has shown that this nerve functor, defined on $\langle \mathbf{m\text{-}Cat} \rangle$ and viewed as landing in the category of simplicial sets with distinguished "hollow" (or "thin") elements, is fully faithful. A good deal of progress has been made by Michael Zaks and Dominic Verity on the characterization (up to isomorphism) of these nerves; but at the time of writing (November 1992), the conjecture of John Roberts (see [S5]) remains unproved.

Finally, we remark that categorical structures can be considered inside categories whose objects are more geometric than sets. Nerves then are simplicial geometric objects whose "geometric realizations" are "classifying spaces" [Sg].

## Acknowledgement

While this paper is really neither a survey paper nor a joint paper, it contains my version of ideas from several collaborations and discussions with Iain Aitchison, Samuel Eilenberg, Michael Johnson, and Steve Schanuel.

## Added in proof

This paper was completed in November 1992. The references have been updated during proofreading and [Sl, S8, S9] have been added. We point to [S9] as suitable for further reading in the area.

There have been two notable developments in the last three years. In July 1993, Dominic Verity completed the proof of the Roberts conjecture (see the end of Section 10). Also, Verity and the author have developed the use of surface diagrams for tricategories generalising the use of string diagrams for bicategories.

## References

[SGA] M. Artin, A. Grothendieck and J.L. Verdier (eds), *Théorie des Topos et Cohomologie Étale des Schémas*, SLNM 269, Springer, Berlin (1972).

[A1] I. Aitchison, *The geometry of oriented cubes*, Macquarie Mathematics Reports 86-002 (December 1986).

[A2] I. Aitchison, *String diagrams for non-abelian cocycle conditions*, Handwritten notes, Talk presented at Louvain-la-Neuve, Belgium (1987).

[AS] I. Aitchison and R. Street, *The algebra of oriented cubes*, Handwritten notes.

[ASn] F.A. Al-Agl and R. Steiner, *Nerves of multiple categories*, Proc. London Math. Soc. **66** (1993), 92–128.

[BW] M. Barr and C. Wells, *Toposes, Triples and Theories*, Springer, Berlin (1985).

[Bc] J. Beck, *Distributive laws*, Seminar on Triples and Categorical Homology Theory, SLNM 80, Springer, Berlin (1969), 119–140.

[Bn1] J. Bénabou, *Introduction to bicategories*, Reports of the Midwest Category Seminar, SLNM 47, Springer, Berlin (1967), 1–77.

[Bn2] J. Bénabou, *Les distributeurs*, Seminaires de Math. Pure, Univ. Catholique de Louvain, Rapport No. 33 (1973).

[Bns] D.B. Benson, *The basic algebraic structures in categories of derivations*, Inform. and Control **28** (1975), 1–29.

[BKP] R. Blackwell, G.M. Kelly and A.J. Power, *Two-dimensional monad theory*, J. Pure Appl. Algebra **59** (1989), 1–41.

[BG1] C. Brown and D. Gurr, *Refinement and simulation of nets – a categorical characterisation*, Proc. Application and Theory of Petri Nets, K. Jensen, ed., Lecture Notes in Comput. Sci. vol. 616, Springer, Berlin (1992), 76–92.

[BG2] C. Brown and Doug Gurr, *Timing Petri nets categorically*, Proc. ICALP 1992, W. Kuich, ed., Lecture Notes in Comput. Sci. vol. 623, Springer, Berlin (1992), 571–582.

[BH] R. Brown and P.J. Higgins, *The equivalence of crossed complexes and $\infty$-groupoids*, Cahiers Topologie Géom. Différentielle Catégoriques **22** (1981), 371–386.

[BS] R. Brown and C.B. Spencer, *G-Groupoids, crossed modules and the fundamental groupoid of a topological group*, Proc. Kon. Nederl. Akad. Wetensch. Ser. A, **79** (1976), 296–302.

[C] D. Conduché, *Modules croisés généralisés de longeur* 2, J. Pure Appl. Algebra **34** (1984), 155–178.

[Da1] B.J. Day, *On closed categories of functors*, Midwest Category Seminar Reports IV, SLNM 137, Springer, Berlin (1970), 1–38.

[Da2] B.J. Day, *A reflection theorem for closed categories*, J. Pure Appl. Algebra **2** (1972), 1–11.

[Da3] B.J. Day, *On closed categories of functors, II*, Category Seminar Sydney 1972/73, SLNM 420, Springer, Berlin (1974), 21–54.

[DM] P. Deligne and J.S. Milne, *Tannakian categories, Hodge cocycles, motives and Shimura varieties*, SLNM 900, Springer, Berlin (1982), 101–228.

[Dr1] V.G. Drinfel'd, *Quantum groups*, Proceedings of the International Congress of Mathematicians at Berkeley, California, USA 1986 (1987), 798–820.

[Dr2] V.G. Drinfel'd, *Quasi-Hopf algebras and Knizhnik–Zamolodchikov equations*, Problems in Modern Quantum Field Theory, A.A. Belavin, A.U. Klimyk and A.B. Zamolodchikov, eds, Research Reports in Physics, Springer, Berlin (1989).

[DS] J. Duskin and R. Street, *Non-abelian cocycles and nerves of $n$-categories*, In preparation.

[Eh1] C. Ehresmann, *Catégories structurées*, Ann. Sci. École Norm. Sup. **80** (1963), 349–425.

[Eh2] C. Ehresmann, *Catégories et Structures*, Dunod, Paris (1965).

[E] S. Eilenberg, *On normal forms*, Proceedings of the Fourth Biennial Meeting of SEAMS on Modern Applications of Mathematics, S. Nualtarance and Y. Kemprasit, eds, Bangkok (1978), 1–9.

[EK] S. Eilenberg and G.M. Kelly, *Closed categories*, Proc. Conf. Categorical Algebra at La Jolla 1965, Springer, Berlin (1966), 421–562.

[ES] S. Eilenberg and R.H. Street, *Rewrite systems, algebraic structures, and higher-order categories*, Handwritten manuscripts.

[F] P. Freyd, *Abelian Categories*, Harper and Row (1964).

[FY1] P. Freyd and D. Yetter, *Braided compact closed categories with applications to low dimensional topology*, Adv. Math. **77** (1989), 156–182.

[FY2] P. Freyd and D. Yetter, *Coherence theorems via knot theory*, J. Pure Appl. Algebra **78** (1992), 49–76.

[GZ] P. Gabriel and M. Zisman, *Calculus of Fractions and Homotopy Theory*, Springer, Berlin (1967).

[Gd] J. Giraud, *Cohomologie non Abélienne*, Springer, Berlin (1971).

[Gt] R. Godement, *Topologie Algébrique et Théorie des Faisceaux*, Publications de L'Institut de Math de L'Université de Strasbourg XIII, Hermann, Paris (1964).

[GPS] R. Gordon, A.J. Power and R.H. Street, *Coherence for tricategories*, Mem. Amer. Math. Soc., to appear.

[Gy1] J.W. Gray, *Formal Category Theory: Adjointness for* 2*-Categories*, SLNM 391, Springer, Berlin (1974).

[Gy2] J.W. Gray, *Coherence for the tensor product of* 2*-categories, and braid groups*, Algebra, Topology, and Category Theory (a collection of papers in honour of Samuel Eilenberg), Academic Press, New York (1976), 63–76.

[H] G. Huet, *Confluent reductions: abstract properties and applications to term rewriting systems*, J. Assoc. Comput. Mach. **27** (1980), 797–821.

[J] M. Johnson, *Pasting Diagrams in $n$-Categories with Applications to Coherence Theorems and Categories of Paths*, PhD Thesis, University of Sydney, October 1987.

[JW] M. Johnson and R. Walters, *On the nerve of an $n$-category*, Cahiers Topologie Géom. Différentielle Catégoriques **28** (1987), 257–282.

[Jt] P.T. Johnstone, *Topos Theory*, Academic Press, New York (1978).

[Joy] A. Joyal, *Foncteurs analytiques et espècies de structures*, Combinatoire Énumérative, Proceedings, Montréal, Québec 1985, SLNM 1234, Springer, Berlin (1991), 126–159.

[JS1] A. Joyal and R. Street, *Braided monoidal categories*, Macquarie Math. Reports #850067 (Dec. 1985); Revised #860081 (Nov. 1986).

[JS2] A. Joyal and R. Street, *The geometry of tensor calculus, I*, Adv. Math. **88** (1991), 55–112.

[JS3] A. Joyal and R. Street, *Tortile Yang–Baxter operators in tensor categories*, J. Pure Appl. Algebra **71** (1991), 43–51.

[JS4] A. Joyal and R. Street, *An introduction to Tannaka duality and quantum groups*, Category Theory, Proceedings, Como 1990; Part II of SLNM 1488, Springer, Berlin (1991), 411–492.

[JS5] A. Joyal and R. Street, *Braided tensor categories*, Adv. Math. **102** (1993), 20–78.

[JT] A. Joyal and M. Tierney, *Algebraic homotopy types*, In preparation.

[KV1] M.M. Kapranov and V.A. Voevodsky, *Combinatorial-geometric aspects of polycategory theory: pasting schemes and higher Bruhat orders (List of results)*, Cahiers Topologie Géom. Différentielle Catégoriques **32** (1991), 11–27.

[KV2] M.M. Kapranov and V.A. Voevodsky, $\infty$-groupoids and homotopy types, Preprint (1990).

[KV3] M.M. Kapranov and V.A. Voevodsky, *2-categories and Zamolodchikov tetrahedra equations*, Proc. Sympos. Pure Math. vol. 56 (1994), 177–259.

[K2] G.M. Kelly, *On MacLane's conditions for coherence of natural associativities, commutativities, etc.*, J. Algebra **1** (1964), 397–402.

[KS] G.M. Kelly and R.H. Street, *Review of the elements of 2-categories*, Category Seminar Sydney 1972/73, SLNM 420, Springer, Berlin (1974), 75–103.

[L] F.W. Lawvere, *Metric spaces, generalized logic, and closed categories*, Rend. Sem. Mat. Fis. Milano **43** (1973), 135–166.

[Lyu] V.V. Lyubashenko, *Tensor categories and RCFT, I. Modular transformations; II. Hopf algebras in rigid categories*, Academy of Sciences of the Ukrainian SSR, Preprints ITP-90-30E and 59E.

[ML1] S. MacLane, *Natural associativity and commutativity*, Rice Univ. Studies **49** (1963), 28–46.

[ML2] S. MacLane, *Categories for the Working Mathematician*, Springer, Berlin (1971).

[MLM] S. MacLane and I. Moerdijk, *Sheaves in Geometry and Logic: A First Introduction to Topos Theory*, Springer, Berlin (1992).

[MP] S. MacLane and R. Paré, *Coherence for bicategories and indexed categories*, J. Pure App. Algebra **37** (1985), 59–80.

[Mj1] S. Majid, *Physics for algebraists: non-commutative and non-cocommutative Hopf algebras by a bicrossproduct construction*, J. Algebra **130** (1990), 17–64.

[MN1] J.M. Maillet and F. Nijhoff, *Multidimensional integrable lattice models, quantum groups, and the d-simplex equations*, Proceedings of the Kiev Conference on Nonlinear and Turbulent Processes in Physics (1989).

[MN2] J.M. Maillet and F. Nijhoff, *The tetrahedron equation and the four-simplex equation*, Phys. Lett. A **134**(4) (2 Jan 1989), 221–228.

[MS] Yu.I. Manin and V.V. Schechtman, *Arrangements of hyperplanes, higher braid groups and higher Bruhat orders*, Advanced Studies in Pure Mathematics vol. 17 (1989), 289–308.

[M] R. Moore, *Apple Macintosh software implementing the excision of extremals algorithm for simplexes and the like*, Macquarie University (1989).

[MS] R. Moore and N. Seiberg, *Classical and quantum conformal field theory*, Comm. Math. Phys. **123**(2) (1989), 177–254.

[PR] R. Penrose and R. Rindler, *Spinors and Space-Time*, Vol. 1, Cambridge Univ. Press, Cambridge, UK (1984).

[Pw1] A.J. Power, *A 2-categorical pasting theorem*, J. Algebra **129** (1990), 439–445.

[Pw2] A.J. Power, *An $n$-categorical pasting theorem*, Category Theory, Proceedings, Como 1990, SLNM 1488, Springer, Berlin (1991), 326–358.

[P] V. Pratt, *Modeling concurrency with geometry*, Preprint, Stanford University (August 1990).

[RT] N.Yu. Reshetikhin and V.G. Turaev, *Ribbon graphs and their invariants derived from quantum groups*, Comm. Math. Phys. **127**(1) (1990), 1–26.

[SR] N. Saavedra Rivano, *Catégories Tannakiennes*, SLNM 265, Springer, Berlin (1972).

[R1] J.E. Roberts, *Mathematical aspects of local cohomology*, Proc. Colloquium on Operator Algebras and Their Application to Mathematical Physics, Marseille (1977).

[R2] J.E. Roberts, *Complicial sets*, Handwritten notes (1978).

[Sch] H. Schubert, *Categories*, Springer, Berlin (1972).

[Sg] G. Segal, *Classifying spaces and spectral sequences*, Inst. Hautes Études Sci. Publ. Math. **34** (1968), 105–112.

[SMC] Shum Mei Chee, *Tortile Tensor Categories*, PhD Thesis, Macquarie University (November 1989); J. Pure Appl. Algebra **93** (1994), 57–110.

[St] J.D. Stasheff, *Homotopy associativity of H-spaces I, II*, Trans. Amer. Math. Soc. **108** (1963), 275–312.

[Sn] R. Steiner, *The algebra of directed complexes*, Applied Categorical Structures **1** (1993), 247–284.

[Sl] J.G. Stell, *Modelling term rewriting systems by sesqui-categories*, Technical Report TR94-02, Dept. Computer Science, Keele University (January 1994).

[S0] R. Street, *Two constructions on lax functors*, Cahiers Topologie Géom. Différentielle Catégoriques **13** (1972), 217–264.

[S1] R. Street, *The formal theory of monads*, J. Pure Appl. Algebra **2** (1972), 149–168.

[S2] R. Street, *Limits indexed by category-valued 2-functors*, J. Pure Appl. Algebra **8** (1976), 148–181.

[S3] R. Street, *Fibrations in bicategories*, Cahiers Topologie Géom. Différentielle Catégoriques **21** (1980), 111–160; **28** (1987), 53–56.

[S4] R. Street, *Higher-dimensional nerves*, Handwritten notes, April–May 1982.

[S5] R. Street, *The algebra of oriented simplexes*, J. Pure Appl. Algebra **49** (1987), 283–335.

[S6] R. Street, *Parity complexes*, Cahiers Topologie Géom. Différentielle Catégoriques **32** (1991), 315–343.

[S7] R. Street, *Fillers for nerves*, SLNM 1348, Springer, Berlin (1988), 337–341.

[S8] R. Street, *Parity complexes: corrigenda*, Cahiers Topologie Géom. Différentielle Catégoriques **35** (1994), 359–362.

[S9] R. Street, *Higher categories, strings, cubes and simplex equations*, Applied Categorical Structures **3** (1995), 29–77.

[T] V.G. Turaev, *The Yang–Baxter equation and invariants of links*, Invent. Math. **92** (1988), 527–553.

[V] D. Verity, *Higher-dimensional nerves*, Handwritten notes, 1991.

[Y] D.N. Yetter, *Quantum groups and representations of monoidal categories*, Math. Proc. Cambridge Phil. Soc. **108** (1990), 261–290.

[Z] A.B. Zamolodchikov, *Tetrahedra equations and integrable systems in three-dimensional space*, Soviet Phys. JETP **52** (1980), 325; Comm. Math. Phys. **70** (1981), 489.

# Section 2B
# Homological Algebra. Cohomology. Cohomological Methods in Algebra. Homotopical Algebra

# The Cohomology of Groups

Jon F. Carlson

*Department of Mathematics, University of Georgia, Athens, Georgia 30602, USA*

*Contents*



HANDBOOK OF ALGEBRA, VOL. 1
Edited by M. Hazewinkel

## 1. Introduction

The cohomology of groups is one of those branches of mathematics which is regarded by many, even some of its most enthusiastic proponents, as a tool for other areas of study. Indeed part of the mystique is that it is a meeting point for so many different subjects. It has had applications in homotopy theory, class field theory, representation theory, $K$-theory and a host of other fields.

The origins and roots of the subject lie both in algebra and topology. First in algebra during the early part of the century, the low dimensional cohomology groups were used to classify objects such as projective representations [Scu] and group extension [Sce]. As a "theory", the cohomology of groups was born in the attempt to understand geometric/topological phenomena. In the mid 1930's, Hurewicz defined the higher homotopy groups $\pi_n(X)$ for $n \geqslant 2$. In 1936 [Hur], he considered aspherical spaces, spaces $X$ with $\pi_n(X) = 0$ for $n \geqslant 1$. He showed that the homotopy type, and hence also the homology and cohomology groups of such a space are completely determined by the fundamental group $G = \pi_1(X)$, assuming that the space is path connected. Hurewicz did not find the actual relationship between $G$ and the (co)homology of $X$. The first step in this direction was taken by Hopf [Hop]. The formulas which Hopf devised and the geometry of Hurewicz served as an inspiration to write an algebraic definition of group cohomology in terms of projective resolutions. More historical information can be found in the article by MacLane [McL2].

The modern topologist associates, to any group $G$, a classifying space $BG$, which is a $K(G, 1)$, an Eilenberg–MacLane space, a CW-complex with no homotopy in dimensions above one and hence aspherical. The cellular chain complex of the universal cover, $EG$ of $BG$, is a free $\mathbb{Z}G$-resolution of $\mathbb{Z}$. Hence the (co)homology of $BG = EG/G$ is the same as that of $G = \pi_1(BG)$. Such spaces have long been a source of motivation and examples in topology.

Yet the clearest indication of the connection between algebraic topology and cohomology of groups is expressed in the more recent theorem of Kan and Thurston [KaT]. Roughly it says that given any connected space $X$ there is a group $G$ with the property that $X$ and $G$ have the same homology and cohomology. Thus groups have a certain universality with respect to homology of spaces. It should be said that the applications of group cohomology and representation theory to topology are not limited to the computation of homology. See [Cas] and [Lan] for just two examples.

From an algebraic viewpoint, the cohomology of groups is really two subjects which share a common set of techniques and interests. For infinite groups, the cohomology theory is very much a part of group theory itself. Groups are often classified according to their homological properties such as cohomological dimension. By contrast, the cohomology of finite groups is much more closely associated to modular and integral representation theory. Here too it can be used as a classification device, but more often for modules than for groups.

For reasons of space this survey will concentrate on the algebraic techniques and applications of the subject. We will not discuss the connections to topology, $K$-theory or other areas beyond what has already been said. We will also not discuss noncommutative cohomology, connections with varieties of groups or other cohomology theories such as those of Lie algebras, algebraic groups or relative theories. Many of these topics will be treated elsewhere in the handbook.

The bibliography at the end of this article is not meant to be exhaustive. We have tried to list, for each subject, a few recent papers from which other references can be found. A lot of material on cohomology of groups can be found in the standard texts on homological algebra such as [CaE, McL1] and [HlS]. Some texts such as [Gru, Lag, Stm] and [Thm3] are aimed at specific aspects of group cohomology. The most modern overall references are the books [Brw1] and [Ben2]. Even at 10 years old Brown's text is close to being up to date particularly for the sections on infinite groups. For finite groups and the connections to representation theory the best text is definitely that of Benson.

## 2. Basic definitions and structures

Throughout this essay it will be assumed that the reader is familiar with fundamental techniques from homological algebra. Background can be found in any of the standard texts on the subject.

**2.1.** *Projective resolutions and homology.* Let $\mathbb{Z}$ denote the ordinary integers and let $G$ be a group. We let $\mathbb{Z}$ also stand for the trivial $\mathbb{Z}G$-module on which the group elements act by multiplication by 1. To define the cohomology groups $H^n(G, \mathbb{Z})$ we begin by taking a projective resolution $(P_*, \partial)$ of $\mathbb{Z}$. This is an exact sequence

$$\cdots \to P_2 \xrightarrow{\partial_2} P_1 \xrightarrow{\partial_1} P_0 \xrightarrow{\varepsilon} \mathbb{Z} \to 0$$

in which each $P_i$ is a projective $\mathbb{Z}G$-module. If $M$ is a $\mathbb{Z}G$-module then the cohomology of $G$ with coefficients in $M$ are the groups

$$H^n(G, M) = H^n\big(\mathrm{Hom}_{\mathbb{Z}G}(P_*, M)\big) = \ker \partial_{n+1}^*/\mathrm{Image}\, \partial_n^*$$

of the complex

$$\cdots \longleftarrow \mathrm{Hom}_{\mathbb{Z}G}(P_1, M) \xleftarrow{\partial_1^*} \mathrm{Hom}_{\mathbb{Z}G}(P_0, M) \longleftarrow 0.$$

Likewise the homology of $G$ is the homology

$$H_n(G, M) = H_n(P_* \otimes_{\mathbb{Z}G} M) = \ker(\partial_n \otimes 1)/\mathrm{Image}(\partial_{n+1} \otimes 1)$$

of the complex

$$\cdots \to P_1 \otimes_{\mathbb{Z}G} M \to P_0 \otimes_{\mathbb{Z}G} M \to 0.$$

The $n$-cycles, $n$-boundaries, $n$-cocycles and $n$-coboundaries are respectively elements of the groups $\ker(\partial_n \otimes 1)$, $\mathrm{Image}(\partial_{n+1} \otimes 1)$, $\ker \partial_{n+1}^*$ and $\mathrm{Image}\, \partial_n^*$.

From the left exactness of the $\mathrm{Hom}_{\mathbb{Z}G}$ functor it can be seen that

$$H^0(G, M) \cong \mathrm{Hom}_{\mathbb{Z}G}(\mathbb{Z}, M) = M^G$$

where

$$M^G = \{m \in M \mid gm = m \text{ for all } g \in G\}$$

is the set of $G$-fixed points of $M$. Likewise $\otimes_{\mathbb{Z}G} M$ is right exact and $H_0(G, M) \cong \mathbb{Z} \otimes_{\mathbb{Z}G} M$, the set of cofixed points. In the terminology of category theory, $H^n(G, -)$ is the $n$-th derived functor of the fixed point functor (see [HlS]).

**2.2.** *Functoriality.* It follows from standard arguments that the homology and cohomology of $G$ are independent of the choice of the projective resolution $(P_*, \partial)$. The map $\varepsilon$: $\mathbb{Z}G \to \mathbb{Z}$ with $\varepsilon(g) = 1$ for all $g \in G$ is called the augmentation map. It is usual to assume that $P_0 = \mathbb{Z}G$ in any projective resolution.

The constructions are easily seen to be functorial. In particular, if $\alpha$: $M \to N$ is a $\mathbb{Z}G$-homomorphism then the chains

$$\alpha_* = \mathrm{Hom}_{\mathbb{Z}G}(P_*, M) \to \mathrm{Hom}_{\mathbb{Z}G}(P_*, N),$$

obtained by composing with $\alpha$, and $1 \otimes \alpha$: $P_* \otimes_{\mathbb{Z}G} M \to P_* \otimes_{\mathbb{Z}G} N$ induce a map $\alpha_*$: $H^n(G, M) \to H^n(G, N)$ and $\alpha_*$: $H_n(G, M) \to H_n(G, N)$.

In particular if

$$0 \to L \xrightarrow{\alpha} M \xrightarrow{\beta} N \to 0$$

is an exact sequence of $\mathbb{Z}G$-modules, then we have a long exact sequence

$$\begin{aligned} 0 \to H^0(G, L) &\xrightarrow{\alpha_*} H^0(G, M) \xrightarrow{\beta_*} H^0(G, N) \xrightarrow{\delta} H^1(G, L) \to \cdots \\ \cdots \to H^n(G, M) &\xrightarrow{\beta_*} H^n(G, N) \xrightarrow{\delta} H^{n+1}(G, L) \xrightarrow{\alpha_*} \cdots. \end{aligned}$$

There is a similar sequence for homology. The maps marked $\delta$ are called the connecting homomorphisms.

It is also true that the homology and cohomology are functorial in $G$, the group variable. However the properties of this functoriality are much more subtle. The restriction and inflation maps are involved. See Section 4.

**2.3.** Ext *and* Tor. The connection between group cohomology and module theory can be seen from the construction. That is, the isomorphisms $H^n(G, M) \cong \mathrm{Ext}^n_{\mathbb{Z}G}(\mathbb{Z}, M)$ and $H_n(G, M) \cong \mathrm{Tor}^{\mathbb{Z}G}_n(\mathbb{Z}, M)$ are obvious from the definitions of Ext and Tor. But further, $M \otimes_{\mathbb{Z}} N$ and $\mathrm{Hom}_{\mathbb{Z}}(M, N)$ can be made into $\mathbb{Z}G$-modules by defining

$$g(m \otimes n) = gm \otimes gn \quad \text{and} \quad gf(m) = g \cdot f(g^{-1}m)$$

for all $g \in G$, $m \in M$, $n \in N$, $f \in \mathrm{Hom}_{\mathbb{Z}}(M, N)$. As such it can be seen that

$$\mathrm{Hom}_{\mathbb{Z}G}(L \otimes_{\mathbb{Z}} M, N) \cong \mathrm{Hom}_{\mathbb{Z}G}\big(L, \mathrm{Hom}_{\mathbb{Z}}(M, N)\big)$$

by the homomorphism which sends $f\colon L \otimes_{\mathbb{Z}} M \to N$ to $g$ where $(g(\ell))(m) = f(\ell \otimes n)$. The extension to cohomology assures that

$$\mathrm{Ext}^n_{\mathbb{Z}G}(M, N) \cong \mathrm{Ext}^n_{\mathbb{Z}G}(\mathbb{Z} \otimes_{\mathbb{Z}} M, N) \cong H^n\big(G, \mathrm{Hom}_{\mathbb{Z}}(M, N)\big).$$

**2.4.** *Example: Finite cyclic group.* Suppose that $G = \langle g \mid g^n = 1\rangle$ is a cyclic group of order $n$. A projective resolution of $\mathbb{Z}$ is given by

$$\cdots \to \mathbb{Z}G \xrightarrow{\partial_2} \mathbb{Z}G \xrightarrow{\partial_1} \mathbb{Z}G \xrightarrow{\varepsilon} \mathbb{Z} \to 0$$

where $\partial_i(1) = g - 1$ if $i$ is an odd integer and $\partial_i(1) = 1 + g + \cdots + g^{n-1}$ if $i$ is an even integer. Now $\mathrm{Hom}_{\mathbb{Z}G}(\mathbb{Z}G, \mathbb{Z}) \cong \mathbb{Z}$ is generated by the augmentation map $\varepsilon$. So we have that $\partial_i^*\colon \mathbb{Z} \to \mathbb{Z}$ is the zero map if $i$ is odd and is multiplication by $n$ if $i$ is even. Hence $H^m(G, \mathbb{Z}) = 0$ if $m$ is odd and $H^m(G, \mathbb{Z}) = \mathbb{Z}/n\mathbb{Z}$ if $m$ is even. Of course $H^0(G, \mathbb{Z}) \cong \mathbb{Z}$. Likewise $\mathbb{Z}G \otimes_{\mathbb{Z}G} \mathbb{Z} \cong \mathbb{Z}$ and we have a similar result on homology.

If $k$ is a field of characteristic $p > 0$ and we regard $k$ as a $\mathbb{Z}G$-module with trivial $G$-action then, assuming $p \mid n$, $H^m(G, k) \cong H_m(G, k) \cong k$ for all $m \geqslant 0$.

**2.5.** *Example: Infinite cyclic groups.* Suppose that $G$ is an infinite cyclic group with generator $g$. Then the homomorphism $\theta\colon \mathbb{Z}G \to \mathbb{Z}G$ which sends 1 to $g - 1$ can be seen to be injective. So the sequence

$$0 \to \mathbb{Z}G \xrightarrow{\theta} \mathbb{Z}G \xrightarrow{\varepsilon} \mathbb{Z} \to 0$$

is exact. It follows that $H^m(G, \mathbb{Z}) \cong H_m(G, \mathbb{Z}) \cong \mathbb{Z}$ if $m = 0$ or 1 and is zero otherwise.

**2.6.** *Direct products.* Let $G_1$ and $G_2$ be groups and $G = G_1 \times G_2$ the direct product. Then $\mathbb{Z}G \cong \mathbb{Z}G_1 \otimes_{\mathbb{Z}} \mathbb{Z}G_2$ with the obvious multiplication. Suppose that we are given projective resolutions $(X_*, \partial')$ of $\mathbb{Z}$ as a $\mathbb{Z}G_1$-module and $(Y_*, \partial'')$ of $\mathbb{Z}$ as a $\mathbb{Z}G_2$-module. Then by the Künneth Theorem, the tensor product (over $\mathbb{Z}$) of the two resolution is a $\mathbb{Z}G$-projective resolution of $\mathbb{Z}$. The tensor product has the form $(P_*, \partial)$ where

$$P_n = \sum_{i+j=n} X_i \otimes_{\mathbb{Z}} Y_j$$

and for $x \in X_i$, $y \in Y_j$,

$$\partial(x \otimes y) = \partial' x \otimes y + (-1)^i x \otimes \partial'' y.$$

It is now possible to construct some of the cohomology of $G$ from that of $G_1$ and $G_2$. For example, if $\alpha\colon X_m \to M$ and $\beta\colon Y_n \to N$ are cocycles for $M$ a $\mathbb{Z}G_1$-module and

$N$ a $\mathbb{Z}G_2$-module, then the homomorphism $\alpha \otimes \beta$ may be composed with the projection $P_{m+n} \to X_m \otimes Y_n$ to yield a $\mathbb{Z}G$-cocycle $P_{m+n} \to M \otimes N$.

**2.7.** *Abelian groups.* If $G = \langle z_1, \dots, z_n \rangle$ is a free abelian group of rank $n$ then $G = \langle z_1 \rangle \times \cdots \times \langle z_n \rangle$. It can be deduced from the above analysis that

$$H^\ell(G, \mathbb{Z}) = \sum_{i_1 + \cdots + i_n = \ell} H^{i_1}(\langle z_1 \rangle, \mathbb{Z}) \otimes_{\mathbb{Z}} \cdots \otimes_{\mathbb{Z}} H^{i_n}(\langle z_n \rangle, \mathbb{Z}).$$

Now suppose that $E = G/\langle z_1^p, \dots, z_n^p \rangle$ is an elementary abelian group of order $p^n$, $p$ a prime. Then $E = \langle x_1 \rangle \times \cdots \times \langle x_n \rangle$ where $x_i$ is the coset of $z_i$. Each $\langle x_i \rangle$ is cyclic of order $p$. It can be seen that

$$H^*(E, k) = H^*(\langle x_1 \rangle, k) \otimes_{\mathbb{Z}} \cdots \otimes_{\mathbb{Z}} H^*(\langle x_n \rangle, k)$$

for any field $k$ of characteristic $p$. The computation of $H^*(E, \mathbb{Z})$ is a bit more tricky because there are $G$-cocycles which cannot be written as products as in (2.6). Nevertheless it can be accomplished with the same sort of analysis (e.g., see [Chp1]).

**2.8.** *Low dimensional cohomology.* In low dimensions the homology and cohomology groups can be expressed in terms of the pieces of a presentation. Notice that if $S \subseteq G$ is a set of generators for $G$ then the augmentation ideal $I(G) = \ker \varepsilon$ is generated as a $\mathbb{Z}G$-module by the set $\{s - 1 \mid s \in S\}$. The proof requires only induction and the observation that if $g = g's$ for $g, g' \in G$, $s \in S$ then $g - 1 = g'(s - 1) + g' - 1$. It follows that the term $P_1$ in a projective resolution $(P_*, \partial)$ of $\mathbb{Z}$ can be taken to be a free $\mathbb{Z}G$-module with a basis $T = \{f_s \mid s \in S\}$ indexed by the set of generators. Then assuming that $P_0 = \mathbb{Z}G$, $\partial_1 \colon P_1 \to P_0$ is given by $\partial_1(f_s) = s - 1$. Of course if $F$ is the free group on the set $T$, then we have a presentation

$$1 \to R \to F \xrightarrow{\theta} G \to 1$$

where $\theta(f_s) = s$ and $R$ is the kernel of $\theta$. With some argument in this direction, it is possible to prove the formula of Hopf [Hop]:

$$H_2(G, \mathbb{Z}) \cong R \cap [F, F]/[R, F].$$

Here $[A, B]$ means the subgroup generated by all commutators $[a, b] = aba^{-1}b^{-1}$ for $a \in A$, $b \in B$. Similar formulas exist for cohomology groups and for homology groups in other dimensions (e.g., see Chapter 3 of [Gru]). In particular we mention the well known formula $H_1(G, \mathbb{Z}) \cong G/[G, G]$. Several other generalizations of the Hopf formula can be found in the literature. The ones by Stöhr [Sto] and by Brown and Ellis [BrE] seem to be the most comprehensive.

**2.9.** *Universal coefficients.* The cohomology of groups has a universal coefficient theorem as does the cohomology of spaces. It says that for $n \geqslant 1$ and any $\mathbb{Z}G$-module $M$ with trivial $G$-action

$$H^n(G, M) \cong \operatorname{Hom}_{\mathbb{Z}}\big(H_n(G, \mathbb{Z}), M\big) \oplus \operatorname{Ext}_{\mathbb{Z}}\big(H_{n-1}(G, \mathbb{Z}), M\big).$$

## 3. Some applications in low dimensions

The early algebraic applications of group cohomology were invented without the benefit of a larger accompanying theory. They can be easily connected to the later-developed theory by considering the standard (bar) resolution for computing the cohomology. As noted earlier, the definitions are independent of the resolutions. So it makes sense to use a specific resolution to fit a task at hand.

**3.1.** *The standard resolution.* Given the group $G$, the standard resolution of $\mathbb{Z}$ as a $\mathbb{Z}G$-module is given by $(P_*, \partial)$ where for each $i \geqslant 0$, $P_i = \mathbb{Z}G \otimes_{\mathbb{Z}} \cdots \otimes_{\mathbb{Z}} \mathbb{Z}G$, $i+1$ factors. The action of $G$ on $P_i$ is defined as multiplication on the left most factor, i.e.

$$g(\alpha_0 \otimes \cdots \otimes \alpha_i) = (g\alpha_0) \otimes \alpha_1 \otimes \cdots \otimes \alpha_i.$$

Hence $P_i$ is a free $\mathbb{Z}G$-module having as $\mathbb{Z}G$-basis the set

$$\{1 \otimes g_1 \otimes \cdots \otimes g_i \mid g_i, \ldots, g_i \in G\}.$$

So $P_0 \cong \mathbb{Z}G$ and $\varepsilon$: $P_0 \to \mathbb{Z}$ is the augmentation. The boundary map $\partial_m$: $P_m \to P_{m-1}$ is given by

$$\begin{aligned} \partial_m(g_0 \otimes g_1 \otimes \cdots \otimes g_m) = & \sum_{j=0}^{m-1} (-1)^j g_0 \otimes \cdots \otimes g_j g_{j+1} \otimes \cdots \otimes g_m \\ & + (-1)^m g_0 \otimes \cdots \otimes g_{m-1}. \end{aligned}$$

So for $m = 1$, $\partial(g \otimes h) = gh - g = g(h-1) \in I(G)$, the proof that the resolution is exact can be accomplished by showing that the boundary homomorphisms have partial inverses as $\mathbb{Z}$-modules.

**3.2.** *Projective representations.* Schur's earliest work on the subject of cohomology concerned projective representations of finite groups [Scu]. Suppose that we are given a vector space $V$ over a field $k$. A projective representation of a finite group $G$ on $V$ is a homomorphism

$$\rho\colon G \to PGL(V) \cong GL(V)/Z\big(GL(V)\big).$$

Here $Z(GL(V)) \cong k^{\times}$ is the center of $GL(V)$. So $PGL(V)$ is the group of invertible linear transformations on the projective space of $V$. Schur's discovery was that the obstruction to lifting a projective representation $\rho$ to an ordinary representation is an element of $H^2(G, k^\times)$ where $k^\times = Z(GL(V))$ has trivial $G$-action.

The reasoning behind Schur's discovery works out as follows. Suppose that $\theta\colon G \to GL(V)$ is a section of $\rho$. That is, $\pi\theta = \rho$ where $\pi\colon GL(V) \to PGL(V)$ is the quotient map. Of course, $\theta$ may not be a homomorphism. However, because $\rho$ is a homomorphism, for every pair $g, h \in G$ there must exist $f(g,h) \in Z(GL(V))$ such that $\theta(g)\theta(h) = f(g,h)\theta(gh)$. Now we define a homomorphism $\psi\colon P_2 \to Z(GL(V)) = k^\times$ by $\psi(1 \otimes g \otimes h) = f(g,h)$. The associative law $[(\theta(g)\theta(h))\theta(j) = \theta(g)(\theta(h)\theta(j))]$ implies that

$$f(g,h)\cdot f(gh,j) = f(g,hj)\cdot f(h,j).$$

The relation says precisely that $\psi \circ \partial_3 = 0$ or that $\psi$ is a 2-cocycle. (Notice that $\psi$ maps an additive group to a multiplicative group.) Hence $\psi$ defines a cohomology class $\mathrm{cls}(\psi) \in H^2(G, k^\times)$.

On the other hand, suppose that we have two sections $\theta_1$ and $\theta_2$, giving maps $f_1$ and $f_2$. Then for some function $\gamma\colon G \to k^\times$, $\theta_1(g) = \gamma(g)\cdot\theta_2(g)$ for all $g \in G$. So for $g, h \in G$

$$\begin{aligned}\theta_1(g)\theta_1(h) &= \gamma(g)\gamma(h)\theta_2(g)\theta_2(h)\\ &= \gamma(g)\gamma(h)f_2(g,h)\theta_2(gh)\\ &= \gamma(g)\gamma(h)\gamma(gh)^{-1}f_2(g,h)\theta_1(gh).\end{aligned}$$

Hence $f_1(g,h) = \gamma(g)\gamma(h)\gamma(gh)^{-1}f_2(g,h)$. Let $\mu\colon P_1 \to k^\times$ be defined by $\mu(1\otimes g) = \gamma(g)$. Then $\mu(\partial(1\otimes g\otimes h)) = \gamma(g)\gamma(h)\gamma(gh)^{-1}$. Hence $\psi_1$ and $\psi_2$, defined by $\theta_1$ and $\theta_2$, differ by the coboundary $\mu$ and must represent the same cohomology class. Now if the representation $\rho$ lifts to an ordinary representation, then there is some $\theta_2$ which is a homomorphism. The corresponding maps $\psi_2$ and $f_2$ are zero, and any cocycle $\psi$ defined as above is in the zero class.

In addition to all of this Schur proved that for any element $\zeta \in H^2(G, k^\times)$, there exists a $k$-space $V$ and a projective representation on $V$ which has $\zeta$ as its cohomology class. Although his methods work for any field, Schur was actually only concerned with the case $k = \mathbb{C}$, the complex numbers. The group $H^2(G, \mathbb{C}^\times)$ is known as the Schur multiplier. It was also proved that there exists an extension

$$0 \to A \xrightarrow{i} E \to G \to 1$$

where $i(A)$ is in the center of $E$, $A \cong H^2(G, \mathbb{C}^\times)$, and every projective representation of $E$ lifts to an ordinary representation. The group $E$ is called a representation group of $G$. The functions $f\colon G \times G \to k^\times$ are called factor sets and they arise again in the following.

**3.3.** *Classifying extensions.* An extension of $G$ by $A$ is a group $E$ such that there is an exact sequence

$$\gamma\colon\ 0 \to A \xrightarrow{j} E \xrightarrow{\pi} G \to 1. \tag{3.4}$$

Here $A$ is a $G$-module, an additive group, with the $G$-action defining the conjugation action in $E$. That is, for $a \in A$ and $s \in E$ with $\pi(s) = g$ then $sj(a)s^{-1} = j(ga)$. Here $G$ can be any group. Two sequences, $\gamma$ and $\gamma'$, are equivalent if there is a commutative diagram

$$\begin{array}{lccccccccc} \gamma\colon & 0 & \longrightarrow & A & \longrightarrow & E & \xrightarrow{\pi} & G & \longrightarrow & 1 \\ & & & \| & & \downarrow{\scriptstyle\theta} & & \| & & \\ \gamma'\colon & 0 & \longrightarrow & A & \longrightarrow & E' & \longrightarrow & G & \longrightarrow & 1 \end{array}$$

where the vertical maps on the ends are identities. The split extension is the one for which there is a map $\sigma\colon\ G \to E$ with $\pi\sigma = Id_G$. Notice however that in the split extension, $\sigma(G)$ is not normal in $E$ unless the action of $G$ on $A$ is trivial.

The classification of such extensions by $H^2(G, A)$ goes back to Schreier [Sce]. It is very similar to the development by Schur given above. Given an extension $\gamma$ as in (3.4), let $\sigma\colon\ G \to E$ be a section, i.e. a function with $\pi\sigma = Id_G$. Because $\sigma$ may not be a homomorphism, there is a factor set $f\colon\ G \times G \to A$ defined by the equation $\sigma(g)\sigma(h) = f(g, h)\sigma(gh)$. The associative law $(\sigma(g)\sigma(h))\sigma(k) = \sigma(g)(\sigma(h)\sigma(k))$ implies that (changing multiplication to addition)

$$gf(h, k) + f(g, hk) = f(g, h) + f(gh, k). \tag{3.5}$$

Hence the function $\psi\colon\ P_2 \to A$ by $\psi(1 \otimes g \otimes h) = f(g, h)$ is a cocycle. It can be checked that two such cocycles differ by a coboundary if and only if they come from equivalent extensions. Finally, suppose that we are given a factor set $f\colon\ G \times G \to A$, defined by a cocycle $\psi$, $\psi(1 \otimes g \otimes h) = f(g, h)$. Then, on the set $E = A \times G$, the operation

$$(a_1, g_1) \cdot (a_2, g_2) = \big(a_1 + a_2 + f(g_1, g_2),\ g_1 g_2\big)$$

makes $E$ into a group. The projection $\pi\colon\ E \to G$ makes $E$ an extension of $G$ by $A$, and the section $\sigma\colon\ G \to E$, given by $\sigma(g) = (0, g)$ has $f$ as the corresponding factor set.

There is an extensive literature on Schur multipliers. The lecture notes of Beyl and Tappe [BeT] are reasonably up to date and give a picture of recent developments.

**3.4.** *Other applications and higher cohomology.* Some similar interpretations have been found for the homology groups in degrees one and two and for the group cohomology in degree one. A few interesting applications can be found in [AsG] and [Gur]. Computer programs have been developed for calculating some of these groups [Hol3, Hol4].

Eilenberg and MacLane worked out a group theoretic interpretation of $H^3(G, A)$ in terms of the obstruction to certain types of group extension of $G$ by a nonabelian group with center $A$. See MacLane's book [McL1] for an account. Although several improvements were made on this work, there was no adequate interpretation of the higher cohomology groups until the late 1970's. At that time several people independently showed that the higher cohomology groups classified crossed extensions. The first of these results is likely due to Huebschmann [Hue2], but see also [Hol1, Hil] and, for a simplified version [Con]. A brief historical account is given in [McL3].

Briefly we review the case with $n = 3$. Suppose that we are given a group $G$ and $\mathbb{Z}G$-module $A$. A triple $(B, C, \theta)$ is a crossed module in this context, if there is an exact sequence of groups

$$0 \to A \xrightarrow{j} B \xrightarrow{\beta} C \to G \to 1.$$

We are assuming that the action of $C$ on $A$ is induced by the given action of $G$ on $A$. The definition of the crossed module $(B, C, \theta)$ requires that $C$ act on $B$ and that the action be completely compatible with the homomorphism $\theta$: $B \to C$. Two crossed modules are said to be equivalent if there is a commutative diagram

$$\begin{array}{ccccccccc} 0 & \longrightarrow & A & \longrightarrow & B & \xrightarrow{\theta} & C & \longrightarrow & G & \longrightarrow & 1 \\ & & \| & & \downarrow & & \downarrow & & \| & & \\ 0 & \longrightarrow & A & \longrightarrow & B' & \xrightarrow{\theta'} & C' & \longrightarrow & G & \longrightarrow & 1 \end{array}$$

The actual equivalence relation is the least equivalence relation which contains this one. Note that the vertical maps on $B$ and $C$ need not be either injective or surjective, so there is no hope of finding an inverse to the given morphism of exact sequences.

The theorem is that the set of equivalence classes of crossed modules is in one-to-one correspondence with the elements of $H^3(G, A)$. The idea of the proof is sketched as follows. To the extension of $G$ by $B/j(A)$, we may associate a factor set

$$f\colon\ G \times G \to B/j(A).$$

We must be careful here as $B/j(A)$ may not be abelian. To lift the factor set to all of $B$, we choose a section $\sigma$: $B/j(A) \to B$. Let $F = \sigma \circ f$ and consider the cocycle relation as in (3.5) for $F$. The two sides of the relation differ by an element $\mu(g, h, k) \in A$ which happens to be a 3-cocycle. This represents the corresponding cohomology class.

The interpretation of the higher dimensional cohomology groups is similar. It is very reminiscent of the Yoneda definition of $\mathrm{Ext}^n_R(C, A)$ as the set of equivalence classes of $n$-fold extensions

$$0 \to A \to B_{n-1} \to \cdots \to B_0 \to C \to 0$$

of $R$-modules. However the noncommutativity in the group case necessitates many complications.

## 4. Some methods and their consequences

As we mentioned in Section 2, the cohomology of groups $H^*(-, M)$ enjoys some functorial properties in the group variables. Any homomorphism $\psi$: $G_1 \to G_2$ of groups can be factored as a surjection $G_1 \to G_1/\ker\psi \cong \psi(G_1)$ followed by an injection $\psi(G_1) \to G_2$. If $M$ is a $\mathbb{Z}G_2$-module then the corresponding homomorphism $H^n(G_2, M) \to H^n(G_1, M)$ is the composition of a restriction map followed by an inflation map. Before defining these maps, we should state that there is no left or right exactness of cohomology with respect to these maps in the group variable. Consequently long exact sequences such as those in (2.2) do not exist. In low degrees there are some substitute exact sequences which can best be explained using a spectral sequence.

**4.1.** *Restriction.* Let $H$ be a subgroup of $G$. The left module $\mathbb{Z}G$ when viewed as a $\mathbb{Z}H$-module is a free module. A basis can be taken as any set of representatives of the left cosets of $H$ in $G$. Hence any projective $\mathbb{Z}G$-module is also a projective $\mathbb{Z}H$-module by restriction. In particular a projective $\mathbb{Z}G$-resolution $(P, \partial)$ of $\mathbb{Z}$ is also a projective $\mathbb{Z}H$-resolution. If $\zeta\colon P_n \to M$ is a $\mathbb{Z}G$-cocycle then it is likewise a $\mathbb{Z}H$-cocycle, and similarly $\mathbb{Z}G$-coboundaries are $\mathbb{Z}H$-coboundaries. Notice that it is possible for a $\mathbb{Z}G$-cocycle to be a $\mathbb{Z}H$-coboundary even when it isn't a $\mathbb{Z}G$-coboundary. Consequently the restriction map which sends the class of a $\mathbb{Z}G$-cocycle to its class as a $\mathbb{Z}H$-cocycle is well defined but not necessarily injective. We denote the restriction map by

$$\mathrm{res}_{G,H}\colon H^n(G, M) \to H^n(H, M).$$

**4.2.** *Inflation.* Let $N$ be a normal subgroup of $G$. Any $G/N$-module $M$ is easily made into a $G$-module by defining $g \cdot m = (gN) \cdot m$ for $g \in G$, $m \in M$. This is the essence of the inflation homomorphism. For the details we need to observe that if $(Q, \partial')$ is a projective $\mathbb{Z}(G/N)$-resolution of $\mathbb{Z}$ and if $(P, \partial)$ is a projective $\mathbb{Z}G$-resolution, then there is a chain map $\mu\colon (P, \partial) \to (Q, \partial')$ lifting the identity on $\mathbb{Z}$ as in the commutative diagram

$$\begin{array}{ccccccccc}
\cdots & \longrightarrow & P_1 & \longrightarrow & P_0 & \xrightarrow{\varepsilon} & \mathbb{Z} & \longrightarrow & 0 \\
 & & \downarrow{\scriptstyle \mu_1} & & \downarrow{\scriptstyle \mu_0} & & \| & & \\
\cdots & \longrightarrow & Q_1 & \longrightarrow & Q_0 & \xrightarrow{\varepsilon'} & \mathbb{Z} & \longrightarrow & 0
\end{array}$$

The inflation map

$$\mathrm{inf}_{G/N,G}\colon H^n(G/N, M) \to H^n(G, M)$$

sends the class of a $G/N$-cocycle $\zeta\colon Q_n \to M$ to the class of the cocycle $\zeta\mu_n$. The fact that $\zeta\mu_n$ is a cocycle follows from the commutativity. It is standard homological algebra to check that the inflation is independent of the resolutions and of the choice of $\mu$.

**4.3.** *Transfer.* There is one other standard homomorphism of this type, called the transfer or corestriction. The transfer dates back to Schur and is an idea inherited from the theory of finite groups. To define it we must assume that $H$ is a subgroup of finite index $|G : H|$ in $G$. Most texts present it in terms of induction of modules but the definition can be stated more simply. We begin by choosing a $\mathbb{Z}G$-resolution $(P, \partial)$ of $\mathbb{Z}$ and make it into a $\mathbb{Z}H$-resolution by restriction. Let $x_1, \ldots, x_r$ be any set of left coset representatives of $H$ in $G$, $r = |G : H|$. Let $f\colon P_n \to M$ be a $\mathbb{Z}H$-cocycle. Then the map

$$tr_H^G(f)\colon P_n \to M, \quad \text{defined by } tr_H^G(f)(u) = \sum_{i=1}^{r} x_i f\left(x_i^{-1} u\right),$$

$u \in P_n$, is a $\mathbb{Z}G$-homomorphism and a $\mathbb{Z}G$-cocycle. It is independent of the choice of the coset representatives, and if $f = g\partial_n$ for $g \in \operatorname{Hom}_{\mathbb{Z}H}(P_{n-1}, M)$ then $tr_H^G(f) = tr_H^G(g) \circ \partial_n$ is a $\mathbb{Z}G$-coboundary. Therefore the map which sends $cls(f) \in H^n(H, M)$ to $cls(tr_H^G(f)) \in H^n(G, M)$ is well defined. This is the transfer homomorphisms and we denote it by $tr_H^G$.

**4.4.** *Restriction-transfer applications.* Notice that if $f\colon P_n \to M$ is a $\mathbb{Z}G$-homomorphism then $tr_H^G(f)(u) = |G : H| f(u)$, $u \in P_n$. Consequently if $f$ is a $G$-cocycle then

$$tr_H^G\left(\operatorname{res}_{G,H}(cls(f))\right) = |G : H| \cdot cls(f).$$

More generally, $tr_H^G \circ \operatorname{res}_{G,H} = |G : H|$. Two applications of this fact follow.

Suppose that $G$ is a finite group with order $|G|$. Let $E = \{1\} \subseteq G$ be the identity subgroup. Then $H^n(E, M) = 0$ for all $n > 0$ and any $\mathbb{Z}E$-module $M$. So if $M$ is a $\mathbb{Z}G$-module, then

$$|G| \cdot H^n(G, M) = |G : H| \cdot H^n(G, M) = tr_E^G\left(\operatorname{res}_{G,E}\left(H^n(G, M)\right)\right) = 0.$$

The second application shows that if $|G : H|$ is not divisible by a prime number $p$ and if $H^n(G, M)_p$ denotes the $p$-torsion in $H^n(G, M)$, then $\operatorname{res}_{G,H}$ is injective on $H^n(G, M)_p$. This is a simple result of the fact that $tr_H^G \circ \operatorname{res}_{G,H}$ is invertible on $H^n(G, M)_p$. In particular if $k$ is a field of characteristic $p$ and if $P$ is a Sylow $p$-subgroup of a finite group $G$ then $\operatorname{res}_{G,P}$ is injective on $H^n(G, k)$, because $H^n(G, k)$ is a $k$-vector space. The same also holds if $k$ is replaced by a $kG$-module $M$, as coefficients.

**4.5.** *Induction of modules.* One other relation between the cohomology of $G$ and a subgroup is given by induction on the coefficient module. Let $M$ be a $\mathbb{Z}H$-module. The induced and coinduced modules of $M$ are defined to be

$$\operatorname{Ind}_H^G(M) = \mathbb{Z}G \otimes_{\mathbb{Z}H} M \quad \text{and} \quad \operatorname{Coind}_H^G(M) = \operatorname{Hom}_{\mathbb{Z}H}(\mathbb{Z}G, M)$$

respectively. They are made into $\mathbb{Z}G$-modules by defining $x(g \otimes m) = xg \otimes m$ and $(xf)(g) = f(gx)$, for $x, g \in G$, $m \in M$, $f \in \operatorname{Hom}_{\mathbb{Z}H}(\mathbb{Z}G, M)$. If $G$ is finite then the induced and coinduced modules of $M$ are isomorphic. The so-called "Shapiro-Lemma" says that

$$H_*(H, M) \cong H_*\left(G, \operatorname{Ind}_H^G(M)\right) \quad \text{and} \quad H^*(H, M) \cong H^*\left(G, \operatorname{Coind}_H^G(M)\right).$$

In the cohomology case the isomorphism is induced from the isomorphism

$$\psi\colon \operatorname{Hom}_{\mathbb{Z}G}\big(P_n, \operatorname{Hom}_{\mathbb{Z}H}(\mathbb{Z}G, M)\big) \to \operatorname{Hom}_{\mathbb{Z}H}(P_n, M),$$

given by $\psi(f)(u) = (f(u))(1)$ for $u \in P_n$. The inverse is given by $(\psi^{-1}(f)(u))(x) = f(xu)$ for $x \in \mathbb{Z}G$. These isomorphisms respect cup products (see below).

**4.6.** *Product structures.* The cup product on group cohomology can be defined in any of several equivalent ways depending on the situation with the coefficients. In $H^*(G, \mathbb{Z})$ a product can be given by the Yoneda splice on exact sequences representing cohomology elements, or if we regard cohomology elements as chain maps on projective resolutions, then the product can be defined by composition of the chain maps. For other coefficients it is easiest first to define an "outer" cup product.

Given a $\mathbb{Z}G$-projective resolution $(P, \partial)$ of $\mathbb{Z}$, we may form the tensor product $(P \otimes_{\mathbb{Z}} P, \partial)$ which is then a projective resolution of $\mathbb{Z} \otimes \mathbb{Z} \cong \mathbb{Z}$ (see (2.6)). Then there is a chain map $\mu\colon (P, \partial) \to (P \otimes P, \partial)$ which lifts the identity on $\mathbb{Z}$. The chain map $\mu$ is sometimes called a diagonal approximation. Suppose that $M$ and $N$ are $kG$-modules and $\alpha\colon P_m \to M$, $\beta\colon P_n \to N$ are cocycles. The tensor product of the maps $\alpha \otimes \beta\colon P_m \otimes P_n \to M \otimes N$ when composed with the projection $\rho_{m,n}\colon (P \otimes P)_{m+n} \to P_m \otimes P_n$ and the chain map $\mu_{m+n}\colon P_{m+n} \to (P \otimes P)_{m+n}$ is a cocycle. It is easy to check that it is a coboundary if either $\alpha$ or $\beta$ is a coboundary. So the product

$$H^m(G, M) \otimes H^n(G, N) \to H^{m+n}(G, M \otimes_{\mathbb{Z}} N)$$

is defined by $cls(\alpha) \otimes cls(\beta) \to cls((\alpha \otimes \beta) \circ \rho_{m,n} \circ \mu_{m+n})$. If $M = N = R$ is some ring on which $G$ acts by automorphisms, then the multiplication $R \otimes_{\mathbb{Z}} R \to R$ gives a product

$$H^m(G, R) \otimes H^n(G, R) \to H^{m+n}(G, R \otimes_{\mathbb{Z}} R) \to H^{m+n}(G, R).$$

The associativity of the product is guaranteed by the coassociativity of the diagonal approximation. That is, the chain maps $(\mu \otimes 1) \circ \mu$ and $(1 \otimes \mu) \circ \mu$ which map $(P, \partial)$ to $(P \otimes P \otimes P, \partial)$ are chain homotopic since they both lift the identity on $\mathbb{Z}$. Consequently they induce the same map on cohomology.

In the case that $R$ is a commutative ring of coefficients with trivial $G$-action, the cup product satisfies the commutative law $\zeta\gamma = (-1)^{mn}\gamma\zeta$ for $m = \deg(\zeta)$, $n = \deg(\gamma)$. So if $\zeta \in H^m(G, R)$ and $m$ is an odd integer then $\zeta^2 = (-1)^{m^2}\zeta^2 = -\zeta^2$. Therefore either $\zeta^2 = 0$ or $\zeta^2$ has (additive) order 2.

**4.7.** *Examples of cohomology rings.* Let $G = \langle z_1, \ldots, z_n \rangle$ be the free abelian group of rank $n$ (see (2.7)). For each $i$, we have that $H^*(\langle z_i \rangle, \mathbb{Z}) = \mathbb{Z}[\zeta_i]/(\zeta_i^2)$. That is, $H^*(\langle z_i \rangle, \mathbb{Z})$ is a free $\mathbb{Z}$-module of rank 2 with generators 1 in degree 0 and $\zeta_i$ in degree 1. Also $\zeta_i^2 = 0$. Then, as in (2.7),

$$H^*(G, \mathbb{Z}) = H^*\big(\langle z_1 \rangle, \mathbb{Z}\big) \otimes \cdots \otimes H^*\big(\langle z_n \rangle, \mathbb{Z}\big).$$

Let $\nu_i = 1 \otimes \cdots \otimes \zeta_i \otimes \cdots \otimes 1$ ($\zeta_i$ in $i$-th spot). The commutativity relations says that $\nu_i \nu_j = -\nu_j \nu_i$. Otherwise the multiplication respects the tensor product so that $\nu_i^2 = 0$. It follows that

$$H^*(G, \mathbb{Z}) = \Lambda_{\mathbb{Z}}(\nu_1, \ldots, \nu_n),$$

is the $\mathbb{Z}$-exterior algebra on $\nu_1, \ldots, \nu_n$.

Next suppose that $G = \langle x \mid x^p = 1 \rangle$ as in (2.4). Assume that $p$ is a prime and that $k$ is a field of characteristic $p > 0$. Then $H^n(G, k) \cong k$ for all $n \geqslant 0$. Let $\gamma_n$ be a generator for $H^n(G, k)$. We have two possibilities.

a) If $p = 2$, then it can be shown that $\gamma^n \neq 0$ and hence by adjusting scalars we have that $H^*(G, k) = k[\gamma_1]$ is a polynomial ring in $\gamma_1$.

b) If $p > 2$ then $\gamma_1^2 = 0$ since $\gamma_1^2$ must have additive order $p \neq 2$. However it can be shown that $\gamma_2^n \neq 0$ and $\gamma_1 \gamma_2^n \neq 0$ for all $n > 0$. So we have $H^*(G, k) = k[\gamma_1, \gamma_2]/(\gamma_1^2)$.

Finally let $G = \langle x_1, \ldots, x_n \rangle$ be an elementary abelian $p$-group ($x_i^p = 1$, $x_i x_j = x_j x_i$) and let $k$ be a field of characteristic $p > 0$. Let

$$\eta_i = 1 \otimes \cdots \otimes \gamma_1 \otimes \cdots \otimes 1$$

where $\gamma_1 \in H^1(\langle x_i \rangle, k)$ appears in the $i$-th position in the factorization

$$H^*(G, k) = H^*\big(\langle x_1 \rangle, k\big) \otimes \cdots \otimes H^*\big(\langle x_n \rangle, k\big)$$

(see (2.7)). If $p = 2$, then $H^*(G, k) = k[\eta_1, \ldots, \eta_n]$. On the other hand if $p \neq 2$, then $\eta_i^2 = 0$. So let $\zeta_i = 1 \otimes \cdots \otimes \gamma_2 \otimes \cdots \otimes 1$ ($i$-th position). Using case (b) above, we see that

$$H^*(G, k) = k[\zeta_1, \ldots, \zeta_n] \otimes \Lambda_k(\eta_1, \ldots, \eta_n).$$

**4.8.** *Spectral sequences.* A spectral sequence is a sequence of complexes which, by taking successive (co)homologies converges to the (co)homology of a given complex or to some graded version thereof. Any of the standard texts on homological algebra or cohomology of groups contains an account of the theory of spectral sequences. Probably the most complete listing of the standard sequences is in [McC]. The most commonly used spectral sequence in group cohomology is the Lyndon–Hochschild–Serre (LHS) spectral sequence. One method of constructing the LHS sequence is outlined in the following.

**4.9.** *The LHS spectral sequence.* Suppose that $H$ is a normal subgroup of $G$. Let $(P, \partial)$ and $(Q, \partial')$ be, respectively, a $\mathbb{Z}G$- and $\mathbb{Z}(G/H)$-projective resolution of $\mathbb{Z}$. We regard $(Q, \partial')$ as a complex of $\mathbb{Z}G$-modules on which $H$ acts trivially. The tensor product $(Q \otimes P, \partial)$ is a projective $\mathbb{Z}G$-resolution of $\mathbb{Z}$. This resolution provides a filtration of the cohomology of $G$ with coefficients in any $\mathbb{Z}G$-module $M$. If $\zeta \in H^m(G, M)$, then $\zeta$ is represented by a cocycle

$$f\colon\ (Q \otimes P)_m = \bigoplus_{j=0}^{m} Q_j \otimes P_{m-j} \to M.$$

We say that $\zeta$ is in the $i$-th filtration $\mathcal{F}_i(H^m(G,M))$ if there exists a cocycle $f$ representing $\zeta$ such that $f$ is supported on

$$\sum_{j=i}^{m} Q_j \otimes P_{m-j}.$$

That is, $f(Q_j \otimes P_{m-j}) = 0$ if $j < i$.

Now let $E_0^{r,s} = \mathrm{Hom}_{\mathbb{Z}G}(Q_r \otimes P_s, M)$. This is a double cochain complex with two different boundary homomorphisms induced from the boundary homomorphisms on $Q$ and $P$. The cohomology of the total complex is, of course, $H^*(G,M)$. We proceed to the first page of the spectral sequence by taking the cohomology with respect to $(1 \otimes \partial)^*$, the coboundary homomorphism induced from that on $(P, \partial)$. Before doing this we should recognize that

$$E_0^{r,s} = \mathrm{Hom}_{\mathbb{Z}G}(Q_r \otimes P_s, M) \cong \mathrm{Hom}_{\mathbb{Z}(G/H)}\big(Q_r, \mathrm{Hom}_{\mathbb{Z}H}(P_s, M)\big),$$

where the isomorphism sends $f$ to $\theta(f)$ such that $\theta(f)(u))(v) = f(u \otimes v)$ for $u \in Q_r$, $v \in P_s$. Consequently the homology with respect to $(1 \otimes \partial)^*$ is

$$E_1^{r,s} = \mathrm{Hom}_{\mathbb{Z}(G/H)}\big(Q_r, H^s(H,M)\big).$$

Next take the cohomology with respect to the boundary homomorphism of $(Q, \partial)$. This is the $d_1$-differential and it maps $d_1$: $E_1^{r,s} \to E_1^{r+1,s}$. Its cohomology gives the terms of the $E_2$ page and we can see that $E_2^{r,s} \cong H^r(G/H, H^s(H,M))$. The differential on the $E_2$ page is $d_2$: $E_2^{r,s} \to E_2^{r+2,s-1}$, and the cohomology gives the $E_3$-page. In general, the $E_t$ page has a boundary homomorphism $d_t$: $E_t^{r,s} \to E_t^{r+t,s-t+1}$ and the cohomology gives the terms of the $E_{t+1}$ page. The spectral sequence converges because for any pair $r, s$ there are only a finite number of values of $t$ for which either $d_t$: $E_t^{r,s} \to E_t^{r+t,s-t+1}$ or $d_t$: $E^{r-t,s+t-1} \to E_t^{r,s}$ are not zero. That is $E_t^{a,b} = 0$ if either $a < 0$ or $b < 0$. In particular, $E_t^{r,s} = E_{t+1}^{r,s} = \cdots = E_\infty^{r,s}$ if $t > r + s$.

For $t > n + 1$, the terms of the $E_t$ page of the spectral sequence are precisely the factors in the filtration $\mathcal{F}$ on $H^n(G,M)$. That is, we have

$$0 \subseteq \mathcal{F}_n\big(H^n(G,M)\big) \subseteq \cdots \subseteq \mathcal{F}_0\big(H^n(G,M)\big) = H^n(G,M)$$

where

$$\mathcal{F}_i\big(H^n(G,M)\big)/\mathcal{F}_{i+1}\big(H^n(G,M)\big) \cong E_t^{i,n-i} = E_\infty^{i,n-i}.$$

**4.10.** *The edge homomorphisms.* On the $E_2$ page of the LHS spectral sequence we have

$$E_2^{r,0} = H^r\big(G/H, H^0(H,M)\big) \cong H^r\big(G/H, M^H\big)$$

where $M^H$ is the $G/H$-module of $H$-fixed points on $M$. For $t \geqslant 2$ the boundary $d_t$ is zero on $E_t^{r,0}$. Hence $E_\infty^{r,0}$ is a quotient of $E_2^{r,0}$. The edge homomorphism

$$E_2^{r,0} = H^r(G/H, M^H) \to E_\infty^{r,0} \subseteq H^r(G, M)$$

is the inflation $H^r(G/H, M^H) \to H^r(G, M^H)$ composed with the map induced by the inclusion $M^H \subseteq M$. So

$$\mathcal{F}_n(H^n(G, M)) = \inf_{G/H,G}(H^n(G/H, M^H)).$$

On the other edge we have that $E_2^{0,s} = H^0(G/H, H^s(G, M))$ is the set of fixed points of $H^s(H, M)$ under the action of the factor group $G/H$. Moreover we have that

$$E_\infty^{0,s} \subseteq E_{s+1}^{0,s} \subseteq \cdots \subseteq E_2^{0,s}$$

is a subgroup of $H^s(H, M)$. Of course, $E_\infty^{0,s} = H^s(G, M)/\mathcal{F}_1(H^s(G, M))$. The quotient map $H^s(G, M) \to E_\infty^{0,s} \subseteq H^s(H, M)$ is, in fact, the restriction homomorphism $\mathrm{res}_{G,H}$.

**4.11.** *The Five term sequence.* From the spectral sequence it is immediate to deduce the standard five term exact sequence for low dimensional group cohomology. The exact sequence is

$$\begin{aligned} 0 \to H^1(G/H, M^H) \xrightarrow{\alpha} H^1(G, M) \xrightarrow{\beta} H^1(H, M)^{G/H} \\ \xrightarrow{\gamma} H^2(G/H, M^H) \xrightarrow{\delta} H^2(G, M). \end{aligned}$$

Here $\alpha$ is the inclusion $E_\infty^{1,0} \subseteq E_2^{1,0}$ into $H^1(G, M)$, $\beta$ is the quotient map by $\mathcal{F}_1(H^1(G, M))$ followed by the inclusion into $H^1(G, M)^{G/H} \cong E_2^{0,1}$, $\gamma$ is $d_2$, and $\delta$ is the quotient by the image of $d_2$ followed by the inclusion into $H^2(G, M)$. Several extension and other versions of the sequence have been given as for example in [BrL] and [Hue3] (see also [ElR] and [ChW]).

**4.12.** *Remarks on LHS.* In the original paper of Hochschild and Serre [HoS] the spectral sequence was defined using the standard (bar) resolution. The construction certainly duplicated some of the ideas of Lyndon [Lyn], though it is not clear that Lyndon had a spectral sequence. Hochschild and Serre actually defined at least two spectral sequences. It was long assumed that the constructions gave equivalent objects, but no proof was offered until that of Evens [Eve2]. The equivalence has been further generalized by Beyl [Bey], but the reader should see [Bar] for the most thorough treatment.

**4.13.** *Other spectral sequences.* There are several other spectral sequences which apply to group cohomology. For example the Eilenberg–Moore sequence was used very effectively in the calculations of Rusin [Rus2]. The hyper-cohomology spectral sequence was useful in [BeC3] and might yet prove useful with constructions such as that in [Web1]. The Bockstein spectral sequence was used in the calculations [LPS] and [HaK]. It should also be mentioned that the LHS spectral sequence as well as some of the others have multiplicative structures.

**4.14.** *The norm map.* When $H$ is a subgroup of finite index in $G$, it is possible to define a multiplicative induction from $\mathbb{Z}H$-modules to $\mathbb{Z}G$-modules. Just as ordinary induction can be used to define the transfer map, which is an additive homomorphism on cohomology, multiplicative induction defines a multiplicative map, the (Evens) norm map

$$H^{2n}(H, \mathbb{Z}) \to H^{2n|G:H|}(G, \mathbb{Z})$$

(see [Eve1]). The norm map can be defined on $H^{2n}(H, M)$ for any $\mathbb{Z}H$-module $M$, but the definition becomes very complicated whenever $M$ is not a commutative ring with trivial $H$-action.

An outline of the construction starts with the fact that, because $|G : H|$ is finite, there is an embedding of $G$ into the wreath product $\widehat{G} = G/H \wr H$. An element of $H^{2n}(H, \mathbb{Z}) \cong \mathrm{Ext}^{2n}_{\mathbb{Z}H}(\mathbb{Z}, \mathbb{Z})$ is represented by an exact sequence $E$ of length $2n$, that begins and ends with $\mathbb{Z}$. Taking the tensor product of $|G : H|$ copies of this sequence (or more precisely, of the complexes formed by truncating the terminal copy of $\mathbb{Z}$) we get an exact sequence of length $2n|G:H|$ on which $G/H$ acts by permuting the copies. Hence we may regard it as a sequence of $\mathbb{Z}\widehat{G}$-modules. Now restrict the sequence to the subgroup isomorphic to $G$. The cohomology class of the restriction, as an element of $H^{2n|G:H|}(G, \mathbb{Z})$ is defined to be the image under the norm map. The norm was originally devised by Evens to prove the finite generation of cohomology rings of finite groups. However it has been very useful in several other ways.

**4.15.** *The Steenrod operations.* The Steenrod operations were invented in a topological setting as operations on the cohomology of spaces. The operations form an algebra and the cohomology of any space or group is an algebra over the Steenrod algebra. Moreover the action of the algebra is natural and commutes with such constructions as restriction, inflation and spectral sequences. See [BeC4] for one example of applications of the Steenrod algebra. Unfortunately, even a list of the properties of the Steenrod operations is too long to include here. See [Ben1] for a condensed list of properties without proof. An algebraic development of the Steenrod operations can be found in [Ben2].

## 5. Topics in finite groups

In recent years there has been a great deal of activity in the area of cohomology of finite groups. Much of it has been motivated by applications to the module theory for group algebras and to topology. Unlike the case in many of the classical applications, the relevant structures have been the more general ring and module theoretic constructions. The methods have included some algebraic geometry and commutative ring theory, as well as simplicial geometry and topology. In several cases the important calculations have first been made or theorems first been proved in the mod-$p$ case for $p$ a prime dividing the order of the group.

**5.1.** *Varieties and cohomology rings.* The primary ingredient which is necessary to begin a theory of cohomology rings is the finite generation theorem of Evens [Eve1] (see also [Ven]). It says that $H^*(G, \mathbb{Z})$ and $H^*(G, k)$, for $k$ a field, are finitely generated. Moreover if $M$ is any finitely generated $\mathbb{Z}G$-module ($kG$-module), then $H^*(G, M)$ is a finitely generated module over $H^*(G, \mathbb{Z})$ (respectively, $H^*(G, k)$). The results were proved by reducing to the case of a $p$-group, using the norm map, and then applying induction on the group order. So when $k$ is a field of characteristic $p$, the $k$-algebra $H^*(G, k)$ has an associated affine variety, $V_G(k)$, its maximal ideal spectrum. Notice that if $p$ is odd then $H^*(G, k)$ is not commutative. However this does not affect the spectrum, because only elements of odd degree fail to commute and they are all nilpotent.

It was Quillen [Qun1] who showed that the dimension of $V_G(k)$ is equal to the $p$-rank of $G$. The components of $V_G(k)$ correspond to the conjugacy classes of maximal elementary abelian $p$-subgroups by way of the restriction maps. In particular, the intersection of the kernels of the maps $\mathrm{res}_{G,E}$: $H^*(G, k) \to H^*(E, k)$, for $E$ an elementary abelian $p$-group, is a nilpotent ideal, and the map on varieties $V_E(k) \to V_G(k)$ is always finite-to-one. The Dimension Theorem is a consequence of the fact that $V_E(k)$ is affine $k$-space, $k^n$, if $k$ is algebraically closed and $E$ has $p$-rank $n$ (see (4.7)). An algebraic proof of the theorem is given in [QuV]. In [Qun2] it was shown that $V_G(k)$ is stratified according to the action of $G$ on its elementary abelian $p$-subgroups.

**5.2.** *Varieties and modules.* Most of the results mentioned above have been extended to the support varieties for finitely generated $kG$-modules. If $M$ is such a module, then let $J(M)$ denote the ideal in $H^*(G, k)$, which is the annihilator of

$$H^*(G, \mathrm{Hom}_k(M, M)) \cong \mathrm{Ext}^*_{kG}(M, M).$$

Let $V_G(M)$ be the subvariety of $V_G(k)$ corresponding to $J(M)$. The first attempt at generalizing Quillen's Dimension Theorem was set in the context of the complexity of $M$ [AlE1]. Roughly speaking, the complexity of $M$ is the polynomial rate of growth of the ring $\mathrm{Ext}^*_{kG}(M, M)$ and is equal to the dimension of $V_G(M)$. Subsequent refinements [AlE2, Avr] proved that

$$V_G(M) = \bigcup \mathrm{res}^*_{G,E}(V_E(M))$$

where the union is taken over the maximal elementary abelian $p$-subgroups of $G$.

In the case of an elementary abelian group $G = E = \langle x_1, \ldots, x_n \rangle$ of order $p^n$, the variety $V_E(M)$ can be computed directly from the structure of $M$. For this, suppose that $k$ is algebraically closed. For $\alpha = (\alpha_1, \ldots, \alpha_n) \in k^n$, let

$$u_\alpha = 1 + \sum \alpha_i(x_i - 1) \in kE.$$

Note that $u_\alpha$ is a unit of order $p$ in $kE$. Let $U_\alpha = \langle u_\alpha \rangle$, and let

$$V_E^r(M) = \{0\} \cup \{\alpha \in k^n \mid M \text{ is not free as a } kU\text{-module}\}.$$

Then $V_E^r(M)$ is called the rank variety of $M$ [Car2]. Under proper identification it is equal to the cohomological variety $V_G(M)$ [AvS]. The varieties of modules have several interesting properties. A couple of the most significant are that $V_G(M) = \{0\}$ if and only if $M$ is projective and that $V_G(M \otimes_k N) = V_G(M) \cap V_G(N)$. Hence it is possible to discover whether the tensor product of two modules is projective without computing the product. See [BeC2] for one application of this fact.

**5.3.** *Depth and systems of parameters.* Several recent investigations have looked at regular sequences and systems of parameters for cohomology rings. It had been a folk theorem (proved using LHS) that any element in $H^{2n}(G,k)$ whose restriction to a cyclic subgroup in the center of a Sylow $p$-subgroup of $G$ is nonzero, must be a regular element. Duflot [Dufl] has extended the result to show that $H^*(G,k)$ has a regular sequence whose length is at least equal to the rank of the center of the Sylow $p$-subgroup of $G$. Landweber and Stong [LaS] have conjectured that the Dickson invariants, taken in proper order, should provide a regular sequence of maximal length. In [BeC3] the authors investigate the case in which the cohomology ring is Cohen–Macaulay and also offer some speculation on the ring structure when the depth is smaller then the $p$-rank.

The cohomology rings of modules, $\operatorname{Ext}^*_{kG}(M,M)$ seem to be more problematic in that they may not be graded commutative. However even here, some progress has been made on the maximal idea structure [Niw].

**5.4.** *Irreducible modules.* One question which has attracted some attention is the role of irreducible modules in the cohomology of groups. In particular, there is an old conjecture that if $M$ is a simple $kG$-module in the principal block of $kG$, then $H^*(G,M) \neq 0$. In the last decade, Linnell [Lin] and Linnell and Stammbach [LiS1, LiS2] have succeeded in proving the statement true under the assumption that $G$ is $p$-solvable or $p$-constrained. The proofs rely on the natural occurrence of the simple modules in things like the composition series of the group itself. Of course, this does not happen in general and the conjecture remains open. The question of when an arbitrary module in the principal block has vanishing cohomology was investigated in [BCR]. However this may or may not be applicable to the question about simple modules.

**5.5.** *Chern rings.* Some recent work has focused on subrings of the cohomology ring $H^*(G,\mathbb{Z})$ or $H^*(G,k)$. The Chern ring $Ch(G) \subseteq H^*(G,\mathbb{Z})$ is the subring generated by the Chern classes coming from all complex representations of $G$. Given a representation $\rho$ of $G$ over $\mathbb{C}$, we may assume that $\rho$: $G \to U(n)$ the unitary group of $n \times n$ matrices. Now $H^*(U(n),\mathbb{Z})$ is a polynomial ring in classes in degrees $2,4,\ldots,2n$. The pull-backs of these classes under $\rho$ are called the Chern classes of $G$ for this representation. If the representation is faithful then $H^*(G,\mathbb{Z})$ is finitely generated as a module over $H^*(BU(n),\mathbb{Z})$. For more information see Thomas' book [Thm3] and the papers [Alz, CaL, Lea, Thm1] and [Thm2].

**5.6.** *Calculations.* Recent years have witnessed a great many interesting and sometimes very impressive calculations of cohomology groups and rings over both the ordinary integers and fields of finite characteristic. Almost all of the computations make use of some sort of spectral sequence. Some, such as [Rus2] have used computer technology. The computations [AMM2] and [AdM] have used the work of Webb on the Brown complex [Web2]. Others employed diagrammatic methods from representation theory [BeC1, BCo]. Some of the computations which have been done are the following.

Extraspecial $p$-groups: [Die, BeC4, HaK, Lea, Lew] and [Qun3];

Other $p$-groups: [EvP1, MiM, Rus1] and [Rus2];

Classical simple groups: [AMM1, AMM2, Car2, Chp3, FiP, Hun1, Hun2, Kle, Qun4, Tez, TeY1, TeY2] and [TeY3];

Sporadic simple groups: [AdM, Chp2] and [Lea].

Several other studies such as [Car2] and [CPS] have considered the cohomology with coefficients in simple modules.

**5.7.** *Other investigations.* Finally we list a few of the other studies which are worth mentioning. The exponents of integral cohomology have been investigated by a few authors (e.g., [Ade1, Car3]). The ring of universally stable elements in the image of every restriction map were studied in [EvP2]. Varieties have been defined and studied for many of the standard constructions such as the image of the transfer map in [EvF]. A great deal of impressive work has been done on the connection between the cohomology of finite groups and that of algebraic groups. See [Fre] for one example. Quillen proved several of the results on varieties for compact Lie groups. For further results in this direction see [Fes1] and [Fes2]. Adem has noted that many of the results on varieties apply to groups with finite virtual cohomological dimension [Qun1, Ade2, Ade3].

## 6. Topics in infinite groups

For infinite groups the cohomology theory is a primary device for classification. It is not surprising that the theory has been mixed with many methods from geometry and topology. However the study of group actions on spaces and geometric objects such as trees could be the subject of another whole essay. We mention some parts of it only briefly here.

**6.1.** *Cohomological dimension.* From the early stages of homological algebra it was natural to ask the question of what groups had finite cohomology or had cohomology in only finitely many degrees. The cohomological dimension of a group $G$ ($cd(G)$) is the smallest natural number $n$ for which there is a $\mathbb{Z}G$-projective resolution $(P, \partial)$ of $\mathbb{Z}$ with $P_i = 0$ for all $i > n$. It is also the smallest $n$ such that $H^i(G, M) = 0$ for all $i > n$ and all $\mathbb{Z}G$-modules $M$. From the definition it is clear that $cd(H) < cd(G)$ whenever $H$ is a subgroup of $G$. Because nontrivial finite groups do not have finite cohomological dimension, all groups with finite cohomological dimension must be torsion free. Serre [Ser] has shown that if $H \subseteq G$ and $G$ is torsion free then $cd(H) = cd(G)$ provided $|G : H|$ is finite.

An easy topological argument proves that free groups have cohomological dimension one. For example if $G$ is free on $n$ generators then the wedge of $n$ circles is a $K(G,1)$. The converse of the statement for $G$ finitely generated was proved in a celebrated paper of Stallings [Sta2]. Swan [Swa] extended the work to show that $cd(G) = 1$ implies that the group $G$ is free. Dunwoody has pushed the result even further. Let $cd_R(G) = n$ if the coefficient ring $R$ as a trivial $RG$-module has a projective resolution of length $n$. In [Dun] it was shown that $cd_R(G) \leqslant 1$ if and only if $G$ is the fundamental group of a graph of groups in which no vertex group has a finite subgroup whose order fails to be invertible in $R$.

**6.2.** *Other finiteness conditions.* A group $G$ is said to be of type $FP_n$ if there exists a projective $\mathbb{Z}G$-resolution

$$\cdots \to P_1 \to P_0 \to \mathbb{Z} \to 0$$

such that $P_i$ is finitely generated for $i < n$. It can be shown that the word "projective" can be replaced by "free" with no loss. Recently Abels and Brown [AbB] (see also [Brw3]) gave examples of groups $G_n$ such that $G_n$ is of type $FP_{n-1}$ but not of type $FP_n$ for $n \geqslant 3$. The examples had been previously known. Abels had proved that they were not of type $FP_1$ and Bieri had shown that $G_n$ was not of type $FP_n$. See the survey by Bieri [Bie2] for more background.

A group $G$ is of type $FP$ if $cd(G) < \infty$ and also $G$ is of type $FP_\infty$. It can be shown that for such a group

$$cd(G) = \max\{n \mid H^n(G, \mathbb{Z}G) \neq 0\}.$$

Brown and Geoghegan have found an example $G$ with $H^n(G, \mathbb{Z}G) = 0$ for $n > 0$, and $G$ torsion free, but $G$ not of type $FP$. In this case $G$ is of type $FP_\infty$ but not of type $FP$. For other work in this direction see [Abe, CuV, Kro1] and [Rat1].

Other groups of interest arise from the sort of combinatorics associated to computers. Automatic groups were introduced in [CEHPT] where it was shown that they are of type $FP_\infty$. The Anick–Groves–Squier Theorem states that a group with a finite complete rewriting system is of type $FP_\infty$ (see [Gro]). A more topological approach is given by Brown [Brw4].

**6.3.** *Duality groups.* A group $G$ of type $FP$ is a duality group if there exists a $\mathbb{Z}G$-module $D$ and a positive integer $n$ such that $H^i(G, M) \cong H_{n-i}(G, D \otimes_{\mathbb{Z}} M)$ for all $i$ and all $\mathbb{Z}G$-modules $M$. If $G$ is a duality group with $n = cd(G)$ then it can be assumed that $D = H^n(G, \mathbb{Z}G)$ with the right-handed $G$-action. Also $D$ must be torsion free as an abelian group. See Brown's book [Brw1] for details. The group $G$ is said to be a Poincaré duality group if in addition $D \cong \mathbb{Z}$. Examples include finitely generated free abelian groups. Finitely generated free groups, knot groups (which always have cohomological dimension 2) and arithmetic groups are duality groups but not Poincaré duality groups.

It is still an open question as to whether there must exist a finite dimensional $K(G,1)$ if $G$ is a Poincaré duality group. Eckmann, Müller and Linnell solve the problem in dimension 2 by showing that a Poincaré duality group of dimension 2 must be the fundamental group of a surface (see the survey article by Eckmann [Eck2]). Partial results in dimension 3 have been given by Hillman [Hlm]. For other recent work on duality groups see [KrR] and [Rot].

**6.4.** *Other results on finite cohomological dimension.* Groups of cohomological dimension 2 have been much studied but have yielded no spectacular results as for those of dimension one. Background on this problem can be found in the notes [Bie1] and [Bie2]. A classical theorem of Lyndon says that torsion free one-relator groups have cohomological dimension two or less. No such theorem is valid for two-relator groups [How], but see also [Hue1]. Other related results include [Gil, HoS] and [Rat2].

Another notion which is of particular interest in knot theory is that of the ends of groups. The ends of a group are defined topologically, but the number of ends is equal to $1 + \mathrm{rank}(H^1(G, \mathbb{Z}G))$. It is well known that a group has one, two or infinitely many ends. Stallings [Sta1] has settled some of the structure of groups with more than one end. For other recent work on ends of groups see [Hol2] and [GeM].

**6.5.** *Solvable and nilpotent groups.* Kropholler [Kro2] has recently finished a problem on solvable groups with finite cohomological dimension. He showed that for such a group $G$ the following are equivalent:

(i) $G$ is constructible,

(ii) $G$ is a duality group,

(iii) $G$ is of type $FP$,

(iv) $cd_{\mathbb{Z}}(G) =$ torsion free rank of $G$,

and

(v) $cd_{\mathbb{Q}}(G) =$ torsion free rank of $G$.

All but the implications (v) $\Rightarrow$ (i) had been proved by Baumslag and Bieri. Kropholler used the methods of [GiS] to finish this part.

A rank function has been defined for torsion free nilpotent groups. It is the sum of the ordinary ranks of the quotients $G_{i-1}/G_i$ where

$$1 = G_n \subseteq G_{n-1} \subseteq \cdots \subseteq G_0 = G$$

is a central series. This rank is called the Hirsh number and is denoted $hG$. It is easy to show that $cd(G) = hG$. Kropholler has recently announced a proof that the rank of any solvable group of type $FP_\infty$ is finite [Kro4]. Several recent results have dealt with the homology and cohomology of nilpotent groups [Rob2, Rob3]. See [Rob1, Rob4] for other references. In addition, the results of Huebschmann [Hue4, Hue5] provide nice answers for groups of class 2. See also [BaD, BDG, Kro3] and [Lor] for other interesting results.

**6.6.** *Virtual ideas.* A group is said to have a virtual property if it has a subgroup of finite index with that property. The property may be something like finite cohomological dimension or duality. In the case of a group $G$ with finite virtual cohomological dimension, there is the notion of Farrell cohomology, $\widehat{H}^n(G, \mathbb{Z})$, which is defined for all values of $n$ [Far]. Farrell cohomology is a generalization of the Tate cohomology for finite groups. It has all of the usual properties of ordinary cohomology with which it coincides in large degrees. It vanishes if the group is torsion free, but it seems to depend not only on the finite subgroups of $G$ but also on the way in which they are embedded in $G$ [Ade2, AdC]. See Brown's book [Brw1] for a full account. For linear groups of finite virtual cohomological dimension see [AlS]. Eckmann and Müller have extended their work on Poincaré duality groups of dimension 2 to virtual $PD^2$-groups [EMu]. For other similar considerations see [GeG].

**6.7.** *Ranks and Euler characteristics.* The Euler characteristic is another major invariant for groups which are virtually $FP$. This notion coincides with the topological Euler characteristic if the group $G$ has a finite $K(G,1)$ (which requires that $G$ be torsion free). The Hattori–Stallings rank (see [Brw1]) makes it possible to define Euler characteristic using complexes of projective modules rather than free modules. Thus it is defined for any group of type $FP$. If $G$ is of virtual type $FP$ then $G$ has a subgroup $H$ with $|G:H| < \infty$ and with torsion free of type $FP$. So the Euler characteristic of $G$ is defined as

$$\chi(G) = \chi(H)/|G:H|.$$

It need not be integral or even positive. In fact if $G$ is finite then $\chi(G) = 1/|G|$, while $\chi(SL_2(\mathbb{Z})) = -1/12$. Brown has shown that the denominator of $\chi(G)$ divides the least common multiple of the orders of the finite subgroups. For more recent results see [Brw2, Dye] and [SmV].

**6.8.** *Relation modules and related objects.* There has been a lot of recent activity concerned with the calculation of cohomology and structures related to the presentations of a group $G$. If

$$1 \to N \to F \to G \to 1$$

is an exact sequence and $F$ is a free group, then the relation module for $G$ is the abelian group $M = N/[N,N]$ made into a $\mathbb{Z}G$-module by the conjugation action. The corresponding extension

$$1 \to M \to F/[N,N] \to G \to 1$$

is called the free abelianized extension. Several investigations have looked at the homology and cohomology of this extension. Gupta [Gup] has shown that the homology can have torsion even when $G$ is torsion free. See [HaS, KKS, Kuz] and [PrS1] for other references.

**6.9.** *Miscellaneous results.* Several results of interest do not fit neatly into the other categories. They include the work of Mislin [Mis] and Eckmann and Mislin [EMi] on Chern classes and the stable range results for congruence subgroups by Charney [Cha] and Arlettaz [Arl]. We should mention also the calculations [PrS2] and [ScV]. Finally Baumslag, Dyer and Miller [BDM] have considered the inverse problem of finding a group $G$ whose $n$-th homology group is isomorphic to a previously given group $A = H_n(G, \mathbb{Z})$.

## References


[AbB] H. Abels and K.S. Brown, *Finiteness property for soluble S-arithmetic groups: an example*, J. Pure Appl. Algebra **44** (1987), 77–83.

[Abe] H. Åberg, *Bieri–Strebel valuations (of finite rank)*, Proc. London Math. Soc. (3) **52** (1986), 269–304.

[Ade1] A. Adem, *Cohomological exponents of $\mathbb{Z}G$-lattices*, J. Pure Appl. Algebra **58** (1989), 1–5.

[Ade2] A. Adem, *On the exponent of cohomology of discrete groups*, Bull. London Math. Soc. **21** (1989), 585–590.

[Ade3] A. Adem, *Euler characteristic and cohomology of p-local discrete groups*, J. Algebra **149** (1992), 183–196.

[AdC] A. Adem and J.F. Carlson, *Discrete groups with large exponents in cohomology*, J. Pure Appl. Algebra **66** (1990), 111–120.

[AMM1] A. Adem, J. Maginnis and R.J. Milgram, *Symmetric invariants and cohomology of groups*, Math. Ann. **278** (1990), 391–411.

[AMM2] A. Adem, J. Maginnis and R.J. Milgram, *The geometry and cohomology of the Mathieu group, $M_{12}$*, J. Algebra **139** (1991), 90–133.

[AdM] A. Adem and R.J. Milgram, *$A_5$-invariants, the cohomology of $L_3(4)$ and related extensions*, Proc. London Math. Soc. (3) **66** (1993), 187–224.

[AlE1] J.L. Alperin and L. Evens, *Representations, resolutions and Quillen's dimension theorem*, J. Pure Appl. Algebra **22** (1981), 1–9.

[AlE2] J.L. Alperin and L. Evens, *Varieties and elementary abelian subgroups*, J. Pure Appl. Algebra **26** (1982), 221–227.

[AlS] R.C. Alperin and P.B. Shalen, *Linear group of finite cohomological dimension*, Invent. Math. **66** (1982), 89–98.

[Alz] K. Alzubaidy, *Rank 2 p-groups, $p > 3$, and Chern classes*, Pacific J. Math. **103** (1982), 259–267.

[Arl] D. Arlettaz, *On the homology and cohomology of congruence subgroups*, J. Pure Appl. Algebra **44** (1987), 3–12.

[AsG] M. Aschbacher and R. Guralnick, *Some applications of the first cohomology group*, J. Algebra **90** (1984), 446–460.

[Avr] G.S. Avrunin, *Annihilators of cohomology modules*, J. Algebra **69** (1981), 150–154.

[AvS] G.S. Avrunin and L. Scott, *Quillen stratification for modules*, Invent. Math. **66** (1982), 227–286.

[Bar] D.W. Barnes, *Spectral sequence constructors in algebra and topology*, Mem. Amer. Math. Soc. **53**, no. 317, (1985).

[BaD] G. Baumslag and E. Dyer, *The integral homology of finitely generated metabelian groups, I*, Amer. J. Math. **104** (1982), 173–182.

[BDG] G. Baumslag, E. Dyer and J.R.J. Groves, *The integral homology of finitely generated metabelian groups, II*, Amer. J. Math. **109** (1987), 133–155.

[BDM] G. Baumslag, E. Dyer and C.F. Miller, III, *On the integral homology of finitely presented groups*, Topology **22** (1983), 27–46.

[Ben1] D. Benson, *Modular Representation Theory: New Trends and Methods*, SLNM 1081, Springer, New York (1984).

[Ben2] D.J. Benson, *Representations and Cohomology, I and II*, Cambridge Studies in Advanced Mathematics, vols 30, 31, Cambridge Univ. Press, New York (1991).

[BeC1] D.J. Benson and J.F. Carlson, *Diagrammatic methods for modular representations and cohomology*, Comm. Algebra **15** (1987), 53–121.

[BeC2] D.J. Benson and J.F. Carlson, *Complexity and multiple complexes*, Math. Z. **195** (1987), 221–238.

[BeC3] D.J. Benson and J.F. Carlson, *Projective resolutions and Poincaré duality complexes*, Trans. Amer. Math. Soc. **342** (1992), 447–488.

[BeC4] D.J. Benson and J.F. Carlson, *The cohomology of extraspecial groups*, Bull. London Math. Soc. **24** (1992), 209–235.

[BCR] D.J. Benson, J.F. Carlson and G. Robinson, *On the vanishing of cohomology*, J. Algebra **131** (1990), 40–73.

[BCo] D.J. Benson and F.R. Cohen, *Mapping class groups of low genus and their cohomology*, Mem. Amer. Math. Soc. **90**, no. 443, (1991).

[Bey] F.R. Beyl, *The spectral sequence of a group extension*, Bull. Sci. Math. (2) **105** (1981), 417–434.

[BeT] F.R. Beyl and J. Tappe, *Group Extensions, Representations, and the Schur Multiplicator*, SLNM 958, Springer, Berlin (1982).

[Bie1] R. Bieri, *On groups of cohomological dimension* 2, Topology and Algebra, Monograph Enseign. Math. 26, Univ. Genieve, Geneva (1978), 55–62.

[Bie2] R. Bieri, *Homological dimension of discrete groups*, 2nd ed., Queen Mary College Mathematics Notes, London (1981).

[Brw1] K.S. Brown, *Cohomology of Groups*, Springer, New York (1982).

[Brw2] K.S. Brown, *Complete Euler characteristics and fixed point theory*, J. Pure Appl. Algebra **24** (1982), 103–121.

[Brw3] K.S. Brown, *Finiteness properties of groups*, J. Pure Appl. Algebra **44** (1987), 45–75.

[Brw4] K.S. Brown, *The geometry of rewriting systems. A proof of the Anick–Groves–Squier Theorem*, Proceedings of the MSRI Workshop on Algorithms, Word Problems and Classification in Combinatorial Group Theory (to appear).

[BrG1] K.S. Brown and R. Geoghegan, *Cohomology with free coefficients of the fundamental group of a graph of groups*, Comment. Math. Helv. **60** (1985), 31–45.

[BrG2] K.S. Brown and R. Geoghegan, *A infinite-dimensional torsion free $FP_\infty$ group*, Invent. Math., 367–381.

[BrE] R. Brown and G.J. Ellis, *Hopf formula for the higher homology of a group*, Bull. London Math. Soc. **20** (1988), 124–128.

[BrL] R. Brown and J.L. Loday, *Excision homotopique en basse dimension*, C. R. Acad. Sci. Paris Ser. I Math. **298** (1984), 353–356.

[CEHPT] J.W. Cannon, D.B.A. Epstein, D.F. Holt, M.S. Paterson and W.P. Thurston, *Word processing in group theory*.

[CaL] H. Cardenas and E. Lluis, *On the Chern classes of representations of the symmetric groups*, Group Theory, de Gruyter, Berlin (1989), 333–345.

[Car1] J.F. Carlson, *The varieties and the cohomology ring of a module*, J. Algebra **85** (1983), 104–143.

[Car2] J.F. Carlson, *The cohomology of irreducible modules over $SL(2,p^n)$*, Proc. London Math. Soc. (3) **47** (1983), 480–492.

[Car3] J.F. Carlson, *Exponents of modules and maps*, Invent. Math. **95** (1989), 13–24.

[Cas] G. Carlsson, *G.B. Segal's Burnside ring conjecture for $(\mathbb{Z}/2)^k$*, Topology **22** (1983), 83–103.

[CaE] H. Cartan and S. Eilenberg, *Homological Algebra*, Princeton Univ. Press, Princeton (1956).

[Chp1] G.R. Chapman, *The cohomology ring of a finite abelian group*, Proc. London Math. Soc. (3) **45** (1982), 564–576.

[Chp2] G.R. Chapman, *Generators and relations for the cohomology ring of Janko's first group in the first twenty-one dimensions*, Groups-St. Andrews 1981, London Math. Soc. Lecture Note Ser. vol. 71, Cambridge Univ. Press, Cambridge (1982), 201–206.

[Chp3] G.R. Chapman, *A class of invariant polynomials and an application in group cohomology*, Publ. Sec. Math. Univ. Autònoma Barcelona **28** (1984), 5–17.

[Cha] R. Charney, *On the problem of homology stability for congruence subgroups*, Comm. Algebra **12** (1984), 2081–2123.

[ChW] C.C. Cheng and Y.C. Wu, *An eight term exact sequence associated with a group extension*, Michigan Math. J. **28** (1981), 323–340.

[CPS] E. Cline, B. Parshall and L. Scott, *Cohomology of finite groups of Lie type, I*, Inst. Hautes Études Sci. Publ. Math. **45** (1975), 169–191.

[Con] B. Conrad, *Crossed $n$-fold extensions of groups, $n$-fold extensions of modules, and higher multipliers*, J. Pure Appl. Algebra **36** (1985), 225–235.

[CuV] M. Culler and K. Vogtmann, *Modulii of graphs and automorphisms of free groups*, Invent. Math. (1986), 91–119.

[Die] T. Diethelm, *The mod p cohomology rings of the nonabelian split metacyclic p-groups*, Arch. Math. **44** (1985), 29–38.

[Duf1] J. Duflot, *Depth and equivariant cohomology*, Comment. Math. Helv. **56** (1981), 627–637.

[Duf2] J. Duflot, *A Hopf algebra associated to the cohomology of the symmetric groups*, The Arcata Conference on Representations of Finite Groups, Proc. Symp. Pure Math. vol. 47, Part 2, Amer. Math. Soc., Providence, RI (1987), 171–186.

[Dun] M.J. Dunwoody, *Accessibility and groups of cohomological dimension one*, Proc. London Math. Soc. (3) **38** (1979), 193–215.

[Dye] M.N. Dyer, *Euler characteristics of groups*, Quart. J. Math. Oxford (2) **38** (1987), 35–44.

[Eck1] B. Eckmann, *Poincaré duality groups of dimension two are surface groups*, Combinatorial Group Theory and Topology, Ann. of Math. Stud. vol. 111, Princeton Univ. Press, Princeton, NJ (1987), 35–51.

[Eck2] B. Eckmann, *Some recent developments in the homology theory of groups (groups of finite and virtually finite dimension)*, J. Pure Appl. Algebra **19** (1980), 61–75.

[EMi] B. Eckmann and G. Mislin, *Chern classes of group representations over a number field*, Compositio Math. **44** (1981), 41–65.

[EMu] B. Eckmann and H. Müller, *Plane motion groups and virtual Poincaré duality of dimension two*, Invent. Math. **69** (1982), 293–310.

[ElR] G.J. Ellis and C. Rodriguez-Fernandez, *An exterior product for the homology of groups with integral coefficients modulo p*, Cahiers Topologie Géom. Différentielle Categoriques **30** (1989), 339–343.

[Eve1] L. Evens, *The cohomology ring of a finite group*, Trans. Amer. Math. Soc. **101** (1961), 224–239.

[Eve2] L. Evens, *The spectral sequence of a finite group extension stops*, Trans. Amer. Math. Soc. **212** (1975), 269–277.

[EvF] L. Evens and M. Feshbach, *Carlson's theorem on varieties and transfers*, J. Pure Appl. Algebra **57** (1989), 39–45.

[EvP1] L. Evens and S. Priddy, *The cohomology of the semidihedral group*, Conference on Algebraic Topology in Honor of Peter Hilton, Contemp. Math. vol. 37, Amer. Math. Soc., Providence, RI (1985), 61–72.

[EvP2] L. Evens and S. Priddy, *The ring of universally stable elements*, Quart. J. Math. Oxford Ser (2) **40** (1989), 339–407.

[Far] F.T. Farrell, *An extension of Tate cohomology to a class of infinite groups*, J. Pure and Appl. Algebra **10** (1977), 153–161.

[Fes1] M. Feshbach, *Some general theorems on the cohomology of classifying spaces of compact Lie groups*, Trans. Amer. Math. Soc. **264** (1981), 49–58.

[Fes2] M. Feshbach, *p-subgroups of compact Lie groups and torsion of infinite height in $H^*(BG)$, II*, Michigan Math. J. **29** (1982), 299–306.

[FiP] Z. Fiedorowicz and S. Priddy, *Homology of classical groups over finite fields and their associated infinite loop spaces*, SLNM 674, Springer, Berlin (1978).

[Fre] E.M. Friedlander, *Multiplicative stability for the cohomology of finite Chevalley groups*, Comment. Math. Helv. **63** (1988), 108–113.

[GeG] T.V. Gedrich and K.W. Gruenberg, *Complete cohomological functors on groups*, Topology Appl. **25** (1987), 203–223.

[GeM] R. Geoghengan and M.L. Mihalik, *Free abelian cohomology of groups and ends of universal covers*, J. Pure Appl. Algebra **36** (1985), 123–137.

[Gil] D. Gildenhuys, *Classification of solvable groups of cohomological dimension two*, Math. Z. **166** (1979), 21–25.

[GiS] D. Gildenhuys and R. Strebel, *On the cohomology of soluble groups, II*, J. Pure Appl. Algebra **26** (1982), 293–323.

[Gro] J.R.J. Groves, *Rewriting systems and homology of groups*, Groups-Canberra 1989, SLNM 1456, Springer, Berlin (1991), 114–141.

[Gru] K.W. Gruenberg, *Cohomological Topics in Group Theory*, SLNM 143, Springer, New York (1970).

[Gup] C.K. Gupta, *The free centre-by-metabelian groups*, J. Austral. Math. Soc. **16** (1973), 294–200.

[Gur] R.M. Guralnick, *Generation of simple groups*, J. Algebra **103** (1986), 381–401.

[HaK] M. Harada and A. Kono, *On the integral cohomology of extraspecial 2-groups*, Proceedings of the Northwestern Conference on Cohomology of Groups, J. Pure Appl. Algebra **44** (1987), 215–219.

[HaS] B. Hartley and R. Stöhr, *Homology of higher relation modules and torsion in free central extensions of groups*, Proc. London Math. Soc. (3) **62** (1991), 325–352.

[Hil] R.O. Hill, Jr., *A natural algebraic interpretation of the group cohomology group* $H^n(Q, A), n \geqslant 4$, Notices Amer. Math. Soc. **25** (1978), A–351.

[Hlm] J.A. Hillman, *Seifert fiber-spaces and Poincaré duality groups*, Math. Z. **190** (1985), 365–369.

[HlS] P.J. Hilton and U. Stammbach, *A Course in Homological Algebra*, Springer, New York (1971).

[HoS] G. Hochschild and J.-P. Serre, *Cohomology of group extensions*, Trans. Amer. Math. Soc. **74** (1953), 110–134.

[Hol1] D.F. Holt, *An interpretation of the cohomology groups* $H^n(G, M)$, J. Algebra **60** (1979), 307–320.

[Hol2] D.F. Holt, *Uncountable locally finite groups have one end*, Bull. London Math. Soc. **13** (1981), 557–560.

[Hol3] D.F. Holt, *The mechanical computation of first and second cohomology groups*, J. Symbolic Comput. **1** (1985), 351–361.

[Hol4] D.F. Holt, *A computer program for the calculation of a covering group of a finite group*, J. Pure Appl. Algebra **35** (1985), 287–295.

[Hop] H. Hopf, *Fundamental gruppe und zweite Bettische Gruppe*, Comment. Math. Helv. **14** (1942), 257–309.

[How] J. Howie, *Two-relator groups with prescribed cohomological dimension*, Proc. Amer. Math. Soc. **100** (1987), 393–394.

[HoS] J. Howie and H.R. Schneebeli, *Groups of finite quasiprojective dimension*, Comment. Math. Helv. **54** (1979), 615–628.

[Hue1] J. Huebschmann, *Cohomology theory of aspherical groups and of small cancellation groups*, J. Pure Appl. Algebra **14** (1979), 137–143.

[Hue2] J. Huebschmann, *Crossed $n$-fold extensions of group cohomology*, Comment. Math. Helv. **55** (1980), 203–313.

[Hue3] J. Huebschmann, *Group extensions, crossed pairs and an eight term exact sequence*, J. Reine Angew. Math. **321** (1981), 150–172.

[Hue4] J. Huebschmann, *Cohomology of nilpotent groups of class* 2, J. Algebra **126** (1989), 400–450.

[Hue5] J. Huebschmann, *Perturbation theory and free resolutions for nilpotent groups of class* 2, J. Algebra **126** (1989), 348–399.

[Hur] W. Hurewicz, *Beiträge zur Topologie der Deformationen IV, Asphärische Räume*, Nederl. Akad. Wetensch. Proc. **39** (1936), 215–224.

[Hun1] Pham Viet Hung, *The* mod 2 *cohomology algebra of a Sylow* 2*-subgroup of the general linear group* $GL(4, \mathbb{Z}_2)$, Acta Math. Vietnam **11** (1986), 136–155.

[Hun2] Pham Viet Hung, *The algebra* $H^*(GL(4,\mathbb{Z}_2),\mathbb{Z}_2)$, Acta Math. Vietnam **12** (1987), 51–60.

[KaT] D.M. Kan and W.P. Thurston, *Every connected space has the homology of a* $K(\pi, 1)$, Topology **15** (1976), 253–258.

[Kle] S.N. Kleinerman, *Cohomology of Chevalley groups of exceptional Lie type*, Mem. Amer. Math. Soc., no. 268, (1982).

[KKS] L.G. Kovacs, Yu.V. Kuz′min and R. Stöhr, *Homology of free abelianized extensions of groups*, Mat. Sb. **182**(4) (1991), 526–542.

[Kro1] P.H. Kropholler, *A note on the cohomology of metabelian groups*, Math. Proc. Cambridge Philos. Soc. **98** (1985), 437–445.

[Kro2] P.H. Kropholler, *Cohomological dimension of soluble groups*, J. Pure Appl. Algebra **43** (1986), 281–287.

[Kro3] P.H. Kropholler, *The cohomology of soluble groups of finite rank*, Proc. London Math. Soc. (3) **53** (1986), 453–473.

[Kro4] P.H. Kropholler, *Soluble groups of type $FP_\infty$ have finite torsion free rank*, Bull. London Math. Soc. **25** (1993), 558–566.

[KrR] P.H. Kropholler and M.A. Roller, *Splittings of Poincaré duality groups, III*, J. London Math. Soc. (2) (1989), 271–284.

[Kuz] Yu.V. Kuz′min, *Homology theory of free abelianized extensions*, Comm. Algebra **16** (1988), 2447–2533.

[LaS] P.S. Landweber and R.E. Stong, *The depth of rings of invariants over finite fields*, Number Theory (New York 1984–1985), SLNM 1240, Springer, Berlin (1987), 259–274.

[Lag] S. Lang, *Rapport sur la Cohomologie des Groupes*, W.A. Benjamin, New York (1966).

[Lan] J. Lannes, *Sur la cohomologie modulo p des p-groupes abeliens elementaires*, Homotopy Theory (Durham, 1985), London Math. Soc. Lecture Notes No. 117, Cambridge Univ. Press, New York (1987), 97–116.

[Lea] I. Leary, *The cohomology of certain finite groups*, Ph.D. Thesis, Cambridge (1990).

[Lew] G. Lewis, *The integral cohomology rings of group of order $p^3$*, Trans. Amer. Math. Soc. (1968), 501–529.

[Lin] P.A. Linnell, *Cohomology of finite soluble groups*, J. Algebra **107** (1987), 53–62.

[LiS1] P.A. Linnell and U. Stammbach, *On the cohomology of p-constrained groups*, The Arcata Conference on Representations of Finite Groups, Proc. Symp. Pure Math. vol. 47, Part 1, Amer. Math. Soc., Providence, RI (1987), 467–469.

[LiS2] P.A. Linnell and U. Stammbach, *The cohomology of p-constrained groups*, J. Pure Appl. Algebra **49** (1987), 273–279.

[LiS3] P.A. Linnell and U. Stammbach, *The block structure of Ext of p-soluble groups*, J. Algebra **108** (1987), 280–282.

[Lor] M. Lorenz, *On the cohomology of polycyclic-by-finite groups*, J. Pure Appl. Algebra **40** (1986), 87–98.

[Lyn] R.C. Lyndon, *The cohomology theory of group extensions*, Duke Math. J. **15** (1948), 271–292.

[McL1] S. MacLane, *Homology*, Springer, New York (1963).

[McL2] S. MacLane, *Origins of the cohomology of groups*, Enseign. Math. (2) **24** (1982), 1–29.

[McL3] S. MacLane, *Historical note*, J. Algebra **60** (1979), 319–320.

[McC] J. McCleary, *User's guide to spectral sequences*, Mathematics Lecture Series vol. 12, Publish or Perish, Inc., Delaware (1985).

[MiM] Phan Anh Minh and Huỳnh Mui, *The* mod *p cohomology algebra of the group $M(p^n)$*, Acta Math. Vietnam **7** (1982), 17–26.

[Mis] G. Mislin, *Classes characteristiques pour les représentations de groupes discrets*, Paul Dubreil and Marie-Paul Malliavin Algebra Seminar, SLNM 294, Springer, Berlin (1982), 296–309.

[Niw] T. Niwasaki, *On Carlson's conjecture for cohomology rings of modules*, J. Pure Appl. Algebra **59** (1989), 265–277.

[PrS1] S.J. Pride and R. Stöhr, *Relation modules with presentations in which each relator involves exactly two types of generators*, J. London Math. Soc. (2) **38** (1988), 99–111.

[PrS2] S.J. Pride and R. Stöhr, *The (co)homology of aspherical Coxeter groups*, J. London Math. Soc. **42** (1990), 49–63.

[LPS] E. Lluis Puebla and V. Snaith, *On the integral homology of the symmetric group*, Bol. Soc. Mat. Mexicana (2) **27** (1982), 51–55.

[Qun1] D. Quillen, *The spectrum of an equivariant cohomology ring I, II*, Ann. Math. (2) **94** (1971), 549–602.

[Qun2] D. Quillen, *A cohomological criterion for p-nilpotence*, J. Pure Appl. Algebra **1** (1971), 361–372.

[Qun3] D. Quillen, *The* mod-2 *cohomology ring of extra-special* 2-*groups and the spinor groups*, Math. Ann. **194** (1971), 197–212.

[Qun4] D. Quillen, *On the cohomology and $K$-theory of the general linear group over a finite field*, Ann. Math. **96** (1972), 552–556.

[QuV] D. Quillen and B.B. Venkov, *Cohomology of finite groups and elementary abelian subgroups*, Topology **11** (1972), 317–318.

[Rat1] J.G. Ratcliffe, *Finiteness conditions for groups*, J. Pure Appl. Algebra **27** (1983), 173–185.

[Rat2] J.G. Ratcliffe, *The cohomology ring of a one-relator group*, Contributions to Group Theory, Contemp. Math. vol. 33, Amer. Math. Soc., Providence, RI, (1984), 455–466

[Rob1] D.J.S. Robinson, *Applications of cohomology to the theory of groups*, Groups-St. Andrews 1981, London Math. Soc. Lecture Notes Ser. 71, Cambridge Univ. Press, Cambridge (1982), 46–80.

[Rob2] D.J.S. Robinson, *Cohomology of locally nilpotent groups*, J. Pure Appl. Algebra **48** (1987), 281–300.

[Rob3] D.J.S. Robinson, *Homology and cohomology of locally supersolvable groups*, Math. Proc. Cambridge Philos. Soc. **102** (1987), 233–250.

[Rob4] D.J.S. Robinson, *Cohomology in infinite group theory*, Group Theory, de Gruyter, Berlin (1989), 29–53.

[Rot] J. Rotman, *A remark on integral duality*, Abelian Group Theory, SLNM 1006, Springer, Berlin (1983), 711–719.

[Rus1] D.J. Rusin, *The* mod-2 *cohomology of metacyclic* 2 *groups*, J. Pure Appl. Math. **44** (1987), 315–327.

[Rus2] D.J. Rusin, *The cohomology of the groups of order* 32, Math. Comp. **53** (1989), 359–385.

[Sce] O. Schreier, *Über die Erweiterungen von Gruppen, I*, Monatsh. Math. Phys. **34** (1926), 165–180.

[Scu] I. Schur, *Über die Darstellung der endlichen Gruppen durch gebrochene lineare Substitutionen*, J. Reine Angew. Math. **127** (1904), 20–50.

[ScV] J. Schwermer and K. Vogtmann, *The integral homology of $SL_2$ and $PSL_2$ of Euclidean imaginary quadratic integers*, Comment. Math. Helv. **58** (1983), 573–598.

[Ser] J.-P. Serre, *Cohomologie des groupes discrets*, Ann. of Math. Stud. vol. 70 (1971), 77–169.

[SmV] J. Smillie and K. Vogtmann, *Automorphisms of graphs, p-subgroups of* $\mathrm{Out}(F_n)$ *and the Euler characteristic of* $\mathrm{Out}(F_n)$, J. Pure Appl. Algebra **49** (1987), 187–200.

[Sta1] J.R. Stallings, *On torsion-free groups with infinitely many ends*, Ann. Math. **88** (1968), 312–334.

[Sta2] J. Stallings, *Groups of cohomological dimension* 1, Application of Categorical Algebra, Proc. Sympos. Pure Math. vol. 17, Amer. Math. Soc., Providence, RI (1970), 124–128.

[Stm] U. Stammbach, *Homology in Group Theory*, SLNM 359, Springer, Berlin (1973).

[StE] N.E. Steenrod and D.B.A. Epstein, *Cohomology Operations*, Ann. of Math. Stud. vol. 50, Princeton Univ. Press, Princeton (1962).

[Sto] R. Stöhr, *A generalized Hopf formula for higher homology groups*, Comment. Math. Helv. **64** (1989), 187–199.

[Swa] R.G. Swan, *Groups of cohomological dimension one*, J. Algebra **12** (1969), 585–601.

[Tez] M. Tezuka, *The cohomology of $SL_2(\mathbb{F}_p)$ and the Hecke algebra actions*, Kodai Math. J. **9** (1986), 440–455.

[TeY1] M. Tezuka and N. Yagita, *The cohomology of subgroups of $GL_n(F_q)$*, Proceedings of the Northwestern Homotopy Theory Conference, Contemp. Math., Amer. Math. Soc., Providence, RI (1983), 379–396.

[TeY2] M. Tezuka and N. Yagita, *The mod-p cohomology ring of $GL_3(F_p)$*, J. Algebra **81** (1983), 295–303.

[TeY3] M. Tezuka and N. Yagita, *The varieties of mod p cohomology rings of extra special p-groups for an odd prime p*, Math. Proc. Cambridge Philos. Soc. **94** (1983), 449–459.

[Thm1] C.B. Thomas, *Filtrations on the representation ring of a finite group*, Proceedings of the Northwestern Homotopy Theory Conference, Contemp. Math. vol. 19, Amer. Math. Soc., Providence, RI (1983), 297–405.

[Thm2] C.B. Thomas, *Chern classes of representations*, Bull. London Math. Soc. **18** (1986), 225–240.

[Thm3] C.B. Thomas, *Characteristic Classes and the Cohomology of Finite Groups*, Cambridge Univ. Press, New York (1986).

[Ven] B.B. Venkov, *Cohomology algebras for some classifying spaces*, Dokl. Akad. Nauk SSSR **127** (1959), 943–944.

[Wal] C.T.C. Wall (ed.), *List of problems, Homological Group Theory*, London Math. Soc. Lecture Notes 36, Cambridge Univ. Press, Cambridge (1979), 369–394.

[Web1] P.J. Webb, *Complexes, group cohomology, and induction theorems for the Green ring*, J. Algebra **104** (1986), 351–357.

[Web2] P.J. Webb, *A local method in group cohomology*, Comment. Math. Helv. **62** (1987), 135–167.

# Relative Homological Algebra. Cohomology of Categories, Posets and Coalgebras

A.I. Generalov

*Sankt Petersburg University, Bibliotechnaya pl. 2, Sankt Petersburg, Russia*

*Contents*



HANDBOOK OF ALGEBRA, VOL. 1
Edited by M. Hazewinkel

## Introduction

The origins of relative homological algebra can be found in different branches of algebra but mainly in the theory of abelian groups and in the representation theory of finite groups. Prüfer introduced in 1923 the notion of purity which nowadays is one of the most important notions of abelian group theory [F]. Generalizations of purity in the category of abelian groups and in module categories have many applications and are really tools of homological algebra.

In representation theory we have also the important notions of relative projectives and relative injectives, and the analysis of their properties has led Hochschild in 1956 to the discovery of "relative homological algebra" [Ho]. It is worth noting that ideas of relative homological algebra were contained "internally" in the "Homological algebra" of Cartan and Eilenberg [CE].

Finally, Buchsbaum [Bc] and others (see [Ma1]) have given axioms for a "proper class" of short exact sequences in any abelian category and MacLane has rewritten in his "Homology" [Ma1] a part of homological algebra from the point of view of relative homological algebra.

The first years after that have yielded many interesting examples of proper classes which have been used for proving "relative" versions of "absolute" theorems. Thus in representation theory Lam and Reiner [LR] discovered the relative Grothendieck group, Warfield [W] introduced the notion of Cohn-purity in a module category which generalized Prüfer purity and he used it for researching the interesting class of algebraically compact modules, Eilenberg and Moore [EM] generalized the notion of proper class starting from a triple and this allows us to develop relative homological algebra in categories more general than the abelian.

Mishina and Skornyakov in 1969 [MS] (see also the expanded English translation in 1976) and then Sklyarenko in 1978 [Sk] have given good surveys of the development of relative homological algebra at that time.

In this article we do not intend to give a review of all contributions to relative homological algebra but we only present the main ideas of the theory and also some recent advances in it.

In Section 1 we introduce the notion of a proper class of cokernels in a pre-abelian category. This generalization of the usual notion of a proper class of short exact sequences will be used in Section 2 for defining relative derived categories. This construction was made in [G7] to give a unified approach to homological algebra in pre-abelian categories which allows to include the approaches in [RW1, Y] in the framework of a single theory.

Section 3 is devoted to relative homological algebra in module categories and we discuss recent results on the classification of inductively closed proper classes which are closely related with algebraically compact modules. Some results concern the structure of such modules. We also discuss in this section the so-called "group of relations" of relative Grothendieck groups.

The language of relative homological algebra is useful in defining the cohomology of small categories (see [HiS]), and the corresponding theory is presented in Section 4. Moreover, we discuss there a new cohomology introduced by Baues and Wirsching [BaW] which generalizes the Hochschild–Mitchell cohomology [Mt1].

Section 5 contains some applications of the results of the preceding sections to the cohomology of partially ordered sets (= posets).

Section 6 contains the cohomology theory of coalgebras (including the relative case). For simplicity we restrict ourselves to the case where the base commutative ring is a field. Note that the general case has been considered in [J], which uses the relative homological algebra developed in [EM].

## 1. Proper classes in preabelian categories

**1.1.** Let $\mathbb{A}$ be a preabelian category, i.e. $\mathbb{A}$ is an additive category in which every morphism has a kernel and a cokernel. Any morphism $f\colon X \to Y$ in $\mathbb{A}$ admits a canonical decomposition

$$f\colon X \xrightarrow{\text{coim } f} \text{Coim } f \xrightarrow{\bar{f}} \text{Im } f \xrightarrow{\text{im } f} Y \tag{1}$$

where coim $f = \text{coker}(\ker f)$ is the coimage of $f$, im $f = \ker(\text{coker } f)$ is the image of $f$ (cf., e.g., [BD]). Recall that an abelian category is a preabelian category such that for any morphism $f$ the morphism $\bar{f}$ in (1) is an isomorphism.

A morphism is called a kernel (respectively a cokernel) if it is a kernel (respectively a cokernel) of some morphism. A morphism $f\colon A \to B$ is called a retraction if there is a morphism $g\colon B \to A$ with $fg = 1_B$.

A sequence

$$A \xrightarrow{i} B \xrightarrow{\sigma} C \tag{2}$$

is called a short exact sequence if $i = \ker \sigma$ and $\sigma = \text{coker}\, i$.

**1.2.** A class $\omega$ of cokernels in a preabelian category $\mathbb{A}$ is said to be a proper class (in short, p.c.) if the following axioms are satisfied:

P0. Every retraction in $\mathbb{A}$ belongs to $\omega$.

P1. If $\sigma, \tau \in \omega$ and $\sigma\tau$ exists then $\sigma\tau \in \omega$.

P2. For any pullback

$$\begin{array}{ccc} A' & \xrightarrow{\sigma'} & B' \\ {\scriptstyle f'}\downarrow & & \downarrow{\scriptstyle f} \\ A & \xrightarrow{\sigma} & B \end{array} \tag{3}$$

if $\sigma \in \omega$ then $\sigma' \in \omega$.

P3. If $\sigma\tau, \tau \in \omega$ then $\sigma \in \omega$.

A short exact sequence (2) is called $\omega$-proper if $\sigma \in \omega$.

From the definition we derive the following statement.

**1.3.** PROPOSITION [G7]. *Let $\omega$ be a p.c. in a preabelian category $\mathbb{A}$. Then:*

a) *$\omega$ is closed with respect to (finite) direct sums of morphisms;*

b) *for any morphism $\tau$ if $\sigma\tau \in \omega$ then $\sigma \in \omega$.*

We can define dually a p.c. of kernels. Usually we do not formulate (but freely use) the dual statements concerning this notion.

**1.4.** EXAMPLES.

a) A cokernel $\sigma$ is a preabelian category $\mathbb{A}$ is called semistable if for any pullback (3) the morphism $\sigma'$ is a cokernel. A semistable kernel is defined dually. A short exact sequence (2) is called stable if $i$ is a semistable kernel and $\sigma$ is a semistable cokernel. In this situation the morphisms $i$ and $\sigma$ are called a stable kernel and a stable cokernel respectively. The class of all stable cokernels (respectively stable kernels) is proper [RW1].

The class of all semistable cokernels (respectively semistable kernels) is also a p.c. [G7]. Note that axiom P2 implies that every cokernel $\sigma$ in a p.c. $\omega$ is semistable, and so the class of semistable cokernels is the largest p.c. of cokernels.

b) Richman and Walker [RW1, RW2] have investigated stable and semistable morphisms is specific preabelian categories such as: a) the category of topological (Hausdorff) modules; 2) the category of valuated groups, and others.

c) The class of all retractions in a preabelian category $\mathbb{A}$ is a p.c.

Further examples of p.c.s (in abelian categories) are discussed below (especially for module categories see Section 3).

**1.5.** REMARKS.

1) In general the class of all cokernels in a preabelian category is not proper.

2) If $\omega$ is a p.c. of cokernels then the class $\{\ker\sigma \mid \sigma \in \omega\}$ does not need to be a p.c. of kernels.

**1.6.** Let $\omega$ be a p.c. of cokernels in a preabelian category $\mathbb{A}$. An object $P \in \mathbb{A}$ is called $\omega$-projective if $P$ is projective w.r.t. all $\sigma \in \omega$, i.e. for any $\sigma$: $M \to N$ in $\omega$ the induced homomorphism $\sigma_*$: $\mathrm{Hom}_{\mathbb{A}}(P, M) \to \mathrm{Hom}_{\mathbb{A}}(P, N)$ is surjective. If $\omega$ is a p.c. of kernels then the notion of an $\omega$-injective object is defined dually.

Let $\mathcal{M}$ be a class of objects in $\mathbb{A}$. Denote by $\omega(\mathcal{M})$ the class of cokernels $\sigma$ in $\mathbb{A}$ such that all objects $M \in \mathcal{M}$ are projective w.r.t. $\sigma$. Plainly $\omega(\mathcal{M})$ is a p.c., and we say that $\omega(\mathcal{M})$ is projectively generated (by $\mathcal{M}$).

A p.c. $\omega$ of cokernels is said to be projective if for any object $A \in \mathbb{A}$ there is a cokernel $\sigma$: $P \to A$ in $\omega$ with $P$ $\omega$-projective.

**1.7.** PROPOSITION. *Let $\omega$ be a projective p.c. Then $\omega$ is projectively generated by the class $\mathcal{M}_\omega$ of $\omega$-projective objects.*

PROOF. Plainly, we have $\omega \subseteq \omega(\mathcal{M}_\omega)$. Now if $\sigma$: $A \to B$ is in $\omega(\mathcal{M}_\omega)$ consider a cokernel $\tau$: $P \to B$ in $\omega$ with $P$ $\omega$-projective. By definition there exists $\tau'$: $P \to A$ such that $\sigma\tau' = \tau$, and by 1.3b we have $\sigma \in \omega$. □

**1.8.** An object $P \in \mathbb{A}$ is said to be cokernel-projective if $P$ is projective w.r.t. all cokernels in $\mathbb{A}$. We say that a preabelian category $\mathbb{A}$ has enough cokernel-projective objects if for any $A \in \mathbb{A}$ there exists a cokernel $\sigma\colon P \to A$ with $P$ cokernel-projective.

**1.9.** PROPOSITION [G7]. *If a preabelian category $\mathbb{A}$ has enough cokernel-projectives then any cokernel in $\mathbb{A}$ is semistable.*

Hence in the situation of the proposition the class of all cokernels in $\mathbb{A}$ is proper and additionally it is projectively generated by the class of cokernel-projectives (cf. 1.7).

**1.10.** The notion of a proper class is more familiar for abelian categories. A p.c. $\omega$ of cokernels in an abelian category $\mathbb{A}$ is uniquely determined by the class of kernels $\{\ker \sigma \mid \sigma \in \omega\}$ which satisfies the dual axioms to P0–P3; we denote by $\mathrm{P0^{op}}, \ldots, \mathrm{P3^{op}}$ these duals respectively. So one can define a proper class of short exact sequences in an abelian category $\mathbb{A}$ using only some pairs of the dual axioms in this list. For example, in [Ma1] the axioms P0, $\mathrm{P0^{op}}$, P1, $\mathrm{P1^{op}}$, P3, $\mathrm{P3^{op}}$ (with slight modifications) are used. Also the axioms of an exact category in the sense of Quillen [Q] are the reformulation of the definition of a p.c. of short exact sequences.

**1.11.** Let $T\colon \mathbb{A} \to \mathbb{B}$ be a functor between abelian categories which reflects epimorphisms (i.e. $T$ is faithful) and suppose that it has a left adjoint $S\colon \mathbb{B} \to \mathbb{A}$, i.e. there are natural transformations $\alpha\colon \mathrm{Id}_{\mathbb{B}} \to TS$ and $\beta\colon ST \to \mathrm{Id}_{\mathbb{A}}$ such that $(\beta S)(S\alpha) = \mathrm{id}_S$, $(T\beta)(\alpha T) = \mathrm{id}_T$. Let $\omega$ be a p.c. of cokernels in $\mathbb{B}$. Define $T^{-1}(\omega)$ as a class of morphisms $f$ in $\mathbb{A}$ with $T(f) \in \omega$. By assumption $\omega' = T^{-1}(\omega)$ consists of cokernels in $\mathbb{A}$.

**1.12.** THEOREM [EM, HiS]. *Under the hypotheses above, suppose additionally that $\omega$ is a projective p.c. Then $\omega'$ is a projective p.c. of cokernels in $\mathbb{A}$.*

Slightly modifying the proof of this theorem we obtain the following result.

**1.13.** THEOREM. *Under the hypotheses above* (*see* 1.11) *let $\omega$ be a p.c. projectively generated by a class of objects $\mathcal{M} \subseteq \mathrm{Ob}(\mathbb{B})$. Then $\omega' = T^{-1}(\omega)$ is a p.c. projectively generated by the class $S(\mathcal{M}) = \{S(M) \mid M \in \mathcal{M}\}$.*

## 2. Relative derived categories

**2.1.** Fix any preabelian category $\mathbb{A}$ and any p.c. $\omega$ of cokernels in $\mathbb{A}$. Denote by $K(\mathbb{A})$ the homotopy category of the category $\mathbb{A}$, i.e. the quotient category of the category $\mathrm{Kom}(\mathbb{A})$ of (cochain) complexes over $\mathbb{A}$ modulo homotopy equivalence. On the category $\mathrm{Kom}(\mathbb{A})$ (and also $K(\mathbb{A})$) there is defined the shift operator $M \mapsto M[1]$ where $(M[1])^n = M^{n+1}$, $d_{M[1]} = -d_M$. The inverse of this operator we denote as follows: $M \mapsto M[-1]$.

A sequence in $\mathrm{Kom}\,(\mathbb{A})$

$$K^{\cdot} \xrightarrow{f} L^{\cdot} \xrightarrow{g} M^{\cdot} \tag{1}$$

is called a short exact sequence if for every $n \in \mathbb{Z}$ there is a short exact sequence in $\mathbb{A}$: $K^n \xrightarrow{f^n} L^n \xrightarrow{g^n} M^n$. The short exact sequence (1) is called $\omega$-proper if $g^n \in \omega$ for every $n$.

A complex $M^{\cdot} = (M^n, d_M^n)$ is called $\omega$-acyclic if for any $n \in \mathbb{Z}$ we have $d_M^n = \mu^n \nu^n$ with $\nu^n \in \omega$ and $\mu^n = \ker d_M^{n+1}$. Denote by $\mathcal{C}(\mathbb{A})$ the full subcategory of $K(\mathbb{A})$ which consists of $\omega$-acyclic complexes.

**2.2.** THEOREM [G7]. *Given an $\omega$-proper sequence* (1) *in* Kom($\mathbb{A}$) *such that some two complexes in it lie in $\mathcal{C}(\mathbb{A})$ the third also lies in $\mathcal{C}(\mathbb{A})$.*

If the category $\mathbb{A}$ is abelian this theorem can be derived from the long exact cohomology sequence which corresponds to the sequence (1). We have no such cohomology sequence in our general context, and the role of the above theorem is to replace it (when we can do so).

**2.3.** COROLLARY. *Given a commutative diagram in* $\mathbb{A}$

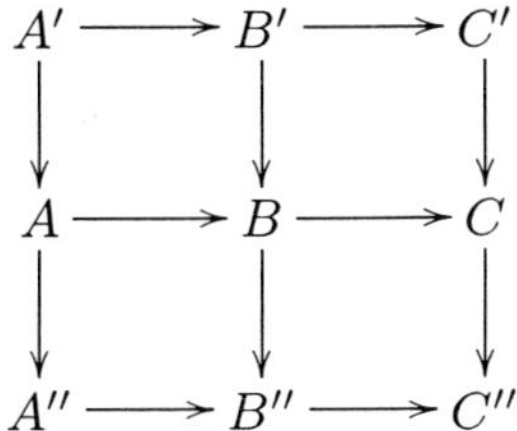

*if all rows and any two columns in it are $\omega$-proper short exact sequences then the third column is also $\omega$-proper provided the composition of the morphisms in the middle column is zero.*

This statement is a generalization of the well-known $3 \times 3$-lemma for abelian categories.

**2.4.** The mapping cone of a morphism $f\colon M^{\cdot} \to N^{\cdot}$ in Kom($\mathbb{A}$) is defined as the complex

$$C(f) = \left(M^{n+1} \oplus N^n, d_{C(f)}\right), \quad d_{C(f)} = \begin{pmatrix} -d_M & 0 \\ f & d_N \end{pmatrix}.$$

The cylinder Cyl($f$) of $f$ is defined as the mapping cone of the morphism

$$\begin{pmatrix} -1 \\ f \end{pmatrix}\colon M^{\cdot} \to M^{\cdot} \oplus N^{\cdot}.$$

We have the following sequence of complexes

$$M^{\cdot} \xrightarrow{f} N^{\cdot} \xrightarrow{g} C(f) \xrightarrow{h} M[1]$$

where

$$N^n \xrightarrow{g^n} C(f)^n = M^{n+1} \oplus N^n$$

and

$$C(f)^n \xrightarrow{h^n} M^{n+1}$$

are the canonical injection and projection respectively. The family of such sequences (up to isomorphism the so-called "distinguished triangles") defines the structure of a triangulated category on the homotopy category $K(\mathbb{A})$ (see the details in [Ha, V] or [GeM]).

**2.5.** PROPOSITION. *$\mathcal{C}(\mathbb{A})$ is a triangulated subcategory of the triangulated category $K(\mathbb{A})$.*

PROOF (*sketch*). We need only to prove that for any morphism $f\colon M^\bullet \to N^\bullet$ in $\mathcal{C}(\mathbb{A})$ the mapping cone $C(f)$ lies in $\mathcal{C}(\mathbb{A})$. We have the short exact sequence:

$$M^\bullet \to \mathrm{Cyl}(f) \to C(f),$$

and as $\mathrm{Cyl}(f) \cong N^\bullet \oplus C(-1_M)$ and $C(-1_M) \sim 0$ the desired statement follows from 2.2. □

**2.6.** A full triangulated subcategory $C$ of a triangulated category $\mathcal{D}$ is said to be épaisse [V] if the following condition is satisfied: given a composition in $\mathcal{D}$ $f\colon X \to V \to Y$ with $V \in \mathcal{C}$ and such that in the distinguished triangle containing $f$

$$X \xrightarrow{f} Y \longrightarrow Z \longrightarrow X[1]$$

$Z \in \mathcal{C}$ we have also $X, Y \in \mathcal{C}$.

The following important fact allows simplification of the recognition of épaisse subcategories.

**2.7.** THEOREM [Ri]. *The full triangulated subcategory $\mathcal{C}$ of a triangulated category $\mathcal{D}$ is épaisse iff any direct summand (in $\mathcal{D}$) of an object $V \in \mathcal{C}$ also lies in $\mathcal{C}$.*

Using this general fact we can prove the following result.

**2.8.** THEOREM [G7]. *Let $\mathbb{A}$ be a preabelian category, $\omega$ be a p.c. of cokernels in $\mathbb{A}$. Then the full subcategory $\mathcal{C}(\mathbb{A})$ of $\omega$-acyclic complexes over $\mathbb{A}$ is an épaisse subcategory of the homotopy category $K(\mathbb{A})$.*

Sketch of the proof. Let $Y \in \mathcal{C}(\mathbb{A})$ and $X$ be a direct summand of $Y$ in $K(\mathbb{A})$, i.e. we have the morphisms $f\colon X \to Y$, $g\colon Y \to X$ such that $gf \sim 1_X$. Consider the morphism $\varphi\colon Y \to C(gf)$ such that

$$\varphi^n = \begin{pmatrix} O \\ g^n \end{pmatrix}\colon\ Y^n \to X^{n+1} \oplus X^n.$$

As $C(gf) \cong C(1_x) \sim 0$, $C(gf) \in \mathcal{C}(\mathbb{A})$ and so by 2.5 $C(\varphi) \in \mathcal{C}(\mathbb{A})$. As is easily seen we have in $\mathrm{Kom}(\mathbb{A})$: $C(\varphi) \cong X[1] \oplus C(g)$, and consequently $X \in \mathcal{C}(\mathbb{A})$. Now the theorem follows from 2.7.

**2.9.** The significance of the notion of an épaisse subcategory is explained by the fact that one can construct the category fractions of a triangulated category w.r.t. an épaisse subcategory (cf. [V]). We discuss this construction for our situation.

**2.10.** A class $S$ of morphisms in a category $\mathcal{D}$ is said to be right localizing ([GaZ]) if the following conditions are satisfied:

a) $1_X \in \mathcal{S}$ for any object $X \in \mathcal{D}$, and if $s, t \in \mathcal{S}$ then $st \in \mathcal{S}$ (provided $st$ exists).

b) for any morphism $f\colon X \to Y$ in $\mathcal{D}$ and any $s\colon Z \to Y$ in $\mathcal{S}$ there exists a morphism $t \in \mathcal{S}$ and a morphism $g$ such that $ft = sg$.

c) if $tf = 0$ for a given morphism $f$, with $t \in \mathcal{S}$, then there exists an $s \in \mathcal{S}$ such that $fs = 0$.

A left localizing class is defined dually. A class of morphisms that is both right and left localizing is called localizing.

For any right localizing class $S$ of morphisms in $\mathcal{D}$ we can form the category of fractions $\mathcal{D}\,[\mathcal{S}^{-1}]$ (or the localization of $\mathcal{D}$ w.r.t. $\mathcal{S}$): it has the same objects as $\mathcal{D}$, and morphisms from $X$ to $Y$ in $\mathcal{D}\,[\mathcal{S}^{-1}]$ are represented by diagrams of the form

$$X \xleftarrow{s} Z \xrightarrow{f} Y \quad \text{with } s \in \mathcal{S}$$

(cf. [GaZ]).

**2.11.** A morphism $f\colon X \to Y$ in $K(\mathbb{A})$ is called an $\omega$-quasi-isomorphism if its mapping cone $C(f)$ is $\omega$-acyclic. We denote by $\mathcal{S}_\omega$ the class of all $\omega$-quasi-isomorphisms in $K(\mathbb{A})$.

Theorem 2.8 implies now the following result.

**2.12.** PROPOSITION. *The class of $\omega$-quasi-isomorphisms $\mathcal{S}_\omega$ in $K(\mathbb{A})$ is localizing.*

**2.13.** By Proposition 2.12 we can construct the localization of $K(\mathbb{A})$ w.r.t. $\mathcal{S}_\omega$, and we define the (relative) derived category $D(\mathbb{A})$ of $\mathbb{A}$ as this localization: $D(\mathbb{A}) = K(\mathbb{A})\,[\mathcal{S}_\omega^{-1}]$. Note that $D(\mathbb{A})$ inherits the structure of a triangulated category from $K(\mathbb{A})$.

If we start with the homotopy category $K^+(\mathbb{A})$ (respectively $K^-(\mathbb{A})$ or $K^b(\mathbb{A})$) of complexes bounded from below (respectively bounded from above or bounded complexes) then we get in a similar way the derived categories $D^+(\mathbb{A})$ (respectively $D^-(\mathbb{A})$ or $D^b(\mathbb{A})$).

**2.14.** If we associate to every object $A \in \mathbb{A}$ the complex $\cdots 0 \to A \to 0 \cdots$ concentrated at zero degree we obtain the functor $I\colon \mathbb{A} \to D(\mathbb{A})$ which is a full embedding.

Now we have the opportunity to introduce the "relative groups of extensions" as follows:

$$\omega\,\mathrm{Ext}^n_{\mathbb{A}}(A, B) = \mathrm{Hom}_{D(\mathbb{A})}\big(I(A), I(B)[n]\big) \quad \text{where } A, B \in \mathbb{A}.$$

**2.15.** THEOREM. *Given an $\omega$-proper sequence in $\mathbb{A}$:*

$$E\colon\ A \xrightarrow{i} B \xrightarrow{\sigma} C$$

*and an object $X \in \mathbb{A}$ the following sequences of abelian groups are exact:*

a) $\cdots \longrightarrow \omega\operatorname{Ext}^n_{\mathbb{A}}\ (C,X) \xrightarrow{\sigma^*} \omega\operatorname{Ext}^n_{\mathbb{A}}\ (B,X) \xrightarrow{i^*} \omega\operatorname{Ext}^n_{\mathbb{A}}\ (A,X) \longrightarrow \omega\operatorname{Ext}^{n+1}_{\mathbb{A}}$ $(C,X) \longrightarrow \cdots$

b) $\cdots \longrightarrow \omega\operatorname{Ext}^n_{\mathbb{A}}\ (X,A) \xrightarrow{i_*} \omega\operatorname{Ext}^n_{\mathbb{A}}\ (X,B) \xrightarrow{\sigma_*} \omega\operatorname{Ext}^n_{\mathbb{A}}\ (X,C) \longrightarrow \omega\operatorname{Ext}^{n+1}_{\mathbb{A}}$ $(X,A) \longrightarrow \cdots$

PROOF. The statement follows from the fact that the functors $\operatorname{Hom}_{D(\mathbb{A})}(U,-)$ and $\operatorname{Hom}_{D(\mathbb{A})}(-,U)$, where $U \in D(\mathbb{A})$, are cohomological (see [GeM]). □

**2.16.** When $\omega = \omega_{st}$ is the class of all stable sequences in a preabelian category $\mathbb{A}$ (or even a subclass of $\omega_{st}$) Theorem 2.15 was established in [RW1]. This result is well-known in the case where $\mathbb{A}$ is an abelian category [Ma1]. Moreover, similarly to the case of abelian categories, the elements of $\omega\operatorname{Ext}^n_{\mathbb{A}}(A,B,)$ $n \geqslant 1$, with $\omega \subseteq \omega_{st}$ are represented by $\omega$-acyclic complexes of the form (see [G7]):

$$0 \longrightarrow B \longrightarrow M^{-n+1} \longrightarrow M^{-n+2} \longrightarrow \cdots \longrightarrow M^0 \longrightarrow A \longrightarrow 0.$$

So, in the case $\omega \subseteq \omega_{st}$ the groups $\omega\operatorname{Ext}^n_{\mathbb{A}}(A,B)$ can be defined alternatively using "Baer addition" on the set of equivalence classes of such complexes [RW1].

**2.17.** Let $\omega$ be a projective p.c. in a preabelian category $\mathbb{A}$. For any object $A \in \mathbb{A}$ we can construct an $\omega$-projective resolution, i.e. the $\omega$-acyclic complex

$$\cdots \longrightarrow P_n \xrightarrow{d_n} P_{n-1} \longrightarrow \cdots \xrightarrow{d_1} P_0 \xrightarrow{\varepsilon} A \longrightarrow 0 \tag{2}$$

with all $P_n$ $\omega$-projective. The morphism $\varepsilon\colon P_0 \to A$ induces obviously a morphism of complexes $\varepsilon\colon P. = (P_n, d_n) \to A$ (here $A$ denotes the complex concentrated in zero degree). The complex $P. = (P_n, d_n)$ is also-called an $\omega$-projective resolution. Any two such $\omega$-projective resolutions of $A$ are homotopic. Moreover, the complex (2) is a mapping cone of the morphism $\varepsilon\colon P. \to A$, and so $\varepsilon$ is an $\omega$-quasi-isomorphism.

Let $K^-(\mathcal{P}_\omega)$ be the homotopy category of complexes bounded from above over the full subcategory $\mathcal{P}_\omega$ of $\mathbb{A}$ consisting of $\omega$-projectives. There is a natural functor $\Phi\colon K^-(\mathcal{P}_\omega) \to D^-_\omega(\mathbb{A})$ which takes $P \in K^-(\mathcal{P}_\omega) \subset K^-(\mathbb{A})$ to its image in $D^-_\omega(\mathbb{A})$ under localization.

**2.18.** THEOREM [G7]. *If $\omega$ is a projective p.c. in a preabelian category $\mathbb{A}$ then the functor $\Phi\colon K^-(\mathcal{P}_\omega) \to D^-_\omega(\mathbb{A})$ is an equivalence of categories.*

Thus, every object in $D^-_\omega(\mathbb{A})$ is represented by a complex bounded on the right over $\mathcal{P}_\omega$. Certainly one has also the dual result for an injective p.c. of kernels in $\mathbb{A}$ (with $D^-_\omega(\mathbb{A})$ being replaced by $D^+_\omega(\mathbb{A})$).

**2.19.** Let $F$: $\mathbb{A} \to \mathbb{C}$ be an (additive covariant) functor from a preabelian category $\mathbb{A}$ to an abelian category $\mathbb{C}$. Define the functor $K^-(F)$: $K^-(\mathcal{P}_\omega) \to K^-(\mathbb{C})$ by $K^-(F)(P_n, d_n) = (F(P_n), F(d_n))$; this definition is correct since homotopic complexes are taken to homotopic ones. Now we get an induced functor

$$D^-(F) = Q_{\mathbb{C}} \circ K^-(F) \circ \Psi\colon\ D^-_\omega(A) \to D^-(\mathbb{C})$$

where $Q_{\mathbb{C}}$: $K^-(\mathbb{C}) \to D^-(\mathbb{C})$ is the corresponding localization functor and $\Psi$ is the equivalence of categories quasi-inverse to $\Phi$. One can prove that $D^-(F)$ is exact (as a functor between triangulated categories, i.e. $D^-(F)$ takes distinguished triangles to the distinguished triangles). Moreover, $D^-(F)$ solves some universal problem defining a "left derived functor" (cf. [V]). Then we can get the classical (relative) left derived functors as the restriction of $D^-(F)$ to $\mathbb{A}$ composed with the cohomology object functors $H^n$:

$$L^n_\omega F = H^n \circ D^-(F)|_{\mathbb{A}}\colon\ \mathbb{A} \to \mathbb{C}.$$

It is easily proved that for $n > 0$ $L^n_\omega F = 0$, and so it is natural to introduce the notation $L^\omega_n F = L^{-n}_\omega F$ for $n \geqslant 0$.

A functor $F$: $\mathbb{A} \to \mathbb{C}$ (where $\mathbb{C}$ is abelian) is called right $\omega$-exact if for any $\omega$-proper sequence in $\mathbb{A}$ $A \to B \to C$ the induced sequence $F(A) \to F(B) \to F(C) \to 0$ is exact.

**2.20.** THEOREM [G7]. *Let $\omega$ be a projective p.c. in a preabelian category $\mathbb{A}$, $F$: $\mathbb{A} \to \mathbb{C}$ be a right $\omega$-exact additive functor to an abelian category $\mathbb{C}$. Given an $\omega$-proper sequence in $\mathbb{A}$ $A \to B \to C$ the following sequence is exact:*

$$\cdots \longrightarrow L^\omega_n F(A) \longrightarrow L^\omega_n F(B) \longrightarrow L^\omega_n F(C) \longrightarrow L^\omega_{n-1}(A) \longrightarrow \cdots$$

$$\cdots \longrightarrow L^\omega_1 F(C) \longrightarrow F(A) \longrightarrow F(B) \longrightarrow F(C) \longrightarrow 0.$$

Essentially this result is a consequence on the exactness of the functor $D^-(F)$.

**2.21.** REMARKS.

a) The (relative) right derived functors $R^n_\omega F$ of an additive covariant functor $F$ with $\omega$ being an injective p.c. of kernels are defined similarly. The right and left derived functors of contravariant functors can be defined by the duality. In particular we can prove that for the contravariant functor $F = \mathrm{Hom}_{\mathbb{A}}(-, X)$: $\mathbb{A} \to \mathcal{A}b$ we have $R^n_\omega F = \omega\,\mathrm{Ext}^n_{\mathbb{A}}(-, X)$ (where $\omega$ is a projective p.c.). Also we may state a similar result for the functor $\mathrm{Hom}_{\mathbb{A}}(X, -)$.

b) In view of 1.9 the main result of [Y] is contained in 2.20.

**2.22.** (Relative) derived categories can be constructed in a more general context. Namely, let $\mathbb{A}$ be an additive category. A class $\omega$ of cokernels in $\mathbb{A}$ is called proper if $\omega$ satisfies the axioms P0–P3 from 1.1. The pair $(\mathbb{A}, \omega)$ is said to be an epi-exact category. This notion generalizes both the exact categories of Quillen and preabelian categories with fixed p.c. $\omega$. In this setting we can repeat the definitions and constructions above which lead us to the derived categories of bounded types, i.e. $D^{\pm}(\mathbb{A})$ and $D^{b}(\mathbb{A})$ [G8]. If $\mathbb{A}$ satisfies additionally the condition of splitting of all idempotents then we can also construct the unbounded derived category $D(\mathbb{A})$ [N, G8].

**2.23.** REMARK. In [N] the derived categories of bounded types of an exact category are constructed under the additional condition that "every weakly split idempotent splits", but as easily seen any epi-exact (hence exact) category satisfies this condition [G8].

## 3. Relative homological algebra in module categories

**3.1.** Let $R$ be an associative ring with identity. Denote by $\operatorname{Mod} R$ (respectively $\operatorname{mod} R$) the category of all (respectively finitely generated) right (unital) $R$-modules. As we pointed out in Section 1 a p.c. $\omega$ in $\operatorname{Mod} R$ may be given by a class of $\omega$-proper sequences or by the class of corresponding $\omega$-proper epimorphisms or else by the class of corresponding $\omega$-proper monomorphisms.

**3.2.** We begin with one of the most famous p.c.s, namely, purity in the category $\mathcal{A}b$ of abelian groups. Recall that a subgroup $H$ of an abelian group $G$ is said to be pure if for any natural number $n$ we have $H \cap nG = nH$. A short exact sequence $0 \to H \xrightarrow{i} G \xrightarrow{\sigma} K \to 0$ in $\mathcal{A}b$ is called pure if $i(H)$ is the pure subgroup of $G$. The class of the pure sequences is proper; usually it is called the (classical) purity. It has many important applications in abelian group theory (see, e.g., [F]). This class is projectively generated by the class of finite abelian groups and it is injectively generated by the same class of groups [F, MS]. The $\omega$-projective (respectively $\omega$-injective) groups for the purity $\omega$ usually are called pure-projective (respectively pure-injective) groups. The class of pure-injective groups coincides with the class of so-called algebraically compact abelian groups [F].

It easily follows from the definition that a monomorphism $H \to G$ is pure iff for any abelian group $X$ the induced homomorphism $H \otimes X \to G \otimes X$ is a monomorphism.

**3.3.** Let $\mathcal{M}$ be a class of left $R$-modules, $\omega$ be the class of monomorphisms $i\colon A \to B$ in $\operatorname{Mod} R$ such that for any $M \in \mathcal{M}$ the induced homomorphism $i \otimes M\colon A \otimes M \to B \otimes M$ is injective. One can readily show that $\omega$ is a p.c., and we call $\omega$ the p.c. flatly generated (by $\mathcal{M}$).

If we take for $\mathcal{M}$ the class of all modules (or even only finitely generated modules) then we get the p.c. which usually is called the Cohn purity. This p.c. was profoundly investigated in [W] and many properties of the classical purity in $\mathcal{A}b$ were extended there to this general context. In particular, the Cohn purity $\omega_P$ is projective and injective, moreover it is projectively generated by the class of finitely presented right $R$-modules. The $\omega_P$-projective (respectively $\omega_P$-injective) modules are called pure-projective (respectively pure-injective).

**3.4.** From the properties of the tensor product, it is easy to see that any flatly generated p.c. $\omega$ is inductively closed, i.e. for any direct system $\{E_i \mid i \in I\}$ of $\omega$-proper sequences in Mod $R$ over a directed set $I$ the colimit $E = \varinjlim_I E_i$ is also an $\omega$-proper sequence. The converse is not true in general, but it is true in the category $\mathcal{A}b$ of abelian groups [Mn, Ku]. More completely, we have the following result over a bounded HNP-ring $R$. Recall that a ring $R$ is said to be bounded if every essential onesided ideal of $R$ contains a nonzero (two-sided) ideal; here "HNP-ring" means "hereditary noetherian prime ring".

**3.5.** THEOREM [G1]. *Let $R$ be a bounded HNP-ring. Every inductively closed p.c. in* Mod $R$ *is flatly generated iff $R$ is a Dedekind prime ring (i.e. $R$ is a HNP-ring without nontrivial idempotent ideals).*

Moreover, in [G1, G6] the classification of all inductively closed p.c.s is given for the cases where $R$ is either a bounded HNP-ring or a time hereditary finite dimensional algebra over a field. These results have interest in view of the still open question (see [MS]): to describe all p.c.s in the category $\mathcal{A}b$ of abelian groups.

Also note that over a bounded HNP-ring $R$ every inductively closed p.c. in Mod $R$ is uniquely determined by its restriction to the subcategory mod $R$ of finitely generated modules [G1], but the similar statement for the case where $R$ is a tame hereditary algebra is false [G6].

Since every (Cohn-)pure sequence in Mod $R$, $R$ being any ring, is the colimit of a direct system of split exact sequences, every inductively closed p.c. contains the Cohn purity [Sk]; so the Cohn purity is the least inductively closed p.c.

By Kaplansky's theorem we have a good description of pure-injective (= algebraically compact) abelian groups [F]. Generalizations of this theorem are proved in [G1, G6] in the cases where $R$ is either a bounded HNP-ring or a tame hereditary algebra.

Part of the results concerning the Cohn purity were generalized in [St] to the category of functors from a small additive category into $\mathcal{A}b$ (see also [Ca]).

**3.6.** Note that we can introduce the notion of a "relative" global dimension of a ring (or even of a category) w.r.t. a projective p.c. $\omega$ similarly to the absolute case (cf. [Ma1]). The notion of the pure global dimension (i.e. w.r.t. the Cohn-purity) is the most interesting among these global dimensions and has been investigated by many authors (see, e.g., [Si, GrJ, BaL]).

**3.7.** Let $\gamma$: $S \to R$ be a (unital) ring morphism and $\gamma^*$: Mod $R \to$ Mod $S$ be a corresponding "forgetful" functor. Let $\omega$ be a p.c. of short exact sequences in Mod $S$ and let $\omega' = (\gamma^*)^{-1}(\omega)$ be the class of short exact sequences in Mod $R$ which being considered over $S$ are $\omega$-proper. Plainly $\omega'$ is a p.c.; we call it the p.c. induced by $\omega$.

**3.8.** PROPOSITION.

a) *If $\omega$ is a p.c. projectively generated by a class $\mathcal{M}$ of $S$-modules then the induced p.c. $\omega'$ is projectively generated by the class of $R$-modules $\{M \otimes_S R \mid M \in \mathcal{M}\}$.*

b) *If $\omega$ is a p.c. injectively generated by a class $\mathcal{M}$ of $S$-modules then the induced p.c. $\omega'$ is injectively generated by the class of $R$-modules $\{\mathrm{Hom}_S(R, M) \mid M \in \mathcal{M}\}$.*

This result follows immediately from 1.13 (and its dual) since the forgetful functor $\gamma^*$ has the functor $- \otimes_S R$ (respectively $\mathrm{Hom}_S(R, -)$) as a left (respectively right) adjoint.

**3.9.** If in the preceding construction $\omega = \omega_0$ is the p.c. in $\operatorname{Mod} S$ consisting of split sequences then the short exact sequences in $\omega_0' = (\gamma^*)^{-1}(\omega_0)$ are called $(R, S)$-proper. When $S$ is a (unital) subring of a ring $R$ and $\gamma\colon\ S \to R$ is the inclusion the p.c. $\omega_0'$ was used by Hochschild [Ho] in the construction of the relative cohomology theory of associative rings (cf. 4.7 below). As a consequence of 3.8 and 1.12 we have the following.

**3.10.** PROPOSITION [Ho]. *Let $\gamma\colon\ S \to R$ be a ring morphism and $\omega$ be the class of $(R, S)$-proper sequences. Then:*

a) *a right $R$-module $M$ is $\omega$-projective (respectively $\omega$-injective) iff $M$ is a direct summand of an $R$-module of the form $X \otimes_S R$ (respectively, $\operatorname{Hom}_S(R, X)$) with an $S$-module $X$;*

b) *$\omega$ is a projective and injective p.c.*

**3.11.** Let $\omega$ be a p.c. of short exact sequences in a (small) abelian category $\mathbb{A}$. Let $F$ be the free abelian group on the set of (representatives of) isomorphism classes $[M]$ of objects $M$ in $\mathbb{A}$. The relative Grothendieck group $K_0(\mathbb{A}, \omega)$ is defined as the quotient of the group $F$ modulo the subgroup $H$ of $F$ generated by elements of the form $r(E) = [A] - [B] + [C]$ which correspond to $\omega$-proper sequences $E\colon\ 0 \to A \to B \to C \to 0$. If $\omega_0$ (respectively, $\omega_1$) is the p.c. of all split exact (respectively exact) sequences then we denote $K_0(\mathbb{A}, \omega_0)$ (respectively $K_0(\mathbb{A}, \omega_1)$) as $K_0(\mathbb{A}, 0)$ (respectively, $K_0(\mathbb{A})$). Plainly $K_0(\mathbb{A})$ is the (absolute) Grothendieck group of the category $\mathbb{A}$.

Consider the short exact sequence

$$0 \to E(\mathbb{A}, \omega) \to K_0(\mathbb{A}, 0) \xrightarrow{\pi} K_0(\mathbb{A}, \omega) \to 0 \tag{1}$$

where $\pi$ is the natural homomorphism. If the Krull–Schmidt theorem is satisfied in the category $\mathbb{A}$ the group $K_0(\mathbb{A}, 0)$ is free and has as a set of free generators the (isomorphism classes of) indecomposable objects in $\mathbb{A}$. In this context the group $E(\mathbb{A}, \omega)$ in (1) is called usually “the group of relations” of the group $K_0(\mathbb{A}, \omega)$ (indeed we have a presentation of this group in terms of generators and relations). Denote also with $E(\mathbb{A}) = E(\mathbb{A}, \omega_1)$, the group of relations of $K_0(\mathbb{A})$.

If $R$ is a right noetherian ring the category $\operatorname{mod} R$ of finitely generated right $R$-modules is abelian, and then we denote the group $K_0(\operatorname{mod} R, \omega)$ (respectively, $E(\operatorname{mod} T, \omega)$ and $E(\operatorname{mod} R)$) by $K_0(R, \omega)$ (respectively, $E(R, \omega)$ and $E(R)$).

**3.12.** Historically the first example of a relative Grothendieck group was considered by Lam and Reiner [LR]. Let $G$ be a finite group, $H$ be a subgroup of $G$, $k$ be a (commutative) field. Denote by $R = k[G]$ and $S = k[H]$ the corresponding group algebras. Let $\omega$ be the class of $(R, S)$-proper short exact sequences in $\operatorname{mod} R$ (3.9) (it will be convenient to call these sequences also $(G, H)$-proper). Then the group $K_0(R, \omega)$ and its relations with $K_0(R, 0)$ and $K_0(R)$ was investigated in [LR].

**3.13.** When $R$ is an artin algebra the groups of relations of the groups $K_0(R, \omega)$ is related with so-called almost split sequences [Bu, A]. Recall that a ring $R$ is said to be an artin algebra if its center $C$ is an artinian ring and $R$ is a finitely generated $C$-module. A ring

$R$ is said to be of finite representation type if it has only a finite number of isomorphism classes of indecomposable modules. A nonsplit short exact sequence

$$E\colon 0 \longrightarrow A \longrightarrow B \xrightarrow{g} C \longrightarrow 0$$

is called an almost split sequence if $A$ and $C$ are indecomposable modules and for any homomorphism $h\colon X \to C$ which is not a split epimorphism there exists a homomorphism $h'\colon X \to B$ such that $gh' = h$.

**3.14.** THEOREM. *Let $R$ be a ring of finite representation type, and let $\omega$ be the p.c. of short exact sequences in* $\operatorname{mod} R$. *Then the group $E(R,\omega)$ if freely generated by elements of the form* $r(E)$ (see 3.11) *where $E$ runs through all $\omega$-proper almost split sequences.*

The absolute case of the theorem is proved for an artin algebra $R$ in [Bu, A], and the general case is proved in [G3].

Note also that over a ring $R$ of finite representation type any inductively closed p.c. in $\operatorname{Mod} R$ is uniquely determined by its restriction to $\operatorname{mod} R$ [G2] (cf. 3.5).

**3.15.** A similar problem is solved in [G4] for tame hereditary finite dimensional algebras $R$. In this situation we add to the almost split sequences new classes of short exact sequences in $\operatorname{mod} R$ and using them we obtain the set of free generators of the group $E(R,\omega)$. Note that these new short exact sequences are also used in the classification of all p.c.s in $\operatorname{mod} R$ (cf. 3.5).

Another set of free generators of absolute group $E(R)$ for (tame or wild) hereditary finite dimensional algebra $R$ was introduced in [Gei].

By the same methods the description of sets of free generators of $E(R,\omega)$ is given for chain right noetherian rings $R$ in [G5].

**3.16.** Some additional information on relative homological algebra in a module category can be found in the excellent review by Sklyarenko [Sk]. For brevity we do not discuss the "relative" spectral sequences considered in [Sk] (see also [Kh1]).

## 4. Cohomology of small categories

**4.1.** Let $J\colon \mathbb{C} \to \mathbb{D}$ be a functor between small categories. We have the induced functor $J^*\colon \mathcal{A}b^{\mathbb{D}} \to \mathcal{A}b^{\mathbb{C}}$, $J^*(T) = T \cdot J$ for a functor $T\colon \mathbb{D} \to \mathcal{A}b$. As the category $\mathcal{A}b$ is complete (i.e. every functor $F\colon \mathbb{C} \to \mathcal{A}b$ has a limit) the functor $J^*$ has a right adjoint $\tilde{J}\colon \mathcal{A}b^{\mathbb{C}} \to \mathcal{A}b^{\mathbb{D}}$; it is called the right Kan extension [Ma2]. If $\omega$ is a projective p.c. in $\mathcal{A}b^{\mathbb{C}}$ then we can construct the (relative) right derived functors $R^n_\omega \tilde{J}\colon \mathcal{A}b^{\mathbb{C}} \to \mathcal{A}b^{\mathbb{D}}$ of the functor $\tilde{J}$. Define the (relative) cohomology $H^*_\omega(J,T)$ of $J$ with coefficients in a functor $T\colon \mathbb{C} \to \mathcal{A}b$ as follows:

$$H^n_\omega(J,T) = R^n_\omega \tilde{J}(T).$$

In this context any functor $T$ in $\mathcal{A}b^{\mathbb{C}}$ is usually called a $\mathbb{C}$-module.

**4.2.** Consider the obvious functor $I\colon \mathbb{C}_d \to \mathbb{C}$ where $\mathbb{C}_d$ is the discrete category corresponding to a category $\mathbb{C}$ (i.e. $\mathrm{Ob}(\mathbb{C}_d) = \mathrm{Ob}(\mathbb{C})$, $\mathbb{C}_d(A, B)$ is empty if $A \neq B$, and $\mathbb{C}_d(A, A) = \{1_A\}$). The induced functor $I^*\colon \mathcal{A}b^{\mathbb{C}} \to \mathcal{A}b^{\mathbb{C}_d}$ is faithful and has a left adjoint $\tilde{I}\colon \mathcal{A}b^{\mathbb{C}_d} \to \mathcal{A}b^{\mathbb{C}}$ (since $\mathcal{A}b$ is a cocomplete category). The category $\mathcal{A}b^{\mathbb{C}_d}$ is a product of copies of $\mathcal{A}b$ and then the class $\omega_d$ of all epimorphisms in $\mathcal{A}b^{\mathbb{C}_d}$ is a projective p.c. Clearly, $(I^*)^{-1}(\omega_d)$ is a class of all epimorphisms in $\mathcal{A}b^{\mathbb{C}}$, and by 1.12 it is a projective p.c. If we take in 4.1 $\omega = (I^*)^{-1}(\omega_d)$ we get the absolute cohomology $H^*(J, T)$ defined in [HiS].

**4.3.** Let now $J\colon \mathbb{C} \to \mathbf{1}$ be the obvious functor where $\mathbf{1}$ is the category with one object and only one morphism. In this situation

$$H^*_\omega(\mathbb{C}, T) = H^*_\omega(J, T) \tag{1}$$

is called the (relative) cohomology of the category $\mathbb{C}$ with coefficients in $T\colon \mathbb{C} \to \mathcal{A}b$. One gets the absolute cohomology $H^*(\mathbb{C}, T)$ if one takes $\omega$ as the p.c. of all epimorphisms in $\mathcal{A}b^{\mathbb{C}}$.

We can identify $J^*\colon \mathcal{A}b = \mathcal{A}b^{\mathbf{1}} \to \mathcal{A}b^{\mathbb{C}}$ with the functor which associated to an abelian group $A$ the corresponding constant functor $\bar{A}\colon \mathbb{C} \to \mathcal{A}b$ (i.e. $\bar{A}(X) = A$ for all objects $X$ in $\mathbb{C}$).

**4.4.** Proposition. *The right adjoint $\tilde{J}\colon \mathcal{A}b^{\mathbb{C}} \to \mathcal{A}b$ of the functor $J^*$ coincides (up to isomorphism) with the functor $\mathcal{A}b^{\mathbb{C}}(\overline{\mathbb{Z}}, -)$.*

Sketch of the proof. Let

$$\sigma\colon \mathcal{A}b^{\mathbb{C}}(\bar{A}, F) \to \mathcal{A}b(A, \mathcal{A}b^{\mathbb{C}}(\overline{\mathbb{Z}}, F))$$

be a map such that for a natural transformation $\eta\colon \bar{A} \to F$ the natural transformation $\sigma(\eta)(a)\colon \overline{\mathbb{Z}} \to F$ is defined as follows: $(\sigma(\eta))_X(1) = \eta_X(a)$ where $X \in \mathrm{Ob}(\mathbb{C})$, $a \in A$ (1 is a generator of $\mathbb{Z}$). The inverse $\tau = \sigma^{-1}$ is a map which to a natural transformation $\zeta\colon A \to \mathcal{A}b^{\mathbb{C}}(\overline{\mathbb{Z}}, F)$ associates the natural transformation $\tau(\zeta)\colon \bar{A} \to F$ such that $\tau(\zeta)_Y(x) = \zeta(x)_Y(1)$.

**4.5.** Proposition. $H^n(\mathbb{C}, T) = \mathrm{Ext}^n_{\mathcal{A}b^{\mathbb{C}}}(\overline{\mathbb{Z}}, T)$.

This statement follows from 4.3 and 4.4. Note that if $P.\colon \cdots \to P_1 \to P_0 \to 0$ is an $\omega$-projective resolution of $\overline{\mathbb{Z}}$ (in $\mathcal{A}b^{\mathbb{C}}$) then we can compute $H^n_\omega(\mathbb{C}, T)$ as the cohomology of the complex $\mathcal{A}b^{\mathbb{C}}(P., T)$.

**4.6.** If we consider a finite group $G$ as a category $\mathbb{G}$ with one object then we get the well-known result (or rather a definition!):

$$H^n(G, T) = H^n(\mathbb{G}, T) = \mathrm{Ext}^n_{\mathbb{Z}G}(\mathbb{Z}, T)$$

where $\mathbb{Z}$ is a trivial $G$-module, $T$ is a $G$-module (cf. [HiS, Ma1]).

If $H$ is a subgroup of the group $G$ and $\omega$ is the p.c. of $(G,H)$-proper sequences (see 3.12) then

$$H^n_\omega(\mathbb{G},T) = \omega\,\mathrm{Ext}^n_{\mathbb{Z}G}(\mathbb{Z},T),$$

that is the relative cohomology $H^n_\omega(\mathbb{G},T)$ coincides with the group $H^n(G,H,T)$ introduced in [Ho].

So we can compute the cohomology $H^n(G,T)$ (respectively, $H^n_\omega(G,T)$) using the standard (or bar) projective (respectively $(G,H)$-projective) resolution $P.$ of $\mathbb{Z}$ (see [Ho]), namely:

$$H^n(G,T) = H^n\big(\mathrm{Hom}_{\mathbb{Z}G}(P.,T)\big)$$

(and similarly in the relative case).

**4.7.** Another example is given by the relative Hochschild cohomology of algebras. Let $\gamma\colon A \to B$ be a $K$-algebra homomorphism where $K$ is a commutative ring. Define

$$S = A \otimes_K B^{\mathrm{op}}, \quad R = B \otimes_K B^{\mathrm{op}}, \quad \bar{\gamma} = \gamma \otimes 1\colon S \to R.$$

Let $\omega$ be a $(R,S)$-proper class (w.r.t. $\bar{\gamma}$) (see 3.9). We have the natural ring homomorphism

$$J\colon R \to B, \quad J(b_1 \otimes b_2) = b_1 b_2.$$

The ring $R$ (respectively $B$) can be considered as an additive category $\mathbb{R}$ (respectively $\mathbb{B}$) with one object [Mt1] and then $J$ is a functor between these categories. Following 4.1 we get the functor

$$J^*\colon \mathcal{A}b^{\mathbb{B}} = B\text{-Mod} \to R\text{-Mod} = \mathcal{A}b^{\mathbb{R}}$$

and its right adjoint $\tilde{J}\colon R\text{-Mod} \to B\text{-Mod}$. As in 4.4 we have $\tilde{J}(-) = \mathrm{Hom}_R(B,-)$, and hence for any $B$-bimodule ($= R$-module) $T$

$$H^n_\omega(J,T) = R^n_\omega \tilde{J}(T) = \omega\,\mathrm{Ext}^n_R(B,T) = \mathrm{Ext}^n_{(R,S)}(B,T).$$

Consequently these cohomology groups coincide with the groups $H^n(B,A,T)$ defined in [Ho].

For computation of this cohomology we may use the standard $(R,S)$-projective resolution of $B$ (see [Ho]):

$$\cdots \longrightarrow X_n \xrightarrow{d_n} X_{n-1} \xrightarrow{d_{n-1}} \cdots \longrightarrow X_0 \xrightarrow{d_0} B \longrightarrow 0$$

where $X_n = B \otimes_A B \otimes_A \ldots \otimes_A B$ ($n+2$ times),

$$d_n(b_0 \otimes \cdots \otimes b_{n+1}) = \sum_{k=1}^{n} (-1)^k\,(b_0 \otimes \cdots \otimes b_k b_{k+1} \otimes \cdots \otimes b_{n+1}).$$

So we have

$$H^*_\omega(J,T) = H^*\big(\mathrm{Hom}_R(X.,T)\big).$$

**4.8.** It is easy to see that for any $\mathbb{C}$-module $T\colon \mathbb{C} \to \mathcal{A}b$ $\varprojlim T \cong \mathcal{A}b^{\mathbb{C}}(\overline{\mathbb{Z}},T)$ and hence we get the following

**4.9.** PROPOSITION. *$H^n(\mathbb{C},T) = \varprojlim^n T$, the $n$-th derived functor of $\varprojlim$.*

**4.10.** Now we discuss a further generalization of the cohomology of small categories introduced in [BaW], namely, cohomology with coefficients in a natural system. Let $\mathbb{C}$ be a small category. The category of factorizations in $\mathbb{C}$ (denoted by $F(\mathbb{C})$) is defined as follows. The objects of $F(\mathbb{C})$ are morphisms in $\mathbb{C}$ and a morphism $f \to g$ is a pair $(\alpha,\beta)$ or morphisms in $\mathbb{C}$ such that $\alpha f\beta = g$; composition is defined naturally.

A natural system (of abelian groups) on $\mathbb{C}$ is a functor $D\colon F(\mathbb{C}) \to \mathcal{A}b$.

The cohomology of a category $\mathbb{C}$ with coefficients in a natural system $D$ is defined as

$$H^*(\mathbb{C},D) = H^*\big(F(\mathbb{C}),D\big)$$

where the right side is the cohomology from (1) in 4.3.

**4.11.** Any functor $M\colon \mathbb{C}^{\mathrm{op}} \times \mathbb{C} \to \mathcal{A}b$ is called a $\mathbb{C}$-bimodule. We have a forgetful functor $\pi\colon F(\mathbb{C}) \to \mathbb{C}^{\mathrm{op}} \times \mathbb{C}$ with associates to a morphism $\sigma\colon X \to Y \in \mathrm{Ob}(F(\mathbb{C}))$ the pair $(X,Y)$. Define the cohomology of the category $\mathbb{C}$ with coefficients in a $\mathbb{C}$-bimodule $M$ as

$$H^n(\mathbb{C},M) = H^n\big(\mathbb{C},\pi^*M\big).$$

This cohomology (usually called Hochschild–Mitchell cohomology) was introduced by Mitchell in [Mt1].

**4.12.** If $\mathbb{C}$ is a small category and $K$ is a commutative ring then $K\mathbb{C}$ is defined as the category whose objects are those of $\mathbb{C}$ and $K\mathbb{C}(A,B)$ is the free $K$-module on the set of morphisms $\mathbb{C}(A,B)$. Composition in $K\mathbb{C}$ is defined naturally so that we obtain an additive category. It is easy to see that there exists an isomorphism of categories $K\mathbb{C} \cong (\mathrm{Mod}\,K)^{\mathbb{C}}$. Moreover, $K\mathbb{C}$ can be considered as a functor $K\mathbb{C}\colon \mathbb{C}^{\mathrm{op}} \times \mathbb{C} \to \mathcal{A}b$.

The case where $K = \mathbb{Z}$ has particular interest.

**4.13.** THEOREM [BaW]. *For any $\mathbb{C}$-bimodule $M$*

$$H^n(\mathbb{C},M) = \mathrm{Ext}^n_{\mathcal{A}b^{F(\mathbb{C})}}\big(\overline{\mathbb{Z}},\pi^*M\big) = \mathrm{Ext}^n_{\mathcal{A}b^{\mathbb{C}^{\mathrm{op}}\times\mathbb{C}}}(\mathbb{Z}\mathbb{C},M).$$

If $p\colon \mathbb{C}^{\mathrm{op}} \times \mathbb{C} \to \mathbb{C}$ is the natural forgetful functor then we have for any $\mathbb{C}$-module $T$ that

$$\mathrm{Ext}^n_{\mathcal{A}b^{F(\mathbb{C})}}\big(\overline{\mathbb{Z}},p^*\pi^*T\big) = \mathrm{Ext}^n_{\mathcal{A}b^{\mathbb{C}}}\big(\overline{\mathbb{Z}},T\big) \quad \text{[BaW]}$$

and hence the definition in 4.10 is indeed a generalization of the definition of the cohomology in 4.3 (cf. 4.5).

**4.14.** Let $\mathbb{C}$ be a category and $N\mathbb{C} = \{N_n(\mathbb{C})\}$ be the nerve of $\mathbb{C}$ (see [GeM]) (for example, $N_n(\mathbb{C})$, $n \geqslant 1$, is the set of sequences $(a_1, \dots, a_n)$ of $n$ composable morphisms $A_0 \xleftarrow{a_1} A_1 \xleftarrow{a_2} \cdots \xleftarrow{a_n} A_n$; $N_0(\mathbb{C}) = \mathrm{Ob}(\mathbb{C})$). Let $D$ be a natural system on $\mathbb{C}$. Denote $a_* = D(a,1)$, $b^* = D(1,b)$ for any morphisms $a$, $b$ in $\mathbb{C}$. Now we construct the following cochain complex $F^{\cdot} = \{F^n, \delta\}$: $F^n = F^n(\mathbb{C}, D)$ is the abelian group of all functions

$$f\colon N_n(\mathbb{C}) \to \dot{\bigcup_{g\in \mathrm{Mor}\,\mathbb{C}}} D(g)$$

(here $\dot{\bigcup}$ denotes a disjoint union) such that $f(a_1, \dots, a_n) \in D(a_1 \cdot a_2 \cdot \dots \cdot a_n)$; the addition in $F^n$ is given "componentwise" in the abelian groups $D(g)$; the coboundary $\delta\colon F^{n-1} \to F^n$ is defined by the formula:

$$(\delta f)(a_1, \dots, a_n) = (a_1)_* f(a_2, \dots, a_n) + \sum_{i=1}^{n-1} (-1)^n f(a_1, \dots, a_i a_{i+1}, \dots, a_n)$$
$$+ (-1)^n a_n^* f(a_1, \dots, a_{n-1}).$$

**4.15.** THEOREM [BaW]. $H^n(\mathbb{C}, D) = H^n(F^{\cdot})$.

This result is proved in [BaW] with the use of a generalized bar resolution $B_* = \{B_n, d\}$ of $\overline{\mathbb{Z}}$ in $\mathcal{A}b^{F(\mathbb{C})}$ for which we have $F^n \cong \mathcal{A}b^{F(\mathbb{C})}(B_n, D)$, and so

$$H^n(F^{\cdot}) = H^n\big(\mathcal{A}b^{F(\mathbb{C})}(B_n, D)\big) = \mathrm{Ext}^n_{\mathcal{A}b^{F(\mathbb{C})}}(\overline{\mathbb{Z}}, D) = H^n(\mathbb{C}, D)$$

(see 4.4).

**4.16.** A function $d \in F^1(\mathbb{C}, D)$ such that $d(xy) = x_*(dy) + y^*(dx)$ is called a derivation (from the category $\mathbb{C}$ to the natural system $D$). An inner derivation is a function $d$ for which there exists an $a = F^0(\mathbb{C}, D)$ such that $d(x) = x_* a(A) - x^* a(B)$ where $x\colon A \to B$. Denote by $\mathrm{Der}\,(\mathbb{C}, D)$ and $\mathrm{Ider}\,(\mathbb{C}, D)$ the abelian groups of all derivations and all inner derivations, respectively.

**4.17.** PROPOSITION [BaW]. $H^1(\mathbb{C}, D) = \mathrm{Der}\,(\mathbb{C}, D)/\mathrm{Ider}\,(\mathbb{C}, D)$.

This statement follows immediately from 4.15 since derivations (respectively inner derivations) are cocycles (respectively coboundaries) in $F^1$.

**4.18.** REMARK. As in the cohomology theory of groups the second cohomology $H^2(\mathbb{C}, D)$ can be described in terms of so-called linear extensions of a category $\mathbb{C}$ by a natural system $D$ [BaW]. An extension of this description is made in [Go] for any $n$-th cohomology $H^n(\mathbb{C}, D)$.

**4.19.** Let $\mathbb{C}$ be a small category and $R$ be a ring. Define the $R$-cohomological dimension of $\mathbb{C}$ by

$$\mathrm{cd}_R\,\mathbb{C} = \sup\left\{k \mid \mathrm{Ext}^k_{R\mathbb{C}}(\overline{R}, -) \neq 0\right\}$$

where for brevity $R\mathbb{C} = (R-\mathrm{Mod})^{\mathbb{C}}$ (see 4.12), and $\overline{R}$ is the constant $R$-valued functor in $R\mathbb{C}$. When $R = \mathbb{Z}$ we write simply $\mathrm{cd}\,\mathbb{C}$.

It follows immediately from the definition that (see [Mt3]):

a) if $F\colon\ \mathbb{C} \to \mathbb{D}$ is any functor (between small categories) then $\mathrm{cd}_R\,\mathbb{C} \leqslant \mathrm{cd}_R\,\mathbb{D}$;

b) if $\gamma\colon\ R \to S$ is a ring homomorphism then $\mathrm{cd}_S\,\mathbb{C} \leqslant \mathrm{cd}_R\,\mathbb{C}$; in particular $\mathrm{cd}_R\,\mathbb{C} \leqslant \mathrm{cd}\,\mathbb{C}$;

c) if $\mathbb{C}$ has an initial object then $\mathrm{cd}_R\,\mathbb{C} = 0$; the converse is true if additionally $\mathbb{C}$ is a connected category in which all idempotents split.

**4.20.** The Hochschild–Mitchell $K$-dimension $\dim_K\mathbb{C}$ of a small category $\mathbb{C}$ (where $K$ is a commutative ring) is defined as the projective dimension of $K\mathbb{C}$ (see 4.12) in the category $\mathcal{A}b^{\mathbb{C}^{\mathrm{op}}\times\mathbb{C}}$, i.e.

$$\dim_K\mathbb{C} = \sup\left\{n \mid \mathrm{Ext}^n_{\mathcal{A}b^{\mathbb{C}^{\mathrm{op}}\times\mathbb{C}}}(K\mathbb{C}, -) \neq 0\right\}.$$

When $K = \mathbb{Z}$ we write simply $\dim\mathbb{C}$. By 4.13 $\dim\mathbb{C}$ can be described as

$$\dim\mathbb{C} = \sup\left\{n \mid H^n(\mathbb{C}, M) \neq 0 \text{ for some } \mathbb{C}\text{-bimodule } M\right\}.$$

The Hochschild–Mitchell dimension is related to another dimensions. For example, we have the following.

**4.21.** THEOREM [Mt1, Mt3]. *Let $\mathbb{C}$ be a small category, $\mathbb{A}$ be an abelian $K$-category (i.e. for any pair $(X,Y)$ of objects in $\mathbb{A}$ $\mathbb{A}(X,Y)$ is a $K$-module and the composition in $\mathbb{A}$ is $K$-bilinear). If $\mathbb{A}$ has exact coproduct then for all $T \in \mathbb{A}^{\mathbb{C}}$*

$$\text{pr.dim.}\,T \leqslant \dim_K\mathbb{C} + \sup\left\{\text{pr.dim.}\,T(X) \mid X \in \mathrm{Ob}(\mathbb{C})\right\}.$$

*Consequently,*

$$\text{gl.dim.}\,\mathbb{A}^{\mathbb{C}} \leqslant \dim_K\mathbb{C} + \text{gl.dim.}\,\mathbb{A}.$$

If we take $\mathbb{A} = \mathrm{Mod}\,K$ and $T = \overline{K}\colon\ K\mathbb{C} \to \mathcal{A}b$, a constant functor, we obtain

**4.22.** PROPOSITION [Mt3]. $\mathrm{cd}_K\mathbb{C} \leqslant \dim_K\mathbb{C}$.

Note that in general $\mathrm{cd}_K\,\mathbb{C} \neq \dim_K\mathbb{C}$, but for example if $\mathbb{C}$ is a group (cf. 4.6) the inequality in 4.22 is an equality [CE].

**4.23.** THEOREM [Mt3]. *Let $\mathbb{C}$ be a small category such that the only idempotents in $\mathbb{C}$ are identities. Then $\dim\mathbb{C} = 0$ iff $\mathbb{C}$ is equivalent to a discrete category.*

We do not discuss other cases where $\mathbb{C}$ has a low Hochschild–Mitchell dimension and note only that many interesting classes of categories with $\dim \mathbb{C} \leqslant 2$ are investigated in [Mt1, Mt3] (cf. 5.10).

**4.24.** The Baues–Wirsching dimension $\operatorname{Dim} \mathbb{C}$ of a small category $\mathbb{C}$ is defined as $\operatorname{Dim} \mathbb{C} = \operatorname{cd} \mathcal{F}(\mathbb{C})$ where $\mathcal{F}(\mathbb{C})$ is the category of factorizations in $\mathbb{C}$ (4.10), i.e. $\operatorname{Dim} \mathbb{C}$ is a projective dimension of the constant natural system $\overline{\mathbb{Z}}$. It easily follows from definition and 4.13 that $\dim \mathbb{C} \leqslant \operatorname{Dim} \mathbb{C}$ (cf. 5.8 below).

**4.25.** THEOREM [BaW].

a) *If* $\mathbb{C}$ *is a free category then* $\operatorname{Dim} \mathbb{C} \leqslant 1$.

b) *If* $\mathbb{C}$ *is a small category such that* $\operatorname{Dim} \mathbb{C} \leqslant 1$ *and* $\Sigma^{-1}\mathbb{C}$ *is a localization of* $\mathbb{C}$ *with respect to a subset* $\Sigma$ *of morphisms in* $\mathbb{C}$ *then* $\operatorname{Dim} \Sigma^{-1}\mathbb{C} \leqslant 1$.

This result corresponds (and partly generalizes) a similar result for the Hochschild–Mitchell dimension (see [CWM]).

## 5. Cohomology of posets

**5.1.** We can apply the results on the cohomology of small categories (see Section 4) if we consider any poset (= partially ordered set) as a category $o$ such that $\operatorname{Ob}(o) = I$ and $o(i,j)$ consists of only one morphism if $i \leqslant j$; $o(i,j) = \varnothing$ otherwise.

Recall that the cohomological $R$-dimension $\operatorname{cd}_R I$ of a poset $I$ where $R$ is a ring is the projective dimension of the constant functor $\overline{R}$: $I \to \mathcal{A}b$ which has a value $R$ on every object in $I$, particularly $\operatorname{cd} I = \operatorname{pr.dim.} \overline{\mathbb{Z}}$ (see 4.19).

**5.2.** As for any functor $T$: $I \to \mathcal{A}b$ we have $H^n(I,T) = \operatorname{Ext}^n_{\mathcal{A}b^I}(\overline{\mathbb{Z}}, T) = \varprojlim{}^n(T)$ (4.9) the groups $H^n(I,T)$ are isomorphic to the cohomology groups of the Roos complex (cf. [Ro]):

$$0 \longrightarrow \prod_{c_0} T(c_0) \xrightarrow{d^0} \prod_{c_0<c_1} T(c_1) \xrightarrow{d^1} \cdots \longrightarrow \prod_{c_0<c_1<\cdots<c_n} T(c_n) \xrightarrow{d^n} \cdots$$

where the differential $d^n$ is defined by the formula:

$$\begin{aligned}(d^n f)(c_0 < \cdots < c_{n+1}) = {} & \sum_{i=0}^{n} (-1)^i f(c_0 < \cdots < \hat{c}_i < \cdots < c_{n+1}) \\ & + (-1)^{n+1} T(c_n < c_{n+1}) f(c_0 < \cdots < c_n).\end{aligned}$$

The following result is an immediate consequence of this observation.

**5.3.** PROPOSITION. *Let* $I$ *be a poset for which there exists a natural number* $n$ *such that every chain of the form* $c_0 < c_1 < \cdots < c_m$ *has the length* $m \leqslant n$. *Then* $\operatorname{cd} I \leqslant n$.

**5.4.** Let $M_n$ be the poset defined as follows: as a set

$$M_n = \{a_1, \ldots, a_n\} \dot{\cup} \{b_1, \ldots, b_n\} \dot{\cup} \{1\}$$

and a partial order on $M_n$ is generated by the relations: $b_i \leqslant a_i$ and $a_i \leqslant 1$ for all $i$; $b_i \leqslant a_{i+1}$ for $i = 1, \ldots, n-1$; $b_n \leqslant a_1$.

**5.5.** THEOREM [CM]. *Let $I$ be a poset with dcc and let $R$ be a ring. Then* $\operatorname{cd}_R I \leqslant 1$ *iff $I$ does not contain $M_n$ as a retract for any $n \geqslant 2$.*

Recall that a subset $J \subseteq I$ is said to be a retract if there exists a morphism of posets $\eta$: $I \to J$ such that $\eta \mid_J = \mathrm{Id}_J$.

We notice additionally that an algorithm was given in [Ch] for determining when a finite poset $I$ has $\operatorname{cd}_R I \leqslant 1$.

**5.6.** THEOREM [Mt2]. *If $I^{\mathrm{op}}$ is a directed set and $R$ is a ring then* $\operatorname{cd}_R I = n + 1$ *where $\aleph_n$ is the smallest cardinal number of a cofinal subset of $I^{\mathrm{op}}$.*

**5.7.** In general there exist finite posets for which $\operatorname{cd}_R I$ is dependent on the ring $R$. In fact, one can make the difference $\operatorname{cd}_R I - \operatorname{cd}_S I$ as large as one likes for suitable $I$, $R$ and $S$ [Mt1]. A similar result holds for $\operatorname{gl\,dim} KI$, the global dimension of the incidence algebra $KI$ of a poset $I$ [Mt1, IZ].

**5.8.** Let $\dim I$ (respectively $\operatorname{Dim} I$) be the Hochschild–Mitchell dimension (respectively Baues–Wirsching dimension) of a poset $I$ (cf. 4.20 and 4.24). In contrast to the inequality $\dim \mathbb{C} \leqslant \operatorname{Dim} \mathbb{C}$ (4.24) we have the following.

**5.9.** PROPOSITION [Kh3]. *If $I$ is a poset then* $\dim I = \operatorname{Dim} I$.

**5.10.** THEOREM [Mt1]. *Let $I$ be a poset and let $K$ be a commutative ring. Then* $\dim_K I \leqslant 1$ *iff $I$ is the free category generated by an oriented graph.*

In particular if $I = \mathbb{Z}$ is a poset of integers then $\dim \mathbb{Z} = 1$.

Note that Mitchell [Mt1] gives also a description of some class of categories with $\dim_K \mathbb{C} \leqslant 2$ including all locally finite posets.

**5.11.** We discuss now the case of totally ordered sets. In this case the Hochschild–Mitchell dimension is a monotone function: if $I$ is a totally ordered set and $J \subseteq I$ then $\dim J \leqslant \dim I$ [Kh3].

**5.12.** THEOREM [Mt3]. *If $I$ is a totally ordered set whose closed intervals all have cardinal numbers at most $\aleph_n$ then* $\dim I \leqslant n + 2$.

This inequality is best possible in view of the following fact: if $I$ contains $\aleph_n + 1$ or $(\aleph_n + 1)^{\mathrm{op}}$ then $\dim_K I \geqslant n + 2$ for any commutative ring $K$ [Mt1]. In particular, if $I = \mathbb{Q}$ is the poset of rational numbers then $\dim \mathbb{Q} = 2$.

**5.13.** The problem of determining the precise Hochschild–Mitchell dimension of any totally ordered set $I$ is still open (see [Mt3]). In the interesting case where $I = \mathbb{R}$ is the poset of real numbers Mitchell proved that $2 \leqslant \dim \mathbb{R} \leqslant 3$ (the latter inequality was proved under assumption of the continuum hypothesis) and supposed that $\dim \mathbb{R}$ depends on the continuum hypothesis [Mt1, Mt3]. This problem was solved by Balcerzyk [Bar] (independent solution was given by Khusainov [Kh2, Kh3]).

**5.14.** THEOREM [Bar, Kh2, Kh3]. $\dim \mathbb{R} = 3$.

So $\dim \mathbb{R}$ does not depend on the continuum hypothesis. In turn Khusainov [Kh3] discusses a new hypothesis about connections between the Hochschild–Mitchell dimension and the continuum hypothesis.

## 6. Cohomology of coalgebras

**6.1.** Let $k$ be a fixed field. We denote the tensor product $\otimes_k$ over $k$ simply as $\otimes$. A coalgebra $(C, \Delta, \varepsilon)$ over a field $k$ is a $k$-vector space $C$ together with $k$-linear maps $\Delta\colon C \to C \otimes C$ and $\varepsilon\colon C \to k$ such that the diagrams

$$\begin{array}{ccc} C & \xrightarrow{\Delta} & C \otimes C \\ \Delta\downarrow & & \downarrow \Delta\otimes 1 \\ C \otimes C & \xrightarrow{1\otimes\Delta} & C \otimes C \otimes C \end{array}, \qquad \begin{array}{ccccc} C & = & C & = & C \\ \cong\downarrow & & \downarrow\Delta & & \downarrow\cong \\ k \otimes C & \xleftarrow{\varepsilon\otimes 1} & C \otimes C & \xrightarrow{1\otimes\varepsilon} & C \otimes k \end{array}$$

are commutative (here $C \cong k \otimes C$ etc. is the canonical isomorphism). $\Delta$ is called the comultiplication and $\varepsilon$ the counit of the coalgebra $C$. We usually simplify the notation as $C = (C, \Delta, \varepsilon)$.

A pair $(M, m)$ (or simply $M$) where $M$ is a $k$-vector space and $m$ is a $k$-linear map is called a left $C$-comodule if the following diagrams are commutative:

$$\begin{array}{ccc} M & \xrightarrow{m} & C \otimes M \\ m\downarrow & & \downarrow \Delta\otimes 1 \\ C \otimes M & \xrightarrow{1\otimes m} & C \otimes C \otimes M \end{array}, \qquad \begin{array}{ccc} M & = & M \\ m\downarrow & & \downarrow\cong \\ C \otimes M & \xrightarrow{\varepsilon\otimes 1} & k \otimes M \end{array}.$$

A right $C$-comodule is defined similarly. Every right $C$-comodule can be considered as a left $C^{\mathrm{op}}$-comodule where $C^{\mathrm{op}}$ is the opposite coalgebra of $C$. Morphisms of left (or right) $C$-comodules and morphisms of coalgebras are defined naturally. The category of left (respectively right) $C$-comodules will be denoted by $C$-Comod (respectively Comod-$C$). It is well known that the category $C$-Comod is abelian and has enough injectives (see, e.g., [D]).

Let $C$, $D$ be two $k$-coalgebras. $M$ is called $(C, D)$-bicomodule if there exist $k$-linear maps $m_l\colon M \to C \otimes M$ and $m_r\colon M \to M \otimes C$ such that $(M, m_l)$ (respectively $(M, m_r)$) is a left $C$-comodule (respectively right $D$-comodule) and $(1 \otimes m_r)m_l = (m_l \otimes 1)m_r$.

Every $(C,C)$-bicomodule is naturally a left $C^e$-comodule where $C^e = C \otimes C^{\mathrm{op}}$ is the enveloping coalgebra of $C$. In particular $C$ is a left $C^e$-comodule.

**6.2.** Define the cohomology of a coalgebra $C$ with coefficients in a $(C,C)$-bicomodule $N$ as

$$H^n(N,C) = \mathrm{Ext}^n_{C^e}(N,C)$$

(see [D]). So if $X$ is an injective resolution of $C$ as a left $C^e$-comodule then

$$H^n(N,C) = H^n\big(\mathrm{Hom}_{C^e}(N,X^{\bullet})\big).$$

**6.3.** We shall construct a standard complex which is similar to the standard complex used in the computation of the Hochschild cohomology (cf. 5.7). Let $K^n = C\otimes\cdots\otimes C$ ($n+2$ times), $n \geqslant 0$, with the following $(C,C)$-bicomodule structure:

$$k_1(c_0\otimes\cdots\otimes c_{n+1}) = \Delta(c_0)\otimes c_1\otimes\cdots\otimes c_{n+1},$$

$$k_r(c_0\otimes\cdots\otimes c_{n+1}) = c_0\otimes\cdots\otimes c_n\otimes\Delta(c_{n+1}),$$

and then define the differential $d^n\colon K^n \to K^{n+1}$ by

$$d^n(c_0\otimes\cdots\otimes c_{n+1}) = \sum_{i=0}^{n+1}(-1)^i c_0\otimes\cdots\otimes\Delta(c_i)\otimes\cdots\otimes c_{n+1}.$$

It is easy to see that

$$C \xrightarrow{\Delta} K^0 \xrightarrow{d^0} K^1 \xrightarrow{d^1} \cdots$$

is an injective resolution of $C$ as a left $C^e$-comodule. As a consequence we have

$$H^n(N,C) = \mathrm{Ext}^n_{C^e}(N,C) = H^n\big(\mathrm{Hom}_{C^e}(N,K^{\bullet})\big).$$

**6.4.** We use the identifications in the standard complex $K^n \cong C^e\otimes\overline{K}^n$ where $\overline{K}^n$ is the $n$-fold tensor product of $C$ ($\overline{K}^0 = k$), and then

$$\mathrm{Hom}_{C^e}(N,K^n) = \mathrm{Hom}_{C^e}\big(N,C^e\otimes\overline{K}^n\big) \cong \mathrm{Hom}_k\big(N,\overline{K}^n\big).$$

Hence the $H^n(N,C)$ are the cohomology groups of the complex

$$\big\{\mathrm{Hom}_k\big(N,\overline{K}^n\big),\delta\big\}_{n\geqslant 0} \quad \text{where } \delta^n\colon \mathrm{Hom}_k\big(N,\overline{K}^n\big) \to \mathrm{Hom}_k\big(N,\overline{K}^{n+1}\big)$$

is given by

$$\delta^n(f) = (1\otimes f)n_1 - (\Delta\otimes 1\otimes\cdots\otimes 1)f + (1\otimes\Delta\otimes\cdots\otimes 1)f - \cdots$$
$$+(-1)^n(1\otimes\cdots\otimes 1\otimes\Delta)f + (-1)^{n+1}(f\otimes 1)n_r.$$

**6.5.** A $k$-linear map $f\colon N \to C$ from a $(C, C)$-bicomodule $N$ into $C$ is called a coderivation if it satisfies the property $\Delta f = (1 \otimes f)n_1 + (f \otimes 1)n_r$. The coderivation $f\colon N \to C$ is called inner if there exists a $k$-linear map $\gamma\colon N \to k$ such that

$$f = (1 \otimes \gamma)n_1 - (\gamma \otimes 1)n_r.$$

Denote by $\mathrm{Coder}(N, C)$ (respectively $\mathrm{Incoder}(N, C)$) the $k$-space of all (respectively inner) coderivations from $N$ into $C$.

**6.6.** PROPOSITION. *For any $(C, C)$-bicomodule $N$ we have:*
a) $H^0(N, C) \cong \mathrm{Hom}_{C^e}(N, C)$;
b) $H^1(N, C) \cong \mathrm{Coder}(N, C)/\,\mathrm{Incoder}(N, C)$.

This follows immediately from the description of $H^n(N, C)$ in 6.4.

**6.7.** We note that Doi [D] establishes one-to-one correspondence between $H^2(N, C)$ and the set of equivalence classes of so-called "extensions over $C$ with cokernel $N$", hence the second cohomology group of a coalgebra is described in a manner which is similar to the classical case of algebras.

**6.8.** A coalgebra $C$ is called coseparable if there exists a $(C, C)$-bicomodule map $\pi\colon C \otimes C \to C$ such that $\pi\Delta = 1_C$.

**6.9.** THEOREM [D]. *Let $C$ be a $k$-coalgebra. The following statements are equivalent:*
a) *$C$ is coseparable;*
b) *for every $(C, C)$-bicomodule $N$ we have $H^n(N, C) = 0$ for all $n \geqslant 1$;*
c) *every coderivation from any $(C, C)$-bicomodule $N$ into $C$ is an inner coderivation.*

**6.10.** Let $M$ and $N$ be $(C, D)$-bicomodules. A (left) cointegration from $M$ into $N$ is a $D$-map $F\colon M \to C \otimes N$ which satisfies the property $(\Delta \otimes 1)f = (1 \otimes f)m_1 - (1 \otimes n_1)f$. A cointegration is called inner if there exists a $D$-map $\phi\colon M \to N$ such that $f = (1 \otimes \phi)m_1 - n_1\phi$. $k$-spaces of cointegrations and of inner cointegrations from $M$ into $N$ will be denoted by $\mathrm{Coint}(M, N)$ and $\mathrm{Incoint}(M, N)$ respectively.

We have a connection between cointegrations and coderivations in the case where $N = C$.

**6.11.** PROPOSITION [Gu]. *If $M$ is $(C, C)$-bicomodule that there exists a natural isomorphism $\alpha$: $\mathrm{Coint}(M, C) \to \mathrm{Coder}(M, C)$ which restricts to a natural isomorphism $\alpha$: $\mathrm{Incoint}(M, C) \to \mathrm{Incoder}(M, C)$.*

PROOF. This isomorphism is given by the relation $\alpha(f) = (1 \otimes \varepsilon)f$ where $f \in \mathrm{Coint}(M, C)$. The inverse map is given by $\beta(g) = (g \otimes 1)m_r$ where $g \in \mathrm{Coder}(M, C)$. □

**6.12.** Consider the short exact sequence of $(C, D)$-bicomodules

$$0 \longrightarrow M \xrightarrow{m_1} C \otimes M \longrightarrow \Omega(M) \longrightarrow 0,$$

i.e. $\Omega(M) = \operatorname{Coker}(m_1)$.

**6.13.** PROPOSITION [Gu]. *There exists a natural isomorphism of abelian groups* $\operatorname{Hom}_{C-D}(N, \Omega(M)) = \operatorname{Coint}(N, M)$.

**6.14.** Let $\omega$ be p.c. of short exact sequences in the category $C$-Comod-$D$ of $(C, D)$-bicomodules which split over $D$. It is easy to see that the forgetful functor

$$S\colon C\text{-Comod-}D \to \text{Comod-}D$$

has as right adjoint

$$T = C \otimes -\colon \text{Comod-}D \to C\text{-Comod-}D.$$

So if $\omega_0$ is the p.c. consisting of split exact sequences in Comod-$D$ then $\omega = S^{-1}(\omega_0)$ (cf. 1.11). Using the dual versions of 1.12 and 1.13 we obtain

**6.15.** PROPOSITION. *The p.c.* $\omega$ *in* 6.14 *is injective and is injectively generated by the class of comodules* $\{C \otimes X \mid X \in \text{Comod-}D\}$.

**6.16.** Let $\omega\operatorname{Ext}^n_{C-D}(N, -)$ (respectively $R^n_\omega \operatorname{Coint}(N, -)$) be the relative right derived functors of the functor $\operatorname{Hom}_{C-D}(N, -)$ (respectively $\operatorname{Coint}(N, -)$) (cf. 2.19 and 2.21).

**6.17.** THEOREM [Gu]. *For any* $(C - D)$*-bicomodules* $M, N$ *there exist natural isomorphisms of abelian groups:*

$$R^n_\omega \operatorname{Coint}(N, M) \cong \omega\operatorname{Ext}^{n+1}_{C-D}(N, M), \quad n \geqslant 1.$$

PROOF. As $C \otimes M$ is $\omega$-injective (6.15) we deduce from 6.12 that

$$\omega\operatorname{Ext}^n_{C-D}\big(N, \Omega(M)\big) \cong \omega\operatorname{Ext}^{n+1}_{C-D}(N, M), \quad n \geqslant 1.$$

Now we see from 6.13 that

$$\omega\operatorname{Ext}^n_{C-D}\big(N, \Omega(M)\big) \cong R^n_\omega \operatorname{Coint}(N, M), \quad n \geqslant 1,$$

and the statement follows. □

**6.18.** PROPOSITION [Gu]. *Let* $R^n_\omega \operatorname{Coder}(M, -)$ *be the relative right derived functor of the functor* $\operatorname{Coder}(M, -)$ *where* $M$ *is* $(C, C)$*-bicomodule. We have the natural isomorphisms of abelian groups:*

$$R^n_\omega \operatorname{Coder}(M, C) \cong \omega\operatorname{Ext}^{n+1}_{C^e}(M, C) \cong H^{n+1}(M, C), \quad n \geqslant 1.$$

This proposition follows from 6.17 and 6.11.

**6.19.** REMARK. Some applications of the developed theory to the cohomology of Hopf algebras can be found in [D].

**6.20.** If we replace the base field $k$ by a commutative ring then the category Comod $C$ of comodules over $k$-coalgebra $C$ is not in general an abelian category (and is even not preabelian). Nevertheless the cohomology of such coalgebras was introduced in [J] using the relative homological algebra developed in [EM]. The corresponding generalizations of some results mentioned above and further investigations can be found in [Gu].

**6.21.** A different approach is presented in [Gr] to the development of cohomology theory of commutative coalgebras. This theory is in many respects dual to the André–Quillen cohomology theory of commutative rings.

## References


[A] M. Auslander, *Relations of Grothendieck group of artin algebra*, Proc. Amer. Math. Soc. **91** (1984), 336–340.

[BaL] D. Baer and H. Lenzing, *A homological approach to representations of algebras, I: the wild case*, J. Pure Appl. Algebra **24** (1982), 227–233.

[Bar] S. Balcerzyk, *The cohomological dimension of the ordered set of real numbers equals three*, Fund. Math. **111** (1981), 37–44.

[BaW] H.-J. Baues and G. Wirsching, *Cohomology of small categories*, J. Pure Appl. Algebra **35** (1985), 187–211.

[Bc] D. Buchsbaum, *A note on homology in categories*, Ann. Math. **69** (1959), 66–74.

[BD] I. Bucur and A. Deleanu, *Introduction to the Theory of Categories and Functors*, Wiley, New York (1968).

[Bu] M.C.R. Butler, *Grothendieck groups and almost split sequences*, SLNM 882, Springer, Berlin (1981), 357–368.

[Ca] B. Calvo, *Pureté des catégories relatives*, C. R. Acad. Sci. Paris Ser. I **303** (1986), 379–382.

[CE] H. Cartan and S. Eilenberg, *Homological Algebra*, Princeton Univ. Press, Princeton, NJ (1956).

[Ch] C. Cheng, *Finite partially ordered set of cohomological dimension one*, J. Algebra **40** (1976), 340–347.

[CM] C. Cheng and B. Mitchell, *DCC posets of cohomological dimension one*, J. Pure Appl. Algebra **13** (1978), 125–137.

[CWM] C. Cheng, Y.-C. Wu and B. Mitchell, *Categories of fractions preserve dimension one*, Comm. Algebra **8** (1980), 927–939.

[D] Y. Doi, *Homological coalgebra*, J. Math. Soc. Japan **33** (1981), 31–50.

[EM] S. Eilenberg and J.C. Moore, *Foundations of relative homological algebra*, Mem. Amer. Math. Soc. **55** (1965).

[F] L. Fuchs, *Infinite Abelian Groups*, vols 1, 2, Academic Press, New York (1970, 1973).

[GaZ] P. Gabriel and M. Zisman, *Calculus of Fractions and Homotopy Theory*, Springer, Berlin (1967).

[Gei] W. Geigle, *Grothendieck groups and exact sequences for hereditary artin algebras*, J. London Math. Soc. **31** (1985), 231–236.

[GeM] S.I. Gel'fand and Yu.I. Manin, *Methods of Homological Algebra*, vol. 1, Nauka, Moscow (1988) (in Russian).

[G1] A.I. Generalov, *Inductively closed proper classes over bounded hnp-rings*, Algebra i Logika **25** (1986), 384–404. English transl. in Algebra and Logic **25**(4) (1986).

[G2] A.I. Generalov, *Inductive purities over rings of finite representation type*, Abelian Groups and Modules, no. 7, Tomsk (1988), 31–41 (in Russian).

[G3] A.I. Generalov, *Butler–Auslander theorem and Hom-matrix over rings of finite representation type*, Algebraic Systems, Volgograd (1990), 60–71 (in Russian).

[G4] A.I. Generalov, *Relative Grothendieck groups and exact sequences over tame hereditary algebras*, Algebra i Analiz **2** (1990), 47–72 (in Russian).

[G5] A.I. Generalov, *Grothendieck groups and short exact sequences over noetherian rings*, Abelian groups and modules, no. 9, Tomsk (1990), 7–30 (in Russian).

[G6] A.I. Generalov, *Algebraically compact modules and relative homological algebra over tame hereditary algebras*, Algebra i Logika **30** (1991), 259–292 (in Russian).

[G7] A.I. Generalov, *Relative homological algebra in preabelian categories, I: Derived categories*, Algebra i Analiz **4**(1) (1992), 98–119. English transl. in St. Petersburg Math. J. **4**(1) (1993).

[G8] A.I. Generalov, *Derived categories of an additive category*, Algebra i Analiz **4**(5) (1992), 91–103 (in Russian).

[Go] M. Golasinski, *$n$-fold extensions and cohomologies of small categories*, Math. Rev. D'Analyse Num. Theor. Approxim. Acad. Rep. Soc. Roumanie, Fil. Cluj-Napoca **31** (1989), 53–59.

[Gr] L. Grunenfelder, *(Co)-homology of commutative coalgebras*, Comm. Algebra **16** (1988), 541–576.

[GrJ] L. Gruson and C.U. Jensen, *Dimensions cohomologiques reliées aux foncteurs* $\varprojlim^{(n)}$, SLNM 867, Springer, Berlin (1981), 234–294.

[Gu] F. Guzman, *Cointegrations, relative cohomology for comodules and coseparable corings*, J. Algebra **126** (1989), 211–224.

[Ha] R. Hartshorne, *Residues and Duality*, SLNM 20, Springer, Berlin (1966).

[HiS] P.J. Hilton and U. Stammbach, *A Course in Homological Algebra*, Springer, Berlin (1971).

[Ho] G. Hochschild, *Relative homological algebra*, Trans. Amer. Math. Soc. **82** (1956), 246–269.

[IZ] K. Igusa and D. Zacharia, *On the cohomology of incidence algebras of partially ordered sets*, Comm. Algebra **18** (1990), 873–887.

[J] D.W. Jonah, *Cohomology of coalgebras*, Mem. Amer. Math. Soc. **82** (1968).

[Kh1] A.A. Khusainov, *On extension groups in a category of abelian diagrams*, Sibirsk. Mat. Zh. **33**(1), (1992), 179–185 (in Russian).

[Kh2] A.A. Khusainov, *On the Hochschild–Mitchell dimension of the set of real numbers*, Dokl. Akad. Nauk (Russia) **322** (1992), 259–261 (in Russian).

[Kh3] A.A. Khusainov, *The Hochschild–Mitchell dimension of reals is equal to* 3, Sibirsk. Mat. Zh. **34**(4) (1993), 217–227 (in Russian).

[Ku] V.I. Kuz'minov, *On groups of pure extensions of abelian groups*, Sibirsk. Mat. Zh. **17** (1976), 1308–1320 (in Russian).

[LR] T.Y. Lam and I. Reiner, *Relative Grothendieck groups*, J. Algebra **11** (1969), 213–242.

[Ma1] S. MacLane, *Homology*, Springer, Berlin (1963).

[Ma2] S. MacLane, *Categories for the Working Mathematician*, Springer, Berlin (1971).

[Mn] A.A. Manovtsev, *Inductive purities in abelian groups*, Mat. Sb. **96** (1975), 416–446 (in Russian).

[MS] A.P. Mishina and L.A. Skornyakov, *Abelian Groups and Modules*, Nauka, Moscow (1969). English transl. in Amer. Math. Soc. Transl. Ser. 2 vol. 107 (1976).

[Mt1] B. Mitchell, *Rings with several objects*, Adv. Math. **8** (1972), 1–161.

[Mt2] B. Mitchell, *The cohomological dimension of a directed set*, Canad. J. Math. **25** (1973), 233–238.

[Mt3] B. Mitchell, *Some applications of module theory to functor categories*, Bull. Amer. Math. Soc. **84** (1978), 867–885.

[N] A. Neeman, *The derived category of an exact category*, J. Algebra **135** (1990), 388–394.

[Q] D. Quillen, *Higher algebraic $K$-theory, I*, SLNM 341, Springer, Berlin (1973), 85–147.

[RW1] F. Richman and E.A. Walker, *Ext in pre-abelian categories*, Pacific J. Math. **71** (1977), 521–535.

[RW2] F. Richman and E.A. Walker, *Valuated groups*, J. Algebra **56** (1979), 145–167.

[Ri] J. Rickard, *Derived categories and stable equivalence*, J. Pure Appl. Algebra **61** (1989), 303–317.

[Ro] J.-E. Roos, *Sur les foncteurs dérivés de* $\varprojlim$. *Applications*, C. R. Acad. Sci. Paris **252** (1961), 3702–3704.

[Si] D. Simson, *On pure global dimension of locally finitely presented Grothendieck categories*, Fund. Math. **96** (1977), 91–116.

[Sk] E.G. Sklyarenko, *Relative homological algebra in categories of modules*, Uspekhi Mat. Nauk **33**(3) (1978), 85–120. English transl. in Russian Math. Surveys **33** (1978).

[St] B. Stenstrom, *Purity in functor categories*, J. Algebra **8** (1968), 352–361.

[V] J.-L. Verdier, *Catégories derivées*, SLNM 569, Springer, Berlin (1977), 262–311.

[W] R.B. Warfield, *Purity and algebraic compactness for modules*, Pacific J. Math. **28** (1969), 699–719.

[Y] A.V. Yakovlev, *Homological algebra in pre-abelian categories*, Zap. Nauchn. Sem. Leningrad. Otdel. Mat. Inst. Steklov (LOMI) **94** (1979), 131–141. English transl. in J. Soviet Math. **19**(1) (1982).

# Homotopy and Homotopical Algebra

J.F. Jardine[1]

*Mathematics Department, University of Western Ontario, London, Ontario N6A 5B7, Canada*
*e-mail: jardine@uwo.ca*

*Contents*



[1]Supported by NSERC.

HANDBOOK OF ALGEBRA, VOL. 1
Edited by M. Hazewinkel

## 1. Combinatorial homotopy theory

The homotopy theory of simplicial sets and homotopical algebra have their roots in a mistake of Poincaré [60, 61]. He made something of a mess of the proof of his famous duality theorem; this theorem asserts a relation

$$b_p = b_{n-p}$$

between the Betti numbers of a compact connected oriented manifold of dimension $n$. The problem was that there was no good definition of Betti numbers – such things were only talked about intuitively at the time. To correct the error, Poincaré introduced the concept of polyhedron, which to us is a finite CW-complex $X$ with enough information about the incidence relations between cells such that a chain complex $C_*X$ and its associated homology groups $H_*X$ can be formed (he specialized to ordinary simplicial complexes later). Then, for $X$, $b_p$ is the rank of $H_pX$. He fixed his proof, and invented the theory of chain complexes and homology in the process, although apparently the homology groups themselves were not introduced until much later [15, 51] at a suggestion of Emmy Noether.

One of the bothersome details left over was the question of whether $H_*X$ was independent of the given triangulation of $X$. This was settled by Alexander, using the simplicial approximation theorem of Brouwer. The theorem says, in modern terms, that if $f\colon |K| \to |L|$ is a continuous map between the realizations of finite simplicial complexes $K$ and $L$, then, after sufficiently many subdivisions of $K$ and $L$, $f$ may be replaced up to homotopy by the realization of a simplicial map. One of the three ways the Alexander proves his result is by putting this theorem together with Poincaré's "observation" that there is an isomorphism

$$sd\colon H_*(X) \xrightarrow{\cong} H_*(sdX)$$

associated to barycentric subdivision.

This was the birth of a field of mathematics that came to be known as Combinatorial Topology. Explicitly, this was the study of manifolds, triangulations of such, and associated chain complexes and homology invariants. The subject languished under this name until Lefschetz rechristened it "Algebraic Topology" in his book of 1942 [48], on the grounds that, by this time, there was more algebra than combinatorics to be found in the subject. The assumption was false, but the name stuck.

Eilenberg's introduction of singular homology theory in his Annals paper of 1944 [19] was a decisive leap forward. I like to write

$$|\Delta^n| = \left\{(t_0,\dots,t_n) \in \mathbb{R}^{n+1} \,\middle|\, \sum t_i = 1,\ t_i \geqslant 0\right\}$$

for the topological standard $n$-simplex. There are maps

$$d^i\colon |\Delta^{n-1}| \to |\Delta^n|, \quad 0 \leqslant i \leqslant n,$$

defined by

$$(t_0, \ldots, t_{n-1}) \mapsto (t_0, \ldots, \overset{i}{0}, \ldots, t_{n-1})$$

which are now called *cofaces*, and which satisfy the fundamental relation

$$d^j d^i = d^i d^{j-1} \quad \text{if } i < j. \tag{1}$$

A singular $n$-simplex of a topological space $Y$ is a continuous map $\sigma\colon |\Delta^n| \to Y$, and one defines the group $C_n(Y;\mathbb{Z})$ of integral singular $n$-chains to be the free abelian group on the set $S_n(Y)$ of singular $n$-simplices. A boundary operator

$$\partial\colon C_n(Y;\mathbb{Z}) \to C_{n-1}(Y;\mathbb{Z})$$

is defined on generators by

$$\partial(\sigma) = \sum_{i=0}^{n} (-1)^i (\sigma \circ d^i),$$

and it follows from (1) above that $\partial^2 = 0$. The resulting homology groups are the integral singular homology groups $H_*(Y;\mathbb{Z})$ of the space $Y$.

This theory has important advantages: it is manifestly functorial, and in no way depends on a triangulation of the space $Y$.

The structure $S(Y)$, consisting of $S_n(Y)$, $n \geqslant 0$, and all face maps

$$d_i\colon S_n(Y) \to S_{n-1}(Y)$$

defined by

$$d_i(\sigma) = \sigma \circ d^i, \quad 0 \leqslant i \leqslant n,$$

is a type of generalized simplicial complex. Eilenberg and Zilber [20] call it a *semi-simplicial complex.*

Recall that a finite oriented simplicial complex $K$ consists of a totally ordered set $\{v_0, \ldots, v_N\}$ of vertices, together with simplices $\sigma = [v_{i_0}, \ldots, v_{i_k}]$, which are elements of the power set of $\{v_0, \ldots, v_N\}$, such that

(1) each $v_i$ is a simplex of $K$, and

(2) if $\tau \subset \sigma$ and $\sigma$ is a simplex of $K$, then so is $\tau$.

Examples include:

$$\Delta^n = \text{all nonempty subsets of the ordinal number } \mathbf{n} = \{0, \ldots, n\} \quad \text{(standard } n\text{-simplex)}, \tag{2}$$

$$\partial\Delta^n = \Delta^n - [0, \ldots, n] \quad (\text{boundary of } \Delta^n), \text{ and} \tag{3}$$

$$\Lambda^n_k = \text{subcomplex of } \Delta^n \text{ generated by all faces } [0, \dots, i-1, i+1, \dots, n]$$
$$\text{except the } k\text{-th face (the } k\text{-th horn of } \Delta^n). \tag{4}$$

A simplicial complex $K$ may be identified with a semi-simplicial complex by putting in the empty set in degrees above its dimension; the face maps are defined by

$$d_j[v_{i_0}, \dots, v_{i_k}] = [v_{i_0}, \dots, \hat{v_{ij}}, \dots, v_{i_k}] \quad \text{(i.e. remove the vertex } v_{i_j}).$$

Each such $K$ has a canonical barycentric realization $|K|$; this is the topological space defined by

$$|K| = \Big\{(t_0, \dots, t_N) \in \mathbb{R}^{N+1} \mid \{v_i \mid t_i \neq 0\} \text{ is a simplex of } K \text{ and}$$
$$\sum t_i = 1,\ t_i \geqslant 0\Big\}.$$

There is a natural map of semi-simplicial complexes $K \to S|K|$, defined by

$$\sigma = [v_{i_0}, \dots, v_{i_k}] \mapsto |\Delta^k| \xrightarrow{|\sigma|} |K|.$$

Eilenberg showed that this map induces a homology isomorphism, but the most efficient proof uses the fact that the standard $k$-simplex $\Delta^k$ has a simplicial contracting homotopy

$$\Delta^n \times \Delta^1 \to \Delta^n,$$

which can be viewed in categorical terms as the diagram

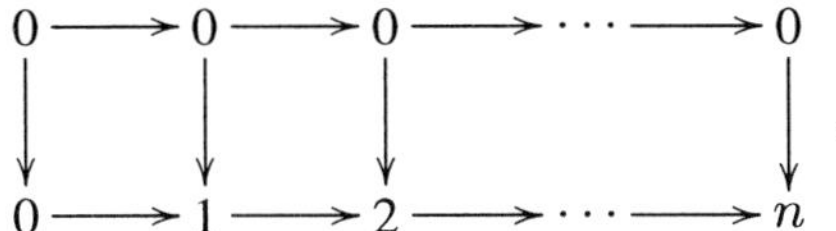


A map $f\colon K \to L$ of oriented simplicial complexes is defined to be an order-preserving function on the vertices such that $f(\sigma)$ is a simplex of $L$ whenever $\sigma$ is a simplex of $K$. Such an $f$ induces a map

$$f_*\colon C_*(K) \to C_*(L),$$

which is given by

$$f_*(\sigma) = \begin{cases} f(\sigma) & \text{if } \dim(\sigma) = \dim(f(\sigma)), \\ 0 & \text{if } \dim(\sigma) > \dim(f(\sigma)). \end{cases}$$

The simplices of $K \times \Delta^1$ are defined to be sequences of pairs of the form

$$(v_{i_0}, \dots, v_{i_n}) \times 0, \quad (v_{i_0}, \dots, v_{i_n}) \times 1,$$

or

$$h_j(v_{i_0}, \dots, v_{i_n}) = \big((v_{i_0}, 0), \dots, (v_{i_j}, 0), (v_{i_j}, 1), \dots, (v_{i_n}, 1)\big),$$

where, in all cases, $[v_{i_0}, \dots, v_{i_n}]$ is a simplex of $K$.

Now let $f, g\colon K \to L$ be maps of simplicial complexes. A simplicial homotopy from $f$ to $g$ is defined to be a commutative diagram of maps of simplicial complexes of the form

$$\begin{array}{ccc} K & & \\ {\scriptstyle i_0}\downarrow & \searrow^{f} & \\ K \times \Delta^1 & \xrightarrow{h} & L, \\ {\scriptstyle i_1}\uparrow & \nearrow_{g} & \\ K & & \end{array} \tag{5}$$

where $i_0$ sends $(v_{i_0}, \dots, v_{i_n})$ to $(v_{i_0}, \dots, v_{i_n}) \times 0$, and $i_1$ sends $(v_{i_0}, \dots, v_{i_n})$ to $(v_{i_0}, \dots, v_{i_n}) \times 1$. Then $h$ determines a chain homotopy from $f_*$ to $g_*$ defined by

$$h_*[v_{i_0}, \dots, v_{i_n}] = \sum_{i=0}^{n} (-1)^i h_i[v_{i_0}, \dots, v_{i_n}].$$

Most of these definitions appear, either in [19] or in the subsequent paper [20]. With the appearance of [20], the foundations of singular homology theory are mostly in place.

In the contracting homotopy for $\Delta^n$, the images of the $h_i[0, \dots, n]$ want to be $(n+1)$-simplices of the form

$$\big[0, \dots, \overset{i}{0}, i, \dots, n\big].$$

It is convenient to keep such "degenerate" simplices (having repeats) around, together with the means of producing them; the definition of

$$f_*\colon C_*(K) \to C_*(L)$$

above was, after all, quite artificial, as was the description of the $h_i$ operators. A *simplicial set* $X$ ("complete semi-simplicial set" in the language of [20]) is often defined to be a sequence of sets $X_n$, $n \geqslant 0$, together with functions

$$d_i\colon X_n \to X_{n-1}, \quad 0 \leqslant i \leqslant n, \text{ (face maps)}$$

and

$$s_j\colon X_n \to X_{n+1}, \quad 0 \leqslant j \leqslant n, \text{ (degeneracies)}$$

such that the following *simplicial identities* hold:

$$d_i d_j = d_{j-1} d_i \quad \text{if } i < j, \tag{6}$$

$$s_i s_j = s_{j+1} s_i \quad \text{if } i \leqslant j,$$

$$d_i s_j = \begin{cases} s_{j-1} d_i & \text{if } i < j, \\ 1 & \text{if } i = j, j+1, \\ s_j d_{i-1} & \text{if } i > j+1. \end{cases}$$

The *singular complex* $S(Y)$ of a topological space $Y$ is a standard example; the $s_j$ are induced by precomposition with the continuous maps

$$s^j\colon \left|\Delta^{n+1}\right| \to \left|\Delta^n\right|,$$

defined by

$$(t_0, \ldots, t_{n+1}) \mapsto (t_0, \ldots, t_j + t_{j+1}, \ldots, t_{n+1}).$$

Any oriented simplicial complex $K$ can be enlarged to a simplicial set in a canonical way, by allowing symbols $[v_{i_0}, \ldots, v_{i_n}]$ with repeats and decreeing that $s_j$ puts in a repeat in the $j$-th place. We shall keep the same notation for the simplicial sets $\Delta^n$, $\partial\Delta^n$ and $\Lambda^n_k$.

One can (and should) also think of a simplicial set as a contravariant functor $X\colon \boldsymbol{\Delta}^{op} \to \mathbf{Sets}$, defined on the category $\boldsymbol{\Delta}$ of finite ordinal numbers $\mathbf{n} = \{0, 1, \ldots, n\}$ and the functors between them. This is a consequence of the presentation of $\boldsymbol{\Delta}$ in terms of generators $d^i$, $s^j$ and relations dual to the simplicial identities which is given by MacLane in [50]. From this point of view $\Delta^n$ is the contravariant functor which is represented by the ordinal number $\mathbf{n}$. Furthermore, and perhaps most importantly, maps of simplicial sets are just natural transformations.

More generally, any category $C$ has associated to it a simplicial set $BC$, called its *nerve*, where $BC_n$ is defined to be the set of functors from $\mathbf{n}$ to $C$ (we shall agree to suppress set-theoretic conundrums here and now). The "space" $BG$ associated to the category with one object canonically associated to a group $G$ is the standard model for the Eilenberg–MacLane space $K(G, 1)$.

One obtains the space in quotes by realizing the simplicial set $BG$. Explicitly, the *realization* $|X|$ of a simplicial set $X$ is defined, most properly in the category of compactly generated Hausdorff spaces, by

$$|X| = \varinjlim_{\Delta^n \xrightarrow{\sigma} X} \left|\Delta^n\right|,$$

where the colimit is taken over the simplex category $\boldsymbol{\Delta} \downarrow X$. The objects of the simplex category are the simplicial set maps

$$\Delta^n \xrightarrow{\sigma} X$$

(aka. simplices of $X$, by the Yoneda Lemma), and the morphisms of $\mathbf{\Delta} \downarrow X$ are the commutative triangles

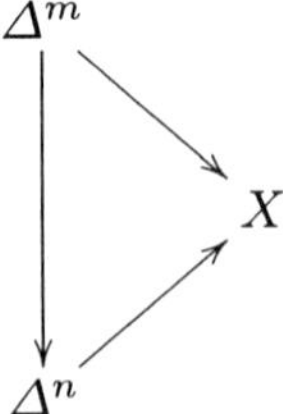

of simplicial set maps.

The realization of a simplicial set was first constructed by Milnor (see [53]) although not in this form. The definition given here appears in [26]. This functor is left adjoint to the singular functor; this was the first discovered example of an adjoint pair of functors, due to Kan [45], who formulated the concept.

This definition of the realization functor immediately implies that the realization of the standard $n$-simplex $\Delta^n$ is canonically homeomorphic to the topological standard $n$-simplex $|\Delta^n|$. Similarly, $|\partial\Delta^n|$ is a triangulated $(n-1)$-sphere, and $|\Lambda^n_k|$ is what is left of this sphere after cutting out the interior of one of the top cells (it's also a cone on an $(n-2)$-sphere). Furthermore (see [26] again), the cell structure of $\Delta^n \times \Delta^1$ can be used to show that the canonical projection map

$$|\Delta^n \times \Delta^1| \to |\Delta^n| \times |\Delta^1|$$

is an isomorphism, so that $|X \times \Delta^1|$ is canonically homeomorphic to $|X| \times |\Delta^1|$, for any simplicial set $X$. Finally, since simplicial sets are unions of skeleta in an obvious sense, the realization of any simplicial set is a CW-complex.

In particular, since $|\Delta^1|$ is a copy of the unit interval, any simplicial homotopy (5) gives an ordinary homotopy of maps of CW-complexes when the realization functor is applied.

Kan showed, in the mid 1950's [41–44], that it is possible to construct a "homotopy theory" which is completely internal to the category $\mathbf{S}$ of simplicial sets. Let me try to explain what might have been the intuition behind the theory, albeit from a modern point of view. Note that all of the following constructions have topological analogues which arise from applying the realization functor to everything in sight.

Everyone agrees that the homotopy extension property is a good thing. To see what it means in the simplicial set world, suppose that $L$ is a subcomplex of a simplicial set $K$, and let $X$ be a simplicial set. The homotopy extension property for the data consisting of the diagrams

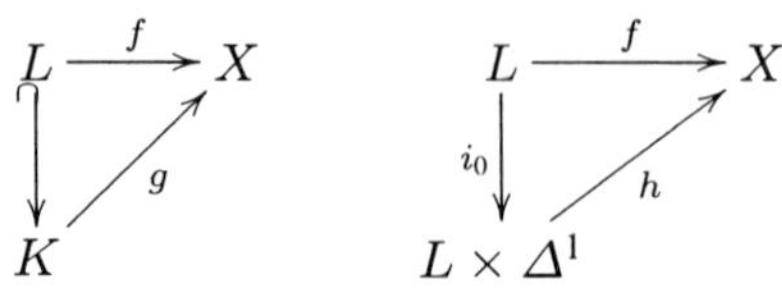

consists of the existence of a dotted arrow $H$ which makes the following diagram commute:

$$\begin{array}{ccc} (K \times 0) \cup_{(L\times 0)} (L \times \Delta^1) & \xrightarrow{(g,h)} & X \\ \big\downarrow & \nearrow_{H} & \\ K \times \Delta^1 & & \end{array}$$

In other words, $H$ is a homotopy which restricts to $h$ on $L \times \Delta^1$ and starts out at $g$. $X$ has the homotopy extension property for all such maps $g$ and inclusions $i$ if and only if the dotted arrow exists making the diagram commute in all diagrams of the following form:

$$\begin{array}{ccc} (\Delta^n \times 0) \cup_{(\partial\Delta^n\times 0)} (\partial\Delta^n \times \Delta^1) & \xrightarrow{\theta} & X \\ \big\downarrow & \nearrow & \\ \Delta^n \times \Delta^1 & & \end{array}$$

In the one-dimensional case ($n = 1$) this requirement can be described using the following picture of $\Delta^1 \times \Delta^1$:

$$\begin{array}{ccc} (0,0) & \xrightarrow{y} & (1,0) \\ {\scriptstyle x}\big\downarrow & \searrow & \big\downarrow \\ (0,1) & \xrightarrow[z]{} & (1,1) \end{array}.$$

We have a simplicial map $\theta$ taking values in $X$ defined on the 1-simplices $x$, $y$, and $z$, and which agrees on their common vertices. We want to extend the map $\theta$ to all of $\Delta^1 \times \Delta^1$. This amounts to solving two extension problems, namely

$$\Lambda^2_1 \subset \Delta^2: \quad \begin{array}{ccc} (0,0) & & \\ \big\downarrow & & \\ (0,1) & \longrightarrow & (1,1) \end{array} \quad \subset \quad \begin{array}{ccc} (0,0) & & \\ \big\downarrow & \searrow & \\ (0,1) & \longrightarrow & (1,1) \end{array} \tag{7}$$

and

$$\Lambda^2_0 \subset \Delta^2: \quad \begin{array}{ccc} (0,0) & \longrightarrow & (1,0) \\ & \searrow & \\ & & (1,1) \end{array} \quad \subset \quad \begin{array}{ccc} (0,0) & \longrightarrow & (1,0) \\ & \searrow & \big\downarrow \\ & & (1,1) \end{array} \tag{8}$$

in that order.

More generally, if one can find the indicated extension in every diagram of the form

$$\begin{array}{ccc} \Lambda^n_k & \longrightarrow & X \\ \cap\downarrow & \nearrow & \\ \Delta^n & & \end{array}, \tag{9}$$

then $X$ has the homotopy extension property with respect to all inclusions of simplicial sets.

Kan calls the property (9) the *extension condition*. Most homotopy theorists are in the habit of saying that a simplicial set which satisfies the extension condition is a *Kan complex*.

There is a rather dramatic source of examples of such, in that the singular set $S(Y)$ of every topological space $Y$ is a Kan complex. In effect, $|\Lambda^n_k|$ is a strong deformation retract of $|\Delta^n|$. The classifying object $BG$ of a group $G$ is also a Kan complex, since the category $G$ is a groupoid.

In general, however, the nerve $BC$ of a category $C$ is not a Kan complex. In particular, $\Delta^n$ is not a Kan complex for $n \geqslant 2$: try finding the dotted arrow in the category **2** which makes the diagram

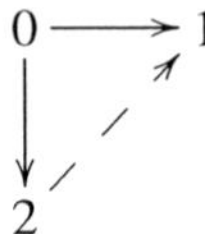

commute.

Kan's early work on cubical complexes suggested the possibility of defining homotopy groups for a Kan complex directly, without passing to the associated realization [46]. This was taken up by Moore in [58]. Let $X$ be a Kan complex, and let $*$ denote a choice of base point (vertex or 0-simplex) and all of its degeneracies. Initially, the symbol $\pi^s_n(X,*)$ stands for the set of simplicial homotopy classes of maps of pairs $(\Delta^n, \partial\Delta^n) \to (X,*)$ in the category of simplicial sets. This set carries a group structure of $n \geqslant 1$, namely the one which makes the homotopy addition theorem work (see [13, 53], for example), and $\pi^s_n(X,*)$ is the *$n$-th simplicial homotopy group*, based at $*$. Of course, $\pi^s_n(X,*)$ is abelian for $n \geqslant 2$, for the standard reason that the multiplication can be defined in two ways satisfying an interchange law in those degrees.

An obvious adjointness argument implies that there is a group isomorphism

$$\pi^s_n\big(S(Y),*\big) \cong \pi_n(Y,*)$$

for all topological spaces $Y$, and so the simplicial homotopy groups coincide with ordinary topological homotopy groups. It is a result of Milnor's (see [53] again) that the natural unit map $\eta$: $X \to S|X|$ induces an isomorphism in all simplicial homotopy groups for all Kan complexes $X$. On the other hand, the counit map $\varepsilon$: $|SY| \to Y$ was well-known to be a homotopy equivalence for CW-complexes $Y$. Furthermore, $S$ and

$|\cdot|$ preserve the respective homotopy relations by adjointness. These results imply that the homotopy category of CW-complexes is equivalent to the homotopy category of Kan complexes. This observation appears in [44].

## 2. Homotopical algebra

The late 1950's is the classical period of simplicial homotopy theory. Moore constructed the natural Postnikov tower of a Kan complex in [58], and the theories of twisted cartesian products and of simplicial fibre bundles were developed at that time [2].

Complaints remained about simplicial sets, however. The most serious (still being heard in some circles) was that this homotopy theory only talked about Kan complexes. Secondly, while the theory clearly demonstrated that homotopy theory in the large was a combinatorial affair, the theory itself was *too* combinatorial from a practical point of view. For example, Kan expressed the extension condition by saying that $X$ satisfies the condition if there is an $n$-simplex $x$ such that $d_i x = x_i$, $i \neq k$, for every $n$-tuple $x_0, \dots, x_{k-1}, x_{k+1}, \dots, x_n$ of $(n-1)$-simplices such that $d_i x_j = d_{j-1} x_i$ if $i < j$ and $i, j \neq k$. The whole of the theory was phrased in similar terms, and everything was proved by using "prismatic" arguments, which involved repeated explicit solution of extension problems. This approach created special nastiness, for example, in discussions of simplicial function spaces.

Both problems were addressed in the 1960's, with the axiomatization of the theory.

The process began with the introduction, by Gabriel and Zisman [26], of anodyne extensions. They define a class $\mathcal{A}$ of simplicial set monomorphisms to be *saturated* if:

(1) all isomorphisms belong to $\mathcal{A}$,
(2) $\mathcal{A}$ is closed under pushout,
(3) $\mathcal{A}$ is closed under retracts, and
(4) $\mathcal{A}$ is closed under countable composition and arbitrary disjoint unions.

An *anodyne extension* is a member of the smallest saturated class of monomorphisms which contains the maps

$$\Lambda_k^n \subset \Delta^n, \quad n \geqslant 1,\ 0 \leqslant k \leqslant n.$$

The inclusions

$$(\Delta^1 \times K) \cup (e \times L) \subset (\Delta^1 \times L), \quad e = 0, 1,$$

associated to an arbitrary inclusion $K \subset L$ of simplicial sets, are also anodyne extensions, by the classical argument.

Now let $\mathcal{B}$ be another class of simplicial set maps. Consider the class $\mathcal{A}'$ of all simplicial set maps which have the *left lifting property* with respect to all members of $\mathcal{B}$. This means that a map $i$ is a member of $\mathcal{A}'$ if and only if, for every solid arrow commutative diagram

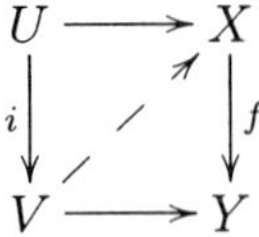

of simplicial set maps with $f \in \mathcal{B}$, the dotted arrow exists making the diagram commute (one also says in this case that $f$ has the *right lifting property* with respect to $i$). The key point is that this class $\mathcal{A}'$ is saturated. Thus, if $\mathcal{B}$ consists of maps having the right lifting property with respect to all inclusions $\Lambda^n_k \subset \Delta^n$, then the class $\mathcal{A}'$ contains all anodyne extensions.

Most of the combinatorics (aka. obstruction theory) in the subject can be compressed into this last observation. The *simplicial function space* (or *function complex*) $\mathbf{hom}(U, X)$ for simplicial sets $U$ and $X$ is defined by

$$\mathbf{hom}(U, X)_n = \text{simplicial set maps from } U \times \Delta^n \text{ to } X.$$

The functor $X \mapsto \mathbf{hom}(U, X)$ is right adjoint to the functor $Z \mapsto Z \times U$, and it is now a formality to show that, if $i$: $U \to V$ is a monomorphism of simplicial sets and $p$: $X \to Y$ has the right lifting property with respect to all anodyne extensions, then the induced map

$$\mathbf{hom}(V, X) \xrightarrow{(i^*, p_*)} \mathbf{hom}(U, X) \times_{\mathbf{hom}(U,Y)} \mathbf{hom}(V, Y) \tag{10}$$

has the same lifting property. This is one of the more powerful properties of simplicial function spaces.

It is now common to say that *Kan fibration* $p$: $X \to Y$ is a map which has the right lifting property with respect to all inclusions of the form $\Lambda^n_k \subset \Delta^n$, and hence with respect to all anodyne extensions. A simplicial set $X$ is a *Kan complex* if and only if the unique map $X \to \Delta^0$ is a Kan fibration. Kan originally defined such fibrations in terms of a relative extension condition.

Some readers will notice that I have started to use the language that Quillen introduced in [62]. Let's just say that a Kan fibration is a *fibration*. A *weak equivalence* of simplicial sets is a map $f$: $X \to Y$ whose realization $|f|$: $|X| \to |Y|$ is a homotopy equivalence of CW-complexes. A *cofibration* is a monomorphism of simplicial sets. Finally, a *trivial fibration* is a map which is both a fibration and a weak equivalence; there is a corresponding (dual) notion of *trivial cofibration*. The following theorem, due to Quillen [62], is now the fundamental result of simplicial homotopy theory:

THEOREM 1. *With the definitions given above, the category* **S** *of simplicial sets is a closed model category in the sense that the following axioms are satisfied:*

**CM1:** **S** *is closed under finite direct and inverse limits.*

**CM2:** *Suppose given maps* $X \xrightarrow{f} Y \xrightarrow{g} Z$ *of simplicial sets. If any two of* $f$, $g$ *or* $g \circ f$ *are weak equivalences, then so is the third.*

**CM3:** *The classes of fibrations, cofibrations and weak equivalences are closed under retraction.*
**CM4:** *Suppose given a commutative solid arrow diagram*

$$\begin{array}{ccc} U & \longrightarrow & X \\ {\scriptstyle i}\downarrow & \nearrow & \downarrow{\scriptstyle p} \\ V & \longrightarrow & Y \end{array}$$

*where $i$ is a cofibration and $p$ is a fibration. If either $i$ or $p$ is trivial, then the dotted arrow exists making the diagram commute.*
**CM5:** *Any map $f$ of simplicial sets may be factored as*
(a) *$f = pi$, where $p$ is a fibration and $i$ is a trivial cofibration, and*
(b) *$f = qj$, where $q$ is a trivial fibration and $j$ is a cofibration.*

In fact, more is true: these days, one most commonly summarizes the situation by saying that the category of simplicial sets is a proper closed simplicial model category (see [3]). The "simplicial" part means that there is a notion of function space (i.e. the function complexes discussed above) which has good adjointness properties and satisfies Quillen's axiom **SM7**. This axiom says that the map (10) is a fibration if $i$ is a cofibration and $p$ is a fibration, and that this map is a trivial fibration if either $i$ or $p$ is trivial. It is proved for simplicial sets by using the theory of anodyne extensions. The word "proper" means, most succinctly, that weak equivalences are stable under base change by fibrations and cobase change by cofibrations. In particular, if

$$\begin{array}{ccc} Z \times_Y X & \xrightarrow{f_*} & X \\ \downarrow & & \downarrow{\scriptstyle p} \\ Z & \xrightarrow[f]{} & Y \end{array}$$

is a pullback diagram in which $p$ is a fibration and $f$ is a weak equivalence, then $f_*$ is a weak equivalence. The second part of the statement deals with pushouts of weak equivalences by cofibrations. Such things are usually proved by applying the realization functor. For this purpose (as well as for many others, including the proof of Theorem 1), it is critical to know another result of Quillen [63]:

THEOREM 2. *The realization of a Kan fibration is a Serre fibration.*

The proof of Theorem 2 is quite subtle, since it depends on the theory of minimal fibrations (which goes back at least to [20]). All extant proofs of Theorem 1 use minimal fibrations as well.

The *homotopy category* $Ho(\mathbf{S})$ is obtained from the category $\mathbf{S}$ of simplicial sets by formally inverting the weak equivalences. A homotopy category $Ho(\mathcal{C})$ may similarly be constructed by formally inverting the weak equivalences in any closed model category $\mathcal{C}$. This construction is essentially incidental to the theory, since associated homotopy categories can almost never be studied directly. Theorem 1 and its relatives have a list of easy corollaries [62], whose proofs display the interplay between cofibrations, fibrations and weak equivalences (and model categories of various descriptions) that homotopy theory is really all about. The part of it that is valid in an arbitrary closed model category is often called *homotopical algebra*.

To illustrate, the Whitehead theorem asserts that any weak equivalence $f\colon X \to Y$ is a homotopy equivalence if $X$ and $Y$ are both fibrant (Kan complexes) and cofibrant (note that all simplicial sets are cofibrant). In view of the factorization axiom, it is enough to assume that $f$ is either a trivial cofibration or a trivial fibration. We shall assume that $f$ is a trivial fibration; the other case follows by duality.

But then the result is proved by finding, in succession, maps $\theta$ and $\gamma$ making the following diagram commute:

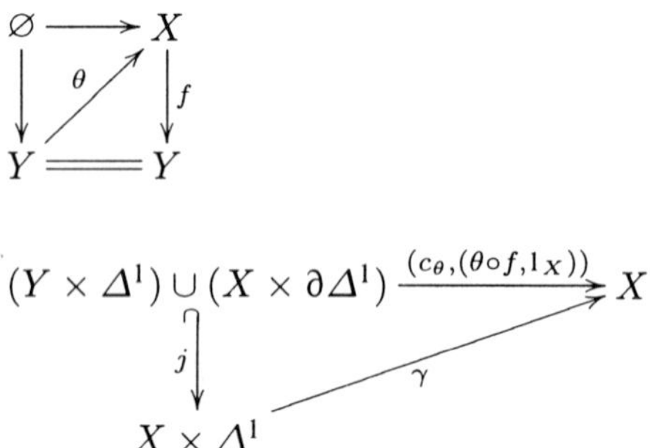


where $c_\theta$ is the constant homotopy

$$Y \times \Delta^1 \xrightarrow{pr} Y \xrightarrow{\theta} X$$

for the map $\theta$. The point is that $j$ is a trivial cofibration, since $\theta \times 1_{\partial\Delta^1}$ is a trivial cofibration, and trivial cofibrations are closed under pushout (this needs a separate proof).

Note as well that CW-complexes are topological spaces which are fibrant (since every topological space $X$ sits in a Serre fibration $X \to *$) and cofibrant, so the Whitehead Theorem given above specializes to the standard topological result.

There is also the original meaning of the word "closed":

LEMMA 3.

(1) *A map $i\colon U \to V$ is a cofibration of and only if $i$ has the left lifting property with respect to all trivial cofibrations.*

(2) *The map $i$ is a trivial cofibration if and only if it has the right lifting property with respect to all fibrations.*

(3) *A map $p\colon X \to Y$ is a fibration if and only if it has the right lifting property with respect to all trivial cofibrations.*

(4) *The map $p$ is a trivial fibration if and only if it has the right lifting property with respect to all cofibrations.*

The point of Lemma 3 is that the various species of cofibrations and fibrations determine each other via lifting properties.

For the first statement, suppose that $i$ is a cofibration, $p$ is a trivial fibration, and that there is a commutative diagram

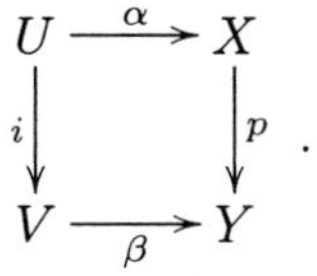

$$\begin{array}{ccc} U & \xrightarrow{\alpha} & X \\ {\scriptstyle i}\downarrow & & \downarrow{\scriptstyle p} \\ V & \xrightarrow[\beta]{} & Y \end{array} . \tag{11}$$

Then there is a map $\theta$: $V \to X$ such that $p\theta = \beta$ and $\theta i = \alpha$, by **CM4**. Conversely suppose that $i$: $U \to V$ is a map which has the left lifting property with respect to all trivial fibrations. By **CM5**, $i$ has a factorization

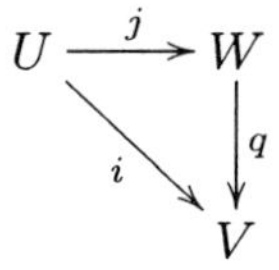

$$\begin{array}{ccc} U & \xrightarrow{j} & W \\ & {\scriptstyle i}\searrow & \downarrow{\scriptstyle q} \\ & & V \end{array}$$

where $j$ is a cofibration and $q$ is a trivial fibration. But then, there is a commutative diagram of the form

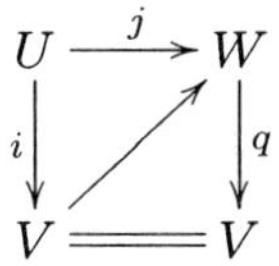

$$\begin{array}{ccc} U & \xrightarrow{j} & W \\ {\scriptstyle i}\downarrow & \nearrow & \downarrow{\scriptstyle q} \\ V & = & V \end{array}$$

and so $i$ is a retract of $j$. **CM3** then implies that $i$ is a cofibration.

The remaining statements of Lemma 3 are proved the same way. The proof of the lemma is as important as its statement: it contains one of the standard tricks that is used to prove that the closed model axioms are satisfied in other settings.

It's hard to tell from this vantage point whether or not it was one of the original goals of the theory, but it's major feature of homotopical algebra that the notion of homotopy itself "explodes" in this context. Quillen defines a *cylinder object* for an object $A$ in a closed model category $\mathcal{C}$ to be a commutative triangle of the form

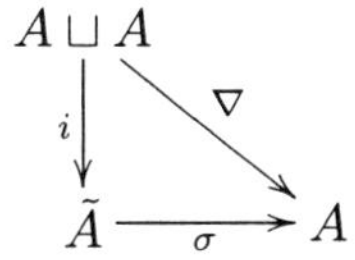

$$\begin{array}{ccc} A \sqcup A & & \\ {\scriptstyle i}\downarrow & \searrow{\scriptstyle \nabla} & \\ \tilde{A} & \xrightarrow[\sigma]{} & A \end{array} \tag{12}$$

where $\nabla$: $A \sqcup A \to A$ is the canonical fold map which is defined to be the identity on $A$ on each summand, $i$ is a cofibration, and $\sigma$ is a weak equivalence. Then a *left homotopy* of maps, $f, g\colon A \to B$ is a commutative diagram

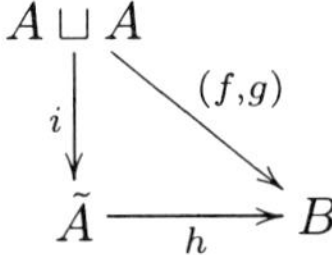


where $(f, g)$ is the map on the disjoint union which is defined by $f$ on one summand and $g$ on the other, and the data consisting of

$$i = (i_0, i_1)\colon\ A \sqcup A \to \tilde{A}$$

comes from *some choice* of cylinder object for $A$.

$X \times \Delta^1$ is a cylinder object for the simplicial set $X$, and $Y \times I$ is a cylinder object for each CW-complex $Y$, so that simplicial homotopy and ordinary homotopy of continuous maps are examples of this phenomenon, but left homotopy is inherently much more flexible, since there are many more cylinder objects around. Any factorization of $\nabla\colon A \sqcup A \to A$ into a cofibration followed by a trivial fibration that one might get out of **CM5** does the trick. In general, though, not much can be said about left homotopy unless the source object is cofibrant [62]:

LEMMA 4.

(1) *Suppose that $A$ is cofibrant, and that* (12) *is a cylinder object for $A$. Then the maps $i_0, i_1\colon A \to \tilde{A}$ are trivial cofibrations.*

(2) *Left homotopy of maps $A \to B$ is an equivalence relation if $A$ is cofibrant.*

Dually, a *path object* for $B$ is a commutative triangle of the form

$$\begin{array}{ccc} & & \hat{B} \\ & \overset{s}{\nearrow} & \big\downarrow{\scriptstyle p=(p_0,p_1)} \\ B & \xrightarrow[\Delta]{} & B \times B \end{array} \tag{13}$$

where $\Delta$ is the diagonal map, $s$ is a weak equivalence, and $p$ (which is given by $p_0$ on one factor and by $p_1$ on the other) is a fibration.

Once again, the factorization axiom **CM5** dictates that there is an ample supply of path objects. If the simplicial set $X$ is a Kan complex, then the function complex $\mathbf{hom}(\Delta^1, X)$ is a path object for $X$, by Quillen's axiom **SM7**.

There is a notion of right homotopy which corresponds to path objects: two maps $f, g\colon A \to B$ are said to be *right homotopic* if there is a diagram

$$\begin{array}{ccc} & & \hat{B} \\ & \overset{h}{\nearrow} & \big\downarrow{\scriptstyle (p_0,p_1)} \\ A & \xrightarrow[(f,g)]{} & B \times B \end{array}$$

where the map $(p_0, p_1)$ arises from some path object (13), and $(f, g)$ is the map which projects to $f$ on the left hand factor and $g$ on the right hand factor.

LEMMA 5.

(1) *Suppose that $B$ is fibrant and that $\hat{B}$ is a path object for $B$ as in* (13). *Then the maps $p_0$ and $p_1$ are trivial fibrations.*

(2) *Right homotopy of maps $A \to B$ is an equivalence relation if $B$ is fibrant.*

Lemma 5 is *dual* to Lemma 4 in a precise sense. If $\mathcal{C}$ is a closed model category, then its opposite $\mathcal{C}^{op}$ is a closed model category whose cofibrations (respectively fibrations) are the opposites of the fibrations (respectively cofibrations) in $\mathcal{C}$. A map in $\mathcal{C}^{op}$ is a weak equivalence for this structure if and only if its opposite is a weak equivalence in $\mathcal{C}$. Then Lemma 5 is an immediate consequence of the instance of Lemma 4 which occurs in $\mathcal{C}^{op}$. This sort of duality is ubiquitous in the theory, and there are several veiled references to it above.

Left and right homotopies are linked by the following result:

PROPOSITION 6. *Suppose that $A$ is cofibrant. Suppose further that*

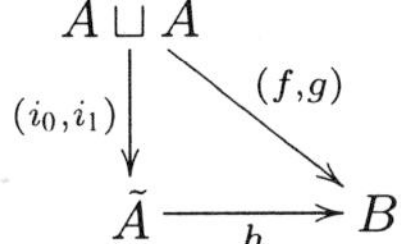


*is a left homotopy between maps $f, g\colon A \to B$, and that*

$$\begin{array}{ccc} & & \hat{B} \\ & \overset{s}{\nearrow} & \Big\downarrow{\scriptstyle p=(p_0,p_1)} \\ B & \xrightarrow[\Delta]{} & B \times B \end{array}$$

*is a fixed choice of path object for $B$. Then there is a homotopy of the form*

$$\begin{array}{ccc} & & \hat{B} \\ & \overset{H}{\nearrow} & \Big\downarrow{\scriptstyle (p_0,p_1)} \\ A & \xrightarrow[(f,g)]{} & B \times B \end{array}.$$

This result has a dual, which the reader should be able to formulate independently. Proposition 6 and its dual together imply

COROLLARY 7. *Suppose given maps $f, g\colon A \to B$, where $A$ is cofibrant and $B$ is fibrant. Then the following are equivalent:*

(1) *$f$ and $g$ are left homotopic.*

(2) *$f$ and $g$ are right homotopic with respect to a fixed choice of path object.*

(3) *$f$ and $g$ are right homotopic.*
(4) *$f$ and $g$ are left homotopic with respect to a fixed choice of cylinder object.*

In other words, all possible definitions of homotopy of maps $A \to B$ are the same if $A$ is cofibrant and $B$ is fibrant.

Quillen's proof of Proposition 6 is an elegant little miracle: $i_0$ is a trivial cofibration since $A$ is cofibrant, and $(p_0, p_1)$ is a fibration, so that there is a commutative diagram of the form

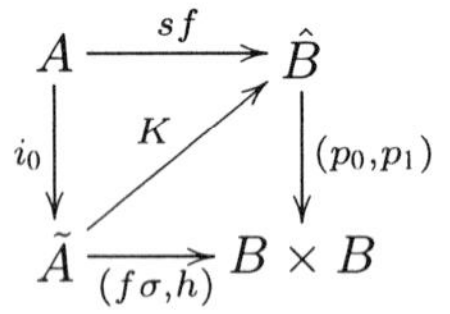

$$\begin{array}{ccc} A & \xrightarrow{sf} & \hat{B} \\ {\scriptstyle i_0}\downarrow & \overset{K}{\nearrow} & \downarrow{\scriptstyle (p_0,p_1)} \\ \tilde{A} & \xrightarrow[(f\sigma,h)]{} & B \times B \end{array}$$

for some choice of lifting $K$. Then the composite $K \circ i_1$ is the desired right homotopy.

We can now, unambiguously, speak of homotopy classes of maps between objects $C$ and $D$ which are both fibrant and cofibrant. Quillen denotes the corresponding set of equivalence classes by $\pi(C, D)$. There is a category $\pi\mathcal{C}_{cf}$ associated to any closed model $\mathcal{C}$: the objects are the cofibrant and fibrant objects of $\mathcal{C}$, and the morphisms from $C$ to $D$ in $\pi\mathcal{C}_{cf}$ are the elements of the set $\pi(C, D)$.

For each object $X$ of $\mathcal{C}$, use **CM5** to choose, in succession, maps of the form

$$* \xrightarrow{i_X} QX \xrightarrow{p_X} X$$

and

$$QX \xrightarrow{j_X} RQX \xrightarrow{q_X} *,$$

where $i_X$ is a cofibration, $p_X$ is a trivial fibration, $j_X$ is a trivial cofibration, and $q_X$ is a fibration. Then $RQX$ is an object which is both fibrant and cofibrant, and $RQX$ is weakly equivalent to $X$, via the maps $p_X$ and $j_X$. Any map $f\colon X \to Y$ lifts to a map $Qf\colon QX \to QY$, and then $Qf$ extends to a map $RQf\colon RQX \to RQY$. Furthermore, the assignment $f \mapsto RQf$ is well-defined up to homotopy and so a functor

$$RQ\colon \mathcal{C} \to \pi\mathcal{C}_{cf}$$

is defined. This functor $RQ$ is universal up to isomorphism for functors $\mathcal{C} \to \mathcal{D}$ which invert weak equivalences, proving:

THEOREM 8. *For any closed model category $\mathcal{C}$, the category $\pi\mathcal{C}_{cf}$ of homotopy classes of maps between fibrant-cofibrant objects is equivalent to the homotopy category $Ho(\mathcal{C})$ which is obtained by formally inverting the weak equivalences of $\mathcal{C}$.*

This result is an elaboration of various old stories: the category of homotopy classes of maps between CW-complexes is equivalent to the full homotopy category of topological spaces, and the homotopy category of simplicial sets is equivalent to the category if simplicial homotopy classes of maps between Kan complexes.

Much of modern homotopy theory is based on Quillen's theorem, and its relatives. The closed model structure on the category of cosimplicial spaces is one of the basic technical devices underlying the Bousfield–Kan theory of homotopy inverse limits [5]. The category of bisimplicial sets carries several different closed model structures (see [5, 3, 57]), which are used, variously, for the theory of homotopy colimits, and the proofs of Quillen's Theorem B and the group completion theorem [35, 65]. Some of the early applications of the theory were in rational homotopy theory [4, 64]. More recently, the closed model structure on the category of supplemented commutative graded algebras over a field was an important technical device in Miller's proof of Sullivan's conjecture on maps from classifying spaces to finite complexes [55]. Stable homotopy theory can be axiomatized this way as well [3].

## 3. Simplicial presheaves

There is a growing list of applications of either the closed model theory or the axiomatic point of view in subjects related to algebraic $K$-theory.

There is a long-standing (and still unsolved) conjecture of Friedlander and Milnor [22] that asserts that the canonical map

$$BGl_n(\mathbb{C}) \to BGl_n(\mathbb{C})^{top}$$

of simplicial spaces induces an isomorphism in cohomology with torsion coefficients. $BGl_n(\mathbb{C})$ is endowed, in each degree, with the discrete topology, and so it may be identified with the corresponding simplicial set. $BGl_n(\mathbb{C})^{top}$ can be identified with a bisimplicial set by applying the singular functor in each degree. The unitary group $U_n$ is the maximal compact subgroup of $Gl_n(\mathbb{C})$, so that $BGl(\mathbb{C})^{top}$ has the homotopy type of the space $BU$ of complex $K$-theory.

One way or another (use the Riemann existence theorem [56]), it is possible to show that

$$H^*\big(BGl_n(\mathbb{C})^{top}, \mathbb{Z}/\ell\big)$$

is isomorphic to the étale cohomology

$$H^*_{et}(BGl_{n,\mathbb{C}}, \mathbb{Z}/\ell)$$

of the simplicial scheme $BGl_{n,\mathbb{C}}$. Since the integral group-scheme $Gl_{n,\mathbb{Z}}$ is cohomologically proper [23],

$$H^*_{et}(BGl_{n,\mathbb{C}}; \mathbb{Z}/\ell)$$

may be identified up to isomorphism with the ring

$$H^*_{et}(BGl_{n,k}; \mathbb{Z}/\ell)$$

associated to $Gl_{n,k}$ defined over any algebraically closed field $k$ of characteristic prime to $\ell$. Finally, the induced map in cohomology

$$H^*\big(BGl_n(\mathbb{C})^{top}; \mathbb{Z}/l\big) \to H^*\big(BGl_n(\mathbb{C}); \mathbb{Z}/l\big)$$

is a special case of a comparison map

$$\varepsilon^*\colon\ H^*_{et}(BG_k; \mathbb{Z}/\ell) \to H^*\big(BG(k); \mathbb{Z}/\ell\big)$$

associated to any reductive algebraic group $G_k$ defined over $k$. The *generalized isomorphism conjecture* [22] asserts that this map $\varepsilon^*$ is an isomorphism.

A proper description of $\varepsilon^*$ requires simplicial sheaves, and a homotopy theory thereof. The theorem of faithfully flat descent implies that the simplicial scheme $BG_k$ represents a simplicial sheaf on the big étale site $(Sch|_k)_{et}$ of schemes over $k$ which are locally of finite type. The simplicial set $BG(k)$ is the simplicial set of global sections of $BG_k$, and the counit of the global sections – constant sheaf adjunction is a map of simplicial sheaves of the form

$$\varepsilon\colon\ \Gamma^* BG(k) \to BG_k.$$

The category $\mathbf{SShv}(Sch|_k)_{et}$ of simplicial sheaves on $(Sch|_k)_{et}$ carries a *category of fibrant objects* structure that I call the *local theory* [9, 32, 33]. A *local fibration* is a map of simplicial sheaves which, for this example, induces Kan fibrations in each stalk. *Weak equivalences* between locally fibrant objects are also defined stalkwise via sheaves of simplicial homotopy groups (one has to be careful with this, because we are not assuming that every section of a locally fibrant object is a Kan complex). These two classes of maps satisfy a list of axioms which is essentially half of a closed model structure, but which implies that the associated homotopy category $Ho(\mathbf{SShv}(Sch|_k)_{et})$ may be constructed via a calculus of fractions. We capture sheaf cohomology in the process: it follows from the Illusie conjecture (later, Van Osdol's theorem) [30, 59, 32] that there is an isomorphism

$$H^n_{et}(BG_k; \mathbb{Z}/\ell) \cong \big[BG, K(\mathbb{Z}/\ell, n)\big],$$

where the square brackets indicate morphisms in the homotopy category and

$$K(\mathbb{Z}/\ell, n)$$

is the constant simplicial sheaf associated to the obvious choice of Eilenberg–MacLane space.

There is a corresponding local homotopy theory for simplicial presheaves on arbitrary Grothendieck sites [32, 33]. The Illusie conjecture asserts, in its most useful form, that every weak equivalence of locally fibrant simplicial presheaves induces an isomorphism on all homology sheaves. It is trivial to prove in cases where the associated Grothendieck topos has enough points; in the general case, it reduces to showing that local fibrations which are weak equivalences induce isomorphisms on homology sheaves, but then this follows from the fact that such maps have a local right lifting property with respect to all inclusions of finite simplicial sets.

Illusie's conjecture [59] can also be proved by using the method of Boolean localization, which is a general trick that faithfully imbeds an arbitrary Grothendieck topos into one that satisfies the axiom of choice (this is "Barr's Theorem", [52], p. 513).

If $A$ is a presheaf of abelian groups, the standard model for the Eilenberg–MacLane object $K(A,n)$ is a presheaf of simplicial abelian groups. The Illusie conjecture implies the existence of a natural isomorphism

$$\big[X, K(A,n)\big] \cong \big[\mathbb{Z}(X), A[n]\big],$$

valid for all simplicial presheaves $X$ between morphisms in the homotopy category of simplicial presheaves, and morphisms in the derived category from the presheaf of Moore chain complexes $\mathbb{Z}(X)$ to the complex $A[n]$ consisting of $A$ concentrated in degree $n$. The latter is identified with a hypercohomology group of $X$ with coefficients in $A$. If $X$ is represented by a simplicial scheme in any of the standard geometric toposes, then such hypercohomology groups can be identified with the standard cohomology groups of $X$ [21, 32] which arise from sheaves on the site fibred over $X$. Sheaf cohomology can therefore always be identified with morphisms in a homotopy category.

The calculus of fractions approach to the construction of the set of morphisms $[X, K(A,n)]$ in the homotopy category implies that there is an isomorphism of the form

$$\big[X, K(A,n)\big] \cong \varinjlim_{Y \xrightarrow{[\pi]} X} \pi\big(\mathbb{Z}(Y), A[n]\big), \tag{14}$$

where the (filtered) colimit is indexed over simplicial homotopy classes of maps $[\pi]$ which are represented by maps

$$\pi\colon Y \to X$$

which are local fibrations and weak equivalences. This isomorphism (14) is a generalized form of the Verdier hypercovering theorem [1, 9, 32], and maps of the form $\pi$ are now called *hypercovers*. Note that when I say simplicial homotopy classes in this context, it means equivalence classes of maps for the relation which is generated by ordinary simplicial homotopy, since simplicial homotopy of maps between locally fibrant simplicial presheaves is not an equivalence relation in general.

The field $k$ is algebraically closed, and is therefore a point in the eyes of étale cohomology. It follows, by an adjointness argument, that there is an isomorphism

$$\big[\Gamma^* BG(k), K(\mathbb{Z}/\ell, n)\big] \cong H^n\big(BG(k); \mathbb{Z}/\ell\big),$$

and so the map $\varepsilon^*$ above is just induced by precomposition with $\varepsilon$ in the homotopy category.

The map $\varepsilon^*$ has been proved to be an isomorphism when the group $G$ is the general linear group $Gl$ [32], as well as for the infinite orthogonal group $O$ and the infinite symplectic group $Sp$ [47], and in general when the underlying field $k$ is the algebraic closure of a finite field [22, 31]. The finite field case reduces, via a Künneth formula, to a study of the interplay between the Lang isomorphism and the Hopf algebra structure of $H^*_{et}(BG_k; \mathbb{Z}/\ell)$. The stable results are consequences of the *rigidity theorem* of Gabber, Gillet and Thomason [25, 27], one form of which says that the mod $\ell$ algebraic $K$-theory sheaf on $(Sch|_k)_{et}$ is weakly equivalent to the constant simplicial sheaf on the mod $\ell$ $K$-theory space associated to the field $k$ (Suslin's results [72, 73] calculate the $K$-groups $K_*(k; \mathbb{Z}/\ell)$, so we know what the sheaves of homotopy groups look like). The rigidity theorem implies – think about what it means stalkwise – that the adjunction map

$$\varepsilon\colon \Gamma^* BGL(k) \to BGl$$

induces an isomorphism in mod $\ell$ homology sheaves, and so it's a cohomology isomorphism as well.

Some readers might feel abused at this point by the fact that the general linear group $Gl$ is not exactly an algebraic group over $k$. It is most correctly viewed as a presheaf of groups on the big étale site for $k$, given by the filtered colimit

$$Gl = \varinjlim Gl_n$$

in the presheaf category. Then the simplicial presheaf $BGl$ is a filtered colimit of the representable objects $BGl_n$, and so the cohomology of $BGl$ may be computed from that of the $BGl_n$'s by a $\varprojlim^1$ exact sequence.

The étale cohomology rings of the classifying simplicial schemes $BG$ of various group-schemes $G$ have been calculated. Schechtman [68] has shown that, if $S$ is a scheme whose ring of global sections contains a primitive $\ell$-th root of unity, then there is an $A_S$-algebra isomorphism

$$H^*_{et}(BGl_{n,S}, \mathbb{Z}/\ell) \cong A_S[c_1, \ldots, c_n]$$

where $A_S$ denotes the étale cohomology ring $H^*_{et}(S, \mathbb{Z}/\ell)$ of the base, and the degree of the Chern class $c_i$ is $2i$. If $\ell = 2$ and the ring of global sections of $S$ contains $1/2$, then the cohomology of the classifying object $BO(n)_S$ can also be calculated [37]: there is an $A_S$-algebra isomorphism

$$H^*_{et}(BO(n)_S, \mathbb{Z}/2) \cong A_S[HW_1, \ldots, HW_n],$$

where $deg(HW_i) = i$. The classes $HW_i$ are called universal Hasse–Witt classes. If the base scheme $S$ is the spectrum of an algebraically closed field, then these classes specialize the Stiefel–Whitney classes. Thus, you get what you would expect from classical calculations for the cohomology of both $BGl_{n,S}$ and $BO(n)_S$.

Suppose that $S = Sp(A)$ is an affine scheme such that $A$ contains a primitive $\ell$-th root of unity. Schechtman's result and the vanishing of certain $\varprojlim^1$ terms together imply that there is an $A_S$-algebra isomorphism of the form

$$H^*(BGl, \mathbb{Z}/\ell) \cong A_S[c_1, c_2, \ldots],$$

where the degree of $c_j$ is $2j$. Any element

$$\alpha \in H_i(BGl(A), \mathbb{Z}/\ell)$$

can be represented by an ordinary chain map

$$\mathbb{Z}/\ell[i] \xrightarrow{\alpha} \mathbb{Z}/\ell(BGl(A)),$$

and so applying the constant presheaf functor gives a map gives rise to a composite

$$\Gamma^*\mathbb{Z}/\ell[i] \xrightarrow{\Gamma^*\alpha} \Gamma^*\mathbb{Z}/\ell(BGl(A)) \xrightarrow{\varepsilon} \mathbb{Z}/\ell(BGl)$$

of maps of presheaves of chain complexes. The Chern class $c_j \in H^{2j}(BGl/\mathbb{Z}/\ell)$ can be identified with a map in the derived category (i.e. homotopy category of presheaves of chain complexes) of the form $c_j\colon \mathbb{Z}/\ell(BGl) \to \Gamma^*\mathbb{Z}/\ell[2j]$, and so the composite $c_j \cdot \varepsilon \cdot \Gamma^*\alpha$ represents an element of $H_{et}^{2j-i}(A, \mathbb{Z}/\ell)$. The Chern class $c_j$ therefore determines a generalized cap product homomorphism

$$(c_j)_*\colon\ H_i(BGl(A), \mathbb{Z}/\ell) \to H_{et}^{2j-i}(A, \mathbb{Z}/\ell).$$

The composite of $(c_j)_*$ with the mod $\ell$ Hurewicz map

$$K_i(A, \mathbb{Z}/\ell) \to H_i(BGl(A), \mathbb{Z}/\ell)$$

is Soulé's $K$-theory Chern class map

$$c_{i,j}\colon\ K_i(A, \mathbb{Z}/l) \to H_{et}^{2j-i}(A, \mathbb{Z}/\ell)$$

for mod $\ell$ étale cohomology. Bruno Kahn has observed that the universal Hasse–Witt classes $HW_j \in H_{et}^i(BO_n, \mathbb{Z}/2)$ induce $L$-theory Stiefel–Whitney classes

$$_{-1}L_i(B, \mathbb{Z}/2) \to H_{et}^{j-i}(B, \mathbb{Z}/2)$$

for rings $B$ containing $1/2$, by exactly the same procedure.

Schechtman's calculation is more or less "motivic" in that it works for any decent cohomology theory with coefficients in a suitable collection of sheaves $\mathcal{F}^\cdot$ [67]. In that case, the Chern class maps, after twisting by certain factorial constants, induce a Chern character map

$$ch\colon\ K_i(A) \to \bigoplus_{j \geqslant 0} H^{2j-i}(A, \mathcal{F}^\cdot),$$

which then induces the Beilinson regulator on the eigenspaces of the Adams operations on $K_i(A)$.

I want to explain the name and the utility of the classes $HW_i$, but it involves a foray into the homotopical description of nonabelian $H^1$. Suppose, quite generally, that $G$ is a sheaf of groups on an arbitrary Grothendieck site $\mathcal{C}$. Then the nonabelian cohomology object $H^1(\mathcal{C}, G)$ has a homotopical classification, given by

THEOREM 9. *There is a natural isomorphism*

$$H^1(\mathcal{C}, G) \cong [*, BG],$$

*where $[*, BG]$ denotes morphisms in the simplicial sheaf homotopy category associated to $\mathcal{C}$ from the terminal object $*$ to the classifying simplicial sheaf $BG$.*

Recall that $H^1(\mathcal{C}, G)$ may also be identified with the set of isomorphism classes of $G$-torsors over the point $*$.

Theorem 9 is proved in [36]. The essential idea is that the homotopy invariant

$$[*, BG] = \varinjlim_{U \xrightarrow{[\pi]} *} \pi(U, BG)$$

coincides with the Čech cohomology object

$$\varinjlim_{V \twoheadrightarrow *} \check{H}^1(V, G),$$

and the latter is well known to coincide with $H^1(\mathcal{C}, G)$.

Elements of $\check{H}^1(V, G)$ can be identified with simplicial homotopy classes of maps $cosk_0(V) \to BG$, where the 0-th coskeleton $cosk_0(V)$ is just the Čech resolution

$$V \leftleftarrows V \times V \leftleftarrows\!\!\!\leftarrow V \times V \times V \quad \cdots$$

of the object $*$. The simplicial sheaf map $cosk_0(V) \to *$ is a hypercover of the terminal object, so there's an obvious comparison map

$$\varinjlim_{V \twoheadrightarrow *} \check{H}^1(V, G) \to \varinjlim_{U \to *} \pi(U, BG). \tag{15}$$

Finally, the map (15) is a bijection, since the set of simplicial homotopy classes $\pi(U, BG)$ can be identified with natural isomorphism classes of functors from the sheaf of fundamental groupoids of $U$ to the groupoid $G$, and the sheaf of fundamental groupoids for $U$ is the trivial groupoid associated to the covering $U_0 \twoheadrightarrow *$ of the terminal object by the sheaf of 0-simplices of $U$. Note that the nerve of this last groupoid is just $cosk_0(U_0)$.

To give a concrete example of Theorem 9 at work, suppose that $K$ is a field of characteristic not equal to 2, and let $\beta$ be a nondegenerate symmetric bilinear form of rank $n$ over $K$. The form $\beta$ trivializes over some finite Galois extension $L/K$, say via a form isomorphism $A\colon \beta \to 1_n$. But, if $G$ denotes the Galois group of $L/K$ and $g$ is a member of $G$, then $g(A)$ is also a form isomorphism, so that the composite

$$1_n \xrightarrow{A^{-1}} \beta \xrightarrow{g(A)} 1_n$$

is an automorphism of the trivial form of rank $n$ over $L$, and hence an element of the orthogonal group $O_n(L)$. The upshot of this construction is that a form $\beta$ gives rise to cocycles

$$f_\beta\colon G \to O_n(L),$$

defined by $g \mapsto g(A)A^{-1}$, for suitable choices of Galois extensions. Each such cocycle can be identified with a map of simplicial presheaves on your choice of étale sites for $K$ of the form

$$f_\beta\colon EG \times_G Sp(L) \to BO_n,$$

where $EG \times_G Sp(L)$ is just notation for the Čech resolution associated to the étale covering $Sp(L) \to Sp(K)$. The fact that this resolution actually is a Borel construction in the homotopy theoretic sense for the action of $G$ on $Sp(L)$ is just undergraduate field theory. The choices that have been made are invariant up to homotopy and refinement, and it's well known that the resulting map from the set of isomorphism classes of nondegenerate symmetric bilinear forms of rank $n$ over $K$ to the nonabelian cohomology object $H^1_{et}(K, O_n)$ is a bijection. Theorem 9 therefore implies that isomorphism classes of such forms may be identified with homotopy classes of maps, i.e. with elements of the set $[*, BO_n]$.

From this point of view, any form $\beta$ and any element of the étale cohomology of $BO_{n,K}$ together give rise to elements of the Galois (or étale) cohomology of $K$, via the induced map

$$\beta^*\colon H^*_{et}(BO_{n,K}, \mathbb{Z}/2) \to H^*_{et}(K, \mathbb{Z}/2).$$

Thus, in the tradition of characteristic class theory in Algebraic Topology, the universal classes

$$HW_i \in H^i_{et}(BO_{n,K}, \mathbb{Z}/2)$$

determine classes $HW_i(\beta) = \beta^*(HW_i)$ in the mod 2 Galois cohomology of $K$, that I call Hasse–Witt classes of the form $\beta$.

The Hasse–Witt classes $HW_i(\beta)$ coincide with the Delzant Stiefel–Whitney classes of $\beta$, by a homotopical reinterpretation of the fact that every nondegenerate symmetric bilinear from over $K$ diagonalizes over $K$. Furthermore, $HW_2(\beta)$ is the classical Hasse–Witt invariant of the form $\beta$, whence the name for the classes $HW_i(\beta)$.

Let $\beta$ and its trivialization $A$ be as above, and recall that the group $O_\beta(K)$ of automorphisms of $\beta$ is the group of $K$-points of a group-scheme $O_\beta$; $O_n$ is a more tractable notation for the group-scheme of automorphisms $O_{1_n}$ of the trivial form $1_n$, from this point of view. Let $G$ be the Galois group of $L/K$, as before, and let $\overline{K}$ be the algebraic closure of $K$. Every representation

$$\rho\colon G \to O_\beta(K)$$

of $G$ gives rise to two sets of Galois cohomological invariants, namely the Stiefel–Whitney classes associated to the composite representation

$$G \xrightarrow{\rho} O_\beta(K) \hookrightarrow O_\beta(L) \cong O_n(L) \hookrightarrow O_n(\overline{K}),$$

and the Hasse–Witt invariants for the Fröhlich twisted form, given by the cocycle

$$g \mapsto g(A)\rho(g)A^{-1}. \tag{16}$$

There has been some activity in recent years relating these invariants to each other, and to the Hasse–Witt classes of the underlying form $\beta$, as well as to the classical spinor norm [24, 37, 40, 70]. It started with Serre's formula for the Hasse–Witt invariant of the trace form, followed by the various proofs of a complementary result of Fröhlich, which was a general formula for the $HW_2$ class of the twisted form (16). There is a similar formula for the $HW_3$ class of the twisted form, which can be obtained from the decomposability of the spinor class, coupled with knowing that the Steenrod algebra [6, 37] acts on everything in sight, and obeys the Wu formulae when applied to Stiefel–Whitney and Hasse–Witt classes in particular.

One of the deep qualitative statements in the area is that the Stiefel–Whitney classes of representations of the form $\rho$ are decomposable; it can be proved by introducing a total Steenrod squaring operation in mod 2 simplicial presheaf cohomology [37], or by a Brauer lift argument that starts with Kahn's calculations [40]. Of course, if anybody ever succeeds in proving the Milnor conjecture that the norm residue homomorphism

$$K_*^M(K) \otimes \mathbb{Z}/2 \to H_{et}^*(K, \mathbb{Z}/2)$$

defines an isomorphism from the mod 2 Milnor $K$-theory to the mod 2 Galois cohomology of $K$, then this result about Stiefel–Whitney classes won't seem so interesting. Apparently, Merkurjev has recently shown that this conjecture follows if one can show that the Steenrod squaring operation $Sq^1$ acts on $H_{et}^*(K, \mathbb{Z})$ as though it were decomposable.

Torsors are a sheaf-theoretic analogue of principal fibrations over some base, and there is a classification theory for principal fibrations of simplicial sets. Theorem 9 could lead one to expect a homotopical classification theory of principal fibrations of simplicial sheaves under a sheaf $H$ of simplicial groups. Such a theory almost certainly doesn't exist, however, since it's not at all clear that a weak equivalence of simplicial sheaves induces a bijection on the level of isomorphism classes of principal $H$-bundles. This property has been singled out as a condition [39] to be placed on $H$ for such a classification theory to exist. The difficulty of importing the principal fibration theory from simplicial sets to simplicial sheaves can be explained by observing that the simplicial set theory involves pseudo cross-sections [53]. These are constructed by solving infinite lists of extension problems, so that the Axiom of Choice is implicitly invoked, and this isn't legal in an arbitrary Grothendieck topos.

This is one of the principal difficulties underlying the work on the characteristic classes associated to symmetric bilinear forms and representations of Galois groups. Fröhlich and his followers proved the formula for the Hasse–Witt invariant of a twisted form, by using an explicit 2-cocycle associated to the central extension

$$\mathbb{Z}/2 \to Pin_n \to O_n.$$

This central extension classifies the Hasse–Witt invariant $HW_2$ in some sense, while $HW_1$ can be recovered from the determinant homomorphism $O_n \to \mathbb{Z}/2$. There is no known corresponding geometric representation for any of the higher Hasse–Witt classes $HW_1$, $i \geqslant 3$, and hence no explicit cocycles that represent them. Finding geometric interpretations of the standard higher characteristic classes is a major open problem in many contexts (see [12], for example). Breen's recent work on nonabelian $H^2$ and $H^3$ [7, 8] can be viewed as progress in this direction, but it has not yet led to explicit formulae. One might also hope for an analogue of Loday's work on $n$-types [49] in the simplicial presheaf setting.

The failure of the Axiom of Choice on the topos level is serious business from a homotopical point for view, since it forces the bifurcation of the ordinary homotopy theory of simplicial sets into a local and a global theory. The local theory was partially described above – recall that it does not arise from a closed model structure. The global theory is due to Joyal [38, 33]; there is an antecedent in work of Brown and Gersten [11]. In the category of simplicial sheaves on a site, a *cofibration* is a monomorphism, a *weak equivalence* is a map which induces an isomorphism in all sheaves of homotopy groups, and a *global fibration* is a map which has the right lifting property with respect to all trivial cofibrations. Joyal's result is that, with these definitions, the category of simplicial sheaves on an arbitrary Grothendieck site has the structure of a closed model category.

Global fibrations are local fibrations. The converse is not true, essentially because sheaf cohomology is nontrivial. Globally fibrant simplicial presheaves $X$ have the *cohomological descent property*

$$\pi_n X(*) = [*, \Omega^n X],$$

meaning that the $n$-th homotopy group of the simplicial set of global sections of a globally fibrant simplicial sheaf $X$ coincides with a certain set of morphisms in the homotopy category of simplicial sheaves. Thus, for example, choose a sheaf $A$ of abelian groups on an étale site for a scheme $S$, and suppose that the map

$$i\colon\ K(A, n) \to GK(A, n)$$

is a globally fibrant model for the Eilenberg–MacLane presheaf $K(A, n)$ in the sense that $i$ is a (stalkwise) weak equivalence and $GK(A,n)$ is globally fibrant. Then

$$\pi_s GK(A,n)(S) \cong [*, \Omega^s GK(A,n)] \cong [*, K(A, n-s)],$$

so that

$$\pi_s GK(A,n)(S) = H_{et}^{n-s}(S, A)$$

for $0 \leqslant s \leqslant n$, and is 0 otherwise, whereas $K(A,n)(S)$ has only one nontrivial homotopy group, namely the global sections group $A(S)$ in degree $n$. Thus, if every locally fibrant object were globally fibrant, then sheaf cohomology theory would be trivial.

Joyal's result was extended to simplicial presheaves in [33]. Any topology on a given Grothendieck site gives rise to a closed model structure and a homotopy theory for diagrams (presheaves) of simplicial sets on that site. This holds in particular for the chaotic topology (meaning, no topology at all), so there is a sensible notion of global fibration and so on for arbitrary diagrams of simplicial sets: cofibrations and weak equivalences are defined pointwise, and global fibrations are defined by the lifting property (compare [5, 16, 62]).

Similar results hold for presheaves of spectra [34], leading, for example, to generalized étale cohomology theories. Such theories are represented by presheaves of spectra in the homotopy categories associated to étale sites of schemes, and can be calculated as stable homotopy groups of global sections of globally fibrant models. Étale $K$-theory is an example [33]. Subject to the existence of certain bounds on cohomological dimension, these groups can be recovered from the cohomology of the underlying site with coefficients in sheaves of stable homotopy groups. This is achieved with the cohomological descent spectral sequence [33, 34, 75], the existence of which is now a formal consequence of the ambient homotopy theory.

The Lichtenbaum–Quillen conjecture can be formulated in the language of globally fibrant models of presheaves of spectra. Suppose, for example, that $L$ is a field of characteristic not equal to $\ell$, which contains a primitive $\ell$-th root of unity and has Galois cohomological dimension $d < \infty$ with respect to $\ell$-torsion sheaves. There is such a thing as the mod $\ell$ $K$-theory presheaf of spectra $\mathbf{K}/\ell$ on the étale site for $L$, with

$$\pi_j \mathbf{K}/\ell(N) = K_j(N, \mathbb{Z}/\ell)$$

for all finite Galois extensions $N/L$. Now take a globally fibrant model

$$\mathbf{K}/\ell \xrightarrow{i} G\mathbf{K}/\ell$$

in the category of presheaves of spectra, for the étale topology. Then the Lichtenbaum–Quillen conjecture, in this setting, asserts that the induced map

$$\pi_j \mathbf{K}/\ell(L) \xrightarrow{i_*} \pi_j G\mathbf{K}/\ell(L)$$

is an isomorphism in ordinary stable homotopy groups for $j \geqslant M$, for some $M$ thought to vary linearly with $d$. It must be emphasized that the mod $\ell$ $K$-theory presheaf of spectra itself could never be globally fibrant if $d \geqslant 1$: the spectrum $\mathbf{K}/\ell(L)$ is connective, so it's missing some homotopy groups in negative degrees which its globally fibrant model certainly has, on account of the descent spectral sequence.

The Lichtenbaum–Quillen conjecture would imply that the torsion part of the algebraic $K$-theory of $L$ could be recovered in high degrees from its Galois cohomology, via the cohomological descent spectral sequence. The torsion $K$-theory of $L$ would therefore be periodic and effectively computable in all but finitely many degrees. The strongest result in this area is Thomason's theorem [75], which asserts that if the field $L$ has a finite Tate–Tsen filtration, then the mod $\ell$ $K$-theory presheaf of spectra on $et|_L$ has the cohomological descent property, after the Bott element has been formally inverted. This conjecture is the focus of much of the current research in algebraic $K$-theory.

## References

[1] M. Artin and B. Mazur, *Étale Homotopy Theory*, SLNM 100, Springer, Berlin (1969).

[2] M. Barratt, V.K.A.M. Guggenheim and J.C. Moore, *On semi-simplicial fibre bundles*, Amer. J. Math. **81** (1959), 639–657.

[3] A.K. Bousfield and E.M. Friedlander, *Homotopy theory of $\Gamma$-spaces, spectra and bisimplicial sets*, SLNM 658, Springer, Berlin (1978), 80–150.

[4] A.K. Bousfield and V.K.A.M. Guggenheim, *On PL de Rham theory and rational homotopy type*, Mem. Amer. Math. Soc. **179** (1976).

[5] A.K. Bousfield and D.M. Kan, *Homotopy Limits, Completions and Localizations*, SLNM 304, Springer, Berlin (1972).

[6] L. Breen, *Extensions du groupe additif*, Inst. Hautes Études Sci. Publ. Math. **48** (1978), 39–125.

[7] L. Breen, *Bitorseurs et cohomologie non-abélienne*, The Grothendieck Festschrift, I, Progress in Mathematics vol. 86, Birkhäuser (1990), 401–476.

[8] L. Breen, *Théorie du Schreier supérieure*, Ann. Scient. École Norm. Sup. (4) **25** (1992), 465–514.

[9] K.S. Brown, *Abstract homotopy theory and generalized sheaf cohomology*, Trans. Amer. Math. Soc. **186** (1973), 419–458.

[10] K.S. Brown, *Buildings*, Springer, Berlin (1989).

[11] K.S. Brown and S.M. Gersten, *Algebraic $K$-theory as generalized sheaf cohomology*, Algebraic $K$-theory I: Higher $K$-theories, SLNM 341, Springer, Berlin (1973), 266–292.

[12] J.-L. Brylinski, *Loop Spaces, Characteristic Classes and Geometric Quantization*, Progress in Mathematics vol. 107, Birkhäuser (1993).

[13] E.B. Curtis, *Simplicial homotopy theory*, Adv. Math. **6** (1971), 107–209.

[14] J. Dieudonné, *Les débuts de la topologie algébrique*, Exposition. Math. **3** (1985), 347–357.

[15] J. Dieudonné, *A History of Algebraic and Differential Topology 1900–1960*, Birkhäuser, Basel (1989).

[16] E. Dror-Farjoun, *Homotopy theories for diagrams of spaces*, Proc. Amer. Math. Soc. **101** (1985), 181–189.

[17] W.G. Dwyer and E.M. Friedlander, *Algebraic and étale $K$-theory*, Trans. Amer. Math. Soc. **292** (1985), 247–280.

[18] W.G. Dwyer and D.M. Kan, *Equivalences between homotopy theories of diagrams*, Algebraic Topology and Algebraic $K$-theory, W. Browder, ed., Ann. Math. Stud. vol. 113, Princeton Univ. Press (1987), 180–205.

[19] S. Eilenberg, *Singular homology theory*, Ann. Math. **45** (1944), 407–477.

[20] S. Eilenberg and J.A. Zilber, *On semi-simplicial complexes and singular homology*, Ann. Math. **51** (1950), 409–513.

[21] E. Friedlander, *Étale Homotopy of Simplicial Schemes*, Ann. Math. Stud. vol. 104, Princeton Univ. Press (1982).

[22] E. Friedlander and G. Mislin, *Cohomology of classifying spaces of complex Lie groups and related discrete subgroups*, Comment. Math. Helv. **59** (1984), 347–361.

[23] E. Friedlander and B. Parshall, *Étale cohomology of reductive groups*, SLNM 854, Springer, Berlin (1981), 127–140.

[24] A. Fröhlich, *Orthogonal representations of Galois groups, Stiefel–Whitney classes and Hasse–Witt invariants*, J. Reine Angew. Math. **360** (1985), 85–123.

[25] O. Gabber, *$K$-theory of Henselian local rings and Henselian pairs*, Algebraic $K$-theory, Commutative Algebra, and Algebraic Geometry, Contemp. Math. vol. 126 (1992), 59–70.

[26] P. Gabriel and M. Zisman, *Calculus of Fractions and Homotopy Theory*, Springer, New York (1967).

[27] H. Gillet and R.W. Thomason, *The $K$-theory of strict Hensel local rings and a theorem of Suslin*, J. Pure Appl. Algebra **34** (1984), 241–254.

[28] J. Giraud, *Cohomologie non Abélienne*, Springer, Berlin (1971).

[29] T.E.W. Gunnarsson, *Abstract homotopy theory and related topics*, Ph.D. thesis, Chalmers Tekniska Högskola, Göteborg, Sweden (1987).

[30] L. Illusie, *Complexe Cotangent et Déformations, I*, SLNM 239, Springer, Berlin (1971).

[31] J.F. Jardine, *Cup products in sheaf cohomology*, Canad. Math. Bull. **29** (1986), 469–477.

[32] J.F. Jardine, *Simplicial objects in a Grothendieck topos*, Applications of Algebraic $K$-theory to Algebraic Geometry and Number Theory, I, Contemp. Math. vol. 55(I) (1986), 193–239.

[33] J.F. Jardine, *Simplicial presheaves*, J. Pure Appl. Algebra **47** (1987), 35–87.

[34] J.F. Jardine, *Stable homotopy theory of simplicial presheaves*, Canad. J. Math. **39** (1987), 733–747.

[35] J.F. Jardine, *The homotopical foundations of algebraic $K$-theory*, Algebraic $K$-theory and Algebraic Number Theory, Contemp. Math. vol. 83 (1989), 57–82.

[36] J.F. Jardine, *Universal Hasse–Witt classes*, Algebraic $K$-theory and Algebraic Number Theory, Contemp. Math. vol. 83 (1989), 83–100.

[37] J.F. Jardine, *Higher spinor classes*, Mem. Amer. Math. Soc. **528** (1994).

[38] A. Joyal, *Letter to A. Grothendieck* (1984).

[39] A. Joyal and M. Tierney, *Classifying spaces for sheaves of simplicial groupoids*, J. Pure Appl. Algebra **89** (1991), 135–161.

[40] B. Kahn, *Classes de Stiefel–Whitney de formes quadratiques et de représentations galoisiennes réelles*, Invent. Math. **78** (1984), 223–256.

[41] D.M. Kan, *Abstract homotopy, I*, Proc. Nat. Acad. Sci. USA **41** (1955), 1092–1096.

[42] D.M. Kan, *Abstract homotopy, II*, Proc. Nat. Acad. Sci. USA **42** (1956), 225–228.

[43] D.M. Kan, *Abstract homotopy, III*, Proc. Nat. Acad. Sci. USA **42** (1956), 419–421.

[44] D.M. Kan, *On c.s.s. complexes*, Amer. J. Math. **79** (1957), 449–476.

[45] D.M. Kan, *Adjoint functors*, Trans. Amer. Math. Soc. **87** (1958), 294–329.

[46] D.M. Kan, *A combinatorial definition of homotopy groups*, Ann. Math. **67** (1958), 288–312.

[47] M. Karoubi, *Relations between algebraic $K$-theory and hermitian $K$-theory*, J. Pure Appl. Algebra **34** (1984), 259–263.

[48] S. Lefschetz, *Algebraic Topology*, Amer. Math. Soc., Providence, RI (1942).

[49] J.-L. Loday, *Spaces with finitely many non-trivial homotopy groups*, J. Pure Appl. Algebra **24** (1982), 179–202.

[50] S. MacLane, *Categories for the Working Mathematician*, Springer, Berlin (1971).

[51] S. MacLane, *Origins of the cohomology of groups*, Enseign. Math. **XXIV** (1978), 1–25.

[52] S. MacLane and I. Moerdijk, *Sheaves in Geometry and Logic: A First Introduction to Topos Theory*, Springer, Berlin (1992).

[53] J.P. May, *Simplicial Objects in Algebraic Topology*, Van Nostrand, Princeton (1968).

[54] A.S. Merkurjev and A.A. Suslin, *$k$-cohomology of Severi–Brauer varieties and the norm residue homomorphism*, Math. USSR-Izv. **21** (1983), 307–340.

[55] H.R. Miller, *The Sullivan conjecture on maps from classifying spaces*, Ann. Math. **120** (1984), 39–87.

[56] J.S. Milne, *Étale Cohomology*, Princeton Univ. Press, Princeton (1980).

[57] I. Moerdijk, *Bisimplicial sets and the group-completion theorem*, Algebraic $K$-theory: Connections with Geometry and Topology, NATO ASI Series C vol. 279, Kluwer (1989), 225–240.

[58] J.C. Moore, *Semi-simplicial complexes and Postnikov systems*, Symposium Internationale de Topologica Algebraica (1958), 232–247.

[59] D.H. Van Osdol, *Simplicial homotopy in an exact category*, Amer. J. Math. **99**(6) (1977), 1193–1204.

[60] H. Poincaré, *Analysis situs*, Journal de l'École Polytechnique, Paris (2) **1** (1895), 1–123.
[61] H. Poincaré, *Compléments à l'analysis situs*, Rendiconti, Circolo Matematico, Palermo **13** (1899), 285–343.
[62] D. Quillen, *Homotopical Algebra*, SLNM 43, Springer, Berlin (1967).
[63] D. Quillen, *The geometric realization of a Kan fibration is a Serre fibration*, Proc. Amer. Math. Soc. **19** (1968), 1499–1500.
[64] D. Quillen, *Rational homotopy theory*, Ann. Math. **90** (1969), 205–295.
[65] D. Quillen, *Higher algebraic $K$-theory I*, Algebraic $K$-theory I: Higher $K$-theories, SLNM 341, Springer, Berlin (1973), 85–147.
[66] D. Quillen, *Homotopy properties of the poset of non-trivial p-subgroups of a group*, Adv. Math. **28** (1978), 101–128.
[67] M. Rapoport, N. Schnappacher and P. Schneider (eds), *Beilinson's Conjectures on Special Values of L-functions*, Perspectives in Mathematics vol. 4, Academic Press (1988).
[68] V.V. Schechtman, *On the delooping of Chern character and Adams operations*, SLNM 1289, Springer, Berlin (1987), 265–319.
[69] G. Segal, *Categories and cohomology theories*, Topology **13** (1974), 293–312.
[70] V.P. Snaith, *Topological Methods in Galois Representation Theory*, Wiley, New York (1989).
[71] D. Sullivan, *Infinitesimal computations in algebraic topology*, Inst. Hautes Études Sci. Publ. Math. **47** (1977), 269–332.
[72] A.A. Suslin, *On the $K$-theory of algebraically closed fields*, Invent. Math. **73** (1983), 241–245.
[73] A.A. Suslin, *On the $K$-theory of local fields*, J. Pure Appl. Algebra **34** (1984), 301–318.
[74] A.A. Suslin, *Algebraic $K$-theory of fields*, Proceedings, International Congress of Mathematicians, 1986, vol. I (1987), 222–244.
[75] R. Thomason, *Algebraic $K$-theory and étale cohomology*, Ann. Scient. École Norm. Sup. (4) **18** (1985), 437–552.

# Derived Categories and Their Uses

Bernhard Keller

*U.F.R. de Mathématiques, U.R.A. 748 du CNRS, Université Paris 7, 75251 Paris Cedex 05, France*
*e-mail: keller@mathp7.jussieu.fr*

*Contents*



HANDBOOK OF ALGEBRA, VOL. 1
Edited by M. Hazewinkel

## 1. Introduction

### 1.1. *Historical remarks*

Derived categories are a 'formalism for hyperhomology' [61]. Used at first only by the circle around Grothendieck they have now become wide-spread in a number of subjects beyond algebraic geometry, and have found their way into graduate text books [33, 38, 44, 62].

According to L. Illusie [32], derived categories were invented by A. Grothendieck in the early sixties. He needed them to formulate and prove the extensions of Serre's duality theorem [55] which he had announced [24] at the International Congress in 1958. The essential constructions were worked out by his pupil J.-L. Verdier who, in the course of the year 1963, wrote down a summary of the principal results [56]. Having at his disposal the required foundations Grothendieck exposed the duality theory he had conceived of in a huge manuscript [25], which served as a basis for the seminar [29] that Hartshorne conducted at Harvard in the autumn of the same year.

Derived categories found their first applications in duality theory in the coherent setting [25, 29] and then also in the étale [60, 13] and in the locally compact setting [57–59, 22].

At the beginning of the seventies, Grothendieck–Verdier's methods were adapted to the study of systems of partial differential equations by M. Sato [53] and M. Kashiwara [37]. Derived categories have now become the standard language of microlocal analysis (cf. [38, 46, 52] or [6]). Through Brylinski–Kashiwara's proof of the Kazhdan–Lusztig conjecture [9] they have penetrated the representation theory of Lie groups [4] and finite Chevalley groups [54]. In this theory, a central role is played by certain abelian subcategories of derived categories which are modeled on the category of perverse sheaves [2], which originated in the sheaf-theoretic interpretation [14] of intersection cohomology [20, 21].

In their fundamental papers [1] and [3], Beilinson, and Bernstein and Gelfand used derived categories to establish a beautiful relation between coherent sheaves on projective space and representations of certain finite-dimensional algebras. Their constructions had numerous generalizations [16, 34–36, 10]. They also led D. Happel to a systematic investigation of the derived category of a finite-dimensional algebra [26, 27]. He realized that categories provide the proper setting for the so-called tilting theory [7, 28, 5]. This theory subsequently reached its full scope when it was generalized to 'Morita theory' for derived categories of module categories [48, 50] (cf. also [39–41]). Morita theory has further widened the range of applications of derived categories. Thus, Broué's conjectures on representations of finite groups [8] are typical of the synthesis of precision with generality that can be achieved by the systematic use of this language.

### 1.2. *Motivation of the principal constructions*

Grothendieck's key observation was that the constructions of homological algebra do not barely yield cohomology groups but in fact complexes with a certain indeterminacy. To make this precise, he defined a *quasi-isomorphism* between two complexes over an abelian category $\mathcal{A}$ to be a morphism of complexes $s\colon L \to M$ inducing an isomorphism $\mathrm{H}^n(s)\colon \mathrm{H}^n(L) \to \mathrm{H}^n(M)$ for each $n \in \mathbf{Z}$. The result of a homological construction is then a complex which is 'well defined up to quasi-isomorphism'. To illustrate this point, let us recall the definition of the left derived functor $\mathrm{Tor}_n^A(M, N)$, where $A$ is an associative ring with 1, $N$ a (fixed) left $A$-module and $M$ a right $A$-module. We choose a resolution

$$\cdots \to P^i \to P^{i+1} \to \cdots \to P^{-1} \to P^0 \to M \to 0$$

(i.e. a quasi-isomorphism $P \to M$) with projective right $A$-modules $P^i$. Then we consider the complex $P \otimes_A N$ obtained by applying $? \otimes_A \ N$ to each term $P^i$, and 'define' $\mathrm{Tor}_n^A(M, N)$ to be the $(-n)$-th cohomology group of $P \otimes_A N$. If $P' \to M$ is another resolution, there is a morphism of resolutions $P \to P'$ (i.e. morphism of complexes compatible with the augmentations $P \to M$ and $P' \to M$), which is a homotopy equivalence. The induced morphism $P \otimes_A N \to P' \otimes_A N$ is still a homotopy equivalence and a fortiori a quasi-isomorphism. We thus obtain a system of isomorphisms between the $H^{-n}(P \otimes_A \ N)$, and we can give a more canonical definition of $\mathrm{Tor}_n^A(M, N)$ as the (inverse) limit of this system. Sometimes it is preferable to 'compute' $\mathrm{Tor}_n^A(M, N)$ using flat resolutions. If $F \to M$ is such a resolution, there exists a morphism of resolutions $P \to F$. This is no longer a homotopy equivalence but still induces a quasi-isomorphism $P \otimes_A N \to F \otimes_A N$. So we have $\mathrm{H}^{-n}(P \otimes_A N) \xrightarrow{\sim} \mathrm{H}^{-n}(F \otimes_A N)$. However, the construction yields more, to wit the family of complexes $F \otimes_A N$ indexed by all flat resolutions $F$, which forms a single class with respect to quasi-isomorphism. More precisely, any two such complexes $F \otimes_A N$ and $F' \otimes_A N$ are linked by quasi-isomorphisms $F \otimes_A N \leftarrow P \otimes_A N \to F' \otimes_A N$. The datum of this class is of course richer than that of the $\mathrm{Tor}_n^A(M, N)$. For example, if $A$ is a flat algebra over a commutative ring $k$, it allows us to recover $\mathrm{Tor}_n^A(M, N \otimes_k X)$ for each $k$-module $X$.

Considerations like these must have led Grothendieck to define the *derived category* $\mathbf{D}(\mathcal{A})$ of an abelian category $\mathcal{A}$ by 'formally adjoining inverses of all quasi-isomorphisms' to the *category* $\mathbf{C}(\mathcal{A})$ *of complexes over* $\mathcal{A}$. So the objects of $\mathbf{D}(\mathcal{A})$ are complexes and its morphisms are deduced from morphisms of complexes by 'abstract localization'. The right (resp. left) 'total derived functors' of an additive functor $F\colon \mathcal{A} \to \mathcal{B}$ will then have to be certain 'extensions' of $F$ to a functor $\mathbf{R}F$ (resp. $\mathbf{L}F$) whose composition with $\mathrm{H}^n(?)$ should yield the traditional functors $\mathbf{R}^n F$ (resp. $\mathbf{L}^n F$).

It was Verdier's observation that one obtains a convenient description of the morphisms of $\mathbf{D}(\mathcal{A})$ by a 'calculus of fractions' if, in a first step, one passes to the *homotopy category* $\mathbf{H}(\mathcal{A})$, whose objects are complexes and whose morphisms are homotopy classes of morphisms of complexes. In a second step, the derived category is defined as the localization of $\mathbf{H}(\mathcal{A})$ with respect to all quasi-isomorphisms. The important point is that in $\mathbf{H}(\mathcal{A})$ the (homotopy classes of) quasi-isomorphisms $M' \leftarrow M$ (resp. $L \leftarrow L'$) starting (resp. ending) at a fixed complex form a *filtered* category $\Sigma_M$ (resp. ${}_L\Sigma$). We have

$$\mathrm{Hom}_{\mathbf{D}(\mathcal{A})}(L, M) = \varinjlim_{\Sigma_M} \mathrm{Hom}_{\mathbf{H}(\mathcal{A})}(L, M') = \varinjlim_{{}_L\Sigma} \mathrm{Hom}_{\mathbf{H}(\mathcal{A})}(L', M).$$

The elements of the two right hand members are intuitively interpreted as 'left fractions' $s^{-1}f$ or 'right fractions' $gt^{-1}$ associated with diagrams

$$L \xrightarrow{f} M' \xleftarrow{s} M \quad \text{or} \quad L \xleftarrow{t} L' \xrightarrow{g} M$$

of $\mathbf{H}(\mathcal{A})$. This also leads to a simple definition of the derived functors: Examples suggest that the derived functors $\mathbf{R}F$ and $\mathbf{L}F$ can not, in general, be defined on all of $\mathcal{DA}$. Following Deligne [12], 1.2, we define the *domain* of $\mathbf{R}F$ (resp. $\mathbf{L}F$) to be the full subcategory of $\mathbf{D}(\mathcal{A})$ formed by the complexes $M$ (resp. $L$) such that

$$\varinjlim{}_{\Sigma_M} FM' \quad (\text{resp. } \varprojlim{}_{L^\Sigma} FL')$$

exists in $\mathbf{D}(\mathcal{B})$ and is preserved by *all* functors starting from the category $\mathbf{D}(\mathcal{B})$. For such an $M$ (resp. $L$) we put

$$\mathbf{R}FM := \varinjlim{}_{\Sigma_M} FM' \quad (\text{resp. } \mathbf{L}FL := \varprojlim{}_{L^\Sigma} FL').$$

The functors thus constructed satisfy the universal property by which Grothendieck and Verdier originally [61] defined derived functors. When they exist, they enjoy properties which apparently do not follow directly from the universal property.

### 1.3. *On the use of derived categories*

Any relation formulated in the language of derived categories and functors gives rise to assertions formulated in the more traditional language of cohomology groups, filtrations, spectral sequences .... Of course, these can frequently be proved without explicitly mentioning derived categories so that we may wonder why we should make the effort of using this more abstract language. The answer is that the simplicity of the phenomena, hidden by the notation in the old language, is clearly apparent in the new one. The example of the Künneth relations [61] serves to illustrate this point:

Let $X$ and $Y$ be compact spaces, $R$ a commutative ring with 1, and $\mathcal{F}$ and $\mathcal{G}$ sheaves of $R$-modules on $X$ and $Y$, respectively. If $R = \mathbf{Z}$, and either $\mathcal{F}$ or $\mathcal{G}$ is torsion-free, we have split short exact sequences

$$\begin{aligned} 0 \to \bigoplus_{p+q=n} \mathrm{H}^p(X,\mathcal{F}) \otimes_R \mathrm{H}^q(Y,\mathcal{G}) &\to \mathrm{H}^n(X \times Y, \mathcal{F} \otimes_R \mathcal{G}) \\ \to \bigoplus_{p+q=n+1} \mathrm{Tor}^{\mathbf{Z}}\big(\mathrm{H}^p(X,\mathcal{F}), \mathrm{H}^q(Y,\mathcal{G})\big) &\to 0. \end{aligned} \tag{1}$$

When the ring $R$ is more complicated, for example $R = \mathbf{Z}/l^r\mathbf{Z}$, $l$ prime, $r > 1$, and if we make no hypothesis on $\mathcal{F}$ or $\mathcal{G}$, then in the traditional language we only have two spectral sequences with isomorphic abutments whose initial terms are

$$'K_2^{p,q} = \bigoplus_{r+s=q} \mathrm{Tor}_{-p}^R\big(\mathrm{H}^r(X,\mathcal{F}), \mathrm{H}^s(Y,\mathcal{G})\big), \tag{2}$$

$$''K_2^{p,q} = \mathrm{H}^p\big(X \times Y, \mathrm{Tor}^R_{-q}(\mathcal{F}, \mathcal{G})\big). \tag{3}$$

However, these spectral sequences and the isomorphism between their abutments are just the consequence and the imperfect translation of the following relation in the derived category of $R$-modules

$$\mathbf{R}\Gamma(X, \mathcal{F}) \otimes^{\mathsf{L}}_R \mathbf{R}\Gamma(Y, \mathcal{G}) \xrightarrow{\sim} \mathbf{R}\Gamma\big(X \times Y, \mathcal{F} \otimes^{\mathsf{L}}_R \mathcal{G}\big), \tag{4}$$

where $\mathbf{R}\Gamma(X, ?)$ denotes the right derived functor of the global section functor and $\otimes^{\mathsf{L}}_R$ denotes the left derived functor of the tensor product functor. (We suppose that $X$ and $Y$ are spaces of finite cohomological dimension, for example, finite cell complexes.) The members of (4) are complexes of $R$-modules which are well determined up to quasi-isomorphism, and whose cohomology groups are the abutment of the spectral sequences (2) and (3), respectively. Of course, formula (4) still holds when $\mathcal{F}$ and $\mathcal{G}$ are (suitably bounded) complexes of sheaves on $X$ and $Y$. The formula is easy to work with in practice and also allows us to formulate commutativity and associativity properties when there are several factors.

Extension of scalars leads to an analogous formula in the derived categories: If $S$ is an $R$-algebra and $\mathcal{F}$ a sheaf of $R$-modules on $X$, we have the relation

$$\mathbf{R}\Gamma(X, \mathcal{F}) \otimes^{\mathsf{L}}_R S \xrightarrow{\sim} \mathbf{R}\Gamma\big(X, \mathcal{F} \otimes^{\mathsf{L}}_R S\big).$$

Metaphorically speaking, one can say that *naïve formulas which are false in the traditional language become true in the language of derived categories and functors.*

## 2. Outline of the chapter

The machinery needed to define a derived category in full generality tends to obscure the simplicity of the phenomena. We therefore start in Section 3 with the example of the derived category of a module category. The same construction applies to any abelian category with enough projectives.

The class of abelian categories is not closed under many important constructions. Thus the category of projective objects or the category of filtered objects of an abelian category are no longer abelian in general. This leads us to working with exact categories in the sense of Quillen [47]. We recall their definition and the main examples in Section 4.

Heller's stable categories [30] provide an efficient approach [26] to the homotopy category. They also yield many other important examples of triangulated categories, and, more generally, of suspended categories (cf. Section 7). We give Heller's construction in Section 6. It is functorial in the sense that exact functors give rise to 'stable functors'. The notion of a triangle functor (= $S$-functor [42] = exact functor [12]) appears as the natural axiomatization of this concept. Triangle functors, equivalences and adjoints are presented in Section 8.

In Section 9, we recall basic facts on the localization of categories from [15]. These are then specialized to triangulated categories in Section 10. Proofs for the results of these sections may be found in [29, 6, 38].

In Section 11, we formulate Verdier's definition of the derived category [56] in the context of exact categories.

In Section 12, we give a sufficient condition for an inclusion of exact categories to induce an equivalence of their derived categories. This is a key result since it corresponds to the theorem on the existence and unicity of injective resolutions in classical homological algebra.

In Sections 13, 14, and 15, we develop the theory of derived functors following Deligne. Derived functors are constructed using a 'generalized calculus of fractions'. This approach makes it possible to easily deduce fundamental results on restrictions, adjoints and compositions in the generality they deserve. Proofs for some nontrivial lemmas of these sections may be found in [12].

## 3. The derived category of a module category

For basic module theoretic notions and terminology we refer to [31], I, IV. We shall sometimes write $\mathcal{C}(X, Y)$ for the set of morphisms from $X$ to $Y$ in a category $\mathcal{C}$.

Let $R$ be an associative ring with 1 and denote by $\operatorname{Mod} R$ the category of right $R$-modules. By definition, the *objects* of $\mathbf{D}^b(\operatorname{Mod} R)$, the *derived category of* $\operatorname{Mod} R$, are the chain complexes

$$P = \left( \cdots \to P_n \xrightarrow{d_n^P} P_{n-1} \to \cdots \right)$$

of projective right $R$-modules $P_n$, $n \in \mathbf{Z}$, such that we have $P_n = 0$ for all $n \ll 0$ and $\mathrm{H}_n(P) = 0$ for all $n \gg 0$, where $\mathrm{H}_n(P)$ denotes the $n$-th homology module of $P$. If $P$ and $Q$ are such complexes a *morphism* $P \to Q$ of $\mathbf{D}^b(\operatorname{Mod} R)$ is given by the equivalence class $\bar{f}$ of a morphism of complexes $f$: $P \to Q$ modulo the subgroup of *null-homotopic morphisms*, i.e. those with components of the form

$$d_{n+1}^Q r_n + r_{n-1} d_n^P$$

for some family of $R$-module homomorphisms $r_n$: $P_n \to Q_{n-1}$, $n \in \mathbf{Z}$. The *composition* of morphisms of $\mathbf{D}^b(\operatorname{Mod} R)$ is induced by the composition of morphisms of complexes.

The category of $R$-modules is related to its derived category by a *canonical embedding*: The canonical functor can: $\operatorname{Mod} R \to \mathbf{D}^b(\operatorname{Mod} R)$ sends an $R$-module $M$ to the complex

$$\cdots \to P_{n+1} \to P_n \to \cdots \to P_1 \to P_0 \to 0 \to 0 \to \cdots$$

given by a chosen projective resolution of $M$. If $f$: $M \to N$ is an $R$-module homomorphism, can $f$ is the uniquely determined homotopy class of morphisms of complexes $g$: can $M \to$ can $N$ such that $\mathbf{H}_0(g)$ is identified with $f$.

We endow $\mathbf{D}^b(\operatorname{Mod} R)$ with the endofunctor $S$, called the *suspension functor* (or shift functor), and defined by

$$(SP)_n = P_{n-1}, \quad d_n^{SP} = -d_{n-1}^P,$$

on the objects $P \in \mathbf{D}^b$ (Mod $R$) and by $S\bar{f} = \bar{g}$, $g_n = f_{n-1}$, on morphisms $\bar{f}$.

We omit the symbol can from the notations to state the fundamental formula

$$\mathrm{Hom}_D(M, S^n N) \xrightarrow{\sim} \mathrm{Ext}^n_R(M, N), \quad n \in \mathbf{N}, \tag{5}$$

where $\mathrm{Hom}_D(,)$ denotes morphisms in the derived category and $M, N$ are $R$-modules. This isomorphism is compatible with the product structures in the sense that the composition

$$L \xrightarrow{\bar{g}} S^m M \xrightarrow{S^m \bar{f}} S^{m+n} N$$

corresponds to the 'splicing product' [31], IV, Exercise 9.3, of the $n$-extension determined by $\bar{f}$ with the $m$-extension determined by $\bar{g}$.

EXAMPLE 3.1. *Fields.* Suppose that $R = k$ is a (skew) field. Then it is not hard to see that each $P \in \mathbf{D}^b$ (Mod $k$) is isomorphic to a finite sum of objects $S^n M$, $M \in \mathrm{Mod}\, k$, $n \in \mathbf{Z}$. Moreover, by formula (5) there are no nontrivial morphisms from $S^i M$ to $S^j N$ unless $i = j$, and

$$\mathrm{Hom}_{\mathbf{D}}\left(S^i M, S^i N\right) \xrightarrow{\sim} \mathrm{Hom}_k(M, N).$$

Thus, $\mathbf{D}^b(\mathrm{Mod}\, k)$ is equivalent to the category of $\mathbf{Z}$-graded $k$-vector spaces with finitely many nonzero components. The equivalence is realized by the homology functor $P \mapsto \mathrm{H}_*(P)$.

EXAMPLE 3.2. *Hereditary rings.* Suppose that $R$ is hereditary (i.e. submodules of projective $R$-modules are projective). For example, we can take for $R$ a principal domain or the ring of upper triangular $n \times n$-matrices over a field. Then, as in Example 3.1, each $P \in \mathbf{D}^b$ (Mod $R$) is isomorphic to a finite sum of objects $S^n M$, $M \in \mathrm{Mod}\, R$, $n \in \mathbf{Z}$. Formula (5) shows that

$$\mathrm{Hom}_{\mathbf{D}}\left(S^i M, S^j N\right) = \begin{cases} 0, & j \neq i, i+1, \\ \mathrm{Hom}_R(M, N), & j = i, \\ \mathrm{Ext}^1_R(M, N) & j = i+1. \end{cases}$$

EXAMPLE 3.3. *Dual numbers.* Let $k$ be a commutative field and let $R = k[\delta]/(\delta^2)$ be the ring of dual numbers over $k$. The complexes

$$\cdots \to 0 \to A \xrightarrow{\delta} A \xrightarrow{\delta} \cdots \xrightarrow{\delta} A \xrightarrow{\delta} A \to 0 \to \cdots$$

with nonzero components in degrees $0, \ldots, N$, $N \geqslant 1$, have nonzero homology in degrees 0 and $N$, only, but they do not admit nontrivial decompositions as direct sums in $\mathbf{D}^b(\mathrm{Mod}\, R)$.

## 4. Exact categories

We refer to [45] for basic category theoretic notions and terminology. A category which is equivalent to a small category will be called *svelte*.

A pair of morphisms

$$A \xrightarrow{i} B \xrightarrow{p} C$$

in an additive category is *exact* if $i$ is a kernel of $p$ and $p$ a cokernel of $i$.

An *exact category* [47] is an additive category $\mathcal{A}$ endowed with a class $\mathcal{E}$ of exact pairs closed under isomorphism and satisfying the following axioms Ex0–Ex2$^{\mathrm{op}}$ [39]. The *deflations* (resp. *inflations*) mentioned in the axioms are by definition the morphisms $p$ (resp. $i$) occurring in pairs $(i, p)$ of $\mathcal{E}$. We shall refer to such pairs as *conflations*.

Ex0. The identity morphism of the zero object is a deflation.

Ex1. A composition of two deflations is a deflation.

Ex1$^{\mathrm{op}}$. A composition of two inflations is an inflation.

Ex2. Each diagram

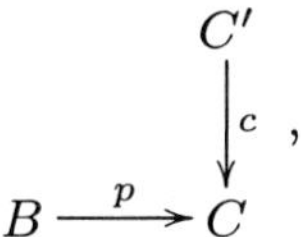

where $p$ is a deflation, may be completed to a cartesian square

$$\begin{array}{ccc} B' & \xrightarrow{p'} & C' \\ {\scriptstyle b}\downarrow & & \downarrow{\scriptstyle c} \\ B & \xrightarrow{p} & C \end{array},$$

where $p'$ is a deflation.

Ex2$^{\mathrm{op}}$. Each diagram

$$\begin{array}{ccc} A & \xrightarrow{i} & B \\ {\scriptstyle a}\downarrow & & \\ A' & & \end{array}$$

where $i$ is an inflation, may be completed to a cocartesian square

$$\begin{array}{ccc} A & \xrightarrow{i} & B \\ {\scriptstyle a}\downarrow & & \downarrow{\scriptstyle b} \\ A' & \xrightarrow{i'} & B' \end{array},$$

where $i'$ is an inflation.

An *abelian category* is an exact category such that each morphism $f$ admits a factorization $f = ip$, where $p$ is a deflation and $i$ an inflation. In this case, the class of conflations coincides with the class of all exact pairs.

If $\mathcal{A}$ and $\mathcal{B}$ are exact categories, an *exact functor* $\mathcal{A} \to \mathcal{B}$ is an additive functor taking conflations of $\mathcal{A}$ to conflations of $\mathcal{B}$.

A *fully exact subcategory* of an exact category $\mathcal{A}$ is a full additive subcategory $\mathcal{B} \subset \mathcal{A}$ which is *closed under extensions*, i.e. if it contains the end terms of a conflation of $\mathcal{A}$, it also contains the middle term. Then $\mathcal{B}$ endowed with the conflations of $\mathcal{A}$ having their terms in $\mathcal{B}$ is an exact category, and the inclusion $\mathcal{B} \subset \mathcal{A}$ is a fully faithful exact functor.

EXAMPLE 4.1. *Module categories and their fully exact subcategories.* Let $R$ be an associative ring with 1. The category $\operatorname{Mod} R$ of right $R$-modules endowed with all short exact sequences is an abelian category. The classes of free, projective, flat, injective, finitely generated, ... modules all form fully exact subcategories of $\operatorname{Mod} R$.

In general, *any svelte exact category may be embedded as a fully exact subcategory of some module category* [47, 39]. As a consequence [39], in any argument involving only a finite diagram and such notions as deflations, inflations, conflations, it is legitimate to suppose that we are operating in a fully exact subcategory of a category of modules.

EXAMPLE 4.2. *Additive categories.* Let $\mathcal{A}$ be an additive category. Endowed with all split short exact sequences $\mathcal{A}$ becomes an exact category.

EXAMPLE 4.3. *The category of complexes.* Let $\mathcal{A}$ be an additive category. Denote by $\mathbf{C}(\mathcal{A})$ the category of differential complexes

$$\cdots \to A^n \xrightarrow{d_A^n} A^{n+1} \to \cdots$$

over $\mathcal{A}$. Endow $\mathbf{C}(\mathcal{A})$ with the class of all pairs $(i, p)$ such that $(i^n, p^n)$ is a split short exact sequence for each $n \in \mathbf{Z}$. Then $\mathbf{C}(\mathcal{A})$ is an exact category.

EXAMPLE 4.4. *$k$-split sequences.* Let $k$ be a commutative ring and $R$ an associative $k$-algebra. Endowed with the sequences whose restrictions to $k$ are split short exact the category $\operatorname{Mod} R$ of Section 3 becomes an exact category.

EXAMPLE 4.5. *Filtered objects.* Let $\mathcal{A}$ be an exact category. The objects of the *filtered category* $\mathbf{F}(\mathcal{A})$ are the sequences of inflations

$$A = \left(\cdots \to A^p \xrightarrow{j_A^p} A^{p+1} \to \cdots\right), \quad p \in \mathbf{Z},$$

of $\mathcal{A}$ such that $A^p = 0$ for $p \ll 0$ and $\operatorname{Cok} j_A^p = 0$ for all $p \gg 0$. The morphisms from $A$ to $B \in \mathbf{F}(\mathcal{A})$ bijectively correspond to sequences $f^p \in \mathcal{A}(A^p, B^p)$ such that $f^{p+1} j_A^p = j_B^p f^p$ for all $p \in \mathbf{Z}$. The sequences whose components are conflations of $\mathcal{A}$ form an exact structure on $\mathbf{F}(\mathcal{A})$. Note that if $\mathcal{A}$ contains a nonzero object, then $\mathbf{F}(\mathcal{A})$ is not abelian (even if $\mathcal{A}$ is).

EXAMPLE 4.6. *Banach spaces.* Let $\mathcal{A}$ be the category of complex Banach spaces. The axioms for an exact structure are satisfied by the sequences which are short exact as sequences of complex vector spaces.

## 5. Exact categories with enough injectives

Let $\mathcal{A}$ be an exact category. An object $I \in \mathcal{A}$ is *injective* (resp. *projective*) if the sequence

$$\mathcal{A}(B,I) \xrightarrow{i^*} \mathcal{A}(A,I) \to 0 \quad \big(\text{resp. } \mathcal{A}(P,B) \xrightarrow{p_*} \mathcal{A}(P,C) \to 0\big)$$

is exact for each conflation $(i,p)$ of $\mathcal{A}$. *We assume from now on that $\mathcal{A}$ has enough injectives*, i.e. that each $A \in \mathcal{A}$ admits a conflation

$$A \xrightarrow{i_A} IA \xrightarrow{p_A} SA$$

with injective $IA$. If $\mathcal{A}$ also has enough projectives (i.e. for each $A \in \mathcal{A}$ there is a deflation $P \to A$ with projective $P$), and the classes of projectives and injectives coincide, we call $\mathcal{A}$ a *Frobenius category.*

EXAMPLE 5.1. *Module categories.* The category of modules over an associative ring $R$ with 1 has enough projectives and injectives. Projectives and injectives coincide for example if $R$ is the group ring of a finite group over a commutative field.

EXAMPLE 5.2. *Additive categories.* In Example 4.2, each object is injective and projective, and we can take $i_A$ to be the identity of $A$ for each $A \in \mathcal{A}$.

EXAMPLE 5.3. *The category of complexes.* In Example 4.3, we define

$$(IA)^n = A^n \oplus A^{n+1}, \quad d^n_{IA} = \begin{bmatrix} 0 & \mathbf{1} \\ 0 & 0 \end{bmatrix}, \quad (i_A)^n = \begin{bmatrix} \mathbf{1} \\ d^n_A \end{bmatrix},$$
$$(SA)^n = A^{n+1}, \quad d^n_{SA} = -d^{n+1}_A, \quad p^n_A = [-d^n_A \; \mathbf{1}].$$

It is easy to see that $IA$ is injective in $\mathbf{C}(\mathcal{A})$. Now the inflation $i_A$ splits iff $A$ is homotopic to zero. Thus, *a complex is injective in $\mathbf{C}(\mathcal{A})$ iff it is homotopic to zero.* Since the complexes $IA$, $A \in \mathbf{C}(\mathcal{A})$, are also projective, $\mathbf{C}(\mathcal{A})$ *is a Frobenius category.*

EXAMPLE 5.4. *$k$-split exact sequences.* In Example 4.4 we can take for $i_M$ the canonical injection

$$M \to \mathrm{Hom}_k(R, M), \quad m \mapsto (r \mapsto rm).$$

If $R = k[G]$ for a finite group $G$, the fully exact subcategory of Mod $R$ formed by finitely generated $k$-free $R$-modules is a Frobenius category.

EXAMPLE 5.5. *Filtered objects.* In Example 4.5 it is not hard to see that $\mathbf{F}(\mathcal{A})$ has enough injectives iff $\mathcal{A}$ has, and in this case the injectives of $\mathbf{F}(\mathcal{A})$ are the filtered objects with injective components [39]. Similarly, $\mathbf{F}(\mathcal{A})$ has enough projectives iff $\mathcal{A}$ has, and in this case the projectives of $\mathbf{F}(\mathcal{A})$ are the filtered objects of $A$ with projective components and such that $j^p_A$ splits for all $p \in \mathbf{Z}$.

EXAMPLE 5.6. *Banach spaces.* As a consequence of the Hahn–Banach theorem, the one-dimensional complex Banach space is injective for the category of Example 4.6. More generally, the space of bounded functions on a discrete topological space is injective. There are enough injectives since each Banach space identifies with a closed subspace of the space of bounded functions on the unit sphere of its dual with the discrete topology.

## 6. Stable categories

Keep the notations and hypotheses of Section 5. The *stable category* $\underline{\mathcal{A}}$ associated with $\mathcal{A}$ has the same objects as $\mathcal{A}$. A *morphism* of $\underline{\mathcal{A}}$ is the equivalence class $\bar{f}$ of a morphism $f\colon A \to B$ of $\mathcal{A}$ modulo the subgroup of morphisms factoring through an injective of $\mathcal{A}$. The *composition* of $\underline{\mathcal{A}}$ is induced by that of $\mathcal{A}$.

EXAMPLE 6.1. *The homotopy category.* The *homotopy category* $\mathbf{H}(\mathcal{A})$ of an additive category $\mathcal{A}$ is by definition the stable category of the category $\mathbf{C}(\mathcal{A})$ of complexes over $\mathcal{A}$ (cf. Example 4.3). So the objects of $\mathbf{H}(\mathcal{A})$ are complexes over $\mathcal{A}$ and the morphisms are homotopy classes of morphisms of complexes, by Example 5.3.

The stable category is an additive category and the projection functor $\mathcal{A} \to \underline{\mathcal{A}}$ is an additive functor. However, in general, $\underline{\mathcal{A}}$ does not carry an exact structure making the projection functor into an exact functor. Nonetheless, in order to keep track of the conflations of $\mathcal{A}$, we can endow $\underline{\mathcal{A}}$ with the following 'less rigid' structure:

First, we complete the assignment $A \mapsto SA$ to an endofunctor of $\underline{\mathcal{A}}$ by putting $S\bar{f} = \bar{h}$, where $h$ is any morphism fitting into a commutative diagram

$$\begin{array}{ccccc} A & \xrightarrow{i_A} & IA & \xrightarrow{p_A} & SA \\ {\scriptstyle f}\downarrow & & \downarrow{\scriptstyle g} & & \downarrow{\scriptstyle h} \\ B & \xrightarrow{i_B} & IB & \xrightarrow{p_B} & SB \end{array} .$$

Indeed, by the injectivity of $IB$, such diagrams exist. Clearly $\bar{h}$ does not depend on the choice of $g$.

Secondly, we associate with each conflation $\varepsilon = (i, p)$ of $\mathcal{A}$ a sequence

$$A \xrightarrow{\bar{i}} B \xrightarrow{\bar{p}} C \xrightarrow{\partial\varepsilon} SA$$

called a *standard triangle* and defined by requiring the existence of a commutative diagram

$$\begin{array}{ccccc} A & \xrightarrow{i} & B & \xrightarrow{p} & C \\ \| & & \downarrow{\scriptstyle g} & & \downarrow{\scriptstyle e} \\ A & \xrightarrow{i_A} & IA & \xrightarrow{p_A} & SA \end{array} , \qquad \partial\varepsilon = \bar{e}.$$

Again, $g$ exists by the injectivity of $IA$, and $\bar{e}$ is independent of the choice of $g$.

If $\mathcal{C}$ is an arbitrary category endowed with an endofunctor $S$: $\mathcal{C} \to \mathcal{C}$, an *$S$-sequence* is a sequence

$$X \xrightarrow{u} Y \xrightarrow{v} Z \xrightarrow{w} SX$$

of $\mathcal{C}$ and a *morphism of $S$-sequences* from $(u, v, w)$ to $(u', v', w')$ is a commutative diagram of the form

$$\begin{array}{ccccccc} X & \xrightarrow{u} & Y & \xrightarrow{v} & Z & \xrightarrow{w} & SX \\ {\scriptstyle x}\downarrow & & \downarrow & & \downarrow & & \downarrow{\scriptstyle Sx} \\ X' & \xrightarrow{u'} & Y' & \xrightarrow{v'} & Z' & \xrightarrow{w'} & SX' \end{array} .$$

With this terminology, we define a *triangle* of $\underline{\mathcal{A}}$ to be an $S$-sequence isomorphic to a standard triangle. A *morphism of triangles* is a morphism of the underlying $S$-sequences. Note that the standard triangle construction defines a *functor* from the category of conflations to the category of triangles.

THEOREM 6.2. *The category $\underline{\mathcal{A}}$ endowed with the suspension functor $S$ and the above triangles satisfies the following axioms* SP0–SP4.

SP0. Each $S$-sequence isomorphic to a triangle is itself a triangle.

SP1. For each object $X$, the $S$-sequence

$$0 \to X \xrightarrow{1_X} X \to S0$$

is a triangle.

SP2. If $(u, v, w)$ is a triangle, then so is $(v, w, -Su)$.

SP3. If $(u, v, w)$ and $(u', v', w')$ are triangles and $x, y$ morphisms such that $yu = u'x$, then there is a morphism $z$ such that

$$zv = v'y \quad \text{and} \quad (Sx)w = w'z.$$

$$\begin{array}{ccccccc} X & \xrightarrow{u} & Y & \xrightarrow{v} & Z & \xrightarrow{w} & SX \\ {\scriptstyle x}\downarrow & & {\scriptstyle y}\downarrow & & {\scriptstyle z}\downarrow & & \downarrow{\scriptstyle Sx} \\ X' & \xrightarrow{u'} & Y' & \xrightarrow{v'} & Z' & \xrightarrow{w'} & SX' \end{array}$$

SP4. For each pair of morphisms

$$X \xrightarrow{u} Y \xrightarrow{v} Z$$

there is a commutative diagram

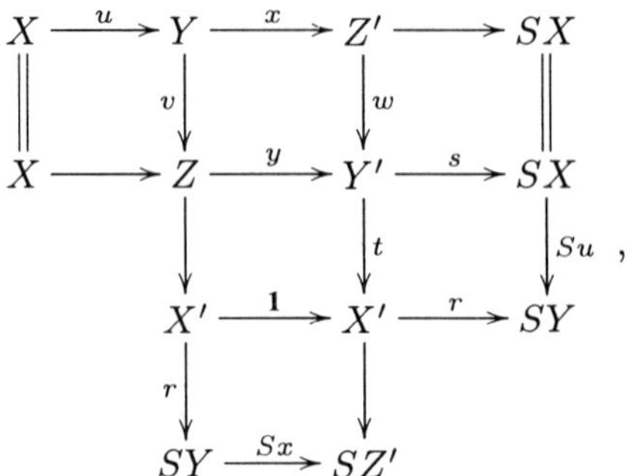


where the first two rows and the two central columns are triangles.

We refer to [26] for a proof of the theorem in the case where $\mathcal{A}$ is a Frobenius category. Property SP4 can be given a more symmetric form if we represent a morphism $X \to SY$ by the symbol

$$X \xrightarrow{+} SY$$

and write a triangle in the form

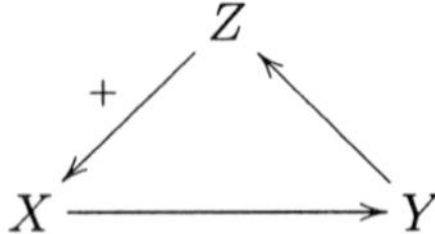


With this notation, the diagram of SP4 can be written as an octahedron in which 4 faces represent triangles. The other 4 as well as two of the 3 squares 'containing the center' are commutative.

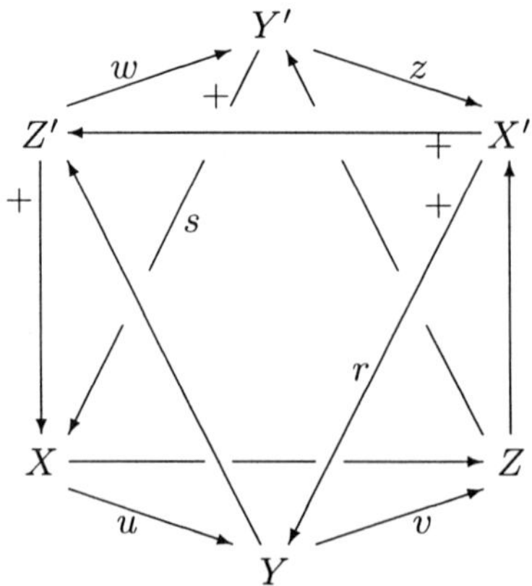

## 7. Suspended categories and triangulated categories

A *suspended category* [42] is an additive category $\mathcal{S}$ with an additive endofunctor $S$: $\mathcal{S} \to \mathcal{S}$ called the *suspension functor* and a class of $S$-sequences called *triangles* and satisfying the axioms SP0–SP4 of Section 6.

A *triangulated category* is a suspended category whose suspension functor is an equivalence.

By Theorem 6.2, the stable category of an exact category $\mathcal{A}$ with enough injectives is a suspended category. If $\mathcal{A}$ is even a Frobenius category, it is easy to see that $\underline{\mathcal{A}}$ is triangulated.

EXAMPLE 7.1. *The mapping cone*. Let $\mathcal{A}$ be an additive category. The homotopy category $\mathbf{H}(\mathcal{A})$ is triangulated (6.1). Here the suspension functor is even an automorphism. Axiom SP4 implies that for each morphism of complexes $f$: $X \to Y$, there is a triangle

$$X \xrightarrow{\bar{f}} Y \xrightarrow{\bar{g}} Z \xrightarrow{\bar{h}} SX.$$

Concretely, we can construct $Z$ as the *mapping cone* $Cf$ over $f$. It is defined as the cokernel of the conflation $[i_X\ f]^t$: $X \to IX \oplus Y$ and hence fits into a diagram

$$\begin{array}{ccccc} X & \xrightarrow{[i_X\ f]^t} & IX \oplus Y & \xrightarrow{[k\ g]} & Cf \\ \| & & \downarrow{\scriptstyle [\mathbf{1}\ 0]} & & \downarrow{\scriptstyle h} \\ X & \xrightarrow{i_X} & IX & \xrightarrow{p_X} & SX \end{array}.$$

The standard triangle provided by this diagram is clearly isomorphic to $(\bar{f}, \bar{g}, \bar{h})$. Explicitly

$$(Cf)^n = X^{n+1} \oplus Y^n, \quad d^n_{Cf} = \begin{bmatrix} -d_X^{n+1} & 0 \\ f^{n+1} & d_Y^n \end{bmatrix},$$

$$g_n = \begin{bmatrix} 0 \\ \mathbf{1} \end{bmatrix}, \quad h^n = [-\mathbf{1}\ 0].$$

The following properties of a suspended category $\mathcal{S}$ are easy consequences of the axioms. Proofs may be found in [29, 6, 38].

a) Each morphism $u$: $X \to Y$ can be embedded into a triangle $(u, v, w)$.

b) For each triangle

$$X \xrightarrow{u} Y \xrightarrow{v} Z \xrightarrow{w} SX$$

and each $V \in \mathcal{S}$, the induced sequence

$$\mathcal{S}(X,V) \leftarrow \mathcal{S}(Y,V) \leftarrow \mathcal{S}(Z,V) \leftarrow \mathcal{S}(SX,V) \leftarrow \mathcal{S}(SY,V) \cdots$$

is exact. In particular, $vu = wv = (Su)w = 0$.

c) If in axiom SP3 the morphisms $x$ and $y$ are invertible, then so is $z$.

d) If $(u, v, w)$ and $(u', v', w')$ are triangles, then so is

$$X \oplus X' \xrightarrow{u \oplus u'} Y \oplus Y' \xrightarrow{v \oplus v'} Z \oplus Z' \xrightarrow{w \oplus w'} S(X \oplus X').$$

e) If

$$X \xrightarrow{u} Y \xrightarrow{v} Z \xrightarrow{w} SX$$

is a triangle, the sequence

$$0 \to Y \xrightarrow{v} Z \xrightarrow{w} SX \to 0$$

is split exact iff $u = 0$.

f) For an arbitrary choice of the triangles starting with $u, v$ and $vu$ in axiom SP4, there are morphisms $w$ and $z$ such that the second central column is a triangle and the whole diagram is commutative.

Now suppose that $\mathcal{S}$ *is a triangulated category*. Then in addition we have

g) If $(v, w, -Su)$ is a triangle of $\mathcal{S}$, then so is $(u, v, w)$.

h) If $(u, v, w)$ and $(u', v', w')$ are triangles and $y, z$ morphisms such that $zv = v'y$, then there is a morphism $x$ such that $yu = u'x$ and $(Sx)w = w'z$.

i) For each triangle

$$X \xrightarrow{u} Y \xrightarrow{v} Z \xrightarrow{w} SX$$

and each $V \in \mathcal{T}$ the induced sequence

$$\mathcal{T}(V, X) \to \mathcal{T}(V, Y) \to \mathcal{T}(V, Z) \to \mathcal{S}(V, SX) \to \mathcal{S}(V, SY) \to \cdots$$

is exact.

This implies in particular that our notion of triangulated category coincides with that of [2], 1.1.

## 8. Triangle functors

We shall denote all suspension functors by the same letter $S$.

Let $\mathcal{S}$ and $\mathcal{T}$ be two suspended categories. A *triangle functor* from $\mathcal{S}$ to $\mathcal{T}$ is a pair consisting of an additive functor $F\colon \mathcal{S} \to \mathcal{T}$ and a morphism of functors $\alpha\colon FS \to SF$ such that

$$FX \xrightarrow{Fu} FY \xrightarrow{Fv} FZ \xrightarrow{(\alpha X)(Fw)} SFX$$

is a triangle of $\mathcal{T}$ for each triangle $(u, v, w)$ of $\mathcal{S}$. This implies that $\alpha$ is invertible, as we see by considering the case $Y = 0$ and using property b) of Section 7.

EXAMPLE 8.1. *Triangle functors induced by exact functors.* Let $\mathcal{A}$ and $\mathcal{B}$ be two exact categories with enough injectives. Let $F$: $\mathcal{A} \to \mathcal{B}$ be an exact functor preserving injectives. Then $F$ induces an additive functor $\underline{F}$: $\underline{\mathcal{A}} \to \underline{\mathcal{B}}$. For each $A \in \mathcal{A}$ define $\alpha A$ to be the class of a morphism $a$ fitting into a commutative diagram

$$\begin{array}{ccccc} FA & \xrightarrow{Fi_A} & FIA & \xrightarrow{Fp_A} & FSA \\ \| & & \downarrow & & \downarrow a \\ FA & \xrightarrow{i_{FA}} & IFA & \xrightarrow{p_{FA}} & SFA \end{array} .$$

Then $(\underline{F}, \alpha)$ is a triangle functor $\underline{\mathcal{A}} \to \underline{\mathcal{B}}$. This construction transforms compositions of exact functors to the compositions of the corresponding triangle functors.

A *morphism of triangle functors* $(F, \alpha) \to (G, \beta)$ is a morphism of functors $\mu$: $F \to G$ such that the square

$$\begin{array}{ccc} FS & \xrightarrow{\alpha} & SF \\ {\scriptstyle \mu S}\downarrow & & \downarrow{\scriptstyle S\mu} \\ GS & \xrightarrow{\beta} & SG \end{array}$$

is commutative. A triangle functor $(F, \alpha)$: $\mathcal{S} \to \mathcal{T}$ is a *triangle equivalence* if there exists a triangle functor $(G, \beta)$: $\mathcal{T} \to \mathcal{S}$ such that the *composed triangle functors* $(GF, (\beta F)(G\alpha))$ and $(FG, (\alpha G)(F\beta))$ are isomorphic to the *identical triangle functors* $(\mathbf{1}_{\mathcal{S}}, \mathbf{1}_S)$ and $(\mathbf{1}_{\mathcal{T}}, \mathbf{1}_S)$, respectively.

LEMMA 8.2. *A triangle functor $(F, \alpha)$ is a triangle equivalence iff $F$ is an equivalence of the underlying categories.*

Let $(R, \rho)$: $\mathcal{S} \to \mathcal{T}$ and $(L, \lambda)$: $\mathcal{T} \to \mathcal{S}$ be two triangle functors such that $L$ is left adjoint to $R$. Let $\Psi$: $\mathbf{1}_{\mathcal{T}} \to RL$ and $\Phi$: $LR \to \mathbf{1}_{\mathcal{S}}$ be two 'compatible' adjunction morphisms, i.e. we have $(\Phi L)(L\Psi) = \mathbf{1}_L$ and $(R\Phi)(\Psi R) = \mathbf{1}_R$. For $X \in \mathcal{T}$ and $Y \in \mathcal{S}$, denote by $\mu(X, Y)$ the canonical bijection

$$\mathcal{S}(LX, Y) \to \mathcal{T}(X, RY), \quad f \mapsto (Rf)(\Psi X).$$

Then it is not hard to see that the following conditions are equivalent

i) $\lambda = (\Phi SL)(L\rho^{-1}L)(LS\Psi)$,

ii) $\rho^{-1} = (RS\Phi)(R\lambda R)(\Psi SR)$,

iii) $\Phi S = (S\Phi)(\lambda R)(L\rho)$,

iv) $S\Psi = (\rho L)(R\lambda)(\Psi S)$,

v) The following diagram is commutative.

$$\begin{array}{ccccc} \mathcal{S}(LX,Y) & \xrightarrow{S} & \mathcal{S}(SLX,SY) & \xrightarrow{\lambda^*} & \mathcal{S}(LSX,SY) \\ \downarrow{\scriptstyle \mu(X,Y)} & & & & \downarrow{\scriptstyle \mu(SX,SY)} \\ \mathcal{T}(X,RY) & \xrightarrow{S} & \mathcal{T}(SX,SRY) & \xleftarrow{\rho_*} & \mathcal{T}(SX,RSY) \end{array} .$$

If they are fulfilled, we say that $\Phi$ and $\Psi$ are *compatible triangle adjunction morphisms* and that $(L, \lambda)$ is a *left triangle adjoint* of $(R, \rho)$.

LEMMA 8.3. *Let $\mathcal{S}$ and $\mathcal{T}$ be triangulated categories, $(R, \rho)$: $\mathcal{S} \to \mathcal{T}$ a triangle functor, $L$ a left adjoint of $R, \Phi$: $LR \to \mathbf{1}_{\mathcal{S}}$ and $\Psi$: $\mathbf{1}_{\mathcal{T}} \to RL$ compatible adjunction morphisms and $\lambda = (\Phi SL)(L\rho^{-1}L)(LS\Psi)$. Then $(L, \lambda)$ is a triangle functor and is a left triangle adjoint of $(R, \rho)$.*

A proof is given in [40], 6.7.

EXAMPLE 8.4. *Infinite sums of triangles.* Let $\mathcal{T}$ be a triangulated category and $I$ a set. Suppose that each family $(X_i)_{i \in I}$ admits a direct sum $\bigoplus_{i \in I} X_i$ in $\mathcal{T}$. This amounts to requiring that the diagonal functor

$$D: \mathcal{T} \to \prod_{i \in I} \mathcal{T},$$

which with each object $X \in \mathcal{T}$ associates the constant family with value $X$, admits a left adjoint. Now the product category admits a canonical triangulated structure with suspension functor $S(X_i) = (SX_i)$, and $(D, \mathbf{1})$ is a triangle functor. Thus, by the lemma,

$$\bigoplus: \prod_{i \in I} \mathcal{T} \to \mathcal{T}$$

can be completed to a triangle functor. Loosely speaking this means that sums of families of triangles indexed by $I$ are still triangles.

## 9. Localization of categories

If $\mathcal{C}$ and $\mathcal{D}$ are categories, we will denote by $\mathrm{Hom}(\mathcal{C}, \mathcal{D})$ the category of functors from $\mathcal{C}$ to $\mathcal{D}$. Note that in general, the morphisms between two functors do not form a set but only a 'class'. A category $\mathcal{C}$ will be called *large* to point out that the morphisms between fixed objects are not assumed to form a set.

Let $\mathcal{C}$ be a category and $\Sigma$ a class of morphisms of $\mathcal{C}$. There always exists [15], I, 1, a large category $\mathcal{C}[\Sigma^{-1}]$ and a functor $Q$: $\mathcal{C} \to \mathcal{C}[\Sigma^{-1}]$ which is 'universal' among the functors making the elements of $\Sigma$ invertible, that is to say that, for each category $\mathcal{D}$, the functor

$$\mathrm{Hom}(Q, \mathcal{D}): \ \mathrm{Hom}\big(\mathcal{C}[\Sigma^{-1}], \mathcal{D}\big) \to \mathrm{Hom}(\mathcal{C}, \mathcal{D})$$

induces an isomorphism onto the full subcategory of functors making the elements of $\Sigma$ invertible.

Now suppose that $\Sigma$ *admits a calculus of left fractions*, i.e.

F1. The identity of each objects is in $\Sigma$.

F2. The composition of two elements of $\Sigma$ belongs to $\Sigma$.

F3. Each diagram

$$X' \xleftarrow{s} X \xrightarrow{f} Y$$

with $s \in \Sigma$ can be completed to a commutative square

$$\begin{array}{ccc} X & \xrightarrow{f} & Y \\ {\scriptstyle s}\downarrow & & \downarrow{\scriptstyle t} \\ X' & \xrightarrow{f'} & Y' \end{array}$$

with $t \in \Sigma$.

F4. If $f, g$ are morphisms and there exists $s \in \Sigma$ such that $fs = gs$, then there exists $t \in \Sigma$ such that $tf = tg$.

Then the category $\mathcal{C}[\Sigma^{-1}]$ admits the following simple description: The *objects* of $\mathcal{C}[\Sigma^{-1}]$ are the objects of $\mathcal{C}$. The *morphisms* $X \to Y$ of $\mathcal{C}[\Sigma^{-1}]$ are the equivalence classes of diagrams

$$X \xrightarrow{f} Y' \xleftarrow{s} Y,$$

where by definition $(s, f)$ is *equivalent* to $(t, g)$ if there exists a commutative diagram

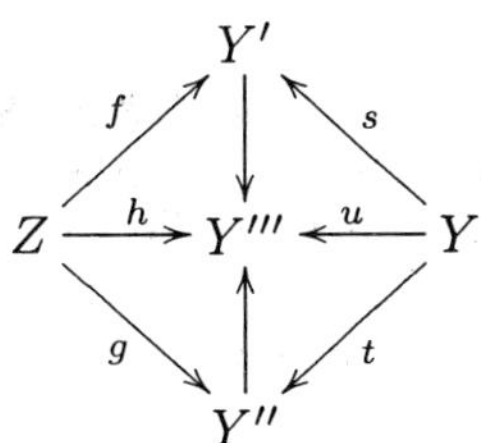


such that $u \in \Sigma$. Let $(s \mid f)$ denote the equivalence class of $(s, f)$. We define the *composition* of $\mathcal{C}[\Sigma^{-1}]$ by

$$(s \mid f) \circ (t \mid g) = (s't \mid g'f),$$

where $s'$ and $g'$ fit into the following commutative diagram, which exists by F3:

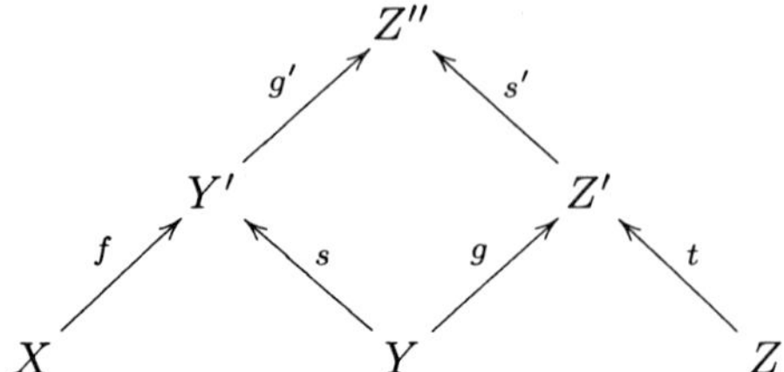

One easily verifies that $\mathcal{C}[\Sigma^{-1}]$ is indeed a category, that the quotient functor

$$Q\colon \mathcal{C} \to \mathcal{C}[\Sigma^{-1}], \quad X \mapsto X, \ f \mapsto (\mathbf{1} \mid f)$$

makes the elements of $\Sigma$ invertible (the inverse of $(\mathbf{1} \mid s)$ is $(s \mid \mathbf{1})$), and that it does have the universal property stated above (cf. [15]).

If $\Sigma$ also *admits a calculus of right fractions* (i.e. the duals of F1–F4 are satisfied), the dual of the above construction yields a category, which, by the universal property, is canonically isomorphic to $\mathcal{C}[\Sigma^{-1}]$.

Now let $\mathcal{B} \subset \mathcal{C}$ be a full subcategory. Denote by $\Sigma \cap \mathcal{B}$ the class of morphisms of $\mathcal{B}$ lying in $\Sigma$. We say that $\mathcal{B}$ *is right cofinal in* $\mathcal{C}$ *with respect to* $\Sigma$, if for each morphism $s\colon X' \to X$ of $\Sigma$ with $X' \in \mathcal{B}$, there is a morphism $m\colon X \to X''$ such that the composition $ms$ belongs to $\Sigma \cap \mathcal{B}$. The *left* variant of this property is defined dually.

LEMMA 9.1. *The class $\Sigma \cap \mathcal{B}$ admits a calculus of left fractions. If $\mathcal{B}$ is right cofinal in $\mathcal{C}$ w.r.t. $\Sigma$, the canonical functor*

$$\mathcal{B}[(\Sigma \cap \mathcal{B})^{-1}] \to \mathcal{C}[\Sigma^{-1}]$$

*is fully faithful.*

## 10. Localization of triangulated categories

Let $\mathcal{T}$ be a triangulated category and $\mathcal{N} \subset \mathcal{T}$ a *full suspended subcategory*, i.e. a full additive subcategory such that $S\mathcal{N} \subset \mathcal{N}$ and $\mathcal{N}$ is *closed under extensions*, i.e. if the terms $X$ and $Z$ of triangle $(X, Y, Z)$ belong to $\mathcal{N}$, then so does $Y$. We say that $\mathcal{N}$ is a *full triangulated subcategory* if we also have $\Sigma^{-1}\mathcal{N} \subset \mathcal{N}$.

Let $\Sigma$ be the class of morphisms $s$ of $\mathcal{T}$ occurring in a triangle

$$N \to X \xrightarrow{s} X' \to SN,$$

with $N \in \mathcal{N}$.

LEMMA 10.1. *The class $\Sigma$ is a multiplicative system with $S\Sigma \subset \Sigma$. Moreover, if, in the setting of axiom* SP3 (*Section* 6), *the morphisms $x$ and $y$ belong to $\Sigma$, then $z$ may be found in $\Sigma$. If $N$ is a full triangulated subcategory of $\mathcal{T}$, we have $S^{-1}\Sigma \subset \Sigma$.*

The localization $\mathcal{T}[\Sigma^{-1}]$ is an additive category and the quotient functor $Q$: $\mathcal{T} \to \mathcal{T}[\Sigma^{-1}]$ an additive functor (by [15], I, 3.3). We endow it with the *suspension functor* $S$ induced by $S$: $\mathcal{T} \to \mathcal{T}$. We declare the *triangles* of $\mathcal{T}[\Sigma^{-1}]$ to be those $S$-sequences which are isomorphic to images of triangles of $\mathcal{T}$ under the quotient functor.

By SP1 and SP2, the morphisms $N \to 0$ with $N \in \mathcal{N}$ belong to $\Sigma$. Thus the quotient functor annihilates $\mathcal{N}$. We define

$$\mathcal{T}/\mathcal{N} := \mathcal{T}[\Sigma^{-1}].$$

If $\mathcal{S}$ is a suspended category, denote by $\mathrm{Hom}_{\mathrm{tria}}(\mathcal{T}, \mathcal{S})$ the large category of triangle functors from $\mathcal{T}$ to $\mathcal{S}$.

PROPOSITION 10.2. *The category $\mathcal{T}/\mathcal{N}$ endowed with the above structure becomes a suspended category and $(Q, \mathbf{1})$: $\mathcal{T} \to \mathcal{T}/\mathcal{N}$ a triangle functor. For each suspended category $\mathcal{S}$, the functor*

$$\mathrm{Hom}_{\mathrm{tria}}(Q, \mathcal{S})\colon \mathrm{Hom}_{\mathrm{tria}}(\mathcal{T}/\mathcal{N}, \mathcal{S}) \to \mathrm{Hom}_{\mathrm{tria}}(\mathcal{T}, \mathcal{S})$$

*induces an isomorphism onto the full subcategory of triangle functors annihilating $\mathcal{N}$. If $\mathcal{N} \subset \mathcal{T}$ is a full triangulated subcategory, then $\mathcal{T}/\mathcal{N}$ is triangulated.*

Let $\mathcal{S} \subset \mathcal{T}$ be a full triangulated subcategory. If

$$N \to X \to X' \to SN$$

is a triangle with $N \in \mathcal{N}$ and $X, X' \in \mathcal{S}$, then $N$ lies in $\mathcal{S} \cap \mathcal{N}$ as an extension of $X$ by $S^{-1}X'$. So the multiplicative system of $\mathcal{S}$ defined by $\mathcal{S} \cap \mathcal{N}$ coincides with $\Sigma \cap \mathcal{S}$.

LEMMA 10.3. *If each morphism $N \to X'$ with $N \in \mathcal{N}$ and $X' \in \mathcal{S}$ admits a factorization $N \to N' \to X'$ with $N' \in \mathcal{N} \cap \mathcal{S}$, then $\mathcal{S}$ is right cofinal w.r.t. $\Sigma$. In particular, the canonical functor $\mathcal{S}/\mathcal{S} \cap \mathcal{N} \to \mathcal{T}/\mathcal{N}$ is fully faithful.*

## 11. Derived categories

Let $\mathcal{A}$ be an exact category (cf. Section 4). A complex $N$ over $\mathcal{A}$ is *acyclic in degree $n$* if $d_N^{n-1}$ factors as

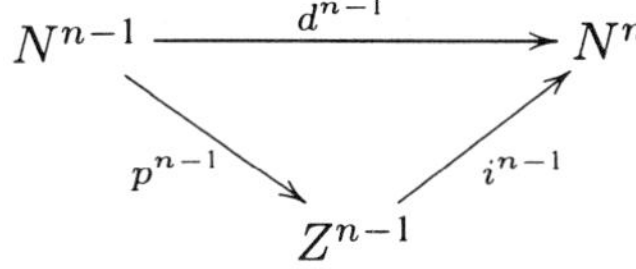


where $p^{n-1}$ is a cokernel for $d^{n-2}$ and a deflation, and $i^{n-1}$ is a kernel for $d^n$ and an inflation. The complex $N$ is *acyclic* if it is acyclic in each degree.

EXAMPLE 11.1. *$\mathcal{A}$ abelian.* Then $N$ is acyclic in degree $n$ iff $\mathrm{H}^n(N) = 0$.

EXAMPLE 11.2. *Null-homotopic complexes.* Let $R$ be an associative ring with 1 and $e \in R$ an idempotent. Let $\mathcal{A}$ be the exact category of free $R$-modules (cf. Example 4.1). The 'periodic' complex

$$\cdots \xrightarrow{1-e} R \xrightarrow{e} R \xrightarrow{1-e} R \xrightarrow{e} \cdots$$

is acyclic iff $\operatorname{Ker} e$ and $\operatorname{Ker}(1-e)$ are free $R$-modules. Note, however, that this complex is always null-homotopic. If $\mathcal{A}$ is any exact category it is easy to see that the following are equivalent

i) Each null-homotopic complex is acyclic.

ii) Idempotents split in $\mathcal{A}$, i.e. $\operatorname{Ker} e$ and $\operatorname{Ker}(1-e)$ exist for each idempotent $e\colon A \to A$ of $\mathcal{A}$.

iii) The class of acyclic complexes is closed under isomorphism in $\mathbf{H}(\mathcal{A})$.

Denote by $\mathcal{N}$ the full subcategory of $\mathbf{H}(\mathcal{A})$ formed by the complexes which are isomorphic to acyclic complexes.

LEMMA 11.3. *$\mathcal{N}$ is a full triangulated subcategory of $\mathbf{H}(\mathcal{A})$.*

The morphisms $\bar{s}$ of $\mathbf{H}(\mathcal{A})$ occurring in triangles $N \to X \xrightarrow{\bar{s}} X' \to SN$ with $N \in \mathcal{N}$ are called *quasi-isomorphisms*. If $\mathcal{A}$ is abelian, a morphism $\bar{s}$ is a quasi-isomorphism if and only if $\mathbf{H}^n(\bar{s})$ is invertible for each $n \in \mathbf{Z}$. By definition (cf. Section 10) the multiplicative system $\Sigma$ associated with $\mathcal{N}$ is formed by all quasi-isomorphisms. The *derived category of $\mathcal{A}$* is the localization (cf. Section 10)

$$\mathbf{D}(\mathcal{A}) := \mathbf{H}(\mathcal{A})/\mathcal{N} = \mathbf{H}(\mathcal{A})\left[\Sigma^{-1}\right].$$

EXAMPLE 11.4. *The abelian case.* If $\mathcal{A}$ is abelian, this definition of $\mathbf{D}(\mathcal{A})$ is identical with Verdier's [56].

EXAMPLE 11.5. *The split case.* If each conflation of $\mathcal{A}$ splits, we have $\mathcal{N} = 0$ and $\mathbf{H}(\mathcal{A}) \xrightarrow{\sim} \mathbf{D}(\mathcal{A})$.

Let

$$\varepsilon\colon X \xrightarrow{i} Y \xrightarrow{p} Z$$

be a sequence of complexes over $\mathcal{A}$ such that $(i^n, p^n)$ is a conflation for each $n \in \mathbf{Z}$. We will associate with $\varepsilon$ a functorial triangle of $\mathbf{D}(\mathcal{A})$ which coincides with the image of $(\bar{\imath}, \bar{p}, \partial\varepsilon)$ if $(i^n, p^n)$ is a split conflation for all $n \in \mathbf{Z}$ (cf. Section 5). Form a commutative diagram

$$\begin{array}{ccccc} X & \xrightarrow{[i_X\ i]^t} & IX \oplus Y & \xrightarrow{[k\ g]} & Ci \\ \| & & \downarrow{\scriptstyle [0\ 1]} & & \downarrow{\scriptstyle s} \\ X & \xrightarrow{i} & Y & \xrightarrow{p} & Z \end{array},$$

where $Ci$ is the mapping cone of Example 7.1. In the notations used there, the triangle determined by $\varepsilon$ is then

$$X \xrightarrow{Q\bar{\iota}} Y \xrightarrow{Q\bar{p}} Z \xrightarrow{Q\bar{h}(Q\bar{s})^{-1}} SX,$$

where $Q$ is the quotient functor and $(Qs)^{-1}$ is well defined by the

LEMMA 11.6. *The morphism $\bar{s}$ is a quasi-isomorphism.*

Let $\mathbf{C}^+(\mathcal{A})$, $\mathbf{C}^-(\mathcal{A})$ and $\mathbf{C}^b(\mathcal{A})$ be the full subcategories of $\mathbf{C}(\mathcal{A})$ formed by the complexes $A$ such that $A^n = 0$ for all $n \ll 0$, resp. $n \gg 0$, resp. all $n \gg 0$ and all $n \ll 0$. Let $\mathbf{H}^+(\mathcal{A})$, $\mathbf{H}^-(\mathcal{A})$ and $\mathbf{H}^b(\mathcal{A})$ be the images of these subcategories in $\mathbf{H}(\mathcal{A})$. Note that these latter subcategories are not closed under isomorphism in $\mathbf{H}(\mathcal{A})$. Nevertheless it is clear that their closures under isomorphism form full suspended subcategories (cf. Section 10) of $\mathbf{H}(\mathcal{A})$. For $* \in \{+,-,b\}$ we put

$$\mathbf{D}^*(\mathcal{A}) := \mathbf{H}^*(\mathcal{A})/\mathbf{H}^*(\mathcal{A}) \cap \mathcal{N}.$$

Note that we have canonical isomorphisms

$$\mathbf{H}^+(\mathcal{A}^{\mathrm{op}}) \xrightarrow{\sim} \mathbf{H}^-(\mathcal{A})^{\mathrm{op}}, \quad \mathbf{D}^+(\mathcal{A}^{\mathrm{op}}) \xrightarrow{\sim} \mathbf{D}^-(\mathcal{A})^{\mathrm{op}}$$

mapping a complex $A$ to the complex $B$ with $B^n = A^{-n}$ and $d_B^n = d_A^{-n-1}$.

LEMMA 11.7. *The canonical functors*

$$\mathbf{D}^*(\mathcal{A}) \to \mathbf{D}(\mathcal{A}), \quad * \in \{+,-,b\},$$

*induce equivalences onto the full subcategories of $\mathbf{D}(\mathcal{A})$ formed by the complexes which are acyclic in degree $n$ for all $n \ll 0$, resp. $n \gg 0$, resp. all $n \gg 0$ and all $n \ll 0$. The subcategory $\mathbf{H}^+(\mathcal{A})$ (resp. $\mathbf{H}^-(\mathcal{A})$) is right (resp. left) cofinal in $\mathbf{H}(\mathcal{A})$ w.r.t. the class of quasi-isomorphisms. The subcategory $\mathbf{H}^b(\mathcal{A})$ is right cofinal in $\mathbf{H}^-(\mathcal{A})$ w.r.t. the class of quasi-isomorphisms.*

## 12. Derived categories of fully exact subcategories

Let $\mathcal{A}$ be an exact category and $\mathcal{B} \subset \mathcal{A}$ a fully exact subcategory (cf. 4). Consider the conditions

C1. For each $A \in \mathcal{A}$ there is a conflation $A \to B \to A'$ with $B \in \mathcal{B}$.

C2. For each conflation $B \to A \to A'$ of $\mathcal{A}$ with $B \in \mathcal{B}$, there is a commutative diagram

$$\begin{array}{ccccc} B & \longrightarrow & A & \longrightarrow & A' \\ \| & & \downarrow & & \downarrow \\ B & \longrightarrow & B' & \longrightarrow & B'' \end{array}$$

whose second row is a conflation of $\mathcal{B}$.

Note that C2 is implied by C1 together with the following stronger condition: For each conflation $B \to B' \to A''$ of $\mathcal{A}$ with $B$ and $B'$ in $\mathcal{B}$, we have $A'' \in \mathcal{B}$.

THEOREM 12.1 (cf. [39], 4.1). a) *Suppose* C1 *holds. Then for each left bounded complex $A$ over $\mathcal{A}$, there is a quasi-isomorphism $A \to B$ for some left bounded complex $B$ over $\mathcal{B}$. In particular, the canonical functor $\mathbf{D}^+(\mathcal{B}) \to \mathbf{D}^+(\mathcal{A})$ is essentially surjective.*

b) *Suppose* C2 *holds. Then the category $\mathbf{H}^+(\mathcal{B})$ is right cofinal in $\mathbf{H}^+(\mathcal{A})$ w.r.t. the class of quasi-isomorphisms. In particular, the canonical functor $\mathbf{D}^+(\mathcal{B}) \to \mathbf{D}^+(\mathcal{A})$ is fully faithful.*

EXAMPLE 12.2. *Injectives.* If $\mathcal{A}$ has enough injectives (cf. 5), conditions C1 and C2 are obviously satisfied for the full subcategory $\mathcal{B} = \mathcal{I}$ formed by the injectives of $\mathcal{A}$ and endowed with the split conflations. Thus we have

$$\mathbf{D}^+(\mathcal{A}) \xleftarrow{\sim} \mathbf{D}^+(\mathcal{I}) \xleftarrow{\sim} \mathbf{H}^+(\mathcal{I}).$$

EXAMPLE 12.3. *Noetherian modules.* Let $R$ be a right noetherian ring and $\operatorname{mod} R$ the category of noetherian $R$-modules. The dual of Condition C2 is clearly satisfied for the fully exact subcategory $\operatorname{mod} R \subset \operatorname{Mod} R$. Thus the functor

$$\mathbf{D}^-(\operatorname{mod} R) \to \mathbf{D}^-(\operatorname{Mod} R)$$

is fully faithful.

EXAMPLE 12.4. *Filtered objects.* Let $\mathcal{E}$ be an exact category and $\mathcal{A}$ the category of sequences

$$A = \left(\cdots \to A^p \xrightarrow{f_A^p} A^{p+1} \to \cdots\right)$$

of morphisms of $\mathcal{E}$ with $A^p = 0$ for all $p \ll 0$ and $f_A^p$ invertible for all $p \gg 0$. Let $\mathcal{B} = \mathbf{F}(\mathcal{E})$ be the category of filtered objects over $\mathcal{E}$ (cf. Example 4.5). It is not hard to prove that $\mathcal{B}$ viewed as a fully exact subcategory of $\mathcal{A}$ satisfies the duals of C1 and C2.

## 13. Derived functors, restrictions, adjoints

Let $\mathcal{S}$ and $\mathcal{T}$ be triangulated categories, and $\mathcal{M} \subset \mathcal{S}$ and $\mathcal{N} \subset \mathcal{T}$ full triangulated subcategories (cf. Section 10). Let

$$(F, \varphi)\colon\ \mathcal{S} \to \mathcal{T}$$

be a triangle functor. We do *not* assume that $F\mathcal{M} \subset \mathcal{N}$. Hence in general, $F$ will not induce a functor $\mathcal{S}/\mathcal{M} \to \mathcal{T}/\mathcal{N}$. Nevertheless there often exists an 'approximation' to such an induced functor, namely a triangle functor $\mathbf{R}F\colon\ \mathcal{S}/\mathcal{M} \to \mathcal{T}/\mathcal{N}$ and a morphism of triangle functors can: $QF \to (\mathbf{R}F)Q$. We follow P. Deligne's approach [12] to the construction of $\mathbf{R}F$.

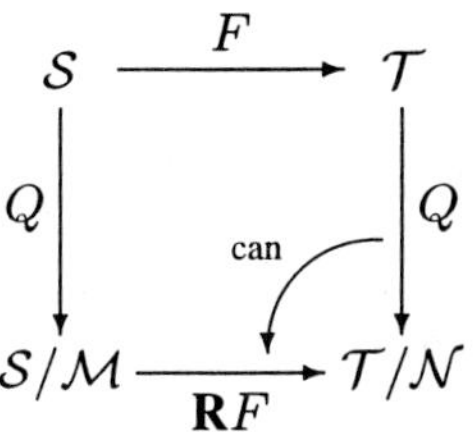


Let $\Sigma$ be the multiplicative system associated with $\mathcal{M}$ (cf. Section 10). Let $Y$ be an object of $\mathcal{S}/\mathcal{M}$. We define a contravariant functor $\mathbf{r}FY$ from $\mathcal{T}/\mathcal{N}$ to the category of abelian groups as follows. The value of $\mathbf{r}FY$ at $X \in \mathcal{T}/\mathcal{N}$ is formed by the equivalence classes $(f \mid s)$ of pairs

$$X \xrightarrow{f} FY', \quad Y' \xleftarrow{s} Y,$$

where $f \in (\mathcal{T}/\mathcal{N})(X, FY')$ and $s \in \Sigma$. Here, two pairs $(f, s)$ and $(g, t)$ are considered *equivalent* if there are commutative diagrams of $\mathcal{T}/\mathcal{N}$ and $\mathcal{S}$

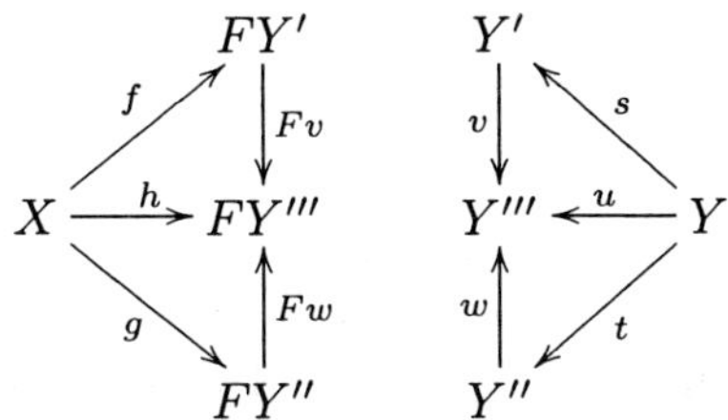


such that $u \in \Sigma$. We say that $\mathbf{R}FY$ is *defined at* $Y$ if $\mathbf{r}FY$ is a representable functor. In this case, we *define* $\mathbf{R}FY$ to be a representative of $\mathbf{r}FY$. So $\mathbf{R}FY$ is an object of $\mathcal{T}/\mathcal{N}$ endowed with an isomorphism

$$(\mathcal{T}/\mathcal{N})(?, \mathbf{R}FY) \xrightarrow{\sim} \mathbf{r}FY.$$

The datum of such an isomorphism is equivalent to the following more explicit data:

- For each $s$: $Y \to Y'$ of $\Sigma$, we have a morphism $\rho_s$: $FY' \to \mathbf{R}FY$ such that $\rho_u(Fv) = \rho_s$ whenever $u = vs$ belongs to $\Sigma$.
- There is some $s_0$: $Y \to Y_0'$ and a morphism $\sigma$: $\mathbf{R}FY \to FY_0'$ such that $(\mathbf{1}_{FY'} \mid s) = (\sigma\rho_s \mid s_0)$ for each $s$: $Y \to Y'$.

In fact, if the isomorphism is given, $\rho_s$ corresponds to the class $(\mathbf{1}_{FY'} \mid s)$ and $\mathbf{1}_{\mathbf{R}FY}$ to $(\sigma \mid s_0)$. Conversely, if the $\rho_s, s_0$, and $\sigma$ are given, the associated isomorphism maps $g$: $X \to (\mathbf{R}F)Y$ to $(\sigma g \mid s_0)$, and its inverse maps $(f \mid s)$ to $\rho_s f$.

If we view $Y$ as an object of $\mathcal{S}$, then, by definition, the *canonical morphism* can: $QFY \to (\mathbf{R}F)QY$ equals $\rho_s$ for $s = \mathbf{1}_Y$.

Let $\alpha = (t \mid g)$ be a morphism $Y \to Z$ of $\mathcal{S}/\mathcal{M}$. We define the *morphism* $\mathbf{r}F\alpha$: $\mathbf{r}FY \to \mathbf{r}FZ$ by

$$\mathbf{r}F\alpha(f \mid s) = \big((Fg')f \mid s't\big),$$

where $s'$ and $g'$ fit into a commutative diagram

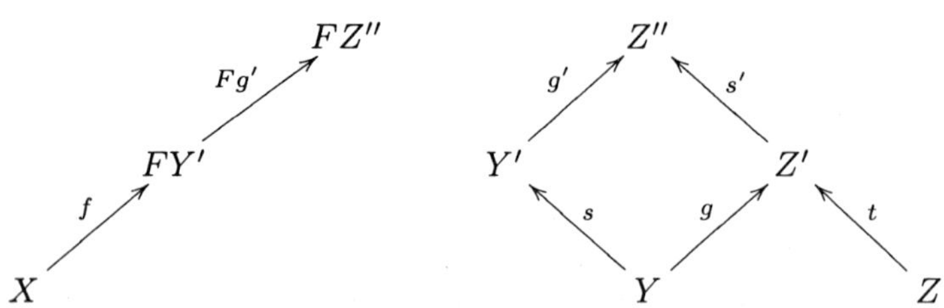


which exists by F3 (cf. Section 9). One easily verifies that this makes $\mathbf{r}F$ into a functor from $\mathcal{S}/\mathcal{M}$ to the category of functors from $\mathcal{T}/\mathcal{N}$ to the category of abelian groups. Now suppose that $\mathbf{R}F$ is defined at $Y$ and $Z$. We define *the morphism* $\mathbf{R}F\alpha$ by the commutative diagram

$$\begin{array}{ccc} (\mathcal{T}/\mathcal{N})(?, \mathbf{r}FY) & \xrightarrow{\sim} & \mathbf{r}FY \\ {\scriptstyle (\mathbf{R}F\alpha)_*}\downarrow & & \downarrow{\scriptstyle \mathbf{r}F\alpha} \\ (\mathcal{T}/\mathcal{N})(?, \mathbf{r}FZ) & \xrightarrow{\sim} & \mathbf{r}FZ \end{array} .$$

Thus $\mathbf{R}F$ becomes a *functor* $\mathcal{U} \to \mathcal{T}/\mathcal{N}$, where $\mathcal{U}$ denotes the full subcategory formed by the objects at which $\mathbf{R}F$ is defined. Suppose that $\mathbf{R}F$ is defined at $Y \in \mathcal{S}/\mathcal{M}$. The following chain of isomorphisms shows that $\mathbf{R}F$ is defined at $SY$ and that $\varphi$: $FS \to SF$ yields a *canonical morphism* $\mathbf{R}\varphi$: $(\mathbf{R}F)S \to S(\mathbf{R}F)$

$$(\mathbf{r}F)(SY) \xleftarrow{\sim} \mathbf{r}(FS)(Y) \xrightarrow{\sim} \mathbf{r}(SF)(Y) \xleftarrow{\sim} (\mathcal{T}/\mathcal{N})(?, S\mathbf{R}FY).$$

PROPOSITION 13.1 (cf. [12], 1.2). *If*

$$X \xrightarrow{u} Y \xrightarrow{v} Z \xrightarrow{w} SX$$

*is a triangle of* $\mathcal{S}/\mathcal{M}$ *and* $\mathbf{R}F$ *is defined at* $X$ *and* $Z$, *then it is defined at* $Y$. *In this case* $(\mathbf{R}Fu, \mathbf{R}Fv, (\mathbf{R}\varphi)(X)\mathbf{R}Fw)$ *is a triangle of* $\mathcal{T}/\mathcal{N}$.

In particular, $\mathcal{U}$ is a triangulated subcategory of $\mathcal{S}/\mathcal{M}$ and $(\mathbf{R}F, \mathbf{R}\varphi)$: $\mathcal{U} \to \mathcal{T}/\mathcal{N}$ is a triangle functor. It is called the *right derived functor* of $(F, \varphi)$ (with respect to $\mathcal{M}$ and $\mathcal{N}$).

Let $I$ denote the inclusion of the preimage of $\mathcal{U}$ in $\mathcal{S}$.

LEMMA 13.2. *The canonical morphism* can: $QFI \to (\mathbf{R}F)QI$ *is a morphism of triangle functors.*

The *left derived functor* $(\mathbf{L}F, \mathbf{L}\varphi)$ *of* $(F, \varphi)$ is defined dually: For $X \in \mathcal{S}/\mathcal{M}$, one defines a covariant functor $\mathbf{l}FX$ whose value at $Y \in \mathcal{T}/\mathcal{N}$ is formed by the equivalence classes $(s \mid f)$ of pairs

$$X \xleftarrow{s} X', \quad FX' \xrightarrow{f} Y,$$

where $f \in (\mathcal{T}/\mathcal{N})(FX', Y)$ and $s \in \Sigma \cdots$. The *canonical morphism* can: $QFX \to \mathbf{L}FQX$ corresponds to the class $(\mathbf{1}_X \mid \mathbf{1}_{FX}) \ldots$.

EXAMPLE 13.3. *Induced functors.* If we have $F\mathcal{M} \subset \mathcal{N}$, then $\mathbf{R}F$ and $\mathbf{L}F$ are isomorphic to the triangle functor $\mathcal{S}/\mathcal{M} \to \mathcal{T}/\mathcal{N}$ induced by $F$, and can: $QF \to \mathbf{R}FQ$ and can: $\mathbf{L}FQ \to QF$ are isomorphisms.

Suppose that $(F', \varphi')$ is another triangle functor and $\mu$: $F \to F'$ a morphism of triangle functors. Then for each $Y \in \mathcal{S}/\mathcal{M}$, the morphism $\mu$ induces a morphism

$$\mathbf{r}\mu\colon \mathbf{r}FY \to \mathbf{r}F'Y$$

and hence a morphism $\mathbf{R}\mu$: $\mathbf{R}FY \to \mathbf{R}F'Y$ if both, $\mathbf{R}F$ and $\mathbf{R}F'$, are defined at $Y$. Note that the assignments $\mu \mapsto \mathbf{r}\mu$ and $\mu \mapsto \mathbf{R}\mu$ are compatible with compositions.

LEMMA 13.4. *The morphism* $\mathbf{R}\mu$ *is a morphism of triangle functors between the restrictions of* $\mathbf{R}F$ *and* $\mathbf{R}F'$ *to the intersection of their domains.*

Keep the above hypotheses and let $\mathcal{U} \subset \mathcal{S}$ be a full triangulated subcategory which is right cofinal in $\mathcal{S}$ with respect to $\Sigma$. Denote by $I$: $\mathcal{U} \to \mathcal{S}$ the inclusion functor. Recall from Lemma 10.3 that the induced functor $\mathbf{R}I$: $\mathcal{U}/(\mathcal{U} \cap \mathcal{M}) \to \mathcal{S}/\mathcal{M}$ is fully faithful.

LEMMA 13.5. *Let* $U \in \mathcal{U}$. *Then* $\mathbf{R}F$ *is defined at* $U$ *if and only if* $\mathbf{R}(FI)$ *is defined at* $U$ *and in this case the canonical morphism*

$$\mathbf{R}(FI)(U) \to \mathbf{R}F\mathbf{R}I(U)$$

*is invertible.*

Keep the above hypotheses. Let $(R, \rho)$: $\mathcal{S} \to \mathcal{T}$ be a triangle functor and $(L, \lambda)$: $\mathcal{T} \to \mathcal{S}$ a left triangle adjoint (cf. Section 8). Let $X \in \mathcal{T}/\mathcal{N}$ and $Y \in \mathcal{S}/\mathcal{M}$ be objects such that $\mathbf{L}L$ is defined at $X$ and $\mathbf{R}R$ is defined at $Y$.

LEMMA 13.6. *We have a canonical isomorphism*

$$\nu(X, Y)\colon\ \mathcal{S}/\mathcal{M}(\mathbf{L}LX, Y) \to \mathcal{T}/\mathcal{N}(X, \mathbf{R}RY).$$

*Moreover the diagram*

$$\begin{array}{ccccc}
\mathcal{S}/\mathcal{M}(\mathbf{L}LX, Y) & \xrightarrow{S} & \mathcal{S}/\mathcal{M}(S\mathbf{L}LX, SY) & \xrightarrow{(L\lambda)^*} & \mathcal{S}/\mathcal{M}(\mathbf{L}LSX, SY) \\
\downarrow{\scriptstyle \nu(X,Y)} & & & & \downarrow{\scriptstyle \nu(SX,SY)} \\
\mathcal{T}/\mathcal{N}(X, \mathbf{R}RY) & \xrightarrow{S} & \mathcal{T}/\mathcal{N}(SX, S\mathbf{R}RY) & \xleftarrow{(R\rho)_*} & \mathcal{T}/\mathcal{N}(SX, \mathbf{R}RSY)
\end{array}$$

*is commutative.*

In particular, if $\mathbf{R}R$ and $\mathbf{L}L$ are defined everywhere, then $\mathbf{L}L$ is a left triangle adjoint of $\mathbf{R}R$ (cf. Section 8).

## 14. Split objects, compositions of derived functors

Keep the hypotheses of Section 13. An object $Y$ of $\mathcal{S}$ is *$F$-split* with respect to $\mathcal{M}$ and $\mathcal{N}$ if $\mathbf{R}F$ is defined at $Y$ and the canonical morphism $FY \to \mathbf{R}FY$ of $\mathcal{T}/\mathcal{N}$ is invertible.

LEMMA 14.1. *The following are equivalent*

i) *$Y$ is $F$-split.*

ii) *For each morphism $s\colon Y \to Y'$ of $\Sigma$, the morphism $QFs$ admits a retraction (= left inverse).*

iii) *For each morphism $f\colon M \to Y$ of $\mathcal{S}$ with $M \in \mathcal{M}$, the morphism $Ff$ factors through an object of $\mathcal{N}$.*

Let $Y_0$ be an object of $\mathcal{S}$. If there is a morphism $s_0\colon Y_0 \to Y$ of $\Sigma$ with $F$-split $Y$, then $\mathbf{R}F$ is defined at $Y_0$ and we have

$$\mathbf{R}FY_0 \xrightarrow{\sim} \mathbf{R}FY \xleftarrow{\sim} FY.$$

Indeed, this is clear since $\mathbf{r}F(s_0 \mid \mathbf{1}_Y)$ provides an isomorphism $\mathbf{r}FY_0 \xrightarrow{\sim} \mathbf{r}FY$.

We say that $\mathcal{S}$ has *enough $F$-split objects* (with respect to $\mathcal{M}$ and $\mathcal{N}$) if, for each $Y_0 \in \mathcal{S}$, there is a morphism $s_0\colon Y_0 \to Y$ of $\Sigma$ with $F$-split $Y$. In this case $\mathbf{R}F$ is defined at each object of $\mathcal{S}/\mathcal{M}$.

Let $\mathcal{R}$ be another triangulated category, $\mathcal{L} \subset \mathcal{R}$ a full triangulated subcategory and $G\colon \mathcal{R} \to \mathcal{S}$ a triangle functor. Suppose that for each object $Z_0$ of $\mathcal{R}$, the multiplicative system defined by $\mathcal{L}$ contains a morphism $Z_0 \to Z$ such that $Z$ is $G$-split and $GZ$ is $F$-split.

LEMMA 14.2. *The functor $\mathbf{R}G$ is defined on $\mathcal{R}/\mathcal{L}$, the functor $\mathbf{R}F$ is defined at each $\mathbf{R}GZ_0$, $Z_0 \in \mathcal{R}/\mathcal{L}$, and we have a canonical isomorphism of triangle functors*

$$\mathbf{R}(GF) \xrightarrow{\sim} \mathbf{R}G\mathbf{R}F.$$

## 15. Derived functors between derived categories

Let $\mathcal{A}$ and $\mathcal{C}$ be exact categories and $F\colon \mathcal{A} \to \mathcal{C}$ an additive (but not necessarily exact) functor. Clearly $F$ induces a triangle functor $\mathbf{H}(\mathcal{A}) \to \mathbf{H}(\mathcal{C})$, which will be denoted by $(F, \varphi)$. The construction of Section 13 then yields the *right derived functor* $(\mathbf{R}F, \mathbf{R}\varphi)$ *of $F$* defined on a full triangulated subcategory of $\mathbf{D}(\mathcal{A})$ and taking values in $\mathbf{D}(\mathcal{C})$. Similarly for the *left derived functor* $(\mathbf{L}F, \mathbf{L}\varphi)$.

If $\mathcal{C}$ is abelian, one defines the $n$-th right (resp. left) derived functor of $F$ by

$$\mathbf{R}^n FX = \mathrm{H}^n(\mathbf{R}FX) \quad (\text{resp. } \mathbf{L}_n FX = \mathrm{H}^{-n}(\mathbf{L}FX)),\ n \in \mathbf{Z}.$$

Typically, $\mathbf{R}F$ is defined on $\mathbf{D}^+(\mathcal{A})$. Lemma 13.5 and Lemma 11 then show that the restriction of $\mathbf{R}F$ to $\mathbf{D}^+(\mathcal{A})$ coincides with the derived functor of the restriction of $F$ to $\mathbf{H}^+(\mathcal{A})$.

An object $A \in \mathcal{A}$ is called *(right) $F$-acyclic* if $A$ viewed as a complex concentrated in degree zero is a (right) $F$-split object of $\mathbf{H}(\mathcal{A})$. The following lemma is often useful for finding acyclic objects.

LEMMA 15.1. *Let $\mathcal{B} \subset \mathcal{A}$ be a fully exact subcategory satisfying condition* C2 *of Section* 12 *and such that the restriction of $F$ to $\mathcal{B}$ is an exact functor. Then $\mathcal{B}$ consists of right $F$-acyclic objects.*

EXAMPLE 15.2. *Injectives.* If $\mathcal{B}$ is the subcategory of the injectives of $\mathcal{A}$, then each conflation of $\mathcal{B}$ splits. So any additive functor restricts to an exact functor on $\mathcal{B}$. Hence an injective object is $F$-acyclic for any additive functor $F$.

Let $\mathcal{A}c \subset \mathcal{A}$ be the full subcategory formed by the $F$-acyclic objects.

LEMMA 15.3. *The category $\mathcal{A}c$ is a fully exact subcategory of $\mathcal{A}$ and satisfies condition* C2 *of Section* 12. *The restriction of $F$ to $\mathcal{A}c$ is an exact functor.*

Now suppose that $\mathcal{A}$ *admits enough (right) $F$-acyclic objects*, i.e. that for each $A \in \mathcal{A}$, there is a conflation

$$A \to B \to A'$$

with $F$-acylic $B$. This means that $\mathcal{A}c$ satisfies condition C1 of Section 12. Hence for each $X \in \mathbf{H}^+(\mathcal{A})$ there is a quasi-isomorphism $X \to X'$ with $X' \in \mathbf{H}^+(\mathcal{A}c)$.

LEMMA 15.4. *The functor $\mathbf{R}F$ is defined on $\mathbf{D}^+(\mathcal{A})$. If $X$ a left bounded complex, we have $\mathbf{R}FX \xrightarrow{\sim} FX'$, where $X \to X'$ is a quasi-isomorphism with $X' \in \mathbf{H}^+(\mathcal{A}c)$. Each left bounded complex over $\mathcal{A}c$ is right $F$-split.*

EXAMPLE 15.5. *Injectives.* If $\mathcal{A}$ has enough injectives, it has enough $F$-acyclic objects for any additive functor $F$. The right derived functor is then computed by evaluating $F$ on an 'injective resolution' $X'$ of the complex $X$ constructed with the aid of Theorem 12.1.

Now let $R$: $\mathcal{A} \to \mathcal{C}$ be an additive functor and $L$: $\mathcal{C} \to \mathcal{A}$ a left adjoint. Suppose that $\mathcal{A}$ admits enough right $R$-acyclic objects and that $\mathcal{C}$ admits enough left $L$-acyclic objects. Then we have well defined derived functors $\mathbf{R}R$: $\mathbf{D}^+(\mathcal{A}) \to \mathbf{D}(\mathcal{C})$ and $\mathbf{L}L$: $\mathbf{D}^-(\mathcal{C}) \to \mathbf{D}(\mathcal{A})$.

LEMMA 15.6. *For $X \in \mathbf{D}^-(\mathcal{C})$ and $Y \in \mathbf{D}^+(\mathcal{A})$, we have a canonical isomorphism*

$$\nu(X,Y)\colon\ \mathbf{D}(\mathcal{A})(\mathbf{L}LX, Y) \xrightarrow{\sim} \mathbf{D}(\mathcal{C})(X, \mathbf{R}RY)$$

*compatible with the suspension functors as in Lemma* 13.6.

## References


[1] A.A. Beilinson, *Coherent sheaves on $\mathbf{P}^n$ and problems of linear algebra*, Funktsional. Anal. i Prilozhen. **12** (1978), 68–69. English translation: Functional Anal. Appl. **12** (1979), 214–216.

[2] A.A. Beilinson, J. Bernstein and P. Deligne, *Faisceaux pervers*, Astérisque **100** (1982).

[3] I.N. Bernstein, I.M. Gelfand and S.I. Gelfand, *Algebraic bundles on $\mathbf{P}^n$ and problems of linear algebra*, Funktsional. Anal. i Prilozhen. **12** (1978), 66–67. English translation: Functional Anal. Appl. **12** (1979), 212–214.

[4] J. Bernstein and V. Lunts, *Equivariant Sheaves and Functors*, SLNM 1578, Springer, Berlin (1994), 1–139.

[5] K. Bongartz, *Tilted algebras*, Representations of Algebras, Puebla 1980, SLNM 903, Springer, Berlin (1981), 26–38.

[6] A. Borel et al., *Algebraic D-modules*, Perspectives in Mathematics vol. 2, Academic Press, Boston (1987).

[7] S. Brenner and M.C.R. Butler, *Generalizations of the Bernstein–Gelfand–Ponomarev reflection functors*, Representation Theory II, Ottawa 1979, SLNM 832, Springer, Berlin (1980), 103–169.

[8] M. Broué, *Blocs, isométries parfaites, catégories dérivées*, C. R. Acad. Sci. Paris **307** (1988), 13–18.

[9] J.-L. Brylinski and M. Kashiwara, *Kazhdan–Lusztig conjecture and holonomic systems*, Invent. Math. **64** (1981), 387–410.

[10] R.-O. Buchweitz, *The Comparison Theorem*, Appendix to R.-O. Buchweitz, D. Eisenbud and J. Herzog, *Cohen–Macaulay modules on quadrics*, Singularities, Representations of Algebras, and Vector Bundles (Lambrecht 1985), SLNM 1273, Springer, Berlin (1987), 96–116.

[11] H. Cartan and S. Eilenberg, *Homological Algebra*, Princeton Univ. Press (1956).

[12] P. Deligne, *Cohomologie à supports propres,* Exposé XVII, SGA 4, SLNM 305, Springer, Berlin (1973), 252–480.

[13] P. Deligne, *La formule de dualité globale*, Exposé XVII in SGA 4, SLNM 305, Springer, Berlin (1973), 481–587.

[14] P. Deligne, Letter to D. Kazhdan and G. Lusztig dated 20 April 1979.

[15] P. Gabriel and M. Zisman, *Calculus of Fractions and Homotopy Theory*, Springer, Berlin (1967).

[16] W. Geigle and H. Lenzing, *A class of weighted projective curves arising in representation theory of finite dimensional algebras*, Singularities, Representations of Algebras, and Vector Bundles (Lambrecht 1985), SLNM 1273, Springer, Berlin (1987), 265–297.

[17] S.I. Gelfand and Yu.I. Manin, *Methods of Homological Algebra*, Vol. 1, Nauka, Moscow (1988) (in Russian), Math. Reviews 90k:18016.

[18] S.I. Gelfand and Yu.I. Manin, *Homological Algebra*, VINITI, Moscow (1989) (in Russian), Math. Reviews 92a:18003a (translated in [44]).

[19] R. Godement, *Topologie Algébrique et Théorie des Faisceaux*, Hermann, Paris (1958).

[20] M. Goresky and R. MacPherson, *Intersection homology theory*, Topology **19** (1980), 135–162.

[21] M. Goresky and R. MacPherson, *Intersection homology, II*, Invent. Math. **72** (1983), 77–130.

[22] P.P. Grivel, *Une démonstration du théorème de dualité de Verdier*, Enseign. Math. **31** (1985), 227–247.

[23] A. Grothendieck, *Sur quelques points d'algèbre homologique*, Tôhoku Math. J. **9** (1957), 119–221.

[24] A. Grothendieck, *The cohomology theory of abstract algebraic varieties*, Proc. Int. Math. Cong. Edinburgh (1958), 103–118.

[25] A. Grothendieck, *Résidus et dualité*, Prénotes pour un séminaire Hartshorne, Manuscript (1963).

[26] D. Happel, *On the derived category of a finite-dimensional algebra*, Comment. Math. Helv. **62** (1987), 339–389.

[27] D. Happel, *Triangulated Categories in the Representation Theory of Finite Dimensional Algebras*, London Math. Soc. Lecture Note Series vol. 119, Cambridge Univ. Press (1988).

[28] D. Happel and C.M. Ringel, *Tilted algebras*, Representations of Algebras, Puebla 1980, Trans. Amer. Math. Soc. **274** (1982), 399–443.

[29] R. Hartshorne, *Residues and Duality*, SLNM 20, Springer, Berlin (1966).

[30] A. Heller, *The loop space functor in homological algebra*, Trans. Amer. Math. Soc. **96** (1960), 382–394.

[31] P.J. Hilton and U. Stammbach, *A Course in Homological Algebra*, Graduate Texts in Math. vol. 4, Springer, New York (1971).

[32] L. Illusie, *Catégories dérivées et dualité: travaux de J.-L. Verdier*, Enseign. Math. (2) **36** (1990), 369–391.
[33] B. Iversen, *Cohomology of Sheaves*, Springer, New York (1986).
[34] M.M. Kapranov, *The derived categories of coherent sheaves on Grassmannians*, Funktsional. Anal. i Prilozhen. **17** (1983), 78–79. English translation: Functional Anal. Appl. **17** (1983), 145–146.
[35] M.M. Kapranov, *The derived category of coherent sheaves on a quadric*, Funktsional. Anal. i Prilozhen. **20** (1986), 67. English translation: Functional Anal. Appl. **20** (1986), 141–142.
[36] M.M. Kapranov, *On the derived categories of coherent sheaves on some homogeneous spaces*, Invent. Math. **92** (1988), 479–508.
[37] M. Kashiwara, *Algebraic study of systems of partial differential equations*, Thesis, University of Tokyo (1970).
[38] M. Kashiwara and P. Schapira, *Sheaves on Manifolds*, Grundlehren Math. Wiss. vol. 292, Springer, Berlin (1990).
[39] B. Keller, *Chain complexes and stable categories*, Manuscripta Math. **67** (1990), 379–417.
[40] B. Keller, *Derived categories and universal problems*, Comm. Algebra **19** (1991), 699–747.
[41] B. Keller, *Deriving DG categories*, Ann. Sci. École Norm. Sup. (4) **27** (1994), 63–102.
[42] B. Keller and D. Vossieck, *Sous les catégories dérivées*, C. R. Acad. Sci. Paris **305** (1987), 225–228.
[43] S. König, *Tilting complexes, perpendicular categories and recollements of derived module categories of rings*, J. Pure Appl. Algebra **73** (1991), 211–232.
[44] A.I. Kostrikin and I.R. Shafarevich (eds), *Algebra V: Homological Algebra*, with contributions by S.I. Gelfand and Yu.I. Manin, Encyclopedia of Math. Sci. vol. 38, Springer, New York (1994).
[45] S. MacLane, *Categories for the Working Mathematician*, Graduate Texts in Math. vol. 5, Springer, New York (1971).
[46] Z. Mebkhout, *Le Formalisme des Six Opérations de Grothendieck pour les $D_X$-modules Cohérents*, Travaux en Cours 35, Hermann, Paris (1989).
[47] D. Quillen, *Higher Algebraic $K$-theory, I*, SLNM 341, Springer, Berlin (1973), 85–147.
[48] J. Rickard, *Morita theory for derived categories*, J. London Math. Soc. (2) **39** (1989), 436–456.
[49] J. Rickard, *Derived categories and stable equivalence*, J. Pure Appl. Alg. **61** (1989), 303–317.
[50] J. Rickard, *Derived equivalences as derived functors*, J. London Math. Soc. (2) **43** (1991), 37–48.
[51] J. Rickard, *Lifting theorems for tilting complexes*, J. Algebra **142** (1991), 383–393.
[52] M. Saito, *On the derived category of mixed Hodge modules*, Proc. Japan Acad. (A) **62** (1986), 364–366.
[53] M. Sato, *Hyperfunctions and partial differential equations*, Proc. Intern. Conference on Functional Analysis and Related Topics, Tokyo 1969, Univ. Tokyo Press (1969), 91–94.
[54] L. Scott, *Simulating algebraic geometry with algebra, I. The algebraic theory of derived categories*, The Arcata Conference on Representations of Finite Groups (Arcata, Calif., 1986), Proc. Sympos. Pure Math. vol. 47 (1987), 271–281.
[55] J.-P. Serre, *Cohomologie et géométrie algébrique*, Proc. Int. Cong. Math. vol. III, Amsterdam (1954), 515–520.
[56] J.-L. Verdier, *Catégories dérivées, état* 0, SGA 4 1/2, SLNM 569, Springer, Berlin (1977), 262–311.
[57] J.-L. Verdier, *Le théorème de dualité de Poincaré*, C. R. Acad. Sci. Paris **256** (1963), 2084–2086.
[58] J.-L. Verdier, *Dualité dans la cohomologie des espaces localement compacts*, Séminaire Bourbaki 65/66 **300**, Benjamin (1966), 300-01–300-13.
[59] J.-L. Verdier, *Théorème de dualité pour la cohomologie des espaces localement compacts*, Dualité de Poincaré, Séminaire Heidelberg–Strasbourg 66/67, Publ. Inst. Rech. Math. Av. **3** (1969), exp. 4.
[60] J.-L. Verdier, *A duality theorem in the étale cohomology of schemes*, Conference on Local Fields, Nuffic Summer School held at Driebergen in 1966, Springer, Berlin (1967), 184–198.
[61] J.-L. Verdier, *Catégories dérivées*, Thèse.
[62] C.A. Weibel, *An Introduction to Homological Algebra*, Cambridge Studies in Advanced Mathematics vol. 38, Cambridge Univ. Press (1994).

# Section 3A
# Commutative Rings and Algebras

# Ideals and Modules

J.P. Lafon

*5, rue André Theuriet, 92340 Bourg La Reine, France*

*Contents*



HANDBOOK OF ALGEBRA, VOL. 1
Edited by M. Hazewinkel

## 1. Why not only ideals?

The theory of ideals in its present form was created by Dedekind [7] from the theory of the so called ideal numbers of Kummer. But, even then, he introduced the more general notion of *fractionary ideals* which are in fact modules and Kronecker [16] already used modules over the ring of polynomials.

Let us say, before giving more precise definitions, that we get modules by using the axioms of vector spaces but allowing the scalars to be in any ring, except that we shall assume here that it is a commutative one.

Modules do appear quite naturally.

Look for instance at Hilbert's *famous theorem* on syzygys:

The data are a (homogeneous) ideal $I$ of the ring $R = k[X_1, \dots, X_n]$ of polynomials where $k$ is a field and a finite system $\{f_1, \dots, f_r\}$ of generators of $I$.

The kernel of the map: $(g_1, \dots, g_r) \mapsto f_1g_1 + \cdots + f_rg_r$ of $R^n$ into $R$ is no longer an ideal but a submodule of the $R$-module $R^n$ called *the first module of syzygys of $I$.*

It is finitely generated. So we can repeat the same procedure with a finite system of generators of it and get *the second module of syzygys of $I$* and so on.

The theorem asserts we get 0 after at most $n$ such operations.

So very often, as in the above example, we need to consider together with an ideal $I$ of a ring $R$ the *$R$-module $R/I$.*

*Another reason to introduce modules is the linearization of some a priori nonlinear notions.*

A very significant example is the example of integral elements [30].

*Let $R$ be a ring and $S$ an overring. An element $x \in S$ is integral over $R$ if it is a root of a monic polynomial*

$$f(X) = X^n + a_{n-1}X^{n-1} + \cdots + a_0 \in R[X]$$

This does not seem at first to be a linear condition but you may translate it into the following linear one: *the $R$-module $R[x]$ is finitely generated.*

Using that condition you very easily get the main properties of integral elements, for instance: sum and product of integral elements are integral, integral dependence is transitive.

*The notion of modules is also important in representation theory.*

- Let $G$ be a group. In its simplest form, a linear complex representation $U$ of $G$ is a group morphism $g \mapsto U(g)$ of $G$ into the additive group of linear transformations of a $\mathbf{C}$-vector space, e.g., $\mathbf{C}^n$.
- It defines a structure of a *group with operators $G$* over $\mathbf{C}^n$; more precisely, we have a map $(g, x) \mapsto U(g)(x)$ from the product $G \times \mathbf{C}^n$ to $\mathbf{C}^n$ with convenient properties.
- Considering the group ring $R = \mathbf{C}[G]$, we get on $G$ the structure of an $R$-module.

*The extension to modules of natural notions or properties of ideals is called modulation.*

This process was quite alive in the sixties, allowing clearest proofs in commutative algebra. Here is an example:

Let $R$ be a noetherian local ring and $S = \widehat{R}$ its completion. If $I$ and $J$ are two ideals of $R$, we get an equality

$$(I \cap J)\widehat{R} = I\widehat{R} \cap J\widehat{R}.$$

This is a *formal* consequence of the fact that $\widehat{R}$ is a *flat* module over $R$. The same process gives the same result for instance when $R$ is the ring of germs of analytic functions and $S$ the ring of germs of indefinitely differentiable functions in a neighborhood of the origin of $R^n$.

We shall see too that *there exist very useful characterizations of rings by the properties of some types of modules over them, e.g., the simplest one: A ring $R$ is a field iff every module over it is free.*

After arguing a strong thesis, we may give a mild antithesis by asserting that *the most important notion is the notion of ideals.*

Some very interesting books or papers do not need and do not use the notion of modules: look, for instance at [23].

It is also worth saying that it is sometimes possible to get properties of a module from properties of an ideal. *Here is the principle of idealization of* Nagata ([22], p. 2).

Let $R$ be a ring and $M$ a module over $R$. On the additive group $S = R \oplus M$, define a product by

$$(r, x)(r', x') = (rr', rx' + r'x).$$

Thus one gets a commutative ring. The map $x \mapsto (0, x)$ identifies $M$ with an ideal of $S$ and every submodule of $M$ with an ideal contained in it.

For instance, M. Nagata gets the primary decomposition of modules from the primary decomposition of ideals which emerged at first in Lasker's works ([22], Exercise 1, p. 24).

## 2. Basic definitions

*Here is a short and nonexhaustive sketch of definitions and basic properties of modules over commutative rings, setting out the category of modules as an abelian category.*

Most of them extend easily to modules over noncommutative rings but some extensions are much more difficult!

Let us recall, for instance, that a ring $R$ is *simple* if it is $\neq 0$ and has no left ideal but 0 and $R$. A commutative ring is simple if it is a field but, if the ring is not commutative, it is a matrix ring over a skew field![1]

*Let $R$ be a commutative ring.*

DEFINITION 2.1. 1. An $R$-module is an abelian group $M$ (in additive notation) with an external product: $(a, x) \mapsto ax$ from $R \times M$ to $M$ such that

- $(\lambda + \mu)x = \lambda x + \mu x, \ \forall \lambda, \mu \in R, \ \forall x \in M.$
- $\lambda(x + y) = \lambda x + \lambda y, \ \forall \lambda \in R, \ \forall x, y \in M.$

[1] You need also, and it is not always difficult, to extend the notions to sheaves of modules over sheaves of rings but everything in its own time!

- $\lambda(\mu x) = (\lambda\mu)x, \ \forall\lambda, \mu \in R, \ \forall x \in M.$
- $1X = x, \forall x \in M.$

2. A morphism of the $R$-module $M$ into the $R$-module $M'$ is a map $f\colon\ M \to M'$ which is linear, i.e. such that

$$f(\lambda x + \mu y) = \lambda f(x) + \mu f(y), \quad \forall\lambda, \mu \in R, \ \forall x, y \in M.$$

3. A submodule of the $R$-module $M$ is an $R$-module $M'$ such that $M' \subset M$ and the natural injection $M' \to M$ is a morphism, i.e. an additive subgroup such that

$$\lambda \in R, \ x \in M' \Longrightarrow \lambda x \in M'.$$

4. We shall use the term overmodule, which is a little unusual, of a module $M'$ to indicate a module $M$ admitting $M'$ as a submodule.

*Examples*

- Let $I$ be a set. The set $R^I$ of maps from $I$ to $R$ has a natural structure of a module over $R$ where $f + g$ and $\lambda f$ are defined by

$$(f + g)(x) = f(x) + g(x), \quad (\lambda f)(x) = \lambda f(x), \quad \forall x \in I.$$

  You may write also

$$R^I = \{(x_i)_{i \in I} \mid x_i \in R\}.$$

- The support of $f \in R^I$ is the subset $\{i \in I \ / \ f(i) \neq 0\}$.
  The subset $R^{(I)}$ of $R^I$ of maps with finite support is a submodule of $R^I$ called the free $R$-module generated by the set $I$.
  If $I$ is finite $R^{(I)} = R^I$.
- A submodule of the $R$-module $R$ is an ideal of the ring $R$.
- The submodule of the $R$-module $M$ *generated by a subset* $X$ of $M$ is the intersection $s(X)$ of all the submodules containing $X$. It is also the submodule of linear combinations of elements of $X$.
- If $s(X) = M$ we say that $X$ is a *set of generators* of $M$.
- If there exists a *finite* set $X$ of generators of $M$, we say that $M$ is *finitely generated.*
  If $I$ is infinite, the $R$-module $R^I$ is not finitely generated.

*Here is a first characterization of rings by properties of modules:*

THEOREM 2.2 (Noetherian rings). *Every submodule of every finitely generated module over R is finitely generated iff every ideal of R is finitely generated.*

Such a ring $R$ is called noetherian.

*Morphisms*
The composite of two morphisms is a morphism. The identity map $1_M$ of a module $M$ is a morphism.

DEFINITION 2.3. A morphism $f\colon M \to M'$ is an isomorphism if there exists a morphism $g$ such that $g \circ f = 1_M$; $f \circ g = 1_{M'}$.

The morphism $f$ is an isomorphism if and only if the map $f$ from the *set* $M$ to the *set* $M'$ is bijective. We have then $g = f^{-1}$.

The composite of two isomorphisms is still an isomorphism. The identity map $1_M$ is an isomorphism.

DEFINITION 2.4 (*Kernel and Image*). Let $f$ be a morphism from the $R$-module $M$ to the $R$-module $M'$.

1. The kernel of $f$ is the submodule $\ker(f)$ of $M$

$$\ker(f) = \{x \in M \mid f(x) = 0\}.$$

2. The image of $f$ is the submodule $\operatorname{im}(f)$ of $M'$

$$\operatorname{im}(f) = \{f(x) \mid x \in M\}.$$

DEFINITIONS 2.5 (*Cokernel and Coimage*).

- Let $N$ be a submodule of the $R$-module $M$. We define on the quotient group $M/N$ a structure of an $R$-module by
  $\lambda\overline{x} = \overline{\lambda x}$ where $x$ is a representative of $\overline{x} \in M/N$.
  It is called the quotient module of $M$ by $N$.
- The cokernel of a morphism $f\colon M \to M'$ is the module $\operatorname{coker}(f) = M'/\operatorname{im}(f)$.
- The coimage of it is the quotient module $\operatorname{coim}(f) = M/\ker(f)$.

The morphism $f$ defines an isomorphism $\xi \mapsto f(x)$ from $\operatorname{coim}(f)$ to $\operatorname{im}(f)$, where $x$ is a representative of $\xi \in \operatorname{coim}(f)$.

DEFINITION 2.6 (*Exact sequences*).

- A sequence $M' \xrightarrow{f} M \xrightarrow{g} M''$ is exact if $\ker(g) = \operatorname{im}(f)$.
  If $M' = 0$, this means that $g$ is one to one.
  If $M'' = 0$, this means that $f$ is onto.
- A sequence of morphisms is exact if every sequence of two consecutive morphisms of it is exact.
- An exact sequence

  $$0 \to M' \xrightarrow{f} M \xrightarrow{g} M'' \to 0,$$

  is called a short exact sequence.

A diagram of morphisms is *commutative* if the composite of morphisms from one module to another does not depend on the path you take.

Here is a typical example of the use of these notions.

PROPOSITION 2.7 (Snake lemma). *Let there be a commutative diagram with exact rows*

$$\begin{array}{ccccc} M' & \xrightarrow{\gamma} & M & \xrightarrow{\delta} & M'' \\ \downarrow{\scriptstyle f} & & \downarrow{\scriptstyle g} & & \downarrow{\scriptstyle h} \\ N' & \xrightarrow{\alpha} & N & \xrightarrow{\beta} & N'' \end{array}$$

*with $\delta$ onto and $\alpha$ one to one. The following sequence is exact:*

$$\ker(f) \xrightarrow{\phi} \ker(g) \xrightarrow{\psi} \ker(h) \xrightarrow{\partial} \operatorname{coker}(f) \xrightarrow{\phi_1} \operatorname{coker}(g) \xrightarrow{\psi_1} \operatorname{coker}(h)$$

*where $\phi, \psi, \phi_1, \psi_1$ are natural and $\partial$ is defined as follows:*

*if $z'' \in \ker(h)$, let $z \in M$ be such that $z'' = \delta(z)$; then $\beta(g(z)) = 0$; let $u' \in N'$ such that $g(z) = \alpha(u')$; then $\partial(z'')$ is the class of $u'$ modulo* $\operatorname{im}(f)$.

The snake lemma has many corollaries. There exist also the *five lemma*, the *four lemma* and others whose proofs use so-called *diagram chasing*. See for instance [5].

### 2.1. *The functor* Hom

*We show here the property of left exactness of the functor* Hom.

*It is very important because from it we can deduce the exactness of some fundamental functors, the so called representable functors, for instance the tensor product.*

*Let $M, N$ be $R$-modules.*

*The set $\operatorname{Hom}_R(M, N)$ of morphisms from $M$ to $N$ has a natural structure of an $R$-module.*

Let us fix $M$. We get a *covariant functor* $F = \operatorname{Hom}_R(M, -)$ from the category of $R$-modules into itself:

- If $N$ is an $R$-module, $F(N) = \operatorname{Hom}_R(M, N)$.
- If $\phi$: $N \to N'$ is a morphism, $F(\phi)$ is the morphism $\psi \mapsto \phi \circ \psi$.

THEOREM 2.8. *The functor* $\operatorname{Hom}_R(M, -)$ *is left exact.*

This means that for every short exact sequence

$$0 \to N' \xrightarrow{f} N \xrightarrow{g} N''$$

the sequence

$$0 \to \operatorname{Hom}_R(M, N') \xrightarrow{\operatorname{Hom}(M,f)} \operatorname{Hom}_R(M, N) \xrightarrow{\operatorname{Hom}(M,g)} \operatorname{Hom}_R(M, N'')$$

is exact.

PROOF.
- As $f$ is one to one, $\mathrm{Hom}(M, f)$ is one to one: $f \circ \phi = 0 \Rightarrow \phi = 0$.
- $\ker(\mathrm{Hom}(M, g)) = \mathrm{im}(\mathrm{Hom}(M, f))$:

$$\phi \in \ker(\mathrm{Hom}(M, g)) \Rightarrow g \circ \phi = 0 \Rightarrow \phi(N) \subset \ker(g) = \mathrm{im}(f)$$
$$\Rightarrow \exists \psi \mid \phi = f \circ \psi.$$

In the same way, if we fix the $R$-module $N$, we get a *contravariant functor* $\mathrm{Hom}_R(-, N)$ which is also left exact, i.e. for every short exact sequence

$$M' \xrightarrow{f} M \xrightarrow{g} M'' \to 0,$$

the sequence

$$0 \to \mathrm{Hom}_R(M'', N) \xrightarrow{\mathrm{Hom}(g,N)} \mathrm{Hom}_R(M, N) \xrightarrow{\mathrm{Hom}(f,N)} \mathrm{Hom}_R(M', N)$$

is exact. □

### 2.2. *Sums and products*

DEFINITIONS 2.9. Let $\{M_i\}_{i \in I}$ be a family of modules over the ring $R$.
- The product

$$P = \prod_{i \in I} M_i$$

  has a natural structure of a module over $R$.
- The direct sum is the submodule of $P$

$$S = \sum_{i \in I} M_i = \{\{x_i\}_{i \in I} \mid x_i = 0 \text{ but for a finite number of } i\}.$$

- A submodule $N$ of $M$ is a *direct summand* if there exists a submodule $P$ of $M$ such that $M \simeq N \times P$.

The modules $P$ with the projections $p_i$: $P \to M_i$ and $S$ with the injections $j_i$: $M_i \to S$ have natural *universal properties*.

If the set $I$ is finite, let us say $I = \{1, \ldots, n\}$, we have $P = S$ and there is a set of natural formulas connecting those projections and injections:

$$\sum_{i=1}^{n} j_i \circ p_i = 1_P,$$
$$p_i \circ j_i = 1_{M_i}, \qquad p_j \circ j_i = 0 \quad \text{if } j \neq i.$$

If for every $i$, $M_i = R$, remark that the product is the module $R^I$ and the sum the module $R^{(I)}$ defined before.

### 2.3. *Direct and inverse limits*

*We introduce now two important notions: the first one, the inverse limit, is connected with the notion of intersection and the second one, the direct limit, with that of union.*

*Let $I$ be a preordered set by a relation $\leqslant$.*

DEFINITION 2.10 (*Inverse system*). An inverse system indexed by $I$ consists of the data:
- A family $\{M_i\}_{i \in I}$ of $R$-modules.
- For every $i, j \in I$ with $i \leqslant j$, of a morphism $f_{ij}$ from $M_j$ to $M_i$ such that
  - if $i \leqslant j \leqslant k$, $f_{ik} = f_{ij} \circ fjk$,
  - $f_{ii} = 1_{M_i}$.

The submodule $M$ of the product

$$\prod_{i \in I} M_i$$

whose elements are the elements $(x_i)_{i \in I}$ such that $i \leqslant j \Rightarrow x_i = f_{ij}(x_j)$ is called *the inverse (or projective) limit of the inverse system*, in notation $M = \varprojlim M_i$.

Let $\alpha_j$ be the restriction to $\varprojlim M_i$ of the projection from the product

$$\prod_{i \in I} M_i$$

to $M_j$.

THEOREM 2.11. *For every $R$-module $N$ we have the following isomorphism:*

$$\mathrm{Hom}_R(N, \varprojlim M_i) \to \varprojlim \mathrm{Hom}_R(N, M_j),$$

$$g \mapsto (\alpha_j o g)_{j \in I}.$$

Here is now the dual notion of direct system and limit.

DEFINITION 2.12. A *direct system* indexed by $I$ of $R$-modules consists of the data:
- A family $\{M_i\}_{i \in I}$ of $R$-modules
- For every $i, j \in I$ with $i \leqslant j$, of a morphism $f_{ji}$ of $M_i$ in $M_j$ such that
  - if $i \leqslant j \leqslant k$, $f_{ki} = f_{kj} \circ f_{ji}$,
  - $f_{ii} = 1_{M_i}$ for every $i \in I$.

The $R$-module

$$M = \left( \bigoplus_{i \in I} M_i \right) / \Phi,$$

where $\Phi$ is the submodule generated by the elements $(y_i)_{i\in I}$ for which there exists $j, k \in I$ with $j \leqslant k$ such that $y_i = 0$ if $i \neq j, k$ and $y_k = f_{kj}(y_j)$, is called the *direct limit (injective limit)* of $(M_i, f_{ji})_{i,j\in I}$, in notation $M = \varinjlim M_i$.

Let $\beta_j$ be the morphism from $M_j$ to $\varinjlim M_i$ which is the composite of the injection of $M_j$ into the direct sum

$$\bigoplus_{i\in I} M_i$$

and the surjection from it to $\varinjlim M_i$.

THEOREM 2.13. *For every $R$-module $N$, the map*

$$g \mapsto (g \circ \beta_j)_{j\in I}$$

*is an isomorphism of* $\mathrm{Hom}_R(\varinjlim M_i, N)$ *onto* $\varprojlim \mathrm{Hom}_R(M_j, N)$.

THEOREM 2.14 (Exactness of inverse and direct limits).
- *The functor inverse limit is left exact.*
- *The functor direct limit is exact.*

Let us explain what exactness means for direct limits.
- A morphism $\phi$ of the direct system $(M_i', f_{ij}')_{i,j\in I}$ into the direct system $(M_i, f_{ij})_{i,j\in I}$ is the data for every $i \in I$ of a morphism $\phi_i$ of $M_i'$ into $M_i$ such that if $i \leqslant j$ $f_{ji} o \phi_i = \phi_j o f_{ji}'$.
- The notion of a short exact sequence

$$0 \to (M_i', f_{ji}') \xrightarrow{\phi=(\phi_i)} (M_i, f_{ji}) \xrightarrow{\psi=(\psi_i)} (M_i'', f_{ji}'') \to 0$$

  of direct systems indexed by $I$ is the evident one.

*Some remarks*

1. If the set $I$ is discretely ordered, i.e. if two distinct elements are incomparable, the inverse limit is the product and the direct limit the direct sum.

2. Of great importance[2] is the case where the ordered set $I$ is directed, which means: $\forall i, j \in I$, there exists $k \in I$ such that $i, j \leqslant k$.

Then we get the equality

$$\ker(\beta_i) = \bigcup_{j\geqslant i} \ker(f_{ji}).$$

3. Mixing case 1 and case 2, you get any inverse or direct limit.

[2] Look for that the study of Grothendieck abelian categories ([13], Axiome AB5).

4. You may look at a module $M$ as the union of its finitely generated submodules but it is often much better to look at it as a direct limit (not the union!) of finitely presented modules, i.e. quotient modules of free finitely generated modules by finitely generated submodules.

*We now leave the general definitions to sink into more subtle notions which came from the theory of abelian groups.*

*This theory of abelian groups goes back to Gauss, who in the Disquisitiones studied the finite abelian group of the classes of quadratic forms with a given discriminant.*

*This was extended to the theory of modules over principal ideal rings and then over more general rings.*

*We shall sometimes introduce a notion for abelian groups and then extend it to modules over any ring.*

## 3. Free modules

### 3.1. *Free modules*

*Elementary linear algebra begins with the study of vector spaces and the main theorem asserts first that every vector space has a basis, maybe empty, and second that the cardinal of such a basis only depends on the space.*

DEFINITION 3.1. An $R$-module $L$ is free if it has a basis, i.e. a subset $\{x_i\}_{i\in I}$ such that every element of $L$ may be written in only one way $\sum_{i\in I}\lambda_i x_i$ with $\lambda_i \in I$ and $\lambda_i = 0$ but for a finite subset of $I$.

The $R$ module $R^{(I)}$ is free with basis $\{x_i\}_{i\in I}$ where $x_i$ is the map $I \to R$ defined by $x_i(i) = 1$ and $x_i(j) = 0$ if $j \neq i$. We shall identify $i$ to $x_i$ and so $I$ to a basis of $R^{(I)}$.

The $R$-module $L$ is free with a basis indexed by $I$ iff it is isomorphic to $R^{(I)}$.

In general, there exist modules which are not free. Here are two simple reasons for that.

- First of all, note that if a module has a nonempty basis it is only cancelled by 0, i.e. if $\lambda \in R$ and $\lambda M = 0$ then $\lambda = 0$.
  If the ring $R$ has an ideal $I \neq (0)$ and $R$, which means that $R$ is neither 0 nor a field, the $R$-module $R/I$ has no basis because it is cancelled by $I$.
  *Hence a ring $R$ is a field iff every $R$-module is free.*
- On the other hand, note that *an ideal of a domain $R$ is a free module iff it is principal, a quite useful characterization of principal!*
  For example the nonprincipal ideal $(3, \sqrt{-5})$ of the ring $\mathbf{Z}[\sqrt{-5}]$ is not free.

THEOREM 3.2. *Two bases of a free module have the same cardinality.*

Proof is easy.

Let $\{x_i\}_{i\in I}$ be a basis of the free $R$-module $M$. If the *commutative* ring $R$ is not 0 it has a maximal ideal $I$. The quotient module $M/IM$ gets a structure of vector-space over the field $R/I$ and, writing $\overline{y}$ for the class of $y \in M$, we see that $\{\overline{x}_i\}_{i\in I}$ is a basis of this vector-space. So the cardinal of $I$ depends only on $M$.

We shall not list here all the extensions to free modules of properties of vector-spaces. Let us give the following one coming from the isomorphism

$$\text{Hom}\left(R^{(I)}, M\right) \simeq (\text{Hom}(R, M))^I \simeq M^I:$$

Suppose that the sequence of $R$-morphisms

$$M' \to M \to M''$$

is exact and that $N$ is a free $R$-module.

Then the transformed sequence

$$\text{Hom}_R(N, M') \to \text{Hom}_R(N, M') \to \text{Hom}_R(N, M'')$$

is exact too.

*Here are some properties connected with free modules.*

- We saw that if the ring $R$ is not a field, there exist modules which are not free. Nevertheless, if $R$ is not a field, any $R$-module $M$ is a quotient of a free $R$-module $L$ for instance $R^{(M)}$.
- The module $M$ is *finitely generated* iff we can chose the free module $L$ with a finite basis.
- We say it is *finitely presented* if it is of type $L/N$ where $L$ is free with finite basis and $N$ is a finitely generated submodule of $L$.
- The ring $R$ is called *coherent* if every finitely generated $R$-module is finitely presented.[3] A noetherian ring is coherent but the converse is wrong: *An infinite product of fields is coherent but is not noetherian.*

### 3.2. *Torsion modules*

If the ring $R$ is an integral domain, an obstruction to an $R$-module $M$ being free is the existence of nonzero elements $\alpha \in R$ and $x \in M$ such that $\alpha x = 0$.

If $M$ is a finitely generated module over a principal ideal domain $R$, for instance a finitely generated abelian group, this is the only obstruction.

To give a more precise structure theorem, let us give some definitions.

DEFINITION 3.3. Let $R$ be an integral domain and $M$ be a module over $R$.

- The subset

$$T(M) = \{x \in M \mid \exists \alpha \in R,\ \alpha \neq 0;\ \alpha x = 0\}$$

is a submodule of $M$ called the torsion submodule of $M$.
- The module $M$ is called torsion if $T(M) = M$ and torsion free if $T(M) = 0$.

[3] This notion is useful in the study of analytic spaces when the dimension is $\infty$.

We remark that the quotient submodule $M/T(M)$ is torsion free and that $T(T(M)) = T(M)$. So you can imagine the axiomatic definition of a general notion of *torsion* and of what is called *torsion theory*: you give the class of modules you want to be *torsion* with natural properties [8, 27].

For instance, if $\Sigma$ is a multiplicative subset of $R$, one can take the class of modules $M$ such that $\Sigma^{-1}M = 0$.

This theory is connected to localization and for instance to the *localizing subcategories* of P. Gabriel [12].

Here is one of the oldest and most famous results.

THEOREM 3.4. *Let $R$ be a principal ideal domain and $M$ be a finitely generated module.*
- *$M$ is the direct sum of its torsion submodule $T(M)$ and a free module.*
- *If $M$ is a torsion module, it may be written in only one way in the form:*

$$M = \sum_{i=1}^{n} R/(d_i)$$

*where $d_i \in R$ and $d_{i+1}$ divides $d_i$ $(i = 1, \ldots, n-1)$.*

The elements $d_i$ are called the *invariant factors* of the torsion module $M$.

*The main example*
*Let $k$ be a field, $V$ a finite dimensional $k$-vector space and $u$ an endomorphism of $V$.*

Define on the additive group $V$ a structure of module $M = V_u$ over the ring $R = k[X]$ of polynomials by setting, if $f(X) \in k[X]$ and $x \in V$,

$$f(X)x = f(u)(x).$$

Then $V_u$ is finitely generated like $V$. It is a torsion module: the Hamilton–Cayley theorem says that, if $\chi_u(X)$ is the characteristic polynomial of $u$, $\chi_u(u) = 0$ and so $\chi_u(X)V_u = (0)$.

## 4. Projective modules

### 4.1. *Projective modules*

If the ring $R$ is principal, any submodule of a free module $L$ and so any *direct summand* $M$ of $L$ is free.

Fortunately, owing to the importance of the problems involved, even the last assertion does not hold for any ring.

*A direct summand of a free module is not necessarily free.*

For instance, let $R$ be the Dedekind ring $\mathbf{Z}[\sqrt{-5}]$ and $M$ be the ideal $(3, 1 + 2\sqrt{-5})$. This ideal is not principal and so the $R$-module $M$ is not free but, as it is a quotient of the free module $R^2$, we shall see below that it is a direct summand of it.

DEFINITION 4.1. A projective module is a direct summand of a free module.

THEOREM 4.2 (Projective ideals). *If the ring $R$ is an integral domain, an ideal $I$ is projective iff it is invertible and then it is finitely generated.*

This means that there exist $\alpha_1, \ldots, \alpha_n, \in I$ and $\beta_1, \ldots, \beta_n, d \neq 0 \in R$ such that, in the field of quotients of $R$:

$$\beta_i I \subset R; \quad 1 = \sum_{i=1}^{n} \alpha_i \left(\beta_i d^{-1}\right).$$

Such an ideal is generated by $\alpha_1, \ldots, \alpha_n$ and so is *finitely generated.*

*Some definitions using projective modules*
First of all let us give this easy result: *every $R$-module is projective if and only if the ring $R$ is a finite product of fields.*

Now we give two important definitions.

DEFINITION 4.3.
- A ring $R$ in which every ideal is projective is called hereditary.
- A ring $R$ in which every finitely generated ideal is projective is called semi-hereditary.

If $R$ is an integral domain, $R$ is hereditary iff it is a Dedekind ring, i.e. a ring in which every proper ideal is a product of prime ideals.

A *semi-hereditary integral domain* is called a Prüfer ring, as is any Bezout domain, i.e. a domain in which every finitely generated ideal is principal.

Here are two well known examples of such rings:
- the ring of entire functions in the complex plane
- the ring of all the algebraic integers.

The following structure theorems of modules over such rings generalize those for principal ideal domains [6].

THEOREM 4.4.
- *If the ring $R$ is semi-hereditary, every finitely generated submodule of a free $R$-module is the direct sum of a finite number of modules each of which is isomorphic with a finitely generated ideal of R.*
- *If the ring $R$ is hereditary, every submodule of a free module is the direct sum of modules each of which is isomorphic with an ideal of $R$* [6].

*Connection between projective and free modules*
The following question is quite natural and important:

Is every projective $R$-module free?

We have to distinguish between the finitely generated and the nonfinitely generated case.

- First of all, let us give the following result of I. Kaplansky [15] improved by C. Walker [31]:
  *Every projective module is a direct sum of countably generated modules.*

- In [2], H. Bass proved that nonfinitely generated modules do behave better than the finitely generated ones!:
  *If $R$ is a connected*[4] *commutative noetherian ring, every not finitely generated projective $R$-module is free.*
- A famous problem of J.P. Serre connected with the problem of triviality of vector bundles over affine spaces was solved independently by D. Quillen [24] and A. Suslin [28] in 1976. They proved:
  *Every projective module over the ring $k[X_1, \ldots, X_n]$ of polynomials over a field $k$ is free.*

### 4.2. *Miscellaneous*

Is any $R$-module (resp. finitely generated module) a quotient of a *minimal* projective module, i.e. a so-called *projective cover?*

If the answer is yes, H. Bass says that the ring $R$ is *perfect* (resp. *semi-perfect*) and he gets structure theorems in the not necessarily commutative case [3].

Here they are in the much simpler commutative case.

- *The ring $R$ is semi-perfect iff it is a finite product of local rings.*
- *The ring $R$ is perfect iff it is a finite product of local rings with $T$-nilpotent maximal ideals. This means that for any sequence $(a_n)_{n\in\mathbf{N}}$ of elements of the maximal ideal, there exists $m$ such that $a_1 \cdots a_m = 0$.*

What about a direct product of projectives? S. Chase proved in 1960:

*Every product of projectives is projective if and only if the ring $R$ is artinian* [9].

## 5. Flat modules

*The notion of flat modules was introduced by J.P. Serre in a famous paper, often named* GAGA, [29], *connecting classical algebraic geometry to so-called analytic geometry, the study of zeros of analytic functions.*

A very important example is that of the ring $\mathbf{C}\{X_1, \ldots, X_n\}$ of convergent power series looked at as a module over the ring $\mathbf{C}[X_1, \ldots, X_n]$ of polynomials.

In the same vein [20] is the example of the ring $\mathcal{E}_n$ of germs of functions of class $C^\infty$ in a neighborhood of the origin in $\mathbf{R}^n$ looked at as a module over the ring of germs of analytic functions.

*Let us first recall very briefly some facts about the tensor product.*

### 5.1. *Tensor product*

*Let $M$ and $N$ be two $R$-modules.*

Consider the set product $M \times N$ and the free $R$-module $L = R^{(M\times N)}$; so we may identify the set $\{(x, y)\}_{x\in M, y\in N}$ to a basis of $L$.

[4] This means without idempotent other than 0 or 1.

Then let $\Phi$ be the submodule of $L$ generated by elements of one of the following types:

- $(x_1 + x_2, y) - (x_1, y) - (x_2, y);\ x_1, x_2 \in M;\ y \in N$
- $(x, y_1 + y_2) - (x, y_1) - (x, y_2);\ x \in M;\ y_1, y_2 \in N$
- $(\lambda x, y) - \lambda(x, y);\ \lambda \in R;\ x \in M;\ y \in N$
- $(x, \lambda y) - \lambda(x, y);\ \lambda \in R;\ x \in M;\ y \in N$

Finally put $M \otimes_R N = L/\Phi$ and write $x \otimes y$ for the class of $(x, y) \in M \times N \subset R^{(M \times N)}$ modulo $\Phi$ and $\alpha$ for the *bilinear* map: $(x, y) \mapsto x \otimes y$ from $M \times N$ to $M \otimes_R N$.

*The R-module $M \otimes_R N$ is called the tensor product of the two modules $M$ and $N$.*[5]

THEOREM 5.1 (Universal property). *For every bilinear map $\psi$ from the product $M \times N$ to an $R$-module $P$ there exists one and only one morphism $\overline{\psi}$ from $M \otimes_R M$ to $P$ such that $\psi = \overline{\psi} \circ \alpha$.*

For a sketchy proof, remark that $\psi$ defines a morphism $\psi_1$ from $L$ to $P$ by $\psi_1((x, y)) = \psi((x, y))$.

The module $\Phi$ has been chosen in such a way that the bilinearity of $\psi$ implies that $\psi_1(\Phi) = 0$. The morphism $\overline{\psi}$ is then defined naturally by quotient.

THEOREM 5.2. *Tensor product is right exact but not always left exact.*

You can prove this using the universal property and the left exactness of the functor Hom [17].

### 5.2. *Flat modules*

DEFINITIONS 5.3. Let $R$ be a ring and $M$ a module over $A$.

1. The module $M$ is flat if, for every exact sequence

$$0 \to N' \xrightarrow{f} N \xrightarrow{g} N'' \to 0, \tag{1}$$

the transformed sequence

$$0 \to M \otimes_R N' \xrightarrow{M \otimes f} M \otimes_R N \xrightarrow{M \otimes g} M \otimes_R N'' \to 0 \tag{2}$$

is exact.

2. It is faithfully flat if the exactness of (2) is equivalent to that of (1).
3. An $R$-algebra $S$ is flat or faithfully flat if the $R$-module $S$ is so.

As the tensor product is right exact, $M$ is flat if every injection remains an injection when tensored by $M$ and faithfully flat if the converse holds.

---

[5] You may define in the same way the tensor product of any number of modules and prove natural properties of associativity.

*Examples*

- The $R$-module $R$ is (faithfully) flat.
- A direct sum of flat modules is flat.
- A direct summand of a flat module is flat.
- A free module is (faithfully) flat.
- A projective module is flat.

The characterization of finitely projective modules as finitely presented modules which become free when localized at any prime ideal of the ring $R$ gives a converse to the last assertion above.

*A finitely presented flat module is projective.* ([5], Chapter II.5 2, Corollary 2 of Theorem 1.)

### 5.3. *Characterization with ideals*

In fact, the special sequences (1) of type $0 \to \mathbf{a} \to R$ with $\mathbf{a}$ ideal of $R$ are enough to give flatness.

THEOREM 5.4. *The R-module M is flat if and only if for every ideal, and even for every finitely generated ideal,* $\mathbf{a}$ *of R the homomorphism*

$$\sum x_i \otimes a_i \mapsto \sum a_i x_i$$

*from* $M \otimes_R \mathbf{a}$ *to* $M$ *is injective.*

As an easy corollary we get:

*If the ring R is a principal ideal domain, the R-module M is flat if and only if it is torsionless.*

The tensor product commutes with direct sums and so *every direct sum of flat modules is still flat.* What about *products* of flat modules?

S. Chase answered that question in [9]:

*The ring R is coherent iff every product of flat modules is flat or iff for every set I the R-module* $R^I$ *is flat.*

### 5.4. *Using linear equations*

*Here is a rather concrete characterization of flat modules by homogeneous linear equations. The characterization of faithfully flat modules is the same but with nonhomogeneous linear equations.*

It is worth applying it to the following examples of flat modules:

- $R$ is a field and the $R$-module $M$ is an overfield.
- $R$ is the ring of germs of analytic functions in a neighborhood of the origin of $\mathbf{R}^n$ and $M$ is the ring $\mathcal{E}_n$ of germs of functions of class $C^\infty$.
- $R$ is a noetherian local ring and $M$ is its completion.

THEOREM 5.5. *Let $j$ be an integer $\geqslant 1$*

*The following assertions are equivalent for an R-module $M$:*

(i) *The R-module $M$ is flat.*

(ii) *For every $a_{ij} \in R$ and $x_i \in M$ $(i = 1, \ldots, r;\ j = 1, \ldots, n)$ such that*

$$\sum_{i=1}^{r} a_{ij} x_i = 0; \quad j = 1, \ldots, n,$$

*there exist*

$$s \in \mathbf{N}^*, \quad b_{ik} \in R, \quad y_k \in M; \quad k = 1, \ldots, s$$

*such that*

$$\sum_{i=1}^{r} a_{ij} b_{ik} = 0; \quad j = 1, \ldots, n \quad \text{and} \quad x_i = \sum_{k=1}^{s} b_{ik} y_k.$$

(iii) *The assertion* (ii) *only for $j = 1$.*

SKETCH OF PROOF ([5]).

(iii) $\Rightarrow$ (i) Let $a_1, \ldots, a_r$ be generators of a finitely generated ideal $I$. The condition

$$\sum_{i=1}^{r} a_i x_i = 0$$

means that the element

$$\sum_{i=1}^{r} x_i \otimes a_i$$

is in the kernel of the morphism

$$\lambda\colon \sum_{i=1}^{r} x_i \otimes a_i \mapsto \sum_{i=1}^{r} a_i x_i$$

from $M \otimes I$ to $M$. The assumption implies that such an element is 0.

By induction on the number of elements of a minimal system of generators, you get: (for instance, [17], p. 243–244).

PROPOSITION 5.6. *A finitely generated module $M$ over a local ring $R$ is flat iff it is free.*

### 5.5. *Lazard's characterization*

*Here now is a very nice result of D. Lazard* [18].

THEOREM 5.7. *Any R flat module $M$ is a filtering direct limit of free R-modules.*

We do not give the proof but only the inverse system of free modules involved.

Let $\Lambda$ be the set of finite subsets of the product $M \times \mathbf{N}$. If $\alpha \in \Lambda$, note $\{e_{(m,n)}\}_{(m,n)\in\alpha}$ the natural basis of the $R$-module $R^{(\alpha)}$.

Let $u_\alpha$ be the morphism from $R^{(\alpha)}$ into $M$ such that $u_\alpha(e_{(m,n)}) = m$.

Let $I$ be the set of elements $(\alpha, N_\alpha)$ with $\alpha \in \Lambda$ and $N_\alpha$ a finite subset of $K_\alpha = \ker(u_\alpha)$ naturally ordered.

For every $i \in I$, let $M_i$ be the finitely presented module $R^{(\alpha)}/s(N_\alpha)$ where $s(X)$ is the module spanned by $X$.

We prove then:

- $M = \varinjlim (M_i)_{i\in I}$.
- The subset $J = \{i \in I \mid M_i \text{ is free}\}$ is cofinal to $I$.
- $M = \varinjlim (M_i)_{i\in J}$.

### 5.6. *Absolutely flat rings*

*It is natural to look at the rings R for which every R-module is flat.*

Those rings are the *regular rings of Von Neumann*. To avoid the possible confusion with the regular rings of commutative algebra, Bourbaki called them absolutely flat and we shall agree with that terminology [5].

DEFINITION 5.8 (*Von Neumann*). The (commutative) ring $R$ is said to be absolutely flat if it satisfies the following equivalent properties where $a$ is any element of $R$: 1. There exists $b \in R$ such that $a = ba^2$. 2. The ideal $aR$ is a direct summand of $R$. 3. The $R$-module $R/aR$ is flat.

Here are some remarks:

- if $a = ba^2$, the idempotent $e_a = ba$ generates the ideal $aR$. So every principal ideal of an absolutely flat ring is generated by an idempotent.
- If $e$ and $f$ are idempotents of a ring $R$, $e + f - ef$ is also one and the ideal $(e, f)$ is equal to the principal ideal $(e + f - ef)$.
- Every finitely generated ideal of an absolutely flat ring $R$ is principal, hence $R$ is a Bezout ring. Such an ideal is generated by an idempotent and is a direct summand of $R$.

THEOREM 5.9. *The ring $R$ is absolutely flat iff every $R$-module is flat.*

## 6. Injective modules

*We introduce here the notion of injective modules, which is in duality with that of projective modules. Some properties may be deduced from those of projective modules using categorical duality but the most important ones, e.g., the fact that every module is a submodule of an injective one, need a specific treatment.*

### 6.1. *Some history*

The notion of *divisible abelian groups* was introduced by R. Baer [1] as a direct summand of every abelian group which admits it as a subgroup.

He also gave the following more explicit definition.

DEFINITION 6.1. A divisible abelian group is a group $G$ such that

$$\forall x \in G,\ \forall n \in \mathbf{N}^*,\ \exists y \in G \mid x = ny.$$

To extend that definition, it is worth translating it in the following way:

*Any homomorphism $f$ from $(n)$ to $G$, given by $x = f(n) \in G$, may be extended to a homomorphism $g$ of $\mathbf{Z}$ into $G$ by $g(1) = y$ with $ny = x$.*

This is illustrated by the following commutative diagram:

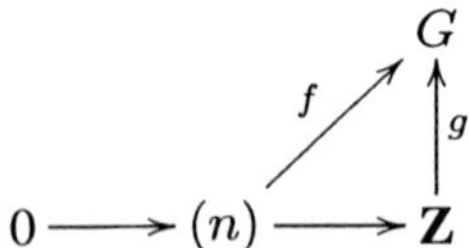

Here are some easy consequences, where group means abelian group:

- A divisible torsionless group $G$ is a vector-space over $\mathbf{Q}$: define $(n/m)x$ with $n, m \in \mathbf{N}^*$, $x \in G$, as $ny$ where $y$ is the only one element such that $my = x$.
- A quotient of a divisible group, for instance a direct summand, is divisible.
- A direct product of divisible groups is divisible.
- A divisible subgroup $H$ of a group $G$ is a direct summand of $G$.

*Here is a sketch of the proof of the last assertion*:

Zorn's lemma gives us a subgroup $K$ of $G$ which is maximal for the property $K \cap H = (0)$. Check then $K + H = G$.

THEOREM 6.2 (Divisible abelian group structure).

- *A divisible abelian group is a direct sum of indecomposable divisible groups.*
- *An indecomposable divisible group is isomorphic to*
  - *the additive group* $\mathbf{Q}$,
  - *the additive group of p-adic integers* $\mathbf{Z}_{p^\infty}$.

Using the property of extensions we get the following generalization of the notion of divisible groups.

DEFINITIONS 6.3. An $R$-module $M$ is injective if it satisfies the following equivalent conditions:

(i) $M$ is a direct summand of any over-module $N$.

(ii) The functor $\mathrm{Hom}_R(-, M)$ is right-exact and so exact.

(iii) Any morphism to $M$ from a submodule $N'$ of a module $N$ may be extended to a morphism of $N$ into $M$.

If the ring $R$ is an integral domain, a torsion free-module is injective iff it is divisible.

The following characterization is noteworthy:

*The integral domain is a Dedekind ring iff every divisible module is an injective one.*

The structure theorem of indecomposable divisible groups was generalized to any injective module over a *noetherian* ring $R$ by E. Matlis.

THEOREM 6.4 (E. Matlis). *Let $R$ be a* noetherian *ring.*

- *Any injective $R$-module is a direct sum of indecomposable injective modules.*
- *Any indecomposable injective module is isomorphic to the completion $\widehat{R_I}$ of $R$ in the $I$-adic topology, i.e. the topology of powers of a prime ideal $I$.*

REMARK. A direct product of injective modules is injective. What about direct sums of injectives?

H. Bass proved in his Ph.D. (1956): *Every direct sum of injectives is injective if and only if the ring $R$ is noetherian* [9].

### 6.2. *Injective hull*

*The notion of injective hulls is dual of that of projective covers, and is quite important in noetherian commutative algebra but also in other parts of mathematics, for instance in algebraic topology.*

*Especially it allows computation of the injective dimension of a module. To convince yourself look at the fundamental paper by H. Bass: On the ubiquity of Gorenstein rings* [4].

*The existence of the injective hull was established first for abelian groups by Baer and then extended by Eckmann and Schopf.*

*Its connection with the primary decomposition of modules was proved by Matlis and extended to abelian categories by P. Gabriel.*

THEOREM 6.5. *For any $R$-module $M$, there exists an injective $R$-module $E(M)$ which is minimal among the injective over-modules of $M$.*

*It is unique up to automorphism and is called the injective hull of $M$.*

SKETCH OF PROOF.

An overmodule $P$ of $M$ is called an *essential extension* of $M$ if for a submodule $N$ of $P$ the two following conditions are equivalent:

- $N = 0$.
- $N \cap M = 0$.

You may also say: for any nonzero element $x$ of $P$ there exists $\lambda \in R$ such that $\lambda x$ is a non zero element of $M$.

The notion of essential extension is transitive.

The following lemma is the crux! But we leave its proof.

LEMMA 6.6. *The $R$-module $P$ is injective iff it has no essential extension other than itself.*

Ordered by inclusion the set of essential extensions of $M$ (except for isomorphism) is inductive: every linearly ordered subset of it has an upper bound.

So Zorn's lemma gives us a maximal essential extension $E(M)$. By transitivity of the notion of essential extension and the characterization of the lemma, we see that the $R$-module $E(M)$ is injective.

It remains to check that it is a minimal injective over-module of $M$ and that it is unique except for isomorphism.

## 7. Some generalizations

The $R$-module $M$ is projective (resp. injective) if the functor $\mathrm{Hom}_R(M, -)$ (resp. $\mathrm{Hom}_R(-, M)$) transforms *every* short exact sequence into another one.

It is difficult to resist the temptation to look at modules $M$ such that $\mathrm{Hom}_R(M, -)$ (resp. $\mathrm{Hom}_R(-, M)$) transforms into an exact sequence every element of a convenient class of short exact sequences.

This gives rise for instance to the notions of *quasi-projective* or *quasi-injective* modules [11, 10].

Here is a significant application [26]. Use the class of short exact sequences of type

$$0 \to \mathbf{a} \to R \to R/\mathbf{a} \to 0.$$

You get the following characterization of *finitely generated* projective modules, dual to that of injective modules using only ideals:

*The* finitely generated *module $P$ is projective iff, for every ideal* **a** *of $R$, every morphism of $M$ into $R/\mathbf{a}$ comes from a morphism of $M$ into $R$.*

## References


[1] R. Baer, *Abelian groups that are direct summands of every containing abelian group*, Bull. Amer. Math. Soc. **46** (1940), 800–806.

[2] H. Bass, *Big projective modules are free*, Ann. Math.

[3] H. Bass, *Finitistic dimension and a homological generalization of semi-primary rings*, Trans. Amer. Math. Soc. **95** (1960), 466–468.

[4] H. Bass, *On the ubiquity of Gorenstein rings*, Math. Z. **82** (1963), 8–28.

[5] N. Bourbaki, *Algèbre Commutative. Chap.* 1 *Modules Plats. Chap.* 2 *Localisation*, Hermann, Paris (1961).

[6] H. Cartan and S. Eilenberg, *Homological Algebra*, Princeton Univ. Press (1956).

[7] R. Dedekind, *Gesammelte Werke*, 3 vols, Brunswick (1930–1932).

[8] Dickson, *A torsion theory for abelian categories*, Trans. Amer. Math. Soc. **121** (1966), 223–235.

[9] S. Chase, *Direct products of Modules*, Trans. Amer. Math. Soc. **97** (1960), 457–473.

[10] G.C. Faith and Y. Utumi, *Quasi-injective modules and their endomorphism rings*, Arch. Math. **15** (1964), 166–174.

[11] K.R. Fuller and D.A. Hill, *On quasi-projective modules via relative projectivity*, Arch. Math. **21** (1970), 369–373.

[12] P. Gabriel, *Des catégories abéliennes*, Bull. Soc. Math France **90** (1962), 323–448.

[13] A. Grothendieck, *Sur quelques points d'algèbre homologique*, Tôhoku Math. J. **9** (1957), 119–221.

[14] I. Kaplansky, *Infinite Abelian Groups*, Univ. of Michigan Press, Ann Arbor (1954).

[15] I. Kaplansky, *Projective modules*, Ann. Math. **68** (1958) 372–377.
[16] L. Kronecker, *Werke*, 6 vols, Leipzig (1895–1930), rééd. 5 vols, New York (1968).
[17] J.P. Lafon, *Les Formalismes Fondamentaux de l'Algèbre Commutative*, Hermann, Paris (1974).
[18] D. Lazard, *Autour de la platitude*, Bull. Soc. Math. France **97** (1968), 81–128.
[19] S. MacLane, *Homology*, Springer, Berlin (1963).
[20] B. Malgrange, *Ideals of differentiable functions*, Oxford Univ. Press (1966).
[21] E. Matlis, *Injective modules over noetherian rings*, Pacific J. Math. **8** (1958), 511–528.
[22] M. Nagata, *Local Rings*, Interscience Publishers (1962).
[23] D.G. Northcott, *Ideal Theory*, Cambridge Univ. Press, Cambridge (1953).
[24] D. Quillen, *Projective modules over polynomial rings*, Invent. Math. **36** (1976), 167–171.
[25] S. Lang, *Algebra*, Addison-Wesley (1965).
[26] E. de Robert, *Projectifs et injectifs*, C. R. Acad. Sci. Paris Sér. A **286** (1969), 361–364.
[27] Stenström, *Ring of Quotients*, Springer, Berlin (1975).
[28] A. Suslin, *Modules projectifs sur un anneau de polynômes*, Dokl. Akad. Nauk SSSR (1976).
[29] J.P. Serre, *Géométrie algébrique et gómétrie analytique*, Ann. Inst. Fourier **6** (1955/56).
[30] B.L. van der Waerden, *Modern Algebra*, I and II, Springer (1940).
[31] C.P. Walker, *Relative homological algebra and abelian groups*, Illinois J. Math **10** (1966), 186–209.
[32] O. Zariski and P. Samuel, *Commutative Algebra*, I and II, Van-Nostrand (1960).

# Section 3B
# Associative Rings and Algebras

# Polynomial and Power Series Rings. Free Algebras, Firs and Semifirs

P.M. Cohn

*University College London, Gower Street, London WC1E 6BT, UK*
*e-mail: pmc@math.ucl.ac.uk*

*Contents*



HANDBOOK OF ALGEBRA, VOL. 1
Edited by M. Hazewinkel

## 0. Introduction

The guiding theme of this article is the Euclidean algorithm. First stated for the integers in Euclid's Elements, Book VII, it was extended to polynomials in one variable by S. Stevin (1585). Its extension to more than one variable is difficult if one allows the variables to commute. But the weak algorithm, described in Section 3, provides a counterpart to the Euclidean algorithm (to which it reduces when commutativity is imposed) and it forms a natural tool for the study of polynomials in several noncommuting indeterminates. Just as in a Euclidean domain every ideal is principal, so the (one-sided) ideals in a ring with weak algorithm are free, as modules over the ring, and this leads to firs and semifirs which form the subject of Section 2. Much of the module theory for firs applies more generally to hereditary rings, and Section 4 summarizes what is known, while Section 5 deals with coproducts, a natural extension of free algebras. In this context it is interesting to note that the weak algorithm was first observed in the coproduct of skew fields (Cohn 1960). I am indebted to M.L. Roberts for his comments on a preliminary version.

## 1. Skew polynomial and power series rings

**1.1.** For any ring $R$ the polynomial ring $R[x]$ is a familiar construction, obtained by adjoining to $R$ an element $x$ subject to the rule

$$ax = xa, \quad \text{for all } a \in R. \tag{1}$$

As is well known, every element of $R[x]$ can be uniquely expressed as a *polynomial* in $x$ with coefficients from $R$:

$$f = a_0 + xa_1 + \cdots + x^n a_n, \quad a_i \in R. \tag{2}$$

The *degree* of $f$, $\deg f$, is defined as $n$ if $a_n \neq 0$; by convention, $\deg 0 = -\infty$.
We observe that

$$\deg(f - g) \leqslant \max\{\deg f, \deg g\}, \tag{3}$$

$$\deg fg \leqslant \deg f + \deg g. \tag{4}$$

If $R$ is an integral domain, then equality always holds in (4) and it follows that in this case $R[x]$ is again an integral domain.

This situation may be generalized by giving up (1) but still demanding that every polynomial can be written in the form (2). We now need to replace (1) by a *commutation rule* which expresses $ax$ in the form (2). If we wish to preserve (4), the most general such rule has the form

$$ax = xa^{\alpha} + a^{\delta}, \tag{5}$$

where $a \mapsto a^\alpha$, $a \mapsto a^\delta$ are well-defined maps of $R$ into itself (by the uniqueness of (2)). The distributive law: $(a+b)x = ax + bx$, shows that

$$x(a+b)^\alpha + (a+b)^\delta = xa^\alpha + a^\delta + xb^\alpha + b^\delta,$$

while the associative law: $(ab)x = a(bx)$ entails

$$x(ab)^\alpha + (ab)^\delta = a\bigl(xb^\alpha + b^\delta\bigr) = \bigl(xa^\alpha + a^\delta\bigr)b^\alpha + ab^\delta.$$

Together with the rule $1.x = x.1 = x$ this yields

$$(a+b)^\alpha = a^\alpha + b^\alpha, \quad (ab)^\alpha = a^\alpha b^\alpha, \quad 1^\alpha = 1, \quad a, b \in R, \tag{6}$$

$$(a+b)^\delta = a^\delta + b^\delta, \quad (ab)^\delta = a^\delta b^\alpha + ab^\delta, \quad 1^\delta = 0, \quad a, b \in R. \tag{7}$$

By (6), $\alpha$ is an endomorphism of $R$; its kernel is a proper ideal of $R$, hence $\alpha$ is injective whenever $R$ is a simple ring, in particular, when $R$ is a field. Equation (7) shows $\delta$ to be a linear map called an $\alpha$-derivation. For a given endomorphism $\alpha$, any element $m$ of $R$ defines an $\alpha$-derivation $\delta_m$ by the rule

$$a^{\delta_m} = am - ma^\alpha;$$

$\delta_m$ is called the *inner* $\alpha$-derivation induced by $m$. An $\alpha$-derivation which is not inner is said to be *outer*.

Given a ring $R$ with an endomorphism $\alpha$ and an $\alpha$-derivation $\delta$, we can define a multiplication on the set of all polynomials (2) by using the commutation rule (5) to 'straighten' products $x^m a x^n b$. Applying (5), we find

$$x^m a x^n b = x^{m+1} a^\alpha x^{n-1} b + x^m a^\delta x^{n-1} b$$

and an induction on $n$ reduces the product to the form (2). The resulting rings is denoted by $R[x; \alpha, \delta]$ and is called the *skew polynomial ring*; it has been defined here by the presentation with defining relations (5) in addition to the relations in $R$. We still need to show that the form (2) for its elements is unique; this follows by letting it act on the right $R$-module $R^{\mathbf{N}}$ consisting of all sequences $(a_i)$ indexed by $\mathbf{N}$, by the rules

$$(a_i)b = (a_i b), \quad (a_i)x = \bigl(a_i^\delta + a_{i-1}^\alpha\bigr), \quad \text{where } a_{-1} = 0.$$

For $f = a_0 + xa_1 + \cdots + x^n a_n$ applied to $(1, 0, 0, \ldots)$ just gives $(a_0, a_1, \ldots)$, so two elements (5) which are equal in $R[x; \alpha, \delta]$ must have the same coefficients. We also note that $\delta$ is now inner, induced by $x$, as we see by writing (5) in the form $a^\delta = ax - xa^\alpha$. If $\delta$ happens to be inner on $R$, induced by $m$, then we can reduce $\delta$ to 0:

$$R[x; \alpha, \delta] = R[x - m; \alpha, 0].$$

In case $\delta$ is zero, one often writes $R[x; \alpha]$ in place of $R[x; \alpha, 0]$.

If all coefficients are written on the left, (5) takes the form $xa = a^{\alpha}x + a^{\delta}$ and we obtain a *left* skew polynomial rings. When $\alpha$ is an automorphism, this is the same as the right skew polynomial ring $R[x; \alpha^{-1}, -\alpha^{-1}\delta]$, but in general the two notions are distinct, i.e. a left skew polynomial ring with a nonsurjective endomorphism cannot always be expressed as a right skew polynomial ring.

If $K$ is a skew field with endomorphism $\alpha$ and $\alpha$-derivation $\delta$, then $K[x; \alpha, \delta]$ is a right Ore domain and so has a skew field of fractions, denoted by $K(x; \alpha, \delta)$. More generally, this holds when $K$ is a right Ore domain (Curtis 1952), cf. (Cohn 1985, p. 54). However, even for a field $K$, $K[x; \alpha, \delta]$ is not left Ore, unless $\alpha$ is an automorphism. For a right Noetherian ring $R$ with an automorphism $\alpha$, the skew polynomial ring $R[x; \alpha, \delta]$ is again right Noetherian (Hilbert basis theorem), but for a general endomorphism this need not be so (Lesieur 1978; Cohn 1985, p. 58, 1990, p. 366).

*Examples of skew polynomial rings*

1. Let $k$ be any field; the *Weyl algebra* $A_1[k]$ is defined as the $k$-algebra generated by $p, q$ over $k$ with defining relation $pq - qp = 1$. If $B = k[p]$, then $A_1[k] = B[q; 1, d/dp]$, where $d/dp$ is the derivation on $B$ mapping $f$ to its derivative $df/dp$. Every element of $A_1[k]$ can be written as a linear combination of terms $p^i q^j$ and $df/dp = fq - qf$; similarly $df/dq = pf - fp$. It is easily checked that $A_1[k]$ is a simple Noetherian domain when char $k = 0$. In prime characteristic $r$ say, $A_1[k]$ is of dimension $r^2$ over its center $k[p^r, q^r]$.

2. Let $k$ be a field containing a primitive $n$-th root of 1, $\omega$ say. Then any cyclic algebra $A$ of degree $n$ over $k$ is generated by elements $u, v$ such that

$$u^n = \alpha, \quad v^n = \beta, \quad \text{where } \alpha, \beta \in k, \quad vu = \omega uv.$$

If $F = k(u)$ and $x^n - \alpha$ is irreducible over $k$, then the Galois group of $F/k$ is generated by $\sigma$: $u \mapsto \omega u$ and $A = F[v, \sigma]/(v^n - \beta)$.

3. The two-dimensional soluble nonabelian Lie algebra over $k$ has a basis $x, y$ with multiplication $[x, y] = y$. Its universal associative envelope may be described as $A[y; \sigma]$, where $A = k[x]$ and $\sigma$ is the shift automorphism, $f(x) \mapsto f(x + 1)$. It has the defining relation $xy = y(x + 1)$ and is also called the *translation ring* over $k$.

4. Let $k$ be a field of prime characteristic $p$ and consider the set $A$ of all polynomials of the form

$$f(x) = a_0 x^{p^m} + a_1 x^{p^{m-1}} + \cdots + a_{m-1}x^p + a_m x,$$

with the usual addition and with substitution as multiplication: $(fg)(x) = f(g(x))$. This set $A$ is a ring under these operations. If $\sigma$: $a \mapsto a^p$ is the Frobenius endomorphism in $k$, then the map $\eta$: $k[x; \sigma] \to A$ defined by $x^m \mapsto x^{p^m}$ and linearity is a homomorphism. This ring was first defined and studied by Ore (1933).

5. If $R$ is a principal ideal domain with an automorphism $\alpha$, then $R[x; \alpha]$ is right Noetherian, but not principal unless $R$ is a skew field. Now Jategaonkar (1969) has observed that when $\alpha$ is an endomorphism of $R$ which maps every nonzero element of $R$ to a unit, then $R[x; \alpha]$ is again right principal, and for a suitably chosen ring he

has shown that this construction can be iterated transfinitely, to produce principal right ideal domains which form a good source of counter-examples: (i) right Noetherian rings whose Jacobson radical is not nilpotent, (ii) right but not left primitive rings, (iii) rings with preassigned (different) left and right global dimensions. Moreover, Lenstra (1974) has shown that the iterated skew polynomial rings constructed by Jategaonkar form the precise class of integral domains with a unique remainder algorithm (cf. Cohn (1985), p. 532ff.). They lead to examples of rings whose set of right ideals is well-ordered, which have been studied by Brungs (1969).

**1.2.** Let $R$ be any ring. If in (2) we allow infinite series, with multiplication rule (1),

$$f = a_0 + xa_1 + x^2a_2 + \cdots, \tag{8}$$

we obtain a ring $R[[x]]$, called the *formal power series ring* in $x$ over $R$. It may be considered entirely formally, or as the completion of the polynomial ring $R[x]$ in the $x$-adic topology obtained by taking the ideals $(x^n)$ consisting of the powers of $x$ as system of neighborhoods of 0. This topological point of view is sometimes useful. Thus if we want to define a skew power series ring, we find that the rule (5) cannot be used when $\delta \neq 0$, because left multiplication by $a \in R$ is not continuous. When $\delta = 0$, we can define the skew power series ring $R[[x;\alpha]]$ as in the polynomial case, using the commutation rule

$$ax = xa^{\alpha}. \tag{9}$$

From $R[[x]]$ we obtain the ring $R((x))$ of *formal Laurent series* by localizing at $x$; the elements are now all Laurent series

$$f = \sum_{i=-N}^{\infty} x^i a_i,$$

with the multiplication defined as before. When $R = K$ is a skew field, $K((x))$ is again a skew field, as is well known. To define skew Laurent series we shall need to assume that $\alpha$ is an automorphism, because the commutation rule (9) now has to be supplemented by

$$ax^{-1} = x^{-1}a^{\alpha^{-1}}.$$

In the context of Laurent series we can deal with derivations by rewriting (5) (with an automorphism $\alpha$) as a commutation rule for $y = x^{-1}$:

$$ya = a^{\alpha}y + ya^{\delta}y.$$

By induction on $n$ we obtain

$$ya = a^{\alpha}y + a^{\delta\alpha}y^2 + a^{\delta^2\alpha}y^3 + \cdots + a^{\delta^{n-1}\alpha}y^n + ya^{\delta^n}y^n. \tag{10}$$

As $n \to \infty$, this expression converges to a series in $y$ (with coefficients on the left!) and this allows us to define the skew field of formal Laurent series in $x^{-1}$. If we think of $x$ as inducing the derivation $\delta$, we see that derivation produces divergence while its inverse produces convergence, a pattern familiar from analysis.

More generally, one can define skew power series over a skew field $K$ for any sequence of mappings $(\delta_n)$ of $K$ such that

$$az = za^{\delta_0} + z^2 a^{\delta_1} + \cdots + z^{n+1} a^{\delta_n} + \cdots,$$

provided that

D.1. The $\delta_i$ are additive maps of $K$ and $\delta_0$ is injective,

D.2.

$$(ab)^{\delta_n} = \sum_{i=0}^{n} a\Delta_i^n b^{\delta_i} \quad (n = 0, 1, 2, \ldots),$$

where $\Delta_i^n$ is the coefficient of $t^{n+1}$ in $(\sum t^{k+1}\delta_k)^{i+1}$ (with a central indeterminate $t$). In particular it follows that $\delta_0$ is an endomorphism of $K$; the sequence $(\delta_0, \delta_1, \ldots)$ is called a *higher* $\delta_0$*-derivation*.

For example, if $\delta$ is an $\alpha$-derivation which is nilpotent: $\delta^{n+1} = 0$, for some $n$, then the sequence $(\alpha, \delta\alpha, \delta^2\alpha, \ldots, \delta^n\alpha, 0, \ldots)$ is a higher $\alpha$-derivation, as follows from (10). Conversely, any higher $\delta_0$-derivation $(\delta_i)$ with $\delta_i = 0$ for $i > n$ is of this form, provided that $\delta_0$ is an automorphism and $\delta_0, \ldots, \delta_n$ are right linearly independent over $K$ (Smits 1968; Cohn 1985, p. 524). In the general case higher derivations are not well understood and there have been a number of studies (Brungs and Törner 1984; Dumas and Vidal 1992).

**1.3.** An interesting generalization of power series, the Malcev–Neumann construction, allows the group algebra of any ordered group to be embedded in a skew field. Let $M$ be a *monoid*, i.e. a semigroup with 1, and $K$ any ring. The *monoid ring* $K[M]$ is the free $K$-module on $M$ as basis, with multiplication

$$au.bv = abuv \quad a, b \in K, \; u, v \in M.$$

The general element of $K[M]$ has the form $\sum a_u u (a_u \in K)$; it could also be described as a family $(a_u)$ indexed by $M$, with almost all components 0. Then the multiplication takes the form

$$(a_u)(b_v) = (c_w), \quad \text{where } c_w = \sum a_u b_{u^{-1}w}. \tag{11}$$

Consider now the set of all series $\sum a_u u$, i.e. all families $(a_u)$. Addition can be defined as before, but we cannot use the multiplication (11), because it leads to infinite sums. To overcome this problem, let us assume that our monoid $M$ is *ordered*, i.e. a total ordering $\leqslant$ is defined, such that

$$x \leqslant x', \; y \leqslant y' \Rightarrow xy \leqslant x'y'.$$

Let $K((M))$ be the set of all series $\sum a_u u$ whose *support* $\{u \in M \mid a_u \neq 0\}$ is well-ordered. It can be shown that this set is a ring in which $\sum a_u u$ is invertible, provided that its leading term $a_s s$ (i.e. the first term in its support) is invertible. In particular, if $K$ is a field and $G$ is an ordered group, then $K((G))$ is a skew field, so the group algebra $K[G]$ has been embedded in a skew field (Malcev 1948; Neumann 1949), cf., e.g., (Cohn 1985, Chapter 8). These series are also called *Malcev–Neumann series*.

Any free group may be totally ordered, e.g., by writing its elements as infinite products of basic commutators and using the lexicographic ordering of its exponents (Hall 1959, Chapter 11). Thus the group algebra of a free group has been embedded in a skew field.

G.M. Bergman (1978) has obtained a useful normal form for series in $K((M))$ under conjugation: If $f = \sum a_u u$ has an invertible leading term $a_s s$ and $s \neq 1$, then there exists $q$ with leading term 1 such that $q^{-1} f q$ has its support in the centralizer of $s$ in $M$. Further, $q$ may be chosen so that its support meets the centralizer of $s$ in 1; under this hypothesis $q$ is unique (cf. (Bergman 1978) or (Cohn 1985, p. 529)). This result is most useful when centralizers are small; e.g., in a free group the centralizer of any element $\neq 1$ is a cyclic subgroup, hence for a free group $F$ any Malcev–Neumann series is conjugate to a Laurent series in a single variable.

## 2. Firs, semifirs and generalizations

**2.1.** Let $R$ be a ring; if all right ideals are projective (i.e. $R$ is *right hereditary*), then every submodule $M$ of a free right $R$-module $F$ is projective. This follows easily by restricting the projections $F \to R$ to $M$ and noting that $M$ splits over the image. Similarly, if all finitely generated right ideals of $R$ are projective (i.e. $R$ is *right semihereditary*), then every finitely generated submodule of a free module is projective.

In homological algebra the global dimension of a ring forms a means of classification, and the hereditary rings, i.e. the rings of global dimension 1 are simplest after the familiar case of global dimension 0 (the semisimple rings). A second mode of classification looks at the form taken by projective modules. Here the simplest class is formed by the *projective-free* rings, in which every finitely generated projective right module is free, of unique rank. By the duality for projective modules: $P^* = \mathrm{Hom}_R(P, R)$, it comes to the same to demand this condition for left modules. We shall be concerned with rings that are hereditary as well as projective-free. They are just the firs, formally defined as follows:

DEFINITION 1. A *right fir* (= free ideal ring) is a ring $R$ with invariant basis number in which every right ideal is free, as right $R$-module. *Left firs* are defined similarly and a left and right fir is called a *fir*.

THEOREM 2.1. *Over a right fir $R$, any submodule of a free right $R$-module is free.*

This follows like the corresponding assertion for hereditary rings mentioned earlier. In the commutative case a fir is just a principal ideal domain (PID), for every ideal is free and its rank cannot exceed 1 because any 2-element set is linearly dependent: $xy - yx = 0$, so $x, y$ cannot form a basis. Hence any nonzero (left or right) ideal is free on a 1-element basis, and this ensures that it is a PID. More generally, a right fir which is also a right Ore domain is a principal right ideal domain. Of course every PID is a fir; as examples of one-sided firs we have right but not left principal ideal domains, but there are also examples of one-sided firs that are not Ore (cf. 3.5 below).

The condition of invariant basis number (IBN) was imposed from the beginning (Cohn 1964) to exclude pathology. *Metafirs*, for which IBN fails, exist in profusion (cf. 3.6 below), but have not really been studied. If we impose finite generation, we obtain a notion which is automatically symmetric:

DEFINITION 2. A *semifir* is a ring $R$ with IBN in which every finitely generated right ideal (or equivalently, every finitely generated left ideal) is free.

To study semifirs, we introduce another condition which is itself symmetric. A relation

$$x_1y_1 + \cdots + x_ny_n = 0 \tag{1}$$

in a ring $R$ is called *trivial* if for each $i = 1, \ldots, n$ either $x_i = 0$ or $y_i = 0$. We shall call (1) an *$n$-term relation* and write it more briefly as

$$x.y = 0,$$

where $x$ is a row and $y$ is a column. If there is an invertible matrix $P$ such that on writing $x' = xP$, $y' = P^{-1}y$, the relation $x'y' = 0$ is trivial, then (1) is said to be *trivializable*. For example, writing $x = (3, -1, 4)$, $y = (2, 10, 1)^T$, we obtain a relation (1) which is not trivial, but which can easily be trivialized over $\mathbf{Z}$.

To prove the symmetry of semifirs, it is convenient to have another definition:

DEFINITION 3. For any integer $n \geqslant 0$, an *$n$-fir* is a ring $\neq 0$ in which every $m$-term relation for $m \leqslant n$ is trivializable.

Thus a 0-fir is a nonzero ring and a 1-fir is a nonzero ring in which $ab = 0$ implies $a = 0$ or $b = 0$; in other words, a 1-fir is just an integral domain.

THEOREM 2.2. *For any nonzero ring $R$ and any integer $n \geqslant 0$ the following conditions are equivalent:*

(a) *every relation of at most $n$ terms can be trivialized,*

(b) *every right ideal of $R$, on at most $n$ generators, is free of unique rank,*

(c) *every submodule on at most $n$ generators of a free right $R$-module is free of unique rank,*

(d) *every matrix relation $XY = 0$, where $X$ has at most $n$ columns, can be trivialized, the left-right analogues of* (a)–(d).

Here a matrix relation $XY = 0$, where $X$ is $r \times m$ and $Y$ is $m \times s$ say, is *trivialized* by an invertible $m \times m$ matrix $P$ if the relation $XP.P^{-1}Y = 0$ is trivial, in the sense that for each $i = 1, \ldots, m$, either the $i$-th column of $XP$ or the $i$-th row $P^{-1}Y$ is 0. A proof of Theorem 2.2 (which is not difficult) may be found in Cohn (1990, p. 427) or (1985, p. 66) or (1995, 1.6, p. 35f.).

If $\mathcal{S}_n$ denotes the class of $n$-firs, we have the inclusions

$$\mathcal{S}_0 \supset \mathcal{S}_1 \supset \cdots, \tag{2}$$

which are all proper (cf. 3.6 below), and $\mathcal{S} = \bigcap \mathcal{S}_n$ is the class of all semifirs, by (b) of Theorem 2.2. Moreover, this theorem shows the symmetry of the definition of a semifir; in fact we may define a semifir symmetrically as a nonzero ring in which every relation is trivializable. By (2) we see that every semifir, in particular, every left or right fir is an integral domain. It can be shown that every semifir can be embedded in a skew field (cf. (Cohn 1971, p. 283, 1985, p. 417, 1995, 4.5, p. 182)). Since every class $\mathcal{S}_n$ is clearly defined by elementary sentences, it follows from (2) by the compactness theorem of logic that embeddability in a skew field cannot be defined by a finite set of elementary sentences (Cohn 1974, 1981, 1995, 6.7, p. 328).

In the commutative case (or more generally, the Ore case) semifirs just reduce to Bezout domains (i.e. integral domains in which every finitely generated ideal is principal). In fact a commutative 2-fir is just a Bezout domain; thus the chain (2) collapses to $\mathcal{S}_0 \supset \mathcal{S}_1 \supset \mathcal{S}_2 = \mathcal{S}$ in the commutative case.

Over a local ring every finitely generated projective module is free (in fact, this holds for any projective module, cf. Kaplansky (1958)), and a local ring always has IBN, hence any local ring which is left (or right) semihereditary is a semifir. Such rings can also be characterized as follows (Cohn 1992b):

THEOREM 2.3. *A local ring $R$ is a semifir if and only if it satisfies the following trivializability condition:*

*Given $a_1, \ldots, a_n \in R$, if there is a nontrivial linear relation*

$$\sum a_i b_i = 0, \quad b_i \in R, \ \textit{not all } 0, \tag{3}$$

*then there exists a relation* (3) *in which one of the $b_i$ is a unit.*

Since the Ore condition holds for all Noetherian domains, it is clear that general firs will not be Noetherian. Nevertheless there is a chain condition satisfied by firs. A module is said to possess the *ascending chain condition on $n$-generator submodules*, ACC$_n$ for short, if any ascending chain of $n$-generator submodules becomes stationary. By a ring with right ACC$_n$ we mean a ring satisfying ACC$_n$ as right module over itself; similarly for left ACC$_n$.

THEOREM 2.4. *Let $R$ be a right fir. Then any free right $R$-module satisfies ACC$_n$ for all $n \geqslant 1$.*

For let $N_1 \subset N_2 \subset \cdots$ be an infinite strictly ascending chain of $n$-generator submodules of a free module $F$. Then the union $N = \bigcup N_i$ is countably but not finitely generated, hence free of countable rank, with basis $u_1, u_2, \ldots$, say. The submodule $P$ generated by $u_1, \ldots, u_{n+1}$ is a direct summand of $N$ and is contained in some $N_{i_0}$, hence a direct summand of $N_{i_0}$, but this contradicts the fact that $N_{i_0}$ is free of rank $n$.

The conclusion of Theorem 2.4 holds more generally for any finitely related $R$-module. Further, the result extends to $\aleph_0$-firs, where a right *$\alpha$-fir* is a ring in which every right ideal with a generating set of cardinal at most $\alpha$ is free, of unique rank (Bergman 1967; Cohn 1967, 1985, p. 72).

**2.2.** Semifirs satisfy a form of Sylvester's law of nullity, once the appropriate notion of rank has been defined. Let $A$ be an $m \times n$ matrix over any ring $R$; there are various ways of writing $A$ as a product of an $m \times r$ by an $r \times n$ matrix:

$$A = PQ, \quad P \in {}^mR^r,\ Q \in {}^rR^n. \tag{4}$$

If the factorization (4) of $R$ is chosen so as to give $r$ its least value, it is called a *rank factorization* of $A$ and $r$ itself is called the *inner rank* or simply the *rank* of $A$, written $r(A)$. For example, an element of $R$, as $1 \times 1$ matrix, has inner rank 1 unless it is 0, for 0 can be written as a product $PQ$, where $P$ is a matrix with no columns and $Q$ a matrix with no rows. When $R$ is a skew field, the above definition agrees with the usual definition of rank; in more general cases the rank is usually not defined, and even when it is (e.g., for commutative rings) it need not coincide with the inner rank.

It is clear that the inner rank of a product of matrices is bounded above by the ranks of the factors: $r(AB) \leqslant \min\{r(A), r(B)\}$. In a semifir we also have a lower bound on the inner rank:

SYLVESTER'S LAW OF NULLITY. *Let $R$ be a semifir and $A \in {}^mR^n$, $B \in {}^nR^p$. Then*

$$r(AB) \geqslant r(A) + r(B) - n. \tag{5}$$

The class of rings satisfying (5) has been studied by Dicks and Sontag (1978) (cf. also (Cohn 1985, 5.5, 1989b)) under the name Sylvester domain. Formally a *Sylvester domain* is defined as a ring $\neq 0$ such that

SD. *For any $A \in {}^mR^n$, $B \in {}^nR^p$, if $AB = 0$, then $r(A) + r(B) \leqslant n$.*

From this special case of (5) the full form can be obtained by taking a rank factorization of $AB$, say $AB = PQ$, where $Q$ has $r(AB)$ rows. Then

$$(A\ P)\begin{pmatrix} -B \\ Q \end{pmatrix} = 0,$$

hence by SD,

$$r(A) + r(B) \leqslant r(A\ P) + r(B\ Q)^T \leqslant n + r(AB),$$

i.e. (5). By taking $A, B$ to be $1 \times 1$, i.e. elements of $R$, we see that a Sylvester domain is indeed an integral domain.

Every Sylvester domain has weak global dimension at most two and is projective-free; in the commutative case the converse also holds, but in general it is not known whether every projective-free ring of weak global dimension at most two is a Sylvester domain. But every projective-free ring of weak global dimension at most 1 which is also right coherent is a semifir (cf. Dicks and Sontag (1978) or Cohn (1985, p. 256)).

A square matrix, say $n \times n$, of inner rank $n$ is said to be *full*. Thus over a skew field, a square matrix is clearly invertible if and only if it is full, and it follows that only full matrices can be inverted under a homomorphism to a skew field. If $\phi$: $R \to K$ is a homomorphism to a skew field which keeps every full matrix full (an *honest* homomorphism), then it is an embedding and the skew field generated by the image is a universal skew field of fractions of $R$, in a sense which can be made precise (cf. Cohn (1985, 7.2)). It can be shown that every semifir has a universal skew field of fractions inverting all full matrices. More generally (Dicks and Sontag 1978; Cohn 1985, p. 417):

THEOREM 2.5. *A ring has an honest homomorphism to a skew field* (*and hence has a universal skew field of fractions*) *if and only if it is a Sylvester domain.*

In a Sylvester domain every full matrix is regular (i.e. a non-zero-divisor); for an Ore domain the converse holds: If in an Ore domain $R$ every full matrix is regular, then $R$ is a Sylvester domain.

As examples of Sylvester domains, apart from semifirs, we have polynomial rings in two variables over a field: $k[x, y]$, but not in more than two variables. For example, in $k[x, y, z]$ the matrix

$$\begin{pmatrix} 0 & z & -y \\ -z & 0 & x \\ y & -x & 0 \end{pmatrix}$$

is full, and so of inner rank 3, but it is a zero-divisor, since it annihilates the row $(x, y, z)$ (Dicks and Sontag 1978; Cohn 1985, 5.5, 1989b). The polynomial ring $\mathbf{Z}[x]$ is a Sylvester domain; more generally, if $A$ is a commutative PID, then the free algebra $A\langle X\rangle$ is a Sylvester domain (Dicks and Sontag 1978; Cohn 1985, p. 260). The same method will show that the group algebra $A[F]$ of a free group $F$ over a commutative PID $A$ is a Sylvester domain; that $A[F]$ is projective-free was shown by Bass (1964).

## 3. The weak algorithm, free rings

**3.1.** Let $R$ be a *filtered ring*, i.e. $R$ has a sequence of additive subgroups

$$R_0 \subseteq R_1 \subseteq \cdots, \quad \bigcup R_h = R,$$

such that $R_i R_j \subseteq R_{i+j}$ and $1 \in R_0$. Then $R_0$ is a subring and each $R_h$ is an $R_0$-bimodule. Moreover, $R$ is an $R_0$-ring; this just means a ring with a homomorphism from $R_0$ into it (in this case an embedding). On $R$ we define a *filtration* $v$ by putting $v(0) = -\infty$ and for $x \neq 0$ defining

$$v(x) = \min\{h \mid x \in R_h\};$$

we shall call $v(x)$ the *degree* of $x$. It is clear that $v(x)$ is an integer-valued function on $R^\times$ satisfying:

V.1. $v(x) \geqslant 0$ for all $x \neq 0$ $(v(0) = -\infty)$,

V.2. $v(x - y) \leqslant \max\{v(x), v(y)\}$,

V.3. $v(xy) \leqslant v(x) + v(y)$,

V.4. $v(1) = 0$.

We shall mainly have to deal with the case where equality holds in V.3; then $v$ is called a *degree function*. Conversely, any $\mathbf{Z}$-valued function satisfying V.1–4 leads to a filtration on $R$. Every ring $R$ is *trivially* filtered by the rule $R = R_0$; the corresponding filtration: $v(x) = 0$ for all $x \neq 0$, is said to be *trivial*.

With a filtered ring $R$ we associate a graded ring $(H_i)$ in the usual way by writing $H_i = R_i/R_{i-1}$ and defining the product of $\alpha \in H_i$ and $\beta \in H_j$ by taking representatives $a \in R_i$ of $\alpha$ and $b \in R_j$ of $\beta$ and putting $\alpha\beta$ equal to the coset $ab + R_{i+j-1}$. Clearly $\alpha\beta$ depends only on $\alpha, \beta$ and not on $a, b$ and with this product $(H_i)$ becomes a graded ring, also written gr $R$. We shall need a notion of linear dependence; this will be defined in terms of the filtration, though it is easily expressed in terms of gr $R$, a task left to the reader.

If $R$ is any filtered ring, with filtration $v$, then a family $(a_i)$ of elements of $R$ is said to be *right $v$-dependent* if $a_i = 0$ for some $i$ or there exist elements $b_i \in R$, almost all 0, such that

$$v\left(\sum a_i b_i\right) < \max_i \{v(a_i) + v(b_i)\}.$$

A *$v$-independent* family is one that is not $v$-dependent. For example, any linearly dependent family is right $v$-dependent, though not conversely, and linear dependence is the special case of $t$-dependence, where $t$ is the trivial filtration.

An element $a \in R$ is said to be *right $v$-dependent* on a family $(a_i)$ if $a = 0$ or if there exist $c_i \in R$, almost all 0, such that

$$v\left(a - \sum a_i c_i\right) < v(a), \quad v(a_i) + v(c_i) \leqslant v(a) \quad \text{for all } i.$$

*Left $v$-dependence* is defined analogously.

Using a degree function we can express the usual division algorithm as follows:

DA. *Given $a, b \in R$, $b \neq 0$, there exist $q, r \in R$ such that*

$$a = bq + r, \quad v(r) < v(b). \tag{1}$$

An equivalent form turns out to be more convenient for us:

DA$'$. *For any $a, b \in R$ the one of $a, b$ of higher* (*or equal*) *degree is right $v$-dependent on the other.*

If DA holds, and $v(b) \leqslant v(a)$ say, then either $b = 0$ or (1) holds, so $v(a - bq) < v(b) \leqslant v(a)$, which shows $a$ to be right $v$-dependent on $b$. Conversely, assume DA$'$ and let $a, b \in R$ be such that $b \neq 0$. Choose $q \in R$ such that $v(a - bq)$ is minimal; to satisfy DA we must show that $v(a - bq) < v(b)$. But if $v(a - bq) \geqslant v(b)$, then by DA′ there exists $q_1 \in R$ such that $v(a - b(q + q_1)) < v(a - bq)$, and this contradicts the definition of $q$. This shows DA and DA′ to be equivalent.

The division algorithm is particularly suited to the study of polynomial rings, for we have

THEOREM 3.1. *Let $R$ be a ring with a degree function $v$. Then $R$ satisfies the division algorithm if and only if either $R$ is a skew field or $R = K[x; \alpha, \delta]$ for a skew field $K$ with endomorphism $\alpha$ and $\alpha$-derivation $\delta$, where $v(x) > 0$* (cf. Jacobson (1934); Cohn (1961, 1985, p. 92)).

For a generalization to several variables it is important *not* to require the variables to commute. The division algorithm is then replaced by a relative dependence condition on finite families:

WA$_n$. *A ring $R$ with a filtration $v$ is said to possess an $n$-term weak algorithm* (*WA$_n$*) *relative to $v$ if in any right $v$-dependent family of at most $n$ elements $a_1, \ldots, a_m$ $(m \leqslant n)$, where*

$$v(a_1) \leqslant \cdots \leqslant v(a_m)$$

*say, some $a_i$ is right $v$-dependent on $a_1, \ldots, a_{i-1}$. If the $n$-term weak algorithm for all $n$ holds in $R$, then $R$ is said to possess a weak algorithm* (*WA*).

To take some simple cases, WA$_1$ states that $v(ab) = v(a) + v(b)$, for $a, b \neq 0$, i.e. $v$ is a degree function; WA$_2$ states that $v$ is a degree function and for any $a, b$ that are right $v$-dependent, where $v(a) \geqslant v(b)$ say, there exists $c \in R$ such that $v(a - bc) < v(b)$. If moreover, $R$ is commutative, then any two elements are linearly dependent, as we have seen, hence right $v$-dependent, so for commutative rings WA$_2$ reduces to the division algorithm in the form DA′. More generally, this holds for any filtered Ore domain with WA$_2$ and it shows that in the commutative (and even Ore) case, WA$_2$ implies WA. By contrast, for general filtered rings the conditions WA$_n$ can be shown to be all distinct (Bergman (1967); see also 3.6 below). In any filtered ring $R$ with WA$_2$ the subring $R_0$ is a skew field; for if $a \in R_0^{\times}$, then $1, a$ are right $v$-dependent, because $1.a = a.1 \neq 0$, and $v(a) = v(1) = 0$, so 1 is right $v$-dependent on $a$: $v(1 - ab) < 0$, hence $ab = 1$. Now $b \in R_0$, so $bc = 1$ for some $c \in R$ and $c = ab.c = a.bc = a$, which shows $a$ to be invertible, as claimed.

By expressing the weak algorithm in terms of the associated graded ring, one can show that WA$_n$ and hence WA is left-right symmetric, i.e. it holds for a ring $R$ if and only if it holds for the opposite ring (Cohn 1961; Bergman 1967; Cohn 1985, 2.3).

A familiar argument shows that any commutative filtered ring with a division algorithm is a PID. Correspondingly one has

THEOREM 3.2. *A filtered ring with $n$-term weak algorithm is an $n$-fir with ACC$_n$; a filtered ring with weak algorithm is a* (*left and right*) *fir.*

To prove the second statement, one writes, for any right ideal $\mathfrak{a}$ of $R$, $\mathfrak{a}_r = \mathfrak{a} \cap R_r$, takes a maximal right $v$-independent subset $B_1$ of $\mathfrak{a}_1$ and defines $B_r$ recursively as a minimal spanning set of $\mathfrak{a}_r$ over $\mathfrak{a}'_r$, the set of elements of $\mathfrak{a}_r$ right $v$-dependent on $\mathfrak{a}_{r-1}$; in effect $B_r$ is an $R_0$-basis for $\mathfrak{a}_r/\mathfrak{a}'_r$. Now $B = \cup B_r$ is a basis for $\mathfrak{a}$ which is therefore free, and the rank is easily verified to be unique, so that IBN holds in $R$. Thus $R$ is a right fir, and by the symmetry of WA it is also a left fir. Now the case WA$_n$ follows easily as a special case, while ACC$_n$ follows by associating with any right ideal generated by $a_1, \ldots, a_m$ the indicator $(v(a_1), \ldots, v(a_m), \infty^{n-m})$ and noting that these indicators are well-ordered (in the lexicographic ordering) (Bergman 1967; Cohn 1985, 2.2).

**3.2.** Let $k$ be a commutative field and $X$ any set. The *free $k$-algebra* on $X$, written $k\langle X\rangle$, is the $k$-algebra generated by $X$ with an empty set of defining relations. Thus $k\langle X\rangle$ consists of all linear combinations of products of elements of $X$ and this expression is unique:

$$f = \sum a_I x_I, \quad a_I \in k, \text{ almost all } 0, \tag{2}$$

where $x_I = x_{i_1} \cdots x_{i_n}$ is a typical product of elements of $X$. The *degree* of $f$, defined as $\max\{|I| \mid a_I \neq 0\}$ provides a degree function on $X$. More generally, we obtain a degree function by assigning arbitrary positive integers as degrees to the elements of $X$. If $X^*$ denotes the free monoid on $X$, then $k\langle X\rangle$ may also be described as $k[X^*]$, the monoid algebra on $X^*$. When $X$ consists of a single element $x$, then $k\langle X\rangle = k[x]$ is just the polynomial ring in $x$, but for $|X| > 1$, $k\langle X\rangle$ is noncommutative. We recall the characteristic property of $k\langle X\rangle$: Given any $k$-algebra $C$ and any mapping $f$: $X \to C$, $x_i \mapsto c_i$, there is a unique homomorphism from $k\langle X\rangle$ to $C$ extending $f$, namely

$$\sum a_I x_{i_1} \cdots x_{i_n} \mapsto \sum a_I c_{i_1} \cdots c_{i_n}.$$

By the uniqueness of the normal form (2) this is well-defined. Since $k$ is a field, it is determined by $k\langle X\rangle$ as the set of all units, together with 0; $X$ cannot be determined in the same way, but its cardinal is an invariant of $k\langle X\rangle$, called the *rank* of the algebra.

Often a more general construction is needed, where the free generators need not commute with the scalars. Let $D$ be a skew field and $k$ a central subfield. Then the *free $D$-ring on a set $X$ over $k$* is defined as the $D$-ring generated by $X$ subject to the defining relations

$$\alpha x = x\alpha, \quad \text{for all } x \in X, \ \alpha \in k. \tag{3}$$

This ring is denoted by $D_k\langle X\rangle$; clearly it reduces to the previous case when $D = k$. It has the universal property that any map of $X$ into a $D$-ring extends to a unique homomorphism; here it is essential for $k$ to be contained in the center of $D$.

The elements of $D_k\langle X\rangle$ do not have quite as good a normal form as in (2) (to get a reasonable form one has to choose a basis $u_\lambda$ for $D$ over $k$ and consider the $k$-linear combinations of products of terms $u_\lambda x_i$), but there is again a degree function, defined as before.

More generally, let $U$ be any $D$-bimodule over $k$; this means that $\alpha u = u\alpha$ for all $u \in U$, $\alpha \in k$. Let us put

$$U^0 = D, \quad U^r = U \otimes U \otimes \cdots \otimes U \quad \text{with } r \text{ factors, } r \geqslant 1,$$

as $D$-bimodule, where the tensor product is taken over $D$. Then the direct sum

$$D\langle U\rangle = U^0 \oplus U^1 \oplus \cdots \tag{4}$$

can be defined as $D$-ring using the $D$-bilinear maps $D^i \otimes D^j \to D^{i+j}$. This ring is called the *tensor $D$-ring* on $U$; confusion with the earlier notation $k\langle X\rangle$ is unlikely, since it is usually clear whether $U$ is a $D$-bimodule or a set. In particular, $D_k\langle X\rangle$ has the form $D\langle U\rangle$, where $U = (D^0 \otimes_k D)^{(X)}$ and $D^0$ is the opposite ring of $D$. We again have a degree function, defined by the natural grading of (4); more generally, $U$ itself may be a direct sum of terms of different positive degrees.

We now show that for any $D$-bimodule $U$ over $k$, $D\langle U\rangle$ is a ring with a weak algorithm:

THEOREM 3.3. *Let $D$ be a skew field with a central subfield $k$ and let $U$ be a $D$-bimodule over $k$. Then the tensor $D$-ring $D\langle U\rangle$ is a ring with weak algorithm relative to the degree function defined by $U$. Hence $D\langle U\rangle$ is a fir.*

PROOF. We put $R = D\langle U\rangle$, decompose $U$ into its homogeneous components if necessary and take a basis $B$ of $U$ as left $D$-space. Then each element of $R$ can be written as a linear combination of monomial terms $au_1 \cdots u_r$, where $a \in D$, $u_i \in B$. Moreover, when two such terms are multiplied:

$$bv_1 \cdots v_s a u_1 \cdots u_r,$$

then the product can be brought to the form of a sum of monomial terms by rearranging $bv_1 \cdots v_s a$. When this has been done, all terms in the sum will clearly end in $u_1 \cdots u_r$.

We fix a particular monomial $u_1 \cdots u_r$ of degree $m$ (usually $m = r$, but this is not essential) and define the *left transduction* for this monomial as the left $D$-linear mapping $f \mapsto f^*$ of $R$ into itself which maps any monomial of the form $wu_1 \cdots u_r$ to $w$ and all others (those not ending in $u_1 \cdots u_r$) to 0. This is a well-defined map, because the different monomials form a $D$-basis of $R$; for any $f \in R$ we have $v(f^*) \leqslant v(f) - m$ and for $f, g \in R$ we have

$$(fg)^* \equiv fg^* \pmod{T_{v(f)-1}}. \tag{5}$$

For when $g$ is a term of degree at least $m$, we actually have equality; when $g$ is a term of degree less than $m$, the right-hand side of (5) vanishes and (5) then holds as a congruence, hence it holds generally by linearity.

We can now verify the weak algorithm for $R$. Assume that $a_1, \ldots, a_n$ is a right $v$-dependent family, i.e. $b_1, \ldots, b_n \in R$ exist such that

$$v\Big(\sum a_i b_i\Big) < d = \max\big\{v(a_i) + v(b_i)\big\}.$$

We may assume that the $a_i$ are numbered so that $v(a_1) \leqslant v(a_2) \leqslant \cdots \leqslant v(a_n)$ and we then have to show that some $a_i$ is right $v$-dependent on $a_1, \ldots, a_{i-1}$. By omitting terms if necessary we may assume that $v(a_i) + v(b_i) = d$ for all $i$; then $v(b_1) \geqslant \cdots \geqslant v(b_n)$.

In terms of our basis $B$ for $U$ let $u_1 \cdots u_r$ be a product of maximal degree $m = v(b_n)$ occurring in $b_n$ with a nonzero coefficient $\alpha$ and write $*$ for the left transduction for $u_1 \cdots u_r$. In $\sum a_i b_i^*$ the $i$-th term differs from $(a_i b_i)^*$ by a term of degree $< v(a_i) \leqslant v(a_n)$. Hence

$$v\Big(\sum a_i b_i^* - \sum (a_i b_i)^*\Big) < v(a_n),$$

while

$$v\Big(\sum (a_i b_i)^*\Big) \leqslant v\Big(\sum a_i b_i\Big) - m < d - m = v(a_n).$$

Thus $v(\sum a_i b_i^*) < v(a_n)$, and this gives a relation of right $v$-dependence of $a_n$ on $a_1, \ldots, a_{n-1}$ because $b_n^* = \alpha \in D^X$. □

This result was first proved for free algebras (Cohn 1961); the above proof is modeled on that of Bergman (1967). Conversely, it can be shown that any filtered $D$-ring with a weak algorithm has a free generating set $X$ for a left ideal complementing $D$ such that every element can be uniquely written in the form (2) (Cohn 1961). More generally, Bergman has given a complete determination of all rings with weak algorithm (Bergman 1967; Cohn 1985, p. 113). We note that Theorem 3.3 shows in particular that $D_k\langle X\rangle$ with the usual degree function (assigning degree 1 to all elements of $X$) has weak algorithm and so is a fir.

**3.3.** Let $R$ be any filtered ring for which $R_0$ is a skew field $K$. If the terms $R_n/R_{n-1}$ of the associated graded ring are finite-dimensional as right $K$-spaces, say $\dim_K(R_n/R_{n-1}) = \alpha_n$, then we can form the formal power series

$$H(R : K) = \sum \alpha_n t^n.$$

It is called the *Hilbert series* (or also Poincaré series) of $R$. For example, in the free algebra $k\langle X\rangle$ let us assign positive degrees to the elements of $X$; if there are $\lambda_d$ elements of degree $d$, we put $H(X) = \sum \lambda_d t^d$. With this notation an easy counting argument shows the truth of

THEOREM 3.4. *The Hilbert series of a free algebra* $k\langle X\rangle$, *where* $X$ *contains* $\lambda_d$ *elements of degree* $d$ *and* $H(X)=\sum\lambda_d t^d$, *is*

$$H(R:k)=\big(1-H(X)\big)^{-1}.$$

In particular, if $X$ consists of $m$ elements, all of degree 1, then $H(R:k)=(1-mt)^{-1}$. This should be compared with the corresponding formula $H(A:k)=(1-t)^{-m}$ for the polynomial ring $A=k[x_1,\dots,x_m]$.

Similar results can be proved more generally for any ring with a weak algorithm (Cohn 1985, p. 107).

Any finitely presented module $M$ over a semifir $R$ has a resolution

$$0\longrightarrow R^m\longrightarrow R^n\longrightarrow M\longrightarrow 0.$$

Here the number $n-m$ depends only on $M$, not on the resolution (by Schanuel's lemma) and it is called the *characteristic* of $M$, $\chi(M)$. If $R=k\langle X\rangle$ as in Theorem 3.4 and $M$ is a finitely presented $R$-module, then it can be shown (Cohn 1985, p. 109) that

$$\chi(M)=\big(1-|X|\big)\dim_k(M). \tag{6}$$

For example, if $R=k\langle x_1,\dots,x_m\rangle$, $\mathfrak{a}$ is a right ideal of rank $r$ and $\dim_k(R/\mathfrak{a})=n$, then by applying (6) to $R/\mathfrak{a}$ we find $\chi(R/\mathfrak{a})=1-r$, hence

$$r-1=(m-1)n. \tag{7}$$

This is the Schreier–Lewin formula (Lewin 1969; Cohn 1969); it is the analogue of Schreier's formula for groups.

**3.4.** In the free algebra $k\langle X\rangle$ we have, besides the degree function, an *order function*, defined as the least degree of terms in the support. This may be regarded as arising from an inverse filtration, satisfying

I.1. $v(x)\in\mathbf{N}$, for $x\neq 0$, $v(0)=\infty$,

I.2. $v(x-y)\geqslant\min\{v(x),v(y)\}$,

I.3. $v(xy)\geqslant v(x)+v(y)$.

An inversely filtered ring $R$ is said to have an *inverse weak algorithm* (IWA) if the corresponding graded ring gr $R$ satisfies the weak algorithm, defined as before. This is not an algorithm ending in a finite number of steps, but in general it has an infinite number of steps and to obtain useful results one has to operate in the completion of $R$ relative to the given filtration. Just as the weak algorithm characterizes free algebras, so the inverse weak algorithm (with completeness) characterizes free power series rings. Here the free power series ring may be defined as the completion of the free algebra $k\langle X\rangle$ in the $X$-adic topology, i.e. the topology defined by the powers of the ideal generated by $X$.

THEOREM 3.5. *A complete inversely filtered ring* $R$ *satisfies an inverse weak algorithm if and only if* $R/R_1$ *is a skew field* $K$ *and* $R_1$ *contains a set* $X$ *such that every element* *of $R$ can be expressed as a convergent series in the products of elements of $X$ with coefficients in a set of representatives of $K$ in $R$.*

We have stated this result somewhat loosely to convey the flavor; for a precise form further definitions are needed (cf. Cohn (1962, 1985, p. 129)).

A ring with IWA is a semifir but not generally a fir; all we can generally say is that it is a 'topological' fir, in the sense that every right ideal $\mathfrak{a}$ has a linearly independent subset generating a right ideal dense in $\mathfrak{a}$.

As for the weak algorithm one can define the inverse $n$-term weak algorithm; in the commutative (or more generally, the Ore) case the IWA is a consequence of the 2-term IWA, and a complete inversely filtered ring with 2-term IWA is a discrete rank 1 valuation ring (or a skew field).

A natural question asks when the $\mathfrak{a}$-adic filtration (by the powers of an ideal $\mathfrak{a}$) defines an IWA. This is answered by

THEOREM 3.6. *Let $R$ be a ring with an ideal $\mathfrak{a}$ such that*
(i) *$R/\mathfrak{a}$ is a skew field,*
(ii) $\mathfrak{a} \otimes_R \mathfrak{a} = \mathfrak{a}^2$,
(iii) $\bigcap \mathfrak{a}^n = 0$.
*Then $R$ has an inverse weak algorithm relative to the $\mathfrak{a}$-adic filtration.*

Condition (ii) holds whenever $\mathfrak{a}$ is flat as right (or left) ideal. Two important conditions where these conditions are satisfied are:

1. $R$ is a fir with an ideal $\mathfrak{a}$ such that $R/\mathfrak{a}$ is a skew field.
2. $R$ is a semihereditary local ring with maximal ideal $\mathfrak{m}$ which is finitely generated as right ideal, and whose powers intersect in zero.

Cf. Cohn (1970, 1985, p. 132, 1992b).

For an inversely filtered ring with IWA there is an important relation with its completion. Let $S$ be a ring and $R$ a subring. Given $a \in S^n$, $b \in {}^nS$, the product

$$ab = \sum a_i b_i$$

is said to lie *trivially* in $R$ if for each $i = 1, \ldots, n$ either $a_i$ and $b_i$ lie in $R$ or $a_i = 0$ or $b_i = 0$. If for any families of rows $(a_\lambda)$ in $S^n$ and columns $(b_\mu)$ in ${}^nS$ such that $a_\lambda b_\mu \in R$ for all $\lambda$, $\mu$ there exists an invertible $n \times n$ matrix $P$ over $S$ such that all the products $a_\lambda P.P^{-1} b_\mu$ lie trivially in $R$, then $R$ is said to be *totally inert* in $S$. If this holds for all finite families $(a_\lambda)$, $(b_\mu)$, $R$ is said to be *inert* in $S$.

To illustrate this notion let us show that every inert embedding is honest. Let $A$ be a full matrix over $R$ and suppose that $A = UV$ is a rank factorization over $S$. Since $R$ is inert in $S$, there exists an invertible matrix $P$ such that all entries of the product $UP.P^{-1}V$ lie trivially in $R$. Now no column of $UP$ and no row of $P^{-1}V$ can vanish, because $UV$ was a rank factorization. Hence all entries of $UP$ and $P^{-1}V$ lie in $R$, and this shows $A$ to be full over $S$, as claimed. Let us state two important cases of inertia:

THEOREM 3.7 (Inertia theorem). *Let $R$ be a fir which is inversely filtered with inverse weak algorithm and let $\widehat{R}$ be its completion. Then $R$ is totally inert in $\widehat{R}$.*

For a proof see Cohn (1985, p. 133). Special cases of this result were obtained by Tarasov (1967) and Bergman (1967). For some purposes the following more elementary result is sufficient:

THEOREM 3.8 (Inertia lemma). *Let $R$ be a semifir. Then for any central indeterminate $t$, $R[[t]]$ is inert in $R((t))$.*

With the help of Amitsur's theorem on generalized polynomial identities this leads to a useful property of full matrices:

THEOREM 3.9 (Specialization lemma). *Let $D$ be a skew field with infinite center $k$ such that $[D : k]$ is infinite. Then any full matrix over $D_k\langle X\rangle$ becomes invertible for some choice of values of $X$ in $D$* (Cohn 1985, p. 285, 1990, p. 429, 1995, 6.2, p. 287).

The condition $[D : k] = \infty$ is clearly necessary; whether it is necessary for $k$ to be infinite is not known, but this is also needed in Amitsur's proof of his theorem on rational identities (Amitsur (1966), or for a proof using the specialization lemma, Cohn (1972)).

**3.5.** Because of its symmetry the weak algorithm is unsuitable for constructing one-sided firs. We therefore try to modify the definition of a degree function. In that definition, V.1–4 of 3.1, we used the addition on **N**. Instead of **N** we shall use the ordinals as values. Thus we consider a function $w$ on a ring $R$ satisfying

T.1. $w$ maps $R^{\times}$ to an initial segment of the ordinals, $w(1) = 0$, $w(0) = -1$,

T.2. $w(a - b) \leqslant \max\{w(a), w(b)\}$,

T.3. $w(ab) \geqslant w(a)$ for any $b \in R^{\times}$.

Such a function $w$ will be called a *transfinite degree function* on $R$. Clearly its existence implies that $R$ is an integral domain (by T.3). We also note that the usual division algorithm satisfies T.1–3. Given a transfinite degree function $w$ on $R$, a family $(a_i)$ in $R$ is called *right $w$-dependent* if $a_i = 0$ for some $i$ or there exist $b_i \in R$ almost all 0, such that

$$w\Bigl(\sum a_i b_i\Bigr) < \max_i \{w(a_i b_i)\},$$

and $a \in R$ is said to be *right $w$-dependent* on a family $(a_i)$ if $a = 0$ or there exist $c_i \in R$ almost all 0 such that

$$w\Bigl(a - \sum a_i c_i\Bigr) < w(a), \quad w(a_i c_i) \leqslant w(a) \quad \text{for all } i.$$

Now $R$ is said to possess a right *transfinite weak algorithm* (TWA) if in any right $w$-dependent family $a_1, \ldots, a_n$ with $w(a_1) \leqslant \cdots \leqslant w(a_n)$, some $a_i$ is right $w$-dependent on $a_1, \ldots, a_{i-1}$. As for the WA one shows that a ring with right TWA for a function $w$ is a right fir and the set $\{x \in R \mid w(x) \leqslant 0\}$ is a skew field.

To find examples of this notion let us turn to monoids. A monoid $M$ is said to be *rigid* if it satisfies cancellation and if $au = bv$, then either $a = bs$ or $b = as$ for some

$s \in M$. Further, if $ab = 1$ implies $a = b = 1$, $M$ is said to be *conical*, and $M$ satisfies right $\mathrm{ACC}_1$ if any ascending chain of principal right ideals $a_1M \subset a_2M \subset \cdots$ breaks off.

On any monoid $M$ we can define a preordering by left divisibility:

$$u \leqslant v \quad \text{if and only if} \quad v = us \text{ for some } s \in M.$$

If $M$ is conical and has cancellation, this is actually a partial ordering, which satisfies the minimum condition precisely when right $\mathrm{ACC}_1$ holds in $M$. Suppose that $M$ is a conical monoid with right $\mathrm{ACC}_1$ which moreover is rigid. Then for any $s \in M$, the lower segment generated by $s$, viz. $\{x \in M \mid x \leqslant s\}$ is totally ordered and so by $\mathrm{ACC}_1$ is an ordinal number, which we shall denote by $w(s)$, and call the *transfinite degree function* defined by left divisibility. It is easily verified that

$$w(u) \leqslant w(v) \Rightarrow w(cu) \leqslant w(cv) \quad \text{for } u, v, c \in M, \tag{8}$$

$$w(b) \leqslant w(c) \Rightarrow w(bu) \leqslant w(cu) \quad \text{for } b, c, u \in M. \tag{9}$$

Let $R = k[M]$ be the monoid algebra of $M$ over a field $k$ and extend $w$ to $R$ by writing

$$w\Bigl(\sum \lambda_s s\Bigr) = \max\bigl\{w(s) \mid \lambda_s \neq 0\bigr\}.$$

Then (8), (9) still hold when $b, c \in R$, $u, v \in M$, and T.1–3 can be verified, as well as the transfinite weak algorithm. Thus we have

THEOREM 3.10. *Let $M$ be a conical rigid monoid with right $ACC_1$. Then the monoid algebra $k[M]$ satisfies the transfinite weak algorithm with respect to the left divisibility ordering of $M$, and hence $k[M]$ is a right fir.*

For details of the proof see Cohn (1985, p. 141). This construction developed from an earlier one (Cohn 1969a) which was suggested by a method of Skornyakov (1965) for constructing one-sided firs.

By way of example consider the monoid $M$ generated by $y, x_i$ $(i \in \mathbf{Z})$ with the defining relations

$$yx_i = x_{i-1}.$$

Every element of $M$ can be written in the form

$$x_{i_1} \ldots x_{i_r} y^m, \quad i_\rho \in \mathbf{Z},\ r, m \geqslant 0,$$

and this form makes it easy to verify that $M$ is a conical rigid monoid and satisfies right $\mathrm{ACC}_1$. Thus the monoid algebra $R = k[M]$ is a right fir, but it is not a left fir, because left $\mathrm{ACC}_1$ fails to hold: $Rx_0 \subset Rx_1 \subset \cdots$ (Cohn 1985, p. 142).

If in Theorem 3.10 we omit right $\mathrm{ACC}_1$, the ring need not even be a 2-fir, as Cedó (1988) has shown by taking the group $G = \langle x, y \mid yxy = x\rangle$. The submonoid $H$ generated by $x$ and $y$ is conical and rigid, but $k[H]$ is not a 2-fir, since the relation

$$(1-x)(1-y) = (1-y)(1+xy)$$

is not trivializable.

To exclude this case, let us define a monoid $M$ to be *irreflexive* if for any $a, b, c \in M$ such that $a$ is a nonunit and $a = bac$, it follows that $b = c = 1$. Then the monoids whose monoid algebra is a right fir can be characterized as rigid irreflexive monoids with right $\mathrm{ACC}_1$ whose group of units is free (Kozhukhov 1982). The monoid algebra of $M$ is a two-sided fir if and only if $M$ is the free product of a free monoid and a free group (Wong 1978).

**3.6.** For any filtered ring $R$ with a degree function $v$ we can define the *dependence number* $\lambda_v(R)$ of $R$ relative to $v$ as the largest number $n$ for which $\mathrm{WA}_n$ holds, or $\infty$ if $\mathrm{WA}_n$ holds for all $n$. The larger $\lambda_v(R)$, the more we can prove about $R$, and it would be desirable to be able to read off $\lambda_v$ from a presentation of $R$. This is too much to expect, but for presentations of a certain prescribed form, basically to ensure that no "short" relations occur, it is possible to establish a lower bound for $\lambda_v(R)$. We shall not describe the precise result here (cf. Cohn (1969b, 1985, p. 145)), but state a simple consequence which is often useful. It requires a basic construction which itself is of independent interest and is not as widely known as it should be.

Let $Rg$ be the category of rings and $Rg_n$ the category of all $n \times n$ matrix rings (over some other ring). Thus an object of $Rg_n$ consists of a ring $R$ with $n^2$ distinguished elements $e_{ij}$, the matrix units, which satisfy the relations

$$e_{ij}e_{kl} = \delta_{jk}e_{il}, \quad \sum e_{ii} = 1. \tag{10}$$

A morphism in $Rg_n$ is just a ring homomorphism preserving the matrix units. We have a functor $M_n\colon Rg \to Rg_n$ which associates with each ring $R$ its $n \times n$ matrix ring $M_n(R)$. This functor has a left adjoint, the *$n$-matrix reduction functor* $W_n$:

$$Rg\big(R, M_n(S)\big) \cong Rg\big(W_n(R), S\big). \tag{11}$$

This functor $W_n$ may also be defined explicitly: Let $F_n(R)$ be the ring generated by $R$ and $n^2$ elements $e_{ij}$ satisfying the equations (10) for matrix units, and in case $R$ is a $k$-algebra, the relations $\alpha e_{ij} = e_{ij}\alpha$ ($\alpha \in k$). In terms of coproducts (Section 5) we have

$$F_n(R) = R \mathbin{*_k} M_n(k).$$

Now $F_n(R)$ contains $n^2$ matrix units, by construction, and so is of the form $F_n(R) = M_n(P)$, where $P$ is the centralizer of all the $e_{ij}$ in $F_n(R)$ (cf., e.g., Cohn (1989a, p. 136)). This ring $P$ is denoted by $W_n(R)$, so that

$$F_n(R) = M_n\big(W_n(R)\big).$$

Intuitively $W_n(R)$ may be described as follows: take the elements of $R$, treat them as $n \times n$ matrices and form the ring consisting of all their entries. Explicitly this means that for each $a \in R$ we have $n^2$ elements $a_{ij}$, given by

$$a_{ij} = \sum_{\nu} e_{\nu i} a e_{j\nu}.$$

By the properties of matrix rings we have

$$\begin{aligned} Rg(R, M_n(S)) &\cong Rg_n(F_n(R), M_n(S)) = Rg_n(M_n(W_n(R)), M_n(S)) \\ &\cong Rg(W_n(R), S). \end{aligned}$$

Thus $W_n(R)$ as defined here satisfies (11), and, as is well known, this determines it up to natural equivalence.

Now the theorem mentioned (but not quoted) above can be used to establish

THEOREM 3.11. *Let $R$ be a nonzero $k$-algebra. Then the matrix reduction $W_n(R)$ has a filtration $v$ for which the $(n-1)$-term weak algorithm holds. Hence $W_n(R)$ is an $(n-1)$-fir.*

For a proof see Bergman (1974b, p. 57) or Cohn (1985, p. 148, or 1995, 5.7, p. 247). More generally this method shows that if $R$ is an $(r-1)$-fir but not an $r$-fir, then $W_n(R)$ is an $(nr-1)$-fir, but not an $nr$-fir. Another possible generalization consists in replacing the elements of $R$ by rectangular matrices such that for any product $ab$ occurring in a defining relation the number of columns of $a$ and rows of $b$ is $\nu$, where $\nu \geqslant n$.

As an application we construct, for any $n \geqslant 1$, $(n-1)$-firs that are not $n$-firs. Let $R$ be generated by $a, b$ with defining relation $ab = 1$. Then $T = W_n(R)$ is an $(n-1)$-fir, but there are two $n \times n$ matrices $A$, $B$ over $T$ such that $AB = I$, $BA \neq I$. Explicitly, this means that $T^n \cong T^n \oplus K$ as right $T$-modules, where $K \neq 0$, and it shows that $T$ is not an $n$-fir. Similarly one can show (using the remark after Theorem 3.11) that the $k$-algebra with $2mn$ generators $a_{ir}, b_{ri}$ $(i = 1, \dots, m,\ r = 1, \dots, n)$ satisfying the relations (in matrix form, writing $A = (a_{ir})$, $B = (b_{ri})$),

$$AB = I_m, \quad BA = I_n$$

is an $r$-fir, where $r = \min(m, n) - 1$. For $m < n$ say, this algebra $R$ is an $(m-1)$-fir, but it is not an $m$-fir, since $R^m \cong R^n$. However, $R$ can be shown to be a metafir, using the methods of Bergman (1974a, 1974b), see Cohn (1995), 5.7, p. 246ff.

Another example illustrates the construction of Sylvester domains (cf. 2.2 above). For any integers $m$, $n$, $r$ denote by $R(m, r, n)$ the $k$-algebra generated by $r(m+n)$ elements forming the entries of an $m \times r$ matrix $A$ and an $r \times n$ matrix $B$, with defining relations (in matrix form) $AB = 0$. By the remark after Theorem 3.11, $R$ is an $(r-1)$-fir. Moreover, Dicks and Sontag (1978) show (using results from Bergman (1974b)) that $R(m, r, n)$

(i) is a Sylvester domain if and only if $r \geqslant m + n$,

(ii) has every full matrix regular if and only if $r > \max(m, n)$,

(iii) has every full matrix left regular if and only if $r > n$.

There is another result for estimating the dependence number, due to Hedges (1987). Let $R$ be any ring and $n \geqslant 1$; an element $c \in R$ is said to be *$n$-irreducible* if it is not zero or a unit and in any representation as a sum of terms

$$c = \sum_1^n a_i b_i$$

either $\sum a_i R = R$ or $\sum R b_i = R$. For example, in an integral domain an element is 1-irreducible precisely when it is an atom (i.e. unfactorable). Now Hedges (1987) proves the following result relating to algebras with a single homogeneous defining relation:

THEOREM 3.12. *Let $F = k\langle X\rangle$ be the free $k$-algebra on a graded set $X$ and let $c$ be an element of $F$ which is homogeneous for the given grading and $n$-irreducible. Then the 1-relator graded algebra $R = F/FcF$ satisfies the $n$-term weak algorithm relative to the degree function induced from $F$. In particular, if $c$ is an atom, then $R$ is an integral domain.*

More precisely, Hedges shows that the $n$-irreducibility of $c$ is necessary as well as sufficient for $\mathrm{WA}_n$ and both are equivalent to the condition that $R$ is an $n$-fir.

## 4. Modules over firs and semifirs

**4.1.** Let $R$ be any ring. An $R$-module $M$ is said to be *bound* if

$$M^* = \mathrm{Hom}_R(M, R) = 0,$$

*unbound* if it contains no bound submodule apart from 0. The pair of classes consisting of all bound and all unbound modules form the torsion theory cogenerated by $R$ (cf., e.g., Stenström (1975, p. 219)). For example when $R = \mathbf{Z}$, the finitely generated bound modules reduce to torsion modules and generally bound modules play a similar role to torsion modules (though the term 'torsion module' will in general be used here in a specific way, defined below). Over a semihereditary ring any finitely generated projective module is unbound; Theorem 4.1 below gives conditions under which the converse holds.

Given any $R$-module $M$ (for any ring $R$), we can form the sum of all bound submodules of $M$ to obtain a submodule $M_b$, the *bound component* of $M$. It is characterized as the largest bound submodule of $M$, or also as the smallest submodule such that the quotient $M/M_b$ is unbound. To state conditions for $M$ to split over $M_b$ we shall need a finiteness condition (Cohn 1992a). A ring $R$ is said to possess *bounded decomposition type* (BDT) if for each $n \geqslant 1$ there exists $r = r(n)$ such that $R^n$ cannot be written as a direct sum of more than $r$ terms. For example, any projective-free ring has BDT (with $r(n) = n$); this includes all semifirs.

THEOREM 4.1. *Let $R$ be a right semihereditary ring with bounded decomposition type and let $M$ be a finitely generated right $R$-module with bound component $M_b$. Then there is a decomposition*

$$M = M_b \oplus P, \tag{1}$$

*where $P$ is projective.*

For if $M$ is unbound, then an induction on the number of generators of $M$ shows $M$ to be isomorphic to a direct sum of right ideals, hence projective. For general $M$ this shows $M/M_b$ to be projective and this leads to the splitting (1).

The characteristic of a finitely presented module has already been defined in 3.3. More generally, for any semihereditary ring $R$ with an embedding in a skew field $K$ we can define a rank function on projective modules by taking $r(P)$ to be the dimension over $K$ of $P \otimes_R K$. This rank function is integer-valued and faithful (i.e. $r(P) \neq 0$ for $P \neq 0$) and it satisfies Sylvester's law of nullity ((5) of 2.2). Conversely, a faithful integer-valued rank function (with $r(R) = 1$) for which Sylvester's law of nullity holds, leads to an embedding in a skew field (Schofield 1985, p. 106). In terms of such a rank function we can again, on a semihereditary ring, define a characteristic $\chi(M)$ for finitely presented modules. In addition we put $\chi(M) = -\infty$ if $M$ is finitely generated but not finitely related, and $\chi(M) = \infty$ if $M$ is not finitely generated. The characteristic so defined is non-negative for commutative rings, but in general it may take negative values, e.g., when $R = k\langle x, y\rangle$, $M = R/(xR + yxR + y^2xR)$, then $\chi(M) = -2$.

Over a semihereditary ring $R$ with such a rank function we define a *torsion module* as an $R$-module $M$ such that $\chi(M) = 0$ and $\chi(M') \geqslant 0$ for all submodules $M'$ of $M$. Over a semifir every finitely presented module is defined by a matrix and here $M$ is a torsion module precisely when the defining matrix is full (cf. 2.2 above). Generally, in any homomorphism between torsion modules, $f$: $M \to N$, the kernel and cokernel are again torsion, by the exact sequence

$$0 \longrightarrow \ker f \longrightarrow M \xrightarrow{f} N \longrightarrow \operatorname{coker} f \longrightarrow 0,$$

which leads to the inequalities $0 \leqslant \chi(\ker f) = \chi(\operatorname{coker} f) \leqslant 0$.

For any semihereditary ring the transpose functor $\operatorname{Tr}(M) = \operatorname{Ext}^1_R(M, R)$ establishes a duality between left and right torsion modules and $\operatorname{Tr}^2(M) \cong M$ (in fact, Tr can be defined more generally and plays a role in the representation theory of algebras). For hereditary rings the torsion modules satisfy both chain conditions (cf. Theorem 2.4 for the case of firs, Cohn (1985, p. 231) for the general case). This leads to the following description of the category of torsion modules (Cohn 1992a).

THEOREM 4.2. *Let $R$ be a semihereditary ring with an embedding in a skew field. Then the category of $\mathcal{T}_R$ of right torsion modules is an abelian category closed under extensions and the functor* $\operatorname{Tr}(M) = \operatorname{Ext}^1_R(M, R)$ *establishes a duality between $\mathcal{T}_R$ and ${}_R\mathcal{T}$, the category of left torsion modules. If $R$ is hereditary, all the objects of $\mathcal{T}_R$ and ${}_R\mathcal{T}$ are of finite length.*

In the special case of firs, Theorem 4.2 can be translated into matrix language and it then yields a theorem on the factorization of matrices. Two matrices $A$, $B$ (not necessarily square) over a ring $R$ are said to be *associated* if there exist invertible matrices $P$, $Q$ over $R$ such that $A = PBQ$. If $A \oplus I$ is associated to $B \oplus I$, where the unit matrices need not be of the same size, then $A$ and $B$ are *stably associated*. Any $m \times n$ matrix $A$ defines a mapping of free left $R$-modules $f$: $R^m \to R^n$ or of free right $R$-modules ${}^nR \to {}^mR$. The map $f$ is injective precisely when $A$ is left regular, i.e. $XA = 0$ implies $X = 0$, and the left $R$-module defined by $A$ is bound if and only if $A$ is right regular. Two regular matrices define isomorphic modules precisely when they are stably associated (Cohn 1985, p. 27f.). Another equivalent condition can be stated in term of comaximal relations. Two matrices $A, B$ each with $m$ rows are *right comaximal* if their columns span the free right $R$-module ${}^mR$; 'left comaximal' is defined similarly and a matrix relation $AB' = BA'$ is called *comaximal* if $A, B$ are right and $A', B'$ left comaximal. We recall that a ring $R$ is *weakly finite* if $R^n \cong R^n \oplus K$ (as left $R$-modules) implies $K = 0$; in terms of matrices this means that $AB = I$ implies $BA = I$ for any square matrices $A$, $B$. For example, any semifir is clearly weakly finite. Now the isomorphism of modules defined by matrices is described in the following theorem (cf. Cohn (1985, p. 28)):

THEOREM 4.3. *Let $R$ be a weakly finite ring and $A$, $A'$ any matrices over $R$. Then* (a), (b) *below are equivalent and imply* (c):

(a) *$A$ and $A'$ satisfy a comaximal relation $AB' = BA'$,*

(b) *$A$ and $A'$ are stably associated,*

(c) *the left modules defined by $A$ and $A'$ are isomorphic.*

*When $A$, $A'$ are left regular, all three conditions are equivalent.*

To translate Theorem 4.2 in the case of firs into matrix language, let us define an *atomic* matrix or *atom* as a square matrix which is not a unit and which cannot be written as a product of (square) nonunits. In particular, such a matrix must be full, since otherwise we could write it as a product of matrices with zero rows and columns.

THEOREM 4.4. *Let $R$ be a fir. Then any full matrix is either a unit or its admits a factorization into atoms. Any two such factorizations have the same number of terms; if they are $C = A_1 \cdots A_r = B_1 \cdots B_r$, then for some permutation $i \mapsto i'$ of $1, \dots, r$, $A_{i'}$ is stably associated to $B_i$.*

This result can be proved more generally for full $n \times n$ matrices over a $2n$-fir with left and right $\mathrm{ACC}_n$ (Cohn 1985, p. 168).

For a PID there is a stronger conclusion; in this case the full matrices are just the regular square matrices, and the inner rank can then be defined as the order of the largest regular square submatrix. We shall say that $a$ is a *total divisor* of $b$, $a\|b$, if $aR \supseteq Rc = cR \supseteq bR$ for some $c \in R$.

THEOREM 4.5. *Let $R$ be a principal ideal domain and $A$ an $n \times n$ matrix of rank $r$ over $R$. Then there exist invertible $n \times n$ matrices $P$, $Q$ over $T$ such that*

$$PAQ = \mathrm{diag}(a_1, \dots, a_r, 0, \dots, 0), \quad \textit{where } a_i \| a_{i+1}. \tag{2}$$

*Moreover, this form is unique up to association for $r \geqslant 2$, and unique up to stable association for $r = 1$.*

The existence (in a weak form) was proved for Euclidean domains by Wedderburn (1932) and in the full form by Jacobson (1937). The step to PID's was taken by Teichmüller (1937), and Nakayama (1938) showed that the form (2) is unique up to stable association, but the question remained whether any two forms (2) are associated. This was answered affirmatively for $r \geqslant 2$ by Guralnick, Levy and Odenthal (1987), who also give examples to show that 'stable' cannot be omitted for $r = 1$.

In terms of modules Theorem 4.5 shows that any finitely generated module over a PID is a direct sum of a finite number of cyclic modules: $M = C_1 \oplus \cdots \oplus C_n$, where $C_{i-1}$ is a homomorphic image of $C_i$. For Euclidean domains the matrices $P$, $Q$ in (2) can be taken to be products of elementary matrices, but this may not be possible in general PID's, e.g., the ring of integers in $\mathbf{Q}(\sqrt{-19})$ (cf. Cohn (1966)).

**4.2.** We can use modules to embed firs in skew fields. Let $R$ be a fir; an $R$-module will be called *prime* if it has characteristic 1 and any nonzero submodule has positive characteristic; e.g., $R$ itself is prime. It is easily checked that for any prime module $M$, $\mathrm{End}_R(M)$ is an integral domain. Now let $\mathcal{L}$ be the category whose objects are homomorphisms $R \to M$ ($M$ a prime left $R$-module) and whose morphisms are maps $M \to N$ forming a commutative triangle with the canonical maps. Between any two objects of $\mathcal{L}$ there can be at most one morphism, thus $\mathcal{L}$ is a preordering, which moreover, can be shown to be directed. Its direct limit $U$ is a left $R$-module with $R$ as submodule and the $R$-endomorphisms of $U$ are transitive on the nonzero elements of $U$, hence $E = \mathrm{End}_R(U)$ is a skew field with $R$ as subring. This proof is due to Bergman (1984) (cf. Cohn (1985, p. 243)).

## 5. Coproducts of rings

The customary notion of a $K$-algebra (with unit element) may be briefly described as a ring $R$ with a homomorphism from $K$ to the center of $R$. Often a more general concept is needed and we define a *$K$-ring* as a ring $R$ with a homomorphism $K \to R$. For example, the free $D$-ring $D_k\langle X\rangle$ defined in 3.2 is a $D$-ring in this sense. A $K$-ring $R$ is *faithful* if the canonical map $K \to R$ is injective.

Let $R_1$, $R_2$ by any $K$-rings; their *coproduct* over $K$, $R_1 \underset{K}{*} R_2$, is defined as the pushout of the diagram with the canonical maps $\lambda_i$:

$$\begin{array}{ccc} K & \xrightarrow{\lambda_1} & R_1 \\ \downarrow{\scriptstyle \lambda_2} & & \downarrow \\ R_2 & \longrightarrow & P \end{array}$$

To see that it exists we simply take presentations of $R_1$, $R_2$ and add the relations

$$\gamma\lambda_1 = \gamma\lambda_2 \quad \text{for all } \gamma \in K.$$

Let us take two faithful $K$-rings $R_1$, $R_2$ and form their coproduct $P = R_1 \underset{K}{*} R_2$. The first questions are:

Q.1. Is $P$ faithful as $K$-ring?

Q.2. Are the natural mappings of $R_1$, $R_2$ in $P$ embeddings?

Q.3. If $R_1$, $R_2$ are embedded in $P$, is their intersection in $P$ equal to $K$?

It is easy to see that in general the answers are negative, e.g., an element of $K$ may have an inverse in $R_1$ and be a zero-divisor in $R_2$. If Q.2 has an affirmative answer, the coproduct $P$ is called *faithful*; if Q.3 has an affirmative answer, $P$ is called *separating* (in Cohn (1959) a faithful separating coproduct was called a 'free product' by analogy with the group case, where these conditions always hold). A $K$-ring $R$ is called *left faithfully flat* if $R$ is a faithful $K$-ring and, identifying $K$ with its image in $R$, we have $R/K$ flat as left $K$-module. The following existence theorem was proved in Cohn (1959):

THEOREM 5.1. *Let $R_\lambda$ be a family of $K$-rings. If each $R_\lambda$ is left faithfully flat, then their coproduct over $K$ is left faithfully flat and separating.*

In particular, this theorem can be applied when $K$ is a skew field or more generally a semisimple ring. In the skew field case there is a direct (nonhomological) proof, obtained by taking left $K$-bases for each $R_\lambda$ adapted to the inclusion $K \subseteq R_\lambda$. This fact can be used to show that the coproduct of a family of skew fields over a common subfield is a fir (Cohn 1960). This proof was greatly generalized by Bergman (1974a) to obtain estimates for the global dimension and information on projective modules for a coproduct over a semisimple ring. For any ring $R$ we denote by $\mathcal{P}(R)$ the 'monoid of projectives', i.e. the set consisting of all isomorphism types $[Q]$ of finitely generated projective left $R$-modules $Q$, which forms a commutative monoid for the operation $[P]+[Q] = [P \oplus Q]$. Each ring homomorphism $R \to S$ induces a homomorphism $\mathcal{P}(R) \to \mathcal{P}(S)$ by the rule $P \mapsto S \otimes_R P$.

THEOREM 5.2 (Bergman's coproduct theorem). *Let $K$ be a semisimple ring, $(R_\lambda)$ a family of faithful $K$-rings and $P = \underset{K}{*} R_\lambda$ their coproduct over $K$. Then*

$$\text{r.gl.dim.}P = \begin{cases} \sup(\text{r.gl.dim.}R_\lambda) & \textit{if this is positive,} \\ \leqslant 1 & \textit{if all } R_\lambda \textit{ have gl.dim.zero.} \end{cases}$$

*Secondly, the monoid $\mathcal{P}(P)$ of projectives is the pushout of the maps $\mathcal{P}(K) \to \mathcal{P}(R_\lambda)$ induced by the canonical maps $K \to R_\lambda$.*

For a proof see Bergman (1974a) or Cohn (1995, 5.3, p. 218ff.) (cf. also Schofield (1985, Ch. 2)). These results were further extended to direct limits of rings by Dicks (1977).

Theorem 5.2 was applied by Bergman (1974b) to construct a hereditary ring $R$ with prescribed monoid of projectives; here $\mathcal{P}(R)$ can be any finitely generated abelian monoid which is conical with order-unit, i.e. an element $E$ such that for each $x \in \mathcal{P}(R)$ there exists $y \in \mathcal{P}(R)$ and an integer $n$ such that $x + y = nE$.

## References


Amitsur, S.A. (1966). *Rational identities, and applications to algebra and geometry*, J. Algebra **3**, 304–359.

Bass, H. (1964). *Projective modules over free groups are free*, J. Algebra **1**, 367–373.

Bergman, G.M. (1967). *Commuting elements in free algebras and related topics in ring theory*, Thesis, Harvard University.

Bergman, G.M. (1974a). *Modules over coproducts of rings*, Trans. Amer. Math. Soc. **200**, 1–32.

Bergman, G.M. (1974b). *Coproducts and some universal ring constructions*, Trans. Amer. Math. Soc. **200**, 33–88.

Bergman, G.M. (1978). *Conjugates and $n$-th roots in Hahn–Laurent group rings*, Bull. Malaysian Math. Soc. (2) **1**, 29–41. *Historical addendum*, ibid. **2** (1979), 41–42.

Bergman, G.M. (1984). *Dependence relations and rank functions on free modules*, Preprint 19.

Brungs, H.-H. (1969). *Generalized discrete valuation rings*, Canad. J. Math. **21**, 1404–1408.

Brungs, H.-H. and G. Törner (1984). *Skew power series rings and derivations*, J. Algebra **87**, 368–379.

Cedó, F. (1988). *A question of Cohn on semifir monoid rings*, Comm. Algebra **16**(6), 1187–1189.

Cohn, P.M. (1959). *On the free product of associative rings*, Math. Z. **71**, 380–398.

Cohn, P.M. (1960). *On the free product of associative rings, II. The case of skew fields*, Math. Z. **73**, 433–456.

Cohn, P.M. (1961). *On a generalization of the Euclidean algorithm*, Proc. Cambridge Phil. Soc. **57**, 18–30.

Cohn, P.M. (1962). *Factorization in non-commutative power series*, Proc. Cambridge Phil. Soc. **58**, 452–464.

Cohn, P.M. (1964). *Free ideal rings*, J. Algebra **1**, 47–69.

Cohn, P.M. (1966). *On the structure of the $GL_2$ of a ring*, Inst. Hautes Études Sci. Publ. Math. **30**, 5–53.

Cohn, P.M. (1967). *Torsion modules over free ideal rings*, Proc. London Math. Soc. (3) **17**, 577–599.

Cohn, P.M. (1969). *Free associative algebras*, Bull. London Math. Soc. **1**, 1–39.

Cohn, P.M. (1969a). *Rings with a transfinite weak algorithm*, Bull. London Math. Soc. **1**, 55–59.

Cohn, P.M. (1969b). *Dependence in rings, II. The dependence number*, Trans. Amer. Math. Soc. **135**, 267–279.

Cohn, P.M. (1970). *On a class of rings with inverse weak algorithm*, Math. Z. **117**, 1–6.

Cohn, P.M. (1971). *Free Rings and Their Relations*, LMS Monographs no. 2, Academic Press, London and New York.

Cohn, P.M. (1972). *Generalized rational identities*, Proc. Conf. in Ring Theory at Park City, Utah, 1971, R. Gordon, ed., Academic Press, New York, 107–115.

Cohn, P.M. (1974). *The class of rings embeddable in skew fields*, Bull. London Math. Soc. **6**, 147–148.

Cohn, P.M. (1981). *Universal Algebra*, 2nd ed., Reidel, Dordrecht.

Cohn, P.M. (1985). *Free Rings and Their Relations*, 2nd ed., LMS Monographs no. 19, Academic Press, London and New York.

Cohn, P.M. (1989a). *Algebra*, 2nd ed., vol. 2, Wiley, Chichester.

Cohn, P.M. (1989b). *Around Sylvester's law of nullity*, Math. Sci. **14**, 73–83.

Cohn, P.M. (1990). *Algebra*, 2nd ed., vol. 3, Wiley, Chichester.

Cohn, P.M. (1992a). *Modules over hereditary rings*, Contemp. Math. **130**, 111–119.

Cohn, P.M. (1992b). *A remark on power series rings*, Publ. Math. **36**, 481–484.

Cohn, P.M. (1995). *Skew Fields*, Encyclopedia of Mathematics and its Applications vol. 57, Cambridge Univ. Press, Cambridge.

Curtis, C.W. (1952). *A note on non-commutative polynomial rings*, Proc. Amer. Math. Soc. **3**, 965–969.

Dicks, W. (1977). *Meyer–Vietoris presentations over colimits of rings*, Proc. London Math. Soc. (3) **34**, 557–576.

Dicks, W. and E.D. Sontag (1978). *Sylvester domains*, J. Pure Appl. Algebra **13**, 243–275.

Dumas, F. and R. Vidal (1992). *Dérivations et hautes dérivations dans certains corps gauches de séries de Laurent*, Pacif. J. Math. **153**, 277–288.

Guralnick, R.M., L.S. Levy and C. Odenthal (1988). *Elementary divisor theorem for noncommutative pids*, Proc. Amer. Math. Soc. **103**, 1003–1011.

Hall, M., Jr. (1959). *The Theory of Groups*, Macmillan, New York.

Hedges, M.C. (1987). *The Freiheitssatz for graded algebras*, J. London Math. Soc. (2) **35**, 395–405.

Jacobson, N. (1934). *A note on non-commutative polynomials*, Ann. Math. **35**, 209–210.

Jacobson, N. (1937). *Pseudo-linear transformations*, Ann. Math. **38**, 484–507.

Jategaonkar, A.V. (1969). *A counter-example in ring theory and homological algebra*, J. Algebra **12**, 418–440.
Kaplansky, I. (1958). *Projective modules*, Ann. Math. **68**, 372–377.
Kozhukhov, I.B. (1982). *Free left ideal semigroup rings*, Algebra i Logika **21**(1), 37–59 (in Russian).
Lenstra, H.W., Jr. (1974). *Lectures on Euclidean rings*, Universität Bielefeld.
Lesieur, L. (1978). *Conditions Noethériennes dans l'anneau de pôlynomes de Ore* $A[X,\sigma,\delta]$, Sém. D'Alg. P. Dubreil 30ème année (Paris, 1976–77), SLNM 641, Springer, Berlin, 220–234.
Lewin, J. (1969). *Free modules over free algebras and free group algebras: the Schreier technique*, Trans. Amer. Math. Soc. **145**, 455–465.
Malcev, A.I. (1948). *On the embedding of group algebras in division algebras*, Dokl. Akad. Nauk SSSR **60**, 1499–1501 (in Russian). (Coll. Works, I, 211–213.)
Nakayama, T. (1938). *A note on the elementary divisor theory in non-commutative domains*, Bull. Amer. Math. Soc. **44**, 719–723.
Neumann, B.H. (1949). *On ordered division rings*, Trans. Amer. Math. Soc. **66**, 202–252.
Ore, O. (1933). *On a special class of polynomials*, Trans. Amer. Math. Soc. **35**, 559–584.
Schofield, A.H. (1985). *Representations of rings over skew fields*, LMS Lecture Notes no. 92, Cambridge Univ. Press, Cambridge.
Skornyakov, L.A. (1965). *On Cohn rings*, Algebra i Logika **4**(3), 5–30 (in Russian).
Smits, T.H.M. (1968). *Nilpotent S-derivations*, Indag. Math. **30**, 72–86.
Stenström, B. (1975). *Rings of Quotients*, Grundl. Math. Wiss. vol. 217, Springer, Berlin.
Stevin, S. (1585). *Arithmétique*, Antwerp. (Vol. II of Coll. Works, 1958.)
Tarasov, G.V. (1967). *On free associative algebras*, Algebra i Logika **6**(4), 93–105 (in Russian).
Teichmüller, O. (1937). *Der Elementarteilersatz für nichtkommutative Ringe*, S.-Ber. Preuss. Akad. Wiss., 169–177.
Wedderburn, J.H.M. (1932). *Noncommutative domains of integrity*, J. Reine Angew. Math. **167**, 129–141.
Wong, R. (1978). *Free ideal monoid rings*, J. Algebra **53**, 21–35.

# Simple, Prime and Semiprime Rings

V.K. Kharchenko[1]

*Institute of Mathematics, Academy of Sciences, University prosp. 4, 630090 Novosibirsk, Russia*
*e-mail: kharchen@math.nsk.su*
*vlad@servidor.dgsca.unam.mx*

*Contents*



[1]The paper was done under the support of International Science Foundation grants RPS 000 and RPS 300.

HANDBOOK OF ALGEBRA, VOL. 1
Edited by M. Hazewinkel

## 1. Simple rings

We shall start with the definition of a simple ring and examples of simple rings.

DEFINITION. A ring $R$ is called *simple* if it has no proper two sided ideals and $R^2 \neq 0$.

First of all any field and any skew field are simple rings. The following lemma shows that the ring of $n$ by $n$ matrices over a (skew) field is simple.

**1.1.** LEMMA. *The ring of $n \times n$ matrices over a ring $R$ is simple if and only if $R$ is simple.*

PROOF. Let $R$ be simple and $I$ a nonzero ideal of the ring of $n$ by $n$ matrices $R_n$. Let us denote for an element $r \in R$ by $r_{ij}$ the matrix whose $(i,j)$ coefficient is $r$ and all others are zero. For a matrix $M \in R_n$ we denote by $M_{ij}$ its $(i,j)$ coefficient.

If $M$ is a nonzero matrix from $I$, $M_{ij} \neq 0$ then $RM_{ij}R$ is a two sided ideal of $R$. This is a nonzero ideal. Indeed, the left annihilator $L = \{l \in R \mid lR = 0\}$ is a two sided ideal which is not equal to $R$ (because $R^2 \neq 0$) and so $L = 0$. If $RM_{ij}R = 0$ then $M_{ij}R \in L = 0$. Analogously, the right annihilator of $R$ is zero, and $M_{ij} = 0$. Thus $RM_{ij}R = R$.

If $N$ is an arbitrary matrix, then $N_{tv} \in RM_{ij}R$, or in detail

$$N_{tv} = \sum_k r^k_{tv} M_{ij} s^k_{tv},$$

and in matrix form

$$N = \sum_{t,v,k} \left(r^k_{tv}\right)_{ti} \cdot M \cdot \left(s^k_{tv}\right)_{jv} \in I.$$

Inversely, if $R$ is not simple and $A$ is proper ideal of it then the set $A_n$ of all matrices with coefficients from $A$ is a proper ideal of $R_n$. □

**1.2.** *Rings of infinite matrices and simple rings.* For a given ring $R$ and infinite set $U$ two types of rings of $U \times U$ matrices can be defined: the ring of column-finite matrices $Lin_c(R)$ and the ring of row-finite matrices $Lin_r(R)$.

If $M, N$ are infinite matrices with coefficients $M_{uv}, N_{uv},\ u, v \in U$, respectively, then addition and multiplication are defined by the usual formulae

$$(M \pm N)_{uv} = M_{uv} \pm N_{uv}, \tag{1}$$

$$(MN)_{uv} = \sum_k M_{uk} N_{kv}. \tag{2}$$

The summation in the second formula would be well defined if $k$ runs over a finite set. It will be so if any of matrices $M, N$ has a finite number of nonzero coefficients in any column $M_k = \{M_{uk} \mid u \in U\}$ (respectively $N_k = \{N_{uk} \mid u \in U\}$). Thus formulae (1), (2) define the ring of column-finite matrices

$$Lin_c(R) = \{M \mid \forall k \text{ the set } \{u \mid M_{uk} \neq 0\} \text{ is finite}\}.$$

In the same way if both of matrices $M, N$ has only a finite number of nonzero coefficients in any row $M^u = \{M_{uk} \mid k \in U\}$ then this formula is also well defined and we obtain the ring of row-finite matrices

$$Lin_r(R) = \{M \mid \forall u \text{ the set } \{k \mid M_{uk} \neq 0\} \text{ is finite}\}.$$

Evidently the intersection $L(R) = Lin_c(R) \cap Lin_r(R)$ is also a ring. If $R$ has a unit element 1 then each of these three rings has a unit element – this is the infinite matrix $E$ with $E_{uu} = 1$, $E_{uv} = 0$, $u \neq v$.

None of these rings is simple. The set $S_c(R)$ of all matrices $M$ each of which has only a finite number of nonzero rows is a two sided ideal in the ring $Lin_c(R)$. Analogously the set $S_r(R)$ of all matrices $M$ each of which has only a finite number of nonzero columns is a two sided ideal of the ring $Lin_r(R)$. Evidently $S_c(R) \cap S_r(R) \lhd L(R)$. More generally, if $\alpha$ is an infinite cardinal then we can define the ideal $S_c^\alpha(R)$ and, symmetrically, the ideal $S_r^\alpha(R)$ of all matrices $M$ such that a cardinality of the set of all nonzero rows (respectively columns) of $M$ is less than $\alpha$.

**1.2.1.** THEOREM. *If $R$ is a simple ring with* 1 *then $S_c(R)$ and $S_r(R)$ are simple rings.*

PROOF. Let $I$ be a nonzero ideal of $S_c(R)$ and $0 \neq M \in I$. If, for instance, $M_{ij} \neq 0$ then $RM_{ij}R = R$ and we can find a presentation of the unit element

$$1 = \sum_k r^{(k)} M_{jt} s^{(k)}.$$

It follows

$$1_{tt} = \sum_k \left(r^{(k)}\right)_{ti} M \left(s^{(k)}\right)_{jt} \in I$$

for all $t \in U$. If $N$ is an element of $S_c(R)$ and $n$ is a number of nonzero rows of $N$ then

$$N = \left(\sum_{t=1}^{n} e_{tt}\right) N \in IN \subseteq I.$$

□

In particular if $R = F$ is a field (or, more generally, skew field) then the rings $S_c(R)$ and $S_r(R)$ are two more examples of simple rings. If the set $U$ is infinite then these rings have no unit elements.

**1.2.2.** If $R = F$ is a (skew) field then the ring $Lin_r(R)$ can be characterized as the complete ring of linear transformations $End_F V$ of a left linear space $V$ of dimension equal to the cardinality of the set $U$ (the space of rows of length $|U|$ with a finite number of nonzero elements). Respectively, the ring $L_c(R)$ can be identified with the complete ring of linear transformations $End\, V_F$ of a right linear space of the same dimension (the space of columns of depth $|U|$ having a finite number of nonzero coefficients).

Indeed, for any linear transformation $f$ of $V$ and a fixed basis $\{v^u \mid u \in U\}$ we can define a finite-column matrix $M^{(f)}$ whose coefficients $M^{(f)}_{uk} = \mu_{uk}$ are uniquely defined by the following formula

$$(v^u)f = \sum_{k \in U} \mu_{uk} v^k. \tag{3}$$

Inversely, any finite-column matrix $M$ formula (3) with $\mu_{uk} = M_{uk}$ defines a linear transformation $f^{(M)}$. The correspondences $f \mapsto M^{(f)}$, $M \mapsto f^{(M)}$ define the isomorphism which gives the desired characterization.

Formula (3) shows immediately that under this characterization the ideal $S^\alpha_c(F)$ is identified with the set of all linear transformations $f$ which have images $im\, f$ of dimension $< \alpha$ (note that here $\alpha$ is an *infinite* cardinal). Recall that the dimension of $im\, f$ is called the *rank* of $f$. So $S^\alpha_c(F)$ is the ideal of all transformations of rank less than $\alpha$.

This characterization allows one to construct another important example of simple rings.

**1.2.3.** THEOREM. *Let $F$ be a (skew) field and $U$ be any set of cardinality $\alpha$. Then $S^\alpha_c(F)$ is a maximal ideal of $L_c(F)$. In particular the factor ring $L_c(F)/S^\alpha_c(F)$ is a simple ring with unit element* $1 + S^\alpha_c(F)$.

PROOF. For a subspace $W$ of $V$ let us denote by $W^\perp$ a direct supplement of $W$ in $V$, i.e. a subspace such that $W + W^\perp = V$, $W \cap W^\perp = 0$. If $W$ is of dimension $\alpha$ then it is isomorphic to $V$ and we can define a linear transformation $p_W\colon V \to V$ such that $im\, p_W = W$ and it is an isomorphism of $V$ onto $W$. So there exists an inverse linear map $p_W^{-1}\colon W \to V$. Let us define $p^*_W = p_W^{-1}$ on $W$ and $p^*_W = 0$ on $W^\perp$. Then $p^*_W \in End_F(V) = Lin_c(F)$.

A transformation $f$ does not belong to $S^\alpha_c(F)$ iff it has rank $\alpha$. Let $W = (ker\, f)^\perp$. The restriction of $f$ to $W$ is an isomorphism of $W$ with $im\, f$ and it is possible to define a transformation $f^*$ which is equal to $f^{-1}$ on $im\, f$ and zero on $(im\, f)^\perp$. We have $1 = p_W \cdot f \cdot f^* \cdot p^*_W$

It follows that if an ideal $I$ contains an element $f$ of rank $\alpha$ then it contains 1 and therefore it is equal to $Lin_c(F)$. □

NOTE. Of course the same statement is true for the ring $Lin_r(F)$.

**1.3.** *Algebra of differential operators* (*Weyl algebra*). Let $F$ be any field and $A = F[t_1, \dots, t_n]$ be the algebra of polynomials in $n$ variables. For any element $a \in A$ we define an operator of (right) multiplication $r_a\colon A \to A$. This is the linear transformation acting by the formula $(f)r_a = fa$. Any variable $t_i$ defines a partial derivative $\partial/\partial t_i$ which is also a linear transformation, $\partial/\partial t_i \in End\, A$.

The subalgebra of $End\, A$ generated by these two types of transformations is called the Weyl algebra or the algebra of differential operators $A_n$.

Let us denote by $x_i$ the multiplication by $t_i$ and by $y_i$ the operator $\partial/\partial t_i$. We have

$$\begin{aligned}(f)(x_i y_j) &= ((f)x_i)y_j = (ft_i)y_j = (ft_i)'_{t_j} \\ &= (f)'_{t_j} \cdot t_i + f \cdot (t_i)'_{t_j} = (f)(y_j x_i + \delta^i_j).\end{aligned}$$

Therefore $x_i y_j = y_j x_i$ if $i \neq j$ and $x_i y_i = y_i x_i + 1$. Evidently $y_i y_j = y_j y_i$ and $x_i x_j = x_j x_i$ by definition. Thus the algebra $A_n$ is generated by the elements $x_1, \ldots, x_n$, $y_1, \ldots, y_n$ with the following relations

$$x_i y_j = y_j x_i, \quad i \neq j, \qquad x_i y_i = y_i x_i + 1,$$

$$x_i x_j = x_j x_i, \qquad y_i y_j = y_j y_i. \tag{4}$$

**1.3.1.** THEOREM. *If the field $F$ has zero characteristic then $A_n$ is a simple ring.*

PROOF. Let $I$ be a nonzero ideal of $A_n$. By using relations (4) one can transform any element $u$ of $A_n$ to the form

$$u = \sum_{i,j} \alpha_{i,j} y_1^{i_1} \cdots y_n^{i_n} x_1^{j_1} \cdots x_n^{j_n}$$

where $i = (i_1, \ldots, i_n)$, $j = (j_1, \ldots, j_n)$ are multi-indices. Let $u \in I$ be a nonzero element with the smallest possible degree, where the degree is the maximum of the sums $i_1 + \cdots + i_n + j_1 + \cdots + j_n$ with $\alpha_{i,j} \neq 0$. If $u$ has zero degree then $u = \alpha_{0,0} \neq 0$ and $1 = u\alpha_{0,0}^{-1} \in I$; therefore $I = A_n$.

Finally, the monomial $u_{i,j} = y_1^{i_1} \cdots y_n^{i_n} x_1^{j_1} \cdots x_n^{j_n}$ commutes with $x_k$ and $y_k$ according to the formulae

$$u_{i,j} x_k - x_k u_{i,j} = i_k \cdot y_1^{i_1} \cdots y_k^{i_k - 1} \cdots y_n^{i_n} x_1^{j_1} \cdots x_n^{j_n},$$

$$u_{i,j} y_k - y_k u_{i,j} = j_k \cdot y_1^{i_1} \cdots y_n^{i_n} x_1^{j_1} \cdots x_k^{j_k - 1} \cdots x_n^{j_n}.$$

This implies that if $u$ has nonzero degree then one of the commutators $[u, x_k], [u, y_k]$ is nonzero and has lower degree. □

**1.4.** *General structure.* Now we are going to consider a structure of an arbitrary simple ring. The first steps show in fact that any simple ring is an algebra over a field.

**1.4.1.** LEMMA. *A center of a simple ring is either a field or zero. In the former case a unit of the center is a unit of the ring.*

PROOF. Recall that the *center* $Z(R)$ of a ring $R$ is the set $\{z \in R \mid \forall r \in R\ (zr = rz)\}$. If $0 \neq z \in Z$ then $zR$ is a nonzero ideal of $R$, so $zR = R$ and for any $z_1 \in Z$ there exists an element $r \in R$ such that $zr = z_1$. It is enough to show that $r \in Z$. We have $0 = [z_1, x] = [zr, x] = z[r, x]$ and therefore $zR[r, x]R = 0$. Here $[r, s]$ is short for $rs - sr$. So either $R[r, x]R = 0$ and $r \in Z$ or $R[r, x]R = R$ and $z = 0$.

Finally, if $e$ is a unit of the center then $e(ex - x) = 0$ and $ReR(ex - x) = 0$, so $R(ex - x) = 0$ and $ex - x = 0$. □

This lemma shows in particular that any simple ring with a unit can be considered as an algebra over a field (its center). If the center of a simple ring is zero then it still has an algebra structure over a field – its centroid. Recall that the *centroid* $C(R)$ of a ring $R$ is defined as the subring of the endomorphism ring of the abelian group $(R, +)$, which commutes with left and right multiplications:

$$C(R) = \{\xi \in End(R, +) \mid \xi(axb) = a\xi(x)b,\ a, b \in R \cup \{1\}\}.$$

It is easy to see that if a ring has a unit then the centroid is equal to the center.

**1.4.2.** LEMMA. *The centroid of a simple ring is a field.*

PROOF. First of all the centroid has a unit 1 which is the identity endomorphism $1(x) = x$. If $0 \neq \xi \in C(R)$ then $\xi(R)$ is a nonzero ideal of $R$. Therefore $\xi(R) = R$. Moreover if $\xi(r) = 0$ then $\xi(RrR) = 0$ which implies that $r = 0$ and $ker\,\xi = 0$. Thus there exists an inverse map $\xi^{-1}\colon R \to R$, which is also an element from the centroid: $\xi^{-1}(axb) = \xi^{-1}(a\xi(\xi^{-1}(x))b) = \xi^{-1}\xi(a\xi^{-1}(x)b) = a\xi^{-1}(x)b$. □

Any simple ring becomes an algebra over its centroid with linear space structure $\xi x = \xi(x)$. Therefore without loss of generality in general theory one can consider only simple algebras over a field.

DEFINITION. A simple algebra is called *central* if the base field is equal to its centroid.

The following result is one of the fundamental facts in the theory of simple algebras.

**1.4.3.** THEOREM. *The tensor product of a central simple algebra with a simple one is simple. If both algebras are central then the product is central as well.*

A proof of this fact can be obtained with the help of the following useful property of linearly independent sets in central simple algebras.

**1.4.4.** LEMMA. *If $d_1, \dots, d_n$ are linearly independent elements of a central simple algebra $R$ and $a$ is an arbitrary element of this algebra then there exist elements $s_1, \dots, s_m, t_1, \dots, t_m$, such that*

$$\sum s_i d_1 t_i = a, \quad \sum s_i d_2 t_i = 0, \quad \dots, \quad \sum s_i d_n t_i = 0. \tag{5}$$

This lemma easily follows from the Jacobson–Chevalley density theorem: the algebra $R$ should be considered as an irreducible module over a tensor product $L = R^{op} \otimes R$, with the action $d \cdot \beta = \sum s_i d t_i$, where $\beta = \sum s_i \otimes t_i$ and $R^{op}$ is the $R$-opposite algebra, i.e. the linear space $R$ with a new multiplication $s * s_1 = s_1 s$. In this case the statement of the lemma means that $R^{op} \otimes R$ is a dense subring in the ring $End\, R$ of linear transformations of the space $R$.

**1.5.** *Embeddings.* In this section we will see, in particular, that a general structure of a simple ring can be quite complicated.

**1.5.1.** THEOREM. *Any algebra over a field can be embedded into a simple ring.*

PROOF. Let $A$ be an algebra of dimension $d$ over a field $F$. Let $V$ be a linear space of the polynomial algebra $R = A[X]$, where $X$ is an infinite set of cardinality $\alpha \geqslant d$. We can embed $A$ into the algebra of linear transformations $End\, V$ by right multiplications $\xi\colon a \mapsto r_a$, where $(f)r_a = fa$. The rank of any right multiplication $r_a, a \in A$ is equal to $\alpha$ because $im\, r_a$ contains the linearly independent set $\{xa,\ x \in X\}$. This implies that the composition

$$A \to Lin_c(F) \to Lin_c(F)/S_c^{\alpha}(F)$$

has zero kernel and by Theorem 1.2.3 we are done. □

This theorem was first obtained by L.A. Bokut' [Bo63]. His method is a little bit complicated and more general. It is based on the Shirshov composition lemma (or the *diamond* lemma). This lemma allows one to find a basis of an algebra $R\langle X \parallel U\rangle$ generated by a set of variables $X$ with a set of relations $U$ of the type $w_i = f_i(X)$, where $w_i$ is a monomial in $X$ and $f_i$ is a polynomial in $X$ with all monomials less then $w_i$ with respect to the natural ordering of monomials. The composition lemma gives sufficient conditions for a set of all monomials $T$ which have no submonomials $w_i$ to be linearly independent in the algebra $R$ (see details in [Bo76] or in [Bo77]).

By means of this method L.A. Bokut' obtained, in particular, the following unexpected results.

**1.5.2.** THEOREM. *Any algebra $R$ can be embedded in a simple algebra $A$ which is a sum of four subalgebras $A_1, A_2, A_3, A_4$ with zero multiplication $A_i^2 = 0$, i.e. any element $a$ of $A$ has a form $a = a_1 + a_2 + a_3 + a_4$, with $a_i \in A_i$.*

**1.5.3.** THEOREM. *Any countable algebra can be embedded in a simple algebra with three generators.*

**1.5.4.** THEOREM. *Any algebra can be embedded in a simple algebra which is a sum of three nilpotent subalgebras.*

This result is completed by the following theorem of O. Kegel [Ke63].

**1.5.5.** THEOREM. *Any algebra which is a sum of two nilpotent subalgebras is nilpotent itself.*

Needless to say that no nilpotent algebra is simple.

The following result of P.M. Cohn [Co58] can also be obtained with the help of the composition lemma.

**1.5.6.** THEOREM. *Any algebra with no zero divisors can be embedded in a simple algebra with no zero divisors.*

**1.6.** *Radicals and simple rings.* Recall that a radical is a map $\rho$ which associates to every ring $R$ an ideal $\rho(R)$ with the following properties

(1) If $\xi$: $R \to A$ is a homomorphism, then $\xi(\rho(R)) \subseteq \rho(\xi(A))$.

(2) The ideal $\rho(R)$ is the biggest ideal with the property $\rho(I) = I$.

(3) For any $R$ the relation $\rho(R/\rho(R)) = 0$ is valid.

If $R = \rho(R)$ then $R$ is called a *$\rho$-radical ring.* If $\rho(R) = 0$ then the ring $R$ is called a *$\rho$-semi-simple ring.* Evidently any simple ring $R$ is either $\rho$-radical or $\rho$-semi-simple. Therefore it is a question of interest to describe all simple $\rho$-radical rings.

To any ring property can be associated the map which associates to a ring the biggest ideal (if any) obeying as a ring this property. If such a map proves to be a radical, then the property is called a *radical* property.

We are going to consider a problem of existence of simple $\rho$-radical rings for the three most important radicals: the Levitzki locally nilpotent radical, the Koethe upper nil-radical and the Jacobson one. The corresponding three questions can be formulated in the following way. Does there exist a simple locally nilpotent ring (i.e. a simple ring such that every finitely generated subring $S \subseteq R$ is nilpotent $S^n = 0$)? Does there exist a simple nil ring (i.e. a simple ring such that each element $a$ is nilpotent $a^n = 0$)? Does there exist a simple Jacobson radical ring (i.e. a simple ring such that for any element $a$ of it there exists an $x$ such that $ax + a + x = 0$)?

The answers are very different: no, nobody knows, yes.

**1.6.1.** THEOREM. *There exist no locally nilpotent simple rings.*

PROOF. It is enough to note that in a locally nilpotent ring $R$ every nonzero element $a$ does not belong to the ideal $RaR$. This would imply that in a simple locally nilpotent ring $RaR = 0$ and $R^3 = 0$, which contradicts the fact that $R = R^2 = R^3$ in any simple ring.

Thus, let

$$a = r_1 a t_1 + \cdots + r_n a t_n. \tag{6}$$

The subring $S$ generated by the elements $r_1, \ldots, r_n$ is nilpotent $S^m = 0$, i.e. $r_{i1} r_{i2} \cdots r_{im} = 0$ for any $1 \leqslant i1, \ldots, im \leqslant n$.

By iterating equality (6) we have

$$\begin{aligned} a &= \sum_{i1=1}^{n} r_{i1} a t_{i1} = \sum_{1 \leqslant i1, i2 \leqslant n} r_{i1} r_{i2} a t_{i2} t_{i1} \\ &= \sum_{1 \leqslant i1, \ldots, in = m \leqslant n} r_{i1} r_{i2} \cdots r_{im} a t_{im} \cdots t_{i2} t_{i1} = 0. \end{aligned}$$

□

**1.6.2.** *Does there exist a simple nil-ring?* This is an open problem. By 1.6.1 a simple nil-ring cannot be locally nilpotent. The existence problem of a nil-ring which in not locally nilpotent was set by J. Levitzki in 1945. This problem and the Kurosh problem [Ku41] concerning the existence of an algebraic algebra which is not locally finite was solved only in 1964 by E.S. Golod in a famous paper [Go64]. So far there is not another known way to construct a nil-ring which is not locally nilpotent. Therefore for a construction of a simple nil-ring one has to either find a new solution of the Kurosh–Levitzki problem or investigate deeply the Golod construction. We will make here only a first small step.

**1.6.3.** LEMMA. *There exists a simple nil-ring if and only if there exists a nil-ring with an element $a$ such that $0 \neq a \in RaRaR$.*

PROOF. Of course if $R$ is a simple nil-ring, then $RaRaR$ is a nonzero (if $a \neq 0$) ideal and therefore $R = RaRaR$, so $a \in RaRaR$.

Inversely, let $0 \neq a \in RaRaR = S$ and $R$ is a nil-ring. The set $\mathbf{M}$ of all ideals $I \triangleleft R$ such that $a \notin I$ and $I \subseteq S$ is not empty, $0 \in \mathbf{M}$. This set is inductive, i.e. for every chain of any cardinality

$$I_1 \subseteq I_2 \subseteq \cdots \subseteq I_\alpha \subseteq \cdots$$

of its elements the union $\bigcup I_\alpha$ belongs to $\mathbf{M}$. Therefore we can apply the Zorn lemma to $\mathbf{M}$, which says that $\mathbf{M}$ has a maximal element $U$. In this case the factor-ring $\widehat{S} = S/U$ is a simple nil-ring.

Indeed, it is nil as any subring of a nil-ring is nil and any factor-ring of a nil ring is nil. As $a \notin U$ then $\hat{a} \neq 0$, where $\hat{a} = a + U$ in the factor-ring. Also we have $a \in RaRaR \subseteq RSRSR \subseteq S^2$ and therefore $\hat{a} \in \widehat{S}^2$. This implies $\widehat{S}^2 \neq 0$.

If $\widehat{W}$ is a proper ideal of $\widehat{S}$, then one can find a set $W \supseteq U$ such that $\widehat{W} = W + U$ in the factor-ring $\widehat{S}$. Now $SWS$ is an ideal of $R$ as so is $S$. Moreover $SWS \subseteq W$ as $\widehat{W}$ is an ideal of $\widehat{S}$. By the choice of $U$ either $a \in SWS + U$ or $SWS \subseteq U$. In the former case $S = RaRaR \subseteq SWS + U \subseteq W$ – a contradiction.

In the last case we consider the set $\ell = \{x \in S \mid xS \subseteq U\}$, which is evidently an ideal of $R$, containing both $SW$ and $U$. If $a \in \ell$ then $aS \subseteq U$ and $a \in RaRSR \subseteq aS \subseteq U$ – a contradiction. This implies $a \notin \ell$ and therefore $\ell \in \mathbf{M}$. By the choice of $U$ we have $\ell = U$. In particular $SW \subseteq U$.

Finally, the set $\ell' = \{x \in S \mid Sx \subseteq U\}$ is an ideal of $R$, which contains both $W$ and $U$. If $a \in \ell'$, then $Sa \subseteq U$ and $a \in RSRaR \subseteq Sa \subseteq U$ – a contradiction. So $\ell' \in \mathbf{M}$ and by the maximality of $U$ we have $\ell' = U$, which is impossible because $\ell' \supseteq W \neq U$. Thus $\widehat{S}$ is a simple nil-ring. □

By this lemma all we need for a solution of the problem is to add the one additional relation

$$x = x_1 x x_2 x x_3 + x_4 x x_5 x x_6$$

to the Golod system of relations, which gives a nil but not nilpotent finitely generated algebra $R = \langle x, x_1, x_2, x_3, x_4, x_5, x_6 \rangle$, and prove that in the result $x$ is not zero. Nobody has proved this to date. The A.Z. Anan'in paper [An85] has something to do with this matter. In this paper there is constructed a finitely generated nil algebra $A$ over a field, such that $A \otimes A$ is not equal to its Jacobson radical, and the intersection

$$\bigcap_{n=1}^{\infty} A^n$$

is not zero. This is a step in the right direction, because for a simple nil-ring $R$ we would have $R \otimes R^{op}$ is primitive, in particular its Jacobson radical is zero, and

$$\bigcap_{n=1}^{\infty} R^n = R.$$

(Note that the Anan'in example $A$ has an involution and therefore it is isomorphic to $A^{op}$.)

**1.6.4.** *Simple Jacobson-radical rings.* The first example of a simple ring which coincides with its Jacobson radical was found by E. Sąsiada in 1961. It was published in a joint paper with P.M. Cohn [SC67] in a simplified and slightly generalized form. This example is based on the following lemma.

**1.6.5.** LEMMA. *Let $R$ be the ring of formal power series in two noncommuting indeterminates $x$ and $y$ over a field $F$. Denote by $I$ the ideal of $R$ generated by $x - yx^2y$; then $x \notin I$.*

This lemma shows that in the factor-ring $A = R'/I$ of the ring of formal power series $R'$ with zero constant terms by the ideal $I$, the element $x$ is nonzero and $x = yx^2y$. The ring $A$ is radical because this is the case for $R'$. If we note that Lemma 1.6.3 is also valid for the Jacobson radical (in the proof we have used only the fact that an ideal of a radical ring is radical), then the relations $0 \neq x = yx^2y = y^2x^2y^2x^2y^2 \in AxAxA$ would imply the existence of a simple Jacobson radical ring.

After the construction of a simple radical ring there arises a natural problem of embedding of an arbitrary radical algebra in a simple radical ring. Attempts towards a positive solution of this problem led P.M. Cohn [Co73] to the following embedding theorem.

**1.6.6.** THEOREM. *Any radical algebra over a field embeddable in a division algebra is embeddable in a simple radical ring.*

A complete solution of this problem was obtained by A.I. Valitskas [Va88] by construction of a big collection of radical algebras which cannot be embedded in simple rings. He also found necessary and sufficient conditions for an algebra to be embedded in a radical algebra.

Recently new examples of simple radical algebras were found by N.I. Dubrovin [Du80]. His examples have a number of additional properties. In particular he has constructed a simple radical chain Ore domain. This construction is based on the old D.M. Smirnov example [Sm66] of a right ordered group.

**1.7.** *Subdirect decompositions and simple rings.* Recall that a ring $R$ is called a *subdirect product* of a family of rings $\{R_\alpha,\ \alpha \in A\}$ if there exists an embedding $\pi$ of $R$ into the direct product

$$\prod_{\alpha\in A} R_\alpha,$$

such that its compositions with the natural projections

$$p_\alpha\colon \prod R_\alpha \to R_\alpha$$

result in epimorphisms. In this case we write $R = S_{\alpha\in A} R_\alpha$.

Notice that the subdirect product is not uniquely defined by the family $\{R_\alpha,\ \alpha \in A\}$. For instance, any intermediate subring $S$,

$$\sum R_\alpha \subseteq S \subseteq \prod R_\alpha,$$

is a subdirect product of $\{R_\alpha,\ \alpha \in A\}$. Here $\sum R_\alpha$ is the subring of all elements $f \in \prod R_\alpha$ which have only a finite number of nonzero components

$$\sum R_\alpha = \left\{ f = (\ldots, f_\alpha, \ldots) \in \prod R_\alpha\colon |\{\alpha,\ f_\alpha \neq 0\}| < \infty \right\}.$$

**1.7.1.** DEFINITION. A ring $R$ is said to be *approximated* by a family of rings $\{R_\alpha,\ \alpha \in A\}$ if there exists a family $\{I_\alpha,\ \alpha \in A\}$ of ideals of $R$ such that

$$\bigcap_{\alpha\in A} I_\alpha = 0, \quad R/I_\alpha \cong R_\alpha.$$

**1.7.2.** PROPOSITION. *A ring $R$ is approximated by the family $\{R_\alpha,\ \alpha \in A\}$ iff $R = S_{\alpha\in A} R_\alpha$.*

PROOF. If $R = S_{\alpha\in A} R_\alpha$ then the family of all kernels $\{ker\, \pi p_\alpha\}$ is the required family of ideals. Inversely the map $\pi\colon\ r \to (\ldots, r + I_\alpha, \ldots)$ is a homomorphism of $R$ to $\prod_{\alpha\in A} R_\alpha$. Its kernel is equal to $\bigcap_A I_\alpha = 0$. So it is an embedding. Evidently,

$$r \cdot \pi p_\alpha = (\ldots, r + I_\alpha, \ldots) = r + I_\alpha$$

and therefore $im\, \pi p_\alpha = R/I_\alpha$. □

**1.7.3.** DEFINITION. A decomposition $R = S_{\alpha\in A} R_\alpha$ is called *trivial* if one of the projections $\pi p_\alpha$ is an isomorphism. A ring $R$ is said to be *subdirectly indecomposable* if it has no nontrivial decomposition in a subdirect product. Proposition 1.7.2 implies immediately that a ring $R$ is subdirectly indecomposable iff the intersection of all its nonzero ideals is nonzero, or, equivalently, the ring $R$ has a smallest nonzero ideal. This ideal is called a *heart* of the ring $R$.

**1.7.4.** THEOREM. *Any ring is a subdirect product of subdirectly indecomposable rings.*

PROOF. For any nonzero element $a$ let us denote by $M_a$ the set of all ideals $I \lhd R$ such that $a \notin I$. This set is inductive. By the Zorn lemma there exists a maximal element $I_a$ in $M_a$. Evidently $\bigcap_{0 \neq a \in R} I_a = 0$ (because $a \notin I_a$). So, it is enough to note that the factor-ring $R_a = R/I_a$ is subdirectly indecomposable.

Any nonzero ideal $\bar{J}$ of this factor-ring is defined by an ideal $J$ of $R$, properly containing the ideal $I_a$. As $I_a$ is a maximal ideal from $M_a$, then $J \notin M_a$, i.e. $a \in J$. It implies that $\bar{a} = a + I_a$ is a nonzero element in the heart of the ring $R_a$. □

Finally, the following lemma of Andrunakievič gives the connection between subdirect decompositions and simple rings.

**1.7.5.** LEMMA. *The heart $H$ of any subdirectly indecomposable ring is either simple or it has zero multiplication $H^2 = 0$.*

PROOF. Let $H^2 \neq 0$. For any ideal $I$ of the ring $H$ the set $HIH$ is an ideal of $R$, contained in $H$. Therefore either $HIH = 0$ or $I \supseteq HIH = H$. In the former case the left annihilator $l(H) = \{h \in H \mid hH = 0\}$ is an ideal of $R$, properly contained in $H$ (as $H^2 \neq 0$) and therefore it is equal to zero. We have $HI \subseteq l(H) = 0$. Analogously $r(H) = \{h \in H \mid Hh = 0\}$ is equal to zero and $I \subseteq r(H) = 0$. □

**1.8.** *Simple artinian rings.* Recall that a ring $R$ is called *left* (*right*) *artinian* if it satisfies the following Descending Chain Condition (DCC) for left (right) ideals: any strictly descending chain of left (right) ideals

$$I_1 \supset I_2 \supset \cdots \supset I_n \supset \cdots$$

has finite length.

Simple left (right) artinian rings are characterized by the classical Wedderburn–Artin theorem, which plays a fundamental role in ring theory.

**1.8.1.** WEDDERBURN–ARTIN THEOREM. *A simple ring $R$ is left* (*right*) *artinian iff it is isomorphic to a ring of $n \times n$ matrices over a skew field. The number $n$ and the skew field are uniquely determined* (*up to isomorphism*).

The modern proof of this theorem is based on the Schur lemma and the Jacobson–Chevalley density theorem. First of all the DCC allows one to find a minimal nonzero left ideal $V$. The Schur lemma says that a ring $D$ of all endomorphisms of the left $R$-module $V$ is a skew field. Therefore $V$ can be considered as a right vector space over $D$. With the help of DCC one now proves that this space has finite dimension $n$. Then, any element $r \in R$ acts like linear transformation on this space by left multiplication $r(v) = rv$. So we have a homomorphism $\xi\colon R \to End_D V$. The kernel of this homomorphism is an ideal of the simple ring $R$. This means that either $\xi$ is an embedding or $RV = 0$. In the latter case the right annihilator $\rho$ of $R$ is an ideal containing $V$ and, by simplicity, $\rho = R$, which implies $R^2 = 0$ in contradiction with the definition of a simple ring. Thus

$\xi$ is an embedding of $R$ into the ring $End_D V$. The last is isomorphic to the ring of $n \times n$ matrices over the skew field $D$. Finally the Jacobson–Chevalley density theorem shows that the image of $\xi$ is equal to $End_D V$, which completes the main part of the proof.

The uniqueness part in the Wedderburn–Artin theorem is no more than the statement that an isomorphism $D_n \cong \Delta_m$ implies $D \cong \Delta$ and $n = m$, which can be proved in a number of ways.

The following fundamental Skolem–Noether theorem also plays an outstanding role in ring theory as well as in the theory of presentations of finite groups.

**1.8.2.** SKOLEM–NOETHER THEOREM. *If $f, g$ are two unitary (i.e. taking the unit element to the unit element) homomorphisms of a simple finite dimensional algebra $B$ to a central simple artinian algebra $A$, then there exists an invertible element $a \in A$, such that $g(x) = a^{-1}f(x)a$.*

This theorem implies a number of important properties of simple finite dimensional algebras. For example any two abstractly isomorphic simple finite dimensional unitary subalgebras in an artinian central simple algebra are conjugate in this algebra. Another corollary says that any automorphism $\xi$ of a central simple finite dimensional algebra is inner, i.e. it has a form $\xi(x) = a^{-1}xa$.

Note that the unitarity of the homomorphisms as well as the centrality of the algebra $A$ are essential in the Skolem–Noether theorem. For instance any ring of $n \times n$ matrices over a (skew) field $D$ has the following embeddings in the ring of $2n \times 2n$ matrices: $f\colon M \to diag(M,\ M)$ and $g\colon M \to diag(M,\ 0)$. For any invertible $a \in D_{2n}$ we have $a^{-1}f(1_n)a = a^{-1}1_{2n}a = 1_{2n} \neq g(1_n)$. Also, if $A, B$ are fields not equal to the base field (so they are noncentral simple algebras) then evidently any two different embeddings $f, g\colon B \to A$ are not conjugate in $A$.

**1.9.** *Simple finite dimensional algebras, the Brauer group.* The most important examples of simple artinian algebras are finite dimensional simple algebras. By the Wedderburn–Artin theorem such an algebra $A$ has a form $D_n$, where $D$ is a skew field of finite dimension over the base field or, in other words, a finite dimensional division algebra (which is called a *component* of the algebra $A$). Therefore, the theory of finite dimensional simple algebras is closely related to classical field theory and, in particular, to Galois theory. We see that the structure of a simple finite dimensional algebra essentially reduces to that of finite dimensional division algebra. It shows that the following definition is rather natural.

**1.9.1.** DEFINITION. Two central simple finite dimensional algebras $A, B$ are called *similar*, $A \sim B$, if their division algebras are isomorphic, i.e. $A \cong D_n,\ B \cong D_m$.

We have seen (1.4.3) that a tensor product of central simple algebras is central and simple. Moreover, if $A, A_1$ are similar $A \cong D_n,\ A_1 \cong D_m$ and $B, B_1$ are similar $B \cong \Delta_s,\ B_1 \cong \Delta_r$ then $A \otimes B \sim A_1 \otimes B_1$. Indeed,

$$A \otimes B \cong (D \otimes \Delta)_{ns} \sim (D \otimes \Delta)_{mr} \cong A_1 \otimes B_1.$$

Therefore we can define a well-defined operation on the set of classes of similar central simple finite dimensional algebras (over a fixed field $F$): $[A] + [B] = [A \otimes B]$, where $[A]$ is the set of all central simple finite dimensional algebras similar to $A$.

**1.9.2.** THEOREM. *The set $Br\, F$ of classes of similar central simple finite dimensional algebras over a field $F$ is an abelian torsion group under the addition just defined.*

This group is called the *Brauer group*, or the *class group* of the field $F$.

PROOF. It is evident that $[F] + [A] = [A]$, which implies that $[F]$ is the zero element. For any $A$ let us consider the opposite algebra $A^{op}$, which has the same linear structure and the opposite multiplication $x * y = yx$. Let us show that $[A^{op}] + [A] = [F]$. For this it is enough to show that $A^{op} \otimes A \cong F_n$.

The dimension of the algebra $A^{op} \otimes A$ is $n^2$. Let $L$ be the algebra of all linear transformations of the space $A$ over $F$. This algebra has dimension $n^2$ also and it is similar to zero $L \cong F_n \sim F$. Any element

$$v = \sum a_i \otimes b_i \in A^{op} \otimes A$$

defines a linear transformation

$$\rho(v)\colon\ x \to \sum a_i x b_i.$$

It is easy to see that $\rho$ is a homomorphism:

$$\begin{aligned} X\rho\Big(\sum a_i \otimes b_i\Big)\rho\Big(\sum c_j \otimes d_j\Big) &= \Big(\sum a_i X b_i\Big)\rho\Big(\sum c_j \otimes d_j\Big) \\ &= \sum c_j\Big(\sum a_i X b_i\Big)d_i = \sum (a_i * c_j)X(b_i d_i) \\ &= X\rho\Big(\Big(\sum a_i \otimes b_i\Big)\cdot\Big(\sum c_j \otimes d_j\Big)\Big). \end{aligned} \tag{7}$$

Its kernel is an ideal of a simple algebra, therefore $\rho$ is an embedding. The image of $\rho$ has dimension $n^2$ in the space $L$ of the same dimension. It means that $\rho$ is an isomorphism.

It can be shown that the order of the element $[A]$ divides $\sqrt{dim_F\, D}$, i.e.

$$[A]\sqrt{dim_F\, D} = 0$$

in the Brauer group, where, as usual, $A \cong D_n$. □

Now we are going to consider the most important construction of central simple finite dimensional algebras.

**1.9.3.** *Crossed products.* Let $G$ be a finite group of automorphisms of a field $K$, and $F$ be the subfield of fixed elements $F = \{a \in K \mid \forall g \in G \quad a^g = a\}$. In other words

$K$ is a normal separable extension of the field $F$ with Galois group $G$. Let us consider a linear space $(K, G)$ over the field $F$ with basis $\{u_g \mid g \in G\}$, where the $u_g$ are new symbols. This means that any element of $(K, G)$ has a form

$$\sum_{g \in G} u_g \alpha_g,$$

where $\alpha_g \in K$. The following formulae, with the help of the distributive rule, define a multiplication on the space $(K, G)$:

$$\alpha u_g = u_g \alpha^g, \tag{8}$$

$$u_g u_h = u_{gh} \xi(g, h), \tag{9}$$

where $\xi\colon G \times G \to K^*$ is a function of two variables on the group $G$ with the values in the group of nonzero elements $K^*$ of the field $K$. In such a way we obtain an algebra $(K, G, \xi)$, in general nonassociative, over the field $F$ with the multiplication

$$\Bigl(\sum_{g \in G} u_g \alpha_g\Bigr)\Bigl(\sum_{h \in G} u_h \beta_h\Bigr) = \sum_{g,h \in G} u_{gh} \xi(g, h) \alpha_g^h \beta_h.$$

For the associative rule to hold it is necessary and sufficient that the equations $(u_g u_h) u_f = u_g(u_h u_f)$ are valid. By using formulae (8) and (9) we have that the following conditions on the function $\xi$ are equivalent to the associativity of the algebra $(K, G, \xi)$:

$$\xi^f(g, h)\xi(gh, f) = \xi(g, hf)\xi(h, f), \tag{10}$$

where by definition $\xi^f(g, h) = (\xi(g, h))^f$.

A function $\xi\colon G \times G \to K^*$ which satisfies formula (10) is called a $(G, K)$-*factor set*. The associated algebra $(K, G, \xi)$ is called the *crossed product* of the field $F$ with the group $G$ and factor set $\xi$. The following theorem is very important.

**1.9.4.** THEOREM. *The crossed product $(K, G, \xi)$ is a central simple algebra over the field $F = K^G$.*

PROOF. The proof of this fundamental result is based on the classical linear independence of automorphisms theorem:

*if $g_1, \dots, g_n$ are different automorphisms of a field $K$ then no linear combination $k_1 g_1 + \dots + k_n g_n$ over $K$ is zero but the one with $k_1 = \dots = k_n = 0$.*

Due to this fact it is easy to see that the crossed product with the trivial factor set $\xi_0(h, g) = 1$ is isomorphic to the matrix algebra $F_n$. Indeed, the algebra $(K, G, \xi_0)$ has dimension $n^2$ over the field $F$. The correspondence

$$\mu\colon \sum u_g \alpha_g \to \sum \alpha_g g$$

is a homomorphism of $(K, G, \xi_0)$ to the algebra of linear transformations of $K$ over $F$ (which also has the dimension $n^2$):

$$x(kg \cdot k_1 g_1) = \left(kx^g\right)(k_1 g_1) = k_1 \left(kx^g\right)^{g_1} = k_1 k^{g_1} x^{gg_1} = k^{g_1} k_1 x^{gg_1}.$$

By the linear independence of automorphisms theorem the kernel of this homomorphism is zero. Thus, $\mu$ is an isomorphism $(K, G, \xi_0) \cong F_n$.

Another fact from Galois theory we need is that there exist elements $a_1, \ldots, a_m, b_1, \ldots, b_m \in K$, such that

$$a_1^g b_1 + \cdots + a_m^g b_m = \delta_1^g, \tag{11}$$

where $\delta_1^g$ is the Kronecker delta: $\delta_1^g = 0$ if $g \neq 1$, and $\delta_1^g = 1$ if $g = 1$. This can be proved in two small steps.

a) Let us consider the linear map $\mu$: $K \otimes_F K \to K^n$ defined by the formula

$$\mu\left(\sum a_i \otimes b_i\right) = \left(\ldots, \sum_i a_i^{g_k} b_i, \ldots\right).$$

All we need is to show that there exists a tensor $v \in K \otimes_F K$, such that $\mu(v) = (1, 0, \ldots, 0)$. We suppose here that $\{g_1 = 1, g_2, \ldots, g_k, \ldots, g_n\}$ is the group $G$. As the dimension of $K \otimes_F K$ and that of $K^n$ are both equal to $n^2$, it is enough to prove that the kernel of $\mu$ is zero. If

$$\mu(v) = 0, \quad v = \sum a_i \otimes b_i,$$

then in the algebra $(K, G, \xi_0)$ for any $u_g$ we have

$$0 = u_g \sum a_i^g b_i = \sum a_i u_g b_i.$$

As $\{u_g\}$ is a basis of the algebra $(K, G, \xi_0)$ then $\sum a_i X b_i = 0$ for any $X \in (K, G, \xi_0)$. We have seen that the algebra $(K, G, \xi_0)$ is isomorphic to the matrix algebra $F_n$ and therefore it is enough to prove the second step:

b) *If $a_1, \ldots, a_m$ are $n \times n$ matrices over $F$, such that $\sum a_i X b_i = 0$ for any $n \times n$ matrix $X$, then $\sum a_i \otimes b_i = 0$ in the tensor product $F_n \otimes F_n$.*

Indeed, we can define a linear map $\rho$: $F_n \otimes F_n \to Lin_F(F_n)$ by the formula

$$X\rho\left(\sum c_i \otimes d_i\right) = \sum c_i^t X d_i,$$

where $c^t$ is a matrix transposed to $c$. This is a homomorphism: see (7). The tensor product $F_n \otimes F_n$ is a simple algebra (it is the algebra of $n^2 \times n^2$ matrices), therefore the kernel of $\rho$ is zero, i.e.

$$\sum a_i^t \otimes b_i = 0.$$

As

$$\sum a_i^t \otimes b_i \to \sum a_i \otimes b_i$$

is an isomorphism of linear spaces, the second step is done and the existence of the $a_i$s and $b_i$s in (11) is proved.

Let, finally,

$$v = \sum u_g \alpha_g$$

be a nonzero element in an ideal $I$ of $(K, G, \xi)$. By multiplying if necessary with $u_{g^{-1}}$ we can suppose that $\alpha_1 \neq 0$. For the elements defined in (11), we have

$$\sum a_i v b_i = u_1 \alpha_1,$$

so $u_1 \in I$ and $u_g = u_1 u_g \xi(1, g)^{-1} \in I$ for all $g \in G$. This means that $I = (K, G, \xi)$ and $(K, G, \xi)$ is a simple algebra.

The center of $(K, G, \xi)$ is equal to $u_1 F \cong F$. Indeed, if

$$v = \sum u_g \alpha_g$$

is a central element then for an arbitrary $k \in K$, we have

$$vk - kv = \sum u_g (k - k^g) \alpha_g = 0.$$

So $\alpha_g = 0$ if $g \neq 1$. Analogously, $0 = vu_g - u_g v = u_g(\alpha_1 - \alpha_1^g)$ and therefore $\alpha_1 = \alpha_1^g$ for all $g \in G$, which implies $\alpha_1 \in F = K^G$. Thus, the crossed product $(K, G, \xi)$ is a central simple algebra. □

**1.9.5.** THEOREM. *Any finite dimensional central simple algebra is similar to a crossed product.*

This theorem shows the importance of a crossed product construction but it surely doesn't means that any finite dimensional central simple algebra is isomorphic to a crossed product. In fact it was an open problem for a long time whether any finite dimensional central simple algebra is isomorphic to a crossed product. The answer is "yes" in many important cases. The following theorem is an achievement of algebraic number theory. Recall that a field $F$ is called a *field of algebraic numbers* if it is a finite extension of the field $\mathbf{Q}$ of rational numbers.

**1.9.6.** THEOREM. *Any finite dimensional central simple algebra over a field of algebraic numbers is isomorphic to a crossed product with a cyclic Galois group.*

The first example of a finite dimensional central division algebra which is not a crossed product with cyclic Galois group was found by A.A. Albert. This division algebra is of dimension 16. Nevertheless any division algebra of dimension 16 is a crossed product. The final solution of the problem was found by S.A. Amitsur [Am72] with the help of a generic matrix construction.

Let $X_i = ||x^i_{jk}||$ be an $n \times n$ matrix with coefficients $x^i_{jk}$ that are algebraically independent over **Q**. S.A. Amitsur proved that the *algebra of generic matrices* $\mathbf{Q}[X_1, \ldots, X_m]$ has no zero divisors and satisfies the Ore condition. It follows that this algebra has a classical ring of quotients $D^{(n)} \cong \mathbf{Q}(X_1, \ldots, X_m)$ which is a division algebra. This division algebra has finite dimension over its center (this can be proved by PI-theory methods). Finally, if $n$ is divisible neither by 8 nor by any square of a prime number then $D^{(n)}$ is not a crossed product. Note that in this example the structure of the base field (this is the center of $D^{(n)}$) is not absolutely clear.

Details of the theory of finite dimensional simple algebras can be found in the classical monographs [Al61, Pi82]. Relations of this theory to the theory of $PI$-rings are considered in the books [Ja75, Ro80, Pr73]. In addition there are a lot of relations between classical problems and modern algebraic $K$-theory and number theory (see volume 1 and 2 of this Handbook of Algebra).

We will finish this section with two old open problems concerning the structure of skew fields. First of all there is the Kurosh problem for division algebras.

**1.9.7.** KUROSH PROBLEM. *Is any algebraic division algebra locally finite?*

Recall that an algebra is said to be *algebraic* if any of its element is a root of some polynomial with coefficients from the base field. To date there is known only one way to construct algebraic nonlocally finite algebras. That is the Golod method. How can one modify it in order to make the result a skew field? This is the problem.

There have been a number of positive results with this problem. First of all the solution essentially depends on the base field $F$. Indeed, if for instance $F$ has only finite algebraic field extensions (like the field of real numbers) then an algebraic division algebra over it has bounded degree. This means that there exists a number $n$ such that any element is a root of a polynomial of degree $n$. In this case the well-known Levitzki–Shirshov theorem answers the problem in the affirmative.

**1.9.8.** THEOREM. *Any algebraic algebra of bounded degree is locally finite.*

Another problem is concerned with the well-known fact that any commutative finitely generated ring (not algebra over a field) which is a field is finite.

**1.9.9.** PROBLEM. *Is a skew field which is finitely generated as a ring, commutative?*

**1.10.** *Simple noetherian rings.* Recall that a ring $R$ is called *left* (*right*) *noetherian* if it satisfies the following Ascending Chain Condition (ACC) for left (right) ideals: any strictly ascending chain of left (right) ideals

$$I_1 \subset I_2 \subset \cdots \subset I_n \subset \cdots$$

has a finite length.

Any artinian ring with a unit element is noetherian (this is an old result of Hopkins [Ho39]). In particular, any simple artinian ring is noetherian, which is completely clear as a matrix ring over a skew field has a finite dimension over this skew field as a left linear space. We have already constructed an example of a simple noetherian ring which is not artinian. It is the Weyl algebra – algebra of differential operators (see 1.3). Passing to matrices preserves both ACC and simplicity. Moreover these properties are stable with respect to Morita-equivalence.

Recall that two unitary rings $R, S$ are called *Morita-equivalent* if the categories of left modules $L(R), L(S)$ over these rings are equivalent.

The following Zalesskii–Neroslavskii construction of a simple noetherian ring results in a ring which is not Morita-equivalent to a domain (i.e. a ring with no zero divisors).

**1.10.1.** THEOREM. *Let $R$ be a commutative noetherian domain and $\alpha$ be an automorphism such that $I^\alpha \neq I$ for any proper ideal $I$ of $R$. If $R$ is not a field then $A = R[x, x^{-1}, \alpha]$ is a simple noetherian domain.*

Recall that $A = R[x, x^{-1}, \alpha]$ is the ring of polynomials in $x, x^{-1}$ with the relations $xx^{-1} = x^{-1}x = 1$, $rx = xr^\alpha$, $rx^{-1} = x^{-1}r^{\alpha^{-1}}$, $r \in R$.

Let $F$ be a field of characteristic 2, $K = F(t)$ the field of rational functions, $R = K[y, y^{-1}]$ be the localization of the polynomial ring $K[y]$ with respect to $y$. If $\alpha$ is the automorphism of $R$ defined by $y^\alpha = ty$ then Theorem 1.10.1 can be applied to $R$ and therefore the ring $A = R[x, x^{-1}, \alpha]$ is a simple noetherian domain. Let finally $h$ be the automorphism of $A$ taking $x$ to $x^{-1}$ and $y$ to $y^{-1}$. The automorphism $h$ generates a group $H$ with two elements. We can define a trivial crossed product (or in other words, a skew group ring) $A\langle H\rangle = u_1 A + u_h A$ where the multiplication is defined by formulae (8), (9) with $\xi(g, h) = 1$.

**1.10.2.** THEOREM. *The crossed product $ZN = A\langle H\rangle$ is a simple noetherian ring which is not Morita-equivalent to a domain.*

Simple left (right) noetherian rings have no complete characterization like artinian rings. The Zalesskii–Neroslavskii example shows that the situation is much more complicated than in the artinian case. Nevertheless a number of structural facts are known. First of all like any prime left Goldie ring, a simple left noetherian ring $R$ is a left order in a simple artinian ring. This means that the set of all left regular elements $T = \{t \in R \mid \forall x \in R \ \ xt = 0 \Rightarrow x = 0\}$ is an Ore set and the classical left ring of quotients $T^{-1}R$ is simple artinian.

In the theory of simple noetherian rings the notions of Morita-equivalence, global dimension, Krull dimension and Goldie dimension play an important role.

The following theorem of C. Faith and G.O. Michler gives a sufficient condition for a simple noetherian ring to be Morita-equivalent to a domain.

**1.10.3.** THEOREM. *Any simple noetherian ring of global dimension $\leqslant 2$ is Morita-equivalent to a simple noetherian domain.*

Another interesting result in the characterization of simple noetherian rings up to Morita-equivalence is due to J.T. Stafford.

**1.10.4.** THEOREM. *Let $R$ be a simple noetherian ring of finite global dimension. If the left Krull dimension of $R$ is less than a natural number $n$ then $R$ is Morita-equivalent to a simple noetherian ring with Goldie dimension less than $n$.*

It should be noted that any noetherian ring has a Krull dimension (possibly infinite) and that the Zalesskii–Neroslavskii example has infinite global dimension.

Details of the structure theory of simple noetherian rings can be found in the special monograph [CF75], which contains lots of examples. The Zalesskii–Neroslavskii examples can be found in the original papers [ZN75, ZN77] or in a monograph by D. Passman [Pa89] on crossed products. Rings with different chain conditions are considered in [CH80].

## 2. Prime rings

**2.0.** *Baer radical and prime and semiprime rings.* The Baer radical is defined by the smallest radical class, which contains all rings $A$ with zero multiplication, $A^2 = 0$. The Baer radical $\mathbf{B}(R)$ of a ring $R$ is the union of the transfinite chain of ideals

$$(0) = N_0 \subseteq N_1 \subseteq N_2 \subseteq \cdots N_\alpha \subseteq \cdots,$$

where the ideal $N_{\alpha+1}$ is chosen in such a way that the sum of all the nilpotent ideals of the factor-ring $R/N_\alpha$ is $N_{\alpha+1}/N_\alpha$ and the equality

$$N_\alpha = \bigcup_{\beta<\alpha} N_\beta$$

holds for limit ordinals $\alpha$.

Recall that an ideal $I$ is called *nilpotent* if there exists a number $n$ such that $x_1 x_2 \cdots x_n = 0$ for all $x_1, x_2, \ldots, x_n \in I$.

**2.0.1.** DEFINITION. A ring is called *semiprime* if it has no nonzero ideals with zero multiplication, i.e., for an ideal $I$ the equality $I^2 = 0$ implies $I = 0$. It is easy to see that a semiprime ring has no nonzero nilpotent ideals as well. Indeed, if $I^n = 0$ then $I^{n-1}$ is an ideal with zero multiplication.

**2.0.2.** DEFINITION. An ideal $I$ of the ring $R$ is called *semiprime* if the factor-ring $R/I$ is semiprime.

We see that the Baer radical of a ring is a semiprime ideal: if $\mathbf{B}(R) = N_\alpha$ then $N_\alpha = N_{\alpha+1}$, which means that the factor-ring $R/\mathbf{B}(R)$ has no nonzero nilpotent ideals.

The Baer radical can be defined as the smallest semiprime ideal of a ring as well. To see this we can use transfinite induction. If $I$ is a semiprime ideal then $N_0 \subseteq I$, which gives a basis for the induction. The inclusion $N_\alpha \subseteq I$ implies that any nilpotent modulo $N_\alpha$ ideal is nilpotent modulo $I$ and due to the semiprimeness of $I$ the inclusion

$N_{\alpha+1} \subseteq I$ holds. Finally if $\alpha$ is a limit ordinal and $N_\beta \subseteq I$ for any $\beta < \alpha$, then evidently $N_\alpha = \bigcup N_\beta \subseteq I$.

**2.0.3.** DEFINITION. A ring $R$ is called *prime* if the product of any two of its nonzero ideals is nonzero. Accordingly, an ideal $I$ is called *prime* if the factor-ring $R/I$ is prime. Therefore, the ring $R$ is prime iff $(0)$ is a prime ideal. The following characterization of prime and semiprime rings in terms of the ring elements often proves to be useful.

**2.0.4.** LEMMA. *A ring $R$ is prime iff for any nonzero $a, b \in R$ there exists an $x \in R$, such that $axb \neq 0$.*

*A ring $R$ is semiprime iff for any nonzero element $a \in R$ there exists $x \in R$, such that $axa \neq 0$.*

PROOF. If in a prime ring $axb = 0$ for each $x \in R$, then the ideal $(a) = a\mathbf{Z} + aR + Ra + RaR$ generated by $a$ and the ideal $J = Rb + RbR$ satisfy $(a) \cdot J = 0$, therefore either $(a) = 0$ or $J = 0$. In the former case $a = 0$; in the latter case $R \cdot (b) = 0$ and $b = 0$. Inversely, if $AB = 0$ for two nonzero ideals, then $aRb = 0$ for nonzero elements $a \in A$, $b \in B$.

If in a semiprime ring $aRa = 0$, then $(Ra + RaR)^2 = 0$, so $Ra = 0$; in particular $a\mathbf{Z}a\mathbf{Z} = 0$ and $(a)^2 = 0$. Inversely, if $A^2 = 0$ and $a \in A$, then $aRa \subseteq A^2 = 0$. □

The most important fact which connects the Baer radical and prime rings is the following theorem.

**2.0.5.** THEOREM. *Any semiprime ring is a subdirect product of prime rings.*

PROOF. Let $a = a_0$ be a nonzero element of the ring $R$. According to Lemma 2.0.4, we can find an element $x_1 \in R$, such that $a_1 = ax_1a \neq 0$. Using the element $a_1$ we find an element $x_2$ such that $a_2 = a_1x_2a_1 \neq 0$. Continuing this process, we can construct a countable sequence of nonzero elements $a_0, a_1, \ldots, a_n, \ldots$ such that $a_{n+1} = a_nx_{n+1}a_n$ for certain elements $x_1, x_2, \ldots, x_n, \ldots$ of the ring $R$.

Let us consider the set $\mathbf{M}$ of all ideals containing no elements of the sequence thus constructed and ordered by inclusion. This set is not empty since it contains the zero ideal as an element. Moreover, the set is *directed,* thus, according to the Zorn lemma, the set $\mathbf{M}$ contains maximal elements. Let $P_a$ be one of them. In this case the ideal $P_a$ does not intersect the sequence $a_0, a_1, \ldots, a_n, \ldots$ but any ideal strictly containing $P_a$ has a nonempty intersection with this sequence. Since $a \notin P_a$,

$$\bigcap_{0 \neq a \in R} P_a = 0$$

and it remains to show that $P_a$ is a prime ideal.

Let $A$ and $B$ be ideals of the ring $R$ not contained in $P_a$. Then $A_1 = A + P_a \supset P_a$, $B_1 = B + P_a \supset P_a$, and, since $P_a$ is maximal in the set $\mathbf{M}$, the ideals $A_1$ and $B_1$ do not belong to $\mathbf{M}$, i.e. there can be found natural $n, m$, such that $a_n \in A_1$, $a_m \in B_1$. Let, for instance, $n \geqslant m$. In this case, since $a_{k+1} = a_kx_{k+1}a_k \in (a_k)$, $a_n \in (a_m)$. Then we get:

$$a_{n+1} = a_nx_{n+1}a_n \in A_1B_1 \subseteq AB + P_a.$$

Therefore, the ideal $AB + P_a$ is not contained in $P_a$ and, hence, $AB$ is not zero modulo $P_a$. □

It is a question of interest how far the Baer radical can be away from being nilpotent because of the transfinite process of its construction. To a certain extent this question is answered by the following theorem.

**2.0.6.** THEOREM. *The Baer radical of a ring is locally nilpotent.*

PROOF. Let us consider the inductive construction. We have $N_0$ is locally nilpotent. Let $\alpha$ be a limit ordinal,

$$N_\alpha = \bigcup_{\beta<\alpha} N_\beta$$

and let us suppose that all the ideals $N_\beta$ are locally nilpotent. If $a_1, \ldots, a_n$ are elements from $N_\alpha$, then there exist transfinite numbers $\beta_1, \ldots, \beta_n$ which are less than $\alpha$ and such that $a_1 \in N_{\beta_1}, \ldots, a_n \in N_{\beta_n}$. Let $\beta$ be the largest from the numbers $\beta_1, \ldots, \beta_n$, then, since the ideals $\{N_\gamma\}$ form a chain, we get $a_1, \ldots, a_n \in N_\beta$ and, due to the local nilpotency of $N_\beta$, these elements generate a nilpotent subring.

If $\alpha$ is not a limit ordinal, $\alpha = \beta + 1$ and the ideal $N_\beta$ is locally nilpotent, then by the definition $N_\alpha$ is a sum of nilpotent modulo $N_\beta$ ideals. As a sum of any finite number of nilpotent ideals is nilpotent we have that any finite set $s_1, \ldots, s_n$ of elements of the ideal $N_\alpha$ generates a subring $S$, such that $S^m \subseteq N_\beta$ . However, the subring $S^m$ is generated by a finite set of elements $\{s_{i_1} s_{i_2} s_{i_3} \cdots s_{i_m} \mid 1 \leqslant i_j \leqslant n\}$ and, hence, $S^m$ and, consequently, also $S$, are nilpotent. □

**2.1.** *General structure, Martindale quotient ring.* Let $R$ be a prime ring and consider the set of all left $R$-module homomorphisms $f\colon I \to R$ where $I$ ranges over all nonzero two-sided ideals of $R$. Two such homomorphisms are said to be *equivalent* if they agree on their common domain, which is a nonzero ideal since the intersection of two nonzero ideals in a prime ring contains their product and therefore is nonzero. It is easy to see that this is an equivalence relation. Indeed, what is needed here is the observation that if $f\colon I \to R$ with $If = 0$ and if $f$ is defined on $r \in R$, then $rf = 0$. This follows since $Ir \subseteq I$ so $0 = (Ir)f = I(rf)$ and hence $rf = 0$ in this prime ring. We let $\overline{f}$ denote the equivalence class of $f$ and let $R_F$ be the set of all such equivalence classes.

The arithmetic of $R_F$ is defined in a fairly obvious manner. Suppose $f\colon I \to R$ and $g\colon J \to R$. Then $\overline{f}+\overline{g}$ is the class of $f+h\colon I \cap J \to R$ and $\overline{f}\overline{g}$ is the class of the product function $fg\colon JI \to R$. It is easy to see that these definitions make sense and that they respect the equivalence relation. Furthermore, the ring axioms are surely satisfied so $R_F$ is a ring with 1. Finally let $r_\rho\colon R \to R$ denote right multiplication by $r \in R$. Then the map $r \to \overline{r_\rho}$ is easily seen to be a ring homomorphism from $R$ into $R_F$. Moreover, if $r \neq 0$ then $Rr_\rho \neq 0$ and hence $\overline{r_\rho} \neq 0$ by the observation of the preceding paragraph. We conclude therefore that $R$ is embedded isomorphically in $R_F$ and we will view $R_F$ as an overring of $R$. It is the *left Martindale ring of quotients* of $R$.

Suppose $f\colon I \to R$ and $a \in I$. Then $a_\rho f$ is defined on $R$ and for all $r \in R$ we have

$$r(a_\rho f) = (ra)f = r(af) = r(af)_\rho.$$

Hence $\overline{a_\rho}\ \overline{f} = \overline{(af)_\rho}$ and the map $f$ translates in $R_F$ to right multiplication by $\overline{f}$. With this observation, the following theorem is an elementary exercise. It is very important that the properties of the quotient ring described in this theorem define the left Martindale quotient ring uniquely up to isomorphism over $R$.

**2.1.1.** THEOREM.

(a) $R \subseteq R_F$.

(b) *If $q \in R_F$ and $Iq = 0$ for some nonzero ideal $I$ of $R$, then $q = 0$.*

(c) *If $q_1, q_2, \ldots, q_n \in R_F$, then there exists a nonzero ideal $I$ of $R$ with $Iq_1$, $Iq_2, \ldots, Iq_n \subseteq R$.*

(d) *If $I$ is a nonzero ideal of $R$ and $\xi\colon I \to R$ is a homomorphism of left $R$-modules, then there exists an element $q \in R_F$ such that $\xi(a) = aq$ for all $a \in I$.*

These properties can be considered as axioms for a left Martindale ring of quotients. Using these axioms one can prove the following useful facts.

**2.1.2.** LEMMA.

(a) *The ring $R_F$ is prime.*

(b) *The center of $R_F$ is a field.*

(c) *The left Martindale ring of quotient of any nonzero ideal $I$ of $R$ is equal to that of $R$, i.e. $I_F = R_F$.*

The *extended centroid* $C = C(R)$ of a prime ring $R$ is defined as the center of $R_F$. The *central closure* $CR$ of the ring $R$ is the linear space over the extended centroid generated by $R$. In the case of a ring without 1 it is natural to consider the *unitary central closure* $S = C + CR$.

Another important subring in $R_F$ is the *symmetrical* quotient ring

$$Q(R) = \{q \in R_F \mid qI \subseteq R \text{ for some nonzero ideal } I \text{ of } R\}.$$

In some sense it is the intersection of the left and right Martindale quotient rings. Here the right Martindale quotient ring is defined in an obvious way by changing left to right in the construction.

The extended centroid of a prime ring plays in the theory of prime rings the same role the centroid does in the theory of simple rings. This role is defined by the following statement which is analogous to the important Lemma 1.4.4.

**2.1.3.** LEMMA. *If $d_1, \ldots, d_n$ are linearly independent over $C$ elements of $R_F$ then there exist elements $s_1, \ldots, s_m, t_1, \ldots, t_m \in R$, such that*

$$\sum s_i d_1 t_i \neq 0, \ \sum s_i d_2 t_i = 0, \ \ldots, \ \sum s_i d_n t_i = 0. \tag{12}$$

**2.1.4.** EXAMPLES. Now we are going to consider a number of examples. Let $R$ be a simple ring with 1. In this case the collection of all nonzero ideals has only one element, $R$. Any left module homomorphism $\xi\colon R \to R$ is defined by the right multiplication by $\xi(1)$: $\xi(r) = \xi(r1) = r\xi(1)$, which implies $R_F = R$. We have seen an example of a simple ring with a unit element; it is the Weyl algebra $A_n(F)$. This algebra has a classical quotient division ring, while the Martindale quotient ring is equal to $A_n(F)$.

If a simple ring $R$ has no unit element then surely $R_F \neq R$ as the quotient ring has a unit. In this case $R$ is a right ideal of $R_F$ and a two-sided ideal of $Q(R)$. Indeed, the collection of all nonzero ideals of $R$ still has only one element, $R$. Therefore, for any element $q \in R_F$ we have $Rq \subseteq R$; if additionally $q \in Q(R)$, then also $qR \subseteq R$. Moreover both of the rings $R_F$ and $Q(R)$ are subdirectly irreducible prime rings. Evidently the heart of $Q(R)$ is $R$, while the heart of $R_F$ is the two-sided ideal $R_F R$ generated by $R$: any nonzero ideal $I$ of $R_F$ contains $RI \triangleleft R$ and therefore $R \subseteq I$. The following example shows that the heart of $R_F$ might not be equal to $R$.

Let $R$ be a ring of all infinite matrices over a field $K$ having only a finite number of nonzero elements. This is a simple subring in both the row-finite, $Lin_r(K)$, and column-finite, $Lin_c(K)$, rings of matrices. It is easy to see that $R$ is a right ideal of $Lin_r(K)$ and a left ideal of $Lin_c(K)$, which implies at least that its left Martindale quotient ring contains $Lin_r(K)$ and the right one contains $Lin_c(K)$. Moreover $R$ is not a two-sided ideal either in $Lin_r(K)$ or in $Lin_c(K)$; so $R$ cannot be an ideal either in the left Martindale quotient ring or in the right one. Note that it can be shown that $R_F = Lin_r(K)$ and $Q(R) = Lin_r(K) \cap Lin_c(K)$.

Another useful example is given by a free algebra. If $K\langle X\rangle$ is the free algebra generated by a set $X$ in two or more elements then its symmetrical quotient ring is equal to $K\langle X\rangle$ (this is not a trivial fact), while the left Martindale quotient ring is quite large. For instance it contains zero divisors. Note here that the symmetrical quotient ring of a ring with no zero divisors has no zero divisors: if $qs = 0$, then for suitable ideals $Iq \subseteq R$, $sJ \subseteq R$ and so $IqsJ = 0$ in the ring $R$, which implies $Iq = 0$ or $sJ = 0$, i.e. $q = 0$ or $s = 0$.

**2.2.** *Generalized polynomial identities.* A *generalized polynomial* over a prime ring $R$ is a polynomial in noncommutative variables with coefficients in $R$ (or more generally in $R_F$).

By this definition, any generalized polynomial $f$ can be presented as a sum of monomials

$$f = \sum_i d_1^{(i)} x_{i1} d_2^{(i)} x_{i2} \cdots d_n^{(i)} x_{in} d_{n+1}^{(i)}, \tag{13}$$

where $d_k^{(i)} \in R_F$, $x_{ij} \in X$. Let $a_1, \ldots, a_n$ be some elements from $R_F$. Then the value of the polynomial $f(x_1, \ldots, x_n)$ at $x_1 = a_1, \ldots, x_n = a_n$ is the element of ring $R_F$ obtained by replacing $x_i$ with $a_i$ in formula (13).

**2.2.1.** DEFINITION. A generalized polynomial $f = f(x_1, \ldots, x_n)$ is said to be a *generalized identity* of the ring $R$ if $f(r_1, \ldots, r_n) = 0$ for all $r_1, \ldots, r_n \in R$. The identity

$f$ is called *trivial* if it is a consequence of the identities defined by the ring properties (distributivity, associativity, etc.) and the identities of the type $xc = cx$ where $c$ is an arbitrary element of the generalized centroid $C$.

The definition of a trivial identity can be stated in terms of free products: a generalized polynomial $f$ is trivial if it is zero as an element of the free product $R_F * C\langle X\rangle$ of algebras over $C$, where $C\langle X\rangle$ is the free algebra with free generators $\{x_1, x_2, \dots, x_n, \dots\} = X$.

By this definition any generalized polynomial can be reduced modulo the trivial ones to the form (13), where the $d_k^{(i)}$ are elements of a given basis of $R_F$ over the field $C$. In this case $f$ is the trivial identity iff in this reduced form all the monomials cancel.

Another very important criterion of nontriviality for a *multilinear* generalized polynomial is based on the observation that all the simplest trivial identities do not change the order of the variables in the monomials. It follows that a multilinear generalized polynomial is trivial iff all its, so called, *generalized monomials* are trivial. Recall that any multilinear $f(x_1, \dots, x_n)$ is a sum of $n!$ generalized monomials $f = \sum f_\pi$, where $\pi$ ranges over all permutations of $\{1, \dots, n\}$ and a generalized monomial $f_\pi$ is a sum of all monomials of $f$ which have the fixed order of variables $x_{\pi(1)}, x_{\pi(2)}, \dots, x_{\pi(n)}$. The criterion is based on the following

**2.2.2.** LEMMA. *A generalized monomial is an identity if and only if it is a trivial identity.*

**2.2.3.** Thus we have the criterion: *A multilinear generalized polynomial is nontrivial iff one of its generalized monomials has a nonzero value.*

The proof of Lemma 2.2.2 follows from Lemma 2.1.3 by an easy induction on degree. Indeed, any generalized monomial with the order of variables $x_1, x_2, \dots, x_n$ can be written in the form

$$f = \sum d_j x_1 b_j g_j(x_2, \dots, x_n),$$

where the $d_1, \dots, d_k$ are linearly independent over $C$ elements of $R_F$. By induction we can suppose that $b_1 g_1$ is not an identity. Lemma 2.1.3 says that there exist elements $s_1, \dots, s_m, t_1, \dots, t_m \in R$, such that

$$a_1 = \sum s_i d_1 t_i \neq 0, \quad \sum s_i d_2 t_i = 0, \quad \dots, \quad \sum s_i d_n t_i = 0$$

we have that the generalized monomial

$$\sum_i s_i f(t_i x_1, x_2, \dots, x_n) = a_1 x_1 b_1 g_1(x_2, \dots, x_n)$$

is identically zero on $R$ (if $f$ is). As $b_1 g_1$ is not an identity, we can find $a_2, \dots, a_n \in R$, such that $v = b_1 g_1(a_2, \dots, a_n) \neq 0$. This implies $a_1 R v = 0$, which is impossible in a prime ring.

**2.2.4.** MARTINDALE THEOREM. *If a prime ring $R$ satisfies a nontrivial generalized identity, then its central closure $RC$ has an idempotent $e$, such that $eRCe$ is a finite-dimensional sfield over $C$.*

Recall that the "sfield" is the short for skew field (= division ring = quasifield). With the help of Lemma 2.1.3 we will prove a number of important corollaries.

**2.2.5.** LEMMA. *A center of the sfield $T = eRCe$ is equal to $Ce$.*

PROOF. Let $t$ be a central element of $T$. Then for any $x \in R$ we have

$$f(x) \equiv txe - exet = 0.$$

If the elements $e$ and $t$ are linearly independent over $C$ in the ring $R_F$, then, by Lemma 2.1.3 there can be found elements $v_i, r_i \in R$, such that

$$0 = \sum v_i e r_i \neq \sum v_i t r_i = a.$$

Then

$$0 = \sum v_i f(r_i x) = axe$$

for all $x$, which is impossible. Therefore, $t = ce$. □

**2.2.6.** LEMMA. *For any linear over $C$ transformation $l$: $T \to T$ there exist elements $a_i, b_i \in T$, such that*

$$l(x) = \sum a_i x b_i.$$

PROOF. If $n$ is the dimension of the sfield $T$ over the center $Ce$, then the dimension of the space of all linear transformations is $n^2$. On the other hand, the $n^2$ linear transformations $l_{ij}$: $x \to a_i x a_j$, where $a_1, \ldots, a_n$ is a basis of $T$ over the center, are linearly independent. Indeed, if

$$f(x) \equiv \sum c_{ij} a_i x a_j = 0, \quad x \in T,$$

then

$$\sum_i a_i x \Big( \sum_j c_{ij} a_j \Big) = 0$$

and, applying Lemma 2.1.3 to $T$ and the system of its linearly independent elements $d_1 = a_1, \ldots, d_n = a_n$, we get $c_{ij} = 0$. □

**2.2.7.** THEOREM. *If a ring with no zero divisors satisfies a nontrivial generalized identity, then its center is nonzero and the central closure is a finite-dimensional sfield.*

PROOF. According to the Martindale theorem, $RC$ has a primitive idempotent, but the ring $Q(R)$ with no zero divisors can have only one nonzero idempotent; it is 1. Therefore, $RC = 1 \cdot RC \cdot 1 = T$ is a finite dimensional sfield over $C$.

Let us consider a linear over $C$ projection $l\colon T \overset{\text{on}}{\to} C$ and assume that $l(T) \cap R = 0$. Then, by Lemma 2.2.6 there are elements $a_i, b_i \in T$, such that

$$l(x) = \sum a_i x b_i$$

for all $x \in T$. For the elements $a_i, b_i$ one can find a nonzero element $q \in R$, such that $a_i q, q b_i \in R$, since $T \subseteq Q(R)$. We have $l(qRq) \subseteq R$ and, therefore, $C = l(T) = l(qTq) = l(qRCq) = l(qRq)C \subseteq (l(T) \cap R)C = 0$. This contradiction proves the corollary. □

**2.2.8.** The Martindale quotient ring arose as a tool for the investigation of prime rings in the original paper [Ma69] where he proves Theorem 2.2.4 as a generalization of a theorem of Amitsur [Am65]. The construction of the quotient ring as a direct limit is quite old (see, for example the survey [El73]). Criterion 2.2.3 and Lemma 2.2.2 are due to L. Rowen [Ro75].

**2.3.** *Prime rings with a primitive idempotent.* In this paragraph we describe the structure of the rings arising in the Martindale theorem. A *primitive idempotent* is an idempotent $e = e^2 \neq 0$ such that $eRe$ is a sfield. In the literature prime rings with a primitive idempotent are called primitive with a nonzero one-sided ideal or primitive with a nonzero socle or quite primitive rings. The structure of such rings was investigated in detail by N. Jacobson in his classical papers and monographs. Here we are going to present a small piece of the subject.

The *socle* $Soc(R)$ of a prime ring with a primitive idempotent $R$ is the two-sided ideal generated by all its primitive idempotents.

**2.3.1.** LEMMA. *The socle $Soc(R)$ of a quite primitive ring $R$ is equal to its heart. In particular, the socle is generated by any primitive idempotent and is a simple ring.*

PROOF. If $I$ is a nonzero ideal and e is an arbitrary primitive idempotent, then $0 \neq eIe \subseteq I \cap eRe$. As a sfield has no proper ideals, $I \cap eRe \ni e$ and, hence, $I \supseteq Soc(R)$. □

It should be recalled that a module $M$ over the ring $R$ is called *irreducible* if it contains no proper submodules (i.e. submodules other than (0) or $M$).

**2.3.2.** LEMMA. *Let $e, f$ be primitive idempotents of a prime ring $R$. Then the sfields $eRe$ and $fRf$ are isomorphic. The right ideals, $fR$ and $eR$, as well as the left ideals, $Re$ and $Rf$, are mutually isomorphic as modules over $R$ and they are irreducible.*

PROOF. Since the ring $R$ is prime, there can be found an element $u$, such that $fue \neq 0$. By the same reason, the set $fueRf$ forms a nonzero right ideal of the sfield $fRf$, i.e. there exists an element $u'$, such that $fue \cdot eu'f = f$. Squaring both parts of the latter equality, we see that $\xi = eu'f\, fue \neq 0$. Therefore, $\xi$ is a nonzero idempotent ($\xi^2 = eu'f(fue \cdot eu'f)\, fue = \xi$) lying in the sfield $eRe$. Thus, $eu'f \cdot fue = e$. It is now absolutely clear that the mappings

$$ere \mapsto fuereu'f, \qquad frf \mapsto eu'frfue$$

give the sought isomorphism of sfields.

Let $N$ be a nonzero right ideal contained in $eR$. Then $eN = N$ and, hence, since the ring $R$ is prime, $eNe = Ne \neq 0$. However, $eNe$ is a right ideal of the sfield $eRe$ and, therefore, $e \in eNe \subseteq N$ and $eR = N$ so that $eR$ (and, analogously, $Re$) is an irreducible module.

Let us now consider the mapping $\varphi$: $eR \to fR$ given by the rule $er \to fuer$, where $u$ is the element determined above. As $fue \neq 0$, $\varphi$ is a nonzero homomorphism of right $R$-modules. As the kernel of $\varphi$ is a submodule of $eR$, then due to irreducibility, $\varphi$ is an embedding. The image of $\varphi$ is a nonzero submodule in $fR$, coinciding, due to its irreducibility, with $fR$. Thus, $\varphi$ is an isomorphism of modules. The lemma is proved. □

The lemma proved above makes it possible to determine the *sfields* of a quite primitive ring $R$ as the sfield $T$, which is isomorphic to $eRe$ for a certain primitive idempotent $e$. Moreover, the lemma states that the module $V = eR$ is also independent of the choice of a primitive idempotent. Since $eRe \cdot eR \subseteq eR$, this module can be viewed as a left vector space over the sfield $T = eRe$. In this case the elements of the ring $R$ turn to transformations of the left vector space $V$ over the sfield $T$. Indeed, the element $r$ is identified with the mapping $v \mapsto vr$. This presentation is exact, since $Vr = 0$ implies $r = 0$, as $R$ is prime. There now naturally arises the ring $Lin_r(T)$ of all linear transformations of the left space $V$ over the sfield $T$ (see 1.2.2).

The embedding of a quite primitive ring $R$ into the ring $Lin_r(T)$ is, by itself, not enough information. The most essential is the fact that at such an embedding $R$ proves to be a *dense* subring in $Lin_r(T)$.

A subring $S \subseteq Lin_r(T)$ is called *dense,* if for any finite-dimensional subspace $W \subseteq V$ and any linear transformation $l \in Lin_r(T)$, there exists an element $s \in S$ which coincides with $l$ on $W$. In terms of matrices this is equivalent to the fact that for any finite subset $J \subset U$ and any $J \times J$-matrix $l = \|l_{\alpha\beta}\|$ there exists a matrix $\|s_{\alpha\beta}\| \in S$, extending $l$, i.e. such that $s_{\alpha\beta} = l_{\alpha\beta}$ at $\alpha, \beta \in J$.

It should also be added that on the ring $Lin_r(T)$ there is a natural topology, called the *finite topology*, such that a subspace is dense in this topology if and only if it is dense in the sense defined above. Namely, in the finite topology the basis of zero neighborhoods are the sets $W^{\perp} = \{l \in Lin_r(T) \mid Wl = 0\}$ where $W$ ranges over all finite-dimensional subspaces of $V$.

**2.3.3.** THEOREM. *A quite primitive ring $R$ is dense in $Lin_r(T)$. A set of all finite rank transformations $l \in R$ coincides with $Soc(R)$.*

With the help of this theorem it is possible to obtain the result of M. Hacque ([Ha82], Lemma 11, and [Ha87], Remark 4.9) on quotient rings of a prime ring with a primitive idempotent.

**2.3.4.** THEOREM. *The left Martindale quotient ring of a prime ring with a primitive idempotent coincides with the ring $Lin_r(T)$.*

**2.3.5.** *The symmetrical quotient ring.* Let us now make a small diversion into general topology. Let us consider an arbitrary set $X$ and a set $F$ of its transformations. Of special interest are the topologies for which $F$ proves to consist of continuous transformations. It is evident that if $\{\tau_\alpha,\ \alpha \in A\}$ is the class of such topologies (a topology is considered to be given by a set of open sets), then their intersection

$$\bigcap_A \tau_\alpha$$

will also be a topology for which all the functions $f \in F$ are continuous.

Recall that the linear space $V$ (supplied with a topology) over a topological sfield $T$ is called a *topological linear space,* provided the linear operations (addition and scalar multiplication $T \times V \to V$), as well as the mappings $tv \mapsto t$ for all $v \neq 0$ are continuous (here $t$ ranges over $T$).

Let us consider $T$ as a topological sfield with the discrete topology.

**2.3.6.** THEOREM.

(a) *There is a weakest topology on $V$, which turns it into a topological linear space over the sfield $T$, such that $R$ consists of continuous transformations.*

(b) *A set of all linear continuous transformations in this topology equals $Q(R)$.*

(c) *A set of all continuous finite rank transformations coincides with the socle of the ring $R$.*

**2.4.** *Essential polynomial identities, prime PI-rings.* The criterion 2.2.3 of nontriviality of a generalized identity inspires the following notion. A multilinear generalized identity $f$ of a prime ring $R$ is called *essential* if the ideal of $R_F$ generated by all values of all its generalized monomials, while the variables range over $R_F$, contains 1 (i.e. it is equal to $R_F$).

**2.4.1.** THEOREM. *If a prime ring $R$ satisfies an essential generalized identity, then its central closure $RC$ is a finite-dimensional central simple algebra over $C$. The center $Z$ of $R$ is nonzero. The extended centroid $C$ is the quotient field of $Z$.*

PROOF. By the criterion 2.2.3 and the Martindale Theorem 2.2.4, $RC$ is a prime ring with a primitive idempotent. Theorem 2.3.3 shows that this ring is a dense subring of the row-finite matrix ring $L = Lin_r(T)$ over a finite-dimensional central division algebra $T$. As the ring operations are continuous in the finite topology, all generalized identities can be extended from $RC$ to $L$. Moreover any multilinear identity of $R$ is also valid on $RC$. Indeed,

$$\begin{aligned} &f(x_1, \dots, c_1x_i + c_2y_i, \dots, x_n) \\ &\quad \equiv c_1 f(x_1, \dots, x_i, \dots, x_n) + c_2 f(x_1, \dots, y_i, \dots, x_n) = 0. \end{aligned}$$

Thus the ring $L$ also satisfies the essential identity $f$. Let $\alpha$ be the dimension of the space $V$. We will prove that $\alpha$ is a finite number.

If $\alpha$ is not finite then the factor-ring $R_1 = L/S_r^\alpha(T)$ is a simple ring with 1 (see 1.2.3). The generalized polynomial $f$ induces a generalized polynomial $\bar{f}$ over the ring $R_1$, which evidently is an identity of $R_1$. By the criterion 2.2.3 this is a nontrivial identity because the ideal generated by all values of its generalized monomials is the homomorphic image of the ideal generated by all values of generalized monomials of $f$ and the last one contains 1. So we can apply the Martindale theorem to $R_1$. This is a simple ring with 1 and therefore it coincides with its left Martindale quotient ring. Therefore it is a prime ring with a primitive idempotent and by 2.3.4, $R_1 = Lin_r(T_1)$. The row-finite matrix ring is simple only if the cardinality $\beta$ of matrix size is finite, as in the opposite case $S_r^\beta(T_1)$ is a proper ideal. Thus we have proved that $R_1$ is the ring of $n \times n$ matrices over a sfield, which is a contradiction as it is easy to see that for an infinite $\alpha$ the factor-ring $L_r^\alpha(T)/S_r^\alpha(T)$ is not artinian.

So the number $\alpha = n$ is finite and $RC$ is a dense subring in the $n \times n$ matrix ring over the skew field $T$. The finite topology is discrete in this case and therefore $RC = T_n$.

Finally, the considerations in statements 2.2.5–2.2.7 show that the center $Z$ of $R$ is nonzero (one should replace $T$ with $RC$ in the proofs). If $c \in C$ is an arbitrary element, then $cI$ is a nonzero ideal of $R$ for a suitable $I \triangleleft R$. The center of $I$ is nonzero as this ring satisfies the same identity and $I_F = R_F$. As $Z(I) \subseteq Z(R)$ we can find nonzero central elements $z_1, z_2$, such that $cz_1 = z_2$. □

**2.4.2.** DEFINITION. A ring $R$ is called a *PI-ring* if it satisfies a polynomial identity with integer coefficients

$$\sum_i \alpha_i x_{i1} x_{i2} \cdots x_{in} = 0 \tag{14}$$

with one of the coefficients $\alpha_i$ equal to 1.

**2.4.3.** THEOREM. *The center $Z$ of a prime PI-ring $R$ is nonzero. The quotient ring $RZ^{-1}$ is a finite-dimensional central simple algebra over the quotient field $C = ZZ^{-1}$.*

PROOF. The standard linearization process allows one to find a multilinear identity of the type (14). This identity considered as a generalized polynomial identity is essential: the generalized monomial $\alpha_i x_{i(1)} x_{i(2)} \cdots x_{i(n)}$ at $x_{i(j)} = 1$ is equal to $\alpha_i$. Due to Theorem 2.4.1 we are done. □

Note that the fact that the center of a prime PI-ring is nonzero is very important for the structure theory of PI-rings. This fact was firstly discovered independently by E. Formanek and Ju. Razmyslov with the help of so called *central polynomials*. These are polynomials with only central values, and which are not identities. Theorem 2.4.3 in fact was obtained by L. Rowen [Ro75]. He proved that any ring with 1 (not necessarily prime) satisfying an essential identity is a PI-ring. This fact is also valid for identities with automorphisms [Kh75] and even for differential identities with automorphisms under some additional restrictions (see [Kh91]).

**2.5.** *Galois theory of prime rings.* The Galois theory of noncommutative rings is a natural outgrowth of the classical Galois theory of fields. Let $G$ be the group of automorphisms of a ring $R$. Then we are concerned with the relationship between $R$ and the fixed ring $R^G = \{r \in R \mid r^g = r \text{ for all } g \in G\}$ and with the relationship between the subgroups of $G$ and intermediate rings $R \supset S \supseteq R^G$. Let

$$A(S) = \{g \in Aut\, R \mid s^g = s \text{ for all } s \in S\}.$$

The ring $R^G$ is called the Galois subring of a group $G$ and the group $A(S)$ the Galois group of a subring $S$. The correspondences $G \mapsto R^G$ and $S \mapsto A(S)$ invert the inclusion relations, i.e. if $G_1 \subseteq G_2$, then $R^{G_1} \supseteq R^{G_2}$ and if $S_1 \subseteq S_2$, then $A(S_1) \supseteq A(S_2)$. We also have $A(R^G) \supseteq G$, $R^{A(S)} \supseteq S$, which immediately yield $R^{A(R^G)} = R^G$, $A(R^{A(S)}) = A(S)$. Therefore, the mappings under discussion set up a one-to-one correspondence between Galois groups and Galois subrings.

In order to prove the correspondence theorem in a class of rings $\Sigma$ it is necessary (and sufficient) to answer the following questions:

(1) Under what conditions does a Galois subring of a group $G$ of automorphisms of a ring $R \in \Sigma$ belong to $\Sigma$?

(2) When will a group $G$ which satisfies condition (1) be a Galois group?

(3) Under what conditions an intermediate ring $S \in \Sigma$, $R^G \subseteq S \subseteq R$, is a Galois subring?

**2.5.1.** *Basic notions.* Any automorphism $g$ of a prime ring $R$ has a unique extension to the symmetrical quotient ring $Q(R)$, therefore we can suppose that all the automorphisms are defined on $Q(R)$. This is a pleasant fact. An unpleasant one is that in this situation there arises the problem of how to distinguish automorphisms of $R$ in the group $Aut\, Q(R)$. The straightforward answer: "by the property $R^g = R$", is not good enough. The point is that from the $Q(R)$-point of view the ring $R$ is in no way better than any of its nonzero ideals as $Q(I) = Q(R)$, while the automorphism groups can be essentially different. This leads us to the following notion

$$\mathcal{A}(R) = \{g \in Aut\, Q(R) \mid \exists I, J \lhd R,\ I \neq 0 \neq J,\ J \subseteq I^g \subseteq R\}.$$

One can prove that $\mathcal{A}(R)$ is a group and $\mathcal{A}(R) = \mathcal{A}(I)$ for any nonzero ideal $I$. We will look at elements of the group $\mathcal{A}(R)$ as the automorphisms under investigation. This approach involves only a little difficulty with the proofs while the results become general enough to be applied to noninvariant nonzero ideals without any trouble.

*Algebra of a group.* The main special effect in noncommutative Galois theory is that there arise so-called *inner* automorphisms. If $b$ is an invertible element then $\hat{b}$: $x \to b^{-1}xb$ is an automorphism. It is easy to see that the group $\mathcal{A}(R)$ contains all inner automorphisms of $Q(R)$. The *algebra of a group* $G \subseteq \mathcal{A}(R)$ is defined as the linear space over the extended centroid $C$ generated by all elements $b$ invertible in $Q(R)$ such that $\hat{b} \in G$. This space is a subalgebra of $Q(R)$. We will denote the algebra of a group $G$ by $\mathcal{B}(G)$. It is an inner part of the group in a ring form.

If $G$ is a finite group, then its algebra $\mathcal{B}(G)$ will be finite-dimensional over $C$. Of the finite order will also be the factor-group $G/G_{inn}$, where $G_{inn}$ is the normal subgroup of all inner for $Q$ automorphisms (such automorphisms are sometimes called $X$-*inner*).

*Reduced order.* A group $G$ is called *reduced-finite* if its algebra $\mathcal{B}(G)$ is finite-dimensional, while the factor-group $G/G_{inn}$ is finite. In this case the number $\dim_C \mathcal{B}(G) \cdot |G/G_{inn}|$ is called the *reduced order* of the group $G$.

*Noether groups.* Let $G$ be a Galois group, $G = A(S)$. If $b = \sum c_i b_i$, $\hat{b}_i \in G$, is an invertible element of $\mathcal{B}(G)$ then it commutes with all elements of $S$ as do the $b_i$s and therefore $\hat{b} \in A(S) = G$. This means that any Galois group is an $N$-group in the sense of the following definition. A group $G$ is called an *$N$-group* (*a Noether group*) if any inner automorphism of the ring $Q$ corresponding to an invertible element of $\mathcal{B}(G)$ belongs to $G$.

*Regular groups.* A reduced-finite group $G$ is called an *$M$-group* (*Maschke group*) if its algebra $\mathcal{B}(G)$ is semisimple. With the help of the well known Maschke theorem it is easy to see that any finite group of order invertible in $C$ is an $M$-group.

A Maschke group $G$ is called *regular* if it is also a Noether group.

A reduced-finite $N$-group with simple algebra is called *quite regular.*

We need the notion of a finite-dimensional quasi-Frobenius algebra (see the paper "Frobenius algebras" by K. Yamagata in this volume).

**2.5.2.** THEOREM. *The following statements on a finite-dimensional algebra $B$ with* 1 *are equivalent:*

(1) *the left $B$-module $B$ is injective;*

(2) *the right $B$-module $B$ is injective;*

(3) *for every left ideal $\lambda$ and right ideal $\rho$ of the algebra $B$ the following equalities are valid:*

$$ann_l(ann_r(\lambda)) = \lambda, \quad ann_r(ann_l(\rho)) = \rho;$$

(4) *the sum of all right ideals conjugate with left ones coincides with the whole algebra $B$;*

(5) *the sum of all left ideals conjugate with right ones coincides with the whole algebra $B$.*

Here $ann_l(\rho)$ is the left annihilator of $\rho$, and $ann_r(\lambda)$ is the right annihilator of $\lambda$. Recall also that left and right modules $\lambda, \rho$ are conjugate if there exists a bilinear nondegenerate associative form $\rho \times \lambda \to C$. It is easy to see that all simple and all semisimple finite-dimensional algebras are quasi-Frobenius (even Frobenius). In particular all regular and any quite regular groups have quasi-Frobenius algebras.

**2.5.3.** THEOREM. *Any reduced-finite $N$-group with quasi-Frobenius algebra is a Galois group.*

The proof of this theorem is based on two important considerations. One of them is a generalization of 2.1.3, which is a noncommutative analog of the automorphism-independence theorem.

**2.5.4.** PROPOSITION. *Let $d_1, d_2, \ldots, d_n \in R_F$ be linearly independent over $C$ elements, and let $s_1, s_2, \ldots, s_m \in R_F$ be arbitrary elements. If none of the automorphisms $g_1, g_2, \ldots, g_m \in \mathcal{A}(R)$ is inner for $Q(R)$ then there exists $a_1, \ldots, a_k; b_1, \ldots, b_k \in R$, such that*

$$\sum_i a_i d_1 b_i \neq 0, \quad \sum_i a_i d_j b_i = 0, \quad \sum_i a_i s_t^{g_t} b_i = 0,$$

*where $j = 2, \ldots, n$; $t = 1, 2, \ldots, m$.*

Another fundamental consideration is the construction of trace forms. Let $T$ be a transversal for $G_{inn}$ in a reduced finite group $G$. If $\lambda, \rho$ are conjugate left and right ideals of $\mathcal{B}(G)$, with basis $a_1, \ldots, a_n$ of $\lambda$ and dual basis $a_1^*, \ldots, a_n^*$, of $\rho$ then the form

$$\tau_{\lambda,\rho}(x) = \sum_{g \in T} \sum_{i=1}^{n} (a_i x a_i^*)^g$$

is invariant, i.e. $\tau_{\lambda,\rho}(x) \in (Q(R))^G$ for all $x \in Q(R)$. Moreover there exists a nonzero ideal $I \triangleleft R$, such that $0 \neq \tau_{\lambda,\rho}(I) \subseteq R^G$. Such a form is called a *trace form.*

The above theorem implies that all regular and all quite regular groups are Galois. An important example of $M$-groups is concerned with the case when the ring $R$ has no zero divisors. Indeed, the algebra of any reduced-finite group $G$ is contained in the ring $Q(R)$, which also has no zero divisors and therefore $\mathcal{B}(G)$ is a division ring. Thus we have a corollary.

**2.5.5.** COROLLARY. *If a ring $R$ has no zero divisors then the Galois closure of a reduced-finite group $G$ is the group $G \cdot \mathcal{B}^*$ generated by $G$ and the inner automorphisms corresponding to invertible elements of $\mathcal{B}(G)$.*

The following theorem with 2.5.3 gives us the solution of questions (1) and (2) above.

**2.5.6.** THEOREM. *Let $G$ be a reduced-finite group with quasi-Frobenius algebra, then*

(a) *the Galois subring $R^G$ is semiprime if and only if $G$ is an $M$-group.*

(b) *the Galois subring $R^G$ is prime if and only if the algebra $\mathcal{B}(G)$ has no proper $G$ invariant ideals.*

Note that the group $G$ naturally acts on $\mathcal{B}(G)$ because an inner automorphism defined by $b^g$ equals $g^{-1}\hat{b}g$.

**2.5.7.** THEOREM. *Let $G$ be a regular group. An intermediate ring $S$, $R \supseteq S \supseteq R^G$, is a Galois subring of an $M$-subgroup of $G$ if and only if it satisfies the following properties* **BM, RC, SI**.

**BM** (*Bimodule property*). Let $e \in \mathcal{B}(G)$ be an idempotent, such that $se = ese$ for all $s \in S$. Then there exists (an idempotent) $f$, such that $ef = f$, $fe = e$ and $fs = sf$ for all $s \in S$.

**RC** (*Rational completeness*). If A is a two-sided ideal of $S$ with zero annihilators in $S$ and $Ar \subseteq S$ for a certain $r \in R$, then $r \in S$.

(This condition is equivalent to $S_F \cap R = S$.)

**SI** (*Sufficiency of invertible elements*).The centralizer

$$Z = \{z \in \mathcal{B}(G) \mid \forall s \in S \ zs = sz\}$$

is generated by its invertible elements and if for an automorphism $g \in G$ there is a nonzero element $b \in \mathcal{B}(G)$, such that $sb = bs^g$ for all $s \in S$, then there exists an invertible element with the same property.

In relation to the first part of condition **SI**, we should at once remark that if a field $C$ (the generalized centroid of $R$) contains at least three elements, then any finite-dimensional algebra with 1 over $C$ is generated by its invertible elements. Therefore, this part of the condition is restrictive only in the case when $C$ is a two-element field. The second part of the condition is automatically satisfied if $R$ has no zero divisors or, generally, if the algebra of the group $G$ is simple.

We can now formulate the traditional Galois theory correspondence theorems.

**2.5.8.** THEOREM. *Let $G \subseteq \mathcal{A}(R)$ be a regular group of automorphisms of a prime ring $R$. Then the mappings $H \mapsto R^H$, $S \mapsto A(S)$ set up a one-to-one correspondence between all regular subgroups of the group G and all intermediate subrings obeying conditions* **BM**, **RC** *and* **SI**.

**2.5.9.** THEOREM. *Let $G \subseteq \mathcal{A}(R)$ be a finite group of X-outer automorphisms of a prime ring $R$. Then the mappings $H \mapsto R^H$, $S \mapsto A(S)$ set up a one-to-one correspondence between all subgroups of the group G and all intermediate rationally-complete subrings of $R$.*

**2.5.10.** THEOREM. *Let $G \subseteq \mathcal{A}(R)$ be a reduced-finite N-group of automorphisms of a domain $R$. Then the mappings $H \mapsto R^H$, $S \mapsto A(S)$ set a one-to-one correspondence between all N-subgroups of the group $G$ and all the intermediate rationally-complete subrings.*

In addition there is a somewhat unexpected Galois correspondence theorem, which is based on the fact that the symmetrical quotient ring of a free algebra coincides with this algebra.

**2.5.11.** THEOREM. *Let $G$ be a finite group of homogeneous automorphisms of a free algebra $F\langle X\rangle$ in more than one variable. Then the mappings $H \mapsto R^H$, $S \mapsto A(S)$ set a one-to-one correspondence between all subgroups of the group G and all intermediate free subalgebras.*

A traditional problem of the Galois theory is that of finding criteria for an intermediate subring $S$ to be a Galois extension over $R^G$. The conditions for a general case being quite complex, we are not going to discuss them here. The considerations of this problem are based on the theorem of extension of isomorphisms. Let us begin with a simple example which shows the extension of isomorphisms over $R^G$ between intermediate subrings to be not always possible for arbitrary $M$-groups.

**2.5.12.** EXAMPLE (*D.S. Passman*). Let $R$ be the ring of all matrices of the order four over a field $F \neq GF(2)$ and let $G$ be the group of all (inner) automorphisms of this algebra. Let us set $S = \{diag(a, a, a, b) \mid a, b \in F\}$ and let $S_1 = \{diag(a, a, b, b) \mid a, b \in F\}$. Then $S \cong S_1$, and the corresponding isomorphism is the identity on $R^G = \{diag(a, a, a, a)\}$. Both rings are Galois subrings, since their centralizers $Z, Z_1$ are generated by invertible elements. At the same time, the isomorphism between $S$, $S_1$ cannot be extended to an automorphism of $R$, as $Z$ is not isomorphic to $Z_1$.

The situation is better under the assumption that $S, S_1$ are Galois subrings of groups with simple algebras.

**2.5.13.** THEOREM. *Let $G$ be a regular group of automorphisms of a prime ring, $S, S_1$ be intermediate Galois subrings of quite regular groups. Then any isomorphism $\varphi$: $S \to S_1$, which is the identity on $R^G$, can be extended to an automorphism $\bar{\varphi} \in G$.*

**2.5.14.** COROLLARY. *Let $G$ be a reduced-finite group of automorphisms of a domain. Then any isomorphism over $R^G$ between intermediate subrings can be extended to an automorphism $g \in \mathcal{A}(R)$.*

**2.5.15.** COROLLARY. *Let $G$ be a finite group of outer (for $Q$) automorphisms of a prime ring $R$. Then any isomorphism over $R^G$ between intermediate subrings can be extended to an automorphism $g \in G$.*

**2.5.16.** Work on noncommutative Galois theory was begun by E. Noether [No33] in her study of inner automorphisms of central simple algebras. This was continued in the 1940s and 1950s where the work still concerned rather special rings $R$. For example the Galois theory of division rings was initiated by N. Jacobson [Ja40] and [Ja47], H. Cartan [Ca47], and G. Hochschild [Ho49]. Complete rings of linear transformations were investigated by T. Nakayama and G. Azumaya [NA47] and J. Dieudonné [Di48], and somewhat later A. Rosenberg and D. Zelinsky [RZ55] studied continuous transformation rings. Much of this can be found in Jacobson's book [Ja56]. In addition, simple artinian rings were considered by G. Hochschild [Ho50], T. Nakayama [Na52] and in a long series of papers by H. Tominaga and T. Nagahara leading to their monograph [TN70].

In the 1960s a great deal of work was done on the Galois theory of separable algebras. Among the many papers on this subject, we note in particular [Mi66] by Miyashita, [CM67] by L.N. Childs and F.R. DeMeyer, [VZ69] by Villamayor and D. Zelinsky and [Kr70] by Kreimer.

The results presented here are basically due to the author. They were first proved in general form for semiprime rings in a series of papers [Kh75, Kh77, Kh78]. The case of prime rings was revised with new achievements by S. Montgomery and D.S. Passman in [MP84] from where we obtained this history information. A somewhat new point of view of the subject is presented in the book [Kh91].

## 3. Semiprime rings

Recall that a ring is called *semiprime* if it has no nonzero ideals with zero multiplication. Evidently a semiprime ring contains no nilpotent ideals: if $I^n = 0$ then $I^{n-1}$ has a zero multiplication. Moreover a semiprime ring has no one-sided nilpotent ideals: if for instance $L$ is a left nilpotent ideal, $L^n = 0$, then $(L + LR)^{n+1} = 0$.

Another important consequence of the definition is that any ideal has zero intersection with its annihilator $I \cap ann\, I = 0$, while left and right annihilators of a two-sided ideal coincide. In this case the sum $I + ann\, I$ is direct and this sum has zero annihilator, in particular it intersects any nonzero ideal, i.e. it is an essential ideal in the sense of the following definition.

DEFINITION. An ideal $I$ is called *essential* if it has nonzero intersection with any nonzero ideal. Respectively a left (right) ideal is called *essential* if it has nonzero intersection with any nonzero left (right) ideal.

The Martindale quotient ring of a semiprime ring is defined in the same way as that of a prime ring, where instead of nonzero ideals one should consider essential two-sided ideals. This quotient ring as well as the symmetrical one have an axiomatic definition in the spirit of Theorem 2.1.1.

**3.0.1.** THEOREM. *Let $R$ be a semiprime ring. There exists a unique (up to isomorphism over $R$) ring $R_F$ such that:*

(a) $R \subseteq R_F$.

(b) *If $q \in R_F$ and $Iq = 0$ for some essential ideal $I$ of $R$, then $q = 0$.*

(c) *If $q_1, q_2, \ldots, q_n \in R_F$, then there exists an essential ideal $I$ of $R$ with $Iq_1$, $Iq_2, \ldots, Iq_n \subseteq R$.*

(d) *If $I$ is an essential ideal of $R$ and $\xi$: $I \to R$ is a homomorphism of left $R$-modules, then there exists an element $q \in R_F$ such that $\xi(a) = aq$ for all $a \in I$.*

Using these axioms one can prove the following useful facts.

**3.0.2.** THEOREM.

(a) *The ring $R_F$ is semiprime.*

(b) *The center of $R_F$ is a commutative self injective von Neumann regular ring.*

(c) *The left Martindale ring of quotients of any essential ideal $I$ of $R$ is equal to that of $R$, i.e. $I_F = R_F$.*

The *extended centroid* $C = C(R)$ of a semiprime ring is also defined as the center of $R_F$, and the *symmetrical* quotient ring

$$Q(R) = \{q \in R_F \mid qI \subseteq R \text{ for some essential ideal } I \text{ of } R\}.$$

Lemma 2.1.3 also remains valid, but instead of linear independence one should take the condition $d_1 \notin Cd_2 + Cd_3 + \cdots + Cd_n$.

**3.1.** *Goldie and prime dimensions.* The *prime dimension* of a ring $R$ is the largest number $n$, such that $R$ contains a direct sum of $n$ nonzero two-sided ideals.

The prime dimension of a prime ring is equal to one. A semiprime ring can have either finite or infinite prime dimension. However, if this dimension equals one, then the ring is prime: if $IJ = 0$, then the sum $I + J$ is direct.

**3.1.1.** THEOREM. *The following statements are equivalent for a semiprime ring $R$.*

(1) *The prime dimension of the ring $R$ is $n$.*

(2) *The ring $R$ contains an essential direct sum of ideals $I_1 \oplus \cdots \oplus I_n$, each of which is a nonzero prime subring.*

(3) *The ring of quotients $R_F$ is isomorphic to a direct sum of $n$ nonzero prime rings.*

(4) *The ring of quotients $Q(R)$ is isomorphic to a direct sum of $n$ nonzero prime rings.*

(5) *The generalized centroid of $R$ is isomorphic to a direct sum of $n$ fields.*

(6) *The ring $R_F$ has exactly $2^n$ different central idempotents.*

We have seen in 2.0.5 that any semiprime ring is a subdirect product of prime rings. It can be proved that a semiprime ring is presented as a subdirect product of $n$ prime rings if and only if its prime dimension is less than or equal to $n$. In this case the following uniqueness theorem is valid.

**3.1.2.** THEOREM. *If the prime dimension of a semiprime ring $R$ equals $n$, then:*

(a) *Any irreducible presentation of the ring $R$ as a subdirect product of prime rings contains exactly $n$ factors.*

(b) *A presentation of $R$ as a subdirect product of $n$ prime rings is unique, i.e. if $R = S^n_{i=1} R_i = S^n_{i=1} R'_i$, then there exists a permutation $\sigma$ such that $R'_i \cong R_{\sigma(i)}$ and $\ker \pi'_i = \ker \pi_{\sigma(i)}$, where $\pi_i, \pi'_i$ are the approximating projections.*

(c) *The ring $R$ has exactly $n$ minimal prime ideals and their intersection equals zero.*

The *left Goldie* dimension of a left module $V$ is the biggest number $n$, such that $V$ contains a direct sum of $n$ nonzero submodules. This dimension is also often called *uniform dimension.*

One can prove (it is not obvious) that if a module does not contain a direct sum of an infinite number of nonzero submodules, then it has finite Goldie dimension.

A ring $R$ of finite Goldie dimension is called a *left Goldie ring* if it satisfies the Ascending Chain Condition for left annihilator ideals.

It is obvious that a semiprime left Goldie ring has finite prime dimension. In this case the prime subdirect factors of the irreducible presentation are prime Goldie rings.

The following statement presents a very important property of semiprime Goldie rings.

**3.1.3.** PROPOSITION. *A left ideal of a semiprime left Goldie ring is essential if and only if it has a regular element.*

It should be recalled that an element $r \in R$ is called *regular* if $sr \neq 0$, $rs \neq 0$ for all $s \in R$, $s \neq 0$.

Most important is the connection of semiprime Goldie rings with the classical quotient ring construction.

**3.1.4.** DEFINITION. Let $R$ be a subring of a ring $S$. The ring $S$ is called a *classical quotient ring* of the ring $R$, and the ring $R$ is called a *left order* in $S$ iff the following conditions are met:

(1) all regular elements of the ring $R$ are invertible in the ring $S$;

(2) all elements of the ring $S$ have the form $a^{-1}b$, where $a, b \in R$ and $a$ is a regular element of the ring $R$.

An arbitrary ring does not always have a classical left quotient ring. Indeed, if a ring $R$ has a classical quotient ring and $a, b \in R$ with $b$ is regular, then $ab^{-1} = c^{-1}d$, where $c$, $d \in R$ and $c$ is regular. Hence, $ca = db$ and we come to the necessity of the following *left Ore* condition.

*For any elements $a, b \in R$, where $b$ is a regular element, one can find elements $c, d \in R$ where $c$ is a regular element, such that $ca = db$.*

This condition is known to be sufficient for the existence of a left classical quotient ring (the Ore theorem). Moreover, the Ore condition guarantees the uniqueness of the left classical quotient ring, this ring being denoted by $Q_{cl}(R)$.

**3.1.5.** GOLDIE THEOREM. (a) *A ring $R$ is a left order in a simple artinian ring if and only if $R$ is a prime left Goldie ring.*

(b) *A ring $R$ is a left order in a semisimple artinian ring if and only if $R$ is a semiprime left Goldie ring.*

Semiprime left Goldie rings have some interesting characterizations. One of them replaces the ACC with a nonsingularity condition. Recall that a ring $R$ is called *left nonsingular* if every essential left ideal has a zero right annihilator.

**3.1.6.** JOHNSON THEOREM. *A semiprime ring $R$ is a left Goldie ring if and only if it is nonsingular and contains no infinite direct sums of nonzero left ideals.*

**3.2.** *A topology on a semiprime ring of infinite prime dimension.* In this paragraph we are going to define a topology on a semiprime ring. This topology becomes discrete if the ring has finite prime dimension. Therefore this construction as well as the constructions in the next paragraph are special tools for the investigation of semiprime rings with infinite prime dimension.

Recall that a partially ordered set $A$ is called *directed* if for any two elements $\alpha_1, \alpha_2 \in A$ there exists an upper bound $\beta \in A$: $\beta \geqslant \alpha_1, \alpha_2$.

The set of all central idempotents of the ring $Q(R)$ is partially ordered: $e_1 \leqslant e_2$ iff $e_1 e_2 = e_1$. Moreover, it can be easily seen (and it is important) that any subset $E_1$ has an exact upper bound $\sup E_1$. Let $E$ be a subring of $C$ generated by all its idempotents. In this case the rings $R_F$, $Q(R)$, $C$, $End_{\mathbf{Z}}(R_F)$ are modules over $E$. Here $End_{\mathbf{Z}} R_F$ is the ring of endomorphisms of the additive group of the ring $R_F$. This ring is a right module over $C$, but it may be not an algebra over it. The action of the central elements is defined by the formula $x(\varphi c) = (x\varphi)c$.

**3.2.1.** DEFINITION. Let $M$ be a module over the ring $E$ and $A$ be a directed partially ordered set. Let us call an element $m \in M$ a *limit* of the family $\{m_\alpha \in M,\ \alpha \in A\}$ if there exists a directed family of idempotents $\{e_\alpha,\ \alpha \in A\}$ such that

(a) $e_\alpha \leqslant e_\beta$ for $\alpha \leqslant \beta$,

(b) $\sup e_\alpha = 1$,

(c) for all $\alpha \in A$ the equalities $me_\alpha = m_\alpha e_\alpha$ are valid.

In this case we shall write

$$m = \lim_A m_\alpha.$$

Accordingly, a set $T \subseteq M$ is called *closed* if any limit of any family of elements from $T$ belongs to $T$. The *closure of a set* is the least closed set containing the given one. Therefore, the operation of closure determines a certain topology on the $E$-module $M$.

Let now $\varphi\colon M_1 \to M$ be a certain mapping of $E$-modules. We shall call $\varphi$ *quite continuous* if the equality

$$m = \lim_A m_\alpha$$

yields

$$\varphi(m) = \lim_A \varphi(m_\alpha).$$

It should be remarked that any quite continuous mapping $\varphi$ is continuous. Indeed, in this case the total preimage of a closed set is closed and, hence, $\varphi$ is a continuous mapping. It is evident that if $\varphi$ preserves the operators of multiplication by the central idempotents: $\varphi(me) = \varphi(m)e$, then $\varphi$ is quite continuous. In particular *any $E$-module as well as $C$-module homomorphism is continuous, while an isomorphism is a homeomorphism.*

It follows that the operators of left and right multiplication $x \mapsto rx$, $x \mapsto xr$ are continuous transformations of the ring $R_F$. Moreover it can be seen that the transformations $\Pi_a\colon x \mapsto x + a$, as well as all automorphisms and derivations of the ring $R_F$ are quite continuous. It is also important that the closure of any subring of $R_F$ is a subring, as well as that the closure of an ideal of a subring $S$ is an ideal of the closure of $S$.

The above definition does not imply that the directed family of elements $\{m_\alpha\}$ cannot have more than one limit.

**3.2.2.** LEMMA. *Any directed family of elements of a module $M$ over $C$ has at most one limit iff $M$ is a nonsingular module.*

The main examples of nonsingular modules are $R_F$, $End_{\mathbf{Z}}(R_F)$ and all their submodules.

**3.2.3.** DEFINITION. A family $\{m_\alpha \mid \alpha \in A\}$ is said to be self consistent if there exists a directed family of idempotents $\{e_\alpha\}$, such that

(a) $\sup e_\alpha = 1$,

(b) if $\alpha \geqslant \beta$ then the relations $m_\alpha e_\beta = m_\beta e_\beta$, and $e_\alpha \geqslant e_\beta$ are valid.

A subset $S \subseteq M$ is called *complete* if any self consistent family of its elements has a limit.

It is very important that the modules $R_F$, $Q(R)$, $C$ and $End_{\mathbf{Z}}(R_F)$ are complete. Evidently any closed subset of a complete module is complete. Moreover any factor-module of a complete module by a closed submodule is complete.

**3.2.4.** THEOREM. *Any complete nonsingular module over $C$ is injective and, vice versa, any injective module over $C$ is complete.*

This theorem immediately implies that the closure of any $C$-submodule of $R_F$ as well as that of any submodule of a complete nonsingular module is equal to its injective hull.

**3.3.** *Canonical sheaf.* In this paragraph we shall present the closure $\widehat{RE}$ in the topology described above as a ring of global sections of a certain sheaf over the structure space of the extended centroid. Such a presentation is useful since it enables one to see clearly the function-theoretic intuition employed when studying a semiprime ring. It is worthwhile adding that in the case of a prime ring all these constructions degenerate and their essence and meaning is to reduce the process of studying semiprime rings to that of studying prime rings (i.e. stalks of a canonical sheaf).

Roughly speaking, the ring of global sections of a sheaf is a set of all continuous functions on a topological space, the only difference being that at every point these functions assume values in their own (local) rings. Let us give the exact definitions.

**3.3.1.** DEFINITION. Let there be given a topological space $X$, and for any open set $U$ let there be given a ring (group or, more generally, an object of a certain fixed category) $\mathfrak{R}(U)$, and let for any two open sets $U \subset V$ there be given a homomorphism $\rho_U^V\colon \mathfrak{R}(V) \to \mathfrak{R}(U)$.

This system is called a *presheaf* of rings, provided the following conditions are met:

(1) if $U$ is empty, then $\mathfrak{R}(U)$ is the zero ring;

(2) $\rho_U^U$ is the identity mapping;

(3) for any open sets $U \subseteq V \subseteq W$ we have $\rho_U^W = \rho_U^V \rho_V^W$.

Such a presheaf will be denoted by one letter, $\mathfrak{R}$.

The simplest example of a presheaf is the presheaf of all functions on $X$ with the values in a ring $A$. In this case $\mathfrak{R}(U)$ consists of all functions on $U$ with values in $A$, and for $U \subset V$ the homomorphism $\rho_U^V$ is the restriction of the function determined on $V$ to the subset $U$.

In order to extend the intuition of this example to the case of any presheaf, the homomorphisms $\rho_U^V$ are called *restriction homomorphisms.* The elements of (the ring) $\mathfrak{R}(U)$ are called the *sections* of the presheaf $\mathfrak{R}$ over $U$. Sections of $\mathfrak{R}$ over $X$ are called *global.* Thus, $\mathfrak{R}(U)$ is a ring of sections of the presheaf $\mathfrak{R}$ over $U$; $\mathfrak{R}(X)$ is a ring of global sections.

Returning to the example of the presheaf of all functions on $X$, let us assume the topological space $X$ to be the union of open sets $U_\alpha$. Then any function on $X$ is uniquely determined by its restrictions to the sets $U_\alpha$. Moreover, if on every $U_\alpha$ a function $f_\alpha$ is given, such that the restrictions of $f_\alpha$ and $f_\beta$ coincide on $U_\alpha \cap U_\beta$, then there exists a function on $X$, such that every $f_\alpha$ is its restriction to $U_\alpha$.

These properties can be formulated for any presheaf and they single out an extremely important class of presheaves.

**3.3.2.** DEFINITION. A presheaf $\mathfrak{R}$ on a topological space $X$ is called a *sheaf* if for each open set $U \subset X$ and every open cover $U = \bigcup U_\alpha$ the following conditions are met:

(1) if $\rho^U_{U_\alpha}(s_1) = \rho^U_{U_\alpha}(s_2)$ for $s_1, s_2 \in \mathfrak{R}(U)$ and all $\alpha$, then $s_1 = s_2$;

(2) if $s_\alpha \in \mathfrak{R}(U_\alpha)$ are such that for any $\alpha, \beta$ the restrictions of $s_\alpha$ and $s_\beta$ on $U_\alpha \cap U_\beta$ coincide, then there exists an element $s \in \mathfrak{R}(U)$, whose restriction to $U_\alpha$ is equal to $s_\alpha$ for all $\alpha$.

Let us try to consider an arbitrary sheaf as a sheaf of functions on the space $X$. To this end it is necessary to determine the value of $s(x)$ for the section $s \in \mathfrak{R}(U)$ at any point $x \in U$. We have the elements $s(V) = \rho^U_V(s)$ for all open neighborhoods $V$ of the point $x$ which are contained in $U$. Thus it is natural to consider their 'limit', and so we have to introduce the direct limit

$$\mathfrak{R}_x = \varinjlim_{x \in V} \mathfrak{R}(V)$$

with respect to the system of homomorphisms

$$\rho^U_V : \mathfrak{R}(U) \to \mathfrak{R}(V).$$

This limit is called the *stalk* of the sheaf (or a presheaf) $\mathfrak{R}$ at the point $x$.

Thus, a stalk element at the point $x$ is determined by any section over an open neighborhood of $x$. And two sections, $u, v \in \mathfrak{R}(U)$, define the same element of the stalk at $x$ if their restrictions to some open neighborhood of $x$ coincide.

For any open set $U \ni x$ this gives a natural homomorphism $\rho^U_x\colon \mathfrak{R}(U) \to \mathfrak{R}_x$, which maps a section to the stalk element determined by it. Thus we can define the *value of a section* $s$ at the point $x$ as $\rho^U_x(s)$.

**3.3.3.** *The construction.* Let us now go over to constructing the canonical sheaf. Let $C$ be a generalized centroid of a semiprime ring $R$. Let us denote by $X$ the set of all its prime ideals but $C$. This set is called a *spectrum* of the ring $C$ and is often denoted by *Spec* $C$. The elements of the spectrum are called *points of the spectrum* or simply *points.*

In order to put a topology on $X$, it is necessary to define the operation of closure. For a set $A \subset X$ let us define the closure $\bar{A}$ as a set of all points containing the intersection

$$\bigcap_{p \in A} p.$$

The topology obtained in this way is called the *spectral topology.*

In order to construct the sheaf over $X$, it is useful to know the structure of open sets in $X$ (they are also called *domains*).

If $e \in E$ is a central idempotent, then by $U(e)$ we denote a set of all points $p$, such that $e \notin p$.

Allowing for the fact that the product $e(1-e) = 0$ belongs to any prime ideal, we see any point of the spectrum contains either $e$ or $1-e$, but not both simultaneously. Therefore,

$$U(e) \cup U(1-e) = X, \qquad U(e) \cap U(1-e) = \varnothing. \tag{15}$$

On the other hand, the closure of $U(e)$ contains only points $q$ containing the intersection

$$\bigcap_{e\notin p} p = \bigcap_{1-e\in p} p.$$

The latter intersection contains the element $1 - e$ and, hence, $1 - e \in q$, which implies $e \notin q$, i.e. by definition, $q \in U(e)$.

Now relations (15) show $U(e)$ to be an open and closed set simultaneously. Such a set is called a *clopen* set.

**3.3.4.** LEMMA. *Each clopen subset of $X$ has the form $U(e)$ for suitable idempotent $e \in E$.*

**3.3.5.** LEMMA. *The sets $U(e)$, $e \in E$ form a fundamental system of open neighborhoods of $X$, i.e. any open set is presented as a union of sets of the type $U(e)$.*

Lemma 3.3.4 shows the set of central idempotents to be in a one-to-one correspondence with the set of closed domains of $X$. One can easily prove that this correspondence preserves the lattice operations and order relation:

$$U(e_1e_2) = U(e_1) \cap U(e_2); \quad U(e_1 + e_2 - e_1e_2) = U(e_1) \cup U(e_2),$$

$$U(e_1) \subseteq U(e_2) \Leftrightarrow e_1 \leqslant e_2. \tag{16}$$

This circumstance allows one in a number of cases to identify central idempotents with closed domains and consider idempotents as objects consisting of points, which makes many considerations extremely vivid. The correspondence under discussion also preserves the exact upper bounds

$$U\Bigl(\sup_\alpha\{e_\alpha\}\Bigr) = \overline{\bigcup_\alpha U(e_\alpha)},$$

where on the right we have the closure of the union of the domains $U(e_\alpha)$.

**3.3.6.** THEOREM. *The space $X = \operatorname{Spec} C$ is an extremely disconnected, compact and Hausdorff topological space.*

Recall that a topological space is called *extremely disconnected* if the closure of each open set is open (and of course closed).

**3.3.7.** DEFINITION. Now we are completely ready to determine the *canonical sheaf* $\Gamma = \Gamma(R)$. Let $U$ be an open set. By formula (15), its closure $\overline{U}$ is open and has the form $U(e)$ for some central idempotent. Let us set

$$\Gamma(U) \stackrel{df}{=} \Gamma(\overline{U}) \stackrel{df}{=} e\widehat{RE}.$$

Since the inclusion $W \subseteq U$ implies $\overline{W} \subseteq \overline{U}$, and this inclusion by formula (16) gives the inequality $f \leqslant e$ for the corresponding idempotents $U(f) = \overline{W}$, $U(e) = \overline{U}$, then the homomorphism $x \to xf$ acting from $e\widehat{RE}$ onto $f\widehat{RE}$ is defined, which will be viewed as the restriction homomorphism $\rho_W^U$.

The validity of axioms (1)–(3) of a presheaf for $\Gamma$ is absolutely obvious.

**3.3.8.** THEOREM. *The presheaf $\Gamma$ determined above is a sheaf.*

So, we have achieved our goal: the ring $\widehat{RE}$ is presented as the ring of global sections of a sheaf $\Gamma$. This sheaf is called the *canonical sheaf* of the ring $R$.

This sheaf satisfies another important condition: any section can be extended to a global one (sheaves satisfying this condition are called *flabby*) and, moreover, the restriction homomorphisms are retractions, so that the ring of global sections naturally contains all the rings of local sections $e\widehat{RE} \subseteq \widehat{RE}$.

**3.3.9.** DEFINITION. The *support* of a global section $s$, or more generally, the support of a subset $S \subseteq \widehat{RE}$ of global sections is defined as the difference $1 - f$, where $f$ is the largest central idempotent with $Sf = 0$.

From the function point of view the support of $s$ is a set of all points where the function $s$ has nonzero values.

This definition allows one to describe the stalks of the canonical sheaf.

**3.3.10.** THEOREM. *A section $s$ belongs to the kernel of the natural homomorphism $\rho_p$ iff the support of the element $s$ belongs to $p$ or, which is equivalent, $s \in p\widehat{RE}$. The stalk $\Gamma_p$ is a prime ring, isomorphic to the factor-ring*

$$\widehat{RE}/\widehat{RE} \cap p\widehat{RE}.$$

**3.4.** *A metatheorem.* The metatheorem is a theorem about theorems, which can be transferred from prime rings to semiprime ones using a canonical sheaf. In fact almost all logical theorems deal with classes of statements and theorems. Therefore our goal is a logical theorem which describes a possibly widest class of properties, which can be transferred from stalks of a canonical sheaf to a ring of global sections. Let us recall briefly the basic definitions of the language of elementary logic.

**3.4.1.** DEFINITION. An *$n$-ary predicate* on a set $A$ is a mapping $P$ of the Cartesian power $A^n$ to a two-element set $\{T, F\}$ ( $T \leftrightarrow$ 'true', $F \leftrightarrow$ 'false') and an *$n$-ary operation* is a mapping $f$ from $A^n$ to $A$. Predicates of the same name (i.e., those with one name, $P$, and the same arity, $n$) or operations of the same name can be defined on different sets. In this case we speak about the *values* of the same predicate, $P$ (operation, $f$) on different sets. A *signature* is a set $\Omega$ of names of predicates and operations put in correspondence with arities. An *algebraic system* of signature $\Omega$ is a set with given values of the predicates and operations from $\Omega$ of the corresponding arity. Sometimes both zero-ary operations and zero-ary predicates are considered. A zero-ary operation on set $A$ is a fixed element of this set, while a zero-ary predicate is either true or false.

The predicates and operations from $\Omega$ defined on an algebraic system of this signature are called *principal* or *basic*. With their help one can construct new predicates: *formula predicates*.

The simplest formula predicates are those given by *atomic formulas*

$$P(x_1,\dots,x_n), \quad x_1 = x_2, \quad F(x_1,\dots,x_n) = x_{n+1}, \tag{17}$$

where $P, F$ are basic predicates and functions.

A formula predicate is given by a formula of the elementary language, i.e. it is 'obtained' from the simplest formula predicates by employing logical connectives, $\&, \vee, \neg, \rightarrow$ and by quantification by $\exists, \forall$ of subject variables $x_i$.

Let us finally describe the class that is most important for us: that of *Horn formulas*. The simplest Horn formulas are the following ones:

$$A_1 \& A_2 \& \cdots \& A_p \rightarrow A_{p+1}; \quad A_i; \quad \neg A_1 \vee \neg A_2 \vee \cdots \vee \neg A_p,$$

where the $A_i$ are formulas of type (17). An arbitrary Horn formula is a quantification of a conjunction of simplest Horn formulas. The predicate defined by a Horn formula is called a *Horn predicate*.

**3.4.2.** DEFINITION. A flabby sheaf $\mathfrak{R}$ over an extremely disconnected space $X$ is called *correct* if $\mathfrak{R}(U) = \mathfrak{R}(\overline{U})$ for any domain $U$, i.e. the mappings $\rho_U^{\overline{U}}$ are isomorphisms.

According to the construction the canonical sheaf $\Gamma$ is correct. Let us now fix a correct sheaf $\mathfrak{R}$ of algebraic systems of a signature $\Omega$ (the reader can view $\mathfrak{R}$ as a sheaf of rings in an extended signature).

**3.4.3.** DEFINITION. Let us call an $n$-ary predicate $P$ a *sheaf predicate* provided its values are given on all rings of sections and on all stalks of $\mathfrak{R}$ and the following conditions are met.

(a) If $P(\bar{s}_1,\dots,\bar{s}_n) = T$ for $\bar{s}_i \in R_t$, then there exists a neighborhood $W$ of the point $t$ and preimages $s_i \in \mathfrak{R}(W)$, such that $P(\rho_U^W(s_1),\dots,\rho_U^W(s_n)) = T$ for every domain $U \subseteq W$.

(b) If $U = \bigcup U_\alpha$ and $s_1,\dots,s_n \in \mathfrak{R}(U)$, in which case $P(\rho_{V_\alpha}^U(s_1),\dots,\rho_{V_\alpha}^U(s_n)) = T$ for all domains $V_\alpha \subseteq U_\alpha$, then $P(s_1,\dots,s_n) = T$.

The following statement shows that it is sheaf predicates that are of primary interest to us.

**3.4.4.** PROPOSITION. *Let $P$ be a sheaf predicate, $s_1,\dots,s_n$ be sections over a domain $U$. Then, if the set of points $t \in U$, for which $P(\rho_t(s_1),\dots,\rho_t(s_n)) = T$ is dense in $U$, then $P(s_1,\dots,s_n) = T$. In particular, if a certain theorem is given by a zero-ary sheaf predicate and is valid in all (or almost all) stalks, then it is also valid in the ring of global sections.*

The proof can be directly obtained by first applying condition (a) to the stalks on which predicate $P$ is true, followed by employing condition (b).

**3.4.5.** METATHEOREM. *Any Horn predicate is a sheaf predicate. In particular, if the formulation of a theorem can be presented as a Horn formula, then the truth of this theorem in almost all stalks of the canonical sheaf implies its truth in the ring of global sections of this sheaf.*

The importance of this theorem is strengthened by the fact that in the class of prime rings *every* elementary formula is equivalent to a Horn one. Indeed, when constructing Horn formulas, it is only forbidden to use disjunctions, but in the class of prime rings $f = 0 \vee g = 0$ is equivalent to $\forall x\ fxg = 0$. In more detail, it is necessary to reduce the quantorless part of the given formula to a conjunctive normal form, and then every subformula of the type

$$f \neq 0 \vee f_2 \neq 0 \vee \cdots \vee f_k \neq 0 \vee g_1 = 0 \vee \cdots \vee g_n = 0$$

should be replaced with the Horn formula

$$\forall x_1 x_2 \ldots x_{n-1} (f_1 = 0 \& \ \cdots \ \& f_n = 0 \rightarrow g_1 x_1 g_2 \ldots x_{n-1} g_n = 0).$$

The following two facts are also useful. The first allows one to reduce theorems from the ring of global sections to theorems about the stalks and the second is a consequence of the compactness of the spectrum.

**3.4.6.** PROPOSITION. *Let $P$ be a sheaf predicate, and $Q$ be a strict sheaf predicate, and let us assume that the predicate $P \rightarrow Q$ is true on all algebraic systems (rings) of sections. Then this predicate is true on all the stalks as well.*

Recall that a predicate $P$ is said to be *strict sheaf* if its values are given on all algebraic systems (rings) of sections and stalks of the sheaf $\mathfrak{R}$, so that $\mathfrak{R}$ remains a sheaf in the category of algebraic systems with the additional predicate $P$.

**3.4.7.** PROPOSITION. *Let $P_1, P_2, \ldots, P_n, \ldots$ be a sequence of $m$-ary Horn predicates, such that the implications $P_n \rightarrow P_{n+1}$ are true on certain global sections $s_1, \ldots, s_m$ of a canonical sheaf $\Gamma$. Then, if for every point $p$ one of the predicates $P_i$ is true on $\bar{s}_1 = \rho_p(s_1), \ldots, \bar{s}_m = \rho_p(s_m)$, then one of the predicates $P_i$ is true on $s_1, \ldots, s_m$.*

The following theorem concerning Martindale quotient rings can be easily proved with the help of the metatheorem.

**3.4.8.** PROPOSITION. *Let $p$ be an arbitrary point of the spectrum. Then the following inclusions are valid:*

$$\Gamma_p(R) \subseteq \Gamma_p(Q) \subseteq \Gamma_p(R_F) \subseteq \big(\Gamma_p(R)\big)_F, \qquad \Gamma_p(Q) \subseteq Q\big(\Gamma_p(R)\big).$$

**3.5.** *Galois theory.* We start with an example, vividly illustrating variations in the basic notions of Galois theory when going over to a semiprime case.

Let

$$R = \prod_{\alpha \in A} F_\alpha$$

be a direct product of isomorphic fields $F_\alpha \cong F$ and $H$ be a finite group of automorphisms of the field $F$. Let us define the action of $H$ on the product $R$ in a componentwise manner, $f^h(\alpha) = (f(\alpha))^h$, i.e. $(\ldots, f_\alpha, \ldots)^h = (\ldots, f_\alpha^h, \ldots)$. On the other hand, on $R$ one can naturally define the action of the direct product

$$G = \prod_{\alpha \in A} H_\alpha$$

of the groups $H_\alpha$ isomorphic to $H$: $f^g(\alpha) = f(\alpha)^{g(\alpha)}$, i.e.

$$(\ldots, f_\alpha, \ldots)^{(\ldots, g_\alpha, \ldots)} = (\ldots, f_\alpha^{g_\alpha}, \ldots).$$

In this case the subrings of fixed elements for the group $H$ and for the group $G$ coincide and are isomorphic to the direct product of copies of $F^H$.

**3.5.1.** *Conjugation modules.* For an automorphism $g \in \mathcal{A}(R)$[1] of a semiprime ring $R$ the *conjugation module* is defined by the formula $\Phi_g = \{a \in Q \mid \forall x \in R \;\; xa = ax^g\}$. It can be proved that this module is a cyclic $C$-submodule $\Phi_g = \varphi_g C$ and that the ring $Q$ has a direct decomposition $Q = i\,(g)\,Q \oplus f\,Q$, where $i\,(g), f$ are central idempotents, such that the action of the automorphism $g$ on $i\,(g)\,Q$ coincides with the conjugation by the invertible (in $i\,(g)\,Q$) element $\varphi_g$. Moreover, for any $e \leqslant i\,(g)$ the automorphism $g$ is inner on $eQ$, which, in particular, implies that $G_e = \{g \in G \mid i(g) \geqslant e\}$ is a subgroup of the group $G$ (for a given e).

**3.5.2.** *The algebra of a group* of automorphisms $G \subseteq \mathcal{A}(R)$ is equal to the sum of all conjugation modules

$$\mathbf{B}(G) = \sum_{g \in G} \Phi_g.$$

It is clear that it is an algebra over the extended centroid $C$.

**3.5.3.** *Reduced finiteness.* The most natural and direct transfer of this notion is as follows: the module $\mathbf{B}(G)$ is finitely generated over $C$ and a factor-group $G/G_{inn}$ is finite. Such a definition, however, excludes from consideration in the above example the group $G$, which is the Galois closure of the finite group $H$. For this reason we have to pass to a "local" variation of this notion.

---

[1] For a semiprime ring $R$ the group $\mathcal{A}(R)$ is defined by the formula $\mathcal{A}(R) = \{g \in Aut\,Q(R) \mid \exists I, J \lhd R,$ $ann_R\, I = ann_R\, J = 0,\;\; J \subseteq I^g \subseteq R\}$, see 2.5.1 for prime rings.

A group $G \subseteq \mathcal{A}(R)$ is called *reduced-finite* if its algebra $\mathbf{B}(G)$ is a finitely generated $C$-module and $\sup\{e \mid |G : G_e| < \infty\} = 1$.

**3.5.4.** *Closure of a group.* Returning to the starting example, we should remark that the action of every $g \in G$ on the factor $F_\alpha$ coincides with that of a certain $h \in H$, which is $\alpha$-dependent. It is this peculiarity that results in the coincidence of the invariants of $G$ and those of $H$. Under general conditions we, by analogy, come to the notion of a local belonging to a group: the automorphism $g$ *locally belongs* to a group $H$, provided there exists a dense family of idempotents $\{e_\alpha \in C \mid \alpha \in A\}$, such that the action of $g$ on $e_\alpha Q$ coincides with that of a certain $h_\alpha \in H$, i.e. $\sup\{e\colon\ g|_{eQ} \in H|_{eQ}\} = 1$.

A group $G$ is called *closed* if any automorphism locally belonging to the group $G$ lies in $G$.

**3.5.5.** LEMMA. *Any Galois group is closed.*

**3.5.6.** EXAMPLE. We are going to consider an example which illustrates the notion of the closure operation in the case of rings with finite prime dimension. Let

$$Q = \underbrace{Q_1 \oplus \cdots \oplus Q_n}_{n},$$

where $Q_i \cong Q_0$ is a prime ring.

If a group $H$ acts on $Q_0$, then on $Q$ we have, first, the product $H^n = H \times \cdots \times H$ acting in a componentwise manner and, second, the group of permutations $S_n$, rearranging the summands

$$(q_1 \oplus \cdots \oplus q_n)^\pi = q_{\pi^{-1}(1)} \oplus \cdots \oplus q_{\pi^{-1}(n)}.$$

Therefore, an action of the semidirect product $G = H^n \propto S_n$ is defined on $Q$. It is obvious that $G$ is a closed group. The inverse statement is also valid.

**3.5.7.** LEMMA. *Let a closed group $G \subseteq \operatorname{Aut} Q$ act transitively on the components of $Q$. In this case the rings $Q_i$ can be identified in such a way that $G = H^n \propto S_n$.*

It is obvious that in the preceding lemma the fixed ring $Q^G$ is isomorphic to $Q_0^H$, where the isomorphism is defined by a diagonal mapping $q \to q^{\sigma_1} + \cdots + q^{\sigma_n}$ with the help of identification isomorphisms $\sigma_1, \ldots, \sigma_n$.

**3.5.8.** *Noether groups ($N$-groups).* This notion remains unchanged: a group $G \subseteq \mathcal{A}(R)$ is called an *$N$-group* provided every invertible element of its algebra $\mathbf{B}(G)$ determines an automorphism from $G$. It is easy to see now that any Galois group is a closed $N$-group.

**3.5.9.** *Maschke groups ($M$-groups).* The definition is preserved: a reduced-finite group $G$ is called an *$M$-group* provided its algebra is semiprime.

**3.5.10.** *Regular groups.* If an $M$-group $H$ is given, then we can extend it to an $N$-group by adding all inner automorphisms corresponding to invertible elements from $\mathbf{B}(H)$. The obtained group will have the same algebra $\mathbf{B}(G) = \mathbf{B}(H)$ and will, therefore, be reduced-finite, the fixed ring $Q^G = Q^H$ remaining unchanged. We can now extend $G$ to a closed group $\widehat{G}$ by adding all the automorphisms locally belonging to $G$. In this case we also have $Q^{\widehat{G}} = Q^H$ and $\mathbf{B}(\widehat{G}) = \mathbf{B}(H)$, but the group $\widehat{G}$ can be not reduced-finite.

An $N$-group $G$ is called *regular* if it is a closure of an $M$-group.

It can be proved that each closed $N$-subgroup of a regular group is a closure of a certain reduced-finite group with the same algebra.

**3.5.11.** THEOREM. *An automorphism $h$ belongs to the Galois closure of an $MN$-group $G$ iff $h$ locally belongs to $G$, i.e. $A(R^G) = \widehat{G}$.*

The proof of this theorem is based on the metatheorem. The canonical sheaf cannot be used here directly because the group acts nontrivially on the space $Spec\,C$ and therefore there is no natural action of $G$ on the stalks. Instead of the canonical sheaf one should consider the so-called *invariant sheaf.* It is defined over the space of orbits $Spec\,C/G$ of the spectrum. If $\pi$ is a map which takes a point $p$ to its orbit $\bar{p} = \{p^g \mid g \in G\}$, then any open set of $Spec\,C/G$ has the form $\pi(V)$, where $V$ is an open set of $Spec\,C$, which allows one to define rings of sections and restriction homomorphisms of the invariant sheaf respectively as $\Gamma(\pi^{-1}(W))$ and $\rho^{\pi^{-1}(U)}_{\pi^{-1}(W)}$. In this way we obtain a correct (see 3.4.2) sheaf as the space of orbits is extremely disconnected. Thus one can apply the metatheorem to the invariant sheaf. For this purpose the structure of stalks of the invariant sheaf is important.

**3.5.12.** THEOREM. *Almost all stalks of the invariant sheaf for a reduced-finite group $G$ have a decomposition*

$$\Gamma_{\pi(p)} = \Gamma_1 + \Gamma_2 + \cdots + \Gamma_n,$$

*where the $\Gamma_i$ are prime rings, isomorphic to the stalk $\Gamma_p$ of the canonical sheaf, and the group induced by the closure of $G$ has the form $H^n \propto S_n$, (see Example 3.5.6) where $H$ is a reduced finite group of automorphisms of the prime ring $\Gamma_p$. If $G$ is respectively a Maschke, Noether or regular group then so is $H$.*

**3.5.13.** *Galois subrings.* Let $G$ be a regular group of automorphisms of a semiprime ring $R$, and let $S$ be an intermediate ring, $R \supseteq S \supseteq R^G$.

It can be shown that the centralizer of $S$ in the ring $R_F$ is contained in the algebra $\mathbf{B}(G)$ of the group $G$. This centralizer will be denoted by $Z$.

Let us reformulate the conditions on an intermediate ring arising in the prime ring case in the semiprime ring situation.

**BM.** Let $e$ be an idempotent from $\mathbf{B}(G)$, such that $se = ese$ for any $s \in S$. Then there is an (idempotent) $f \in Z$, such that $ef = f$, $fe = e$.

**SI.** The $C$-algebra $Z$ is generated by its invertible elements and if for an automorphism $g \in G$ there is an element $b \in \mathbf{B}(G)$, such that $sb = bs^g$ for all $s \in S$, then there is an invertible in $e(b)Q$ element with the same property. Here $e(b)$ is the support of $e$.

**RC.** If $A$ is an essential ideal of $S$, and $Ar \subseteq S$ for a certain $r \in R$, then $r \in S$.

**3.5.14.** THEOREM. *Any intermediate Galois subring of an $M$-subgroup of the group $G$ satisfies conditions* **BM**, **SI** *and* **RC**.

**3.5.15.** THEOREM. *Let $G$ be a regular group. Then any intermediate ring satisfying conditions* **BM**, **SI** *and* **RC** *is a Galois subring of a regular subgroup of the group $G$.*

The proofs of these theorems are also based on the metatheorem for the invariant sheaf.

**3.5.16.** EXTENSION THEOREM. *Let $G$ be a regular group of automorphisms of a semiprime ring $R$ and let $S'$, $S''$ be intermediate Galois subrings of $M$-subgroups. If $\varphi$: $S' \to S''$ is an isomorphism that is the identity on $R^G$, and the ring $S'$ satisfies condition* **SI** *for all mappings from $\varphi G$, then $\varphi$ can be extended to an isomorphism from $G$.*

It is interesting to note that the proof of this theorem can be easily obtained as a corollary of the correspondence theorem. In this case one should consider the semiprime ring $\overline{R} = R \oplus R$ with the group $\overline{G} = G^2 \propto S_2$. Then $S = \{s \oplus s^{\varphi},\ s \in S'\}$ will be an intermediate subring and one can apply Theorem 3.5.15.

**3.6.** The topology considered in 3.2 was introduced in paper [Kh79] as a tool for the investigation of derivations in semiprime rings. The metatheorem was discovered independently in three different forms by K.I. Beidar and A.V. Mikhalev [BM85], S. Burris and H. Werner [BW79], and V.A. Lyubetzky and E.I. Gordon [LG82, Ly86] at approximately the same time. The first approach was based on the notion of orthogonal completeness and was oriented to applications in ring theory. The second one was concerned with investigations of sheaves of algebraic systems independent of applications in ring theory. The last approach is based on nonstandard analysis. Within this framework a semiprime orthogonal complete ring (or a closed ring in the topology defined in 3.2) can be viewed as a nonstandard prime ring. The formulation of the metatheorem here is taken from the book [Kh91] where the background intuition of all three approaches is presented. In this book the Galois theory for semiprime rings is considered in detail from the modern point of view.

## References


[Al61] A.A. Albert, *Structure of Algebras,* Amer. Math. Soc., Providence, RI (1961), pp. 205.

[Am65] S.A. Amitsur, *Generalized polynomial identities and pivotal monomials,* Trans. Amer. Math. Soc. **114**(1) (1965), 210–226.

[Am72] S.A. Amitsur, *On central division algebras,* Israel J. Math. **12**(4) (1972), 408–420.

[An85] A.Z. Anan'in, *Nil-algebras with nonradical tensor square,* Sibirsk. Mat. Zh. **26**(2) (1985), 192–194.

[BM85] K.I. Beidar and A.V. Mikhalev, *Orthogonal completeness and algebraic systems,* Uspekhi Mat. Nauk **40**(6) (246) (1985), 79–115.

[Bo63] L.A. Bokut', *Some embedding theorems for rings and semigroups,* Sibirsk. Mat. Zh. **4**(3) (1963), 500–518; *II*, Sibirsk. Mat. Zh. **3**(5) (1963), 729–743.

[Bo76] L.A. Bokut', *Embeddings in simple associative rings,* Algebra i Logika **15**(2) (1976), 215–246.

[Bo77] L.A. Bokut', *Associative Rings I, II,* Novosibirsk, Novosibirsk State University (1977, 1981).

[BW79] S. Burris and H. Werner, *Sheaf constructions and their elementary properties,* Trans. Amer. Math. Soc. **248**(2) (1979), 269–309.

[Ca47] H. Cartan, *Théorie de Galois pour les corps non commutatifs,* Ann. Sci. École Norm. Sup. (3) **64** (1947), 59–77.

[CF75] J. Cozzens and C. Faith, *Simple Noetherian Rings,* Cambridge Univ. Press, Cambridge (1975), pp. 135.

[CH80] A.W. Catters and C.R. Hajarnavis, *Rings with Chain Conditions,* Pitman, Boston (1980), pp. 198.

[CM67] L.N. Childs and F.R. DeMeyer, *On automorphisms of separable algebras,* Pacific J. Math. **23**(1) (1967), 25–34.

[Co58] P. Cohn, *On a class of simple rings,* Matematika **5**(10) (1958), 103–117.

[Co73] P. Cohn, *The embedding of radical rings in simple radical rings,* Bull. London Math. Soc. **3**(8) (1971), 185–188; *Correction,* ibid. **4** (1972), 54; *2nd correction,* ibid. **5** (1973), 322.

[Di48] J. Dieudonné, *La théorie de Galois des anneaux simples et semi-simples,* Comment. Math. Helv. **21**(2) (1948), 154–184.

[Du80] N.I. Dubrovin, *Chain domains,* Vestnik MGU, Matem., Mekh., no. 2 (1980), 51–54.

[El73] V.P. Elizarov, *Strong pretorsions and strong filters, quotient rings and modules,* Sibirsk. Mat. Zh. **4**(3) (1973), 549–559.

[FO80] J.W. Fisher and J. Osterburg, *Finite group actions on noncommutative rings: a survey since* 1970, Ring Theory and Algebra III, Dekker, New York (1980), 357–393.

[Go64] E.S. Golod, *On nil-algebras and finitely approximated p-groups,* Izv. Akad. Nauk SSSR, Ser. Mat. **28** (1964), 273–276.

[Ha82] M. Hacque, *Anneaux fidelment représentés sur leur socle droit,* Comm. Algebra **10** (1982), 1027–1072.

[Ha87] M. Hacque, *Theorie de Galois des anneaux presque-simples,* J. Algebra **108**(2) (1987), 534–577.

[Ho49] G. Hochschild, *Double vector spaces over division rings,* Amer. J. Math. **71**(2) (1949), 443–460.

[Ho50] G. Hochschild, *Automorphisms of simple algebras,* Trans. Amer. Math. Soc. **69** (1950), 292–301.

[Ho39] C. Hopkins, *Rings with minimal condition for left ideals,* Ann. Math. **40**(3) (1939), 712–730.

[Ja40] N. Jacobson, *The fundamental theorem of the Galois theory for quasifields,* Ann. Math. **41**(1) (1940), 1–7.

[Ja47] N. Jacobson, *A note on division rings,* Amer. J. Math. **69**(1) (1947), 27–36.

[Ja56] N. Jacobson, *Structure of Rings,* AMS Colloq. Publ. vol. 37, Amer. Math. Soc., Providence, RI (1956, revised 1964).

[Ja75] N. Jacobson, *PI-algebras. An Introduction,* SLNM 441, Springer, Berlin (1975), pp. 115.

[Ke63] O. Kegel, *Zur nilpotent gewissen assoziativer Ringe,* Math. Ann. **149**(3) (1963), 258–260.

[Kh75] V.K. Kharchenko, *Generalized identities with automorphisms of associative rings with a unite,* Algebra i Logika **14**(6) (1975), 681–696.

[Kh77] V.K. Kharchenko, *Galois theory of a semiprime ring,* Algebra i Logika **16**(3) (1977), 313–363.

[Kh78] V.K. Kharchenko, *Algebras of invariants of free algebras,* Algebra i Logika **17**(4) (1978), 478–487.

[Kh79] V.K. Kharchenko, *Differential identities of semiprime rings,* Algebra i Logika **18**(1) (1979), 86–119.

[Kh91] V.K. Kharchenko, *Automorphisms and Derivations of Associative Rings,* Kluwer, Dordrecht (1991), pp. 385.

[Kr70] H.F. Kreimer, *On the Galois theory of separable algebras,* Pacific J. Math. **34**(3) (1970), 729–740.

[Ku41] A.G. Kurosh, *Ring theory problems connected with Burnside problem on periodic groups,* Izv. Akad. Nauk SSSR, Ser. Mat. **5** (1941), 233–241.

[Le45] J. Levitzki, *On three problems concerning mil-rings,* Bull. Amer. Math. Soc. **51**(12) (1945), 913–919.

[LG82] V.A. Lyubetsky and E.I. Gordon, *Imbedding of sheaves in a Geiting-sign universum,* Dep. VINITI, n. 4782-82, Moscow (1982), 1–29.

[Ly86] V.A. Lyubetsky, *Some applications of topos theory in investigations of algebraic systems,* Appendix in the translation to Russian of the book: P.T. Johnstone, *Topos Theory,* Moscow, Nauka (1986), p. 440.

[Ma69] W.S. Martindale, *Prime rings satisfying a generalized polynomial identity,* J. Algebra **12**(4) (1969), 567–584.

[Mi66] Y. Miyashita, *Finite outer Galois theory of noncommutative rings,* J. Fac. Sci. Hokkaido Univ. Ser. I, **19** (1966), 115–134.

[MP84] S. Montgomery and D. Passman, *Galois theory of prime rings,* J. Pure Appl. Algebra **31** (1984), 139–184.

[Na52] T. Nakayama, *Galois theory of simple rings,* Trans. Amer. Math. Soc. **73**(2) (1952), 276–292.

[NA47] T. Nakayama and G. Azumaya, *On irreducible rings,* Ann. Math. **48**(4) (1947), 949–965.

[No33] E. Noether, *Nichtkommutative algebra,* Math. Z. **37** (1933), 514–541.

[Pa89] D. Passman, *Infinite Crossed Products,* Academic Press, Boston (1989), pp. 468.

[Pi82] R.S. Pierce, *Associative Algebras,* Springer, Berlin (1982), pp. 436.

[Pr73] C. Procesi, *Rings with Polynomial Identities,* Dekker, New York (1973), pp. 190.

[Ro75] L. Rowen, *Generalized Polynomial Identities,* J. Algebra **34**(3) (1975), 458–480.

[Ro80] L. Rowen, *Polynomial Identities in Ring Theory,* Academic Press, New York (1980), pp. 364.

[RZ55] A. Rosenberg and D. Zelinsky, *Galois theory of continuous linear transformation rings,* Trans. Amer. Math. Soc. **79**(2) (1955), 429–452.

[SC67] E. Sąsiada and P.M. Cohn, *An example of a simple radical ring,* J. Algebra **5**(3) (1967), 373–377.

[Sm66] D.M. Smirnov, *Right ordered groups,* Algebra i Logika **5**(5) (1966), 41–59.

[TN70] H. Tominaga and T. Nagahara, *Galois Theory of Simple Rings,* Okayama Math. Lectures, Okayama University (1970).

[Va88] A.I. Valitskas, *Examples of radical rings not embeddable in simple radical rings,* Dokl. Akad. Nauk SSSR **294**(4) (1988), 790–792.

[VZ69] O.E. Villamayor and D. Zelinsky, *Galois theory with infinitely many idempotents,* Nagoya Math. J. **35** (1969), 83–98.

[ZN75] A.E. Zalesskii and O.M. Neroslavskii, *On simple noetherian rings,* Izv. Akad. Nauk BSSR, Ser. Fiz.-Mat. **5** (1975), 38–42.

[ZN77] A.E. Zalesskii and O.M. Neroslavskii, *There exist simple noetherian rings with zero divisors and without idempotents,* Comm. Algebra **5**(3) (1977), 231–244.

# Algebraic Microlocalization and Modules with Regular Singularities over Filtered Rings

A. van den Essen

*Department of Mathematics, Catholic University of Nijmegen, Toernooiveld, 6525 ED Nijmegen, The Netherlands*
*e-mail: essen@sci.kun.nl*

*Contents*



HANDBOOK OF ALGEBRA, VOL. 1
Edited by M. Hazewinkel

## Introduction

During the seventies a new branch of mathematics was born: the theory of $\mathcal{D}$-modules. Roughly speaking it is an algebraic frame work in which systems of linear partial differential equations can be studied. The theory grew rapidly, in particular the study of modules with regular singularities got much attention and culminated in the so-called Riemann–Hilbert correspondence (a generalization of Deligne's solution of Hilbert's 21st problem) ([Me1, Me2] and [KK]). Application of the theory to several parts of mathematics, [Be1, Be2, BeiBe, Br1, Br2, Br3, BrK, Sai1, Sai2, Sai3], stimulated many people to study the theory. Some nice survey papers are [v.D, LeMe] and [Od]. Also several books on $\mathcal{D}$-modules appeared [Bj1, Bo, P, Sch] and [Me3]. However the theory is not as algebraic as an algebraist would like, since at several places analysis and in particular micro-local analysis is used to obtain deep results (see, for example, [KO] and [KK]). One of the origins of the theory is the highly nontrivial fact that the characteristic variety of a $\mathcal{D}$-module is involutive. This result was first proved in [SKK] by micro-local analysis. Later in 1981 O. Gabber gave a purely algebraic proof of this important result [Ga]. His paper was the starting point of a purely algebraic study of $\mathcal{D}$-modules: in fact it was Springer who, after reading Gabber's paper introduced the algebraic counter-part of the analytic microlocalization, the so-called algebraic microlocalization ([Sp]): let $R$ be a filtered ring such that the associated graded ring $grR$ is commutative and $S$ a multiplicatively closed subset of $R$. Then it is not always possible to localize at $S$. However if $\sigma(S)$ is a multiplicatively closed set of $grR$ ($\sigma$ denotes the principal symbol map), then it is possible to lift the localization of $grR$ at $\sigma(S)$ to a kind of localization of $R$ at $S$. This result has been generalized by the author in [v.E1] to arbitrary filtered rings (i.e. $grR$ need no longer be commutative) in order to generalize several of the results of the analytic $\mathcal{D}$-module theory to a large class of filtered rings ([v.E2]). This chapter is an account on algebraic microlocalization and its application to the study of (holonomic) modules with regular singularities.

The efforts to generalize results from the $\mathcal{D}$-module theory to filtered rings has revealed some new approach to filtered rings. This is exposed in Section 0 where we introduce two principles for studying filtered rings and their modules (we also refer to the papers [Gi] and [ABO] where a similar kind of approach is described).

In Section 1 algebraic microlocalization is introduced and several of its properties are given (universal property, the graded ring of microlocalization, flatness of microlocalization, ...). In Section 2 we introduce holonomic $R$-modules, based on the involutiveness of the characteristic variety. Furthermore we define $R$-modules with regular singularities and give several equivalent descriptions of this notion. One of them is the existence of a very good filtration, which makes the link with results obtained in [KK]. The main result, Theorem 2.6.1 gives a local-global description of regular singularities in terms of the minimal prime components of the characteristic variety. This theorem should be considered as the algebraic analogue of Theorem 4.1 of Deligne in [D]. Finally in Section 3 we mention some other applications of algebraic microlocalization.

## 0. Preliminaries

All rings are associative with identity. A module is always a left module, unless mentioned otherwise.

### 0.1. *Filtered rings and modules*

A ring $R$ is called a *filtered ring* if there exists an ascending chain $FR$:

$$\cdots \subset F_{-1}R \subset F_0R \subset F_1R \subset \cdots$$

of additive subgroups of $R$ such that $1 \in F_0R$ and $F_nRF_mR \subset F_{n+m}R$ for all $n, m \in \mathbb{Z}$. The chain $FR$ is called a *filtration of* $R$. Observe that $F_0R$ is a subring of $R$. A left $R$-module $M$ is called a *filtered module* if there exists an ascending chain $FM$:

$$\cdots \subset F_{-1}M \subset F_0M \subset F_1M \subset \cdots$$

of additive subgroups of $M$ such that $F_nRF_mM \subset F_{n+m}M$ for all $n, m \in \mathbb{Z}$. The chain $FM$ is called a *filtration of* $M$. Throughout this paper all filtrations considered will be exhaustive, i.e.

$$\bigcup F_nR = R \quad \text{and} \quad \bigcup F_nM = M.$$

For a filtered ring $R$ one defines the so-called *associated graded ring*, denoted $grR$, by

$$grR = \bigoplus gr_nR, \quad \text{where } gr_nR = F_nR/F_{n-1}R.$$

The ring structure on $grR$ is given by the formula

$$(r + F_{n-1}R) \cdot (r' + F_{m-1}R) = rr' + F_{n+m-1}R.$$

We define the symbol map $\sigma\colon R \to grR$ by

$$\sigma(r) = 0 \quad \text{if } r \in \bigcap F_nR,$$
$$\sigma(r) = r + F_{n-1}R \quad \text{if } r \in F_nR \backslash F_{n-1}R.$$

Similarly, if $M$ is a filtered $R$-module, replacing $R$ by $M$ in the definitions above, we can define $grM = \bigoplus gr_nM$ where $gr_nM = F_nM/F_{n-1}M$ and also the symbol map $\sigma\colon M \to grM$. The formula

$$(r + F_{n-1}M)(m + F_{p-1}M) = rm + F_{n+p-1}M$$

equips $grM$ with the structure of a graded $grR$-module. The importance of the associated graded ring, module and the symbol map $\sigma$, comes from the following principle.

LIFTING PRINCIPLE. It is often possible to lift a property of the associated graded module (resp. ring) to the filtered module (resp. ring) via the symbol map $\sigma$.

EXAMPLE 0.1.1. Suppose

$$\bigcap F_n R = \{0\}.$$

If $grR$ has no zero-divisors, then $R$ has no zero-divisors.

PROOF. Let $rr' = 0$ for some $r, r' \in R$ not both zero. Say $r \in F_n R \backslash F_{n-1} R$ and $r' \in F_m R \backslash F_{m-1} R$. Then $\sigma(r)\sigma(r') = rr' + F_{n+m-1} R = 0$. So $grR$ has zero-divisors, a contradiction. □

COROLLARY 0.1.2. *The ring of differential operators with polynomial coefficients*

$$A_n = \mathbb{C}[x_1, \dots, x_n, \partial_1, \dots, \partial_n],$$

*the $n$-th Weyl algebra, has no zero-divisors (since $grA_n \simeq \mathbb{C}[x_1, \dots, x_n, \zeta_1, \dots, \zeta_n]$, a polynomial ring in $2n$ variables, where the filtration on $A_n$ is the usual $\partial$-filtration, see* [Bj1], *Chapter* 3, *for more details).*

EXAMPLE 0.1.3. Let $FR$ be complete and separated (see 0.2 for the definitions) and $grR$ left noetherian. Then $R$ is left noetherian.

This is a special case of

EXAMPLE 0.1.4. Let $FR$ be complete and $M$ a filtered $R$-module with $FM$ separated (i.e. $\bigcap_n F_n M = \{0\}$). If $grM$ is a finitely generated $grR$-module, then $M$ is a finitely generated $R$-module. More precisely, there exist a finite number of elements $m_i$ in $M$ and $v_i \in \mathbb{Z}$ such that $F_n M = \Sigma F_{n-v_i} R m_i$ (cf. [v.E1], Proposition 6.5).

The type of filtrations described in Example 0.1.4 play a crucial role in the theory of filtered modules:

DEFINITION 0.1.5. Let $R$ be an arbitrary filtered ring and $M$ a filtered $R$-module. A filtration $FM$ on $M$ is called *good* if there exist a finite number of elements $m_i$ in $M$ and elements $v_i \in \mathbb{Z}$ such that $F_n M = \Sigma F_{n-v_i} R m_i$ for all $n \in \mathbb{Z}$.

An $R$-module possesses a good filtration if and only if it is finitely generated over $R$. Furthermore, if $FM$ and $F'M$ are two good filtrations on $M$, then one easily verifies that they are equivalent, i.e.

DEFINITION 0.1.6. Let $FM$ and $F'M$ be two arbitrary filtrations on a filtered $R$-module $M$. Then they are called *equivalent* if there exists an integer $c \in \mathbb{N}$ such that

$$F_{n-c} M \subset F'_n M \subset F_{n+c} M,$$

for all $n \in \mathbb{Z}$.

### 0.2. *The topology defined by a filtration*

Let $R$ be a filtered ring with filtration $FR$ and $M$ a filtered $R$-module with filtration $FM$. On $M$ we have an *order function* $v$: $M \to \mathbb{Z} \cup \{-\infty\}$ defined by $v(m) = -\infty$ if $m \in \bigcap F_n M$, $v(m) = n$ if $m \in F_n M \backslash F_{n-1} M$.

This gives a nonarchimedean pseudo-norm on $M$ (called the *associated pseudo-norm*) defined by $|m| = 2^{v(m)}$.

The topology induced on $M$ by this pseudo-norm is called *the topology defined by the filtration* $FM$. The filtration $FM$ is called *discrete*, resp. *separable*, resp. *complete* if the topology defined by $FM$ has the respectively mentioned properties. So more concretely, $FM$ is discrete if there exists an integer $N$ such that $F_n M = 0$ for all $n < N$, $FM$ is separated if $\bigcap F_n M = \{0\}$ and $FM$ is complete if $FM$ is separated and all Cauchy sequences in the $FM$-topology converge in $M$. Finally observe that $FM$ is separated if and only if $|\ |$ is a norm on $M$.

### 0.3. *Strongly filtered rings*

DEFINITION 0.3.1. A filtered ring $R$ with filtration $FR$ is called *strongly filtered* if there exists an element $s \in F_1 R \backslash F_0 R$ invertible in $R$ such that $s^{-1} \in F_{-1} R$.

EXAMPLE 0.3.2. The localized polynomial ring $\mathbb{Z}[X, X^{-1}]$ with the natural filtration defined by $v(X^{-n} a) = \deg a - n$ $(a \in \mathbb{Z}[X],\ n \in \mathbb{Z})$ is a strongly filtered ring; just take $s = X$.

EXAMPLE 0.3.3. Let $X$ be a complex analytic manifold and $p \in T^*X \backslash T^*_X X$. Then the ring $E_p$ of germs of microlocal differential operators is a strongly filtered ring. ([Bj1], Chapter 4, Theorem 3.5.)

Filtrations on modules over strongly filtered rings are easy to describe. To see this let $R$ be a strongly filtered ring. Then we have the following (easy to verify) formulas:

$$F_n R = s^n F_0 R = F_0 R s^n \quad \text{for all } n \in \mathbb{Z}.$$

Furthermore, if $M$ is a filtered $R$-module with filtration $FM$, then

$$F_n M = s^n F_0 M = F_n R F_0 M \quad \text{for all } n \in \mathbb{Z}.$$

So $M = RF_0 M$, since $M = \bigcup F_n M$ and $R = \bigcup F_n R$. From these formulas we see that a filtration $FM$ on $M$ is completely determined by the $F_0 R$-module $F_0 M$. Conversely, an arbitrary $F_0 R$-submodule $M_0$ of $M$ satisfying $RM_0 = M$ gives rise to a filtration $FM$ on $M$ by putting $F_n M := F_n R M_0$. Summarizing: we get a bijection $\mathcal{F}$ between the set of $F_0 R$-submodules $M_0$ of $M$ generating $M$, i.e. $RM_0 = M$ and the set of filtrations of $M$, given by

$$\mathcal{F}\colon\ M_0 \rightsquigarrow \mathcal{F}(M_0) := (F_n R M_0)_{n \in \mathbb{Z}}.$$

One easily verifies that restriction of $\mathcal{F}$ to the set of $F_0R$-submodules of $M$ of finite type gives a bijection with the good filtrations of $M$ (see Definition 0.1.5).

### 0.4. *Strongly filtered rings associated to filtered rings*

The importance of strongly filtered rings comes from the fact that to every filtered ring one can associate a strongly filtered ring which can be used to reduce problems over arbitrary filtered rings to problems over strongly filtered rings (see the reduction principle below).

Let $R$ be an arbitrary filtered ring with filtration $FR$ and $X$ an indeterminate. Consider the localized polynomial ring $R_X := R[X, X^{-1}]$ which becomes a filtered ring by putting

$$F_nR_X = \{\Sigma r_iX^i \mid v(r_i) + i \leqslant n \text{ for all } i \in \mathbb{Z}\}.$$

One readily verifies that $R_X$ is a strongly filtered ring and call it *the strongly filtered ring associated to* $R$. Similarly, if $M$ is a filtered $R$-module with filtration $FM$ we can form $M_X := M[X, X^{-1}]$ which is a filtered $R_X$-module with filtration

$$F_nM_X = \{\Sigma m_iX^i \mid v(m_i) + i \leqslant n \text{ for all } i \in \mathbb{Z}\}.$$

The associated strongly filtered ring plays a crucial role in describing and finding properties of arbitrary filtered rings and modules.

This is expressed in the following principle.

REDUCTION PRINCIPLE. To find properties for arbitrary filtered rings (resp. modules over filtered rings) first find them for strongly filtered rings (resp. modules over strongly filtered rings) and then rewrite the result for the strongly filtered ring $R_X$ (resp. for modules over $R_X$) in terms of the filtered ring $R$ (resp. in terms of $R$-modules).

Let us first illustrate the reduction principle with two simple examples

EXAMPLE 0.4.1. Good filtrations.

Suppose we only defined good filtrations for strongly filtered rings by the condition that $F_0M$ is a finitely generated $F_0R$-module. According to the reduction principle the corresponding notion of a good filtration for modules over an arbitrary filtered ring $R$ would be $F_0M_X$ is a finitely generated $F_0R_X$-modules. Then rewriting this condition in terms of the filtration $FM$ of $M$ would give us exactly the condition formulated in Definition 0.1.5.

EXAMPLE 0.4.2. Equivalent filtrations.

Suppose we only defined the notion of equivalent filtrations for strongly filtered rings by the condition that their induced topologies are the same. Then one readily verifies that this is equivalent to the existence of an integer $c \in \mathbb{N}$ satisfying $F_{-c}M \subset F'_0M \subset F_cM$. According to the reduction principle the corresponding notion of equivalent filtrations on a module over an arbitrary filtered ring would be that $F_{-c}M_X \subset F'_0M_X \subset F_cM_X$

for some $c \in \mathbb{N}$. Rewriting this condition in terms of the filtrations $FM$ and $F'M$ of $M$ would give us exactly the condition formulated in Definition 0.1.6.

REMARK. For an arbitrary filtered ring the condition that two filtrations are equivalent is stronger then asking that their topologies are the same.

Now we give two more results obtained by using the reduction principle (these results will be very useful in the sequel, cf. Section 2).

Let $R$ be an arbitrary filtered ring and $M$ an $R$-module with filtration $FM$. Let $M'$ be an $R$-submodule of $M$. On $M'$ we have the *induced filtration* defined by $F_nM' = M' \cap F_nM$ and the *quotient filtration* defined by $F_nM/M' = M' + F_nM/M'$. One readily verifies that if $FM$ is good on $M$ the quotient filtration is good on $M/M'$. However in general the induced filtration of a good filtration need not be good. So we can ask

QUESTION 1. Under which conditions on $FR$ good filtrations on $M$ induces good filtrations on submodules $M'$ of $M$?

To solve this question we apply the reduction principle. So first assume that $R$ is strongly filtered and $F_0M$ is a finitely generated $F_0R$-module. Then Question 1 amounts to ask: under which conditions on $FR$ is the $F_0R$-module $M' \cap F_0M$ a finitely generated $F_0R$-module?

A natural condition is to assume that $F_0R$ is a (left) noetherian ring, since then obviously $M' \cap F_0M$ is finitely generated over $F_0R$. Now assume that $R$ is an arbitrary filtered ring. Then according to the reduction principle we consider the condition: $F_0R_X$ is a (left) noetherian ring. Indeed, if $F_0R_X$ is left noetherian then using the formula $F_nM'_X = M'_X \cap F_nM_X$ we in particular have that $F_0M'_X = M'_X \cap F_0M_X$, so $F_0M'_X$ is a finitely generated left $F_0R_X$-module and hence $F_0M'$ is good by Example 0.4.1.

The ring $F_0R_X = \Sigma F_nRX^{-n} = \Sigma F_nRT^n$, where $T = X^{-1}$, is a graded ring with the obvious $T$-grading. It is called the *Rees ring* of $R$ and denoted $\widetilde{R}$. So the previous arguments show that $\widetilde{R}$ is left noetherian is a sufficient condition for Question 1. In a completely analogous way one can show that the condition $\widetilde{R}$ is left noetherian is also a sufficient condition for

QUESTION 2. Under which conditions on $FR$ filtrations equivalent with good filtrations are good?

So summarizing we have

PROPOSITION 0.4.3. *If $\widetilde{R}$ is left noetherian then good filtrations induces good filtrations on submodules and filtrations equivalent with good filtrations are good.*

REMARK 0.4.4. In [ABO, LiO] and [Li1] an extensive study is made of filtered rings and their interplay with both the Rees ring and the associated graded ring (observe that $grR \simeq \Sigma F_nRX^{-n}/\Sigma F_{n-1}RX^{-n} = F_0R_X/F_{-1}R_X = gr_0R_X$). In particular Proposition 0.4.3 was obtained in [LiO], Proposition 2.1.

Using Proposition 0.4.3 we get a nice correspondence between the good filtrations on $M$ and the good filtrations on $M_X$: if $FM$ is a filtration on $M$ let $\mathcal{F}_X(FM)$ denote the filtration $FM_X$ defined above on $M_X$. Conversely, if $F$ denotes a filtration on $M_X$ $G(F)$ denotes the filtration on $M$ defined by $F_nG(F) = M \cap F_n$. Observe that $G\mathcal{F}_X(FM) = FM$ and that $FM$ is good on $M$ if and only if $\mathcal{F}_X(FM)$ is good on $M_X$ (see Example 0.4.1). For the applications in Section 2 we need

PROPOSITION 0.4.5. *Let $\widetilde{R}$ be left noetherian and $M$ a finitely generated left $R$-module. If $F = (F_n)_{n\in\mathbb{Z}}$ is a good filtration on $M_X$ then $G(F)$ is a good filtration on $M$.*

PROOF. Choose a good filtration $FM$ on $M$. By Proposition 0.4.3 it suffices to prove that $G(F)$ is equivalent with $FM$. Since $\mathcal{F}_X(FM)$ is good on $M_X$ there exists an integer $c \in \mathbb{N}$ such that $F_{n-c}M_X \subset F_n \subset F_{n+c}M_X$, for all $n \in \mathbb{Z}$.

Since $F_nM_X \cap M = F_nM$ for all $n \in \mathbb{Z}$, we get $F_{n-c}M \subset F_nG(F) \subset F_{n+c}M$ for all $n \in \mathbb{Z}$, i.e. $G(F)$ is equivalent with $FM$, as desired. □

## 1. Algebraic microlocalization

### 1.1. *Introduction*

Let $R$ be any ring (containing 1) and $S \subset R$ a multiplicatively closed subset of $R$. Then one can not always form the left ring of fractions. This is only possible if $S$ is a left Ore set[1] (cf. [Ste]). However, if $R$ is a filtered ring with filtration $FR$ and $S$ is a multiplicatively closed subset of $R$ such that $\sigma(S)$ is an Ore set in $grR$ (i.e. we can localize at the graded level), then it turns out that there exist a filtered ring, denoted $E_S(R)$ and called the left algebraic microlocalization of $R$ with respect to $S$, and a filtered morphism $\varphi$: $R \to E_S(R)$ such that each element $\varphi(s)$, $s \in S$ is invertible in $E_S(R)$. So roughly speaking, we can lift the localization at the graded level to a microlocalization at the ring level (the lifting principle!).

Algebraic microlocalization was introduced by T.A. Springer in his seminar on $\mathcal{D}$-modules in the autumn of 1982. His construction assumed that $grR$ was commutative ([Sp]). Other constructions of microlocalizations using the assumption that $grR$ is commutative, where given in [La] and [Gi]. In [v.E1] the construction of Springer was extended to the general case, i.e. without any condition on the filtered ring $R$. More constructions of algebraic microlocalizations where given in [WK] and [ABO]. This last construction generalizes ideas of [Gi]. The results of [v.E1] where used in [v.E2] to develop a purely algebraic theory of modules with regular singularities over a large class of filtered rings (see Section 2 for details), including the rings of differential and micro-local differential operators considered in [KO, KK] and [SKK].

### 1.2. *Definition of algebraic microlocalization and some properties*

In the remainder of Section 1 we assume:

[1] A multiplication subset $S$ is an associative ring $R$ is called a left Ore set if for all $s \in S$, $r \in R$ then exist $s' \in S$, $r' \in R$ such that $s'r = r's$.

$R$ is a filtered ring with filtration $FR$, $S$ a multiplicatively closed subset of $R$ such that $\sigma(S)$ is a multiplicatively closed subset of $grR$ with $0 \notin \sigma(S)$ and satisfying the (left) Ore conditions. Furthermore $M$ denotes a filtered (left) $R$-module with filtration $M$.

THEOREM 1.2.1 ([v.E1], Theorems I and II). *There exist $E_S(R)$, resp. $E_S(M)$, a complete separated filtered ring, resp. a complete separated filtered $E_S(R)$-module, and a canonical morphism of filtered rings $\varphi_R$: $R \to E_S(R)$ satisfying $\varphi_R(s)$ is invertible in $E_S(R)$ for all $s \in S$ and $\varphi_R(s)^{-1} \in F_{-n}E_S(R)$ if $\sigma(s) \in gr_n R$, resp. a morphism of filtered $R$-modules $\varphi_M$: $M \to E_S(M)$, having the following universal property: for every filtered morphism $h_R$: $R \to R'$, resp. for every filtered morphism of filtered $R$-modules $h_M$: $M \to M'$, such that $FR'$, resp. $FM'$, is complete and separated and such that for every $s \in S$ $h_R(s)$ is invertible in $R'$ with $h_R(s)^{-1} \in F_{-n}R'$ if $\sigma(s) \in gr_n R$, there exist a unique morphism of filtered rings $\mathcal{X}_R$: $E_S(R) \to R'$ satisfying $\mathcal{X}_R \circ \varphi_R = h_R$, resp. a unique morphism of $R$-modules $\mathcal{X}_M$: $E_S(M) \to M'$ satisfying $\mathcal{X}_M \circ \varphi_M = h_M$.*

The ring $E_S(R)$ resp. the module $E_S(M)$ is called the *left algebraic microlocalization of $R$ resp. $M$ with respect to $S$*. One easily verifies that the properties stated in the theorem characterize the algebraic microlocalizations with respect to $S$.

REMARK. The module $E_S(M)$ depends on the filtration $FM$. So it would be better to write $E_S(M, FM)$ instead. However if no confusion is possible we write $E_S(M)$.

To get a more concrete description of microlocalizations and their filtrations, we list the following properties ([v.E1]):

**1.2.2.** The norm on $E_S(M)$ defined by the filtration on $E_S(M)$ satisfies

$$\|\varphi_R(s)^{-1}\varphi_M(m)\| = |s|^{-1}|m|_S, \quad \text{for all } s \in S,\ m \in M,$$

where $|\ |_S$ denotes the localized pseudo-norm on $M$ (cf. [v.E1]) defined by

$$|m|_S := \inf_{\rho \in S} |\rho|^{-1}|\rho m|$$

(it is shown in [v.E1], Proposition 3.2, that $|sm|_S = |s|_S|m|_S$ and $|s|_S = |s|$ for all $m \in M$ and all $s \in S$).

**1.2.3.** From 1.2.2 we can describe the kernel of $\varphi_M$: $M \to E_S(M)$:

$$\varphi_M(m) = 0, \text{ if and only if } \|\varphi_M(m)\| = 0 \text{ if and only if } \inf_{\rho \in S} |\rho|^{-1}|\rho m| = 0.$$

**1.2.4.** The elements $\varphi_R(s)^{-1}\varphi_M(m)$ form a dense subset in $E_S(M)$ with respect to the $\|\ \|$-topology.

**1.3.** *Examples*

1° Let $S$ be a left Ore set in $R$. Then $E_S(R) \simeq \widehat{S^{-1}R^S}$, where $\wedge_S$ denotes the completion of $S^{-1}R$ with respect to the following pseudo-norm on $S^{-1}R$: $|s^{-1}r|_S = |s|^{-1}|r|_S$, all $r \in R$, $s \in S$ where $|\ |_S$ is as in 1.2.2. (The proof follows readily from the universal property of microlocalization. See also [ABO], Remark 3.11.3.)

2° *Completion.* Taking $S = \{1\}$ in 1° we get $E_{\{1\}}(R) \simeq \widehat{R}$ (the $FR$-completion of $R$).

3° *Noncommutative localization.* If $S$ is a left Ore set of $R$ and $FR$ is trivial (i.e. $F_nR = R$ for all $n \geqslant 0$ and $F_nR = 0$ for all $n < 0$) then $E_S(R) = S^{-1}R$ (this follows from 1° by observing that the filtration on $S^{-1}R$ is trivial and hence complete).

**1.4.** *More properties of microlocalizations*

PROPOSITION 1.4.1 ([v.E1], Proposition 5.24). *There exists an isomorphism $\psi_R$ of graded rings from $\sigma(S)^{-1}grR$ to $grE_S(R)$ defined by*

$$\psi_R\big(\sigma(s)^{-1}\sigma(r)\big) = \varphi_R(s)^{-1}\varphi_R(r) + F_{n-1}E_S(R),$$

*for all $\sigma(s)^{-1}\sigma(r) \in (\sigma(S)^{-1}grR)(n)$.*

*More generally: there exists an isomorphism $\psi_M$ of graded $\sigma(S)^{-1}grR$-modules from $\sigma(S)^{-1}grM$ to $grE_S(M)$ defined by*

$$\psi_M\big(\sigma(s)^{-1}\sigma(m)\big) = \varphi_R(s)^{-1}\varphi_M(m) + F_{n-1}E_S(M),$$

*for all $\sigma(s)^{-1}\sigma(m) \in (\sigma(S)^{-1}grM)(n)$.*

An immediate consequence of this proposition and the Examples 0.1.3 and 0.1.4 is

COROLLARY 1.4.2. 1) *If $grR$ is left noetherian, then $grE_S(R)$ is left noetherian.*

2) *If $grM$ is a finitely generated $grR$-module, then $grE_S(M)$ is a finitely generated $grE_S(R)$-module.*

3) *If $FM$ is good on $M$, then $FE_S(M)$ is good on $E_S(M)$.*

PROPOSITION 1.4.3 ([ABO], Corollary 3.20). *Let $\widetilde{R}$ be left noetherian. Then*

1) *The functor $E_S(R)\otimes_R$-preserves strict maps and is exact on $R$-modules.*

2) *If $FM$ is good, then $E_S(R) \otimes_R M \simeq E_S(M)$, as filtered $R$-modules.*

REMARK 1.4.4. If $FM$ and $F'M$ are equivalent filtrations on $M$, then $E_S(M, FM) = E_S(M, F'M)$ ([v.E1], Proposition 6.3). Consequently if $M$ is an $R$-module of finite type we can take any good filtration on it (since they are all equivalent) and apply Proposition 1.4.3 to the filtered module obtained in this way, to get an isomorphism of $R$-modules:

$$E_S(R) \otimes_R M \simeq E_S(M).$$

**1.5.** *Algebraic microlocalization as a completion of a localization*

Finally we mention a result (Corollary 1.5.3 below) obtained in [WK] and [ABO] which shows that in many practical cases microlocalization is just a suitable completion of a localization at some Ore set of $R$, i.e. the example 1° in 1.3 is almost always the general case.

We say that the filtration $FR$ satisfies the (left) *comparison condition* if for any finitely generated (left) ideal $I$ in $R$, say

$$I = \sum_{j=1}^{s} Rr_j$$

there exists an integer $c$ such that

$$F_nR \cap I \subset \sum_{j=1}^{s} F_{n+c}r_j, \quad \text{for all } n \in \mathbb{Z}.$$

Let $S$ be as stated in the beginning of Section 1. Since $0 \notin \sigma(S)$ we get $\sigma(s)\sigma(s') = \sigma(ss')$ for all $s, s' \in S$ ([v.E1], Corollary 1.11i)). Then we have the following beautiful example of the lifting principle

PROPOSITION 1.5.1 ([WK], Proposition 17). *If $R$ satisfies the left comparison condition and $S = \sigma^{-1}(\sigma(S))$ (i.e. $S$ is saturated) then $S$ is a left Ore set in $R$.*

LEMMA 1.5.2. *Put $S_{\text{sat}} = \sigma^{-1}(\sigma(S))$. Then $E_S(R) = E_{S_{sat}}(R)$.*

PROOF. By the universal property it suffices to prove that $\varphi_R(\underline{s})$ is invertible in $E_S(R)$ for every $\underline{s} \in S_{\text{sat}}$. Write $\underline{S} = S_{\text{sat}}$. So let $\underline{s} \in \underline{S}$. Since $\sigma(\underline{s}) = \sigma(s)$ for some $s \in S$, say $v(s) = n$, we have that $\underline{s} - s \in F_{n-1}R$. Hence $\varphi_R(\underline{s}) - \varphi_R(s) \in F_{n-1}E_S(R)$. Since $v(\varphi_R(s)^{-1}) = -n$ we get $z := \varphi_R(s)^{-1}\varphi_R(\underline{s}) - 1 \in F_{-1}E_S(R)$. It is well known that the completeness of $E_S(R)$ implies that each element of the form $1 + z$, with $z \in F_{-1}E_S(R)$ is invertible in $E_S(R)$. So $\varphi_R(s)^{-1}\varphi_R(\underline{s})$ is invertible in $E_S(R)$, so $\varphi_R(\underline{s})$ is invertible in $E_S(R)$, implying the lemma. □

Now combining the example 1° of 1.3 with Proposition 1.5.1 and Lemma 1.5.2 we obtain

COROLLARY 1.5.3. *If $R$ satisfies the left comparison condition, then for any multiplicatively closed subset $S$ of $R$ (as in the introduction of Section 1)*

$$E_S(R) = \widehat{S_{\text{sat}}^{-1}R}^{S_{\text{sat}}},$$

*i.e. microlocalization is just a suitable completion at some Ore set of $R$.*

REMARK 1.5.4. This Corollary was obtained in [ABO] under the assumption that $\widetilde{R}$ is left noetherian. However the condition $\widetilde{R}$ is left noetherian implies the comparison condition but is not equivalent with it (see [Li1], Theorem 3.5.4 and 3.6(D)).

## 2. Modules with regular singularities over filtered rings

In this section we develop an algebraic theory of modules with regular singularities for a large class of filtered rings, including the rings of micro-local differential operators and differential operators over the formal and convergent power series in several variables over $\mathbb{C}$. Modules with regular singularities will be introduced in steps. First we define regular singularities for strongly filtered rings and then we deduce the definition for the general case by using the reduction principle.

In this section we assume: $R$ is a filtered ring with filtration $FR$ such that $grR$ is a *commutative* ring.

### 2.0. *The characteristic ideal*

Let $M$ be a finitely generated filtered $R$-module with filtration $F = FM$. Let $I_F(M)$ denote the annihilator of the $grR$-module $grM$ and let $J_F$ be the radical of $I_F$. Suppose that $F'$ is any filtration on $M$ equivalent with $F$, then one can show that $J_F = J_{F'}$ (see [Ga]). Consequently if $F$ is a good filtration on $M$ then $J_F$ does not depend on the choice of the good filtration (since all good filtrations are equivalent). We therefore may denote this ideal by $J(M)$: it is called *the characteristic ideal of* $M$.

### 2.1. $\mathcal{D}_1$*-modules with regular singularities*

To motivate Definition 2.2.1 below we first recall the classical definition of a $\mathcal{D}_1$-module with regular singularities, also called a Fuchsian $\mathcal{D}_1$-module (cf. [Ma]), where $\mathcal{D}_1$ is the ring of differential operators over the convergent power series in one variable $t$. So $\mathcal{D}_1 = \mathcal{O}[\partial]$, where $\mathcal{O} = \mathbb{C}\{t\}$ and $\partial = \frac{\mathrm{d}}{\mathrm{d}t}$.

An ordinary differential equation is called *Fuchsian* or an equation with regular singularities at 0 if it is of the form

$$\begin{aligned}&\big((t\partial)^r + a_{r-1}(t\partial)^{r-1} + \cdots + a_0\big)y = 0,\\ &\quad\text{for some } r \in \mathbb{N} \text{ and } a_i \in \mathcal{O}.\end{aligned} \tag{2.1.1}$$

Let $M$ be a left $\mathcal{D}_1$-module. Then, following [Ma], $M$ is called a *Fuchsian* $\mathcal{D}_1$-module or a $\mathcal{D}_1$*-module with regular singularities* if every $m \in M$ satisfies an equation of the form (2.1.1). One easily verifies that it is equivalent to say that for each $m \in M$ the $\mathcal{O}$-module

$$\sum_{i=0}^{\infty} \mathcal{O}(t\partial)^i m$$

is finitely generated.

Assume now that $M$ is a finitely generated (left) $\mathcal{D}_1$-module, say

$$M = \sum_{j=1}^{q} \mathcal{D}_1 m_j.$$

Let

$$M_0 = \sum_{j=1}^{q} \sum_{i=0}^{\infty} \mathcal{O}(t\partial)^i m_j.$$

Then one readily verifies that $\mathcal{D}_1 M_0 = M$ and $t\partial M_0 \subset M_0$. Furthermore, if $M$ has regular singularities then $M_0$ is a finitely generated $\mathcal{O}$-module. So if $M$ has regular singularities it contains an $\mathcal{O}$-submodule of finite type $M_0$ which generates $M$ as a $\mathcal{D}$-module and which is stable under the action of $t\partial$. It is not difficult to prove that the converse also holds. The condition $t\partial M_0 \subset M_0$ can be expressed in the following way: put

$$\mathcal{J} = \{\tau \in \operatorname{Der}_{\mathbb{C}} \mathcal{O} \mid \tau(0t) \subset \mathcal{O}t\}.$$

Then $\mathcal{J} = \mathcal{O}t\partial$. So $t\partial M_0 \subset M_0$ if and only if $\mathcal{J} M_0 \subset M_0$. Summing up

PROPOSITION 2.1.2. *Let $M$ be a finitely generated left $\mathcal{D}_1$-module. Then $M$ has regular singularities if and only if their exists an $\mathcal{O}$-submodule $M_0$ of $M$ of finite type such that $\mathcal{D} M_0 = M$ and $\mathcal{J} M_0 \subset M_0$.*

### 2.2. *Modules with regular singularities over strongly filtered rings*

Let $X$ be a complex analytic manifold and $\underline{E}$ the sheaf of micro-local differential operators on $X$. In [KO] and [KK] $\underline{E}$-modules with regular singularities were introduced and extensively studied. If $p \in T^*X \backslash T^*_X X$ then $E_p$ is a strongly filtered ring (Example 0.3.3). An $E_p$-module of finite type $M$ is said to have regular singularities along $J(M)$ if there exists a finitely generated $F_0 E_p$-submodule $M_0$ of $M$ generating $M$ as a left $F_0 E_p$-module and which is stable under the action of the elements $P \in F_1 E_p$ such that $\sigma_1(P)$ vanishes on $V(J(M))$, the characteristic variety of $M$.

Now let $R$ be a strongly filtered ring and $M$ a finitely generated left $R$-module. Generalizing the discussion above we put

$$\mathcal{J} = \mathcal{J}(M) = \{\tau \in F_1 R \mid \sigma_1(\tau) \in J(M)\}.$$

DEFINITION 2.2.1. $M$ has *regular singularities along* $J(M)$ (abbreviated R.S) if and only if there exists a finitely generated $F_0 R$-submodule $M_0$ of $M$ such that $RM_0 = M$ and $\mathcal{J} M_0 \subset M_0$.

**2.3.** *Modules with regular singularities over filtered rings*

Now let again $R$ be an arbitrary filtered ring with $grR$ commutative. We use the reduction principle to find the definition of $R$-modules with regular singularities. Therefore we need to rewrite "$M_X$ is an $R_X$-module with regular singularities along $J(M_X)$" in terms of the $R$-module $M$. The result is

PROPOSITION 2.3.1. *Suppose $\widetilde{R}$ is left noetherian. Then $M_X$ is an $R_X$-module with regular singularities along $J(M_X)$ if and only if $M$ possesses a very good filtration, i.e. a good filtration such that $AnngrM$ is a radical ideal, i.e. $AnngrM = J(M)$.*

PROOF. i) Suppose $M_X$ is an $R_X$-module with R.S. So there exists an $F_0R_X$-submodule of finite type $M_0$ of $M_X$ generating $M_X$ which is stable under $\mathcal{J}(M_X)$. So as observed in 0.3 the filtration $\mathcal{F}(M_0)$ is good on $M_X$ and hence by Proposition 0.4.5 the filtration $G := G(\mathcal{F}(M_0))$ is good on $M$. We claim that this filtration is very good. Therefore let $\sigma(r) \in J(M) \cap gr_kR$ and $m \in F_nG$, i.e. $m \in M \cap F_nR_XM_0$. We must show that $rm \in F_{n+r-1}G$, i.e. we must show that $rm \in F_{n+r-1}R_XM_0$. Obviously we may assume $v(r) = k$. Now observe that $\sigma(r) \in J(M) \subset J(M_X)$. So $X^{-(k-1)}r \in \mathcal{J}(M_X)$. Consequently $X^{-(k-1)}rM_0 \subset M_0$. Finally since $X^{-n}m \in M_0$ we get $rm \in X^{n+k-1}M_0 \subset F_{n+k-1}R_XM_0$, as desired.

ii) Conversely, let $FM$ be a very good filtration on $M$. Then $FM_X$ is a good filtration on $M_X$. So it remains to show that $AnngrM_X$ is a radical ideal. Therefore we first observe that $grR_X = grR[\overline{X}, \overline{X}^{-1}]$ the external homogenization of $grR$ (see [NO]), where $\overline{X}$ denotes the class of $X$ in $grR_X$. Then one easily verifies that

$$AnngrM_X = grR[\overline{X}, \overline{X}^{-1}]AnngrM$$

and this is a radical ideal in $grR_X$ since $AnngrM$ is a radical ideal in $grR$ ($FM$ is very good). □

DEFINITION 2.3.2. Let $M$ be a finitely generated left $R$-module. Then $M$ *has regular singularities along* $J(M)$ (abbreviated $M$ has R.S) if $M$ possesses a very good filtration.

REMARK 2.3.3. It is not difficult to show that in case $R$ is a strongly filtered ring, Definition 2.2.1 coincides with Definition 2.3.2 ([v.E2], Proposition 5.4). More precisely an $F_0R$-submodule $M_0$ of $M$ satisfies the conditions of Definition 2.2.1 if and only if the corresponding filtration $\mathcal{F}(M_0)$ on $M$ is very good.

An almost immediate consequence of the definition is that R.S is preserved under microlocalizations: let $S$ be a multiplicatively closed set such that $\sigma(S)$ is a multiplicatively closed subset of $grR$ not containing 0. Then

PROPOSITION 2.3.4. *If $M$ is an $R$-module with R.S, then $E_S(M)$ is an $E_S(R)$-module with R.S.*

PROOF. Let $FM$ be a very good filtration on $M$. Then $FE_S(M)$ is a good filtration on $E_S(M)$ (Corollary 1.4.2). So $J(E_S(M)) = r(AnngrE_S(M)) = r(\sigma(S)^{-1}AnngrM)$

(by Proposition 1.4.1) $= \sigma(S)^{-1}r(AnngrM) = \sigma(S)^{-1}AnngrM$ ($FM$ is very good) $= AnngrE_S(M)$ (by Proposition 1.4.1). So $FE_SM$ is very good, as desired. □

## 2.4. *The characteristic ideal is involutive*

Since $grR$ is commutative we have $[r,r'] := rr' - r'r \in F_{n+m-1}R$ for all $r \in F_nR$ and $r' \in F_mR$. This enables us to equip $grR$ with the structure of a Lie-ring by putting

$$\{r + F_{n-1}R, r' + F_{m-1}R\} := [r,r'] + F_{n+m-2}R.$$

So for every $n, m \in \mathbb{Z}$ we get a $\mathbb{Z}$-bilinear map $\{,\}$: $gr_nR \times gr_mR \to gr_{n+m}R$ which can be extended to a $\mathbb{Z}$-bilinear map $\{,\}$: $grR \times grR \to grR$, called the *Poisson-product.* In fact $\{,\}$ is a bi-derivation, i.e. it is a derivation in the first variable if the second is kept fixed (the same holds for the second variable). An ideal $I \subset grR$ is called *involutive* if $\{r,r'\} \in I$ for all $r, r' \in I$.

EXAMPLE 2.4.1. If $I$ is any ideal in $grR$, then $I^2$ (or more general $I^n$ for $n \geqslant 2$) is involutive since $\{ab, cd\} = a\{b,c\}d + b\{a,c\}d + a\{b,d\}c + b\{a,d\}c \in I^2$ for all $a, b, c, d \in I$.

EXAMPLE 2.4.2. Let $M$ be a finitely generated $R$-module with good filtration $FM$. Then one easily verifies that $AnngrM$ is an involutive ideal.

EXAMPLE 2.4.3. The radical of an involutive ideal is in general not involutive. For example take $R = \mathbb{C}[x, \partial]$ the first Weyl algebra with the usual $\partial$-filtration. Then $I = (x, \xi)$ is not an involutive ideal in $\mathbb{C}[x,\xi](= grR)$. So by Example 2.4.1, $I^2$ is an involutive ideal whose radical is not involutive.

The more surprising is the following result

THEOREM 2.4.4 ([Ga]). *If $grR$ is a noetherian $\mathbb{Q}$-algebra, then $J(M)$ is involutive.*

REMARK 2.4.5. Suppose $R$ is a strongly filtered ring. Using the special element $s$ it is not difficult to verify that $J(M)$ is involutive if and only if $\mathcal{J}(M)$ (see 2.2) is a Lie-algebra.

## 2.5. *Holonomic R-modules*

In this section we want to introduce holonomic $R$-modules for a large class of filtered rings $R$. Therefore we first consider the classical case that $R = \mathcal{D}_n$, the ring of differential operators over the convergent power series over $\mathbb{C}$. Let $\mathcal{I}^*$ be the set of involutive prime ideals of $gr\mathcal{D}_n$. Then it is well known that the height of each element of $\mathcal{I}^*$ is $\leqslant n$ and that sup $htp$, $p \in \mathcal{I}^*$, equals $n$.[2] A $\mathcal{D}_n$-module $M$ of finite type is called holonomic

[2] The height of a prime ideal $p$ is a commutative ring $A$ is the largest number $h$ such that there exists a chain of different prime ideals $p_0 \subset p_1 \subset \cdots \subset p_h = p$; i.e. $ht(p) = \dim(A_p)$.

if the minimal prime components of $J(M)$, which are all involutive by Theorem 2.4.4 have the maximal height $n$.

Let now $R$ be a filtered ring satisfying the following conditions

a) $grR$ is a commutative noetherian $\mathbb{Q}$-algebra.

b) all good filtrations are separable.

Furthermore we put $\mathcal{I}$ (resp. $\mathcal{I}^*$) the set of all involutive (resp. homogeneous involutive) prime ideals of $grR$ and $\nu_R = \sup htp,\ p \in \mathcal{I}$ (resp. $\mu_R = \sup htp,\ p \in \mathcal{I}^*$).

DEFINITION 2.5.1. A finitely generated $R$-module $M \neq 0$ is called *holonomic* if $htp = \mu_R$ for all minimal prime components of $J(M)$. Also $M = 0$ is called holonomic.

REMARK 2.5.2. Condition b) implies that for any finitely generated (left) $R$-module $M$ with good filtration $FM$, $grM = 0$ if and only if $M = 0$ and hence that $J(M) \neq grR$. From condition a) we conclude that $J(M)$ is involutive and hence so are all its minimal prime components. In particular $\mathcal{I}^*$ is not empty since for any finitely $R$-module $M \neq 0$ the minimal prime components of $J(M)$ belong to $\mathcal{I}^*$.

In order to have the following proposition we put one more condition on $R$:

c) good filtrations induces good filtrations on submodules.

PROPOSITION 2.5.3. *Let $0 \to M_1 \to M \to M_2 \to 0$ be an exact sequence of R-modules of finite type. Then M is holonomic if and only if $M_1$ and $M_2$ are holonomic.*

PROOF. Let $FM$ be a good filtration on $M$. Then the quotient filtration is good on $M_2$ and the induced filtration is good on $M_1$ (by c)). It follows that $J(M) = J(M_1) \cap J(M_2)$, which implies the proposition. □

REMARK 2.5.4. A filtered ring such that $grR$ is left noetherian, good filtrations induces good filtrations on submodules and good filtrations are separated, is called a *Zariskian ring* and $FR$ a *Zariskian filtration* ([Li1], Definition 3.5.5, or [LiO], Theorem 3.3). So the conditions a), b), c) above imply that $R$ is a Zariskian ring. It is proved in [Li1], Theorem 3.5.4, or [LiO], Theorem 3.3, that $R$ is a Zariskian ring if and only if $\widetilde{R}$ is left noetherian and $F_{-1}R \subset J(F_0R)$, where $J(F_0R)$ denotes the Jacobson radical of $F_0R$. A complete filtered ring is an example of a Zariskian ring, so in particular all discrete rings and all algebraic microlocalizations are examples of Zariskian rings. For more details on Zariskian rings the reader is referred to [LiO] and [Li1]. The last reference is an enormous source of facts concerning Zariskian rings.

In order to be able to use the reduction principle we need that $M$ is a holonomic $R$-module if and only if $M_X$ is a holonomic $R_X$-module. (observe that $R_X$ satisfies the conditions a) and b): a) is obvious. To see b) take any good filtration on $M$. The corresponding filtration on $M_X$ is good and separated. Since any good filtration on $M_X$ is equivalent to this good and separated filtration, condition b) is satisfied.)

PROPOSITION 2.5.5 ([v.E2], Corollary 6.10 and Proposition 6.11). *There is equivalence between*

1) *$M$ is a holonomic $R$-module if and only if $M_X$ is a holonomic $R_X$-module.*
2) $\mu_R = \nu_R$.

Therefore we put on $R$ the condition
d) $\mu_R = \nu_R$.

REMARK 2.5.6. It is shown in [v.E2], §8, that condition d) is satisfied if $R = \mathcal{D}(B)$, the ring of differential operators with the usual filtration when $B$ is either

i) the ring $\mathcal{O}_n$ of formal or convergent power series over a field of characteristic zero (a complete field in the convergent case) or

ii) $A(V)$ the coordinate ring of an irreducible nonsingular affine variety of dimension $n$.

It is shown that $\mu_{\mathcal{D}(B)} = \nu_{\mathcal{D}(B)} = gl \dim \mathcal{D}(B) = n$. Furthermore it is proved that the notion of a holonomic $\mathcal{D}(B)$-module as defined above, coincides with the one defined in the literature, i.e. $\mathrm{Ext}^v_{\mathcal{D}(B)}(M, \mathcal{D}(B)) = 0$ for all $v \neq n$.

### 2.6. *A local-global theorem for modules with regular singularities*

From now on (unless mentioned otherwise) we assume: $R$ is a filtered ring satisfying the conditions a), b), c) and d) above.

Let $p$ be a prime ideal of $grR$. Put $S_p = \{r \in R \mid \sigma(r) \notin p\}$. Then $S_p$ (resp. $\sigma(S_p)$) is a multiplicatively closed subset of $R$ (resp. $grR$) and since $grR$ is commutative $\sigma(S_p)$ satisfies the Ore conditions. Furthermore $0 \notin \sigma(S_p)$. So we can define $E_p(R) := E_{S_p}(R)$ and $E_p(M, FM) := E_{S_p}(M, FM)$ for any filtered $R$-module $M$ with filtration $FM$. If $FM$ is good on $M$ we write $E_p(M)$ instead of $E_p(M, FM)$ and by Remark 1.4.4 and Remark 2.5.4 we have $E_p(M) \simeq E_p(R) \otimes_R M$. Now we are able to formulate the main result of this paper concerning modules with regular singularities (which should be considered as the micro-local analogue of Deligne's Theorem 4.1 in [D]).

THEOREM 2.6.1 ([v.E2], Theorem 7.3). *Let $M$ be a holonomic $R$-module. Then there is equivalence between*

1) *$M$ is an $R$-module with R.S.*

2) *$E_p(M)$ is an $E_p(R)$-module with R.S for every prime ideal $p$ of $grR$.*

3) *$E_p(M)$ is an $E_p(R)$-module with R.S for every minimal prime component $p$ of $J(M)$.*

PROOF (*Sketch*). 1) $\to$ 2) follows from Proposition 2.3.4. 2) $\to$ 3) is obvious. So it remains to prove 3) $\to$ 1). By using results like $M$ has R.S if and only if $M_X$ has R.S, $E_p(M)$ has R.S if and only if $E_p(M)_X$ has R.S, and Proposition 2.5.5 and some technical lemmas, we can reduce the proof of this theorem to the case that $R$ is a strongly filtered ring (cf. [v.E2], §7, for details or [v.E4]). The remainder of Section 2 is devoted to the proof of this theorem in the strongly filtered ring case.

PROPOSITION 2.6.2. *Let $M$ be a finitely generated (left) $R$-module. Then $E_p(M) \neq 0$ if and only if $p \supset J(M)$.*

PROOF. $E_p(M) = 0$ if and only if $grE_p(M) = 0$ (since $FE_p(M)$ is separated) if and only if $\sigma(S_p)^{-1}grM = 0$ (Proposition 1.4.1) if and only if $\sigma(S_p) \cap AnngrM \neq \varnothing$ if and only if $p \not\supset J(M)$. □

REMARK 2.6.3. This proposition is the algebraic analogue of the well-known fact that the characteristic variety of a coherent sheave of $\underline{\mathcal{D}}$ modules $\underline{\mathcal{M}}$ equals the support of the sheaf of $\underline{E}$-modules $\underline{E} \otimes_{\underline{\mathcal{D}}} \underline{\mathcal{M}}$.

## 2.7. *An important involutiveness result for strongly filtered rings*

Let $R$ be a strongly filtered ring with special element $s \in F_1R \backslash F_0R$, invertible in $R$ and with $s^{-1} \in F_{-1}R$. On $grR$ we have defined the Poisson product. Since $grR \simeq gr_0R[X, X^{-1}]$ ([v.E2], Proposition 4.3) all crucial information of $grR$ is already contained in $gr_0R = F_0R/F_{-1}R$. Therefore we bring the Poisson product, which we have on $grR$, over to $gr_0R$, by putting

$$\{r_0 + F_{-1}R, r_0' + F_{-1}R\} := s[r_0, r_0'] + F_{-1}R.$$

One checks again that this Poisson product on $gr_0R$ is a bi-derivation. An ideal $I$ in $gr_0R$ is called *involutive* if $\{a, b\} \in I$ for all $a, b$ in $I$. The following lemma is not difficult to verify

LEMMA 2.7.1. *Let $p_0$ be a prime ideal of $gr_0R$. Put $p = grRp_0$.*
i) *$p$ is a homogeneous prime ideal of $grR$.*
ii) *$p$ is involutive in $grR$ if and only if $p_0$ is involutive in $gr_0R$.*
([v.E2], Proposition 4.3, Proposition 9.9 and Proposition 4.7).

As before we assume that $R$ satisfies the conditions a) and c). We don't need d) for Theorem 2.7.2 below. Since $grR = \bigoplus gr_nR$ is noetherian it follows easily that $gr_0R$ is noetherian and furthermore that $R$ and $F_0R$ are noetherian ([v.E2], Corollary 4.5). Let $M$ be a finitely generated left $R$-module and $N$ an $F_0R$ submodule of $M$. In the sequel we will need a criterion to decide if $N$ is a finitely generated $F_0R$-module. Therefore choose a good filtration $FM$ on $M$. So $F_0M$ is a finitely generated $F_0R$-module (by 0.3) and since $F_nM = s^nF_0M$ it follows that each $F_nM$ is finitely generated over $F_0R$. Since $F_0R$ is noetherian and $M = \bigcup F_nM$ we therefore get:

$N$ is finitely generated over $F_0R$ if and only if $N \subset F_{n_0}M$ for some $n_0$ if and only if

$$Q(n, N) := F_nM \cap N/F_{n-1}M \cap N = 0$$

for all $n \geqslant n_0$. Observe that $s^{-1}(F_nM \cap N) \subset F_{n-1}M \cap N$, so $Q(n, N)$ is a $gr_0R = F_0R/F_{-1}R$-module. It is straightforward to prove that the left multiplication by $s^{-1}$ induces a $gr_0R$ linear map from $Q(n+1, N)$ into $Q(n, N)$. So if we put $I(n) = Ann_{gr_0(R)}Q(n, N)$ and $J(n) = r(I(n))$ then we get ascending chains $I(1) \subset I(2) \subset \cdots \subset gr_0R$ and $J(1) \subset J(2) \subset \cdots \subset gr_0R$. Since $gr_0R$ is noetherian there exists an integer $n_0 \geqslant 0$ such that $I(n) = I(n_0)$ for all $n \geqslant n_0$ and hence $J(n) = J(n_0)$ for all $n \geqslant n_0$. So $J(n_0) = \bigcup J(n)$. We denote this ideal in $gr_0 R$ by $J$.

THEOREM 2.7.2. *$J$ is an involutive ideal in $gr_0 R$.*

This result was presented by O. Gabber during the Luminy Congress on $\mathcal{D}$-modules, July 1983 (see [Bj2] for a proof).

## 2.8. *A local-global finiteness result for modules over strongly filtered rings*

The notations are the same as in 2.7. So $N$ is an $F_0R$-submodule of $M$, which is a finitely generated $R$-module with a good filtration $FM$ and $R$ is a strongly filtered ring satisfying a), b) and c). For each prime ideal $p$ of $grR$ we define $N(p)$ to be the $F_0E_p(R)$-submodule of $E_p(M)$ generated by the elements $\varphi_p(n)$, with $n \in N$ and we put

$$Q(n, N(p)) = F_nE_p(M) \cap N(p)/F_{n-1}E_p(M) \cap N(p).$$

Now we show the following remarkable finiteness result.

THEOREM 2.8.1. *Let $M$ be a holonomic $R$-module. Then $N$ is a finitely generated $F_0R$-module if and only if $N(p)$ is a finitely generated $F_0E_p(R)$-module for each minimal prime component $p$ of $J(M)$.*

PROOF. If $n_1, \ldots, n_s$ generate $N$ as an $F_0R$-module, then $\varphi_p(n_1), \ldots, \varphi_p(n_s)$ generate $N(p)$ as an $F_0E_p(R)$-module. Conversely, suppose $N$ is not finitely generated over $F_0R$ but $N(p)$ is finitely generated over $F_0E_p(R)$ for every minimal prime component of $J(M)$. From the observations in 2.7 we deduce that $1 \notin J$. So we can choose a minimal prime component $p_0$ of $J$. Then

$$p_0 \supset J \supset J(n) \supset I(n) \quad \text{for all } n \in \mathbb{N},$$

whence $Q(n, N)_{p_0} \neq 0$ for all $n \in \mathbb{N}$. By Lemma 2.7.1 $p := grRp_0$ is a prime ideal in $grR$ and from Lemma 2.8.2 below we obtain that

$$Q(n, N(p)) \neq 0 \quad \text{for all } n \in \mathbb{N} \tag{$*$}$$

implying that $E_p(M) \neq 0$. So $p \supset J(M)$ by Proposition 2.6.2. Since $M$ is holonomic it follows that $p$ contains some minimal prime component $p'$ of $J(M)$ with $htp' = \mu_R$. So $htp \geqslant \mu_R$. However by Theorem 2.7.2 $J$ is involutive, hence so is $p_0$ and this implies that $p$ is involutive (Lemma 2.7.1), so $htp \leqslant \mu_R$. Consequently $htp = \mu_R$ implying $p = p'$. So $p$ is a minimal prime component of $J(M)$. By our hypothesis it follows that $N(p)$ is finitely generated over $F_0E_p(R)$ and hence there exists some $n_0 \in \mathbb{N}$ such that $Q(n, N(p)) = 0$ for all $n \geqslant n_0$ (arguing as in 2.7). But this contradicts $(*)$. So is $N$ finitely generated over $F_0R$, as desired. □

LEMMA 2.8.2. *Let* $\mathfrak{p}_0$ *be a prime ideal in* $gr_0R$ *and* $\mathfrak{p} = grR\mathfrak{p}_0$. *Then* $\mathfrak{p}$ *is a prime ideal in* $grR$ *and the canonical morphism of* $gr_0R$*-modules* $\mathcal{X}$: $Q(n,N) \to Q(n,N(\mathfrak{p}))$ *can be extended uniquely to an injective morphism of* $(gr_0R)_{\mathfrak{p}_0}$*-modules*

$$\widetilde{\mathcal{X}}\colon\ Q(n,N)_{\mathfrak{p}_0} \to Q\big(n,N(\mathfrak{p})\big).$$

PROOF. i) Let $r_0 + F_{-1}R \in gr_0R\backslash\mathfrak{p}_0$. Then $r_0 \in S_\mathfrak{p}$ and since $\|\varphi_\mathfrak{p}(r_0)\| = |r_0| = 1$ $\varphi_\mathfrak{p}(r_0) + F_{-1}E_\mathfrak{p}(R)$ is invertible in $gr_0E_\mathfrak{p}(R)$. Hence the canonical map $gr_0R \to gr_0E_\mathfrak{p}(R)$ extends to a ringhomomorphism $\psi$: $(gr_0R)_{\mathfrak{p}_0} \to gr_0E_\mathfrak{p}(R)$. Since by $\psi$ $Q(n,N(\mathfrak{p}))$ is a left $(gr_0R)_{\mathfrak{p}_0}$-module, $\mathcal{X}$ extends uniquely to a $(gr_0R)_{\mathfrak{p}_0}$-module homomorphism $\widetilde{\mathcal{X}}$: $Q(n,N)_{\mathfrak{p}_0} \to Q(n,N(\mathfrak{p}))$.

ii) It remains to verify that $\widetilde{\mathcal{X}}$ is injective. Let $m \in F_nM \cap N$ and suppose $\varphi_\mathfrak{p}(m) \in F_{n-1}E_\mathfrak{p}(M)$. Then $|m|_\mathfrak{p} = |\varphi_\mathfrak{p}(m)| \leqslant 2^{n-1}$. So by 1.2.2 there exists $t \in S_\mathfrak{p}$, say $v(t) = k$, with $|t|^{-1}|tm| \leqslant 2^{n-1}$. Observe that for the special element $s \in F_1R\backslash F_0R$ with $s^{-1} \in F_{-1}R$, $\sigma(s)$ is a unit in $grR$ with inverse $\sigma(s^{-1})$. In particular $\sigma(s)$ and $\sigma(s^{-1})$ do not belong to $\mathfrak{p}$, so $s$ and $s^{-1}$ belong to $S_\mathfrak{p}$. Consequently $s^r \in S_\mathfrak{p}$ and $\sigma(s^r)$ is a unit in $grR$ for all $r \in \mathbb{Z}$. Now put $\rho = s^{-k}t$. So $\rho \in S_\mathfrak{p}$. If $tm \neq 0$ then $\sigma(tm) \neq 0$ and hence $\sigma(s^{-k})\sigma(tm) \neq 0$ (since $\sigma(s^{-k})$ is a unit) whence $|s^{-k}tm| = |s^{-k}|\,|tm|$. So $|\rho|^{-1}|\rho m| = |t|^{-1}|s|^k|s|^{-k}|tm| \leqslant 2^{n-1}$, i.e. $|\rho m| \leqslant 2^{n-1}$ since $|\rho| = 1$ (obviously if $tm = 0$ then also $|\rho m| = 0 \leqslant 2^{n-1}$). So in any case we have $\sigma(\rho)\overline{m} = 0$ in $Q(n,N)$ implying that $\overline{m} = 0$ in $Q(n,N)_{\mathfrak{p}_0}$ since $\sigma(\rho) \in gr_0R\backslash\mathfrak{p}_0$. So $\widetilde{\mathcal{X}}$ is injective. □

## 2.9. *The proof of Theorem* 2.6.1 *for strongly filtered rings*

Notations as in 2.8. To prove the implication 3) → 1) of Theorem 2.6.1 we use one more characterization of modules with regular singularities based on the involutiveness of $J(M)$. Recall (2.2)

$$\mathcal{J} = \mathcal{J}(M) = \big\{r \in F_1R \mid \sigma_1(r) \in J(M)\big\}.$$

Since $F_0R$ is noetherian and $\mathcal{J} \subset F_1R = F_0Rs$, $\mathcal{J}$ is a finitely generated $F_0R$-module. Furthermore $\mathcal{J}$ is a Lie-algebra since

LEMMA 2.9.1. $J(M)$ *is involutive if and only if* $\mathcal{J}$ *is a Lie-algebra.*

PROOF. Using the special element $s$ it is not difficult to verify that the result follows from

$$\sigma_1\big([r,r']\big) = \big\{\sigma_1(r),\sigma_1(r')\big\} = \big\{\sigma(r),\sigma(r')\big\} \quad \text{if } r,r' \in F_1R\backslash F_0R.$$

□

COROLLARY 2.9.2. $M$ *has* R.S *if and only if*

$$\sum_{i=0}^{\infty} F_0R\tau^i m$$

*is a finitely generated $F_0R$-module for each $m \in M$ and each $\tau \in \mathcal{J}$.*

PROOF. i) Suppose $M$ has R.S. So there exists a finitely generated $F_0R$-submodule $M_0$ of $M$ such that $RM_0 = M$ and $\mathcal{J}M_0 \subset M_0$. Let $\tau \in \mathcal{J}$ and $m \in M$ and put

$$N = \sum_{i=0}^{\infty} F_0R\tau^i m.$$

Since $RM_0 = M$ there exists an integer $k$ such that $m \in F_kRM_0$. Furthermore since $\tau r m = r\tau m + [\tau, r]m$ for all $r \in F_kR$ and $[\tau, r] \in F_kR$ it follows that $\tau F_kRM_0 \subset F_kRM_0$. Consequently $N \subset F_kRM_0 = s^kM_0$. Since $F_0R$ is noetherian and $M_0$ is finitely generated over $F_0R$, hence so is $N$.

ii) Conversely, since $\mathcal{J}$ is finitely generated over $F_0R$ we have

$$\mathcal{J} = \sum_{i=0}^{d} F_0R\tau_i$$

for some $\tau_i \in \mathcal{J}$. Finally

$$M = \sum_{j=1}^{q} Rm_j$$

for some $m_j \in M$. By the hypothesis we can find a positive integer $k$ such that

$$\tau_i^k m_j \in \sum_{p=0}^{k-1} F_0R\tau_i^p m_j$$

for each $1 \leqslant i \leqslant d$, $1 \leqslant j \leqslant q$. Then using that $\mathcal{J}$ is a Lie-algebra one readily verifies that the $F_0R$-module generated by the elements $\tau_1^{i_1} \cdots \tau_d^{i_d} m_j$ with $0 \leqslant i_1, \ldots, i_d \leqslant k-1$ and $1 \leqslant j \leqslant q$ is stable under $\mathcal{J}$ (and of course generate $M$ as an $R$-module). □

REMARK 2.9.3. In the proof of Corollary 2.9.2 we only used that $F_0R$ is noetherian and that $grR$ is a commutative noetherian $\mathbb{Q}$-algebra.

PROOF OF THEOREM 2.6.1 (*for strongly filtered rings*). It remains to prove 3) $\to$ 1). Let $m \in M$ and $\tau \in \mathcal{J}$. By Corollary 2.9.2 it suffices to prove that $N = \Sigma F_0R\tau^i m$ is a finitely generated $F_0R$-module. We want to apply Theorem 2.8.1. So let $p$ be a minimal prime component of $J(M)$. Then $N(p) = \Sigma F_0E_p(R)\varphi_p(\tau)^i\varphi_p(m)$. Since

$$\varphi_p(\tau) \in \mathcal{J}\big(E_p(M)\big)$$

it follows from Corollary 2.9.2 and the hypothesis that $N(p)$ is a finitely generated $F_0E_p(R)$-module, whence $N$ is a finitely generated $F_0R$-module by Theorem 2.8.1.

**2.10.** *Modules with regular singularities and short exact sequences*

Let $R$ be a filtered ring satisfying the conditions a), b), c) and d). Let $M$ be an $R$-module with R.S. So it possesses a very good filtration. Let $M'$ be an $R$-submodule of $M$. Then the induced filtration and the quotient filtration are again good, however they need not be very good. Nevertheless we have

THEOREM 2.10.1. *Let $0 \to M' \to M \to M'' \to 0$ be an exact sequence of holonomic $R$-modules. Then $M$ has R.S if and only if $M'$ and $M''$ have R.S.*

PROOF. i) We use the reduction principle. The sequence

$$0 \to M'_X \to M_X \to M''_X \to 0$$

is an exact sequence of holonomic $R_X$-modules (Proposition 2.5.5). So by Proposition 2.3.1 and Definition 2.3.2 we may assume that $R$ is a strongly filtered ring.

ii) Assume first that $M$ has R.S. We show that $M'$ has R.S by showing that $E_p(M')$ has R.S for every minimal prime component of $J(M')$ (Theorem 2.6.1). So let $p$ be a minimal prime component of $J(M')$. Then $p$ is also a minimal prime component of $J(M)$ (since each minimal prime component of $J(M')$ and each of $J(M)$ has height $\mu_R$). Using that microlocalization is exact we obtain the exact sequence

$$0 \to E_p(M') \to E_p(M) \to E_p(M'') \to 0.$$

If $p$ is not a minimal prime component of $J(M'')$ then $E_p(M'') = 0$ (by Proposition 2.6.2) hence $E_p(M') \simeq E_p(M)$, so $E_p(M')$ has R.S since $E_p(M)$ has R.S by Proposition 2.3.4. So we may assume that $p$ is also a minimal prime component of $J(M'')$. Now observe that

$$J(E_p(M)) = \sigma(S_p)^{-1}J(M) = \sigma(S_p)^{-1}p$$

and similarly

$$J(E_p(M')) = J(E_p(M'')) = \sigma(S_p)^{-1}p.$$

So

$$\mathcal{J}(E_p(M)) = \mathcal{J}(E_p(M')) = \mathcal{J}(E_p(M'')).$$

Then the result follows easily from Corollary 2.9.2 and Lemma 2.10.2 below. The proof that $M''$ has R.S is similar. Finally if both $M'$ and $M''$ have R.S then by arguing in a similar way we can show that $E_p(M)$ has R.S along $J(E_p(M))$ for every minimal prime component of $J(M)$. Then apply Theorem 2.6.1.

LEMMA 2.10.2. *Let $R$ be a strongly filtered ring such that $F_0R$ is noetherian and $\tau \in F_1R$. Let*

$$0 \to M' \to M \to M'' \to 0$$

*be an exact sequence of R-modules. Then $\Sigma F_0 R\tau^i m$ is a finitely generated $F_0R$-module for all $m \in M$ if and only if $\Sigma F_0 R\tau^i m'$ and $\Sigma F_0 R\tau^i m''$ are finitely generated $F_0R$-modules for all $m' \in M'$ and $m'' \in M''$.*

The proof of this elementary lemma is left to the reader.

**2.11.** *How to construct very good filtrations from good filtrations?*

The main result of this section (Theorem 2.11.1) gives a new characterization of $R$-modules with R.S which enable us to construct very good filtrations from generators of the $R$-module $M$ and generators of its characteristic ideal. In fact the result follows from Corollary 2.9.2 using the reduction principle.

In this section we only assume: $R$ is a filtered ring such that $\widetilde{R}$ is noetherian and $grR$ is a commutative $\mathbb{Q}$-algebra (it follows from $\widetilde{R}$ noetherian that $grR$ is noetherian too). It follows that $F_0R_X$ is noetherian and that $grR_X$ is a commutative noetherian $\mathbb{Q}$-algebra.

THEOREM 2.11.1 ([v.E3], Theorem 2). *There is equivalence between*

1) *$M$ has R.S.*

2) *For every $m \in M$ and every $D \in R$ with $\sigma(D) \in J(M)$ there exists a positive integer $\rho \in \mathbb{N}$ such that*

$$D^\rho m \in \sum_{i=0}^{\rho-1} F_{(\rho-i)(r-1)} R D^i m, \quad \text{where } r = v(D). \tag{2.11.2}$$

*More precisely, let $m_1, \dots, m_\ell$ generate the R-module $M$ and $\sigma(D_1), \dots, \sigma(D_q)$ the ideal $J(M)$. If 2) is satisfied then there exists $\rho \in \mathbb{N}$ such that (2.11.2) holds for all $D_j$, all $m_t$ and the filtration $FM$ on $M$ defined by*

$$F_n M = \sum_{i=1}^{\ell} \sum_{j=1}^{q} \sum_{i_j=0}^{\rho-1} F_{n-(i_1(r_1-1)+\cdots+i_q(r_q-1))} R D_1^{i_1} \cdots D_q^{i_q} m_t$$

*is a very good filtration on $M$.*

PROOF. 1) $\to$ 2). If $M$ has R.S then $M_X$ has R.S. If $D \in R$ with $\sigma(D) \in J(M)$, say $v(D) = r$, then

$$\tau := X^{-(r-1)} D \in \mathcal{J}(M_X).$$

So if $m \in M$ then by Corollary 2.9.2 there exists an integer $\rho \in \mathbb{N}$ with

$$\tau^\rho m \in \sum_{i=0}^{\rho-1} F_0 R_X \tau^i m.$$

Multiply this equation with $X^{\rho(r-1)}$. Then

$$D^\rho m = X^{\rho(r-1)}\tau^\rho m \in \sum_{i=0}^{\rho-1} X^{(\rho-i)(r-1)} F_0 R_X D^i m \cap M$$

which implies formula (2.11.2).

2) $\to$ 1) (sketch). Since $J(M) = (\sigma(D_1), \ldots, \sigma(D_q))$ we get

$$J(M_X) = \big(\sigma(\tau_1), \ldots, \sigma(\tau_q)\big)$$

with $\tau_j = X^{-(r_j-1)} D_j$ and $r_j = v(D_j)$. Hence

$$\mathcal{J}\big(M_X\big) = F_0 R_X \tau_1 + \cdots + F_0 R_X \tau_q + F_0 R_X.$$

Furthermore

$$M_X = \sum_{t=1}^{\ell} R_X m_t.$$

Then from the proof of Corollary 2.9.2 we know that

$$M_0 := \sum_{t=1}^{\ell} \sum_{j=1}^{q} \sum_{i_j=0}^{\rho-1} F_0 R_X \tau_1^{i_1} \cdots \tau_q^{i_q} m_t$$

is stable under $\mathcal{J}(M_X)$ and hence the corresponding filtration $(F_n R_X M_0)_{n\in\mathbb{Z}}$ is very good in $M_X$ (Remark 2.3.3). Then one readily verifies that the filtration $(M \cap F_n R_X M_0)_{n\in\mathbb{Z}}$ on $M$ has the form described above and is very good. □

## 2.12. *An application: Modules with regular singularities on a curve*

Let $\mathcal{O} = k\langle t\rangle$ be the ring of formal or convergent power series in one variable $t$ over a field $k$ of characteristic zero, $K$ its quotient field on $A$ a subring of $\mathcal{O}$ such that $\dim_k \mathcal{O}/A$ is finite. By $\mathcal{D}(K)$ we denote the set of meromorphic differential operators of the form $\Sigma a_i \partial^i$ with $a_i \in K$ and by $\mathcal{D}(A)$ the set of $D \in \mathcal{D}(K)$ satisfying $D(A) \subset A$. It is shown in [St Sm] that $\mathcal{D}(A)$ and $\mathcal{D} = \mathcal{D}(\mathcal{O})$ are Morita equivalent. The equivalence is given by the functor $P \otimes_{\mathcal{D}}$-, where $P$ is the set of $D \in \mathcal{D}(K)$ satisfying $D(\mathcal{O}) \subset A$. Using Theorem 2.11.1 it is shown in [v.E5] that under the Morita equivalence holonomic $\mathcal{D}$-modules with R.S correspond with holonomic $\mathcal{D}(A)$-modules with R.S. Furthermore the following remarkable theorem is proved

THEOREM 2.12.1 ([v.E5], Theorem 3.10). *Let $N$ be a finite $\mathcal{D}(A)$-module and $D$ in $\mathcal{D}(A)$ satisfying $\sigma(D) = t\xi^n$ where $n = \operatorname{ord} D$. Then $N$ is a holonomic $\mathcal{D}(A)$-module with R.S if and only if $N$ is $D$-regular (i.e. each element $m \in N$ satisfies an equation of the form* (2.11.2)).

## 3. Final remarks

REMARK 3.1. In [Gi] algebraic microlocalization is used to give a new proof of the Gabber–Kashiwara theorem ([Gi], Theorem V8). In fact the key point of that proof is to show that the well-known Kashiwara filtration $D_0M \subset D_1M \subset \cdots \subset D_dM = M$, $d = d(M)$ is compatible with microlocalization, i.e. $E_S(R) \otimes_R D_jM \simeq D_j(E_S(R) \otimes_R M)$ ([Gi], Proposition V9). For more details on pure modules we refer to [EkH, Bj3, Ek1, Ek2].

REMARK 3.2. It was shown in [C] that if $R$ is a separated filtered ring such that $grR$ is an Ore domain then $R$ can be embedded in a skew field. More generally one can show: if $R$ is a separated filtered ring and $S$ a multiplicatively closed subset of $R$, not containing 0, such that the elements of $\sigma(S)$ form a regular Ore set in $grR$, then the map $\varphi_R$: $R \to E_S(R)$ is injective. Furthermore, if $grR$ is a domain and $\sigma(R\backslash\{0\})$ is an Ore set in $grR$, then $E_{R\backslash\{0\}}(R)$ is a skew field. It has been shown in [Li2] that in case $L$ is any Lie-algebra then the skew field $D$ constructed in [C] (in which $U(L)$ can be embedded) is isomorphic to $E_{U(L)\backslash\{0\}}(U(L))$.

REMARK 3.3. Historically the first paper containing microlocalization goes back to I. Schur in 1905 [Schu]: let $K$ be the algebraic closure of $\mathbb{C}((X))$, the field of formal Laurent series (so $K = \bigcup_{p\geqslant 1} \mathbb{C}((X^{1/p})))$ and let $R = K[\partial]$, where $\partial = \frac{\mathrm{d}}{\mathrm{d}x}$. Then Schur constructs the algebraic microlocalization $E_S(R)$, where $S = \{\partial^n \mid n \in \mathbb{N}\}$ in $R$. Using this ring he shows that if $P \in R$ then any two elements $Q_1, Q_2 \in R$ which commute with $P$ commute with each other, i.e. $[Q_1, Q_2] = 0$.

## References


[ABO] J. Asensio, M.v.d. Bergh and F.v. Oystaeyen, *A new approach to microlocalization of filtered rings*, Trans. Amer. Math. Soc. **316**(2) (1989), 537–553.

[Be1] I.N. Bernstein, *Modules over rings of differential operators. Study of fundamental solutions of equations with constant coefficients*, Funct. Anal. Appl. **5** (1971), 89–101.

[Be2] I.N. Bernstein, *The analytic continuation of generalized functions with respect to a parameter*, Funct. Anal. Appl. **6** (1972), 273–285.

[BeiBe] A.A. Beilinson and I.N. Bernstein, *Localisation de g-modules*, C. R. Acad. Sci. Paris **292** (1981), 15–18.

[Bj1] J.-E. Björk, *Rings of Differential Operators*, North-Holland, Amsterdam (1979).

[Bj2] J.-E. Björk, Notes distributed at the Autumn School Hamburg Univ. Oct. 1–7 (1984).

[Bj3] J.-E. Björk, *The Auslander condition on noetherian rings*, Sém. Dubreil–Malliavin, SLNM 1404, Springer, Berlin (1987–88), 137–173.

[Bo] A. Borel et al., *Algebraic $\mathcal{D}$-modules*, Perspectives in Math. vol. 2, Academic Press, New York (1987).

[Br1] J.-L. Brylinski, *Differential operators on the flag manifold, Tableaux de Young et foncteurs de Schur en algèbre et géométrie*, Astérisque **87–88** (1981), 43–60.

[Br2] J.-L. Brylinski, *Modules holonomes à singularités régulières et filtration de Hodge, I*, Algebraic Geometry (Proc. La Rábida), SLNM 961, Springer, Berlin (1982), 1–21.

[Br3] J.-L. Brylinski, *Modules holonomes à singularités régulières et filtration de Hodge, II*, Analyse et Topologie sur les Espaces Singuliers (II–III), Astérisque **101–102** (1983), 75–117.

[BrK] J.-L. Brylinski and M. Kashiwara, *Kazhdan–Lusztig conjecture and holonomic systems*, Invent. Math. **64** (1981), 387–410.

[C] P.M. Cohn, *On the embedding of rings in skew fields*, Proc. London Math. Soc. (3) **11** (1961), 511–530.

[D] P. Deligne, *Equations Différentielles à Points Singuliers Réguliers*, SLNM 163, Springer, Berlin (1970).

[v.D] R. van Doorn, *Classification of regular holonomic $\mathcal{D}$-modules*, Thesis, Catholic Univ., Nijmegen (1987).

[EkH] E.K. Ekström and Ho Dinh Duan, *Purity and equidimensionality of modules for a commutative Noetherian ring with finite dimension*, Report Univ. Stockholm 18 (1986).

[Ek1] E.K. Ekström, *Pure modules over Auslander–Gorenstein rings*, Preprint.

[Ek2] E.K. Ekström, *The Auslander condition on graded and filtered Noetherian rings*, Sém. d'Algèbre, Dubreil–Malliavin, SLNM 1404, Springer, Berlin (1987–88), 220–245.

[v.E1] A. van den Essen, *Algebraic micro-localization*, Comm. Algebra **14** (1986), 971–1000.

[v.E2] A. van den Essen, *Modules with regular singularities over filtered rings*, Publ. Res. Inst. Math. Sci. **22** (1986), 849–887.

[v.E3] A. van den Essen, *Une construction de filtrations réduites pour les modules réguliers sur un anneau filtré*, C. R. Acad. Sci. Paris **303** (1986), 741–743.

[v.E4] A. van den Essen, *Modules with regular singularities over filtered rings and algebraic micro-localization*, Sém. Dubreil–Malliavin, SLNM 1296, Springer, Berlin (1986), 125–157.

[v.E5] A. van den Essen, *Modules with regular singularities on a curve*, J. London Math. Soc. (2) **40** (1989), 193–205.

[Ga] O. Gabber, *The integrability of characteristic varieties*, Amer. J. Math. **103** (1981), 445–468.

[Gi] V. Ginsburg, *Characteristic varieties and vanishing cycles*, Invent. Math. **84** (1986), 327–402.

[KK] M. Kashiwara and T. Kawai, *On holonomic systems with regular singularities, III*, Publ. Res. Inst. Math. Sci. **17** (1981), 813–879.

[KO] M. Kashiwara and T. Oshima, *Systems of differential equations with regular singularities and their boundary value problems*, Ann. Math. **106** (1977), 145–200.

[La] G. Laumon, *Transformations canoniques et spécialisations pour les D-modules filtrés*, Systèms Différentiels et Singularités, Astérisque **130** (1985), 56–129.

[LeMe] D.T. Lê and Z. Mebkhout, *Introduction to linear differential systems, singularities*, Proc. of Symposia in Pure Math. vol. 40, part 2, Amer. Math. Soc., Providence, RI (1983), 31–63.

[Li1] Li HuiShi, *Non-commutative Zariskian rings*, Thesis, Univ. Antwerpen (1989).

[Li2] Li HuiShi, *Note on microlocalizations of filtered rings and the embedding of rings in skew fields*, Preprint (1990).

[LiO] Li HuiShi and F.v. Oystaeyen, *Zariskian filtrations*, Comm. Algebra **17**(12) (1989), 2945–2970.

[LiOWK] Li HuiShi, F.v. Oystaeyen and E. Wexler-Kreindler, *Microlocalisation, platitude et théorie de torsion*, Comm. Algebra **16** (1988), 1813–1852.

[Ma] Y. Manin, *Moduli Fuchsiani*, Ann. Scuola Norm. Pisa (3) (1965), 113–126.

[Me1] Z. Mebkhout, *Une équivalence de catégories*, Compositio Math. **51** (1984), 51–62.

[Me2] Z. Mebkhout, *Une autre équivalence de catégories*, Compositio Math. **51** (1984), 63–88.

[Me3] Z. Mebkhout, *Le formalisme des six opérations de Grothendieck pour les coefficients de de Rham*, Sém. de Plans-sur-Bex, march 1984, Travaux en Cours, Hermann, Paris (1986).

[NO] C.V. Năstàsescu and F.v. Oystaeyen, *Graded Ring Theory*, North-Holland Publ. Comp. 28.

[Od] T. Oda, *Introduction to algebraic analysis on complex manifolds, Algebraic varieties and analytic varieties*, Adv. Studies in Pure Math. vol. 1, North-Holland, Amsterdam (1983), 29–48.

[P] F. Pham, *Singularités des systèmes différentiels de Gauss–Manin*, Prog. in Math. vol. 2, Birkhäuser, Basel (1979).

[Sai1] M. Saito, *Gauss–Manin systems and mixed Hodge structure*, Proc. Japan Acad., Ser. A **58**(1) (1982), 29–32.

[Sai2] M. Saito, *Hodge filtrations on Gauss–Manin systems, I*, J. Fac. Sci. Univ. Tokyo, Sect. IA **30**, 489–498.

[Sai3] M. Saito, *On the derived category of mixed Hodge modules*, Proc. Japan Acad., Ser. A (9) **62** (1986), 364–366.

[SKK] M. Sato, M. Kashiwara and T. Kawai, *Microfunctions and pseudodifferential equations, Hyperfunctions and pseudodifferential equations*, SLNM 287, Springer, Berlin (1973), 265–529.

[Sch] P. Schapira, *Microdifferential Systems in the Complex Domain*, Grundlehren der Math. Wissenschaften 269, Springer, Berlin (1985).

[Schu] I. Schur, *Über vertauschbare lineare Differentialausdrücke*, Sitzungsber. Berliner Math. Gesellsch. (1905).

[Sp] T.A. Springer, *Micro-localisation algébrique*, Sém. d'Algèbre Dubreil–Malliavin, SLNM 1146, Springer, Berlin (1985).

[StSm] J.T. Stafford and S.P. Smith, *Differential operators on an affine curve*, Proc. London Math. Soc. (3) **56** (1988), 229–259.

[Ste] B. Stenström, *Rings of Quotients*, Springer, Berlin (1975).

[WK] E. Wexler-Kreindler, *Microlocalisation, platitude et théorie de torsion*, Comm. Algebra **16** (1988), 1813–1852.

# Frobenius Algebras

Kunio Yamagata

*Institute of Mathematics, University of Tsukuba, Tsukuba, Ibaraki 305, Japan*
*e-mail: yamagata@sakura.cc.tsukuba.ac.jp*

*Contents*



HANDBOOK OF ALGEBRA, VOL. 1
Edited by M. Hazewinkel

All algebras in this article are finite dimensional associative algebras over a field $k$, unless otherwise stated. In 1903, Frobenius [F03] studied algebras for which the left and the right regular representations are equivalent, and gave a necessary and sufficient condition for this equivalence. As a natural generalization of group algebras, Brauer and Nesbitt [BNe37, Ne38] pointed out the importance of the algebras studied by Frobenius and named them Frobenius algebras. They gave a simple characterization of Frobenius algebras in the following form: an algebra $A$ is Frobenius if and only if there is a $k$-linear map $A \to k$ whose kernel contains no nonzero one sided ideals. Nakayama [N3941] defined quasi-Frobenius algebras, and studied the structure of Frobenius and of quasi-Frobenius algebras. Moreover, in his paper [N58] he conjectured that a $k$-algebra $A$ is quasi-Frobenius if it has the infinite dominant dimension, namely there is an infinite exact sequence

$$0 \to A \to I_1 \to I_2 \to \cdots$$

where all $I_n$ are injective and projective modules over the enveloping algebra

$$A^e := A \bigotimes_k A^{op}.$$

Generalizations of quasi-Frobenius algebras were proposed by Thrall [Th48] by introducing the three kinds of algebras called QF-1, QF-2 and QF-3 algebras, and Morita studied them extensively [Mo58a, Mo58b]. Among them, QF-3 algebras were studied deeply in connection with the above conjecture by Nakayama. Tachikawa [Ta64] considered another definition of dominant dimension by substituting a given algebra $A$ for the enveloping algebra $A^e$ in the Nakayama's definition. However, B. Müller [Mu68a] pointed out that the two dimensions are the same, and proved that the conjecture is equivalent to say that a generator-cogenerator $M$ over an algebra $A$ is projective when $\mathrm{Ext}^i_A(M, M) = 0$ for all $i > 0$. QF-1 algebras seem to be more difficult algebras. The QF-1 algebras with squared zero radical or with faithful serial modules are characterized ideal-theoretically by Ringel [Rin73] and Makino [Ma91]. But the problem posed by Thrall to characterize the QF-1 algebras ideal-theoretically is not yet solved completely, and, because of the difficulty probably, the study now seems to be not so popular. An important class of QF-2 algebras, surprisingly, appeared in the representation theory of vector space categories which were introduced by Nazarova and Rojter to prove the second Brauer–Thrall conjecture. This was discovered and developed by Simson [Si85a, Si85b]. On the other hand, generalizations to Artinian or Noetherian, or general rings without chain conditions were one of the main themes in ring theory in the 1960's and in the early 1970's. See [Os66a] and [F76]. Results before 1973 on the generalizations by Thrall and on the Nakayama conjecture were propagated by Tachikawa in his lecture notes [Ta73]. In particular, he proposed two conjectures by dividing Müller's restatement above of the Nakayama conjecture. Another homological conjecture was proposed by Auslander and Reiten [AR75b], called the generalized Nakayama conjecture, which is apparently stronger than the Nakayama conjecture. Those conjectures are implied by the finitistic dimension conjecture. In the representation theory of algebras, Riedtmann's work, starting with her paper [Rie80a], put quasi-Frobenius algebras on the stage as a main theme in the representation theory of algebras, and the quasi-Frobenius algebras of finite representation type were classified. In the representation theory of finite groups, the defect groups of blocks were determined for tame representation type (including finite representation type) by Higman [Hi54], Bondarenko and Drozd [BD77] and Brenner [B70]. The work of Dade, Janusz and Kupisch enabled us to treat the group algebras of finite representation type as algebras independent of the given groups [G73]. In fact, Gabriel and Riedtmann [GR79] classified by quivers and relations the symmetric algebras with stable Auslander–Reiten quivers of the form $\mathbb{Z}A_m/n$ for arbitrary integers $m$ and $n$, and showed that the blocks of finite representation type belong to the class of these symmetric algebras. The shape of stable Auslander–Reiten components of blocks was studied mainly by Webb [We82], using the ideas of Riedtmann [Rie80a] and Happel, Preiser and Ringel [HPR80]. By making use of Webb's theorem and methods of the representation theory of algebras, Erdmann [Er90] described by quivers and relations some symmetric algebras including all blocks of tame infinite representation type. Moreover she characterized stable Auslander–Reiten components of wild blocks [Er94].

The aim of this article is to survey recent developments in quasi-Frobenius algebras over a field, but we devote ourselves to the general ring theory of these algebras. For developments in the representation theory, we refer to survey papers by Gabriel [G79] for algebras of finite representation type, Skowroński [Sk90] for algebras of infinite representation type, and lecture notes by Erdmann [Er90] and Benson [Be84] for group algebras.

This article consists of three sections. In the first preparatory section, we recall Morita theory and some homological definitions for modules. The second section includes the original definition of (quasi-)Frobenius algebras and classical structure theory. We emphasize the Nakayama automorphisms to clarify the difference between Frobenius algebras and symmetric algebras. Trivial extension algebras are known as an important class of quasi-Frobenius algebras. The study of the trivial extensions of hereditary algebras of finite representation type was proposed by Tachikawa, and the Auslander–Reiten quivers were studied by the author [Y81a82] for the trivial extensions of arbitrary hereditary algebras of finite or infinite representation type. These ideas were used by Simson and Skowroński [SiSk81, Sk82] in the study of some classes of QF-3 algebras and their representations. They were developed further by Hughes and Waschbüsch [HW83] where they introduced repetitive algebras (canonical Galois coverings of trivial extension algebras), playing an important role in the representation theory of quasi-Frobenius algebras (e.g., [H87, ASk88, Sk89, PSk91, ErSk92, SkY2]). Happel studied the relation between the category of modules over the repetitive algebra of an algebra $A$ and the derived category of bounded complexes of $A$-modules, and gave a necessary and sufficient condition on the derived category for the global dimension of $A$ to be finite [H87, H89]. Moreover Morita theory for derived categories was established by Rickard [Ric89a, Ric89b, Ric91] and applied to quasi-Frobenius algebras. Happel's theorem on derived categories and Rickard's theorem on quasi-Frobenius algebras are introduced in this section, as well as the main results in [Y81a82]. The Riedtmann's classification theorem of quasi-Frobenius algebras of finite representation type is given [Rie80a]. As an example of quasi-Frobenius algebras appearing in other branches of mathematics, we shall introduce a theorem of

Mikio Sato concerning a characterization of Frobenius algebras in terms of the associated prehomogeneous vector spaces.

The third section is devoted to generalizations of quasi-Frobenius algebras by Thrall. The Nakayama conjecture and some related conjectures are put here, because they are well stated using QF-3 algebras. This section has two main subjects. One is the Nakayama conjecture, but the definition of the dominant dimension is slightly changed. Namely, an algebra $A$ is said to be of dominant dimension $n$ ($\geqslant 0$) when, for a minimal injective resolution of $A$ as a left $A$-module

$$0 \to A \to I_0 \to I_1 \to \cdots \to I_{n-1} \to I_n \to \cdots,$$

$I_i$ is *nonzero* projective for $0 \leqslant i < n$ and $I_n$ is zero or nonprojective. Then a quasi-Frobenius algebra has the dominant dimension one, and the Nakayama conjecture means that *every algebra has finite dominant dimension.* Moreover we can state a finitistic dimension conjecture for dominant dimensions of algebras with a given number of simple modules. First we shall show a characterization by Morita [Mo58a] of the endomorphism ring of a generator-cogenerator and, via Müller's theorems on dominant dimension [Mu68a], we shall show some relations between the Nakayama conjecture and the other conjectures stated above. Ideas or sketches of the proofs will be given to all results related to those conjectures. This section is based on my talk at Bielefeld in 1989, and it may serve as a selfcontained and shorter introduction to both research workers and graduate students interesting in the Nakayama conjecture. The second aim is to introduce the work of Ringel and Makino on QF-1 algebras. Usually it is not easy to know whether a given module has the double centralizer property, but there is an important criterion by Morita [Mo58b] which will be given with a sketch of the proof.

In some examples, we shall consider algebras defined by quivers and relations. See Gabriel [G80] or Ringel [Rin84] for the definition of quivers with relations.

The author wishes to express his appreciation to H. Tachikawa for his encouragement and suggestions, to D. Simson and A. Skowroński for reading the manuscript and their helpful suggestions, to R. Makino for valuable discussions; and to J. Rickard for his permission to include his unpublished result.

## 1. Preliminaries (Morita theory)

Throughout this article all rings are finite dimensional associative algebras with unit element over a (commutative) field $k$, unless otherwise stated. We denote by $\operatorname{Mod} A$ the category of left modules over a ring $A$ and by $\operatorname{mod} A$ the category of finitely generated left $A$-modules. By $M^n$ we denote the direct sum of $n$ copies of a module $M$. For a module $M$, by $\operatorname{Add}(M)$ (resp. $\operatorname{add}(M)$) we denote the family of modules isomorphic to direct summands of the direct sums of copies (resp. finitely many copies) of $M$. Morphisms of a module operate from the left on the module, and for a ring $A$ the opposite ring is denoted by $A^{op}$. Usually, by a module we understand a *finitely generated left* module.

Let $A$ be an algebra and $M$ a nonzero (finitely generated) left $A$-module. The Jacobson radical of $M$ is denoted by $\operatorname{rad} M$. By $\operatorname{soc} M$ (the *socle of $M$*) we denote the sum of all simple submodules of $M$ and, by $\operatorname{top} M$ (the *top of* $M$) the factor module $M/\operatorname{rad} M$. In particular, $\operatorname{soc}({}_AA)$ is called the left socle of the ring $A$, and $\operatorname{soc}(A_A)$ the right socle. An idempotent $e_1$ of $A$ is *isomorphic to* $e_2$ when $Ae_1$ is isomorphic to $Ae_2$ ($Ae_1 \simeq Ae_2$), or equivalently $e_1A \simeq e_2A$. By $I(M)$ or an embedding $M \hookrightarrow I(M)$ we denote the *injective hull* (envelope) of $M$. Since $A$ is finite dimensional over $k$, $I(M)$ is the module of minimal composition length among the injective modules having submodules isomorphic to $M$. The *projective cover* of $M$ is denoted by $P(M)$ or by an epimorphism $p$: $P(M) \to M$, that is, $P(M)$ has a minimal composition length among the projective modules having factor modules isomorphic to $M$. $I(M)$ and $P(M)$ are uniquely determined by $M$ up to isomorphism. Let $\{S_i\}_{1\leqslant i\leqslant n}$ be a set of all nonisomorphic simple $A$-modules. Then the sets $\{P(S_i)\}_i$ and $\{I(S_i)\}_i$ give all nonisomorphic indecomposable projective $A$-modules and nonisomorphic indecomposable injective $A$-modules, respectively. Clearly, $\operatorname{top} P(S_i) \simeq S_i \simeq \operatorname{soc} I(S_i)$, and

$$I(S_i) \simeq \operatorname{Hom}\left(\operatorname{Hom}_A\left(P(S_i), A\right), k\right).$$

A module $M$ is *faithful* if $aM \neq 0$ for any $0 \neq a \in A$. This is equivalent to say that the canonical ring morphism $\varphi$: $A \to \operatorname{End}_{\mathbb{Z}}(M)$ with $(\varphi(a))(x) = ax$ for $a \in A$, $x \in M$, is monomorphic, where $\operatorname{End}_{\mathbb{Z}}(M)$ denotes the endomorphism ring of $M$ as an additive group. For a finitely generated faithful left $A$-module $M$, since the right $A$-module $\operatorname{Hom}(M,k)$ is faithful, there is a monomorphism $A \xrightarrow{u} \operatorname{Hom}(M,k)^n$ for some integer $n$, and we have an epimorphism $\operatorname{Hom}(u,k)$: $M^n \to \operatorname{Hom}(A,k)$. An $A$-module $M$ is called a *generator* if there is an epimorphism $M^n \to A$ for some $n$, or equivalently, $M^n \simeq A \oplus X$ as left $A$-modules for some $A$-module $X$. A module $M$ is a *cogenerator* if for any finitely generated injective module $X$ there is a monomorphism $X \to M^n$ for some $n$. Any generator and any cogenerator is faithful. The endomorphism ring of an indecomposable module is a local ring (i.e. it has a unique maximal left or right ideal). The Krull–Schmidt theorem holds for indecomposable decompositions of finitely generated modules.

For a finitely generated module $M$ over an algebra $A$, the kernel of the projective cover of $M$ is denoted by $\Omega_A(M)$ or $\Omega(M)$ simply, and $\Omega_A^-(M)$ or $\Omega^-(M)$ denotes the cokernel of an injective hull $I(M)$;

$$0 \to \Omega(M) \to P(M) \to M \to 0,$$

$$0 \to M \to I(M) \to \Omega^-(M) \to 0.$$

Let $\Omega^0(M) = M$, $\Omega^1(M) = \Omega(M)$, $\Omega^{-1}(M) = \Omega^-(M)$. Inductively we define

$$\Omega^{n+1}(M) = \Omega\big(\Omega^n(M)\big) \quad \text{and} \quad \Omega^{-n-1}(M) = \Omega^-\left(\Omega^{-n}(M)\right)$$

for all non-negative integers $n$. The operation $\Omega_A$ is called the *Heller function* and plays an important role in the category $\operatorname{mod} A$ over a *self-injective* algebra $A$. Here, an arbitrary ring (associative with 1) $A$ is said to be *left self-injective* when ${}_AA$ is injective. The self-injective for an algebra is left-right symmetric (Theorem 2.2.1). Now assume that $A$ is a left self-injective algebra and let

$$0 \to Y \xrightarrow{u} P \xrightarrow{v} X \to 0$$

be a short exact sequence in $\operatorname{mod} A$. Then $X$ *is nonprojective indecomposable and* $v$ *is a projective cover if and only if* $Y$ *is noninjective indecomposable and* $u$ *is an injective hull* (Heller [He61]).

We say that two rings $A$ and $B$ are *Morita equivalent* when there is a category isomorphism between $\operatorname{Mod} A$ and $\operatorname{Mod} B$, denoted by $\operatorname{Mod} A \approx \operatorname{Mod} B$.

MORITA EQUIVALENCE THEOREM.

(1) *Let* $P$ *be a left module over an algebra* $A$ *and* $B = \operatorname{End}_A(P)^{op}$. *Then* (i) $P_B$ *is a generator if* ${}_AP$ *is finitely generated projective*; $P_B$ *is finitely generated projective and* $A = \operatorname{End}_B(P)$ *if* ${}_AP$ *is a generator.* (ii) *If* ${}_AP$ *is a finitely generated projective generator, then so is* $P_B$ *and* $A \simeq \operatorname{End}_B(P)$ *canonically.*

(2) *Two algebras* $A$ *and* $B$ *are Morita equivalent if and only if there is a finitely generated projective generator* ${}_AP$ *and* $B \simeq \operatorname{End}_A(P)^{op}$. *In this case, the equivalence functor* $S$: $\operatorname{Mod} A \to \operatorname{Mod} B$ *is isomorphic to*

$$\operatorname{Hom}_A(P, -) \simeq \operatorname{Hom}_A(P, A) \otimes_B -.$$

A typical example is given by basic algebras. Let

$$1 = \sum_{i=1}^{m} e_i$$

be the sum of orthogonal primitive idempotents of $A$ and let $e$ be a sum of all nonisomorphic idempotents from $\{e_i\}$. Then the algebra $A$ is Morita equivalent to $eAe$; the projective generator $Ae$ defines the equivalence. The idempotent $e$ is called a *basic idempotent* and the algebra $eAe$ is said to be a basic subalgebra of $A$, which is uniquely determined up to algebra isomorphism. In case $e = 1_A$, the algebra $A$ is called a *basic algebra.* A *basic module* $M$ is by definition a direct sum of nonisomorphic indecomposable submodules. A property of modules is said to be *Morita invariant* if it is preserved by any category equivalence functor.

EXAMPLE. For an algebra $A$, every matrix ring $(A)_n$ is Morita equivalent to $A$. (Take ${}_AM = {}_AA^n$.) More generally, let $M$ and $N$ be $A$-modules such that $M$ is *similar* to $N$, i.e. $\operatorname{add}(M) = \operatorname{add}(N)$. Then

$$B := \operatorname{End}(M)^{op} \quad \text{and} \quad C := \operatorname{End}(N)^{op}$$

are Morita equivalent. In this case, under the equivalence $\operatorname{Mod} B^{op} \approx \operatorname{Mod} C^{op}$, the right $B$-module $M$ corresponds to the right $C$-module $N$. Thus we have that $\operatorname{End}_B(M) \simeq \operatorname{End}_C(N)$ as algebras.

MORITA DUALITY THEOREM.

(1) *Let $Q$ be a left $A$-module and $B = \mathrm{End}(Q)^{op}$. Then* (i) *$Q_B$ is a cogenerator if $_AQ$ is finitely generated injective; $Q_B$ is finitely generated injective and $A \simeq \mathrm{End}_B(Q)$ if $_AQ$ is a cogenerator.* (ii) *If $_AQ$ is a finitely generated injective cogenerator, then so is $Q_B$ and $A \simeq \mathrm{End}_B(Q)$ canonically.*

(2) *Let $D_1$: $\mathrm{mod}\, A \to \mathrm{mod}\, B^{op}$ and $D_2$: $\mathrm{mod}\, B^{op} \to \mathrm{mod}\, A$ be contravariant functors. Then the functors define a duality (i.e. $D_1D_2 \simeq \mathbf{1}_{\mathrm{mod}\, B^{op}}$ and $D_2D_1 \simeq \mathbf{1}_{\mathrm{mod}\, A}$) if and only if there is an $(A, B)$-bimodule $_AQ_B$ satisfying the following two conditions*: (a) *$D_1 \simeq \mathrm{Hom}_A(-, Q)$, $D_2 \simeq \mathrm{Hom}_B(-, Q)$, and* (b) *$_AQ$ and $Q_B$ are finitely generated injective cogenerators, and $B \simeq \mathrm{End}_A(Q), A \simeq \mathrm{End}_B(Q)$ canonically.*

A finitely generated injective module $I$ is a cogenerator in $\mathrm{mod}\, A$ if and only if the injective left $A$-module $\mathrm{Hom}(eA, k)$ is isomorphic to a submodule of $I$ where $e$ is a basic idempotent of $A$; an injective module isomorphic to the module $_A\mathrm{Hom}(eA, k)$ is called a *minimal injective cogenerator*. A typical example of a Morita duality is defined by $\mathrm{Hom}(-, k)$, and by the Morita duality theorem it should be induced by an $A$-bimodule $Q$. In fact, we have that $\mathrm{Hom}(-, k) \simeq \mathrm{Hom}_A(-, \mathrm{Hom}(A, k))$ on $\mathrm{mod}\, A$ and $\mathrm{mod}\, A^{op}$, cf. Proposition 2.4.2. See Morita [Mo58a] and Azumaya [Az59] for further results on duality.

## 2. Frobenius algebras

### 2.1. *Frobenius algebras*

Let $A$ be an algebra and $\{u_1, \dots, u_n\}$ a $k$-basis of $A$. Then for each $a \in A$ we have the two matrices $L(a)$ and $R(a)$ over $k$ such that

$$a(u_1, \dots, u_n) = (u_1, \dots, u_n)L(a), \quad (u_1, \dots, u_n)^T a = R(a)(u_1, \dots, u_n)^T.$$

The correspondences $L$: $A \to (k)_n$ and $R$: $A \to (k)_n$ are $k$-algebra homomorphisms called the left and the right regular representations. By linear algebra, there is an $A$-bilinear form $\beta$: $A \times A \to k$ (i.e. $\beta(ab, c) = \beta(a, bc)$ for $a, b, c \in A$) exactly when there is a matrix $P$ such that $R(a)P = PL(a)$ for all $a \in A$. ($P$ is the matrix with $(i, j)$-entry $\beta(u_i, u_j)$.) Moreover, in this case, $\beta$ is nondegenerate if and only if $P$ is nonsingular. For example, the group algebra $kG$ of a finite group $G$ with coefficients in $k$ has a nondegenerate $kG$-bilinear form $\beta$ (let $\beta(x, y) = \delta_{x,y^{-1}}$ (Kronecker $\delta$) for $x, y \in G$).

Now the canonical $A$-bimodule isomorphism

$$\mathrm{Hom}(A, k) \xrightarrow{\sim} \mathrm{Hom}(A \otimes_A A, k)$$

induces an $A$-bimodule isomorphism

$$\theta\colon \mathrm{Hom}(A, k) \xrightarrow{\sim} \mathrm{Hom}(A \times_A A, k),$$

where $\mathrm{Hom}(A \times_A A, k)$ is the set of all $A$-bilinear forms. We denote by $\theta(\lambda) = \beta_\lambda$, $\theta^{-1}(\beta) = \lambda_\beta$ for any $\lambda \in \mathrm{Hom}(A, k)$ and $\beta \in \mathrm{Hom}(A \times_A A, k)$, respectively.

Then clearly $\beta(x,y) = \lambda_\beta(xy)$, and $\beta$ is nondegenerate if and only if $\lambda_\beta(xA)$ is not zero for any $0 \neq x \in A$, if and only if $\lambda_\beta(Ax)$ is not zero for any $0 \neq x \in A$. Moreover an $A$-bilinear form $\beta$: $A \times A \to k$ defines two $A$-homomorphisms

$$\beta_l\colon\ {}_AA \to {}_A\operatorname{Hom}(A,k), \qquad \beta_r\colon\ A_A \to \operatorname{Hom}(A,k)_A$$

such that $\beta_l(x) = \beta(-,x)$, $\beta_r(x) = \beta(x,-)$. Thus $\lambda_\beta = \beta(-,1) = \beta(1,-)$ and $\beta_l(x) = x\lambda_\beta$, $\beta_r(x) = \lambda_\beta x$ for any $x \in A$. Taking account of these observations, we have the following equivalent properties, where we note that the associated $A$-bilinear form $\beta_\lambda$ is nondegenerate for any element $\lambda \in \operatorname{Hom}(A,k)$ such that $\operatorname{Hom}(A,k) = A\lambda$ or $\lambda A$.

DEFINITION 2.1.1. An algebra $A$ is called a *Frobenius algebra* if the following equivalent conditions hold.

(1) There is a nondegenerate $A$-bilinear form.
(2) There is an isomorphism $A \simeq \operatorname{Hom}(A,k)$ as left $A$-modules.
(3) There is an isomorphism $A \simeq \operatorname{Hom}(A,k)$ as right $A$-modules.
(4) $\operatorname{Hom}(A,k)$ is a cyclic module as a left or right $A$-module.

For a subset $X$ of an algebra $A$, by $\ell_A(X)$ we denote the left annihilator set of $X$ in $A$, that is,

$$\ell_A(X) = \{a \in A \mid ax = 0\}.$$

Similarly, $r_A(X)$ denotes the right annihilator set of $X$ in $A$. For a Frobenius algebra $A$ with a nondegenerate $A$-bilinear form $\beta$: $A \times A \to k$ and $\lambda := \lambda_\beta$, we denote by ${}^\perp I$ the set $\{a \in A \mid \beta(a,I) = 0\}$ and by $I^\perp$ the set $\{a \in A \mid \beta(I,a) = 0\}$ for a subset $I \subset A$. Then, by definition of $\lambda$,

$$r_A(I) = \{a \in A \mid \lambda(Ia) = 0\} = I^\perp$$

for a left ideal $I$. Similarly, ${}^\perp J = \ell_A(J)$ for a right ideal $J$. Moreover we have that $I = \ell_A r_A(I)$ and $\dim_k A = \dim_k I + \dim_k r_A(I)$ for any left ideal $I$. The following theorem is a main theorem in Nakayama [N3941], I. Cf. Theorem 2.2.3.

THEOREM 2.1.1 (Nakayama). *An algebra $A$ is Frobenius if and only if, for a left ideal $I$ and a right ideal $J$, $\ell_A r_A(I) = I$, $r_A\ell_A(J) = J$ and*

$$\dim_k A = \dim_k I + \dim_k r_A(I) = \dim_k J + \dim_k \ell_A(J).$$

*Moreover, in this case, $r_A(I) = I^\perp$ and $\ell_A(J) = {}^\perp J$ for the defining $A$-bilinear form $\beta$.*

Let $A$ be a Frobenius algebra with a nondegenerate $A$-bilinear form $\beta$: $A \times A \to k$. Let $\bar\beta$: $A \times A \to k$ be the bilinear form such that $\bar\beta(x,y) = \beta(y,x)$ for $x, y \in A$. Then, since $\bar\beta$ is a nondegenerate $k$-bilinear form, there is a $k$-linear map $\nu$: $A \to A$ such that $\bar\beta(y,-) = \beta(y^\nu,-)$ for any $y \in A$. Observe that $\nu$ is an algebra isomorphism.

Thus there is an algebra automorphism $\nu$ depending on $\beta$ such that $\beta(x,y) = \beta(y^\nu, x)$, which is called a *Nakayama automorphism* of $A$. See (2.4) for another definition. The next theorem shows that such an automorphism is uniquely determined up to an inner automorphism.

THEOREM 2.1.2. *Let $A$ be a Frobenius algebra with a nondegenerate $A$-bilinear form $\beta$ with Nakayama automorphism $\nu$. Then for any nondegenerate $A$-bilinear form $\beta'$ with Nakayama automorphism $\nu'$, there is an invertible element $c$ of $A$ such that $\nu' \circ \nu^{-1} = \mathbf{c}$, where $x^{\mathbf{c}} := c^{-1}xc$ for any $x \in A$.*

This follows, because there is an elements $c$ such that $\beta'(-,1) = \beta(-,c)$ (linear algebra) and then $c$ has no nonzero annihilator in $A$ since $\beta'$ is nondegenerate. Observe that any element without nonzero annihilator in $A$ is invertible and that

$$\beta'\left(x^{\nu'}, y\right) = \beta'\left(\left(c^{-1}xc\right)^{\nu}, y\right)$$

for $x, y \in A$.

Nakayama proved furthermore that *an algebra is uniserial if and only if each of its factor algebras is Frobenius*. Here we recall the definition of the uniserial algebra. A module is said to be *uniserial* when it has a unique composition series, and a *serial module* is a direct sum of uniserial modules. In particular, an algebra $A$ is said to be *left serial* in case ${}_AA$ is serial, and is said to be *serial* if $A$ is left and right serial. A *uniserial algebra* is by definition a direct product of finitely many local serial algebras.

## 2.2. *Quasi-Frobenius algebras*

The defining properties in Definition 2.1.1 of Frobenius algebras are not Morita invariant, but they imply the injectivity of $A$ as an $A$-module, which is Morita invariant.

DEFINITION 2.2.1. An algebra $A$ is said to be *quasi-Frobenius* when ${}_AA$ is injective.

A basic quasi-Frobenius algebra is Frobenius obviously. The theorem below shows that this definition is independent of the side of operation of $A$ on itself [N3941]. The permutation $\pi$ in the theorem is called the *Nakayama permutation* of $A$.

THEOREM 2.2.1 (Nakayama). *For an algebra $A$ with basic idempotent $e = \sum_{i=1}^n e_i$, the following statements are equivalent and $\pi' = \pi^{-1}$.*

(1) *${}_AA$ is injective.*
(2) *$Ae \simeq \mathrm{Hom}(eA, k)$ as left $A$-modules.*
(3) *There is a permutation $\pi$ on $\{1, \ldots, n\}$ such that* $\mathrm{soc}(Ae_i) \simeq \mathrm{top}(Ae_{\pi(i)})$.
(4) *There is a permutation $\pi$ on $\{1, \ldots, n\}$ such that* ${}_AAe_i \simeq {}_A\mathrm{Hom}(e_{\pi(i)}A, k)$.
(1′) *$A_A$ is injective.*
(2′) *$eA \simeq \mathrm{Hom}(Ae, k)$ as right $A$-modules.*
(3′) *There is a permutation $\pi'$ on $\{1, \ldots, n\}$ such that* $\mathrm{soc}(e_iA) \simeq \mathrm{top}(e_{\pi'(i)}A)$.
(4′) *There is a permutation $\pi'$ on $\{1, \ldots, n\}$ such that* $e_iA_A \simeq \mathrm{Hom}(Ae_{\pi'(i)}, k)_A$.

In fact, (4) $\Rightarrow$ (2) is trivial, and (4) $\Leftrightarrow$ (4′) and (2) $\Rightarrow$ (1) follow from the duality $\mathrm{Hom}(-, k)$. The implications (1) $\Rightarrow$ (4) $\Rightarrow$ (3) are easy (see Preliminaries). Thus it suffices to show (3) $\Rightarrow$ (4). Now assume that $A = eAe$ and suppose (3). Then, since

$$\mathrm{soc}(Ae_i) \simeq \mathrm{top}(Ae_{\pi(i)}) \simeq \mathrm{soc}\,\mathrm{Hom}(e_{\pi(i)}A, k),$$

there is a monomorphism from $\mathrm{soc}(Ae_i)$ to $\mathrm{Hom}(e_{\pi(i)}A, k)$, which is extended to a monomorphism

$$f_i\colon\ {}_AAe_i \to {}_A\mathrm{Hom}(e_{\pi(i)}A, k)$$

because ${}_A\mathrm{Hom}(e_{\pi(i)}A, k)$ is injective. Hence we have a monomorphism

$$f = \bigoplus_i f_i\colon\ {}_AA \doteq \bigoplus_i Ae_i \to \bigoplus_i \mathrm{Hom}(e_{\pi(i)}A, k) = {}_A\mathrm{Hom}(A, k),$$

which implies that, by comparing the dimensions, $f$ and so each $f_i$ is isomorphic.

THEOREM 2.2.2. *An algebra $A$ is quasi-Frobenius if and only if* $\mathrm{Hom}_A(-, A)$ *is a duality functor between* $\mathrm{mod}\,A$ *and* $\mathrm{mod}\,A^{\mathrm{op}}$, *if and only if every left and every right simple $A$-module is reflective.*

The first statement of the theorem follows from Theorem 2.2.1 and the Morita duality theorem; see Morita [Mo58a] or Azumaya [Az59] for the second statement. Here, a module ${}_AM$ is said to be *reflexive* when the canonical morphism $M \to \mathrm{Hom}_A(\mathrm{Hom}_A(M, A), A)$ is an isomorphism.

Now, if an algebra $A$ satisfies that $r_A\ell_A(J) = J$ for any right ideal $J$, then we know that $\mathrm{Hom}_A(A/I, A)$ is a simple right $A$-module for a maximal left ideal $I$ (because $\mathrm{Hom}_A(A/I, A) \xrightarrow{\sim} r_A(I)$ naturally). On the other hand, in case $A$ is quasi-Frobenius, applying $\mathrm{Hom}_A(-, A)$ twice to the exact sequence

$$0 \to I \to A \to A/I \to 0,$$

we have the canonical morphism

$$f\colon\ I \to \mathrm{Hom}_A\big(A/r(I), A\big) = \ell_A r_A(I)$$

and

$$g\colon\ A/I \to \mathrm{Hom}_A\big(\mathrm{Hom}_A(A/I, A), A\big),$$

each of which is an isomorphism if the other is. Thus, thanks to Theorem 2.2.2, we have the following theorem which is one of the main theorems in [N3941].

THEOREM 2.2.3 (Nakayama). *A necessary and sufficient condition for an algebra $A$ to be quasi-Frobenius is that it satisfies the following annihilator condition: $\ell_A r_A(I) = I$ for any left ideal $I$ and $r_A\ell_A(J) = J$ for any right ideal $J$.*

NOTES.

(1) Since every injective (not necessarily finitely generated) module over an algebra $A$ is a direct sum of indecomposable injective modules (E. Matlis), *$A$ is quasi-Frobenius if and only if every injective module in* $\operatorname{Mod} A$ *is projective* (Faith and Walker [FW67]).

(2) Suppose that $A$ is an arbitrary (not necessarily finite dimensional) ring (associative with 1). Then $A$ is left and right Artinian, and right self-injective if it is left or right Noetherian, and left self-injective. (See Faith [F76].) A self-injective Artinian ring is said to be *quasi-Frobenius*. But, in case $A$ is neither left Noetherian nor right Noetherian, self-injectivity is not necessarily left-right symmetric. For example, *the endomorphism ring of any nonfinitely generated free left module over a quasi-Frobenius $k$-algebra $A$ (e.g., $A = k$) is left self-injective but not right self-injective* (Osofsky [Os66b], Sandomierski [San70]). See Kambara [K87] for another example. Goodearl [Go74] gave examples of simple, left and right self-injective rings which are not Artinian. We refer to [Mo58a] and [Az59] for other homological characterizations of quasi-Frobenius Artinian rings, and [Os66a, F76] for some generalizations.

(3) It is still open whether a left or a right self-injective semi-primary ring is quasi-Frobenius ([Os66a], cf. [F90]). Note that a left and right self-injective semi-primary ring is quasi-Frobenius.

(4) The problem whether a self-injective cogenerator ring on the left side (left PF-ring) is a right PF-ring was unsolved until the middle of the 1980's (see Azumaya [Az66], Faith [F76]). Dischinger and W. Müller [DM86] gave a counterexample in 1986 to this problem.

### 2.3. *Symmetric algebras*

A Frobenius algebra $A$ with a symmetric $A$-bilinear form $\beta\colon\ A \times A \to A$ is called a *symmetric* algebra, i.e. there is a symmetric nonsingular matrix $P$ such that $R(a) = P^{-1}L(a)P$ for any $a \in A$ [BNe37]. The bilinear form $\beta$ defined in 2.1 for a group algebra $kG$ is symmetric. We refer to Kupisch [Ku6570] for the structure of symmetric algebras.

The following theorem clarifies a homological difference between Frobenius algebras and symmetric algebras (see Theorem 2.4.1 and Nakayama [N3941], Theorem 12).

THEOREM 2.3.1. *An algebra $A$ is symmetric if and only if $A$ is isomorphic to* $\operatorname{Hom}(A, k)$ *as an $A$-bimodule.*

In fact, assume that $A$ is symmetric and $\beta$ is a symmetric $A$-bilinear form. There is an isomorphism $f\colon\ A \to \operatorname{Hom}(A, k)$ as left $A$-modules such that $f(a) = a\lambda$ for $a \in A$, where $\lambda \in \operatorname{Hom}(A, k)$ and $\beta(x, y) = \lambda(xy)$. Then it suffices to show that $a\lambda = \lambda a$ for $a \in A$: easy. Conversely assume that there is an $A$-bimodule isomorphism

$$f\colon\ A \to \operatorname{Hom}(A, k)$$

and let $f(1) = \lambda$. Obviously $a\lambda = \lambda a$ for any $a \in A$. Then the corresponding $A$-bilinear form $\beta_\lambda$ is clearly nondegenerate and symmetric.

EXAMPLE. A semi-simple algebra is symmetric [BNe37, Ne38, EN55].

In fact, assume that $A$ is a simple algebra and let $C$ be the center of $A$. Then $A \otimes_C A^{op}$ is a simple $C$-algebra which has a unique simple left module ${}_{A\otimes A^{op}}A$ (with canonical operation), and a unique simple right module $A_{A\otimes A^{op}}$. Hence, there is an isomorphism $A \xrightarrow{\sim} \operatorname{Hom}_C(A, C)$ as left $A \otimes_C A^{op}$-modules or as $A$-bimodules. Moreover, since $k$ is a subfield of $C$, the canonical map $\operatorname{Hom}_k(A, k) \to \operatorname{Hom}_C(A, C)$ is clearly an $A$-bimodule isomorphism. Thus we have that an $A$-bimodule isomorphism $A \xrightarrow{\sim} \operatorname{Hom}_k(A, k)$.

Now, consider an $A$-bimodule isomorphism $f\colon A \to \operatorname{Hom}(A, k)$ for a symmetric algebra $A$. Then, for a basic idempotent

$$e = \sum_{i=1}^{n} e_i$$

of $A$, we have that

$$\begin{aligned}{}_A Ae_i &\xrightarrow{\sim} {}_A \operatorname{Hom}(A, k)e_i \xrightarrow{\sim} {}_A \operatorname{Hom}(e_i A, k)\\ &\xrightarrow{\sim} {}_A \operatorname{Hom}\big(\operatorname{Hom}(Ae_{\pi^{-1}(i)}, k), k\big) \xrightarrow{\sim} {}_A Ae_{\pi^{-1}(i)}.\end{aligned}$$

This implies that $i = \pi(i)$ for any $i$, hence $\pi$ is identity. A quasi-Frobenius $k$-algebra $A$ with identity Nakayama permutation is said to be *weakly symmetric*. In [NN38] there is an example of nonsymmetric and weakly symmetric algebra. Another examples will be given in 2.5.

Rickard developed the Morita theory (equivalence) for derived categories of bounded complexes of modules [Ric89a, Ric89b, Ric91]. In particular, he characterized the symmetric algebras in terms of derived categories:

THEOREM 2.3.2 (Rickard). *An algebra derived equivalent to a symmetric algebra is itself symmetric.*

Here two algebras $A, B$ are said to be *derived equivalent* if $D^b(\operatorname{mod} A) \approx D^b(\operatorname{mod} B)$, where $D^b(\operatorname{mod} R)$ is the derived category of bounded complexes of $R$-modules in $\operatorname{mod} R$.

PROPOSITION 2.3.1. *For a symmetric k-algebra A, $\Omega^2(M) \simeq D\operatorname{Tr}(M)$ and $\Omega^{-2}(M) \simeq \operatorname{Tr} D(M)$ for any indecomposable nonprojective module M.*

Here, $\operatorname{Tr}(M)$ is the transpose of $M$, i.e. $\operatorname{Tr}(M) := \operatorname{Coker}(\operatorname{Hom}_A(f, A))$ for a minimal projective resolution

$$P_1 \xrightarrow{f} P_0 \to M \to 0.$$

Hence there is an exact sequence

$$0 \to D\operatorname{Tr}(M) \to \operatorname{Hom}\big(\operatorname{Hom}(P_1, A), k\big) \xrightarrow{\bar{f}} \operatorname{Hom}\big(\operatorname{Hom}(P_0, A), k\big),$$

where $\bar{f} = \mathrm{Hom}(\mathrm{Hom}_A(f, A), k)$. But, since $A \simeq \mathrm{Hom}(A, k)$ as $A$-bimodules by Theorem 2.3.1, we have an exact sequence

$$0 \to D\,\mathrm{Tr}(M) \to P_1 \xrightarrow{f} P_0 \to M \to 0$$

and hence $\Omega^2(M) \simeq D\,\mathrm{Tr}(M)$.

A module $M$ over a quasi-Frobenius algebra $A$ is said to be *bounded* if the lengths of $\Omega^n_A(M)$ $(n > 0)$ have a common upper bound, and *periodic* if $M$ is isomorphic to $\Omega^n_A(M) \oplus P$ for some $n > 0$ and a projective $A$-module $P$. For the group algebra $kG$ of a finite group $G$, Alperin [Al77] proved that every bounded $kG$-module is periodic in case $k$ is algebraic over its prime field, and Eisenbud [Ei80] proved it for any field $k$.

THEOREM 2.3.3 (Alperin–Eisenbud). *Let $kG$ be the group algebra of a finite group $G$ with coefficients in a field $k$. Then every bounded $kG$-module is periodic.*

Thanks to Proposition 2.3.1, a nonprojective indecomposable periodic $kG$-module $M$ is $D\,\mathrm{Tr}$-periodic, i.e. $M \simeq (D\,\mathrm{Tr})^n(M)$ for some $n > 0$ and $D = \mathrm{Hom}(-, k)$. $D\,\mathrm{Tr}$-periodic modules over an algebra are very important in the representation theory of algebras. See Happel, Preiser and Ringel [HPR80], Ringel [Rin84], and Auslander, Platzeck and Reiten [APR77]. Schulz [Schu86], using an idea of Evens [Ev61], gave a condition for a bounded module over a quasi-Frobenius algebra to be periodic. We refer to [Al77, Ei80] and Carlson [Ca79a, Ca79b] for further results, problems and examples of periodic modules. See also Okuyama [Ok87], Erdmann and Skowroński [ErSk92, LS93, AR94].

As to Frobenius algebras appearing in other branches of mathematics, see Humphreys [Hu82] for Lie algebras, Pareigis [Pa71] for Hopf algebras, and Kimura [Ki86] for Mikio Sato's theory of the prehomogeneous vector spaces associated with finite dimensional algebras. From there let us discuss Sato's theorem on Frobenius algebras from around 1962.

Assume that $k$ is an algebraically closed field of characteristic zero, and let the dual space of a $k$-vector space $V$ be denoted by $V^*$. Let $G$ be a connected linear algebraic group over $k$ and $\rho$ a rational representation of $G$ on a finite-dimensional vector space $V$. Suppose that the triplet $(G, \rho, V)$ is a *prehomogeneous vector space* (simply P.V.), i.e. $V$ has a Zariski-dense $G$-orbit $Y = V - S$. Let $g = Lie(G)$ and $g_1 = Lie(G_1)$, where $G_1$ is the subgroup of $G$ generated by the commutator subgroup $[G, G]$ and a generic isotropy subgroup $G_x (x \in Y)$. Then it is known that for $\omega \in g^*$ there is a rational map $\varphi_\omega \colon Y \to V^*$ such that

$$\varphi_\omega\big(\rho(g)x\big) = \rho^*(g)\varphi_\omega(x) \quad \text{for } g \in G,\ x \in Y.$$

Moreover, in this case

$$\big(\varphi_\omega(x)\big)\big(\mathrm{d}\rho(a)x\big) = \omega(a)$$

for all $x \in Y$ and $a \in g$ if and only if $\omega(g_1) = 0$, where $\mathrm{d}\rho$ is the infinitesimal representation of $\rho$ (hence $\frac{\mathrm{d}}{\mathrm{d}t}\rho(\exp ta)x|_{t=0} = \mathrm{d}\rho(a)x$). A P.V. $(G, \rho, V)$ is said to be *regular* if it has a nondegenerate relative invariant (a relative invariant $f(x)$ is said to be nondegenerate if its Hessian $Hess_f(x) = \det(\partial^2 f/\partial x_i \partial x_j)$ is not identically zero), and the P.V. is *quasi-regular* if there exists $\omega \in (g/g_1)^*$ such that $\varphi_\omega\colon Y \to V^*$ is dominant. Note that a regular P.V. is quasi-regular.

Now let $A$ be a $k$-algebra and $A^\times$ the multiplicative group of all invertible elements of $A$. Then $G := A^\times$ acts on $V := A$ by $\rho(g)v = gv$ for $g \in G, v \in V$ so that the triplet $(G, \rho, V)$ is a P.V. which is called the *P.V. of A*. Sato's theorem is now stated as follows.

THEOREM 2.3.4 (Mikio Sato). *Let $(G, \rho, V)$ be the P.V. of an algebra A.*

(1) *The algebra A is Frobenius if and only if dual triplet $(G, \rho^*, V^*)$ is a P.V.*

(2) *The algebra A is symmetric (resp. semisimple) if and only if its P.V. $(G, \rho, V)$ is quasi-regular (resp. regular).*

## 2.4. *Nakayama automorphisms*

For an automorphism $\sigma$ of an algebra $A$ and a left $A$-module $M$, ${}_\sigma M$ is understood to be the $A$-module whose underlying additive group is $M$ with the operation of $A$: $a \cdot m = a^\sigma m$ for $a \in A$, $m \in M$. Similarly $N_\sigma$ is defined for a right $A$-module $N$. For an $A$-bimodule $M$ and $a \in A$, $a_L\colon M \to M$ and $a_R\colon M \to M$ stand for the left and the right multiplication of $a$: $a_L(x) = ax$, $a_R(x) = xa$ for $x \in M$.

Let $A$ be a Frobenius algebra with a nondegenerate $A$-bilinear form $\beta$ and let $\nu$ be the Nakayama automorphism associated with $\beta$; $\beta(x, y) = \beta(y^\nu, x)$ for $x, y \in A$. We know that $\beta$ induces (one-sided) $A$-isomorphisms $\beta_\ell$ and $\beta_r$ from $A$ to $\mathrm{Hom}(A, k)$ (2.1). Moreover it is easy to see that they are (after twisting) in fact $A$-bimodule isomorphisms, i.e.

$$\beta_\ell\colon A \xrightarrow{\sim} \mathrm{Hom}(A, k)_\nu \quad \text{and} \quad \beta_r\colon A \xrightarrow{\sim} {}_{\nu^{-1}}\mathrm{Hom}(A, k).$$

Since the right $A$-module $\mathrm{Hom}(A, k)$ is faithful, the automorphism $\nu$ is uniquely determined with respect to the property that $\beta_\ell\colon A \to \mathrm{Hom}(A, k)_\nu$ is an isomorphism of right $A$-modules. Thus we have the following theorem, generalizing Theorem 2.3.1, which follows from Theorem 2.1.1 and the fact that

$$\mathrm{Hom}\left(A/J^\nu, k\right) \simeq \left\{a \in A \mid \beta(J^\nu, a) = 0\right\} = \left\{a \in A \mid \beta(a, J) = 0\right\} = {}^\perp J.$$

THEOREM 2.4.1. *For a Frobenius algebra A, an automorphism $\nu$ of A is Nakayama if and only if there is an A-bimodule isomorphism*

$$A \xrightarrow{\sim} \mathrm{Hom}(A, k)_\nu, \quad \text{or} \quad A \xrightarrow{\sim} {}_{\nu^{-1}}\mathrm{Hom}(A, k).$$

*Moreover, in this case,*

$$A/J^\nu_A \simeq \mathrm{Hom}\left(\ell_A(J), k\right)_A, \qquad {}_A A/I^{\nu^{-1}} \simeq {}_A\mathrm{Hom}\left(r_A(I), k\right)$$

*for any left ideal I and any right ideal J. In particular, A is symmetric if and only if $\nu$ is inner.*

As an application we can prove the following Skolem–Noether Theorem.

EXAMPLE. An automorphism of a central simple $k$-algebra is an inner automorphism.

PROOF. For an automorphism $\varphi$ of $A$, $\overline{\varphi} = \varphi \otimes 1$ is an automorphism of the central simple algebra $A \otimes_k A^{op}$. Since $A$ (resp. $A_{\overline{\varphi}}$) is the unique simple left (resp. right) $A \otimes_k A^{op}$-module (up to isomorphism), we have an $A$-bimodule isomorphism

$$A \to \mathrm{Hom}({}_{\varphi}A, k) = \mathrm{Hom}(A, k)_{\varphi}.$$

As $A$ is symmetric (2.3), it follows from the above theorem that $\varphi$ is inner. □

THEOREM 2.4.2. *Let $A$ be a Frobenius algebra and $\nu$ its Nakayama automorphism. Let*

$$e = \sum_{i=1}^{n} e_i$$

*be a basic idempotent of $A$. Then $Ae_{\pi(i)} \simeq A(e_i)^{\nu}$ for any $i$.*

PROOF. Let $e_i' = 1 - e_i$. Since $A = e_i^{\nu}A \oplus (e_i')^{\nu}A$, it follows from Theorems 2.4.1 and 2.2.1 that

$$e_i^{\nu}A = A/(e_i')^{\nu}A = A/(e_i'A)^{\nu} \simeq \mathrm{Hom}\left(\ell_A(e_i'A), k\right) = \mathrm{Hom}(Ae_i, k) \simeq e_{\pi(i)}A.$$

□

THEOREM 2.4.3. *Let $A$ be a Frobenius algebra with Nakayama automorphism $\nu$. Then*

$$\mathrm{soc}({}_AA) = \mathrm{soc}(A_A) =: \mathrm{soc}(A) \quad \textit{and} \quad \mathrm{soc}(A) \simeq \mathrm{top}(A)_{\nu}$$

*as $A$-bimodules. In particular* $\mathrm{soc}(A)$ *is a cyclic module as a left and a right $A$-module.*

PROOF. Let $\beta$ be an $A$-bilinear form defining $\nu$. Then it follows from Theorem 2.1.1 that

$$\begin{aligned}\mathrm{soc}({}_AA) &= r_A(\mathrm{rad}\,A) = (\mathrm{rad}\,A)^{\perp} = \left((\mathrm{rad}\,A)^{\nu}\right)^{\perp}\\ &= \left\{x \in A \mid \beta\left((\mathrm{rad}\,A)^{\nu}, x\right) = 0\right\} = \left\{x \in A \mid \beta(x, \mathrm{rad}\,A) = 0\right\}\\ &= {}^{\perp}(\mathrm{rad}\,A) = \ell_A(\mathrm{rad}\,A) = \mathrm{soc}(A_A).\end{aligned}$$

Moreover $\beta$ induces isomorphisms

$$\mathrm{soc}({}_AA_A) \simeq \mathrm{Hom}\left(\mathrm{top}(A_A), k\right)_{\nu} \simeq \mathrm{Hom}\left(\mathrm{top}({}_{\nu}A_A), k\right) \simeq \mathrm{top}({}_AA)_{\nu}.$$

□

The Nakayama automorphism first appeared in [N3941], II, as well as the Nakayama permutation. The importance of the automorphisms is realized mainly in representation theory (e.g., Skowroński [Sk89], Skowroński and Yamagata [SkY1, SkY2]). The following fact was quite recently pointed out in [SkY2].

PROPOSITION 2.4.1. *Let $\beta$: $A \times A \to k$ be a nondegenerate $A$-bilinear form of a $k$-algebra $A$ with Nakayama automorphism $\nu$. Then $\nu = \beta_r^{-1} \cdot \beta_\ell$: $A \xrightarrow{\sim} {}_\nu A_\nu$ is an $A$-bimodule isomorphism.*

PROOF. First note that $\beta_r = D\beta_\ell \cdot (\,)^{**}$, where $D = \mathrm{Hom}(-,k)$ and $(\,)^{**}$: $A \to D^2A$ is the canonical map $((\,)^{**}(a))(f) = f(a)$ for $a \in A$ and $f \in DA$. Let $a^{**} = (\,)^{**}(a)$. Then $\beta_r(1) = 1^{**} \cdot \beta_\ell$, and we have that for any $a \in A$,

$$\big(\beta_r(1)\big)(a) = 1^{**}\big(\beta_\ell(a)\big) = 1^{**}\big(a\beta_\ell(1)\big) = \big(a\beta_\ell(1)\big)(1) = \big(\beta_\ell(1)\big)(a)$$

and hence $\beta_r^{-1}\beta_\ell(1) = 1$. Hence

$$\big(\beta_r^{-1} \cdot \beta_\ell\big)(a) = \beta_r^{-1}\big(\beta_\ell(a)\big) = \beta_r^{-1}\big(a\beta_\ell(1)\big) = a^\nu\big(\beta_r^{-1}\beta_\ell(1)\big) = a^\nu.$$

Therefore we have that $\nu = \beta_r^{-1} \cdot \beta_\ell$. □

Now we describe the Nakayama automorphisms of Frobenius algebras as a special case of general relations between the modules defining Morita dualities and the outer automorphism group (Morita [Mo58a]; cf. Yamagata [Y88], Cohn [Co66]).

Let $A$ be a basic $k$-algebra and

$$1 = \sum_{i=1}^{n} e_i$$

a sum of orthogonal primitive idempotents. By the Morita duality theorem, a duality $D$: $\mathrm{mod}\, A \to \mathrm{mod}\, A^{op}$ is defined by the following $A$-bimodule $Q$: ${}_AQ$ and $Q_A$ are minimal injective cogenerators and $\mathrm{End}({}_AQ)^{op} \simeq A$, $\mathrm{End}(Q_A) \simeq A$ naturally. We shall call such a module $Q$ a *duality module* which is denoted by $DA$ for a self-duality $D$, namely $D = \mathrm{Hom}_A(-, DA)$. Two self-dualities $D_1, D_2$ are isomorphic if and only if $D_1A \simeq D_2A$ as $A$-bimodules. The *Nakayama permutation* $\pi$ of a given $A$-bimodule $DA$ (or of $D$) is by definition a permutation of the index set $\Omega = \{1, \ldots, n\}$ such that $\mathrm{soc}(DA)e_i \simeq \mathrm{top}\, Ae_{\pi(i)}$ for $i \in \Omega$. The module $\mathrm{Hom}(A,k)$ is a typical duality module with identity permutation. Since there is an isomorphism $\omega$: ${}_ADA \to {}_A\mathrm{Hom}(A,k)$ as left $A$-modules, for any $a \in A$ there is a unique element $a^\nu$ of $A$ such that

$$\mathrm{Hom}\left(\left(a^\nu\right)_L, k\right) = \omega \cdot a_R \cdot \omega^{-1}\colon\ {}_A\mathrm{Hom}(A,k) \to {}_A\mathrm{Hom}(A,k).$$

It is easy to see that the correspondence $\nu$: $A \to A$ is an algebra automorphism so that $\omega$: $DA \to \mathrm{Hom}(A,k)_\nu$ is an $A$-bimodule isomorphism. In particular, $\nu$ is inner if and only if $DA$ is isomorphic to $\mathrm{Hom}(A,k)$ as an $A$-bimodule. We call such an automorphism $\nu$ the *Nakayama automorphism* of $A$ defined by $\omega$ (or by $DA$ or $D$). In case $A$ is Frobenius, the permutations and the automorphisms coincide with those in 2.1 and 2.2.

PROPOSITION 2.4.2. *The isomorphism classes of duality modules over a $k$-algebra $A$ correspond bijectively to the outer automorphism group* $\mathrm{Aut}(A)/\mathrm{Inn}(A)$: $DA \mapsto \nu$ *under the condition that* $DA \simeq \mathrm{Hom}(A,k)_\nu$ *as $A$-bimodules, where* $\mathrm{Aut}(A)$ *and* $\mathrm{Inn}(A)$ *denote the automorphism group and the inner automorphism group of $A$, respectively.*

Let $\nu$ be the Nakayama automorphism associated with a duality module $DA$ over an algebra $A$. It induces naturally a Morita equivalence $\tilde{\nu}$: $\mathrm{Mod}\, A \to \mathrm{Mod}\, A$ such that $\tilde{\nu}(M) = {}_\nu M$, with inverse $\tilde{\nu}^{-1}(M) = {}_{\nu^{-1}}M$. Then it follows from the proposition above that $\tilde{\nu} \simeq \mathrm{Hom}(DA,k) \otimes_A -$ and $\tilde{\nu}^{-1} \simeq \mathrm{Hom}_A(\mathrm{Hom}(DA,k), -)$. The functor $\tilde{\nu}$ is isomorphic to the identity if and only if $\nu$ is inner. Moreover, $\tilde{\nu}$ is isomorphic to $\mathrm{Hom}(A,k) \otimes_A -$ (which is called the *Nakayama functor* (Gabriel [G80])) if and only if $A$ is Frobenius with Nakayama automorphism $\nu$.

EXAMPLE. Let $A$ and $T$ be $k$-algebras with the same quiver and the following relations:

$$1 \underset{\beta}{\overset{\alpha}{\rightleftarrows}} 2; \qquad A\colon\ \alpha\beta = \beta\alpha = 0, \quad T\colon\ (\alpha\beta)^2 = (\beta\alpha)^2 = 0.$$

Let $DA$ be the ideal $\mathrm{rad}^2 T$. Then $A \simeq T/DA$ as algebras and $DA$ is a duality $A$-module with Nakayama permutation (12). This is also the Nakayama permutation of $T$. The Nakayama automorphism $\nu$ of $A$ by $DA$ is the automorphism given by $\nu(\alpha) = \beta$, $\nu(\beta) = \alpha$.

### 2.5. *Hochschild extension algebras*

An important class of quasi-Frobenius algebras is given by extension algebras of algebras by duality modules. Let $A$ be an algebra and $DA$ a duality module. Then an algebra $T$ is called an extension algebra of $A$ by $DA$ if there is an exact sequence

$$0 \to DA \xrightarrow{\kappa} T \xrightarrow{\rho} A \to 0$$

such that $\rho$ is an algebra morphism and $\kappa$ is a $T$-bimodule monomorphism from ${}_\rho(DA)_\rho$ to $T$. If we identify $DA$ with $\mathrm{Ker}\,\rho$, then $DA$ is a $T$-ideal with $(DA)^2 = 0$ and the factor algebra $T/DA$ is isomorphic to $A$ by $\rho$. Since $DA$ is nilpotent in $T$, a set of orthogonal idempotents $\{e_i\}$ with

$$1_A = \sum_{i=1}^{n} e_i$$

can be lifted to a set of orthogonal idempotents of $T$, which are denoted by $\{\mathbf{e}_i\}$, so that

$$1_T = \sum_{i=1}^{n} \mathbf{e}_i$$

and $e_i = \rho(\mathbf{e}_i)$ for any $i$. Hence the Nakayama permutations $\pi_A, \pi_T$ are the same permutations of the set $\{1, 2, \ldots, n\}$. For $\sigma_A \in \mathrm{Aut}(A)$ and $\sigma_T \in \mathrm{Aut}(T)$, $\sigma_T$ is called an extension of $\sigma_A$ ($\sigma_A$ is a restriction of $\sigma_T$) if $\sigma_A \rho = \rho \sigma_T$. An extension of $A$ by $DA$ corresponds to a 2-cocycle $\alpha$: $A \times A \to DA$, namely, $\alpha$ satisfies the relation

$$a\alpha(b, c) + \alpha(a, bc) = \alpha(ab, c) + \alpha(a, b)c$$

for all $a, b, c \in A$ and it defines associative multiplication on the $k$-space $A \oplus DA$ by the rule:

$$(a, x)(b, y) = \big(ab, ay + xb + \alpha(a, b)\big) \quad \text{for } (a, x), (b, y) \in A \oplus DA.$$

Two extensions

$$0 \to DA \xrightarrow{\kappa} T \xrightarrow{\rho} A \to 0, \qquad 0 \to DA \xrightarrow{\kappa'} T \xrightarrow{\rho'} A \to 0$$

are equivalent if there is an algebra morphism $\tau$: $T \to T$ such that $\kappa' = \tau\kappa$ and $\rho = \rho'\tau$ (so $\tau$ is an automorphism of $T$). The equivalence classes of extensions of $A$ by $DA$ form the Hochschild cohomology group $\mathrm{H}^2(A, DA)$, in which the zero element is the class of splittable extensions (an extension

$$0 \to DA \xrightarrow{\kappa} T \xrightarrow{\rho} A \to 0$$

is said to be splittable if there is an algebra morphism $\rho'$: $A \to T$ with $\rho\rho' = 1 \in \mathrm{Aut}(A)$). (See [CE56].) Every $A$-module $M$ is naturally considered as a $T$-module by $\rho$: $tm = \rho(t)m$ for $t \in T$, $m \in M$. Hence $\mathrm{Mod}\, A$ is naturally isomorphic to the full subcategory of $\mathrm{Mod}\, T$ whose objects are annihilated by $DA$. The full subcategory (of $\mathrm{mod}\, T$) of $T$-modules which are not annihilated by $DA$ is denoted by $\mathrm{mod}(T \backslash A)$. An essential property of extension algebras is the following [Y81a82].

PROPOSITION 2.5.1. *Let*

$$0 \to DA \xrightarrow{\kappa} T \xrightarrow{\rho} A \to 0$$

*be an extension. Then, for any idempotent* $\mathbf{e}$ *of* $T$ *and* $e = \rho(\mathbf{e})$, $\rho$ *canonically induces isomorphisms*:

$$T\mathbf{e}/(DA)\mathbf{e} \simeq Ae, \qquad \mathrm{soc}({}_T T\mathbf{e}) = \mathrm{soc}\big({}_T(DA)\mathbf{e}\big) \simeq \mathrm{soc}_A\big((DA)e\big),$$

$$\mathrm{top}(T\mathbf{e}) \simeq \mathrm{top}(Ae).$$

*Moreover,* $T$ *is a quasi-Frobenius algebra with the same Nakayama permutation as the Nakayama permutation of* $DA$, *and the Nakayama automorphism of* $A$ *defined by* $DA$ *extends to* $T$.

We shall show briefly how to prove the last statement on automorphisms (the others are easy). Let $\omega$: $DA \to \operatorname{Hom}(A,k)_\nu$ be an $A$-bimodule isomorphism with Nakayama automorphism $\nu$ of $A$. Since ${}_TT$ is an injective hull of $DA$ by the first assertion, the morphism

$$\operatorname{Hom}(\rho,k)\omega\colon\ {}_TDA \to {}_T\operatorname{Hom}(T,k)$$

extends to an isomorphism

$$\omega'\colon\ {}_TT \xrightarrow{\sim} {}_T\operatorname{Hom}(T,k)$$

along $\kappa$, i.e. $\operatorname{Hom}(\rho,k)\omega = \omega'\kappa$. Hence, for the automorphism $\nu'$ of $T$ such that

$$\omega'\colon\ T \xrightarrow{\sim} \operatorname{Hom}(T,k)_{\nu'}$$

as $T$-bimodules, the morphism $\omega$ induces an isomorphism $DA \to \operatorname{Hom}(A,k)_{\nu'}$ as $T$-bimodules. Consequently, $\nu$ is the restriction of $\nu'$ by the definition of $\nu$.

Every splittable extension is equivalent to the trivial extension

$$0 \to DA \xrightarrow{\kappa} A \ltimes DA \xrightarrow{\rho} A \to 0,$$

where $\kappa$ and $\rho$ are the canonical injection and projection, respectively. The *trivial extension algebra* $A \ltimes DA$ is the $k$-vector space $A \oplus DA$ with multiplication

$$(a,x)(b,y) = (ab, ay + xb)$$

for $a,b \in A$ and $x,y \in DA$. Note that $T(A) := A \ltimes \operatorname{Hom}(A,k)$ is a symmetric algebra because the map

$$\varphi\colon\ T(A) \to \operatorname{Hom}\big(T(A),k\big), \quad \big(\varphi(a,f)\big)(b,g) = f(b) + g(a),$$

is clearly a $T(A)$-bimodule isomorphism [IW80]. Moreover, by the following theorem [Y88], symmetric extension algebras of $A$ by $DA$ exist only in the case where the duality module $DA$ is isomorphic to $\operatorname{Hom}(A,k)$ as an $A$-bimodule.

THEOREM 2.5.1. *Let $\nu$ be an automorphism of a k-algebra A. Then there is a symmetric extension algebra T of A by* $\operatorname{Hom}(A,k)_\nu$ *if and only if* $A \ltimes \operatorname{Hom}(A,k)_\nu$ *is a symmetric algebra, if and only if $\nu$ is inner.*

For example, let $K/k$ be a finite normal separable field extension with Galois group $G \neq \{1\}$, and let $A = K$ and take $\nu(\neq 1) \in G$. Then the above theorem implies that $K \ltimes \operatorname{Hom}(K,k)_\nu$ is a nonsymmetric and weakly symmetric algebra.

To state some results on the module categories of extension algebras, let us recall the definition of the Auslander–Reiten quiver [AR75a77]. A nonsplittable short exact sequence

$$0 \to Z \xrightarrow{v} Y \xrightarrow{u} X \to 0$$

in mod $A$, where $X$ and $Z$ are indecomposable, is called an *almost split sequence* if it satisfies the condition: for any nonsplittable morphism $u''\colon W \to X$ in mod $A$, there is a morphism $u'\colon W \to Y$ such that $u'' = uu'$, or equivalently, for any nonsplittable morphism $v''\colon Z \to W$ in mod $A$ there is a morphism $v'\colon Y \to W$ such that $v'' = v'v$. Auslander and Reiten proved that *any nonprojective (resp. noninjective) indecomposable module $M$ has a unique almost split sequence (up to isomorphism) of the form* $0 \to \tau(M) \to X \to M \to 0$ (*resp.* $0 \to M \to Y \to \tau^{-}(M) \to 0$) where $\tau(M) = \mathrm{Hom}_A(\mathrm{Tr}(M), k)$ and $\tau^{-}(M) = \mathrm{Tr}(\mathrm{Hom}_A(M, k))$ [AR75a77]. A morphism $f\colon X \to Y$ between indecomposable modules is said to be *irreducible* if for any decomposition of

$$f\colon X \xrightarrow{u} W \xrightarrow{v} Y$$

in mod $A$, $u$ is a splittable monomorphism or $v$ is a splittable epimorphism. Thus $f$ is irreducible if and only if $f \in \mathrm{rad}\,\mathrm{Hom}_A(X, Y)$ but not in $\mathrm{rad}^2\,\mathrm{Hom}_A(X, Y)$, where $\mathrm{rad}\,\mathrm{Hom}_A(X, Y)$ denotes the set of nonisomorphisms from $X$ to $Y$, and $\mathrm{rad}^2\,\mathrm{Hom}_A(X, Y)$ is the set of all $f \in \mathrm{Hom}_A(X, Y)$ with $f = f''f'$, where $f' \in \mathrm{rad}\,\mathrm{Hom}_A(X, M)$, $f'' \in \mathrm{rad}\,\mathrm{Hom}_A(M, Y)$ for some $A$-module $M$. The *Auslander–Reiten quiver* $\Gamma(A)$ is the oriented graph whose vertices are the isomorphism classes of indecomposable $A$-modules, and the number of arrows from $[X] \to [Y]$ is $a(X, Y)$, where $a(X, Y)$ is the dimension of $\mathrm{rad}\,\mathrm{Hom}(X, Y)/\mathrm{rad}^2\,\mathrm{Hom}(X, Y)$ over $F(X) := \mathrm{End}(X)/\mathrm{rad}\,\mathrm{End}(X)$ or equivalently over $F(Y)$ (see [Rin84]). The stable subquiver $\Gamma_s(A)$ denotes the full subquiver (of $\Gamma(A)$) of those vertices $[X]$ such that $\tau^n X$ is neither projective nor injective for any integer $n$. A *regular component* is a component without any projective modules and any injective modules. The following was proved by Auslander and Reiten [AR75a77], V.

PROPOSITION 2.5.2. *Let*

$$0 \to Z \xrightarrow{v} Y \oplus P \xrightarrow{u} X \to 0$$

*be an almost split sequence over a quasi-Frobenius algebra A, where $Y$ has no nonzero projective direct summands and $P$ is projective. Then there is an almost split sequence of the form*

$$0 \to \tau(Z) \xrightarrow{\tau(v)} \tau(Y) \oplus Q \xrightarrow{\tau(u)} \tau(X) \to 0,$$

*where $Q$ is injective.*

Modules over a trivial extension algebra $T(A) := A \ltimes \mathrm{Hom}(A, k)$ with $\mathrm{rad}^2 A = 0$ were first studied by W. Müller [MuW74] (and Green and Reiten [GrRe76]). Tachikawa asked if the trivial extension algebra $T(A)$ is of *finite representation type* (i.e. the number of isomorphism classes of indecomposable modules is finite) for a hereditary $k$-algebra $A$ of finite representation type. As to this problem, the theorem below, concerning a correspondence of indecomposable modules in mod $A$ and in $\mathrm{mod}(T(A)\backslash A)$ and concerning the Auslander–Reiten quiver $\Gamma(T(A))$, was first proved by the author

[Y81a82], I. Then Tachikawa gave another proof of (1) in the theorem and announced it at a meeting (Tsukuba, October 1978) and reported in [Ta80] with additional (2), by using the description of $T(A)$-modules by Fossum, Griffith and Reiten [FGR75] (namely, the $T(A)$-module is characterized as the pair $(M, \varphi)$ with an $A$-module $M$ and an $A$-homomorphism $\varphi\colon \operatorname{Hom}(A,k)\otimes_A M \to M$ such that $\varphi\cdot(\operatorname{Hom}(A,k)\otimes\varphi)=0$). It should be noted that the theorem below treats the Auslander–Reiten quivers of hereditary algebras of *any* representation type (cf. the inaccurate introduction of Section 4 in [Ta80]), and that the proof is valid without any change for not only trivial extensions but also all extensions of hereditary algebras by a duality module.

For an algebra $A$, $\operatorname{ind} A$ (resp. $\operatorname{proj} A$, $\operatorname{inj} A$) denotes the isomorphism classes of indecomposable (resp. indecomposable projective, indecomposable injective) $A$-modules, and $\operatorname{ind} T\setminus \operatorname{ind} A$ is denoted by $\operatorname{ind}(T\setminus A)$ for an extension algebra $T$ of $A$ by a module $DA$.

THEOREM 2.5.2 (Yamagata). *Let $A$ be a hereditary $k$-algebra and let $DA = \operatorname{Hom}(A,k)$, $T = A \ltimes DA$ and $\Omega = \Omega_T$. Then*

(1) *$\Omega$ and $\Omega^-$ induce the following bijective correspondences:*

$$\Omega\colon\ \operatorname{proj} A \to \operatorname{inj} A,\qquad \Omega\colon\ \operatorname{ind} A\setminus \operatorname{proj} A \to \operatorname{ind}(T\setminus A)\setminus \operatorname{proj} T,$$

$$\Omega^-\colon\ \operatorname{inj} A \to \operatorname{proj} A,\qquad \Omega^-\colon\ \operatorname{ind} A\setminus \operatorname{inj} A \to \operatorname{ind}(T\setminus A)\setminus \operatorname{inj} T.$$

(2) *The irreducible maps and the almost split sequences in* $\operatorname{mod} A$ *remain so in* $\operatorname{mod} T$. *The Auslander–Reiten quiver $\Gamma(T)$ is obtained from $\Gamma(A)$ by using $\Omega$. In particular, in case $A$ is basic and of infinite representation type, any regular component of $\Gamma(A)$ is a component of $\Gamma(T)$, and*

$$\Gamma_s(T) = \Gamma_s(A) \cup \Omega\big(\Gamma_s(A)\big) \cup \mathbf{P}_s \cup \mathbf{I}_s \quad (\textit{disjoint union}),$$

*where* $\mathbf{P}$ (*resp.* $\mathbf{I}$) *is the component containing all projective* (*resp. injective*) *$A$-modules.*

The proof was given in [Y81a82], I, Theorems 2.12, 4.1 and pp. 425–426; the statements on the Auslander–Reiten quivers and the facts that $\Omega(\Gamma_s(A)) = \Omega^{-1}(\Gamma_s(A))$ are trivial consequences of the first statement in (2) and Proposition 2.5.2 (also see [Y88], Remark on p. 39). Moreover, $\mathbf{P}$ and $\mathbf{I}$ are the components of $\Gamma(T)$ obtained from $\mathbf{p}(A)\cup\Omega(\mathbf{i}(A))$ and $\Omega(\mathbf{p}(A))\cup\mathbf{i}(A)$ by locating $T$-projective modules, where $\mathbf{p}(A)$ (resp. $\mathbf{i}(A)$) is the component of $\Gamma(A)$ containing all projective (resp. injective) $A$-modules (i.e. the *preprojective* (resp. *preinjective*) component, where one needs to remember that $A$ is supposed to be basic and hereditary). See Ringel [Rin86], p. 68, for a generalization of the statements in (2).

An important feature of trivial extension algebras is that it implies a relation between the finiteness of global dimension of an algebra $A$ and the derived category $D^b(\operatorname{mod} A)$ of bounded complexes of $A$-modules (Happel [H87, H89]). Consider the trivial extension algebra $T(A)$ as a $\mathbb{Z}$-graded algebra whose elements $(a,0)$ and $(0,f)(a\in A,\ f\in \operatorname{Hom}(A,k))$ are of degree 0 and 1, respectively. By $\operatorname{mod}^{\mathbb{Z}} T(A)$ we denote the category of finitely generated $\mathbb{Z}$-graded $T(A)$-modules with morphisms of degree zero.

THEOREM 2.5.3 (Happel). *An algebra $A$ is of finite global dimension if and only if $D^b(\operatorname{mod} A)$ is equivalent to $\mathsf{mod}^{\mathbb{Z}}\, T(A)$ as a triangulated category.*

Here, the *stable category of* $\operatorname{mod} A$, denoted by $\operatorname{mod} A$, has the same objects as $\operatorname{mod} A$. For $X, Y \in \mathsf{mod}\, A$, the set of morphisms $\mathsf{Hom}(X, Y)$ in $\mathsf{mod}\, A$ is the factor group of $\operatorname{Hom}_A(X, Y)$ by the morphisms factoring through projective $A$-modules. The category $\mathsf{mod}^{\mathbb{Z}}\, T(A)$ above is similarly defined and becomes a triangulated category in the following way: For a module $X$, let

$$0 \to X \xrightarrow{u} I(X) \to \Omega^{-1}(X) \to 0$$

be the injective hull of $X$. For a morphism $f\colon X \to Y$, form the pushout

$$X \xrightarrow{f} Y \xrightarrow{g} Z = X \xrightarrow{u} I(X) \to Z$$

and the corresponding short exact sequence

$$0 \to Y \xrightarrow{g} Z \xrightarrow{h} \Omega^{-1}(X) \to 0.$$

Then $\Omega^{-1}$ is an automorphism of $\mathsf{mod}^{\mathbb{Z}}\, T(A)$ and

$$X \xrightarrow{f} Y \xrightarrow{g} Z \xrightarrow{h} \Omega^{-1}(X)$$

is a triangle in $\mathsf{mod}^{\mathbb{Z}}\, T(A)$. The algebras $A, B$ are *stably equivalent* when there is a category equivalence $\mathsf{mod}\, A \approx \mathsf{mod}\, B$. Obviously, a Morita equivalence induces both a derived equivalence and a stable equivalence, but the converse is not true in general. Morita theory for stable categories is not known yet, but, as mentioned in 2.3, Rickard developed Morita theory for derived categories [Ric89b] and proved the following.

THEOREM 2.5.4 (Rickard). *Derived equivalent quasi-Frobenius algebras are stably equivalent. Moreover, in case $A$ and $B$ are derived equivalent algebras, $T(A)$ and $T(B)$ are derived equivalent.*

Thus we know that $T(A)$ and $T(B)$ are stably equivalent if $A$ and $B$ are derived equivalent algebras. By the following theorem, for example, we know when two algebras have stably equivalent trivial extension algebras [TaW87].

THEOREM 2.5.5 (Tachikawa–Wakamatsu). *For two algebras $A$ and $B$, $T(A)$ and $T(B)$ are stably equivalent if there is a tilting $(A, B)$-bimodule.*

Here a finitely generated left $A$-module $M$ is called a *tilting* module when it satisfies the following conditions:

(1) $\operatorname{proj.dim}_A M \leqslant 1$,
(2) $\operatorname{Ext}^1_A(M, M) = 0$, and
(3) there is an exact sequence $0 \to {}_AA \to M_0 \to M_1 \to 0$ with $M_0, M_1 \in \operatorname{add}(M)$.

If ${}_AM$ is tilting and $B = \mathrm{End}_A(M)^{op}$, then $M_B$ is also a tilting module and $A = \mathrm{End}_B(M)$ (Brenner and Butler [BB80], Happel and Ringel [HR82]); ${}_AM_B$ is called a tilting $(A, B)$-bimodule. The above theorem is also obtained as a corollary of Rickard's theory of derived categories [Ric89b]. We refer to Happel [H88] for the derived categories for algebras, Reiten [R76] and Martínez Villa [Mar89] for algebras stably equivalent to quasi-Frobenius algebras, and to Wakamatsu [Wak90, Wak93] for more general discussion and further results.

Quasi-Frobenius algebras of finite representation type are closely related to trivial extension algebras. For this, see Hughes and Waschbüsch [HW83] and Bretscher, Läser and Riedtmann [BrLR81]. This fact is based on tilting module theory and the classification theorem by Riedtmann [Rie80a]: Let $A$ be a quasi-Frobenius algebra of finite representation type. For the stable Auslander–Reiten quiver $\Gamma_s(A)$ let $x^\tau$ denote the $\tau$-orbit of a vertex $x$. Let $\Delta$ be the graph whose vertices are the $\tau$-orbits of $\Gamma_s(A)$, and there is an edge $x^\tau - y^\tau$ in $\Delta$ when there is an arrow $a \to b$ in $\Gamma_s(A)$ such that $x^\tau = a^\tau$, $y^\tau = b^\tau$ or $x^\tau = b^\tau$, $y^\tau = a^\tau$. Then $\Delta$ is called the type of $\Gamma_s(A)$.

THEOREM 2.5.6 (Riedtmann). *Let $A$ be a quasi-Frobenius algebra over an algebraically closed field of finite representation type. Then the type of $\Gamma_s(A)$ is a disjoint union of Dynkin diagrams.*

For example, let $A$ be a hereditary algebra whose quiver is of Dynkin type $\Delta$. Then the trivial (or any) extension algebra $T$ of $A$ by $\mathrm{Hom}(A, k)$ has a stable Auslander–Reiten quiver of type $\Delta$ (Theorem 2.5.2), and so has $T(B)$ for an algebras $B$ with a tilting $(A, B)$-bimodule (Theorem 2.5.5). We refer to [Rie80a, G79] for details and for further references, and Webb [We82], Linnel [Li85], Okuyama [Ok87], Erdmann [Er91, Er94], and Erdmann and Skowroński [ErSk92] for the graph structure of stable Auslander–Reiten components of group algebras. See Riedtmann [Rie80b, Rie83] and Waschbüsch [Was80, Was81] for the classification of quasi-Frobenius algebras of finite representation type, and Kupisch [Ku6570, Ku75]. For quasi-Frobenius algebras of infinite representation type, see Skowroński [Sk89, ErSk92], Erdmann [Er90] and [ANS89, Neh89, NehS89].

EXAMPLES.

(1) A purely separable extension field $K$ of $k$ has a nonsplittable extension [Y81a82]:

$$0 \to \mathrm{Hom}(K, k) \to T \to K \to 0.$$

More generally, there is a hereditary algebra $A$, with an arbitrary quiver, which has a nonsplittable extension of $A$ by $\mathrm{Hom}(A, k)$. A concrete example will appear in [SkY1]. (A hereditary algebra $A$ over an algebraically closed field $k$ has no nonsplittable extensions over $\mathrm{Hom}(A, k)$.)

(2) Let $A$ be a factor algebra of a hereditary algebra (i.e. an algebra without oriented cycles in its quiver) and let $DA$ be any duality $A$-module. Then any two extension algebras of $A$ by $DA$ are stably equivalent [Y88].

(3) Let $T_0$ and $T_1$ be the algebras with the same quiver and the following relations;

$$1 \underset{\beta}{\overset{\alpha}{\rightleftarrows}} 2 \circlearrowleft \rho; \qquad \begin{array}{l} T_0\colon\ \alpha\beta = \rho^2,\ \beta\alpha = 0, \\ T_1\colon\ \alpha\beta = \rho^2,\ \rho^4 = 0,\ \beta\alpha = \beta\rho\alpha. \end{array}$$

Then, in case $\operatorname{char} k = 2$, the algebras $T_0$ and $T_1$ are not isomorphic (Riedtmann). Moreover, for the algebra $A$ with the same quiver as the $T_i$ $(i = 0, 1)$ but with the relation that the composite of any two arrows is zero, we have that $A \simeq T_0/\operatorname{rad}^2 T_0 \simeq T_1/\operatorname{rad}^2 T_1$ naturally as algebras, and, by these isomorphisms, $\operatorname{rad}^2 T_0$ and $\operatorname{rad}^2 T_1$ become $A$-bimodules that are isomorphic to each other. Let $DA$ be a duality $A$-bimodule isomorphic to $\operatorname{rad}^2 T_0 (\simeq \operatorname{rad}^2 T_1)$ as $A$-bimodules. Then $T_0$ and $T_1$ both are nonsplittable extension algebras of $A$ over $DA$. It is not difficult to see that $T_0$ and $T_1$ are symmetric algebras (e.g., use Theorem 2.3.1). Hence, in case $\operatorname{char} k = 2$, there are at least two nonisomorphic symmetric nonsplittable extensions of $A$ by $DA$.

(4) Let $\lambda$ be a nonzero element of a field $k$ and let $A(\lambda)$ be the local $k$-algebra defined by two arrows $\alpha, \beta$ with the relation: $\alpha^2 = \beta^2 = \alpha\beta + \lambda\beta\alpha = 0$. Every $A(\lambda)$ is then a local Frobenius algebra on four dimensional $k$-vector space and a nonsplittable extension algebra of the Frobenius $k$-algebra $k[x]/(x^2)$ by itself (with indeterminate $x$). In the case where $\operatorname{char}(k) = 2$, $A(1)$ is the group algebra of Klein's 4-group with coefficients in $k$. It seems that, as a natural degeneration of $A(1)$, these algebras $A(\lambda)$ were first observed by Nakayama [N3941], I, where he showed that $A(\lambda)$ is symmetric exactly when $\lambda = 1$ by checking all ideals (in fact, $A := A(\lambda), \operatorname{rad} A, \operatorname{rad}^2 A, 0$ and $Aa = aA = k1_A + ka$ for $a \in k\alpha + k\beta$ are the only ideals). As to the Nakayama automorphisms of $A(\lambda)$, let $\nu(\alpha) = \lambda\alpha$ and $\nu(\beta) = \lambda^{-1}\beta$. Then $\nu$ induces an automorphism of $A(\lambda)$ naturally, so that there is an $A(\lambda)$-bimodule isomorphism $A(\lambda) \xrightarrow{\sim} \operatorname{Hom}(A(\lambda), k)_\nu$. It is easily seen that $\nu$ is inner if and only if $\lambda = 1$. Hence we know again that $A(\lambda)$ is symmetric exactly when $\lambda = 1$ (Theorem 2.4.1). Moreover we have the following interesting fact which was discovered by Rickard (unpublished).

THEOREM (Rickard). *$A(\lambda)$ and $A(\lambda')$ are stably equivalent if and only if $\lambda' = \lambda$ or $\lambda' = \lambda^{-1}$, in which case $A(\lambda)$ and $A(\lambda')$ are isomorphic.*

## 3. Generalizations and the Nakayama conjecture

The Nakayama conjecture is a characterization problem of quasi-Frobenius algebras and is deeply related to a class of QF-3 algebras introduced by Thrall [Th48].

### 3.1. *Thrall's generalizations*

Let $A$ be a $k$-algebra, $M$ an $A$-module and let $B = \operatorname{End}(M)^{op}$. Then $M$ is said to be *balanced* (or have the double centralizer property) if the canonical algebra morphism $\varphi_M\colon A \to \operatorname{End}_B(M)$ is epimorphic. A module $M$ is said to be *minimal faithful* if it is faithful and any proper direct summand of $M$ is not faithful. Note that ${}_AA$ has projective minimal faithful submodules and ${}_A\operatorname{Hom}(A,k)$ has injective minimal faithful submodules.

EXAMPLE. A module $M$ is balanced if $M^m$ contains ${}_AA$ as a direct summand for some $m > 0$ (Nesbitt and Thrall (1946); see Section 1, Example, and Morita [Mo58a], Theorem 16.5).

DEFINITION 3.1.1.

(1) A is a left QF-1 algebra if every faithful left $A$-module is balanced.

(2) $A$ is a left QF-2 algebra if every indecomposable projective left $A$-module has a simple socle.

(3) A is a left QF-3 algebra if there is a unique minimal faithful left $A$-module up to isomorphism.

The definition of QF-1, QF-3 algebras does not depend on the side of modules. Because, faithful left $A$-modules are in a 1–1 correspondence by $\operatorname{Hom}(-,k)$ with faithful right $A$-modules. (See Harada [Ha66], Morita [Mo69] for the side problem for QF-3 Artinian rings, see also Masaike [Mas92].) It is clear that any quasi-Frobenius algebra is QF-1 (use the above example), QF-2 and QF-3. Conversely, Floyd [Fl68] proved that *commutative QF-*1 *algebras are quasi-Frobenius*. (Cf. [DF70, C70, Rin74].) See Morita [Mo69] and Fuller [Fu69] for the Morita duality induced by an arbitrary QF-3 algebra.

The following characterization was proved by Thrall [Th48], Theorem 5 (see Jans [J59]).

PROPOSITION 3.1.1. *An algebra A is QF-*3 *if and only if there is a projective, injective and faithful left A-module, and if and only if the injective hull $I({}_AA)$ of ${}_AA$ is projective.*

PROOF. Assume that $A$ is QF-3, and let $P$ and $Q$ be minimal faithful summands of ${}_AA$ and ${}_A\operatorname{Hom}(A,k)$, respectively. Then, by uniqueness, $P \simeq Q$ as $A$-modules which implies that $P$ is injective. Conversely, assume that $P$ is a projective injective faithful module, and let $M$ be a faithful module. There is a monomorphism $\varphi\colon A \to M^{(n)}$ for some $n$. Clearly we may assume that $P$ is basic, so that $P \simeq Ae$ for some $e = e^2 \in A$. Since $P$ is injective, the restriction $\varphi|_P\colon P \hookrightarrow M^{(n)}$ is splittable and hence $P$ is isomorphic to a summand of $M$ because $P$ is basic. Thus the basic module $P$ is a unique minimal faithful module. The last statement is obviously equivalent to the second statement.

Now assume that $A$ is left and right QF-2. Then there are minimal faithful modules ${}_AP$ and ${}_AQ$, which are projective and injective respectively, such that any indecomposable summand has both a simple top and a simple socle. Let ${}_AP = \bigoplus_i P_i$ and ${}_AQ = \bigoplus_j Q_j$ be indecomposable decompositions. Then every $P_i$ is embedded into some $Q_{\sigma(i)}$, because $Q$ is faithful and $\operatorname{soc} P_i$ is simple, so that $Q_{\sigma(i)}$ is projective (note: the projective cover of $Q_{\sigma(i)}$ has a simple socle because $A$ is left QF-2). Hence we have that the module ${}_AQ$ is projective, injective and faithful. Thus we know that *every left and right QF-*2 *algebra is QF-*3. □

NOTES. QF-2 algebras play an important role in the representation theory of vector space categories [Si85a, Si85b]. We refer to Simson [Si92] for a general discussion of vector space categories and their representations. See also [Ha82, Ha83].

**3.2.** *A construction of QF-3 algebras*

There is an important construction of QF-3 algebras as the endomorphism rings of generator-cogenerators over an algebra $A$. This was first studied by Morita [Mo58a].

Let $M$ be a module over an algebra $A$. Recall that $M$ is a generator if and only if ${}_AA \in \operatorname{add}(M)$, and $M$ is a cogenerator if and only if ${}_A\operatorname{Hom}(A,k) \in \operatorname{add}(M)$. For a generator-cogenerator $M$, decompose it as $M = M_0 \oplus M_1 \oplus M_2 \oplus M_3$ such that $M_0$ has neither nonzero projective summands nor nonzero injective summands, $M_1$ is projective without nonzero injective summands, $M_2$ is projective and injective, and $M_3$ is injective without nonzero projective summands. Let $M(\rho) = M_1 \oplus M_2$ and $M(\kappa) = M_2 \oplus M_3$. Let $B = \operatorname{End}_A(M)^{op}$ and let $e_M(\rho)\colon M \to M(\rho) \hookrightarrow M$ and $e_M(\kappa)\colon M \to M(\kappa) \hookrightarrow M$ be idempotents of $B$ which are composites of a canonical projection and a canonical injection. Now assume that $A$ and ${}_AM$ are basic. Then, since ${}_AM$ is supposed to be a generator-cogenerator, we have that ${}_AM(\rho) \simeq {}_AA$, ${}_AM(\kappa) \simeq {}_A\operatorname{Hom}(A,k)$, and $e_M(\rho)B_B \simeq M_B$, $A \simeq e_M(\rho)Be_M(\rho)$ as algebras. Hence $e_M(\rho)B_B$ is a projective, injective and faithful right $B$-module, because so is $M_B$ by Morita theory. On the other hand, $B$ is naturally identified with $\operatorname{End}_A(M^*)$ and then

$$e_M(\kappa) = e_M(\kappa)^*\colon\ M^* \to M(\kappa)^* \to M^*,$$

where $(\ )^* := \operatorname{Hom}(-,k)$. Moreover,

$$M^*(\rho) = M(\kappa)^*_A \simeq A_A,$$

$$M^*(\kappa) = M(\rho)^*_A \simeq \operatorname{Hom}(A,k)_A, \qquad e_{M^*}(\rho) = e_M(\kappa)^*,$$

and

$${}_BBe_M(\kappa) = {}_BBe_{M^*}(\rho) \simeq {}_BM^*.$$

This implies that ${}_BBe_M(\kappa)$ is also projective, injective and faithful. Thus we know that $B$ is a QF-3 algebra with minimal faithful $B$-modules $Be_M(\kappa)$ and $e_M(\rho)B$, which proves the next theorem.

THEOREM 3.2.1 (Morita). *Let $M$ be a generator-cogenerator over an algebra $A$ and $B = \operatorname{End}_A(M)^{op}$. Then $B$ is a QF-3 algebra with projective, injective and faithful modules $Be_M(\kappa)$ and $e_M(\rho)B$. Moreover, $A$ and $e_M(\rho)Be_M(\rho)$ are Morita equivalent.*

COROLLARY 3.2.1. $\operatorname{End}_A(A \oplus \operatorname{Hom}(A,k))$ *is a QF-3 algebra for any algebra $A$.*

Another important example of endomorphism algebras in the above theorem is given by Auslander algebras. Here an *Auslander algebra* is by definition the endomorphism algebra of an $A$-module $M$ such that $\operatorname{add} M = \operatorname{add} M(A)$, where $A$ is an algebra of finite representation type and $M(A)$ is the direct sum of all nonisomorphic indecomposable $A$-modules. In this case, obviously ${}_AM$ is a generator-cogenerator, and we know that gl dim $B \leqslant 2$ for $B = \mathrm{End}_A(M)^{op}$. Indeed, for this it suffices to show that $\mathrm{Ker}\, h$ is projective for any morphism $h\colon P_1 \to P_0$ where $P_0$ and $P_1$ are projective $B$-modules. Let $F = {}_B\mathrm{Hom}_A(M, -)$ and $G = {}_AM \otimes_B -$. Then, since $\mathrm{Ker}\, G(h) \in \mathrm{add}({}_AM)$ (by the definition of ${}_AM$) and $FG \simeq \mathbf{1}$ on the subcategory of projective $B$-modules, we have that

$$\mathrm{Ker}\, h \simeq F\big(\mathrm{Ker}\, G(h)\big) \in \mathrm{add}\, F(M) = \mathrm{add}({}_BB).$$

This implies that $\mathrm{Ker}\, h$ is projective as desired. A characterization of Auslander algebras is given in Theorem 3.3.2 below, which is proved by Auslander [A71, A74]. Ringel and Tachikawa proved it for Artinian rings [RinT75]. Tachikawa related Auslander algebras with the problem when the category of projective modules is abelian with generators and products [Ta73].

### 3.3. *QF-3 maximal quotient algebras*

Not every QF-3 algebra is the endomorphism ring of a generator-cogenerator. In fact, for $A$, $M$ and $B$ as in Theorem 3.2.1, let $0 \to M \to I_0 \to I_1$ be a minimal injective presentation of the $A$-module $M$. Then $I_0$ and $I_1$ belong to $\mathrm{add}({}_A\mathrm{Hom}(A, k))$. Here note that ${}_A\mathrm{Hom}(A, k)$ is in $\mathrm{add}({}_AM)$. Applying $\mathrm{Hom}_A(M, -)$ to the sequence, we have an exact sequence of left $B$-modules such that $0 \to {}_BB \to P_0 \to P_1$ where $P_i = \mathrm{Hom}_A(M, I_i)$ belongs to $\mathrm{add}({}_BBe(\kappa))$. Hence both $P_0$ and $P_1$ are projective, injective and belong to $\mathrm{add}(I({}_BB))$, where $I({}_BB)$ is the injective hull of ${}_BB$. Thus $B$ is a left maximal quotient algebra. Here recall that the *left maximal quotient ring* $Q_\ell(R)$ of a (*general*) ring $R$ is the endomorphism ring of the right $\mathrm{End}_R(I({}_RR))^{op}$-module $I({}_RR)$ (i.e. $Q_\ell(R)$) is the double centralizer of $I({}_RR)$). Since each element of $R$ defines an endomorphism (as a left multiplication) of $\mathrm{End}_R(I({}_RR))^{op}$-module $I({}_RR)$, there is a canonical ring homomorphism $\varphi_\ell\colon R \to Q_\ell(R)$. A ring $R$ is called a *left maximal quotient ring* if the canonical morphism $\varphi_\ell$ is an isomorphism. A right maximal quotient ring $Q_r(R)$ is defined similarly. A ring $R$ is said to be *maximal quotient* when it is left and right maximal quotient, i.e. both $\varphi_\ell$ and $\varphi_r$ are isomorphisms. This definition is due to Lambek (originally defined by Utsumi, Osaka Math. J. **8** (1956)) and he proved the following effective characterization [L86], §4.3, Proposition 1, which is a special case of Theorem 3.5.1.

PROPOSITION 3.3.1 (Lambek). *A ring $R$ is a left maximal quotient ring if and only if there is an exact sequence $0 \to {}_RR \to I_0 \to I_1$ of $R$-modules such that $I_0$ and $I_1$ are direct summands of a direct product of copies of $I({}_RR)$. In case $R$ is a finite dimensional algebra over a field we can take as $I_0$ and $I_1$ direct summands of $I({}_RR)^m$ for some $m$. (Note: we can take $I({}_RR)$ as $I_0$.)*

Now we are in a position to state a characterization of the endomorphism algebra of a generator-cogenerator by Müller [Mu68a], Theorem 2, and Morita [Mo69], Theorem 1.2.

THEOREM 3.3.1. *The following statements for an algebra $A$ are equivalent.*

(1) *$A$ is the endomorphism algebra of a generator-cogenerator over an algebra.*

(2) *$A$ is a QF-3 maximal quotient algebra.*

(3) *For a minimal injective presentation $0 \to {}_AA \to I_0 \to I_1$, both $I_0$ and $I_1$ are projective.*

This follows from Theorem 3.2.1, Propositions 3.1.1, 3.3.1, and the following lemma.

LEMMA 3.3.1. *Let $A$ be a QF-3 algebra, and let ${}_AAe$ and $fA_A$ be unique minimal faithful $A$-modules. Then $A$ is left maximal quotient if and only if it is right maximal quotient. Moreover, in this case, $Ae_{eAe}$ and ${}_{fAf}fA$ are generator-cogenerators, $A \simeq \mathrm{End}_{eAe}(Ae) \simeq \mathrm{End}_{fAf}(fA)^{op}$ as algebras, and $\mathrm{Hom}_{eAe}(Ae, -) \otimes_A Ae \simeq \mathbf{1}$ on $\mathrm{add}(Ae_{eAe})$.*

PROOF. $I({}_AA)$ and $Ae$ are similar, so $Q_\ell \simeq \mathrm{End}_{eAe}(Ae)$ as algebras (Section 1, Example). Similarly, $Q_r(A) \simeq \mathrm{End}_{fAf}(fA)^{op}$. On the other hand, since $fA \simeq \mathrm{Hom}(Ae, k)$ as right $A$-modules, their double centralizers are isomorphic, i.e.

$$\mathrm{End}_{fAf}(fA)^{op} \simeq \mathrm{End}_{eAe}(Ae).$$

Hence $\varphi_\ell\colon A \to Q_\ell(A)$ is isomorphic if and only if so is $\varphi_r\colon A \to Q_r(A)$. In particular, by Morita theory, $Ae_{eAe}$ is a generator-cogenerator because ${}_AAe$ is projective and injective. □

PROPOSITION 3.3.2. *Let $A$ be the endomorphism algebra of a generator-cogenerator ${}_BM$ over an algebra $B$. Then the following assertions hold.*

(1) $\mathrm{Ext}^i_B(M, M) = 0$ *for some $i$ if and only if* $\mathrm{Ext}^i_{fAf}(fA, fA) = 0$, *where $fA$ is a unique minimal faithful right $A$-module.*

(2) *${}_BM$ is projective if and only if $A$ is quasi-Frobenius.*

This is an easy consequence of Morita Theory. First note that $M_A$ is projective, injective faithful. (1) By Theorem 3.3.1 there is a unique minimal faithful right $A$-module $fA$. Then $M_A$ and $fA_A$ are similar. This implies that $B(= \mathrm{End}_A(M))$ and $fAf(= \mathrm{End}_A(fA))$ are Morita equivalent, so the assertion (1) follows (Section 1, Example). (2) Assume that ${}_BM$ is projective. Then $M_A$ is a generator by the Morita equivalence theorem. Hence $A_A$ is a summand of $M^n_A$ for some $n > 0$ and so $A_A$ is injective. Conversely, if $A$ is quasi-Frobenius, an embedding $A_A \hookrightarrow M^m_A$ for some $m > 0$ is splittable because $A_A$ is injective. Hence $M_A$ is a generator and so, by the Morita equivalence theorem again, ${}_BM$ is projective.

As an application of Theorem 3.3.1 and Lemma 3.3.1 we can prove a characterization of Auslander algebras as mentioned in 3.2, where one direction is clear from Theorem 3.3.1 and the observation after Corollary 3.2.1.

THEOREM 3.3.2 (Auslander). *An algebra $A$ is an Auslander algebra if and only if $\mathrm{gl\,dim}\, A \leqslant 2$ and there is an injective presentation $0 \to {}_AA \to I_0 \to I_1$ such that $I_0, I_1$ are projective $A$-modules.*

PROOF. Assume that gl dim $A \leqslant 2$ and $A$ satisfies the equivalent conditions in Theorem 3.3.1. Let $Ae$ be a unique minimal faithful $A$-module, and $F = \mathrm{Hom}_{eAe}(Ae, -)$ and $G = - \otimes_A Ae$. Then, to show that $A$ is an Auslander algebra it suffices to show that any indecomposable right $eAe$-module $M$ belongs to $\mathrm{add}(Ae_{eAe})$ (Lemma 3.3.1). Now, since $Ae_{eAe}$ is a cogenerator, there is an exact sequence of right $eAe$-modules;

$$0 \to M \to Ae^m \xrightarrow{h} Ae^n$$

for some integers $m, n$. Then we have that $\mathrm{Ker}\, F(h)$ is a projective right $A$-module, i.e. $\mathrm{Ker}\, F(h) \in \mathrm{add}(A_A)$, because $F(Ae)_A \simeq A_A$ (Lemma 3.3.1) and gl dim $A \leqslant 2$. Hence

$$\mathrm{Ker}\, GF(h) \simeq G\big(\mathrm{Ker}\, F(h)\big) \in \mathrm{add}\,\big(G(A_A)\big) = \mathrm{add}(Ae_{eAe}).$$

On the other hand, since $\mathbf{1} \simeq GF$ on $\mathrm{add}(Ae_{eAe})$ by Lemma 3.3.1, $h$ is isomorphic to $GF(h)$, so that $\mathrm{Ker}\, h \simeq \mathrm{Ker}\, GF(h)$. In consequence, we have that $M \in \mathrm{add}(Ae_{eAe})$ as desired. □

NOTES. For an arbitrary ring $A$, $Q_\ell(A)$ is a left self-injective ring if and only if $Q_\ell(A)$ is an injective left $A$-module ([L86], Proposition 3).

(1) Masaike [Mas71] gave an ideal theoretical characterization on $A$ for $Q_\ell(A)$ to be quasi-Frobenius: *$Q_\ell(A)$ is quasi-Frobenius if and only if A satisfies the ascending chain condition for left $I({}_AA)$-annihilators and $r_A(L) = 0$ for any left ideal L of A such that there is an A-homomorphism from L to A which is not extended properly to any left ideals*. For example, the ring of triangular matrices over a quasi-Frobenius ring is a left and right QF-3 Artinian ring whose maximal left and maximal right quotient rings are quasi-Frobenius. Sumioka [Su75] proved the converse under some additional condition.

(2) Kambara [K90] characterized a (von Neumann) regular ring $A$ with $Q_\ell(A)$ being left and right self-injective. In fact, in this case, $Q_\ell(A)$ is a regular and maximal right quotient ring. This answers a question posed by Goodearl and Handelman [GoH75].

(3) B. Müller [Mu68b] proved a structure theorem for QF-3 algebras: *An algebra A is a QF-3 algebra if and only if A is a subalgebra of a QF-3 maximal quotient algebra R such that A contains suitable minimal faithful ideals ${}_RRe, fR_R$ and the unit 1 of R.* See [RinT75] for a generalization to arbitrary QF-3 rings.

### 3.4. *The Nakayama conjecture*

Nakayama [N58] conjectured that an algebra $A$ is quasi-Frobenius if there exists an infinite exact sequence

$$0 \to A \to X_1 \to \cdots \to X_n \to \cdots$$

of projective and injective $A$-bimodules $X_i$, and he proved this for serial algebras (= generalized uniserial algebras, see 2.1). Tachikawa [Ta64] considered similar sequences where the $X_i$ are projective and injective left $A$-modules. He defined the *left dominant dimension* of $A$ to be greater than or equal to $n$, $\text{l.dom.dim}\, A \geqslant n$, when there exists an exact sequence

$$0 \to A \to X_1 \to \cdots \to X_n$$

of projective and injective left $A$-modules $X_i$ $(1 \leqslant i \leqslant n)$. B. Müller [Mu68a] proved that the Nakayama conjecture is equivalent to say that $A$ is quasi-Frobenius if $\text{l.dom.dim}\, A = \infty$ (i.e. $\geqslant n$ for any $n$). However, compared to the usual homological dimensions (e.g., projective dimension, injective dimension), we define it in a slightly different way.

DEFINITION 3.4.1. The *left dominant dimension* of an algebra $A$, $\text{l.dom.dim}\, A$, is $n+1$ if for a minimal injective resolution of ${}_AA$

$$0 \to {}_AA \to I_0 \to I_1 \to \cdots \to I_n \to I_{n+1} \to \cdots,$$

$I_i$ $(0 \leqslant i \leqslant n)$ are nonzero projective but $I_{n+1}$ is not. Similarly $\text{r.dom.dim}\, A$ is defined and $\text{l.dom.dim}\, A = 0$ if $I({}_AA)$ is not projective.

Then, $\text{l.dom.dim}\, A > 0$ means just that $A$ is left QF-3, and $A$ is a QF-3 maximal quotient algebra if and only if $\text{l.dom.dim}\, A > 1$ (Theorem 2.3.1). Since the dominant dimension of quasi-Frobenius algebras is 1, the Nakayama conjecture can be stated as follows:

NAKAYAMA CONJECTURE (NC). *Every algebra has finite dominant dimension.*

A crucial theorem for dominant dimensions was first proved by Müller [Mu68a], Lemma 3, where $D = \text{Hom}(-, k)$.

THEOREM 3.4.1 (B. Müller). *Suppose that* $\text{l.dom.dim}\, A > 1$, *and let* ${}_AAe, fA_A$ *be unique minimal faithful modules. Then* $\text{l.dom.dim}\, A > n+1$ *if and only if*

$$\text{Ext}^i_{fAf}(fA, fA) = 0 \quad \textit{for } 0 < i < n+1,$$

*if and only if*

$$\text{Ext}^i_{eAe}\big(D(Ae), D(Ae)\big) = 0 \quad \textit{for } 0 < i < n+1.$$

Now suppose that $\text{l.dom.dim}\, A > 1$ as in the theorem. Then it follows from Theorem 3.3.1 and Lemma 3.3.1 that

$$A = \text{End}_{fAf}(fA)^{op} = \text{End}_{eAe}(Ae),$$

and ${}_{fAf}fA_A \simeq {}_{eAe}D(Ae)_A$ as bimodules via the canonical isomorphism

$$fAf = \text{End}_A(fA) \simeq \text{End}_A\big(D(Ae)\big) = eAe.$$

Thus the above theorem is obtained immediately from the first assertion in the next lemma which is used essentially in [Mu68a]. For a module $M$, a minimal injective (resp. projective) resolution is denoted by $I_\bullet(M)$: $0 \to M \to I_0(M) \to I_1(M) \to \cdots$ (resp. $P_\bullet(M)$: $\cdots \to P_1(M) \to P_0(M) \to M \to 0$).

LEMMA 3.4.1. *Let $A$ be an algebra $A$ and $M$ an $A$-module.*

(1) *The following statements are equivalent for a projective $A$-module $P_A$ and $B = \mathrm{End}_A(P)$.*

(a) $I_i(M) \in \mathrm{add}({}_A\mathrm{Hom}(P,k))$ *for* $0 \leqslant i \leqslant n$.

(b) $M \simeq \mathrm{Hom}_B(P, P \otimes_A M)$ *canonically as $A$-modules and*

$$\mathrm{Ext}^i_B(P, P \otimes_A M) = 0 \quad \textit{for } 0 < i < n.$$

(2) *The following statements are equivalent for a projective $A$-module ${}_AQ$ and $B = \mathrm{End}_A(Q)^{op}$.*

(a) $P_i(M) \in \mathrm{add}({}_AQ)$ *for* $0 \leqslant i \leqslant n$.

(b) $M \simeq Q \otimes_B \mathrm{Hom}_A(Q, M)$ *canonically as $A$-modules and*

$$\mathrm{Tor}^B_i\big(Q, \mathrm{Hom}_A(Q, M)\big) = 0 \quad \textit{for } 0 < i < n.$$

It is enough to show (1) because of the dual argument, and it is a routine task: Let $\varphi_X$: $X \to \mathrm{Hom}_B(P, P \otimes_A X)$ be a canonical $A$-morphism for an $A$-module $X$.

(a) $\Rightarrow$ (b): Since ${}_A\mathrm{Hom}(P,k)$ is injective and $\varphi_X$ is an isomorphism for $X = \mathrm{Hom}(P,k)$ and so for $X \in \mathrm{add}_A(\mathrm{Hom}(P,k))$, it follows that $\mathrm{Hom}_B(P, P \otimes_A I_\bullet(M))$ is isomorphic to $I_\bullet(M)$ up to $n$.

(b) $\Rightarrow$ (a): Any injective $B$-module $J$ is a direct summand of $\mathrm{Hom}(B,k)^m$ for some $m > 0$. Hence, as $A$-modules, $\mathrm{Hom}_B(P, J)$ is a direct summand of $\mathrm{Hom}_B(P, \mathrm{Hom}(B,k))^m$ and $\mathrm{Hom}_B(P, \mathrm{Hom}(B,k)) \simeq \mathrm{Hom}(P,k)$. Thus it follows from the assumption in (b) that $\mathrm{Hom}_B(P, I_i({}_BP \bigotimes_A M))$ contains $I_i(M)$ as a direct summand for $0 \leqslant i \leqslant n$.

EXAMPLE 1. For any $k$-algebra $A$, there is an idempotent $f$ such that

$$\mathrm{add}_A \mathrm{Hom}(fA, k) = \mathrm{add}\left(\bigoplus_{i=1}^{n(A)} I_i({}_AA)\right)$$

where $n(A)$ is a natural number such that

$$I_j({}_AA) \in \mathrm{add}\left(\bigoplus_{i=1}^{n(A)} I_i({}_AA)\right)$$

for all $j \geqslant 0$. Then it follows from the Morita equivalence theorem and Lemma 3.4.1(1) that ${}_{fAf}fA$ is a generator, $A \simeq \mathrm{End}_{fAf}(fA)$ and $\mathrm{Ext}^i_{fAf}(fA, fA) = 0$ for all $i > 0$.

We know that l.dom.dim $A > 0$ if and only if r.dom.dim $A > 0$ (Theorem 3.3.1). But, in this case, it is true that both dimensions are equal [Mu68b].

PROPOSITION 3.4.1. *If* l.dom.dim $A > 0$ *and* r.dom.dim $A > 0$, *then* l.dom.dim $A =$ r.dom.dim $A$.

PROOF. Assume that l.dom.dim $A > n + 1$ for a non-negative integer $n$. We have only to show that $P_i(D(A_A)) \in \operatorname{add}(Ae)$ for $0 \leqslant i \leqslant n + 1$, where $Ae$ is a unique minimal faithful $A$-module. Hence, by Lemma 3.4.1, it suffices to show that $\operatorname{Tor}_i^{eAe}(Ae, e(DA)) = 0$ for $0 < i < n + 1$, because applying $D := \operatorname{Hom}(-, k)$ to the algebra isomorphism $A \simeq \operatorname{Hom}_{eAe}(Ae, Ae)$ we have that ${}_ADA \simeq Ae \otimes_{eAe} e(DA)$. But this follows from Theorem 3.4.1 and the isomorphism

$$D\operatorname{Tor}_i^{eAe}\bigl(Ae, e(DA)\bigr) \simeq \operatorname{Ext}_{eAe}^i\bigl(D(Ae), D(Ae)\bigr).$$

□

The following is a direct consequence of Proposition 3.3.2 and Theorem 3.4.1.

THEOREM 3.4.2 (B. Müller). *Let* ${}_BM$ *be a nonprojective generator-cogenerator over an algebra* $B$, *and* $A = \operatorname{End}_B(M)^{op}$. *Then* l.dom.dim $A < \infty$ *if and only if*

$$\operatorname{Ext}_B^i(M, M) \neq 0 \quad \text{for some } i > 0.$$

By using this theorem, the Nakayama conjecture is restated as follows (B. Müller).

(NC-M) *A generator-cogenerator* $M$ *over an algebra* $A$ *is projective if*

$$\operatorname{Ext}_A^i(M, M) = 0 \quad \text{for all } i > 0.$$

There are other several conjectures related to the Nakayama conjecture.

(FDC) The finitistic dimension conjecture. *For an algebra* $A$, *there is a bound for the finite projective dimensions of finitely generated* $A$*-modules.*

See [Zi92] for a brief history and the recent development of the conjecture.

(SNC) Strong Nakayama conjecture [CF90]. *For any nonzero finitely generated left module* $M$ *over an algebra* $A$, *there is an integer* $n \geqslant 0$ *such that* $\operatorname{Ext}_A^n(M, A) \neq 0$.

(GNC) Generalized Nakayama conjecture [AR75b]. *For an algebra* $A$, *any indecomposable injective* $A$*-module belongs to*

$$\bigcup_{i \geqslant 0} \operatorname{add} I_i({}_AA),$$

*where* $0 \to {}_AA \to I_0({}_AA) \to I_1({}_AA) \to \cdots$ *is a minimal injective resolution of* ${}_AA$.

This is equivalent to the next statement because $\mathrm{Ext}^i_A(S,A) \neq 0$ for a simple module $S$ if and only if $S \subset I_i({}_AA)$:

(GNC′) *For any simple module $S$ over an algebra $A$, there is an integer $n \geqslant 0$ such that* $\mathrm{Ext}^n_A(S,A) \neq 0$.

The following is an analogue of NC-M, which is also equivalent to GNC [AR75b].

(GNC-M) *A generator $M$ over an algebra $A$ is projective if* $\mathrm{Ext}^i_A(M,M) = 0$ *for all* $i > 0$.

Tachikawa divided Müller's restatement NC-M into the following two statements, where $DA = \mathrm{Hom}(A,k)$, and by NC-T we understand the two statements NC-T1, T2 together.

(NC-T1) *An algebra $A$ is quasi-Frobenius provided that* $\mathrm{Ext}^i_A(DA,A) = 0$ *for all* $i > 0$.

(NC-T2) *A finitely generated module $M$ over a quasi-Frobenius algebra $A$ is projective if* $\mathrm{Ext}^i_A(M,M) = 0$ *for all* $i > 0$.

Now our aim is to show the following implications among those statements, where the implication FDC$\Rightarrow$NC was first noted by Müller and the last equivalence is in [Ta73].

THEOREM 3.4.3. FDC $\Rightarrow$ SNC $\Rightarrow$ GNC $\Leftrightarrow$ GNC-M $\Rightarrow$ NC $\Leftrightarrow$ NC-M $\Leftrightarrow$ NC-T.

FDC$\Rightarrow$SNC: Let $n$ be a bound for the finite projective dimensions, and assume that $\mathrm{Ext}^i_A(M,A) = 0$ for some nonzero $A$-module $M$ and all $i \geqslant 0$. Let

$$\cdots \xrightarrow{f_2} P_1 \xrightarrow{f_1} P_0 \xrightarrow{f_0} M \to 0$$

be a minimal projective resolution of $M$ and $F = \mathrm{Hom}_A(-,A)$. Then, by assumption we have the exact sequence

$$0 \to F(P_0) \xrightarrow{F(f_1)} F(P_1) \xrightarrow{F(f_2)} \cdots,$$

where all $F(P_i)$ $(i \geqslant 0)$ are projective right $A$-modules. Hence we have a projective resolution of $\mathrm{Im}\, F(f_{n+2})$:

$$0 \to F(P_0) \xrightarrow{F(f_1)} F(P_1) \to \cdots \to F(P_{n+1}) \to \mathrm{Im}\, F(f_{n+2}) \to 0.$$

Since the projective dimension of $\mathrm{Im}\, F(f_{n+2})$ is bounded by $n$ by assumption, $F(f_1)$ and so $f_1 = F(F(f_1))$ must be splittable. Hence ${}_AM$ is projective and $\mathrm{Hom}(M,A) \neq 0$, a contradiction.

SNC$\Rightarrow$GNC is trivial (use GNC$=$GNC′).

GNC$\Rightarrow$GNC-M: Let ${}_AM$ be a generator such that $\mathrm{Ext}^i_A(M,M) = 0$ for all $i > 0$ and $B = \mathrm{End}_A(M)^{op}$. By the Morita equivalence theorem, $M_B$ is projective and $A = \mathrm{End}_B(M)$. Take an idempotent $f$ in $B$ such that $\mathrm{add}(M_B) = \mathrm{add}(fB_B)$. Then $A$ and $fBf$ are Morita equivalent, under which ${}_AM$ corresponds to ${}_{fBf}fB$ (Section 1,

Example). Hence it suffices to show that ${}_{fBf}fB$ is projective. Now $\mathrm{Ext}^i_{fBf}(fB, fB) = 0$ for all $i > 0$ because of the assumption for ${}_AM$, which implies that

$$I_i({}_BB) \in \mathrm{add}_B \operatorname{Hom}(fB, k)$$

for all $i \geqslant 0$ (Lemma 3.4.1(1)). Consequently, ${}_B \operatorname{Hom}(fB, k)$ is a cogenerator by GNC for $B$, i.e. $fB_B$ is a generator and so ${}_{fBf}fB$ is projective by the Morita equivalence theorem.

GNC-M $\Rightarrow$ GNC: For an algebra $A$ take an idempotent $f$ of $A$ as in Example 1 above. Then ${}_{fAf}fA$ is projective by GNC-M and hence $fA_A$ is a generator by the Morita equivalence theorem because $A \simeq \mathrm{End}_{fAf}(fA)$. Hence ${}_A \operatorname{Hom}(fA, k)$ is a cogenerator.

GNC $\Rightarrow$ NC: By GNC any indecomposable injective module ${}_AI$ appears in some $I_i({}_AA)$ as a summand. Hence, if $\mathrm{dom.dim}\, A = \infty$, then every ${}_AI$ is projective and so ${}_AA$ is injective, i.e. $\mathrm{dom.dim}\, A = 1$, a contradiction. (Note: GNC-M $\Rightarrow$ NC-M is trivial.)

NC-M $\Rightarrow$ NC-(T1+T2): First, assume that $\mathrm{Ext}^i_A(DA, A) = 0$ for all $i > 0$, and let $M = A \oplus DA$. Since $M$ is a generator-cogenerator and

$$\mathrm{Ext}^i_A(M, M) = \mathrm{Ext}^i_A(DA, A) = 0,$$

it follows from NC-M that ${}_AM$ is projective and so ${}_ADA$ is projective, i.e. $A_A$ is injective. Second, let $A$ be quasi-Frobenius and $M$ an $A$-module such that $\mathrm{Ext}^i_A(M, M) = 0$ for $i > 0$. Then

$$\mathrm{Ext}^i_A(M \oplus A, M \oplus A) = 0$$

for all $i > 0$ because ${}_AA$ is injective. Hence, by NC-M, ${}_A(M \oplus A)$ and so $M$ are projective.

NC-(T1+T2) $\Rightarrow$ NC-M: Let ${}_AM$ be a generator-cogenerator such that $\mathrm{Ext}^i_A(M, M) = 0$ for $i > 0$. We shall show that ${}_AM$ is projective. Since $M$ is a generator-cogenerator, there is a summand ${}_AN$ such that $\mathrm{add}(N) = \mathrm{add}(A \oplus DA)$. Then $\mathrm{Ext}^i_A(N, N) = 0$ for $i > 0$ implies that

$$0 = \mathrm{Ext}^i_A(A \oplus DA, A \oplus DA) = \mathrm{Ext}^i_A(DA, A)$$

for $i > 0$. Hence, by NC-T1, $A$ is quasi-Frobenius and so, by NC-T2, ${}_AM$ is projective.

REMARK. All implications, except GNC $\Leftrightarrow$ GNC-M, in Theorem 3.4.3 are true for a given algebra $A$. But the proof of the equivalence GNC $\Leftrightarrow$ GNC-M involves the conjectures for the endomorphism algebras of some generators over a given algebra $A$.

EXAMPLE 2. There is a simple construction of algebras with large dominant dimension [Y90] (cf. Theorem 3.4.1). Let $\mathcal{D}$ be the class of $A$-modules $M$ such that $I_i(M)$ $(i = 0, 1)$ is projective, and $D = \operatorname{Hom}(-, k)$. Assume the following two conditions:

(i) ${}_AA = P_1 \oplus Q_1$ *as left $A$-modules and* $A_A = P_2 \oplus Q_2$ *as right $A$-modules, where every submodule of $P_i$ is projective noninjective and $Q_i$ is injective.*

(ii) $\mathrm{Hom}(X, M) = 0$ *for any* $M \in \mathcal{D}$ *and any indecomposable nonprojective injective* $A$*-module* $X$.

Then, for $B = \mathrm{End}_A({}_AA \oplus_A DA)^{op}$, we have that

$$\text{l.dom.}\dim A + 1 \leqslant \text{l.dom.}\dim B \leqslant \text{l.dom.}\dim A + 2,$$

$$\text{gl}\dim A + 1 \leqslant \text{gl}\dim B \leqslant \text{gl}\dim A + 2,$$

and

$$0 \leqslant \text{gl}\dim B - \text{l.dom.}\dim B \leqslant (\text{gl}\dim A - \text{l.dom.}\dim A) + 1.$$

For example, taking a hereditary algebra $A$ as $A_1$ and the endomorphism algebras

$$A_{i+1} := \mathrm{End}_{A_i}(A_i \oplus DA_i)$$

consecutively, we have then that

$$\text{l.dom.}\dim A_1 < \cdots < \text{l.dom.dim}\, A_n < \cdots.$$

By this method, the number of simple modules is increasing. However, there is not known any class of algebras with arbitrarily large dominant dimension and with the same number of simple modules. We conjecture that *there is an upper bound for the dominant dimensions of algebras having finite dominant dimension and with a given number of simple modules*. This is still open for the class of algebras (having finite dominant dimension) with only two simple modules, see [KK90] and [Zi93]. We state a similar problem related to GNC.

PROBLEM. Find a number $n(A)$ for an algebra $A$ such that

$$I_j({}_AA) \in \mathrm{add}\left(\bigoplus_{i=0}^{n(A)} I_i({}_AA)\right)$$

for all $j > 0$. If we take $n(A)$ minimal, is there an upper bound among those numbers $n(A)$ of algebras $A$ with a given number of simple modules?

Green and Zimmermann-Huisgen [GZ91] proved FDC for algebras $A$ with $\mathrm{rad}^3 A = 0$, hence by Theorem 3.4.3 (see the above Remark) these algebras also satisfy GNC and NC. This was partially generalized recently by Dräxler and Happel [DH92] as follows:

PROPOSITION 3.4.2. *An algebra $A$ satisfies the generalized Nakayama conjecture provided that, for some natural number $n$, $\mathrm{rad}^{2n+1} A = 0$ and $A/\mathrm{rad}^n A$ is representation finite.*

NOTES. Wilson [Wi86] proved that positively graded algebras satisfy the generalized Nakayama conjecture. See Auslander and Reiten, and Solberg [AR91, AR92, AS92], Fuller and Zimmermann-Huisgen [FuZ86], and Martínez Villa [Mar92] for general discussion on GNC. It is known that NC-T2 holds for the group algebra $kG$ of a finite group over a field $k$. This is a consequence of the work of Evens [Ev61] and Alperin and Evens [AlE81]. It was first proved for finite $p$-groups by Tachikawa [Ta73]; see also Schulz [Schu86] and Donovan [Do88]. See also Hoshino [Ho82] for NC-T2 for the trivial extension algebras considered in Theorem 2.5.2.

### 3.5. *QF*-1 *algebras*

There is a homological characterization of balanced modules by Morita [Mo71] and Suzuki [Suz71]. Cf. [Mo70], Theorem 3.4.

THEOREM 3.5.1. *Let ${}_AM$ be a (not necessarily finitely generated) faithful A-module, and let $B = \mathrm{End}_A(M)^{\mathrm{op}}$ and $C = \mathrm{End}_B(M)$. Then ${}_AM$ is balanced if and only if $M$ satisfies the following two conditions*:

(a) *the canonical morphism ${}_CM_B \to {}_C\mathrm{Hom}_A({}_AC_C, {}_AM_B)$ is an isomorphism,*

(b) *there is an exact sequence $0 \to {}_AA \to M_0 \to M_1$, where each $M_i$ is a direct product of copies of ${}_AM$ (in case $M$ is finitely generated, we can take both $M_i \in \mathrm{add}({}_AM)$).*

In fact, if ${}_AM$ is balanced, then (a) is trivial and (b) follows by applying $\mathrm{Hom}_B(-, M_B)$ to a projective presentation of $M_B$:

$$\bigoplus_{j\in J} B_B \to \bigoplus_{i\in I} B_B \to M_B \to 0.$$

To prove the converse, first observe that $\varphi^{-1}(f) = fu$ for $f \in \mathrm{Hom}_A(C, M)$ and the inclusion $u\colon A \hookrightarrow C$, where $\varphi\colon M \to \mathrm{Hom}_A(C, M)$ is the canonical isomorphism. Then, applying $\mathrm{Hom}_A(C, -)$ to the exact sequence in (b), we have an isomorphism

$$\mathrm{Hom}(C, A) \to A\colon\ f \mapsto fu,$$

in particular, the composite identity

$${}_AA \overset{u}{\hookrightarrow} {}_AC \xrightarrow{f} {}_AA$$

for some $f$. Let

$${}_AC = {}_AA \oplus {}_AC'.$$

Then $\mathrm{Hom}_A(C', M) = 0$ by (a), which forces that $C' = 0$ because $C'M \subset M$ and ${}_CM$ is faithful. Thus we know that $u$ is an isomorphism.

Although the theorem does characterize balanced faithful modules, to study QF-1 algebras there is more effective criterion by Morita [Mo58b] to check if a module is balanced. Up to this time, it is the most important tool to study QF-1 algebras. In fact, the main results mentioned below were obtained by using the criterion effectively.

MORITA'S CRITERION. *Let $M$ be a faithful balanced $A$-module. Then, for an indecomposable $A$-module $N$, $M \oplus N$ is balanced if and only if $M$ either generates or cogenerates $N$ (i.e. there is an epimorphism $M^m \to N$ or a monomorphism $N \to M^m$ for some $m$).*

PROOF. Let

$$_AL = M \oplus N, \qquad B = \mathrm{End}_A(L)^{op},$$

and

$$e_M\colon L \xrightarrow{\pi_M} M \overset{inc}{\hookrightarrow} L, \qquad e_N\colon L \xrightarrow{\pi_N} N \overset{inc}{\hookrightarrow} L$$

be the composites of a canonical projection and a canonical injection. Let

$$N_1 := \sum \{\mathrm{Im}\, f \mid f \in \mathrm{Hom}_A(M, N)\},$$

and

$$N_0 := \bigcap \{\mathrm{Ker}\, g \mid g \in \mathrm{Hom}_A(N, M)\}.$$

Note that $N_1 = M(e_M B e_N) \subseteq MB$. Since $N$ is indecomposable, $D := \mathrm{End}_A(N)$ is a local ring. Let $\overline{D} = D/\operatorname{rad} D$, $N_1^* = N_1 + N \operatorname{rad} D$ and

$$N_0^* = \ell_{N_0}(\operatorname{rad} D)(= \{x \in N_0 \mid x(\operatorname{rad} D) = 0\}) \neq 0.$$

Now assume that $N_1 \neq N$ and $N_0 \neq 0$. Then $N/N_1^*$ and $N_0^*$ are nonzero $(A, \overline{D})$-bimodules. Since $\overline{D}$ is a division ring, there is a nonzero $\overline{D}$- morphism from $N/N_1^*$ to $N_0^*$, so that we have a nonzero $D$-homomorphism $\varphi\colon N \to N$ such that $\varphi(N_1^*) = 0$ and $\varphi(N) \subseteq N_0^*$. Let

$$\Phi\colon L \xrightarrow{\pi_N} N \xrightarrow{\varphi} N \overset{inc}{\hookrightarrow} L$$

be the composite of $\varphi$ and canonical morphisms. Observe that $\Phi$ is a $B$-homomorphism. Moreover, since $\Phi(M) = 0$, $\Phi$ is not a left multiplication of any element of $A$ because $aM \neq 0$ for any $0 \neq a \in A$. Hence we know that $_AL$ is not balanced. Conversely, take any $B$-homomorphism $\Phi\colon L_B \to L_B$, and assume that $N_1 = N$ or $N_0 = 0$. Since $\Phi(M) \subset M$, the restriction $\Phi|_M\colon M \to M$ is a $C$-monomorphism, where

$$C := e_M B e_M \simeq \mathrm{End}_A(M)$$

canonically. Then $\Phi|_M = a_L$ (left multiplication) for some $a \in A$, because ${}_AM$ is balanced. Let

$$\Phi' = \Phi - a_L\colon\ L_B \to L_B.$$

Then $\Phi'(M) = 0$, and it suffices to show that $\Phi'(N) = 0$. In fact, in case $N = N_1$, $N \subset MB$ and so $\Phi'(N) \subset \Phi'(M)B = 0$. In case $N_1 = 0$, there are $b_1, \ldots, b_m \in B$ such that

$$(b_1, \ldots, b_m)\colon\ N \to M^m$$

is a monomorphism for some $m > 0$. If $\Phi'(N) \neq 0$, then there is some $b_i$ such that $\Phi'(N)b_i \neq 0$ because $\Phi'(N) \subset N$ (note that $\Phi'$ is a $B$-homomorphism). But, in this case,

$$\Phi'(N)b_i = \Phi'(Nb_i) \subset \Phi'(M) = 0,$$

a contradiction. □

COROLLARY 3.5.1. *An algebra $A$ is QF-1 if and only if the following two conditions hold:*

(a) *every minimal faithful $A$-module is balanced,*

(b) *for any minimal faithful $A$-module $M$, every indecomposable $A$-module is either generated or cogenerated by $M$.*

Thus the condition for all faithful modules in the definition of a QF-1 algebra is reduced to some conditions for the minimal faithful modules. Obviously every quasi-Frobenius algebra has a unique minimal faithful module. Floyd [Fl68] conjectured that QF-1 algebras have at most finitely many indecomposable faithful modules. But, in 1986, Makino [Ma86] constructed QF-1 algebras having infinitely many minimal faithful balanced modules. Before introducing his example, we mention some results on QF-1 algebras related to the problem of Floyd. A general structure theorem of QF-1 algebras was first proved by Ringel [Rin73].

THEOREM 3.5.2 (Ringel). *Assume that $A$ is a QF-1 algebras and let $e$ and $f$ be primitive idempotents with $f(\mathrm{soc}({}_AA) \cup \mathrm{soc}(A_A))e \neq 0$. Then*

(1) *either* $|\mathrm{soc}(A_A)e| = 1$ *or* $|f\ \mathrm{soc}({}_AA)| = 1$,

(2) $|\mathrm{soc}({}_AA)e| \times |f\ \mathrm{soc}(A_A)| \leqslant 2$, *and*

(3) $|\mathrm{soc}({}_AA)e| = 2$ *implies that* $\mathrm{soc}(A_A)e \subset \mathrm{soc}({}_AA)e$.

Note that, for a primitive idempotent $e$ of an algebra $A$, $\mathrm{soc}(A_A)e$ is not zero if and only if $Ae$ is a direct summand of a minimal faithful projective $A$-module. (This is used in several papers, e.g., Fuller [Fu70], Makino [Ma86]). Indeed, let $Af$ be a minimal faithful projective module with an idempotent $f$, and let $J = \mathrm{soc}(A_A)$. Clearly $Je \neq 0$ if and only if $JeAf \neq 0$, because $Af$ is faithful. But $JeAf \neq 0$ just means that $eAf$ is not contained in $\mathrm{rad}\,A$, or equivalently $Ae$ is a direct summand of $Af$.

By using the above theorem, Ringel answered Floyd's question as follows.

THEOREM 3.5.3. *A $QF$-1 algebra having an indecomposable faithful module is Morita equivalent to a local quasi-Frobenius algebra.*

Thus QF-1 algebras have at most one indecomposable faithful module, and *any $QF$-1 algebra which is not quasi-Frobenius has no indecomposable faithful module.*

The structure of QF-1 algebras and of QF-1 serial algebras are known as follows:

THEOREM 3.5.4 (Ringel). *Let $A$ be an algebra such that* $\operatorname{rad}^2 A = 0$, *and let* $L = \operatorname{soc}({}_A A)$ *and* $J = \operatorname{soc}(A_A)$. *Then the following conditions are equivalent:*

(1) *$A$ is a $QF$-1 algebra.*

(2) *$A$ satisfies the following conditions:*

(i) *for primitive idempotents $e$ and $f$ with $f(L \cap J)e \neq 0$, we have that* (a) $|Je| = 1$ *or* $|fL| = 1$, *and* (b) $|Le| \times |fJ| \leqslant 2$,

(ii) *$Je \subset Le$ for every primitive idempotent $e$ with $|Le| = 2$,*

(iii) *$fL \subset fJ$ for every primitive idempotent $f$ with $|fJ| = 2$.*

(3) *Every indecomposable $A$-module has either a simple top or a simple socle, and $A$ is a left maximal and a right maximal quotient algebra.*

Fuller [Fu68] gave a necessary and sufficient condition for a serial algebra to be QF-1. The following theorem by Ringel and Tachikawa [RinT75] implies a construction of serial QF-1 algebras.

THEOREM 3.5.5. *An algebra $A$ over a field $k$ is a serial $QF$-1 algebra if and only if $A$ is isomorphic to the endomorphism algebra of a module $M$ over a serial $k$-algebra $B$ such that*

(a) $\operatorname{add}(M) = \operatorname{add}(B \oplus \operatorname{Hom}(B, k))$ *and*

(b) $\operatorname{Hom}(X, Y) = 0$ *for any indecomposable nonprojective and any indecomposable noninjective direct summand $X$ and $Y$ of $M$, respectively.*

Moreover, every indecomposable left module over a left serial QF-1 algebra has a simple socle (Tachikawa [Ta75]). It should be noted that both algebras with squared zero radical and left serial algebras have faithful serial modules. Thus the class of algebras in the following theorem contains the algebras in Theorem 3.5.4 and left serial QF-1 algebras.

THEOREM 3.5.6 (Makino). *Let $A$ be an algebra over an algebraically closed field. If $A$ is a $QF$-1 algebra with faithful serial modules, then every indecomposable $A$-module has either a simple top or a simple socle.*

Moreover, in fact, he gave a necessary and sufficient condition for an algebra to be a QF-1 algebra with a faithful serial module (see Makino [Ma91], Theorem II).

Now we shall give an example of an algebra with infinitely many minimal faithful balanced modules (see [Ma86] for further information).

Let $A$ be the algebra over a field $k$ with the following quiver and relations:

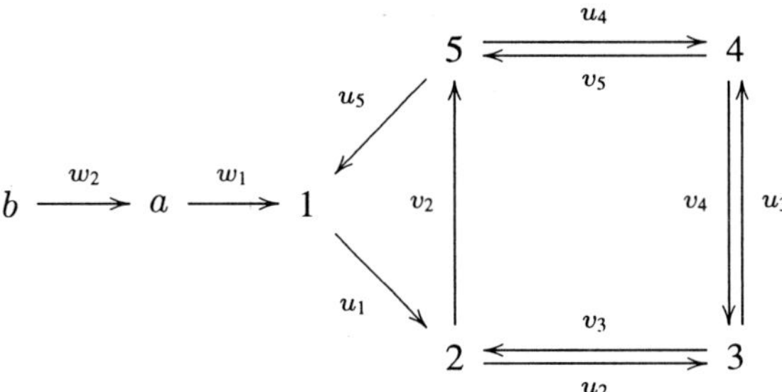


$$\begin{aligned}
&u_5u_4,\ u_4u_3,\ u_3u_2,\ u_2u_1\\
&v_5v_2,\ v_2v_3,\ v_3v_4,\ v_4v_5\\
&v_3u_2u_1,\ u_1w_1,\ w_1w_2\\
&v_3u_2 - u_1u_5v_2,\ u_2v_3 - v_4u_3\\
&u_3v_4 - v_5u_4,\ u_4v_5 - v_2u_1u_5.
\end{aligned}$$

Then

$$A = A_{e_b} \oplus Ae_a \oplus Ae_1 \oplus Ae_2 \oplus \cdots \oplus Ae_5,$$

where $Ae_b$, $Ae_i$ $(2 \leqslant i \leqslant 5)$ are injective, and $Ae_a$, $Ae_1$ are not injective. Hence, every faithful $A$-module should have a summand isomorphic to

$$Ae_b \oplus \left(\bigoplus_{i=2}^{5} Ae_i\right).$$

Moreover, a module of the form

$$Ae_b \oplus \left(\bigoplus_{i=1}^{5} Ae_i\right) \oplus X$$

with noninjective indecomposable module $X$ is minimal faithful if and only if $Ae_a$ is isomorphic to a submodule of $X$. Now, taking account of this fact, let $X_n$ be the indecomposable module below and consider the module

$$M_n = Ae_b \oplus Ae_1 \oplus \cdots \oplus Ae_5 \oplus X_n \quad \text{for } n > 0.$$

Then we have that every $M_n$ is minimal faithful because $Ae_a \hookrightarrow X_n$.

THEOREM 3.5.7 (Makino). *The algebra $A$ is QF-1, and every $M_n$ $(n > 0)$ is a minimal faithful balanced module.*

$X_n$:

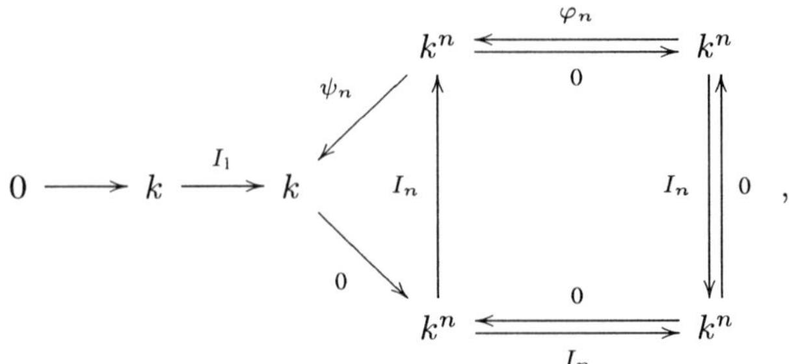


*where*

$$\varphi_n = \begin{pmatrix} 0 & & & 0 \\ 1 & \ddots & \ddots & \\ & \ddots & \ddots & \\ 0 & & 1 & 0 \end{pmatrix},$$

$\psi_n = (10\ldots0)$ *and* $I_n$ *is the* $n \times n$ *identity matrix.*

The following list of papers is *not* intended as a comprehensive bibliography of quasi-Frobenius algebras. I have selected those articles most relevant to the areas covered in this article, which were published after 1970 basically. Articles before 1970, can be found in various books or lecture notes, e.g., Curtis and Reiner [CR62, CR8187], Lambek [L86], Faith [F76] or Tachikawa [Ta73], Erdmann [Er90], etc. For articles on the representation theory of quasi-Frobenius algebras, see Gabriel [G79] and Skowroński [Sk90].

## References


[Al77] J.L. Alperin, *Periodicity in groups*, Illinois J. Math. **21** (1977), 776–783.

[AlE81] J.L. Alperin and L. Evens, *Representations, resolutions and Quillen's dimension theorem*, J. Pure Appl. Algebra **22** (1981), 1–9.

[ANS89] I. Assem, J. Nehring and A. Skowroński, *Domestic trivial extension of simply connected algebras*, Tsukuba J. Math. **13** (1989), 31–72.

[ASk88] I. Assem, and A. Skowroński, *Algebras with cycle-finite derived categories*, Math. Ann. **280** (1988), 441–463.

[A71] M. Auslander, *Representation dimension of Artin algebras*, Queen Mary Coll. Notes, London (1971).

[A74] M. Auslander, *Representation theory of Artin algebras, I*, Comm. Algebra **1** (1974), 177–268; *II*, ibid. (1974), 269–310.

[APR77] M. Auslander, M.I. Platzeck and I. Reiten, *Periodic modules over weakly symmetric algebras*, J. Pure Appl. Algebra **11** (1977), 279–291.

[AR75a77] M. Auslander and I. Reiten, *Representation theory of Artin algebras, III*, Comm. Algebra **3** (1975), 239–294; *IV*, ibid. **5** (1977), 519–554; *V*, ibid. **5** (1977), 519–554.

[AR75b] M. Auslander and I. Reiten, *On a generalized version of the Nakayama conjecture*, Proc. Amer. Math. Soc. **52** (1975), 69–74.

[AR91] M. Auslander and I. Reiten, *Applications of contravariantly finite subcategories*, Adv. Math. **86** (1991), 111–152.

[AR92] M. Auslander and I. Reiten, *k-Gorenstein algebras and syzygy modules*, Preprint, Mathematics No. 12/1992, The University of Trondheim (1992).

[AR94] M. Auslander and I. Reiten, *D* Tr*-periodic modules and functors*, Preprint, Mathematics No. 16/1994, The University of Trondheim (1994).

[AS92] M. Auslander and Ø. Solberg, *Gorenstein algebras and algebras with dominant dimension at least* 2, Preprint, Mathematics No. 14/1992, The University of Trondheim (1992).

[Az59] G. Azumaya, *A duality theory for injective modules (Theory of quasi-Frobenius modules)*, Amer. J. Math. **81** (1959), 249–278.

[Az66] G. Azumaya, *Completely faithful modules and self-injective rings*, Nagoya Math. J. **27** (1966), 697–708.

[Be84] D.J. Benson, *Modular Representation Theory: New Trends and Methods*, SLNM 1081, Springer, Berlin (1984).

[BD77] V.M. Bondarenko and J.A. Drozd, *The representation type of finite groups*, Zap. Nauchn. Sem. Leningrad. Otdel. Mat. Inst. Steklov. **57** (1977), 24–41. English translation: J. Soviet Mat. **20** (1982), 2525–2528.

[BNe37] R. Brauer and C. Nesbitt, *On the regular representations of algebras*, Proc. Nat. Acad. Sci. USA **23** (1937), 236–240.

[B70] S. Brenner, *Modular representations of p-groups*, J. Algebra **15** (1970), 89–102.

[BB80] S. Brenner and M.C.R. Butler, *Generalizations of the Bernstein–Gelfand–Ponomarev reflection functors, Representation Theory, II*, SLNM 832, Springer, Berlin (1980), 103–169.

[BrLR81] O. Bretscher, C. Läser and C. Riedtmann, *Self-injective and simply connected algebras*, Manuscripta Math. **36** (1981), 253–307.

[C70] V.P. Camillo, *Balanced rings and a problem of Thrall*, Trans. Amer. Math. Soc. **149** (1970), 143–153.

[CF72] V.P. Camillo and K.R. Fuller, *Balanced and QF-*1 *algebras*, Proc. Amer. Math. Soc. **34** (1972), 373–378.

[Ca79a] J.F. Carlson, *Periodic modules with large periods*, Proc. Amer. Math. Soc. **76** (1979), 209–215.

[Ca79b] J.F. Carlson, *The dimensions of periodic modules over modular group algebras*, Illinois J. Math. **23** (1979), 295–306.

[CE56] H. Cartan and S. Eilenberg, *Homological Algebra*, Princeton Univ. Press, Princeton (1956).

[Co66] P.M. Cohn, *Morita Equivalence and Duality*, University of London, Queen Mary College, London (1966).

[CF90] R.R. Colby and K.R. Fuller, *A note on the Nakayama conjectures*, Tsukuba J. Math. **14** (1990), 343–352.

[CR62] C.W. Curtis and I. Reiner, *Representation Theory of Finite Groups and Associative Algebras*, Interscience, New York (1962).

[CR8187] C.W. Curtis and I. Reiner, *Methods of Representation Theory, I, II*, Wiley, New York (1981, 1987).

[DF70] S.E. Dickson and K.R. Fuller, *Commutative QF-*1 *Artinian rings are QF*, Proc. Amer. Math. Soc. **24** (1970), 667–670.

[DM86] F. Dischinger and W. Müller, *Left PF is not right PF*, Comm. Algebra **14** (1986), 1223–1227.

[DR73] V. Dlab and C.M. Ringel, *The structure of balanced rings*, Proc. London Math. Soc. **26** (1972), 446–462.

[DR72a] V. Dlab and C.M. Ringel, *Rings with the double centralizer property*, J. Algebra **22** (1972), 480–501.

[DR72b] V. Dlab and C.M. Ringel, *Balanced rings*, SLNM 246, Springer, Berlin (1972), 74–143.

[DR72c] V. Dlab and C.M. Ringel, *A class of balanced non-uniserial rings*, Math. Ann. **195** (1972), 279–291.

[Do88] P.W. Donovan, *A criterion for a modular representation to be projective*, J. Algebra **117** (1988), 434–436.

[DH92] P. Dräxler and D. Happel, *A proof of the generalized Nakayama conjecture for algebras with* $J^{2\ell+1} = 0$ *and* $A/J^{\ell}$ *representation finite*, J. Pure Appl. Algebra **78** (1992), 161–164.

[EN55] S. Eilenberg and T. Nakayama, *On the dimension of modules and algebras, II (Frobenius algebras and quasi-Frobenius rings)*, Nagoya Math. J. **9** (1955), 1–16.

[Ei80] D. Eisenbud, *Homological algebra on a complete intersection, with an application to group representations*, Trans. Amer. Math. Soc. **260** (1980), 35–64.

[Er90] K. Erdmann, *Blocks of Tame Representation Type and Related Algebras*, SLNM 1428, Springer, Berlin (1990).

[Er91] K. Erdmann, *On Auslander–Reiten components for wild blocks*, Progr. Math. vol. 95 (1991), 371–387.

[Er94] K. Erdmann, *On Auslander–Reiten components for group algebras*, Preprint, Oxford (1994).

[ErSk92] K. Erdmann and A. Skowroński, *On Auslander–Reiten components of blocks and self-injective biserial algebras*, Trans. Amer. Math. Soc. **330** (1992), 169–189.

[Ev61] L. Evens, *The cohomology ring of a finite group*, Trans. Amer. Math. Soc. **101** (1961), 224–239.

[F76] C. Faith, *Algebra, II. Ring Theory*, Springer, Berlin (1976).

[F90] C. Faith, *When self-injective rings are QF: A report on a problem*, Preprint, Univ. Autónoma Barcelona (1990).

[FW67] C. Faith and E.A. Walker, *Direct sum representations of injective modules*, J. Algebra **5** (1967), 203–221.

[Fl68] C.R. Floyd, *On QF-1 algebras*, Pacific J. Math. **27** (1968), 81–94.

[FGR75] R.M. Fossum, P.A. Griffith and I. Reiten, *Trivial Extensions of Abelian Categories*, SLNM 456, Springer, Berlin (1975).

[F03] G. Von Frobenius, *Theorie der hyperkomplexen Größen*, Sitzung der phys.-math. Kl. (1903), 504–538, 634–645.

[Fu68] K.R. Fuller, *Generalized uniserial rings and their Kupisch series*, Math. Z. **106** (1968), 248–260.

[Fu69] K.R. Fuller, *On indecomposable injective over Artinian rings*, Pacific J. Math. **29** (1969), 115–135.

[Fu70] K.R. Fuller, *Double centralizers of injectives and projectives over Artinian rings*, Illinois J. Math. **14** (1970), 658–664.

[FuZ86] K.R. Fuller and B. Zimmermann-Huisgen, *On the generalized Nakayama conjecture and the Cartan determinant problem*, Trans. Amer. Math. Soc. **294** (1986), 679–691.

[G72] P. Gabriel, *Unzerlegbare Darstellungen, I*, Manuscripta Math. **6** (1972), 71–193.

[G73] P. Gabriel, *Unzerlegbare Darstellungen, II*, Symp. Math. Ist. Naz. Alta Mat. (1973), 81–104.

[G79] P. Gabriel, *Algébre auto-injectives de représentation finie (d'aprés Christine Riendtmann)*, Séminaire Bourbaki, 32e année (1979/80), no. 545.

[G80] P. Gabriel, *Auslander–Reiten sequences and representation-finite algebras*, SLNM 831, Springer, Berlin (1980), 1–71.

[GR79] P. Gabriel and C. Riedtmann, *Group representations without groups*, Comment. Math. Helv. **54** (1979), 240–287.

[Go74] K.R. Goodearl, *Simple self-injective rings need not be Artinian*, Comm. Algebra **2** (1974), 83–89.

[GoH75] K.R. Goodearl and D. Handelman, *Simple self-injective rings*, Comm. Algebra **3** (1975), 797–834.

[GrRe76] E.L. Green and I. Reiten, *On the construction of ring extensions*, Glasgow Math. J. **17** (1976), 1–11.

[GZ91] E.L. Green and B. Zimmermann-Huisgen, *Finitistic dimension of artinian rings with vanishing radical cube*, Math. Z. **206** (1991), 505–526.

[H87] D. Happel, *On the derived category of a finite-dimensional algebras*, Comment. Math. Helv. **62** (1987), 339–389.

[H88] D. Happel, *Triangulated categories in the representation theory of finite-dimensional algebras*, **119** Cambridge Univ. Press, Cambridge–New York (1988).

[H89] D. Happel, *Auslander–Reiten triangles in derived categories of finite-dimensional algebras*, Proc. Amer. Math. Soc. **112** (1991), 641–648.

[HPR80] D. Happel, U. Preiser and C.M. Ringel, *Vinberg's characterization of Dynkin diagrams using subadditive functions with applications to DTr-periodic modules*, SLNM 832, Springer, Berlin (1980), 280–294.

[HR82] D. Happel and C.M. Ringel, *Tilted algebras*, Trans. Amer. Math. Soc. **274** (1982), 399–443.

[Ha66] M. Harada, *QF-3 and semiprimary PP-rings, II*, Osaka J. Math. **3** (1966), 21–27.

[Ha82] M. Harada, *Self mini-injective rings*, Osaka J. Math. **19** (1982), 587–597.

[Ha83] M. Harada, *A characterization of QF-algebras*, Osaka J. Math. **20** (1983), 1–4.

[He61] A. Heller, *Indecomposable representations and the loop-space operation*, Proc. Amer. Math. Soc. **12** (1961), 640–643.

[Hi54] D. Higman, *Indecomposable representations at characteristic $p$*, Duke J. Math. **21** (1954), 377–381.

[Ho82] M. Hoshino, *Modules without self-extensions and Nakayama's conjecture*, Arch. Math. **43** (1984), 493–500.

[HW83] D. Hughes and J. Waschbüsch, *Trivial extensions of tilted algebras*, Proc. London Math. Soc. **46** (1983), 347–364.

[Hu82] J.E. Humphreys, *Restricted Lie algebras (and beyond)*, Contemp. Math. vol. 13, Amer. Math. Soc. (1982), 91–98.

[IW80] Y. Iwanaga and T. Wakamatsu, *Trivial extension of Artin algebras*, SLNM 832, Springer, Berlin (1980), 295–301.

[J59] J.P. Jans, *Projective injective modules*, Pacific J. Math. **9** (1959), 1103–1108.

[K87] H. Kambara, *Examples of a directly finite self-injective regular rings which is not left self-injective*, Proc. the 20th Symp. Ring Theory, Okayama (1987), 141–145.

[K90] H. Kambara, *On directly finite regular rings*, Osaka J. Math. **27** (1990), 629–654.

[Ki86] T. Kimura, *A classification theory of prehomogeneous vector spaces*, Adv. Stud. Pure Math. **14** (1986), 223–256.

[KK90] E. Kirkman and J. Kuzmanovich, *Algebras with large homological dimensions*, Proc. Amer. Math. Soc. **109** (1990), 903–906.

[Ku6570] H. Kupisch, *Symmetrische Algebren mit endlich vielen unzerlegbaren Darstellungen, I,* J. Reine Angew. Math. **219** (1965), 1–25; *II*, ibid. **245** (1970), 1–13.

[Ku75] H. Kupisch, *Quasi-Frobenius algebras of finite representation type*, SLNM 488, Springer, Berlin (1975), 184–200.

[L86] J. Lambek, *Rings and Modules*, Chelsea, New York (1986).

[Li85] P.A. Linnell, *The Auslander–Reiten quiver of a finite group*, Arch. Mat. **45** (1985), 289–295.

[LS93] S. Liu and R. Schulz, *The existence of bounded infinite $D\,\mathrm{Tr}$-orbits*, Research Report No. 552/1993, National University of Singapore (1993).

[Ma74] R. Makino, *Balanced conditions for direct sums of serial modules*, Sci. Rep. Tokyo Kyoiku Daigaku Sec. A **12** (1974), 181–201.

[Ma85a] R. Makino, *QF-1 algebras of local-colocal type*, Math. Z. **189** (1985), 571–592.

[Ma85b] R. Makino, *Double centralizers of minimal faithful modules over left serial algebras*, J. Algebra **96** (1985), 18–34.

[Ma86] R. Makino, *QF-1 algebras which have infinitely many minimal faithful modules*, Comm. Algebra **14** (1986), 193–210.

[Ma91] R. Makino, *QF-1 algebras with faithful direct sums of uniserial modules*, J. Algebra **136** (1991), 175–189.

[Mar89] R. Martínez Villa, *The stable group of a selfinjective Nakayama algebras*, Monogr. Inst. Mat. U.N.A.M. (1989), 192.

[Mar92] R. Martínez Villa, *Modules of dominant and codominant dimension*, Comm. Algebra **20** (1992), 3515–3540.

[Mas71] K. Masaike, *Quasi-Frobenius maximal quotient rings*, Sci. Rep. Tokyo Kyoiku Daigaku Sec. A **11** (1971), 1–5.

[Mas92] K. Masaike, *Reflexive modules over QF-3 rings*, Canad. Math. Bull. **35** (1992), 247–251.

[Mo58a] K. Morita, *Duality for modules and its applications of the theory of rings with minimum condition*, Sci. Rep. Tokyo Kyoiku Daigaku Sec. A **6** (1958), 83–142.

[Mo58b] K. Morita, *On algebras for which every faithful representation is its own second commutator*, Math. Z. **69** (1958), 429–434.

[Mo69] K. Morita, *Duality in QF-3 rings*, Math. Z. **108** (1969), 237–252.

[Mo70] K. Morita, *Localizations in categories of modules, I,* Math. Z. **114** (1970), 121–144.

[Mo71] K. Morita, *Flat modules, injective modules and quotient rings*, Math. Z. **120** (1971), 25–40.

[Mu68a] B. Müller, *The classification of algebras by dominant dimension*, Canad. J. Math. **20** (1968), 398–409.

[Mu68b] B. Müller, *On algebras of dominant dimension one*, Nagoya Math. J. **31** (1968), 173–183.

[Mu68c] B. Müller, *Dominant dimension of semi-primary rings*, J. Reine Angew. Math. **232** (1968), 173–179.

[MuW74] W. Müller, *Unzerlegbare Moduln über artinschen Ringen*, Math. Z. **137** (1974), 197–226.

[N3941] T. Nakayama, *On Frobeniusean algebras, I*, Ann. Math. **40** (1939), 611–633; *II*, ibid. **42** (1941), 1–21.

[N58] T. Nakayama, *On algebras with complete homology*, Abh. Math. Sem. Univ. Hamburg **22** (1958), 300–307.

[NN38] T. Nakayama and C. Nesbitt, *Note on symmetric algebras*, Ann. Math. **39** (1938), 659–668.

[Neh89] J. Nehring, *Polynomial growth trivial extensions of non-simply connected algebras*, Bull. Polish Acad. Sci. 9/10 (1989).

[NehS89] J. Nehring and A. Skowroński, *Polynomial growth trivial extensions of simply connected algebras*, Fund. Math. **132** (1989), 117–134.

[Ne38] J. Nesbitt, *Regular representations of algebras*, Ann. Math. **39** (1938), 634–658.

[Ok87] T. Okuyama, *On the Auslander–Reiten quiver of a finite group*, J. Algebra **110** (1987), 425–430.

[Os66a] B.L. Osofsky, *A generalization of quasi-Frobenius rings*, J. Algebra **4** (1966), 373–387; *Erratum* 9 (1968), 120.

[Os66b] B.L. Osofsky, *Cyclic injective modules of full linear rings*, Proc. Amer. Math. Soc. **17** (1966), 247–253.

[Pa71] B. Pareigis, *When Hopf algebras are Frobenius algebras*, J. Algebra **18** (1971), 588–596.

[PSk91] Z. Pogorzały and A. Skowroński, *Selfinjective biserial standard algebras*, J. Algebra **138** (1991), 491–504.

[R76] I. Reiten, *Stable equivalence of self-injective algebras*, J. Algebra **40** (1976), 63–74.

[Ric89a] J. Rickard, *Morita theory for derived categories*, J. London Math. Soc. **39** (1989), 436–456.

[Ric89b] J. Rickard, *Derived categories and stable equivalence*, J. Pure Appl. Algebra **61** (1989), 303–317.

[Ric91] J. Rickard, *Derived equivalences as derived functors*, J. London Math. Soc. **43** (1991), 37–48.

[Rie80a] C. Riedtmann, *Algebren, Darstellungsköcher, Überlagerungen und zurück*, Comment. Math. Helv. **55** (1980), 199–224.

[Rie80b] C. Riedtmann, *Representation-finite selfinjective algebras of type $A_n$*, SLNM 832, Springer, Berlin (1980), 449–520.

[Rie82] C. Riedtmann, *Preprojective partitions for selfinjective algebras*, J. Algebra **76** (1982), 532–539.

[Rie83] C. Riedtmann, *Representation-finite selfinjective algebras of type $D_n$*, Compositio Math. **49** (1983), 231–282.

[Rin73] C.M. Ringel, *Socle conditions for QF-1 rings*, Pacific J. Math. **41** (1973), 309–336.

[Rin74] C.M. Ringel, *Commutative QF-1 rings*, Proc. Amer. Math. Soc. **42** (1974), 365–368.

[Rin84] C.M. Ringel, *Tame Algebras and Integral Quadratic Forms*, SLNM 1099, Springer, Berlin (1985).

[Rin86] C.M. Ringel, *Representation theory of finite-dimensional algebras*, London Math. Soc. Lecture Notes 116, Cambridge Univ. Press, Cambridge–New York (1986), 7–79.

[RinT75] C.M. Ringel and H. Tachikawa, *QF-3 rings*, J. Reine Angew. Math. **272** (1975), 49–72.

[San70] F.L. Sandomierski, *Some examples of right self-injective rings which are not left self-injective*, Proc. Amer. Math. Soc. **26** (1970), 244–245.

[Sam87] M. Sato, *On simply connected QF-3 algebras and their construction*, J. Algebra **106** (1987), 206–220.

[Schu86] R. Schulz, *Boundness and periodicity of modules over QF rings*, J. Algebra **101** (1986), 450–469.

[Si85a] D. Simson, *Special schurian vector space categories and $\ell$-hereditary right QF-2 rings*, Comment. Math. **25** (1985), 135–147.

[Si85b] D. Simson, *Vector space categories, right peak rings and their socle projective modules*, J. Algebra **92** (1985), 532–571.

[Si92] D. Simson, *Linear Representations of Partially Ordered Sets and Vector Space Categories*, Algebra, Logic and Applications 4, Gordon and Breach Sci. Pub., London (1992).

[SiSk81] D. Simson and A. Skowroński, *Extensions of artinian rings by hereditary injective modules*, SLNM 903, Springer, Berlin (1981), 315–330.

[Sk82] A. Skowroński, *A characterization of a new class of Artin algebras*, J. London Math. Soc. **26** (1982), 53–63.

[Sk89] A. Skowroński, *Selfinjective algebras of polynomial growth*, Math. Ann. **285** (1989), 177–199.

[Sk90] A. Skowroński, *Algebras of polynomial growth*, Topics in Algebra, Banach Center Publications vol. 26, part 1, (1990), 535–568.

[SkY1] A. Skowroński and K. Yamagata, *Socle deformations of selfinjective algebras*, Proc. London Math. Soc., to appear.

[SkY2] A. Skowroński and K. Yamagata, *Selfinjective algebras and generalized standard components*, In preparation.

[Su75] T. Sumioka, *A characterization of the triangular matrix rings over QF-rings*, Osaka J. Math. **12** (1975), 449–456.

[Suz71] Y. Suzuki, *Dominant dimension of double centralizers*, Math. Z. **122** (1971), 53–56.

[Ta64] H. Tachikawa, *On dominant dimensions of QF-3 algebras*, Trans. Amer. Math. Soc. **112** (1964), 249–266.

[Ta73] H. Tachikawa, *Quasi-Frobenius Rings and Generalizations*, SLNM 351, Springer, Berlin (1973).

[Ta75] H. Tachikawa, *Balancedness and left serial of finite type*, SLNM 488, Springer, Berlin (1975), 351–378.

[Ta80] H. Tachikawa, *Representations of trivial extensions of hereditary algebras*, SLNM 832, Springer, Berlin (1980), 579–599.

[TaW86] H. Tachikawa and T. Wakamatsu, *Applications of reflection functors for selfinjective algebras*, SLNM 1177, Springer, Berlin (1986), 308–327.

[TaW87] H. Tachikawa and T. Wakamatsu, *Tilting functors and stable equivalence for selfinjective algebras* J. Algebra **109** (1987), 138–165.

[Th48] R.M. Thrall, *Some generalizations of quasi-Frobenius algebras*, Trans. Amer. Math. Soc. **64** (1948), 173–183.

[Wak90] T. Wakamatsu, *Stable equivalence of selfinjective algebras and a generalization of tilting modules*, J. Algebra **134** (1990), 298–325.

[Wak93] T. Wakamatsu, *Tilting theory and selfinjective algebras*, Preprint, Saitama University (1993).

[Was80] J. Waschbüsch, *A class of self-injective algebras and their indecomposable modules*, SLNM 832, Springer, Berlin (1980), 632–647.

[Was81] J. Waschbüsch, *Symmetrische Algebren von endlichen Modultyp*, J. Reine Angew. Math. **321** (1981), 78–98.

[We82] P. Webb, *The Auslander–Reiten quiver of a finite group*, Math. Z. **179** (1982), 79–121.

[Wi86] G. Wilson, *The Cartan map on categories of graded modules*, J. Algebra **86** (1983), 390–398.

[Y80] K. Yamagata, *On extensions over Artinian rings with self-dualities*, Tsukuba J. Math. **4** (1980), 67–75.

[Y81a82] K. Yamagata, *Extensions over hereditary Artinian rings with self-dualities, I*, J. Algebra **73** (1981), 386–433; *II*, J. London Math. Soc. **26** (1982), 28–36.

[Y81b85] K. Yamagata, *On algebras whose trivial extensions are of finite representation type, I*, SLNM 903, Springer, Berlin (1981), 364–371; *II*, J. London Math. Soc. **32** (1985), 203–216.

[Y88] K. Yamagata, *Representations of nonsplittable extension algebras*, J. Algebra **115** (1988), 32–45.

[Y90] K. Yamagata, *Dominant dimension of algebras*, Colloquium talk at Paderborn University, January 1990.

[Zi92] B. Zimmermann-Huisgen, *Homological domino effects and the first Finitistic Dimension Conjecture*, Invent. Math. **108** (1992), 369–383.

[Zi93] B. Zimmermann-Huisgen, *Bounds on global and finitistic dimension for finite dimensional algebras with vanishing radical cube*, J. Algebra **161** (1993), 47–68.

# Subject Index